Engineering Drawing and Design
THIRD EDITION

CECIL JENSEN
Former Technical Director
R. S. McLaughlin Collegiate and
Vocational Institute
Oshawa, Ontario, Canada

JAY D. HELSEL
Professor
Department of Industrial Arts
and Technology
California University of Pennsylvania
California, Pennsylvania

GREGG DIVISION
McGRAW-HILL BOOK COMPANY

New York   Atlanta   Dallas   St. Louis   San Francisco   Auckland   Bogotá
Guatemala   Hamburg   Johannesburg   Lisbon   London   Madrid   Mexico   Montreal
New Delhi   Panama   Paris   San Juan   São Paulo   Singapore   Sydney   Tokyo   Toronto

Sponsoring Editor: D. Eugene Gilmore
Editing Supervisor: Alfred Bernardi
Design and Art Supervisor: Patricia Lowy
Production Supervisor: Frank Bellantoni

Cover Photographs: Computervision and Evans & Sutherland

Library of Congress Cataloging in Publication Data

Jensen, Cecil Howard, (date)
    Engineering drawing and design.

    Includes bibliographical references and index.
    1. Mechanical drawing.   2. Engineering design.
I. Helsel, Jay D.   II. Title.
T353.J47 1985   604.2   83-19896
ISBN 0-07-032533-2

Engineering Drawing and Design, Third Edition

1 2 3 4 5 6 7 8 9 0   VHVH   8 9 10 9 8 7 6 5 4

ISBN 0-07-032533-2

CONTENTS

PREFACE

Engineering Drawing and Design, Third Edition, is prepared for a two-semester course in engineering drawing. The contents are consistent with the trends and practices currently used in the preparation of engineering drawings.

Technical drafting, like all technical areas, is constantly changing. The computer has revolutionized the way in which drawings are prepared. For this reason, three new topics have been introduced in the third edition of *Engineering Drawing and Design*—computer-aided drafting (CAD), computer-aided manufacturing (CAM), and electrical and electronics drafting. In this new edition, the authors have made every effort to translate the most current technical information available into the most usable form from the standpoint of both teacher and student. The latest developments and current practices in all areas of graphic communication, computer-aided drafting (CAD), electronics drafting, functional drafting, materials representation, shop processes, numerical control, true positioning, geometric tolerancing, and metrication have been incorporated into this text in a manner that synthesizes, simplifies, and converts complex drafting standards and procedures into understandable instructional units. Extensive author research and visits to drafting rooms throughout the country have resulted in a combination of current drafting practices and practical pedagogical techniques that produces the most efficient learning system yet designed for the instruction of engineering drawing.

A new Chapter 32, "Computer-Aided Design and Drafting," explains the basic concepts that a drafter or student drafter needs to know about CAD. It provides an excellent introduction to this topic.

Chapter 25, "Electrical and Electronics Drawings," introduces the student to the new state of the art—the use of computer chips and logic diagrams. The authors are indebted to Robert Chadwick, Technical Director at the McLaughlin C.V.I., Oshawa, Canada, for assisting in the selection of topics and projects for this chapter.

Additional problems and the clustering of existing problems provides greater choice of material. Every chapter in the text is divided into a number of single-concept units

each with its own objectives, instruction, examples, review, and assignments. This organization provides the student with a logical sequence of experiences which can be adjusted to individual needs and also provides for maximum efficiency in learning essential concepts. Development of each unit is from the simple to the complex and from the familiar to the unfamiliar. Checkpoints are included to provide maximum reinforcement at each level.

Although the adoption of the metric system for drawings by smaller industries is not keeping pace with the large international companies, it is increasing in use. For that reason ANSI Y14.5M-1982 Dimensioning and Tolerancing, ANSI B4.2-1978 Preferred Metric Limits and Fits, and ANSI Y14.36-1978 Surface Texture Symbols have all been published in the metric units of size and measurement. In order to prepare our students for gainful employment upon graduation, it is recommended that both the International System of Units (SI) and the U.S. Customary System (USCS) units of measurement be included in all technical drawing programs.

Both SI and USCS units are used throughout the text and in all problems. Thus, the text may be used in a completely metric-oriented course, or in a course which utilizes both metric and customary systems. The teacher may also customize the course by selecting appropriate problems or materials to emphasize or deemphasize any degree of metrication. The dual dimensions shown in this book, especially in the assignment sections, are neither hard nor soft conversions. Instead, the sizes are those that would be most commonly used in the particular dimensioning units and so are only approximately equal. Dual dimensioning in this way avoids awkward amounts and allows instructor and student to be confident that a drawing using either set of dimensions will be no more difficult to work than one dimensioned exclusively in either dimensioning system.

Two sets of A4- or B-size worksheets are available separately for the completion of the problems. The worksheets include the problem in metric form on one side and in customary form on the reverse side. They are preprinted with light lines to provide the student with a beginning to each problem. Using these worksheets eliminates some of the initial work such as preparing borders, legends, data lines, and so forth. The worksheets also provide the student with the positioning of the drawing on each sheet, thus enabling the student to concentrate on the solution to the problem rather than on the mechanics of beginning the drawing. This focuses attention specifically on the concept under consideration and eliminates time wasted in nonessential aspects of the lesson. In earlier units, a certain amount of the work is completed for the student; in later units, however, fewer lines are provided.

A complete solutions manual to most graphic problems found in the text is also available from the publisher.

The authors would like to thank the many users of the previous edition of the textbook for their thoughtful and useful comments. In addition, the help of Hal Lindquist, Donald Voisinet, and John Nee is appreciated and gratefully acknowledged.

Cecil Jensen
Jay D. Helsel

ABOUT THE AUTHORS

CECIL JENSEN is the author or coauthor of many successful technical books, including *Engineering Drawing and Design, Fundamentals of Engineering Drawing, Drafting Fundamentals, Interpreting Engineering Drawings, Architectural Drawing and Design for Residential Construction, Home Planning and Design,* and *Interior Design.* Some of these books are printed in three languages and are used in many countries.

He has twenty-seven years of teaching experience in mechanical and architectural drafting and was a technical director for a large vocational school in Canada.

Before entering the teaching profession, Mr. Jensen gained several years of design experience in the industry. He has also been responsible for the supervision of the teaching of technical courses for General Motors apprentices in Oshawa, Canada.

He is a member of the Canadian Standards Committee (CSA) on Technical Drawings (which includes both mechanical and architectural drawing) and is chairman of the Committee on Dimensioning and Tolerancing. Mr. Jensen is Canada's representative on the American (ANSI) Standards for Dimensioning and Tolerancing and has recently represented Canada at two world (ISO) conferences in Oslo (Norway) and Paris on the standardization of technical drawings.

He took an early retirement from the teaching profession in order to devote his full attention to writing.

JAY D. HELSEL is a professor of industrial arts and technology at California University of Pennsylvania. He completed his undergraduate work in industrial arts at California State College and was awarded a master's degree from Pennsylvania State University. He has done advanced graduate work at West Virginia and at the University of Pittsburgh, where he completed a doctoral degree in educational communications and technology. In addition, Dr. Helsel holds a certificate in airbrush techniques and technical illustration from the Pittsburgh Art Institute.

He has worked in industry and has taught drafting, metalworking, woodworking, and a variety of laboratory and professional courses at both the secondary and college levels. During the past twenty years, he has also worked as a free-lance artist and illustrator. His work appears in many technical publications.

Dr. Helsel is coauthor of *Engineering Drawing and Design, Fundamentals of Engineering Drawing, Programmed Blueprint Reading,* and *Mechanical Drawing.* He is also the author of a series of *Mechanical Drawing Film Loops.*

DRAFTING UPDATE (PRESENT DRAWING PRACTICES)

FEATURE		SYMBOL	ANSI PUBLICATION Y14.5M−1982 (EXCEPT WHERE NOTED, REFER TO CLAUSE NO.)
LINES (THREE LINE WIDTHS NOW REPLACED BY TWO LINE WIDTHS.)		▬▬▬ THICK ——— THIN	Y14.2M−1979
LETTERING (TWO APPROVED STYLES. HEIGHT OF LETTERING DEPENDENT ON DRAWING SIZE.)			Y14.2M−1979
MILLIMETER DIMENSIONING PRACTICES			CLAUSE 1.6.1
METRIC LIMITS AND FITS			B 4.2−1978
DIAMETER SYMBOL (NOW PRECEDES THE DIAMETER VALUE. THE SYMBOL REPLACES THE ABBREVIATION DIA)		Ø	CLAUSE 1.8.1
RADIUS SYMBOL (NOW PRECEDES THE RADIUS VALUE)		R	CLAUSE 1.8.2
REFERENCE DIMENSION		(8.6)	CLAUSE 1.7.6
SURFACE TEXTURE SYMBOL		√ √ √ √⌐	Y14.36−1978
SPECIFYING REPETITIVE FEATURES		X	CLAUSE 1.9.5
COUNTERBORE OR SPOTFACE		⌴	CLAUSE 3.3.10
COUNTERSINK		⌵	CLAUSE 3.3.11
DEPTH		↧	CLAUSE 3.3.12
CONICAL TAPER		▷ 0.2 : 1	CLAUSE 2.13
FLAT TAPER		◁ 0.15 : 1	CLAUSE 2.14
SYMMETRICAL OUTLINES			CLAUSE 1.8.8
ALL AROUND		○	CLAUSE 3.4.2.3
DIMENSIONING CHORDS, ANGLES, AND ARCS		50 50͡ 60°	CLAUSE 1.8.3
NOT TO SCALE DIMENSION		◄— 120 —►	CLAUSE 1.7.9
FEATURE CONTROL FRAME (FORMERLY CALLED FEATURE CONTROL SYMBOL. ORDER OF SEQUENCE CHANGED)		⊕ | Ø 0.1 | A	CLAUSE 3.4.2
DATUM TARGET SYMBOL		Ø10 / A3	CLAUSE 3.3.3 AND 4.5.1
GEOMETRIC CHARACTERISTIC SYMBOLS	STRAIGHTNESS	——	CLAUSE 6.4.1
	FLATNESS	▱	CLAUSE 6.4.2
	PARALLELISM	//	CLAUSE 6.6.3
	SYMMETRY	⌖	CLAUSE 5.12
	CIRCULAR RUNOUT	↗	CLAUSE 6.7.2.1
	TOTAL RUNOUT	↗↗	CLAUSE 6.7.2.2

PART 1 Basic Drawing Design

CHAPTER 1

The Language of Industry

UNIT 1-1
The Language of Industry

Since earliest times people have used drawings to communicate and record ideas so that they would not be forgotten. The earliest forms of writing, such as the Egyptian hieroglyphics, were picture forms.

The word *graphic* means dealing with the expression of ideas by lines or marks impressed on a surface. A drawing is a graphic representation of a real thing. Drafting, therefore, is a graphic language, because it uses pictures to communicate thoughts and ideas. Because these pictures are understood by people of different nations, drafting is referred to as a "universal language."

Drawing has developed along two distinct lines, with each form having a different purpose. On the one hand, artistic drawing is concerned mainly with the expression of real or imagined ideas of a cultural nature. Technical drawing, on the other hand, is concerned with the expression of technical ideas or ideas of a practical nature, and it is the method used in all branches of technical industry.

Even highly developed word languages are inadequate for describing the size, shape, and relationship of physical objects. For every manufac-

tured object there are drawings that describe its physical shape completely and accurately, communicating engineering concepts to manufacturing. For this reason, drafting is referred to as the "language of industry."

Drafters translate the ideas, rough sketches, specifications, and calculations of engineers, architects, and designers into working plans which are used in making a product. See Figs. 1-1-1 through 1-1-7. Drafters may calculate the strength, reliability, and cost of materials. In their drawings and specifications, they describe exactly what materials workers are to use on a particular job. To prepare their drawings, drafters use instruments such as compasses, dividers, protractors, templates, and triangles, as well as drafting machines that combine the functions of several devices. They also may use engineering handbooks, tables, calculators, and computers to assist in solving technical problems.

Drafters are often classified according to their type of work or their level of responsibility. Senior drafters (designers) take the preliminary information provided by engineers and architects to prepare design "layouts" (drawings made to scale of the object to be built). Detailers (junior drafters) make drawings of each part shown on the layout, giving dimensions, material, and any other information necessary to make the detailed drawing clear and complete. Checkers carefully

examine drawings for errors in computing or recording dimensions and specifications.

Drafters also may specialize in a particular field of work, such as mechanical, electrical, electronic, aeronautical, structural, or architectural drafting.

DRAWING STANDARDS

Throughout the long history of drafting, many drawing conventions, terms, abbreviations, and practices have come into common use. It is essential that different drafters use the same practices if drafting is to serve as a reliable means of communicating technical theories and ideas.

In the interest of efficient communication, the American National Standards Institute (ANSI) has adopted a set of drafting standards which are recommended for drawing practice in all fields of engineering and are used and explained throughout this text. These standards apply primarily to end product drawings, which usually consist of detail or part drawings and assembly or subassembly drawings, and are not intended to fully cover other supplementary drawings such as checklists, parts lists, schematic diagrams, electrical wiring diagrams, flowcharts, installation drawings, process drawings, architectural drafting, and pictorial drawing.

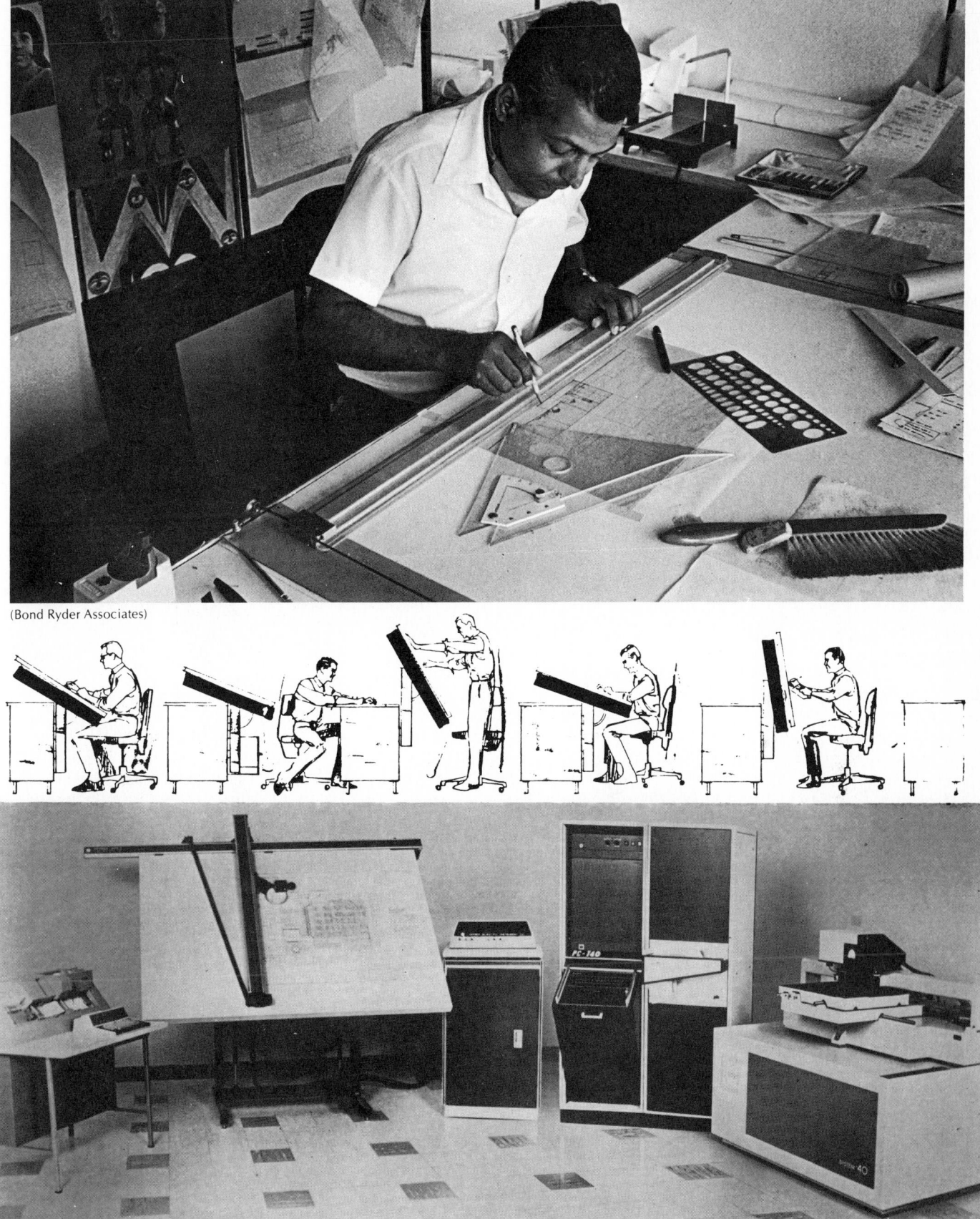

(Bond Ryder Associates)

Fig. 1-1-1 Drafting—today and tomorrow. (Auto-Trol Corporation.)

ROTO HAMMER

Fig. 1-1-2 **Pictorial drawings.** (Skil Corp.)

OIL TANK VENT

OIL TANK FILL LINE

FUEL OIL TANK

OIL TO FUEL OIL PUMPS

PRESSURE RELIEF RETURN TO OIL TANK

PRESSURE RELIEF VALVE

SUCTION TWIN OIL STRAINER

BYPASS

MOTOR DRIVEN FUEL OIL PUMP

PRESSURE RELIEF VALVE

AUXILIARY STANDBY FUEL OIL PUMP

Fig. 1-1-4 **Pipe drawings.** (Jenkins Bros., Ltd.)

Fig. 1-1-3 **Structural drawings.** (American Institute of Steel Construction.)

Fig. 1-1-5 **Machine drawings.**

Fig. 1-1-6　Architectural drawings.

PLAN

AREA: 1350 sq ft

(EXCLUDING CARPORT AND OUTSIDE STORAGE)

Fig. 1-1-7　Electrical drawings.

substantially to the improved quality of photographically reproduced engineering drawings.

CHANGING TIMES[1]

Fifty years have brought great changes to the drafting room. Its physical appearance, furnishings, even its drafters and engineers have moved quickly from their battered domain of old into the Space Age.

These changes were brought about largely by the recognition of many factors that affect the performances of working people. Because designing and drafting are specialized technical fields today that require a high level of precision, personnel efficiency in these areas has been closely linked to the working atmosphere.

A constant reappraisal of this atmosphere should be a prime responsibility of all chief engineers and chief drafters. With an eye to improving working conditions, thereby increasing efficiency and bettering performance, they should reevaluate periodically the tables, boards, seating arrangements, drafting machines and tools, lighting, reference materials, and file units assigned to their department.

Drafting room technology has progressed at the same rapid pace as the economy of our country. Many changes have taken place in the modern drafting room as compared to a typical drafting room scene before the turn of the century, shown in Fig. 1-1-8. Not only are there far more tools, but they are of much higher quality. From automated drafting machines to computer-aided drafting systems and from combination reference tables with adjustable drawing boards to drawing media that contain all the desired qualities for reproduction. Noteworthy progress has been made and continues to be made as our expanding technology takes giant steps forward in this modern age.

PLACES OF EMPLOYMENT[2]

There are over 400 000 people working in drafting positions in the United States. Approximately 4 percent are women. About 9 out of 10 drafters are employed in private industry. Manufacturing industries that employ large

The information and illustrations shown have been revised to reflect current industrial practices in the preparation and handling of engineering documents. The increased use of reduced-size copies of engineering drawings made from microfilm and the reading of microfilm require the proper preparation of the original engineering document. All future drawings should be prepared for eventual photographic reduction or reproduction. The observance of the drafting practices described in this text will contribute

Fig. 1-1-8 The drafting office at the turn of the century. (Bettman Archive, Inc.)

numbers are those making machinery, electrical equipment, transportation equipment, and fabricated metal products. Nonmanufacturing industries employing large numbers are engineering and architectural consulting firms, construction companies, and public utilities.

Over 25 000 drafters work for the government; the majority work for the armed services. Drafters employed by state and local governments work chiefly for highway and public works departments. Several thousand drafters are employed by colleges and universities and by nonprofit organizations.

TRAINING, QUALIFICATIONS, ADVANCEMENT

Young people interested in becoming drafters can acquire the necessary training from a number of sources, including technical institutes, junior and community colleges, extension divisions of universities, vocational and technical high schools, and correspondence schools. Others may qualify for drafting positions through on-the-job training programs combined with part-time schooling or through 3- or 4-year apprenticeship programs.

The prospective drafter's training, whether obtained in high school or post-high school drafting programs, should include courses in mathematics and physical sciences, as well as in mechanical drawing and drafting.

Studying shop practices and learning some shop skills also are helpful, since many higher-level drafting jobs require knowledge of manufacturing or construction methods. Many technical schools offer courses in structural design, strength of materials, and physical metallurgy.

Young people having only high school drafting training usually start out as tracers, or detailers. Those having some formal post-high school technical training can often qualify as junior drafters. As drafters gain skill and experience, they may advance to higher-level positions as checkers, detailers, senior drafters, designers, or supervisors of other drafters. See Fig. 1-1-9. Drafters who take courses in engineering and mathematics are sometimes able to transfer to engineering positions.

Qualifications for success as a drafter may include the ability both to visualize objects in three dimensions and to do freehand drawing. Although artistic ability is not generally required, it may be helpful in some specialized fields.

Drafting work also requires good eyesight (corrected or uncorrected), eye-hand coordination, and manual dexterity.

EMPLOYMENT OUTLOOK

Employment opportunities for drafters are expected to be favorable in the future. Prospects will be best for those having post-high school drafting training as many industries now regard the 2-year post-high school program as a prerequisite for their drafters. Well-qualified high school graduates who have had only high school drafting, however, also will be in demand.

Employment of drafters is expected to rise rapidly as a result of the increasingly complex design problems of modern products and processes. In addition, as engineering and scientific occupations continue to grow, more drafters will be needed as support personnel. On the other hand, photo-reproduction of drawings and expanding use of electronic drafting equipment and computers are eliminating some routine tasks done by drafters. This development will probably reduce the need for some less skilled drafters.

References
1. Charles Bruning Co.
2. *Occupational Outlook Handbook.*

Fig. 1-1-9 Positions within the drafting office.

CHAPTER 2

Drafting Skills and Drawing Office Practices

UNIT 2-1
The Drafting Office

The drafting office is the starting point for all engineering work. Its product, the *engineering drawing*, is the main method of communication between all people concerned with the design and manufacture of parts. Therefore the drafting office must provide accommodations and equipment for the drafters, from designer and checker to detailer or tracer; for the personnel who make copies of the drawings and file the originals; and for the secretarial staff who assist in the preparation of the drawings (Fig. 2-1-1).

Most engineering departments still rely on manual drafting needs. In the majority of cases, this is all that is necessary. Equipment for manual drafting is varied and is steadily being improved. Where a high volume of finished or repetitive work is not necessary, this equipment does the job adequately and inexpensively, and most designers are accustomed to working with it.

A growing number of companies have turned to automated drafting. The reason is not simply to speed the drafting process. Automated drafting can serve as a full partner in the design process, enabling the designer to do jobs that are simply not possible or feasible with manual equipment.

Computer-aided-drafting, normally referred to as CAD, and drawings for numerical control are covered in detail in Chaps. 32 and 16, respectively.

UNIT 2-2
Manual Drafting Equipment and Supplies

Over the years, the designer's chair and drafting table have evolved into a drafting station which provides a comfortable, integrated work area. Yet much of the equipment and supplies employed years ago are still in use today, although they have been vastly improved.

DRAFTING FURNITURE

Special tables and desks are manufactured for use in single-station or multistation design offices. Typical are desks with attached drafting boards (Fig. 2-2-1). The boards may be used by the occupant of the desk to which it is attached, in which case it may swing out of the way when not in use, or may be reversed for use by the person in the adjoining station.

In addition to such special work stations, a variety of individual desks, chairs, tracing tables, filing cabinets, and special storage devices for drawings are available (Fig. 2-2-2).

The simplest manually adjustable tables typically consist of a hinged surface riding on a vertical rod secured by a setscrew. The setscrew is loosened,

Fig. 2-1-1 Drafting Office. (Vemco Corp.)

7

Fig. 2-2-1   Drafting tables are available in a variety of sizes and styles. (Bruning Division, Addressograph Multigraph Corp.)

Drafting Machines

In the well-equipped engineering department, where the designer is expected to do accurate drafting, the T square has been replaced largely by the drafting machine. This device, which combines the functions of T square, triangles, scale, and protractor, is estimated to save up to 50 percent of the user's time. All positioning is done with one hand, while the other hand is free to draw.

Drafting machines may be attached to any drafting board or table. Two types are currently available. In the track type, a vertical beam carrying the drafting instruments rides along a horizontal beam fastened to the top of the table. In the arm, or elbow type (Fig. 2-2-5), two arms pivot from the top of the machine and are relative to each other.

Fig. 2-2-2   Drafting work stations. (Teledyne Post.)

(Bruning.)

the top is set at the desired angle, and the setscrew is retightened.

DRAFTING EQUIPMENT

See Fig. 2-2-3 for a variety of drafting equipment.

Drawing Boards

The drawing sheet is attached directly to the surface of a drafting table or a portable drawing board (Fig. 2-2-4). Drafting boards are used in schools and for home use and generally have a smaller work surface than what is found on drafting tables. They are designed to stay flat and have straight guiding edges.

Fig. 2-2-3   Drafting equipment. (Staedtler-Mars.)

Fig. 2-2-4 **Drawing boards.** (Teledyne Post.)

Fig. 2-2-6 **Drafting table with parallel slide.**
(Addressograph Multigraph Corp.)

(A) WOOD OR PLASTIC HEAD WITH
PLASTIC EDGE BLADE

(B) STEEL BLADE

(C) ADJUSTABLE HEAD WITH PLASTIC
EDGE BLADE

Fig. 2-2-7 **T squares are available in various styles and materials.** (AM Bruning International.)

Fig. 2-2-5 **Arm type drafting machine.** (Keuffel and Esser Co.)

Track-type drafting machines are especially suitable for long-line work and large drawings.

Parallel Slide

The parallel slide is used in drawing horizontal lines and for supporting triangles, when vertical and sloping lines are being drawn. (See Fig. 2-2-6.) It is fastened on each end to cords, which pass over pulleys. This arrangement permits movement up and down the board while maintaining the parallel slide in a horizontal position.

T Squares

The T square (Fig. 2-2-7) performs the same function as the parallel slide. T squares are made of various materials, the more popular being plastic-edged wood blades with heads made from wood or plastic.

To check the accuracy of a T square draw a sharp line along the drawing edge of the T square on a sheet of paper. Turn the T square upside down and using the same drawing edge check the line for error. If the drawing edge and the pencil line do not match, the T square is not accurate.

The head of the T square is placed on the left side of a drawing board for use by right-handed people and on the right side of the drawing board for use by left-handed people.

Triangles

Triangles are used together with the parallel straightedge or T square when you are drawing vertical and sloping lines (Fig. 2-2-8). The triangles most commonly used are the 30/60° and the 45° triangles. Singly or in combination, these triangles can be used to form angles in multiples of 15°. For other

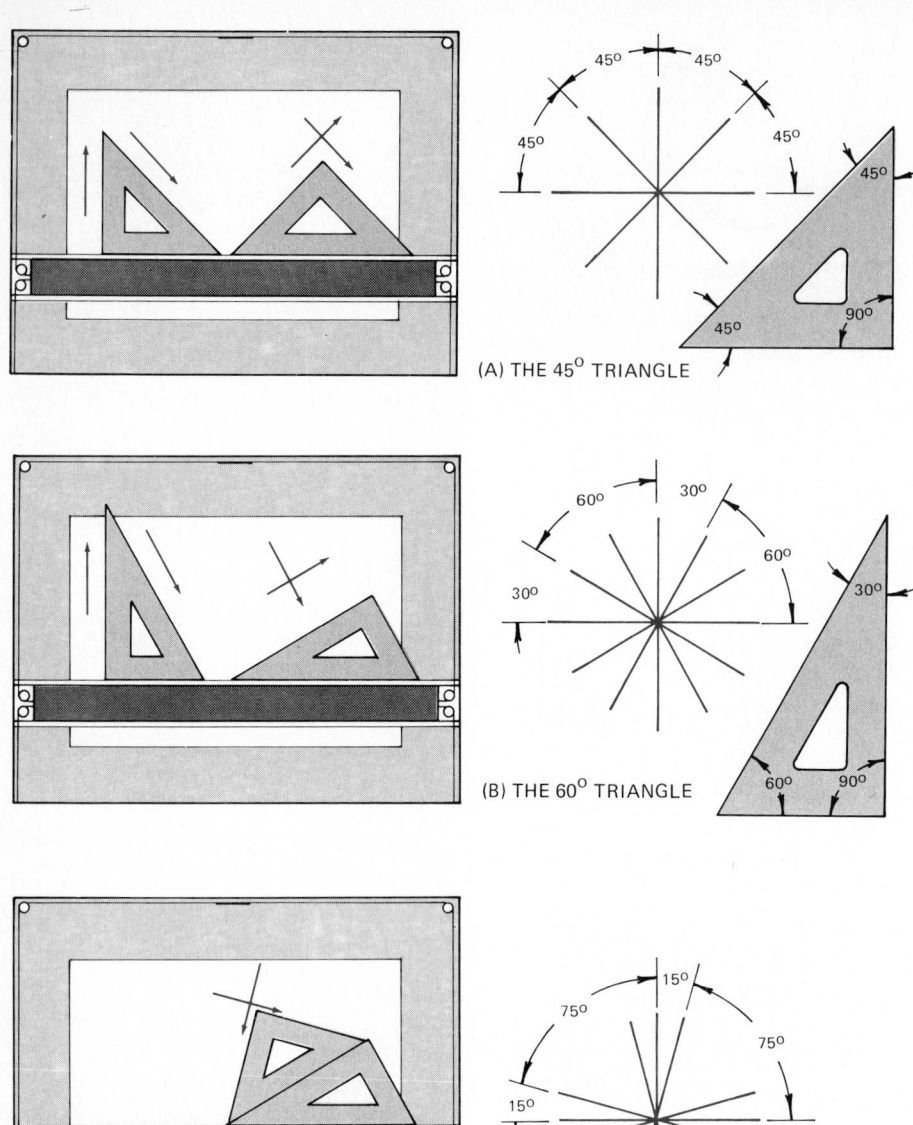

(A) THE 45° TRIANGLE

(B) THE 60° TRIANGLE

(C) THE TRIANGLES IN COMBINATION

Fig. 2-2-8   The triangles.

Fig. 2-2-9   A protractor is used to lay out, or measure, angles.

angles, the protractor (Fig. 2-2-9) is used. All angles can be drawn with the adjustable triangle (Fig. 2-2-10); this instrument replaces the two common triangles and the protractor.

Fig. 2-2-10   Adjustable triangle. (Charles Bruning Co.)

Scales

Shown in Fig. 2-2-11 are the common shapes of scales used by drafters to make measurements on their drawings. Scales are used only for measuring and are not to be used as a straightedge when drawing lines. It is important that drafters draw accurately to scale. The scale to which the drawing is made must be given in the title block or strip.

REGULAR   RELIEVED FACET
TRIANGULAR SCALES

DOUBLE BEVEL   OPPOSITE BEVEL   FLAT BEVEL

FLAT SCALES

Fig. 2-2-11   End view shapes of scales.

When objects are drawn at their actual size, the drawing is called *full scale* or *scale 1:1*. Many objects, however, such as buildings, ships, or airplanes, are too large to be drawn full scale, so they must be drawn to a reduced scale. An example would be the drawing of a house to a scale of ¼ in. = 1 ft.

Frequently, objects such as small watch parts are drawn larger than their actual size so that their shape can be seen clearly. Such a drawing has been drawn to an enlarged scale. The minute hand of a wristwatch, for example, could be drawn to a scale of 5:1.

Many mechanical parts are drawn to half scale, 1:2, and quarter scale, 1:4, or nearest metric scale, 1:5. Notice that the scale is expressed as an equation. The left side of the equation represents a unit of the size drawn; on the right side, a unit of the actual drawing equals five units of measurement of the actual object.

Scales are made with a variety of combined scales marked on their surfaces. This combination of scales spares the drafter the necessity of calculating the sizes to be drawn when working to a scale other than full size.

Metric Scales

The linear unit of measurement for mechanical drawings is the millimeter. Scale multipliers and divisors of 2 and 5 are recommended, which give the scales shown in Fig. 2-2-12.

The numbers shown indicate the difference in size between the drawing and the actual part. For example, the ratio 10:1 shown on the drawing means that the drawing is 10 times the actual size of the part, whereas a ratio of 1:5 on the drawing means the object is 5 times as large as it is shown on the drawing.

The units of measurement for architectural drawings are the meter and millimeter. The same scale multipliers and divisors as used for mechanical drawings are used for architectural drawings.

U.S. Customary Scales

Inch Scales There are three types of scales which show various values that are equal to 1 inch (in.) (Fig. 2-2-13). They are the decimal inch scale, the fractional inch scale, and the scale which has divisions of 10, 20, 30, 40, 50, 60, and 80 parts to the inch. The last scale is known as the civil engineer's scale. It is used for making maps and charts. The divisions or parts of an inch can be used to represent feet, yards, rods, or miles. This scale is also useful in mechanical drawing when the drafter is dealing with decimal dimensions.

On fractional inch scales, multipliers or divisors of 2, 4, 8, and 16 are used, offering such scales as full size, half size, quarter size, etc.

1:1 SCALE (1 mm DIVISIONS)

1:2 SCALE (2 mm DIVISIONS)

1:5 SCALE (5 mm DIVISIONS)

1:50 SCALE (50 mm DIVISIONS)

ENLARGED	SIZE AS	REDUCED
1000 : 1	1 : 1	1 : 2
500 : 1		1 : 5
200 : 1		1 : 10
100 : 1		1 : 20
50 : 1		1 : 50
20 : 1		1 : 100
10 : 1		1 : 200
5 : 1		1 : 500
2 : 1		1 : 1000

Fig. 2-2-12   Metric scales.

DECIMAL INCH SCALE (FULL SIZE)

DECIMAL INCH SCALE (HALF SIZE)

FRACTIONAL INCH SCALE (FULL SIZE)

FRACTIONAL INCH SCALE (HALF SIZE)

CIVIL ENGINEER SCALE (10 DIVISIONS)

CIVIL ENGINEER SCALE (30 DIVISIONS)

Fig. 2-2-13   Inch scales.

Foot Scales These scales are used mostly in architectural work. See Fig. 2-2-14. They differ from the inch scales in that each major division represents a foot, not an inch, and the end units are subdivided into inches or parts of an inch. The more common scales are ⅛ in. = 1 ft, ¼ in. = 1 ft, 1 in. = 1 ft, and 3 in. = 1 ft. The most commonly used inch and foot scales are shown in Fig. 2-2-15.

1″ = 1′– 0″ SCALE

1/4″ = 1′– 0″ SCALE

Fig. 2-2-14 Foot scales.

DECIMALLY DIMENSIONED DRAWINGS	FRACTIONALLY DIMENSIONED DRAWINGS	DIMENSIONED IN FEET AND INCHES	
		SCALE	EQUIVALENT RATIO
10 : 1	8 : 1	6 IN.= 1 FT	1 : 2
5 : 1	4 : 1	3 IN.= 1 FT	1 : 4
2 : 1	2 : 1	1½ IN.= 1 FT	1 : 8
1 : 1	1 : 1	1 IN.= 1 FT	1 : 12
1 : 2	1 : 2	¾ IN.= 1 FT	1 : 16
1 : 5	1 : 4	½ IN.= 1 FT	1 : 24
1 : 10	1 : 8	⅜ IN.= 1 FT	1 : 32
1 : 20	1 : 16	¼ IN.= 1 FT	1 : 48
ETC.	ETC.	3/16 IN.= 1 FT	1 : 64
		⅛ IN.= 1 FT	1 : 96
		1/16 IN.= 1 FT	1:192

Fig. 2-2-15 Commonly used foot and inch scales.

Compasses

The compass is used for drawing circles and arcs. Several basic types and sizes are available (Fig. 2-2-16).

- *Friction head compass*, standard in most drafting sets.
- *Bow compass*, which operates on the jackscrew or ratchet principle by turning a large knurled nut.
- *Drop bow compass*, mostly used for drawing small circles. The center rod contains the needle point and remains stationary while the pencil or pen leg revolves around it.
- *Beam compass*, a bar with an adjustable needle and pencil-and-pen attachment for drawing large arcs or circles.
- *Circuit scribing instrument*, a modified drop bow compass, used to cut terminal pads and prepare printed-circuit layouts on scribe coat film.

The bow compass is adjusted by turning a screw whose knurled head is located either in the center or to one side. The bow compass can be used and adjusted with one hand as shown in Fig. 2-2-17. The proper technique is:

1. Adjust the compass to the correct radius.
2. Hold the compass between the thumb and finger.

Fig. 2-2-16 Compasses. (Keuffel & Esser Co.)

(A)   SETTING TO THE RADIUS

(B)   START

(C)

Fig. 2-2-17 Adjusting the radius for the bow pencil compass.

STATIONARY ROD OR PIN

TUBE CARRIES PEN OR PENCIL AND REVOLVES AROUND ROD

SPRING

ADJUSTING SCREW

DROP TUBE BODY AND REVOLVE TO DRAW CIRCLE

Fig. 2-2-18 The drop-spring bow compass is used for drawing very small circles, especially where there are many to be drawn. (Keuffel & Esser Co.)

3. With greater pressure on the leg with the needle located on the intersection of the center lines rotate the compass in a clockwise direction. The compass should be slightly tipped in the direction of motion.

The drop-spring bow compass shown in Fig. 2-2-18 is used for drawing very small circles.

Dividers

Lines are divided and distances transferred (moved from one place to another) with dividers. The basic types of dividers are shown in Fig. 2-2-19.

Dividers have a steel pin insert in each leg and come in a variety of sizes and designs, similar to the compasses. A compass can be used as a divider by replacing its lead point with a steel pin.

Drawing Instrument Sets

Many drafters have a complete drawing set, which usually includes several compasses and dividers with extension attachments for making inked drawings (Fig. 2-2-20).

FRICTION   BOW   PROPORTIONAL

(A) TYPES OF DIVIDERS

TO ADJUST

SECOND CENTER

FIRST CENTER   START HERE

m n

(B) DIVIDERS ARE USED TO DIVIDE AND TO TRANSFER DISTANCES

Fig. 2-2-19 Dividers. (Keuffel & Esser Co.)

RULING PEN

BOX WITH LEADS

BOX WITH REPAIR PARTS
AND NEEDLE POINTS

FRICTION DIVIDERS

PEN FOR COMPASS
SCREWDRIVER

PEN HANDLE

FRICTION COMPASS

LENGTHENING BAR
FOR COMPASS

RULING PEN

BOW PENCIL

CENTER TACK

BOW PEN

BOW DIVIDERS

Fig. 2-2-20   A three-bow set of drawing instruments. (Keuffel & Esser Co.)

Drafting Leads and Pencils

Leads Because of the drawing media used and the type of reproduction required, pencil manufacturers have marketed three types of lead for the preparation of engineering drawings.
Graphite Lead This is the conventional type of lead which has been used for years. It is made from graphite, clay, and resin. It is available in a variety of grades or hardnesses—9H, 8H, 7H, and 6H (hard); 5H and 4H (medium hard); 3H and 2H (medium); H and F (medium soft); and HB, B, 2B, 3B, 4B, 5B, and 6B (very soft), the latter not being recommended for drafting. The selection of the proper grade of lead is important. A hard lead might penetrate the drawing while a soft lead will smear. The next two types of drafting leads were developed as a result of the introduction of film as a drawing medium. A limited number of grades are available in these leads, and they do not correspond to the grades used for graphite lead.
Plastic Lead This type of lead is designed for use on film only. It has good microform reproduction characteristics.
Plastic-Graphite Lead As the name implies, this lead is made of plastic and graphite. There are two basic types: fired and extruded. They are similar in material content to plastic fired lead,

but they are processed differently. They are designed for use on film only, erase well, do not readily smear, and produce a good opaque line which is suitable for microform reproduction. The main drawback with this type of lead is that it does not hold a point well.

Drafting Pencils The leads are held either in the conventional wood-bonded cases known as wooden pencils or in metal or plastic cases known as mechanical pencils. See Figs. 2-2-21 and 2-2-22. With the latter, the lead is ejected to the desired length of projection from the clamping chuck and then pointed in the same manner as the wood-bonded pencil. Recently, dis-

posable mechanical pencils became available. These operate just as any mechanical pencil, but they are discarded after the lead has been used.

Lead Pointers

A fast, convenient means of putting a clean drafter's point on mechanical or wood-cased pencils is not only desirable, but necessary. Mechanical sharpeners (Fig. 2-2-23A) are made with special drafter's cutters that remove the wood as shown. The required point shape is then formed by hand sanding or by a special pointer.

When hand sanding (Fig. 2-2-23B) rub the lead back and forth on a sandpaper block or a fine file, while turning

CONICAL   WEDGE OR CHISEL   BEVEL

Fig. 2-2-21   Pencil point shapes.

WOOD-BONDED CASE

GRADE MARK

(A) WOODEN PENCILS

STANDARD SIZE LEAD

STANDARD LEAD HOLDER

METAL OR PLASTIC CASE

THIN LEAD (REQUIRES
NO SHARPENING)

THIN LEAD HOLDER

(B) MECHANICAL PENCILS

Fig. 2-2-22   Drafting pencils.

Fig. 2-2-23 Lead pointers. (A) A drafter's pencil sharpener cuts the wood, not the lead. (B) Shaping the lead by hand sanding. (C) Shaping the lead with a lead pointer.

(A) (B) (C)

it slowly to form the point. Some drafters prefer to use a chisel or bevel point. Keep the sandpaper block at hand so that you can sharpen the pencil often.

Special pointers are used for shaping the lead, as shown in Fig. 2-2-23C. Such devices may be hand-operated or electrically powered.

Erasers and Cleaners

Erasers A variety of erasers have been designed to do special jobs—remove surface dirt, minimize surface damage on film or vellum, and remove ink or pencil lines. See Fig. 2-2-24.

Fig. 2-2-24 Erasers.

Erasing Machines Hardly any pressure is required when using the electrically powered erasing machine because the high-speed rotation of the shaft actually does the clean-up job rapidly and flawlessly. These machines make erasures with pinpoint accuracy, and no drawing damage can occur because the motor is designed to stall when too much erasing pressure is applied (Fig. 2-2-25).

Fig. 2-2-25 Erasing machine. (Keuffel & Esser Co.)

Cleaners An easy way to clean tracings is to sprinkle them lightly with gum eraser particles while working. Then triangles, scales, etc., stay spotless and clean the surface automatically as they are moved back and forth. The particles contain no grit or abrasive, and will actually improve the ink-taking quality of the drafting surface.

Inking Powders Surfaces of tracing cloths and prints made on reproduction papers require preparation before they are drawn on with ink. This is done by sprinkling an inking powder

on the surface and rubbing it with a felt pad. The powder is then completely removed from the drawing surface.

Erasing Shields These thin pieces of metal or plastic have a variety of openings to permit the erasure of fine detail lines or lettering without disturbing nearby work that is to be left on the drawing. See Fig. 2-2-26. Through the use of this device, erasures can be performed quickly and accurately.

Fig. 2-2-26 Erasing shield. (Charles Bruning Co.)

KEEPING DRAWINGS CLEAN

It goes without saying that the end product, the drawing, must be clean and sharp for reproduction purposes. Dirt from the drafter's hands and instruments, graphite from the pencils, and dirt from the air are the main contributors to dirty drawings. Preventive maintenance goes a long way in keeping drawings clean.

The drafter's hands should always be clean. Dirty, oily, or perspiring hands should never come in contact with drawings.

Drafting instruments, such as triangles, scales, parallel slides, and the blade of the T square should be cleaned at least daily.

The drawing surface of the board must be brushed frequently to remove the dirt particles built up by dirt in the air, from erasing, and drawing lines.

Other recommendations are:

- Keep away from the drawings when sharpening the lead of pencils and compasses. Wipe the lead with a cleaning tissue to remove the loose graphite.
- Always use a brush when cleaning the drawing. Never wipe a drawing with your hands.
- Roll up your shirt sleeves as the cloth and buttons may damage or smudge your drawings.

Brushes

A light brush (Fig. 2-2-27) is used to keep the drawing area clean. By using a brush to remove eraser particles and any accumulated dirt, the drafter avoids smudging the drawing.

Fig. 2-2-27 Drafter's brush. (Charles Bruning Co.)

Lettering Aids

Lettering sets or guides (Fig. 2-2-28) are also used when it is desirable to have more uniform and accurate letters and numerals than can be obtained by the freehand method. Lettering sets contain a number of guide templates that give a variety of letter shapes and sizes, as well as different slope angles.

Dry transfer lettering is a product which offers a wide variety of lettering of good quality and can be applied speedily. It adheres firmly to paper, wood, glass, and metal and is available in different colors. In case of errors, letters can be removed with cellophane tape or a pencil eraser.

Lettering typewriters have been used in drafting offices for some time for the lettering of bills of material and typing on appliques. But now small, movable typewriters can letter any-

(A)

(B)

Fig. 2-2-28 Lettering aids. (A) Mechanical lettering. (Addressograph Multigraph Corp.) **(B) Appliques.** (Letraset.)

where on the drawing without the need for removing the drawing from the board.

Templates

To save time, many drafters now use templates for drawing small circles and arcs. Templates are also available for drawing standard square, hexagonal, triangular, and elliptical shapes and standard electrical and architectural symbols. See Fig. 2-2-29.

Irregular Curves

For drawing curved lines in which, unlike circular arcs, the radius of curvature is not constant, a tool known as an irregular or French curve (Fig. 2-2-30) is used. The patterns for these curves are based on various combinations of ellipses, spirals, and other mathematical curves. The curves are available in a variety of shapes and sizes. Generally, the drafter plots a series of points of intersection along the desired path and then uses the French curve to join these points so that a smooth-flowing curve results.

Fig. 2-2-29 Templates. (Teledyne-Post.)

Fig. 2-2-30 Irregular curves. (Teledyne Post.)

Curved Rules and Splines

Curved rules and splines (Fig. 2-2-31) solve the problem of ruling a smooth curve through a given set of points. They lie flat on the board and are as easy to use as a triangle; yet they can be bent to fit any contour to a 3 in. (75 mm) minimum radius and will hold the position without support. A clear, plastic ruling edge stands away from the board just far enough to prevent ink lines from smearing.

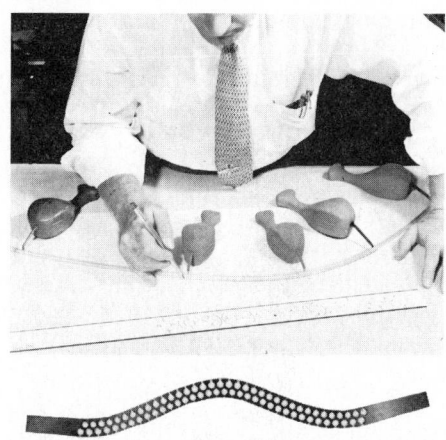

Fig. 2-2-31 Curved rules and splines. (Keuffel & Esser Co.)

Inking Equipment

Although most production drawings are drawn with pencil, in the last few years the number of ink drawings have been on the increase. The use of this type of drawing for technical illustrations and the demand for good, clear drawings for microform reproduction have brought about the introduction of new and improved inking methods and techniques. Typical inking equipment is shown in Fig. 2-2-32.

Inking Pens Two types of pens are used to produce ink lines. The ruling pen with an adjustable blade for drawing different-width lines has been avail-

Fig. 2-2-32 Inking instruments. (A) Blade type. (Charles Bruning Co.) (B) Needle-in-tube-type pen. (J. S. Staedtler, Inc.)

able for many years. It is filled by a dropper cap, squeeze bottle, or cartridge tube. Inking compasses are available with permanent or detachable blades, which interchange with the lead attachment part of the compass. Specially designed pens for drawing curved, multiple broken, or hidden lines are also available.

The second type, called a *technical fountain pen*, a needle-in-tube type pen, is relatively new in comparison with the ruling pen, and has gained wide acceptance with drafters because the line widths are fixed. Also it is suitable for drawing both lines and letters. Different-size needle points are available which produce different-width lines. Several types of technical fountain pens and other needle-in-tube type pens now provide compass attachments so that these may be clamped to, or inserted on, a standard compass leg, thus providing the advantages of the compass while eliminating the inconvenience of the blade pen.

The use of a technical drawing pen with a circular template has, in many instances, replaced the drawing of circles and arcs by means of a compass. For best results, the pen should be perpendicular to the paper.

Calculators

Calculators, such as those shown in Fig. 2-2-33, are used by drafters to

Fig. 2-2-33 Calculators.

make fast mathematical calculations using division, multiplication, and extractions of square roots, and to solve problems involving areas, volumes, masses, strengths of materials, pressures, etc. Because of cost, time, accuracy, and ease of operation, the calculator has replaced the slide rule in the engineering office.

BASIC EQUIPMENT

Figure 2-2-34 shows the basic drafting equipment often found in a student drafting kit. This equipment along with other common items is listed below.

Drawing board
T square, parallel-ruling straightedge (parallel slide), or drafting machine
Drawing sheets (paper or film)
Drafting tape
Drafting pencils
Pencil sharpener
Lead pointer or sandpaper
Eraser
Erasing shield
Triangles, 45° and 30/60° (not required with drafting machine)
Scales
Irregular curve
Drawing instrument set
Black drawing ink
Technical fountain pens
Brush
Protractor
Cleaning powder

Fig. 2-2-34 Student drafting kit. (AM Bruning International.)

Your drafting instructor can tell you exactly what equipment will be needed for your course.

ASSIGNMENTS

See Assignments 1 through 4 for Unit 2-2 on page 38.

UNIT 2-3
Automated Drafting

Most automatic drafting machines (ADM) are digitally controlled. See Fig. 2-3-1. They were developed from, and resembled in their basic mode of operation, numerically controlled machine tools. The last few years have seen many earth-shaking developments in automatic drafting equipment. The price/performance ratio has improved considerably, and high-quality, relatively inexpensive equipment has become available. Manufacturers have introduced modular systems which can be easily upgraded.

Most automatic drafting is done with dot-matrix plotters, such as electrostatic machines.

All machines can be made to automatically generate common shapes, such as squares, circles, ellipses, and other second-degree curves. Many can generate complex shapes, can do high-speed contouring and curve fitting, and can carry out interpolations, extrapolations, and fairing. Some systems can rotate the shape about any axis, scale down or blow up, and produce mirror-image drawings.

Any symbol or shape—circuit mechanism, rivet, screw, etc.—can be stored in memory and called out at any time. It is then drawn at the specified point. When a drawing must be updated, the ADM can make a new drawing, leaving blank those sections which must be revised; changes can be entered manually through the keyboard.

Now systems are available that can take a designer's rough sketch and transform it into finished drawings (with a person operating the system), photo-ready artwork, numerical control (NC) tapes, and various other output forms. These systems are simple to use and require virtually no programming skill.

One of the heaviest and most successful uses of automated drafting has been in electric-circuit design. This is mainly because circuits are highly repetitive. A relatively small number of standard symbols is used over and over; these symbols, stored digitally in the machine's memory, can be called out quickly.

In addition to drawing electrical symbols where the designer indicates, the automatic plotter can route wiring properly and place components in their optimum relationship. This can be extremely useful, especially in designing multilayer printed-circuit boards (Fig. 2-3-2). In fact, as circuits get smaller and smaller, such tasks become nearly impossible except through computerized methods. After

Fig. 2-3-1 Automated drafting equipment. (Gerber Scientific Instrument Co.)

Fig. 2-3-2 Printed circuit produced by automated drafting machine in 25 minutes. (Gerber Scientific Instrument Co.)

the circuit has been designed, the plotter can (assuming that it is sufficiently accurate) draw the artwork masters necessary to produce the printed circuits.

Under certain conditions, automated drafting is justified for certain types of mechanical drawing where a number of standard components such as standard shapes—screws, bolts, letters, dimensions, and other symbols—are involved. Such applications are adaptable to automated drafting, as are cases where the machine's logic can be used to position components, calculate dimensions, or do some other time-consuming task.

Interactive drafting, as shown in Fig. 2-3-3, allows the user to see the final version of the sketch at any point. Typically, the user enters the rough sketch and calls for all or part of it to be

Fig. 2-3-3 Interactive drafting. (Systems Engineering Laboratories.)

displayed. He or she can then make any changes by light pen, digitizer, or teletypewriter. The user can also manipulate the sketch or zoom in on any particular part of it for a closer look. Once satisfied, the user can instruct the drafting machine to make a finished drawing.

Computer-aided drafting (CAD) is covered in detail in Chapter 32.

Reference
1. *Machine Design*, July 1971.

UNIT 2-4
Drawing and Layout Form

STANDARD DRAWING SIZES

Inches Drawing sizes in the inch system are based on dimensions of commercial letterheads, 8.5 × 11 in., and standard rolls of paper or film 36 and 42 in. wide. They can be cut from these standard rolls with a minimum of waste. See Fig. 2-4-1.

Fig. 2-4-1 Standard drawing paper sizes. (ANSI.)

INCH DRAWING SIZES		
DRAWING SIZE	BORDER SIZE	OVERALL PAPER SIZE
A	8.00 X 10.50	8.50 X 11.00
B	10.50 X 16.50	11.00 X 17.00
C	16.25 X 21.25	17.00 X 22.00
D	21.00 X 33.00	22.00 X 34.00
E	33.00 X 43.00	34.00 X 44.00

METRIC DRAWING SIZES (MILLIMETERS)		
DRAWING SIZE	BORDER SIZE	OVERALL PAPER SIZE
A4	190 X 267	210 X 297
A3	277 X 390	297 X 420
A2	400 X 564	420 X 594
A1	574 X 811	594 X 841
A0	821 X 1159	841 X 1189

Fig. 2-4-2 Metric drawing paper.

Metric Metric drawing sizes are based on the A0 size, having an area of 1 square meter (m^2) and a length-to-width ratio of $1:\sqrt{2}$. Each smaller size has an area half of the preceding size, and the length-to-width ratio remains constant. See Fig. 2-4-2.

DRAWING FORMAT

A general format for drawings is shown in Fig. 2-4-3, which illustrates a drawing trimmed to size. It is recommended that preprinted drawing forms be made to the trimmed size and have rounded corners, as shown, to minimize dog-ears and tears.

Zoning System

Drawings larger than B size may be zoned for easy reference, by dividing the space between the trimmed size and the inside border into zones measuring 4.25×5.50 in. These zones are numbered horizontally and lettered vertically, with uppercase letters, from the lower RH (right-hand) corner, as in Fig. 2-4-3, so that any area of the drawing can be identified by a letter and a number, such as B3.

Marginal Marking[1]

In addition to zone identification, the margin may also carry fold marks to enable folding and a graphical scale to facilitate reproduction to a specific size. In the process of microforming, it is necessary to center the drawing within rather close limits in order to meet standards. To facilitate this operation, it has become common practice to put a centering arrow or mark on at least three sides of the drawing. Most practices include the arrows on each of the four sides. If three sides are used, the arrows should be on the two sides and on the bottom. This helps the camera operator to align the drawing properly since the copyboard usually contains cross hairs through the center of the board at right angles. With any three arrows aligned on the cross hairs, centering is automatic. The arrows should be on the center of the border which outlines the information area of the drawing, not at the edge of the sheet on which the drawing is made.

TITLE BLOCKS AND TABLES

Title Block

The title block is located in the lower right-hand corner. The arrangement and size of the title block are optional, but the following information should be included:

1. Drawing number
2. Name of firm or organization
3. Title or description
4. Scale

Provision may also be made within the title block for the date of issue, signatures, approvals, sheet number, drawing size, job, order, or contract number; references to this or other documents; and standard notes such as tolerances or finishes. An example of a typical title block is shown in Fig. 2-4-4. In classrooms, a title strip is often used on A- and B-size drawings, such as shown in Fig. 2-4-5.

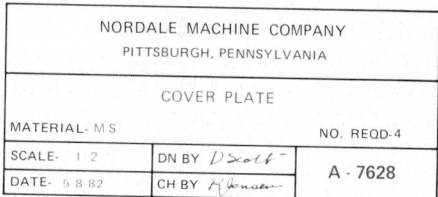

Fig. 2-4-4 Title block.

Material List and Order Table

The whole space above the title block, with the exception of the auxiliary number block, should be reserved for tabulating materials, change of order, and revision; drawing in this space should be avoided. On preprinted forms, the right-hand inner border may be graduated to facilitate ruling, as shown in Fig. 2-4-6.

Change or Revision Table

All drawings should carry a change or revision table, either down the right-hand side or across the bottom of the drawing. In addition to the description of drawing changes, provision may be made for recording a revision symbol zone location, issue number, date, and approval of the change. Typical revision tables are shown in Fig. 2-4-7.

Fig. 2-4-3 Drawing paper format.

Fig. 2-4-5 Title strip.

AMT	DET	STOCK SIZE	MAT.

NORALE MACHINE CO.
PITTSBURGH, PENNSYLVANIA

MODEL _____
PART NAME_____

PART NO._____
_____
OPERATION_____

FOR USE ON_____
METAL_____ DIE CLEARANCE_____
TOLERANCE: ±0.5mm UNLESS OTHERWISE SPECIFIED   METRIC

LAYOUT_____ DRAWN_____
CHECKED_____ APPROVED_____
SCALE_____ DATE_____
FROM B. P._____ DATED_____

SHEETS	SHEET	**NO.**

Fig. 2-4-6 Combined title block, order table, and material list.

Auxiliary Number Blocks

An auxiliary number block, approximately 2 × .25 in. (50 × 10 mm) is placed above the title block so that when prints are folded, the number will appear close to the top RH corner of the print, as in Fig. 2-4-3. This is done to facilitate identification when the folded prints are filed on edge.

REVISIONS		
SYMBOL	DESCRIPTION	DATE & APPROVAL

(A) VERTICAL REVISION TABLE

ZONE OR CHANGE SYMBOL
DATE
APPROVAL

REVISION TABLE	1	12/4/82			
	DESCRIPTION				

(B) HORIZONTAL REVISION TABLE

Fig. 2-4-7 Revision tables.

Auxiliary number blocks are usually placed within the inside border, but they may be placed in the margin outside the border line if space permits.

Reference
1. National Microfilm.

UNIT 2-5
Drawing Media[1]

The families of drafting and tracing materials—*paper and film*—differ sufficiently between and within themselves to provide a wide choice of qualities and characteristics for selection of the perfect material for drawing requirements.

PAPER

Drafting papers come with a wide range of qualities—strength, erasability, permanence, translucency, etc. The distinguishing feature between drawing and tracing paper is translucency.

Opaque papers are used primarily as drawing papers. Because of a change

(C) APPLICATION

in drafting practices, the need for drawing papers today is limited to the two extremes in the scale of quality—the very highest grade, permanent drawing papers with the best possible erasing quality for maps or master drawings, which are later photographed, and the inexpensive school type of papers for the educational market. Since master drawings often must be revised and corrected, the major consideration in high-grade drawing papers is erasability.

The more translucent papers are used as tracing papers. Formerly, they were used almost exclusively for the ink tracing of pencil drawings made on opaque papers. Translucency was the prime requirement.

Today, the usual drafting practice is to develop the master drawing directly on translucent paper from which reproductions can be made, thereby saving the time, expense, and checking involved in the tracing process. This practice means that, in addition to good translucency, the modern high-grade tracing paper must be able to withstand considerable handling. The paper should retain these qualities for a long time to avoid the eventual necessity of redrawing, either manually or photographically.

FILM[2]

The most recently developed drafting medium is film. The advantages of polyester film as a drawing material are many. Raw polyester has natural dimensional stability, great tearing strength, high transparency, age and heat resistance, nonsolubility, and waterproofness. The outstanding virtue of film over any other drafting medium is that film is almost indestructible. Its amazing permanence safeguards the important investment in engineering drawings and records. It is permanently translucent, waterproof, unaffected by aging, superb for pencil drafting, ink work, and typewriting, and has unequaled erasability.

Polyester materials, however, present some problems. The material must be sufficiently dense to avoid reflection from the copyboard in microforming, but translucent enough for backlighting or contact printing.

PREPRINTED GRID SHEETS[3]

This type of drafting medium makes the job of preparing engineering and freehand drawings easier and quicker. Cross-sectional guidelines can be pre-printed directly on the paper or used as a liner beneath the drawing paper to provide an accurate guide for all drawing work. These cross-sectional guides are available in several grid sizes. The squared and pictorial styles (isometric, perspective, and oblique) are the more common preprinted grid papers used by drafters. These grids, when applied directly on the paper, as shown in Fig. 2-5-1, or on film with a special non-reproducible ink, will not appear on the prints when the drawing is reproduced by diazo and photograph methods.

For freehand work, whether rough sketches in the field or in the drafting room, whether these are preliminary or to be used as the finished drawings, grid-line papers are an invaluable aid. The cross-sectional patterns serve as ready-made guides for base lines, dimensioning, and angles. See Fig. 2-5-2.

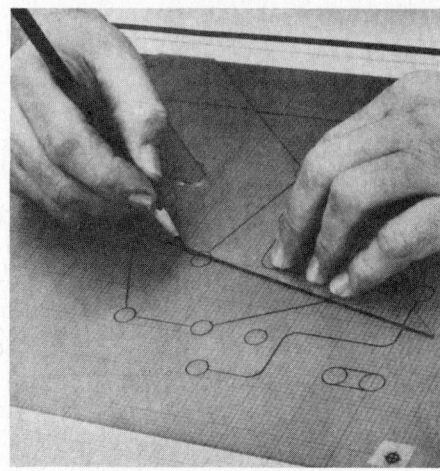

Fig. 2-5-2 Preprinted grid sheet being used as an underlay. (Keuffel & Esser Co.)

References
1. Keuffel and Esser Co.
2. *Machine Design* and National Microfilm
3. Eastman Kodak Company.

(A) DRAWING ON GRID SHEET.

(B) PRINT OF DRAWING SHOWN IN (A).

Fig. 2-5-1 Drawing directly on preprinted grid sheet.

UNIT 2-6
Basic Drafting Skills

LETTERING

Single-Stroke Gothic Lettering

The most important requirements for lettering are legibility, reproducibility, and ease of execution. These are particularly important because of the increased use of microforming which requires optimum clarity and adequate size of all details and lettering. It is recommended that all drawings be made to conform to these requirements and that particular attention be paid to avoid the following common faults:

1. Unnecessarily fine detail
2. Poor spacing of details
3. Carelessly drawn figures and letters
4. Inconsistent delineation
5. Incompleted erasures that leave ghost images
6. Use of differing densities, such as pencil, ink, and typescript on the same drawing

These requirements are met in the recommended single-stroke Gothic

ABCDEFGHIJKLMNOP
QRSTUVWXYZ&
1234567890

INCLINED LETTERS

ABCDEFGHIJKLMNOP
QRSTUVWXYZ&
1234567890

VERTICAL LETTERS

Fig. 2-6-1 Approved lettering for engineering drawings.

ABCDEFGHIJKLMNO
PQRSTUVWXYZ
1234567890

Fig. 2-6-2 Microfont letters. (National Microfilm Assoc.)

GOOD SPACING OF
CHARACTERS AND EVEN
LINE WEIGHT PRODUCE
CONSISTENTLY GOOD
RESULTS ON MICROFILM

UNDESIRABLE CRAMPED LETTERING

POORLY SPACED AND
FORMED, OR CRAMPED
LETTERING MEANS POOR
RESULTS IN MICROFILMING

Fig. 2-6-4 Spacing of lettering. (National Microfilm Assoc.)

characters shown in Fig. 2-6-1, or adaptations thereof, which improve reproduction legibility. One such adaptation by the National Microfilm Association is the vertical Gothic-style Microfont alphabet (Fig. 2-6-2) intended for general usage.

Either inclined or vertical lettering is permissible, but only one style of lettering should be used throughout a drawing. The preferred slope for the inclined characters is 2 in 5, or approximately 68° with the horizontal.

Uppercase letters should be used for all lettering on drawings unless lowercase letters are required to conform with other established standards, equipment nomenclature, or marking.

Lettering for titles, subtitles, drawing numbers, and other uses may be made freehand, by typewriter, or with the aid of mechanical lettering devices such as templates and lettering machines. Regardless of the method used, all characters are to conform, in general, with the recommended Gothic style and must be legible in full- or reduced-size copy by any accepted method of reproduction.

The recommended minimum freehand and mechanical letter heights for various applications are given in Fig. 2-6-3. So that lettering will be uniform and of proper height, light guidelines, properly spaced, are drawn first and then the lettering is drawn between these lines.

Letters in words should be spaced so that the background areas between the letters are approximately equal, and words are to be clearly separated by a space equal to the height of the lettering. See Fig. 2-6-4. The vertical space between lines of lettering should be no more than the height of the lettering and no less than half the height of the lettering.

Notes should be placed horizontally on drawings and separated vertically by spaces at least equal to double the height of the character size used, to maintain the identity of each note.

Decimal points must be uniform, dense, and large enough to be clearly visible on approved types of reduced copy. Decimal points should be placed in line with the bottom of the associated digits and be given adequate space.

Lettering should not be underlined except when special emphasis is required. The underlining should not be less than .06 in. (1.5 mm) below the lettering.

When drawings are being made for microforming, the size of the lettering

USE	INCH		METRIC mm		DRAWING SIZE
	FREEHAND	MECHANICAL	FREEHAND	MECHANICAL	
DRAWING NUMBER IN	0.250	0.240	7	7	UP TO AND INCLUDING 17 x 22 INCHES
TITLE BLOCK	0.312	0.290			LARGER THAN 17 x 22 INCHES
DRAWING TITLE	0.250	0.240	7	7	ALL
SECTION AND TABULATION LETTERS	0.250	0.240	7	7	
ZONE LETTERS AND NUMERALS IN BORDER	0.188	0.175	5	5	
DIMENSION, TOLERANCE, LIMITS, NOTES, SUBTITLES FOR SPECIAL VIEWS, TABLES, REVISIONS, AND ZONE LETTERS FOR THE BODY OF THE DRAWING	0.125	0.120	3.5	3.5	UP TO AND INCLUDING 17 x 22 INCHES
	0.156	0.140	5	5	LARGER THAN 17 x 22 INCHES

THIS IS AN EXAMPLE OF .125 IN. LETTERING

THIS IS AN EXAMPLE OF .188 IN. LETTERING

THIS IS AN EXAMPLE OF .250 IN. LETTERING

Fig. 2-6-3 Recommended lettering heights. (ANSI Y14.2M 1979.)

is an important consideration. A drawing may be reduced to half size when microformed at 30X reduction and blown back at 15X magnification. (Most microform engineering readers and blowback equipment have a magnification of 15X. If a drawing is microformed at 30X reduction, the enlarged blown-back image is 50 percent, at 24X, it is 62 percent of its original size.)

Standards generally do not allow characters smaller than .12 in. (3 mm) for drawings to be reduced 30X, and the trend is toward larger characters. Figure 2-6-5 shows the proportionate size of letters after reduction and enlargement.

The lettering heights, spacing, and proportions in Fig. 2-6-5 normally provide acceptable reproduction or camera reduction and blowback. However, manually, mechanically, optimechanically, or electromechanically applied lettering (typewriter, etc.) with heights, spacing, and proportions less than those recommended are acceptable when the reproducibility requirements of the accepted industry or military reproduction specifications are met.

LINE WORK

The various lines used in drawing form the "alphabet" of the drafting language: like letters of the alphabet, they

Fig. 2-6-6   Application of lines. (ANSI Y14.2M 1975.)

Fig. 2-6-5   Proportionate size of letters after reduction and enlargement. (National Microfilm Assoc.)

are different in appearance. See Figs. 2-6-6 and 2-6-10. The distinctive features of all lines that form a permanent part of the drawing are the differences in their width and construction. Lines must be clearly visible and stand out in sharp contrast to one another. This line contrast is necessary if the drawing is to be clear and easily understood.

The drafter first draws very light construction lines, setting out the main shape of the object in various views. Since these first lines are very light, they can be erased easily should changes or corrections be necessary. When the drafter is satisfied that the layout is accurate, the construction lines are then changed to their proper type, according to the alphabet of lines. Guidelines, used to ensure uniform lettering, are also drawn very lightly.

Line Widths

Two widths of lines, thick and thin, as shown in Fig. 2-6-7, are recommended for use on drawings. Thick lines are .030 to .038 in. (.05 to 0.8 mm) wide, thin lines between .015 and .022 in. (0.3 to 0.5 mm) wide. The actual width of each line is governed by the size and style of the drawing and the smallest size to which it is to be reduced. All lines of the same type should be uniform throughout the drawing. Spacing between parallel lines should be such that there is no fill-in when the copy is reproduced by available photographic methods. Spacing of no less than .12 in. (3 mm) normally meets reproduction requirements.

THICK
WIDTH .032 IN. (0.7mm)

THIN
WIDTH .016 IN. (0.35mm)

Fig. 2-6-7 Line widths.

All lines should be clean-cut, opaque, uniform, and properly spaced for legible reproduction by all commonly used methods, including microforming in accordance with industry and government requirements. There should be a distinct contrast between the two widths of lines.

Types of Lines

The types of lines used on engineering drawings are illustrated in Figs. 2-6-6 and 2-6-10. All lines must be clear and dense to obtain good reproduction. When additions or revisions are made to existing drawings, the line widths and density should match the original work.

Visible Lines The visible lines should be used for representing visible edges or contours of objects. Visible lines should be drawn so that the views they outline clearly stand out on the drawing with a definite contrast between these lines and secondary lines.

Hidden Lines Hidden lines consist of short, evenly spaced dashes and are used to show the hidden features of an object. See Fig. 2-6-8. The lengths of the dashes may vary slightly in relation to the size of the drawing. Hidden lines should always begin and end with a dash, in contrast with the visible or

ENLARGED DETAILS

Fig. 2-6-8 Hidden-line technique.

hidden line from which they start, except when such a dash would form a continuation of a visible line. Dashes should join at corners, and arcs should start with dashes at tangent points. Hidden lines should be omitted when they are not required for the clarity of the drawing.

Although features located behind transparent materials may be visible, they should be treated as concealed features and shown with hidden lines.

Center Lines Center lines consist of alternating long and short dashes (Fig. 2-6-9). They are used to represent the axis of symmetrical parts and features, bolt circles, and paths of motion. The long dashes of the center lines may vary in length, depending upon the size of the drawing. Center lines should start and end with long dashes and should not intersect at the spaces between dashes. Center lines should extend uniformly and distinctly a short distance beyond the object or feature of the drawing unless a longer extension is required for dimensioning or for some other purpose. They should not terminate at other lines of the drawing, nor should they extend through the space between views. Very short center lines may be unbroken if no confusion results with other lines.

CENTER LINE NOT BROKEN WHEN EXTENDED BEYOND OBJECT

USE TWO SHORT DASHES AT POINT OF INTERSECTION

Fig. 2-6-9 Center-line technique.

Extension and Dimension Lines These lines are used when dimensioning a drawing and are explained in detail in Chap. 5.

Leaders Leaders are used to indicate the part of a drawing to which a note refers. See Chap. 5 for further details.

Cutting-Plane Lines Cutting-plane lines are used to show the location of cutting planes for sectional views and are explained in detail in Chap. 6.

Section Lining Section lining is used to show the surface in the section view imagined to be cut along the cutting-plane line. Refer to Chap. 6 for further details.

Symmetry Lines Symmetry lines are center lines used as an axis of symmetry for partial views. The line of symmetry is identified by two thick, short parallel lines drawn at right angles to the center line. Symmetry lines are used when representing partially drawn views and partial sections of symmetrical parts. See Fig. 2-6-8. Symmetrical view visible and hidden lines may extend past the symmetrical line if clarity will be improved.

Viewing-Plane Lines Viewing-plane lines are used to locate the viewing position for removed partial views.

Break Lines Break lines are shown in Fig. 2-6-11. They are used to shorten the view of long uniform sections or when only a partial view is required. Such lines are used on both detail and assembly drawings. The straight thin

TYPE OF LINE	APPLICATION	DESCRIPTION
VISIBLE LINE ——————————— THICK		THE VISIBLE LINE IS USED TO INDICATE ALL VISIBLE EDGES OF AN OBJECT. THEY SHOULD STAND OUT CLEARLY IN CONTRAST TO OTHER LINES SO THAT THE SHAPE OF AN OBJECT IS APPARENT TO THE EYE.
HIDDEN LINE — — — — — — — THIN		THE HIDDEN OBJECT LINE IS USED TO SHOW SURFACES, EDGES, OR CORNERS OF AN OBJECT THAT ARE HIDDEN FROM VIEW
CENTER LINE THIN ALTERNATE LINE AND SHORT DASHES		CENTER LINES ARE USED TO SHOW THE CENTER OF HOLES AND SYMMETRICAL FEATURES.
SYMMETRY LINE CENTER LINE THICK SHORT LINES		SYMMETRY LINES ARE USED WHEN PARTIAL VIEWS OF SYMMETRICAL PARTS ARE DRAWN. IT IS A CENTER LINE WITH TWO THICK SHORT PARALLEL LINES DRAWN AT RIGHT ANGLES TO IT AT BOTH ENDS.
EXTENSION AND DIMENSION LINES THIN DIMENSION LINE EXTENSION LINE		EXTENSION AND DIMENSION LINES ARE USED WHEN DIMENSIONING AN OBJECT.
LEADERS DOT ARROW THIN		LEADERS ARE USED TO INDICATE THE PART OF THE DRAWING TO WHICH A NOTE REFERS. ARROWHEADS TOUCH THE OBJECT LINES WHILE THE DOT RESTS ON A SURFACE.
BREAK LINES THIN LONG BREAK THICK SHORT BREAK		BREAK LINES ARE USED WHEN IT IS DESIRABLE TO SHORTEN THE VIEW OF A LONG PART.

Fig. 2-6-10   Types of line.

TYPE OF LINE	APPLICATION	DESCRIPTION
CUTTING-PLANE LINE THICK OR		THE CUTTING-PLANE LINE IS USED TO DESIGNATE WHERE AN IMAGINARY CUTTING TOOK PLACE.
SECTION LINES THIN LINES		SECTION LINING IS USED TO INDICATE THE SURFACE IN THE SECTION VIEW IMAGINED TO HAVE BEEN CUT ALONG THE CUTTING PLANE LINE.
VIEWING-PLANE LINE THICK OR	OR	THE VIEWING-PLANE LINE IS USED TO INDICATE DIRECTION OF SIGHT WHEN A PARTIAL VIEW IS USED.
PHANTOM LINE THIN		PHANTOM LINES ARE USED TO INDICATE ALTERNATE POSITION OF MOVING PARTS, ADJACENT POSITION OF MOVING PARTS, ADJACENT POSITION OF RELATED PARTS, AND REPETITIVE DETAIL. FOR PHANTOM LINE APPLICATIONS SEE FIGURE 2-6-12.
STITCH LINE THIN OR SMALL DOTS	OR OR	STITCH LINES ARE USED FOR INDICATING A SEWING OR STITCHING PROCESS.
CHAIN LINE THICK		CHAIN LINES ARE USED TO INDICATE THAT A SURFACE OR ZONE IS TO RECEIVE ADDITIONAL TREATMENT OR CONSIDERATIONS.

Fig. 2-6-10 (Continued)

(A) SHORT BREAK — ALL SHAPES
DRAWN FREEHAND — THICK

(B) LONG BREAK — ALL SHAPES
DRAWN FREEHAND — THIN

(C) SOLID SQUARE

(D) SOLID CYLINDER
DRAWN FREEHAND OR WITH INSTRUMENTS
THICK

(E) TUBULAR (SQUARE)

(F) TUBULAR (ROUND)

(G) WOOD
DRAWN FREEHAND — THIN

Fig. 2-6-11 Conventional break lines.

EXISTING COLUMN
NEW GIRDER

(A)

INDICATION OF ADJACENT PARTS

(B)

(C)

(D)

INDICATION OF REPEATED DETAIL
Fig. 2-6-12 Phantom-line application.

line with freehand zigzags is recommended for long breaks, the thick freehand line for short breaks, and the jagged line for wood parts.

The special breaks shown for cylindrical and tubular parts are useful when an end view is not shown; otherwise, the thick break line is adequate.

Phantom Lines Phantom lines consist of long dashes separated by pairs of short dashes. See Fig. 2-6-12. The long dashes may vary in length, depending on the size of the drawing. Phantom lines are used to indicate alternate positions of moving parts, adjacent positions of related parts, and repeated detail. These lines are also used for features such as bosses and lugs (later removed), for delineating machining stock and blanking developments, for piece parts in jigs and fixtures, and for mold lines on drawings or formed metal parts. Phantom lines should start and end with long dashes.

Stitch Lines Two forms of stitch lines are approved for general use as follows:

1. Short thin dashes and spaces of equal lengths.
2. Dots approximately .016 in. (0.35 mm), .12 in. (3 mm) apart.

Stitch lines are used for indicating a sewing or stitching process. See Fig. 2-6-10.

Chain Lines Chain lines, as shown in Figs. 2-6-6 and 2-6-10 consist of thick alternating long and short dashes. This line is used to show that a surface or surface zone is to receive additional treatment or considerations within limits specified on a drawing.

DRAWING LINES

Straight Lines
When using a T square to draw horizontal lines (Fig. 2-6-13 and 2-6-14), hold the head of the T square against the edge of the drawing board and slide the T square either up or down to the desired position. Firmly press down on the blade of the T square to prevent it from moving, then proceed to draw the line. When drawing vertical lines, a triangle, which rests on the top side of the T square, is moved to the desired position and both the blade of the T square and the triangle are held firmly to the drawing board with the hand not holding the pencil.

ROTATE PENCIL
MOTION
ABOUT 15°

Fig. 2-6-13 Drawing pencil lines.

When a parallel slide is used, as in Fig. 2-6-15, it will always be in a horizontal position as the wire and rollers in the slide move both ends of the slide simultaneously and at the same speed.

A general rule to follow when drawing straight lines is this: Lean the pencil in the direction of the line which you are about to draw. A right-handed person would lean the pencil to the right and draw horizontal lines from

(A) DRAWING A HORIZONTAL LINE

(B) DRAWING A VERTICAL LINE

Fig. 2-6-14   Drawing lines with the aid of a T square.

Fig. 2-6-15   Drawing sloped lines.

left to right. The left-handed person would reverse this procedure. When drawing vertical lines, lean the pencil away from yourself, toward the top of the drafting board, and draw lines from bottom to top. Lines sloping from the bottom to the top right are drawn from bottom to top; lines sloping from the bottom to the top left are drawn from top to bottom. This procedure for sloping lines would be reversed for a left-handed person.

When using a conical-shaped lead, rotate the pencil slowly between your thumb and your forefinger when drawing lines. This keeps the lines uniform in width and the pencil sharp. Do not rotate a pencil having a bevel or wedge-shaped lead.

Circles and Arcs

When drawing circles and arcs with a compass, it is recommended that the compass lead be softer and blacker than the pencil lead being used on the

same drawing. For example, if you are drawing with a 2H or 3H pencil, use an H compass lead. This will produce a drawing having similar line work since it is necessary to compensate for the weaker impression left on the drawing medium by the compass lead as compared with the stronger direct pressure of the pencil point. For drawing circles and arcs, see Figs. 2-6-16 and 2-6-18.

It is essential that the compass lead be reasonably sharp at all times in order to ensure proper line width. The compass lead should be sharpened to a bevel point, with the top rounded off as shown in Fig. 2-6-17. The pencil lead is slightly shorter than the needle

point, and the needle point must be readjusted after each sharpening.

Drafters find it much easier and faster to use plastic templates. There are sets which contain all common sizes and shapes of holes that most drafters are ever called upon to draw.

Fig. 2-6-17   Sharpening and setting the compass lead.

Fig. 2-6-16   Drawing circles and arcs.

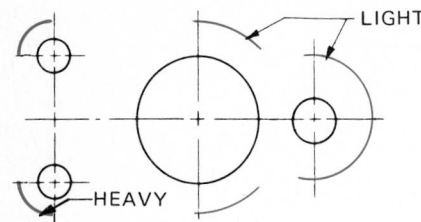

(A) ESTABLISH CENTER LINES AND
RADII MARKS

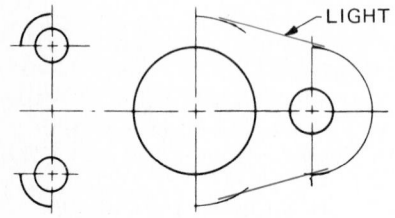

(B) DRAW CIRCLES AND ARCS

(C) DRAW TANGENT LINES

FIRST POSITION

SECOND POSITION

THIRD POSITION

Fig. 2-6-19 Drawing a curved line.

(D) COMPLETE OBJECT LINES

Fig. 2-6-18 Sequence of steps for drawing a view having circles and arcs.

Curved Lines

Curved lines may be drawn with the aid of irregular curves, flexible curves, and elliptical templates (Fig. 2-6-19). After you have established the points through which the curved line passes, draw a light freehand line through these points. Next fit the irregular curve or other instrument by trial against a part of the curved line and draw a portion of the line. Move the curve to match the next portion, and so forth. Each new position should fit enough of the part just drawn (overlap) to ensure continuing a smooth line. It is very important to notice whether the radius of the curved line is increasing or decreasing and to place the irregular curve in the same way. If the curved line is symmetrical about the axis, the position of the axis may be marked on the irregular curve with a pencil for one side and then reversed to match and draw the other side.

Erasing Techniques[1]

Revision or change practice is inherent in the method of making engineering drawings. It is much more economical to introduce changes or additions on an original drawing than to redraw the entire drawing. Consequently, erasing has become a science all its own. Proper erasing is extremely important since some drawings are revised a great number of times. Consequently,

good techniques and materials must be used which permit repeated erasures on the same area. Some recommendations follow.

1. Avoid damaging the surface of the drawing medium by selecting the proper eraser.
2. Lines not thoroughly erased produce ghostlike images on prints, resulting in reduced legibility.
3. A hard, smooth surface, such as a triangle, placed under the lines being removed makes erasing easier.
4. Using an erasing shield protects the adjacent lines and lettering and also eliminates wrinkling. See Fig. 2-6-20.
5. Use an eraser on the back of the drawing medium as well as on the drawing side.
6. Be sure to completely remove erasure debris from the drawing surface.
7. When extensive changes are required, it may be more economical to cut and paste or make an intermediate drawing.
8. When erasing, use no more pressure than necessary.
9. The drawing quality of the drawing medium which may have been damaged by erasing may be improved by sprinkling an inking powder on the surface and rubbing it with a cloth.

Fig. 2-6-20 Using the eraser shield.

In addition to these suggestions, it is necessary to match the density of the surrounding background when erasures are made. Often, the erased area is much cleaner than the rest of the drawing. If the change is made on this clean area, the contrast between line and background is different and that area presents a problem in reproduction.

It is usual practice to "smudge" the erased area so that it looks about the same shade as the surrounding area.

Removing Lines on Film[2]

Erasers Lines on photoreproduction film fall into two classifications: photographic lines and pencil-and-ink lines. All these lines can be removed easily so that the erased area can be used for further drafting. Here are some tips for removing lines.

There are three basic types of erasers: rubber, plastic, and liquid. Rubber and plastic erasers may tend to cause a shine on the drafting surface. This is not necessarily detrimental. Good drafting lines can be drawn easily over areas from which lines have been erased many times. A good general rule to follow is to use a soft, nonabrasive eraser and only enough pressure to remove the line. See Fig. 2-6-21.

Fig. 2-6-21   Erasing. (Keuffel & Esser Co.)

Liquid erasers do not put a shine on the drafting surface and can be used to make many erasures in the same place. When plastic erasers are used, the shiny appearance actually may be transparentizing because of the plasticizers used in their manufacture rather than wear on the drafting surface. The transparentizing effect is not detrimental since it does not reduce the ability of the surface to take a pencil line.

When the drafting surface is affected by excessive erasures, it can be repaired by rolling a regular typewriter eraser across the smooth area or by rubbing a small amount of drafting powder into the area with a finger.

Fig. 2-6-22   Positioning the paper on the board.

Pencil Lines Pencil lines can be removed from all film with a soft, nonabrasive rubber or plastic eraser or with liquid eraser. To keep the drafting surface from becoming too shiny, avoid excessive pressure.

FASTENING PAPER TO THE BOARD

The most common method of holding the drawing paper to the drafting board is with masking tape.

When fastening the paper to the board, line up the bottom or top edge of the paper with the top horizontal edge of the T square, parallel straightedge, or horizontal scale of the drafting machine. See Fig. 2-6-22. When refastening a partially completed drawing, use lines on the drawing rather than the edge of the paper for alignment.

SKETCHING

Freehand sketching is a necessary part of drafting because the drafter in industry frequently sketches ideas and designs prior to making instrumental drawings.

The drafter may also use sketches to explain thoughts and ideas to other people in discussions of mechanical parts and mechanisms. Sketching, therefore, is an important method of communication. Practice in sketching helps the student to develop a good sense of proportion and accuracy of observation.

A fairly soft (HB, F, or H) pencil should be used for preliminary practice. Many types of graph or ordinate paper are available and can be used to advantage when close accuracy to scale or proportion is desirable. Freehand sketching of lines, circles, and arcs is illustrated in Fig. 2-6-23.

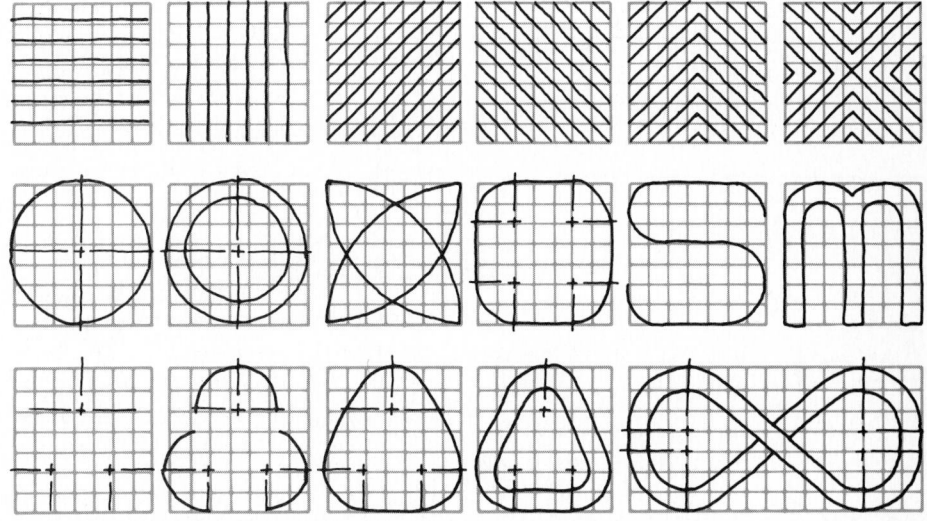

Fig. 2-6-23   Sketching lines, circles, and arcs.

Since the shapes of objects are made up of flat and curved surfaces, the lines forming views of objects will be both straight and curved. Do not attempt to draw long lines with one continuous stroke. First plot points along the desired line path; then connect these points with a series of light strokes.

When you are sketching a view (or views), first lightly sketch the overall size as a rectangular or square shape, estimating its proportions carefully. Then add lines for the details of the shape, and thicken all lines forming the view. See Fig. 2-6-24.

Figure 2-6-25 shows two methods of sketching circles. Figure 2-6-26 illustrates, both pictorially and orthographically, the use of graph paper for the sketching of a machine part. Coordinate and isometric sketching paper are shown in Fig. 2-6-27.

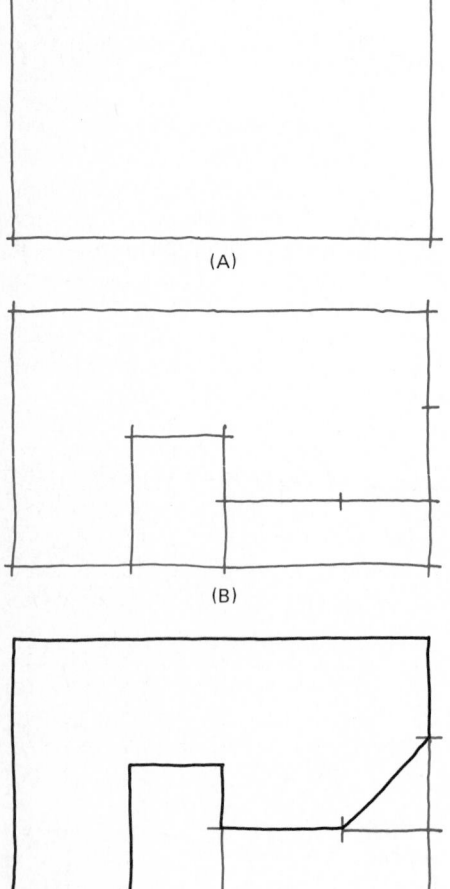

Fig. 2-6-24 Sketching a view having straight lines.

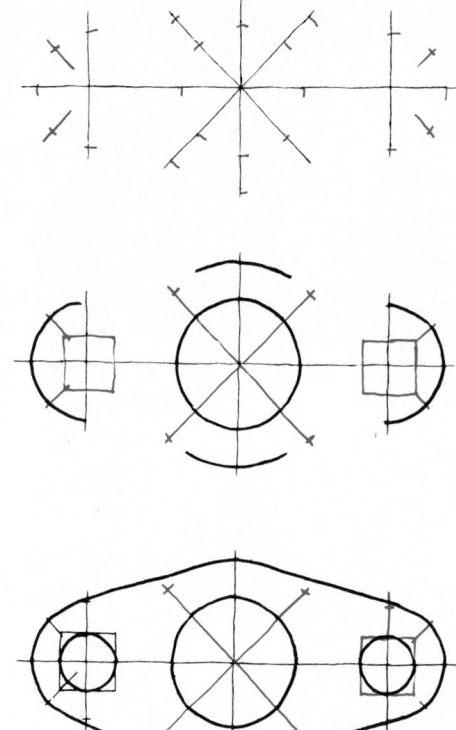

Fig. 2-6-25 Sketching a figure having circles and arcs.

A

B

Fig. 2-6-26 Usual procedure for sketching three views.

(A) COORDINATE SKETCHING PAPER

(B) ISOMETRIC SKETCHING PAPER

Fig. 2-6-27 Sketching paper.

INKING[3]

In inking, technique is an important factor. The pen should be moved as it touches the film or paper at the beginning of a line and as it leaves the film or paper at the end of the line. This will minimize belling, or spreading of ink, at the line ends.

There are two inking pens on the market today, the needle-in-tube type and the blade type. Needle-in-tube type pens should be held nearly vertical and moved with a light touch across the drawing medium. See Fig. 2-6-28. Blade-type ruling pens should be inclined at an angle up to 15° from the vertical in the direction of motion.

There are several inks designed specially for use on drawing media. They can be used with ruling pens, capillary-type pens, lettering pens, or pen nibs. When inking on film, blade-type pens should be filled about half-way, as is customary for drawing on vellum.

Normal handling of drawing media is bound to soil them. Ink lines applied over soiled areas do not adhere well and may be chipped off or flake in

(A) INKING STRAIGHT LINES WITH A BLADE RULING PEN

(B) INKING STRAIGHT LINES WITH A NEEDLE-IN-TUBE PEN

Fig. 2-6-28   Drawing ink lines.

Fig. 2-6-30   Scribing. (Keuffel & Esser Co.)

Fig. 2-6-29   Inking. (J. S. Staedtler, Inc.)

time. It is always good practice to keep the drawing paper or film clean. Soiled areas can be cleaned effectively with a cleaner.

The use of a needle-in-tube pen with a circular template has, in many instances, replaced the drawing of circles by means of the blade-type pen compass, which must be frequently refilled with ink (Fig. 2-6-29). When you are inking circles by template, the only thing to look out for is to keep the template above the drawing media. This is accomplished by inserting another template or other thin material between the template and the drawing surface in order to avoid blotting through capillary underflow of ink.

SCRIBING

Scribing is a drafting technique which in some areas has replaced pen-and-pencil drafting. It has already done so for some types of close-tolerance work. With regular drafting, using pencil or pen, draw on top of the work. With scribing, you incise, or cut, lines into a special surface with scribing tools, making lines that are sharp and clean, never vary in width, and can't be smudged. Scribe lines also produce the sharpest prints, direct from the scribed original. See Fig. 2-6-30.

Scribing is used for a variety of applications—as a drafting medium for the tools and templates from which fabricated parts are produced; in aircraft lofting and automobile layouts; in mapping; and for close-tolerance electronic circuitry and microcircuit layout.

In the past, glass, steel, and aluminum were considered the only stable base materials for scribing because of their extremely high dimensional stability. These materials were adequate but difficult to handle and file. In subsequent years, glass cloth was developed, and it became quite popular as a stable base material. At present, the most useful and stable drafting and reproduction medium is polyester film. It is available in a wide range of surfaces for the precision drafting and reproduction requirements of such industries as mapping and electronics. There are specially prepared scribe surfaces for the critical line work of mapping and undimensioned drafting—coatings which can be cut with a blade and then peeled back to the transparent film base.

Fig. 2-7-1 Drawing reproduction. (Eastman Kodak Co.)

Fig. 2-7-2 A whiteprint machine. (*Bruning.*)

ASSIGNMENTS

See Assignments 5 through 10 for Unit 2-6 on page 40.

References

1. National Microfilming Association and Keuffel & Esser Co.
2. Keuffel & Esser Co.
3. Eastman Kodak Company.

UNIT 2-7
Drawing Reproduction[1]

A revolution in reproduction technologies and methods began in the 1940s and 1950s. It brought with it new equipment and supplies which have made quick copying commonplace. The new technologies make it possible to apply improved systems approaches and new information handling techniques to all types of files ranging from small documents to large engineering drawings. See Fig. 2-7-1.

The pressures on business and government for greater efficiency, space savings, cost reductions, lower investment costs, and equally important factors provided a fertile field for the new reproduction technologies. There is no reason to believe that such pressures will diminish; in fact, as the years go by, it is certain that more and more improvements will occur, newer and better reproduction and information handling equipment and methods will be discovered, and the advantages which they offer will find ever-widening application.

REPRODUCTION EQUIPMENT[2]

Studies of reproduction facilities, existing or proposed, should first consider the nature of the demand for this service, then the processes which best satisfy the demand, and finally the particular machines which employ the processes. Factors to consider at these stages of study include:

- *Input originals*—sizes, paper mass, color, artwork
- *Quality of output copies*—depending on expected use and degree of legibility required
- *Size of copies*—same size, enlarged, reduced
- *Color*—copy paper and ink
- *Registration*—in multiple-color work
- *Volume*—numbers of orders and copies per order
- *Speed*—machine productivity, convenient start-stop, and load-unload
- *Cost*—direct labor, direct material, overhead
- *Future requirements*

Two general kinds of reproduction are recognized: copying and duplicating.

Copying machines are suitable for both line work and pictorials, often on large sheets. They operate at slow speeds, and so the cost per copy is relatively high.

In contrast, duplicating processes are characterized by high speed, high volume, and low cost per copy. They can be used with a wide variety of papers in many sizes, masses, and finishes. Although duplicators—spirit, stencil, and offset—are used in conjunction with copying machines in the engineering and drafting offices, they will not be covered in this text.

Copiers

The principal kinds of copiers are diazo, electrostatic, thermographic, and photographic devices.

Blueprinting was the classical process for copying engineering drawings. It provided white lines on a blue background. Prints were made on a continuous roll-fed machine and were not intended for reproduction. The process was wet, and the quality of prints was poor. In recent years, blueprinting has been replaced, mainly by diazo.

Diazo (Whiteprint)

In this process (Figs. 2-7-2 and 2-7-3), paper or film coated with a photosensitive diazonium salt is exposed to light passing through an original of translucent paper or film. The exposed coated sheet is then developed by an alkaline

DEVELOPED PRINT COMES OUT HERE

RISING VAPOR

AMMONIA

PRINT FACE UP

TRACING

TRACING

BRIGHT LIGHT

PRINT PAPER CHEMICAL SIDE UP

GLASS CYLINDER

ROLLERS MOVE THE TRACING AND PRINT AROUND THE LIGHT, AND MOVE THE PRINT PAST THE RISING AMMONIA VAPOR.

Fig. 2-7-3   The diazo printing process.

agent such as ammonia vapor. Where the light passes through the clear areas of the master, it decomposes the diazonium salt, leaving a clear area on the copy. Where markings on the original block the light, the ammonia and the unexposed coating produce an opaque dye image of the original markings. A positive original makes a positive copy, and a negative original makes a negative copy. Therefore, the polarity is said to be *nonreversing*. The three diazo processes currently used differ mainly in the way the ammonia is introduced to the diazo coat. These are ammonia vapor, moist developing, and pressure developing.

Perhaps the most significant characteristic of the diazo process is that it allows reproduction of fine detail. Diazo is a high-contrast process and thus ideal for document reproduction.

Photographic

Photography is the process of creating latent images on light-sensitive silver halide material by exposure to light. The images are made visible and permanent by developing and fixing techniques. A camera provides for enlargement or reduction of the image size.

Contact printing and projection printing are the two principal methods of making photographic prints—contact for prints the same size as originals and projection for reduced or enlarged prints.

Electrostatic

Electrostatic reproduction, one form of which is xerography, is a dry copying process which uses electrostatic force to deposit dry powder on copy paper.

Electrostatic transfer enables printing on plain paper, offset paper masters, or transparent materials. Some machines copy only at the same size as the originals, while others can reduce or enlarge. See Fig. 2-7-4.

MICROFORM

Microforming of engineering drawings is now an established practice in many drafting offices. See Fig. 2-7-5. This has come about because of the primary savings in lower transportation, labor, and storage costs of microfilm.

Microform prints are B size, regardless of their original size, and are much easier to handle and store. From the drafting room, drawings are taken to a camera, photographed, and stored in rolls or on cards.

Forms of Film

One way to classify microfilm is according to the physical forms, called *microforms*, in which it is used.

Roll Film This is the form of the film after it has been removed from the camera and developed. Microfilm

Fig. 2-7-4   Photoreproduction process. This machine will reproduce, reduce, fold, and sort prints. (Xerox Corp.)

References
1. National Microfilming Association.
2. *Machine Design*, July 1971.

Fig. 2-7-5 Microforming. (3M Co. and American Motors Co.)

comes in four different widths—16, 35, 70, and 105 mm—and is stored in magazines.

Aperture Cards Perhaps the simplest of the flat microforms is finished roll film cut into separate frames, each mounted on a card having a rectangular hole as shown in Fig. 2-7-6. Aperture cards are available in many sizes.

Jackets Jackets are made of thin, clear plastic and have channels into which short strips of microfilm are inserted. They come in a variety of film channel combinations for 16- and/or 35-mm microfilm. Like aperture cards, jackets can be viewed easily.

Microfiche A microfiche is a sheet of clear film containing a number of microimages arranged in rows. A common size, 100 × 150 mm, frequently is arranged to contain 98 images. Microfiches are especially well suited for quantity distribution of standard information, parts, and service lists.

Readers and Viewers

Microform readers magnify film images large enough to be read and project the images onto a translucent or opaque screen. Some readers accommodate only one microform (rolls, jackets, microfiches, or aperture cards), while others can be used with two or more. Scanning-type readers, having a variable-type magnification, are used when frames containing a large drawing are viewed. Generally, only parts of the drawing can be viewed at one time.

Reader-Printers

Two kinds of equipment are used to make enlarged prints from microform: reader-printers and enlarger-printers. The reader-printer, as illustrated in Fig. 2-7-7, is a reader which incorporates a means of making hard copy from the projected image. The enlarger-printer is designed only for copying and does not include the means for reading.

Fig. 2-7-6 Aperture cards.
(Eastman Kodak Co.)

Fig. 2-7-7 Reader-printer.
(Eastman Kodak Co.)

UNIT 2-8
Filing Systems[1]

One of the most common and difficult problems facing an engineering department is how to set up and maintain an efficient engineering filing area. Normal office file methods are not considered satisfactory for engineering drawings. To properly serve its function, an engineering filing area must meet two important criteria: accessibility of information and protection of valuable documentation.

For this kind of system to be effective, drawings must be readily accessible. The degree of accessibility is dependent on whether drawings are considered active, semiactive, or inactive.

FILING ORIGINAL DRAWINGS

Unless a company has developed a full microforming system, the original drawings which the drafter produced must be kept and filed for future use or reference. Unlike the prints, the originals must *not* be folded to avoid crease lines, which would appear in copies. They are filed in either a flat or rolled position. See Fig. 2-8-1.

In determining what type of equipment to use for engineering files, it should be remembered that different types of drawings require different kinds of files. Also, in planning a filing system, keep in mind that filing requirements are always increasing; unlike normal office files that can be purged each year, the more drawings produced, the more need to be stored. Therefore, any filing system must have the flexibility of being easily expanded, and generally in a minimum of space.

Microform Filing Systems

It seems logical that reducing drawings to tiny images on film would make them more difficult to locate. However, this is not the case, for while they are reduced in size, they are made more uniform. This results in improved file arrangements.

Roll Film Roll film can be coded in several ways for visual or automatic retrieval. The more common methods used are flash cards, code lines, sequential numbering image control, and binary code patterns.

Aperture Cards In many respects, aperture cards have the same filing and retrieval capabilities as jackets and microfiches. However, there is one important difference. It is possible to use aperture cards in machine records handling systems. They can be printed, punched, and sorted by machine.

Jackets and Microfiches Both these microforms are basically the same with respect to retrieval. See Fig. 2-8-2. Each has a title header for identifying numbers and titles. Each jacket or microfiche contains a group of images arranged in a logical sequence so that the particular images can be readily found.

FOLDING OF PRINTS

To facilitate handling, mailing, and filing, prints should be folded to letter size, 8.5 × 11 in. (210 × 297 mm), in such a way that the title block and auxiliary number always appear on the front face and the last fold is always at the top. In filing, this prevents other drawings from being pushed into the folds of filed prints.

Recommended methods of folding standard-size prints are illustrated in Fig. 2-8-3.

On preprinted forms, it is recommended that fold marks be included in the margin of the drawings on size B and larger and be identified by number, for example, "fold 1," "fold 2." In zoned prints, the fold lines will coincide with zone boundaries; but they should, nevertheless, be identified.

To avoid loss of clarity by frequent folding, important details should not be placed close to fold areas.

Reference
1. Eastman Kodak Company and "Setting Up and Maintaining an Effective Drafting Filing System," *Reprographics*, March 1975.

Fig. 2-8-1 Filing systems. (A) Vertical filing. (Ulrick Plan File.) **(B) Horizontal filing.** (AM Bruning.)

Fig. 2-8-2 Microfilm jacket. (Eastman Kodak Co.)

Fig. 2-8-3 Folding prints.

ASSIGNMENTS for Chapter 2

Assignments for Unit 2-2, Manual Drafting Equipment and Supplies

1. Using the scales shown in Fig. 2-2-A determine lengths A to K.
2. Inch measurement assignment. With reference to Fig. 2-2-B and using the scale:
   Half size decimal inch scale measure distances A-F
   Half size fractional inch scale measure distances G-M
   Full size decimal inch scale measure distances N-T
   Full size fractional inch scale measure distances U-Z

3. Metric measurement assignment. With reference to Fig. 2-2-B and using the scale:
   1:1 measure distances A-E
   1:2 measure distances F-K
   1:5 measure distances L-P
   1:10 measure distances Q-U
   1:50 measure distances V-Z

4. Foot and inch measurement assignment. With reference to Fig. 2-2-B and using the scale:
   1″ = 1′ − 0″ measure distances A-F
   3″ = 1′ − 0″ measure distances G-M
   ¼″ = 1′ − 0″ measure distances N-T
   ⅜″ = 1′ − 0″ measure distances U-Z

Fig. 2-2-A Test in reading drafting scales.

Fig. 2-2-B   Scale measurement assignment.

Assignments for Unit 2-6, Basic Drafting Skills

5. Lettering assignment. Set up a B- or A3-size sheet similar to that shown in Fig. 2-6-A. Using vertical uppercase Gothic lettering shown in Fig. 2-6-2 complete each line. Each letter and number is to be drawn several times to the three recommended lettering heights shown. Very light guidelines must be drawn first. The bracketed dimensions are millimeters.

6. Lettering assignment. Same as assignment 5 except use inclined lettering.

7. On a B- or A3-size sheet draw the template shown in Fig. 2-6-B. Scale full (1:1). Do not dimension.

8. On an A3- or B-size sheet draw the template shown in Fig. 2-6-C. Scale 1:1. Do not dimension.

9. On a B- or A3-size sheet draw any four of the six inlay designs shown in Fig. 2-6-D. Scale full (1:1). Do not dimension.

10. On an A3- or B-size sheet draw the four structural steel shapes shown in Fig. 2-6-E. The fillets and radii are one-half the material thickness. Scale 1:1. Do not dimension.

Fig. 2-6-A   Lettering assignment.

NOTE – DIMENSIONS NOT ENCLOSED IN BRACKETS ARE IN INCHES. DIMENSIONS ENCLOSED IN BRACKETS ARE IN MILLIMETERS.

Fig. 2-6-B   Template no. 1.

NOTE - DIMENSIONS NOT ENCLOSED IN BRACKETS ARE IN MILLIMETERS. DIMENSIONS ENCLOSED IN BRACKETS ARE IN INCHES.

A Note About Dual Dimensions

The dual dimensions shown in this book, especially in the assignment sections, are neither hard nor soft conversions. Instead, the sizes are those that would be most commonly used in the particular dimensioning units and so are only approximately equal. Dual dimensioning this way avoids awkward amounts and allows instructor and student to be confident that a drawing using either set of dimensions will be no more difficult to work than one dimensioned exclusively in either dimensioning system.

Fig. 2-6-C   Template no. 2.

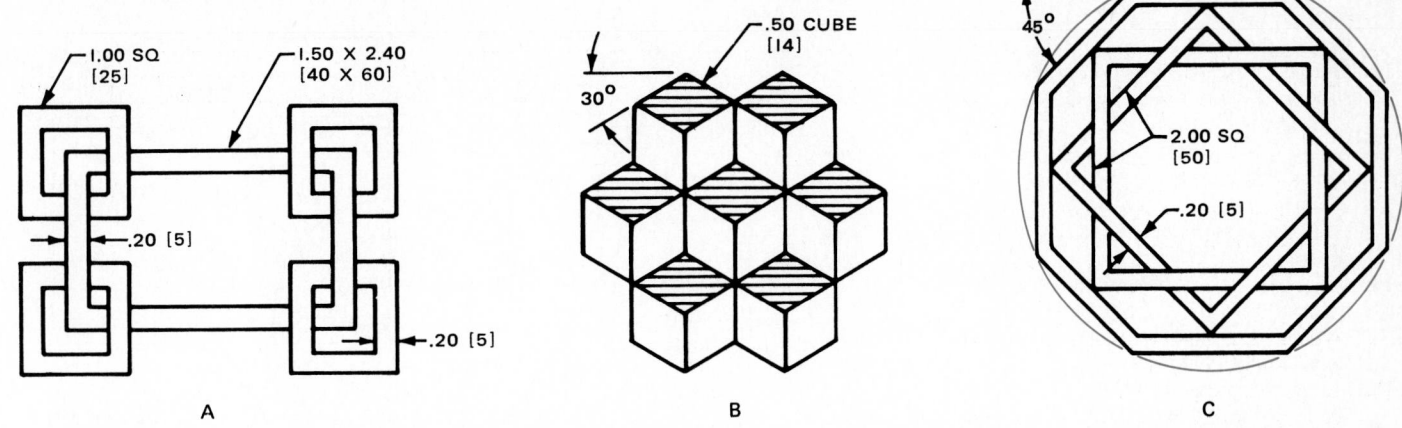

A

B

C

NOTE — DIMENSIONS NOT ENCLOSED IN BRACKETS ARE IN INCHES.
DIMENSIONS IN BRACKETS ARE IN MILLIMETERS.

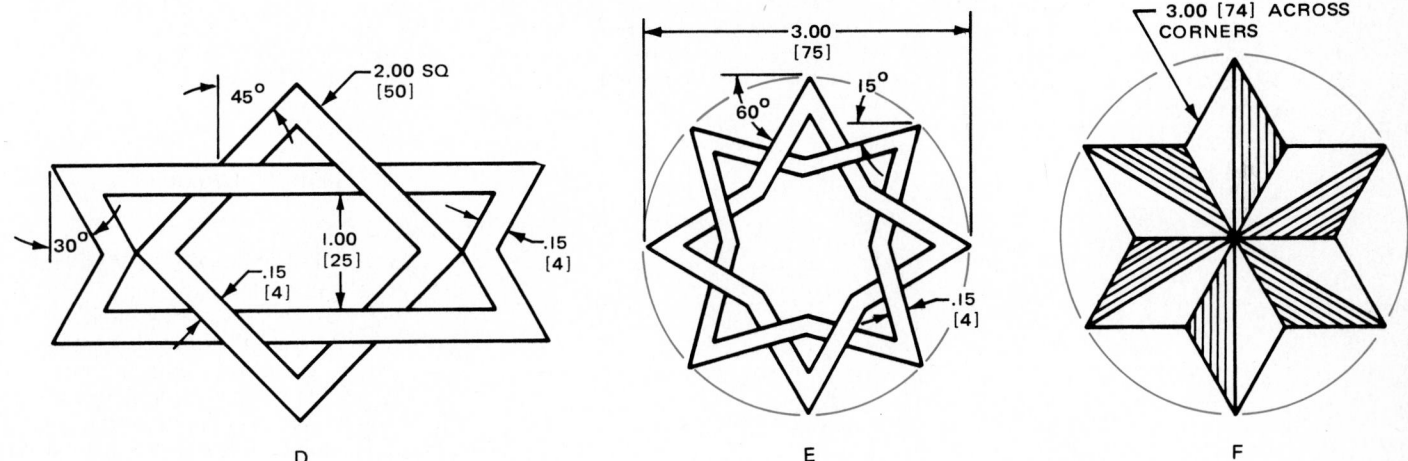

D

E

F

Fig. 2-6-D   Inlay designs.

NOTE – DIMENSIONS NOT ENCLOSED IN BRACKETS ARE IN MILLIMETERS.
DIMENSIONS IN BRACKETS ARE IN INCHES.

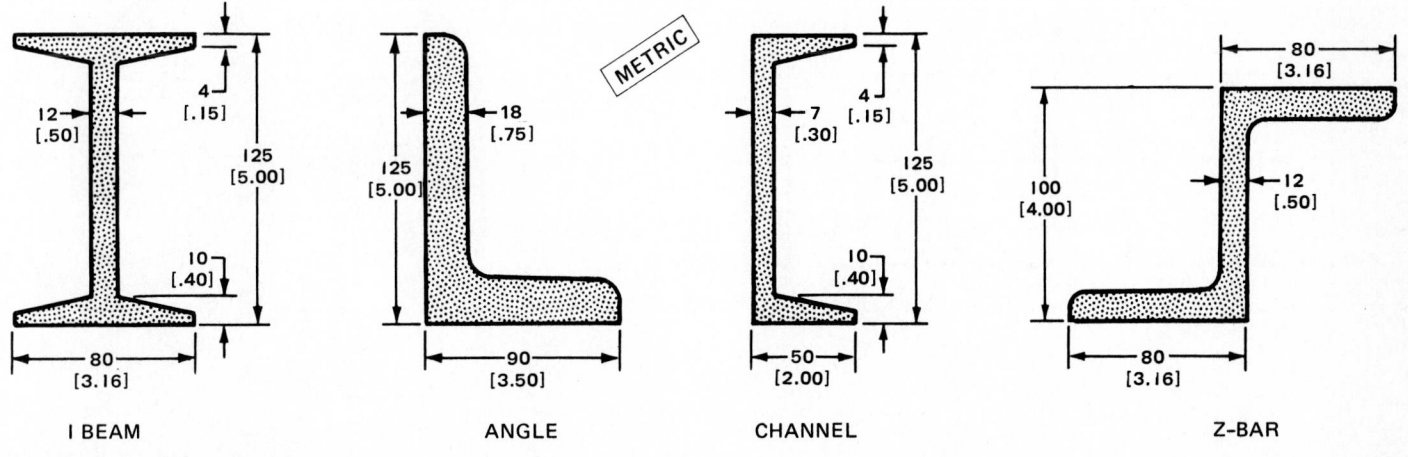

I BEAM   ANGLE   CHANNEL   Z-BAR

Fig. 2-6-E   Structural steel shapes.

11. On a B- or A3-size sheet draw the template shown in Fig. 2-6-F. Scale full (1:1). Do not dimension.

12. On a B- or A3-size sheet draw the shaft support shown in Fig. 2-6-G. Scale full (1:1). Do not dimension.

13. On an A3- or B-size sheet draw the dial indicator shown in Fig. 2-6-H. Scale 2:1. Do not dimension but add the word DEGREES and the degree numbers shown.

14. On an A3- or B-size sheet draw the dart board shown in Fig. 2-6-J. Scale 1:2. Use diagonal line shading and add the numbers. Do not dimension.

15. Using grid paper draw the line graph shown in Fig. 2-6-K using .25 in. or 5 mm grids.

16. On a B- or A3-size sheet layout the pattern for the table leg shown in Fig. 2-6-L to half scale (1:2).

Fig. 2-6-G   Shaft support.

Fig. 2-6-H   Dial indicator.

Fig. 2-6-F   Template.

Fig. 2-6-J   Dart board.

Fig. 2-6-K   Line graph.

1.0 IN. OR 25 mm
SQUARES

2 HOLES
DRILL AND CSK FOR
#10 WOOD SCREW

Fig. 2-6-L   Table leg.

CHAPTER 3

Theory of Shape Description

UNIT 3-1
Theory of Shape Description

Chapter 2 illustrated many simple parts that required only one view to completely describe them. However, in industry, the majority of parts that have to be drawn are more complicated than the ones previously described. More than one view of the object is required to show all the construction features.

Pictorial (three-dimensional) drawings of objects are sometimes used, but the majority of drawings used in mechanical drafting for completely describing an object are multiview drawings as shown in Fig. 3-1-1.

Pictorial projections, such as axonometric, oblique, and perspective projection, are useful for illustrative purposes and are frequently employed in installation and maintenance drawings and design sketches.

As a result of new drawing techniques and equipment, pictorial drawings are becoming a popular form of communication, especially with people not trained to read engineering drawings. Practically all drawings of do-it-yourself projects for the general public or of assembly-line instructions for nontechnical personnel are done in pictorial form.

SHAPE DESCRIPTION BY VIEWS

When looking at objects, we normally see them as three-dimensional, having *width*, *depth*, and *height*, or *length*, *width*, and *height*. The choice of terms used depends on the shape and proportions of the object.

Spherical shapes, such as a basketball, are described as having a certain *diameter* (one term).

Cylindrical shapes, such as a baseball bat, have *diameter* and *length*. However, a hockey puck would have *diameter* and *thickness* (two terms).

Objects which are not spherical or cylindrical require three terms to describe their overall shape. The terms used for a car would probably be length, width, and height; for a filing cabinet, width, height, and depth; for a sheet of drawing paper, length, width, and thickness. These terms are used interchangeably according to the *proportions* of the object being described, and the *position* it is in when being viewed. For example, a telephone pole lying on the ground would be described as having diameter and length, but when placed in a vertical position, its dimensions would be diameter and height.

In general, distances from left to right are referred to as width or length, distances from front to back as depth or width, and vertical distances (except when very small in proportion to the others) as height. On drawings, the multidimensional shape is represented by a view or views on the flat surface of the drawing paper.

PICTORIAL VIEWS

Pictorial drawings represent the shape with just one view, and are frequently used for illustrative purposes, for installation and maintenance drawings, and do-it-yourself projects for the general public. However, the majority of parts manufactured in industry are

ORTHOGRAPHIC PROJECTION   ISOMETRIC   OBLIQUE   PERSPECTIVE

PICTORIAL DRAWINGS

Fig. 3-1-1   Types of projection used in drafting.

too complicated in shape and detail to be described successfully by a pictorial view.

ORTHOGRAPHIC PROJECTION

The drafter must represent the part which appears as three-dimensional (width, height, depth) to the eye on the flat plane of the drawing paper. Different views of the object—front, side,

(A) PICTORIAL DRAWING (ISOMETRIC)

THICK SOLID LINE USED TO INDICATE VISIBLE OBJECT LINES

(B) ORTHOGRAPHIC PROJECTION DRAWING (THIRD ANGLE)

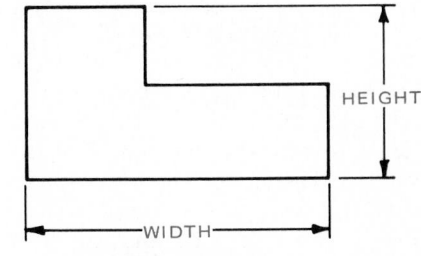

(C) ORTHOGRAPHIC VIEW

Fig. 3-1-2 A simple object shown in pictorial and orthographic projection.

Fig. 3-1-3 Systematic arrangement of views.

and top views—are systematically arranged on the drawing paper to convey the necessary information to the reader (Fig. 3-1-2B and 3-1-3). Features are projected from one view to another. This type of drawing is called an *orthographic projection*. The word *orthographic* is derived from two Greek words: *orthos*, meaning straight, correct, at right angles to; and *graphikos*, meaning to write or describe by drawing lines.

An orthographic view is what you would see looking directly at one side or "face" of the object. When looking directly at the front face, you would see width and height (two dimensions) but not the third dimension, depth. Each orthographic view gives two of the three major dimensions.

Orthographic Systems

The principles of orthographic projection can be applied in four different "angles" or systems: first-, second-, third-, and fourth-angle projection (Fig. 3-1-4).

However, only two systems—first- and third-angle projection—are used. Third-angle projection is used in the United States, Canada, and many other countries throughout the world. First-angle projection, which will be

described in detail in Unit 3-7, is used mainly in many European and Asiatic countries. As world trade has brought about the exchange of engineering drawings as well as the end products, drafters are now called upon to communicate in, as well as understand, both types of orthographic projection.

ISO Projection Symbol

With two types of projection being used on engineering drawings, a method of identifying the type of projection is necessary. The International

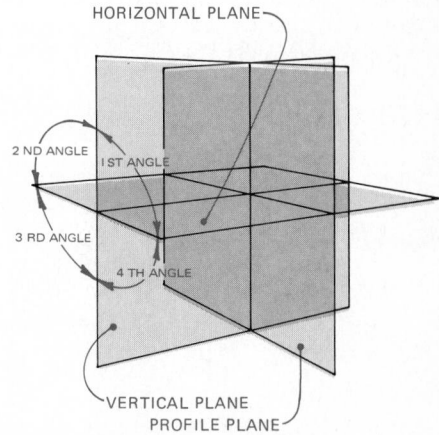

Fig. 3-1-4 The three planes used in orthographic projection.

Standards Organization, known as ISO, has recommended that the symbol shown in Fig. 3-1-5 be shown on all drawings and located preferably in the lower right-hand corner of the drawing, adjacent to the title block (Fig. 3-1-6). To aid the reader in learning the language of industry, many objects throughout this text have been drawn in first- as well as third-angle projection. The ISO symbol will indicate the type of projection used.

Third-Angle Projection

In third-angle projection, the object is positioned in the third-angle quadrant, as shown in Fig. 3-1-7. The person viewing the object does so from six different positions, namely, from the

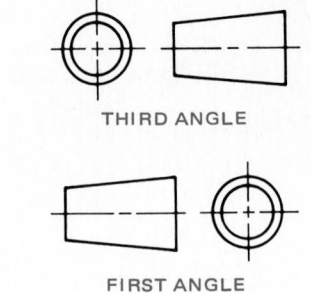

THIRD ANGLE

FIRST ANGLE

Fig. 3-1-5   ISO projection symbol.

TITLE BLOCK

Fig. 3-1-6   Locating ISO symbol on drawing paper.

REAR VIEW

TOP VIEW

LEFT-SIDE VIEW

RIGHT-SIDE VIEW

FRONT VIEW

BOTTOM VIEW

Fig. 3-1-7   Relationship of object with viewing planes in third-angle projection.

VIEWED FROM TOP

VIEWED FROM LEFT SIDE

VIEWED FROM REAR

FRONT

VIEWED FROM RIGHT SIDE

VIEWED FROM BOTTOM

VIEWING THE OBJECT FROM ALL SIX SIDES

TOP

LEFT

FRONT

RIGHT

REAR

BOTTOM

UNFOLDING GLASS BOX TO GIVE THIRD ANGLE LAYOUT OF VIEWS

B

A

D

OBJECT ENCLOSED IN GLASS BOX

Fig. 3-1-8   Systematic arrangement of views.

TOP VIEW

FRONT

LEFT-SIDE VIEW

FRONT VIEW

RIGHT-SIDE VIEW

REAR VIEW

BOTTOM VIEW

top, front, right side, left side, rear, and bottom. The views or pictures seen from these positions are then recorded or drawn on the plane located between the viewer and the object. These six viewing planes are then rotated or positioned so that they lie in a single plane, as shown in Fig. 3-1-8. Rarely are all six views used. Only the views which are necessary to fully describe the object are drawn. Simple objects, such as a gasket, can be described sufficiently by one view alone. However, in mechanical drafting two- or three-view drawings of objects are more common, the rear, bottom, and one of the two side views being rarely used. Figure 3-1-9 shows simple objects drawn in orthographic and pictorial form.

To fully appreciate the shape and detail of views drawn in third-angle orthographic projection, the units for this chapter have been designed according to the types of surfaces generally found on objects. These surfaces can be divided into flat surfaces parallel to the viewing planes with and with-

out hidden features; flat surfaces which appear inclined in one plane and parallel to the other two principal reference planes (called *inclined* surfaces); flat surfaces which are inclined in all three reference planes (called *oblique* surfaces); and surfaces which have diameters or radii. These drawings are so designed that only the top, front, and right side views are required.

All Surfaces Parallel to the Viewing Planes and All Edges and Lines Visible When a surface is parallel to the viewing planes, that surface will show as a surface on one view and a line on the other views. The lengths of these lines are the same as the lines shown on the surface view. The drawing has been made showing each side to represent the exact shape and size of the object and the relationship of the three views to one another.

ASSIGNMENTS

See Assignments 1 and 2 for Unit 3-1 on page 56.

UNIT 3-2
Spacing of Views and Miter Lines

SPACING THE VIEWS

It is important for clarity and good appearance that the views be well balanced on the drawing paper, whether the drawing shows one view, two views, three views, or more. The drafter must anticipate the approximate space required. This is determined from the size of the object to be drawn, the number of views, the scale used, and the space between views. Ample space should be provided between views to permit placement of dimensions on the drawing without crowding. Space should also be allotted so that notes can be added without crowding. However, space between views should not be excessive. Once the size of paper, scale, and number of views are established, the balancing of the three views is relatively simple. A

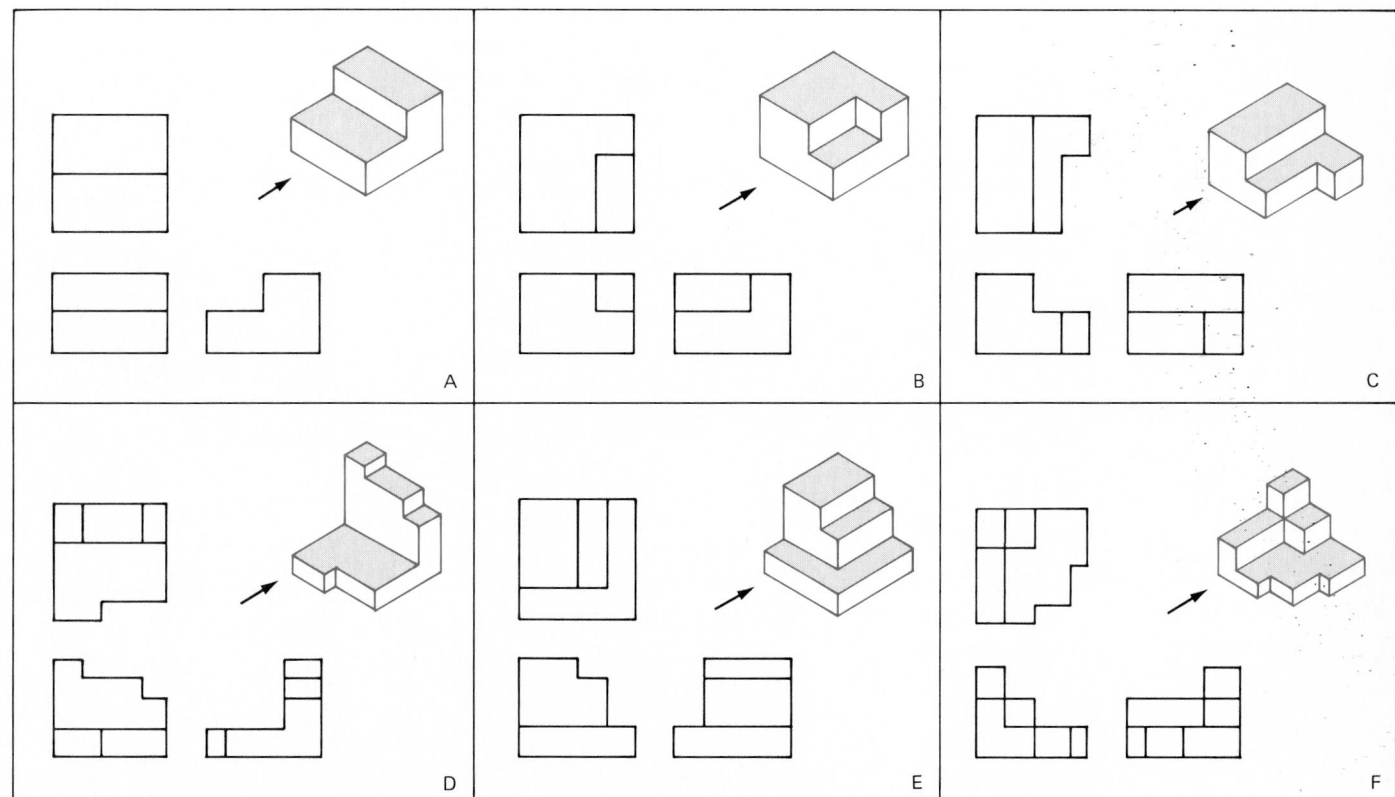

NOTE: ARROWS INDICATE DIRECTION OF SIGHT WHEN LOOKING AT THE FRONT VIEW.

Fig. 3-1-9   Illustrations of objects drawn in third-angle orthographic projection.

simple method of positioning the views on the drawing paper is shown in Fig. 3-2-1. In this example, a distance of 40 mm (1.50 in.) is left between views. For the beginning drafter, between 30 and 40 mm (1.20 to 1.50 in.) is recommended for the distance between views.

(A) DECIDING THE VIEWS TO BE DRAWN AND THE SCALE TO BE USED

(B) CALCULATING DISTANCES A AND B

(C) ESTABLISHING LOCATION OF PLANES 1 AND 2

Fig. 3-2-1 Balancing the drawing on the drawing paper.

USE OF A MITER LINE

The use of a miter line provides a fast and accurate method of constructing the third view once two views are established (Fig. 3-2-2).

Using a Miter Line to Construct the Right Side View

1. Given the top and front views, project lines to the right of the top view.
2. Establish how far from the front view the side view is to be drawn (distance D).
3. Construct the miter line at 45° to the horizon.
4. Where the horizontal projection lines of the top view intersect the miter line, drop vertical projection lines.
5. Project horizontal lines to the right of the front view and complete the side view.

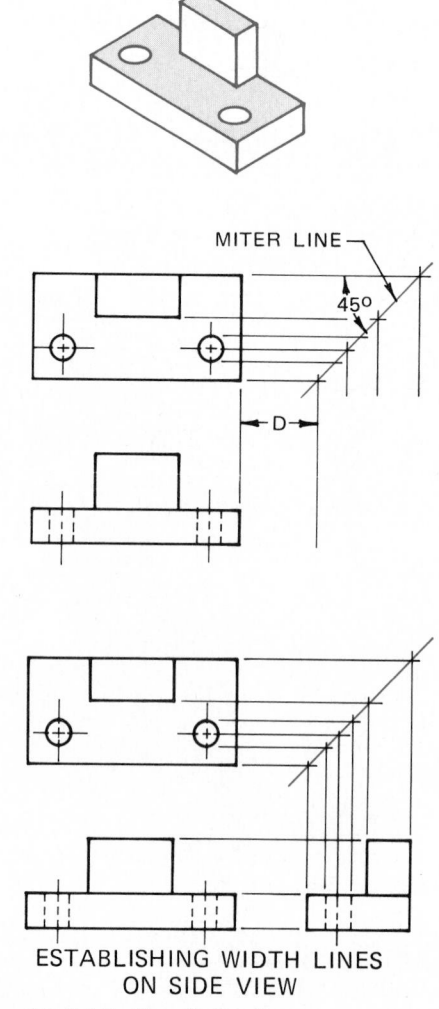

ESTABLISHING WIDTH LINES ON SIDE VIEW

Fig. 3-2-2 Use of miter line.

Using a Miter Line to Construct the Top View

1. Given the front and side views, project vertical lines up from the side view.
2. Establish how far away from the front view the top view is to be drawn (distance D).
3. Construct the miter line at 45° to the horizon.
4. Where the vertical projection lines of the side view intersect the miter line, project horizontal lines to the left.
5. Project vertical lines up from the front view and complete the top view.

ASSIGNMENTS

See Assignments 3 through 6 for Unit 3-2 on page 56.

ESTABLISHING WIDTH LINES ON TOP VIEW

UNIT 3-3
All Surfaces Parallel to the Viewing Planes with Some Edges and Surfaces Hidden

Most objects drawn in engineering offices are more complicated than the ones shown in Fig. 3-3-1. Many features (lines, holes, etc.) cannot be seen when viewed from outside the piece. These hidden edges are shown with *hidden lines* and are normally required

on the drawing to show the true shape of the object. Figure 3-3-2 shows additional examples of objects requiring hidden lines.

ASSIGNMENTS

See Assignments 7 through 10 for Unit 3-3 on page 57.

UNIT 3-4
Inclined Surfaces

If the surfaces of an object lie in either a horizontal or a vertical position, the surfaces appear in their true shapes in one of the three views, and these surfaces appear as a line in the other two views.

When a surface is inclined or sloped in only one direction, then that surface is not seen in its true shape in the top, front, or side view. It is, however, seen in two views as a distorted surface. On the third view it appears as a line.

The true length of surfaces A and B in Fig. 3-4-1 is seen in the front view only. In the top and side views, only the width of surfaces A and B appears in its true size. The length of these surfaces is foreshortened.

Where an inclined surface has important features that must be shown clearly and without distortion, an *auxiliary* or helper view must be used. This type of view will be discussed in

detail in Chapter 13. Figure 3-4-2 shows additional examples of objects having inclined surfaces.

ASSIGNMENTS

See Assignments 11 through 14 for Unit 3-4 on page 60.

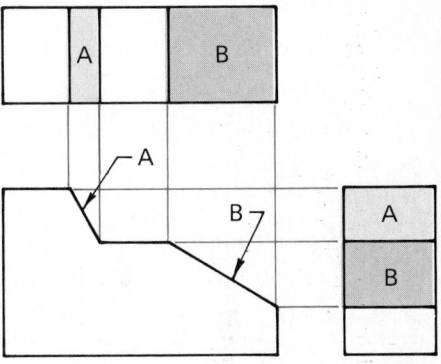

NOTE: THE TRUE SHAPE OF SURFACES A AND B DO NOT APPEAR ON THE TOP OR SIDE VIEWS.

Fig. 3-4-1   Sloping surfaces.

HIDDEN EDGE LINES SHOWN IN FRONT VIEW

HIDDEN EDGE LINE

Fig. 3-3-1   Hidden lines.

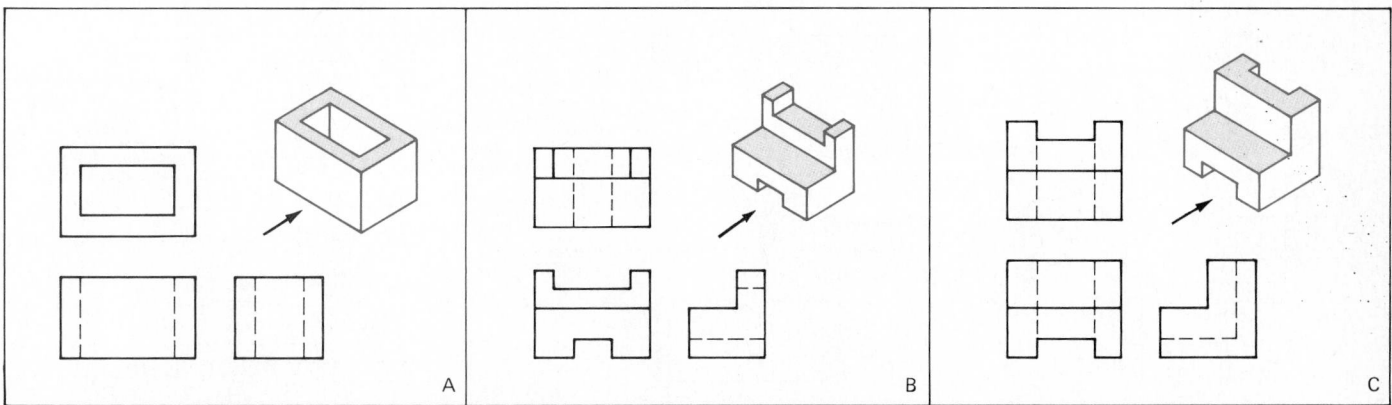

Fig. 3-3-2   Illustrations of objects having hidden features.

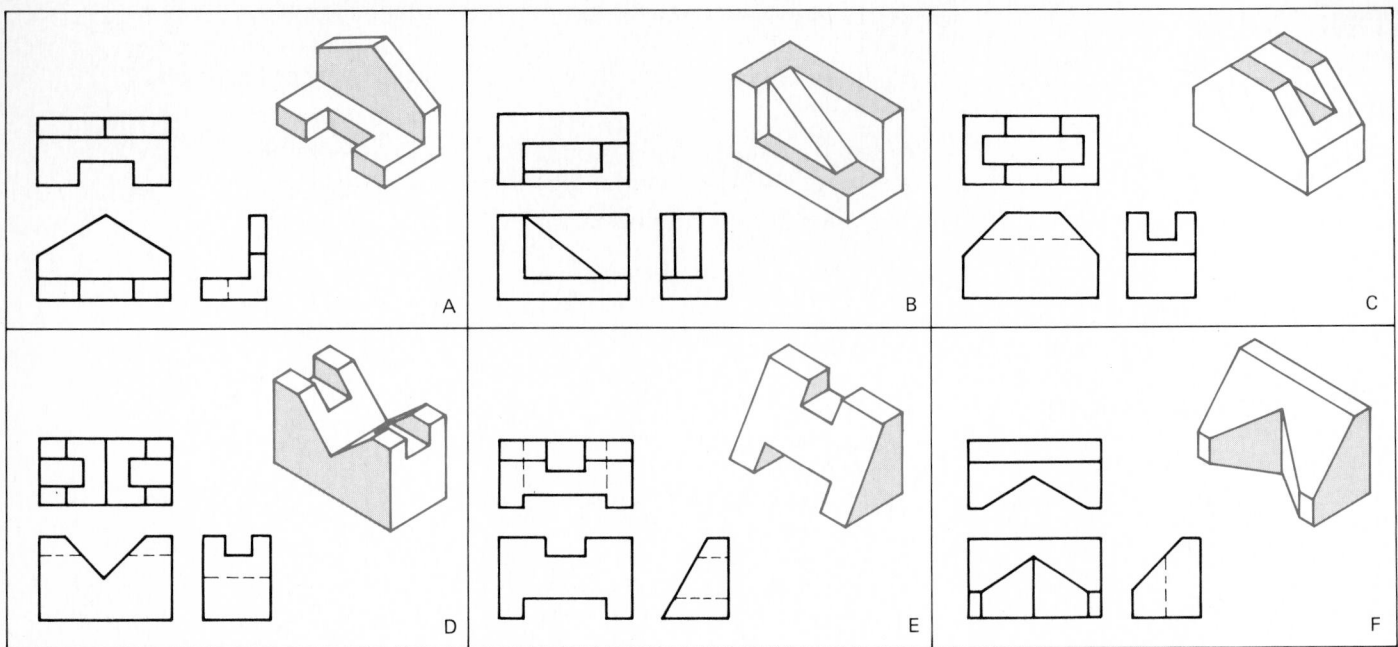

Fig. 3-4-2   Illustrations of objects having sloping surfaces.

UNIT 3-5
Circular Features

Typical parts with circular features are illustrated in Fig. 3-5-1. Note that the circular feature appears circular in one view only and that no line is used to show where a curved surface joins a flat surface. Hidden circles, like hidden flat surfaces, are represented on drawings by a hidden line.

Center Lines
A center line is drawn as a thin, broken line of long and short dashes, spaced alternately. Such lines may be used to locate center points, axes of cylindrical parts, and axes of symmetry, as shown in Fig. 3-5-2. Solid center lines are often used when the circular features are small. Center lines should project for a short distance beyond the outline of the part or feature to which they refer. They must be extended for use as extension lines for dimensioning purposes, but in this case the extended portion is not broken.

On views showing the circular features, the point of intersection of the two center lines is shown by the two intersecting short dashes.

Fig. 3-5-1   Illustrations of objects having circular features.

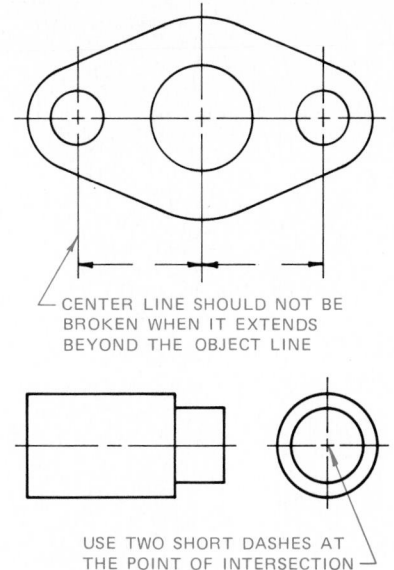

CENTER LINE SHOULD NOT BE
BROKEN WHEN IT EXTENDS
BEYOND THE OBJECT LINE

USE TWO SHORT DASHES AT
THE POINT OF INTERSECTION

Fig. 3-5-2   Center line application.

ASSIGNMENTS

See Assignments 15 through 19 for Unit 3-5 on page 64.

UNIT 3-6
Oblique Surfaces

When a surface is sloped so that it is not perpendicular to any of the three viewing planes, it will appear as a surface in all three views but never in its true shape. This is referred to as an *oblique surface* (Fig. 3-6-1). Since the oblique surface is not perpendicular to the viewing planes, it cannot be parallel to them and consequently appears foreshortened. If a true view is required for this surface, two auxiliary views—a primary and a secondary view—need to be drawn. This is discussed in detail under Secondary Auxiliary Views in Unit 13-4. Figure 3-6-2 shows additional examples of objects having oblique surfaces.

ASSIGNMENTS

See Assignments 20 and 21 for Unit 3-6 on page 67.

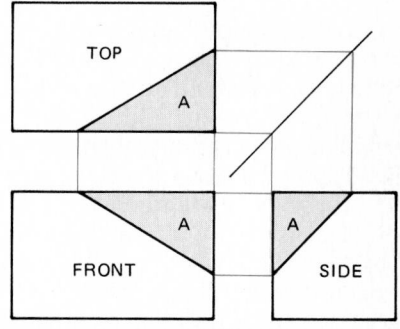

Fig. 3-6-1   Oblique surface A not true shape in any of the three views.

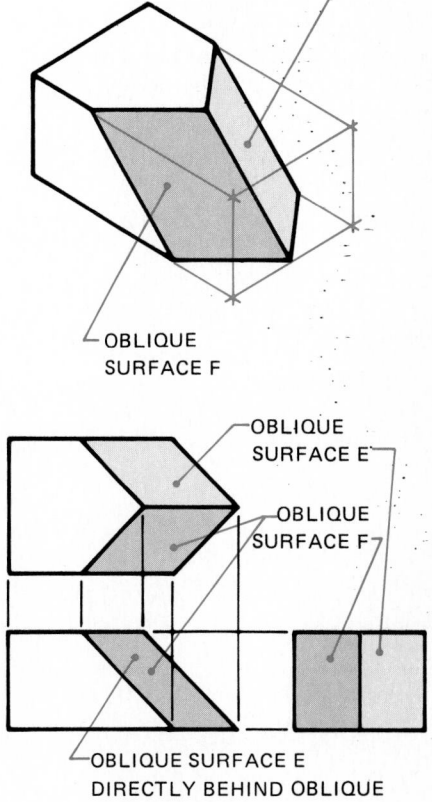

Fig. 3-6-2   Examples of objects having oblique surfaces.

UNIT 3-7
First-Angle Orthographic Projection

As mentioned previously, first-angle orthographic projection is used in many countries throughout the world. Today with global marketing and the interchange of drawings with different countries, drafters are called upon to prepare and interpret drawings in both first- and third-angle projection. In first-angle projection, all the views are projected onto the planes located behind the objects rather than onto the planes lying between the objects and the viewer, as in third-angle projection. This is shown in Fig. 3-7-2. The unfolding and positioning of the views in one plane are shown in Fig. 3-7-3. Note that the views are on opposite sides of the front view with the exception of the rear view. A comparison between the views of first- and third-angle projections is shown in Figs.

3-7-1 and 3-7-4. Remember that the views are identical in shape and detail, and only their location in reference to the front view has changed.

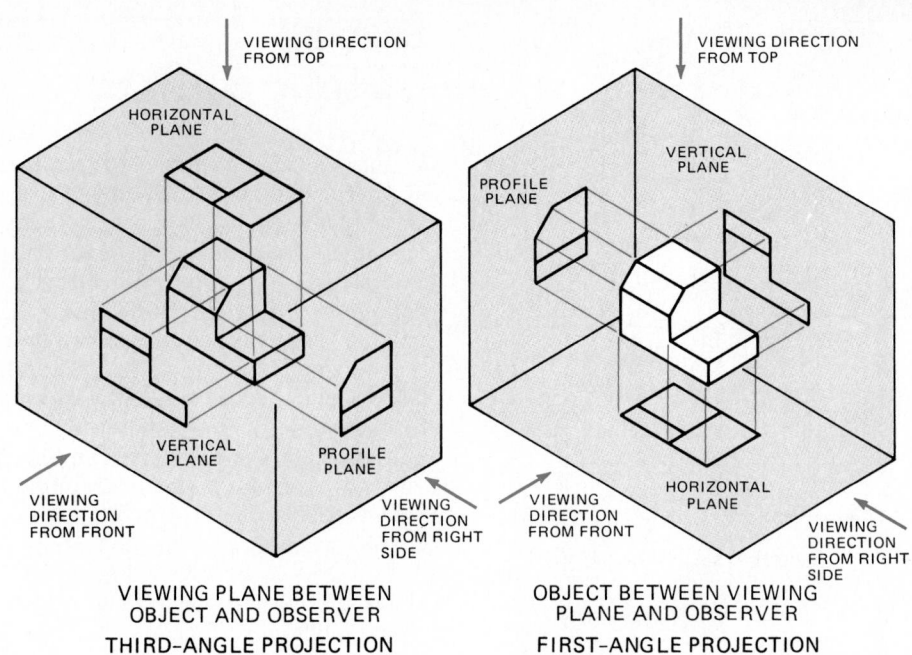

VIEWING PLANE BETWEEN OBJECT AND OBSERVER
THIRD-ANGLE PROJECTION

OBJECT BETWEEN VIEWING PLANE AND OBSERVER
FIRST-ANGLE PROJECTION

Fig. 3-7-1   A comparison between third- and first-angle projection.

ASSIGNMENTS

See Assignments 22 and 23 for Unit 3-7 on page 68.

Review for Assignment
Unit 2-6   Sketching
Unit 3-1   ISO Projection Symbol

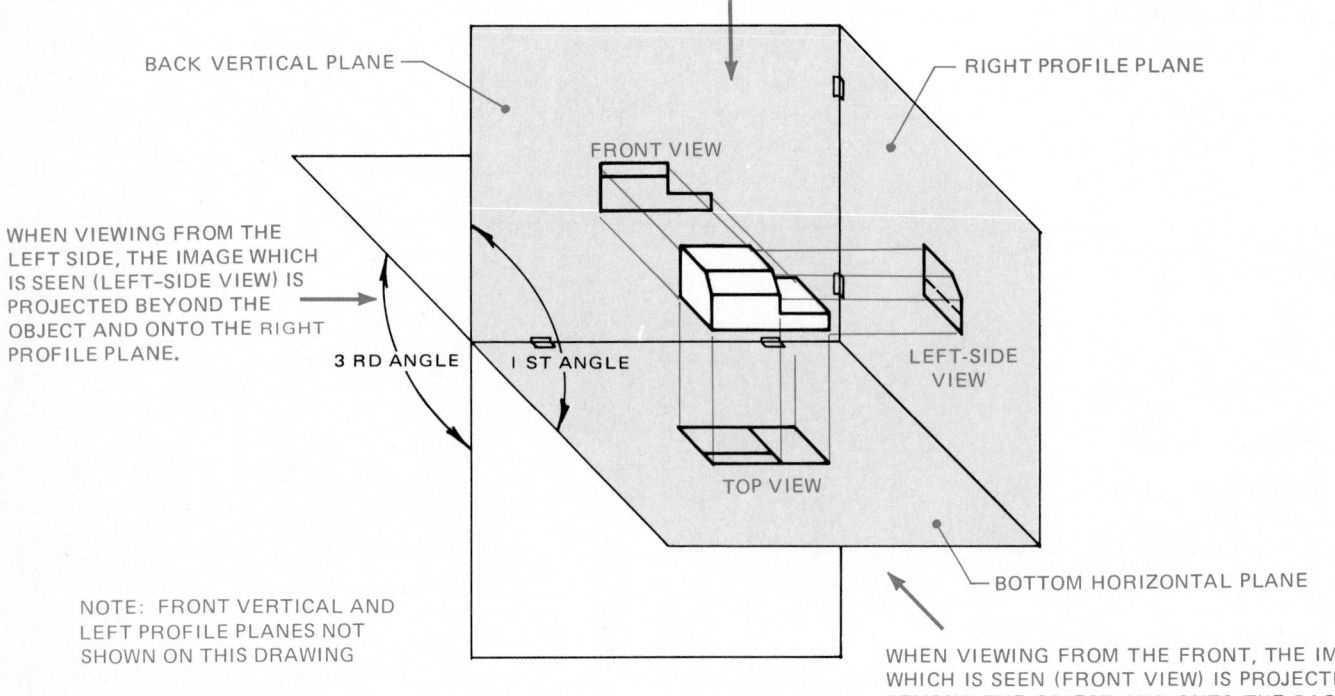

Fig. 3-7-2   Relationship of object with viewing planes in first-angle projection.

Fig. 3-7-3 First-angle orthographic projection.

(A) PICTORIAL DRAWING
(ISOMETRIC)

(B) THIRD–ANGLE PROJECTION

(C) FIRST–ANGLE PROJECTION

Fig. 3-7-4 A simple object shown in pictorial and orthographic form.

UNIT 3-8
One- and Two-View Drawings

VIEW SELECTION

Views should be chosen that will best describe the object to be shown. Only the minimum number of views that will completely portray the size and shape of the part should be used. They should also be chosen to avoid hidden feature lines whenever possible, as shown in Fig. 3-8-1.

Except for complex objects of irregular shape, it is seldom necessary to

THIS END VIEW AVOIDED

THIS END VIEW PREFERRED

Fig. 3-8-1 Avoidance of hidden-line features.

draw more than three views. For representing simple parts, one- or two-view drawings will often be adequate.

ONE-VIEW DRAWINGS

In one-view drawings, the third dimension, such as thickness, may be

expressed by a note or by descriptive words or abbreviations, such as DIA, Ø, or HEXAGON ACROSS FLATS. Square sections may be indicated by light crossed diagonal lines. This applies whether the face is parallel or inclined to the drawing plane. These are illustrated in Fig. 3-8-2.

When cylindrical-shaped surfaces include special features such as a keyseat, a side view (often called an *end view*) is required.

TWO-VIEW DRAWINGS

Frequently the drafter will decide that only two views are necessary to

.06 THICK

(A) FLAT PART

HEX 1.60 AF

□ 1.00

Ø1.00

.84 TWO FLATS
DIAMETRICALLY OPPOSITE

1.000 – 8 UNC – 2A

□ .62

(B) TURNED PART

Fig. 3-8-2   One-view drawings.

explain fully the shape of an object (Fig. 3-8-3). For this reason, some drawings consist of two adjacent views, such as the top and front views only, or front and right side views only. Two views are usually sufficient to explain fully the shape of cylindrical objects; if three views were used, two of them would be identical, depending on the detail structure of the part.

ASSIGNMENT

See Assignment 24 for Unit 3-8 on page 69.

Review for Assignments
Unit 2-6   Line Work and Drawing
     Lines

(A) SIDE VIEW NOT REQUIRED

SYMMETRY LINE

(A) WITH HALF VIEW

LEFT SIDE ONLY

(B) TOP VIEW NOT REQUIRED

Fig. 3-8-3   Two-view drawings.

Fig. 3-9-1   Partial views.

UNIT 3-9
Partial Views

Symmetrical objects may often be adequately portrayed by half views (Fig. 3-9-1A). A center line is used to show the axis of symmetry. Two short thick lines, above and below the view of the object, are drawn at right angles to and on the center line to indicate the line of symmetry.

Partial views, which show only a limited portion of the object with remote details omitted, should be used, when necessary, to clarify the meaning of the drawing (Fig. 3-9-1B). Such views are used to avoid the necessity of drawing many hidden features.

On drawings of objects where two side views can be used to better advantage than one, each need not be complete if together they depict the shape. Show only the hidden lines of features immediately behind the view (Fig. 3-9-1C).

ASSIGNMENT

See Assignment 25 for Unit 3-9 on page 70.

Review for Assignments
Unit 2-6   Line Work and Drawing
     Lines

VIEWING-PLANE LINE (THICK)

A

VIEW A-A

A

(B) PARTIAL VIEW WITH A VIEWING-PLANE
LINE USED TO INDICATE DIRECTION

RIGHT SIDE ONLY

(C) PARTIAL SIDE VIEWS

UNIT 3-10
Rear Views and Enlarged Views

PLACEMENT OF VIEWS

When views are placed in the relative positions shown in Fig. 3-1-8, it is rarely necessary to identify them. When they are placed in other than the regular projected position, the removed view must be clearly identified.

Whenever appropriate, the orientation of the main view on a detail drawing should be the same as on the assembly drawing. To avoid the crowding of dimensions and notes, ample space must be provided between views.

SEE VIEW A

VIEW A
SCALE 3:1

(A) ENLARGED VIEW OF FEATURE

DETAIL A
SCALE 5:1

METRIC

SEE DETAIL A
SCALE SHOWN ON DRAWING

(B) ENLARGED VIEW OF ASSEMBLY

B

VIEW B
SCALE 5:1

(C) ENLARGED REMOVED VIEW

Fig. 3-10-1 Enlarged views.

ENLARGED VIEWS

Enlarged views are used when it is desirable to show a feature in greater detail or to eliminate the crowding of details or dimensions (Fig. 3-10-1). The enlarged view should be oriented in the same manner as the main view. However, if an enlarged view is rotated, state the direction and the amount of rotation of the detail. The scale of enlargement must be shown, and both views should be identified by one of the three methods shown.

REAR VIEWS

Rear views are normally projected to the right or left. When this projection is not practical, because of the length of the part, particularly for panels and mounting plates, the rear view must not be projected up or down. Doing so would result in the part being shown upside down. Instead, the view should be drawn as if it were projected sideways but located in some other position, and it should be clearly labeled REAR VIEW REMOVED (Fig. 3-10-2).

ASSIGNMENT

See Assignment 26 for Unit 3-10 on page 71.

Review for Assignments
Unit 12-4   Kinds of Plastics
Unit 12-5   Forming Methods
Unit 12-6   Design Consideration—Parts

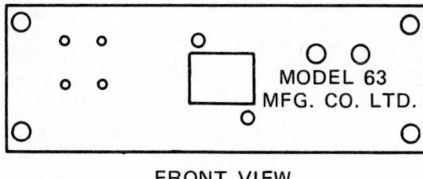

MODEL 63
MFG. CO. LTD.

FRONT VIEW

1   2
POS. A

1   2

3   4
POS. C

POS. B

REAR VIEW REMOVED

Fig. 3-10-2 Removed rear view.

UNIT 3-11
Opposite-Hand Views and Key Plans

OPPOSITE-HAND VIEWS

Where parts are symmetrically opposite, such as for right- and left-hand usage, one part is drawn in detail and the other is described by a note such as PART B SAME EXCEPT OPPOSITE HAND. It is preferable to show both part numbers on the same drawing (Fig. 3-11-1).

PT 1

PT 2

(A) TWO DRAWINGS

PT 1 AS SHOWN
PT 2 OPPOSITE HAND

(B) ONE DRAWING REPLACES TWO VIEWS

Fig. 3-11-1 Opposite hand views.

KEY PLAN

A method particularly applicable to structural work is to include on each sheet of a series of drawings a small key plan drawn in bold lines the relationship of the detail on that sheet to the whole work, as in Fig. 3-11-2.

Fig. 3-11-2 Key plan.

ASSIGNMENT

See Assignment 27 for Unit 3-11 on page 71.

ASSIGNMENTS for Chapter 3

Assignments for Unit 3-1, Theory of Shape Description

1. On two A-size sheets of preprinted grid paper (.25-in. or 10-mm grids) sketch three views of each of the objects shown in Figs. 3-1-A and 3-1-B. Draw three objects on each sheet. Each square shown on the objects represents one square on the grid paper. Allow one grid space between views and a minimum of two grid spaces between objects. Identify the type of projection used by placing the appropriate ISO projection symbol at the bottom of the drawing.

2. On an A- or A4 sheet draw three views of one of the parts shown in Figs. 3-1-C to 3-1-F. Allow 1 in. or 25 mm between views. Scale full or 1:1. Do not dimension.

Assignments for Unit 3-2, Spacing of Views and Miter Lines

3. Make a sketch similar to Figs. 3-2-1B and C, and establish the distance between plane 1 and the left border line, and between plane 2 and the bottom border line, given the following:
Scale full
Drawing space 7.00 × 11.00 in.
Size of object to be drawn:
    width 4.00, depth 2.20, height 1.30
Space between views 1.5 in. (40 mm).

4. Make a sketch similar to Figs. 3-2-1B and C, and establish the distance between plane 1 and the left border line, and between plane 2 and the bottom border line, given the following:
Scale 1:2
Drawing space 180 × 280 mm
Size of object to be drawn:
    width 240, depth 60, height 80
Space between views 40 mm (1.5 in.)

5. Spacer 1, Fig. 3-2-A, sheet size A, scale full. Make a three-view drawing using a miter line to complete the right side view. Space between views to be 1.5 in. (40 mm).

6. Bracket 2, Fig. 3-2-B, sheet size A4, scale 1:1. Make a three-view drawing using a miter line to complete the top view. Space between views to be 40 mm (1.5 in.).

Fig. 3-1-A   Sketching assignment.

Fig. 3-1-B   Sketching assignment.

Fig. 3-1-C   Step support.

Fig. 3-1-D   Corner block.

Fig. 3-1-E T bracket.

Fig. 3-1-F Angle step bracket.

Fig. 3-2-A Spacer.

Fig. 3-2-B Bracket.

Assignments for Unit 3-3, All Surfaces Parallel to the Viewing Plane with Some Edges and Surfaces Hidden

7. On two A- or A4-size sheets of pre-printed grid paper (.25-in. or 10-mm grids) sketch three views of each of the objects shown in Figs. 3-3-A and 3-3-B. Draw three objects on each sheet. Each square shown on the objects represents one square on the grid paper. Allow one grid space between views and a minimum of two spaces between objects. Identify the type of projection by placing the ISO projection symbol at the bottom of the drawing.

Fig. 3-3-A Drawing assignment.

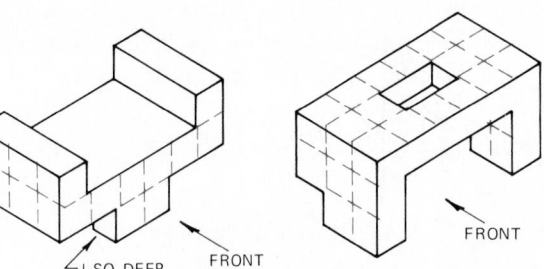

Fig. 3-3-B Sketching assignment.

8. Sketching assignment. Make three-view sketches of the parts shown in Fig. 3-3-C.

9. On an A- or A4-size sheet make a three-view instrument drawing of one of the parts shown in Figs. 3-3-D to 3-3-G. Scale full or 1:1. Allow 1.2 in. (30 mm) between views.

10. Matching test. Match the pictorial drawings to the orthographic drawings shown in Fig. 3-3-H.

FRONT

2 FRONT

3 FRONT

4 FRONT

5 FRONT

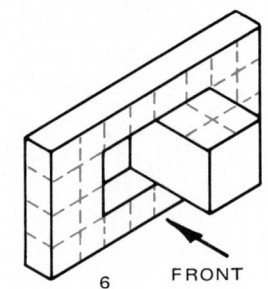
6 FRONT

Fig. 3-3-C Sketching assignment.

Fig. 3-3-D Adapter.

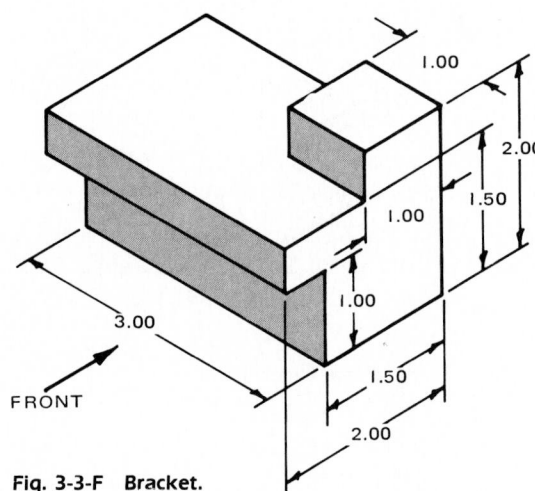

Fig. 3-3-F Bracket.

Fig. 3-3-E Link.

Fig. 3-3-G Guide block.

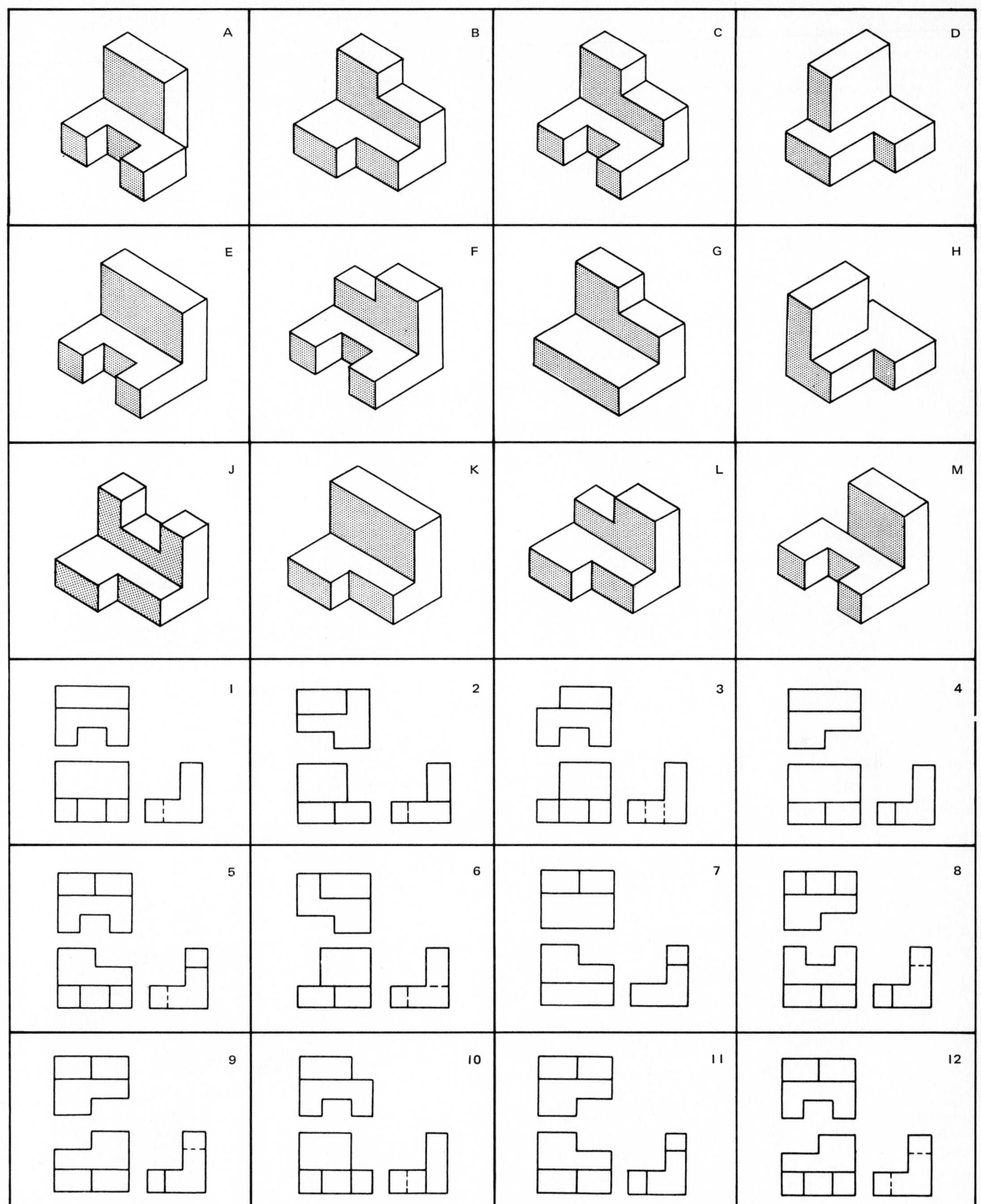

Fig. 3-3-H   Match pictorial drawings A through M with orthographic drawings.

Assignments for Unit 3-4, Inclined Surfaces

11. On two A- or A4-size sheets of pre-printed grid paper (.25 in. or 10 mm grids) sketch three views of each of the objects shown in Figs. 3-4-A and 3-4-B. Draw three objects on each sheet. Each square shown on the objects represents one square on the grid paper. Allow one grid space between views and a minimum of two grid spaces between objects. The sloped (inclined) surfaces on each of the three objects are identified by a letter. Identify the sloped surfaces on each of the three views with a corresponding letter. Also identify the type of projection used by placing the appropriate ISO symbol at the bottom of the drawing.

12. On a B- or A3-size sheet, make a three-view drawing of one of the parts shown in Figs. 3-4-C to 3-4-F. Allow 1.2 in. (30 mm) between views. Do not dimension. Scale full or 1:1.

13. Sketching assignment. Make three-view sketches of the parts shown in Figs. 3-4-G to 3-4-K.

14. Matching test. Match the pictorial drawings to the orthographic drawings shown in Fig. 3-4-L.

Fig. 3-4-A   Sketching assignment.

Fig. 3-4-B   Sketching assignment.

Fig. 3-4-C   Slide bar.

Fig. 3-4-E   Adjusting guide.

Fig. 3-4-D   Flanged support.

Fig. 3-4-F   Separator.

1

FRONT

2

FRONT

FRONT   3

4

FRONT

Fig. 3-4-G   Sketching assignment.

5

FRONT

FRONT   6

1

FRONT

2

FRONT

FRONT

3

FRONT   4

Fig. 3-4-H   Sketching assignment.

FRONT   5

FRONT   6

FRONT

FRONT 2

FRONT 3

FRONT 4

Fig. 3-4-J Sketching assignment.

FRONT 5

FRONT 6

FRONT

FRONT 2

FRONT 3

FRONT 4

Fig. 3-4-K Sketching assignment.

FRONT 5

FRONT 6

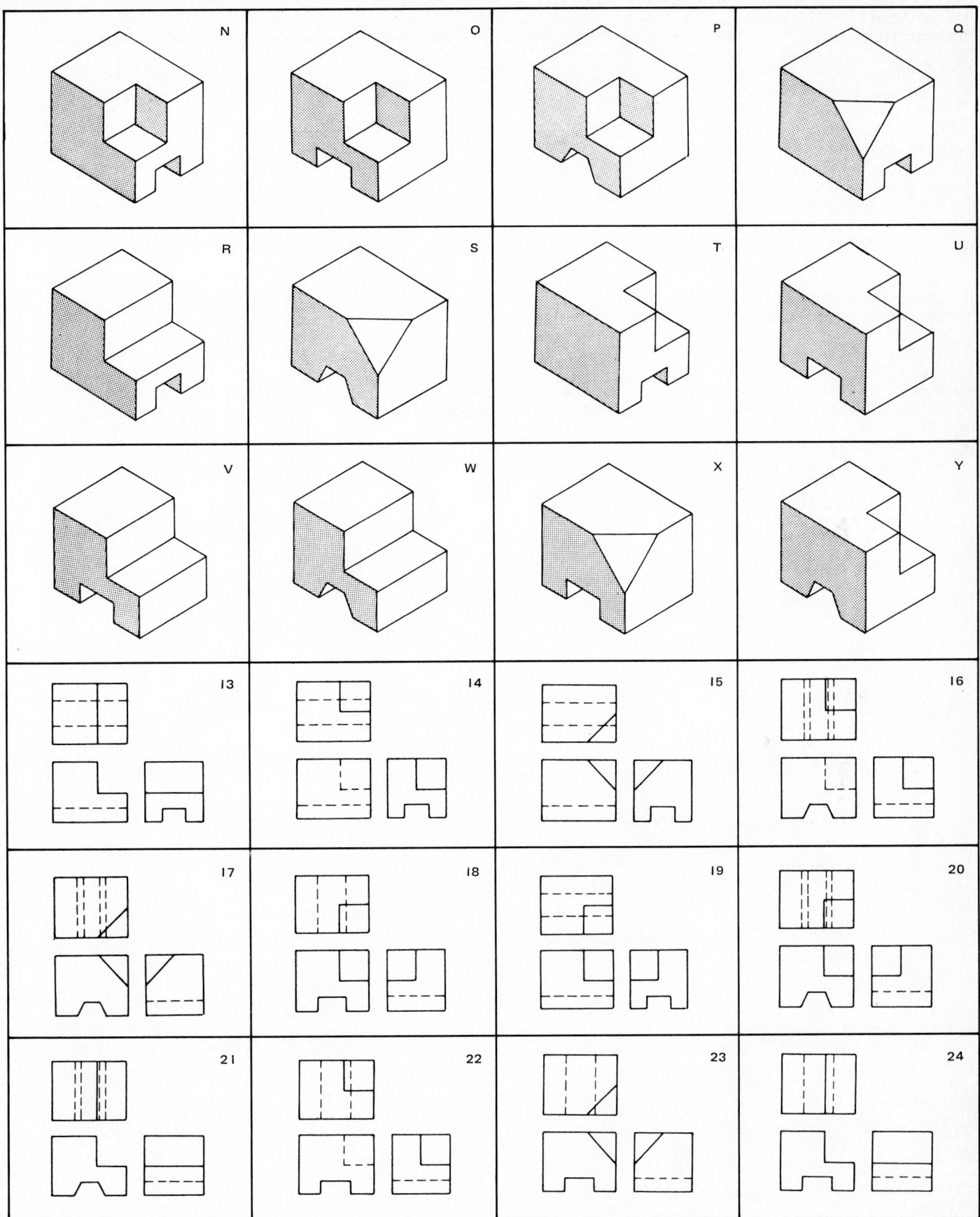

Fig. 3-4-L   Matching test.

Assignments for Unit 3-5, Circular Features

15. On two A- or A4-size sheets of pre-printed grid paper (.25 in. or 10 mm grids) sketch three views of each of the objects shown in Figs. 3-5-A and 3-5-B. Draw three objects on each sheet. Each square shown on the objects represents one square on the grid paper. Allow one gird space between views and a minimum of two grid spaces between objects. Identify the type of projection used by placing the appropriate ISO projection symbol at the bottom of the drawing.

16. On a B- or A4-size sheet, make a three-view drawing of one of the parts shown in Figs. 3-5-C to 3-5-F. Allow 1.2 in. (30 mm) between views. Do not dimension. Scale full or 1:1.

17. Sketching assignment. Make three-view sketches of the parts shown in Fig. 3-5-G.

18. Sketching assignment. Sketch the views needed for a multiview drawing for the parts shown in Fig. 3-5-H. Choose your own sizes and estimate proportions.

19. Completion test. See Fig. 3-5-J.

Fig. 3-5-A   Sketching assignment.

Fig. 3-5-B   Sketching assignment.

Fig. 3-5-C   Rod support.

Fig. 3-5-D   Pillow block.

Fig. 3-5-E   Cradle support.

Fig. 3-5-F   Rocker arm.

FRONT

1

2

FRONT

3

FRONT

4

FRONT

FRONT

5

FRONT

6

Fig. 3-5-G   Sketching assignment.

1

2

3

4

5

6

Fig. 3-5-H   Sketching assignment.

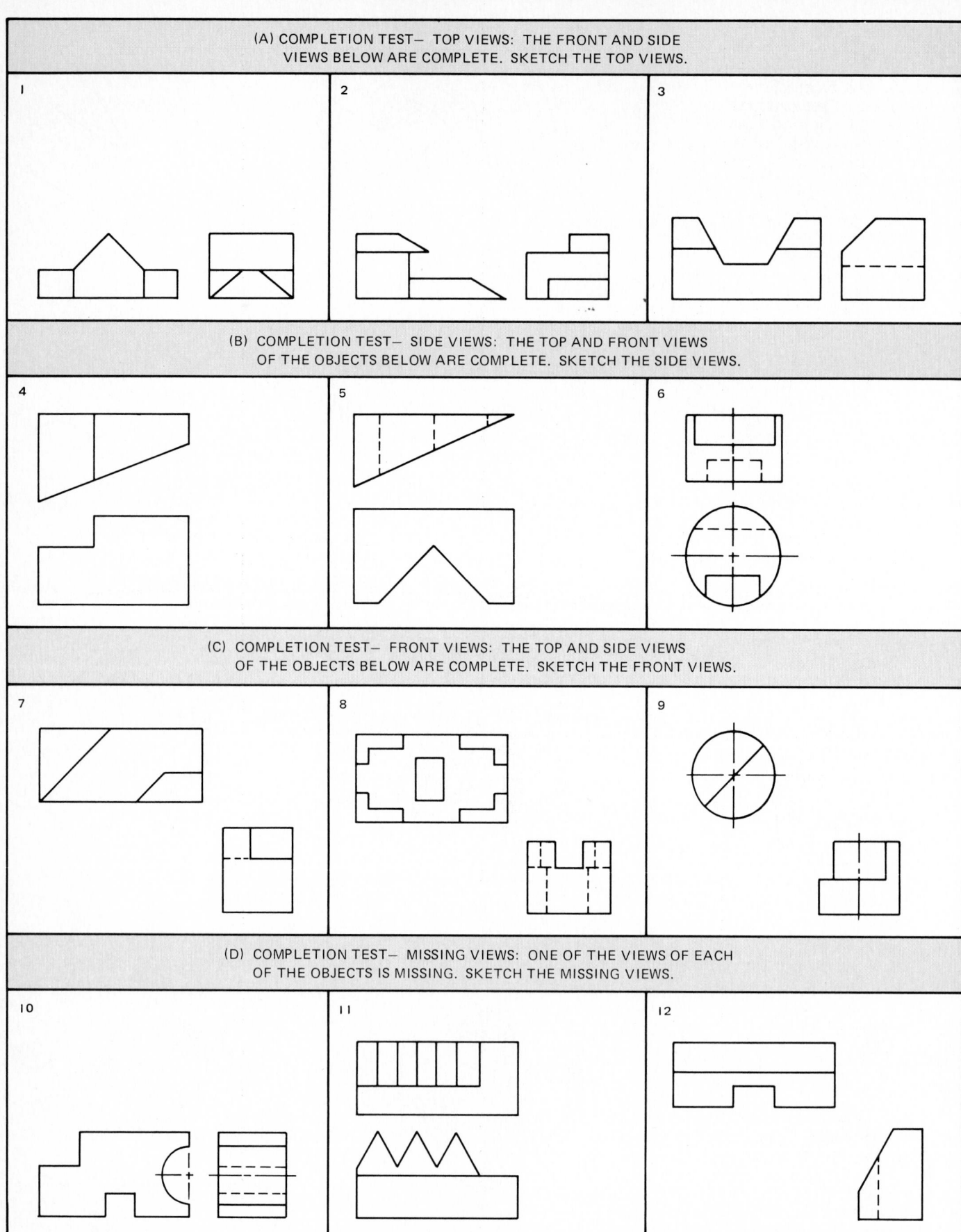

(A) COMPLETION TEST— TOP VIEWS: THE FRONT AND SIDE
VIEWS BELOW ARE COMPLETE. SKETCH THE TOP VIEWS.

1

2

3

(B) COMPLETION TEST— SIDE VIEWS: THE TOP AND FRONT VIEWS
OF THE OBJECTS BELOW ARE COMPLETE. SKETCH THE SIDE VIEWS.

4

5

6

(C) COMPLETION TEST— FRONT VIEWS: THE TOP AND SIDE VIEWS
OF THE OBJECTS BELOW ARE COMPLETE. SKETCH THE FRONT VIEWS.

7

8

9

(D) COMPLETION TEST— MISSING VIEWS: ONE OF THE VIEWS OF EACH
OF THE OBJECTS IS MISSING. SKETCH THE MISSING VIEWS.

10

11

12

Fig. 3-5-J Completion tests.

Assignments for Unit 3-6, Oblique Surfaces

20. On two A- or A4-size sheets of pre-printed grid paper (.25 in. or 10 mm grids) sketch three views of each of the objects shown in Figs. 3-6-A to 3-6-C. Draw three objects on each sheet. Each square on the objects represents one square on the grid paper. Allow one grid space between views and a minimum of two grid spaces between objects. The oblique surfaces on the objects are identified by a letter. Identify the oblique surfaces on each of the three views with a corresponding letter. Also identify the type of projection used by placing the appropriate ISO symbol at the bottom of the drawing.

21. On a B- or A4-size sheet, make a three-view drawing of one of the parts shown in Figs. 3-6-D to 3-6-G. Allow 1.20 in. (30 mm) between views. Do not dimension.

Fig. 3-6-A   Sketching assignment.

Fig. 3-6-B   Sketching assignment.

Fig. 3-6-C   Sketching assignment.

Fig. 3-6-D   Base plate.

Fig. 3-6-F   Support.

Fig. 3-6-E   Angle brace.

Fig. 3-6-G   Locking base.

Assignments for Unit 3-7, First-Angle Orthographic Projection

22. On two A- or A4-size sheets of pre-printed grid paper (.25 in. or 10 mm grids) sketch three views in first-angle orthographic projection of each of the objects shown in Figs. 3-7-A and 3-7-B. Draw three objects on each sheet. Each square on the objects represents one square on the grid paper. Allow one grid space between views and a minimum of two grid spaces between objects. Identify the type of projection used by placing the ISO projection symbol at the bottom of the drawing.

23. On a B- or A3-size sheet, make a three-view drawing in first-angle projection of one of the parts shown in Figs. 3-7-C to 3-7-F. Allow 1.20 in. (30 mm) between views.

Fig. 3-7-A   Sketching assignment.

Fig. 3-7-B   Sketching assignment.

Fig. 3-7-C   Spacer.

Fig. 3-7-E   Angle slide.

Fig. 3-7-D   Spacer.

Fig. 3-7-F   Bracket.

Assignment for Unit 3-8, One- and Two-View Drawings

24. On a B- or A3-size sheet, select any four of the objects shown in Fig. 3-8-A or 3-8-B and draw only the necessary views in orthographic third-angle projection which will completely describe each part. Use symbols or abbreviations where possible. The drawings need not be to scale but should be drawn in proportion to the illustrations shown.

PART I

PART 2

PART 3

PART 4

PART 5

PART 6

Fig. 3-8-A   Drawing assignment.

PART I

PART 2

PART 3

PART 4

PART 5

PART 6

Fig. 3-8-B   Drawing assignment.

Assignment for Unit 3-9, Partial Views

25. On a B- or A3-size sheet, select any one of the objects shown in Figs. 3-9-A to 3-9-D and draw only the necessary views (full and partial) which will completely describe each part. Add dimensions and machining symbols when required.

Fig. 3-9-A   Round flange.

Fig. 3-9-B   Square flange.

ROUNDS AND FILLETS R 2
MATL — CI

Fig. 3-9-C   Flanged coupling.

ROUNDS AND FILLETS R .06
MATL — CI

Fig. 3-9-D   Flanged adapter.

Assignment for Unit 3-10, Rear Views and Enlarged Views

26. On a B- or A3-size sheet, select one of the panels shown in Fig. 3-10-A or 3-10-B and make a detail drawing of the part. Enlarged views are recommended. Panels such as these, where labeling is used to identify the terminals, are used extensively in the electrical and electronics industry.

In addition to the detail drawing, you are required to select a plastic material for the part. Design requirements dictate that the part be strong, have good electrical properties, be opaque, and, because of the large quantity required, be capable of being molded. Prepare a report to accompany the drawing which lists a minimum of three choices of materials that could be specified.

Assignment for Unit 3-11, Opposite-Hand Views and Key Plans

27. With most truss drawings, the scale used on the overall assembly is such that intricate detail cannot be clearly shown. As a result, enlarged detail views are added. With this type of assembly, many parts are opposite-hand to their counterparts.

The primary problem in the assembly of large-span multimember space structures has been to find some simple, inexpensive, and repetitive way to connect many members into and through a typical joint. The approach shown in this assignment is to shop-fabricate as much as possible, to ship the subassemblies to the site, and finally to complete the assembly by bolting the subassemblies together. The number, size, and strength of bolts are calculated by using the loading requirements at each connection.

Benefits of shop prefabrication of large units include minimization of field erection time and less possible error in the field resulting from the greater tolerance control in the shop.

On a B-size sheet, draw the enlarged views of the gusset assemblies shown in Fig. 3-11-A to a scale of 1 in. = 1 ft. On a second sheet, draw up the complete bill of materials necessary to make one complete truss. The bolts used to hold the truss to the walls are not to be included. Show the half truss assembly on this sheet.

Shop Bolting Data. All structural members will be bolted to the gusset plate with four .375 in. high-strength bolts. Spacing is 1.5 in. from end and 3.0 in. center to center.

Field Bolting Data. All connections are to be made with five .375 in. high-strength bolts. Spacing is 1.5 in. from end and 3.0 in. center to center.

Fig. 3-10-A Radio cover plate.

Fig. 3-10-B Transceiver cover plate.

Fig. 3-11--A Crescent truss.

CHAPTER 4

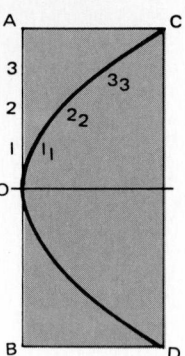

Applied Geometry

UNIT 4-1
Straight Lines

Most of the lines forming the views on mechanical drawings can be drawn using the instruments and equipment described in Chap. 2. However, geometric constructions have important uses, both in making drawings and in solving problems with graphs and diagrams. Sometimes it is necessary to use geometric constructions, particularly if the drafter does not have the advantages afforded by a drafting machine, an adjustable triangle, or templates for drawing hexagonal and elliptical shapes.

To Draw a Line or Lines Parallel to and at a Given Distance from an Oblique Line

1. Given line *AB* (Fig. 4-1-1), erect a perpendicular *CD* to *AB*.

2. Space the given distance from the line *AB* by scale measurement or by an arc along line *CD*.

3. Position a triangle, using a second triangle or a T square as base, so that one side of the triangle is parallel with the given line.

4. Slide this triangle along the base to the point at the desired distance from the given line, and draw the required line.

To Draw a Straight Line Tangent to Two Circles

Place a T square or straightedge so that the top edge just touches the edges of the circles, and draw the tangent line (Fig. 4-1-2). Perpendiculars to this line from the centers of the circles give the tangent points T_1 and T_2.

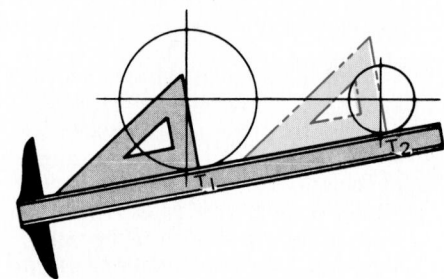

Fig. 4-1-2   Drawing a straight line tangent to two circles.

To Bisect a Straight Line

1. Given line *AB* (Fig. 4-1-3), set the compass to a radius greater than ½ *AB*.

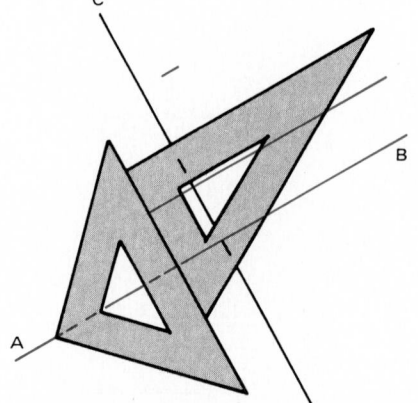

Fig. 4-1-1   Drawing parallel lines with the sides of triangles.

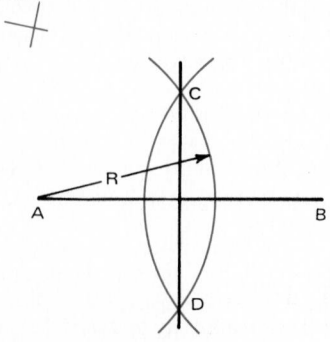

Fig. 4-1-3   Bisecting a line.

2. Using centers at *A* and *B*, draw intersecting arcs above and below line *AB*. A line *CD* drawn through the intersections will divide *AB* into two equal parts and will be perpendicular to line *AB*.

To Bisect an Arc

1. Given arc *AB* (Fig. 4-1-4), set the compass to a radius greater than ½ *AB*.
2. Using points *A* and *B* as centers, draw intersecting arcs above and below arc *AB*. A line drawn through the intersections *C* and *D* will divide the arc *AB* into two equal parts.

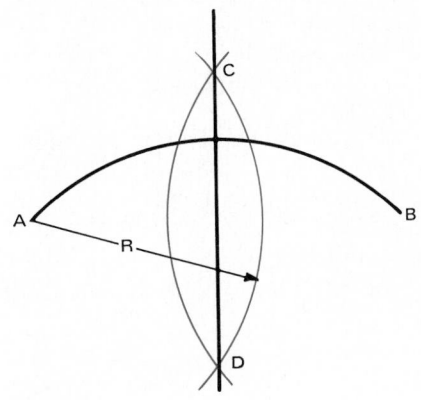

Fig. 4-1-4 Bisecting an arc.

To Bisect an Angle

1. Given angle *ABC*, with center *B* and a suitable radius (Fig. 4-1-5) draw an arc to cut *BC* at *D* and *BA* at *E*.
2. With centers *D* and *E* and equal radii, draw arcs to intersect at *F*.
3. Join *B* and *F* and extend to *G*. Line *BG* is the required bisector.

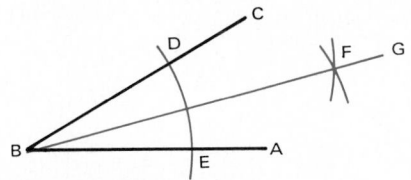

Fig. 4-1-5 Bisecting an angle.

To Divide a Line into a Given Number of Equal Parts

1. Given line *AB* and the number of equal divisions desired (12, for example), draw a perpendicular from *A*.

Fig. 4-1-6 Dividing a straight line into equal parts.

2. Place the scale so that the desired number of equal divisions is conveniently included between *B* and the perpendicular. Then mark these divisions, using short vertical marks from the scale divisions as shown in Fig. 4-1-6.
3. Draw perpendiculars to line *AB* through the points marked, dividing the line *AB* as required.

ASSIGNMENT

See Assignment 1 for Unit 4-1 on page 78.

UNIT 4-2
Arcs and Circles

To Draw an Arc Tangent to Two Lines at Right Angles to Each Other

Given radius *R* of the arc (Fig. 4-2-1).

1. Draw an arc having radius *R* with center at *B*, cutting the lines *AB* and *BC* at *D* and *E*, respectively.
2. With *D* and *E* as centers and with the same radius *R*, draw arcs intersecting at *O*.
3. With center *O*, draw the required arc. The tangent points are *D* and *E*.

Fig. 4-2-1 Arc tangent to two lines at right angles to each other.

To Draw an Arc Tangent to the Sides of an Acute Angle

Given radius *R* of the arc (Fig. 4-2-2).

1. Draw lines inside the angle, parallel to the given lines, at distance *R* away from the given lines. The center of the arc will be at *C*.
2. Set the compass to radius *R*, and with center *C* draw the arc tangent to the given sides. The tangent points *A* and *B* arc found by drawing perpendiculars through point *C* to the given lines.

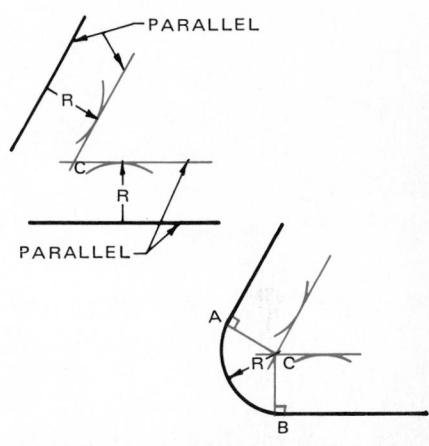

Fig. 4-2-2 Drawing an arc tangent to the sides of an acute angle.

To Draw an Arc Tangent to Two Sides of an Obtuse Angle

Follow the same procedure as for an acute angle (Fig. 4-2-3).

Fig. 4-2-3 Drawing an arc tangent to the sides of an obtuse angle.

To Draw a Circle on a Regular Polygon

1. Given the size of the polygon (Fig. 4-2-4), bisect any two sides; for example, *BC* and *DE*. The center of the polygon is where bisectors *FO* and *GO* intersect at point *O*.
2. The inner circle radius is *OH*, and the outer circle radius is *OA*.

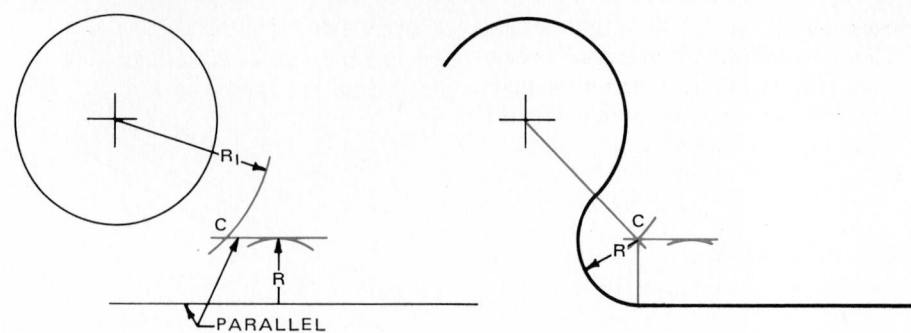

Fig. 4-2-6 Drawing an arc tangent to a circle and a straight line.

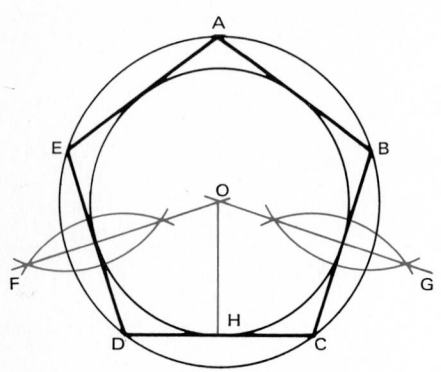

Fig. 4-2-4 Drawing a circle on a regular polygon.

To Draw an Arc Tangent to a Given Circle and Straight Line

1. Given *R*, the radius of the arc (Fig. 4-2-6), draw a line parallel to the given straight line between the circle and the line at distance *R* away from the given line.
2. With the center of the circle as center and radius R_1 (radius of the circle plus *R*), draw an arc to cut the parallel straight line at *C*.
3. With center *C* and radius *R*, draw the required arc tangent to the circle and the straight line.

To Draw an Arc Tangent to Two Circles (Fig. 4-2-7)

Figure 4-2-7A

1. Given the radius of arc *R*, with the center of circle *A* as center and radius R_2 (radius of circle *A* plus *R*), draw an arc in the area between the circles.
2. With the center of circle *B* as center and radius R_3 (radius of circle *B* plus *R*), draw an arc to cut the other arc at *C*.
3. With center *C* and radius *R*, draw the required arc tangent to the given circles.

To Draw a Reverse, or Ogee, Curve Connecting Two Parallel Lines

1. Given two parallel lines *AB* and *CD* and distances *X* and *Y* (Fig. 4-2-5), join points *B* and *C* with a line.
2. Erect a perpendicular to *AB* and *CD* from points *B* and *C*, respectively.
3. Select point *E* on line *BC* where the curves are to meet.
4. Bisect *BE* and *EC*.
5. Points *F* and *G* where the perpendiculars and bisectors meet are the centers for the arcs forming the ogee curve.

(A)

(B)

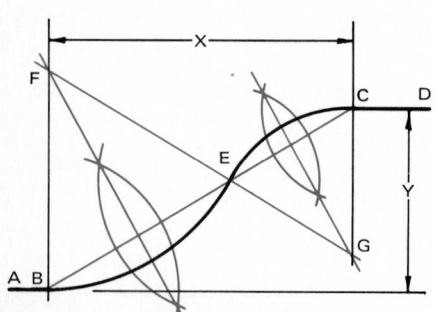

Fig. 4-2-5 Drawing a reverse (ogee) curve connecting two parallel planes.

Fig. 4-2-7 Drawing an arc tangent to two circles.

Figure 4-2-7B

1. Given radius of arc R, with the center of circle A as center and radius $R - R_2$, draw an arc in the area between the circles.
2. With the center of circle B as center and radius $R - R_3$, draw an arc to cut the other arc at C.
3. With center C and radius R, draw the required arc tangent to the given circles.

To Draw an Arc or Circle Through Three Points Not in a Straight Line

1. Given points A, B, and C (Fig. 4-2-8), join points A, B, and C as shown.

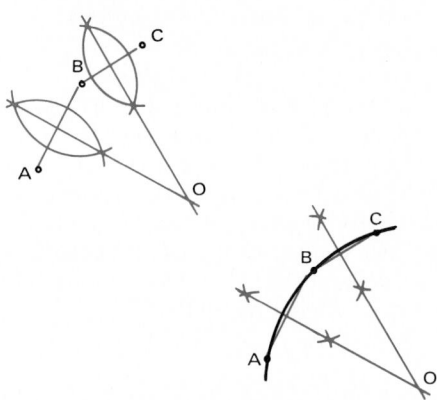

Fig. 4-2-8 Drawing an arc or circle through three points not in a straight line.

2. Bisect lines AB and BC and extend bisecting lines to intersect at O. Point O is the center of the required circle or arc.
3. With center O and radius OA draw an arc.

ASSIGNMENT

See Assignment 2 for Unit 4-2 on page 78.

UNIT 4-3
Polygons

A regular polygon is a plane figure bounded by straight lines of equal length and contains angles of equal size.

To Draw a Hexagon (Fig. 4-3-1), Given the Distance Across the Flats

1. Establish horizontal and vertical center lines for the hexagon.
2. Using the intersection of these lines as center, with radius one-half the distance across the flats, draw a light construction circle.
3. Using the 60° triangle, draw six straight lines, equally spaced, passing through the center of the circle.
4. Draw tangents to these lines at their intersection with the circle.

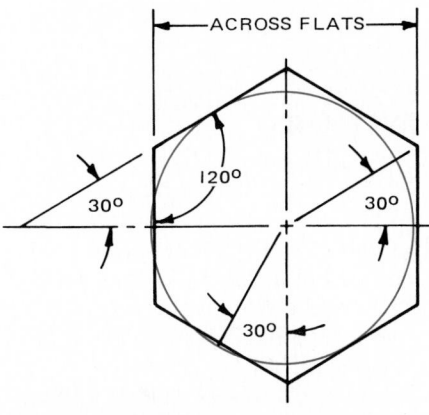

Fig. 4-3-1 Constructing a hexagon, given distance across flats.

To Draw a Hexagon, Given the Distance Across the Corners (Fig. 4-3-2)

1. Establish horizontal and vertical center lines, and draw a light construction circle with radius one-half the distance across the corners.
2. With a 60° triangle, establish points on the circumference 60° apart.
3. Draw straight lines connecting these points.

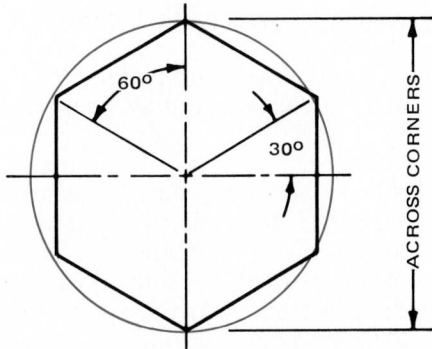

Fig. 4-3-2 Constructing a hexagon, given distance across corners.

To Draw an Octagon, Given the Distance Across the Flats (Fig. 4-3-3)

1. Establish horizontal and vertical center lines and draw a light construction circle with radius one-half the distance across the flats.
2. Draw horizontal and vertical lines tangent to the circle.
3. Using the 45° triangle, draw lines tangent to the circle at a 45° angle from the horizontal.

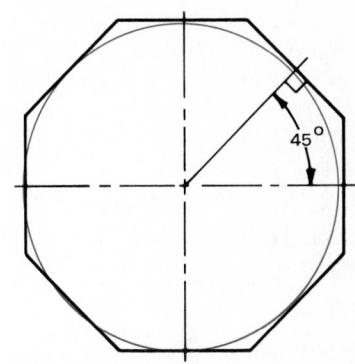

Fig. 4-3-3 Constructing an octagon, given distance across flats.

To Draw an Octagon, Given the Distance Across the Corners (Fig. 4-3-4)

1. Establish horizontal and vertical center lines and draw a light construction circle with radius one-half the distance across the corners.
2. With the 45° triangle, establish points on the circumference between the horizontal and vertical center lines.
3. Draw straight lines connecting these points to the points where the center lines cross the circumference.

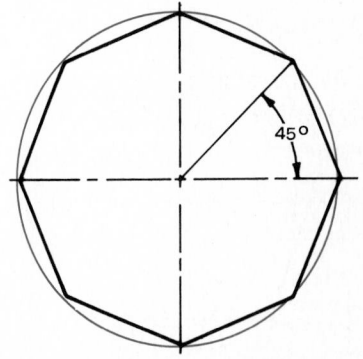

Fig. 4-3-4 Constructing an octagon, given distance across corners.

To Draw a Regular Polygon, Given the Length of the Sides

As an example, let a polygon have seven sides.

1. Given the length of side AB (Fig. 4-3-5), with radius AB and A as center, draw a semicircle and divide it into seven equal parts using a protractor.
2. Through the second division from the left, draw radial line A2.
3. Through points 3, 4, 5, and 6 extend radial lines as shown.
4. With AB as radius and B as center, cut line A6 at C. With the same radius and C as center, cut line A5 at D. Repeat at E and F.
5. Connect these points with straight lines.

These steps can be followed in drawing a regular polygon with any number of sides.

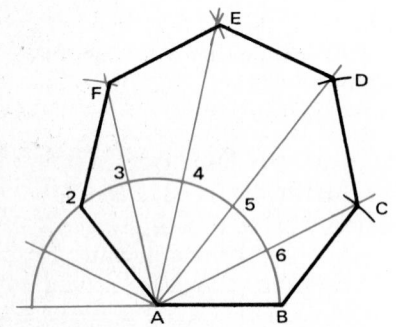

Fig. 4-3-5   Constructing a regular polygon, given length of one side.

To Inscribe a Regular Pentagon in a Given Circle

1. Given circle with center O (Fig. 4-3-6), draw the circle with diameter AB.

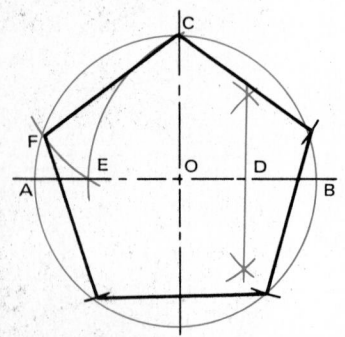

Fig. 4-3-6   Inscribing a regular pentagon in a given circle.

2. Bisect line OB at D.
3. With center D and radius DC, draw arc CE to cut the diameter at E.
4. With C as center and radius CE, draw arc CF to cut the circumference at F. Distance CF is one side of the pentagon.
5. With radius CF as a chord, mark off the remaining points on the circle. Connect the points with straight lines.

ASSIGNMENT

See Assignment 3 for Unit 4-3 on page 78.

UNIT 4-4
The Ellipse

The *ellipse* is the plane curve generated by a point moving so that the sum of the distances from any point on the curve to two fixed points, called *foci*, is a constant.

Often a drafter is called upon to draw oblique and inclined holes and surfaces which take the form of an ellipse. Several methods, true and approximate, are used for its construction. The terms *major diameter* and *minor diameter* will be used in place of *major axis* and *minor axis* so the reader won't become confused with the mathematical X and Y axes.

To Draw an Ellipse—Two-Circle Method

1. Given the major and minor diameters (Fig. 4-4-1), construct two concentric circles with diameters equal to AB and CD.
2. Divide the circles into a convenient number of equal parts. Figure 4-4-1 shows 12.
3. Where the radial lines intersect the outer circle, as at 1, draw lines parallel to line CD inside the outer circle.
4. Where the same radial line intersects the inner circle, as at 2, draw a line parallel to axis AB away from the inner circle. The intersection of these lines, as at 3, gives points on the ellipse.
5. Draw a smooth curve through these points.

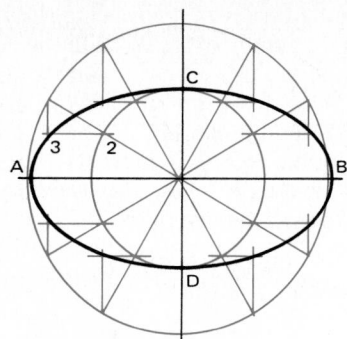

Fig. 4-4-1   Drawing an ellipse—two circle method.

To Draw an Ellipse—Four-Center Method

1. Given the major diameter CD and the minor diameter AB (Fig. 4-4-2), join points A and C with a line.
2. Lay off AE equal to CO-AO.
3. Draw the right bisector of CE, locating point G on line CO and point F on line AB. (Line AB may have to be extended.)
4. Make OF equal to OF_1 and OG equal to OG_1.
5. Points F, F_1, G, and G_1 are the centers for the two large and two small arcs forming the ellipse.

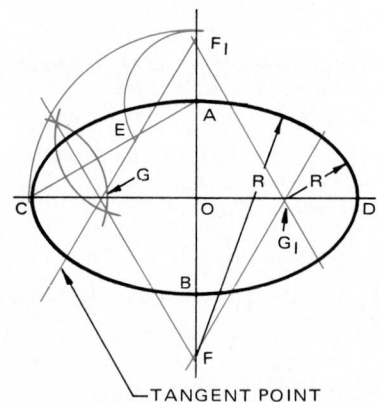

Fig. 4-4-2   Drawing an ellipse—four center method.

To Draw an Ellipse—Parallelogram Method

1. Given the major diameter CD and minor diameter AB (Fig. 4-4-3), construct a parallelogram.
2. Divide CO into a number of equal parts. Divide CE into the same number of equal parts. Number the points from C.
3. Draw a line from B to point 1 on line CE. Draw a line from A through

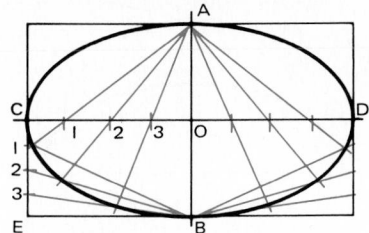

Fig. 4-4-3 Drawing an ellipse—parallelogram method.

point 1 on *CO*, intersecting the previous line. The point of intersection will be one point on the ellipse.

4. Proceed in the same manner to find other points on the ellipse.
5. Draw a smooth curve through these points.

ASSIGNMENT

See Assignment 4 for Unit 4-4 on page 78.

UNIT 4-5
Helix and Parabola

HELIX

The *helix* is the curve generated by a point that revolves uniformly around and up or down the surface of a cylinder. The *lead* is the vertical distance that the point rises or drops in one complete revolution.

To Draw a Helix

1. Given the diameter of the cylinder and the lead (Fig. 4-5-1), draw the top and front views.
2. Divide the circumference (top view) into a convenient number of parts (use 12) and label them.
3. Project lines down to the front view.
4. Divide the lead into the same number of equal parts and label them as shown in Fig. 4-5-1.
5. The points of intersection of lines with corresponding numbers lie on the helix. *Note:* Since points 8 to 12 lie on the back portion of the cylinder, the helix curve starting at point 7 and passing through points 8, 9, 10, 11, 12 to point 1 will appear as a hidden line.

6. If the development of the cylinder is drawn, the helix will appear as a straight line on the development.

PARABOLA

The *parabola* is a plane curve generated by a point that moves along a path equidistant from a fixed line (directrix) and a fixed point (focus).

To Construct a Parabola—Parallelogram Method

1. Given the sizes of the enclosing rectangle, distances *AB* and *AC* (Fig. 4-5-2A), construct a parallelogram.
2. Divide *AC* into a number of equal parts. Divide *AO* into the same number of equal parts. Number the points as shown.
3. Draw a line from *O* to point 1 on line *AC*. Draw a line parallel to the axis through point 1 on line *AO*, intersecting the previous line *O*-1. The point of intersection will be one point on the parabola.
4. Proceed in the same manner to find other points on the parabola.
5. Connect the points using an irregular curve.

To Construct a Parabola—Offset Method

1. Given the sizes of the enclosing rectangle, distances *AB* and *AC* (Fig. 4-5-2B), construct a parallelogram.

2. Divide *OA* into four equal parts.
3. The offsets vary in length as the square of their distances from *O*. Since *OA* is divided into four equal parts, distance *AC* will be divided into 4^2, or 16, equal divisions. Thus since *O*1 is one-fourth the length of *OA*, the length of line 1-1$_1$ will be $(1/4)^2$, or $1/16$, the length of *AC*.
4. Since distance *O*2 is one-half the length of *OA*, the length of line 2-2$_1$ will be $(1/2)^2$, or $1/4$, the length of *AC*.
5. Since distance *O*3 is three-fourths the length of *0A*, the length of line 3-3$_1$ will be $(3/4)^2$, or $9/16$, the length of *AC*.
6. Complete the parabola by joining the points with an irregular curve.

ASSIGNMENT

See Assignment 5 for Unit 4-5 on page 78.

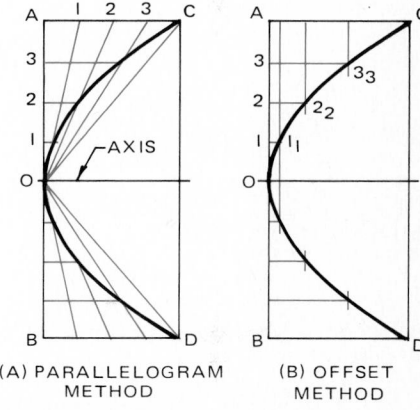

(A) PARALLELOGRAM METHOD   **(B) OFFSET METHOD**

Fig. 4-5-2 Common methods used to construct a parabola.

Fig. 4-5-1 Drawing a cylindrical helix.

ASSIGNMENTS for Chapter 4

Assignment for Unit 4-1, Straight Lines

1. Divide a B- or A3-size sheet as shown in Fig. 4-1-A. In the designated areas draw the geometric constructions.

Assignment for Unit 4-2, Arcs and Circles

2. Divide a B- or A3-size sheet as shown in Fig. 4-2-A. In the designated areas draw the geometric constructions.

Assignment for Unit 4-3, Polygons

3. Divide an A3- or B-size sheet as shown in Fig. 4-3-A. In the designated areas, draw the geometric constructions.

Assignment for Unit 4-4, The Ellipse

4. Divide a B- or A3-size sheet as shown in Fig. 4-4-A. In the designated areas draw the geometric constructions.

Assignment for Unit 4-5, Helix and Parabola

5. Divide an A3- or B-size sheet, as shown in Fig. 4-5-A. In the designated areas, draw the geometric constructions.

Review Assignments

6. On an A3- or B-size sheet draw one of the parts shown in Figs. 4-6-A to 4-6-D. Do not erase construction lines. Scale full or 1:1. Do not dimension.

Fig. 4-1-A Straight-line construction.

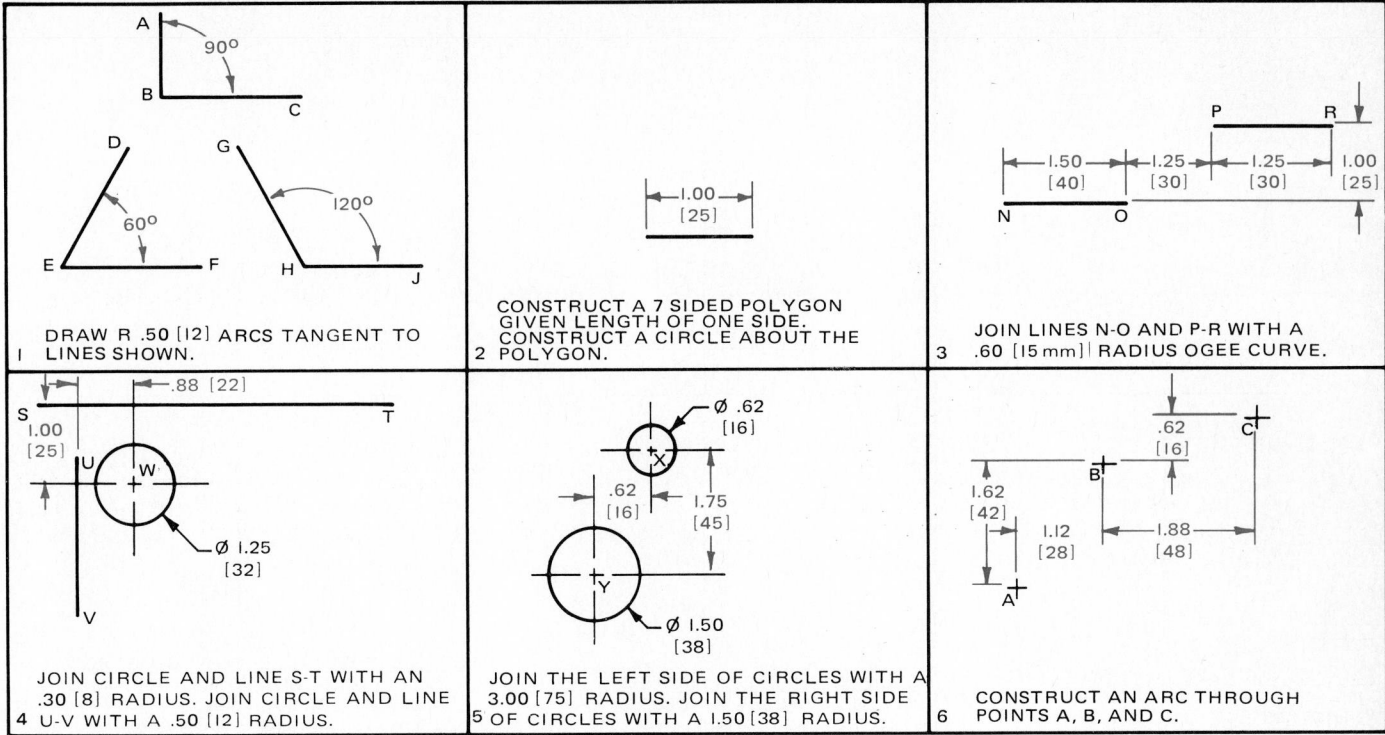

1 DRAW R .50 [12] ARCS TANGENT TO LINES SHOWN.

2 CONSTRUCT A 7 SIDED POLYGON GIVEN LENGTH OF ONE SIDE. CONSTRUCT A CIRCLE ABOUT THE POLYGON.

3 JOIN LINES N-O AND P-R WITH A .60 [15 mm] RADIUS OGEE CURVE.

4 JOIN CIRCLE AND LINE S-T WITH AN .30 [8] RADIUS. JOIN CIRCLE AND LINE U-V WITH A .50 [12] RADIUS.

5 JOIN THE LEFT SIDE OF CIRCLES WITH A 3.00 [75] RADIUS. JOIN THE RIGHT SIDE OF CIRCLES WITH A 1.50 [38] RADIUS.

6 CONSTRUCT AN ARC THROUGH POINTS A, B, AND C.

Fig. 4-2-A Drawing assignment.

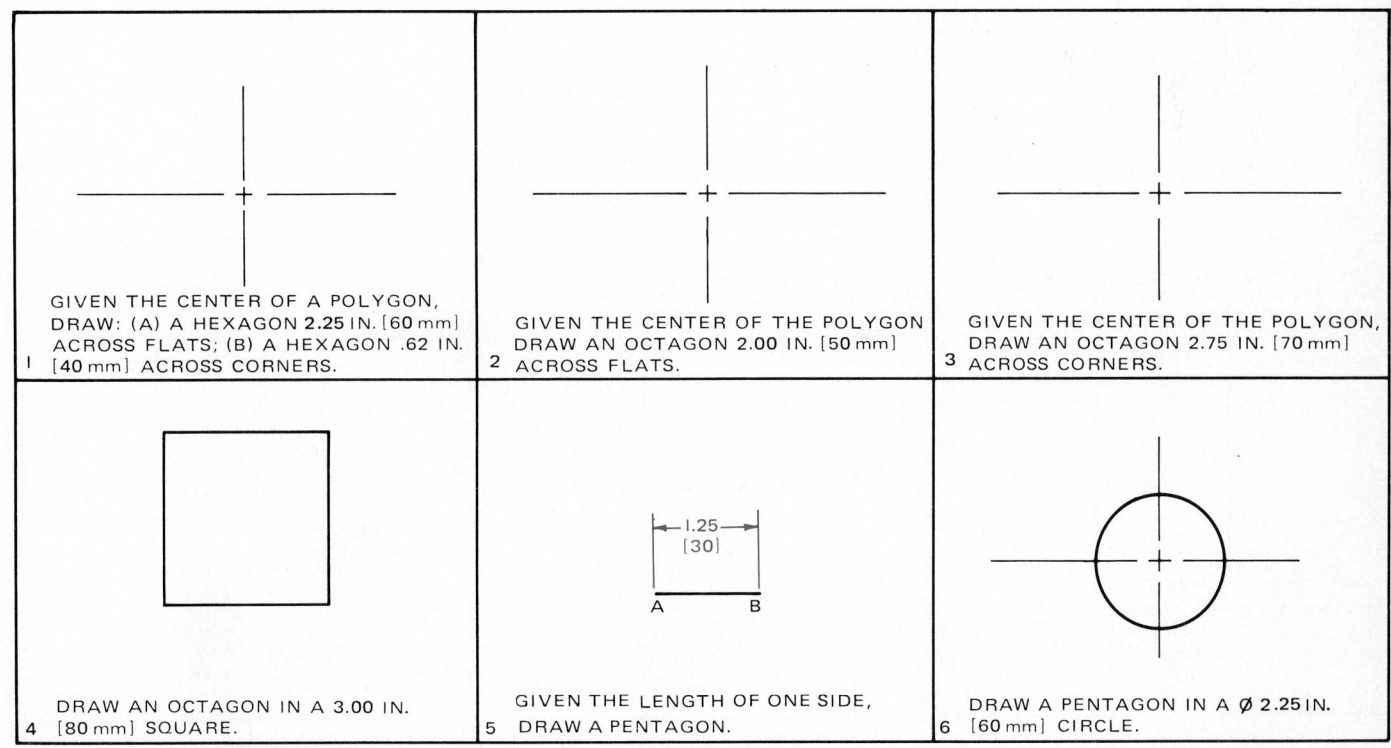

1 GIVEN THE CENTER OF A POLYGON, DRAW: (A) A HEXAGON 2.25 IN. [60 mm] ACROSS FLATS; (B) A HEXAGON .62 IN. [40 mm] ACROSS CORNERS.

2 GIVEN THE CENTER OF THE POLYGON DRAW AN OCTAGON 2.00 IN. [50 mm] ACROSS FLATS.

3 GIVEN THE CENTER OF THE POLYGON, DRAW AN OCTAGON 2.75 IN. [70 mm] ACROSS CORNERS.

4 DRAW AN OCTAGON IN A 3.00 IN. [80 mm] SQUARE.

5 GIVEN THE LENGTH OF ONE SIDE, DRAW A PENTAGON.

6 DRAW A PENTAGON IN A Ø 2.25 IN. [60 mm] CIRCLE.

Fig. 4-3-A Drawing assignment.

ELLIPSE 2 CIRCLE METHOD
GIVEN 2 CIRCLES OF Ø 2.00 [50]
AND Ø 4.00 [100], DRAW AN
ELLIPSE.

ELLIPSE 4 CENTER METHOD
GIVEN 2 CIRCLES OF Ø 2.75 [70]
AND 4.25 [110], DRAW AN ELLIPSE.

ELLIPSE PARALLELOGRAM METHOD
GIVEN MAJOR DIA OF 4.80 [120] AND
MINOR DIA OF 2.40 [60]
DRAW AN ELLIPSE.

DIMENSIONS IN INCHES

Fig. 4-4-A   Drawing assignment.

GIVEN DIA AND LEAD
CONSTRUCT A HELIX
STARTING AT POINT A.

OFFSET METHOD
GIVEN A RECTANGLE,
CONSTRUCT A PARABOLA.

PARALLELOGRAM METHOD
GIVEN A RECTANGLE
CONSTRUCT A PARABOLA.

Fig. 4-5-A   Drawing assignment.

Fig. 4-6-A   Hex wrench.

Fig. 4-6-B   Template.

Fig. 4-6-C   Adjustable fork.

Fig. 4-6-D   Fan base.

CHAPTER 5

Basic Dimensioning

UNIT 5-1
Basic Dimensioning

A working drawing is one from which a tradesperson can produce a part. The drawing must be a complete set of instructions, so that it will not be necessary to give further information to the people making the object. A working drawing, then, consists of the views necessary to explain the shape, the dimensions needed by the tradesperson, and required specifications, such as material and quantity needed. The latter information may be found in the notes on the drawing, or it may be located in the title block.

DIMENSIONING

Dimensions are indicated on drawings by extension lines, dimension lines, leaders, arrowheads, figures, notes, and symbols. They define geometric characteristics such as lengths, diameters, angles, and locations. See Fig. 5-1-1. The lines used in dimensioning are thin in contrast to the outline of the object. The dimensions must be clear and concise and permit only one interpretation. In general, each surface, line, or point is located by only one set of dimensions. Deviations from the approved rules for dimensioning should be made only in exceptional cases, when they will improve the clar-

Fig. 5-1-1 Basic dimensioning elements.

ity of the dimensions. An exception to these rules is for arrowless and tabular dimensioning, which is discussed in Unit 5-4.

In general, each surface, line, or point is located by only one set of dimensions. These dimensions are not duplicated in other views, except for the purpose of identification, the improvement of clarity, or both.

Drawings for industry requires some form of tolerancing on dimensions so that components can be properly assembled and manufacturing and production requirements can be met. This chapter deals only with basic dimensioning and tolerancing techniques. Modern engineering tolerancing, such as true positioning and tolerance of form, is covered in detail in Chap. 30.

Dimension and Extension Lines

Dimension lines are used to determine the extent and direction of dimensions, and they are terminated by neatly made, uniform arrowheads, as shown in Fig. 5-1-2. Arrowheads are usually drawn freehand, and the recommended length and width should be in a ratio of 3:1 (Fig. 5-1-3B). The length of the arrowhead should be equal to the height of the dimension numerals. Where space is limited, a small circular dot may be used in lieu of an arrowhead (Fig. 5-1-3D). Normally, a break is made near the center of the dimension line for the insertion of the dimension which indicates the distance between the extension lines. When several dimension lines are directly above or next to one another, it is good practice to stagger the dimensions in order to improve the clarity of the drawing. In special cases, such as dual dimensioning that combines inch and metric units of measure on the same engineering drawing, or as a simplified drafting practice, the dimension line may be unbroken. The spacing suitable for most drawings between parallel dimension lines is .25 in. (8 mm), and the spacing between the outline of the object and the nearest dimension line should be approximately .50 in. (10 mm). When the space between the extension lines is too small to permit the placing of the dimension line complete with arrowheads and dimension, then the alternate method of placing the dimension line, dimension, or both outside the extension lines are used. See Fig. 5-1-3D. Center lines should never be used for dimension lines. Every effort should be made to avoid crossing dimension lines by placing the shortest dimension closest to the outline (Fig. 5-1-3E).

Dimension lines should be placed outside the view where possible and should extend to extension lines rather than visible lines. However, when readability is improved by avoiding either extra-long extension lines (Fig. 5-1-4) or the crowding of dimensions, placing of dimensions on views is permissible. Avoid dimensioning to hidden lines. In order to do so, it may be necessary to use a sectional view or a broken-out section. When the termination for a dimension is not included, as when used on partial views, a double arrow is used. For symmetrical fea-

(A) PLACEMENT OF DIMENSIONS

(B) ARROWHEAD SIZE

A SMALL CIRCULAR DOT MAY BE USED IN LIEU OF ARROWHEADS WHERE SPACE IS RESTRICTED.

(C) OBLIQUE DIMENSIONING

(E) STAGGER DIMENSIONS FOR CLARITY

(D) DIMENSIONING IN RESTRICTED AREAS

Fig. 5-1-3 Dimension lines.

(F) USE OF DOUBLE ARROWHEADS

Fig. 5-1-2 Dimension and extension lines.

(A) IMPROVING READABILITY OF DRAWING

(B) AVOIDING LONG EXTENSION LINES

Fig. 5-1-4 Placing dimensions on view.

tures the dimension line should extend beyond the center line before the double arrowheads are added (Fig. 5-1-3F).

Extension (*projection*) lines are used to indicate the point or line on the drawing to which the dimension applies. See Fig. 5-1-5. A small gap is left between the extension line and the outline to which it refers, and the extension line extends about .12 in. (3 mm) beyond the outermost dimension line. However, when extension lines refer to points, as in Fig. 5-1-5E, they should extend through the points. Extension lines are usually drawn perpendicular to dimension lines. However, to improve clarity or when there is overcrowding, extension lines may be drawn at an oblique angle as long as clarity is maintained.

Center lines may be used as extension lines in dimensioning. The portion of the center line extending past the circle is not broken.

Where extension lines cross other extension lines, dimension lines, or visible lines, they are not broken. However, if extension lines cross arrowheads or dimension lines close to arrowheads, a break in the extension line is recommended.

Leaders

Leaders are used to direct notes, dimensions, symbols, item numbers, or part numbers to features on the

drawing. See Fig. 5-1-6. A leader should generally be a single straight inclined line (not vertical or horizontal) except for a short horizontal portion extending to the center of the height of the first or last letter or digit of the note. The leader is terminated by an arrowhead or a dot of at least .06 in. (1.5 mm) in diameter. Arrowheads should always terminate on a line; dots should be used within the outline of the object and rest on a surface. Leaders should not be bent in any way unless it

Fig. 5-1-6 Leaders.

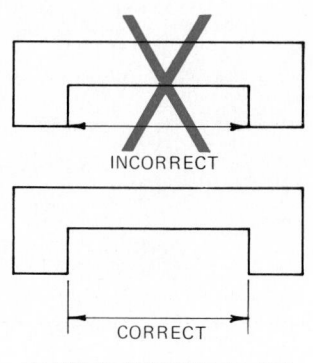

(A) USE OF EXTENSION LINES

(B) CENTER LINE USED AS EXTENSION LINE

(C) BREAK IN EXTENSION LINES

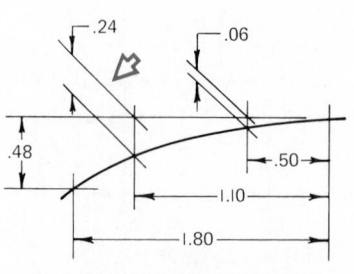

(D) OBLIQUE EXTENSION LINES

Fig. 5-1-5 Extension lines.

(E) EXTENSION LINE FROM POINTS

(F) EXTENSION LINE FROM POINTS

is unavoidable. Leaders should not cross one another, and two or more leaders adjacent to one another should be drawn parallel if practicable. It is better to repeat dimensions or references than to use long leaders.

Where a leader is directed to a circle or circular arc, its direction should point to the center of the arc or circle. Regardless of the reading direction used, aligned or unidirectional, all notes and dimensions used with leaders are placed in a horizontal position.

Notes

Notes are used to simplify or complement dimensioning by giving information on a drawing in a condensed and systematic manner. They may be general or local notes, and should be in the present or future tense.

General Notes These refer to the part or the drawing as a whole. They should be shown in a central position below the view to which they apply or placed in a general note column. Typical examples of this type of note are

- FINISH ALL OVER
- ROUNDS AND FILLETS R .06
- REMOVE ALL SHARP EDGES

LOCAL NOTES These apply to local requirements only and are connected by a leader to the point to which the note applies. Typical examples are

- ⌀ 6, 4 HOLES
- 2 × 45°
- ⌀ 3
  ⌀ 11.5 × 86° CSK
- M 12 × 1.25

UNITS OF MEASUREMENTS

Although the metric system of dimensioning is becoming the offical standard of measurement, many drawings in current use are still dimensioned in inches or feet and inches. For this reason, drafters should be familiar with all the dimensioning systems which they may encounter. The dimensions used in this book are primarily decimal inch. However, metric and dual dimensions are also used very frequently.

Inch Units of Measurement

Decimal-Inch System Parts are designed in basic decimal increments, prefera-

(A) DECIMAL INCH

(B) FEET AND INCHES

(C) MILLIMETERS

Fig. 5-1-7 Dimensioning units.

bly .02 in., and are expressed with a minimum of two figures to the right of the decimal point. See Fig. 5-1-7. Using the .02 in. module, the second decimal place (hundredths) is an even number or zero. By using the design modules having an even number for the last digit, dimensions can be halved for center distances without increasing the number of decimal places. Decimal dimensions which are not multiples of .02, such as .01, .03, and .15, should be used only when it is essential to meet design requirements such as to provide clearance, strength, smooth curves, etc. When greater accuracy is required, sizes are expressed as three- or

four-place decimal numbers, for example, 1.875.

Whole dimensions will show a minimum of two zeros to the right of the decimal point.

24.00 *not* 24

An inch value of less than 1 is shown without a zero to the left of the decimal point.

.44 *not* 0.44

In cases where parts have to be aligned with existing parts or commercial products, which are dimensioned in fractions, it may be necessary to use decimal equivalents of fractional dimensions.

Fractional-Inch System In this system, parts are designed in basic units of common fractions down to 1/64 in. Decimal dimensions are used when finer divisions than 1/64 in. must be made. Common fractions are used for specifying the size of holes that may be produced by drills ordinarily stocked in fraction sizes and for the sizes of standard screw threads.

When common fractions are used on drawings, the fraction bar must not be omitted and should be horizontal except when applied with a typewriting machine which does not have a horizontal fraction bar.

When a dimension intermediate between 1/64 increments is necessary, it is expressed in decimals, such as .30, .257, or .2575 in.

The inch marks (″) should not be shown with the dimensions. A note such as

DIMENSIONS ARE IN INCHES

should be clearly shown on the drawing. The exception is when the dimension "1 in." is shown on the drawing. The 1 should then be followed by the inch marks—1″, *not* 1.

Foot and Inch System Feet and inches are often used for installation drawings, drawings of large objects, and floor plans associated with architectural work. In this case, all dimensions 12 in. or greater are specified in feet and inches. For example, 24 in. is expressed as 2′–0, and 27 in. is expressed as 2′–3. Parts of an inch are usually expressed as common fractions, rather than as decimals.

The inch marks (″) are not shown except on architectural drawings. The drawing should carry a note such as

DIMENSIONS ARE IN FEET AND INCHES UNLESS OTHERWISE SPECIFIED

A dash and space should be left between the foot and inch values. For example, 1′–3, not 1′3.

SI Metric Units of Measurement

The standard metric units on engineering drawings are the millimeter (mm) for linear measure and micrometer (μm) for surface roughness. See Fig. 5-1-7. For architectural drawings, meter and millimeter units are used.

Whole numbers from 1 to 9 are shown without a zero to the left of the number or a zero to the right of the decimal point.

2 *not* 02 or 2.0

A millimeter value of less than 1 is shown with a zero to the left of the decimal point.

0.2 *not* .2 or .20
0.26 *not* .26

Decimal points should be uniform and large enough to be clearly visible on reduced-size prints. They should be placed in line with the bottom of the associated numbers and be given adequate space.

Commas should not be used to separate groups of three numbers in either inch or metric values. A space should be used in place of the comma.

32 541 *not* 32,541
2.562 827 6 *not* 2.562827

Identification A metric drawing should include a general note, such as

UNLESS OTHERWISE SPECIFIED DIMENSIONS ARE IN MILLIMETERS

and be identified by the word METRIC prominently displayed near the title block.

Units Common to Either System

Some measurements can be stated so that the callout will satisfy the units of both systems. For example, tapers

such as .006 in. per inch and 0.006 mm per millimeter can both be expressed simply as the ratio 0.006:1 or in a note: TAPER 0.006:1. Angular dimensions are also specified the same in both inch and metric systems.

Standard Items

Fasteners and Threads Either inch or metric fasteners and threads may be used. Refer to the Appendix and Chap. 8 for additional information.

Hole Sizes Tables showing standard inch and metric drill sizes are shown in the Appendix.

DUAL DIMENSIONING

With the great exchange of drawings taking place between the United States and the rest of the world, at one time it became advantageous to show drawings in both inches and millimeters. As a result, many companies adopted a dual system of dimensioning. Today, however, this type of dimensioning should be avoided if possible. However, when it is necessary or desirable

(A) POSITION METHOD

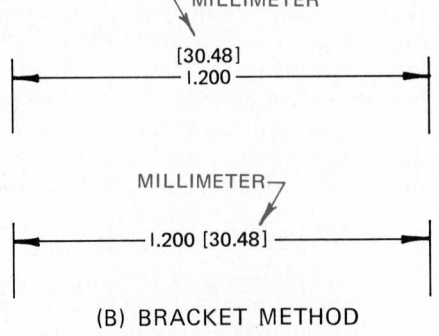

(B) BRACKET METHOD

Fig. 5-1-8 Dual dimensioning.

to give dimensions in both inches and millimeters on the same drawing, the following guidelines, as illustrated in Fig. 5-1-8, should be observed.

Show the millimeter dimension above the inch dimension separated by a horizontal line, or to the left of the inch dimension separated by a slash (oblique) line, or by enclosing the inch dimension in brackets.

When a note with a leader is used, the units of measurement are separated by an oblique line or by enclosing the latter dimension in brackets. The converted values are not shown below the measurements.

The method(s) used on the drawing should be clearly stated on the drawing.

A note or illustration should be located near the title block or strip to identify the inch and millimeter dimensions.

$$\frac{\text{MILLIMETER}}{\text{INCH}}$$

MILLIMETER/INCH

MILLIMETER [INCH]

Examples of dual-dimensioned drawings are shown in Figs. 5-1-9 and 5-1-10.

Dual dimensioning, as described above, has some disadvantages. More space is required to place both sets of dimensions, errors may occur when dual dimensions are manually converted and placed on the drawing, and the time spent by the drafter to calculate the dual dimensions may be costly. Some drawing offices have added the dual dimensions to the drawing in chart form. The drawing is dimensioned to the units (either inches or millimeters) by the drafter. Then it is given to an assistant who feeds all the dimensions into a computer which accurately tabulates the dual dimensions. The results are printed on an adhesive translucent material which is attached to the drawing.

ANGULAR UNITS

Angles are measured in degrees. The decimal degree is now preferred over the use of degrees, minutes, and seconds. For example, the use of 60.5° is preferred to the use of 60°30′. Where only minutes or seconds are specified,

Fig. 5-1-9 Dual dimensioned drawing.

Fig. 5-1-10 Drawing with millimeter conversion chart.

the number of minutes or seconds is preceded by 0°, or 0°0', as applicable. Some examples follow.

Decimal Degree	Degrees, Minutes, and Seconds
10° ± 0.5°	10° ± 0°30'
0.75°	0°45'
0.004°	0°0'15"
90° ± 1.0°	90° ± 1°
25.6° ± 0.2°	25°36' ± 0°12'
25.51°	25°30'36"

The dimension line of an angle is an arc drawn with the apex of the angle as the center point for the arc, wherever practicable. The position of the dimension varies according to the size of the angle and appears in a horizontal position. Recommended arrangements are shown in Fig. 5-1-11.

Fig. 5-1-11 Angular units.

READING DIRECTION

Dimensions and notes should be placed to be read from the bottom of the drawing. The exception to this rule is for architectural and structural drawings, where the aligned system of dimensioning is preferred.

In both methods, angular dimensions and dimensions and notes shown with leaders should be aligned with the bottom of the drawing. See Fig. 5-1-12.

Fig. 5-1-12 Reading direction of dimensions.

BASIC RULES FOR DIMENSIONING

Refer to Fig. 5-1-13.

- Place dimensions between the views when possible.
- Place the dimension line for the shortest length, width, or height nearest the outline of the object. Parallel dimension lines are placed in order of their size, making the longest dimension line the outermost.
- Place dimensions near the view that best shows the characteristic contour or shape of the object. In following this rule, dimensions will not always be between views.
- On large views, dimensions can be placed on the view to improve clarity.
- Use only one system of dimensions, either the unidirectional or the aligned, on any one drawing.
- Dimensions should not be duplicated in other views.

(A) PLACE DIMENSIONS BETWEEN VIEWS

(B) PLACE SMALLEST DIMENSION NEAREST THE VIEW BEING DIMENSIONED

(C) DIMENSION THE VIEW THAT BEST SHOWS THE SHAPE

Fig. 5-1-13 Basic dimensioning rules.

- Dimensions should be selected so that it will not be necessary to add or subtract dimensions in order to define or locate a feature.

SYMMETRICAL OUTLINES

Partial views are often drawn for the sake of economy. When only one-half the outline of a symmetrically shaped part is drawn, symmetry is indicated by applying the symmetry symbol to the center line on both sides of the part. See Fig. 5-1-14.

REFERENCE DIMENSIONS

A reference dimension is shown for information only, and it is not required for manufacturing or inspection purposes. It is enclosed in parentheses, as

Fig. 5-1-14 Dimensioning symmetrical outlines or features.

Fig. 5-1-15 Reference dimensions.

shown in Fig. 5-1-15. Formerly the abbreviation REF was used to indicate a reference dimension.

NOT-TO-SCALE DIMENSIONS

When a dimension on a drawing is altered, making it not to scale, it should be underlined with a straight, thick line (Fig. 5-1-16), except when the condition is clearly shown by break lines.

Fig. 5-1-16 Not to scale dimensions.

OPERATIONAL NAMES

The use of operational names with dimensions, such as turn, bore, grind, ream, tap, and thread, should be avoided. While the drafter should be aware of the methods by which a part can be produced, the method of manufacture is better left to the producer. If the completed part is adequately dimensioned and has surface texture symbols showing finish quality desired, it remains a shop problem to meet the drawing specifications.

ABBREVIATIONS

Abbreviations and symbols are used on drawings to conserve space and time, but used only where their meanings are quite clear. Therefore, only commonly accepted abbreviations such as those shown in the Appendix should be used on drawings.

References
1. ANSI Y14.5, *Dimensioning and Tolerancing*.

ASSIGNMENTS

See Assignments 1 through 3 for Unit 5-1 on page 114.

Review for Assignment
Unit 2-6   Drafting Skills

UNIT 5-2
Dimensioning Circular Features

DIAMETERS

Where the diameter of a single feature or the diameters of a number of concentric cylindrical features are to be specified, it is recommended that they be shown on the longitudinal view. Use the diameter symbol ∅. See Fig. 5-2-1.

Where the circular view is dimensioned, the dimension for the diameter is also preceded by the symbol ∅.

Where space is restricted or when only a partial view is used, diameters may be dimensioned as illustrated in Fig. 5-2-2.

(A) TWO-VIEW DRAWING

(B) ONE-VIEW DRAWING

Fig. 5-2-1   Diameters.

(C) DIMENSIONING DIAMETERS ON END VIEW

Fig. 5-2-2   Dimensioning diameters where space is restricted.

RADII

The general method of dimensioning a circular arc is by giving its radius. A radius dimension line passes through, or is in line with, the radius center and terminates with an arrowhead touching the arc. See Fig. 5-2-3. An arrowhead is never used at the radius center. The size of the dimension is preceded by the abbreviation R for both customary and metric dimensioning. Where space is limited, as for a small radius, the radial dimension line may extend through the radius center. Where it is inconvenient to place the arrowhead between the radius center

(A) RADII WHICH NEED NOT HAVE THEIR CENTERS LOCATED

(B) LOCATING RADIUS CENTER

(C) RADII WITH COMMON TANGENT POINTS

(D) FORESHORTENED RADII

(E) RADII LOCATED BY TANGENTS

Fig. 5-2-3   Radii.

and the arc, it may be placed outside the arc, or a leader may be used (Fig. 5-2-3A).

Where a dimension is given to the center of the radius, a small cross should be drawn at the center (Fig. 5-2-3B). Extension lines and dimension lines are used to locate the center. Where the location of the center is unimportant, a radial arc may be located by tangent lines (Fig. 5-2-3E).

Where the center of a radius is outside the drawing or interferes with another view, the radius dimension line may be foreshortened (Fig. 5-2-3D). The portion of the dimension line next to the arrowhead should be radial relative to the curved line. Where the radius dimension line is foreshortened and the center is located by coordinate dimensions, the dimensions locating the center should be shown as foreshortened or the dimension shown as not to scale.

Simple fillet and corner radii may also be dimensioned by use of a general note, such as

ALL ROUNDS AND FILLETS
UNLESS OTHERWISE SPECIFIED
R .20 or ALL RADII R 5

Where a radius is dimensioned in a view that does not show the true shape of the radius, TRUE R is added before the radius dimension, as illustrated in Fig. 5-2-4.

Fig. 5-2-4   Indicating true radius.

Rounded Ends

Overall dimensions should be used for parts or features having rounded ends. For fully rounded ends, the radius (R) is shown but not dimensioned (Fig. 5-2-5A). For parts with partially rounded ends, the radius is dimensioned (Fig. 5-2-5B). Where a hole and radius have the same center and the hole location is more critical than the location of a radius, then either the

Fig. 5-2-5   External surfaces with rounded ends.

radius or the overall length should be shown as a reference dimension (Fig. 5-2-5C).

Dimensioning Chords, Arcs, and Angles

The difference in dimensioning chords, arcs, and angles is shown in Fig. 5-2-6.

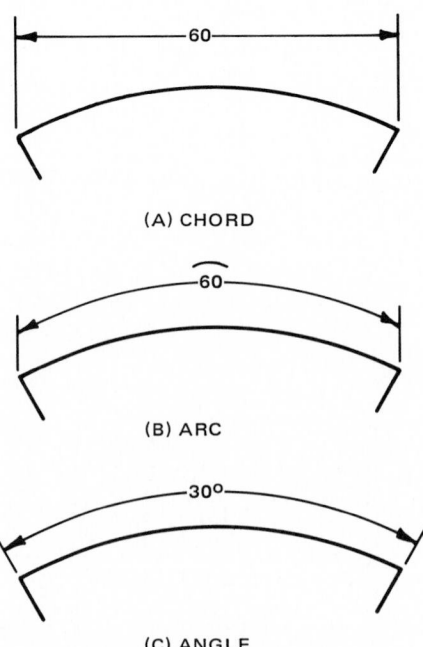

Fig. 5-2-6   Dimensioning chords, arcs, and angles.

Spherical Features

Spherical surfaces may be dimensioned as diameters or radii, but the dimension should be used with the abbreviations SR or S∅. See Fig. 5-2-7.

Fig. 5-2-7   Spherical surfaces.

Cylindrical Holes

Plain, round holes are dimensioned in various ways, depending upon design and manufacturing requirements (Fig. 5-2-8). However, the leader is the method most commonly used. When a leader is used to specify diameter sizes, as with small holes, the dimension is identified as a diameter by preceding the numerical value with the diameter symbol ∅.

The size, quantity, and depth may be shown on a single line, or on several lines if preferable. For through holes, the abbreviation THRU should follow the dimension if the drawing does not make this clear. The depth dimension of a blind hole is the depth of the full diameter and is normally included as part of the dimensioning note.

When more than one hole of a size is required, the number of holes should be specified. However, care must be taken to avoid placing the hole size and quantity values together without ade-

(A) DIMENSIONING ONE HOLE

Ø .75
1.00 DEEP

PLAN VIEW NOT SHOWN

OR

Ø .75

1.00

(B) DIMENSIONING A BLIND HOLE

Ø .62 THRU

(C) DIMENSIONING A THROUGH HOLE
WHICH IS NOT SHOWN IN A
LONGITUDINAL VIEW

Ø .168 4 HOLES

Ø .50
4 HOLES EQ SP

Ø 2.50

(D) DIMENSIONING A GROUP OF HOLES

Fig. 5-2-8 Cylindrical holes.

quate spacing. It may be better to show the note on two or more lines than to use a line note which might be misread (Fig. 5-2-8D).

Minimizing Leaders If too many leaders would impair the legibility of a drawing, letters or symbols as shown in Fig. 5-2-9 should be used to identify the features.

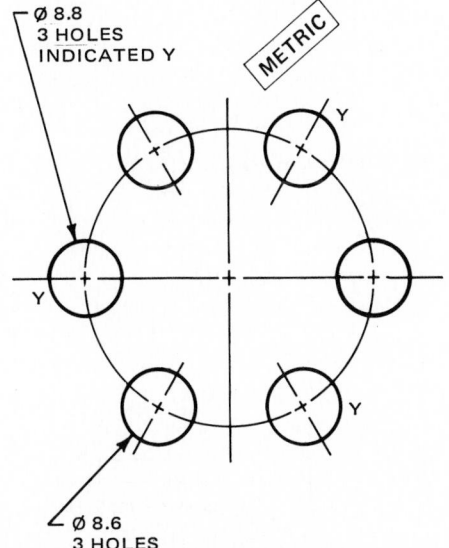

Ø 8.8
3 HOLES
INDICATED Y

METRIC

Ø 8.6
3 HOLES

Fig. 5-2-9 Minimizing leaders.

3.86

2.00

.90

.14

.60

.40

R
2 PLACES

(A)

1.00

1.25

.60

.40 X .84

R
2 PLACES

(B)

.56

.50

1.00

.38

.54

Ø.62

.32

R
2 PLACES

(C)

Fig. 5-2-10 Slotted holes.

Slotted Holes

Elongated holes and slots are used to compensate for inaccuracies in manufacturing and to provide for adjustment. See Fig. 5-2-10. The method selected to locate the slot would depend on how the slot was made. The method shown in Fig. 5-2-10B is used when the slot is punched out and the location of the punch is given. Figure 5-2-10A shows the dimensioning method used when the slot is machined out.

Countersinks, Counterbores, and Spotfaces

The abbreviations CSK, CBORE, and SF for countersink, counterbore, and spotface, respectively, indicate the form of the surface only and do not restrict the methods used to produce that form. The dimensions for them are usually given as a note, preceded by the size of the through hole. See Fig. 5-2-11.

A countersink is an angular-sided recess to accommodate the head of flathead screws, rivets, and similar items. The diameter at the surface and the included angle are given. When the depth of the tapered section of the countersink is critical, this depth is specified in the note or by dimension. For counterdrilled holes, the diameter, depth, and included angle of the counterdrill are given.

A counterbore is a flat-bottomed, cylindrical recess which permits the head of a fastening device, such as a bolt, to lie recessed into the part. The diameter, depth, and corner radius are specified in a note. In some cases, the thickness of the remaining stock may be dimensioned rather than the depth of the counterbore.

A spotface is an area where the surface is machined just enough to provide smooth, level seating for a bolt head, nut, or washer.

The diameter of the faced area and either the depth or the remaining thickness are given. A spotface may be specified by a note only, and not delineated on the drawing. If no depth or remaining thickness is specified, it is implied that the spot-facing is the minimum depth necessary to clean up the surface to the specified diameter.

The symbols for counterbore or

Ø .41
Ø .8I CDRILL
.19 DEEP

Ø .28
⌄ Ø .62 X 82°

OR

118°

COUNTERDRILL   COUNTERSINK

(A) COUNTERDRILLS AND
COUNTERSINKS

Ø .28
Ø .41
⊽ .12

Ø .28
⊔ Ø .56

OR

R .03   R .03

.31

(B) COUNTERBORES

Ø .28
⊔ Ø .75
4 HOLES

Ø .75

.06

.62 MIN   Ø .28

(C) SPOT FACES

Fig. 5-2-11 Counterdrills, countersinks, counterbores, and spotfaces.

spotface, countersink, and depth are shown in Fig. 5-2-12. In each case the symbol precedes the dimension.

References and Source Materials
1. ANSI Y14.5, *Dimensioning and Tolerancing.*

ASSIGNMENTS

See Assignments 4 through 6 for Unit 5-2 on page 116.

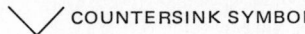

∨ COUNTERSINK SYMBOL

⊔ COUNTERBORE OR
SPOTFACE SYMBOL

⊽ DEPTH SYMBOL

(A) SYMBOLS

Ø .28 THRU
∨ Ø .40 X 82°

COUNTERSINK SYMBOL

Ø .28 THRU
⊔ Ø .56

COUNTERBORE OR SPOTFACE SYMBOL

.24 – .26

Ø .371 – .376
⊽ .75

DEPTH SYMBOL

(B) APPLICATION

Fig. 5-2-12   Hole symbols.

UNIT 5-3
Dimensioning Common Features

Tapers

Circular Tapers Tapered shanks are used on many small tools, such as drills, reamers, counterbores, and spotfaces, to hold them accurately in the machine spindle. See Fig. 5-3-1. *Taper* means the difference in diameter or width in a given length. There are many standard tapers; the Morse taper and the Brown and Sharpe taper are the most common.

The following dimensions may be used, in suitable combinations, to define the size and form of tapered features:

- The diameter (or width) at one end of the tapered feature
- The length of the tapered feature
- The rate of taper
- The included angle
- The taper ratio

In dimensioning a taper by means of the taper ratio, the taper symbol ▷ should precede the ratio figures.

1.20

Ø 1.00   Ø .60

$$TAPER = \frac{1.00 - .60}{1.20} = \frac{.40}{1.20} = 1 : 3$$

1.40

Ø 1.00

4°

0.2 : 1

Ø 1.00

Fig. 5-3-1   Dimensioning circular tapers.

Flat Tapers Flat tapers are used as locking devices, such as taper keys and adjusting shims. The methods recommended for dimensioning flat tapers are shown in Fig. 5-3-2.

Fig. 5-3-2 Dimensioning flat tapers.

Chamfers

The process of chamfering, that is, cutting away the inside or outside piece, is done to facilitate assembly. Chamfers are normally dimensioned by giving their angle and length. See Fig. 5-3-3. When the chamfer is 45°, it may be specified as a note.

When a very small chamfer is permissible, primarily to break a sharp corner, it may be dimensioned but not drawn, as in Fig. 5-3-3C. If not otherwise specified, an angle of 45° is understood.

Internal chamfers may be dimensioned in the same manner, but it is often desirable to give the diameter over the chamfer. The angle may also be given as the included angle if this is a design requirement. This type of dimensioning is generally necessary for larger diameters, especially those over 2 in. (50 mm), whereas chamfers on small holes are usually expressed as countersinks. Chamfers are never measured along the angular surface.

Fig. 5-3-3 Dimensioning chamfers.

Knurls

Knurling is specified in terms of type, pitch, and diameter before and after knurling. See Fig. 5-3-4. The letter P precedes the pitch number. Where control is not required, the diameter after knurling is omitted. Where only portions of a feature require knurling, axial dimensions must be provided. Where required to provide a press fit between parts, knurling is specified by a note on the drawing which includes the type of knurl required, the pitch, the toleranced diameter of the feature prior to knurling, and the minimum acceptable diameter after knurling. Commonly used types are straight, diagonal, spiral, convex, raised diamond, depressed diamond, and radial. The pitch is usually expressed in terms of teeth per inch or millimeter and may be the straight pitch, circular pitch, or diametral pitch. For cylindrical surfaces, the latter is preferred.

The knurling symbol is optional and is used only to improve clarity on working drawings.

Fig. 5-3-4 Dimensioning knurls.

Formed Parts

In dimensioning formed parts, the inside radius is usually specified, rather than the outside radius, but all forming dimensions should be shown on the same side if possible. Dimensions apply to the side on which the dimensions are shown unless otherwise specified. See Fig. 5-3-5.

Undercuts

The operation of undercutting or necking, that is, cutting a recess in a diameter, is done to permit two parts to come together, as illustrated in Fig. 5-3-6A. It is indicated on the drawing by a note listing the width first and then the diameter. If the radius is shown at the bottom of the undercut, it will be assumed that the radius is equal to one-half the width unless otherwise specified, and the diameter will apply to the center of the undercut. When the size of the undercut is unimportant, the dimension may be left off the drawing.

Fig. 5-3-5 Dimensioning theoretical points of intersection.

(A)

Fig. 5-3-7 Dimensioning symmetrical features.

(B)

PART CANNOT FIT FLUSH IN
HOLE BECAUSE OF SHOULDER

CHAMFER ADDED TO
HOLE TO ACCEPT
SHOULDER OF PART

SAME PART WITH
UNDERCUT ADDED
PERMITS PART TO
FIT FLUSH

(A) CHAMFER AND UNDERCUT
APPLICATION

3 X Ø 9

(B) PLAIN UNDERCUT

3 X Ø 9

(C) RADIUSED UNDERCUT

Fig. 5-3-6 Dimensioning undercuts.

Symmetrical Outlines

Symmetrical outlines may be dimensioned on one side of the axis of symmetry only. See Fig. 5-3-7. Where only part of the outline is shown, because of functional drafting procedures, the size of the part, or space limitations, symmetrical shapes may be shown by only one-half of the outline, and the symmetry is indicated by applying the symbol for part symmetry to the center line. In such cases, the outline of the part should extend slightly beyond the center line and terminate with a break line. Note the dimensioning method of extending the dimension lines to act as extension lines for the perpendicular dimensions.

USING DESCRIPTIVE NOTES	USING " NUMBER OF TIMES" SYMBOL

.40   16 SPACES AT .60

.30

Ø .258 17 HOLES

.40   16 X .60   .60

.30

17X Ø .258

.34   4.00

.28

Ø .28
5 HOLES EQ SP

Ø .281
8 HOLES EQ SP ON Ø 2.34

8X Ø .281
EQ SP ON Ø 2.34

.40   .16

22.5°

4 SLOTS
EQUALLY SPACED

Ø .25
8 HOLES EQUALLY
SPACED ON Ø 1.60

4X .40   4X .16

22.5°

XX

8X 45°

8X Ø .25

Fig. 5-3-8 Dimensioning repetitive detail.

Repetitive Features and Dimensions

Repetitive features and dimensions may be specified on a drawing by the use of an "X" in conjunction with the numeral to indicate the "number of times" or "places" they are required. A space is shown between the "X" and the dimension.

An "X" which means "by" is often used between coordinate dimensions specified in note form. Where both are used on a drawing, care must be taken to ensure each is clear. See Fig. 5-3-8.

Wire, Sheet Metal, and Drill Rod

Wire, sheet metal, and drill rod, which are manufactured to gage or code sizes, should be shown by their decimal dimensions; but gage numbers, drill letters, etc., may be shown in parentheses following those dimensions.

EXAMPLES

Sheet —.141 (No. 10 USS GA)
        —.081 (No. 12 B & S GA)

Source Materials

1. ANSI Y 14.5, *Dimensioning and Tolerancing*.

ASSIGNMENTS

See Assignments 7 through 11 for Unit 5-3 on page 119.

Review for Assignment

Unit 2-6   Drafting Skills

UNIT 5-4
Dimensioning Methods

The choice of the most suitable dimensions and dimensioning methods will depend, to some extent, on how the part will be produced and whether the drawings are intended for unit or mass production.

Unit production refers to cases where each part is to be made separately, using general-purpose tools and machines.

Fig. 5-4-1   Rectangular coordinate dimensioning.

Mass production refers to parts produced in quantity, where special tools and gages are usually provided.

Either linear or angular dimensions may locate features with respect to one another (point-to-point) or from a datum. Point-to-point dimensions may be adequate for describing simple parts. Dimensions from a datum may be necessary if a part with more than one critical dimension must mate with another part.

The following systems of dimensioning are used more commonly for engineering drawings.

Rectangular Coordinate Dimensioning

This is a method for indicating distance, location, and size by means of linear dimensions measured parallel or perpendicular to reference axes or datum planes that are perpendicular to one another.

Coordinate dimensioning with dimension lines must clearly identify the datum features from which the dimensions originate. See Fig. 5-4-1.

Rectangular Coordinate Dimensioning Without Dimension Lines Dimensions may be shown on extension lines without the use of dimension lines or arrowheads. The base lines may be zero coordinates or they may be labeled as X, Y, and Z. See Figs. 5-4-2 and 5-4-3.

Tabular Dimensioning Tabular dimensioning is a type of coordinate dimensioning in which dimensions from mutually perpendicular planes are listed in a table on the drawing rather than on the pictorial delineation. This method is used on drawings which require the location of a large number of similarly shaped features. See Fig. 5-4-3.

Polar Coordinate Dimensioning

Polar coordinate dimensioning is commonly used in circular planes or circular configurations of features. It is a method of indicating the position of a point, line, or surface by means of a linear dimension and an angle, other

HOLE SYMBOL	HOLE SIZE
A	.246
B	.189
C	.154
D	.125

Fig. 5-4-2   Rectangular coordinate dimensioning (arrowless dimensioning).

HOLE DIA	HOLE SYMBOL	LOCATION		
		X	Y	Z
5.6	A₁	60	40	18
4.8	B₁	10	40	THRU
	B₂	75	40	THRU
	B₃	60	16	THRU
	B₄	80	16	THRU
4	C₁	18	40	THRU
	C₂	55	40	THRU
	C₃	10	20	THRU
	C₄	30	20	THRU
	C₅	75	20	THRU
	C₆	18	16	THRU
3.2	D₁	55	8	12

Fig. 5-4-3 Tabular dimensioning.

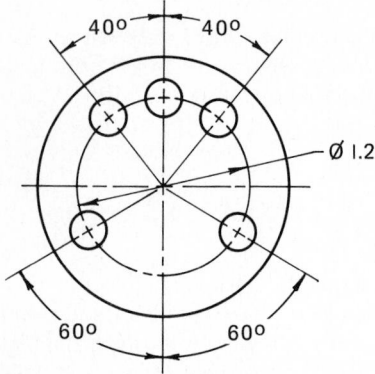

Fig. 5-4-4 Polar coordinate dimensioning.

than 90°, that is implied by the vertical and horizontal center lines. See Fig. 5-4-4.

Chordal Dimensioning

The chordal dimensioning system may also be used for the spacing of points on the circumference of a circle relative to a datum, where manufacturing methods indicate that this will be more convenient. See Fig. 5-4-5.

Fig. 5-4-5 Chordal dimensioning.

True Position Dimensioning

True position dimensioning has many advantages over the coordinate dimensioning system. See Fig. 5-4-6. Because of its scope, it is covered as a complete topic in Chap. 30.

Fig. 5-4-6 True position dimensioning.

(A) CHAIN DIMENSIONING

(B) DATUM OR COMMON-POINT DIMENSIONING

Fig. 5-4-7 A comparison between chain and datum dimensioning.

Chain Dimensioning

When a series of dimensions is applied on a point-to-point basis, it is called *chain dimensioning*. See Fig. 5-4-7. A possible disadvantage of this system is that it may result in an undesirable accumulation of tolerances between individual features. See Unit 5-5.

Datum or Common-Point Dimensioning

When several dimensions emanate from a common reference point or line, the method is called *common-point* or *datum dimensioning*.

References

1. ANSI, Y14.5, *Dimensioning and Tolerancing*.

ASSIGNMENTS

See Assignments 12 and 13 for Unit 5-4 on page 120.

Review for Assignment
Unit 2-6 Drafting Skills

UNIT 5-5
Limits and Tolerances

In the 6000 years of the history of technical drawing as a means for the communication of engineering information, it seems inconceivable that such an elementary practice as the tolerancing of dimensions, which we take so much for granted today, was introduced for the first time about 70 years ago.

Apparently, engineers and workers came in a very gradual manner to the realization that exact dimensions and shapes could not be attained in the manufacture of materials and products.

The skilled tradespeople of old prided themselves on being able to work to exact dimensions. What they really meant was that they dimensioned objects with a degree of accuracy greater than that with which they could measure. The use of modern measuring instruments would have shown the deviations from the sizes which they called exact.

Note: LC1 and LC2 locational clearance fits may also be used as sliding fits with greater tolerances.

RC3 Precision Running Fit This fit is about the closest fit which can be expected to run freely, and is intended for precision work for oil-lubricated bearings at slow speeds and light journal pressures, but is not suitable where appreciable temperature differences are likely to be encountered.

RC4 Close Running Fit This fit is intended chiefly as a running fit for grease or oil-lubricated bearings on accurate machinery with moderate surface speeds and journal pressures, where accurate location and minimum play are desired.

RC5 and RC6 Medium Running Fits These fits are intended for higher running speeds and/or where temperature variations are likely to be encountered.

RC7 Free Running Fit This fit is intended for use where accuracy is not essential, and/or where large temperature variations are likely to be encountered.

RC8 and RC9 Loose Running Fits These fits are intended for use where materials made to commercial tolerances are involved such as cold-rolled shafting, tubing, etc.

Locational Clearance Fits

Locational clearance fits are intended for parts which are normally stationary, but which can be freely assembled or disassembled. They run from snug fits for parts requiring accuracy of location, through the medium clearance fits for parts such as spigots, etc., to the looser fastener fits where freedom of assembly is of prime importance.

These are classified as follows:

LC1 to LC4 These fits have a minimum zero clearance, but in practice the probability is that the fit will always have a clearance. These fits are suitable for location of nonrunning parts and spigots, although classes LC1 and LC2 may also be used for sliding fits.

LC5 and LC6 These fits have a small minimum clearance, intended for close location fits for nonrunning parts. LC5 can also be used in place of RC2 as a free-slide fit, and LC6 may be used as a medium running fit having greater tolerances than RC5 and RC6.

LC7 to LC11 These fits have progressively larger clearances and tolerances, and are useful for various loose clearances for assembly of bolts and similar parts.

Locational Transition Fits

Locational transition fits are a compromise between clearance and interference fits, for application where accuracy of location is important, but either a small amount of clearance or interference is permissible.

These are classified as follows:

LT1 and LT2 These fits average a slight clearance, giving a light push fit, and

EXAMPLE – Ø .7500 RC2 FIT (BASIC HOLE SYSTEM)

(A) CLEARANCE FIT

EXAMPLE – Ø .7500 LT2 FIT (BASIC HOLE SYSTEM)

(B) TRANSITION FIT

EXAMPLE – Ø .7500 FN2 FIT (BASIC HOLE SYSTEM)

(C) INTERFERENCE FIT

Fig. 5-6-3 Types of inch fits.

are intended for use where the maximum clearance must be less than for the LC1 to LC3 fits, and where slight interference can be tolerated for assembly by pressure or light hammer blows.

LT3 and LT4 These fits average virtually no clearance, and are for use where some interference can be tolerated, for example: to eliminate vibration. These are sometimes referred to as an easy keying fit, and are used for shaft keys and ball race fits. Assembly is generally by pressure or hammer blows.

LT5 and LT6 These fits average a slight interference, although appreciable assembly force will be required when extreme limits are encountered, and selective assembly may be desirable. These fits are useful for heavy keying, for ball race fits subject to heavy duty and vibration, and as light press fits for steel parts.

Locational Interference Fits

Locational interference fits are used where accuracy of location is of prime importance, and for parts requiring rigidity and alignment with no special requirements for bore pressure. Such fits are not intended for parts designed to transmit frictional loads from one part to another by virtue of the tightness of fit, as these conditions are covered by force fits.

These are classified as follows:

LN1 and LN2 These are light press fits, with very small minimum interference, suitable for parts such as dowel pins, which are assembled with an arbor press in steel, cast iron, or brass. Parts can normally be dismantled and reassembled, as the interference is not likely to overstrain the parts, but the interference is too small for satisfactory fits in elastic materials or light alloys.

LN3 This is suitable as a heavy press fit in steel and brass, or a light press fit in more elastic materials and light alloys.

LN4 to LN6 While LN4 can be used for permanent assembly of steel parts, these fits are primarily intended as press fits for more elastic or soft materials, such as light alloys and the more rigid plastics.

Fig. 5-6-4 Applications of international tolerance (IT) grades.

Force or Shrink Fits

Force or shrink fits constitute a special type of interference fit, normally characterized by maintenance of constant bore pressures throughout the range of sizes. The interference therefore varies almost directly with diameter, and the difference between its minimum and maximum values is small to maintain the resulting pressures within reasonable limits.

These fits may be described briefly as follows:

FN1 Light Drive Fit Requires light assembly pressure, and produces more

Fig. 5-6-5 Tolerancing grades for machining processes.

or less permanent assemblies. It is suitable for thin sections or long fits, or in cast-iron external members.

FN2 Medium Drive Fit Suitable for ordinary steel parts, or as a shrink fit on light sections. It is about the tightest fit that can be used with high-grade cast-iron external members.

FN3 Heavy Drive Fit Suitable for heavier steel parts or as a shrink fit in medium sections.

FN4 and FN5 Force Fits Suitable for parts which can be highly stressed, and/or for shrink fits where the heavy pressing forces required are impractical.

Basic Hole System

In the basic hole system, which is recommended for general use, the basic size will be the design size for the hole, and the tolerance will be plus. The design size for the shaft will be the basic size minus the minimum clearance, or plus the maximum interference, and the tolerance will be

Fig. 5-6-6 Tolerance symbol.

minus, as given in the tables in the Appendix. For example, (see table 48) for a 1 in. RC7 fit, values of +.0020, .0025, and −.0012 are given; hence, limits will be

Hole ϕ 1.0000 $\begin{smallmatrix}+.0020\\-.0000\end{smallmatrix}$

Shaft ϕ .9975 $\begin{smallmatrix}+.0000\\-.0012\end{smallmatrix}$

Basic Shaft System

Fits are sometimes required on a basic shaft system, especially in cases where two or more fits are required on the same shaft. This is designated for design purposes by a letter S following the fit symbol; for example, RC7S.

Tolerances for holes and shaft are identical with those for a basic hole system, but the basic size becomes the design size for the shaft and the design size for the hole is found by adding the minimum clearance or subtracting the maximum interference from the basic size.

For example, for a 1-in. RC7S fit, values of +.0020, .0025, and −.0012 are given; therefore, limits will be

Hole ϕ 1.0025 $\begin{smallmatrix}+.0020\\-.0000\end{smallmatrix}$

Shaft ϕ 1.000 $\begin{smallmatrix}+.0000\\-.0012\end{smallmatrix}$

PREFERRED METRIC LIMITS AND FITS

The ISO system of limits and fits for mating parts is approved and adopted for general use in the United States. It establishes the designation symbols used to define specific dimensional limits on drawings.

The general terms "hole" and "shaft" can also be taken as referring to the space containing or contained by two parallel faces of any part, such as the width of slot, or the thickness of a key.

An "International Tolerance grade" establishes the magnitude of the tolerance zone or the amount of part size variation allowed for internal and external dimensions alike. See Table 42, Appendix. There are eighteen tolerance grades which are identified by the prefix IT, such as IT6; IT11; etc. The smaller the grade number the smaller the tolerance zone. For general applications of IT grades see Fig. 5-6-4.

Grades 1 to 4 are very precise grades intended primarily for gage making and similar precision work, although grade 4 can also be used for very precise production work.

Grades 5 to 16 represent a progressive series suitable for cutting operations, such as turning, boring, grinding, milling, and sawing. Grade 5 is the most precise grade, obtainable by fine grinding and lapping, while 16 is the coarsest grade for rough sawing and machining.

Grades 12 to 16 are intended for manufacturing operations such as cold heading, pressing, rolling, and other forming operations.

As a guide to the selection of tolerances, Fig. 5-6-5 has been prepared to show grades which may be expected to be held by various manufacturing processes for work in metals. For work in

EXAMPLE — D9/h9 PREFERRED SHAFT BASIS FIT FOR A ϕ 20 SHAFT
(A) CLEARANCE FIT

EXAMPLE — H7/k6 PREFERRED HOLE BASIS FIT FOR ϕ 20 HOLE
(B) TRANSITION FIT

EXAMPLE — S7/h6 PREFERRED SHAFT BASIS FIT FOR A ϕ 20 SHAFT
(C) INTERFERENCE FIT

Fig. 5-6-7 Types of metric fits.

other materials, such as plastics, it may be necessary to use coarser tolerance grades for the same process.

A fundamental deviation establishes the position of the tolerance zone with respect to the basic size. Fundamental deviations are expressed by "tolerance position letters." Capital letters are used for internal dimensions, and lower case letters for external dimensions.

Symbols

Tolerance Symbol By combining the IT grade number and the tolerance position letter, the tolerance symbol is established which identifies the actual maximum and minimum limits of the part. The toleranced sizes are thus defined by the basic size of the part followed by the symbol composed of a letter and a number. See Fig. 5-6-6 on page 104.

Preferred Tolerance Zones

The preferred tolerance zones are shown in Table 43 of the Appendix. The encircled tolerance zones (13 each) are first choice, the framed tolerance zones are second choice, and the open tolerance zones are third choice.

Hole Basis Fits System

In the hole basis fits system (see Tables 44 and 46 of the Appendix) the basic size will be the minimum size of the hole. For example, for a 25 ⌀ H8/f7 fit, which is a Preferred Hole Basis Clearance Fit, the limits for the hole and shaft will be as follows:

Refer to Tables 44 and 46 of the Appendix.
Hole limits = ⌀ 25.000 − ⌀ 25.033
Shaft limits = ⌀ 24.959 − ⌀ 24.980
Minimum clearance = 0.020
Maximum clearance = 0.074
If a 25 ⌀ H7/s6 Preferred Hole Basis Interference Fit is required, the limits for the hole and shaft will be as follows:
Refer to Tables 44 and 46 of the Appendix.
Hole limits = ⌀ 25.000 − ⌀ 25.021
Shaft limits = ⌀ 25.035 − ⌀ 25.048
Minimum interference = −0.014
Maximum interference = −0.048

Shaft Basis Fits System

Where more than two fits are required on the same shaft, the shaft basis fits system is recommended. Tolerances for holes and shaft are identical with those for a basic hole system. However, the basic size becomes the maximum shaft size. For example, for a 16 ⌀ C11/h11 fit, which is a Preferred Shaft Basis Clearance Fit, the limits for the hole and shaft will be as follows:
Refer to Tables 45 and 47 of the Appendix.
Hole limits = ⌀ 16.095 − ⌀ 16.205
Shaft limits = ⌀ 15.890 − ⌀ 16.000
Minimum clearance 0.095
Maximum clearance 0.315

Preferred Fits

First choice tolerance zones are shown to relative scale in Table 43 of the Appendix. Hole basis fits have a fundamental deviation of "H" on the hole, and shaft basis fits have a fundamental deviation of "h" on the shaft. Normally, the hole basis system is preferred. Figure 5-6-7 on page 105 shows examples of three common fits.

Fit Symbol A fit is indicated by the basic size common to both components, followed by a symbol corresponding to each component, with the internal part symbol preceding the external part symbol. See Fig. 5-6-8.

Fig. 5-6-8 Fit symbol.

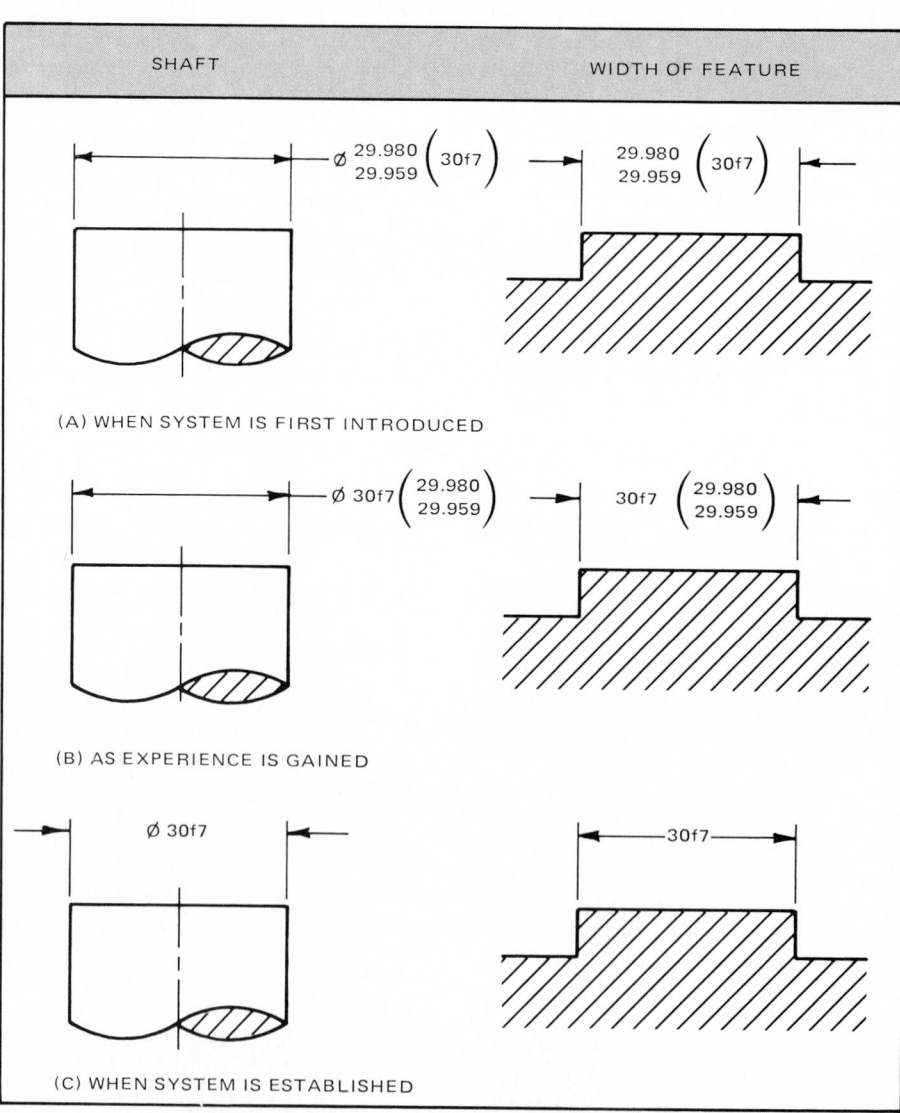

(A) WHEN SYSTEM IS FIRST INTRODUCED

(B) AS EXPERIENCE IS GAINED

(C) WHEN SYSTEM IS ESTABLISHED

Fig. 5-6-9 Metric tolerance symbols.

The limits of size for a hole having a tolerance symbol 40H8 (See Table 46) is

∅ 40.039 MAXIMUM LIMIT
∅ 40.000 MINIMUM LIMIT

The limits of size for the shaft having a tolerance symbol 40f7 (See Table 46) is

∅ 39.975 MAXIMUM LIMIT
∅ 39.950 MINIMUM LIMIT

The method shown in Fig. 5-6-9A is recommended when the system is first introduced. In this case limit dimensions are specified, and the basic size and tolerance symbol are identified as reference.

As experience is gained, the method shown in Fig. 5-6-9B may be used. When the system is established and standard tools, gages, and stock materials are available with size and symbol

identification, the method shown in Fig. 5-6-9C may be used.

This would result in a clearance fit of 0.025 to 0.089 mm. A description of the preferred metric fits is shown in Fig. 5-6-10.

References

1. ANSI B4.2, *Preferred Metric Limits and Fits.*

ASSIGNMENTS

See Assignments 15 through 18 for Unit 5-6 on page 121.

Review for Assignments

Unit 2-6   Drafting Skills

| ISO SYMBOL | | DESCRIPTION |
HOLE BASIS	SHAFT BASIS	
H11/c11	C11/h11	LOOSE RUNNING FIT FOR WIDE COMMERCIAL TOLERANCES OR ALLOWANCES ON EXTERNAL MEMBERS.
H9/d9	D9/h9	FREE RUNNING FIT NOT FOR USE WHERE ACCURACY IS ESSENTIAL, BUT GOOD FOR LARGE TEMPERATURE VARIATIONS, HIGH RUNNING SPEEDS, OR HEAVY JOURNAL PRESSURES.
H8/f7	F8/h7	CLOSE RUNNING FIT FOR RUNNING ON ACCURATE MACHINES AND FOR ACCURATE LOCATION AT MODERATE SPEEDS AND JOURNAL PRESSURES.
H7/g6	G7/h6	SLIDING FIT NOT INTENDED TO RUN FREELY, BUT TO MOVE AND TURN FREELY AND LOCATE ACCURATELY.
H7/h6	H7/h6	LOCATIONAL CLEARANCE FIT PROVIDES SNUG FIT FOR LOCATING STATIONARY PARTS; BUT CAN BE FREELY ASSEMBLED AND DISASSEMBLED.
H7/k6	K7/h6	LOCATIONAL TRANSITION FIT FOR ACCURATE LOCATION, A COMPROMISE BETWEEN CLEARANCE AND INTERFERENCE.
H7/n6	N7/h6	LOCATIONAL TRANSITION FIT FOR MORE ACCURATE LOCATION WHERE GREATER INTERFERENCE IS PERMISSIBLE.
H7/p6	P7/h6	LOCATIONAL INTERFERENCE FIT FOR PARTS REQUIRING RIGIDITY AND ALIGNMENT WITH PRIME ACCURACY OF LOCATION BUT WITHOUT SPECIAL BORE PRESSURE REQUIREMENTS.
H7/s6	S7/h6	MEDIUM DRIVE FIT FOR ORDINARY STEEL PARTS OR SHRINK FITS ON LIGHT SECTIONS, THE TIGHTEST FIT USABLE WITH CAST IRON.
H7/u6	U7/h6	FORCE FIT SUITABLE FOR PARTS WHICH CAN BE HIGHLY STRESSED OR FOR SHRINK FITS WHERE THE HEAVY PRESSING FORCES REQUIRED ARE IMPRACTICAL.

(Left margin labels: CLEARANCE FITS, TRANSITION FITS, INTERFERENCE FITS. Right margin labels: MORE CLEARANCE, MORE INTERFERENCE.)

Fig. 5-6-10   Description of preferred fits.

UNIT 5-7
Surface Texture

Modern development of high-speed machines has resulted in higher loadings and increased speeds of moving parts. To withstand these more severe operating conditions with minimum friction and wear, a particular surface finish is often essential, making it necessary for the designer to accurately describe the required finish to the persons who are actually making the parts.

For accurate machines it is no longer sufficient to indicate the surface finish by various grind marks, such as "g," "f," or "fg." It becomes necessary to define surface finish and take it out of the opinion or guesswork class.

All surface finish control starts in the drafting room. The designer has the responsibility of specifying the right surface to give maximum performance and service life at the lowest cost. In selecting the required surface finish for any particular part, the designer bases her or his decision on past experience with similar parts, on field service data, or on engineering tests. Such factors as size and function of the parts, type of loading, speed and direction of movement, operating conditions, physical characteristics of both materials on contact, whether they are subjected to stress reversals, type and amount of lubricant, contaminants, temperature, etc., influence the choice.

There are two principal reasons for surface finish control:

1. To reduce friction
2. To control wear

Whenever a film of lubricant must be maintained between two moving parts, the surface irregularities must be small enough so they will not penetrate the oil film under the most severe operating conditions. Bearings, journals, cylinder bores, piston pins, bushings, pad bearings, helical and worm gears, seal surfaces, machine ways, and so forth, are examples where this condition must be fulfilled.

Surface finish is also important to the wear of certain pieces which are subject to dry friction, such as machine tool bits, threading dies, stamping dies, rolls, clutch plates, brake drums, etc.

Smooth finishes are essential on certain high-precision pieces. In mechanisms such as injectors and high-pressure cylinders, smoothness and lack of waviness are essential to accuracy and pressure-retaining ability.

Surfaces, in general, are very complex in character. Only the height, width, and direction of surface irregularities will be covered in this section since these are of practical importance in specific applications.

Surface Texture Characteristics

Refer to Fig. 5-7-1.

Microinch A microinch is one millionth of an inch (.000 001 in.). For written specifications or reference to surface roughness requirements, microinches may be abbreviated as μin.

Micrometer A micrometer is one millionth of a meter (0.000 001 m). For written specifications or reference to surface roughness requirements, micrometer may be abbreviated as μm.

Roughness Roughness consists of the finer irregularities in the surface texture usually including those which result from the inherent action of the production process. These are considered to include traverse feed marks and other irregularities within the limits of the roughness-width cutoff.
Roughness-Height Value Roughness-height value is rated as the arithmetic average (AA) deviation expressed in microinches or micrometers measured normal to the center line. ISO and many European countries use the term CLA (center line average) in lieu of AA. Both have the same meaning.

Roughness Spacing Roughness spacing is the distance parallel to the nominal surface between successive peaks or ridges which constitute the predominant pattern of the roughness. Roughness spacing is rated in inches or millimeters.

Roughness-Width Cutoff The greatest spacing of repetitive surface irregularities is included in the measurement of average roughness height. Roughness-width cutoff is rated in inches or millimeters and must always

be greater than the roughness width in order to obtain the total roughness height rating.

Waviness Waviness is the usually widely spaced component of surface texture and is generally of wider spacing than the roughness-width cutoff. Waviness may result from such factors as machine or work deflections, vibration, chatter, heat treatment, or warping strains. Roughness may be considered as superimposed on a "wavy" surface. Although waviness is not currently in ISO Standards, it is included as part of the surface texture symbol to follow present industrial practices in the United States.

Lay The direction of the predominant surface pattern, ordinarily determined by the production method used, is the *lay*. Lay symbols are specified as shown in Fig. 5-7-9.

Flaws Flaws are irregularities which occur at one place or at relatively infrequent or widely varying intervals in a surface. Flaws include such

defects as cracks, blow holes, checks, ridges, scratches, etc. Unless otherwise specified, the effect of flaws is not included in the roughness-height measurements.

Surface Texture Symbol

Surface characteristics of roughness, waviness, and lay may be controlled by applying the desired values to the surface texture symbol, shown in Figs. 5-7-2 and 5-7-3, in a general note, or both. Where only the roughness value is indicated, the horizontal extension line on the symbol may be omitted. The horizontal bar is used whenever any surface characteristics are placed above the bar or to the right of the symbol. The point of the symbol should be located on the line indicating the surface, on an extension line from the surface, or on a leader pointing to the surface or extension line. See Fig. 5-7-4. When numerical values accompany the symbol, the symbol should be in an upright position in order to be readable from the bottom. This means

Fig. 5-7-1 Surface texture characteristics.

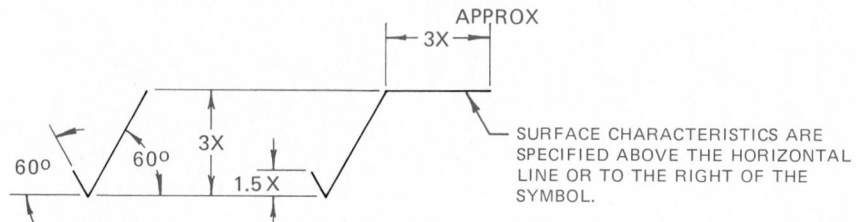

X = FIGURE HEIGHT OF VALUES.
HORIZONTAL EXTENSION BAR REQUIRED
WHEN WAVINESS RATINGS ARE SHOWN.

Fig. 5-7-2 Basic surface texture symbol.

PRESENT SYMBOLS VALUES SHOWN IN CUSTOMARY OR METRIC		FORMER SYMBOLS VALUES SHOWN IN MICROINCHES AND INCHES	
BASIC SURFACE TEXTURE SYMBOL	√	√	BASIC SURFACE TEXTURE SYMBOL
ROUGHNESS HEIGHT RATING IN MICROINCHES OR MICROMETERS AND N SERIES ROUGHNESS NUMBERS	63 √ N8 √	63 √	ROUGHNESS HEIGHT RATING IN MICROINCHES.
MAXIMUM AND MINIMUM ROUGHNESS HEIGHT IN MICROINCHES OR MICROMETERS	63 32 √	63 32 √	MAXIMUM AND MINIMUM ROUGHNESS HEIGHT RATINGS IN MICROINCHES
WAVINESS HEIGHT IN INCHES OR MILLIMETERS (F)	63 32 √ F	63 32 √ .002	WAVINESS HEIGHT IN INCHES
WAVINESS SPACING IN INCHES OR MILLIMETERS (G)	63 32 √ F—G	63 32 √ .002 —1	WAVINESS WIDTH IN INCHES
LAY SYMBOL (D)	63 32 √⊥	63 32 √⊥	LAY SYMBOL
MAXIMUM ROUGHNESS SPACING IN INCHES OR MILLIMETERS (B)	63 32 √⊥ B	√⊥ .002—1 .008	SURFACE ROUGHNESS WIDTH IN INCHES
ROUGHNESS SAMPLING LENGTH OR CUT OFF RATING IN INCHES OR MILLIMETERS (C)	63 32 √⊥ C	√⊥ .002—1 .030 .008	ROUGHNESS WIDTH CUTOFF IN INCHES

Fig. 5-7-3 Location of notes and symbols on surface texture symbols.

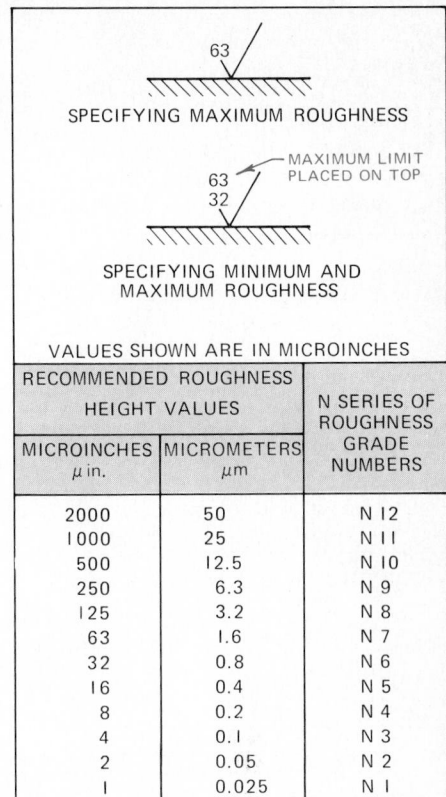

SPECIFYING MAXIMUM ROUGHNESS

SPECIFYING MINIMUM AND MAXIMUM ROUGHNESS

VALUES SHOWN ARE IN MICROINCHES

RECOMMENDED ROUGHNESS HEIGHT VALUES		N SERIES OF ROUGHNESS GRADE NUMBERS
MICROINCHES μ in.	MICROMETERS μm	
2000	50	N 12
1000	25	N 11
500	12.5	N 10
250	6.3	N 9
125	3.2	N 8
63	1.6	N 7
32	0.8	N 6
16	0.4	N 5
8	0.2	N 4
4	0.1	N 3
2	0.05	N 2
1	0.025	N 1

Fig. 5-7-5 Roughness height ratings.

that the long leg and extension line are always on the right. When no numerical values are shown on the symbol, the symbol may also be positioned to be readable from the right side. If necessary, the symbol may be connected to the surface by a leader line terminating in an arrow. The symbol applies to the entire surface, unless otherwise specified. The symbol for the same surface should not be duplicated on other views.

Application

Plain (Unplated or Uncoated) Surfaces Surface texture values specified on plain surfaces apply to the completed surface unless otherwise noted.

Plated or Coated Surfaces Drawings or specifications for plated or coated parts must indicate whether the surface texture value applies before, after, or both before and after plating or coating.

Surface Texture Ratings The roughness value rating is indicated at the left of the long leg of the symbol. See Fig. 5-7-4. The specification of only one rating indicates the maximum value, and any lesser value is acceptable. The specification of two ratings indicates the minimum and maximum values, and anything lying within that range is acceptable. See Fig. 5-7-5. The maximum value is placed over the minimum.

Typical surface roughness-height applications are shown in Fig. 5-7-6.

The surface roughness range for common production methods is shown in Fig. 5-7-7.

ALL SURFACES 6.3 √ UNLESS OTHERWISE SPECIFIED

NOTE: VALUES SHOWN ARE IN MICROMETERS

Fig. 5-7-4 Application of surface texture symbols and notes.

MICROINCHES AA RATING	MICROMETERS AA RATING	APPLICATION
1000	25.2	ROUGH, LOW GRADE SURFACE RESULTING FROM SAND CASTING, TORCH OR SAW CUTTING, CHIPPING OR ROUGH FORGING. MACHINE OPERATIONS ARE NOT REQUIRED AS APPEARANCE IS NOT OBJECTIONABLE. THIS SURFACE, RARELY SPECIFIED, IS SUITABLE FOR UNMACHINED CLEARANCE AREAS ON ROUGH CONSTRUCTION ITEMS.
500	12.5	ROUGH, LOW GRADE SURFACE RESULTING FROM HEAVY CUTS AND COARSE FEEDS IN MILLING TURNING, SHAPING, BORING, AND ROUGH FILING, DISC GRINDING AND SNAGGING. IT IS SUITABLE FOR CLEARANCE AREAS ON MACHINERY, JIGS, AND FIXTURES. SAND CASTING OR ROUGH FORGING PRODUCES THIS SURFACE.
250	6.3	COARSE PRODUCTION SURFACES, FOR UNIMPORTANT CLEARANCE AND CLEANUP OPERATIONS, RESULTING FROM COARSE SURFACE GRIND, ROUGH FILE, DISC GRIND, RAPID FEEDS IN TURNING, MILLING, SHAPING, DRILLING, BORING, GRINDING, ETC., WHERE TOOL MARKS ARE NOT OBJECTIONABLE. THE NATURAL SURFACES OF FORGINGS, PERMANENT MOLD CASTINGS, EXTRUSIONS, AND ROLLED SURFACES ALSO PRODUCE THIS ROUGHNESS. IT CAN BE PRODUCED ECONOMICALLY AND IS USED ON PARTS WHERE STRESS REQUIREMENTS, APPEARANCE, AND CONDITIONS OF OPERATIONS AND DESIGN PERMIT.
125	3.2	THE ROUGHEST SURFACE RECOMMENDED FOR PARTS SUBJECT TO LOADS, VIBRATION, AND HIGH STRESS. IT IS ALSO PERMITTED FOR BEARING SURFACES WHEN MOTION IS SLOW AND LOADS LIGHT OR INFREQUENT. IT IS A MEDIUM COMMERCIAL MACHINE FINISH PRODUCED BY RELATIVELY HIGH SPEEDS AND FINE FEEDS TAKING LIGHT CUTS WITH SHARP TOOLS. IT MAY BE ECONOMICALLY PRODUCED ON LATHES, MILLING MACHINES, SHAPERS, GRINDERS, ETC., OR ON PERMANENT MOLD CASTINGS, DIE CASTINGS, EXTRUSION, AND ROLLED SURFACES.
63	1.6	A GOOD MACHINE FINISH PRODUCED UNDER CONTROLLED CONDITIONS USING RELATIVELY HIGH SPEEDS AND FINE FEEDS TO TAKE LIGHT CUTS WITH SHARP CUTTERS. IT MAY BE SPECIFIED FOR CLOSE FITS AND USED FOR ALL STRESSED PARTS, EXCEPT FAST ROTATING SHAFTS, AXLES, AND PARTS SUBJECT TO SEVERE VIBRATION OR EXTREME TENSION. IT IS SATISFACTORY FOR BEARING SURFACES WHEN MOTION IS SLOW AND LOADS LIGHT OR INFREQUENT. IT MAY ALSO BE OBTAINED ON EXTRUSIONS, ROLLED SURFACES, DIE CASTINGS AND PERMANENT MOLD CASTINGS WHEN RIGIDLY CONTROLLED.
32	0.8	A HIGH-GRADE MACHINE FINISH REQUIRING CLOSE CONTROL WHEN PRODUCED BY LATHES, SHAPERS, MILLING MACHINES, ETC., BUT RELATIVELY EASY TO PRODUCE BY CENTERLESS, CYLINDRICAL OR SURFACE GRINDERS. ALSO, EXTRUDING, ROLLING, OR DIE CASTING MAY PRODUCE A COMPARABLE SURFACE WHEN RIGIDLY CONTROLLED. THIS SURFACE MAY BE SPECIFIED IN PARTS WHERE STRESS CONCENTRATION IS PRESENT. IT IS USED FOR BEARINGS WHEN MOTION IS NOT CONTINUOUS AND LOADS ARE LIGHT. WHEN FINER FINISHES ARE SPECIFIED, PRODUCTION COSTS RISE RAPIDLY; THEREFORE, SUCH FINISHES MUST BE ANALYZED CAREFULLY.
16	0.4	A HIGH QUALITY SURFACE PRODUCED BY FINE CYLINDRICAL GRINDING, EMERY BUFFING, COARSE HONING OR LAPPING. IT IS SPECIFIED WHERE SMOOTHNESS IS OF PRIMARY IMPORTANCE, SUCH AS RAPIDLY ROTATING SHAFT BEARINGS, HEAVILY LOADED BEARINGS AND EXTREME TENSION MEMBERS.
8	0.2	A FINE SURFACE PRODUCED BY HONING, LAPPING, OR BUFFING. IT IS SPECIFIED WHERE PACKINGS AND RINGS MUST SLIDE ACROSS THE DIRECTION OF THE SURFACE GRAIN, MAINTAINING OR WITHSTANDING PRESSURES, OR FOR INTERIOR HONED SURFACES OF HYDRAULIC CYLINDERS. IT MAY ALSO BE REQUIRED IN PRECISION GAGES AND INSTRUMENT WORK, OR SENSITIVE VALUE SURFACES, OR ON RAPIDLY ROTATING SHAFTS AND ON BEARINGS WHERE LUBRICATION IS NOT DEPENDABLE.
4	0.1	A COSTLY REFINED SURFACE PRODUCED BY HONING, LAPPING, AND BUFFING. IT IS SPECIFIED ONLY WHEN THE REQUIREMENTS OF DESIGN MAKE IT MANDATORY. IT IS REQUIRED IN INSTRUMENT WORK, GAGE WORK, AND WHERE PACKINGS AND RINGS MUST SLIDE ACROSS THE DIRECTION OF SURFACE GRAIN SUCH AS ON CHROME-PLATED PISTON RODS, ETC., WHERE LUBRICATION IS NOT DEPENDABLE.
2 1	0.05 0.025	COSTLY REFINED SURFACES PRODUCED ONLY BY THE FINEST OF MODERN HONING, LAPPING, BUFFING, AND SUPERFINISHING EQUIPMENT. THESE SURFACES MAY HAVE A SATIN OR HIGHLY POLISHED APPEARANCE DEPENDING ON THE FINISHING OPERATION AND MATERIAL. THESE SURFACES ARE SPECIFIED ONLY WHEN DESIGN REQUIREMENTS MAKE IT MANDATORY. THEY ARE SPECIFIED ON FINE OR SENSITIVE INSTRUMENT PARTS OR OTHER LABORATORY ITEMS, AND CERTAIN GAGE SURFACES, SUCH AS ON PRECISION GAGE BLOCKS.

Fig. 5-7-6 Typical surface roughness height applications.

SURFACE ROUGHNESS AVERAGE OBTAINABLE BY COMMON PRODUCTION METHODS

PROCESS	ROUGHNESS HEIGHT RATING MICROMETERS, μm (MICROINCHES, μ in.) AA
(μm)	50 25 12.5 6.3 3.2 1.6 0.8 0.4 0.2 0.1 0.05 0.025 0.012
(μ in.)	(2000) (1000) (500) (250) (125) (63) (32) (16) (8) (4) (2) (1) (0.5)
FLAME CUTTING	
SNAGGING	
SAWING	
PLANING, SHAPING	
DRILLING	
CHEMICAL MILLING	
ELECT. DISCHARGE MACH.	
MILLING	
BROACHING	
REAMING	
ELECTRON BEAM	
LASER	
ELECTROCHEMICAL	
BORING, TURNING	
BARREL FINISHING	
ELECTROLYTIC GRINDING	
ROLLER BURNISHING	
GRINDING	
HONING	
ELECTROPOLISH	
POLISHING	
LAPPING	
SUPERFINISHING	
SAND CASTING	
HOT ROLLING	
FORGING	
PERM MOLD CASTING	
INVESTMENT CASTING	
EXTRUDING	
COLD ROLLING, DRAWING	
DIE CASTING	

TYPICAL APPLICATION

- VERY ROUGH SURFACE. EQUIV TO SAND CASTING.
- ROUGH SURFACE. RARELY USED.
- COARSE FINISH, EQUIV TO ROLLED SURFACES AND FORGINGS.
- MEDIUM FINISH COMMONLY USED. REASONABLE APPEAR.
- GOOD FOR CLOSE FITS, UNSUITABLE FOR FAST ROTATING MEMBERS.
- USED ON SHAFTS AND BEARINGS WITH LIGHT LOADS AND MODERATE SPEEDS.
- USED ON HIGH SPEED-SHAFTS AND BEARINGS.
- USED ON PRECISION GAGE AND INSTRUMENT WORK. COSTLY.
- REFINED FINISH. COSTLY TO PRODUCE.
- SUPER FINISH. COSTLY. SELDOM USED.

THE RANGES SHOWN ABOVE ARE TYPICAL OF THE PROCESSES LISTED. HIGHER OR LOWER VALUES MAY BE OBTAINED UNDER SPECIAL CONDITIONS

KEY	AVERAGE APPLICATION ▬▬	LESS FREQUENT APPLICATION ═══

Fig. 5-7-7 Surface roughness range for common production.

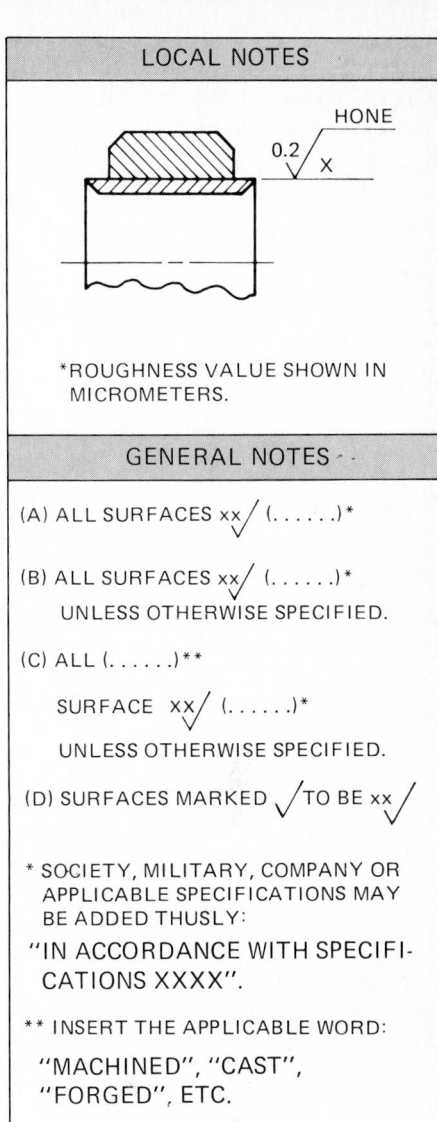

LOCAL NOTES

*ROUGHNESS VALUE SHOWN IN MICROMETERS.

GENERAL NOTES

(A) ALL SURFACES xx/ (......)*

(B) ALL SURFACES xx/ (......)*
UNLESS OTHERWISE SPECIFIED.

(C) ALL (......)**
SURFACE xx/ (......)*
UNLESS OTHERWISE SPECIFIED.

(D) SURFACES MARKED / TO BE xx/

* SOCIETY, MILITARY, COMPANY OR APPLICABLE SPECIFICATIONS MAY BE ADDED THUSLY:
"IN ACCORDANCE WITH SPECIFICATIONS XXXX".

** INSERT THE APPLICABLE WORD:
"MACHINED", "CAST", "FORGED", ETC.

Fig. 5-7-8 Surface texture notes.

the symbol should be supplemented by the words NO LEAD.

Roughness sampling length or cutoff rating is in inches or millimeters and is located below the horizontal extension (Fig. 5-7-3). Unless otherwise specified, roughness sampling length is .03 in. (0.8 mm). See Fig. 5-7-10.

Notes

Notes relating to surface roughness can be local or general.

General Note Normally, a general note is used where a given roughness requirement applies to the whole part or the major portion. Any exceptions to the general note are given in a local note. See Fig. 5-7-8.

Waviness-height rating is specified in inches or millimeters and is located above the horizontal extension of the symbol (Fig. 5-7-3). Any lesser value is acceptable.

Waviness spacing is indicated in inches or millimeters and is located above the horizontal extension and to the right, separated from the waviness-height rating by a dash (Fig. 5-7-3). Any lesser value is acceptable. If the waviness value is a minimum, the abbreviation MIN should be placed after the value.

Lay symbols which indicate the directional pattern of the surface texture are shown in Fig. 5-7-9. The symbol is located to the right of the long leg of the symbol. On surfaces having parallel or perpendicular lay designated, the lead resulting from machine feeds may be objectionable. In these cases,

SYMBOL	DESIGNATION	EXAMPLE
=	LAY PARALLEL TO THE LINE REPRE-SENTING THE SURFACE TO WHICH THE SYMBOL IS APPLIED.	DIRECTION OF TOOL MARKS
⊥	LAY PERPENDICULAR TO THE LINE REPRESENTING THE SURFACE TO WHICH THE SYMBOL IS APPLIED.	DIRECTION OF TOOL MARKS
X	LAY ANGULAR IN BOTH DIRECTIONS TO LINE REPRESENTING THE SURFACE TO WHICH SYMBOL IS APPLIED.	DIRECTION OF TOOL MARKS
M	LAY MULTIDIRECTIONAL.	M
C	LAY APPROXIMATELY CIRCULAR RELATIVE TO THE CENTER OF THE SURFACE TO WHICH THE SYMBOL IS APPLIED.	C
R	LAY APPROXIMATELY RADIAL RELA-TIVE TO THE CENTER OF THE SUR-FACE TO WHICH THE SYMBOL IS APPLIED.	R
P	LAY NONDIRECTIONAL, PITTED OR PROTUBERANT.	P

Fig. 5-7-9 Lay symbols.

PRESENT SYMBOL			FORMER SYMBOL		
LAY SYMBOLS					
16 (.300) C	63 (.300) ⊥	125 (1.000) M	16 0.030 C	63 .030 ⊥ .002	32 .002 .010 M

STANDARD ROUGHNESS SAMPLING LENGTH VALUES	
INCHES	MILLIMETERS
.003	0.08
.010	0.25
.030	0.8
.100	2.54
.300	8
1.000	25.4

Fig. 5-7-10 Lay and roughness sampling length specifications.

REMOVAL OF MATERIAL BY MACHINING IS

Fig. 5-7-11 Indicating the removal of material on the surface texture symbol.

MACHINED SURFACES

In preparing working drawings or parts to be cast, molded, or forged, the drafter must indicate the surfaces on the drawing which will require machining or finishing. The symbol √ identifies those surfaces which are produced by machining operations. See Fig. 5-7-11. It indicates that material is to be provided for removal by machining. Where all the surfaces are to be machined, a general note such as FINISH ALL OVER may be used, and the symbols on the drawing may be omitted. Where space is restricted, the machining symbol may be placed on an extension line.

Machining symbols, like dimensions, are not normally duplicated. They should be used on the same view as the dimensions that give the size or location of the surfaces concerned. The symbol is placed on the line representing the surface or, where desirable, on the extension line locating the surface. Figures 5-7-12 and 5-7-13 show examples of the use of machining symbols.

Fig. 5-7-12 Application of machining symbols.

Fig. 5-7-13 Extra metal allowance for machined surfaces.

Fig. 5-7-15 Indicating machining allowance on drawings.

Fig. 5-7-14 Indication of machining allowance.

Fig. 5-7-16 Symbol for removal of material not permitted.

Material Removal Allowance

When it is desirable to indicate the amount of material to be removed, the amount of material in inches or millimeters is shown to the left of the symbol. Illustrations showing material removal allowance are shown in Figs. 5-7-14 and 5-7-15.

Material Removal Prohibited

When it is necessary to indicate that a surface must be produced without material removal, the machining prohibited symbol shown in Fig. 5-7-16 must be used.

Former Machining Symbols

Former machining symbols, as shown in Fig. 5-7-17, may be found on many drawings in use today. When called upon to make changes or revisions to a drawing already in existence, a drafter must adhere to the drawing conventions shown on that drawing.

References and Source Materials
1. ANSI Y14.36, *Surface Texture Symbols.*
2. GAR.
3. General Motors.

ASSIGNMENTS

See Assignments 19 through 22 for Unit 5-7 on page 125.

Fig. 5-7-17 Former machining symbols.

Review for Assignments

ASSIGNMENTS for Chapter 5

Fig. 5-1-A   Template no. 1.

MATL — SAE 1020
.10 THICK

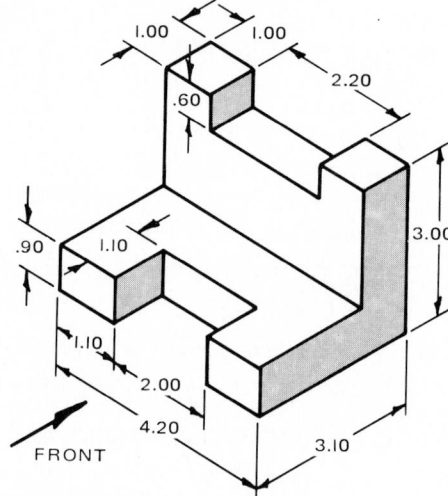

Fig. 5-1-B   Template no. 2.

MATL — SAE 1010
1.0 THICK

METRIC

Assignments for Unit 5-1, Basic Dimensioning

1. Select one of the template drawings (Fig. 5-1-A or 5-1-B) and on a B- or A3-size sheet make a one-view drawing, complete with dimensions, of the part. Scale is full or 1:1.

2. Select one of the parts shown in Figs. 5-1-C to 5-1-F and on a B- or A3-size sheet make a three-view drawing, complete with dimensions, of the part. Scale is full or 1:1.

3. Select one of the parts shown in Figs. 5-1-G to 5-1-I and on a B- or A3-size sheet make a three-view drawing, complete with dimensions, of the part. Place the dimensions on the view which best shows the shape of the part or feature. Scale is full or 1:1.

Fig. 5-1-D   Notched block.

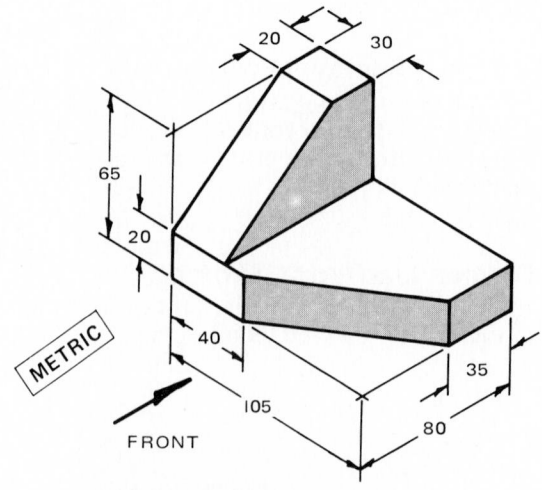

Fig. 5-1-C   Cross slide.

Fig. 5-1-E   Angle plate.

METRIC

Fig. 5-1-F Guide stop.

Fig. 5-1-G Base.

Fig. 5-1-H Separator.

PARTIAL RIGHT
SIDE VIEW

METRIC

Fig. 5-1-I Stand.

Assignments for Unit 5-2, Dimensioning Circular Features

4. Select one of the problems shown in Figs. 5-2-A and 5-2-B. On a B- or A3-size sheet make a one-view drawing, complete with dimensions, of the part. Scale is full or 1:1.

5. Select one of the parts shown in Figs. 5-2-C to 5-2-G and on a B- or A3-size sheet make a three-view drawing, complete with dimensions, of the part. Scale is full or 1:1.

6. Select one of the parts shown in Fig. 5-2-H and on a B- or A3-size sheet redraw the part and add dimensions. Select one of the scales shown to scale the drawing. Scale is full or 1:1.

Fig. 5-2-A   Gasket.

Fig. 5-2-B   Dial indicator.

Fig. 5-2-C   Guide block.

Fig. 5-2-D   Bracket.

Fig. 5-2-F   Offset plate.

Fig. 5-2-E   Shaft support.

Fig. 5-2-G   Yoke.

Fig. 5-2-H Problems in dimensioning practice.

Fig. 5-3-A Handle.

Fig. 5-3-B Selector shaft.

Assignments for Unit 5-3, Dimensioning Common Features

7. Redraw the handle shown in Fig. 5-3-A on a B- or A3-size sheet. Scale is full or 1:1. The following features are to be added and dimensioned.

   (a) .12 × 45° chamfer
   (b) 33P diamond knurl for 1.20 in. starting .80 in. from left end
   (c) 1:8 circular taper for 1.20 in. length on right end of ⌀1.25
   (d) .16 × ⌀.54 in. undercut on ⌀.75
   (e) ⌀.189 × .25 in. DEEP, 4 holes equally spaced
   (f) .10 × 30° chamfer. The .10 in. dimension taken horizontally along the shaft.

8. On a B- or A3-size sheet redraw the selector shaft shown in Fig. 5-3-B and dimension. Scale the drawing for sizes.

9. On an A3- or B-size sheet make a one-view drawing (plus a partial view of the blade), with dimensions, of the screwdriver shown in Fig. 5-3-C. Scale is 1:1.

10. On a B- or A3-size sheet make a one-view drawing with dimensions of the indicator rod shown in Fig. 5-3-D. Scale is full or 1:1.

11. On an A3- or B-size sheet make a half-view drawing of one of the parts shown in Fig. 5-3-E or 5-3-F. Add the symmetry symbol to the drawing and dimension. Scale is 1:1 for 5-3-E and 10:1 for 5-3-F.

Fig. 5-3-C Screwdriver.

Fig. 5-3-D Indicator rod.

Fig. 5-3-E   Gasket.

Fig. 5-3-F   Adjusting locking plate.

Assignments for Unit 5-4, Dimensioning Methods

12. Select one of the problems shown in Figs. 5-4-A and 5-4-B, and on an A3- or B-size sheet make a working drawing of the part. The arrowless dimensioning shown is to be replaced with rectangular coordinate dimensioning and has the following dimensioning changes.

For Fig. 5-4-A
- Holes A, E, and D are located from the zero coordinates.
- Holes B are located from center of hole E.
- Hole C is located from center of hole D

For Fig. 5-4-B
- Holes E and D are located from left and bottom edges.
- Holes A and C are located from center of hole D.
- Hole B is located from center of hole E.

For the sake of clarity, some dimensions may best be shown on the part. Scale for drawings is 1:1.

13. Divide a B- or A3-size sheet into four quadrants by bisecting the vertical and horizontal sides. In each quadrant draw the adapter plate shown in Fig. 5-4-C. Different methods of dimensioning are to be used for each drawing. The methods are rectangular coordinate, chordal, arrowless, and tabular. Scale is full or 1:1.

HOLE	SIZE
A	8
B	4
C	5
D	76
E	12

MATL — SAE 1006   3 THICK

Fig. 5-4-A   Cover plate.

HOLE	SIZE
A	.30
B	.16
C	.24
D	.40 X 2.75
E	.50

MATL — SAE 1008
.12 THICK

Fig. 5-4-B   Transmission cover.

Fig. 5-4-C   Adapter plate.

	A	B	C	D	E			F	G	H	J	K			L	M	N	P	R
BASIC SIZE					⊠		BASIC SIZE					⊠		BASIC SIZE					⊠
TOLERANCE					⊠		TOLERANCE					⊠		TOLERANCE					⊠
LIMITS OF SIZE — MAX							LIMITS OF SIZE — MAX							LIMITS OF SIZE — MAX					
LIMITS OF SIZE — MIN							LIMITS OF SIZE — MIN							LIMITS OF SIZE — MIN					

Fig. 5-5-A   Inch limits and tolerances.

Assignment for Unit 5-5, Limits and Tolerances

14. Calculate the sizes and tolerances for one of the drawings shown in Fig. 5-5-A or 5-5-B.

Assignments for Unit 5-6, Fits and Allowances

15. Using the tables of fits located in the Appendix, calculate the missing dimensions in any of the four charts shown in Figs. 5-6-A to 5-6-D.

	A	B	C	D	E			F	G	H	J	K			L	M	N	P	R
BASIC SIZE					⊠		BASIC SIZE					⊠		BASIC SIZE					⊠
TOLERANCE					⊠		TOLERANCE					⊠		TOLERANCE					⊠
LIMITS OF SIZE — MAX							LIMITS OF SIZE — MAX							LIMITS OF SIZE — MAX					
LIMITS OF SIZE — MIN							LIMITS OF SIZE — MIN							LIMITS OF SIZE — MIN					

Fig. 5-5-B   Metric limits and tolerances.

CLEARANCE FITS	INTERFERENCE FITS	CLEARANCE AND INTERFERENCE FITS

Q1. DIMENSION SHAFT (J) TO HAVE A TOLERANCE OF 0.05 AND A MINIMUM CLEARANCE OF 0.02.

Q2. DIMENSION BUSHING (K) TO HAVE A TOLERANCE OF 0.07 AND A MAXIMUM INTERFERENCE OF 0.22.

Q3. DIMENSION SHAFT (J) TO HAVE A TOLERANCE OF 0.02 AND A MINIMUM CLEARANCE OF 0.05.

Q4. DIMENSION BUSHING (K) TO HAVE A TOLERANCE OF 0.07 AND A MAXIMUM INTERFERENCE OF 0.25.

COMPLETE THIS CHART FROM THE INFORMATION GIVEN ABOVE

	A	B	C	D
TOLERANCE ON HOLE	✕	✕	✕	✕
TOLERANCE ON SHAFT				
MINIMUM CLEARANCE				
MAXIMUM CLEARANCE				

COMPLETE THIS CHART FROM THE INFORMATION GIVEN ABOVE

	E	F	G	H
TOLERANCE ON PART	✕	✕	✕	✕
TOLERANCE ON SLOT				
MINIMUM INTERFERENCE				
MAXIMUM INTERFERENCE				

COMPLETE THIS CHART FROM THE INFORMATION GIVEN ABOVE

	J	K
Q1		✕
Q2	✕	
Q3		✕
Q4	✕	

Fig. 5-6-A   Inch fits.

CLEARANCE FITS (RUNNING OR SLIDING)	TRANSITION FITS (LOCATIONAL)	INTERFERENCE FITS (FORCE OR SHRINK)

G7/h6 SLIDING FIT

H9/d9 RUNNING FIT

H7/h6 LOCATIONAL CLEARANCE

K7/h6 LOCATIONAL TRANSITION

H7/p6 LOCATIONAL INTERFERENCE

U7/h6 FORCE FIT

H7/s6 SHRINK FIT

COMPLETE THIS CHART USING THE PROPER LIMIT AND FIT TABLES

DIA	H7/h6	K7/h6	H7/p6
Ø E		✕	✕
Ø F		✕	✕
Ø G	✕	✕	✕
Ø H	✕		
Ø J	✕	✕	
Ø K	✕	✕	

COMPLETE THIS CHART USING THE PROPER LIMIT AND FIT TABLES

FIT	Ø A	Ø B	Ø C	Ø D
G7/h6			✕	✕
H9/d9	✕	✕		

COMPLETE THIS CHART USING THE PROPER LIMIT AND FIT TABLES

FIT	Ø L	Ø M	Ø N	Ø P
			✕	✕
	✕	✕		

Fig. 5-6-B   Metric fits.

Fig. 5-6-C   Metric fits.

CLEARANCE FITS					INTERFERENCE FITS					CLEARANCE AND INTERFERENCE FITS		

Q1. DIMENSION SHAFT (J) TO HAVE A TOLERANCE OF 0.05 AND A MINIMUM CLEARANCE OF 0.02.
Q2. DIMENSION BUSHING (K) TO HAVE A TOLERANCE OF 0.07 AND A MAXIMUM INTERFERENCE OF 0.22.
Q3. DIMENSION SHAFT (J) TO HAVE A TOLERANCE OF 0.02 AND A MINIMUM CLEARANCE OF 0.05.
Q4. DIMENSION BUSHING (K) TO HAVE A TOLERANCE OF 0.07 AND A MAXIMUM INTERFERENCE OF 0.25.

COMPLETE THIS CHART FROM THE INFORMATION GIVEN ABOVE		A	B	C	D	COMPLETE THIS CHART FROM THE INFORMATION GIVEN ABOVE		E	F	G	H	COMPLETE THIS CHART FROM THE INFORMATION GIVEN ABOVE		J	K
TOLERANCE ON HOLE		✕				TOLERANCE ON PART		✕				Q1			
TOLERANCE ON SHAFT		✕				TOLERANCE ON SLOT		✕				Q2			
MINIMUM CLEARANCE						MINIMUM INTERFERENCE						Q3			
MAXIMUM CLEARANCE						MAXIMUM INTERFERENCE						Q4			

Fig. 5-6-C   Metric fits.

CLEARANCE FITS (RUNNING OR SLIDING)				TRANSITION FITS (LOCATIONAL)				INTERFERENCE FITS (FORCE OR SHRINK)			

RC2 SLIDING FIT

RC5 RUNNING FIT

LC2 LOCATIONAL CLEARANCE

LT3 LOCATIONAL TRANSITION

LN2 LOCATIONAL INTERFERENCE

FN1 FORCE FIT

FN4 SHRINK FIT

COMPLETE THIS CHART USING THE PROPER LIMIT AND FIT TABLES			
DIA	LC2	LT3	LN2
Ø E			
Ø F			
Ø G			
Ø H			
Ø J			
Ø K			

COMPLETE THIS CHART USING THE PROPER LIMIT AND FIT TABLES			
FIT	Ø A	Ø B	Ø C	Ø D
RC2				
RC5				

COMPLETE THIS CHART USING THE PROPER LIMIT AND FIT TABLES			
FIT	Ø L	Ø M	Ø N	Ø P
FN1				
FN4				

Fig. 5-6-D   Inch fits.

16. Using the fit tables in the Appendix, complete the table shown in Fig. 5-6-E.

(A) SHAFT IN BUSHED HOLE

(B) GEAR AND SHAFT IN BUSHED BEARING

(C) CONNECTING-ROD BOLT

(D) LINK PIN (SHAFT BASIS FITS)

(E) CRANK PIN IN CAST IRON

INCH FITS

(A) SHAFT IN BUSHED HOLE

(B) GEAR AND SHAFT IN BUSHED BEARING

(C) CONNECTING-ROD BOLT

(D) LINK PIN (SHAFT BASIS FITS)

(E) CRANK PIN IN CAST IRON

METRIC FITS

DESIGN SKETCH	BASIC DIAMETER SIZE	SYMBOL	BASIS	FEATURE	LIMITS OF SIZE		CLEARANCE OR INTERFERENCE	
					MAX	MIN	MAX	MIN
A	.375 (10)		HOLE	HOLE				
				SHAFT				
A	.250 [6]		HOLE	HOLE				
				SHAFT				
B	.500 [12]		HOLE	HOLE				
				SHAFT				
B	.625 [16]		HOLE	HOLE				
				SHAFT				
B	.750 [20]		HOLE	HOLE				
				SHAFT				
C	.312 [8]		SHAFT	HOLE				
				SHAFT				
D	.188 [5]		HOLE	HOLE				
				SHAFT				
D	.312 [8]		SHAFT	HOLE				
				SHAFT				
E	.812 [18]		HOLE	HOLE				
				SHAFT				

Fig. 5-6-E   Fit problems.

17. On an A- or A4-size sheet make a detail drawing of the roller guide base shown in Fig. 5-6-F. Scale the drawing for sizes. Other considerations are:

(a) Keyseat to be for standard square key and limits on the hole controlled by either an H9/d9 (metric) or RC6 (inch) fit.

(b) Control critical machine surfaces to 0.8 μm or 32 μin.

(c) Dimension in metric or decimal inch.

18. On an A- or A4-size sheet make a detail drawing of the spindle shown in Fig. 5-6-G. Scale the drawing for sizes. Other considerations are:

(a) "A" diameter to have an LC3 (inch) or H7/h6 (metric) fit.

(b) "B" diameter requires a 96 diamond knurl or its equivalent.

(c) "C" diameter to have an LT3 (inch) or H7/K6 (metric) fit.

(d) "D" diameter to be a minimum relief (undercut).

(e) "E" to be a standard No. 807 Woodruff key in center of segment and the diameter to be controlled by an RC3 (inch) or H7/g6 (metric) fit.

(f) "F" to be undercut for a standard retaining ring and controlled form to manufacturer's specifications.

(g) Dimensions in decimal inch or metric.

Assignments for Unit 5-7, Surface Texture

19. On an A3- or B-size sheet make a working drawing of the cross slide shown in Fig. 5-7-A. Scale is 1:1. The following surface texture information is to be added to the drawing:

The dovetail slot is to have a maximum roughness value of 3.2 μm and a machining allowance of 2 mm.

The ends of the shaft support are to have maximum and minimum roughness values of 1.6 and 0.8 μm and a machining allowance of 2 mm.

The hole is to have an H8 tolerance.

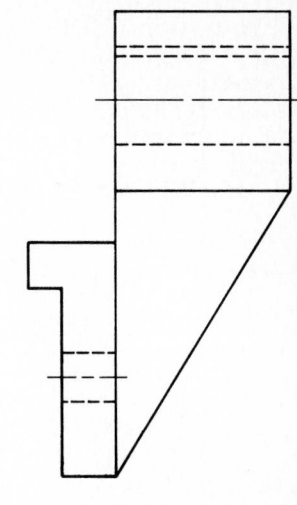

Fig. 5-6-F   Roller guide base.

Fig. 5-6-G   Spindle.

Fig. 5-7-A   Cross slide.

6 X 3 KEYSEAT
Ø 15
Ø 23
RIBS BOTH SIDES
ROUNDS AND FILLETS R2
MATL — MALLEABLE IRON
METRIC
FRONT

METRIC

INCH

20. On a B- or A3-size sheet make a working drawing of the column bracket shown in Fig. 5-7-B. Scale is full. The following surface texture information is to be added to the drawing.

The bottom of the base is to have a maximum roughness value of 125 μin. and a machining allowance of .06 in.

The tops of the bosses are to have a maximum roughness value of 250 μin. and a machining allowance of .04 in.

The end surfaces of the hubs supporting the shafts are to have maximum and minimum roughness values of 63 and 32 μin. and a machining allowance of .04 in.

The large hole is to be dimensioned for an RC4 fit. The small hole is to be dimensioned for an LN3 fit for plain bearings.

ROUNDS AND FILLETS R 2.5
MATL – MALLEABLE IRON

Fig. 5-7-C   Adjustable base plate.

ROUNDS AND FILLETS R .10

MATL – MALLEABLE IRON

FRONT VIEW

Fig. 5-7-B   Column bracket.

21. On an A3- or B-size sheet make a working drawing of the adjustable base plate shown in Fig. 5-7-C. The amount of material to be removed on the surfaces requiring machining is 2 mm. The center hole is to be dimensioned having an H8 tolerance. Scale is 1:1.

22. On a B- or A3-size sheet make a working drawing of the link shown in Fig. 5-7-D. The amount of material to be removed from the end surfaces of the hub is .09 in. and .06 in. on the bosses and bottom of the vertical hub. The two large holes are to have an LN3 fit for journal bearings. Scale is full.

ROUNDS AND FILLETS R .12

Fig. 5-7-D   Link.

CHAPTER 6

Working Drawings

Working Drawings

A *working drawing* is a drawing that supplies information and instructions for the manufacture or construction of machines or structures. Generally, working drawings may be classified into two groups: detail drawings, which provide the necessary information for the manufacture of the parts, and assembly drawings, which supply the necessary information for their assembly.

Since working drawings may be sent to other companies to make or assemble the parts, the drawings should conform with the drawing standards of that company. For this reason, most companies follow the drawing standards of their country. The drawing standards recommended by the American National Standards Institute (ANSI) and the Canadian Standards Association (CSA), which is very similar to ANSI, have been adopted by the majority of industries in North America.

DETAIL DRAWINGS

A detail drawing (Fig. 6-1-1) must supply the complete information for the construction of a part. This information may be classified under three headings: shape description, size description, and specifications.

Shape Description This term refers to the selection and number of views to show or describe the shape of the part. The part may be drawn in either pictorial or orthographic projection, the latter being used more frequently. Sectional views, auxiliary views, and enlarged detail views may be added to the drawing in order to provide a clearer image of the part.

Size Description Dimensions which show the size and location of the shape features are then added to the drawing. The manufacturing process will influence the selection of some dimensions, such as datum features. Tolerances are then selected for each dimension.

Specifications This term refers to general notes, material, heat treatment, finish, general tolerances, and number

Fig. 6-1-1 A simple detail drawing.

required. This information is located on or near the title block or strip.

Additional Drawing Information In addition to the information pertaining to the part, a detail drawing includes additional information such as drawing number, scale, method or projection, date, name of part or parts, and the drafter's name.

The selection of paper size is determined by the number of views selected, the number of general notes required, and the drawing scale used. If the drawing is to be microformed, then the lettering size would be another factor to consider. The drawing number usually carries a prefix or suffix number or letter to indicate the sheet size, such as A-571 or 4-571; the letter A indicates that it is made on an 8.50 × 11.00 in. sheet, and the number 4 indicates that the drawing is made on a 210 × 297 mm sheet.

DRAWING CHECKLIST

As an added precaution against errors occurring on a drawing, many companies have provided checklists for drafters to follow before a drawing is issued to the shop. A typical checklist may be as follows:

1. *Dimensions.* Is the part fully dimensioned, and are the dimensions clearly positioned? Is the drawing dimensioned to avoid unnecessary shop calculations?
2. *Scale.* Is the drawing to scale? Is the scale shown?
3. *Tolerances.* Are the clearances and tolerances specified by the linear and angular dimensions and by local, general, or title block notes suitable for proper functioning? Are they realistic? Can they be liberalized?

Fig. 6-1-2 Manufacturing process influences the shape of the part.

4. *Standards.* Have standard parts, design, materials, processes, or other items been used where possible?
5. *Surface Texture.* Have surface roughness values been shown where required? Are the values shown compatible with overall design requirements?
6. *Material.* Have proper material and heat treatment been specified?

Qualifications of a Detailer

The detailer should have a thorough understanding of materials, shop processes, and operations in order to properly dimension the part and call for the correct finish and material. In addition, the detailer must have a thorough knowledge of how the part functions in order to provide the correct data and tolerances for each dimension.

The detailer may be called upon to work from a complete set of instructions and drawings, or he or she may be required to make working drawings of parts which involve the design of the part. Design considerations are limited in this chapter, but are covered in detail in Chap. 33.

MANUFACTURING METHODS

The type of manufacturing process will influence the selection of material and detailed feature of a part. See Fig. 6-1-2. For example, if the part is to be cast, rounds and fillets will be added. Additional material will also be required where surfaces are to be finished.

The more common manufacturing processes are machining from standard stock; prefabrication which includes welding, riveting, soldering, brazing, and gluing; forming from sheet stock; casting; and forging. The latter two processes can be justified only when large quantities are required and for specially designed parts. All these processes are described in detail in other chapters.

Several drawings may be made for the same part, each one giving only the information necessary for a particular step in the manufacture of the part. A part which is to be produced by forging, for example, may have one draw-

ing showing the original rough forged part and one detail of the finished forged part. See Figs. 6-1-2C and 6-1-2D.

ASSIGNMENTS

See Assignments 1 through 3 for Unit 6-1 on page 133.

Review for Assignments
Unit 5-1   Basic Dimensioning
Unit 5-2   Dimensioning Circular Features
Unit 5-3   Dimensioning Common Features
Unit 10-10 Sand Castings

UNIT 6-2
Multiple Detail Drawings

Detail drawings may be shown on separate sheets, or they may be grouped on one or more large sheets.

Often the detailing of parts is grouped according to the department in which they are made. For example, wood, fiber, and metal parts are used in the assembly of a transformer. Three separate detail sheets—one for wood parts, one for fiber parts, and the third for the metal parts—may be drawn. These parts would be made in the different shops and sent to another area for assembly. In order to facilitate assembly, each part is given an identification part number which is shown on the assembly drawing. A typical detail drawing showing multiple parts is illustrated in Fig. 6-2-1.

If the details are few, the assembly drawing may appear on the same sheet or sheets.

ASSIGNMENT

See Assignment 4 for Unit 6-2 on page 134.

Review for Assignment
Unit 3-2   Shape Description
Chapter 5   Basic Dimensioning

Fig. 6-2-1 **Detail drawing containing many details on one drawing.**

UNIT 6-3
Drawing Revisions

Revisions are made to an existing drawing when manufacturing methods are improved, to reduce cost, to correct errors, and to improve design. A clear record of these revisions must be registered on the drawing.

All drawings must carry a change or revision table, either down the right-hand side or across the bottom of the drawing. In addition to a description of drawing changes, provision may be made for recording a revision symbol, a zone location, an issue number, a date, and the approval of the change. Should the drawing revision cause a dimension or dimensions to be other than the scale indicated on the drawing, then the dimensions that are not to scale should be indicated by the method shown in Fig. 5-1-16. Typical revision tables are shown in Fig. 6-3-1.

At times, when there are a large number of revisions to be made, it may be more economical to make a new drawing. When this is done, the words REDRAWN and REVISED should appear in the revision column of the new drawing. A new date is also shown for updating old prints.

(A) DRAWING REVISIONS

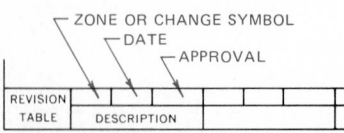

REVISIONS		
SYMBOL	DESCRIPTION	DATE & APPROVAL
1	LENGTH WAS 150	J. Heisel 12-4-76
2	CHAMFER ADDED	F. Newman 2-2-77

(B) VERTICAL REVISION BLOCK

(C) HORIZONTAL REVISION BLOCK

Fig. 6-3-1   Drawing revisions.

References
1. ANSI Y14.5, *Dimensions and Tolerancing*.

ASSIGNMENT

See Assignment 5 for Unit 6-3 on page 136.

Review for Assignment
Unit 5-1   Not-to-Scale Drawings

UNIT 6-4
Assembly Drawings

All machines and mechanisms are composed of numerous parts. A drawing showing the product in its completed state is called an *assembly drawing*.

Assembly drawings vary greatly in the amount and type of information they give, depending on the nature of the machine or mechanism they depict. The primary functions of the assembly drawing are to show the product in its completed shape, to indicate the relationship of its various components, and to designate these components by a part or detail number. Other information that might be given includes overall dimensions, capacity dimensions, relationship dimensions between parts (necessary information for assembly), operating instructions, and data on design characteristics.

Design Assembly Drawings
When a machine is designed, an assembly drawing or a design layout is first drawn to clearly visualize the performance, shape, and clearances of the various parts. From this assembly drawing, the detail drawings are made and each part is given a part number.

To assist in the assembling of the machine, part numbers of the various details are placed on the assembly drawing. Small circles .31 to .50 in. (8 to 12 mm) in diameter, that contain the part number are then attached to the corresponding part with a leader, as illustrated in Fig. 6-4-1. It is important that the detail drawings not use identical numbering schemes when several bills of material are used.

Installation Assembly Drawings
This type of assembly drawing is used when many unskilled people are employed to mass-assemble parts. Since these people are not normally trained to read technical drawings, simplified pictorial assembly drawings similar to the one shown in Fig. 6-4-2 are used.

Assembly Drawings for Catalogs
Special assembly drawings are prepared for company catalogs. These assembly drawings show only pertinent details and dimensions that would interest the potential buyer. Often one drawing, having letter dimensions accompanied by a chart, is used to cover a range of sizes, such as the pillow block shown in Fig. 6-4-3B.

Bills of Material
A *bill of material* is an itemized list of all the components shown on an assembly drawing or a detail drawing. See Fig. 6-4-4. Often, a bill of material is placed on a separate sheet for ease of handling and duplicating. Since the bill of material is used by the purchasing department to order the necessary material for the design, the bill of material should show the raw material size, rather than the finished size of the part.

For castings, a pattern number should appear in the size column in lieu of the physical size of the part.

Standard components, which are purchased rather than fabricated, such as bolts, nuts, and bearings, should have a part number and appear on the bill of material. Information in the descriptive column should be sufficient for the purchasing agent to order these parts.

Parts lists for bills of material placed on the bottom of the drawing should read from bottom to top, while bills of material placed on the top of the drawings should read from top to bottom. This practice allows additions to be made at a later date.

ASSIGNMENT

See Assignment 6 for Unit 6-4 on page 136.

LEADER WITH ARROW TOUCHING PART

CIRCLE .40 TO .50 MIN IN DIAMETER

SAME PART NUMBER AS SHOWN ON DETAIL SHEET

3.00 MAX

Fig. 6-4-1 Identification numbers on assembly drawings.

800 MAX 0 MIN

230

Ø100

350 X 350

2000

1300 MAX

METRIC

280 X 300

840

(A) DRILL PRESS

A SQUARE

B SQUARE

J

W

S

G

E D

X

K

C

d

A CHART IS USED WITH THIS TYPE OF DRAWING TO COVER A RANGE IN SIZES

(B) PILLOW BLOCK

Fig. 6-4-3 Assembly drawings used in catalogs.

R32   R33

TS1

R16

C17

R49   R8   R9

SW3   TS3

C18

R40

C19   C6

R50   C20   C5   SW1

C21

ORANGE

TS2

BLACK   SW2

C16

GREEN

R45

R55   C25

BROWN

BLUE

R59

BROWN

PL1

R56

BLK

J1   J3

J2

R57

ITEM 36A   ITEM 36   R30   SW4   ITEM 29

DETAIL A

Fig. 6-4-2 Installation assembly drawings.

QTY	ITEM	MATL	DESCRIPTION	PT NO.
I	BASE	CI	PATTERN # A3154	I
I	CAP	CI	PATTERN # B7156	2
I	SUPPORT	MS	.38 X 2.00 X 4.38	3
I	BRACE	MS	.25 X 1.00 X 2.00	4
I	COVER	ST	.1345 (# 10 GA USS) X 6.00 X 7.50	5
I	SHAFT	CRS	Ø1.00 X 6.50	6
2	BEARINGS	SKF	RADIAL BALL # 6200Z	7
2	RETAINING CLIP	TRUARC	N5000-725	8
I	KEY	ST	WOODRUFF # 608	9
I	SET SCREW	CUP POINT	HEX SOCKET .25UNC X 1.50	10
4	BOLT–HEX HD–REG	SEMI-FIN	.38UNC X 1.50LG	11
4	NUT–REG HEX	ST	.38UNC	12
4	LOCK WASHER–SPRING	ST	.38 - MED	13
				14

(A) TYPICAL BILL OF MATERIAL   PARTS 7 TO 13 ARE PURCHASED ITEMS

SOME SUGGESTED SIZES

5.24

.50   1.62   .62   2.00   .50

.50

.31

QTY	ITEM	MATL	DESCRIPTION	PT NO.

(B) SAMPLE SIZES

Fig. 6-4-4 Bill of material.

UNIT 6-5
Exploded Assembly Drawings

In many instances parts must be identified or assembled by persons unskilled in the reading of engineering drawings. Examples are found in the appliance-repair industry, which relies on assembly drawings for repair work and for reordering parts. Exploded assembly drawings, like that shown in Fig. 6-5-1 are used extensively in these cases, for they are easier to read. This type of assembly drawing is also used frequently by companies that manufacture do-it-yourself assembly kits, such as model-making kits.

For this type of drawing, the parts are aligned in position. Frequently, shading techniques are used to make the drawings appear more realistic.

ASSIGNMENT

See Assignment 7 for Unit 6-5 on page 137.

Review for Assignment
Unit 14-1 Pictorial Drawings
Unit 14-8 Technical Illustration

UNIT 6-6
Detailed Assembly Drawings

Often these are made for fairly simple objects, such as pieces of furniture, when the parts are few in number and are not intricate in shape. All the dimensions and information necessary for the construction of each part and for the assembly of the parts are given directly on the assembly drawing. Separate views of specific parts, in enlargements showing the fitting together of parts, may also be drawn in addition to the regular assembly drawing. Note that in Fig. 6-6-1 the enlarged views are drawn in picture form, not as regular orthographic views. This method is peculiar to the cabinetmaking trade and is not normally used in mechanical drawing.

ASSIGNMENT

See Assignment 8 for Unit 6-6 on page 138.

Review for Assignment
Unit 6-4 Assembly Drawings
Unit 6-1 Detail Drawings

UNIT 6-7
Subassembly Drawings

Many completely assembled items, such as a car and a television set, are assembled with many preassembled components as well as individual parts. These preassembled units are referred to as *subassemblies*. The assembly drawings of a transmission for an automobile and the transformer

ROTO HAMMER

(A) PICTORIAL EXPLODED ASSEMBLY

Ø NOTES
USE 38849 NAMEPLATE WITH CAT. NO. 38848.
45459 FRICTION PLATE USES 3 CLUTCH DISCS
UNITS WITH 4 CLUTCH DISCS ON FRICTION
PLATE USE 46293.

45276 RETAINING RING
45277 1/4"-20 X 1 1/2" SET SCREW
Ø45459 FRICTION PLATE
45279 ADJUSTING KNOB

45281 CLUTCH PLATE
45282 PRESSURE SPRING
45283 1/4"-20 HEX NUT
45284 LOCK CUP
Ø38551 NAMEPLATE
45286 FIBRE WASHER
45287 WING NUT

34030 1/4" LOCKWASHER (2)
405 1/4"-20 HEX. NUT (2)
45502 MOUNTING HOOK (1/4"-20THD)
45275 BRACKET
32114 WIRE CONNECTOR (2)
45288 FIBRE WASHER
45289 ARBOR SCREW
46271 TURRET SCREW
46276 LOCK WASHER
46274 GROUNDING LUG
46273 KEP NUT
46272 TURRET
46275 CAP NUT
45293 GROMMET
45290 RUBBER STOP
46270 CORD & PLUG ASSEMBLY

PARTS NOT SHOWN
46288 CONNECTION DIAGRAM
45503 MOUNTING HOOK (WOOD SCREW)

ELECTRIC BALANCE REEL

38314 HUBBLE TWIST LOCK - MALE
(OMIT ON 38275)

38313 HUBBLE TWIST LOCK - FEMALE
(OMIT ON 38275)

(B) ORTHOGRAPHIC EXPLODED ASSEMBLY

Fig. 6-5-1 Exploded assembly drawings.

Fig. 6-6-1 Detailed assembly drawing.

for a television set are typical examples of subassembly drawings.

Subassemblies are designed to simplify final assembly as well as permit the item to be either assembled in a more suitable area or purchased from an outside source. This type of drawing shows only those dimensions which would be required for the completed assembly. Examples are size of the mounting holes and their location, shaft locations, and overall sizes. This type of drawing is found frequently in catalogs.

ASSIGNMENT

See Assignment 9 for Unit 6-7 on page 138.

Review for Assignment
Unit 6-4   Assembly Drawings
Unit 2-6   Phantom Lines

ASSIGNMENTS for Chapter 6

Assignment for Unit 6-1, Working Drawings

1. On a B- or A3-size sheet, make a working drawing of one of the parts shown in Fig. 6-1-A or 6-1-B. Select appropriate views and dimensions and add to the drawing the information needed so that the parts can be completely manufactured. Scale is to suit.

2. On a B- or A3-size sheet, select one of the parts shown in Fig. 6-1-C or 6-1-D and make a three-view working drawing. Balance the views on the paper, and with the use of a miter line project all lines and points from one view to the other. Dimensions are to be converted to millimeters if required. Only the dovetail and T slot dimensions are critical and must be taken to an accuracy of one point beyond the decimal point. All other dimensions are to be rounded off to whole numbers.

3. On a B- or A3-size sheet make detail drawings of any of the parts assigned by your instructor from the assembly drawings shown in Fig. 6-1-E or 6-1-F. The scale and selection of views are to be decided by the student.

MATL — MALLEABLE IRON
ROUNDS AND FILLETS R.06

ENLARGED VIEW OF T—SLOT

Fig. 6-1-A   Cross slide.

Ø 13
Ø 24 SFACE
2 HOLES

25

38

15

15

R 19

50

18

10
2 RIBS

32

25

10

125

38

64

13

METRIC

ROUNDS AND FILLETS R 2
MATL – MALLEABLE IRON

Fig. 6-1-B   Guide bracket.

FINISH ALL SURFACES ON THE TEE
AND DOVETAIL SLOTS
ROUNDS AND FILLETS R .06
MATL – CI

Ø .380
2 HOLES

1.40

.90

1.80

1.40

15°

.90

.40

.40

.45

.50

.40

1.20

.30
.40

.25

.50

1.00

.30
.60

60°

.40

.30

4.00

2.20

1.20

2.00

FRONT

3.30

Fig. 6-1-D   Locating stand.

Ø .440

1.90

1.00

1.00

1.80

.40

30°

45°

.40

.40

600°

2.50

.20

.50

.50

1.25

.40

.50

.75

1.50

.50

2.50

1.00

600°

.45

4.00

FRONT

FINISH ALL SURFACES ON THE TEE
AND DOVETAIL SLOTS
ROUNDS AND FILLETS R .06
MATL – CI

Fig. 6-1-C   Cross slide.

Assignment for Unit 6-2, Multiple Detail Drawings

4. On a B- or A3-size sheet, make detail drawings of all the parts shown of one of the assemblies in Fig. 6-2-A or 6-2-B. Since time is money, select only the views necessary to describe each part. Below each part show the following information: part number, name of part, material, number required. Scale is full or 1:1.

Ø 7.5 X 80 DEEP
.125 x 27 NPT

M24 PREVAILING TORQUE NUT

3 RIBS
8mm
THICK
EQUALLY
SPACED

WOODRUFF
KEY #809

25

2

8

R5

Ø 54

40

R5

Ø 28

35

18

5

30°

10

Ø 70

130

Ø 130

Ø 110

Ø 45

90

Ø 35

135

Ø 13 4 HOLES
EQUALLY
SPACED

1 x 8 OIL
GROOVE

480

Ø 26

R10

R40

R10

30

R

22

12

12

R 20

350

Ø 160

Ø 200

8 x 4 KEYSEAT

METRIC

Fig. 6-1-E   Pulley assembly.

Fig. 6-1-F Adjustable pulley.

①	FRAME	⑤	WASHER	
②	SHAFT	⑥	NUT	
③	PULLEY	⑦	BOLT	
④	COLLAR	⑧	WASHER	

Fig. 6-2-A Shaft support.

MATL – SAE 1025

Fig. 6-2-B Shaft pivot support.

METRIC

MATL – SAE 1050

Fig. 6-3-A Gasket.

Ø.50
6 HOLES EQUALLY
SPACED ON Ø4.38

BACK SURFACE

2.38

2.25

Ø 2.88

.75

Ø 5.75

REVISIONS:
1. Ø.50 TO BE Ø.53
2. Ø5.75 TO BE Ø6.00
3. 2.25 TO BE 2.30
4. 2.38 TO BE 2.25

ROUNDS AND FILLETS R.06
MATL — MALLEABLE IRON

Fig. 6-3-B Axle cap.

Ø 8
4 HOLES

88 SQ

64

64

R 5

12

6

13

12

45°

6

2

32

32

28

28

METRIC

REVISIONS:
1. 88 TO BE 92
2. 12 TO BE 14
3. Ø8 TO BE Ø10
4. 28 TO BE 30

MATL — NEOPRENE

Assignment for Unit 6-3, Drawing Revisions

5. Select one of the drawings shown in Fig. 6-3-A or 6-3-B and make appropriate revisions to these drawings, recording the changes in the drawing revision column.

Assignment for Unit 6-4, Assembly Drawings

6. On a B- or A3-size sheet, make a one-view assembly drawing of one of the assemblies shown in Fig. 6-4-A or 6-4-B. For Fig. 6-4-B show a round bar (Ø 1 in.) in phantom being held in position.

Include on the drawing a bill of material and identification part numbers. Scale is full or 1:1.

Assignment for Unit 6-5, Exploded Assembly Drawings

7. On a B- or A3-size sheet, make an exploded assembly drawing in orthographic projection of one of the assemblies shown in Fig. 6-5-A or 6-5-B. Use center lines to align parts and holes. To make the parts appear more realistic, shading techniques are recommended. Scale is 1:1.

.44

.625 SQ

.19

PT 1 — SCREW
MATL — SAE 1112

Ø1.06

R.09

.625–11UNC–2A END OF SCREW
Ø.38 X .10 LG

2.50
TO END

Ø1.25

.75

PT 2 — POST
MATL — CAST STEEL

.50

Ø1.75

.62

PT 4 — RING
MATL — SAE 1020

4.00
LENGTH

R5.00

MED DIAMOND KNURL

.69

R5.00

PT 3 — WEDGE
MATL — SAE 1050

Ø3.25

3.00

.50

.44

.50

Ø1.81

Ø3.50

PT 5 — BLOCK
MATL — SAE 1020

3.69

.50

Ø2.31

.31

Ø2.25

.31

2.75 SQ

Fig. 6-4-A Tool post holder.

Fig. 6-4-B V-block clamp.

PT I – ADJUSTING SCREW
MATL – SAE 1112

P 0.8 DIAMOND KNURL

Ø 22

16 8

Ø 3

NECK 1.5 X 1.5

M 12

76 LENGTH
OF SCREW

END OF SCREW
Ø 10 X 6 LG

19

R 5

PT 2 – YOKE
MATL –
CAST STEEL

R 32

R 22

R 3

6

22

45°

38

10

28

66

54

20

23

5

4

12

20

22

50

14

45°

75

FRONT

PT 3 – BASE
MATL – SAE 1020

1.5 X 45° CHAMFER

4

9

25

43

3

Ø 7.5

Ø 14

Ø 18

PT I – POST
MATL – SAE 1112
I REQD

PEEN AT ASSEMBLY

15

2

40

3

Ø 16

Ø 7.5

0.73

2 X 45°
RIB

Ø 6.4

1.5 X 45° CHAMFER
PT 3 – SHAFT
MATL – SAE 1112
I REQD

R 3

2.5

6

Ø 14.5

22 44

Ø 8
2 HOLES

30

R 10

60

PT 2 – BRACKET
MATL – 2.38 (#13G S GA)
I REQD

PT 6 – RETAINING RING
EXT SERIES 5133
I REQD
MATL – STEEL

R 2

28

Ø 100

Ø 70

3

Ø 30

Ø 14

Ø 14

R I

Ø 8

28

PT 5 – BUSHING
MATL – BRASS
I REQD

PT 4 – WHEEL
MATL – HARD RUBBER
I REQD

METRIC

Fig. 6-5-B Caster.

.12 X Ø .31

.12

Ø .56

.25

1.38

1.06

.438 - 20 NF

Ø .31

PT 3 STUD
4 REQD MATL - STEEL

Ø .750

.438 - 20 NF - 2B

PT I FORK 2 REQD MATL C I

1.50

Ø 1.75

1.50

Ø .750

PT 2 RING

.41

.82

DRILL & REAM FOR NO.4 TAPER PIN

R.62

Ø .56

.12

.38

.12

2.38

R.50

Ø 1.00

1.12

Ø 1.38

2.62

1.62

1.50

.75

R.38

R1.00

BILL OF MATERIAL			
PT	ITEM	QTY	MATL	DESCRIPTION
I	FORK	2	C I	
2	RING	I	STEEL	
3	STUD	4	STEEL	
	NO 4 TAPER PIN	2		PURCHASED

Fig. 6-5-A Universal joint.

Assignment for Unit 6-6, Detailed Assembly Drawings

8. On an A3- or B-size sheet, make a detailed assembly drawing of one of the assemblies shown in Fig. 6-6-A or 6-6-B. Include on the drawing the method of assembly (i.e., nailing, wood screws, doweling, etc.) and a bill of material. Include in the bill of material the assembly materials. Use scale 1:5 for Fig. 6-6-A and 1.50 in. = 1 ft. for Fig. 6-6-B.

Assignment for Unit 6-7, Subassembly Drawings

9. On an A3- or B-size sheet make a one-view subassembly drawing of one of the assemblies shown in Fig. 6-7-A or 6-7-B. A broken-out or partial section view is rec-ommended to show the interior features. Include on the drawing pertinent dimensions, identification numbers on assembly drawing, a bill of material, and a phantom outline of the adjoining part or features. Scale is 1:1 or full.

Fig. 6-6-A   Book rack.

Fig. 6-6-B   Saw horse.

NOTE: WOOD SIZES (THICKNESS AND WIDTH) ARE NOMINAL SIZES

Fig. 6-7-A   Wheel assembly.

Ø 1.90 Ø .625

.638

.12

5.50

PT 2 IDLER PULLEY FRAME

Ø 1.25 Ø .625

4.75

1.00 2.75

Ø .875

.12 .50

.50

2.00

1.00

.75 2.75

4.25

OIL GROOVE .12 WIDE X .03 DEEP
OIL GROOVE
.31 WIDE X .03 DEEP

Ø .750 Ø 1.06

.25

2.68

PT 3 IDLER PULLEY BUSHING

.625 – 11 UNC – 2A
#405 WOODRUFF
KEYSEAT

Ø .328
Ø .328 X 1.75 DEEP
.125 PIPE
THREAD

Ø .625 Ø .750 Ø .875

.62 .38 .40 1.00

2.75

5.25

PT 4 IDLER PULLEY SHAFT

2.68

Ø 3.25 Ø 1.75 Ø 1.06 Ø 2.75
Ø 2.25

R.18

1.09 .50

PT 1 IDLER PULLEY

Fig. 6-7-B   Idler pulley.

BILL OF MATERIAL				
PT	ITEM	MATL	DESCRIPTION	QTY
1	IDLER PULLEY	CI	A – 5432	1
2	IDLER PULLEY FRAME	CI	A – 1734	1
3	IDLER PULLEY BUSHING	BRZ		1
4	IDLER PULLEY SHAFT	CRS		1
5	HEX NUT	STEEL	.625 UNC	1
6	WOODRUFF KEY	STD	NO. 405	1
7	OILER	STD	.125	1

CHAPTER 7

Sections and Conventions

UNIT 7-1
Sectional Views

Sectional views, commonly called *sections*, are used to show interior detail that is too complicated to be shown clearly by regular views containing many hidden lines. For some assembly drawings, they show a difference in materials. A sectional view is obtained by supposing the nearest part of the

Fig. 7-1-1   A full-section drawing.

object to be cut or broken away on an imaginary cutting plane. The exposed or cut surfaces are identified by section lining or cross-hatching. Hidden lines and details behind the cutting-plane line are usually omitted unless they are required for clarity or dimensioning. It should be understood that only in the sectional view is any part of the object shown as having been removed.

A sectional view frequently replaces one of the regular views. For example, a regular front view is replaced by a front view in section, as shown in Fig. 7-1-1.

Whenever practical, except for revolved sections, sectional views should be projected perpendicular to the cutting plane and be placed in the normal position for third-angle projection.

When the preferred placement is not practical, the sectional view may be removed to some other convenient position on the drawing, but it must be clearly identified, usually by two capital letters, and labeled.

Cutting-Plane and Viewing-Plane Lines

Cutting-plane and viewing-plane lines (Fig. 7-1-2) are used to indicate the location of cutting planes for sectional views and the viewing position for removed partial views. Two forms of cutting-plane lines are approved for general use.

The first form consists of evenly spaced, thick dashes with arrowheads (Fig. 7-1-2A). The second form consists of alternating long dashes and pairs of short dashes (Fig. 7-1-2A). The long dashes may vary in length, depending on the size of the drawing.

Both forms of lines should be drawn to stand out clearly on the drawing. The ends of the lines are bent at 90° and terminated by bold arrowheads to indicate the direction of sight for viewing the section.

The cutting-plane line can be omitted when it corresponds to the center line of the part and it is obvious where the cutting plane lies. On drawings with a high density of line work, cutting-plane lines may be modified by omitting the dashes between the line ends for the purpose of obtaining clarity, as shown in Fig. 7-1-2B.

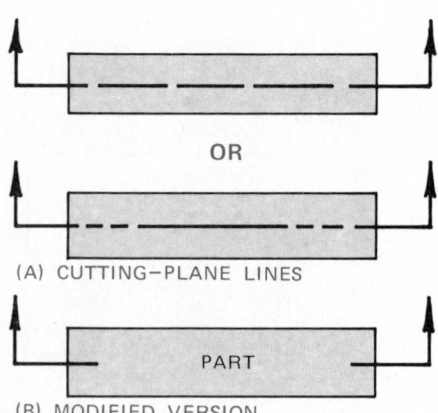

Fig. 7-1-2   Cutting-plane and viewing-plane lines.

(A) SIDE VIEW NOT SECTIONED

CUTTING PLANE

FRONT SECTION REMOVED

CUTTING–PLANE LINE

SECTION VIEW REPLACES
CONVENTIONAL SIDE VIEW

(B) SIDE VIEW IN FULL SECTION

Fig. 7-1-3   Full-section view.

Full Sections

When the cutting plane extends entirely through the object in a straight line and the front half of the object is theoretically removed, a full section is obtained. See Figs. 7-1-3 and 7-1-4. This type of section is used for both detail and assembly drawings. When the section is on an axis of symmetry, it is not necessary to indicate its location. See Fig. 7-1-5. However, it may be identified and indicated in the normal manner to increase clarity, if so desired.

(A) **INCOMPLETE** LINES BEHIND
CUTTING PLANE NOT SHOWN

(B) **POOR PRACTICE** HIDDEN
LINES NOT NECESSARY

(C) **GOOD PRACTICE** HIDDEN
LINES OMITTED, VISIBLE
LINES SHOWN

**Fig. 7-1-4   Visible and hidden lines on
section views.**

Section Lining

Section lining, sometimes referred to as cross-hatching, can serve a double purpose. It indicates the surface that has been theoretically cut and makes it stand out clearly, thus helping the observer to understand the shape of the object. Section lining may also indicate the material from which the object is to be made, when the lining symbols shown in Fig. 7-1-6 are used.

Section Lining for Detail Drawings Since the exact material specifications for a

HIDDEN LINES SHOW
INTERIOR POORLY

OR

**Fig. 7-1-5   Cutting-plane line may be
omitted when it corresponds with the center
line.**

1. IRON AND GENERAL
   PURPOSE USE FOR
   ALL MATERIALS

2. CORK, FELT,
   FABRIC, LEATHER,
   FIBER

3. MARBLE, SLATE,
   PORCELAIN,
   GLASS, ETC.

4. BRONZE, BRASS,
   COPPER, AND
   COMPOSITIONS

5. STEEL

6. EARTH

7. MAGNESIUM,
   ALUMINUM, AND
   ALUMINUM ALLOYS

8. THERMAL
   INSULATION

9. ROCK

10. SOUND INSULATION

11. WHITE METAL,
    ZINC, LEAD,
    BABBITT AND
    ALLOYS

12. RUBBER, PLASTIC,
    ELECTRICAL
    INSULATION

13. SAND

14. CONCRETE

ACROSS GRAIN

WITH GRAIN

15. WOOD

16. WATER AND
    OTHER LIQUIDS

Fig. 7-1-6   Symbolic section lining.

part are usually given elsewhere on the drawing, the general-purpose section lining symbol is recommended for most detail drawings. An exception may be made for wood when it is desirable to show the direction of the grain.

The lines for section lining·are thin and are usually drawn at an angle of 45°

Fig. 7-1-7 Direction of section lining.

Fig. 7-1-8 Outline section lining.

Fig. 7-1-9 Section lining omitted to accommodate dimensions.

Fig. 7-1-10 Thin parts in section.

to the major outline of the object. The same angle is used for the whole "cut" surface of the object. If the part shape would cause section lines to be parallel, or nearly so, to one of the sides of the part, then some angle other than 45° should be chosen. See Fig. 7-1-7. The spacing of the hatching lines should be reasonably uniform to give a good appearance to the drawing. The pitch, or distance between lines, normally varies between .03 and .12 in. (1 and 3 mm) depending on the size of the area to be sectioned.

Large areas shown in section need not be entirely section-lined. See Fig. 7-1-8. Section lining around the outline will usually be sufficient, providing clarity is not sacrificed.

Dimensions or other lettering should not be placed in sectioned areas. When this is unavoidable, the section lining should be omitted for the numerals or lettering. See Fig. 7-1-9.

Sections which are too thin for effective section lining, such as sheet-metal items, packing, and gaskets, may be shown without section lining; or the area may be filled in completely. See Fig. 7-1-10.

ASSIGNMENT

See Assignment 1 for Unit 7-1 on page 153.

Review for Assignment
Unit 10-1   Sand Castings
Unit 5-2   Spot Faces
Unit 2-6   Drawing Circles and Arcs

UNIT 7-2
Two or More Sectional Views on One Drawing

If two or more sections appear on the same drawing, the cutting-plane lines are identified by two identical large, single-stroke, Gothic letters, one at each end of the line, placed behind the arrowhead so that the arrow points away from the letter. The identification letters should not include I, O, Q, and Z. See Fig. 7-2-1.

Sectional view subtitles are given when identification letters are used and appear directly below the view, incorporating the letters at each end of the cutting-plane line thus: SECTION A-A, or abbreviated, SECT. B-B. When the scale is different from the main view, it is stated below the subtitle thus

SECTION A-A
SCALE 1:4

ASSIGNMENT

See Assignment 2 for Unit 7-2 on page 154.

Review for Assignment
Unit 5-2   Countersinks, Counterbores, and Spot Faces
Unit 4-3   Drawing a Hexagon
Unit 5-3   Dimensioning Tapers

Fig. 7-2-1 Detail drawing having two section views.

UNIT 7-3
Half Sections

A *half section* is a view of an assembly or object, usually symmetrical, showing one half of the view in section. See Figs. 7-3-1 and 7-3-2. Two cutting-plane lines, perpendicular to each other, extend halfway through the view, and one quarter of the view is considered removed with the interior exposed to view.

Similar to the practice followed for full-section drawings, the cutting-plane line need not be drawn for half sections when it is obvious where the cutting took place. Instead, center lines may be used. When a cutting plane is used, the common practice is to show only one end of the cutting-plane line, terminating with an arrow to show the direction of sight for viewing the section.

On the sectional view a center line or a visible object line may be used to divide the sectioned half from the unsectioned half of the drawing. This type of sectional drawing is best suited for assembly drawings where both internal and external construction is shown on one view and where only overall and center-to-center dimensions are required. The main disadvantage of using this type of sectional drawing for detail drawings is the difficulty in dimensioning internal features without adding hidden lines. However, hidden lines may be added for dimensioning, as shown in Fig. 7-3-3.

HIDDEN LINES ADDED FOR DIMEN-
SIONING

Fig. 7-3-3   Dimensioning half-section view.

ASSIGNMENT

See Assignment 3 for Unit 7-3 on page 154.

Review for Assignment
Unit 19-1   Vee-Belt Drive
Unit 9-1   Dimensioning of Keyseats
Unit 9-1   Keys
Unit 5-2   Double-Arrow Dimensioning

Fig. 7-3-1   A half-section drawing.

Fig. 7-3-2   Half-section views.

UNIT 7-4
Threads in Section

True representation of a screw thread is seldom provided on working drawings because it would require very laborious and accurate drawing involving repetitious development of the helix curve of the thread. A symbolic representation of threads is now standard practice.

Three types of conventions are in general use for screw thread representation. See Fig. 7-4-1. These are known as pictorial, schematic, and simplified representations. Simplified representation should be used whenever it will clearly portray the requirements. Schematic and pictorial representations require more drafting time, but

EXTERNAL

(A) PICTORIAL REPRESENTATION

INTERNAL

(B) SCHEMATIC REPRESENTATION

CHAMFER CIRCLE

EXTERNAL

END OF FULL THREAD

INTERNAL

(C) SIMPLIFIED REPRESENTATION (ANSI)

EXTERNAL

INTERNAL

(D) SIMPLIFIED REPRESENTATION (ISO)

Fig. 7-4-1   Threads in section.

Fig. 7-4-2   Threaded assembly.

NOTE: EXTERNAL THREADS ARE SHOWN ON THREADED ASSEMBLIES.

BEFORE ASSEMBLY   AT ASSEMBLY

Fig. 7-4-3   Drawing threads in assembly drawings.

views, the externally threaded part is always shown covering the internally threaded part, as illustrated in Fig. 7-4-3.

ASSIGNMENT

See Assignment 4 for Unit 7-4 on page 154.

Review for Assignment
Unit 8-2   Thread Conventions
Unit 8-2   Pipe Threads
Unit 8-1   Threading Considerations
Unit 7-18   Intersections of
            Unfinished Surfaces

are sometimes necessary to avoid confusion with other parallel lines or to more clearly portray particular aspects of the threads.

Threaded Assemblies
Any of the thread conventions shown here may be used for assemblies of threaded parts, and two or more methods may be used on the same drawing, as shown in Fig. 7-4-2. In sectional

UNIT 7-5
Assemblies in Section

SECTION LINING ON ASSEMBLY DRAWINGS

General-purpose section lining is recommended for most assembly drawings, especially if the detail is small. Symbolic section lining is generally not recommended for drawings that will be microformed.

General-purpose section lining should be drawn at an angle of 45° with the main outlines of the view. On adjacent parts, the section lines should be drawn in the opposite direction, as shown in Figs. 7-5-1 and 7-5-2.

PART A   PART C
45°
15
60°
45°
PART B

Fig. 7-5-1   Direction of section lining.

(A) ADJACENT PARTS

(B) SPACING OF SECTION LINING ACCORDING TO SIZE OF AREA TO BE SECTIONED.

Fig. 7-5-2   Arrangement of section lining.

For additional adjacent parts, any suitable angle may be used to make each part stand out separately and clearly. Section lines should not be purposely drawn to meet at common boundaries.

When two or more thin adjacent parts are filled in, a space is left between them, as shown in Fig. 7-5-3.

Symbolic section lining is used on special-purpose assembly drawings such as illustrations for parts catalogs, display assemblies, promotional materials, etc., when it is desirable to distinguish between different materials (Fig. 7-1-6).

All assemblies and subassemblies pertaining to one particular set of drawings should use the same symbolic conventions.

Shafts, Bolts, Pins, Keyseats, and Similar Solid Parts, in Section Shafts, bolts, nuts, rods, rivets, keys, pins, and similar solid parts, the axes of which lie in the cutting plane, should not be sectioned except that a broken-out section of the shaft may be used to describe more clearly the key, keyseat, or pin. See Fig. 7-5-4.

ASSIGNMENT

See Assignment 5 for Unit 7-5 on page 154.

Review for Assignment

STEEL PLATES

Fig. 7-5-3   Assembly of thin parts in section.

GASKETS

Fig. 7-5-4   Section lining used to distinguish between stationary and moving parts.

UNIT 7-6
Offset Sections

In order to include features that are not in a straight line, the cutting plane may be offset or bent, so as to include several planes or curved surfaces. See Figs. 7-6-1 and 7-6-2.

An offset section is similar to a full section in that the cutting-plane line extends through the object from one side to the other. The change in direction of the cutting-plane line is not shown in the sectional view.

ASSIGNMENT

See Assignment 6 for Unit 7-6 on page 156.

Review for Assignment

NOTE: CHANGE IN DIRECTION OF CUTTING-PLANE LINE NOT SHOWN IN SECTION VIEW ¬

Fig. 7-6-1   An offset section.

SECTION B—B

SECTION A—A

Fig. 7-6-2 Positioning offset sections.

ALTERNATE CROSS-HATCHING
AND HIDDEN LINES USED TO
INDICATE RIB

RIB B   RIB A

SECTION D-D

(A) BASE

RIBS C

(B) PULLEY

Fig. 7-7-2 Alternate method of showing ribs in section.

UNIT 7-7
Ribs, Holes, and Lugs in Section

RIBS IN SECTIONS

A true-projection sectional view of a part, such as shown in Fig. 7-7-1, would be misleading when the cutting plane passes longitudinally through the center of the rib. To avoid this impression of solidity, a section not showing the ribs section-lined or cross-hatched is preferred. When there is an odd number of ribs, such as those shown in Fig. 7-7-1B, the top rib is aligned with the bottom rib to show its true relationship with the hub and flange. If the rib is not aligned or revolved, it appears distorted on the sectional view and is therefore misleading.

At times it may be necessary to use an alternative method of identifying ribs in a sectional view. Figure 7-7-2 shows a base and a pulley in section. If rib A of the base were not sectioned as previously mentioned, it would appear exactly like rib B in the sectional view and would be misleading. Similarly, rib C shown on the pulley may be overlooked. To distinguish between the ribs on the base and the ribs and spaces on the pulley, alternate section lining on the ribs is used. The line between the rib and solid portions is shown as a broken line.

HOLES ARE ROTATED TO CUTTING PLANE TO SHOW THEIR
TRUE RELATIONSHIP WITH THE REST OF THE ELEMENT

RIBS ARE NOT SECTIONED

SECTION A-A
PREFERRED

SECTION A-A
TRUE PROJECTION

(A) CUTTING PLANE PASSING THROUGH BOTH RIBS

HOLE AND RIB ARE ROTATED
TO CUTTING PLANE

TRUE PROJECTION GIVES
A DISTORTED IMPRESSION

SECTION B-B
PREFERRED

SECTION B-B
TRUE PROJECTION

(B) CUTTING PLANE PASSING THROUGH ONE RIB AND ONE HOLE

Fig. 7-7-1 Preferred and true projection through ribs and holes.

SECTION B-B

(1) HOLES ALIGNED

SECTION C-C

(2) LUGS ALIGNED AND SECTIONED

SECTION D-D

(3) LUGS ALIGNED AND SECTIONED

SECTION E-E

(4) LUG NOT SECTIONED

Fig. 7-7-3   Aligning holes and lugs in section drawings.

HOLES IN SECTIONS

Holes, like ribs, are aligned as shown in Fig. 7-7-1 to show their true relationship to the rest of the part.

LUGS IN SECTION

Lugs, like ribs and spokes, are also aligned to show their true relationship to the rest of the part, because true projection may be misleading. Figure 7-7-3 shows several examples of lugs in section. Note how the cutting-plane line is bent or offset so that the features may be clearly shown in the sectional view.

Some lugs are shown in section, and some are not. When the cutting plane passes through the lug crosswise, the lug is sectioned; otherwise, the lugs are treated in the same manner as ribs.

ASSIGNMENT

See Assignment 7 for Unit 7-7 on page 156.

Review for Assignment
Unit 10-1   Bosses
Unit 5-1   Dual Dimensioning
Unit 7-17   Foreshortened Projection
Unit 7-18   Intersection of Unfinished Surfaces

UNIT 7-8
Revolved and Removed Sections

Revolved and removed sections are used to show the cross-sectional shape of ribs, spokes, or arms when the shape is not obvious in the regular views. See Figs. 7-8-1 to 7-8-3. Often end views are not needed when a revolved section is used. For a revolved section, draw a center line through the shape on the plane to be described, imagine the part to be rotated 90°, and superimposed on the view the shape that would be seen when rotated. See Figs. 7-8-1 and

(A) END VIEW NOT CLEAR

(B) REVOLVED SECTION

LINE SHOULD NOT GO THROUGH SECTION

AVOID

(C) PARTIAL VIEW SHOWING REVOLVED SECTION

CROSSING LINES TEND TO CONFUSE

AVOID

(D) REVOLVED SECTION WITH MAIN VIEW BROKEN FOR CLARITY

(E) PARTIAL VIEW SHOWING REVOLVED SECTION

Fig. 7-8-1   Revolved sections.

SECTION A–A
DOUBLE SIZE

SECTION B–B
DOUBLE SIZE

SECTION C–C
DOUBLE SIZE

VIEW D–D
DOUBLE SIZE

(A) REMOVED SECTIONS AND REMOVED VIEW

(B) CRANE HOOK

(C) NUT

Fig. 7-8-2   Removed sections.

Fig. 7-8-3   Revolved (superimposed) sections.

Fig. 7-8-4   Reference zone location.

The removed section differs in that the section, instead of being drawn right on the view, is removed to an open area on the drawing (Fig. 7-8-2). Frequently the removed section is drawn to an enlarged scale for clarification and easier dimensioning. Removed sections of symmetrical parts should be placed, whenever possible, on the extension of the center line (Fig. 7-8-2B).

On complicated drawings where the placement of the removed view may be some distance from the cutting plane, auxiliary information, such as the reference zone location (Fig. 7-8-4), may be helpful.

Placement of Sectional Views

Whenever practical, except for revolved sections, sectional views should be projected perpendicular to the cutting plane and be placed in the normal position for third-angle projection. See Fig. 7-8-5.

When the preferred placement is not practical, the sectional view may be removed to some other convenient position on the drawing, but it must be clearly identified, usually by two capital letters, excluding I, O, Q, and Z, and be labeled.

ASSIGNMENT

See Assignment 8 for Unit 7-8 on page 156.

Review for Assignment
Unit 7-14   Conventional Breaks
Unit 7-18   Intersections of
                   Unfinished Surfaces

7-8-3. If the revolved section does not interfere with the view on which it is revolved, then the view is not broken unless it would provide for clearer dimensioning. When the revolved section interferes or passes through lines on the view on which it is revolved, then the general practice is to break the view (Fig. 7-8-3). Often the break is used to shorten the length of the object. In no circumstances should the lines on the view pass through the section. When superimposed on the view, the outline of the revolved section is a thin, continuous line.

Fig. 7-8-5   Placement of sectional views.

UNIT 7-9
Spokes and Arms in Section

A comparison of the true projection of the wheel with spokes and the wheel with a web is made in Figs. 7-9-1A and B. This comparison shows that a preferred section for the wheel and spokes is desirable so that it will not appear to be a wheel with a solid web. In preferred sectioning, any part that is not solid or continuous around the hub is drawn without the section lining, even though the cutting plane passes through the spoke. When there is an odd number of spokes, as shown in Fig. 7-9-1C, the bottom spoke is aligned with the top spoke to show its true relationship to the wheel and to the hub. If the spoke were not revolved or aligned, it would appear distorted in the sectional view.

ASSIGNMENT

See Assignment 9 for Unit 7-9 on page 158.

Review for Assignment
Unit 9-2   Pin Fasteners
Unit 7-18   Intersections of
            Unfinished Surfaces

(A) FLAT PULLEY WITH WEB

SECTION A-A

(C) HANDWHEEL WITH ODD NUMBER OF SPOKES

SECTION C-C PREFERRED   SECTION C-C TRUE PROJECTION

Fig. 7-9-1   Preferred and true projection through spokes.

UNIT 7-10
Partial or Broken-Out Sections

Where a sectional view of only a portion of the object is needed, partial sections may be used. See Fig. 7-10-1. An irregular break line is used to show the extent of the section. With this type of section, a cutting-plane line is not required.

Fig. 7-10-1   Broken-out or partial sections.

(B) HANDWHEEL WITH EVEN NUMBER OF SPOKES

SECTION B-B PREFERRED   SECTION B-B TRUE PROJECTION

(D) HANDWHEEL WITH ODD NUMBER OF OFFSET SPOKES

SECTION D-D PREFERRED   SECTION D-D TRUE PROJECTION

ASSIGNMENT

See Assignment 10 for Unit 7-10 on page 158.

Review for Assignment
Unit 9-3   Retaining Rings

UNIT 7-11
Phantom or Hidden Sections

A phantom section is used to show the typical interior shapes of an object in one view when the part is not truly symmetrical in shape, as well as to show mating parts in an assembly drawing. See Fig. 7-11-1. It is a sectional view superimposed on the regular view without the removal of the front portion of the object. The section lining used for phantom sections consists of thin, evenly spaced, broken lines.

Fig. 7-11-1   Phantom or hidden section.

ASSIGNMENT

See Assignment 11 for Unit 7-11 on page 159.

Review for Assignment
Unit 5-3   Undercuts
Unit 5-6   Fits and Allowances
Unit 6-4   Design Assembly Drawings

UNIT 7-12
Sectional Drawing Review

In Units 7-1 through 7-11 the different types of sectional views have been explained and drawing problems have been assigned with each type of section drawing.

In the drafting office, it is the drafter who must decide which views are required to fully explain the part to be made. In addition, the drafter must select the proper scale(s) which will show the features clearly.

This unit has been designed to review the sectional-view options open to the drafter.

ASSIGNMENT

See Assignment 12 for Unit 7-12 on page 160.

Review for Assignment
Units 7-1 to 7-11   Sectional Drawings
Unit 6-1   Working Drawings

UNIT 7-13
Conventional Representation of Common Features

To simplify the representation of common features, a number of conventional drafting practices are used. Many conventions are deviations from true projection for the purpose of clarity; others are used to save drafting time. These conventions must be executed carefully, for clarity is even more important than speed.

Many drafting conventions such as those used on thread, gear, and spring drawings appear in various chapters throughout the text. Only the conventions not described in those chapters appear here.

Repetitive Details
Repetitive features, such as gear and spline teeth, are shown by drawing a partial view, showing two or three of these features, with a phantom line or lines to indicate the extent of the remaining features. See Fig. 7-13-1A and B. Alternatively, gears and splines may be shown with a solid thick line representing the basic outline of the part and a thin line representing the root of the teeth. This is essentially the same convention that is used for screw threads. The pitch line may be added by using the standard center line.

(A) SERRATED SHAFT

(B) SPLINED SHAFT

(C) DIAMOND KNURLING

(D) STRAIGHT KNURLING

(E) HOLES IN CIRCULAR PITCH

(F) HOLES IN LINEAR PITCH

(G) REPEATED PARTS

(H) REPEATED DETAILS

(I) SQUARE SECTIONS

Fig. 7-13-1   Conventional representation of common features.

Knurls
Knurling is an operation which puts patterned indentations in the surface of a metal part to provide a good finger grip. See Fig. 7-13-1C and D. Commonly used types of knurls are straight, diagonal, spiral, convex, raised diamond, depressed diamond, and radial. The pitch refers to the distance between corresponding indentations, and it may be a straight pitch, a

circular pitch, or a diametral pitch. For cylindrical surfaces, the latter is preferred. The pitch of the teeth for coarse knurls (measured parallel to the axis of the work) is 14 teeth per inch (TPI) or about 2 mm; for medium knurls, 21 TPI or about 1.2 mm; and for fine knurls, 33 TPI or 0.8 mm. The medium-pitch knurl is the most commonly used.

As a time-saver, the knurl symbol is shown on only a part of the surface being knurled.

Holes
A series of similar holes is indicated by drawing one or two holes and showing only the center for the others. See Fig. 7-13-1E and F.

Repetitive Parts
Repetitive parts, or intricate features, are shown by drawing one in detail and the others in simple outline only. A covering note is added to the drawing. See Fig. 7-13-1G and H.

Square Sections
Square sections on shafts and similar parts may be illustrated by thin, crossed, diagonal lines, as shown in Fig. 7-13-1[I].

ASSIGNMENT

See Assignment 13 for Unit 7-13 on page 160.

Review for Assignment
Unit 2-2   Enlarged Scales
Unit 3-7   One- and Two-View Drawings
Unit 5-3   Dimensioning Common Features

UNIT 7-14
Conventional Breaks

Long, simple parts such as shafts, bars, tubes, and arms need not be drawn to their entire length. Conventional breaks located at a convenient position may be used and the true length indicated by a dimension. See Fig. 7-14-1. Often a part can be drawn to a larger scale to produce a clearer drawing if a conventional break is used. The breaks used on circular

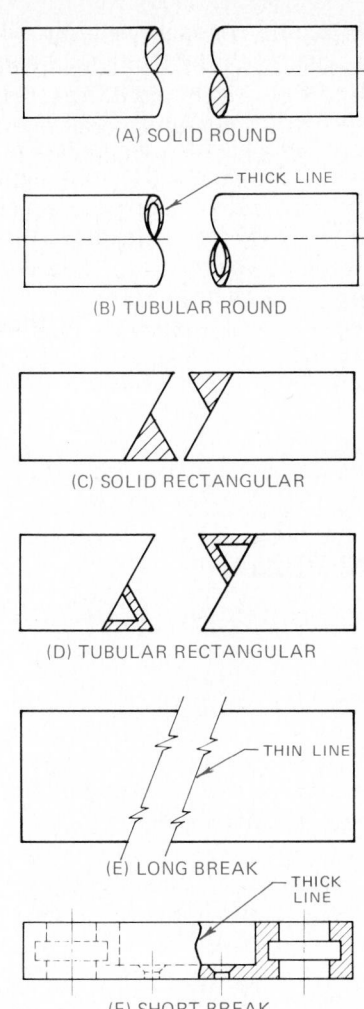

(A) SOLID ROUND

THICK LINE

(B) TUBULAR ROUND

(C) SOLID RECTANGULAR

(D) TUBULAR RECTANGULAR

THIN LINE

(E) LONG BREAK

THICK LINE

(F) SHORT BREAK

Fig. 7-14-1 Conventional breaks.

objects, known as "S breaks," may be drawn freehand, with an irregular curve, or a template or compass.

ASSIGNMENT

See Assignment 14 for Unit 7-14 on page 160.

Review for Assignment
Unit 4-3   Drawing a Hexagon
Unit 3-9   Enlarged Views

UNIT 7-15
Materials of Construction

Symbols used to indicate materials in sectional views are shown in Fig. 7-1-6. Those shown for concrete, wood, and

transparent materials are also suitable for outside views. Other symbols which may be used to indicate areas of different materials are shown in Fig. 7-15-1. It is not necessary to cover the entire area affected with such symbolic lining, as long as the extent of the area is shown on the drawing.

COARSE WIRE MESH

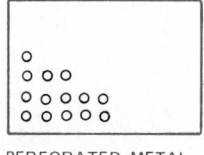

FINE WIRE MESH

MARBLE

PERFORATED METAL

Fig. 7-15-1 Symbols to indicate materials of construction.

Transparent Materials
These should generally be treated in the same manner as opaque materials; i.e., details behind them are shown with hidden lines if such detail is necessary.

ASSIGNMENT

See Assignment 15 for Unit 7-15 on page 161.

Review for Assignment
Unit 2-2   Enlarged Views
Unit 6-6   Detail Assembly Drawings
Unit 7-14   Conventional Breaks
Appendix   Sheet-Metal Gage Sizes

UNIT 7-16
Cylindrical Intersections

The intersections of rectangular and circular contours, unless they are very large, are shown conventionally as in Figs. 7-16-1 and 7-16-2. The same convention may be used to show the intersection of two cylindrical contours, or the curve of intersection may be shown as a circular arc.

(A)

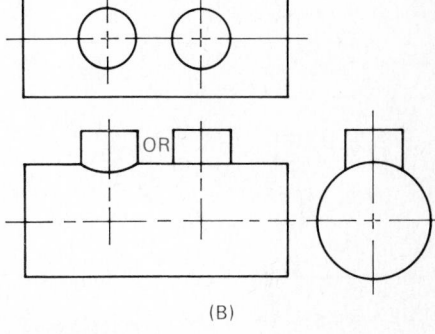

OR

(B)

Fig. 7-16-1 Conventional representation of external intersections.

PREFERRED   TRUE

Fig. 7-16-2 Conventional representation of cylindrical intersections.

ASSIGNMENT

See Assignment 16 for Unit 7-16 on page 162.

Review for Assignment
Unit 5-6   Fits
Unit 5-7   Surface Texture Symbols
Unit 7-10   Partial Views
Unit 7-8   Revolved Sections

UNIT 7-17
Foreshortened Projection

When the true projection of a feature would result in confusing foreshortening, it should be rotated until it is parallel to the line of the section or projection. See Fig. 7-17-1.

(A) ALIGNMENT OF RIB AND HOLES

(B) ALIGNMENT OF PART

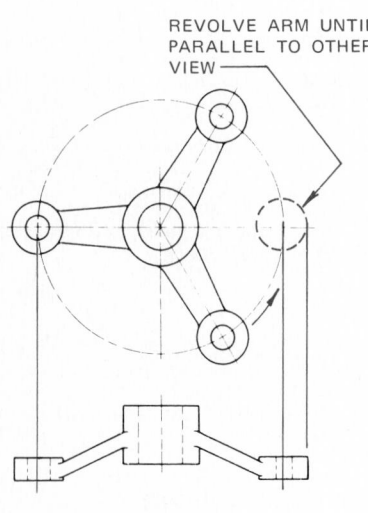

(C) ALIGNMENT OF ARM

Fig. 7-17-1 Alignment of parts and holes to show true relationship.

Holes Revolved to Show True Distance from Center

Drilled flanges in elevation or section should show the holes at their true distance from the center, rather than the true projection.

ASSIGNMENT

See Assignment 17 for Unit 7-17 on page 162.

Review for Assignment
Unit 5-6   Fits
Unit 5-7   Surface Finish
Unit 9-1   Keys
Unit 7-8   Revolved Sections

UNIT 7-18
Intersections of Unfinished Surfaces

The intersections of unfinished surfaces that are rounded or filleted may be indicated conventionally by a line coinciding with the theoretical line of intersection. The need for this convention is demonstrated by the examples shown in Fig. 7-18-1, where the upper top views are shown in true projection. Note that in each example the true projection would be misleading. In the case of the large radius, such as shown in Fig. 7-18-1D, no line is drawn. Members such as ribs and arms that blend into other features terminate in curves called *runouts*. Small runouts are usually drawn freehand. Large runouts are drawn with an irregular curve, template, or compass. See Fig. 7-18-2.

ASSIGNMENT

See Assignment 18 for Unit 7-18 on page 163.

Review for Assignment
Unit 5-2   Dimensioning Circular
           Features
Unit 5-7   Surface Texture Symbols
Unit 3-11   Miter Lines

Fig. 7-18-1 Conventional representation of rounds and fillets.

Fig. 7-18-2   Conventional representation of runouts.

(A)   (B)   (C)   (D)

FLAT
RIB

ROUND
RIB

(E)   (F)   (G)   (H)

ASSIGNMENTS for Chapter 7

Assignment for Unit 7-1, Sectional Views

1. Select one of the problems shown in Fig. 7-1-A or 7-1-B, and on a B- or A3-size sheet make a two-view working drawing of the part, showing one of the views in full section. Scale is full or 1:1.

Ø .500
2 HOLES

R .76   R .50

2.25   .50

Ø .34
Ø .70 SF
6 HOLES

R1.30

R1.80

1.60   .44

ROUNDS AND FILLETS R .10   MATL — MALLEABLE IRON

Fig. 7-1-A   Shaft base.

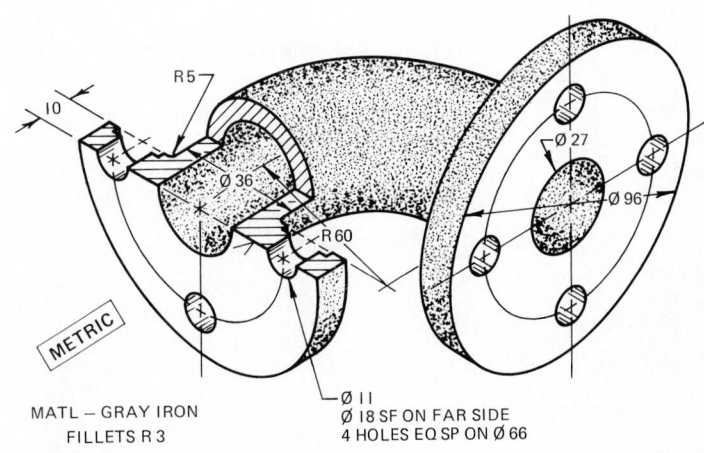

R 5

10

Ø 36

R 60

METRIC

MATL — GRAY IRON
FILLETS R 3

Ø 27

Ø 96

Ø 11
Ø 18 SF ON FAR SIDE
4 HOLES EQ SP ON Ø 66

Fig. 7-1-B   Flanged elbow.

Assignment for Unit 7-2, Two or More Sectional Views on One Drawing

2. Select one of the problems shown in Fig. 7-2-A or 7-2-B and on a B- or A3-size sheet make a working drawing of the part showing the appropriate views in sections. Refer to the Appendix for taper sizes. Scale is 1:1.

Assignment for Unit 7-3, Half Sections

3. Select one of the problems shown in Fig. 7-3-A or 7-3-B. On a B- or A3-size sheet, make a two-view working drawing of the part, showing the side view in half section. Scale is full or 1:1.

Assignment for Unit 7-4, Threads in Section

4. Select one of the problems shown in Fig. 7-4-A or 7-4-B. On a B- or A3-size sheet make a working drawing of the part. Determine the number of views and the best type of section which will clearly describe the part. Scale is full or 1:1.

Assignment for Unit 7-5, Assemblies in Section

5. On an A3- or B-size sheet, make a one-view section assembly drawing of one of the problems shown in Fig. 7-5-A or 7-5-B. Include on your drawing a bill of materials and identify the parts on the assembly. Assuming that this drawing will be used in a catalog, place on the drawing the dimensions and information required by the potential buyer. Scale is 1:1 or full.

Fig. 7-2-A   Casing.

Fig. 7-2-B   Housing.

Fig. 7-3-A   Double-V pulley.

Fig. 7-3-B   Step-V pulley.

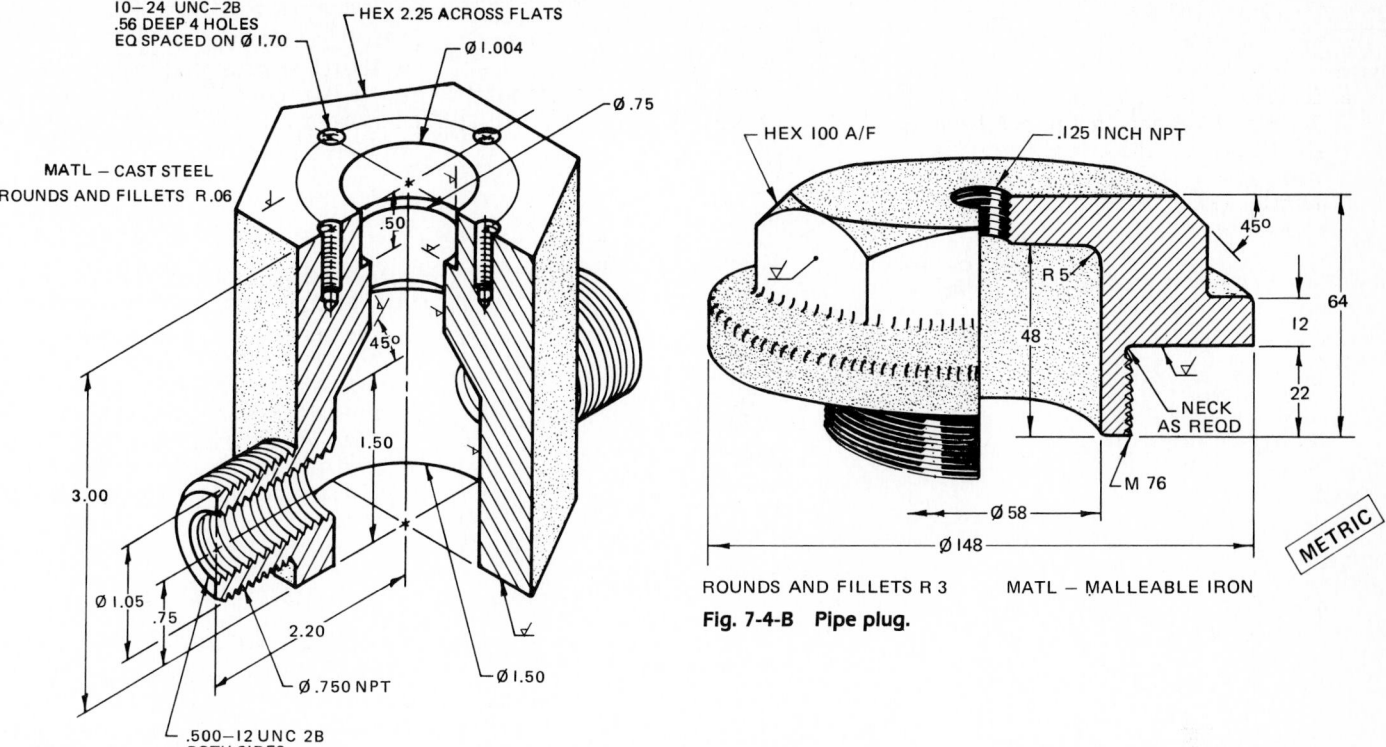

10-24 UNC-2B
.56 DEEP 4 HOLES
EQ SPACED ON Ø 1.70

HEX 2.25 ACROSS FLATS

Ø1.004

Ø.75

MATL – CAST STEEL
ROUNDS AND FILLETS R.06

.50

45°

3.00

1.50

Ø 1.05

.75

2.20

Ø .750 NPT

Ø 1.50

.500–12 UNC 2B
BOTH SIDES

Fig. 7-4-A   Valve body.

HEX 100 A/F

.125 INCH NPT

45°

R 5

48

12

64

22

NECK
AS REQD

M 76

Ø 58

Ø 148

METRIC

ROUNDS AND FILLETS R 3   MATL – MALLEABLE IRON

Fig. 7-4-B   Pipe plug.

PT 3 – AXLE SUPPORT
MATL – MALLEABLE IRON

38

26

12

58   82

10

Ø 16

10

Ø 11
Ø 20 SFACE
2 HOLES

48

R 19

PT I – TOP PLATE

MATL – MALLEABLE IRON

44

R 12

4   16

4

Ø 11
4 HOLES

60

112

58

82

136

Ø 34

Ø 22

12

Ø 22

12

19

PT 5 – BUSHING
MATL – BRONZE

2 X 45°
CHAMFER

Ø 22

Ø 16

40

20   80

PT 4 – AXLE
MATL – SAE 1020

Ø 38

Ø 22

38

10

5

Ø 100

38

3

PT 2 – WHEEL
MATL – MALLEABLE IRON

Fig. 7-5-A   Caster.

METRIC

FIT	BETWEEN PARTS
H8/f7	4 AND 5
H7/p6	3 AND 5
H8/f7	2 AND 4

FASTEN ASSEMBLY TO A 6mm STEEL PLATE BY
FOUR M10 X 50mm LG HEX HD BOLTS, NUTS AND
LOCK WASHERS. SHOW THE STEEL PLATE IN
PHANTOM LINES.

ROUNDS AND FILLETS R 3

PARTIAL DETAIL OF MOUNTING BRACKET

4.25 · 4.00

Ø1.06

CLEARANCE
HOLES FOR
BOLTS

PT 1 – LINK
MATL – MALLEABLE IRON
ROUNDS AND FILLETS R.12

Ø .75 X .12 HIGH BOSS

4.00

Ø 2.50

Ø1.88 (RC 4 FIT WITH BUSHING)

Ø1.06 (LN 3 FIT
WITH BUSHING)

R1.75

1.00

.50

Ø 1.50

3.50

Ø.44 — 4 HOLES EQUALLY
SPACED ON

Ø 1.50 (RC 4 FIT WITH SHAFT)

Ø 1.88 (LN 3 FIT IN LINK)

4.00

PT 3 – BUSHING
MATL – BRONZE

Ø1.50 (RC 4 FIT IN BUSHING)

PT 4 – SHAFT
MATL – SAE 1020 8.00 LG

Ø.750 (RC 4 FIT WITH SHAFT)
Ø1.06 (LN 3 FIT IN LINK)
Ø 1.25

1.00
1.25

PT 2 – BUSHING
MATL – BRONZE

Ø .750
(RC 4 FIT IN
BUSHING)

PT 5 – SHAFT
MATL – SAE 1020 12.00 LG

FASTEN ASSEMBLY TO THE 6 INCH STEEL MOUNTING
BRACKET SHOWN BY FOUR .375 X 1.25 LG HEX HD
BOLTS, NUTS AND LOCK WASHERS. SHOW THE
STEEL PLATE IN PHANTOM LINES.

Fig. 7-5-B   Connecting link.

Assignment for Unit 7-6, Offset Sections

6. Select one of the problems shown in Fig. 7-6-A or 7-6-B. On a B- or A3-size sheet, make a working drawing of the part. Scale is full or 1:1.

Assignment for Unit 7-7, Ribs, Holes, and Lugs in Section

7. Select one of the problems shown in Fig. 7-7-A or 7-7-B. On a B- or A3-size sheet make a three-view working drawing of the part showing the front and side views in section. Both parts are to be used here and abroad, so a dual dimensioning system must be used. Scale is full or 1:1.

Assignment for Unit 7-8, Revolved and Removed Sections

8. Select one of the problems shown in Fig. 7-8-A or 7-8-B and on a B- or A3-size sheet make a working drawing of the part. For clarity it is recommended that an enlarged removed view be used to show the detail of the small hole. Scale is full or 1:1.

DRAWING SET-UP

REPLACE END VIEW WITH
SECTIONS L–L, M–M, AND N–N

ROUNDS AND FILLETS R .12

Fig. 7-6-A   Base plate.

HOLE	HOLE SIZE	LOCATION		
		X	Y	Z
A	.500–13UNC–2B	1.25	1.38	
B	Ø.281 CSK Ø.50 X 82°	2.25	1.94	
C	Ø.281 CBORE Ø.50 X .25 DEEP	1.12	3.50	
D₁	Ø.31	3.50	.75	
D₂	Ø.31	3.50	1.75	
E	.500–13UNC–2B X .75 DEEP	2.62		.75
F₁	Ø.50	.88		1.00
F₂	Ø.50		1.25	1.00
F₃	Ø.50		3.25	1.00
F₄	Ø.50	4.00	3.00	
G	Ø.12 THROUGH		3.00	.75

DRAW TOP, FRONT AND 3 SECTION VIEWS
MATL – MALLEABLE IRON

HOLE	HOLE SIZE	LOCATION		
		X	Y	Z
A₁	Ø 12	16	9	
A₂	Ø 12	100	9	
A₃	Ø 12	30	92	
A₄	Ø 12	87	92	
B₁	Ø 8	38	32	
B₂	Ø 8	80	32	
C₁	M6 X 12 DEEP	12	50	
C₂	M6	104	52	
D	Ø 6 CBORE Ø 12 X 6 DEEP	58	70	
E	Ø 10 X 12 DEEP	58		11
F₁	Ø 6		32	20
F₂	Ø 6		70	20

MATL – MALLEABLE IRON
ROUNDS AND FILLETS R3

REPLACE END VIEW WITH
SECTIONS G–G, H–H, AND J–J.

Fig. 7-6-B   Mounting plate.

MATL – MALLEABLE IRON
ROUNDS AND FILLETS R.10

Fig. 7-7-A   Bracket bearing.

ENLARGED VIEW AT A–A

ROUNDS AND
FILLETS R.12

MATL – CAST STEEL

Fig. 7-8-A   Idler support.

MATL – MALLEABLE IRON

Fig. 7-7-B   Shaft support.

ENLARGED VIEW OF SMALL END

MATL – MALLEABLE IRON
ROUNDS AND FILLETS R3

Fig. 7-8-B   Shaft support.

Assignment for Unit 7-9, Spokes and Arms in Section

9. Select one of the problems shown in Fig. 7-9-A or 7-9-B and on a B- or A3-size sheet make a two-view working drawing. Draw the side view in full section, and show a revolved section of the spoke in the front view. Scale is full or 1:1.

Assignment for Unit 7-10, Partial or Broken-Out Sections

10. Select one of the problems shown in Fig. 7-10-A or 7-10-B and make a two-view working drawing on a B- or A3-size sheet. Use partial sections where clarity of drawing can be achieved. Scale is full or 1:1.

VIEW OF HUB

Fig. 7-9-A   Handwheel.

ROUNDS AND FILLETS R.12
MATL — MALLEABLE IRON

Fig. 7-10-A   Tumble box.

ROUNDS AND FILLETS R3
MATL — CAST STEEL

Fig. 7-9-B   Offset handwheel.

GROOVE FOR N5000—51
INTERNAL RETAINING RING
(SEE APPENDIX)

ROUNDS AND FILLETS R3
MATL — MALLEABLE IRON

Fig. 7-10-B   Hold-down bracket.

Ø 1.25
.25
1.25
Ø .500 RC 5 FIT
FOR SHAFT

.06 X .06 NECK
MATL — BRONZE
BUSHING

Ø .750 LN 2 FIT
FOR HOUSING

R 1.00

VIEW A—A

A

A

Ø 1.25
.25
.35
Ø .40 SLOT

R .40

.25

R 1.00
1.50
Ø .750 LN 2
FIT FOR BUSHING
3.00

.25

Ø .40
.40 DEEP

1.50
.35
R .40
Ø 1.25
Ø 1.00

HOUSING
ROUNDS AND FILLETS R .10
MATL — MALLEABLE IRON

Fig. 7-11-A   Drill-jig assembly.

Assignment for Unit 7-11, Phantom or Hidden Sections

11. On an A3- or B-size sheet, make a two-view assembly drawing of one of the assemblies shown in Fig. 7-11-A or 7-11-B. The front view is to be drawn as a phantom section drawing. Scale is 1:1 or full. Show only the bushing hole sizes and their locations on the drawing.

Ø 20
Ø 48
10
MATL — BRONZE
BUSHING
2 X 2 NECK
Ø 30
46

Ø 64
Ø 30
Ø 48
12
38
MATL — MALLEABLE IRON
HOUSING

38
24
10
24
Ø 24
Ø 16
10
10
Ø 16

METRIC

□76
20
Ø 8

FRONT
ROUNDS AND FILLETS R3
H7/s6 FIT FOR BUSHING IN HOUSING

MATL — BRONZE
BUSHING

Fig. 7-11-B   Housing.

Assignment for Unit 7-12, Sectional Drawing Review

12. On a B- or A3-size sheet, make a three-view working drawing of one of the parts shown in Fig. 7-12-A or 7-12-B. From the information on section drawings found in Units 7-1 to 7-11, select appropriate sectional views which will improve the clarity of the drawing. Scale is full or 1:1.

Assignment for Unit 7-13, Conventional Representation of Common Features

13. On a B- or A3-size sheet, make a working drawing of one of the parts shown in Fig. 7-13-A or 7-13-B. Wherever possible, simplify the drawing by using conventional representation of features.

MATL — ASTM CLASS 30 GREY IRON
ROUNDS AND FILLETS R .12

Fig. 7-12-A   Two-post column base.

ROUNDS AND FILLETS R 3
MATL — MALLEABLE IRON

METRIC

Fig. 7-12-B   Shaft support base.

MATL — SAE 1050

Fig. 7-13-A   Adjustable locking plate.

FILLETS R.50
MATL — SAE 1040

SCALE 10 : 1

METRIC

Fig. 7-13-B   Clock stem.

.12 X 45°
CHAM

.06
45°
.12

14.00

6.00

HEX 1.12 ACROSS
CORNERS

R .50

Ø .75

2.00

1.00

1.00

.12

FINISH — HEAT TREAT
MATL — SAE 1080

Fig. 7-14-A Hand chisel.

Assignment for Unit 7-14, Conventional Breaks

14. On a B- or A3-size sheet, make a working drawing of one of the parts shown in Fig. 7-14-A or 7-14-B. Use conventional breaks to shorten the length of the part. An enlarged view is also recommended where the detail cannot be clearly shown at full scale. Scale is 1:1.

Assignment for Unit 7-15, Materials of Construction

15. On a B- or A3-size sheet, make a detailed assembly drawing of one of the assemblies shown in Fig. 7-15-A or 7-15-B. Enlarged details are recommended for the steel mesh and joints. Use conventional breaks to shorten the length. Scale is 1:4 (inches) or 1:5 (metric).

M 6
4 HOLES

R 6

Ø 10 SLOT

254

76

127

38

50

25

6

SLOT 1.5 WIDE
X 2 DEEP
HEX 12 AF
BOTH
SIDES

25

25

12

Ø 12
2 HOLES

25

354

13

10

26

12

12

METRIC

HEX AND SLOT TO BE ∀
ROUNDS AND FILLETS R 2
MATL — CAST STEEL
FINISH — HEAT TREAT

Fig. 7-14-B Fixture base.

45°

DETAIL OF NOTCH
FOR CORNER BENDS

□ .50
.12 TYP

ENLARGED
VIEW "A"

24.00

TACKWELD 16 USS GA
PERFORATED METAL
TO FRAME

"A"

BUTT WELD SEAM

L 1.50 X 1.50 X .25

12.00

Fig. 7-15-A Barbecue grill.

2 X 12 DEEP
GROOVE

□ 50

25 DEEP
□ 25 X 25 LG

25

2 X 12 DEEP
GROOVE
ON CENTER

□ 50

12.5

25

ENLARGED EXPLODED VIEW AT "B"

PINE FRAME

76

16 USS GA
PERFORATED
METAL

2438

"A"

METRIC

"B"

76

□ 50

□ 20

6

ENLARGED VIEW AT "A"

600

Fig. 7-15-B Room divider.

Assignment for Unit 7-16, Cylindrical Intersections

16. On an A3- or B-size sheet, make a working drawing of one of the parts shown in Fig. 7-16-A or 7-16-B. For 7-16-A a bushing is to be pressed (H7/s6) into the large hole and the stepped smaller hole is to have a running fit (H8/f7) with its respective shaft. These sizes are to be given as limit dimensions. All other finished surfaces are to have a 3.2-μm or equivalent finish. For Fig. 7-16-B an LN3 fit is required for the two large holes. Finished surfaces are to have a 63-μin. finish with an .06-in. material-removal allowance. Use your judgment in selecting the number of views required and deciding whether some form of sectional view would be desirable to improve the readability of the drawing. Scale is 1:1.

Assignment for Unit 7-17, Foreshortened Projection

17. On an A3- or B-size sheet, make a detail drawing of one of the parts shown in Fig. 7-17-A or 7-17-B. All surface finishes are to be 1.6 μm or 63 μin. Keyed holes will have H9/d9 or RC6 fits with shafts. Where required, rotate the features to show their true distances from the centers and edges. To show the true shape of the ribs or arms, a revolved section is recommended. Scale is 1:1.

Fig. 7-16-A Steering knuckle.

Fig. 7-16-B Shift lever.

Fig. 7-17-A Clutch.

Fig. 7-17-B Mounting bracket.

Assignment for Unit 7-18, Intersections of Unfinished Surfaces

18. On an A3- or B-size sheet, make a three-view detail drawing of one of the problems shown in Figs. 7-18-A through 7-18-C. Scale is 1:1. Surface finish requirements are essential for all parts. For Fig. 7-18-A the T slot surfaces should have a maximum roughness of 0.8 μm and a maximum waviness of 0.05 mm for a 25-mm length. The back surface should have a maximum roughness of 3.2 μm with no restrictions on waviness. For Fig. 7-18-B the back surface and notch should have an equivalent control as the T slot in Fig. 7-18-A. The faces on the boss should have a maximum rough-ness of 125 μin. with no restrictions on waviness.

The slots in Fig. 7-18-C are to have a surface finish of 3.2 μm and a machining allowance of 2 mm. The base is to have the same surface finish but with a 3-mm machining allowance.

Ø .40
Ø .75 X 82° CSK
2 HOLES

R .38

Ø .502
2 HOLES

ROUNDS AND FILLETS R.12
MATL — MALLEABLE IRON

Fig. 7-18-A   Cut-off stop.

Ø 7
Ø 12 X 82° CSK
2 HOLES

METRIC

ROUNDS AND FILLETS R3
MATL - MALLEABLE IRON

Fig. 7-18-B   Sparker bracket.

Ø 10
3 HOLES

R 12

Ø 8
2 HOLES

METRIC

Fig. 7-18-C   Pipe vise base.

Ø 8
2 HOLES

VIEW IN DIRECTION OF ARROW A

NOTE: RIB AND WALL THICKNESS 3 mm
EXCEPT WHERE OTHERWISE SHOWN.
MATL — MALLEABLE IRON
ROUNDS AND FILLETS R 2

PART 2

Fasteners, Materials, and Forming Processes

CHAPTER 8

Threaded Fasteners

Thread Forms[1,2]

Fastening devices are important in the construction of manufactured products, in the machines and devices used in manufacturing processes, and in the construction of all types of buildings. Fastening devices are used in the smallest watch to the largest ocean liner. See Fig. 8-1-1.

There are two basic kinds of fasteners: *permanent* and *removable*. Rivets and welds are permanent fasteners. Bolts, screws, studs, nuts, pins, rings, and keys are removable fasteners. As industry progressed, fastening devices became standardized, and they developed definite characteristics and names. A thorough knowledge of the design and graphic representation of the more common fasteners is an essential part of drafting.

The cost of fastening, once considered only incidental, is fast becoming recognized as a critical factor in total product cost. "It's the in-place cost that counts, not the fastener cost" is an old saying of fastener design. The art of holding down fastener cost is not learned simply by scanning a parts catalog. More subtly, it entails weighing such factors as standardization, automatic assembly, tailored fasteners, and joint preparation.

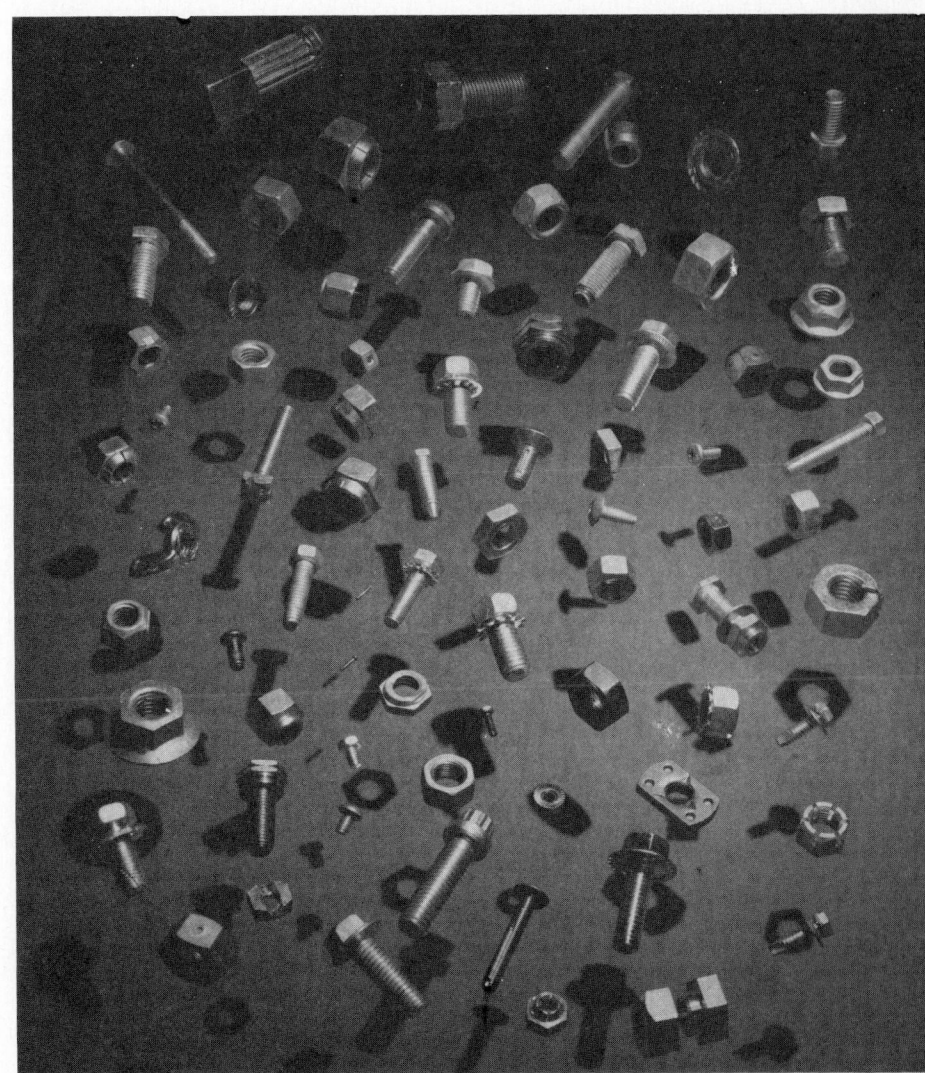

Fig. 8-1-1   Fasteners. (Industrial Fasteners Institute)

Standardization A favorite cost-reducing method, standardization, not only cuts the cost of parts but also reduces paperwork and simplifies inventory and quality control. By standardizing on type and size, it may be possible to reach the level of usage required to make power tools or automatic assembly feasible.

SCREW THREADS

A *screw thread* is a ridge of uniform section in the form of a helix on the external or internal surface of a cylinder. See Fig. 8-1-2.

The *pitch* of a thread *P* is the distance from a point on the thread form to the corresponding point on the next form, measured parallel to the axis. See Fig. 8-1-3. The *lead L* is the distance the threaded part would move parallel to the axis during one complete rotation in relation to a fixed mating part (the distance a screw would enter a threaded hole in one turn).

Fig. 8-1-2   The helix.

THREAD FORMS

Figure 8-1-4 shows some of the more common thread forms in use today. The ISO metric thread shown in Fig. 8-1-4 will eventually replace all the V-shaped metric and inch threads. As for the other thread forms shown, the proportions will be the same for both metric- and inch-size threads.

The knuckle thread is usually rolled or cast. A familiar example of this form is seen on electric light bulbs and sockets. See Fig. 8-1-5. The square and acme forms are designed to transmit motion or power, as on the lead screw of a lathe. The buttress thread takes pressure in only one direction—against the surface perpendicular to the axis.

THREAD REPRESENTATION

True representation of a screw thread is seldom provided on working drawings because it would require very laborious and accurate drawing involving repetitious development of the helix curve of the thread. See Figs. 8-1-3 and 8-1-6. Symbolic representations of threads are now standard practice. There are two basic thread rep-

Fig. 8-1-3   Screw-thread terms.

Fig. 8-1-4   Common thread forms and proportions.

Fig. 8-1-5   Application of a knuckle thread.

Fig. 8-1-6 The helix of a square thread.

Fig. 8-1-7 Steps in drawing detailed representation of screw thread.

resentations used, the detailed and simplified.

DETAILED REPRESENTATION

Detailed Representation of Screw Threads
The detailed representation for V-shaped threads uses the sharp-V profile. Straight lines are used to represent the helices of the crest and root lines.

The order of drawing the screw threads is shown in Fig. 8-1-7. The pitch is seldom drawn to scale; generally it is approximated and drawn to look good. Lay off the pitch *P* and the half-pitch *P*/2, as shown in Fig. 8-1-7A. Adjust the triangle to the slope, and draw the crest lines (if a drafting machine is used, set the ruling arm to the slope of the crest line). At B draw the V profile for one thread, top and bottom, locating the root diameter. Draw light construction lines for the root diameter. At C set the ruling face of the 30° triangle and draw one side of the remaining V's (thread profile). Reverse the triangle and draw the

other side of the V's, completing the thread profile. At D, draw the root lines, which complete the pictorial representation of the threads.

Detailed Representation of Square Threads
The depth of the square thread is one-half the pitch. In Fig. 8-1-8A, lay off spaces equal to *P*/2 along the diameter and add light lines to locate the depth of thread. At B draw the crest lines. At C draw the root lines, as shown. At D

the internal square thread is drawn in section. Note the reverse direction of the lines.

Detailed Representation of Acme Threads
The depth of the acme thread is one-half the pitch. See Fig. 8-1-8E through H. The stages in drawing acme threads are shown at E. The pitch diameter is midway between the outside diameter and the root diameter and locates the pitch line. On the pitch line, lay off

Fig. 8-1-8 Steps in drawing detailed representation of square and acme threads.

half-pitch spaces and draw the root lines to complete the view. The construction shown at F is enlarged.

A sectional view of an internal acme thread is shown at G. Other representations used for internal threads are hidden lines and sections. These are shown at H.

RIGHT- AND LEFT-HAND THREADS

Unless designated otherwise, threads are assumed to be right-hand. A bolt being threaded into a tapped hole

(A) RIGHT-HAND THREAD

(B) LEFT-HAND THREAD

Fig. 8-1-9   Right- and left-hand threads.

would be turned in a right-hand (clockwise) direction. See Fig. 8-1-9. For some special applications, such as turnbuckles, left-hand threads are required. When such a thread is necessary, the letters LH are added after the thread designation.

SINGLE AND MULTIPLE THREADS

Most screws have single threads. It is understood that unless the thread is designated otherwise, it is a single thread. The single thread has a single ridge in the form of a helix. See Fig. 8-1-10. The lead of a thread is the distance traveled parallel to the axis in one rotation of a part in relation to a fixed mating part (the distance a nut would travel along the axis of a bolt with one rotation of the nut). In single threads, the lead is equal to the pitch. A double thread has two ridges, started 180° apart, in the form of helices; and the lead is twice the pitch. A triple thread has three ridges, started 120° apart, in the form of helices; and the lead is 3 times the pitch. Multiple threads are used where fast movement is desired with a minimum number of rotations, such as on threaded mechanisms for opening and closing windows.

THREADED ASSEMBLIES

It is often desirable to show threaded assembly drawings in detailed form, e.g., in presentation or catalog drawings. Hidden lines are normally omitted on these drawings, as they do nothing to add to the clarity of the drawing. See Fig. 8-1-11.

References and Source Material
1. ANSI Y14.6, *Screw Threads*.
2. *Machine Design*, Materials reference issue, March 1981.

ASSIGNMENTS

See Assignments 1 and 2 for Unit 8-1 on page 184.

UNIT 8-2
Simplified Representation of Threads

Two types of conventions are in general use for screw-thread representation. These are known as detailed (as explained in Unit 8-1), and simplified representations. Simplified representation should be used whenever it will clearly portray the requirements. Detailed representation requires more drafting time, but is sometimes necessary to avoid confusion with other parallel lines or to more clearly portray particular aspects of the threads.

SIMPLIFIED REPRESENTATION

Simplified representation, as shown in Fig. 8-2-1B should be used whenever it conveys the required information without confusion, because it requires the least amount of drafting effort. In this system the thread crests, except in hidden views, are represented by a thick outline and the thread roots by a thin broken line. The end of the full-

(A) EXTERIOR VIEW

(B) INTERIOR VIEW

Fig. 8-1-11   Detailed threaded assembly.

LEAD = 1 P

(A) SINGLE THREAD

LEAD = 2 P

(B) DOUBLE THREAD

LEAD = 3 P

(C) TRIPLE THREAD

Fig. 8-1-10   Single and multiple threads.

(A) PICTORIAL REPRESENTATION (FORMERLY KNOWN AS SEMICONVENTIONAL)

USED ON ENLARGED DETAIL AND OTHER SPECIAL APPLICATIONS

CHAMFER CIRCLE →

END OF FULL THREAD →

(B) SIMPLIFIED REPRESENTATION

USED WHENEVER IT CONVEYS THE INFORMATION WITHOUT LOSS OF CLARITY

Fig. 8-2-1 American National Standard Thread conventions.

(A) EXTERNAL THREAD

(B) INTERNAL THREAD

Fig. 8-2-2 Former schematic thread conventions.

Fig. 8-2-3 Simplified and detailed representation of threads in assembly drawings.

NOMINAL DIAMETER
NUMBER OF THREADS PER INCH
THREAD SERIES
.750—10 UNC

(A) BASIC THREAD CALLOUT

CLASS OF THREAD FIT
.750—10 UNC—2A
DESIGNATION FOR EXTERNAL THREAD

EXTERNAL THREAD

CLASS OF THREAD FIT
.750—16 UNF—3B
DESIGNATION FOR INTERNAL THREAD

INTERNAL THREAD

(B) TOLERANCE CALLOUT

.625—11 UNC—2A
.80 DEEP

.80

(C) BLIND HOLE

2.00—2 SQUARE

Ø 2.00

.50

(D) MISCELLANEOUS THREAD FORMS

Fig. 8-2-4 Thread specifications for inch-size threads.

CHAMFER SHOWN AT BEGINNING OF THREAD
CHAMFER SIZE NEED NOT BE SHOWN

RECESS SHOWN AT END OF THREAD
RECESS SIZE NEED NOT BE SHOWN

Fig. 8-2-5 Omission of thread information on detail drawings.

form thread is indicated by a thick line across the part, and imperfect or runout threads are shown beyond this line by running the root line at an angle to meet the crest line. If the length of runout threads is unimportant, this portion of the convention may be omitted.

The former schematic thread representation is shown in Fig. 8-2-2.

THREADED ASSEMBLIES

Either thread convention may be used for assemblies of threaded parts, and both methods may be used on the same drawing, as shown in Fig. 8-2-3. In sectional views, the externally threaded part is always shown covering the internally threaded part.

INCH THREADS

In the United States a great number of threaded assemblies are still designed using inch-sized threads. In this system the *pitch* is equal to

$$\frac{1}{\text{Number of threads per inch}}$$

The number of threads per inch is set for different diameters in what is called a thread *series*. For the Unified National system there is the coarse-thread series and the fine-thread series. See Fig. 8-2-4.

In addition, there is an extra-fine-thread series, UNEF, for use where a small pitch is desirable, such as on thin-walled tubing. For special work and for diameters larger than those specified in the coarse and fine series, the Unified Thread system has three series that provide for the same number of threads per inch regardless of the diameter. These are the 8-thread series, the 12-thread series, and the 16-thread series. These are called constant pitch threads.

Thread Class

Three classes of external thread (classes 1A, 2A, and 3A) and three classes of internal thread (classes 1B, 2B, and 3B) are provided. These classes differ in the amount of allowances and tolerances provided in each class.

The general characteristics and uses of the various classes are as follows.

Classes 1A and 1B These classes produce the loosest fit, that is, the greatest amount of play in assembly. They are useful for work where ease of assembly and disassembly is essential, such as for some ordnance work and for stove bolts and other rough bolts and nuts.

Classes 2A and 2B These classes are designed for the ordinary good grade of commercial products, such as machine screws and fasteners, and for most interchangeble parts.

Classes 3A and 3B These classes are intended for exceptionally high-grade commercial products, where a particularly close or snug fit is essential and the high cost of precision tools and machines is warranted.

Thread Designation

Thread designation for inch threads, whether external or internal, is expressed in this order: diameter (nominal or major diameter), number of threads per inch, thread form and series, and class of fit.

METRIC THREADS

Metric threads are grouped into diameter-pitch combinations distinguished from one another by the pitch applied to specific diameters. See Fig. 8-2-5. The *pitch* for metric threads is the distance between corresponding points on adjacent teeth. In addition to a coarse- and fine-pitch series, a series of constant pitches is available.

Coarse-Thread Series This series is intended for use in general engineering work and commercial applications.
Fine-Thread Series The fine-thread series is for general use where a finer thread than the coarse-thread series is desirable. In comparison with a coarse-thread screw, the fine-thread screw is stronger in both tensile and torsional strength and is less likely to loosen under vibration.

Thread Grades and Classes The *fit* of a screw thread is the amount of clearance between the internal and external threads when they are assembled.

For each of the two main thread elements—pitch diameter and crest diameter—a number of tolerance grades have been established. The number of the tolerance grades reflects the size of the tolerance. For example, grade 4 tolerances are smaller than grade 6 tolerances, and grade 8 tolerances are larger than grade 6 tolerances.

In each case, grade 6 tolerances should be used for medium-quality length-of-engagement applications. The tolerance grades below grade 6 are intended for applications involving fine quality and/or short lengths of engagement. Tolerance grades above grade 6 are intended for coarse quality and/or long lengths of engagement.

In addition to the tolerance grade, a positional tolerance is required. This tolerance defines the maximum-material limits of the pitch and crest diameters of the external and internal threads and indicates their relationship to the basic profile.

In conformance with current coating (or plating) thickness requirements and the demand for ease of assembly, a series of tolerance positions reflecting the application of varying amounts of allowance has been established as follows:

For external threads:

- Tolerance position e (large allowance)
- Tolerance position g (small allowance)
- Tolerance position h (no allowance)

For internal threads:

- Tolerance position G (small allowance)
- Tolerance position H (no allowance)

PITCH MAY BE OMITTED FOR COARSE THREADS

(A) BASIC THREAD CALLOUT

(B) TOLERANCE CALLOUT

(C) INTERNAL THREAD CALLOUT

(D) MISCELLANEOUS THREAD FORMS

Fig. 8-2-6   Thread specifications for metric threads.

Thread Designation

ISO metric screw threads are defined by the nominal size (basic major diameter) and pitch, both expressed in millimeters. An "M" specifying an ISO metric screw thread precedes the nominal size, and an "x" separates the nominal size from the pitch. For the coarse-thread series only, the pitch is not shown unless the dimension for the length of the thread is required. In specifying the length of thread, an "x" is used to separate the length of thread from the rest of the designations. For external threads, the length of thread may be given as a dimension on the drawing.

For example, a 10 mm diameter, 1.25 pitch, fine-thread series is expressed as M10 × 1.25. A 10 mm diameter, 1.5 pitch, coarse-thread series is expressed as M10; the pitch is not shown unless the length of thread is required. If the latter thread were 25 mm long and this information was required on the drawing, then the thread callout would be M10 × 1.5 × 25.

Fig. 8-2-7   Comparison of thread sizes.

A complete designation for an ISO metric screw thread comprises, in addition to the basic designation, an identification for the tolerance class. The tolerance class designation is separated from the basic designation by a dash and includes the symbol for the pitch diameter tolerance followed immediately by the symbol for crest diameter tolerance. Each of these symbols consists of a numeral indicating the grade tolerance followed by a letter representing the tolerance position (a capital letter for internal threads and a lowercase letter for external threads). Where the pitch and crest diameter symbols are identical, the symbol need be given only once.

For external threads, the length of thread is given as a dimension on the drawing. The length given is to be the minimum length of full thread. For threaded holes that go all the way through the part, the term THRU is sometimes added to the note. If no depth is given, the hole is assumed to go all the way through. For threaded holes that do not go all the way through, the depth is given in the note, for example, M12 × 1.75 × 20 DEEP. The depth given is the minimum depth of full thread.

Neither the chamfer shown at the beginning of a thread, nor the relief recess at the end of a thread where a small diameter meets a larger diameter is required to be dimensioned, as shown in Fig. 8-2-6. A comparison of customary and metric thread sizes is shown in Fig. 8-2-7.

PIPE THREADS

The pipe universally used is the inch-sized pipe and, as such, will not be changed. When pipe is ordered, the outside diameter and wall thickness (in inches or millimeters) are given. In calling for the size of thread, the note used is similar to that for screw threads. See Fig. 8-2-8.

EXAMPLE 1
4 × 8NPT

where 4 = nominal diameter of pipe, in inches
8 = number of threads per inch
N = American Standard
P = pipe
T = taper thread

References and Source Material
1. American National Standards Institute.

ASSIGNMENTS
See Assignments 3 through 6 for Unit 8-2 on page 185.

Review for Assignments

UNIT 8-3
Common Threaded Fasteners

FASTENER SELECTION[1]

Fastener manufacturers agree that product selection must begin at the design stage. For it is here, when a product is still a figment of someone's imagination, that the best interests of the designer, production manager, and purchasing agent can be served. Designers, naturally, want optimum performance; production people are interested in the ease and economics of assembly; purchasing agents are keen to minimize initial costs and stocking costs.

The answer, pure and simple, is to determine the objectives of the particular fastening job and then consult fas-

(A) TERMINOLOGY

(B) CONVENTION USED FOR STRAIGHT OR TAPERED THREADS

(C) CONVENTION USED TO SHOW DIRECTION AND TAPER OF THREAD

Fig. 8-2-8   Pipe thread terminology and conventions.

ROUND HEAD   FLAT HEAD   OVAL HEAD   UNDERCUT OVAL HEAD   FILLISTER HEAD   TRUSS HEAD   PAN HEAD   HEXAGON HEAD   HEXAGON WASHER HEAD

(A) SCREWS

HEX HEAD   SQUARE HEAD

(B) BOLTS

DOUBLE-END STUD   (C) STUDS   CONTINUOUS-THREAD STUD

Fig. 8-3-1 Common threaded fasteners.

tener suppliers. These technical experts can often shed light on the situation and then recommend the right item at the best in-place cost.

Machine screws are among the most common fasteners in industry. See Figs. 8-3-1 and 8-3-2. They are the easiest to install and remove. They are also among the least understood. To obtain maximum machine-screw efficiency, thorough knowledge of the properties of both the screw and the materials to be fastened together is required.

PAN HEAD   FLAT HEAD   FILLISTER HEAD

(A) MACHINE SCREWS

(B) CAP SCREWS

(C) BOLTS   (D) STUDS

Fig. 8-3-2 Fastener application.

For a given application, a designer should know the load which the screw must withstand, whether the load is one of tension or shear, and whether the assembly will be subject to impact shock or that perennial nemesis, vibration. Once these factors have been determined, then the size, strength, head shape, and thread type can be selected.

FASTENER DEFINITIONS

Machine Screws *Machine screws* have either fine or coarse threads and are available in a variety of heads. They may be used in tapped holes or with nuts.

Cap Screws A *cap screw* is a threaded fastener which joins two or more parts by passing through a clearance hole in one part and screwing into a tapped hole in the other. A cap screw is tightened or released by torquing the head. Cap screws range in size from .25 in. (6 mm) in diameter and are available in five basic types of head.

Captive Screws *Captive screws* are those that remain attached to the panel or parent material even when disengaged from the mating part. They are used to meet military requirements, to prevent screws from being lost, to speed assembly and disassembly operations, and to prevent damage from loose screws falling into moving parts or electrical circuits.

Tapping Screws *Tapping screws* cut or form a mating thread when driven into preformed holes.

Bolts A *bolt* is a threaded fastener which passes through clearance holes in assembled parts and threads into a nut. Bolts and nuts are available in a variety of shapes and sizes. The square and hexagon head are the two most popular designs.

Studs *Studs* are shafts threaded at both ends, and they are used in assemblies. One end of the stud is threaded into one of the parts being assembled; and the other assembly parts, such as washers and covers, are guided over the studs through clearance holes and are held together by means of a nut which is threaded over the exposed end of the stud.

Explanatory Data

A bolt is designed for assembly with a nut. A screw has features in its design which make it capable of being used in a tapped or other preformed hole in the work. Because of basic design, it is possible to use certain types of screws in combination with a nut. Any externally threaded fastener which has a majority of the design characteristics that assist its proper use in a tapped or other preformed hole is a *screw*, regardless of how it is used in its service application.

THE CHANGE TO METRIC FASTENERS

In the United States, the Industrial Fasteners Institute (IFI) has undertaken a major compilation of standards in its 296-page volume *Metric Fastener Standards*. It covers more than 30 metric fastener standards which have been developed by the IFI over the past few years.

FASTENER CONFIGURATION

Head Styles

Which of the various head configurations to specify depends on the type of driving equipment used (screwdriver, socket wrench, etc.), the type of joint load, and the external appearance desired. The head styles shown in Fig. 8-3-3 can be used for both bolts and screws but are most commonly identified with the fastener category called machine screw or cap screw.

PAN

BINDING

HEX

SQUARE

WASHER (FLANGED)

OVAL

FLAT

FILLISTER

TRUSS

12-POINT

Fig. 8-3-3 Common head styles.

Hex and Square The hex head is the most commonly used head style. The hex head design offers greater strength, ease of torque input, and area than the square head.

Pan This head combines the qualities of the truss, binding, and round head types.

Binding This type of head is commonly used in electrical connections because its undercut prevents fraying of stranded wire.

Washer (flanged) This configuration eliminates the need for a separate assembly step when a washer is required, increases the bearing areas of

HEX CAP   SLOTTED   PHILLIPS[R]   CLUTCH TYPE A

TORQ-SET[R]   TRIPLE SQUARE   MULTI-SPLINE   CLUTCH TYPE G

TRI-WING[R]   TORX[R]   SCRULOX[R]   SLAB HEAD

POZIDRIV[R]   REED & PRINCE (FREARSON)   SQUARE   HEXAGON

Fig. 8-3-4 Drive configurations.

the head, and protects the material finish during assembly.

Oval Characteristics of this head type are similar to those of the flat head but it is sometimes preferred because of its neat appearance.

Flat Available with various head angles, this fastener centers well and provides a flush surface.

Fillister The deep slot and small head allow a high torque to be applied during assembly.

Truss This head covers a large area. It is used where extra holding power is required, holes are oversize, or the material is soft.

12-Point This twelve-sided head is normally used on aircraft-grade fasteners. Multiple sides allow for a very sure grip and high torque during assembly.

OVAL SHOULDER   ROUND SHOULDER

FIN NECK   SQUARE (CARRIAGE) NECK

Fig. 8-3-5 Shoulders and necks.

Drive Configurations

Figure 8-3-4 shows sixteen different driving designs.

Shoulders and Necks

The shoulder of a fastener is the enlarged portion of the body of a threaded fastener or the shank of an unthreaded fastener. See Fig. 8-3-5.

Point Styles

The point of a fastener is the configuration of the end of the shank of a headed fastener, or each end of a headless fastener. Standard point styles are shown in Fig. 8-3-6.

Cup Most widely used when the cutting-in action of point is not objectionable.

Flat Used when frequent resetting of a part is required. Particularly suited for use against hardened steel shafts. This point is preferred where walls are thin or the threaded member is a soft material.

Cone Used for permanent location of parts. Usually spotted in a hole to half its length.

Oval Used when frequent adjustment is necessary or for seating against angular surfaces.

Half Dog Normally applied where permanent location of one part in relation to another is desired. It is spotted in a shaft hole.

PROPERTY CLASSES OF FASTENERS

Inch Fasteners

The strength of customary fasteners for most common uses is determined by the size of the fastener and the material

CUP   FLAT   CONE

OVAL   HALF DOG

Fig. 8-3-6 Point styles.

HEAD DESIGNATION	GRADE				
	GRADES 0, 1, 2	GRADE 3	GRADE 5	GRADE 7	GRADE 8
MINIMUM TENSILE STRENGTH KIPS	0—NO REQUIRE-MENTS 1—55 2—69 64 55	110 100	120 115 105	133	150

Fig. 8-3-7 Mechanical requirements for inch-size threaded fasteners.

from which it is made. Property classes are defined by the Society of Automotive Engineers (SAE), or the American Society for Testing and Materials (ASTM).

Figure 8-3-7 lists the mechanical requirements of inch-sized fasteners and their identification patterns.

Metric Fasteners

For mechanical and material requirements, metric fasteners are classified under a number of property classes. Bolts, screws, and studs have seven property classes of steel suitable for general engineering applications. The property classes are designated by numbers where increasing numbers generally represent increasing tensile strengths. The designation symbol consists of two parts: the first numeral of a two-digit symbol or the first two numerals of a three-digit symbol approximates one-hundredth of the minimum tensile strength in megapascals; and the last numeral approximates one-tenth of the ratio expressed as a percentage of minimum yield strength and minimum tensile strength.

EXAMPLE 1 A property class 4.8 fastener (see Fig. 8-3-8) has a minimum tensile strength of 420 MPa and a minimum yield strength of 340 MPa. One percent of 420 is 4.2. The first digit is 4. The minimum yield strength of 340 MPa is equal to approximately 80 percent of the minimum tensile strength of 420 MPa. One-tenth of 80 percent is 8. The last digit of the property class is 8.

EXAMPLE 2 A property class 10.9 fastener (see Fig. 8-3-8) has a minimum tensile strength of 1040 MPa and a minimum yield strength of 940 MPa. One

percent of 1040 is 10.4. The first two numerals of the three-digit symbol are 10. The minimum yield strength of 940 MPa is equal to approximately 90 percent of the minimum tensile strength of 1040 MPa. One-tenth of 90 percent is 9. The last digit of the property class is 9.

Machine screws are normally available only in classes 4.8 and 9.8; other bolts, screws, and studs are available in all classes within the specified product size limitations given in Fig. 8-3-7.

For guidance purposes only, to assist designers in selecting a property class, the following information may be used.

- Class 4.6 is approximately equivalent to SAE grade 1 and ASTM A 307, grade A.
- Class 5.8 is approximately equivalent to SAE grade 2.
- Class 8.8 is approximately equivalent to SAE grade 5 and ASTM A 449.
- Class 9.8 has properties approximately 9 percent stronger than SAE grade 5 and ASTM A 449.
- Class 10.9 is approximately equivalent to SAE grade 8 and ASTM A 354 grade BD.

PROPERTY CLASS	NOMINAL DIAMETER	MINIMUM TENSILE STRENGTH MPa	MINIMUM YEILD STRENGTH MPa
4.6	M5 THRU M36	400	240
4.8	M1.6 THRU M16	420	340
5.8	M5 THRU M24	520	420
8.8	M16 THRU M36	830	660
9.8	M1.6 THRU M16	900	720
10.9	M5 THRU M36	1040	940
12.9	M1.6 THRU M36	1220	1100

Fig. 8-3-8 Mechanical requirements for metric bolts, screws, and studs.

HEX

SQUARE

COUNTERSUNK

FLANGED

12 – POINT FLANGED

PAN

CARRIAGE

LAG

TRACK

BENT

PLOW

ELEVATOR

Fig. 8-3-9 Common bolts.

BOLT STYLES

The more common types of bolt styles are shown in Fig. 8-3-9. A brief explanation of their uses follows.

Hex The most commonly used of the standard fasteners. The hex head design offers greater strength and ease of torque input than the square head.

Square This bolt is supplied in two strength grades and in popular sizes.

Countersunk For flush mounting of assemblies.

Flanged This head provides a large bearing area; it often eliminates the need for a separate washer.

12-Point Flanged Also referred to as *aircraft-type*.

Pan Presents a smooth, attractive external appearance. It is tightened by torquing the mated nut. Recommended by the IFI to replace the round head.

Carriage Normally made with a round head for an attractive external appearance; has ribs or flats on the shank to prevent turning when the bolt is tightened.

Lag A square-headed fastener with a threaded conical point. Normally used in wood or in concrete with an expansion anchor.

Track One of the family of bolts designed specifically for use with railroad tracks. The version shown has an elliptical neck to prevent rotation.

Bent A threaded rod with an end formed in the shape of an eye or right angle bend.

Plow Usually made for flush mounting, this bolt has a square countersunk head which may also include a key to prevent turning.

Elevator This large diameter flat-head bolt provides a nearly flush surface and large bearing area for use with softer materials. Its square neck prevents rotation.

Fastener Markings

Slotted and crossed recessed screws of all sizes and other screws and bolts of sizes .25 in. or M4 and smaller need not be marked. All other bolts and

PROPERTY CLASS	IDENTIFICATION SYMBOL	
	BOLTS, SCREWS AND STUDS	STUDS SMALLER THAN M12
4.6	4.6	–
4.8	4.8	–
5.8	5.8	–
8.8 (I)	8.8	○
9.8 (I)	9.8	+
10.9 (I)	10.9	□
12.9	12.9	△

NOTE I: PRODUCTS MADE OF LOW CARBON MARTENSITE STEEL SHALL BE ADDITIONALLY IDENTIFIED BY UNDERLINING THE NUMERALS.

Fig. 8-3-10 Metric property class identification symbols for bolts, screws, and studs.

screws of sizes .25 in. or M5 and larger are marked to identify their strength. The property class symbols for metric fasteners are shown in Fig. 8-3-10. The symbol is located on the top of the bolt head or screw. Alternatively, for hex-head products, the markings may be indented on the side of the head.

All studs of size .25 in. or M5 and larger are identified by the property class symbol. The marking is located on the extreme end of the stud. For

Fig. 8-3-11 Hex-nut styles.

PROPERTY CLASS	NOMINAL NUT SIZE	SUGGESTED PROPERTY CLASS OF MATING BOLT, SCREW OR STUD
5	M5 THRU M36	4.6, 4.8, 5.8
9	M5 THRU M16 M20 THRU M36	5.8, 9.8 5.8, 8.8
10	M6.3 THRU M36	10.9

Fig. 8-3-12 Metric nut selection for bolts, screws, and studs.

studs with an interference fit thread the markings are located at the nut end. Studs smaller than .50 in. or M12 use different identification symbols.

Nuts

The customary terms *regular* and *thick* for describing nut thicknesses have been replaced by the terms *style 1* and *style 2* for metric nuts. The design of style 1 and 2 steel nuts shown in Fig. 8-3-11 is based on providing sufficient nut strength to reduce the possibility of thread stripping. There are three property classes of steel nuts available. See Fig. 8-3-12.

Hex Flanged Nuts These nuts are intended for general use in applications requiring a large bearing contact area, such as where the nut face is to bear directly on the work of overslotted or oversized clearance holes or against material of relatively low compressive strength, or where the conditions might normally necessitate the use of a flat washer under a hex nut. The two styles of flanged hex nuts differ dimensionally in thickness only. The standard property classes for hex flanged nuts are identical to the hex nuts. All metric nuts are marked to identify their property class.

DRAWING A BOLT AND NUT

Bolts and nuts are not normally drawn on detail drawings unless they are of a special size or have been modified slightly. On some assembly drawings it may be necessary to draw the nut and bolt. Approximate nut and bolt sizes are shown in Fig. 8-3-13. Actual sizes are found in the Appendix. Nut and bolt templates are also available and are recommended as a cost-saving device. Conventional drawing practice is to draw the nuts and bolt heads in the across-corners position in all views.

STUDS

Studs, as shown in Fig. 8-3-14, are still used in large quantities to best fulfill the needs of certain design functions and for overall economy.

Fig. 8-3-13 Approximate head proportions for hex-head cap screws, bolts, and nuts.

(A) CAP SCREW

(B) HEX BOLT

(C) 12 SPLINE FLANGE SCREW

STYLE 1   STYLE 2

(D) HEX NUTS

(A)   DOUBLE   END

(B)   CONTINUOUS   THREAD

Fig. 8-3-14   Studs.

Stud Standards

There are four basic types of studs.

Type 1: Unfinished These studs have an unfinished body with no specific body tolerances.

Type 2: Finished, Full or Undersize Body These studs have a maximum body diameter equal to the basic major

diameter of the thread and a minimum body diameter equal to the rolled thread blank size.

Type 3: Finished, Full Body These studs have a maximum body diameter equal to basic major diameter of the thread and a minimum body diameter equal to the specified minimum major diameter of the thread.

Type 4: Finished, Close Body These studs have body diameter tolerances as specified by the purchaser.

Designation

Double-End Studs To avoid possible misunderstanding in specifying double-end studs, it is recommended that they be designated in the following sequence: Type and name; nominal

size; thread information; stud length; material, including grade identification; and finish (plating or coating) if required.

EXAMPLE

TYPE 2 DOUBLE-END STUD, .500—13 UNC—2A × 4.00, CADMIUM PLATED

Continuous-Thread Studs To avoid possible misunderstanding in specifying continuous-thread studs, it is recommended that they be designated in the following sequence: Product name, nominal size, thread information, stud length, material and finish (plating or coating) if required.

EXAMPLE

TYPE 3 CONTINUOUS THREAD STUD, M24 × 3 × 200, STEEL CLASS 8.8, ZINC PHOSPHATE AND OIL

WASHERS

Washers are one of the most common forms of hardware and perform many varied functions in mechanically fastened assemblies. They may only be required to span an oversize clearance hole, to give better bearing for nuts or screw faces, or to distribute loads over a greater area. Often, they serve as locking devices for threaded fasteners. They are also used to maintain a spring-resistance pressure, to guard surfaces against marring, and to provide a seal.

Classification of Washers

Washers are commonly the elements which are added to screw systems to keep them tight, but not all washers are locking types. Many washers serve other functions, such as calibrated and uncalibrated load distribution, surface protection, insulation, sealing, electrical connection, and spring-tension take-up devices.

Flat Washers Plain, or flat, washers are used primarily to provide a bearing surface for a nut or a screw head, to cover large clearance holes, and to distribute fastener loads over a larger area—particularly on soft materials such as aluminum or wood. See Fig. 8-3-15.

Conical Washers These washers are used with screws to effectively add spring take-up to the screw elongation.

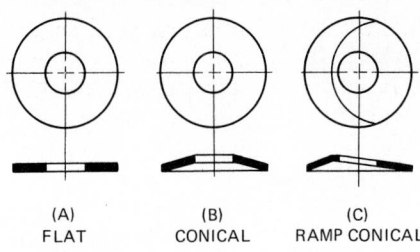

(A)
FLAT

(B)
CONICAL

(C)
RAMP CONICAL

Fig. 8-3-15 Flat and conical washers.

(A) PLAIN

(B) NONLINK POSITIVE

Fig. 8-3-16 Helical spring washers.

They are usually made of hardened spring steel.

Belleville is a term commonly, but incorrectly, applied to coned (conical) washers. Cone washers do not have any auxiliary locking features other than friction, and in the flattened position, they are the equivalent of any flat washer as far as locking action is concerned.

Helical Spring Washers These washers are made of slightly trapezoidal wire formed into a helix of one coil so that the free height is approximately twice the thickness of the washer section. See Fig. 8-3-16.

Tooth Lock Washers Made of hardened carbon steel, a tooth lock washer has teeth that are twisted or bent out of the plane of the washer face so that sharp cutting edges are presented to both the workpiece and the bearing face of the screw head or nut. See Fig. 8-3-17.

Spring Washers There are no standard designs for spring washers. See Fig. 8-3-18. They are made in a great variety of sizes and shapes and are usually selected from a manufacturer's catalog for some specific purpose.

Special-Purpose Washers Molded or stamped nonmetallic washers are available in many materials and may be used as seals, electrical insulators, or for protection of the surface of assembled parts.

EXTERNAL TYPE

INTERNAL TYPE

HEAVY DUTY
INTERNAL TYPE

COUNTERSUNK TYPE

EXTERNAL–INTERNAL
TYPE

DOME TYPE

DISHED TYPE

PYRAMIDAL TYPE

Fig. 8-3-17 Tooth lock washers.

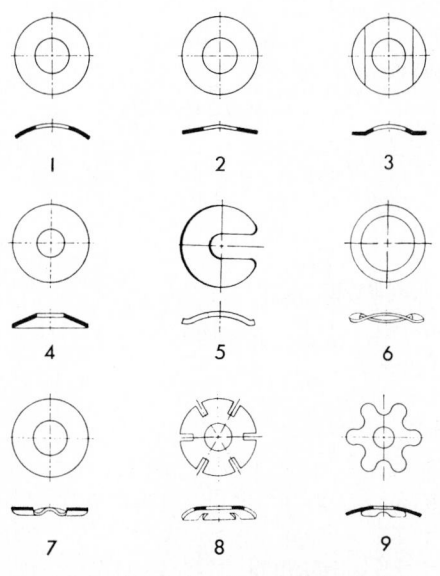

1   2   3

4   5   6

7   8   9

Fig. 8-3-18 Typical spring washers design.

Many plain, cone, or tooth washers are available with special mastic sealing compounds firmly attached to the washer. These washers are used for sealing and vibration isolation in high-production industries.

Terms Related to Threaded Fasteners

The *tap drill size* for a threaded (tapped) hole is a diameter equal to the minor diameter of the thread. The *clearance drill size*, which permits the free passage of a bolt, is a diameter slightly greater than the major diameter of the bolt. See Fig. 8-3-19. A *counterbored hole* is a circular, flat-bottomed recess that permits the head of a bolt or cap screw to rest below the surface of the part. A *countersunk hole* is an angular-sided recess that accommodates the shape of a flat-head cap screw or machine screw or an oval-head machine screw. *Spot-facing* is a machine operation that provides a smooth, flat surface where a bolt head or a nut will rest.

SPECIFYING FASTENERS

In order for the purchasing department to properly order the fastening device which has been selected in the design, the following information is required. (*Note:* The information listed will not apply to all types of fasteners):

1. Type of fastener
2. Thread specifications
3. Fastener length
4. Material
5. Head style
6. Type of driving recess
7. Point type (setscrews only)
8. Property class
9. Finish

EXAMPLES

.375—16 UNC—2A × 4.00
HEX BOLT, ZINC PLATED

M10 × 1.5 × 50, 9.8 12-SPLINE
FLANGE SCREW, CADMIUM
PLATED

TYPE 2 DOUBLE-END STUD, M10
× 1.5 × 100, STEEL CLASS 9.8,
CADMIUM PLATED

NUT, HEX, STYLE 1, .500 UNC
STEEL

(A) THREADED ASSEMBLIES

(B) DIMENSIONING HOLES

(C) DESCRIPTION OF FASTENERS

Fig. 8-3-19 Specifying threaded fasteners and holes.

MACH SCREW, PHILLIPS ROUND HD, 8—32 UNC × 1.00 LG, BRASS

WASHER, FLAT 8.4 ID × 17 OD × 2 THK, STEEL LOCKWASHER.

References and Source Material
1. *Machine Design*, Fastening and joining reference issue, Nov. 1981.
2. "Fastening Technology '77," *Design Engineering* and Staff of Stelco's "Fastener Facts."

ASSIGNMENTS

See Assignments 7 and 8 for Unit 8-3 on page 186.

Review for Assignments

SETSCREWS

Setscrews are used as semipermanent fasteners to hold a collar, sheave, or gear on a shaft against rotational or translational forces. In contrast to most fastening devices, the setscrew is essentially a compression device. Forces developed by the screw point on tightening produce a strong clamping action that resists relative motion between assembled parts. The basic problem in setscrew selection is to find the best combination of setscrew form, size, and point style that provides the required holding power.

Setscrews can be categorized in two ways: by their forms and by the point style desired. See Fig. 8-4-1. Each of the standardized setscrew forms is available in any one of six point styles.

Conventional approach to setscrew selection is usually based on a rule-of-thumb procedure: The setscrew diameter should be roughly equal to one-half the shaft diameter. This rule of thumb often gives satisfactory results, but its range of usefulness is limited.

Setscrews and Keyseats When a setscrew is used in combination with a key, the screw diameter should be equal to the width of the key. In this combination the setscrew is locating the parts in an axial direction only. The torsional load on the parts is carried by the key.

The key should be tight-fitting, so that no motion is transmitted to the screw. Under high reversing or alternating loads, a poorly fitted key will cause the screw to back out and lose its clamping force.

KEEPING FASTENERS TIGHT

Fasteners are inexpensive, but the cost of installing them can be substantial. Probably the simplest way to cut assembly costs is to make sure that, once installed, fasteners stay tight.

The American National Standards Institute has identified three basic locking methods: free-spinning, prevailing-torque, and chemical locking. Each has its own advantages and disadvantages. See Fig. 8-4-2.

STANDARD POINTS	
CUP	Most generally used. Suitable for quick and semi-permanent location of parts on soft shafts, where cutting in of edges of cup shape on shaft is not objectionable.
FLAT	Used where frequent resetting is required, on hard steel shafts, and where minimum damage to shafts is necessary. Flat is usually ground on shaft for better contact.
CONICAL	For setting machine parts permanently on shaft, which should be spotted to receive cone point. Also used as a pivot or hanger.
SPHERICAL	Should be used against shafts spotted, splined or grooved to receive it. Sometimes substituted for cup point.
HALF DOG	For permanent location of machine parts, although cone point is usually preferred for this purpose. Point should fit closely to dia. of drilled hole in shaft. Sometimes used in place of a dowel pin.

STANDARD HEADS	
HEXAGON SOCKET	Standard size range: No. 0 to 1.0 in. (2 to 24mm), threaded entire length of screw in .06 in. (2mm) increments from .25 to .62 in. (6 to 16mm), .12 in. (3mm) increments from .62 to 1.0 in. (16 to 24mm). Coarse or fine thread series.
SLOTTED	Standard size range: No. 5 to .75 in. (3 to 20mm) threaded entire length of screw. Coarse or fine thread series.
FLUTED SOCKET	Same as hexagon socket. No. 0 and 1 (2 to 3mm) have four flutes. All others have six flutes.
SQUARE HEAD	Standard size range: No. 10 to 1.50 in. (5 to 36mm). Entire body is threaded. Coarse or fine-thread series. Sizes .25 in. (6mm) and larger are normally available in coarse threads only.

Fig. 8-4-1   Setscrews.

Free-spinning devices include toothed and spring lockwashers and screws and bolts with washerlike heads. With these arrangements, the fasteners spin free in the clamping direction, which makes them easy to assemble, and the break-loose torque is greater than the seating torque. However, once break-loose torque is exceeded, free-spinning washers have no prevailing torque to prevent further loosening.

Prevailing-torque methods make use of increased friction between nut

(A) FREE-SPINNING

(B) PREVAILING TORQUE

Fig. 8-4-2   Typical setscrew installation locking fasteners.

and bolt. Metallic types usually have deformed threads or contoured thread profiles that jam the threads on assembly. Nonmetallic types make use of nylon or polyester insert elements that produce interference fits on assembly.

Chemical locking is achieved by coating the fastener with an adhesive.

LOCKNUTS

A *locknut* is a nut with special internal means for gripping a threaded fastener or connected material to prevent rotation in use. Generally it has the dimensions, mechanical requirements, and other specifications of a standard nut, but with a locking feature added.

Locknuts are divided into two general classifications: prevailing-torque and free-spinning types. These are shown in Fig. 8-4-3.

Prevailing-Torque Locknuts

Prevailing-torque locknuts spin freely for a few turns, and then must be wrenched to final position. The maximum holding and locking power is reached as soon as the threads and the

locking feature are engaged. Locking action is maintained until the nut is removed. Prevailing-torque locknuts are classified by basic design principles:

1. Thread deflection causes friction to develop when the threads are mated; thus the nut resists loosening.
2. The out-of-round top portion of the tapped nut grips the bolt threads and resists rotation.
3. The slotted section of locknut is pressed inward to provide a spring frictional grip on the bolt.
4. Inserts, either nonmetallic or of soft metal, are plastically deformed by the bolt threads to produce a frictional interference fit.
5. A spring wire or pin engages the bolt threads to produce a wedging or ratchet-locking action.

Free-Spinning Locknuts

Free-spinning locknuts are free to spin on the bolt until seated. Additional tightening locks the nuts.

Free-spinning locknuts are often specified when long travel of nut on bolt is unavoidable. Since most free-spinning locknuts depend on clamping force for their locking action, they are usually not recommended for joints that might relax through plastic deformation or for fastening materials that might crack or crumble.

Other Locknut Types

Jam nuts are thin nuts used under full-sized nuts to develop locking action. The large nut has sufficient strength to elastically deform the lead threads of the bolt and jam nut. Thus, a considerable resistance against loosening is built up. The use of jam nuts is decreasing; a one-piece, prevailing-torque locknut usually is used instead at a savings in assembled cost.

The jam nut is considered ideal for assemblies where long travel of nut on bolt under load is necessary to bring mating parts into position.

Slotted and castle nuts have slots which receive a cotter pin that passes through a drilled hole in the bolt and thus serves as the locking member. These nuts are essentially free-spinning nuts with the locking feature

Nonmetallic collar clamped in the top of this nut produces locking action.

Threaded elliptical spring-steel insert grips the bolt and prevents turning.

Slotted section of this prevailing-torque nut forms beams which are deflected inward and grip the bolt.

Three sectors of tapered cone, preformed inwardly, are elastically returned to circu-form when the nut is applied.

(A) PREVAILING TORQUE LOCKNUTS

Deformed bearing surface. Teeth on the bearing surface "bite" into work to provide a ratchet locking action.

Nylon insert flows around the bolt rather than being cut by the bolt threads to provide locking action and an effective seal.

Jam nut, applied under a large regular nut, is elastically deformed against bolt threads when the large nut is tightened.

Nut with a captive-toothed washer. When tightened, the captive washer provides the locking means with spring action between the nut and working surface.

(B) FREE-SPINNING LOCKNUTS

Slotted nut uses a cotter pin through a hole in the bolt for locking action.

Single-thread locknut, which is speedily applied, lock by grip of arched prongs when bolt or screw is tightened.

(C) OTHER TYPES

Fig. 8-4-3 Locknuts.

added after the preload condition is developed. Castle nuts differ from slotted nuts in that they have a circular crown of a reduced diameter.

Single-thread engaging locknuts are spring steel fasteners which may be speedily applied. Locking action is provided by the grip of the thread-engaging prongs and the reaction of the

For holding a motor mounting securely in position.

Use of locknut on a spring clamp.

Use of locknut on a bolted connection that requires pre-determined play.

For an extruded part assembly.

Use of locknut for tubular fastening.

For rubber-insulated and cushion mountings where the nut must remain stationary.

For spring-mounted connections where the nut must remain stationary or is subject to adjustment.

Use of locknut where assembly is subjected to vibratory or cyclic motions that could cause loosening.

Fig. 8-4-4 Typical locknut applications.

arched base. Their use is limited to nonstructural assemblies and usually to screw sizes below 6 mm in diameter. See Fig. 8-4-4 for typical uses of locknuts.

CAPTIVE OR SELF-RETAINING NUTS

Self-retained or captive nuts provide a permanent, strong, multiple-thread fastener in many types of thin materials. See Fig. 8-4-5. They are especially good where there are blind locations, and they can generally be attached without damaging finishes. Methods of attaching these types of nuts vary and tools required for assembly are generally uncomplicated and inexpensive. In this section, the self-retained nuts are grouped according to four means of attachment:

1. Plate or anchor nuts: These nuts have mounting lugs which can be riveted, welded, or screwed to the part.
2. Caged nuts: A spring-steel cage retains a standard nut. The cage snaps into a hole or clips over an edge to hold the nut in position.

3. Clinch nuts: They are specially designed nuts with pilot collars which are clinched or staked into the parent part through a precut hole.
4. Self-piercing nuts: A form of clinch nut that cuts its own hole.

SINGLE-THREAD ENGAGING NUTS

Single-thread engaging nuts are formed by stamping a thread-engaging impression in a flat piece of metal. The stamped impression can take a number of shapes; for example, shear-formed helical prongs engage and lock on the screw-thread root diameter (Fig. 8-4-6A), or a protruding truncated cone (Fig. 8-4-6B) stamped into the

metal provides a ramp that the screw climbs as it turns. Another type of impression has the thread-engaging elements spirally formed to match the pitch of the screw threads (Fig. 8-4-6C).

INSERTS

Inserts are a special form of nut designed to serve the function of a tapped hole in blind or through-hole locations. See Fig. 8-4-7. They are sometimes referred to as *solid bushings*. Another basic type of screw-thread insert consists of precision-formed wire, spirally coiled to provide threads of proper form for installation in a tapped hole. This is known as a *wire insert*.

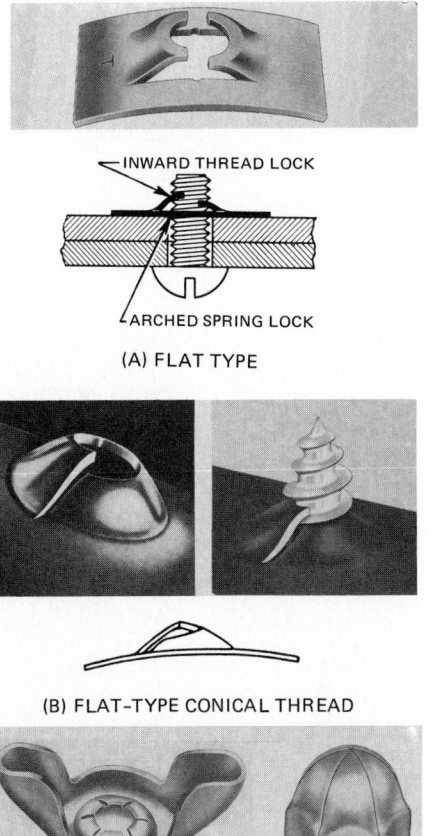

(A) FLAT TYPE

(B) FLAT–TYPE CONICAL THREAD

(C) SPIRAL-FORMED THREAD

Fig. 8-4-6   Single-thread engaging nuts.

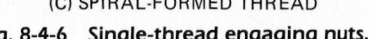

(A) PLATE NUT

(B) CAGED NUT

(C) CLINCH NUT

(I) UNIVERSAL PIERCE NUT

(2) HIGH-STRESS PIERCE NUT

(3) PIERCE NUT WITH CORNERS CLINCHED
(D) PIERCE NUTS

Fig. 8-4-5   Captive or self-retaining nuts.

(A) EXTERNAL-INTERNAL THREADED INSERT

(B) SELF-TAPPING INSERT

(C) PRESSED-IN INSERT

(D) MOLDED-IN INSERT

(E) THIN MATERIAL INSERT

(F) SANDWICH PANEL INSERT

Fig. 8-4-7   Inserts.

SEALING FASTENERS

Fasteners hold two or more parts together, but they can perform other functions as well. One important auxiliary function is that of sealing gases and liquids against leakage.

(A) FASTENERS SEPARATELY SEALED

(B) SEALING ELEMENT CLAMPED IN PLACE

Fig. 8-4-8 Types of sealed-joint construction.

Two types of sealed-joint constructions are possible with fasteners. See Fig. 8-4-8. In one approach, the fasteners enter the sealed medium and are separately sealed.

The second approach uses a separate sealing element which is held in place by the clamping forces produced by conventional fasteners, such as rivets or bolts.

Sealing Fastener Types

There are many methods of obtaining a seal using sealing fasteners, as shown in Fig. 8-4-9.

SEALING SCREWS

MASTIC SEALING COMPOUND

LIQUID PLASTIC COATING

MOLDED RUBBER RING

BRONZE SLEEVE

LEAD WASHER

PREASSEMBLED NEOPRENE WASHER

PREASSEMBLED METAL AND NEOPRENE WASHER

PREASSEMBLED METAL WASHER AND O-RING

PREASSEMBLED NYLON WASHER

O-RING

O-RING WITH TEFLON WASHER

SEALING RIVETS

MOLDED RUBBER RING

SOFT–ALUMINUM WASHER

PLASTIC JACKET

O-RING

O-RING

INTERFERENCE FIT

SEALING NUTS

FLOWED-IN SEALANT

NYLON BODY

COPPER INSERT

NYLON COLLAR

NYLON PELLET

MOLDED RUBBER GASKET OR O-RING

SEALING WASHERS

MOLDED NYLON–SEAL RING

MOLDED RUBBER TOROID

LAMINATED NEOPRENE TO METAL

NYLON SLEEVE

O-RING

FLOWED-IN SEALANT

Fig. 8-4-9 Sealing fasteners.

References and Source Material

1. *Machine Design*, Fastening and joining reference issue, Nov. 1981.

ASSIGNMENTS

See Assignments 9 and 10 for Unit 8-4 on page 186.

Review for Assignments
Unit 7-5   Shafts in Section
Unit 9-2   Pin Fasteners

UNIT 8-5
Fasteners for Light-Gage Metal, Plastic, and Wood

TAPPING SCREWS

Tapping screws cut or form a mating thread when driven into drilled or cored holes. These one-piece fasteners permit rapid installation, since nuts are not used and access is required from only one side of the joint. The mating thread produced by the tapping screw fits the screw threads closely, and no clearance is necessary. This close fit usually keeps the screw tight, even under vibrating conditions. See Fig. 8-5-1.

Tapping screws are practically all case-hardened and, therefore, can be driven tight and have a relatively high ultimate torsional strength. The screws are used in steel, aluminum (cast, extruded, rolled, or die-formed)

TYPE AB

TYPE B

TYPE F

TYPE U

TYPE 21

Fig. 8-5-1 Self-tapping screws.

<table>
<tr><td colspan="2">

HEAVY GAGE SHEET METAL AND
STRUCTURAL STEEL.
USE TYPES B, BP, C, U, D, F, G, T.

</td><td>

LIGHT GAGE SHEET METAL.
USE TYPES AB, B, BP, C.

</td></tr>
<tr><td>

Holes may be drilled or clean-punched.

Use a pierced hole in workpiece if clearance hole is needed in part to be fastened.

</td><td>

Two parts may have pierced holes to nest burrs. This results in a stronger joint.

Extruded hole may also be used in workpiece if clearance hole is needed in fastened part.

</td><td>

Holes may be drilled or clean-punched the same size in both sheet metal parts. For thicker sheet metal and structural steel, a clearance hole should be provided in the part to be fastened. Hole size depends on thickness of the workpiece.
Notes: 1. Use hex-head on Type B and BP screws.
     2. Type C screw for sheet metal only; maximum thickness, .135 in. or 3.5mm.
     3. With Type U screws, material should be thick enough to permit sufficient thread engagement—at least one screw diameter.

</td></tr>
<tr><td colspan="2">

PLASTICS.
USE TYPES B, BP, U, D, F, G, T, BF, BT.

Screw holes may be molded or drilled. If material is brittle or friable, molded holes should be formed with a rounded chamfer, and drilled holes should be machine chamfered. Provide a clearance in the part to be fastened. Depth of penetration should be held within the "minimum and maximum" limits recommended. The hole should be deeper than the screw penetration to allow for chip clearance.

</td><td>

CASTINGS AND FORGINGS.
USE TYPES B, BP, U, D, F, G, T, BF, BT.

Holes may be cored if it is practical to maintain close tolerances. Otherwise blind drill holes to recommended hole size. Provide a clearance hole for screw in the part to be fastened. The hole in the casting, if it is a blind hole, should be deeper than the screw penetration to allow for chip clearance.
Notes: 1. Hole in fastened part may be the same size as workpiece hole for Type U screws.
     2. Types B, BP, BF, BT are only suitable for use in nonferrous castings.

</td></tr>
</table>

Fig. 8-5-2 Tapping-screw application chart.

die castings, cast iron, forgings, plastics, reinforced plastics, asbestos, and resin-impregnated plywood. See Fig. 8-5-2.

Types C, D, F, G, and T tapping screws are available in both coarse- and fine-thread series. Coarse threads should be used with weak materials.

Self-drilling tapping screws, types BSD and CSD, have special points for drilling and then tapping their own holes. See Fig. 8-5-3. These eliminate

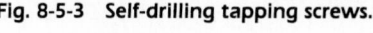

Fig. 8-5-3 Self-drilling tapping screws.

drilling or punching, but they must be driven by a power screwdriver.

Special Tapping Screws
Typical special tapping screws are the self-captive screws and double-thread combinations for limited drive. Self-captive screws combine a coarse-pitched starting thread (similar to type B) with a finer pitch (machine-screw thread) farther along the screw shank.

Sealing tapping screws, with pre-assembled washers or O-rings (Fig. 8-5-4B) can be used to control leaks, squeaks, crazing of enamel, and electrolysis in all types of metal structures and assemblies.

References and Source Material
Machine Design, Fastening and joining reference issue, Nov. 1981.

(A) TAPPING SCREWS WITH PREASSEMBLED WASHERS. THESE ARE AVAILABLE IN A GREAT VARIETY OF THREAD FORMS, HEAD STYLES, AND WASHER CONFIGURATIONS.

(B) TAPPING SCREWS WITH PREASSEMBLED SEALING WASHERS OR COMPOUNDS.

Fig. 8-5-4 Special tapping screws.

ASSIGNMENT
See Assignment 11 for Unit 8-5 on page 187.

Review for Assignments
Unit 7-1   Sectional Drawings

ASSIGNMENTS for Chapter 8

.9680
Ø.9672

.62 ⊢ 1.00

8.75

Ø 1.0310
1.0305

Ø 1.50

.12 X .06 NECK

.06 X 45° CHAMFER

BUTTRESS THREAD
PITCH = .25
LEFT HAND

MATL — SAE 1050

JACK SCREW   DRAW TO FULL SCALE

40

260

Ø 20

12

SHARP V THREAD
PITCH = 2.5
TRIPLE THREAD

MATL — SAE 1112
DRAW TO SCALE 2 : 1

Ø 11.20
11.00

GUIDE ROD

METRIC

45°

KNUCKLE THREAD
PITCH = .125

60°

Ø 1.12

Ø 1.000
Ø .25
Ø .38

.06

.12

.50   .38

1.19

DRAW TO SCALE: 2 X SIZE

FUSE

Fig. 8-1-A   Jack screw and fuse.

HEX 70 AF

Ø 44 X 6 W NECK

SQUARE THREAD
PITCH = 12 SINGLE THREAD

Ø 56

25

95

MATL — SAE 1006

PLUG   DRAW TO SCALE 1 : 1

Fig. 8-1-B   Guide rod and plug.

Assignments for Unit 8-1, Thread Forms

1. On a B- or A3-size sheet draw the two parts shown in Fig. 8-1-A or 8-1-B. Use pictorial representation for the threads. Use conventional breaks to shorten the lengths of the guide rod and jack screw.
2. On a B- or A3-size sheet draw either Figure 8-1-C or 8-1-D showing the connector and supports. The supports are to be drawn in section. Scale is full or 1:1.

Ø 2.00 X .25 WIDE NECK
HEX 3.00 A/F
2.50 - 2 SQUARE THREAD
2.50 - 2 DOUBLE ACME THREAD

2.50   2.50

6.50

1.50   1.50

Fig. 8-1-C   Connector and supports.

END ROD

SQUARE THREAD
PITCH = 12
DOUBLE THD–LH

METRIC

Ø 52 X 6
NECK
BOTH SIDES

Ø64

HEX 76 A/F

ACME THREAD
PITCH = 12
DOUBLE THD

END ROD

Ø64

Ø110

72   72

170

Fig. 8-1-D Connector.

Assignments for Unit 8-2, Simplified Representation of Threads

3. On an A3- or B-size sheet, make a two-view assembly drawing of the parallel clamps shown in Fig. 8-2-A. Use simplified thread connections and include on your drawing a bill of material calling for all the parts. The only dimension required on the drawing is the maximum opening of the jaws. Identify the parts on the assembly. Scale is 1:1.

4. On an A3- or B-size sheet make detail drawings of the parts shown in Fig. 8-2-A. Scale is 1:1. Use your judgment for the number of views required for each part.

5. On a B- or A3-size sheet make a one-view assembly drawing of the turnbuckle shown in Fig. 8-2-B. Draw the assembly in its shortest length and show the maximum position in phantom lines. The only dimensions required are the minimum and maximum distances between the eye centers. Scale is full.

6. On a B- or A3-size sheet make detail drawings of the parts shown in Fig. 8-2-B. Scale is full.

Ø.50

Ø.31

.31 GAP

R.38

3.00

2.00

.312–18 UNC–2A

3.50   .50

.25

.50   .25

.50   .75

.50

.312–18 UNC–2B

Ø 1.00

Ø .62

.312–18 UNC–2B–LH

.312–18 UNC–2A–LH

Fig. 8-2-A Turnbuckle.

PT 2 STATIONARY JAW
I REQD MATL–SAE 1020
AS SHOWN OTHERWISE
SAME AS PT I

5

12

30

80

M8
2 HOLES

12

35

R 6

PT I MOVABLE JAW
I REQD
MATL–SAE 1020

Fig. 8-2-B Parallel clamps.

Ø 12

KNURL P 0.8

1.5 X 1.5 NECK

14

80

M8

4.5

Ø 4.5

PT 3 OUTER SCREW
I REQD MATL–SAE 1112

METRIC

Ø 10

18

M3 X 8 DEEP

Ø 4.8 X 6 DEEP

PT 4 INNER SCREW
I REQD MATL–SAE 1112

60   90

M8

3

14

Ø 9 X 1.5 NECK

Ø 5   5

Ø 12

KNURL P 0.8

PT 5 CLIP
MATL 1.52 (16 USS)

3

R 4.5

9   18

Ø 3.2

R 6

PT 6 MACHINE SCREW RD HD
M3 X 10 LG – I REQD

THREADED FASTENERS 185

Assignments for Unit 8-3, Common Threaded Fasteners

7. On a B- or A3-size sheet draw the four fastener assemblies in full section shown in Fig. 8-3-A. Dimension both the clearance and the threaded holes. A top view may be drawn, if desired. Scale is full.

8. On an A3- or B-size sheet, draw the five standard fastener assemblies in full section shown in Fig. 8-3-B. Use simplified thread symbols. Scale is 1:1. Dimension both the clearance and the threaded holes. If desired, a top view of the fasteners may be drawn.

Assignments for Unit 8-4, Special Fasteners

9. On a B- or A3-size sheet make a one-view assembly drawing of the flexible coupling shown in Fig. 8-4-A. The shafts, which are coupled, are 1.50 in. in diameter and are to be shown in the assembly. They are to extend beyond the coupling for approximately 2.00 in. and end with a conventional break. Show the set-screws and keys in position. Scale is full.

10. On an A3- or B-size sheet, make a one-view assembly drawing of the adjustable shaft support shown in Fig. 8-4-B. Draw the base in full section. A broken-out section is recommended to clearly show the setscrews in the yoke. Add part numbers to the assembly drawing and include a bill of material. Do not dimension. Scale is 1:1.

Fig. 8-3-A   Threaded fasteners.

CONNECTION A
M 10 X 30 LG
HEX HD CAP SCREW

CONNECTION B
M 10 X 40 LG STUD
THREAD EACH END 20 LG
HEX NUT STYLE I AND
SPRING LOCKWASHER

CONNECTION C
M 10 X 30 LG
FL HD CAP SCREW

CONNECTION D
M 10 X 1.25 X 25 LG
SOCKET HEAD CAP SCREW
AND SPRING LOCKWASHER

Fig. 8-3-B   Standard fasteners.

Fig. 8-4-A   Flexible couplings.

MAXIMUM BORE	A	B	C	D	E	F
.9375	3.00	3.75	1.75	.88	1.50	2.38
1.1875	3.50	4.69	2.19	1.06	1.81	2.75
1.4375	4.00	5.62	2.62	1.25	2.12	3.12
1.6875	5.00	6.56	3.06	1.44	2.44	3.50
1.9375	5.50	7.50	3.50	1.50	2.50	4.00
2.1875	6.00	8.44	3.94	1.81	3.06	4.38

DIMENSIONS SHOWN ARE IN INCHES

PT 3 YOKE

MATL—CI I REQD

ROUNDS AND
FILLETS R 3

R 10

M 10
3 HOLES

R 12

PT 6 SET SCREW
M 10 X 30 LG 2 REQD

PT 8 HEX HD JAM NUT
M 10 2 REQD

38
12
12
20
20
34
20
20
10
Ø 38

PT 7 SET SCREW
M 10 X 10 LG
HEX SOCKET
DOG POINT
I REQD

Ø 20 SLIDE FIT
FOR PT 2

PT 5 BEARINGS
MATL—BRONZE 2 REQD

20
20
Ø 20.3
20.0
Ø 25 PRESS FIT
IN PT 4
50
25

CSK Ø6 X 90°
3 HOLES
SPACED AT 90°

Ø 25

Ø 32

PT 4 BEARING HOUSING
MATL—STEEL I REQD

Ø 14
Ø 20
100

PT 2 VERTICAL SHAFT
MATL—STEEL I REQD

Ø 20 SLIDE FIT FOR PT 2
M 10
Ø 38
10
ROUNDS AND FILLETS R 3
20
Ø 8 SLOTS
5
70
100
Ø 60
44
8
120
8

METRIC

Courtesy Boston Gear Works

NOTE—DIAMETERS SHOWN
FOR SLIDE AND PRESS FITS
ARE NOMINAL DIMENSIONS

PT I BASE
MATL—CI I REQD

FRONT

Fig. 8-4-B   Adjustable shaft support.

Assignment for Unit 8-5, Fasteners for Light-Gage Metal, Plastic, and Wood

11. On a B- or A3-size sheet draw the two
    assemblies shown in Fig. 8-5-A. Either
    inch or metric fasteners may be used. In
    each assembly indicate the hole and fas-
    tener size. Scale to suit.

TAPPING SCREW ASSEMBLY
The steel post is fastened to the panel by two
rows of tapping screws. The steel strap is
held to the post by a single tapping screw
which has the equivalent strength (body
area) of at least three of the other tapping
screws.

WOOD FASTENERS
The steel post is held to the wood furring
with no. 8 RHWS. For safety reasons, the
screw holding the steel strap must be flush
with the face of the strap. A no. 10 wood
screw is required.

3.00
.75
1.50
.75

16 GA STEEL
STRAP

13 GA STEEL POST

18 GA STEEL
PANEL

TAPPING SCREWS

METRIC

80
50
15
15

50 X 50
FURRING

4mm STEEL
STRAP

12 GA STEEL POST

WOOD SCREWS

Fig. 8-5-A   Special fastener problems.

CHAPTER 9

Miscellaneous Types of Fasteners

UNIT 9-1
Keys, Splines, and Serrations

KEYS

A *key* is a piece of steel lying partly in a groove in the shaft and extending into another groove in the hub. These grooves are called keyseats and keyways. See boxes 6 and 8 in Fig. 9-1-1. A key is used to secure gears, pulleys, cranks, handles, and similar machine parts to shafts, so that the motion of the part is transmitted to the shaft, or the motion of the shaft to the part, without slippage. The key may also act in a safety capacity; its size is generally calculated so that when overloading takes place, the key will shear or break before the part or shaft breaks.

There are many kinds of keys. The most common types are shown in Fig. 9-1-2. Square and flat keys are widely used in industry. The width of the square and flat key should be approximately one-quarter the shaft diameter, but for proper key selection refer to the Appendix. These keys are also available with a 1:100 taper on their top surfaces and are then known as *square taper* or *flat tapered* keys. The keyway in the hub is tapered to accommodate the taper on the key.

The gibhead key is the same as the square or flat tapered key but has a head added for easy removal.

The Pratt and Whitney key is rec-

tangular with rounded ends. Two-thirds of this key sits in the shaft, one-third sits in the hub.

The Woodruff key is semicircular and fits into a semicircular keyseat in the shaft and a rectangular keyway in the hub. The width of the key should be approximately one-quarter the diameter of the shaft, and its diameter should approximate the diameter of the shaft. Half the width of the key extends above the shaft and into the hub. Refer to the Appendix for exact sizes. Woodruff keys are identified by a number which gives the nominal

dimensions of the key. The numbering system, which originated many years ago, is identified with the fractional-inch system of measurement. The last two digits of the number give the normal diameter in eighths of an inch, and the digits preceding the last two give the nominal width in thirty-seconds of an inch. For example, a No. 1210 Woodruff key indicates a key $12/32 \times 10/8$ in., or a $3/8 \times 1 1/4$ in. key.

In calling up keys on a bill of material, only the information shown in the column "Specifications" in Fig. 9-1-2 need be given.

1 RETAINING COMPOUND JOINT	2 PRESS FIT	3 KNURLED JOINT
4 TAPERED SHAFT	5 SLIDING FIT	6 DRIVEN KEY
7 SPLINES	8 SLIP FIT WITH KEY	9 BRAZED JOINT
10 SETSCREW	11 PINS	12 SPLIT HUB

Fig. 9-1-1   Miscellaneous types of fasteners. (Design Engineering, Oct. 1968.)

TYPE OF KEY	ASSEMBLY SHOWING KEY, SHAFT AND HUB	SPECIFICATION
SQUARE		.25 SQUARE KEY, 1.25 LG OR .25 SQUARE TAPERED KEY, 1.25 LG
FLAT		.188 X .125 FLAT KEY, 1.00 LG OR .188 X .125 FLAT TAPERED KEY, 1.00 LG
GIB-HEAD		.375 SQUARE GIB-HEAD KEY, 2.00 LG
PRATT AND WHITNEY		NO. 15 PRATT AND WHITNEY KEY
WOODRUFF		NO. 1210 WOODRUFF KEY

Fig. 9-1-2   Common keys.

Dimensioning of Keyseats

Keyseats are dimensioned by width, depth, location, and, if required, length. The depth is dimensioned from the opposite side of the shaft or hole. See Fig. 9-1-3.

Tapered Keyseats The depth of tapered keyseats in hubs, which is shown on the drawing, is the nominal depth $H/2$ minus an allowance. This is always the depth at the large end of the tapered keyseat and is indicated on the drawing by the abbreviation LE.

The radii of fillets, when required, must be dimensioned on the drawing, for example, .50 × .25 × R.04.

Since standard milling cutters for Woodruff keys have the same appropriate number, it is possible to call for a Woodruff keyseat by the number only.

Where it is desirable to detail Woodruff keyseats on a drawing, all dimensions are given in the form of a note in the following order: width, depth, and radius of cutter.

Woodruff keyseats may alternately be dimensioned in the same manner as

for square and flat keys, specifying first the width and then the depth. See Fig. 9-1-4.

SPLINES AND SERRATIONS

A *splined shaft* is a shaft having multiple grooves, or keyseats, cut around its circumference for a portion of its length, in order that a sliding engagement may be made with corresponding internal grooves of a mating part.

Splines are capable of carrying heavier loads than keys, permit lateral movement of a part while maintaining positive rotation, and allow the attached part to be indexed or changed to another angular position.

Splines have either straight-sided teeth or curved-sided teeth. The latter type is known as an *involute spline*.

Involute Splines These splines are similar in shape to involute gear teeth but have pressure angles of 30, 37.5, or 45°. There are two types of fits, the *side fit* and the *major-diameter fit*. See Fig. 9-1-5.

Parallel-Side Splines The most popular are the SAE parallel-side splines, as shown in Fig. 9-1-6. They have been used in many applications in the automotive and machine industry.

Serrations are shallow, involute splines with 45° pressure angles. They are primarily used for holding parts, such as plastic knobs on steel shafts.

Drawing Data

It is essential that a uniform system of drawing and specifying splines and serrations be used on drawings. The conventional method of showing

Fig. 9-1-3   Dimensioning.

Fig. 9-1-4   Alternate method of detailing a woodruff keyseat.

(A) SIDE FIT

(B) MAJOR DIAMETER FIT

Fig. 9-1-5   Involute splines.

Fig. 9-1-6 Sizes of SAE parallel-side splines.

NUMBER OF SPLINES	W FOR ALL FITS	PERMANENT FIT		TO SLIDE WITHOUT LOAD		TO SLIDE UNDER LOAD	
		H	R	H	R	H	R
4	0.241 D	0.075 D	0.85 D	0.125 D	0.75 D		
6	0.250 D	0.050 D	0.90 D	0.075 D	0.85 D	0.100 D	0.80 D
10	0.156 D	0.045 D	0.91 D	0.070 D	0.86 D	0.095 D	0.81 D
16	0.098 D	0.045 D	0.91 D	0.070 D	0.86 D	0.095 D	0.81 D

splines on a drawing is shown in Fig. 9-1-7. The drawing callout is shown in Fig. 9-1-8. Distance *L* does not include the cutter runout. The drawing callout shows the symbol indicating the type of spline followed by the type of fit, the pitch diameter, number of teeth and pitch for involute splines, and number of teeth and outside diameter for straight-sided teeth.

ASSIGNMENT

See Assignment 1 for Unit 9-1 on page 204.

Review for Assignment
Unit 7-5   Assemblies in Section

(A) EXTERNAL SPLINE

(B) INTERNAL SPLINE

(C) ASSEMBLY DRAWINGS

Fig. 9-1-7 Representing splines on drawings.

SYMBOLS USED IN CALLOUT WHEN NECESSARY TO SHOW DIFFERENCE FROM A SCREW THREAD.

(A) INVOLUTE SPLINE

(B) STRAIGHT-SIDED TEETH

Fig. 9-1-8 Drawing callout for splines.

UNIT 9-2
Pin Fasteners

Pin fasteners are an inexpensive and effective approach to assembly where loading is primarily in shear. They can be separated into two groups: *semipermanent* and *quick-release*.

SEMIPERMANENT PINS

Semipermanent pin fasteners require application of pressure or the aid of tools for installation or removal. The two basic types are machine pins and radial-locking pins.

The following general design rules apply to all types of semipermanent pins:

- Avoid conditions where the direction of vibration parallels the axis of the pin.
- Keep the shear plane of the pin a minimum distance of 1 diameter from the end of the pin.
- In applications where engaged length is at a minimum and appear-

ance is not critical, allow pins to protrude the length of the chamfer at each end for maximum locking effect.

Machine Pins

Four types are generally considered to be most important: *hardened and ground dowel pins and commercial straight pins*, *taper pins*, *clevis pins*, and *standard cotter pins*. Descriptive data and recommended assembly practices for these four traditional types of machine pins are presented in Fig. 9-2-1. For proper size selection of cotter pins, refer to Fig. 9-2-2.

Radial Locking Pins

Two basic pin forms are employed: *solid with grooved surfaces* and *hollow spring pins*, which may be either slotted or spiral-wrapped.

Grooved Straight Pins

Locking action of the grooved pin is provided by parallel, longitudinal grooves uniformly spaced around the pin surface. Rolled or pressed into solid pin stock, the grooves expand the effective pin diameter. When the pin is driven into a drilled hole corresponding in size to nominal pin diameter, elastic deformation of the raised groove edges produces a secure force-fit with the hole wall. Figure 9-2-3 shows six of the grooved-pin constructions that have been standardized. For typical grooved pin applications and size selection, refer to Figs. 9-2-4 and 9-2-6.

Hollow Spring Pins

Resilience of hollow cylinder walls under radial compression forces is the principle of spiral-wrapped and slotted tubular pins. Both pin forms are made to controlled diameters greater than the holes into which they are pressed. Compressed when driven into the hole, the pins exert spring pressure against the hole wall along their entire engaged length to develop locking action.

Standard slotted tubular pins are designed so that several sizes can be used inside one another. In such combinations, shear strengths of the individual pins are additive. For spring pin application refer to Fig. 9-2-5.

HARDENED AND GROUND DOWEL PIN	TAPER PIN	CLEVIS PIN	COTTER PIN
Standardized in nominal diameters ranging from .12 to .88 (3 to 22 mm). 1. Holding laminated sections together with surfaces either drawn up tight or separated in some fixed rea some fixed relationship. 2. fastening machine parts where accuracy of alignment is a primary consideration. 3. locking components on shafts, in the form of transverse pin key.	Standard pins have a taper of 1:48 measured on the diameter. Basic dimension is the diameter of the large end. Used for light duty service in the attachment of wheels, levers and similar components to shafts. Torque capacity is determined on the basis of double shear, using the average diameter along the tapered section in the shaft for area calculations.	Standard nominal diameters for clevis pins range from .19 to 1.00 (5 to 25mm). Basic function of the clevis pin is to connect mating yoke, or fork, and eye members in knuckle-joint assemblies. Held in place by a small cotter pin or other fastening means, it provides a mobile joint construction, which can be readily disconnected for adjustment or maintenance.	Sizes have been standardized in nominal diameters ranging from .03 to .75 (I to 20mm). Locking device for other fasteners. Used with a castle or slotted nut on bolt, screws, or studs, it provides a convenient, low-cost locknut assembly. Hold standard clevis pins in place. Can be used with or wothout a plain washer as an artificial shoulder to lock parts in position on shafts.

Fig. 9-2-1   Machine pins.

NOMINAL THREAD SIZE		NOMINAL COTTER PIN SIZE		COTTER PIN HOLE		END CLEARANCE*	
in.	(mm)	in.	(mm)	in.	(mm)	in.	(mm)
.250	(6)	.062	(1.5)	.078	(1.9)	.11	(3)
.312	(8)	.078	(2)	.094	(2.4)	.11	(3)
.375	(10)	.094	(2.5)	.109	(2.8)	.14	(4)
.500	(12)	.125	(3)	.141	(3.4)	.17	(5)
.625	(14)	.156	(3)	.172	(3.4)	.23	(5)
.750	(20)	.156	(4)	.172	(4.5)	.27	(7)
1.000	(24)	.188	(5)	.203	(5.6)	.31	(8)
1.125	(27)	.188	(5)	.203	(5.6)	.39	(8)
1.250	(30)	.219	(6)	.234	(6.3)	.41	(10)
1.375	(36)	.219	(6)	.234	(6.3)	.44	(11)
1.500	(42)	.250	(6)	.266	(6.3)	.48	(12)
1.750	(48)	.312	(8)	.312	(8.5)	.55	(14)

* DISTANCE FROM EXTREME POINT OF BOLT OR SCREW TO CENTER OF COTTER PIN HOLE.

Fig. 9-2-2   Recommended cotter pin sizes.

SHAFT DIA		TRANSVERSE KEY			LONGITUDINAL KEY	
		PIN DIA		TAPER PIN NO.	PIN DIA	
in.	(mm)	in.	(mm)		in.	(mm)
.375	(10)	.125	(3)	3/0	.094	(2.5)
.438	(12)	.156	(4)	0	.125	(3)
.500	(14)	.156	(5)	0	.125	(4)
.562	(16)	.188	(5)	2	.156	(4)
.625	(18)	.188	(6)	2	.156	(5)
.750	(20)	.250	(6)	4	.156	(5)
.875	(22)	.250	(6)	4	.219	(6)
1.000	(24)	.312	(8)	6	.250	(6)
1.062	(26)	.312	(8)	6	–	
1.125	(28)	.375	(10)	7	–	
1.188	(30)	.375	(10)	7	–	
1.250	(32)	.375	(10)	7	.312	(8)
1.375	(34)	.438	(11)	7	.375	(10)
1.438	(36)	.438	(11)	7	–	
1.500	(38)	.500	(12)	8	.438	(11)

Fig. 9-2-4   Recommended groove pin sizes.

SOLID WITH GROOVED SURFACES		HOLLOW SPRING PINS
TYPE A Full-length grooves. Used for general purpose fastening. **TYPE B** Grooves extend half length of the pin. Used as a hinge or linkage "bolt" but also can be employed for other functions in through-drilled holes where a locking fit over only part of the pin length is required. **TYPE E** Half-length groove section centered along the pin surface. Used as a cotter pin or in similar functions where an artificial shoulder or a locking fit over the center portion of the pin is required.	**TYPE A3** Full-length grooves with pilot section at one end to facilitate assembly. Expanded dimension of this pin is held to a maximum over the full grooved length to provide uniform locking action. It is recommended for applications subject to severe vibration or shock loads where maximum locking effect is required. **TYPE D** Reverse tapered grooves extend half the pin length. It is the counterpart of the Type B pin for assembly in blind holes. **TYPE U** Full-length grooves with pilot section at both ends for hopper feeding. Same as Type C.	**SPIRAL WRAPPED** **SLOTTED TUBULAR**

Fig. 9-2-3   Radial locking pins.

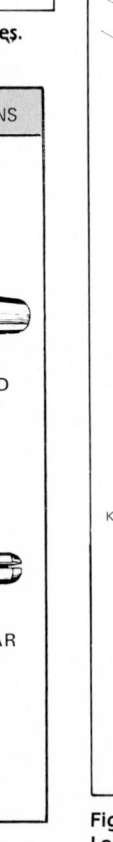

DOWEL APPLICATION

USED AS A SPACER

KEYING PULLEY TO SHAFT

STOP PIN

T HANDLE

TO PREVENT SHAFT ROTATION

COTTER PIN

HINGE IN LIGHT GAGE METAL

Fig. 9-2-5   Spring pin application. (Drive Lok.)

Fig. 9-2-6 Grooved pin application.

QUICK-RELEASE PINS

Commercially available quick-release pins vary widely in head styles, types of locking and release mechanisms, and range of pin lengths. See Fig. 9-2-7.

Quick-release pins may be divided into two basic types: *push-pull* and *positive-locking* pins. The positive-locking pins can be further divided into three categories: heavy-duty cotter pins, single-acting pins, and double-acting pins.

Push-Pull Pins

These pins are made with either a solid or a hollow shank, containing a detent assembly in the form of a locking lug, button, or ball backed up by some type of resilient core, plug, or spring. The detent member projects from the surface of the pin body until sufficient

(A) COMMON TYPES

DRAW-BAR HITCH PIN

CLEVIS-SHACKLE PIN   RIGID COUPLING PIN

TUBING LOCKPIN   ADJUSTMENT PIN   SWIVEL HINGE PIN

(B) APPLICATIONS

Fig. 9-2-7 Quick-release pins.

force is applied in assembly or removal to cause it to retract against the spring action of the resilient core and release the pin for movement.

Positive-Locking Pins

For some quick-release fasteners, the locking action is independent of insertion and removal forces. As in the case of push-pull pins, these pins are primarily suited for shear-load applications. However, some degree of tension loading usually can be tolerated without affecting the pin function.

Positive-locking pins can be divided into three categories: heavy-duty cotter pins, single-acting pins, and double-acting pins.

Heavy-duty cotter pins employ a forged, high-carbon-steel body to replace the conventional split-cotter construction. Locking action is provided by a tempered-steel snap ring mounted on the head of the pin.

Single-acting pins have locking action controlled by a plunger-actuated locking mechanism. In normal (locked) position, the locking element projects beyond the surface of the pin shank to provide a positive lock. Either ball- or pin-type locking elements may be employed.

Double-acting pins are a modification of single-acting types and have a bidirectional, spring-located plunger. Movement of the plunger in either direction along its barrel releases the locking balls.

Reference and Source Material

1. *Machine Design*, Fastening and joining reference issue, November 1981.

ASSIGNMENTS

See Assignments 2 through 5 for Unit 9-2 on page 204.

Review for Assignment

Unit 8-3   Washers
Unit 7-5   Assemblies in Section

UNIT 9-3
Retaining Rings

Retaining rings, or *snap rings*, are used to provide a removable shoulder to accurately locate, retain, or lock components on shafts and in bores of housings. See Fig. 9-3-1. They are easily installed and removed, and since they are usually made of spring steel, retaining rings have a high shear strength and impact capacity. In addition to fastening and positioning, a number of rings are designed for taking up end play caused by accumulated tolerances or wear in the parts being retained. In general, these devices can be placed into three categories which describe the type and method of fabrication: stamped retaining rings, wire-formed rings, and spiral-wound retaining rings.

STAMPED RETAINING RINGS

Stamped retaining rings, in contrast to wire-formed rings with their uniform cross-sectional area, have a tapered radial width which decreases symmetrically from the center section to the free ends. The tapered construction permits the rings to remain circular when they are expanded for assembly over a shaft or contracted for insertion into a bore or housing. This constant circularity ensures maximum contact surface with the bottom of the groove.

Stamped retaining rings can be classified into three groups: axially

Fig. 9-3-1   Retaining ring application. (Waldes Kohinoor, Inc.)

AXIAL ASSEMBLY RINGS

INTERNAL   EXTERNAL   INTERNAL   EXTERNAL   EXTERNAL   EXTERNAL

BASIC TYPES   INVERTED RINGS   HEAVY-DUTY RINGS

END PLAY RINGS

INTERNAL   EXTERNAL   INTERNAL   EXTERNAL   EXTERNAL   EXTERNAL

BOWED RINGS   BEVELED RINGS   LOCKING-PRONG RADIAL RINGS

SELF-LOCKING RINGS

CIRCULAR EXTERNAL RINGS   CIRCULAR INTERNAL RINGS   GRIP EXTERNAL RINGS   TRIANGULAR RETAINER

RADIAL ASSEMBLY RINGS

EXTERNAL   EXTERNAL   EXTERNAL   EXTERNAL

CRESCENT RING   E-RINGS   REINFORCED E-RING   INTERLOCKING RING

Fig. 9-3-2   Stamped retaining rings. (Machine Design, Vol. 53 No. 26, 1981.)

assembled rings, radially assembled rings, and self-locking rings which do not require grooves. *Axially assembled rings* slip over the ends of shafts or down into bores, while *radially assembled rings* have side openings which permit the rings to be snapped directly into grooves on a shaft.

Commonly used types of stamped retaining rings are illustrated and compared in Fig. 9-3-2.

WIRE-FORMED RETAINING RINGS

The *wire-formed retaining ring* is a split ring formed and cut from spring wire of uniform cross-sectional size and shape. The wire is cold-drawn or rolled into shape from a continuous coil or bar. Then the gap ends are cut into various configurations for ease of application and removal.

Rings are available in many cross-sectional shapes, but the most commonly used are the rectangular and circular, or round, cross sections.

SPIRAL-WOUND RETAINING RINGS

Spiral-wound retaining rings consist of two or more turns of rectangular material, wound on edge to provide a continuous crimped or uncrimped coil.

Reference and Source Material
1. *Machine Design*, Fastening and joining reference issue, Nov. 1981.

ASSIGNMENT

See Assignment 6 for Unit 9-3 on page 206.

Review for Assignment
Unit 7-5   Assemblies in Section

UNIT 9-4
Springs

Springs may be classified into three general groups according to their application.

Controlled Action Springs Controlled action springs have a well-defined function, or a constant range of action for each cycle of operation. Examples are valve, die, and switch springs.

Variable-Action Springs Variable-action springs have a changing range of action because of the variable conditions imposed upon them. Examples are suspension, clutch, and cushion springs.

Static Springs Static springs exert a comparatively constant pressure or tension between parts. Examples are packing or bearing pressure, anti-rattle, and seal springs.

TYPES OF SPRINGS

The type or name of a spring is determined by characteristics such as function, shape of material, application, or design. Figure 9-4-1 illustrates common springs in use. Figure 9-4-2 designates spring nomenclature.

Compression Springs
A *compression spring* is an open-coil, helical spring that offers resistance to a compressive form. It has a wide variety of uses and is made in various forms and from different shapes of wire, depending upon its application.

The most common form of this type is the same diameter through its entire length, and it is known as a *straight spring*. Tapered and cone-shaped springs are used quite extensively, and sometimes a combination of straight and tapered sections works to good advantage.

Compression Spring Ends Figure 9-4-3A shows the ends commonly used on compression springs.

Plain open ends are produced by straight cutoff with no reduction of helix angle. See Fig. 9-4-4. The spring should be guided on a rod or in a hole to operate satisfactorily.

Ground open ends are produced by parallel grinding of open-end coil springs. Advantages of this type of end are improved stability and a larger number of total coils.

Fig. 9-4-2   Spring nomenclature.

Fig. 9-4-1   Types of springs.

PLAIN ENDS

SQUARED AND GROUND ENDS

SQUARED OR CLOSED ENDS NOT GROUND

PLAIN ENDS — GROUND

(A)

MACHINE HALF LOOP OPEN

SHORT TWISTED LOOP

RAISED HOOK

FULL TWISTED LOOP

RECTANGULAR HOOK

FULL LOOP AT SIDE

REDUCED SIDE LOOP

DOUBLE TWISTED LOOP

V — HOOK

MACHINE CUTOFF

THREADED PLUG TO FIT PLAIN END SPRING

(B) END STYLES FOR EXTENSION SPRINGS

SPECIAL ENDS

DOUBLE TORSION

SHORT HOOK END

STRAIGHT OFFSET

HINGE ENDS

STRAIGHT TORSION

(C) END STYLES FOR TORSION SPRINGS

Fig. 9-4-3 End styles for helical springs.

Power Springs

Clock or Motor Type A *flat coil spring*, also known as a clock or motor spring, consists of a strip of tempered steel wound on an arbor and usually confined in a case or drum.

Flat Springs

Flat springs are made of flat material formed in such a manner as to apply force in the desired direction when deflected in the opposite direction.

Leaf Springs A *leaf spring* is composed of a series of flat springs nested together and arranged to provide approximately uniform distribution of stress throughout its length. Springs may be used in multiple arrangements, as shown in Fig. 9-4-5.

Belleville Springs Belleville springs are washer-shaped, made in the form of a short, truncated cone.

Belleville washers may be assembled in series to accommodate greater deflections, in parallel to resist greater forces, or in combination of series and parallel, as shown in Fig. 9-4-6.

SPRING DRAWINGS

On working drawings, a schematic drawing of a helical spring is recommended to save drafting time. See Fig. 9-4-7. As in screw-thread representation, straight lines are used in place of the helical curves. On assembly draw-

Plain closed ends are produced with a straight cutoff and with reduction of helix angle to obtain closed-end coils, resulting in a more stable spring.

Ground closed ends are produced by parallel grinding of closed-end coil springs, resulting in maximum stability.

Extension Springs

An *extension spring* is a close-coiled helical spring that offers resistance to a

pulling force. It is made from round or square wire.

Extension Spring Ends The end of an extension spring is usually the most highly stressed part. Thus, proper consideration should be given to its selection. The types of ends shown in Fig. 9-4-3B are most commonly used on extension springs. Different types of ends can be used on the same spring.

Torsion Springs

Springs exerting pressure along a path which is a circular arc, or, in other words, providing a torque, are called torsion springs, motor springs, power springs, etc. The term *torsion spring* is usually applied to a helical spring of round, square, or rectangular wire, loaded by torque.

The variation in ends used is almost limitless, but a few of the more common types are illustrated in Fig. 9-4-3C.

Torsion Bar Springs A torsion bar spring is a relatively straight bar anchored at one end, on which a torque may be exerted at the other end, thus tending to twist it about its axis. A torsion bar is sometimes favored because of its high efficiency in material utilization.

OPEN-END PLAIN

6.5

ONE COIL

LEFT-HAND HELIX

6.5 TOTAL COILS

6 TOTAL COILS

4 ACTIVE COILS

LEFT-HAND HELIX

CLOSED-END PLAIN

Fig. 9-4-4 Coil definitions.

Fig. 9-4-5 Leaf springs—multiple arrangements.

SERIES

PARALLEL

Fig. 9-4-6 Belleville spring installation.

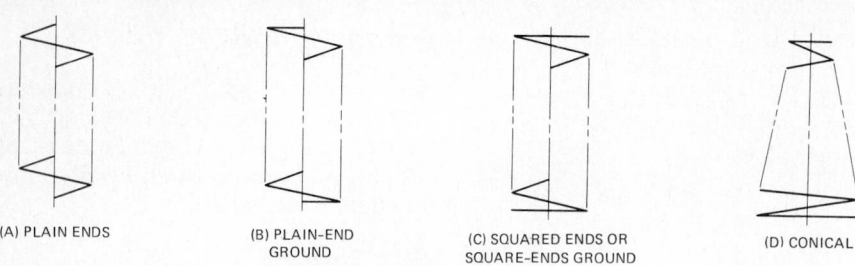

<table>
<tr><td>(A) PLAIN ENDS</td><td>(B) PLAIN-END
GROUND</td><td>(C) SQUARED ENDS OR
SQUARE-ENDS GROUND</td><td>(D) CONICAL</td></tr>
</table>

Fig. 9-4-7 Schematic drawing of compression springs.

ings, springs are normally shown in section, and either cross-hatching lines or solid black shading is recommended, depending on the size of the wire's diameter. See Fig. 9-4-8.

Dimensioning Springs The following information should be given on a drawing of a spring:

- Size, shape, and kind of material used in the spring
- Diameter (outside or inside)
- Pitch or number of coils
- Shape of ends
- Length
- Load and rate (not covered in this text)

EXAMPLE

ONE HELICAL TENSION SPRING 3.00 LG (OR NUMBER OF COILS), .50 ID, PITCH .25, 18 B & S GA SPRING BRASS WIRE

In using single-line representation, the dimensions should state the applicable size of material required to ensure correct interpretation on such features as inside diameter, outside diameter, and end loops. See Fig. 9-4-9.

SPRING CLIPS

Spring clips are a relatively new class of industrial fasteners. They perform multiple functions, eliminate the handling of several small parts, and thus reduce assembly costs. See Fig. 9-4-10.

The *spring clip* is generally self-retaining, requiring only a flange, panel edge, or mounting hole to clip to.

Basically, spring clips are light-duty fasteners and serve the same function as small bolts and nuts, self-tapping screws, clamps, spot welding, and formed retaining plates.

Dart-Type Spring Clips

Dart-shaped panel retaining elements have hips to engage within panel or component holes. The top or arms of the fastener can be formed in any shape to perform unlimited fastening functions.

<table>
<tr><td>(A) COMPRESSION</td><td>(B) EXTENSION</td><td>(C) TORSIONAL</td></tr>
</table>

Fig. 9-4-9 Dimensioning springs.

(A) LARGE SPRINGS

(B) SMALL SPRINGS

Fig. 9-4-8 Showing helical springs on assembly drawings.

(A) DART-TYPE SPRING CLIPS

(B) STUD RECEIVER CLIPS

(C) CABLE, WIRE AND TUBE CLIPS

(D) SPRING MOLDING CLIPS

(E) U-SHAPED, S-SHAPED AND C-SHAPED CLIPS

Fig. 9-4-10 Spring clips.

Stud Receiver Clips

There are three basic types of stud receivers: push-ons, tubular types, and self-threading fasteners. All are designed to make attachments to unthreaded studs, rivets, pins, or rods of metal or plastic.

Cable, Wire, and Tube Clips

These fasteners incorporate self-retaining elements for engaging panel holes or mounting on panel edges and flanges.

Spring-clip cable, wire, and tubing fasteners are front-mounting devices, requiring no access to the back of the panel.

Spring Molding Clips

Molding retaining clips are formed with legs that hold the clips to a panel and arms that positively engage the flanges of various sizes and shapes of trim molding and pull the molding tightly to the attaching panel.

U-Shaped, S-Shaped, and C-Shaped Spring Clips

These spring clips get their names from their shapes. The fastening function is accomplished by using inward compressive spring force to secure assembly components or provide self-retention after installation.

References and Source Material

1. General Motors Corporation.
2. The Wallace Barnes Company Limited.
3. *Machine Design*, Fastening and joining reference issue, Nov. 1981.

ASSIGNMENT

See Assignment 7 for Unit 9-4 on page 206.

UNIT 9-5
Rivets

STANDARD RIVETS

Riveting is a popular method of fastening and joining, primarily because of its simplicity, dependability, and low cost. A myriad of manufactured products and structures, both small and large, are held together by these fasteners. Rivets are classified as permanent fastenings, as distinguished from removable fasteners, such as bolts and screws.

SINGLE-RIVETED LAP JOINT DOUBLE-RIVETED LAP JOINT

(A) LAP JOINTS

SINGLE-RIVETED BUTT JOINT DOUBLE-RIVETED BUTT JOINT

(B) BUTT JOINTS

Fig. 9-5-1 Common riveted joints.

BUTTON HEAD   HIGH BUTTON HEAD   CONE HEAD

PAN HEAD   FLAT-TOP COUNTERSUNK HEAD   ROUND-TOP COUNTERSUNK HEAD

Fig. 9-5-2 Approximate sizes and types of large rivets .50 in. (12 mm) and up.

Basically, a *rivet* is a ductile metal pin which is inserted through holes in two or more parts, and the ends are formed over to securely hold the parts.

Another important reason for riveting is versatility, with respect to both the properties of rivets as fasteners and the method of clinching.

- Part materials: Rivets can be used to join dissimilar materials, metallic or nonmetallic, in various thicknesses.
- Multiple functions: Rivets can serve as fasteners, pivot shafts, spacers, electric contacts, stops, or inserts.
- Fastening finished parts: Rivets can be used to fasten parts that have already received a final painting or other finishing.

Riveted joints are neither watertight nor airtight, although such a joint may be attained at some added cost by using a sealing compound. The riveted parts cannot be disassembled for maintenance or replacement without knocking the rivet out and clinching a new one in place for reassembly. Common riveted joints are shown in Fig. 9-5-1.

Large Rivets

Large rivets are used in structural work of buildings and bridges. Today, however, high-strength bolts have almost completely replaced rivets in field connections because of cost, strength, and the noise factor. Rivet joints are of two types: butt and lapped. The more common types are shown in Fig. 9-5-2. In order to show the difference between *shop rivets* (rivets that are put in the structure at the shop) and *field rivets* (rivets that are used on the site), two types of symbols are used. In drawing shop rivets, the diameter of the rivet head is shown on the drawings. For field rivets, the shaft diameter is used. Figure 9-5-3 shows the conventional rivet symbols adopted by the American and Canadian Institutes of Steel Construction.

Fig. 9-5-3 Conventional rivet symbols.

Small Rivets

Design of small rivet assemblies is influenced by two major considerations:

1. The joint itself, its strength, appearance, and configuration

2. The final riveting operation, in terms of equipment capabilities and production sequence.

Fig. 9-5-4   Basic types of small rivets.

Types of Small Rivets (Fig. 9-5-4)

Semitubular This is the most widely used type of small rivet. The depth of the hole in the rivet, measured along the wall, does not exceed 112 percent of the mean shank diameter. The hole may be extruded (straight or tapered) or drilled (straight), depending on the manufacturer and/or rivet size.

Full Tubular This rivet has a drilled shank with a hole depth more than 112 percent of the mean shank diameter. It can be used to punch its own hole in fabric, some plastic sheets, and other soft materials, eliminating a preliminary punching or drilling operation.

Bifurcated (Split) The rivet body is sawed or punched to produce a pronged shank that punches its own hole through fiber, wood, or plastic.

Compression This rivet consists of two elements: the solid or blank rivet and the deep-drilled tubular member. Pressed together, these form an interference fit.

Design Recommendations (Fig. 9-5-5)

Select the Right Rivets Basic types are covered in Fig. 9-5-4. Rivet standards for all types but compression rivets have been published by the Tubular and Split Rivet Council.

Rivet Diameters The optimum rivet diameter is determined not by performance requirements but by economics—the costs of the rivet and the labor to install it. A recommended size for most applications is 1.4 times the square root of the thickness of the workpiece. On the other hand, the required rivet shank length is fixed by the amount of rivet material needed for clinching and the total material thickness. The rivet length-to-diameter ratio should not exceed 6:1.

Rivet Positioning The location of the rivet in the assembled product influences both joint strength and clinching requirements. The important dimensions are edge distance and pitch distance.

Edge distance is the interval between the edge of the part and the center line of the rivet.

The recommended edge distance for plastic materials, either solid or laminated, is between 2 and 3 diameters, depending on the thickness and inherent strength of the material.

Pitch distance—the interval between center lines of adjacent rivets—should not be too small. Unnecessarily high stress concentrations in the riveted material and buckling

Fig. 9-5-5   Design types for small rivets.

at adjacent empty holes can result if the pitch distance is less than 3 times the diameter of the largest rivet in the assembly (metal parts) or 5 times the diameter (plastic parts).

BLIND RIVETS

Blind riveting is a technique for setting a rivet without access to the reverse side of the joint. However, blind rivets may also be used in applications in which both sides of the joint are actually accessible.

Blind rivets are classified according to the methods with which they are set: pull-mandrel, drive-pin, and chemically expanded. See Fig. 9-5-6.

Design Considerations (Fig. 9-5-7)

Type of Rivet Selection depends on a number of factors, such as speed of assembly, clamping capacity, available sizes, adaptability to the assembly, ease of removal, cost, and structural integrity of the joint.

Joint Design Factors that must be known include allowable tolerances of rivet length versus assembly thickness, hole clearance, joint configuration, and type of loading.

Speed of Installation The fastest, most efficient installation is done with power tools—air, hydraulic, or electric. Manual tools, such as special pliers, can be used efficiently with practically no training.

In-Place Costs Blind rivets often have lower in-place costs than solid rivets or tapping screws because of low tooling investment, high installation speed, and single-operator requirements.

Loading A blind-rivet joint is usually in compression or shear, both of which the rivets can support somewhat better than tensile loading.

Material Thickness Some rivets can be set in materials as thin as .02 in. (0.5 mm) if the head is properly formed and shank expansion is carefully controlled during setting. If possible, however, it is best to form the blind head against the thicker sheet of a combination. Also, if one component is of compressible material, rivets with extra-large head diameter should be used.

Edge Distance The average recommended edge distance (Fig. 9-5-7, top) is twice the diameter of the rivet.

Spacing Rivet pitch (Fig. 9-5-7, top) should be 3 times the diameter of the rivet. Depending upon the nature of

Fig. 9-5-6   Basic types of blind rivets and methods of setting. (Machine Design, Vol. 53, No. 26, 1981.)

Fig. 9-5-7   Blind rivet design data. (Machine Design, Vol. 53, No. 26, 1981.)

load, it may be desirable to decrease or increase this distance.

Length The amount of length needed (Fig. 9-5-7, top) for clinching action varies greatly and depends on material being fastened, necessary strength, and the method of riveting. Most rivet manufacturers provide data on grip ranges of their rivets to simplify selection for the user.

Back-Up Clearance Full entry of the rivet is essential for tightly clinched joints. Sufficient back-up clearance (Fig. 9-5-7, top) must be provided to accommodate the full length of the unclinched rivet, A.

Blind Holes or Slots A useful application of a blind rivet is in fastening members in a blind hole (Fig. 9-5-7, top). At A, the formed head bears against the side of the hole only. As could be expected, this joint is not as strong as the other two (B and C), but the clinching action of the rivet head provides sufficient strength for light tension loads and moderate shear loads.

Riveted Joints Riveted cleat or batten holds a butt joint, A. The simple lap joint, B, must have sufficient material beyond the hole for strength. Excessive material beyond rivet hole C may curl up or vibrate or cause interference problems, depending on the installation (Fig. 9-5-7, middle).

Flush Joints Generally, flush joints are made by countersinking one of the sections and using a rivet with a countersunk head, A.

Another popular method of providing flush assembly and gaining additional bearing strength is to dimple the sheet by forming a conical projection on the back of the sheets with a die, C.

Weatherproof Joints A hollow-core rivet can be sealed by capping it, A; by plugging it, B; or by using both a cap and a plug, C. To obtain a true seal, however, a gasket or mastic should be used between the sections and perhaps under the rivet head. An ideal solution is to use a closed-end rivet, D (Fig. 9-5-7, middle).

Rubber, Plastic, and Fabric Joints Some plastics, such as reinforced molded Fiberglass, or polystyrene, which are

reasonably rigid, present no problem for most small rivets. However, when the material is very flexible or is a fabric, set the rivet as shown at A or B, with the upset head against the solid member. If this practice is not possible, use a back-up strip as shown at C (Fig. 9-5-7, middle).

Pivoted Joints There are a number of ways of producing a pivoted assembly. Three are shown in Fig. 9-5-7, middle.

Attaching Solid Rod When attaching a rod to other members, the usual practice is to pass the rivet completely through the rod (Fig. 9-5-7, bottom).

Attaching Tubing Attaching tubing is an application for which the blind rivet is ideally suited.

Joining Tubing This tubing joint is a common form of blind riveting, used for both structural and low-cost power transmission assemblies (Fig. 9-5-7, bottom).

Making Use of Pull-Up By judicious positioning of rivets and parts that are to be assembled with rivets, the setting force can sometimes be used to pull together unlike parts (Fig. 9-5-7, bottom).

Honeycomb Sections Inserts should be employed to strengthen the section and provide a strong joint.

References and Source Material
1. *Machine Design*, Fastening and joining reference issue, Nov. 1981.

ASSIGNMENT

See Assignment 8 for Unit 9-5 on page 206.

Review for Assignment
Unit 7-5   Assemblies in Section

UNIT 9-6
Welded Fasteners

The most common forms of welded fasteners are screws and nuts. Welded pins or unthreaded studs generally serve as locating or bearing surfaces, rather than as fasteners. In this sec-

tion, welded fasteners are grouped into *resistance-welded* threaded fasteners and *arc-welded* studs.

RESISTANCE-WELDED FASTENERS

Simply defined, a *resistance-welded fastener* is an externally or internally threaded metal part designed to be fused permanently in place by standard production welding equipment. Two methods of resistance welding are used to attach these fasteners (Fig. 9-6-1): projection welding and spot welding.

With either method, the fusion of the fastener to the metal-part surface is the result of the natural resistance of metal

Fig. 9-6-1   Basic methods for attaching resistance-welded fasteners.

Fig. 9-6-2   Application of resistance-welded fasteners. (Ohio Nut and Bolt Co.)

SPHERICAL   BUTTON   RING

RIB   PYRAMIDAL

Fig. 9-6-3 Basic types of weld projection.

to a controlled current under pressure.

The most common forms of weld fasteners are screws and nuts. Pins are also available but generally serve as a locating or bearing surface when applied as a welded component rather than as a fastener. See Fig. 9-6-2.

Design Considerations

Before fasteners can be used, three basic requirements must be met. See Figs. 9-6-3 and 9-6-4.

1. The materials to be joined, both part and fastener, must be suitable for resistance welding.
2. The parts to be welded must be portable enough to be carried to the welder.
3. Production volume should be great enough to justify tooling costs.

Figure 9-6-5 shows typical resistance-welded fasteners.

ARC-WELDED STUDS

There are two basic stud welding processes: *electric-arc* and *capacitor-discharge*.

Electric-Arc Stud Welding The more widely used stud welding process is a semiautomatic electric-arc process in which the heat for end welding the studs is from a motor, generator or

Use Projection-Weld Fasteners When	Use Spot-Weld Fasteners When
• Suitable projection welding equipment is available.	• Suitable rocker-arm welding equipment is available.
• Appearance is an important consideration. Projection welding does not mark the surface on the opposite side of the weld.	• Appearance of the part surface opposite the weld is not critical. Spot welding leaves a slight indentation from the electrode tips.
• Simultaneous welding of multiple fasteners is required.	• Other spot welds are being performed on parts of the assembly.
• Spacing between fasteners must be kept close.	• Length of production run without maintenance is not too important. Spot-welding electrode tips will mushroom to some extent in production welding. Shorter runs before refacing or redressing must be expected.
• Fasteners must be welded to part sections of varying thicknesses.	
• Fasteners must be welded to parts of unusual shape or a watertight weld joint is required.	• Dissimilar materials, such as aluminum, copper, or magnesium, are being welded.
• Welding fixtures can be used for easier locating or automatic feeding.	• Shape, size, or space requirements do not permit use of projection welded fasteners.
• Length of production run without maintenance is critical.	

Fig. 9-6-4 Guide to weld fastener selection.

Application Factors	A	B	C	D	E	F	G	H	I	J	K	L	M
Flat Surfaces	A Y	A Y	A Y	A Y	A Y	A Y	A Y	A Y	A Y	A Y	A Y	A Y	A
Curved Surfaces (concave)	A Y	A Y	A Y		A Y	A Y	A Y			A Y			
Round Surfaces (convex)	A Y	A Y	A Y	Y	A Y	A Y	Y A	A Y	A Y		Y	Y A	A Y
Tubing	A Y	A Y	A Y		A	A Y			Y A	A Y			Y
Channels	A Y	A Y	A Y	A	A Y	A	A	A				A	
Narrow Flanges	A Y	A Y	A Y	A	A Y	A	A Y	A				A	
Offset	A Y	A Y	A Y		A Y		A					A	
Wall Corners	A	A	A		A		A						
Blind Hole								A	A			A	
Wire	A Y	A Y		Y				Y					A
Through Hole	A Y	A Y	A Y	A Y	A	A	A Y	A	A	Y	Y A		A
Tension Against Weld		A			A Y		A	A	A Y	A	A	A Y	A
Hermetic Seal								A					A
Right Angle		Y		Y		Y			A Y				
Extra Thread										A	A	A	A
Bridging			A	A	A		A						
Dual Tapped				A									
Self Locating Pilot	A	A	A	A	A	A	A	A					
No Hole Required in Sheet		Y	Y	Y		Y	Y		A Y	A	A	Y A	A Y
Used with Keyhole Slot				Y									
Pilotless									A		A	A	

Spotweld Nuts — Projection Weld Nuts

Spotweld Screws and Pins (N O P Q R S T U V W X Y) — Projection Weld Screws and Pins

LEGEND

A Single Tab	F Button Projection	J Blind-Hole Flange	O Right-Angle Spade Pin	R Blind-Hole	V Button, Right-Angle Spade
B Targeted	G Four-Button Projection	K Through-Hole	P Keyhole-Slot, Right-Angle Spade Pin	S Spade	W Through-Hole Pin
C Double Tab	H Pilotless	L Tee-Shape	Q Through-Hole	T Hermetic Seal	X Blind-Hole Pin
D Dual Tapped	I Right-Angle Bracket	M Hermetic Seal		U Button-Projection, Blind-Hole	Y Spade Pin
E Dual Projection		N Right-Angle Spade			

Fig. 9-6-5 Resistance-welded fasteners.

transformer-rectifier supplying dc current which passes through an arc from the stud (electrode) to the plate (work).

Electric-arc stud welding is used to best advantage when the base plate is heavy enough to support the full strength of the welded fastener; however, lighter-gage materials are often arc-welded. As a general rule, to avoid burn-through, the plate thickness should be at least one-fifth the weld base diameter.

Capacitor-Discharge Stud Welding The second basic stud welding process derives its heat from an arc produced by a rapid discharge of stored electrical energy, with pressure applied during or immediately following the electrical discharge.

Design Considerations
In most instances, the thickness of the plate for stud attachment will determine the stud welding process. Electric-arc stud welding is generally used for fasteners .32 in. (8 mm) and larger. The two capacitor-discharge methods are used for smaller diameters.

References and Source Material
1. *Machine Design*, Fastening and joining reference issue, Nov. 1981.

ASSIGNMENT

See Assignment 9 for Unit 9-6 on page 208.

Review for Assignment
Unit 7-10   Partial Sections

UNIT 9-7
Adhesive Fastenings

Industrial designers and manufacturers are relying on adhesives more than ever before. They allow greater versatility in design, styling, and materials. They can also cut costs. However, as with any engineering tool, there are limitations as well as advantages.

ADHESION VERSUS STRESS

Adhesion is the force that holds materials together.

Fig. 9-7-1   Basic types of stress on bonded joints. (3M Co.)

Stress, on the other hand, is the force pulling materials apart. See Fig. 9-7-1. The basic types of stress in adhesives are:

1. *Tensile*. Pull is exerted equally over the entire joint. Pull direction is straight and away from the adhesive bond. All adhesive contributes to bond strength.
2. *Shear*. Pull direction is across the adhesive bond. The bonded materials are being forced to slide over one another.
3. *Cleavage*. Pull is concentrated at one edge of the joint and exerts a prying force on the bond. The other edge of the joint is theoretically under zero stress.
4. *Peel*. One surface must be flexible. Stress is concentrated along a thin line at the edge of the bond.

Resistance to stress is one reason for the rapid increase in the use of adhesives for product assembly. The following points elaborate on stress resistance and the other advantages of adhesives.

Advantages
1. Adhesives allow uniform distribution of stress over the entire bond area. See Fig. 9-7-2. This eliminates stress concentration caused by rivets, bolts, spot welds, and similar fastening techniques. Lighter, thinner materials can be used without sacrificing strength.
2. Adhesives can effectively bond dissimilar materials. Laminates of dissimilar material can often produce combinations superior in strength and performance to either adherent alone.
3. Continuous contact between mating surfaces effectively bonds and seals against many environmental conditions.

Fig. 9-7-2   Stresses caused by fasteners.

4. Adhesives maintain the integrity of the bonded material. They eliminate holes needed for mechanical fasteners and surface marks resulting from spot welding, brazing, etc.

Limitations
1. Adhesive bonding can be slow or require critical processing. This is particularly true in mass production. Some adhesives require heat and pressure, or special jigs and fixtures, to establish the bond.
2. Adhesives are sensitive to surface conditions. Special surface preparation may be required for optimum bonding results.
3. Some adhesive solvents present hazards. Special ventilation may be required to protect employees from toxic vapors.
4. Environmental conditions can reduce bond strength of some adhesives. Some do not hold well when exposed to low temperatures, high humidity, severe heat, chemicals, water, etc.

JOINT DESIGN

Joints should be specifically designed for use with structural adhesives. This is largely a matter of common sense and experience. Two basic factors should be the design guidelines. First, structural joints should be designed so that all the bonded area shares the load equally. Second, the joint configuration should be designed so that basic stress is primarily in shear or tensile, with cleavage and peel minimized or eliminated.

The following structural joints and their advantages and disadvantages illustrate some typical design alternatives. They are not, of course, the limit of possible adhesive bonded joints. See Fig. 9-7-3.

Lap Joints

Lap joints are most practical and applicable in bonding thin materials. The *simple lap joint* is offset. This can result in cleavage and peel stress under load when thin materials are used. A *tapered single lap joint* is more efficient than a simple lap joint. The tapered edge allows bending of the joint edge under stress. The *joggle lap joint* gives more uniform stress distribution than either the simple or tapered lap joint. The joint can be formed by simple metal-forming operations. The curing pressure is easily applied.

The *double-butt lap joint* gives more uniform stress distribution in the load-bearing area than the above joints. This type of joint, however, requires machining which is not always feasible with thinner-gage metals. *Double-scarf lap joints* have better resistance to bending forces than double-butt joints. This type of joint, however, also requires machining.

Angle Joints

Angle joints give rise to either peel or cleavage stress depending on the gage of the metal. Typical approaches to the reduction of cleavage are illustrated.

Butt Joints

A straight butt joint has poor resistance to cleavage. The following recessed butt joints are recommended: landed scarf tongue and groove, conventional tongue and groove, and scarf tongue and groove. The landed scarf tongue and groove joints act as stops which can control adhesive line thickness. Tongue and groove joints are self-aligning during assembly and act as a reservoir for void-filling-type adhesives.

Cylindrical Joints

The *T joint* and *overlap slip joint* are typical for bonding cylindrical parts such as tubing, bushings, and shafts. With adhesive bonding, all available contact area contributes to carrying the load.

Corner Joints—Sheet Metal

Corner joints can be assembled with adhesives by using simple supplementary attachments. This permits joining and sealing in a single operation. Typical designs are right-angle butt joints, slip joints, and right-angle support joints.

Corner Joints— Rigid Members

Corner joints, as in storm doors or decorative frames, can be adhesive-bonded. End lap joints are the simplest design type, although they require machining. Adhesives requiring pressure during curing may be utilized in such designs. Mortise and tenon joints are excellent from a design standpoint, but they also require machining. The mitered joint with a spline is best if both members are hollow extrusions.

Stiffener Joints

Deflection and flutter of thin metal sheets can be minimized with adhesive-bonded stiffeners. When such assemblies are flexed, peel stresses are exerted on the adhesives. If the flanges on the stiffening section can bend with the sheet, minimum peel stress on the bond will result. Increasing sheet gage or decreasing the gage of the stiffener flange will give equivalent results.

Reference and Source Material
1. 3M Company.

ASSIGNMENT

See Assignment 10 for Unit 9-7 on page 208.

Review for Assignment
Unit 7-5   Assemblies in Section

Fig. 9-7-3   Adhesive joint design. (3M Co.)

ASSIGNMENTS for Chapter 9

Assignment for Unit 9-1, Keys, Splines, and Serrations

1. On a B- or A3-size sheet, lay out the two fastener assemblies shown in Fig. 9-1-A or 9-1-B. The following fasteners are used:

    For Fig. 9-1-A
    - *Assembly A:* flat key
    - *Assembly B:* serrations

    For Fig. 9-1-B
    - *Assembly A:* square key
    - *Assembly B:* Woodruff key

    Refer to the Appendix and manufacturers' catalogs for sizes and use your judgment for dimensions not shown. Show the dimensions for the keyseats and serrations. Scale is full or 1:1.

Assignments for Unit 9-2, Pin Fasteners

2. On a B- or A3-size sheet, complete the pin assemblies shown in Fig. 9-2-A or 9-2-B, given the following information:

    For Fig. 9-2-A
    - *Assembly A.* Slotted tubular spring pins are used to fasten the cap and handle to the shaft. Scale is half size.
    - *Assembly B.* A clevis pin whose area is equal to the four rivets is used to fasten the trailer hitch to the tractor draw bar. Scale is half size.

    For Fig. 9-2-B
    - *Assembly A.* A type E grooved pin holds the roller to the bracket. A washer and cotter pin are used to fasten the bracket to the pushrod. Scale is 1:1.
    - *Assembly B.* A type A3 grooved pin holds the V-belt pulley to the shaft. Scale is 1:1.

Refer to manufacturers' catalogs for pin sizes and provide the complete information to order each fastener. Scale is as shown.

3. Prepare detail drawings of the parts shown in Fig. 9-2-C. Use your judgment for the scale and selection of views. Include a bill of material.

4. On an A3- or B-size sheet, make a two-view assembly drawing of the crane hook shown in Fig. 9-2-D. The hook is to be held to the U-frame with a slotted locknut. A spring pin is inserted through the locknut slots to prevent the nut from turning. A clevis pin with washer and cotter pin holds the pulley to the frame. Include on the drawing a bill of material. Scale is 1:1.

5. Prepare detail drawings of the parts in assignment 4. Use your judgment for the scale and selection of views.

ASSEMBLY A (FLAT KEY)

ASSEMBLY A (SQUARE KEY)

ASSEMBLY B (SERRATIONS)

Fig. 9-1-A   Key and serration fasteners.

ASSEMBLY B (WOODRUFF KEY)

Fig. 9-1-B   Key fasteners.

ASSEMBLY A (CABINET HANDLE)

ASSEMBLY A (CAM FOLLOWER)

Fig. 9-2-A Pin fasteners.

ASSEMBLY B (DRAW BAR HITCH)

ASSEMBLY B (V-BELT PULLEY)

Fig. 9-2-B Pin fasteners.

Fig. 9-2-C Wheel assembly.

Fig. 9-2-D Crane hook.

Assignments for Unit 9-3, Retaining Rings

6. Divide a B- or A3-size sheet into two sections, and draw the assemblies shown in Fig. 9-3-A or 9-3-B. Complete the assemblies by adding suitable retaining rings as per the information supplied below. Refer to manufacturers' catalogs and show on the drawing the catalog number for the retaining ring. Add ring and groove sizes. Scale is full or 1:1. Use your judgment for dimensions not shown.

For Fig. 9-3-A
- *Assembly A.* An external radial retaining ring mounted on the shaft is to act as a shoulder for the shaft support. An external axial retaining ring is required to hold the gear on the shaft.
- *Assembly B.* The plunger and punch are held into the punch holder by internal retaining rings.

For Fig. 9-3-B
- *Assembly A.* External self-locking retaining rings hold the roller shaft in position on the bracket.
- *Assembly B.* An external self-locking ring holds the plastic housing to the viewer case. An internal self-locking ring holds the lens in position.

Assignments for Unit 9-4, Springs

7. On a B- or A3-size sheet, lay out the two assembly drawings as shown in Fig. 9-4-A or 9-4-B. Complete the drawings from the information supplied below, and make detail drawings of the springs in the spaces provided. Use your judgment for sizes not given.

For Fig. 9-4-A
- *Assembly A.* The license plate holder is held to the frame of the car by a hinge. A torsion spring is required to keep the plate holder in position. The torsion spring is slipped over the hinge pin during assembly, and one end of the spring passes through the hole in the bumper. The other end of the spring is locked into the spring-retaining notch in the license plate holder. Scale is half or 1:2.
- *Assembly B.* Flat springs are positioned in openings C and D in the tape deck player. These springs hold the cassette against the bottom, and the locating pin positioned in the left side of the tape deck. Scale is half or 1:2.

For Fig. 9-4-B
- *Assembly A.* An extension spring con-

trols the lever. The spring is fastened to the neck in the pin and through the hole in the lever. Scale is 1:1.
- *Assembly B.* A compression spring mounted on the shaft of the handle provides sufficient pressure to hold the lever in position, thus maintaining the door against the panel. To open the door, the handle is pushed in and turned. This action compresses the spring and forces the lever away from the notch in the panel edge, thus permitting the lever to turn. Scale is 1:1.

Assignment for Unit 9-5, Rivets

8. On a B- or A3-size sheet, draw the two assembly drawings shown in Fig. 9-5-A or 9-5-B. Complete the drawings from the information supplied below. Refer to manufacturers' catalogs for rivet type and sizes, and on each assembly show the callout for the rivets. Use your judgment for sizes not given.

For Fig. 9-5-A
- *Assembly A.* Padlock brackets are riveted to the locker door and door frame with two blind rivets in each bracket. Scale is full.
- *Assembly B.* The roof truss is assembled in the shop with five evenly spaced Ø .50-in. (12-mm) rivets in each angle. Use quarter scale.

For Fig. 9-5-B
- *Assembly A.* The grill is held to the panel by four truss-head full tubular rivets. Scale is 1:1.
- *Assembly B.* The support is held to the plywood panel by drive rivets uniformly spaced on the gage lines. Two rivets hold the bracket to the support. Scale is 1:1.

ASSEMBLY A (EXTERNAL RETAINING RINGS)

ASSEMBLY B (INTERNAL RETAINING RINGS)

Fig. 9-3-A   Retaining ring fasteners.

ASSEMBLY A
(EXTERNAL SELF-LOCKING)

ASSEMBLY B
(EXTERNAL AND INTERNAL SELF-LOCKING)

Fig. 9-3-B   Retaining ring fasteners.

ASSEMBLY A (TORSION SPRING)

BUMPER
LICENSE PLATE HOLDER 10.00 X 5.75
SPRING RETAINING NOTCH
Ø.25 HINGE PIN
SPRING-RETAINING HOLE IN BUMPER
SECTION

Ø.25 LOCATING PIN
CASSETTE 3.50 WD
TAPE DECK PLAYER 7.88 X 2.00
SECTION A–A
SECTION B–B

ASSEMBLY B (FLAT SPRINGS)

Fig. 9-4-A   Spring fasteners.

70
Ø6 PIN
TRAVEL
LEVER 45 X 18

ASSEMBLY A (EXTENSION SPRING)

METRIC

PANEL
Ø17 WASHER
Ø16 SPRING HOLDER
Ø14 INSIDE
Ø26 KNOB
PUSH AND TURN
DOOR
SPACER
RETAINING RING
LEVER 44 X 20
NOTCH IN PANEL EDGE
24

ASSEMBLY B (COMPRESSION SPRING)

Fig. 9-4-B   Spring fasteners.

LOCKER DOOR 14 GA
Ø.25 HOLE
MAX CL .12
DOOR FRAME
PADLOCK BRACKET 12 GA
1.50
.75

ASSEMBLY A (BLIND RIVETS)

2L 4.00 X 4.00 X .38
.44 GUSSET
2L 3.50 X 2.50 X .38

ASSEMBLY B (LARGE STRUCTURAL RIVETS)

Fig. 9-5-A   Rivet fasteners.

PLASTIC GRILL 76 X 76 X 4
PLASTIC PANEL 4 THICK
50

ASSEMBLY A (SMALL RIVETS)

METRIC

PLYWOOD
RIVET GAGE LINES
SUPPORT 10 GA
26
6
38
23
BRACKET 11 GA

ASSEMBLY B (DRIVE RIVETS)

Fig. 9-5-B   Rivet fasteners.

Assignment for Unit 9-6, Welded Fasteners

9. On a B- or A3-size sheet, draw the two assemblies shown in Fig. 9-6-A or 9-6-B. Refer to manufacturers' catalogs and the Appendix for standard fastener components. Complete the drawings from the information supplied below. Use your judgment for sizes not given. Scale is full or 1:1.

For Fig. 9-6-A
- *Assembly A.* Two resistance-welded threaded fasteners, one on each side of the pipe, are required. The bracket drops over the fasteners, and lockwashers and nuts secure the bracket to the pipe.
- *Assembly B.* A leakproof attaching method (stud welding) is required to hold the adapter to the panel.

For Fig. 9-6-B
- *Assembly A.* A spot-weld nut is to be attached to the panel. A hole in the clamp permits a machine screw to fasten the pipe clamp to the nut.
- *Assembly B.* A right-angle bracket is to be fastened (projection welding) to the bottom plate. The vertical plate is secured to the bracket by a machine screw and lockwasher.

Assignment for Unit 9-7, Adhesive Fastenings

10. On a B- or A3-size sheet, draw the two adhesive-bonded assemblies shown in Fig. 9-7-A or 9-7-B. Complete the drawings from the information supplied below and the adhesive chart in the Appendix. List the adhesive product number and state the method of application you would recommend. Use your judgment for sizes not shown, and dimension the joint. Scale is to suit.

For Fig. 9-7-A
- *Assembly A.* The riveted joint shown is to be replaced by a tongue and groove joint. It must be fast-drying.
- *Assembly B.* The sheet-metal corner joint shown is to be replaced by a slip joint. It must be water-resistant.

For Fig. 9-7-B
- *Assembly A.* Three pieces of wood are to be assembled into the shape shown. Joint design has not been shown.
- *Assembly B.* The riveted joint shown is to be replaced by a joggle lap joint. Must meet specification requirements of MMM-A-121.

ASSEMBLY A (PIPE ATTACHMENT)

ASSEMBLY B (LEAKPROOF ATTACHMENT)

Fig. 9-6-A   Welded fasteners.

ASSEMBLY A (TAB ATTACHMENT)

ASSEMBLY B (RIGHT-ANGLE ATTACHMENT)

Fig. 9-6-B   Welded fasteners.

ASSEMBLY A (BUTT JOINT)

ASSEMBLY B (SLIP JOINT)

Fig. 9-7-A   Adhesive fastenings.

ASSEMBLY A (ANGLE JOINT)

ASSEMBLY B (LAP JOINT)

Fig. 9-7-B   Adhesive fastenings.

CHAPTER 10

Forming Processes

UNIT 10-1
Castings

FORMING PROCESSES[1]

When a component of a machine takes shape on the drawing board of the designer, the method of its manufacture may still be entirely open. The number of possible manufacturing processes is increasing day by day, and the optimum process is found only by carefully weighing technological advantages and drawbacks in relation to the economy of production.

The choice of the manufacturing process depends on the size and shape of the component. Manufacturing processes are therefore important to the engineer and drafter in order to properly design a part. They must be familiar with the advantages, disadvantages, costs, and machines necessary for manufacturing. Since the cost of the part is influenced by the production method, such as welding or casting, the designer must be able to choose wisely the method which will reduce the cost. In some cases it may be necessary to recommend the purchase of a new or different machine in order to produce the part at a competitive price.

This means the designer should design the part for the process as well as for the function. Most of all, unnecessarily close tolerances on nonfunctional dimensions should be avoided.

This chapter covers the following manufacturing processes: casting, forging and powder metallurgy. Forming by means of welding and stamping (punches and dies) is covered in Chaps. 11 and 27, respectively.

CASTING PROCESSES[2,3]

Casting is the process whereby parts are produced by pouring molten metal into a mold. A typical cast part is shown in Fig. 10-1-1. Casting processes for metals can be classified by either the type of mold or pattern or the pressure or force used to fill the mold. Conventional sand, shell, and plaster molds utilize a permanent pattern, but the mold is used only once. Permanent molds and die-casting dies are machined in metal or graphite sections and are employed for a large number of castings. Investment casting and the relatively new full-mold process involve both an expendable mold and an expendable pattern.

Fig. 10-1-1   Typical cast part. (General Motors Corp.)

Casting metals are usually alloys or compounds of two or more metals. They are generally classed as ferrous or nonferrous metals. *Ferrous* metals are those which contain iron, the most common being gray iron, steel, and malleable iron. *Nonferrous* alloys, which contain no iron, are those containing metals such as aluminum, magnesium, and copper.

Sand Mold Casting

The most widely used casting process for metals uses a permanent pattern of metal or wood that shapes the mold cavity when loose molding material is compacted around the pattern. This material consists of a relatively fine sand, which serves as the refractory aggregate, plus a binder.

A typical sand mold, with the various provisions for pouring the molten metal and compensating for contraction of the solidifying metal, and a sand core for forming a cavity in the casting are shown in Fig. 10-1-2. Sand molds consist of two or more sections: bottom (*drag*), top (*cope*), and intermediate sections (*cheeks*) when required. The sand is contained in flasks equipped with pins and plates to ensure the alignment of the cope and drag.

Molten metal is poured into the sprue, and connecting runners provide flow channels for the metal to enter the mold cavity through gates. Riser cavities are located over the heavier sections of the casting. A vent is usually added to permit the escape of gases

DRAG HALF OF PATTERN (WITH DOWEL HOLES)
MOLDING SAND
DRAG FLASK
ALIGNMENT PINS
MOLD BOARD

(A) STARTING TO MAKE THE SAND MOLD

PARTING SURFACE
BOTTOM BOARD

(B) AFTER ROLLING OVER THE DRAG

SPRUE PIN
RISER PIN
COPE FLASK
LUG

(C) PREPARING TO RAM MOLDING SAND IN COPE

POURING BASIN
RISER CAVITY
VENT
SPRUE

(D) REMOVING RISER AND GATE SPRUE PINS AND ADDING POURING BASIN

GATES
RUNNER
CORE PRINTS
CORE

(E) PARTING FLASKS TO REMOVE PATTERN AND TO ADD CORE AND RUNNER

(F) SAND MOLD READY FOR POURING

SPRUE
RISER
CORED HOLE
RUNNER

SPRUE, RISER, AND RUNNER TO BE REMOVED FROM CASTING.

(G) CASTING AS REMOVED FROM THE MOLD

Fig. 10-1-2 Sequence in preparing a sand casting.

CHAPLETS
CORE PRINTS
CORE
MOLD

Fig. 10-1-3 Core prints and chaplets. (General Motors Corp.)

SHELL-MOLD HALF
PATTERN PLATE
EJECTOR PINS

Fig. 10-1-4 Shell mold being stripped from pattern.

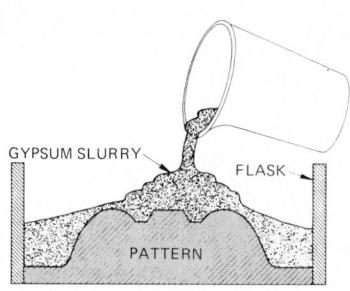

GYPSUM SLURRY
FLASK
PATTERN

Fig. 10-1-5 Pouring slurry over a plaster-mold pattern.

which are formed during the pouring of metal.

When a hollow casting is required, a form called a *core* is usually used. Cores occupy that part of the mold which is intended to be hollow in the casting. Cores, like molds, are formed of sand and placed in the supporting impressions or *core prints* in the molds. The core prints ensure positive location of the core in the mold and, as such, should be placed so that they support the mass of the core uniformly to prevent shifting or sagging. Metal core supports called *chaplets*, which are used in the mold cavity and which fuse into the casting, are sometimes used by the foundry in addition to core prints. See Fig. 10-1-3. Chaplets and their locations are not usually specified on drawings.

In producing sand molds, a metal or wooden pattern must first be made. The pattern, normally made in two parts, is slightly larger in every dimension than the part to be cast, to allow for shrinkage when the casting cools. This is known as *shrinkage allowance*, and the pattern maker allows for it by using a shrink rule for each of the cast metals.

Drafts or slight tapers are also placed on the pattern to allow for easy withdrawal from the sand mold. The parting line location and amount of draft are very important considerations in the design process.

In the construction of patterns for castings in which various points on the surface of the casting must be machined, sufficient excess metal should be provided for all machined

surfaces. Allowance depends on the metal used, the shape and size of the part, the tendency to warp, the machining method, and setup.

The molten metal is poured into the pouring basin and runs down the sprue to a runner and into the mold cavity.

When the metal has hardened, the sand mold is broken and the casting removed. Next the excess metal, gates, and risers are removed and remelted.

Shell Mold Casting[2]

The refractory sand used in shell molding is bonded by a thermostable resin that forms a relatively thin shell mold. A heated, reusable metal pattern plate (Fig. 10-1-4) is used to form each half of the mold by either dumping a sand-resin mixture on top of the heated pattern or by blowing resin-coated sand under air pressure against the pattern.

Plaster Mold Casting

Plaster of Paris and fillers are mixed with water and setting-control agents to form a *slurry*. This slurry is poured around a reusable metal or rubber pattern and sets to form a gypsum mold. See Fig. 10-1-5. The molds are then dried, assembled, and filled with molten (nonferrous) metals. Plaster mold casting is ideal for producing thin, sound walls. As in sand mold casting, a new mold is required for each casting. Castings made by this process have smoother finish, finer detail, and greater dimensional accuracy than sand castings.

Permanent Mold Casting

This process makes use of a metal mold, similar to a die, which is utilized to produce many castings from each mold. See Fig. 10-1-6. It is used to produce some ferrous alloy castings, but due to rapid deterioration of the mold caused by the high pouring temperatures of these alloys, and the attendant high mold cost, the process is confined largely to production of nonferrous alloy castings.

Investment Mold Casting

Investment castings have been better known in the past by the term *lost wax castings*. The term *investment* refers to the refractory material used to encase the wax patterns.

This process uses both an expendable pattern and an expendable mold. Patterns of wax, plaster, or frozen mercury are cast in metal dies. The molds are formed either by pouring a slurry of a refractory material around the pattern positioned in a flask or by building a thick layer of shell refractory on the pattern by repeated dipping into slurries and drying. The arrangement of the wax patterns in the flask method is shown in Fig. 10-1-7.

Full Mold Casting

The characteristic feature of the full mold process is the use of lost patterns made of foamed plastic. These are not

(1) THE DIE

(2) THE WAX PATTERN

(3) THE CLUSTER ASSEMBLY

(4) REFRACTORY MOLD

(5) FIRED MOLD

(6) THE CASTING

Fig. 10-1-7 Investment mold casting.

extracted from the mold, but are vaporized by the molten metal.

The full mold process is suitable for individual castings and for small series of up to five castings. The advantages it offers are obvious: it is very economical and reduces the delivery time required for prototypes, articles urgently needed for repair jobs, or individual large machine parts.

Centrifugal Casting

In the centrifugal casting process, commonly applied to cylindrical casting of either ferrous or nonferrous alloys, a permanent mold is rotated rapidly about the axis of the casting while a measured amount of molten metal is poured into the mold cavity. See Fig. 10-1-8. The centrifugal force is

used to hold the metal against the outer walls of the mold with the volume of metal poured determining the wall thickness of the casting. Rotation speed is rapid enough to form the central hole without a core. Castings made by this method are smooth, sound, and clean on the outside because impurities, being lighter than the metal, work toward the inner surface of the molten metal while rotating. The impurities can be removed by machining.

Continuous Casting

Continuous casting produces semifinished shapes such as uniform section rounds, ovals, squares, rectan-

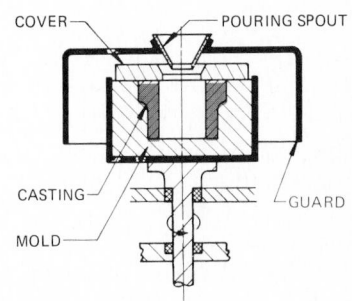

Fig. 10-1-8 Centrifugal mold equipment. (General Motors Corp.)

Fig. 10-1-6 Permanent mold. (General Motors Corp.)

gles, and plates. These shapes are cast from nearly all ferrous and nonferrous metals by continuously pouring the molten metal into a water-jacketed mold. The metal solidifies in the mold, and the solid billet exits continuously into a water spray. These sections are processed further by rolling, drawing, or extruding into smaller, more intricate shapes. Iron bars cast by this process are finished by machining.

Fig. 10-1-9 Die-casting machines.

Die Casting

One of the least expensive, fastest, and most efficient processes used in the production of metal parts is die casting. Die castings are made by forcing molten metal into a die or mold. Large quantities, accurately cast, can be produced with a die-casting die, thus eliminating or reducing machining costs. Many parts are completely finished when taken from a die. Since die castings can be accurate to within .001 in. (0.02 mm) of size, internal and external threads, gear teeth, and lugs can readily be cast.

Die casting has its limitations. Only nonferrous alloys can be die-cast economically because of the lack of a suitable die material to withstand the higher temperatures required for steel and iron.

Die-casting machines are of two types: the *submerged-plunger* type for low-melting alloys containing zinc, tin, lead, etc., and the *cold-chamber* type for high-melting nonferrous alloys containing aluminum and magnesium. See Fig. 10-1-9.

SELECTION OF PROCESS

Selection of the most feasible casting process for a given part requires an evaluation of the type of metal, the number of castings required, their shape and size, the dimensional accuracy required, and the casting finish required. When the casting can be produced by a number of methods, selection of the process is based on the most economical production of the total requirement. Since final cost of the part, rather than price of the rough casting, is the significant factor, the number of finishing operations necessary on the casting is also considered. Those processes that provide the closest dimensions, the best surface finish, and the most intricate detail generally require the smallest number of finishing operations.

A direct comparison of the capabilities, production characteristics, and limitations of several processes is indicated in Fig. 10-1-10.

DESIGN CONSIDERATIONS[3]

The advantages of using castings for engineering components are well appreciated by designers. Of major importance is the fact that they can produce shapes of any degree of complexity and of virtually any size.

PROCESS	METALS CAST	USUAL MASS RANGE	MINIMUM PRODUCTION QUANTITIES	RELATIVE SETUP COST	CASTING DETAIL FEASIBLE	MINIMUM THICKNESS in. (mm)	DIMENSIONAL TOLERANCES in. (mm)	SURFACE FINISH, RMS (μ in.)
SAND (Green, Dry, and Core) CO$_2$ Sand	All ferrous and nonferrous	Less than I lb. (0.5 kg) to several tons	3, without mechanization	Very low to high depending on mechanization	Fair	.12 to .25 (3 to 6)	± .03 (0.8)	350
						.10 to .25 (2.5 to 6)	± .02 (0.5)	250
SHELL	All ferrous and nonferrous	0.5 to 30 lb. (0.2 to 15 kg)	50	Moderate to high depending on mechanization	Fair to good	.03 to .10 (0.8 to 2.5)	± .015 (0.4)	200
PLASTER	Al, Mg, Cu, and Zn alloys	Less than I lb. to 3000 lb. (0.5 to 1350 kg)	1	Moderate	Excellent	.03 to .08 (0.8 to 2)	± .01 (0.2)	100
INVESTMENT	All ferrous and nonferrous	Less than I oz. to 50 lb. (30 g to 25 kg)	25	Moderate	Excellent	0.2 to .06 (0.5 to 1.5)	± .005 (0.1)	80
PERMANENT MOLD Metal Mold	Nonferrous and cast iron	I to 40 lb. (0.5 to 2C kg)	100	Moderate to high	Poor	.18 to .25 (4.5 to 6)	± .02 (0.5)	200
Graphite Mold	Steel	5 to 300 lb. (2 to 150 kg)	100			.25 (6)	± .03 (0.8)	200
DIE	Sn, Pb, Zn, Al, Mg, and Cu alloys	Less than I lb. to 20 lb. (0.5 to 10 kg)	1000	High	Excellent	.05 to .08 (1.2 to 2)	± .002 (0.5)	60

* Values listed are primarily for aluminum alloys, but data applies generally to other metals also.

† Depends on surface area.   Double if dimension is across parting line.

Fig. 10-1-10 General characteristics of casting processes.

Solidification of Metal in a Mold

While this is not the first step in the sequence of events, it is of such fundamental importance that it forms the most logical point to begin understanding the making of a casting.

Consider a few simple shapes transformed into mold cavities and filled with molten metal.

In a sphere, heat dissipates from the surface through the mold while solidification commences from the outside and proceeds progressively inward, in a series of layers. As liquid metal solidifies, it contracts in volume, and unless feed metal is supplied, a shrinkage cavity may form in the center. See Fig. 10-1-11.

The designer must realize that a shrinkage problem exists and that the foundry worker must attach risers to the casting or resort to other means to overcome it.

When the simple sphere has solidified further, it continues to contract in volume, so that the final casting is smaller than the mold cavity.

Consider a shape with a square cross section such as the one shown in Fig. 10-1-12A. Here again, cooling proceeds at right angles to the surface and is necessarily faster at the corners of the casting. Thus, solidification proceeds more rapidly at the corners.

The resulting hot spot prolongs solidification, promoting solidification shrinkage and lack of density in this area. The only logical solution, from the designer's viewpoint, is the provision of very generous fillets or radii at the corners. Additionally, the relative size or shape of the two sections forming the corner is of importance. If they are materially different, as in Fig. 10-1-12D, contraction in the lighter member will occur at a different rate from that in the heavier member. Differential contraction is the major cause of casting stress, warping, and cracking.

General Design Rules

Design for Casting Soundness See Fig. 10-1-13. Most metals and alloys shrink when they solidify. Therefore, the design must be such that all members of the parts increase in dimension progressively to one or more suitable locations where feeder heads can be placed to offset liquid shrinkage.

Fillet All-Sharp Angles See Fig. 10-1-14. Fillets have three functional purposes: to reduce stress concentration in the casting in service; to eliminate cracks, tears, and draws at reentry angles; and to make corners more moldable to eliminate hot spots.

Bring the Minimum Number of Adjoining Sections Together See Fig. 10-1-15. A well-designed casting brings the minimum number of sections together and avoids acute angles.

Fig. 10-1-12 Cooling effect on mold cavities filled with molten metal. (Meehanite Metal Corp.)

Fig. 10-1-14 Fillet all-sharp angles. (Meehanite Metal Corp.)

Fig. 10-1-13 Design members so that all parts increase progressively to feeder risers. (Meehanite Metal Corp.)

Fig. 10-1-15 Bring the minimum number of adjoining sections together. (Meehanite Metal Corp.)

Fig. 10-1-11 Circular mold cavity filled with molten metal.

Design All Sections as Nearly Uniform in Thickness as Possible Shrink defects and casting strains existed in the casting illustrated in Fig. 10-1-16. Redesigning eliminated excessive metal and resulted in a casting that was free from defects, was lighter in mass, and prevented the development of casting strains in the light radial veins.

Avoid Abrupt Section Changes—Eliminate Sharp Corners at Adjoining Sections See Fig. 10-1-17. The difference in the relative thickness of adjoining sections should be a minimum and not exceed a 2:1 ratio.

When a change of thickness must be less than 2:1, it may take the form of a fillet; where the difference must be greater, the form recommended is that of a wedge.

Wedge-shaped changes in wall thickness are to be designated with a taper not exceeding 1 in 4.

Design Ribs for Maximum Effectiveness See Fig. 10-1-18. Ribs have two functions: to increase stiffness and to reduce the mass. If too shallow in depth or too widely spaced, they are ineffectual.

Bosses and Pads Should Not Be Used Unless Absolutely Necessary Bosses and pads increase metal thickness, create hot spots, and cause open grain or draws. Blend these into the casting by tapering or flattening the fillets. Bosses should not be included in casting design when the surface to support bolts, etc., may be obtained by milling or countersinking.

Spoked Wheels See Fig. 10-1-19. A curved spoke is preferred to a straight one. It will tend to straighten slightly, thereby offsetting the dangers of cracking.

Use an Odd Number of Spokes A wheel having an odd number of spokes will not have the same direct tensile stress along the arms as one having an even number and will have more resiliency to casting stresses.

Wall Thicknesses Walls should be of minimum thickness, consistent with good foundry practice, and should provide adequate strength and stiffness. Wall thicknesses for different materials are as follows:

1. Walls of gray-iron castings and aluminum sand castings should not be less than .16 in. (4 mm) thick.
2. Walls of malleable iron and steel castings should not be less than .18 in. (5 mm) thick.
3. Walls of bronze, brass, or magnesium castings should not be less than .10 in. (2.4 mm) thick.

Parting Lines A *parting line* is a line along which the pattern is divided for molding, or along which the sections of a mold separate. Selection of a parting line depends on a number of factors:

- Shape of the casting
- Elimination of machining on draft surfaces
- Method of supporting cores
- Location of gates and feeders

Holes in Castings Small holes usually are drilled and not cored.

ORIGINAL DESIGN INCORRECT

CORRECT DESIGN

Fig. 10-1-16   Design all sections as nearly uniform in thickness as possible. (Meehanite Metal Corp.)

(A) BAD DESIGN

(B) FAIR DESIGN

(C) GOOD DESIGN

(D) BEST IN SOME CASES

L=4 (T−t)

(E) PROPORTIONS FOR CHANGING THICKNESS

Fig. 10-1-17   Avoid abrupt changes. (Meehanite Metal Corp.)

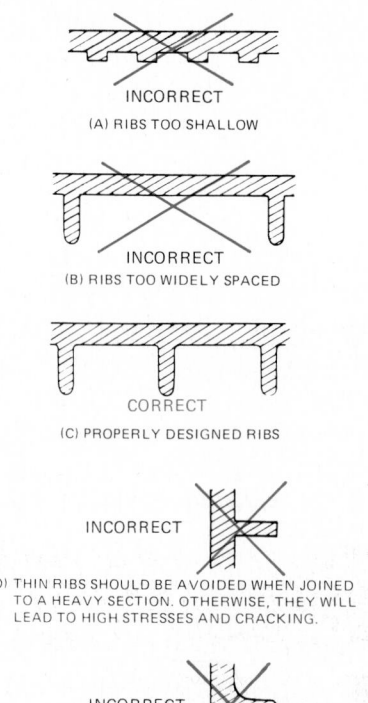

INCORRECT
(A) RIBS TOO SHALLOW

INCORRECT
(B) RIBS TOO WIDELY SPACED

CORRECT
(C) PROPERLY DESIGNED RIBS

INCORRECT
(D) THIN RIBS SHOULD BE AVOIDED WHEN JOINED TO A HEAVY SECTION. OTHERWISE, THEY WILL LEAD TO HIGH STRESSES AND CRACKING.

INCORRECT
(E) AS FAR AS POSSIBLE, JUNCTION BETWEEN RIBS AND MAIN CASTING SHOULD PREVENT ANY LOCAL ACCUMULATION OF METAL.

CORRECT
(F) RIBS SHOULD SOLIDIFY BEFORE THE CASTING SECTION THEY ADJOIN.

(G) T- AND H-SHAPED RIBBED DESIGNS HAVE THE ADVANTAGE OF UNIFORM METAL SECTIONS AND HENCE UNIFORM COOLING.

THICKNESS OF RIBS SHOULD APPROXIMATE 0.8 CASTING THICKNESS.

Fig. 10-1-18   Design ribs for maximum effectiveness. (Meehanite Metal Corp.)

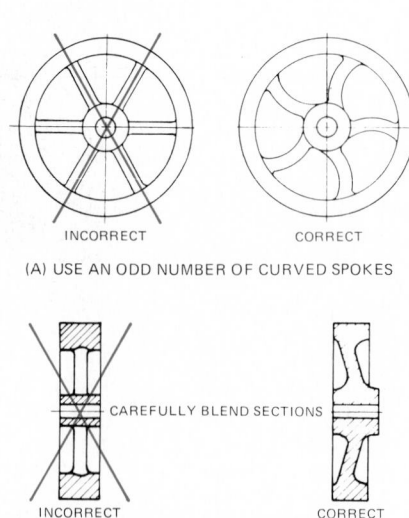

INCORRECT   CORRECT
(A) USE AN ODD NUMBER OF CURVED SPOKES

CAREFULLY BLEND SECTIONS

INCORRECT   CORRECT
(B) AVOID EXCESSIVE SECTION VARIATION

Fig. 10-1-19   Spoked-wheel design. (Meehanite Metal Corp.)

Drafting Practices[2]

It is important that a detail drawing give complete information on all cast parts, e.g.:

- Machining allowances
- Surface texture
- Draft angles
- Limits on cast surfaces that must be controlled
- Locating points
- Parting lines

On small, simple parts all casting information is included on the finished drawing. See Fig. 10-1-20. On more complicated parts, it may be necessary to show additional casting views and sections to completely illustrate the construction of the casting. These additional views should show the rough casting outline in phantom lines and the finished contour in solid lines.

Material In the selection of material for any particular application, the designer is influenced primarily by the physical characteristics such as strength, hardness, density, resistance to wear, mass, antifrictional properties, conductivity, corrosion resistance, shrinkage, and melting point.

Machining Allowance In the construction of patterns for castings in which various points on the surface of the casting must be machined, sufficient excess metal should be provided for all machined surfaces. Unless otherwise specified, Fig. 10-1-21 may be used as a guide to machine finish allowance.

Fillets and Radii Generous fillets and radii (rounds) should be provided on cast corners and specified on the drawing.

Casting Tolerances A great many factors contribute to the dimensional vari-

(A) WORKING DRAWING OF A CAST PART

(B) PATTERN DRAWING FOR THE CAST PART SHOWN IN (A)

Fig. 10-1-20 Cast part drawings.

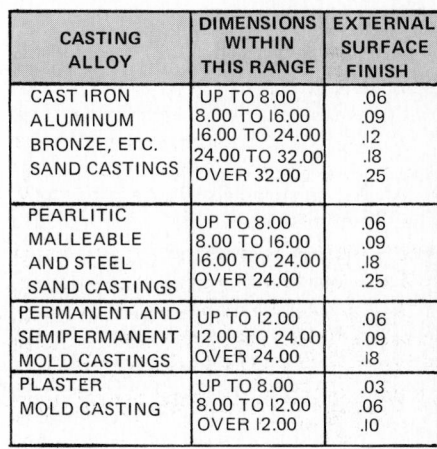

CASTING ALLOY	DIMENSIONS WITHIN THIS RANGE	EXTERNAL SURFACE FINISH
CAST IRON ALUMINUM BRONZE, ETC. SAND CASTINGS	UP TO 8.00 8.00 TO 16.00 16.00 TO 24.00 24.00 TO 32.00 OVER 32.00	.06 .09 .12 .18 .25
PEARLITIC MALLEABLE AND STEEL SAND CASTINGS	UP TO 8.00 8.00 TO 16.00 16.00 TO 24.00 OVER 24.00	.06 .09 .18 .25
PERMANENT AND SEMIPERMANENT MOLD CASTINGS	UP TO 12.00 12.00 TO 24.00 OVER 24.00	.06 .09 .18
PLASTER MOLD CASTING	UP TO 8.00 8.00 TO 12.00 OVER 12.00	.03 .06 .10

Fig. 10-1-21 Guide to machining allowance for castings.

TYPE OF CASTING	DIMENSIONS WITHIN THIS RANGE	STANDARD DRAWING TOLERANCE (±)
IRON AND ALUMINUM SAND CASTINGS	UP TO 8.00 8.00 TO 16.00 16.00 TO 24.00 24.00 TO 32.00 OVER 32.00	.03 .06 .07 .09 .12
PEARLITIC, MALLEABLE IRON AND STEEL SAND CASTINGS	UP TO 8.00 8.00 TO 16.00 16.00 TO 24.00 OVER 24.00	.03 .06 .09 .12
PERMANENT MOLD CASTING (SEMIPERMANENT MOLD CASTING)	UP TO 5.00 5.00 TO 12.00 12.00 TO 24.00 OVER 24.00	.03 .03 .06 .09
PLASTER MOLD CASTING	UP TO 4.00 4.00 TO 8.00 8.00 TO 12.00 OVER 12.00	.02 .02 .03 .06
CENTRIFUGAL PRECISION CASTING	UP TO .50 .50 TO 5.00 OVER 5.00	.02 .02 .02

Fig. 10-1-22 Guide to casting tolerances.

(A) DRAFT ANGLES

EXCESS METAL DUE
TO DRAFT

(B) DRAFT AND MACHINING ALLOWANCE

Fig. 10-1-23   Draft for removal of pattern from mold.

DATUM SURFACE B

DATUM SURFACE C

DATUM SURFACE A

Fig. 10-1-24 Casting datums.

ations of castings. However, the standard drawing tolerances specified in Fig. 10-1-22 can be satisfactorily attained in the production of castings.

Draft All casting methods require a draft or taper on all surfaces perpendicular to the parting line, to facilitate removal of the pattern and ejection of the casting. The permissible draft must be specified on the drawing, in either degrees of taper for each surface, inches of taper per inch of length, or millimeters of taper per millimeter of length.

Suitable draft angles for general use, for both sand and die castings, are 1° for external surfaces and 2° for internal surfaces, as shown in Fig. 10-1-23.

The drawing must always clearly indicate whether the draft should be added to, or subtracted from, the casting dimensions.

CASTING DATUMS[2]

It is recognized that in many cases a drawing is made of the fully machined end product; and casting dimensions, draft, and machining allowances are left entirely to the patternmaker or foundry worker. However, for mass-production purposes it is generally advisable to make a separate casting drawing, with carefully selected datums, to ensure that parts will fit into machining jigs and fixtures and will meet final requirements after machining. Under these circumstances, dimensioning requires the selection of two sets of datum surfaces, lines, or points—one for the casting and one for the machining—to provide common reference points for measuring, machining, and assembly. To select suitable datums, it will be necessary to know how the casting is to be made, where the parting line or lines are to be, and how the part is going to fit into machining jigs and fixtures.

The first step in dimensioning is to select a primary datum surface for the casting, which is sometimes referred to as the *base* surface, and to label it

Fig. 10-1-25   Machined casting drawing illustrating datum lines, set-up points, and finish mark symbols.

DATUM A. See Fig. 10-1-24. This primary datum should be a surface which meets the following criteria as closely as possible:

1. It must be a surface, or datum targets on a surface (see Fig. 10-1-25), which can be used as the basis for measuring the casting and which can later be used for mounting and locating the part in a jig or fixture, for the purpose of machining the finished part.
2. It should be a surface which will not be removed by machining, so that control of material to be removed is not lost, and can be checked at final inspection.
3. It should be parallel with the top of the mold, or parting line; that is, a surface which has no draft or taper.
4. It should be integral with the main

PRIMARY DATUM – PLANE A

SECONDARY DATUM – PLANE B

TERTIARY DATUM – PLANE C

Fig. 10-1-26 Datum planes and datum targets.

body of the casting, so that measurements from it to the main surfaces of the casting will be least affected by cored surfaces, parting lines, or gated surfaces.
5. It should be a surface, or target areas on a surface, on which the part can be clamped without causing any distortion, so that the casting will not be under a distortional stress for the first machining operation.
6. It should be a surface which will provide locating points as far apart as possible, so that the effect of any flatness error will be minimized.

The second step is to select two other planes to serve as secondary and tertiary surfaces. These planes should be at right angles to one another and to the primary datum surface. They probably will not coincide with actual surfaces, because of taper or draft, except at one point, usually a point adjacent to the primary datum surface. These are labeled DATUM B and DATUM C, respectively, as shown in Fig. 10-1-26.

In the case of a circular part, the end-view center lines may be selected as secondary and tertiary datums, as shown in Fig. 10-1-27. In this case, unless otherwise specified, the center lines represent the center of the outside or overall diameter of the part.

MACHINING DATUMS

The first step in dimensioning the machined or finished part is to select a primary datum surface for machining and to label it DATUM D. This surface is the first surface on the casting to be machined and is thereafter used as the datum surface for all other machining operations. It should be selected to meet the following criteria:

Fig. 10-1-27 Datums for circular casting.

1. It is generally preferable, though not essential, that it be a surface which is parallel to the primary casting datum surface.
2. It may be a large, flat, machined surface or several small areas of surfaces in the same or parallel planes.
3. If the primary casting datum surface is smooth and does not require machining, as in die castings, or if suitable target areas have been selected, the same surface may be used as the machining datum surface.
4. If the primary casting datum surface of sand castings appears to be the only suitable surface, it is recommended that three or four pads be provided, which can be machined to form the machining datum surface, as shown in Fig. 10-1-28.
5. When pads or small target areas are selected, they should be placed as far apart as possible and located where the part can be readily clamped in jigs or fixtures without distorting it or interfering with other machining operations.

The second step is to select two other surfaces to serve as secondary and tertiary datums. If these datum surfaces are required only for locating and dimensioning purposes, and not for clamping in a jig or fixture, some suitable datums other than flat, machined surfaces may be chosen. These could be the same datums as used for casting, if the locating point in each case is clearly defined and is not removed in machining. For circular parts, a hole drilled in the center, or a turned diameter other than the outside diameter, may provide suitable center lines for use as secondary datum surfaces, as shown in Fig. 10-1-29.

The third step is to specify the *datum-locating dimension*, that is, the dimension between each casting datum surface and the corresponding machining datum surface. See Fig. 10-1-28. There is never more than one

Fig. 10-1-28 Primary machining datum.

NOTES: 1. DATUMS E AND F ARE CENTER LINES FOR ØG
2. START MACHINING WITH DATUM D TO DIMENSION SHOWN

Fig. 10-1-29 Machining datums for circular parts.

such dimension from each casting datum surface.

Dimensions

When suitable datum surfaces have been selected, with datum-locating dimensions for the machined casting drawing, dimensioning may proceed, with dimensions being specified directly from the datums to all main surfaces. However, where it is necessary to maintain a particular relationship between two or more surfaces or features, regular point-to-point dimensioning is usually the preferred method. This will generally include all such items as thickness of ribs, height of bosses, projections, depth of grooves, most diameters and radii, and center distances between holes and similar features. Whenever possible, specify dimensions to surfaces or surface intersections, rather than to radii centers or nonexistent center lines.

Dimensions given on the casting drawing should not be repeated, except as reference dimensions, on the machined part drawing.

References and Source Material

1. American Iron and Steel Institute, "Principles of Forging Design."
2. General Motors Corporation.
3. Meehanite Metal Corporation.

ASSIGNMENTS

See Assignments 1 through 3 for Unit 10-1 on page 223.

Review for Assignments

Unit 7-8   Revolved and Removed Sections
Unit 7-4   Threads in Section
Unit 6-1   Detail Drawings

UNIT 10-2
Forging and Powder Metallurgy

FORGING

Forging consists of plastically deforming, either by a squeezing pressure or sharp blows, a cast or sintered ingot, a wrought bar or billet, or a powder-metal shape, to produce a desired shape with good mechanical properties. Practically all ductile metals can be forged. See Fig. 10-2-1.

Types of Forgings

Closed-Die Forgings Closed-die forgings are made by hammering or pressing metal until it conforms closely to the shape of the enclosing dies. Grain flow in the closed-die-forged parts can be oriented in the direction requiring greatest strength.

Three-dimensional control of the material to be forged requires a closed die, a simple and common form of which is the impression die.

(A) BILLET

(B) TONGHOLD IS FIRST FORGED

(C) PREFORMED IMPRESSION

(D) PREFORMED IMPRESSION

(E) PREFORMED IMPRESSION

(F) BLOCKING AND FINISHING

(G) AFTER TRIMMING, CRANKS ARE TWISTED INTO POSITION

Fig. 10-2-1 The forging of a crankshaft. (Wyman-Gordon Co.)

In the simplest example of impression-die forging (Fig. 10-2-2) the workpiece is cylindrical and is placed in the bottom-half die. On closing the top-half die, the cylinder undergoes elastic compression until its enlarged sides touch the side walls of the die impression. At this point, a small amount of excess material begins to form the flash between the two die faces. In the further course of die approach, this flash is gradually thinned.

The forging impression die gives control over all three directions, except when the die is similar to that shown in Fig. 10-2-2, and the deforming forging machine tool has an unlimited stroke (e.g., a hammer or hydraulic press). In the latter case, the die must be shaped to allow complete closing of the striking faces at the end of the stroke. See Fig. 10-2-3.

In practice, *closed-die forging* has become the term applied to all forging operations involving three-dimensional control. However, it is seen that the die is only closed by virtue of flash formation.

Forging dies can be divided into three main classes: single-impression, double-impression, and interlocking. See Fig. 10-2-4.

Single-impression dies have the impression of the desired forging entirely in one half of the die.

Fig. 10-2-2 Compression in impression dies.

Fig. 10-2-3 Forging die with flash gutter.

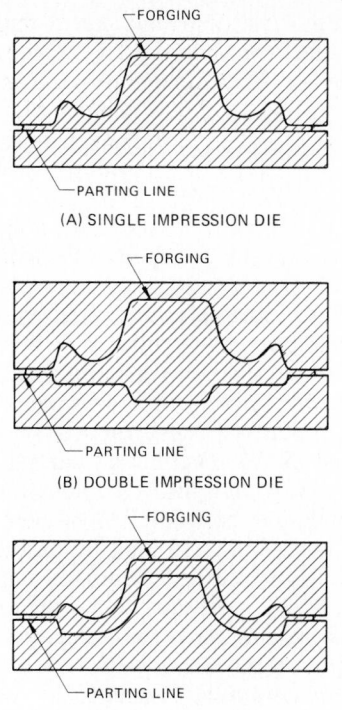

Fig. 10-2-4 Forging dies. (General Motors Corp.)

Double-impression dies have part of the impression of the desired forging sunk in each die in such a manner that no part of the die projects past the parting line into the other die. This type is the most common class of forging.

Trimming

Because the quantity of forging metal is generally in excess of the space in the die cavity, space is provided between the die surfaces for the escape of the excess metal. This space is called the *flash space*, and the excess metal which flows into it is called *flash*. The flash thickness is proportionate to the mass of the forging.

The flash is removed from forgings by trimming dies which are formed to the outline of the part. These dies shear or punch the forging out of the flash, leaving a scar or mark, the width and raggedness of which depends on the quality and state of wear of the dies. See Fig. 10-2-5.

BEFORE TRIMMING FLASH AFTER TRIMMING

Fig. 10-2-5 Flash trimming.

Design Considerations

Corner and Fillet Radii It is important in forging design to use correct radii where two surfaces meet. Corner and fillet radii on forgings should be sufficient to facilitate the flow of metal for sound forgings and to permit economical manufacture.

Stress concentrations resulting from abrupt changes in section thickness or direction are minimized by corner and fillet radii of correct size. Any radius larger than recommended will increase die life. Any radius smaller than recommended will decrease die life. See Fig. 10-2-6 for recommendations.

Sharp fillets cause the formation of cold shuts. In a forging, a *cold shut* is a lap where two surfaces of metal have folded against each other, forming an

H		R
OVER	TO AND INCL	
0	1.00 (25)	.06 (1.5)
1.00 (25)	1.50 (35)	.09 (2.5)
1.50 (35)	2.00 (50)	.12 (3)
2.00 (50)	3.00 (80)	.18 (4.5)

MIN CORNER RADII

H		R
OVER	TO AND INCL	
0	.30 (8)	R = H
.30 (8)	.50 (13)	R = $\frac{3H}{4}$

FILLET RADII FOR SMALL RIBS

FILLET RADII WHEN METAL IS CONFINED

DEPTH OF A FORGED RECESS SHOULD NOT EXCEED 0.67 X DIA

FILLET RADII WHEN METAL IS NOT CONFINED

Fig. 10-2-6 Corner and fillet radii. (General Motors Corp.)

undesirable flow of metal. A cold shut causes a weak spot that may be opened into a crack by heat treatment. Cold shuts are most likely to form at fillets in deep depressions or in deep sections, especially where the metal is confined. See Fig. 10-2-7. In these cases larger fillets are required, as shown in Fig. 10-2-6.

Draft Angle Draft is one of the first factors to be considered in designing a forged part. See Fig. 10-2-8. *Draft* is defined as the angle of taper given to the side walls of the die in order to facilitate removal of the forging. Where little or no draft is allowed, stripper or ejection mechanisms must be used. The usual amount of draft for exterior contours is 7° and for interior contours 10°.

1. STOCK BETWEEN UPPER AND LOWER DIES

2. METAL MOVED TO THE EDGE OF DIE DEPRESSION

3. METAL DROPPING INTO DIE DEPRESSION

4. METAL AT BOTTOM OF DIE DEPRESSION

5. METAL FLOW IS SLOWER AROUND A

6. METAL FLOW AT B CAUSING COLD SHUT

Fig. 10-2-7 Cold shut. (General Motors Corp.)

Fig. 10-2-8 Draft application.

FLAT PARTING

SIMPLE LOCKED PARTING

COMPOUND LOCKED PARTING

Fig. 10-2-10 Parting line application. (General Motors Corp.)

DEPTH OF DRAFT	DRAFT EQUIV FOR ANGLE OF		
	5°	7°	10°
.20 (5)	.018 (0.437)	.024 (0.614)	.035 (0.882)
.40 (10)	.035 (0.875)	.050 (1.228)	.070 (1.763)
.60 (15)	.052 (1.312)	.074 (1.842)	.106 (2.645)
.80 (20)	.070 (1.750)	.100 (2.456)	.140 (3.527)
1.00 (25)	.088 (2.187)	.123 (3.070)	.176 (4.408)

Fig. 10-2-9 Die draft equivalent. (General Motors Corp.)

Die draft equivalent is the amount of offset that results from draft. Figure 10-2-9 shows the draft equivalents for varying angles and depth of draft.

Parting Line The surfaces of dies that meet in forgings are the striking surfaces. The line of meeting is the parting line. The parting line of the forging must be established in order to determine the amount of draft and its location.

The location and the type of parting as applied to simple forgings are shown in Fig. 10-2-10.

Drafting Practices

In preparing forging drawings, it is important to consider drafting practices which may be peculiar to forgings, such as:

- Dimensioning
- Draft angles and parting lines
- Corner and fillet radii
- Forging tolerances
- Allowance for machining
- Material specifications
- Heat treatment
- Location of trademark, part number, and vendor specification

Dimensioning It is generally desirable to apply dimensions to the forged part of the depths of the die impressions. Draft is additive to these dimensions and should be expressed in degrees or linear dimensions.

When the depth of the die impression is located, only one dimension

Fig. 10-2-11 Dimensioning. (General Motors Corp.)

Fig. 10-2-12 Composite forging drawing. (General Motors Corp.)

should originate from the parting line. This surface should then be used to establish other dimensions, as shown in Fig. 10-2-11.

Allowance for Machining When a forging is to be machined, allowance must be made for metal to be removed.

Composite Drawings Generally, a forged part should be shown on one drawing with the forging outline shown in phantom lines, as in Fig. 10-2-12. Forging outlines for machining allowance should not be dimensioned unless the amount of finish cannot be controlled by the machining symbol.

Separate drawings for rough forgings should be made only when the part is complicated and the outline of the rough forging cannot be clearly visualized, or where the outline of the rough forging must be maintained for tooling purposes.

Where both the forging and machining drawings are shown on the same sheet, as in Fig. 10-2-13, place the headings FORGING DRAWING and MACHINING DRAWING directly under the corresponding views.

POWDER METALLURGY

Powder metallurgy is the process of making parts by compressing and sintering various metallic and nonmetallic powders into shape. See Fig. 10-2-14.

Dies and presses known as *briquetting machines* are used to compress the powders into shape. These bri-

MATERIAL XXXX   EST MASS 2.5 LB

ALL DRAFT ANGLES 7°
UNLESS OTHERWISE SPECIFIED

FILLETS AND ROUNDS R.10

(A) FORGING DRAWING

SECTION A-A

TOLERANCES— THICKNESS +.05 −.02
—MISMATCH .02
—DIE WEAR .035

(B) MACHINING DRAWING

Fig. 10-2-13 Separate forging and machining drawings.

Axial Variations Slots having a depth greater than one-fourth the axial length of the part require multiple punch action and result in higher production costs. See Fig. 10-2-15.

Corner Reliefs Corner reliefs can be molded or machined. A molded corner relief will save machining. See Fig. 10-2-16.

Reverse Tapers Reverse tapers cannot be molded. They must be machined. See Fig. 10-2-17.

25 PERCENT AXIAL
LENGTH A OR LESS

MORE THAN 25
PERCENT AXIAL
LENGTH A

PREFERRED   NOT RECOMMENDED

Fig. 10-2-15 Axial variations. (General Motors Corp.)

CAN BE MOLDED   MUST BE MACHINED

PREFERRED   NOT RECOMMENDED

Fig. 10-2-16 Corner relief. (General Motors Corp.)

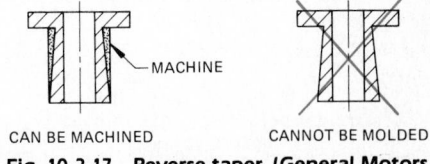

CAN BE MACHINED   CANNOT BE MOLDED

Fig. 10-2-17 Reverse taper. (General Motors Corp.)

quets or compacts are then sintered or heated in an atmosphere-controlled furnace, bonding the powdered materials.

Design Considerations

The following should be considered when powder metal parts are designed in order to realize the maximum benefits from the powder metallurgy process. This process is most applicable to the production of cylindrical, rectangular, or irregular shapes that do not have large variations in cross-sectional dimensions. Surface indentations or projections can be formed on either end or both ends of a part. Splines, gear teeth, axial holes, counterbores, straight knurls, serrations, slots, and keyseats present few problems.

Ejection from the Die The shape of the part must permit ejection from the die. The design requirements for some parts can be achieved only by subsequent machining, as in some corner relief designs, reverse tapers, holes at right angles to the direction of pressing, diamond knurls, and undercuts.

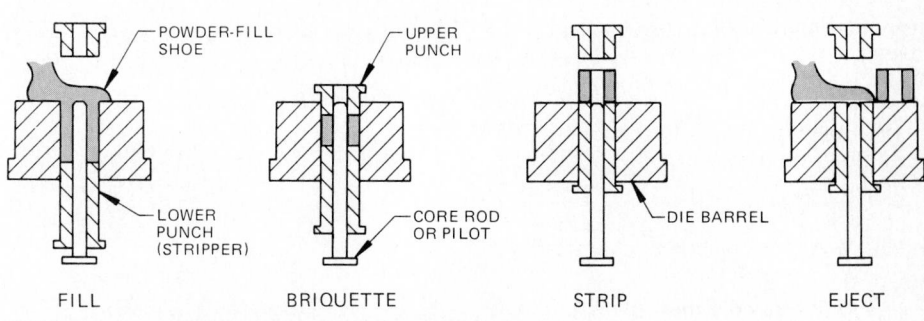

FILL   BRIQUETTE   STRIP   EJECT

Fig. 10-2-14 Compacting sequence. (General Motors Corp.)

Holes at Right Angles to the Direction of Pressing Right-angle holes must be machined. See Fig. 10-2-18.

Knurls Straight knurls can be molded; diamond knurls cannot. See Fig. 10-2-19.

Undercuts Undercuts must be machined. See Fig. 10-2-20.

Wall Thickness In general, sidewalls bordering a depression or hole should be a minimum of .03 in. (0.8 mm) thick. See Fig. 10-2-21.

Corners A fillet radius must be provided under the flange on a flanged part. It allows uniform powder flow in the die and produces a high-strength part. See Fig. 10-2-22.

Flanges A .06-in. (1.5-mm) minimum flange overhang is desired to provide longer tool life. See Fig. 10-2-23.

Blind Holes If a flange is opposite the blind end of the hole, the part must be modified to allow powder to fill in the die. See Fig. 10-2-24.

Changes in Cross Section Large changes in cross section should be avoided because they cause density variation. Warping and cracking are likely to occur during sintering. See Fig. 10-2-25.

Chamfers Care in the design of chamfers minimizes sharp edges on tools and improves tool life. See Fig. 10-2-26.

Holes A variety of odd-shaped holes can be produced economically by the powder metallurgy process. See Fig. 10-2-27.

References and Source Material
1. Frank Burbank, "Forging," *Machine Design*, vol. 37, no. 21, 1965.
2. General Motors Corporation.

ASSIGNMENTS

See Assignments 4 and 5 for Unit 10-2 on page 224.

Review for Assignments
Unit 6-1   Detail Drawings
Unit 7-1   Full Sections

Fig. 10-2-18   Right-angle hole. (General Motors Corp.)

Fig. 10-2-19   Knurls. (General Motors Corp.)

Fig. 10-2-20   Undercuts. (General Motors Corp.)

Fig. 10-2-21   Wall thickness. (General Motors Corp.)

Fig. 10-2-22   Corners. (General Motors Corp.)

Fig. 10-2-23   Flanges. (General Motors Corp.)

Fig. 10-2-24   Blind holes. (General Motors Corp.)

Fig. 10-2-25   Change in cross section. (General Motors Corp.)

Fig. 10-2-26   Chamfers with flat hand. (General Motors Corp.)

Fig. 10-2-27   Holes. (General Motors Corp.)

ASSIGNMENTS for Chapter 10

Assignments for Unit 10-1, Castings

1. On a B- or A3-size sheet, complete the detail drawing of the base for the adjustable shaft support assembly shown in Fig. 10-1-A. Cored holes are to be used for the shaft holes. Scale is full or 1:1.

2. On an A3- or B-size sheet, complete the detail drawing of the fork for the hinged pipe vise assembly shown in Fig. 10-1-B. Use your judgment for dimensions not given. Scale is 1:1.

3. On a B- or A3-size sheet, prepare both the casting and the machining drawings for the connector shown in Fig. 10-1-C. Draw a one-view full section, complete with the necessary dimensions for each drawing. Scale is full or 1:1.

Fig. 10-1-B   Pipe vise.

Fig. 10-1-A   Adjustable shaft support.

Fig. 10-1-C   Connector.

Fig. 10-2-A Bracket.

Assignments for Unit 10-2, Forging, Cold Heading, and Powder Metallurgy

4. On a B- or A3-size sheet, prepare a forging drawing of one of the parts shown in Fig. 10-2-A or 10-2-B. Scale is full or 1:1.

5. On a B- or A3-size sheet prepare a forging drawing for the wrench handle shown in Fig. 10-2-C. Scale is full or 1:1.

Fig. 10-2-B Open-end wrench.

HANDLE DETAIL

Fig. 10-2-C Wrench handle.

WRENCH HANDLE
MATL – FORGED STEEL
I REQUIRED

CHAPTER 11

Welding Drawings

CRANKS AND CRANKSHAFTS	
LINKS AND CLEVISES	
WHEELS	
LEVERS	

Fig. 11-1-1   A variety of weldments. (James F. Lincoln Arc Welding Foundation.)

UNIT 11-1
Designing for Welding[1,2]

The primary importance of welding is to unite various pieces of metal so that they will operate as a unit structure to support the loads to be carried. In order to design such a structure, which will be both economical and efficient, the drafter must have a knowledge of the basic principles of welding practice and an understanding of the advantages and limitations of the process.

In order to produce an economical and pleasing design, the designer should endeavor to utilize the method of construction which is clearly the most advantageous for the application under consideration. This may mean a combination of welding and bolting, or even the incorporation of pressings, forgings, or even castings where they may be advantageous. The possibility of using structural steel shapes and tubes should also be kept in mind. See Figs. 11-1-1 and 11-1-2.

Fig. 11-1-2   Design ideas for fabricated parts. (James F. Lincoln Arc Welding Foundation.)

Of more than 40 welding processes used in industry today, only a few are industrially important. Arc welding, gas welding, and resistance welding are the three most important types of welding.

The workpieces are melted along a common edge or surface so that their molten metal—and usually a filler metal also—is allowed to form a common pool or puddle. The pieces are fused when the puddle solidifies. See Figs. 11-1-3 and 11-1-4.

Gas welding, the most common form of which is oxyacetylene welding, gets its heat from the burning of flammable gases. This process is slow compared to other modern welding methods, so gas welding is normally confined to repair and maintenance work rather than being a major mass-production technique. See Fig. 11-1-5.

The major industrial welding process is arc welding, where heat is generated by an electric arc struck between a welding electrode, or rod, and the workpiece. The arc is quite hot, and melting and subsequent solidification of the weld metal occur very rapidly.

Resistance welding is also widely used, especially in mass-production work. As in arc welding, resistance

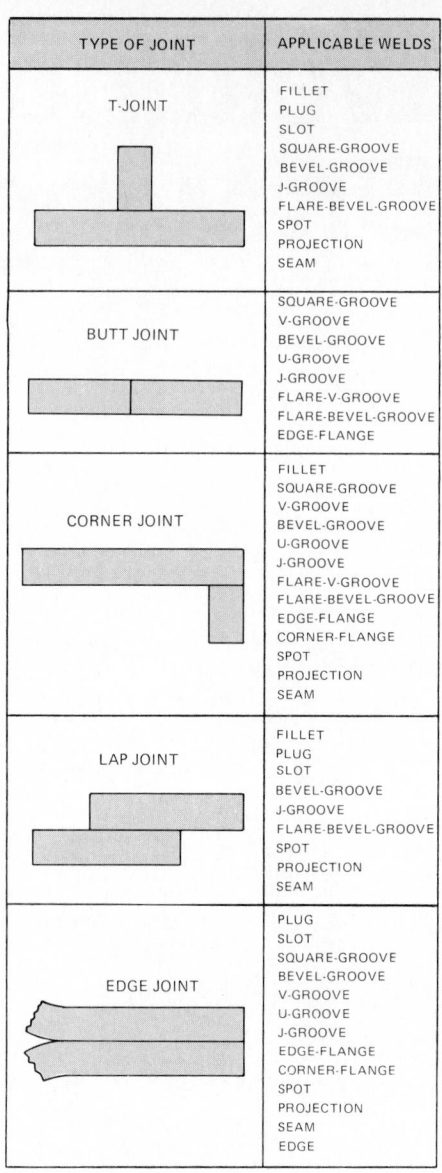

TYPE OF JOINT	APPLICABLE WELDS
T-JOINT	FILLET PLUG SLOT SQUARE-GROOVE BEVEL-GROOVE J-GROOVE FLARE-BEVEL-GROOVE SPOT PROJECTION SEAM
BUTT JOINT	SQUARE-GROOVE V-GROOVE BEVEL-GROOVE U-GROOVE J-GROOVE FLARE-V-GROOVE FLARE-BEVEL-GROOVE EDGE-FLANGE
CORNER JOINT	FILLET SQUARE-GROOVE V-GROOVE BEVEL-GROOVE U-GROOVE J-GROOVE FLARE-V-GROOVE FLARE-BEVEL-GROOVE EDGE-FLANGE CORNER-FLANGE SPOT PROJECTION SEAM
LAP JOINT	FILLET PLUG SLOT BEVEL-GROOVE J-GROOVE FLARE-BEVEL-GROOVE SPOT PROJECTION SEAM
EDGE JOINT	PLUG SLOT SQUARE-GROOVE BEVEL-GROOVE V-GROOVE U-GROOVE J-GROOVE EDGE-FLANGE CORNER-FLANGE SPOT PROJECTION SEAM EDGE

Fig. 11-1-4   Basic welding joints.

(A) BEND WHERE POSSIBLE

AVOID   BEND IF POSSIBLE

AVOID   STOCK T-BAR

(B) USE STANDARD FORMS

AVOID   PREFERRED

(C) AVOID MACHINING OF WELD METAL

AVOID   PREFERRED

(D) USE THICK PADS WHEN MACHINING PADS ARE REQUIRED.

Fig. 11-1-3   Preferred welding design.

METAL OR ALLOY	GAS	ARC
ALUMINUM		
—COMMERCIALLY PURE	X	X
—Al-Mn ALLOY	X	X
BRASS, COMMERCIAL	X	
BRONZE, COMMERCIAL	X	
COPPER (DEOXIDIZED)	X	
IRON		
—GRAY AND ALLOY	X	
—MALLEABLE		
LEAD	X	
MAGNESIUM ALLOYS	X	
NICKEL AND NICKEL ALLOYS	X	X
STEELS, CARBON		
—LOW AND MEDIUM CARBON	X	X
—HIGH CARBON		X
—TOOL STEEL		
STEEL, CAST	X	X
STEELS, STAINLESS		
—CHROMIUM		X
CHROMIUM-NICKEL	X	X

Fig. 11-1-5   Weldability of various metals and alloys.

welding employs electricity. But no arc is generated. Instead, heat is created from resistance losses as a high-amperage current is sent across a joint between two mating surfaces.

DESIGN OF WELDED STRUCTURES

In designing machine frames and similar structures for fabrication by welding, the main considerations, apart from ensuring that the part will fulfill its intended function, are usually confined to the necessity for evolving an article which will be pleasing in appearance and which can be economically produced. In this respect the drafter has more scope for the application of his or her inventiveness than in other branches of welding design. The drafter should try to avoid being unduly influenced by the design principles which have been developed for other methods of construction. For example, in designing machinery parts for fabrication, especially when they are intended to replace or supersede castings and forgings, it is generally essential for the drafter to avoid any tendency to design on the basis of making the weldment look like a casting for forging.

Weight (Mass) Saving

When castings are to be superseded by weldments, the higher labor costs of the weldment must be offset by simplifying the design and reducing the weight or mass. With a casting some extra thickness usually has been provided to allow for defective metal and maybe for shifting cores. With steel there is practically no risk of defective material so that this surplus can be eliminated. Moreover, since cast iron has less than half the tensile strength of steel, the mass or size of a steel part can be reduced proportionately. For example, for the same overall dimensions, because of the higher stresses that can be allowed, a steel section need be no more than half the thickness of a cast-iron one. This is shown in the drawing of a pump base in Fig. 11-1-6.

Conclusion

To a large extent, the ultimate cost of the job is usually the yardstick

(A) ORIGINAL PUMP BASE SUBJECT OF COST STUDY

(B) ALTERNATIVE BASE DESIGNS

Fig. 11-1-6   Design of pump base.

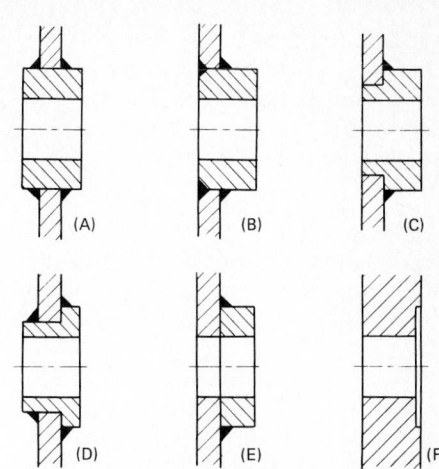

Fig. 11-1-7   Various types of boss attachments.

by which the advantages of any type of construction are measured. The drafter should, therefore, review those factors which are contributory to the cost of a weldment. Although the cost of steel is low compared with that of cast iron or cast steel, and generally it is possible to use less metal in a weldment than in an equivalent casting, it is essential to remember that there are more operations involved in the production of a weldment than there are in the case of a casting. The plate or section must be prepared for welding, the various components must then be assembled and fitted; finally, there is the actual welding which may be followed by stress relieving.

Choice of Raw Materials
In this type of design, the drafter has a wide choice of raw materials, plates, structural shapes, forgings, tubes, castings, etc. Careful consideration of the function of the various components of the structure is desirable in order to enable the most suitable raw material to be selected to ensure efficiency, economy, and pleasing appearance. Steel plates will no doubt provide the basic element in the majority of cases, and by flamecutting there is no limit to the variety of shapes that can be produced.

Steel plate surfaces are usually flat and smooth enough to be used as seating or bolting surfaces without further machining. Moreover, where bearings in a plate are required or should a machined surface be considered desirable for the seating of bolt heads, collars, washers, etc., it is often not essential to weld on bosses such as would ordinarily be employed on a casting. By making the plate a little thicker than normal, the machined areas can be spot-faced into the plate surface. The spot-facing costs about the same as a boss machining operation, but the work in preparation and welding on of bosses is eliminated. See Fig. 11-1-7 for types of boss attachments.

References and Source Material
1. American Welding Society.
2. *Machine Design*, Fastening and joining reference issue, Nov. 1981.

ASSIGNMENT
See Assignment 1 for Unit 11-1 on page 242.

Review for Assignment
Unit 6-4   Bills of Material
Unit 6-6   Detailed Assembly Drawing

UNIT 11-2
Welding Symbols

The introduction of welding symbols enables the designer to indicate clearly the type and size of weld required to meet design requirements, and it is becoming increasingly important for the designer to specify the required type of weld correctly. Points which must be made clear are the type of weld, the joint preparation, the weld size, the root gap (if any), and the degree of penetration required. These points can be clearly indicated on the drawing by the welding symbol. See Fig. 11-2-1.

Welding symbols are a shorthand language. They save time and money and ensure understanding and accuracy. It is desirable that they should be a universal language; and for this reason the symbols of the American Welding Society, already well established, have been adopted. See Figs. 11-2-2 and 11-2-3.

(A) FILLET WELD

(B) GROOVE WELD

in. [mm]
D MIN = T + .30 [8]   D MAX = 2.2 X T_l
WHEN T = $\leq$ .62 [16] T_l = T
WHEN T = $\leq$ .62 [16] T_l MIN = .62 [16] OR T/2
WHICHEVER IS BIGGER

(C) PLUG AND SLOT WELD

Fig. 11-2-2   Basic weld terminology.

The use of the words *far side* and *near side* in the past has led to confusion because when joints are shown in section, all welds are equally distant from the reader, and the words *near* and *far* are meaningless. In the present

Fig. 11-2-1   Welding symbols.

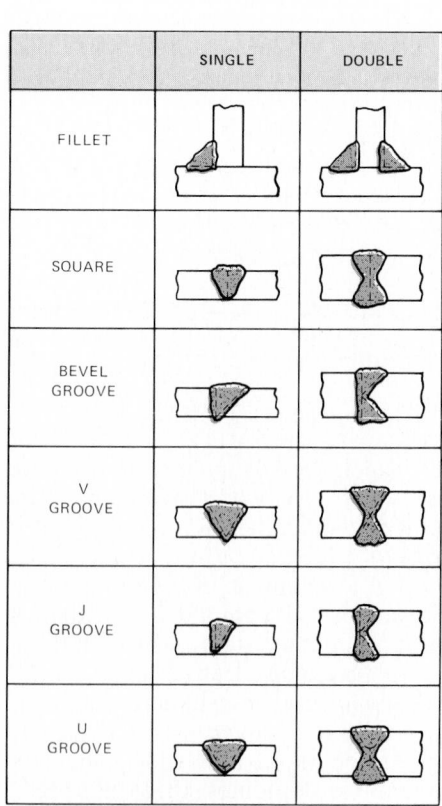

Fig. 11-2-3   Basic types of welds.

Fig. 11-2-4 Identification of arrow side and other side of joint.

system the joint is the basis of reference. Any joint, the welding of which is indicated by a symbol, will always have an arrow side and another side. Accordingly, the words *arrow side*, *other side*, and *both sides* are used here to locate the weld with respect to the joint. See Fig. 11-2-4.

The tail of the symbol is used for designating the welding specifications, procedures, or other supplementary information to be used in the making of the weld. See Fig. 11-2-5. The notation to be placed behind the tail of the symbol is to indicate the process, the type of filler metal to be used, and whether peening or root chipping is required. If notations are not used, the tail of the symbol may be omitted.

The use of letters can designate different welding and cutting processes. See Figs. 11-2-6 and 11-2-7.

(A) REFERENCE   (B) PROCESS

(C) NO SPECIFICATIONS REQUIRED

Fig. 11-2-5 Location of reference and processes on welding symbols.

DESIGNATION	WELDING PROCESS
CAW	Carbon-Arc Welding
CW	Cold Welding
DB	Dip Brazing
DFW	Diffusion Welding
EBW	Electron Beam Welding
EW	Electrostag Welding
EXW	Explosion Welding
FB	Furnace Brazing
FCAW	Flux Cord Arc Welding
FOW	Forge Welding
FRW	Friction Welding
FW	Flash Welding
GMAW	Gas Metal-Arc Welding
GTAW	Gas Tungsten-Arc Welding
IB	Induction Brazing
IRB	Infrared Brazing
IW	Induction Welding
LBW	Laser Beam Welding
OAW	Oxyacetylene Welding
OHW	Oxyhydrogen Welding
PAW	Plasma-Arc Welding
PEW	Percussion Welding
PGW	Pressure Gas Welding
RB	Resistance Brazing
RPW	Projection Welding
RSEW	Resistance Seam Welding
RSW	Resistance Spot Welding
SAW	Submerged Arc Welding
SMAW	Shielded Metal Arc Welding
SW	Stud Welding
TB	Torch Brazing
TW	Thermit Welding
USW	Ultrasonic Welding
UW	Upset Welding

Fig. 11-2-6 Designation of welding processes by letters.

DESIGNATION	CUTTING PROCESS
AAC	Air-Carbon Arc Carbon
AC	Arc Cutting
AOC	Oxygen Arc Cutting
CAC	Carbon Arc Cutting
FOC	Chemical Flux Cutting
MAC	Metal-Arc Cutting
OC	Oxygen Cutting
PAC	Plasma Arc Cutting
POC	Metal Powder Cutting

Fig. 11-2-7 Designation of cutting processes by letters.

Location Significance of Arrow

1. In the case of fillet and groove welding symbols, the arrow connects the welding symbol reference line to one side of the joint, and this side is considered the *arrow side* of the joint. The side opposite the arrow side of the joint is considered the *other side* of the joint.

2. When a joint is depicted by a single line on the drawing and the arrow of a welding symbol is directed to this line, the arrow side of the joint is considered the *near side* of the joint.

3. In the case of plug, slot, arc-spot, arc-seam, resistance-spot, resistance-seam, and projection welding symbols, the arrow connects the welding symbol reference line to the outer surface of one of the members of the joint at the center line of the desired weld. The member to which the arrow points is the *arrow-side* member. The remaining member of the joint is considered the *other member*.

4. When a joint is depicted as an area parallel to the plane of projection in the drawing and the arrow of a welding symbol is directed to that area, the arrow-side member of the joint is considered the *near member* of the joint.

Location of Weld Symbol with Respect to Joint

1. Welds on the arrow side of the joint are shown by placing the weld symbol on the bottom side of the reference line.

2. Welds on the other side of the joint are shown by placing the weld symbol on the top side of the reference line.

3. Welds on both sides of the joint are shown by placing the weld symbol on both sides of the reference line.

Use of Weld-All-Around Symbol

A weld extending completely around a joint is indicated by means of a weld-all-around symbol placed at the intersection of the reference line and the arrow. See Fig. 11-2-8.

Use of Field Weld Symbol

Field welds (welds not made in a shop or at the place of initial construction) are indicated by means of the field weld symbol placed at the intersection of the reference line and the arrow. The flag always points toward the tail of the arrow. See Fig. 11-2-9.

Combined Welding Symbols

For joints having more than one weld, a symbol is shown for each weld. See Fig. 11-2-10.

Finishing of Welds

Finishing of welds, other than cleaning, is indicated by suitable contour and finish symbols. See Fig. 11-2-11.

Fig. 11-2-8 Application of weld all-around symbol.

Fig. 11-2-9 Application of field weld symbol.

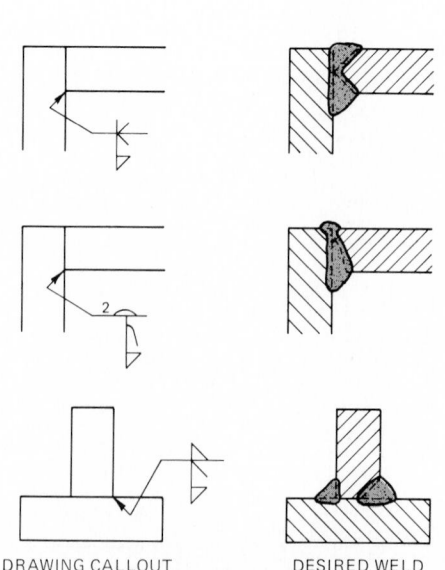

Fig. 11-2-10 Combined welding symbols.

The following finishing symbols indicate the method, not the degree, of finish:

- C = Chipping
- G = Grinding
- M = Machining
- R = Rolling
- H = Hammering

For methods of indicating degree of finish see Unit 5-7.

THE DESIGN OF WELDED JOINTS

Since loads are transferred from one member to another through the welds on a fabricated assembly, the type of joint and weld is specified by the designer. Figure 11-2-4 shows basic joint and weld types. Specifying the joint does not by itself describe the type of weld to be used. Several types of welds may be used for making a joint.

The fillet weld, requiring no groove penetration, is one of the most commonly used welds. Corner welds are also widely used in machine design. The corner-to-corner joint, shown in Fig. 11-2-12A, is difficult to assemble because neither plate can be supported

Fig. 11-2-11 Welding finishing symbols.

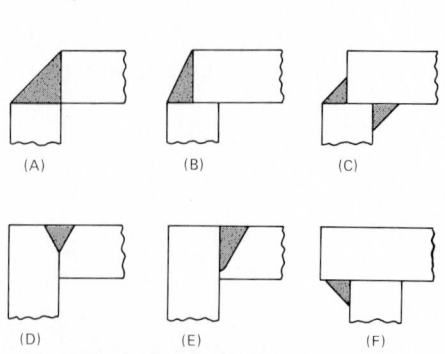

Fig. 11-2-12 Corner joints.

by the other. The joint also requires a larger amount of weld than the other joints illustrated. The corner joint shown in Fig. 11-2-12B is easy to assemble and requires half the amount of weld metal as the joint in Fig. 11-2-12A. However, by using half the weld size, but placing two welds, one outside, as in Fig. 11-2-12C, it is possible to obtain the same total throat as with the first weld. Only half the weld metal is required.

With thick plates, a partial-penetration groove joint, as in Fig. 11-2-12D, is used. This requires beveling. For a deeper joint, a J preparation, as in Fig. 11-2-12E, may be used in preference to a bevel. The fillet weld in Fig. 11-2-12F is out of sight and makes a neat and economical corner.

The size of the weld should always be designed with reference to the size of the thinner member. The joint cannot be made any stronger by using the thicker member for the weld size, and much more weld metal may be required, as illustrated in Fig. 11-2-13.

The designer is frequently faced with the question of whether to use fillet or groove welds. Here cost becomes a major consideration. The fillet welds in Fig. 11-2-14A are easy to apply and require no special plate preparation.

In comparison, the double-bevel groove weld in Fig. 11-2-14B has about one-half the weld area of the fillet welds. However, it requires extra preparation and the use of smaller-diameter electrodes with lower welding currents to place the initial pass without burning through. As plate thickness increases, this initial low-deposition region becomes a less important factor, and the higher cost factor decreases in significance.

Refer to Fig. 11-2-14C. It will be noted that the single-bevel groove weld requires about the same amount of weld metal as the fillet welds deposited in Fig. 11-2-14A. Thus, there is no apparent economic advantage. There are some disadvantages, though. The single-bevel joint requires bevel preparation and initially a lower deposit rate at the root of the joint. From a design standpoint, however, it offers a direct transfer of force through the joint, which means that it is probably better under fatigue loading. Although the illustrated full-strength fillet welds, having leg sizes equal to 75 percent of the plate thickness, would be sufficient, some codes have lower allowable limits for fillet welds and may require a leg size equal to the plate thickness. In this case, the cost of the fillet-welded joint may exceed the cost of a single-bevel groove in thicker plates. Also, if the joint is so positioned that the weld can be made in a flat position, a single-bevel groove

weld would be less expensive than if the fillet welds were specified. As can be seen in Fig. 11-2-15, one of the fillets would have to be made in the overhead position—a costly operation.

References and Source Material
1. American Welding Society.
2. The Lincoln Electric Company.

ASSIGNMENTS

See Assignments 2 and 3 for Unit 11-2 on page 242.

Review for Assignments
Unit 3-9   Enlarged Views
Unit 7-5   Direction of Section Lining

UNIT 11-3
Fillet Welds

FILLET WELD SYMBOLS

Figure 11-3-1 shows the fillet weld symbol and its relative position on the reference line. Figures 11-3-2 and 11-3-3 show applications of the fillet weld and appropriate symbols.

LOCATION SIGNIFICANCE	SYMBOL
ARROW SIDE	
OTHER SIDE	
BOTH SIDES	

Fig. 11-3-1   Fillet weld symbol and its location significance.

1. Dimensions of fillet welds are shown on the same side of the reference line as the weld symbol.

Fig. 11-2-13   Size of weld determined by thinner member.

W = 0.75 T

DOUBLE FILLET	DOUBLE-BEVEL GROOVE	SINGLE-BEVEL GROOVE
(A)	(B)	(C)

Fig. 11-2-14   Comparison between fillet and groove welds.

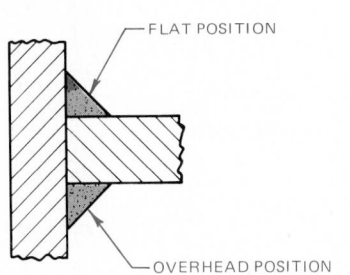

Fig. 11-2-15   In the flat position, a single-groove joint is less expensive than two fillet welds.

2. When both sides of a joint have same-size fillet welds, one or both may be dimensioned.

3. When both sides of a joint have different-size fillet welds, both are dimensioned.

4. When there appears on a drawing a general note governing the dimension of fillet welds, such as ALL FILLET WELDS .25 IN. UNLESS OTHERWISE NOTED, and all the welds have dimensions governed by the note, the dimension need not be shown on the weld symbols.

5. When the dimensions of either arrow side or other side or both welds differ from the dimensions given in the general note, either or both welds are dimensioned.

6. The size of a fillet weld is shown to the left of the weld symbol.

WELDING SYMBOLS

WELD 1

WELD ALL-AROUND ON ONE PLANE

NOTE: WELDING SYMBOL REFERS TO NEAR SIDE

WELDS 2 AND 5

WELD 3

WELD 4

WHEN THE FAR SIDE IS IDENTICAL TO THE NEAR SIDE, THE WELDING SHOWN FOR THE NEAR SIDE SHALL BE DUPLICATED ON THE FAR SIDE. WELDS 2 AND 5 INVOLVE SYMMETRY ABOUT AXIS X–X.

APPLIES TO WELDS 3 AND 4 WHICH ARE SYMMETRICAL ABOUT AXES Y–Y.

Fig. 11-3-2 Application of fillet welds for shaft support.

7. When a fillet weld is required to go completely around a part, a weld-all-around symbol is used.

8. The size of a fillet weld with unequal legs is shown in parentheses to the left of the weld symbol. Weld orientation is not shown by the symbol. It is shown on the drawing when necessary.

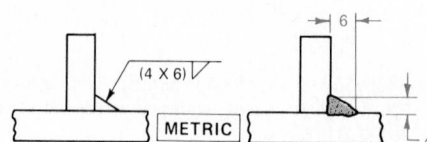

9. The length of a fillet weld, when indicated on the welding symbol, is shown to the right of the weld symbol.

10. Specific lengths of fillet welds may be indicated by symbols in conjunction with dimension lines.

11. The pitch (center-to-center spacing) of intermittent fillet welding is shown as the distance between centers of increments on one side of the joint. It is shown to the right of the length dimension.

12. Staggered intermittent fillet welds are shown with the weld symbols staggered.

13. Fillet welds that are to be welded approximately flat-faced without recourse to any method of finishing are shown by adding the flush-contour symbol to the weld symbol.

Fig. 11-3-3 Welded-steel shaft support.

DIMENSIONS IN INCHES				DIMENSIONS IN MILLIMETERS			
	Strength Design	Rigidity Design			Strength Design	Rigidity Design	
		50% Full-strength Weld	33% of Full-strength Weld			50% of Full-strength Weld	33% of Full-strength Weld
Plate Thickness	Full-strength Weld			Plate Thickness	Full-strength Weld		
Less than .25	.12	.12	.12	6	3	3	3
.25	.19	.19	.19	6	5	5	5
.31	.25	.19	.19	8	6	5	5
.38	.31	.19	.19	10	8	5	5
.44	.38	.19	.19	11	10	5	5
.50	.38	.19	.19	12	10	5	5
.56	.44	.25	.25	14	11	6	6
.62	.50	.25	.25	16	12	6	6
.75	.56	.31	.25	20	14	8	6
.88	.62	.38	.31	22	16	10	8
1.00	.62	.38	.31	25	16	10	8
1.12	.88	.44	.31	28	22	11	8
1.25	1.00	.50	.31	32	25	12	8
1.38	1.00	.50	.38	35	25	12	10
1.50	1.12	.56	.38	38	28	14	10

Fig. 11-3-4 Rule-of-thumb fillet-weld sizes where the strength of the weld metal matches the plate.

SIZE OF FILLET WELDS

Figure 11-3-4 gives the sizing of fillet welds for rigidity designs at various strengths and plate thicknesses, where the strength of the weld metal matches the plate.

In machine design work, where the primary design requirement is rigidity, members are often made with extra-heavy sections, so that improvement under load would be within very close tolerances. The question arises of how to determine the weld size for these types of rigidity designs.

A very practical method is to design the weld for the thinner plate, making it sufficient to carry one-third to one-half the carrying capacity of the plate. This means that if the plate were stressed one-third to one-half its usual value, the weld would be of sufficient size. Most rigidity designs are stressed much below these values. However, any reduction in weld size below one-third the full-strength value would give a weld too small in appearance for general acceptance.

EXAMPLE 1 What size fillet weld is required to match the strength of the fabricated design shown in Fig. 11-3-5A?
Solution With reference to Fig. 11-3-4, a full-strength weld is required. Thinner plate = .31 in. Fillet weld required = .25 in.

EXAMPLE 2 What size fillet weld is required to hold the rib to the plate shown in Fig. 11-3-5B? Weld design is for rigidity only, and only 33 percent of full-strength weld is required.
Solution Thinner plate = .31 in. With reference to Fig. 11-3-4, the weld size under rigidity design, 33 percent opposite .31 in., is .19 in.

Fig. 11-3-5 Calculating fillet-weld size.

References and Source Material
1. American Welding Society.
2. The Lincoln Electric Company.

ASSIGNMENT

See Assignment 4 for Unit 11-3 on page 243.

Review for Assignment

UNIT 11-4
Groove Welds

USE OF BREAK IN ARROW OF BEVEL AND J-GROOVE WELDING SYMBOLS

When a bevel or J-groove weld symbol is used, the arrow points with a definite break toward the member which is to be chamfered. In cases where the member to be chamfered is obvious, the break in the arrow may be omitted. See Figs. 11-4-1 through 11-4-3.

GROOVE WELD SYMBOLS

1. Dimensions of groove welds are shown on the same side of the reference line as the weld symbol.

2. When both sides of a double-groove weld have the same dimensions, one or both may be dimensioned.

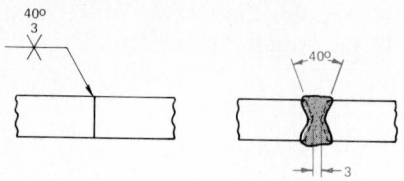

3. When both sides of a double-groove weld differ in dimensions, both are dimensioned.

4. When there appears on the drawing a general note governing the dimensions of groove welds, such as ALL V-GROOVE WELDS ARE TO HAVE A 60° ANGLE UNLESS OTHERWISE NOTED, groove welds need not be dimensioned.

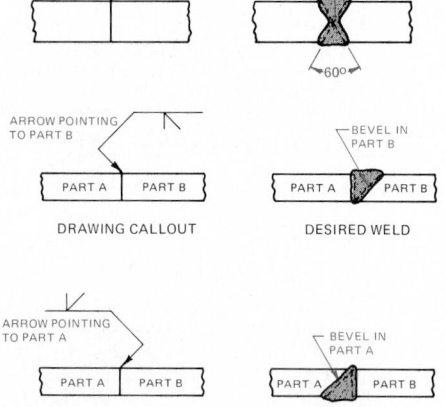

Fig. 11-4-2 Use of break in arrow.

LOCATION SIGNIFICANCE	SQUARE	V	BEVEL	U	J	FLARE-V	FLARE-BEVEL
ARROW SIDE							
OTHER SIDE							
BOTH SIDES							

Fig. 11-4-1 Basic groove welding symbols and their location significance.

NOTE 1 - 45° ALL POSITIONS, 30° FLAT AND OVERHEAD ONLY
NOTE 2 - 45° ALL POSITIONS, 20° FLAT AND OVERHEAD ONLY

Fig. 11-4-3   Spacing and material thickness for common butt joints.

5. For bevel and groove welds, the arrow points with a definite break toward the member being beveled.

6. When the dimensions of one or both welds differ from the dimensions given in the general note, both welds are dimensioned.

7. The size of groove welds is shown to the left of the weld symbol.

8. When the single-groove and symmetrical double-groove welds extend completely through the member or members being joined, the size of the weld need not be shown on the welding symbol.

9. When the groove welds extend only partly through the member being joined, the size of the weld is shown on the welding symbol.

10. The size of groove welds with specified root penetration, except square-groove welds, is given by showing both the depth of chamfering and the root penetration to the left of the weld symbol. The size of the square-groove welds is indicated by showing only the root penetration. The depth of chamfering and the root penetration, enclosed in brackets, are read in that order from left to right along the reference line.

11. The size of flare-groove welds is considered as extending only to the tangent points. The extension beyond the point of tangency is treated as an edge or lap joint. See Fig. 11-4-4.

(A) JOINING ROUNDS

(B) JOINING ROUNDED CORNERS

(C) JOINING ROUND AND FLAT—ONE SIDE

(D) JOINING ROUND AND FLAT—BOTH SIDES

(E) COMBINED WELDS

Fig. 11-4-4   Application of flare-V and flare-bevel welds.

12. Root opening of groove welds is the user's standard unless otherwise indicated. Root opening of groove welds, when not the user's standard, is shown inside the weld symbol.

13. Groove angle of groove welds is the user's standard, unless otherwise indicated. Groove angle of groove welds, when not the user's standard, is shown.

14. Groove radii and root faces of U- and J-groove welds are the user's standard, unless otherwise specified. When groove radii and root faces of U- and J-groove welds are not the user's standard, the weld is shown by a cross section, detail, or other data, with a reference thereto on the welding symbol, observing the usual locational significance.

15. Bead-type back and backing welds of single-groove welds are shown by means of the back or backing weld symbol.

16. Groove welds that are welded approximately flush without recourse to any method of finishing are shown by adding the flush-contour symbol to the weld symbol, observing the usual locational significance.

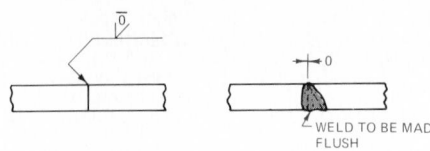

17. Groove welds that are to be made flush by mechanical means are shown by adding both the flush-contour symbol and the user's standard finish symbol to the weld symbol, observing the usual locational significance. C = chipping, G = grinding, M = machining, R = rolling, H = hammering.

GROOVE JOINT DESIGN

Figure 11-4-5 indicates that the root opening R is the separation between the members to be formed. A root opening is used for electrode accessibility to the base or root of the joint. The smaller the angle of bevel, the larger the root opening must be to get good fusion at the root. If the root opening is too small, root fusion is more difficult to obtain and smaller electrodes must be used, thus slowing down the welding process.

Fig. 11-4-5　Root openings.

Figure 11-4-6 indicates how the root opening must be increased as the included angle of the bevel is decreased. Backup strips are used on larger root openings. All three preparations are acceptable; all are conducive to good welding procedure and good weld quality. Selection, therefore, is usually based on cost.

Fig. 11-4-6　Root opening increases as the angle decreases.

Root opening and joint preparation will directly affect weld cost, and selection should be made with this in mind. Joint preparation involves the work required on plate edges prior to welding and includes beveling and providing a root face.

Using a double-groove joint in preference to a single-groove, as in Fig. 11-4-7, halves the amount of welding. This reduces distortion and makes possible alternating the weld passes on each side of the joint, again reducing distortion.

Fig. 11-4-7　Single-V weld uses twice as much weld material as double-V weld.

References and Source Material
1. American Welding Society.
2. Lincoln Electric Company.

ASSIGNMENT
See Assignment 5 for Unit 11-4 on page 244.

Review for Assignment

LOCATION SIGNIFICANCE	PLUG OR SLOT	SPOT OR PROJECTION	SEAM	BACK OR BACKING	SURFACING	FLANGE		MELT THRU
						EDGE	CORNER	
ARROW SIDE				GROOVE WELD SYMBOL				
OTHER SIDE				GROOVE WELD SYMBOL	NOT USED			
BOTH SIDES	NOT USED	NOT USED	NOT USED	NOT USED	NOT USED	NOT USED	NOT USED	NOT USED
NO ARROW SIDE OR OTHER SIDE SIGNIFICANCE	NOT USED			NOT USED	NOT USED	NOT USED	NOT USED	NOT USED

Fig. 11-5-1 Other basic welding symbols and their location significance.

UNIT 11-5
Other Basic Welds

In order for engineers to keep abreast of national and international thinking and to reduce the complexity inherent in providing symbols for a variety of ways of making the same type of weld, new symbols for spot and seam welds have been established. See Fig. 11-5-1. As a result, the old symbols shown in Fig. 11-5-2 should no longer be used. They have been included in this text in order to acquaint the drafter with welding symbols found on existing drawings.

PLUG WELDS

1. Holes in the arrow-side member of a joint for plug welding are specified by placing the weld symbol below the reference line.

DRAWING CALLOUT   DESIRED WELD

2. Holes in the other-side member of a joint for plug welding are indicated by placing the weld symbol above the reference line.

DRAWING CALLOUT   DESIRED WELD

3. The size of a plug weld is shown on the same side and to the left of the weld symbol.

DRAWING CALLOUT   DESIRED WELD

4. The included angle of countersink of plug welds is the user's standard, unless otherwise indicated. Included angle, when not the user's standard, is shown.

DRAWING CALLOUT   DESIRED WELD

5. The depth of filling of plug welds is complete unless otherwise indicated. When the depth of filling is less than complete, the depth of filling, in inches or millimeters, is shown inside the weld symbol.

DRAWING CALLOUT   DESIRED WELD

6. Pitch (center-to-center spacing) of plug welds is shown to the right of the weld symbol.

DRAWING CALLOUT

DESIRED WELD

ARC–SEAM OR ARC–SPOT	RESISTANCE SPOT	PROJECTION	RESISTANCE SEAM	FLASH OR UPSET
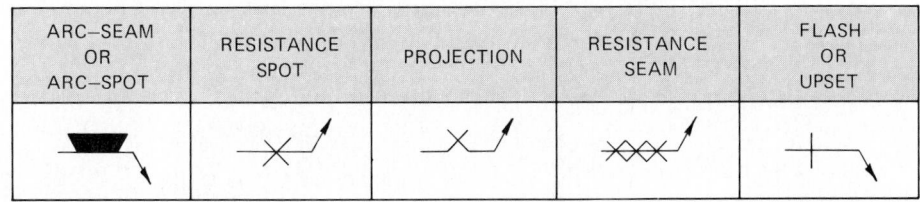				

USE PREFERRED SYMBOL WITH PROCESS REFERENCE IN THE TAIL

Fig. 11-5-2 Former welding symbols.

7. Plug welds that are to be welded approximately flush without recourse to any method of finishing are shown by adding the flush-contour symbol to the weld symbol.

8. Plug welds that are to be made flush by mechanical means are shown by adding both the flush-contour symbol and the user's standard finish symbol to the weld symbol.

DRAWING CALLOUT   DESIRED WELD

SLOT WELDS

1. Slots in the arrow-side member of a joint for slot welding are indicated by placing the weld symbol below the reference line. Slot orientation must be shown on the drawing.

2. Slots in the other-side member of a joint for slot welding are indicated by placing the weld symbol above the reference line.

3. Depth of filling of slot welds is complete unless otherwise indicated. When the depth of filling is less than

complete, the depth of filling, in inches or millimeters, is shown inside the welding symbol.

4. Length, width, spacing, included angle of countersink, orientation, and location of slot welds should be shown on the drawing or by a detail with reference to it on the welding symbol, observing the usual locational significance.

SEE DETAIL A

DETAIL A

DETAIL B

5. Slot welds that are to be welded approximately flush without recourse to any method of finishing are shown by adding the flush-contour symbol to the welding symbol.

6. Slot welds that are to be made flush by mechanical means are shown by adding both the flush-contour symbol and the user's standard finish symbol to the weld symbol.

SPOT OR PROJECTION WELDS

The symbol for all spot or projection welds is a circle, regardless of the welding process used. There is no attempt to provide symbols for different ways of making a spot weld, such as resistance, arc, and electron beam welding.

The symbol for a spot weld is a circle placed:

1. Below the reference line, indicating arrow side.
2. Above the reference line, indicating other side.
3. On the reference line, indicating that there is no arrow or other side.

Spot- or Projection-Weld Application

1. Dimensions of spot welds are shown on the same side of the reference line as the weld symbol, or on either side when the symbol is located astride the reference line and has no arrow-side or other-side significance. They are dimensioned by either the size or the strength. The size is designated as the diameter of the weld and is shown to the left of the weld symbol. The strength of the spot weld is desig-

(A) SPECIFYING DIAMETER OF SPOT

(B) SPECIFYING STRENGTH OF SPOT

nated in pounds (or newtons) per spot and is shown to the left of the weld symbol.

2. The pitch (center-to-center spacing) is shown to the right of the weld symbol.

3. When spot welding extends less than the distance between abrupt changes in the direction of the welding or less than the full length of the joint, the extent is dimensioned.

4. When projection welding is used, the projection-welding process shall be referenced in the tail of the welding symbol. The spot-weld symbol is placed either above or

ARROW SIDE SPOT WELD SYMBOL (GAS TUNGSTEN-ARC SPOT)

ARROW SIDE SPOT WELD SYMBOL (ELECTRON BEAM SPOT)

NO ARROW OR OTHER SIDE REFERENCE OR SIGNIFICANCE (RESISTANCE SPOT)

below (not on) the reference line to designate in which member the embossment is placed.

5. When the exposed surface of one member of a spot-welded joint is to be flush, that surface is indicated by adding the flush-contour symbol to the weld symbol, observing the usual locational significance.

6. When a definite number of spot welds is desired in a certain joint, the number is shown in parentheses either above or below the weld symbol.

METRIC

SEAM WELDS

The symbol for all seam welds is a circle transversed by two horizontal parallel lines. This symbol is used for all seam welds regardless of the way they are made. The seam-weld symbol is placed (1) below the reference line to indicate arrow side, (2) above the reference line to indicate other side, and (3) on the reference line to indicate that there is no arrow or other side.

Seam-Weld Application

1. Dimensions of seam welds are shown on the same side of the reference line as the weld symbol. They are dimensioned by either size or strength. The size of seam welds is designated as the width of the weld and is shown to the left of the weld symbol. The strength of seam welds is designated in pounds per square inch (psi) or newtons per millimeter (N/mm) and is shown to the left of the weld symbols.

(A) SPECIFYING WIDTH OF THE WELD

(B) SPECIFYING STRENGTH OF WELD

2. The length of a seam weld, when indicated on the welding symbol, is shown to the right of the weld symbol. When seam welding extends for the full distance between abrupt changes in the direction of the welding, no length dimension needs to be shown on the welding symbol. When a seam weld extends less than the full length of the joint, the extent of the weld should be shown.

3. The pitch of intermittent seam welds is shown as the distance between centers of the weld increments. The pitch is shown to the right of the length dimension. Unless otherwise indicated, intermittent seam welds are interpreted as having the length and pitch measured parallel to the axis of the weld.

DRAWING CALLOUT

INTERPRETATION

4. When the exposed surface of one member of a seam-welded joint is to be flush, that surface is indicated by adding the flush-contour symbol to

the weld symbol, observing the usual locational significance.

5. When the pitch or length of seam is not parallel to the axis of the weld, then it must be shown on the drawing.

DETAIL B

6. The seam-weld process is referenced in the tail of the welding symbol.

DRAWING CALLOUT

INTERPRETATION

BACK OR BACKING WELDS

The back or backing weld symbol is used to indicate bead-type back or backing welds of single-groove welds.

1. Back or backing welds of single-groove welds are shown by placing a back or backing weld symbol above or below the reference line opposite the groove-weld symbol. Dimensions of back or backing welds are not shown on the welding symbol. If it is desired to specify these dimensions, they are shown on the drawing.

BACK WELD
GROOVE WELD

GROOVE WELD
BACK WELD

2. Back or backing welds that are to be welded approximately flush, without recourse to any method of finishing, are shown by adding the flush-contour symbol to the back or backing weld symbol.

FLUSH

3. Back or backing welds that are made flush by mechanical means are shown by adding both the flush-contour symbol and the user's standard finish symbol to the back or backing weld symbol.

GRIND FLUSH

SURFACING WELDS

1. The surfacing-weld symbol is used to indicate surfaces built up by welding. Surfaces built up by welding, whether by single- or multiple-pass surfacing welds, are shown by the surfacing-weld symbol. The surfacing-weld symbol does not indicate the welding of a joint, and hence has no arrow- or other-side significance. This symbol is drawn below the reference line, and the arrow must point clearly to the surface on which the weld is to be deposited.

2. Dimensions used in conjunction with the surfacing-weld symbol are shown on the same side of the reference line as the weld symbol. The size or height of the surface built up by welding is indicated by showing the minimum height of the weld deposit to the left of the weld symbol. When no specific height of weld deposit is desired, no size dimension need be shown on the welding symbol. When the entire area of a plane or curved surface is to be built up by welding, no dimension other than size (height of deposit) need be shown on the welding symbol.

DRAWING CALLOUT

.12 MINIMUM SURFACE BUILDUP

INTERPRETATION

SURFACE BUILDUP OF A FLAT SURFACE

DRAWING CALLOUT

.12 MINIMUM SURFACE BUILDUP

INTERPRETATION

FLANGED WELDS

The following welding symbols are intended to be used for light-gage metal joints involving the flaring or flanging of the edges to be joined.

1. Edge-flange welds are shown by the edge-flange-weld symbol. This symbol has no both-sides significance.

2. Corner-flange welds are shown by the corner-flange-weld symbol. This symbol has no both-sides significance.

DRAWING CALLOUT   INTERPRETATION

3. Dimensions of flange welds shall be shown on the same side of the reference line as the weld symbol. The radius and the height above the point of tangency are indicated by showing both the radius and the height, separated by a plus mark, and placed to the left of the weld symbol. The radius and the height read in that order from left to right along the reference line.

4. The size of flange welds is shown by a dimension placed outward of the flanged dimensions.

DRAWING CALLOUT   INTERPRETATION

5. Root opening of flange welds is not shown on the welding symbol. If it is desired to specify this dimension, it is shown on the drawing.

6. For flange welds, when one or more pieces are inserted between the two outer pieces, the same welding symbol as for the two outer pieces is used regardless of the number of pieces inserted.

MELT-THROUGH WELDS

1. The melt-through symbol is used where at least 100 percent joint penetration of the weld through the material is required in welds made from one side only. Melt-through welds are shown by placing the melt-through weld symbol on the side of the reference line opposite

the groove- or flange-weld symbol. No dimensions of melt-through, except height of reinforcement, are shown on the welding symbol. If it is desired to specify height of reinforcement, it is shown to the left of the melt-through symbol.

2. Melt-through welds that are to be made flush by mechanical means are shown by adding both the flush-contour symbol and the user's standard finish symbol to the melt-through weld symbol.

3. Melt-through welds that are to be mechanically finished to a convex contour are shown by adding the convex-contour symbol to the melt-through weld symbol.

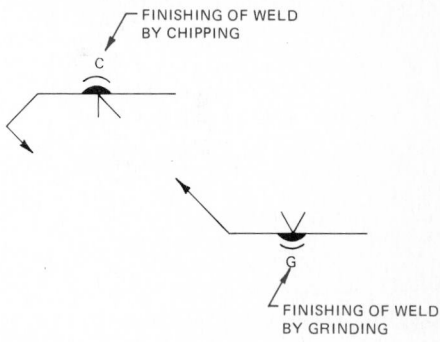

Reference and Source Material
1. American Welding Society.

ASSIGNMENT

See Assignment 6 for Unit 11-5 on page 244.

Review for Assignment
Unit 7-5   Assemblies in Section

ASSIGNMENTS for Chapter 11

Assignment for Unit 11-1, Designing for Welding

1. On a B- or A3-size sheet, redesign one of the cast parts shown in Fig. 11-1-A or 11-1-B for fabrication by welding, using standard sheet sizes and shapes. Make a detail assembly drawing. Welding symbols or sizes are not required. Include on the drawing a bill of material and identify each part on the assembly. Scale is full or 1:1.

Assignments for Unit 11-2, Welding Symbols

2. On a B- or A3-size sheet, complete the enlarged views of the welded joints for the drawing callouts shown in Fig. 11-2-A. Use notes to explain any additional welding requirements.

3. On a B- or A3-size sheet, add the information shown above Fig. 11-2-B to the seven welding symbols shown in this assignment.

Fig. 11-2-A   Showing weld type and proportion on drawings.

Weld	Welding Process Required	Type of Weld	Additional Requirements
1	Carbon-Arc Welding	Bevel	
2	Oxyacetylene Welding	Double Fillet	Both Sides Field Weld
3	Oxyacetylene Welding	Fillet	Both Sides
4	No Specifications Required	J Groove	
5	Carbon-Arc Welding	Fillet	All Around
6	Carbon-Arc Welding	Fillet	All Around Field Weld
7	Gas Metal-Arc Welding	Double V-Groove	

Fig. 11-1-A   Pivot arm.

Fig. 11-1-B   Link.

BEVEL AND FILLET WELDS

FILLET WELD ALL AROUND

J-GROOVE WELD

DOUBLE V-GROOVE WELD

Fig. 11-2-B   Indicating welding symbols on drawings.

Assignment for Unit 11-3, Fillet Welds

4. On a B- or A3-size sheet, select one of the problems shown in Figs. 11-3-A to 11-3-D and make a two-view working drawing complete with dimensions and welding symbols. Include on the drawing a bill of material, and identify each part on the assembly. Use full-strength welds. Scale is full or 1:1.

MATL — AISI C—1040

Fig. 11-3-A   Swing bracket.

Fig. 11-3-C   Slide bracket.

Fig. 11-3-D   Caster frame.

FRONT

Fig. 11-3-B   Step bracket.

MATL — AISI C—1040

Assignment for Unit 11-4, Groove Welds

5. On a B- or A3-size sheet, select one of the problems shown in Figs. 11-4-A through 11-4-D and make a working drawing complete with dimensions and welding symbols. Include on the drawing a bill of material and identify each part on the assembly. Use full-strength welds. Scale is full, 1:1, or as shown. A comparison of a cast and welded steel part is shown above Fig. 11-4-C.

Assignment for Unit 11-5, Other Basic Welds

6. On a B- or A3-size sheet, draw the parts and welds shown in Fig. 11-5-A or 11-5-B and add the weld-size dimensions.

MATERIAL NO. 30 ASTM GREY IRON

MATERIAL—SAE 1032

COMPARISON OF A CAST SHAFT SUPPORT WITH A WELDED-STEEL SHAFT SUPPORT

Fig. 11-4-A   Swing bracket.

TYPE SLF SPRING MOUNTING

Fig. 11-4-B   Connecting bracket.

Fig. 11-4-C   Fan and motor base.

244   FASTENERS, MATERIALS, AND FORMING PROCESSES

Ø 2.18
Ø 1.50

1.75
3.50
.50
C

.62
1.24

3.38
6.76

5.38   4.88
1.31
.62

Ø .40
4 HOLES

.50

8.26
2.88

.69

Ø .312–18UNC–2B
4 HOLES

R .50
1.75

9.24

12.24

.62
2.00
1.50
.25
.50

MATL – ASTM CLASS 50 GREY IRON
ROUNDS AND FILLETS R .12

Fig. 11-4-D   Drill press base.

.31 ⌷ 5

.50 ⌷

.31 ⌷
20°
G

SCALE I:I

PLUG WELDS

.06 ◯ 1.38

SCALE I:I

SPOT OR PROJECTION WELDS

.25 ◯ .75–3.00 ⟍ GTAW

SCALE I:2

SEAM WELD

Fig. 11-5-A   Plug and seam welds.

.25

Ø .750
  .745

NOTE – WORN SHAFT TO BE BUILT
UP AND TURNED TO ORIGINAL
SIZE SHOWN.

1.50

SCALE I:I

SURFACE WELD

Fig. 11-5-B   Spot and surface welds.

CHAPTER 12

Manufacturing Materials

UNIT 12-1
Cast Irons

This chapter is an up-to-date reference on manufacturing materials. It provides the drafter and designer with basic information on materials and their properties to ensure the proper selection of the product material.

FERROUS METALS

Iron and the large family of iron alloys called steel are the most frequently specified metals. Iron is abundant (iron ore constitutes about 5 percent of the earth's crust), easy to convert from ore to a useful form, and iron and steel are sufficiently strong and stable for most engineering applications.

All commercial forms of iron and steel contain carbon, which is an integral part of the metallurgy of iron and steel.

CAST IRON

Because of its low cost, cast iron is often considered a simple metal to produce and to specify. Actually, the metallurgy of cast iron is more complex than that of steel and other familiar design materials. Whereas most other metals are usually specified by a standard chemical analysis, the same analysis of cast iron can produce several entirely different types of iron, depending upon rate of cooling, thickness of the casting, and how long the casting remains in the mold. By con-

trolling these variables, the foundry can produce a variety of irons for heat- or wear-resistant uses, or for high-strength components. See Fig. 12-1-1.

Types of Cast Iron

Ductile, or Nodular, Iron Ductile, or nodular, iron is not as available as gray iron, and it is more difficult to control in production. However, ductile iron can be used where higher ductility or strength is required than is available in gray iron. See Fig. 12-1-2.

Ductile iron, sometimes called nodular iron, is used in applications such as crankshafts because of its good machinability, fatigue strength, and high modulus of elasticity; heavy-duty gears because of its high yield strength

and wear resistance; and automobile door hinges because of its ductility.

Gray Iron Gray iron is a supersaturated solution of carbon in an iron matrix. The excess carbon precipitates out in the form of graphite flakes. Typical applications of gray iron include automotive blocks, flywheels, brake disks and drums, machine bases, and gears. Generally, gray iron serves well in any machinery applications because of its fatigue resistance.

White Iron White iron is produced by a process called *chilling* which prevents graphitic carbon from precipitating out. Either gray or ductile iron can be chilled to produce a surface of white iron. In castings that are white iron throughout, however, the composition

Fig. 12-1-1   Schematic diagram of a blast furnace, hot blast stone, and skiploader. (American Iron and Steel Institute.)

MECHANICAL PROPERTY		DUCTILE				WHITE	GRAY						MALLEABLE						
		50-55-06	60-40-18	100-70-03	120-90-02		20	25	30	40	50	60	32510	35018	40010	45006	50005	70003	90001
Yield strength	10^3 lb/in.²	60–75	45–60	75–90	90–125								32	35	40	45	50	70	90
	MPa	410–520	310–410	520–620	620–860		*	*	*	*	*	*	220	240	275	310	345	485	620
Tensile strength	10^3 lb/in.²	90–110	60–80	100–120	125–150	20–50	20–25	25–30	30–35	40–48	50–57	60–66	50	53	60	65	70	85	105
	MPa	620–760	410–550	690–825	860–1035	140–345	140–170	170–205	205–240	275–330	345–390	415–455	345	365	415	450	480	585	725
Elongation in 2.00 in. (50 mm)	%	3–10	10–25	6–10	2–7	—	1	1	1	0.8	0.5	0.5	10	18	10	6	5	3	1
Modulus of elasticity	10^3 lb/in.²	22–25	22–25	22–25	22–25	—	12	13	15	17	19	20	25	25	26	26	26	26–28	26–28
	10^3 MPa	150–170	150–170	150–170	150–170	8 / —	83	90	103	117	131	138	172	172	180	180	180	180–193	180–193

*Yield strength usually about 65–80% of tensile strength.

Fig. 12-1-2 Mechanical properties of cast iron.

of iron is selected according to part size to ensure that the volume of metal involved can chill rapidly enough to produce white iron.

Because of their extreme hardness, white irons are used primarily for applications requiring wear and abrasion resistance such as mill liners and shot-blasting nozzles. Other uses include railroad brake shoes, rolling-mill rolls, clay-mixing and brick-making equipment, and crushers and pulverizers. Generally, plain (unalloyed) white iron costs less than other cast irons.

The principal disadvantage of white iron is that it is very brittle.

High-Alloy Irons High-alloy irons are ductile, gray, or white irons that contain over 3 percent alloy content. These irons have properties that are significantly different from the unalloyed irons and are usually produced by specialized foundries.

Malleable Iron Malleable iron is white iron that has been converted to a malleable condition by a two-stage heat-treating process.

It is a commercial cast material which is similar to steel in many respects. It is strong and ductile, has good impact and fatigue properties, and has excellent machining characteristics.

The two basic types of malleable iron are ferritic and pearlitic. Ferritic grades are more machinable and ductile, whereas the pearlitic grades are stronger and harder.

References and Source Material
1. *Machine Design*, Materials reference issue, Mar. 1981.

ASSIGNMENT

See Assignment 1 for Unit 12-1 on page 264.

Review for Assignments
Unit 4-2 Arcs Tangent to Two Lines
Unit 5-1 Basic Dimensioning
Unit 6-1 Detail Drawings
Unit 7-8 Revolved Sections
Unit 7-18 Intersection of Unfinished Surfaces

UNIT 12-2
Carbon Steel

Carbon steel is essentially an iron-carbon alloy with small amounts of other elements (either intentionally added or unavoidably present) such as silicon, magnesium, copper, and sulfur. Steels can be either cast to shape or wrought into various mill forms from which finished parts can be machined, forged, formed, stamped, or otherwise generated.

Wrought steel is either poured into ingots or is sand-cast. After solidification the metal is reheated and hot rolled—often in several steps—into the finished wrought form. Hot-rolled steel is characterized by a scaled surface and a decarburized skin.

CARBON AND LOW-ALLOY CAST STEELS

Carbon and low-alloy cast steels lend themselves to the formation of streamlined, intricate parts with high strength and rigidity.

A number of advantages favor steel casting as a method of construction

1. The metallographic structure of steel castings is uniform in all directions. It is free from the directional variations in properties of wrought-steel products.
2. Cast steels are available in a wide range of mechanical properties depending on the compositions and heat treatments.
3. Steel castings can be annealed, normalized, tempered, hardened, or carburized.
4. Steel castings are as easy to machine as wrought steels.
5. Most compositions of carbon and low-alloy cast steels are easily welded because their carbon content is under 0.45 percent.

The making of steel is illustrated in Fig. 12-2-1.

HIGH-ALLOY CAST STEELS

The term *high alloy* is applied arbitrarily to steel castings containing a minimum of 8 percent nickel and/or chromium. Such castings are used mostly to resist corrosion or provide strength at temperatures above 1200°F (650°C).

CARBON STEELS

Carbon steels are the workhorse of product design. They account for over 90 percent of total steel production. More carbon steels are used in product manufacturing than all other metals combined.

A thorough understanding of the selection and specification criteria for all types of steel requires knowledge of what is implied by carbon-steel mill forms, qualities, grades, tempers, finishes, edges, and heat treatments; also how and where these terms relate to dimensions, tolerances, physical and mechanical properties, and manufacturing requirements.

The designer's specification job really begins the instant that molten steel hits the mold. The conditions under which steel solidifies have a significant effect on production and on performance of subsequent mill products.

STEEL SPECIFICATION

Several ways are used to identify a specific steel: by chemical or mechanical properties, by its ability to meet a standard specification or industry-accepted practice, or by its ability to be fabricated into an identified part.

Chemical Composition

The steel producer can be instructed to produce a desired composition in one of three ways

1. By a maximum limit
2. By a minimum limit
3. By an acceptable range

The following are some commonly specified elements.

Carbon The principal hardening element in steel. As carbon content is increased to about 0.85 percent, hardness and tensile strength increase, but ductility and weldability decrease.

Manganese A lesser contributor to hardness and strength. Properties depend on carbon content. Increasing manganese increases the rate of carbon penetration during carburizing.

Phosphorus Large amounts increase strength and hardness but reduce ductility and impact toughness, particu-

Fig. 12-2-1 Flow chart of steelmaking.

larly in the higher-carbon grades. Phosphorus in low-carbon, free-machining steels improves machinability.

Silicon A principal deoxidizer in the steel industry. Silicon increases strength and hardness but to a lesser extent than manganese.

Sulfur Increased sulfur content reduces transverse ductility, notch-impact toughness, and weldability. Sulfur is added to improve machinability of steels.

Copper Improves atmospheric corrosion resistance when present in excess of 0.15 percent.

Fig. 12-2-2 Steel designating system.

TYPE OF CARBON STEEL	NUMBER SYMBOL	PRINCIPAL PROPERTIES	COMMON USES
Plain carbon	10XX		
Low carbon steel (0.06 to 0.20% carbon)	1006 to 1020	Toughness and less strength	Chains, rivets, shafts, and pressed steel products
Medium carbon steel (0.20 to 0.50% carbon)	1020 to 1050	Toughness and strength	Gears, axles, machine parts, forgings, bolts, and nuts
High carbon steel (over 0.50% carbon)	1050 and over	Less toughness and greater hardness	Saws, drills, knives, razors, finishing tools, and music wire
Sulfurized (free-cutting)	11XX	Improves machinability	Threads, splines, and machined parts
Phosphorized	12XX	Increases strength and hardness but reduces ductility	
Manganese steels	13XX	Improves surface finish	

Fig. 12-2-3 Carbon steel designations, properties, and uses.

Lead

Lead Improves the machinability of steel.

Classification Bodies

The specifications covering the composition of iron and steel have been issued by various classification bodies. They serve as a selection guide and provide a means for the buyer to conveniently specify certain known and recognized requirements. The main classification bodies are

SAE—Society of Automotive Engineers

AISI—American Iron and Steel Institute This is an association of steel producers which issues steel specifications for the steel-making industry and cooperates with the SAE in using the same numbers for the same steel.

ASTM—American Society for Testing and Materials This group is interested in materials of all kinds and writes specifications. The ASTM steel specifications for steel plate and structural shapes are used by all steel makers in North America.

The ASTM has several specifications covering structural steel. Both the AISI and AISC (American Institute of Steel Construction) refer to ASTM specifications.

ASME—American Society of Mechanical Engineers This group is interested in the steel used in boilers and other mechanical equipment.

SAE AND AISI—SYSTEMS OF STEEL IDENTIFICATION

The specifications for steel bar are based on a code that specifies the composition of each type of steel covered. They include both plain carbon and alloy steels. The code is a four-number system. See Figs. 12-2-2 and 12-2-3. Each figure in the number has the following specific function; the first or left-side figure represents the major class of steel, the second figure represents a subdivision of the major class; for example, the series having *one* (1) as the left-hand figure covers the carbon steels. The second figure breaks

this class up into normal low-sulfur steels, the high-sulfur free-machining grades, and another grade having higher than normal manganese.

Originally the second figure represented the percentage of the major alloying element present, and this is true of many of the alloy steels. However, this had to be varied in order to account for all the steels that are available.

The third and fourth figures represent carbon content in hundredths of 1 percent, thus the figure xx15 means 0.15 of 1 percent carbon.

Class 1	
Carbon steels	1xxx
Basic open hearth and acid Bessemer carbon steels nonsulfurized and non-phosphorized	10xx
Basic open hearth and acid Bessemer carbon steels, sulfurized but not phosphorized	11xx
Basic open hearth carbon steels phosphorized	12xx

EXAMPLE SAE 2335 is a nickel steel containing 3.5 percent nickel and 0.35 of 1 percent carbon.

Typical properties of rolled carbon steels are shown in Fig. 12-2-4.

Carbon-Steel Sheet

Flat-rolled carbon-steel sheets are made from heated slabs that are progressively reduced in size as they move through a series of rolls.

Hot-Rolled Sheets Hot-rolled sheets are produced into three principal qualities, commercial, drawing, and physical.

MECHANICAL PROPERTY		AISI STEEL											
		1015/1020/1022			1035/1040			1045/1050			1095		
		Hot-Rolled	Cold-Drawn	Annealed	Hot-Rolled	Cold-Drawn	Quenched and Tempered	Hot-Rolled	Cold-Drawn	Quenched and Tempered	Hot-Rolled	Cold-Drawn and Annealed	Quenched and Tempered
Yield strength	10^3 lb/in.^{2}	40	51	42	42	71	63–96	49	84	68–117	66	76	80–152
	MPa	270	350	295	290	440	435–660	335	580	470–800	455	525	580–1050
Tensile strength	10^3 lb/in.^{2}	65	61	60	76	85	96–130	90	100	105–137	130	99	130–216
	MPa	450	420	415	525	585	660–895	620	690	725–945	895	680	895–1490
% Elongation in 2.00 in. (50 mm)		25	15	38	18	12	17–24	15	10	25–15	9	13	10–84

Fig. 12-2-4 Typical mechanical properties of rolled carbon steel.

Cold-Rolled Sheets Cold-rolled sheets are made from hot-rolled coils which are pickled, then cold reduced to the desired thickness. The commercial quality of cold-rolled sheets is normally produced with a matte finish suitable for painting or enameling but not suitable for electroplating.

Carbon-Steel Plates

Carbon-steel plates are produced (in rectangular plates or in coils) by hot rolling directly from the ingot or slab. Plate thickness ranges from .19 in. (4 mm) and thicker for plates up to 48 in. (1200 mm) wide, and from .25 in. (6 mm) and thicker for plates wider than 48 in. (1200 mm). Thickness is specified in millimeters or inches. It can also be specified by weight (lb/ft^2) or mass (kg/m^2).

Carbon-Steel Bars

Hot-Rolled Bars Hot-rolled carbon-steel bars are produced from blooms or billets in a variety of cross sections and sizes. See Figs. 12-2-5 and 12-2-6.

Cold-Finished Bars Cold-finished carbon-steel bars are produced from hot-rolled steel by a cold-finishing process which improves surface finish, dimensional accuracy, and alignment. Cold drawing and cold rolling also increase the yield and tensile strength. For machinability ratings of cold drawn carbon steel see Fig. 12-2-7.

AISI No.	RATING*
12L14	195
1213	137
1215	137
1212	100
1211	91
1117	89
1114	85
1137	72
1141	69
1018	66
1045	55

*Based on 1212 = 100%

Fig. 12-2-7 Machinability rating of cold-drawn carbon steel.

Steel Wire

Steel wire is made from hot-rolled rods produced in continuous-length coils. Most wire is drawn, but some special shapes are rolled.

Pipe and Tubing

Pipe and tubing range from the familiar plumber's black pipe to high-precision mechanical tubing for bearing races. Pipe and tubing may contain fluids, support structures, or be a primary shape from which products are fabricated.

Welded Tubular Products Welded tubular products are made from hot-rolled or cold-rolled flat steel coils.

Pipe Pipe is produced from carbon or alloy steel to nominal dimensions. Nominal pipe sizes are expressed in inch sizes, but in the metric system the outside diameter and the wall thickness are expressed in millimeters. The outside diameter is often much larger than the nominal size. For example, a .75-in. standard-weight pipe has an outside diameter of 1.050 in. (26.7 mm). The outside diameter of nominal size pipe always remains the same and the mass or wall thickness changes. ANSI B36 has developed 10 different wall thicknesses (schedules) of pipe (see Table 56, Appendix).

Nominal pipe size designation stops at 12 in. Pipe 14 in. and over is listed on the basis of outside diameter and wall thickness.

Tubing Tubing is usually specified by a combination of either outside diameter, inside diameter, or wall thickness.

Fig. 12-2-5 Standard stock. (American Iron and Steel Institute.)

ROUND SECTIONS SHAPED BY THREE SYSTEMS OF ROLL PASSING SHOW COMPARATIVE REDUCING ABILITIES

SHAPING AND FINISHING PASSES FOR VARIOUS SECTIONS

OVAL AND SQUARE

DIAMOND AND SQUARE

FLAT AND EDGE

SQUARE

HEXAGON

12 STAND BAR MILL

ROLL PASSES 1 TO 12

CHANNEL

ANGLE

MOST FREQUENTLY USED SYSTEM. HEAVY BUT GOOD REDUCTION.

MODERATELY SEVERE REDUCTION USED MOSTLY FOR MEDIUM BARS.

GENERALLY USED TO ROLL LARGE DIAMETER BARS.

SMALL STRUCTURAL SHAPES MAY BE FORMED BY A WIDE VARIETY OF PASSING PROCEDURES.

Fig. 12-2-6 Bar-mill roll passes. (American Iron and Steel Institute.)

Sizes range from approximately .25 to 5.00 in. (6 to 125 mm), in increments of .12 in. (3 mm). Wall thickness is usually specified in inches, millimeters, or by gage numbers.

Structural-Steel Shapes

A large tonnage of structural-steel shapes goes into manufactured products rather than buildings. The frame of a truck, railroad car, or earth-moving equipment is a structural design problem, just as is a high-rise building.

Size Designations Several ways are used to describe a structural section in a specification, depending primarily on its shape.

1. Beams and channels are measured by the depth of the section in inches (millimeters) and by weight (lbs/ft) or mass (kg/m).
2. Angles are described by length of legs and thickness in inches (millimeters), or more commonly, by lengths of legs and weight (lbs/ft) or mass (kg/m). The longest leg is always stated first.
3. Tees are specified by width of flange, overall depth of stem, and pounds per foot (kilograms per meter) in that order.
4. Zees are specified by width of flange and thickness in inches (millimeters), or by depth, width across flange, and pounds per foot (kilograms per meter).
5. Wide-flange sections are described by depth, width across flange, and pounds per foot (kilograms per meter).

HIGH-STRENGTH LOW-ALLOY STEELS

The properties of high-strength low-alloy (HSLA) steels generally exceed those of conventional carbon structural steels. These low-alloy steels are usually chosen for their high ratios of yield to tensile strength, resistance to puncturing, abrasion resistance, corrosion resistance, and toughness.

ASTM Specifications

ASTM has six specifications covering high-strength low-alloy steels. These are

ASTM A 94 Used primarily for riveted and bolted structures and for special structural purposes.

ASTM A 242 Used primarily for structural members where light weight or low mass and durability are important.

ASTM A 374 Used where high strength is required and where resistance to atmospheric corrosion must be at least equal to that of plain copper-bearing steel.

ASTM A 375 This specification differs slightly from ASTM A 374 in that material can be specified in the annealed or normalized condition.

ASTM A 440 This covers high-strength intermediate-manganese steels for nonwelded applications.

ASTM A 441 This covers the intermediate manganese HSLA steels which are readily weldable when proper welding procedures are used.

LOW- AND MEDIUM-ALLOY STEELS

There are two basic types of alloy steel: *through hardenable* and *surface hardenable*. Each type contains a broad family of steels whose chemical, physical, and mechanical properties make them suitable for specific product applications. See Fig. 12-2-8.

STAINLESS STEELS

Stainless steels have many industrial uses because of their desirable corrosion resistance and strength properties.

Free-Machining Steels

A whole family of free-machining steels has been developed for fast and economical machining. See Fig. 12-2-9. These steels are available in bar stock in various compositions, some standard and some proprietary. When utilized properly, they lower the cost of machining by reducing metal removal time.

TYPE OF STEEL	ALLOY SERIES	APPROXIMATE ALLOY CONTENT (%)	PRINCIPAL PROPERTIES	COMMON USES
Manganese steel	13xx	Mn 1.6–1.9	Improve surface finish	
Molybdenum steels	40xx	Mo 0.15–0.3		
	41xx	Cr 0.4–1.1; Mo 0.08–0.35		Axles, forgings, gears, cams, mechanical parts
	43xx	Ni 1.65–2; Cr 0.4–0.9; Mo 0.2–0.3	High strength	
	44xx	Mo 0.45–0.6		
	46xx	Ni 0.7–2; Mo 0.15–0.3		
	47xx	Ni 0.9–1.2; Cr 0.35–0.55; Mo 0.15–0.4		
	48xx	Ni 3.25–3.75; Mo 0.2–0.3		
Chromium steels	50xx	Cr 0.3–0.5	Hardness, great strength and toughness	Gears, shafts, bearings, springs, connecting rods
	51xx	Cr 0.7–1.15		
	E51100	C 1.0; Cr 0.9–1.15		
	E52100	C 1.0; Cr 0.9–1.15		
Chromium vanadium steel	61xx	Cr 0.5–1.1; V 0.1–0.15	Hardness and strength	Punches and dies, piston rods, gears, axles
Nickel-chromium-molybdenum steels	86xx	Ni 0.4–0.7; Cr 0.4–0.6; Mo 0.15–0.25	Rust resistance hardness, and stength	Food containers, surgical equipment
	87xx	Ni 0.4–0.7; Cr 0.4–0.6; Mo 0.2–0.3		
	88xx	Ni 0.4–0.7; Cr 0.4–0.6; Mo 0.3–0.4		
Silicon-manganese steel	92xx	Si 1.8–2.2	Springiness and elasticity	Springs

Fig. 12-2-8 AISI designation system for alloy steel.

MECHANICAL PROPERTY		12L13/12L14 12L15		1211/1212/ 1213		1117/1118/1119			1137			1141/1144		
		Hot-Rolled	Cold-Drawn	Hot-Rolled	Cold-Drawn	Hot-Rolled	Cold-Drawn	Quenched and Tempered	Hot-Rolled	Cold-Drawn	Quenched and Tempered	Hot-Rolled	Cold-Drawn	Quenched and Tempered
Yield strength	10^3 lb/in.²	34	60–80	33	58	34–46	51–68	50–76	48	82	136	51	90	163
	MPa	235	416–550	225	400	235–315	350–470	345–525	330	565	335	350	620	1120
Tensile strength	10^3 lb/in.²	57	70–90	55	75	62–76	69–78	89–113	88	98	157	94	100	190
	MPa	390	480–620	380	517	425–525	475–535	615–780	605	675	1080	650	690	1310
Elongation in 2.00 in. (50 mm)	%	22	10–18	25	10	23–33	15–20	17–22	15	10	5	15	10	9
Machinability	(B1212 = 100%)	195–296		91–137		89–100			71			69		

Fig. 12-2-9   Typical mechanical properties of free-machining carbon steels.

References and Source Material

1. *Machine Design*, Materials reference issue, Mar. 1981.

ASSIGNMENT

See Assignment 2 for Unit 12-2 on page 264.

Review for Assignments

Unit 5-3   Dimensioning Common Features
Unit 6-1   Detail Drawings
Unit 8-1   Pictorial Representation and Thread Forms
Unit 9-1   Keys

UNIT 12-3
Nonferrous Metals

Although ferrous alloys are specified for more engineering applications than all nonferrous metals combined, the large family of nonferrous metals offers a wider variety of characteristics and mechanical properties. For example, the lightest metal is lithium, .02 lb/in.³ (0.53 g/cm³); the heaviest is osmium with a weight of .81 lb/in.³ (mass of 22.5 g/cm³)—nearly twice the weight of lead. Mercury melts at around −38°F (−39°C), while tungsten, the highest-melting metal, liquefies at 6170°F (3410°C).

Availability, abundance, and the cost to convert the metal into useful forms all play an important part in selecting a nonferrous metal. Although nearly 80 percent of all elements are called "metals," only about two dozen of these are used as structural engineering materials. Of the balance, however, many are used not structurally but as coatings, in electronic devices, as nuclear materials, and as minor constituents in other systems.

One of the most important aspects in selecting a material for a mechanical or structural application is how easily the material can be shaped into the finished part—and how its properties can be either intentionally or inadvertently altered in the process. See Fig. 12-3-1. Frequently, metals are simply cast into the finished part. In other cases, metals are cast into an intermediate form (such as an ingot) then worked or "wrought" by rolling, forging, extruding, or other deformation processes.

MANUFACTURING WITH METALS (Refer to Fig. 12-3-1)

Machining Most metals can be machined. Machinability is best for metals that allow easy chip removal with minimum tool wear.

Powder Metallurgy (PM) Compacting Parts can be made from most metals and alloys by PM compacting, although only a few are economically justified. Iron and iron-copper alloys are most commonly used.

Casting Theoretically, any metal that can be melted and poured can be cast. However, economic limitations usually narrow down the number of ways metals are cast commercially.

Forming method	Aluminum	Copper	Iron	Lead	Magnesium	Nickel	Silver, Gold, Platinum	Molybdenum, Copper, Tantalum, Tungsten	Steel	Tin	Titanium	Zinc
Casting												
Centrifugal	✓	✓	✓			✓			✓			
Continuous	✓	✓	✓	✓					✓			
Ceramic mold	✓	✓	✓		✓	✓			✓			
Investment	✓	✓			✓	✓	✓		✓			
Permanent mold	✓	✓	✓	✓	✓	✓			✓	✓		✓
Sand	✓	✓	✓	✓	✓	✓			✓	✓		✓
Shell mold	✓	✓	✓		✓	✓			✓			
Die casting	✓			✓	✓					✓		✓
Cold heading	✓	✓		✓		✓	✓		✓			
Deep drawing	✓	✓			✓	✓			✓		✓	✓
Extruding	✓	✓		✓				✓	✓	✓	✓	
Forging	✓	✓			✓			✓	✓		✓	
Machining	✓	✓	✓			✓	✓	✓	✓		✓	✓
PM compacting	✓	✓	✓			✓	✓	✓	✓		✓	
Stamping and forming	✓	✓			✓	✓	✓	✓	✓		✓	✓

Fig. 12-3-1   Common methods of forming metals.

Extruding and Forging Metals to be forged or extruded must be ductile and not work-harden at working temperature. Some metals show these characteristics at room temperature and can be cold worked; others must be heated.

Stamping and Forming Most metals, except brittle alloys, can be press worked.

Cold Heading Metals must be ductile and should not work-harden rapidly. Annealing should restore ductility and softness in cold heading alloys.

Deep Drawing Deep drawing involves severe deformation and the metal is usually stretched over the die.

ALUMINUM

The density of aluminum is about one-third that of steel, brass, nickel, or copper. Yet, some alloys of aluminum are stronger than structural steel. Under most service conditions, aluminum has high resistance to corrosion and forms no colored salts which might stain or discolor adjacent components. See Fig. 12-3-2.

COPPER

Copper alloys, approximately 250 of them, are fabricated in rod, sheet, tube, and wire form. Each of these alloys has some property or combination of properties which makes it unique. They can be grouped into several general headings, such as coppers, brasses, leaded brasses, phosphor bronzes, aluminum bronzes, silicon bronzes, beryllium coppers, cupronickels, and nickel silvers. See Fig. 12-3-3.

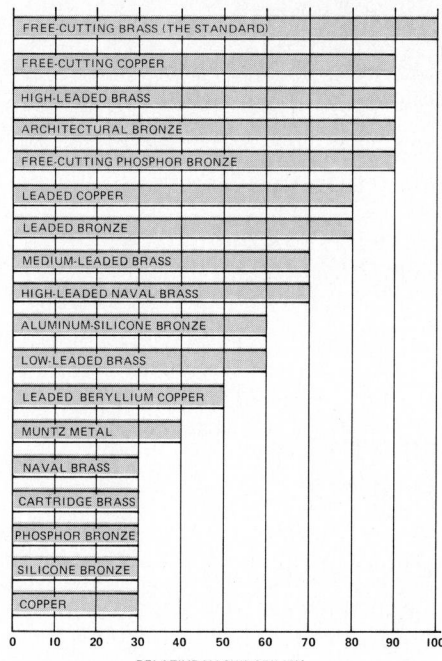

Fig. 12-3-3 Free-machining copper alloys.

MAJOR ALLOYING ELEMENT	DESIGNATION
Aluminum (99% or more)	1xxx
Copper	2xxx
Manganese	3xxx
Silicon	4xxx
Magnesium	5xxx
Magnesium and silicon	6xxx
Zinc	7xxx
Other elements	8xxx
Unused series	9xxx

Fig. 12-3-2 Wrought aluminum alloy designations.

Copper alloys are used where one or more of the following properties is needed: thermal or electrical conductivity, corrosion resistance, strength, ease of forming, ease of joining, and color.

The major alloy usages are

1. Copper in pure form as a conductor in the electrical industry
2. Copper or alloy tubing for water, drainage, air conditioning, and refrigeration lines
3. Brasses, phosphor bronzes, and nickel silvers as springs or in construction of equipment if corrosive conditions are too severe for iron or steel

An advantage of copper and its alloys, offered by no other metals, is the wide range of colors available.

NICKEL

Commercially pure wrought nickel is a grayish-white metal capable of taking a high polish. Because of its combination of attractive mechanical properties, corrosion resistance, and formability, nickel or its alloys is used in a variety of structural applications usually requiring specific corrosion resistance.

MAGNESIUM

Magnesium, with density of only .06 lb/in.^{3} (1.74 g/cm^3), is the world's lightest structural metal. The combination of low density and good mechanical strength makes possible alloys with a high strength-to-weight ratio.

ZINC

Zinc is a relatively inexpensive metal which has moderate strength and toughness and outstanding corrosion resistance in many types of service.

The principal characteristics which influence the selection of zinc alloys for die castings include the dimensional accuracy obtainable, castability of thin sections, smooth surface, dimensional stability, and adaptability to a wide variety of finishes.

TITANIUM

Titanium is a light metal at .16 lb/in.^{3} (4.43 g/cm^3); 60 percent heavier than aluminum but 45 percent lighter than alloy steel. It is the fourth most abundant metallic element in the earth's crust and the ninth most common element.

Titanium-based alloys are much stronger than aluminum alloys and superior in many respects to most alloy steels.

BERYLLIUM

Beryllium has a strength-to-weight ratio comparable to high-strength steel, yet it is lighter than aluminum. Its melting point is 2345°F (1285°C) and it has excellent thermal conductivity. It is nonmagnetic and a good conductor of electricity.

REFRACTORY METALS

Refractory metals are those metals with melting points above 3600°F (2000°C). Among these, the best known and most extensively used are tungsten, tantalum, molybdenum, and niobium.

Refractory metals are characterized by high-temperature strength, corrosion resistance, and high melting points.

Tantalum and Niobium

Tantalum and niobium are usually discussed together, since most of their working operations are identical. Unlike molybdenum and tungsten, tantalum and niobium can be worked at room temperatures. The major differences between tantalum and niobium are in density, nuclear cross section, and corrosion resistance. The density of tantalum is almost twice that of niobium.

Molybdenum

Molybdenum is widely used in missiles, aircraft, industrial furnaces, and nuclear projects. Its melting point is lower than that of tantalum and tungsten. Molybdenum has a high strength-to-weight ratio and a low vapor pressure, is a good conductor of heat and electricity, has a high modulus of elasticity, and a low coefficient of expansion.

Tungsten

Tungsten is the only refractory metal that has the combination of excellent corrosion resistance, good electrical and thermal conductivity, a low coefficient of expansion, and high strength at elevated temperatures.

PRECIOUS METALS

Gold costs over 8000 times more than an equal amount of iron; rhodium costs nearly 32 000 times more than copper. With prices such as these, why are precious metals ever specified?

In some cases, precious metals are used for their unique surface characteristics. They reflect light better than other metals. Gold, for example, is specified as a surface for heat reflectors, insulators, and collectors because of its outstanding ability to reflect ultraviolet radiation.

The family of metals called precious metals can be divided into three subgroups: silver and silver alloys; gold and gold alloys; and the so-called platinum metals, which are platinum, palladium, rhodium, ruthenium, iridium, and osmium.

References and Source Material

1. *Machine Design*, Materials reference issue, Mar. 1981.

See Assignments 3 and 4 for Unit 12-3 on page 264.

Review for Assignments

UNIT 12-4
Plastics

This unit will acquaint drafters with the general characteristics of commercially available plastics so that they may make proper use of plastics in products.

Plastics may be defined as nonmetallic materials capable of being formed or molded with the aid of heat, pressure, chemical reactions, or a combination of these. See Fig. 12-4-1.

Plastics are strong, tough, durable materials which solve many problems in machine and equipment design. Metals, it is true, are hard and rigid. This means that they can be machined, to very close tolerances, into cams, bearings, bushings, and gears, which will work smoothly under heavy loads for long periods. Although some come close, no plastic has the hardness and creep resistance of, say, steel. However, metals have many weaknesses which engineering plastics do not. Metals corrode or rust, they must be lubricated, their working surfaces wear readily, they cannot be used as electrical or thermal insulators, they are opaque and noisy, and where they must flex, metals fatigue rapidly.

Plastics can resolve these weaknesses, though not necessarily all with one material. The engineering plastics are resistant to most chemicals; fluorocarbon is one of the most chemically inert substances known. None of the engineering plastics corrode or rust; acetal resin and fluorocarbon are unaffected even when continuously immersed in water. Engineering plastics can be run at low speeds and loads, and, without lubrication, are among the world's slipperiest solids, being comparable to ice. Engineering plas-

Fig. 12-4-1   A variety of plastic parts. (Automatic Plastics.)

tics are resilient; therefore, they run more quietly and smoothly than equivalent metal ones, and they are able to stand periodic overloads without harmful effects.

Plastics are a family of materials—not a single material—each member of which has its special advantages.

Being manufactured, plastics raw materials are capable of being variously combined to give almost any property desired in an end product. But these are controlled variations unlike those of nature's products. Some thermoplastics can be sterilized.

The widespread and growing use of plastics in almost every phase of modern living can be credited in large part to their unique combinations of advantages. These advantages are light weight, range of color, good physical properties, adaptability to mass-production methods, and, often, lower cost.

Aside from the range of uses attributable to the special qualities of different plastics, these materials achieve still greater variety through the many forms in which they can be produced.

They may be made into definite shapes like dinnerware and electric switchboxes. They may be made into flexible film and sheeting such as shower curtains and upholstery. Plastics may be made into sheets, rods, and tubes that are later shaped or machined into internally lighted signs or airplane blisters. They may be made into filaments for use in household screening, industrial strainers, and sieves. Plastics may be used as a coating on textiles and paper. They may be used to bind such materials as fibers of glass and sheets of paper or wood to form boat hulls, airplane wing tips, and tabletops.

Plastics are usually classified as either *thermoplastic* or *thermosetting*.

THERMOPLASTICS

These materials soften, or liquefy, and flow when heat is applied. Removal of the heat causes these materials to set or solidify. They may be reheated and reformed or reused. In this group fall the acrylics, the cellulosics, nylons (polyamides), polyethylene, polystyrene or styrene, polyfluorocarbons, the vinyls, polyvinylidene, ABS, acetal resin, polypropylene, and polycarbonates. See Fig. 12-4-2.

THERMOPLASTICS			
Name of Plastic	Properties	Forms and Methods of Forming	Uses
ABS (Acrylonitrile Butadiene-Styrene)	Strong, tough, good electrical properties.	Available in powder or granules for injection molding, extrusion, and calendering and as sheet for vacuum forming.	Pipe, wheels, football helmets, battery cases, radio cases, children's skates, tote boxes.
ACETAL RESIN	Rigid without being brittle, tough, resistant to extreme temperatures, good electrical properties.	Produced in powder form for molding and extrusion. Available in rod, bar, tube, strip, slab.	Automobile instrument clusters, gears, bearings, bushings, door handles, plumbing fixtures, threaded fasteners, cams.
ACRYLICS	Exceptional clarity and good light transmission. Strong, rigid, and resistant to sharp blows. Excellent insulator, colorless or full range of transparent, translucent, or opaque colors.	Available in sheet, rod, tube, and molding powders. Plastic products can be produced by fabricating of sheets, rods, and tubes, hot forming of sheets, injection and compression molding of powder, extrusion, casting.	Airplane canopies and windows, television and camera viewing lenses, combs, costume jewelry, salad bowls, trays, lamp bases, scale models, automobile tail lights, outdoor signs.
CELLULOSICS (A) Cellulose Acetate		Available in pellets, sheets, film, rods, tubes, strips, coated cord. Can be made into products by injection, compression molding, extrusion, blow molding, and vacuum forming, or sheets and coating.	Spectacle frames, toys, lamp shades, combs, shoe heels.
(B) Cellulose Acetate Butyrate	Among the toughest of plastics. Retains a lustrous finish under normal wear. Transparent, translucent, or opaque in wide variety of colors and in clear transparent. Good insulators.	Available in pellets, sheets, rods, tubes, strips, and as a coating. Can be made into products by injection, compression molding, extrusion, blowing and drawing of sheet, laminating, coating.	Steering wheels, radio cases, pipe and tubing, tool handles, playing cards.
(C) Cellulose Propionate		Available in pellets for injection extrusion or compression molding.	Appliance housing, telephone hand sets, pens and pencils.
(D) Ethyl Cellulose		Available in granules, flake, sheet, rod, tube, film, or foil. Can be made into finished products by injection, compression molding, extrusion, drawing.	Edge moldings, flashlights, electrical parts.
(E) Cellulose Nitrate		Available in rods, tubes, sheets for machining and as a coating.	Shoe heel covers, fabric coating.

Fig. 12-4-2 Thermoplastics. (Cont'd on page 256). (The Society of Plastics Industry, Inc.)

Name of Plastic	Properties	Forms and Methods of Forming	Uses
FLUOROCARBONS	Low coefficient of friction, resistant to extreme heat and cold. Strong, hard, and good insulators.	Available as powder and granules in resin form. Sheet, rod, tube, film, tape, and dispersions. Molded, extruded, and machined.	Valve seats, gaskets, coatings, linings, tubings.
NYLON (Polyamides)	Resistant to extreme temperatures. Strong and long-wearing range of soft colors.	Available as a molding powder, in sheets, rods, tubes, and filaments. Injection, compression, blow molding, and extrusion.	Tumblers, faucet washers, gears. As a filament, it is used as brush bristles, fishing line.
POLYCARBONATE	High impact strength, resistant to weather, transparent.	Primarily a molding material, may take form of film, extrusion, coatings, fibers, or elastomers.	Parts for aircraft, automobiles, business machines, gages, safety-glass lenses.
POLYETHYLENE	Excellent insulating properties, moisture proof. Clear, transparent, translucent.	Available in pellet, powder, sheet, film, filament, rod, tube, and foamed. Injection, compression, blow molding, extrusion, coating, and casting.	Ice cube trays, tumblers, dishes, bottles, bags, balloons, toys, moisture barriers.
POLYSTYRENE	Clear, transparent, translucent, or opaque. All colors. Water and weather resistant. Resistance to heat or cold.	Available in molding powders or granules, sheets, rods, foamed blocks, liquid solution, coatings, and adhesives. Injection, compression molding, extrusion, laminating, machining.	Kitchen items, food containers, wall tile, toys, instrument panels.
POLYPROPYLENES	Good heat resistance. High resistance to cracking. Light range of color.	Processed by injection molding, blow molding, and extrusion.	Thermal dishware, washing machine agitators, pipe and pipe fittings, wire and cable insulation, battery boxes, packaging film and sheets.
URETHANES	Tough and shock resistant for solid materials. Flexible for foamed material, can be foamed in place.	Solid type—starting two reactants, final article can be extruded, molded, calendered, or cast. Foamed type—can be made by either a prepolymer or one-shot process, in either slab stock or molded form.	Mattresses, cushioning, padding, toys, rug underlays, crashpads, sponges, mats, adhesion, thermal insulation, industrial tires.
VINYLS	Strong and abrasion-resisting. Resistant to heat and cold. Wide color range.	Available in molding powder, sheet, rod, tube, granules, powder. It can be formed by extrusion, casting, calendering, compression, and injection molding.	Raincoats, garment bags, inflatable toys, hose, records, floor and wall tile, shower curtains, draperies, pipe, paneling.

Fig. 12-4-2   (cont'd.)

THERMOSETTING PLASTICS

These materials undergo an irreversible chemical change when heat is applied or when a catalyst or reactant is added. They become hard, insoluble, and infusible, and they do not soften upon reapplication of heat. Thermosetting plastics include phenolics, amino plastics (melamine and urea), cold-molded polyesters, epoxies, silicones, alkyds, allylics, and casein.

SELECTION OF MATERIALS

The properties of various materials influence the shape of the part. The essential properties data needed are static strength values in tension, compression, shear, and flexure and mechanical constants, such as moduli of elasticity and shear. Most thermoplastics have flow properties which cause degradation of fastening torque retention. Composition and fabrication methods influence the mechanical properties of plastics and, therefore, the working stresses.

MACHINING

Practically all thermoplastics and thermosets can be satisfactorily machined on standard equipment with adequate tooling. The nature of the plastic will determine whether heat should be applied, as in some laminates, or avoided, as in buffing some thermoplastics. Standard machining operations can be used, such as turning, drilling, tapping, milling, blanking, and punching.

MATERIAL SELECTION

One of the first decisions a designer makes is the choice of materials. The choice is influenced by many factors, such as the end use of the product and the properties of the selected material. No attempt is made at this point to discuss the engineering approach to selection of materials. This is covered in Chap. 32.

However, a basic examination and selection of a plastic material at this time will help acquaint the drafter with the wide range of plastics available.

References and Source Material
1. The Society of the Plastics Industry, Inc.
2. Crystaplex Plastics.
3. General Motors Corporation.

ASSIGNMENT
See Assignment 5 for Unit 12-4 on page 266.

Review for Assignment
Unit 6-4   Bill of Material

UNIT 12-5
Plastic Forming Methods

The most common methods of forming plastic parts are by the application of heat and pressure, casting, and machining. Heat and pressure may be applied by different methods, the most common of which are compression molding, injection molding, transfer molding, extrusion, blow and vacuum forming, and embossing. Other methods include laminating or layup and cold forming and embossing. See Fig. 12-5-1. The method of forming is governed by the material, part, part design, and cost.

INJECTION MOLDING

Injection is the principal method of forming thermoplastic materials. Modifications of the injection process are sometimes used for thermosetting plastics.

In *injection molding*, plastic material is put into a hopper which feeds into a heating chamber. A plunger pushes the plastic through the long heating chamber where the material is softened to a fluid state. At the end of the chamber is a nozzle which abuts firmly against an opening into a cool, closed mold. The fluid plastic is forced at high pressure through the nozzle into the cold mold. As soon as the plastic cools to a solid state, the mold opens and the finished plastic piece is ejected from the press. See Fig. 12-5-2.

EXTRUDING

Extruding is a process generally used with the thermoplastic materials, although it is applicable to the thermosetting plastics. It lends itself to forming of shapes having uniform sections such as sheets, rods, tubes, and filaments. Extrusion is accomplished by forcing material, which has been softened in a heating chamber, through dies of the desired shape, using the pressure created from a screw or hydraulic ram. Extrusions may be conducted continuously and rapidly with a screw-type machine, as shown in Fig. 12-5-3.

BLOW MOLDING

Blow molding is a method of forming used with thermoplastic materials. Basically, blow molding consists of stretching and then hardening a plastic against a mold. See Fig. 12-5-4.

Fig. 12-5-1 Plastic parts being removed from an injection mold. (Union Carbide Corp.)

PLUNGER TYPE

ROTARY SPREADER TYPE

RECIPROCATING SCREW TYPE

Fig. 12-5-2 Injection molding machines. (General Motors Corp.)

Fig. 12-5-3 Extruding machine. (General Motors Corp.)

EXTRUDED PARISON
MOLD OPEN

MOLD CLOSED
BOTTLE BLOWN

FINISHED BOTTLE

Fig. 12-5-4 Blow molding.

COMPRESSION MOLDING

The molding material is placed in the lower portion of a heated mold cavity having no top and with sides high enough to retain the material. See Fig. 12-5-5. The upper portion of the mold is forced down over the material, usually with high pressure. The combination of heat and pressure causes the molding material to liquefy and flow, filling the mold.

TRANSFER MOLDING

Transfer molding is most generally used for thermosetting plastics. This method is like compression molding in that the plastic is cured into an infusible state in a mold under heat and pressure. It differs from compression molding in that the plastic is heated to a point of plasticity before it reaches the mold and is forced into a closed mold by means of a hydraulically operated plunger. See Fig. 12-5-6.

PULP MOLDING

Thermosetting plastics are used in *pulp molding*. In this process a porous form, approximating the shape of a finished article, is lowered into a tank containing a mixture of pulp, plastic resins, and water. The water is drawn off through the porous form by a vacuum. This causes the pulp and resin mixture to be drawn to the form and adhere to it. When a sufficient thickness of pulp has been drawn onto the form, it is removed and then molded into final shape.

SOLVENT MOLDING

Solvent molding is based on the fact that when a mold is immersed in a solution and withdrawn, or when it is filled with a liquid plastic and then emptied, a layer of plastic film adheres to the sides of the mold. Thermoplastic materials are used in this process of forming.

Some articles thus formed, like a bathing cap or vial, are removed from the molds. Other solvent moldings remain permanently on the form as, for example, a plastic coating on a metal tube.

CASTING

Casting may be employed for both thermoplastic and thermosetting materials in making special shapes, rigid sheets, film and sheeting, rods, and tubes.

The essential difference between casting and molding is that no pressure is used in casting as it is in molding.

In casting, the plastic material is heated to a fluid mass, poured into either open or closed molds, cured at varying temperatures depending on the plastic used, and removed from the molds.

CALENDERING

Calendering can be used to process thermoplastics into film and sheeting and to apply a plastic coating to textiles or other supporting materials. See Fig. 12-5-7. *Film* refers to thicknesses up to and including .01 in. (0.2 mm) while sheeting includes thicknesses over .01 in. (0.2 mm).

COATING

Thermosetting and thermoplastic materials may both be used as a coat-

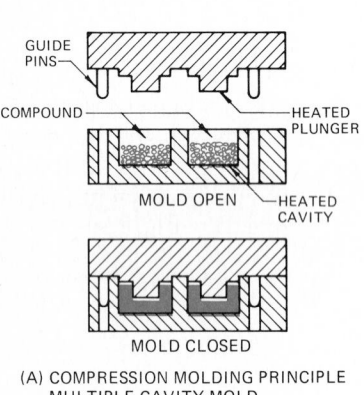

(A) COMPRESSION MOLDING PRINCIPLE MULTIPLE CAVITY MOLD

(B) STRIPPER PLATE MOLD

(C) SPLIT CAVITY MOLD

Fig. 12-5-5   Compression molding. (General Motors Corp.)

LOADING OF MOLD

MOLD CLOSED

MOLD OPEN

Fig. 12-5-6   Transfer molding. (General Motors Corp.)

Fig. 12-5-7   Calendering.

ing. The materials to be coated may be metal, wood, paper, fabric, leather, glass, concrete, ceramics, or other plastics.

Methods of coating are varied and include knife or spread coating, spraying, roller coating, dipping, and brushing. See Fig. 12-5-8.

HIGH-PRESSURE LAMINATING

Thermosetting plastics are most generally used in *high-pressure laminating* which is distinguished by the use of high heat and pressure. See Fig. 12-5-9. These plastics are used to hold together the reinforcing materials that comprise the body of the finished product. The reinforcing materials may be cloth, paper, wood, or fibers of glass.

REINFORCING

Reinforced plastics mostly employ thermoset plastics, though some thermoplastics are used.

Fig. 12-5-8 Coating.

Fig. 12-5-9 High-pressure laminating.

Fig. 12-5-10 Reinforcing.

Reinforced plastics differ from high-pressure laminates in that very low or no pressure is used in the processing. The two methods are alike in that the plastic is used to bind together the cloth, paper, or glass-fiber reinforcing material used for the body of the product. The reinforcing materials may be in sheet or mat form, and their selection depends on the qualities desired in the end product. See Fig. 12-5-10.

FABRICATING

Fabricating covers operations on sheet, rod, tube, sheeting, film, and special shapes to make them into finished products. The materials may be thermosetting or thermoplastic.

Fabricating divides into three broad categories: machining; cutting, sewing, and sealing of film and sheeting; and forming.

FINISHING

The finishing of plastics includes the different methods of adding either decorative or functional surface effects to a plastic product.

References and Source Material
1. The Society of the Plastics Industry, Inc.
2. General Motors Corporation.

ASSIGNMENT

See Assignment 6 for Unit 12-5 on page 266.

Review for Assignment
Unit 12-4 Plastics

UNIT 12-6
Plastic Design Considerations— for Single Parts

The design of molded parts involves several factors not normally encountered with machine-fabricated and assembled parts. It is important that designers take these factors into consideration.

Shrinkage *Shrinkage* is defined as the difference between dimensions of the mold and the corresponding dimensions of the molded part. Normally the mold designer is more concerned with shrinkage than the molded part designer. Shrinkage does, however, affect dimensions, warpage, residual stress, and moldability.

Section Thickness Solidification is a function of heat transfer from or to the mold for both thermoplastics and thermosets. Each material has a fixed rate of heat transfer. Therefore, where section thickness varies, areas within a molded part will solidify at different rates. The varying rates will cause irregular shrinkage, sink marks, additional strain, and warpage. For these reasons, uniform section thickness is important and may be maintained by adding holes or depressions, as shown in Figs. 12-6-1 and 12-6-2.

UNIFORM THICKNESS   VARYING THICKNESS
PREFERRED   NOT RECOMMENDED

Fig. 12-6-1 Section thickness.

MATERIAL	ABSOLUTE MINIMUM		RECOMMENDED MINIMUM	
	(in.)	(mm)	(in.)	(mm)
THERMOPLASTICS				
ABS	.024	0.6	.064	1.6
ACRYLIC	.024	0.6	.100	2.4
CELLULOSIC	.024	0.6	.076	1.9
FLUOROCARBON	.010	0.2	.010	0.3
POLYAMIDE	.016	0.4	.060	1.5
POLYCARBONATE	.024	0.6	.100	2.4
POLYETHYLENE	.036	0.9	.064	1.6
PLOYSTYRENE	.032	0.8	.064	1.6
POLYVINYL CHLORIDE	.100	2.4	.100	2.4
THERMOSETS				
EPOXY	.064	1.6	.130	3.2
POLYESTER				
GLASS FILLED	.040	1.0	.190	4.7
MINERAL FILLED	.040	1.0	.130	3.2
PHENOLICS				
GENERAL PURPOSE	.050	1.3	.130	3.2
FABRIC FILLED	.064	1.6	.190	4.7
MINERAL FILLED	.130	3.2	.190	4.7
UREA AND MELAMINES				
GENERAL PURPOSE	.036	0.9	.100	2.5
FABRIC FILLED	.050	1.3	.130	3.2
MINERAL FILLED	.040	1.0	.190	4.7

Fig. 12-6-2 Section thickness for various plastics. (General Motors Corp.)

Molded Holes A through hole is more advantageous than a blind hole since it is more accurate and economical. Blind holes should not be more than twice as deep as their diameter, as shown in Fig. 12-6-3. Avoid placing holes at angles other than perpendicular to the flash line. If such holes are necessary, consider using a drilled hole to maintain simple molding. See Fig. 12-6-4.

Threads External and internal threads can be easily molded by means of loose-piece inserts and rotating core pins. External threads may be formed by placing the cavity so that the threads are formed in the mold pattern.

Gates Gate location should be anticipated during the design stage. Avoid gating into areas subjected to high stress levels, fatigue, or impact. To optimize molding, locate gates in the heaviest section of the part. See Fig. 12-6-5.

Internal and External Draft Draft is necessary on all rigid molded articles to facilitate removal of the part from the mold. Draft may vary from 0.25 to 4° per side, depending upon the length of the vertical wall, surface area, finish, kind of material, and the mold or method of ejection used.

Parting or Flash Line *Flash* is that portion of the molding material which flows or exudes from the mold parting line during molding. Any mold which is made of two or more parts may produce flash at the line of junction of the mold parts. The thickness of flash usually varies between .002 and .016 in. (0.05 and 0.40 mm), depending upon the accuracy of the mold, type of material, and the process used. See Fig. 12-6-6.

Fillets and Radii The principal functions of fillets and radii (rounds) are to ease the flow of plastic within the mold, to facilitate ejection of the part, and to distribute stress in the part in service. During molding, the material is liquefied, but it is a heavy, viscous liquid which does not easily flow around sharp corners. The liquid tends to bend around corners; therefore, rounded corners permit the liquid plas-

tic to flow smoothly and easily through the mold. For recommended radii, see Fig. 12-6-7.

Undercuts Parts with undercuts should be avoided. Normally, parts with external undercuts cannot be withdrawn from a one-piece mold. Internal undercuts are considered impractical and should be avoided. If an internal undercut is essential, it may be achieved by machining or by use of a flexible mold core material. See Fig. 12-6-9.

Ribs and Bosses Ribs increase rigidity of a molded part without increasing

(A) RIB AND BOSS PROPORTION

(B) RIB SPACING

(C) BOSSES

(D) EJECTOR PIN BOSSES

Fig. 12-6-8   Ribs and bosses. (General Motors Corp.)

Fig. 12-6-3   Blind holes.

Fig. 12-6-4   Avoiding molded holes which are perpendicular to parting line.

Fig. 12-6-5   Gating. (Union Carbide Corp.)

Fig. 12-6-6   Parting or flash line. (General Motors Corp.)

Fig. 12-6-7   Fillets and radii. (General Motors Corp.)

Fig. 12-6-9 Undercuts. (General Motors Corp.)

wall thickness and sometimes facilitate flow during molding. Bosses reinforce small, stressed areas, providing sufficient strength for assembly with inserts or screws. Recommended proportions for ribs and bosses are shown in Fig. 12-6-8A.

References and Source Material
1. General Motors Corporation.
2. General Electric Company.

ASSIGNMENT

See Assignment 7 for Unit 12-6 on page 266.

UNIT 12-7
Plastic Design Considerations— for Assemblies

The design of molded parts which are to be assembled with typical fastening methods involves factors different from those normally encountered with metal.

HOLES AND THREADS

Mechanical fasteners, in general, depend upon a hole of some type. Holes should be designed and located to provide maximum strength and minimum molding problems. Any straight hole, molded or machined, should have between it and an adjacent hole, or side wall, an amount of material equal to or greater than the diameter or width of the hole. Any threaded hole, molded or tapped, should have between it and an adjacent hole, or

side wall, an amount of material at least 3 times the outside diameter of the thread. Spacing may be reduced, however, by proper use of bosses.

Drilled holes are often more accurate and easier to produce than molded holes, even though they require a second operation.

Tapped holes provide an economical means of joining a molded part to its assembly. The designer should avoid threads with a pitch of less than .03 in. (0.8 mm). Holes which are to be tapped should be countersunk to prevent chipping when the tap is inserted.

External and internal threads can be molded integrally with the part. Molded threads are generally more expensive to form than other threads because either a method of unscrewing the part from the mold must be provided or a split mold must be used.

INSERTS

After the molding material has been determined, the insert should be

designed. The molded part should be designed around the insert.

Inserts of round rod stock, coarse diamond knurled and grooved, provide the strongest anchorage under torque and tension. A large single groove with knurling on each end, as in Fig. 12-7-1, is superior to two or more grooves with smaller knurled surface areas. See also examples of inserts in Fig. 12-7-2.

KNURLING DEPTH SHOULD BE ABOUT .01 in. ANGULAR GROOVES GIVE INCREASED AXIAL ANCHORAGE. PLASTIC SHRINKAGE ALONE SHOULD NOT BE RELIED ON TO PROVIDE FIRM SUPPORT FOR INSERTS.

Fig. 12-7-1 Supporting inserts. (General Motors Corp.)

Fig. 12-7-2 Insert application. (General Motors Corp.)

PRESS AND SHRINK FITS

Inserts may be secured by a press fit, or the plastic molding material may be assembled to a larger part by a shrink fit, as shown in Fig. 12-7-3. Both methods rely on shrinkage of the material, which is greatest immediately after removal from the mold.

HEAT FORMING AND HEAT SEALING

Most thermoplastics can be reformed by the application of heat and pressure, as shown in Fig. 12-7-4. This reforming often eliminates the need for other assembly methods, such as adhesive bonding and mechanical fasteners. This method cannot be used with thermosetting materials.

MECHANICAL FASTENING

Various designs of mechanical fasteners are commercially available. Spring-type metal hinges and clips, speed clips or nuts, and expanding rivets are a few of these designs. Design of the parts for assembly requires that molded parts have sufficient sectional strength to withstand the stresses that will be encountered with fasteners. A strengthening of the area which will receive the brunt of these applied stresses is usually required. See Fig. 12-7-5.

RIVETS

Conventional riveting equipment and procedures can be used with plastics. Care must be exercised to minimize stresses induced during the fastening operation. To do this, the rivet head should be 2.5 to 3 times the shank diameter. Also, rivets should be backed with either plates or washers to avoid high localized stresses. See Fig. 12-7-6.

Drilled holes rather than punched holes are preferred for fasteners. If possible, fastener clearance in the hole should be at least .01 in. (0.3 mm) to maintain a plane stress condition at the fastener.

BOSS CAPS

A boss cap is a cup-shaped metal ring which is pressed onto the boss by hand, with an air cylinder, or with a light-duty press. It is designed to reinforce the boss against the expansion force exerted by tapping screws. See Fig. 12-7-7.

ADHESIVE BONDING

When two or more parts are to be joined into an assembly, adhesives permit a strong, durable fastening between similar materials and often are the only fastening method available for joining dissimilar materials. Structural adhesives are made from the same basic resins as many plastics and thus react to their operating environment in a similar manner. In order to provide maximum strength, adhesives must be applied as a liquid to thoroughly wet the surface of the part. The bonding surface must be chemically clean to permit complete wetting. Basic plastics vary in physical properties, so adhesives made from these materials also vary. See Fig. 12-7-8.

PRESS FIT

SHRINK FIT

Fig. 12-7-3 Press and shrink fits. (General Motors Corp.)

BEFORE FORMING

AFTER FORMING IN ASSEMBLY

Fig. 12-7-4 Heat forming.

Fig. 12-7-5 Mechanical fasteners.

PLASTIC OR METAL   PLASTIC

REINFORCING WASHER

NOTE: BREAK ALL SHARP EDGES ON RIVET, WASHER AND HOLES.

Fig. 12-7-6 Recommended riveting procedure.

THREAD SIZE	DIMENSIONS				
	A	B	C	D	R
≠ 6(.138)	.14	.21	.10	.28	.02
≠ 8(.164)	.16	.25	.13	.34	.02
≠ 10(.190)	.18	.29	.17	.40	.02

Fig. 12-7-7 Boss cap design.

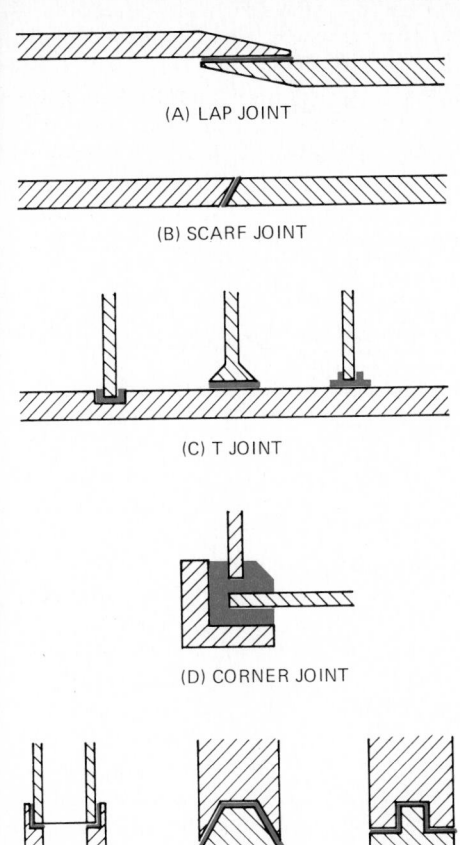

(A) LAP JOINT

(B) SCARF JOINT

(C) T JOINT

(D) CORNER JOINT

(E) BUTT JOINTS

Fig. 12-7-8 Adhesive bonding.

Fig. 12-7-9 Design joints for ultrasonic bonding.

HORN

STUD

METAL

PLASTIC

(A) BEFORE STAKING

(B) AFTER STAKING

Fig. 12-7-10 Typical ultrasonic staking operation.

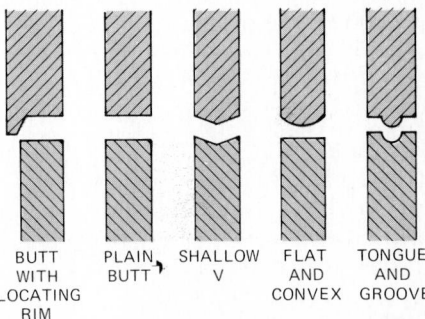

BUTT WITH LOCATING RIM | PLAIN BUTT | SHALLOW V | FLAT AND CONVEX | TONGUE AND GROOVE

Fig. 12-7-11 Joint shapes for spin welding.

ULTRASONIC BONDING

Ultrasonic bonding often is used instead of solvent cementing to bond plastic parts. By using this technique, irregularly shaped parts can be bonded in 2 seconds or less. The bonded parts may be handled and used at reasonable temperatures within minutes after joining.

Only one of the mating parts comes in contact with the horn. The part transmits the ultrasonic vibration to small, hidden bonding areas, resulting in fast, perfect welds. Both mating halves remain cool except at the seam, where the energy is quickly dissipated.

This technique is not recommended where high impact strength is required in the bond area. See Fig. 12-7-9.

ULTRASONIC STAKING

Ultrasonic staking frequently involves the assembly of metal parts. In this technique, a stud molded into the plas-

tic part protrudes through a hole in the metal part. The surface of the stud is vibrated with a horn having high amplitude and a relatively small contact area. The vibration causes the stud to melt and re-form in the configuration of the horn tip. See Fig. 12-7-10.

FRICTION OR SPIN WELDING

This welding technique is limited to parts with circular joints. It is especially useful for large parts where ultrasonic welding or chemical bonding is impractical.

In friction or spin welding, the faces to be joined are pressed together while one part is spun and the other is held fixed. Frictional heat produces a molten zone that becomes a weld when spinning stops. See Fig. 12-7-11.

Drawings

In addition to the usual considerations, the following points should be taken into account when a detail drawing of a plastic part is made

1. Can the part be removed from the mold?
2. Is location of flash line consistent with design requirements?
3. Is section thickness consistent? Are there thick sections? Thin sections? Could greater uniformity of section thickness be maintained?
4. Has the material been correctly specified?
5. Is each feature in accordance with the thinking of competent materials engineers and molders?
6. Have close tolerance requirements been reviewed with responsible engineers?
7. Have marking requirements been specified to inform field service people of the material from which the part is fabricated?

References and Source Material
1. General Motors Corporation.
2. General Electric Company.

ASSIGNMENTS

See Assignments 8 and 9 for Unit 12-7 on page 266.

Review for Assignments
Unit 6-1   Working Drawings
Unit 7-1   Full Sections

ASSIGNMENTS for Chapter 12

Assignment for Unit 12-1, Cast Irons

1. On a B- or A3-size sheet make a two-view working drawing of one of the parts shown in Fig. 12-1-A or 12-1-B. Use a revolved section to show the center section of the arm. Select a suitable cast iron for the part. Scale is full or 1:1.

Assignment for Unit 12-2, Carbon Steel

2. On a B- or A3-size sheet make a working drawing of one of the parts shown in Fig. 12-2-A or 12-2-B. Show the worm threads in pictorial form. Scale is 2× size or 2:1. Select a suitable steel for the part. Conventional breaks may be used to shorten the length of the view.

Assignments for Unit 12-3, Nonferrous Metals

3. On a B- or A3-size sheet make a two-view working drawing of the outboard motor clamp shown in Fig. 12-3-A or 12-3-B. Use lines or surfaces marked A, B, and C as the zero lines and use arrowless dimensioning. Scale is half size or 1:2. Select a suitable material noting that the part must be water resistant, have a painted finish, have moderate strength, and have a light weight or mass.

4. On an A-size sheet recommend the material for each of the parts shown in Figs. 12-3-C and 12-3-D. State your reasons for each of the materials selected.

Fig. 12-1-A   Plug wrench.

Fig. 12-1-B   Door closer arm.

Fig. 12-2-A   Raising bar.

Fig. 12-2-B   Worm for gear jack.

Fig. 12-3-A   Outboard motor clamp.

ROUNDS AND FILLETS R.12

Fig. 12-3-B   Outboard motor clamp.

ROUNDS AND FILLETS R 3

METRIC

KEYSEAT
.188 X .094

.250—20 UNC—2B

Fig. 12-3-C   Coupling.

LINK
ROUNDS AND FILLETS R 3
Ø 19 X 3 HIGH BOSS

Ø 27

R 40

Ø 38

4 HOLES EQUALLY
SPACED ON Ø 58

Ø 8

Ø 64

Ø 48

Ø 48 H7/s6 FIT IN LINK

METRIC

Ø 38

BEARING

Ø 19
Ø 27 H7/s6 FIT IN LINK
Ø 32

BUSHING

Fig. 12-3-D   Connecting link.

Assignment for Unit 12-4, Plastics

5. On a B- or A3-size sheet, prepare a production report for the selection of materials for the parts shown in Fig. 12-4-A or 12-4-B. Convert as many parts as possible to plastic. The crane hook assembly is to be used in a water dip-tank operation. Assume that the production run is such that all forming processes can be considered. Include with your report a bill of material.

Assignment for Unit 12-5, Plastic Forming Methods

6. Prepare a report to your supervisor recommending the method of manufacture for one of the parts shown in Figs. 12-5-A and 12-5-B. The report must be concise and brief. The manner of presentation and the analysis arriving at your conclusion could well be a deciding factor for future promotion within a company.

Assignment for Unit 12-6, Plastic Design Considerations—for Single Parts

7. On a B- or A3-size sheet, prepare a casting drawing for one of the parts shown in Figs. 12-6-A to 12-6-C. Refer to the casting recommendations shown in this unit and indicate the parting line on the drawing. Use your judgment for dimensions not given. Scale is full or 1:1.

Assignments for Unit 12-7, Plastic Design Considerations—for Assemblies

8. On a B- or A3-size sheet, add a threaded insert to one of the parts shown in Fig. 12-7-A or 12-7-B. Use your judgment for dimensions not shown and the type and number of views required. Scale is 2 × size or 2:1.

9. On a B- or A3-size sheet, make an assembly drawing of the parts shown in Fig. 12-7-C. The retaining ring is to be positioned in the center of the part and molded into position. Modification to the retaining ring may be required to prevent the ring from turning in the wheel. Scale is 5× size or 5:1. Show a top view and a full-section view. Dimension the finished assembly.

Fig. 12-4-A   Crane hook assembly.

Fig. 12-4-B   Caster assembly.

Fig. 12-5-A   Coupling.

Fig. 12-5-B   Swing bracket.

NOTE: BASE EXTENDED BEYOND WALLS FOR WELDING PURPOSES ONLY

Fig. 12-6-A   Shaft base.

Fig. 12-6-B   Caster frame.

Fig. 12-6-C   Slide bracket.

MATL – CELLULOSE

Fig. 12-7-A   Connector.

MATL – CELLULOSE

Fig. 12-7-B   Lamp adjusting knob.

MATL – NYLON

NO. 5105 X 31 TRUARC RETAINING RING
TO BE RETAINED IN THE WHEEL, AS AN
INSERT, AT THE CENTER OF THE THICKNESS.

Fig. 12-7-C   Cassette-tape drive wheel.

PART 3

Intermediate Drawing Design

CHAPTER 13

Auxiliary Views

UNIT 13-1
Primary Auxiliary Views

Many machine parts have surfaces that are not perpendicular, or at right angles, to the plane of projection. These are referred to as *sloping* or *inclined* surfaces. In the regular orthographic views, such surfaces appear to be foreshortened, and their true shape is not shown. When an inclined surface has important characteristics that should be shown clearly and without distortion, an auxiliary view is used so that the drawing completely and clearly explains the shape of the object. In many cases, the auxiliary view will replace one of the regular views on the drawing, as illustrated in Fig. 13-1-1.

One of the regular orthographic views will have a line representing the edge of the inclined surface. The auxiliary view is projected from this edge line, at right angles, and is drawn parallel to the edge line.

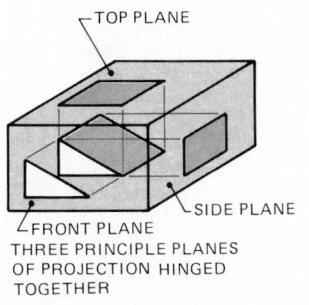

TOP PLANE

FRONT PLANE
SIDE PLANE

THREE PRINCIPLE PLANES OF PROJECTION HINGED TOGETHER

PLANES UNFOLDED

NOT TRUE SHAPE OF SURFACE

PLANES REMOVED SHOWING THREE REGULAR (TOP, FRONT, SIDE) VIEWS

NOTE – IN NONE OF THESE VIEWS DOES THE SLANTED (COLORED) SURFACE APPEAR IN ITS TRUE SHAPE
(A) WEDGED BLOCK SHOWN IN THREE REGULAR VIEWS

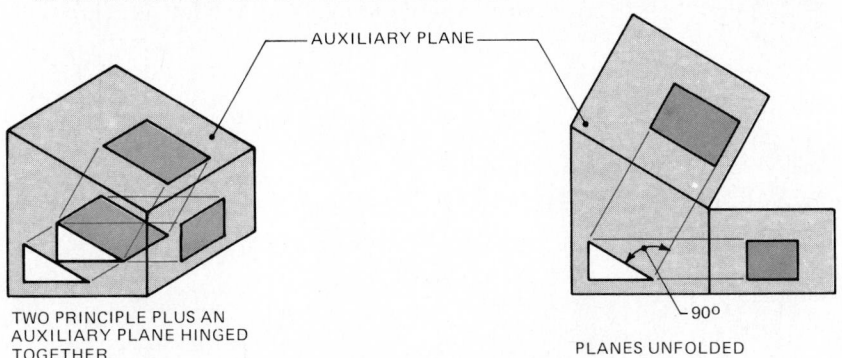

AUXILIARY PLANE

TWO PRINCIPLE PLUS AN AUXILIARY PLANE HINGED TOGETHER

90°

PLANES UNFOLDED

TRUE SHAPE OF COLORED SURFACE

AUXILIARY VIEW

90°

PLANES REMOVED SHOWING FRONT, SIDE, AND AUXILIARY VIEW

NOTE : IN THIS EXAMPLE THE AUXILIARY PLANE REPLACED THE TOP PLANE IN ORDER THAT THE SLANTED (COLORED) SURFACE MAY BE SHOWN IN ITS TRUE SHAPE
(B) REPLACING THE TOP PLANE WITH AN AUXILIARY PLANE

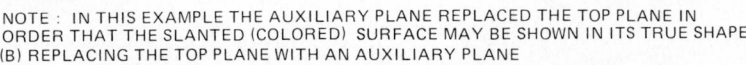

Fig. 13-1-1   Relationship of the auxiliary plane to the three principle planes.

Fig. 13-1-2 Auxiliary views replacing regular views.

Only the true shape features on the views need be drawn, as shown in Fig. 13-1-2. Since the auxiliary view shows only the true shape and detail of the inclined surface or features, a partial auxiliary view is all that is necessary. Likewise, the distorted features on the regular views may be omitted. Hidden lines are usually omitted unless required for clarity. This procedure is recommended for functional and production drafting where drafting costs are an important consideration. However, the drafter may be called upon to draw the complete views of the part. This type of drawing is often used for catalog and standard parts drawings.

Additional examples of auxiliary view drawings are shown in Fig. 13-1-3.

Figure 13-1-4 shows how to make an auxiliary view of a symmetrical object. Figure 13-1-4A shows the object in a pictorial view. In this illustration a center plane is used as a reference plane. In Fig. 13-1-4B the center plane is drawn parallel to the inclined surface shown in the front view. The edge

NOTE: CONVENTIONAL BREAK OR PROJECTED SURFACE ONLY NEED BE SHOWN ON PARTIAL VIEWS

Fig. 13-1-3 Examples of auxiliary view drawings.

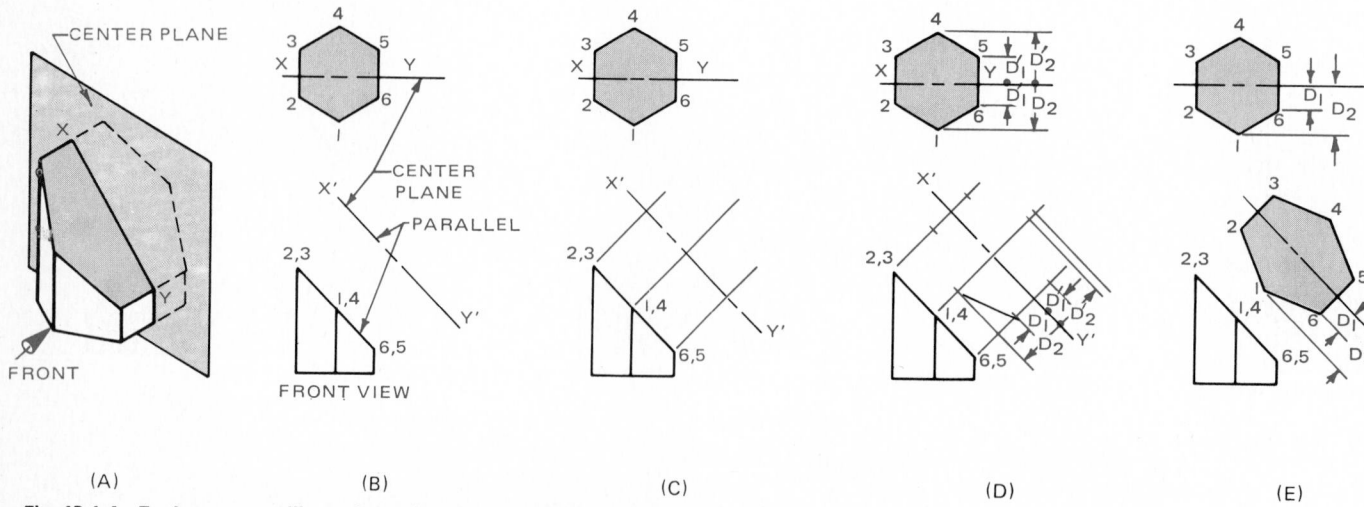

Fig. 13-1-4 To draw an auxiliary view using the center plane reference.

view of this plane appears as a center line, line XY, on the top view. Number the points on the top view. Then transfer these numbers to the edge view of the inclined surface on the front view, as shown. Parallel to this edge view and at a convenient distance from it, draw the line $X'Y'$, as in Fig. 13-1-4C. Now, in the top view, find the distances D_1 and D_2 from the numbered points to the center line. These are the depth measurements. Transfer them onto the corresponding construction lines that you have just drawn, measuring them off on either side of line $X'Y'$, as shown in Fig. 13-1-4D. The result will be a set of points on the construction lines. Connect and number these points, as shown in Fig. 13-1-4E, and the front auxiliary view of the inclined surface results. The remaining portions of the object may also be projected from the center reference plane.

DIMENSIONING AUXILIARY VIEWS

One of the basic rules of dimensioning is to dimension the feature where it can be seen in its true shape and size. Thus the auxiliary view will show only the dimensions pertaining to those features for which the auxiliary view was drawn. The recommended dimensioning method for engineering drawings is the unidirectional system. Not only is this method of dimensioning easier to

prepare and read, but it is readily adaptable to mechanical lettering. See Fig. 13-1-5.

ASSIGNMENTS

See Assignments 1 through 4 for Unit 13-1 on page 274.

Review for Assignments
Unit 7-18   Rounds and Fillets

UNIT 13-2
Circular Features in Auxiliary Projection

As mentioned in Unit 13-1, at times it is necessary to show the complete views of an object. If circular features are involved in auxiliary projection, then the surfaces appear elliptical, not circular, in one of the views.

The method most commonly used to draw the true-shape projection of the curved surface is the plotting of a series of points on the line, the number of points being governed by the accuracy of the curved line required.

Figure 13-2-1 illustrates an auxiliary view of a truncated cylinder. The shape seen in the auxiliary view is an *ellipse*. This shape is drawn by plotting *lines of intersection*. The perimeter of the circle in the top view is divided to

Fig. 13-2-1   Truncated cylinder and auxiliary view.

give a number of equally spaced points—in this case, 12 points, A to M, spaced 30° apart (360°/12 = 30°). These points are projected down to the edge line on the front view, then at right angles to the edge line to the area where the auxiliary view will be drawn. A center line for the auxiliary view is drawn parallel to the edge line, and width settings taken from the top view are transferred to the auxiliary view. Note width setting R for point L. Because the illustration shows a true cylinder and the point divisions in the top view are all equal, the width setting R taken at L is also the correct width

Fig. 13-1-5   Dimensioning auxiliary view drawings.

Fig. 13-2-2   Constructing the true shape of a curved surface by the plotting method.

setting for *C*, *E*, and *J*. Width setting *S* for *B* is also the correct width setting for *F*, *H*, and *M*. When all the width settings have been transferred to the auxiliary view, the resulting points of intersection are connected by means of an irregular curve to give the desired elliptical shape.

It is often necessary to construct the auxiliary view first in order to complete the regular views. This is shown in Fig. 13-2-2.

ASSIGNMENT

See Assignment 6 for Unit 13-2 on page 274.

Review for Assignment
Unit 13-1 Dimensioning Auxiliary Views
Unit 7-18 Intersection of Unfinished Surfaces

Fig. 13-3-1 Auxiliary views added to regular views to show true shape of features.

UNIT 13-3
Multi-Auxiliary-View Drawings

Some objects have more than one surface not perpendicular to the plane of projection. In preparing working drawings of these objects, an auxiliary view may be required for each surface. Naturally, this would depend upon the amount and type of detail lying on these surfaces. This type of drawing is often referred to as a *multi-auxiliary-view* drawing. See Fig. 13-3-1.

One can readily see the advantage of using the unidirectional system of dimensioning for dimensioning an object such as shown in Fig. 13-3-2.

ASSIGNMENT

See Assignment 7 for Unit 13-3 on page 276.

Review for Assignment
Unit 7-18 Intersection of Unfinished Surfaces
Unit 13-1 Dimensioning Auxiliary View

Fig. 13-3-2 Dimensioning a multi-auxiliary-view drawing.

UNIT 13-4
Secondary Auxiliary Views

Some objects, because of their shape, require a secondary auxiliary view to show the true shape of the surface or feature. See Fig. 13-4-1. The surface or feature is usually oblique to the principal planes of projection. In order to draw a secondary auxiliary view, the primary auxiliary view is first constructed so that it is perpendicular to

the inclined surface and to one of the principal planes. The secondary auxiliary view is then projected from the primary auxiliary view, perpendicular to it. Figure 13-4-2 shows the procedure for drawing a secondary auxiliary view.

ASSIGNMENT

See Assignment 8 for Unit 13-4 on page 276.

Review for Assignment
Unit 13-1 Dimensioning Auxiliary Views

Fig. 13-4-1 Secondary auxiliary-view drawing.

Ø.531
2 HOLES

PRIMARY AUXILIARY
VIEW

.62

3.50

2.25

120°

75°

15°

1.25

2.25

2.62

1.50

4.90

4.50   1.00

6.50

HEXAGON
1.50 ACROSS
FLATS

SECONDARY
AUXILIARY
VIEW

.75

NOTE: MANY UNNECESSARY HIDDEN LINES
ARE OMITTED FOR CLARITY.

---

1. DRAW PARTIAL TOP AND FRONT VIEWS

30°

---

2. DRAW PRIMARY AUXILIARY VIEW OUTLINE

120°

PRIMARY
AUXILIARY VIEW

---

3. DRAW SECONDARY AUXILIARY VIEW AND COMPLETE
PRIMARY AUXILIARY VIEW

SECONDARY
AUXILIARY VIEW

---

4. COMPLETE TOP VIEW BY PROJECTING LINES AND POINTS
FROM PRIMARY AUXILIARY VIEW AND TRANSFERRING
DISTANCES FROM SECONDARY AUXILIARY VIEW

W

S

W
2

W

S

W
2

---

5. COMPLETE FRONT VIEW BY PROJECTING LINES AND POINTS
FROM TOP VIEW AND TRANSFERRING DISTANCES FROM
PRIMARY AUXILIARY VIEW

H

L

M

N

H

L

M

N

Fig. 13-4-2 Steps in drawing a secondary auxiliary view.

ASSIGNMENTS for Chapter 13

Assignments for Unit 13-1, Primary Auxiliary Views

1. On a B-size sheet, make a working drawing of the angle bracket shown in Fig. 13-1-A. Replace the top view with an auxiliary view. Draw complete views with hidden lines. Scale is full.

2. On an A3-size sheet, make a working drawing of the cross-slide bracket shown in Fig. 13-1-B. Replace the side view with an auxiliary view. Only partial views need be drawn, and hidden lines may be added to improve clarity. Scale is 1:1.

3. On a B- or A3-size sheet, draw the top, front, and auxiliary views of the angle plate shown in Fig. 13-1-C. Only partial views are required for the auxiliary and front views. Hidden lines may be added to improve the clarity. Scale is full or 1:1.

4. On an A3- or B-size sheet draw the top, front, and partial auxiliary view of each of the two statue bases shown in Fig. 13-1-D. Scale is 1:1.

5. On a B- or A3-size sheet select any two of the truncated prisms shown in Fig. 13-1-E, and draw the top, front, and auxiliary views for each. All views are to be complete with hidden lines shown. Scale is full or 1:1.

Assignment for Unit 13-2, Circular Features in Auxiliary Projection

6. On a B- or A3-size sheet, make a working drawing of one of the parts shown in Fig. 13-2-A or 13-2-B. Add hidden lines when required for clarity. Locational dimensions are to the center of auxiliary views. Scale is full or 1:1.

Fig. 13-1-A Angle bracket.

Fig. 13-1-B Cross-slide bracket.

Fig. 13-1-C Angle plate.

Fig. 13-1-D Statue bases.

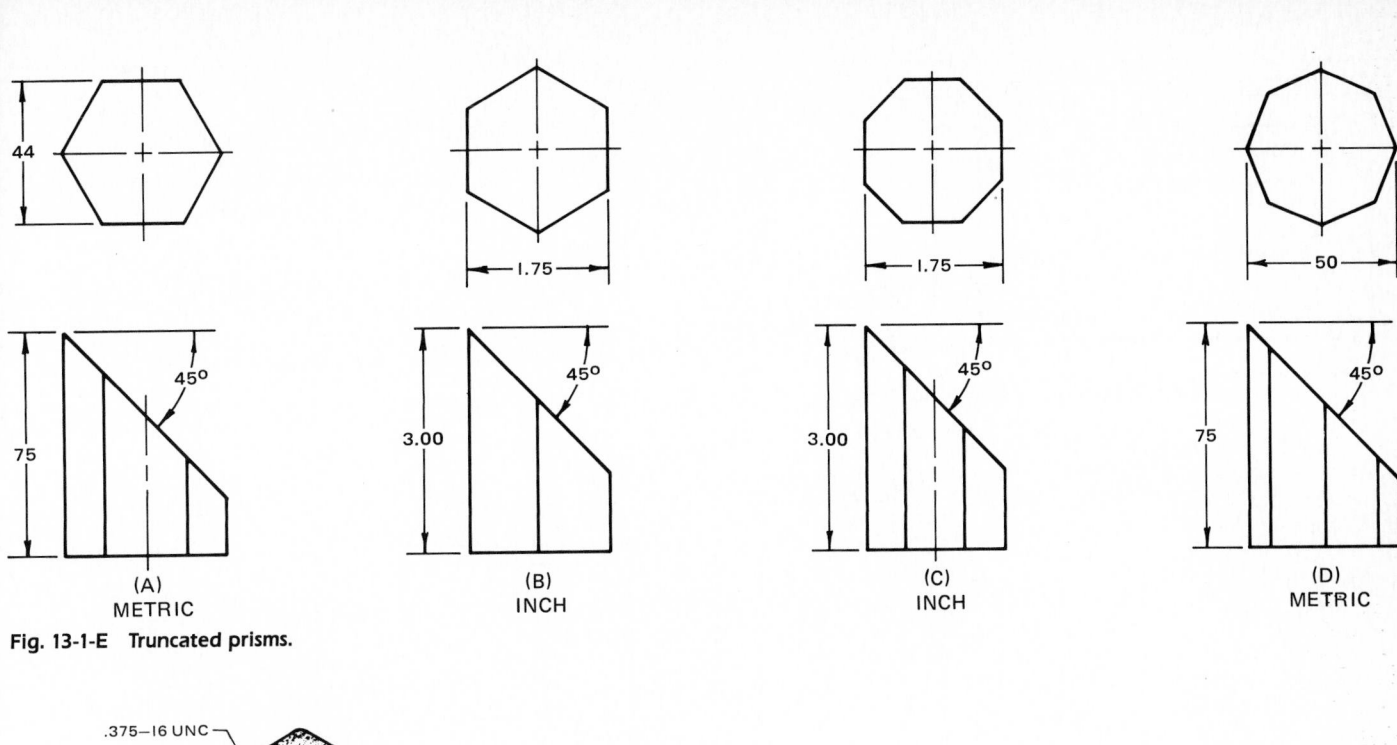

Fig. 13-1-E   Truncated prisms.

(A) METRIC
(B) INCH
(C) INCH
(D) METRIC

.375–16 UNC
1.38
1.00
∅ .875 THRU
R.88
.75
1.12
.44
.75
30°
5.00
.62
.75   .75
30°   1.26
2.76

DOVETAIL BOTH ENDS
ROUNDS AND FILLETS R.12
MATL – CI

Fig. 13-2-A   Control block.

2.00
4.00
4.00
1.50
6.00

DRAWING SET-UP

190
75
90
70

DRAWING SET-UP
"A3" SIZE PAPER

100
80
30°

MATL–CI

Fig. 13-2-B   Link.

∅ 15
R 15
BOTH ARMS
30
50
10
16
15
10
∅ 25
16
∅ 50
30
R 20
R 35
R 8
10
R 23
15
30

METRIC

Fig. 13-3-A Inclined stop.

Assignment for Unit 13-3, Multi-Auxiliary-View Drawings

7. On a B- or A3-size sheet, make a working drawing of one of the parts shown in Figs. 13-3-A to 13-3-C. The selection and placement of views are shown. Only partial views need be drawn except where noted, and hidden lines may be added to improve clarity. Scale is full or 1:1.

Assignment for Unit 13-4, Secondary Auxiliary Views

8. On a B- or A3-size sheet, make a working drawing of one of the parts shown in Fig. 13-4-A or 13-4-B. The selection and placement of views are shown beside the drawing. Only partial auxiliary views need be drawn, and hidden lines may be added to improve clarity. Scale is full or 1:1.

DRAWING SET-UP

Fig. 13-3-B Connecting bar.

Ø .625
2 HOLES

45°

.75
.38

1.75

2.50

1.26

.62

2.00

1.25

1.00

Ø .312
Ø .62 SFACE
4 HOLES

.25
.25

45°

.88

.62

.75

.88   .68   .88   .68

.50

3.00

4.00 SQ

Fig. 13-3-C   Angle slide.

2.50

2.50

.40

2.75

1.75

DRAWING SET-UP

ALL ROUNDS AND FILLETS R .12
MATL — CI

2.00

2.75

2.00

3.38

.75

DRAWING SET-UP

Fig. 13-4-A   Dovetail bracket.

.50

2.40

.30

60°

1.20

2.00

.25

60°

.70   1.20

1.40

60°

HEX .75 A/F

Ø .38
Ø .75 SFACE
4 HOLES

.40

30°

3.50

1.38

4.70

1.56

2.76

.60

.60

ROUNDS & FILLETS R .12
MATL—CI

LOCATION OF Ø 20 HOLE
ON AUXILIARY VIEW

75

60°

85

50

145

25

DRAWING SET-UP

Fig. 13-4-B   Pivot arm.

32

12

30°

30°

120°

32

8

16

10

40

60

90°

22

50

40

Ø 20

10

25

45°

METRIC

90

30°

110

CHAPTER 14

Pictorial Drawings

UNIT 14-1
Pictorial Drawings

Pictorial drawing is the oldest written method of communication known, but the character of pictorial drawing has continually changed with the advance of civilization. In this text only those kinds of pictorial drawings commonly used by the engineer and drafter are considered. Pictorial drawings are useful in design, construction or production, erection or assembly, service or repairs, and sales. They are used to explain complicated engineering drawings to people who do not have the training or ability to read the conventional multiview drawings; to help the designer work out problems in space, including clearances and interferences; to train new employees in the shop; to speed up and clarify the assembly of a machine or the ordering of new parts; to transmit ideas from one person to another, from shop to shop, or from salesperson to purchaser; and as an aid in developing the power of visualization. The type of pictorial drawing used depends on the purpose for which it is drawn.

There are three general types into which pictorial drawings may be divided: axonometric, oblique, and perspective. These three differ from one another in the fundamental scheme of projection, as shown in Fig. 14-1-1.

AXONOMETRIC PROJECTION[1]

A projected view in which the lines of sight are perpendicular to the plane of projection, but in which the three faces of a rectangular object are all inclined to the plane of projection, is called an *axonometric projection*. See Fig. 14-1-2. The projections of the three

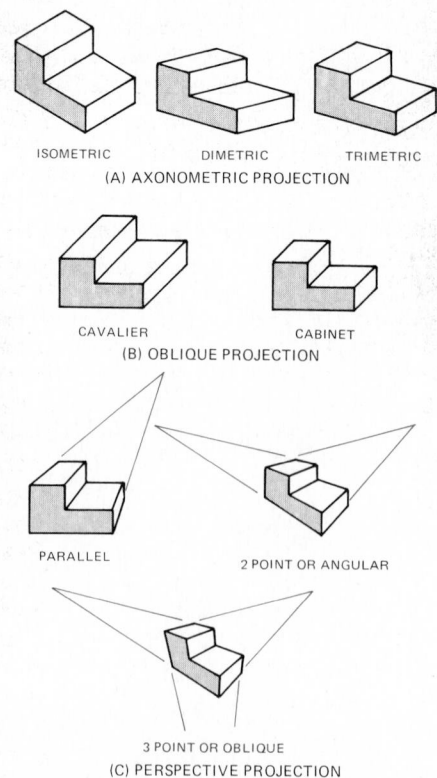

Fig. 14-1-1 Types of pictorial drawings.

principal axes may make any angle with one another except 90°. Axonometric drawings, as shown in Figs. 14-1-3 and 14-1-4, are classified into three forms: *isometric drawings*, where the three principal faces and axes of the object are equally inclined to the plane of projection; *dimetric drawings*, where two of the three principal faces and axes of the object are equally inclined to the plane of projection; and *trimetric drawings*, where all

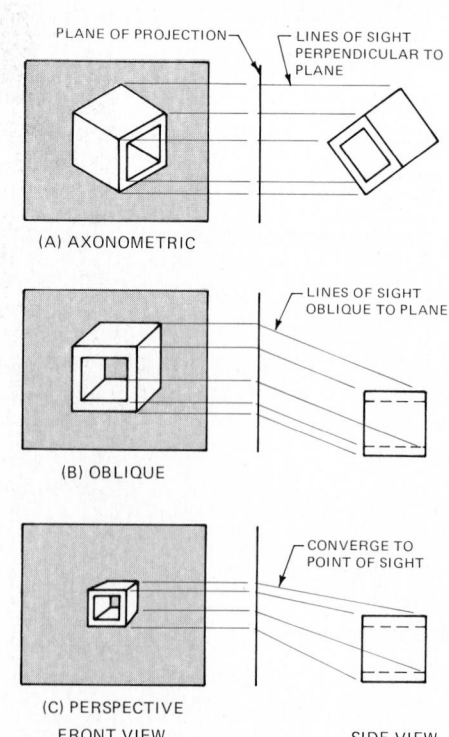

Fig. 14-1-2 Kinds of projection.

(A) ISOMETRIC PROJECTION

(B) DIMETRIC PROJECTION

(C) TRIMETRIC PROJECTION

Fig. 14-1-3   Types of axonometric projection. (Graphic Standard Instruments Co.)

ISOMETRIC

USE OF THIS SET ENABLES YOU TO SHOW THE OBJECT FROM THESE 48 VIEWPOINTS:

DIMETRIC

OBJECT DRAWN UPRIGHT

OBJECT ROTATED 90° CW OR CCW

USE OF THIS SET ENABLES YOU TO SHOW THE OBJECT FROM 144 VIEWPOINTS:

TRIMETRIC

OBJECT DRAWN UPRIGHT

OBJECT ROTATED 90° CW OR CCW

USE OF THIS SET ENABLES YOU TO SHOW THE OBJECT FROM 288 VIEWPOINTS:

Fig. 14-1-4   Axonometric projection. (Graphic Standard Instruments Co.)

(1) REVOLVING THE OBJECT

DRAWING OBJECT TO ACTUAL MEASUREMENTS

NOT TRUE LENGTH APPROX. 0.8 ACTUAL SIZE

(2) TIPPING THE OBJECT

(3) ISOMETRIC PROJECTION

(4) ISOMETRIC DRAWING

(A) ISOMETRIC PROJECTION

ISOMETRIC AXES

(B) ISOMETRIC AXES

TOP

FRONT   SIDE

Fig. 14-1-5   Isometric axes and projections.

three faces and axes of the object make different angles with the plane of projection. The most popular form of axonometric projection is the isometric.

ISOMETRIC DRAWINGS[2]

This method is based on a procedure of revolving the object at an angle of 45° to the horizontal, so that the front corner is toward the viewer, then tipping the object up or down at an angle of 35° 16'. See Fig. 14-1-5. When this is done to a cube, the three faces visible to the viewer appear equal in shape and size, and the side faces are at an angle of 30° to the horizontal. If the isometric view were actually projected from a view of the object in the tipped position, the lines in the isometric view would be foreshortened and would, therefore, not be seen in their true length. To simplify the drawing of an isometric view, the actual measurements of the object are used. Although the object appears slightly larger without the allowance for foreshortening, the proportions are not affected. All isometric drawings are started by constructing the isometric axes, which are a vertical line for height and isometric lines to left and right, at an angle of 30° from the horizontal, for length and width. The three faces seen in the isometric view are the same faces that would be seen in the normal orthographic views: top, front, and side. Figure 14-1-5B illustrates the selection of the front corner (A), the construction of the isometric axes, and the completed isometric view. Note that all lines are drawn to their true length, measured along the isometric axes, and that hidden lines are usually omitted. Vertical edges are represented by vertical lines, and horizontal edges by lines at 30° to the horizontal. Two techniques can be used for making an isometric drawing of an irregularly shaped object, as illustrated in Fig. 14-1-6. In one method, the object is divided mentally into a number of sections and the sections are drawn one at a time in their proper relationship to one another. In the second method, a box is drawn with the maximum height, width, and depth of the object; then the parts of the box that are not part of the object are removed, leaving the pieces that form the total object.

Fig. 14-1-6 Developing an isometric drawing.

(A) DEVELOPMENT BY SECTIONS

(B) BOX CONSTRUCTION

(A) BLOCK IN FEATURES

(B) DARKEN ISOMETRIC LINES

(C) COMPLETE NONISOMETRIC LINES

Fig. 14-1-8 Sequence in drawing an object having nonisometric lines.

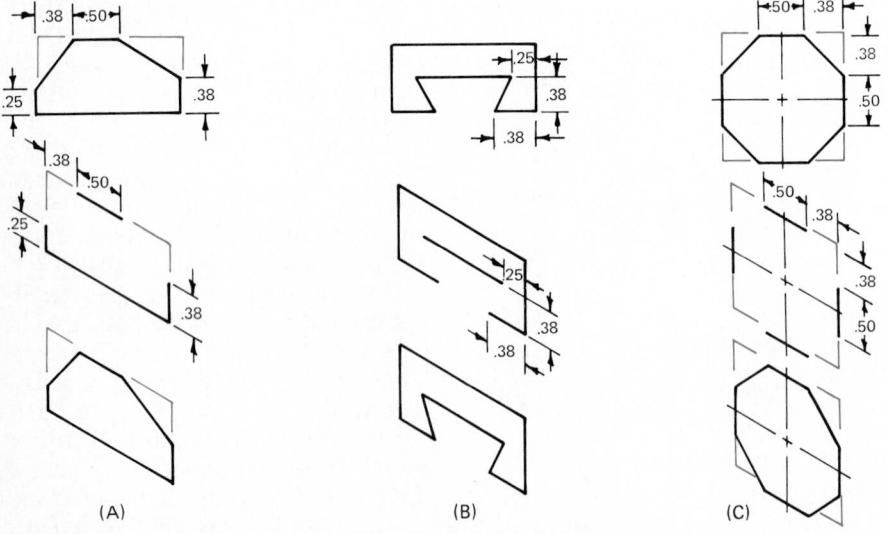

(A)   (B)   (C)

Fig. 14-1-7 Examples in the construction of nonisometric lines.

Nonisometric Lines

Many objects have sloping surfaces that are represented by sloping lines in the orthographic views. In isometric drawing, sloping surfaces appear as *nonisometric* lines. To draw them, locate their endpoints, found on the ends of isometric lines, and join them with a straight line. Figures 14-1-7 and 14-1-8 illustrate examples in the construction of nonisometric lines.

Dimensioning Isometric Drawings

At times, an isometric drawing of a simple object may serve as a working drawing. In such cases, the necessary dimensions and specifications are placed on the drawing.

Dimension lines, extension lines, and the line being dimensioned should be in the same plane. Arrowheads, which should be long and narrow, should be in the plane of the dimension and extension lines. See Fig. 14-1-9.

Fig. 14-1-9 Orienting the dimension line, arrowhead, and extension line.

Fig. 14-1-10 Isometric dimensioning.

Unidirectional dimensioning is the preferred method of dimensioning isometric drawings. The letters and numbers should be vertical and read from the bottom of the sheet. An example of this type of dimensioning is shown in Fig. 14-1-10.

Since the isometric is a one-view drawing, it is not usually possible to avoid placing dimensions on the view or across dimension lines. However, this practice should be avoided whenever possible.

Isometric Grid Paper

Isometric grid sheets are another time-saving device. Designers and engineers frequently use isometric grid paper on which they sketch their ideas and designs. See Fig. 14-1-11. Many companies, such as those which prepare pipe drawings, have large drawing sheets made with nonreproducible isometric grid lines.

Fig. 14-1-11 Isometric grid paper.

References and Source Material

1. Extracted from *American Drafting Standards Manual, Pictorial Drawing* (ANSI Y14.4), with the permission of the publisher. The American Society of Mechanical Engineers, 345 East 47th Street, New York, N.Y. 10017.
2. General Motors Corporation.

ASSIGNMENTS

See Assignments 1 through 3 for Unit 14-1 on page 298.

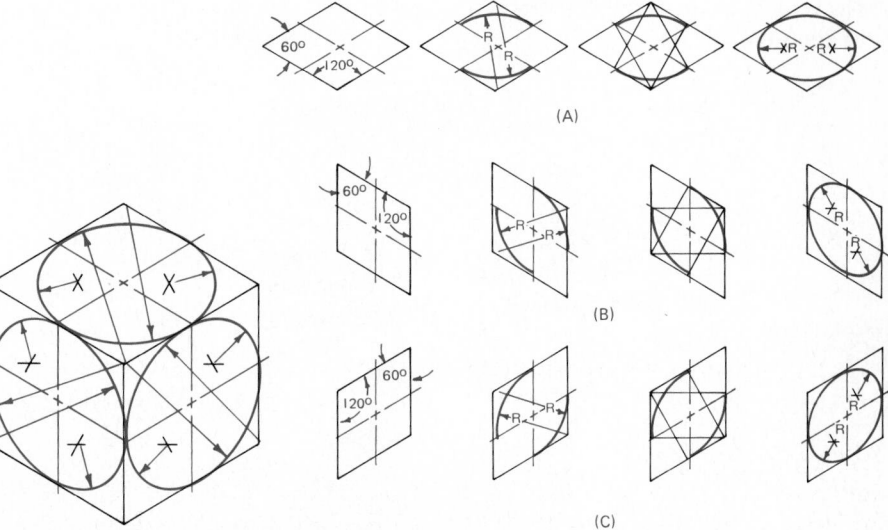

Fig. 14-2-2 Sequence in drawing isometric circles.

UNIT 14-2
Curved Surfaces in Isometric

CIRCLES AND ARCS IN ISOMETRIC

A circle on any of the three faces of an object drawn in isometric has the shape of an ellipse. See Fig. 14-2-1. Figure 14-2-2 illustrates the steps in drawing circular features on isometric drawings.

1. Draw the center lines and a square, with sides equal to the circle diameter, in isometric.
2. Using the obtuse-angled (120°) corners as centers, draw arcs tan-

(A) A SQUARE DRAWN IN THE THREE ISOMETRIC POSITIONS

CIRCLE TOUCHES SQUARE AT MIDPOINT OF EACH SIDE

(B) A CIRCLE PLACED INSIDE A SQUARE AND DRAWN IN THE THREE ISOMETRIC POSITIONS

Fig. 14-2-1 Circles in isometric.

gent to the sides forming the obtuse-angled corners, stopping at the points where the center lines cross the sides of the square.

3. Draw construction lines from these same points to the opposite obtuse-angled corners. The points at which these construction lines intersect are the centers for arcs drawn tangent to the sides forming the acute-angled corners, meeting the first arcs.

In drawing concentric circles, each circle must have its own set of centers for the arcs, as shown in Fig. 14-2-3.

The same technique is used for drawing part-circles (arcs). See Fig. 14-2-4. Construct an isometric square with sides equal to twice the radius, and draw that portion of the ellipse necessary to join the two faces. When these faces are parallel, draw half of an ellipse (one long radius and one short radius); when they are at an obtuse angle (120°), draw one long radius; and when they are at an acute angle (60°), draw one short radius.

ISOMETRIC TEMPLATES

For convenience and time saving, isometric ellipse templates should be used whenever possible. A wide variety of elliptical templates are available. The template shown in Fig. 14-2-5 combines ellipses, scales, and angles. Markings on the ellipses coincide with the center lines of the holes which speeds up drawing circles and arcs. Figure 14-2-6 shows the same part as shown in Fig. 14-2-4 but with the arcs and circles being constructed with a template.

SKETCHING CIRCLES AND ARCS

In sketching circles and arcs on isometric grid paper, locate the center lines first, then lightly sketch in construction boxes (isometric squares) where the circles and arcs should be. See Fig. 14-2-7. Sketch the ellipse (isometric circle) just touching the center of each of the four sides of the square.

DRAWING IRREGULAR CURVES IN ISOMETRIC

To draw curves other than circles or arcs, the plotting method shown in Fig. 14-2-8 is used

1. Draw an orthographic view, and divide the area enclosing the curved line into equal squares.
2. Produce an equivalent area on the isometric drawing, showing the offset squares.
3. Take positions relative to the squares from the orthographic view, and plot them on the corresponding squares on the isometric view.
4. Draw a smooth curve through the established points with the aid of an irregular curve.

ASSIGNMENTS

See Assignments 4 through 6 for Unit 14-2 on page 299.

Review for Assignments
Unit 14-1 Dimensioning Isometric Drawings

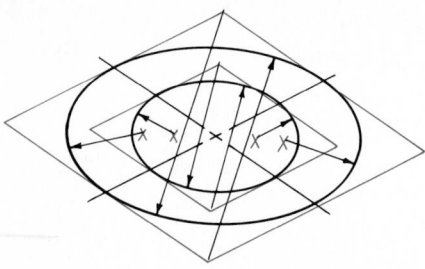
Fig. 14-2-3   Drawing concentric isometric circles.

Fig. 14-2-5   Isometric ellipse template.

Fig. 14-2-6   Circles and arcs drawn with isometric ellipse templates.

Fig. 14-2-4   Drawing isometric arcs.

Fig. 14-2-7   Isometric sketching paper.

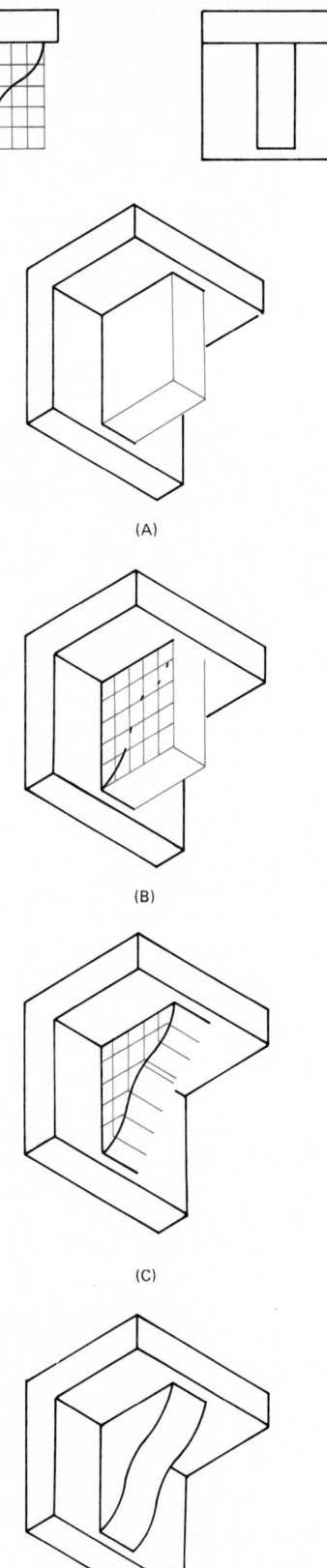

(A)

(B)

(C)

(D)

Fig. 14-2-8 Curves drawn in isometric by means of offset measurements.

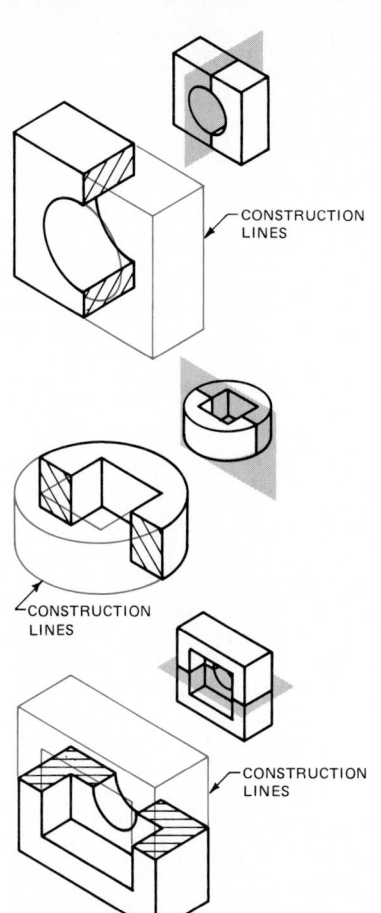

CONSTRUCTION LINES

CONSTRUCTION LINES

CONSTRUCTION LINES

Fig. 14-3-1 Examples of isometric full sections.

UNIT 14-3
Common Features in Isometric

ISOMETRIC SECTIONING

Isometric drawings are generally made as outside views, but sometimes a sectional view is needed. The section is taken on an isometric plane, that is, on a plane parallel to one of the faces of the cube. Figure 14-3-1 shows isometric full sections taken on a different plane for each of three objects. Note the construction lines indicating the part that has been cut away. Isometric half sections are illustrated in Fig. 14-3-2. Notice the outlines of the cut surfaces in A and B. The cut method is to draw the complete outside view and the isometric cutting plane.

When an isometric drawing is sectioned, the section lines are drawn at an angle of 60° with the horizontal or in a horizontal position, depending on where the cutting-plane line is located. In half sections, the section lines are sloped in opposite directions, as shown in Fig. 14-3-2.

FILLETS AND ROUNDS

For most isometric drawings of parts having small fillets and rounds, the

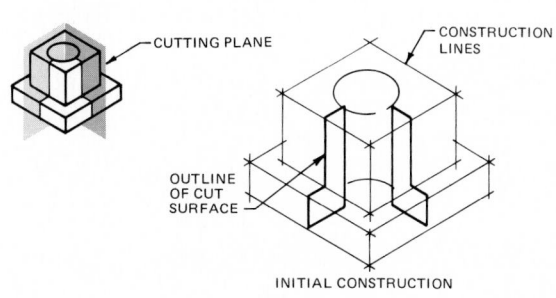

CUTTING PLANE

CONSTRUCTION LINES

OUTLINE OF CUT SURFACE

INITIAL CONSTRUCTION

FINISHED DRAWING

(A) PART I

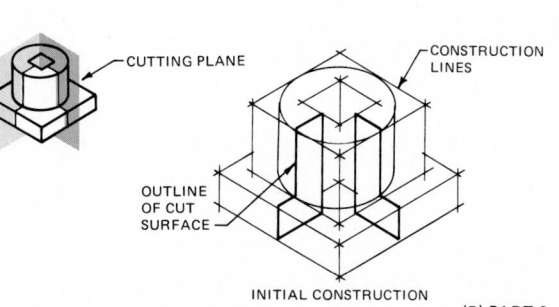

CUTTING PLANE

CONSTRUCTION LINES

OUTLINE OF CUT SURFACE

INITIAL CONSTRUCTION

FINISHED DRAWING

(B) PART 2

Fig. 14-3-2 Examples of isometric half sections.

(A) CURVED LINE

(B) STRAIGHT LINE

Fig. 14-3-3 Representation of fillets and rounds.

ACCEPTABLE

PREFERRED

ACCEPTABLE

Fig. 14-3-5 Conventional breaks.

ACCEPTABLE

PREFERRED

Fig. 14-3-4 Representation of threads in isometric.

Fig. 14-3-6 Isometric assembly drawings.

Ø 5.6
4 HOLES

736 PACKING NUT ASSEMBLY

921 ARM SCREW

938 FILLING PLUG

906 PINION

903 END PLUG

927 REGULATING SCREW ASSEMBLY (2 REQD)

929 REGULATING SCREW PACKING (2 REQD)

70I-N SHELL INCLUDING BEARING

730 SPRING

913-I PISTON

903 END PLUG

Fig. 14-3-7 Exploded isometric assembly drawing. (Yale and Towne Inc.)

adopted practice is to draw the corners as sharp features. However, when it is desirable to represent the part, normally a casting, as having a more realistic appearance, either of the methods shown in Fig. 14-3-3 may be used.

THREADS

The conventional method for showing threads in isometric is shown in Fig. 14-3-4. The threads are represented by a series of ellipses uniformly spaced along the center line of the thread. The spacing of the ellipses need not be the spacing of the actual pitch.

BREAK LINES

For long parts, break lines should be used to shorten the length of the drawing. Freehand breaks are preferred, as shown in Fig. 14-3-5.

ISOMETRIC ASSEMBLY DRAWINGS

Regular or exploded assembly drawings are frequently used in catalogs and sales literature, as illustrated by Figs. 14-3-6 and 14-3-7.

ASSIGNMENTS

See Assignments 7 through 12 for Unit 14-3 on page 300.

Review for Assignments
Unit 6-4   Bill of Materials
Unit 6-4   Assembly Drawings

UNIT 14-4
Oblique Projection

This method of pictorial drawing is based on the procedure of placing the object with one face parallel to the

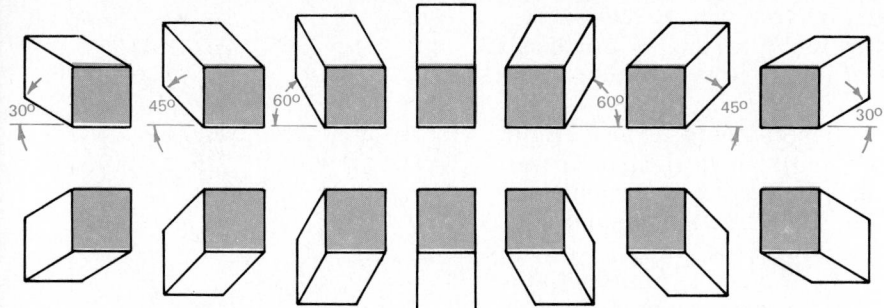

Fig. 14-4-1 Typical positions of receding axes for oblique projections.

Fig. 14-4-2 Two general rules for oblique drawings.

frontal plane and placing the other two faces on oblique (or receding) planes, to left or right, top or bottom, at a convenient angle. The three axes of projection are *vertical*, *horizontal*, and *receding*. Figure 14-4-1 illustrates a cube drawn in typical positions with the receding axis at 60°, 45° and 30°. This form of projection has the advantage of showing one face of the object without distortion. The face with the greatest irregularity of outline or contour, or the face with the greatest number of circular features, or the face with the longest dimension, faces the front. See Fig. 14-4-2.

Two types of oblique projection are used extensively. In *cavalier oblique*, all lines are drawn to their true length, measured on the axes of the projection. In *cabinet oblique*, the lines on the receding axis are shortened by one-half their true length to compensate for

distortion and to approximate more closely what the human eye would see. For this reason, and because of the simplicity of projection, cabinet oblique is a commonly used form of pictorial representation, especially when circles and arcs are to be drawn. Figure 14-4-3 shows a comparison of cavalier and cabinet oblique. Note that hidden lines are omitted unless required for clarity. Many of the drawing techniques for isometric projection apply to oblique projection. Figure 14-4-4 illustrates the construction of an irregularly shaped object by the box method.

INCLINED SURFACES

Angles which are parallel to the picture plane are drawn as their true size. Other angles can be laid off by locating the ends of the inclined line.

CAVALIER PROJECTION   CABINET PROJECTION

Fig. 14-4-3 Types of oblique projection.

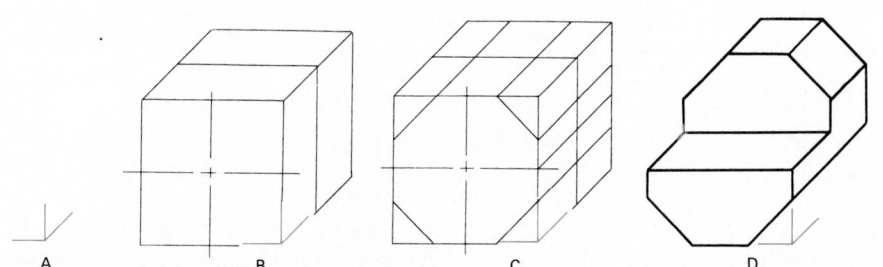

A   B   C   D

Fig. 14-4-4 Oblique construction by the box method.

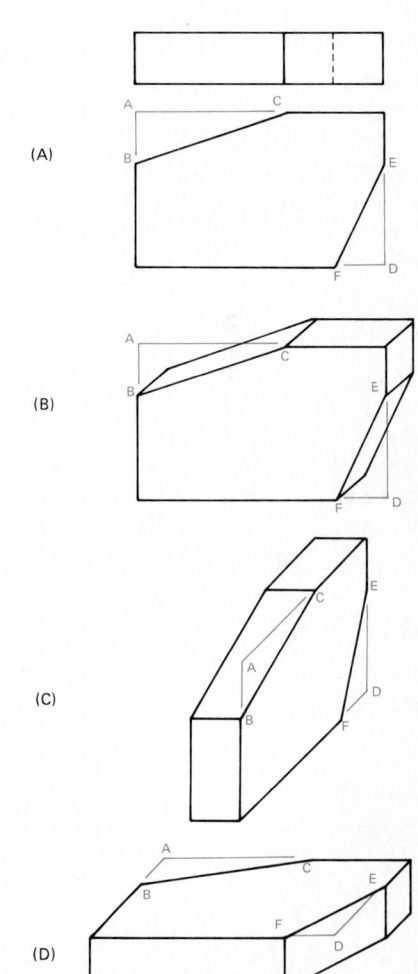

Fig. 14-4-5 Drawing inclined surfaces.

A part with notched corners is shown in Fig. 14-4-5A. An oblique drawing with the angles parallel to the picture plane is shown at Fig. 14-4-5B. In Fig. 14-4-5C the angles are parallel to the profile plane. In each case the angle is laid off by measurement parallel to the oblique axes, as shown by the construction lines. Since the part, in each case, is drawn in cabinet oblique, the receding lines are shortened by one-half their true length.

OBLIQUE SKETCHING

Specially designed oblique sketching paper with 45° lines is available and, like isometric sketching paper, is used extensively by engineers and drafters. See Fig. 14-4-6.

DIMENSIONING AN OBLIQUE DRAWING

Dimension lines are drawn parallel to

Fig. 14-4-6 Oblique sketching paper.

the axes of projection. Extension lines are projected from the horizontal and vertical object lines whenever possible.

The dimensioning of an oblique drawing is similar to that of an isometric drawing. The recommended method is unidirectional dimensioning, which is shown in Fig. 14-4-7. As in isometric dimensioning, usually it is necessary to place some dimensions directly on the view.

ASSIGNMENTS

See Assignments 13 through 15 for Unit 14-4 on pages 301 and 302.

Review for Assignments
Unit 14-1   Dimensioning Isometric Drawings

UNIT 14-5
Common Features in Oblique

CIRCLES AND ARCS

Whenever possible, the face of the object having circles or arcs should be selected as the *front* face, so that such circles or arcs can be easily drawn in their true shape. See Fig. 14-5-1. When circles or arcs must be drawn on one of

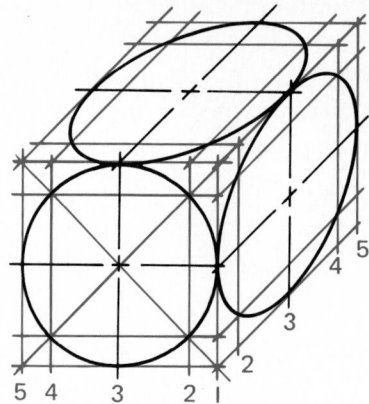

Fig. 14-5-2 Drawing oblique circles by means of offset measurements.

the oblique faces, the offset measurement method illustrated in Fig. 14-5-2 may be used.

1. Draw an oblique square about the center lines, with sides equal to the diameter.
2. Draw a true circle within the oblique square, and establish equally spaced points about its circumference.
3. Project these point positions to the edge of the oblique square, and draw lines on the oblique axis from these positions. Similarly spaced lines are drawn on the other axis, forming offset squares and giving intersection points for the oval shape.

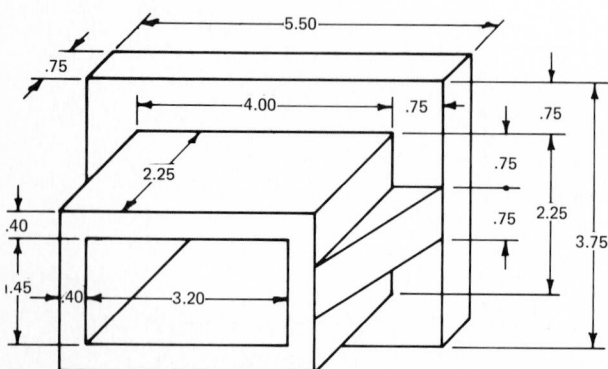

Fig. 14-4-7 Dimensioning an oblique drawing.

SCHEMATIC OF A COMPLETELY AUTOMATIC REGISTRATION CONTROL SYSTEM MAINTAINING THE LOCATION OF CUTOFF ON A CONTINUOUS PRINTED WEB.

Fig. 14-5-1 Application of oblique drawing.

Fig. 14-5-3 Circles parallel to the picture plane are true circles; on other planes, ellipses.

(A) CURVED LINE   (B) STRAIGHT LINE

Fig. 14-5-6 Representing rounds and fillets.

Ø.562, 4 HOLES

2.00
1.25
1.12
2.24
3.24
.50
.50
.50
4.00
5.00
Ø1.006
.250 X .125 KEYSEAT

Fig. 14-5-4 Construction and dimensioning of an oblique object.

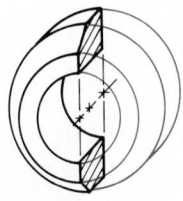

(A) FULL SECTION   (B) HALF SECTION

Fig. 14-5-5 Oblique full section and an oblique half section.

Fig. 14-5-7 Representation of threads in oblique.

Another method used when circles or arcs must be drawn on one of the oblique surfaces is the *four-center method*. In Fig. 14-5-3A a circle is shown as it would be drawn on a front plane, a side plane, and a top plane.

In Fig. 14-5-3B, the oblique drawing has some arcs in a horizontal plane. In Fig. 14-5-3C, the oblique drawing shown has some arcs in a profile plane.

Circles not parallel to the picture plane when drawn by the approximate method are not pleasing but are satisfactory for some purposes. Ellipse templates, when available, should be used because they reduce drawing time and give much better results. If a template is used, the oblique circle should first be blocked in as an oblique square in order to locate the proper position of the circle. Blocking in the circle first also helps the drafter select

the proper size and shape of the ellipse. The construction and dimensioning of an oblique part are shown in Fig. 14-5-4.

OBLIQUE SECTIONING

Oblique drawings are generally made as outside views, but sometimes a sectional view is necessary. The section is taken on a plane parallel to one of the faces of an oblique cube. Figure 14-5-5 shows an oblique full section and an oblique half section. Note the construction lines which show the part that has been cut away.

TREATMENT OF CONVENTIONAL FEATURES

Fillets and Rounds Small fillets and rounds normally are drawn as sharp corners. When it is desirable to show the corners rounded, then either of the methods shown in Fig. 14-5-6 is recommended.

Threads The conventional method of showing threads in oblique is shown in Fig. 14-5-7. The threads are represented by a series of circles uniformly spaced along the center of the thread.

ACCEPTABLE
PREFERRED
ACCEPTABLE
PREFERRED

Fig. 14-5-8 Conventional breaks.

The spacing of the circles need not be the spacing of the pitch.

Breaks Figure 14-5-8 shows the conventional method for representing breaks.

ASSIGNMENTS

See Assignments 16 and 17 for Unit 14-5 on page 302.

(A) PARALLEL PERSPECTIVE

(B) ANGULAR PERSPECTIVE

Fig. 14-6-1   Perspective drawings.

PLAN VIEW

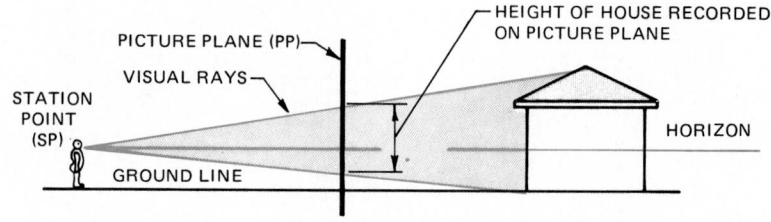

ELEVATION

PICTURE RECORDED ON PICTURE PLANE AS SEEN BY OBSERVER

Fig. 14-6-2   Perspective drawings.

UNIT 14-6
Perspective Projection

Perspective is a method of drawing that depicts a three-dimensional object on a flat plane as it appears to the eye. See Fig. 14-6-1. A pictorial drawing made by the intersection of the picture plane with lines of sight converging from points on the object to the point of sight, which is located at a finite distance from the picture plane, is called a *perspective*. See Fig. 14-6-2.

Perspective drawings are more realistic than axonometric or oblique drawings because the object is shown as the eye would see it. Since they are far more difficult to draw than the other types of pictorial drawings, their use in drafting is limited mainly to production or presentation illustrations and illustrations of proposed structures by architects.

The main elements of a perspective drawing are the *picture plane* (plane of

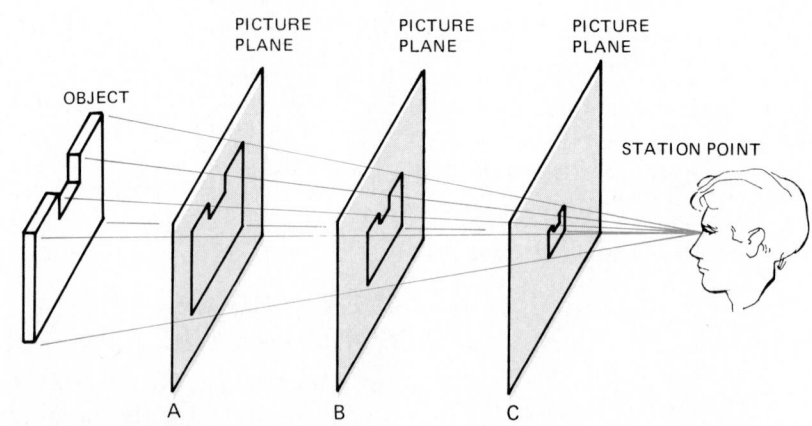

Fig. 14-6-3   Location of the picture plane.

projection), the *station point* (the position of the observer's eye when he or she is viewing the object), the *horizon* (an imaginary horizontal line taken at eye level), the *vanishing point or points* (a point or points on the horizon where all the receding lines converge), and the *ground line* (the base line of the

picture plane and object).

To avoid undue distortion in perspective, the point of sight (station point) should be located so that the cone of rays from the observer's eye has an angle at the apex not greater than 30°. This would place the station point a distance away from the outside

portion of the object of approximately two to two and one-half times the width of the object being viewed. See Figs. 14-6-2 and 14-6-3.

TYPES OF PERSPECTIVE DRAWINGS

There are three types of perspective drawings

1. Parallel: One vanishing point
2. Angular: Two vanishing points
3. Oblique: Three vanishing points

In industry they are normally referred to as one-point, two-point, and three-point perspectives, respectively. See Fig. 14-6-4. Only parallel and angular perspectives are covered in this text.

Parallel, or One-Point, Perspective

Parallel-perspective drawings are similar to oblique drawings, except the

Fig. 14-6-5 Parallel or one-point perspective.

receding lines all converge at one point on the horizon. In drawing a parallel-perspective drawing, one face of the object is placed on the picture-plane line so that it will be drawn in its true size and shape, as shown in Fig. 14-6-5. The *PP* line shown in the top view represents the picture plane line, and point *SP* (station point) is the position of the observer. The lines of the object, which are not on the picture plane, are found by projecting lines down from the top view from the point of intersection of the visual ray and the picture plane, as shown by point *N* in Fig. 14-6-5A(1).

Where the true height of a line or a point does not lie on the picture plane, such as point *P* in Fig. 14-6-5A(2), the

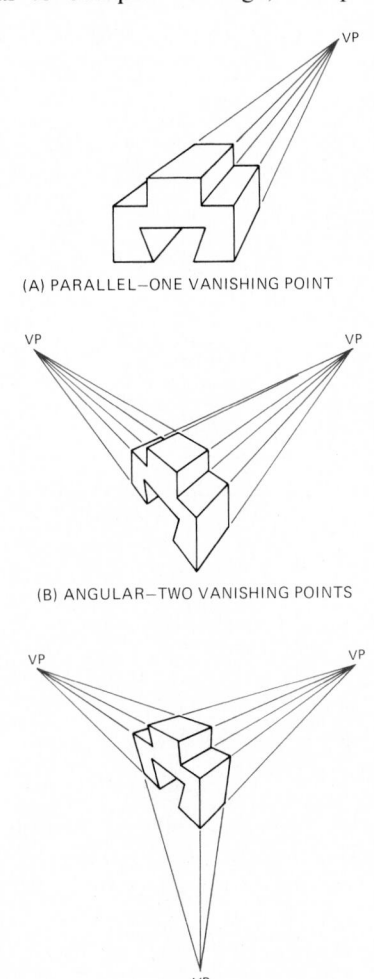

(A) PARALLEL—ONE VANISHING POINT

(B) ANGULAR—TWO VANISHING POINTS

(C) OBLIQUE—THREE VANISHING POINTS

Fig. 14-6-4 Perspective drawings.

Fig. 14-6-6 Construction of a one-point perspective.

PICTURE PLANE

DOOR HEIGHT

HORIZON

VP

GROUND LINE

WINDOW HEIGHT

FIREPLACE HEIGHT

Fig. 14-6-7 Parallel-perspective drawing of an interior of a house.

true height may be found by extending line *PR* to point *S* on the picture plane. Since point *S* lies on the picture plane and is the same height as point *P*, it may readily be found on the perspective drawing. Point *P* will lie on the receding line joining point *S* to the line *VP*.

In drawing a one-point perspective, a side or front view and a top view are normally drawn first—the top view to locate the part with respect to the picture plane and the side or front view to obtain the height of the various features. Figure 14-6-6 shows a simple, one-point perspective drawing with construction lines.

One of the most common uses of a parallel-perspective drawing is for representing the interior of a building. With this type of drawing, the vanishing point is located inside the room and is normally at eye level. See Figs. 14-6-7 and 14-6-10.

(A) EXTERIOR GRID

(B) INTERIOR GRID

Fig. 14-6-8 Parallel-perspective grid types.

.50
1.50
1.00
1.50
4.00
2.00
Ø.50
2.00
1.25
.50

PART

Fig. 14-6-9 Part drawn on parallel-perspective paper—exterior grid.

SCALE: 1 GRID = .50 in.

Fig. 14-6-10   Interior of a room drawn on parallel-perspective paper—interior grid.

ASSIGNMENTS

See Assignments 18 through 20 for Unit 14-6 on page 303.

UNIT 14-7
Angular, or Two-Point, Perspective

Two-point perspective is used quite extensively for architectural and product illustration, as shown in Fig. 14-7-1. *Angular-perspective* drawings are similar to axonometric drawings except that the receding lines converge at two vanishing points located on the

Parallel-Perspective Grid

A variety of perspective grid sheets is available, which enables the drafter to produce perspective drawings in less time than the conventional manner. Using a grid eliminates the tedious effort of establishing and projecting from the vanishing points for each individual feature. It also eliminates the problem of having the vanishing points located, in many instances, beyond the drawing area.

The cube grid, which is most widely used, has two basic variations: an exterior grid and an interior grid. See Fig. 14-6-8. The grid sizes are dependent upon the desired scale of the parts to be drawn. The height and width planes are subdivided into identical increments, each increment representing any convenient size, such as 1½ in., ¼″ = 1′, or 10, 100, or 1000 mm. The plane or surface representing the depth is subdivided into increments which are proportionately foreshortened as it recedes from the picture plane and thus creates the perspective illusion. See Figs. 14-6-9 and 14-6-10.

References and Source Material

1. Extracted from *American Drafting Standards Manual, Pictorial Drawing* (ANSI Y14.4), with the permission of the publisher. The American Society of Mechanical Engineers, 345 East 47th Street, New York, N.Y. 10017.

Fig. 14-7-1   Perspective drawing.

Fig. 14-7-2   Angular-perspective drawing of a prism.

Fig. 14-7-3 Angular-perspective view of an object which does not touch the picture plane.

Fig. 14-7-4 Constructing a circle in angular perspective.

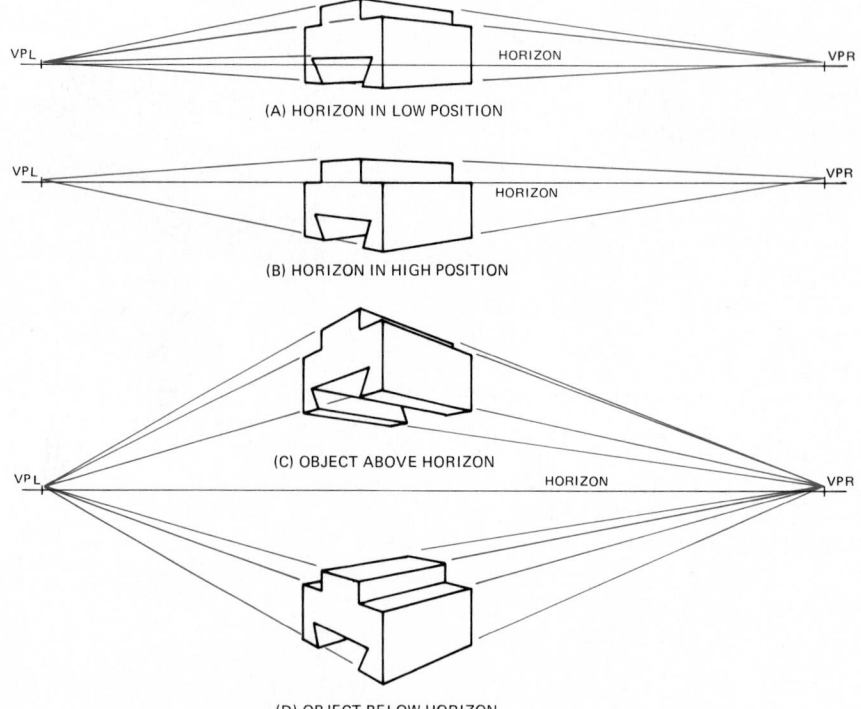

Fig. 14-7-5 Horizon lines.

horizon. Normally the height or vertical lines are parallel to the picture plane, and the length and width lines recede.

The construction for a simple prism is shown in Fig. 14-7-2. Since line 1-2 rests on the picture plane, it will appear as its true height on the perspective drawing and will be located directly below line 1-2 on the top view. The next step is to join points 1 and 2 with light receding lines to both vanishing points. These receding lines represent the width and length lines of the prism; the width lines recede to *VPL* and the length lines recede to *VPR*. Since line 3-4 on the top view does not rest on the picture plane, it will not appear in its true height nor as its true distance from line 1-2 in the perspective. To find its position on the perspective drawing, joint line 3-4, which appears as a point in the top view, to *SP* with a visual ray line. Where this visual ray line intersects the picture plane at *C*, project a vertical line down to the perspective view until it intersects the receding lines 1-*VPR* and 2-*VPR* at points 3 and 4 respectively. Line 5-6 may be found in the same manner. Next join point 3 to *VPL* and point 5 to *VPR* with light receding lines. The intersection of these lines is point 7.

Lines Not Touching on the Picture Plane Figure 14-7-3 illustrates the construction of a perspective drawing where none of the object lines touch the picture plane.

All these lines can be constructed by using the following procedure, which locates the position and size of lines 1-2 and 3-4. Extend line 1-3 in the top view to intersect the picture plane at point *C*. Project a line down from *C* to intersect horizontal lines 1-*D* and 2-*E* at *D* and *E*, respectively. Had line 1-2 been located at *C* in the top view, it would have appeared at its true height and at *D-E* on the perspective. Join points *D* to *VPR* and *E* to *VPR* with light receding lines. Somewhere along these lines are points 1, 2, 3, and 4. Next join lines 1-2 and 3-4 in the top view to *SP* with visual ray lines. Where these visual ray lines intersect the picture plane at *F* and *G*, respectively, project vertical lines down to the perspective view intersecting line *D-VPR* at 1 and 3 and *E-VPR* at 2 and 4.

Construction of Circles and Curves in Perspective Circles and curves may be constructed in perspective, as illustrated in Fig. 14-7-4. Using orthographic projections, oriented with respect to the subject in the plan and side views, plot and label the desired points (using numbers) on the curved surfaces. From the plan view project these points to the picture plane, then vertically down to the perspective view. Project horizontally from the side view to the true-height line in the perspective view, the height of the plotting numbers. The position of the plotting numbers may now be located on the perspective view. Locate the points of intersection of the lines projected down from the picture plane with the visual ray lines receding to the right vanishing point from the appropriate numbers on the true-height line.

Horizon Line Figure 14-7-5 illustrates different effects produced by repositioning the object with respect to the horizon.

ANGULAR-PERSPECTIVE GRIDS

Exterior Grid When the three adjacent exterior planes of the cube are developed, the resultant image is referred to as an *exterior grid*. In using this grid, the points are projected from the top plane downward and from the picture planes away from the observer. See Figs. 14-7-6 and 14-7-7.

Interior Grid When the three adjacent interior planes of the cube are exposed and developed, the resultant image is

(A) EXTERIOR GRID   **(B) INTERIOR GRID**

Fig. 14-7-7   Angular-perspective grids.

HORIZON   HORIZON

(A) BIRDS-EYE VIEW   **(B) WORMS-EYE VIEW**

Fig. 14-7-8   Grid variations.

Fig. 14-7-6   Angular-perspective grid.

Fig. 14-7-9   Angular-perspective grid application.

referred to as an *interior grid*. In using this grid, the points are projected from the base plane upward and from the picture planes toward the observer. The choice of usage of either variation is a matter of individual preference. Each produces the same results.

Two further variations of both the exterior and interior grids are known as the *bird's-eye* and *worm's-eye* grids. These effects are achieved by rotating the vertical plane of the grid about the horizon line.

Objects drawn in the bird's-eye grid appear as if they were being viewed from above the horizon line, as seen in Fig. 14-7-8. Objects drawn in the worm's-eye grid appear as if they were being viewed from below the horizon line.

Grid Increments

The three surfaces or planes of the grid are subdivided into multiple vertical and horizontal increments. Each increment is proportionately foreshortened as it recedes from the picture plane and thus creates the perspective illusion. The grid increments can be any size desired. See Fig. 14-7-9.

Reference and Source Material

1. General Motors Corporation.

ASSIGNMENTS

See Assignments 21 and 22 for Unit 14-7 on page 304.

UNIT 14-8
Technical Illustration

Technical illustrations have an important place in all phases of engineering drawing. They form an essential part of technical manuals and catalogs, as well as illustrations appearing in technical magazines. However, this unit will not cover technical illustration techniques beyond the scope of the general drafting office.

Technical illustration drawings vary from simple sketches to rather extensive shaded drawings. They may be based upon any of the pictorial methods: isometric, oblique, or perspec-

tive, as shown in Fig. 14-8-1. The project being illustrated may be an assembly, a single part, or just a portion of a part. It may take the form of an exterior view, a sectional view, or a phantom view. The purpose in all cases is to provide a clear and easily

understood drawing. See Fig. 14-8-2.

In many cases the drafting department may do the simple illustrations. However, for most purposes the special requirements of such drawings call for work by a professional technical illustrator.

Fig. 14-8-1 A variety of pictorial methods used in technical illustrating. (Graphic Standard Instruments Co.)

Fig. 14-8-2 Technical illustrations.

ISOMETRIC   DIMETRIC   TRIMETRIC

OBLIQUE CAVALIER   OBLIQUE CABINET   SINGLE-POINT (PARALLEL) PERSPECTIVE   TWO-POINT (ANGULAR) PERSPECTIVE

Fig. 14-8-3 V-block in various types of pictorial drawings.

PICTORIAL LINE DRAWINGS

Since nearly all technical illustrations are basically pictorial line drawings, a complete understanding of the various types and their applications is necessary.

While any type of pictorial drawing can be used as the basis for a technical illustration, some types are more suitable than others. This is especially true if the illustration is to be rendered. Figure 14-8-3 shows a V-block drawn in various types of pictorial form. Notice the differences in appearance of each. Isometric is the least natural in appearance; perspective is the most natural. This might suggest, then, that all technical illustrations should be drawn in perspective. This is not necessarily true. While perspective is the most natural in appearance, it takes more time to draw if perspective grid sheets are not used. Thus, it could be more costly.

The shape of the object and how the drawing is to be used also influence the type of pictorial drawing chosen. If the object is circular, it will be easier to draw in oblique rather than isometric if elliptical templates are not available. If the illustration is to be used in a publication such as a journal, operator's manual, technical publication, etc., dimetric, trimetric, or perspective may be the best choice.

LINE APPLICATION

Line thickness in a pictorial drawing has an illusionary meaning. A pictorial drawing created with lines only, that is, without shading, must rely on converging lines and subtle line thicknesses to appear three-dimensional. See Fig. 14-8-4A. Basically, there are three types of lines required to create a good linear pictorial drawing. They are thick, medium, and thin.

Thick Lines Thick or heavy lines are used to emphasize a part in a pictorial drawing. Also, an illusion of depth is created when certain lines are thicker than others. Similarly, a tapered line is used to illustrate depth on a curved surface.

Medium Lines Medium-thickness lines are used as secondary emphasizing lines in the main object of a pictorial. They are also used as depth lines in an unemphasized or background object, as shown in Fig. 14-8-4B.

Thin Lines Thin or fine lines are used as front-edge lines on the surface of the object nearest to the light source. This gives a feeling of light on this edge which further enhances the illusion of perspective. A second use of thin lines is in subduing or deemphasizing an object because it is of secondary importance in the pictorial. Phantom and broken lines are also used to illustrate background or secondary parts.

Figure 14-8-5 illustrates technical illustrations featuring line application. Notice that only the necessary detail is shown and that just enough shading is added to emphasize and give form to the parts.

WING UPPER SURFACE

WIRE OR HEAVY CORD. TYPICAL METHOD OF SUPPORTING ACTUATOR WHEN CONNECTING OR DISCONNECTING LINKS

IDLER LINK (M)

DRIVE LINK (J)

IDLER LINK (H)

PIN (K)

ACTUATOR SUPPORT STRUCTURE

EXAMPLE (A)

PAD

PLATE

PEG

LOOSE PLATE

EXAMPLE (B)

Fig. 14-8-5 Line application.

THICK LINE GIVES ILLUSION OF DEPTH

MEDIUM LINE

THIN LINE

THICK OUTLINE EMPHASIZES PART

TAPERED THICK LINE FOR SHADOW ILLUSION ON CURVES OR CIRCULAR FORMS

(A)

OBJECT DRAWN WITH CONSTANT THICKNESS OF LINE (MEDIUM)

UNEMPHASIZED DEPTH LINE

PHANTOM LINE (THIN)

(B)

Fig. 14-8-4 Line usage.

IDENTIFICATION ILLUSTRATIONS

Pictorial drawings are very useful in identifying parts. They help save time when the parts are manufactured or assembled in place and are useful for illustrating operating instruction manuals and parts catalogs.

Identification illustrations often take the form of exploded views. If parts are few, they can be identified by names and leaders. The identification illustration in Fig. 14-8-6 is an example showing numbers for the parts and a tabulated parts list. This method is especially recommended where identification of parts is desirable and the viewer is not trained to read technical drawings.

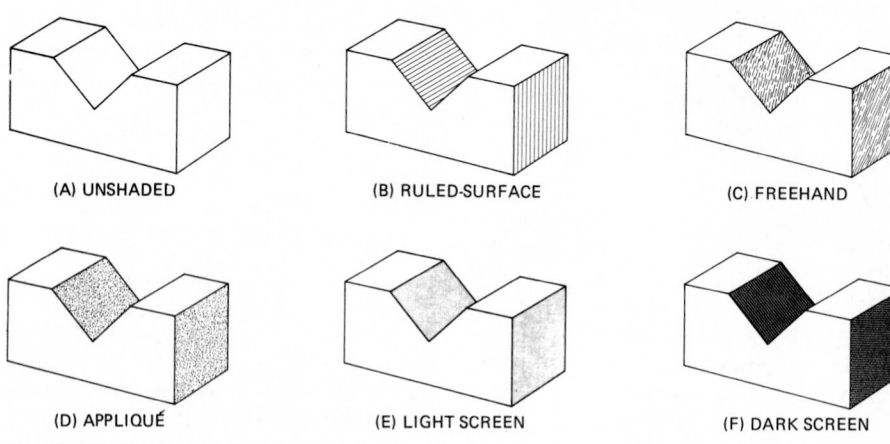

Fig. 14-8-6   Exploded view.

PART NO.	PART NAME	QTY.
403	QUICK CHANGE BOX	1
404	COVER, TOP	1
405	GASKET, COVER	1
406	SCREW, SOCKET HEAD CAP	8
407	SCREW	2
408	SHAFT, SHIFTER	1
409	LINK, SHIFTER	1
410	PIN	1
411	SHOE, SHIFTER	1
412	GASKET (MAKE IN PATTERN SHOP-BOX TO BED)	2
413	"O" RING	2
414	SHAFT, SHIFTER	1
415	PIN, TAPER	2
416	LINK, SHIFTER	1
417	SHOE, SHIFTER	1
418	COVER, SLIP GEAR	1
419	SCREW	4
420	PLUG	2
421	SCREW	3
422	SCREW	1
423	PLUG (NOT USED WITH SCREW REVERSE)	1
424	SCREW	3
425	PIN	2
426	SCREW	6
427	COLLAR	2
428	PLUNGER	2
429	SPRING	2
430	KNOB	2
431	LEVER	2
432	PLATE, FEED-THD.	1
433	PLATE, COMPOUND	1
434	PLATE, ENGLISH INDEX	1
435	COVER	1
436	SCREW	7

RENDERING

For certain purposes or where shapes are difficult to read, surface shading or rendering of some kind may be desirable. For most industrial illustrations, accurate descriptions of shapes and positions are more important than fine artistic effects. Desired results can often be obtained without any shading. In general, surface shading should be limited to the least amount necessary to define the shapes illustrated. Different ways of rendering technical illustrations include line shading, screen tints, and special appliqués and pencil shading (smudge).

Some shaded surfaces are indicated in Fig. 14-8-7. An unshaded view is shown at A for comparison. Ruled-surface shading is shown at B, freehand shading at C, and appliqué shading at D.

Line Shading Line shading is a simple, fast, and effective method of defining form. The technique may use straight or curved lines, as shown in Figs. 14-8-8 and 14-8-9.

With the light rays coming in the usual conventional direction, as in Fig. 14-8-10A, the top and front surfaces would be lighted and the right-hand surface woud be shaded, as in Fig. 14-8-10B. The front surface can have light shading with heavy shading on the right-hand side, as in Fig. 14-8-10C.

Appliqué Shading Commercial products of varied patterns are used to ren-

(A) UNSHADED   (B) RULED-SURFACE   (C) FREEHAND

(D) APPLIQUÉ   (E) LIGHT SCREEN   (F) DARK SCREEN

Fig. 14-8-7   Examples of various kinds of rendering.

STRAIGHT LINE   CURVED LINE

Fig. 14-8-8   Line shading.

Fig. 14-8-9   Application of line shading.

Fig. 14-8-10 Line rendering the faces of a cube.

SCREENS

LINES

Fig. 14-8-11 Appliqués.

PATTERNS

SHADOW SCREEN

(A) CIRCULAR FEATURES

WASHER
SPRING
FERRULE
STUD

.25

LATCH

1.00

1.38

SCREW, WASHER & LOCK WASHER

.50

(B) FLAT FEATURES

SHADOW EFFECTS

(C) MATERIALS

Fig. 14-8-12 Examples of appliqué rendering.

der areas which require distinction. See Fig. 14-8-11. These products are self-adhering and are easily cut out to match the areas to be covered. Shadow effects are achieved by the addition of another layer of the same material or of a contrast pattern. Both methods are illustrated in Fig. 14-8-12.

Stippling consists of dots, short crooked lines, or similar treatment to produce a shaded effect. It is a good method when it is well done.

Screens (measured in the number of dots per area) are available in a great variety of patterns and are very effective. A combination of line and appliqué rendering is shown in Fig. 14-8-13.

Tonal Pencil Rendering Tonal pencil rendering, while in limited use, produces a finished drawing of more professional quality. This method is usually used with diazo reproduction of limited distribution. See Fig. 14-8-14.

SPECIAL FEATURES

Screw Threads Both internal and external threads are conveniently shown by a series of ellipses uniformly spaced along the axis of the thread. The spacing of these lines may be greater than the actual thread pitch to allow room for the effective line shading. See Fig. 14-8-15.

Reference and Source Material
1. General Motors Corporation.

ASSIGNMENTS
See Assignments 23 and 24 for Unit 14-8 on page 305.

Fig. 14-8-13 Combination of line and appliqué rendering. (Holman Bros.)

Fig. 14-8-14 Tonal pencil rendering. (General Motors Corp.)

Fig. 14-8-15 Shaded screw threads.

ASSIGNMENTS for Chapter 14

Assignments for Unit 14-1, Pictorial Drawings

1. On isometric grid paper sketch the four parts shown in Fig. 14-1-A. Do not show hidden lines. Each square shown on the drawing represents one isometric square on the grid.

2. On isometric grid paper sketch the four parts shown in Fig. 14-1-B. Do not show hidden lines. Each square shown on the drawing represents one isometric square on the grid.

3. On a B- or A3-size sheet, make an isometric drawing, complete with dimensions, of one of the parts shown in Figs. 14-1-C to 14-1-F. Scale is full or 1:1.

Fig. 14-1-A   Sketching problems.

Fig. 14-1-B   Sketching problems.

Fig. 14-1-C   Base plate.

Fig. 14-1-E   Base block.

MATL—SAE 1020

METRIC

Fig. 14-1-D   Support bracket.

MATL—SAE 1050

METRIC

Fig. 14-1-F   Step block.

Assignments for Unit 14-2, Curved Surfaces in Isometric

4. On isometric grid paper sketch the four parts shown in Fig. 14-2-A. Each square shown on the figure represents one square on the isometric grid. Hidden lines may be omitted for clarity.

5. On isometric grid paper sketch the four parts shown in Fig. 14-2-B. Each square shown on the figure represents one square on the isometric grid. Hidden lines may be omitted for clarity.

6. On a B- or A3-size sheet, make an isometric drawing complete with dimensions of one of the parts shown in Figs. 14-2-C to 14-2-G. Use half scale for Fig. 14-2-G. For all others the scale is full or 1:1.

Fig. 14-2-B Sketching problems.

Fig. 14-2-A Sketching problems.

Fig. 14-2-F T-guide.

R1.50
Ø2.10
3.25
.90
(1.50)
R1.00
2.40
R.50
.50

Fig. 14-2-C Cradle bracket.

Ø30
12
Ø60
ROUNDS AND FILLETS R2
MATL—CI
Ø19
A
KEYSEAT FOR SQUARE KEY
Ø38
95
A
45
30
12
6
METRIC

Fig. 14-2-D Link.

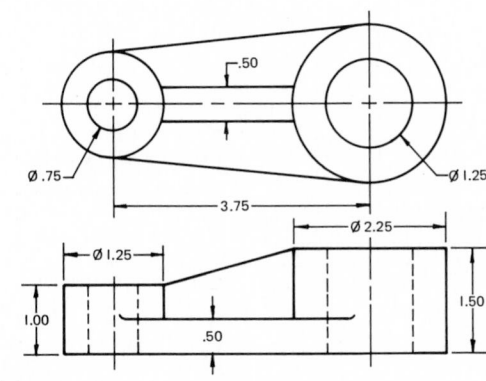

.50
Ø.75
Ø1.25
3.75
Ø2.25
Ø1.25
1.00
.50
1.50

Fig. 14-2-E Link.

Ø1.50 X .50 DEEP
4.00
2.00
3.00
6.00
2.00
2.00
3.00
1.50
5.00
1.25
1.50
.50
1.50
3.00
1.50
1.00
2.00
2.00
1.25
Ø1.25 2 HOLES THRU
1.50
1.00
1.50

Fig. 14-2-G Base.

Assignments for Unit 14-3, Common Features in Isometric

7. On a B- or A3-size sheet draw an isometric half-section view of one of the parts shown in Fig. 14-3-A or 14-3-B. Scale is full or 1:1.

8. On a B- or A3-size sheet draw an isometric full section drawing of one of the parts shown in Fig. 14-3-C or 14-3-D. Scale is full or 1:1.

9. On a B- or A3-size sheet draw an isometric drawing of the shaft shown in Fig. 14-3-E. Use a conventional break to shorten the length. Scale is full or 1:1.

10. On a B- or A3-size sheet, draw an isometric assembly drawing of the two-post die set, model 302, shown in Fig. 14-3-F. Allow 2 in. between the top and base. Use half scale. Do not dimension. Include on the drawing a bill of material. Using part numbers, identify the parts on the assembly.

11. On an A3- or B-size sheet, draw an isometric exploded assembly drawing of the book rack shown in Fig. 14-3-G. Choose either size A or B. Scale is 1:2. Do not dimension. Include on the drawing a bill of material. Using part numbers, identify the parts on the assembly.

12. On a B- or A3-size sheet draw an isometric exploded assembly drawing of the universal joint shown in Fig. 14-3-H. Scale is full or 1:1. Do not dimension. Include on the drawing a bill of material. Using part numbers, identify the parts on the assembly.

Fig. 14-3-B   Base.

MATL – BIRCH

Fig. 14-3-C   Pencil holder.

Fig. 14-3-A   Guide block.

Fig. 14-3-D   Adapter.

Fig. 14-3-E   Shaft.

SIZE	NO. 302 DIE SET	NO. 304 DIE SET
A	11.12	14.00
B	6.00	7.50
C	6.50	8.00
D	1.62	1.75
E	1.50	1.62
F	1.25	1.25
G	1.00	1.12
H	1.25	1.38
J	2.00	2.25
K	2.00	2.25
L	2.00	2.00

Fig. 14-3-F   Two-post die set.

SIZE	NO. 1	NO. 2	NO. 3
A	200	250	300
B	320	370	420

METRIC

Fig. 14-3-G   Book rack.

Assignments for Unit 14-4, Oblique Projection

13. On coordinate grid paper make oblique sketches of the three parts shown in Fig. 14-4-A. Each square shown on the figure represents one square on the grid paper. Hidden lines may be omitted to improve clarity.

14. On coordinate grid paper make oblique sketches of the three parts shown in Fig. 14-4-B. Each square shown on the figure represents one square on the grid paper. Hidden lines may be omitted to improve clarity.

PT 1 – FORK – 2 REQD

PT 2 – RING – 1 REQD

PT 3 – Ø .25 SPRING PIN – 2 REQD

PT 4 – .250 – 20 FHMS – .62 LG – 4 REQD

Fig. 14-3-H   Universal joint.

Fig. 14-4-A   Sketching problems.

Fig. 14-4-B   Sketching problems.

Fig. 14-4-C   V-block rest.

ROUNDS AND FILLETS R .10

Ø .88
Ø 1.50 CSK X 82°
2 HOLES

Ø .60 SLOTS

Ø 1.25

R 2.00

Ø 2.00

Fig. 14-5-A   End bracket.

Ø 1.875

Ø .562
2 HOLES

R .75

Fig. 14-5-C   Bearing support.

Fig. 14-4-D   Dovetail guide.

METRIC

15. On a B- or A3-size sheet, make an oblique drawing, complete with dimensions, of one of the parts shown in Fig. 14-4-C or 14-4-D. Scale is full or 1:1.

Assignments for Unit 14-5, Common Features in Oblique

16. On a B- or A3-size sheet, make an oblique drawing, complete with dimensions, of one of the parts shown in Fig. 14-5-A or 14-5-B. Scale is half or 1:2.

17. On a B- or A3-size sheet, make an oblique drawing, complete with dimensions, of one of the parts shown in Fig. 14-5-C or 14-5-D. Use the straight-line method of showing the rounds and fillets. Scale is half or 1:2 for Fig. 14-5-C and full or 1:1 for Fig. 14-5-D.

M24 X 3 X 30 LG
BOTH ENDS

Ø 40

4 X 12
KEYSEAT

METRIC

Fig. 14-5-B   Shaft.

Ø 90

ROUNDS AND
FILLETS R 3

Ø 10
Ø 30 CBORE X 10 DEEP

Ø 50

Ø 125

METRIC

Fig. 14-5-D   Bushing holder.

Fig. 14-6-A   Vise base.

PARALLEL PERSPECTIVE
OF A DRILL VISE BASE

Fig. 14-6-B   Spacer block.

Fig. 14-6-C   Shaft support.

Assignments for Unit 14-6, Perspective Projection

18. On a B- or A3-size sheet, make a parallel-perspective drawing of the vise base shown in Fig. 14-6-A. Scale is full or 1:1.

19. On a B- or A3-size sheet, make a parallel-perspective drawing of one of the parts shown in Fig. 14-6-B or 14-6-C. Scale is full or 1:1.

20. With the aid of a parallel-perspective grid, make a perspective drawing of the triple bookcase shown in Fig. 14-6-D. Scale is to suit. Each unit is 36 × 12 in. × 6 ft or 900mm W × 300mm D × 1800mm H. Use your judgment for sizes not shown.

Fig. 14-6-D   Bookcase.

Assignments for Unit 14-7, Angular, or Two-Point, Perspective

21. On a B- or A3-size sheet, make an angular-perspective drawing of the planter box or monument shown in Fig. 14-7-A or 14-7-B. Scale is 1:5 or 1:4.

22. On angular-perspective grid paper make a bird's-eye perspective view of the tool holder or cross slide shown in Fig. 14-7-C or 14-7-D. Scale is to suit.

PLACE HORIZON 400 ABOVE GROUND LINE

METRIC

Fig. 14-7-A Planter box.

DWG SET-UP

Fig. 14-7-B Monument.

Fig. 14-7-C Tool holder.

METRIC

Fig. 14-7-D Cross slide.

304 INTERMEDIATE DRAWING DESIGN

Assignments for Unit 14-8, Technical Illustration

23. On a B- or A3-size sheet, prepare two identical pictorial drawings of one of the parts shown in Fig. 14-8-A or 14-8-B. On the first drawing use line shading; on the second, use appliqués. Do not dimension. Select any pictorial method you wish. Scale is full or 1:1.

24. On a B- or A3-size sheet, make an exploded isometric assembly drawing of one of the assemblies shown in Fig. 14-8-C or 14-8-D. Scale is full or 1:1. Do not dimension. Use a rendering technique of your choice to improve the appearance of the drawing.

UNLESS OTHERWISE SHOWN
ROUNDS AND FILLETS R.I0

Fig. 14-8-A   Swing bracket.

Fig. 14-8-B   Stop button.

Fig. 14-8-C   Gear clamp.

Fig. 14-8-D   Bearing bracket.

CHAPTER 15

Functional Drafting

UNIT 15-1
Functional Drafting

Since the basic function of the drafting department is to provide sufficient information to produce or assemble parts, functional drafting must embrace every possible means to communicate this information in the least expensive manner. Functional drafting also applies to any method which would lower the cost of producing the part. New technological developments have provided many new ways of producing drawings at lower costs and/or in less time. This means that the drafting office must be prepared to discard some of the old, traditional methods in favor of these newer means of communication.

There are many ways in which to reduce the drafting time in preparing a drawing. These drawing shortcuts, when collectively used, are of prime importance in an effective drafting system.

These newer techniques cannot be blindly applied, however, but must be carefully evaluated to make certain that the benefits outweigh the potential disadvantages. This evaluation should answer the following questions:

- What is its purpose?
- Is it a personal preference disguised as a project requirement?
- Does it meet contractual requirements?
- Will the shortcut increase costs in other areas such as manufacturing, purchasing, or inspection?
- Is it an effective communication link?
- How much training or education is required to make effective use of it?
- Are facilities available to implement it?
- Does the shortcut bypass a real bottleneck?

As each of these categories is examined, the advantages of the shortcuts will become apparent.

PROCEDURAL SHORTCUTS

There are a number of procedural shortcuts which, if properly applied and carefully managed, can shorten the drawing preparation cycle and result in savings.

Streamlined Approval Requirements It is obvious that the more signatures required on a drawing, the greater the delays in releasing data. The decision as to who will approve drawings and drawing changes must be carefully considered to make certain that all necessary functions have been taken into account (checkers, responsible engineers, important technical specialists, etc.) without imposing undue restrictions. Project ground rules and contractual requirements also play an important part in this decision.

Eliminating the Drawing Check from the Preparation Cycle One of the most common suggested shortcuts, proposed

QTY	PART	MATL	DESCRIPTION	PT NO.
2	CABLE SUPPORT	MAPLE	A–5374 PT 1	1
2	CABLE SUPPORT	MAPLE	A–5374 PT 2	2
3	CABLE SUPPORT	MAPLE	A–5374 PT 4	3

(A) DRAWING CALLOUT

Fig. 15-1-1 Standard tabulated drawings.

PT	A	B	Ø C
4	5.75	2.40	.80
3	4.00	1.60	.80
2	4.00	1.60	1.00
1	5.00	1.60	1.00

(B) STANDARD PART

Fig. 15-1-2 Standard parts drawings stored on microfilm. (Eastman Kodak Co.)

usually when a project is behind schedule or exceeding its budget or when experienced personnel are involved, is to eliminate checking from the drawing preparation cycle.

Using Standard and Existing Drawings Every year numerous drawings of parts are prepared which are repetitions of existing drawings. If the drafter were to incorporate in the new design parts that were already drawn, many drawing hours would be saved. Good drawing application records and an efficient multiple-use drawing system can eliminate a great deal of duplication. Standard tabulated drawings may be used to eliminate hun-

Fig. 15-1-3 General Motors drawing standards manual.

dreds of drawings. See Figs. 15-1-1 and 15-1-2. For additional information on using existing drawings refer to Unit 15-3 on scissors and paste-up drafting.

Standard Drafting Practices Standard drafting practices are obviously the backbone of efficient drafting room operations. The best way to establish and implement these practices is through a good drafting room manual, whose requirements must be strictly observed by all personnel. See Fig. 15-1-3.

The drafting room manual should contain data on the use and preparation of specific types of drawings, drawing and part number requirements, standard and special drafting practices, rules for dimensioning and tolerancing, specifications for associated lists, and company procedures for the preparation, handling, release, and control of drawings.

Team Drafting Many engineering departments have turned out drawings by the method of one drafter to one drawing. Team drafting involves a number of people producing one drawing. While this may seem uneconomical, it is an expeditious means with visible cost savings over the traditional method.

Some firms are using team drafting because it is a better utilization of skill

levels. It is a training program through which drafting skills are taught and semiskilled people are given an opportunity to gain experience.

Data Retrieval The use of microform reader-printers in the drafting room provides quick and ready access to standard drawings and parts. The use of microfiche cards is becoming popular, because they can hold up to 70 pages of information. However, for this method to be effective, a full-time librarian is needed.

Standard Parts and Design Standard Information Encouraging the use of standard parts and standard approaches to design will not only result in drafting time saved, but will also cut costs in areas such as purchasing, material control, manufacturing, etc. The odd-size cutout that requires special tooling, the design that calls for nonstandard hardware, and the equipment that uses a wide variety of fasteners when only one or two would suffice are typical cases where properly applied standards would reduce both time and cost.

Copying Machines One of the most important time-saving devices, which should be available in every drafting area, is a copying machine for reference copies, checking prints of work in preparation, and similar uses. See Fig. 15-1-4. When a drafter needs a copy, work is delayed until the copy is made available. Therefore a good copying machine will soon pay for itself in drawing hours saved.

Training Programs To provide drafters with standard procedures and technical information is not enough; they must be trained in their use. New drafters are frequently overwhelmed by a strange environment, while old

Fig. 15-1-4 Copying machine.

employees fail to keep up with new requirements or properly use the services available. Training programs for the indoctrination of new personnel and the updating of long-service employees are rewarded by more efficient and versatile operation.

DRAFTING EQUIPMENT AND MATERIALS

The quality of the material and supplies used in the preparation of drawings is as important as the quality of the instruments used in fabrication. Leads which break easily, vellum which ghosts or smudges, and inks that crack and chip are some of the many material factors which contribute to increased drawing preparation time and decreased drawing life. Drafting materials must be carefully evaluated before orders are placed.

Numerous timesaving devices are available: templates for every application, "pens" for easier linework, and more application of tape to artwork, transfer-type lettering, etc. Since drafting applications vary so widely, only the drafting supervisor can determine which devices will increase the drafting production.

Drafting aids are designed to facilitate the making of drawings by removing or reducing some of the more tedious aspects of drafting.

Templates Templates, such as shown in Fig. 15-1-5, play an important part in functional drafting, for they save a great deal of time in drawing common shapes of details such as rounds, squares, hexagons, and ellipses. In addition to common shapes, templates have been made for standard parts such as nuts, and bolt heads, for electrical symbols, outlines of tools and equipment, and many other outlines which are often repeated.

Mechanical Lettering The results of a recent survey disclosed that mechanical lettering is generally replacing hand lettering. When mechanical lettering is required, it should be performed, whenever possible, by a subordinate. Mechanical lettering is particularly effective when it is used for general notes, particularly when preprinted standard notes on adhesive material are used.

REDUCING THE NUMBER OF DRAWINGS REQUIRED

The cost of a project is, to some extent, directly related to the number of drawings which must be prepared. Therefore, careful planning to reduce the number of drawings required can result in significant savings. Some ways to reduce the number of drawings are explained below.

Detail Assembly Drawings Detail assembly drawings, in which parts are detailed in place on the assembly (see Fig. 15-1-6), and multidetail assembly drawings, in which there are separate detail views for the assembly and each of its parts, will reduce the number of drawings required. However, they must be used with extreme care. They can easily become too complicated and confusing to be an effective means of communication.

Selecting the Most Suitable Type of Projection to Describe the Part The selection of the type of projection (orthographic, isometric, or oblique) can greatly increase the ease with which some drawings can be read and, in many cases, reduce drafting time. For example, a single-line piping drawing drawn in isometric projection simplifies an otherwise difficult drawing problem in orthographic projection. See Fig. 15-1-7.

References and Source Material
1. George R. Beck, "Reorganization in the Drafting Room," *Graphic Science*, October 1966.
2. Xerox Corp.

ASSIGNMENTS

See Assignments 1 through 3 for Unit 15-1 on page 314.

Review for Assignments
Unit 14-2   Isometric Circles

Fig. 15-1-5   Templates are made for many different uses and save a lot of time.

NOTE: WOOD SIZES ARE NOMINAL

Fig. 15-1-6   Detail assembly drawing of a sawhorse.

(A) ORTHOGRAPHIC PROJECTION   (B) ISOMETRIC PROJECTION

Fig. 15-1-7   Selecting the most suitable type of projection.

UNIT 15-2
Simplified Drafting

The challenge of modern industry is to produce more and better goods at competitive prices. Drafting, like all other branches of industry, must share in the responsibility for making this increased productivity possible. The old concept of drafting—that of producing an elaborate and beautiful drawing, complete with all the lines, projected views, and sections—must give way to a simplified method. The new, simplified method of drafting must embrace many modern economical drafting practices but surrender nothing in either clarity of presentation or accuracy of dimensioning. Drafting stripped of its frills is the new standard. See Fig. 15-2-1.

Although simplified drafting was pursued largely to reduce drafting costs and to improve the drafting product, it had a beneficial side effect when the drawing had to be microformed. Even more important, the worker on the shop floor has an "easier to understand" print. Much of the fine, difficult-to-produce detail, such as unnecessary frills and curlicues, has been eliminated and replaced by simplified or symbolic presentation. Drawings cluttered by repetitive detail are now shown simplified and are more suitable for making quality microform reproduction and reduced-size drawings.

Three most effective practices used in simplified drafting are

• Simplification of dimensioning
• Simplification of detail drawing
• Extensive use of freehand sketching

SIMPLIFICATION OF DIMENSIONING

Simplification of dimensioning not only reduces drafting time but also avoids cluttering a drawing with unnecessary lines, thereby making it easier to read.

No other single factor has more influence on the use of drawings than dimensioning. It is not sufficient to have dimensions numerically correct; it is equally important to dimension a drawing properly so that computation of sizes is unnecessary.

• Use arrowless dimensioning
• Use tabular dimensioning
• Use abbreviations and symbols

Arrowless Dimensioning To avoid having a large number of dimensions extending away from the part, arrowless or ordinate dimensioning may be used. See Fig. 15-2-2. In this system, the "zero" lines represent the vertical and horizontal datum lines, and each of the dimensions without an arrowhead gives the distance from the zero line. There is never more than one zero line in each direction.

Tabular Dimensioning When there is a very large number of holes or repetitive features, such as in a chassis or a printed circuit board, and where the multitude of center lines would make a drawing difficult to read, tabular dimensioning is recommended. See Fig. 15-2-3. In this system each hole or feature is assigned a letter, or a letter with a numeral subscript. The feature dimensions and the feature location along the X and Y axes are given in a table.

Abbreviations and Symbols Abbreviations and symbols are shortened forms of words or expressions used to conserve drafting time and drawing space. Refer to the Appendix for commonly used abbreviations and symbols.

SIMPLIFICATION OF DETAIL DRAWING

1. Complicated parts are best described by means of a drawing. However, explanatory notes can complement the drawing, thereby eliminating views that are time-consuming to draw. See Figs. 15-2-4 and 15-2-5.

CONVENTIONAL DRAWING   SIMPLIFIED DRAWING

Fig. 15-2-1   A comparison between conventional and simplified drafting.

HOLE SYMBOL	HOLE DIA
A	6
B	5
C	4
D	3

Fig. 15-2-2 Arrowless dimensioning.

HOLE DIA	HOLE SYMBOL	LOCATION X→	Y↓	Z
2.50	A1	2.30	1.50	.70
1.94	B1	.25	1.50	THRU
	B2	3.00	1.50	•
	B3	2.34	.50	•
	B4	3.20	.50	•
1.54	C1	.62	1.50	•
	C2	1.90	1.50	•
	C3	.25	.80	•
	C4	1.20	.80	•
	C5	3.00	.80	•
	C6	.62	.50	•
1.16	D1	1.90	.25	.50

Fig. 15-2-3 Tabular dimensioning.

NOTE PT 2 Ø.70 X 2.00LG
Ø.238 HOLE – .62 FROM END

PART DESCRIBED BY A NOTE
EXAMPLE 3

Fig. 15-2-5 Simplified drafting practices for detailed parts.

2. Use simplified drawing practices, as described throughout this text, especially on threads and common features.

3. Avoid unnecessary views. In many cases one or two views are sufficient to explain the part fully.

4. When a large number of holes of similar size are to be made in a part, there is a chance that the person producing the part may misinterpret a conventional drawing. To simplify the drawing and reduce the chance of error, hole symbols such as shown in Figs. 15-2-6 and 15-2-7 are recommended.

Fig. 15-2-4 Comparison between conventional and simplified drawings.

Fig. 15-2-6 Recommended hole symbols in order of preference.

5. The use of the symmetry symbol means that all dimensions are symmetrical about that line.
6. A simplified assembly drawing should be used for assembly purposes only. Some means of simplification are:

   • Standard parts such as nuts, bolts, and washers need not be drawn.
   • Reference part circles and arrowheads on leaders can be omitted.
   • Small fillets and rounds on cast parts need not be shown.
   • Phantom outlines of complicated details can often be used.

7. Use templates where possible.
8. Within limits, a small drawing is made more quickly than a large drawing.
9. Eliminate hidden lines which do not add clarification.
10. Show only partial views of symmetrical objects. See Fig. 15-2-8.
11. Avoid the use of elaborate pictorial and repetitive detail.

QUARTER VIEW

HALF VIEW

Fig. 15-2-8   Partial views.

12. Eliminate repetitive data by use of general notes or phantom lines.
13. Eliminate views where the shape or dimension can be given by description, for example, hex, sq, $\emptyset$, thk, etc.

FREEHAND SKETCHING

Many shops care little whether the drawing is freehand, whether one view is shown, or whether the drawing is to scale, as long as the proportions are approximate. They want the necessary information clearly shown. Freehand sketches and drawings made with instruments can be shown on one sheet. However, it must be clearly understood that the use of freehand sketching does not give the drafter a license to turn out sloppy work.

Savings as high as 30 percent in the preparation of working drawings have been attributed to the use of freehand sketches as opposed to instrument-produced drawings. However, freehand sketching has its limitations. It is highly effective on simple detailed parts, for small radii, such as rounds and fillets, and for small holes. In many cases the term *freehand* is not entirely correct. For instance, templates may be used to draw circles, resistors, or other common features, or a straightedge may be used to produce long lines since it is faster and more accurate than freehand sketching. But short lines are drawn more quickly freehand.

Drawing paper with nonreproducible grid lines is ideal for freehand sketching. See Fig. 15-2-9. For this reason many companies have their drawing paper made with nonreproducible grid lines over the entire drawing area. Other advantages of having the grid lines on the paper are that (1) they may

(A) CONVENTIONAL DRAFTING

(B) SIMPLIFIED DRAFTING

Fig. 15-2-7 Application hole symbols and arrowless dimensioning.

Fig. 15-2-9   Sketching of parts on coordinate paper.

(A) FREEHAND SKETCH

(B) COMBINED FREEHAND AND INSTRUMENT DRAWING

Ø 1.000 / .994

.556 / .554

Ø 1.56

Ø .159 THRU

M 12

MADE FROM M 20 X 120 LG HEX BOLT

---

serve as guidelines in lettering notes and dimensions and (2) they may be used for measuring distances, thereby reducing the number of times the scale is used for measuring.

References and Source Material
1. ANSI Y14.5M, *Dimensioning and Tolerancing.*

ASSIGNMENTS

See Assignments 4 through 7 for Unit 15-2 on pages 314 and 316.

Review for Assignments
Unit 2-6   Sketching
Unit 6-1   Detail Drawings
Appendix   Keys and Keyseats

UNIT 15-3
Reproduction Shortcuts

In the past few years, a number of reproduction techniques have been developed which, if properly used, can greatly reduce drawing preparation time. An understanding of available techniques and their limitations, supported by the close cooperation of a reproduction group familiar with drafting operations can help the drafting supervisor to make significant cost savings.

REPRODUCIBLES FROM EXISTING DRAWINGS

When a new drawing is to be made from an existing drawing with few changes, reproducibles will save a great deal of preparation time. This procedure involves making a translucent or transparent print from the original drawing, removing unwanted material from this print, and adding the new information to the drawing. The main drawback to this method is that the existing drawing may not conform to the latest standard drawing practice.

SCISSORS AND PASTE-UP DRAFTING

No matter how original a design may be, a great number of part features are repetitive. With the aid of modern reproduction methods, drawings can be created by using unchanged portions of existing drawings. Transferring them from one drawing to the next is accomplished by scissors and paste-up drafting. It provides a way of using all or parts of existing drawings, notes, charts, and drawing forms to revise existing drawings and to create new drawings. Through the utilization of existing drawings much valuable drafting time is freed for creative design drafting rather than hand copying.

Finished prints can be made on paper, acetate, or vellum. They can be the same size or reduced to different sizes, depending on the reproduction equipment being used.

Another important advantage of cut-and-paste drafting is that materials copied from existing drawings do not have to be rechecked minutely, as must be done with new drawings. Rechecking time is reduced. See Fig. 15-3-1.

Fig. 15-3-1   Cut and paste drafting. (Xerox Corp.)

Fig. 15-3-2 A variety of shapes and sizes of appliqués. (Graphic Standard Instruments Co.)

(A) TYPICAL APPLIQUES

(B) APPLICATION

Fig. 15-3-4 Electronics appliqués. (Bishop Industries Corp.)

APPLIQUÉS

One of the most successful methods of reducing drawing time is the use of appliqués. When parts, shapes, symbols, or notes are used repeatedly, appliqués should be considered. These pressure-sensitive overlays may be printed on opaque, transparent, or translucent sheets with an adhesive backing.

Appliqués are available in a great variety of standard symbols or patterns (Fig. 15-3-2) and in blank (unprinted) sheets. A matte surface on the blank sheet will accept typewriter copy as well as pencil or ink lines. This material is often used for making corrections on drawings and for adding materials lists or detailed notes which can be typed faster than they can be lettered. Figure 15-3-3 shows a drawing which used many of the appliqués shown in Fig. 15-3-2.

(A) WORM'S EYE VIEW

(B) BIRD'S EYE VIEW

Fig. 15-3-3 Application of appliqués shown in Fig. 15-3-2. (Graphic Standard Instruments Co.)

Appliqués are available in two basic types: *cutout* and *transfer*. Cutout appliqués are applied by positioning the desired image in the correct position on the drawing, burnishing (rubbing) the image area, and cutting around it to remove the portion not wanted. The transfer-type pressure-sensitive appliqué works on a somewhat different principle. The carrier is removed from the translucent image sheet, and the area to be transferred is placed in position on the drawing. The image to be transferred is then rubbed over the top surface of the transfer sheet with a burnishing stick.

The combined use of cut-and-paste drafting and appliqués for new drawings is found extensively in industry, especially in the electronics and piping fields. See Fig. 15-3-4.

References and Source Material
1. George R. Beck, "Reorganization in the Drafting Room," *Graphic Science*, October 1966.
2. Eastman Kodak Company.
3. Xerox.

ASSIGNMENTS

See Assignments 8 and 9 for Unit 15-3 on page 317.

UNIT 15-4
Photodrawings

Photodrawings, that is, engineering drawings into which one photograph, or more, is incorporated, have increased in popularity because they can sometimes present a subject even more clearly than conventional drawings. Photodrawings supplement rather than replace conventional engineering drawings by eliminating much of the tedious and time-consuming effort involved when the subject is difficult to draw. They are particularly useful for assembly drawings, piping diagrams, large machine installations, switchboards, etc., provided, of course, that the subject of the drawings exists so that it may be photographed.

Photodrawings are also a comprehensive means of clearly transmitting technical information; they free the drafter from having to draw things which already exist. See Fig. 15-4-1.

Photodrawings have other advantages. They are easy to make and usually take much less time to prepare than an equivalent amount of conventional drafting.

Fig. 15-4-1 Photodrawing. (Eastman Kodak Co.)

Background Any photodrawing must begin with a photograph of an object, a part or assembly, a building, a model, or whatever else may be the subject of the drawing.

Photography The best photographic angle usually is one which shows the subject in a flat view with as little perspective as possible. (If the situation calls for a perspective, select the angle that best describes the object.) Make certain that all the parts important to the photodrawing are in view of the camera.

Reference and Source Material
1. Eastman Kodak Company.

ASSIGNMENT

See Assignment 10 for Unit 15-4 on page 318.

ASSIGNMENTS for Chapter 15

Assignments for Unit 15-1, Functional Drafting

1. After the number of drawings made over the last 6 months was reviewed, it was discovered that a great number of cable straps, shown in Fig. 15-1-A, were being made which were similar in design. On a B-size sheet, prepare a standard tabulated drawing similar to Fig. 15-1-1, reducing the number of standard parts to 4. Scale is full.

2. On an A3 sheet, the rod guide shown in Fig. 15-1-B is to be drawn twice and the drawing time for each recorded. First, on plain paper make an isometric drawing of the part, using a compass to draw the circles and arcs. Next repeat the draw-ing, only this time use isometric grid paper and a template for drawing the circles and arcs. From the drawing times recorded, state in percentage the time saved by the use of grid paper and templates. Scale is 1:1. Do not dimension.

3. On a B-size sheet, the book rack shown in Fig. 15-1-C is to be drawn twice and the drawing time for each drawing recorded. Using one half of the sheet, make a three-view orthographic projection drawing of the book rack assembly showing only those dimensions and instructions pertinent to the assembly. On the other half of the sheet, prepare detail drawings for the parts required.

Scale to suit. On a second sheet, make an isometric detailed assembly drawing of the book rack showing the dimensions and instructions necessary to completely make and assemble the parts. Scale to suit. From the drawing times recorded, state a percentage of time saved by the use of detailed assembly drawings.

Assignments for Unit 15-2, Simplified Drafting

4. On a B- or A3-size sheet, use simplified drawing practices to make freehand detail sketches of either of the parts shown in Fig. 15-2-A. Scale to suit.

Fig. 15-1-A   Cable straps.

Fig. 15-1-B   Rod guide.

TUBE SUPPORT

Fig. 15-1-C   Book rack.

END DETAIL OF PT 1

TAPER PIN
Ø.25 X 2.00 LG

NOTE: LUMBER SIZES SHOWN
ARE NOMINAL SIZES.

2mm THICK
GASKET

Fig. 15-2-A   Tube support and gasket.

5. On a B-size sheet, redraw the two parts shown in Fig. 15-2-B; use arrowless dimensioning and simplified drawing practices. Use half scale. For the cover plate use the bottom and left-hand edge for the datum surfaces. For the back plate use the bottom and the center of the part for the datum surfaces.

6. On an A3-size sheet, make simplified drawings of the parts shown in Fig. 15-2-C. Scale to suit.

7. On a B-size sheet, redraw the terminal board shown in Fig. 15-2-D using tabular dimensioning and hole symbols. Scale is full.

CHANNEL CLAMP

FLANGED COUPLING

Fig. 15-2-C   Simplification of detail assignment.

COVER PLATE

BACK PLATE

Fig. 15-2-B   Arrowless dimensioning assignment.

TERMINAL BOARD

MATL–.12 THK FIBER

Fig. 15-2-D   Tabular dimensioning and hole symbol assignment.

Assignments for Unit 15-3, Reproduction Shortcuts

8. An exploded isometric assembly drawing of the wheel puller shown in Fig. 15-3-A is urgently required. Time does not permit one drafter to do the entire drawing; thus three drafters will be required to draw the parts. Scale is half or 1:2. On plain paper draw all the parts in isometric. When all the parts have been drawn, cut them from the paper and assemble them in the exploded position on a B- or A3-size sheet. Use glue or rubber cement. Make a suitable print of the exploded assembly on a copying machine.

Fig. 15-3-A   Wheel puller. (Cut and paste)

Fig. 15-3-B   Electronic diagram. (Appliqué)

9. On a B- or A3-size sheet, make the electronic layout drawing shown in Fig. 15-3-B. Appliqués and templates are to be used. If appliqués of the electronic components are not available, make your own by photostating Fig. 15-3-4 then cut them out and glue them to your drawing. There is no scale.

Assignment for Unit 15-4, Photodrawing

10. On a B- or A3-size sheet, make a photostat of the sprocket shown in Fig. 15-4-A. The photostat, which is to serve as a photodrawing, is to replace the two-view drawing. Make a new chart listing metric sizes to replace the existing chart if directed to do so. Leaders, dimensions, and dimension lines, which are to be inked, are to be added to the photodrawing.

HARDENED STEEL BUSHING

SHEAR PIN ADAPTER
SHEAR PIN HUB
STEEL PLATE SPROCKET
SAFETY COLLAR

Hub Bore Range	Shear Pin Assembly Number	Shear Pin		Diameters				Length Thru			Hub Flange Thick-ness	Adapt Flange Thick-ness	Sprocket Seat Width	Bolts	
		Radius	Pin Dia.	Flange	Shear Pin Hub	Adapt Hub & Collar	Sprocket Seat	Shear Pin Hub	Adapt	Collar				Number & Size	Bolt Circle
		R	B	C	D	E	F	G	H	M	J	K	L	N	P
1.00 & under	SP-17	1.80	.25	5.25	1.75	2.50	2.62	2.44	1.38	.38	.56	.56	.44	4-.38	4.00
1.06-1.25	SP-18	2.18	.25	6.00	2.25	3.25	3.38	2.94	1.75	.50	.56	.56	.56	4-.38	4.75
1.30-1.50	SP-19	2.56	.30	6.75	2.75	4.00	4.12	3.56	2.12	.62	.68	.68	.68	4-.50	5.50
1.56-1.75	SP-20	3.00	.38	7.75	3.25	4.75	4.88	4.18	2.50	.75	.80	.80	.68	4-.50	6.25
1.80-2.00	SP-21	3.30	.45	8.75	3.75	5.25	5.38	4.80	2.88	.88	.94	.94	.94	4-.62	7.00
2.06-2.25	SP-22	3.80	.50	9.75	4.25	6.25	6.38	5.18	3.00	1.00	1.06	1.06	1.18	4-.62	8.00
2.30-2.50	SP-23	4.00	.50	10.00	4.50	6.50	6.62	5.68	3.50	1.00	1.06	1.06	1.38	4-.62	8.25
2.56-2.75	SP-24	4.40	.55	11.50	5.00	7.00	7.12	6.30	3.88	1.12	1.18	1.18	1.38	4-.62	9.25
2.80-3.00	SP-25	4.90	.62	12.50	5.50	8.00	8.12	6.94	4.25	1.25	1.30	1.30	1.38	6-.62	10.25

Fig. 15-4-A Bolt-on shear pin sprocket.

CHAPTER 16

Drawing for Numerical Control

UNIT 16-1
Drawing for Numerical Control

Numerical control is a means of automatically directing some or all of the functions of a machine from instructions. The instructions are generally stored on tape and are fed to the controller through a tape reader. The controller interprets the coded instructions and directs the machine through the required operations.

It has been established that because of the consistent high accuracy of numerically controlled machines, and because human errors have been almost entirely eliminated, scrap has been considerably reduced.

Because both setup and tape preparation times are short, numerically controlled machines produce a part faster than manually controlled machines.

When changes become necessary on a part, they can easily be implemented by changing the original tapes. The process takes very little time and expense in comparison to the alteration of a jig or fixture.

Another area where numerically controlled machines are better is in the quality or accuracy of the work. In many cases a numerically controlled machine can produce parts more accurately at no additional cost, resuting in reduced assembly time and better interchangeability of parts. This latter fact is especially important when spare parts are required.

Acceptance of numerical control processes is being accelerated by the development of computer-aided design (CAD) and computer-aided manufacturing (CAM) techniques.

DIMENSIONING FOR NUMERICAL CONTROL

Common guidelines have been established that enable dimensioning and tolerancing practices to be used effectively in delineating parts for both numerical control and conventional fabrication.

COORDINATE SYSTEM

The numerical control concept is based on the system of rectangular or cartesian coordinates in which any position can be described in terms of distance from an origin point along either two or three mutually perpendicular axes. Two dimensional coordinates (X, Y) define points in a plane. See Fig. 16-1-1.

The X axis is horizontal and is considered the first and basic reference axis. Distances to the right of the zero Y axis are considered positive X values and to the left of the zero Y axis as negative X values.

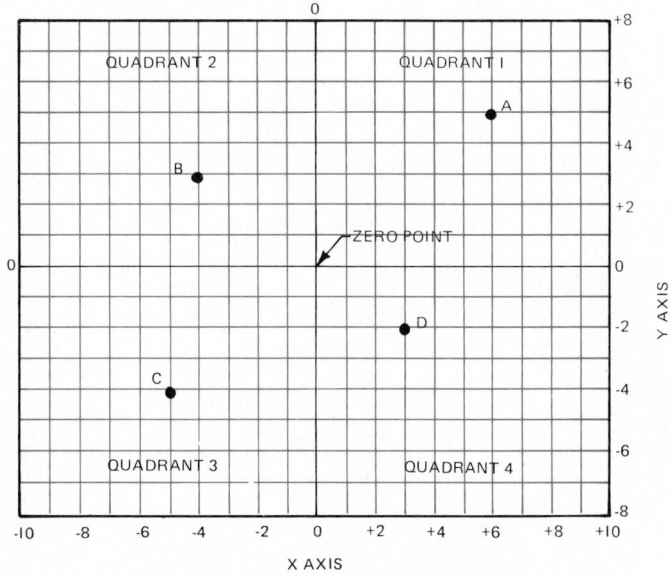

Fig. 16-1-1 Two dimensional coordinates. (X and Y)

The Y axis is vertical and perpendicular to the X axis in the plane of a drawing showing XY relationships. Distances above the zero X axis are considered positive Y values and below the zero X axis as negative Y values. The position where the X and Y axes cross is called the *origin*, or *zero point*.

For example, four points lie in a plane, as shown in Fig. 16-1-1. The plane is divided into four quadrants. Point A lies in quadrant 1 and is located at position (6, 5), with the X coordinate first, followed by the Y coordinate. Point B lies in quadrant 2 and is located at position $(-4, 3)$. Point C lies in quadrant 3 and is located at position $(-5, -4)$. Point D lies in quadrant 4 and is located at position $(3, -2)$.

Designing for numerical control would be greatly simplified if all work were done in the first quadrant because all the values would be positive and the plus and minus signs would not be required. However, any of the four quadrants may be used, and, as such, programming in any of the quadrants should be understood.

Some numerically controlled machines are designed for locating points in only the X and Y directions. These are called *two-axis machines*.

The function of these machines is to move the machine table or tool to a specified position in order to perform work, as shown in Fig. 16-1-2. With the fixed spindle and movable table as shown in Fig. 16-1-2B, hole A is drilled; then the table moves to the left, posi-

tioning point B below the drill. This is the most frequently used method. With the fixed table and movable spindle as shown in Fig. 16-1-2C, hole A is drilled; then the spindle moves to the right, positioning the drill above point B. This changes the direction of the motion, but the movement of the cutter as related to the work remains the same.

Zero Point

As previously mentioned, the zero point is the point where the X and Y axes intersect. It is the point from which all coordinate dimensions are measured. Many machines have a fixed zero point built in.

Two examples of fixed zero points on machine tables are shown in Fig.

(A) FINISHED PART

(B) FIXED SPINDLE, TABLE MOVES

(C) FIXED TABLE, SPINDLE MOVES

Fig. 16-1-2   Positioning the work.

(A) LOCATION OF PART AND ZERO POINT RESULTS IN Ist QUADRANT NC DIMENSIONING.

(B) LOCATION OF PART AND ZERO POINT RESULTS IN 3rd QUADRANT DIMENSIONING.

Fig. 16-1-3   Zero point location.

(A) POINT-TO-POINT DIMENSIONING

(B) BASE LINE DIMENSIONING

Fig. 16-1-4 Dimensioning for N/C.

16-1-3. In Fig. 16-1-3A all points are located in the first quadrant, resulting in positive X and Y values. In Fig. 16-1-3B all points are located in the third quadrant, resulting in negative X and Y values.

Setup Point

The setup point is located on the part or the fixture holding the part. It may be the intersection of two finished surfaces, the center of a previously machined hole in the part, or a feature of the fixture. It must be accurately located in relation to the zero point, as shown in Fig. 16-1-4.

POINT-TO-POINT PROGRAMMING

Point-to-point programming is the most common type of positioning system. With this system each new position is given from the last position. To compute the next position wanted, it is necessary to establish the sequence in which the work is to be done.

An example of this type of dimensioning is shown in Fig. 16-1-4A. The distance between the left edge of the part and hole 1 is given as .75 in. From hole 1 to hole 4 the dimension shown is 4.50 in. (X axis), and from hole 1 to hole 2 the dimension shown is 1.50 in. (Y axis). These dimensions give the distance from the last drilled hole to the next drilled hole. Assume the holes are to be drilled in the sequence shown in the figure. Hole 1 is located (2.75,

2.75) from the zero point. After hole 1 has been drilled, the drill spindle is positioned above the center of hole 2. Hole 2 has the same X-coordinate dimension as hole 1, making the X increment zero. Since the vertical distance between holes 1 and 2 is 1.50 in., the Y increment becomes $+1.50$.

After hole 2 is drilled, the drill spindle is positioned above the center of hole 3. Since the horizontal distance between holes 2 and 3 is 4.50 in., the X increment is $+4.50$. Hole 3 has the same Y-coordinate dimension as hole 2, making the Y increment zero.

From hole 3 the drill spindle is positioned above the center of hole 4. Hole 4 has the same X-coordinate dimension as hole 3, making the X increment zero. Since the vertical distance between hole 3 and hole 4 is 1.50 in., the Y increment is -1.50.

Figure 16-1-5 lists the distance between holes and indicates the direction of motion by plus and minus signs. It can be seen that each pair of coordinates shows the distance between each location in sequence.

COORDINATE PROGRAMMING

Many machines use coordinate programming instead of the point-to-point method of dimensioning. With this type of dimensioning all dimensions are taken from the zero point; as such, base-line dimensioning, as shown in Fig. 16-1-4B, is used. For example,

after hole 1 is drilled, the drill spindle has to be positioned above the center of hole 2. The coordinates for hole 2 are (2.75, 4.25). Figure 16-1-6 shows the coordinate dimensions of the holes shown in Fig. 16-1-4B.

HOLE	X	Y
1	+2.75	+2.75
2	0	+1.50
3	+4.50	0
4	0	-1.50

Fig. 16-1-5 Point-to-point dimensioning of holes shown in Fig. 16-1-4.

HOLE	X	Y
1	+2.75	+2.75
2	+2.75	+4.25
3	+7.25	+4.25
4	+7.25	+2.75

Fig. 16-1-6 Coordinate dimensioning of holes shown in Fig. 16-1-4.

ASSIGNMENTS

See Assignments 1 through 3 for Unit 16-1 on pages 323 through 325.

Review for Assignments

Appendix Trigonometric Functions

UNIT 16-2
Three-Axis Control Systems

Many numerically controlled machines operate in three directions, the table and carriage moving in the X and Y directions, as explained in Unit 16-1, and the tool spindle, such as a turret drill, traveling in an up-and-down direction. A vertical line taken through the center of the machine spindle is referred to as the Z axis and is perpendicular to the plane formed by the X and Y axes. See Fig. 16-2-1.

Thus, a point in space can be described by its X, Y, and Z coordinates. For example, P_1 in Fig. 16-2-2 can be described by its (X, Y, Z) coordinates as (4, 3, 5) and P_2 as (11, 2, 8).

A popular system used on many numerically controlled machines, such as the turret drill, is to establish the Z zero reference plane above the workpiece. Each tool is then adjusted and calibrated to the Z zero reference plane.

For example, Fig. 16-2-3 shows a part requiring three drilled holes. As the center hole is drilled through, the part is raised by gage blocks so that the drill does not touch the machine table.

The height of the gage blocks is determined by the distance the drill passes through the workpiece plus clearance, or .06 in. + 0.3D + .12 in. See Fig. 16-2-4. If a .75 in. drill were used, the gage block height would be .06 + .23 + .12 = .41 in.

If the distance from the top of the workpiece to the Z zero reference plane is set at .75 in., the Z coordinates for the three holes shown are $-(.75 + A)$, $-(.75 + B)$, and $-(.75 + C)$.

DIMENSIONING AND TOLERANCING

Recommended guidelines for dimensioning and tolerancing practices for use in defining parts for numerical-control fabrication are:

1. When the basic coordinate system is established, the setup point should be placed at an appropriate location on the part itself.
2. Any number of subcoordinate systems may be used to define features of a part as long as these systems can be related to the basic coordinate system of the given part.
3. Define part surfaces in relation to three mutually perpendicular reference planes. Establish these planes along part surfaces which parallel the machine axes if these axes can be predetermined.
4. Dimension the part precisely so that the physical shape can be readily determined. Dimension to points on the part surfaces.

Fig. 16-2-3 Calculating Z distance.

Fig. 16-2-4 Determining gage block height.

Fig. 16-2-1 X, Y, and Z axes.

Fig. 16-2-2 Points in space.

5. Regular geometric contours such as ellipses, parabolas, hyperbolas, etc., may be defined on the drawing by mathematical formulas. The numerically controlled machinery can easily be programmed to approximate these curves by linear interpolation, that is, as a series of short, straight lines whose endpoints are close enough together to ensure meeting the required tolerances for contour. In the case of arbitrary curves, the drawing should specify appropriate points on the curve by coordinate dimensions or a table of coordinates. Consideration should be given to the number of points needed to define the curve; however, one should keep in mind the fact that the tighter the tolerance or the smaller the radius of curvature, the closer together the points should be. Such terms as *blend* *smoothly* and *faired curve* are not used. Curves may also be defined by other coordinates, such as polar, spherical, or cylindrical, as applicable.

6. Changes in contour should be unambiguously defined with prime consideration for design intent.

7. Holes in a circular pattern should preferably be located with coordinate dimensions.

8. Where possible, express angular dimensions relative to the X axis in degrees and decimal parts of a degree.

9. Use plus and minus tolerances, not limit dimensions. Preferably, the tolerance should be equally divided bilaterally.

10. Positional tolerancing, form tolerancing, and datum referencing should be used where applicable. Datum features specified on the drawing in proper sequence will clearly indicate their usage for setup.

11. Where profile tolerances are specified, the geometric boundary should be equally disposed bilaterally along the true profile. Avoid profile tolerances applied unilaterally along the true profile. Include no less than four defined points along the profile.

12. Tolerances are specified only on the basis of actual design requirement. The accuracy capability of numerically controlled equipment is not a basis for specifying more restrictive tolerances than are functionally required.

ASSIGNMENTS

See Assignments 4 through 6 for Unit 16-2 on pages 325 through 327.

ASSIGNMENTS for Chapter 16

Assignments for Unit 16-1, Drawing for Numerical Control

1. On a B- or A3-size sheet, prepare two drawings of the cover plate shown in Fig. 16-1-A. One drawing is to use point-to-point dimensioning for the 10 holes; the other drawing is to use coordinate dimensioning. Only the dimensions locating the holes need be shown. The radial and angular dimensions are to be replaced with coordinate dimensions and taken to two decimal places. Below each drawing prepare a chart listing each hole and their X and Y coordinates. The letters shown on the holes show the sequence in which they are to be drilled. The center of the part is the zero point. Scale is full or 1:1.

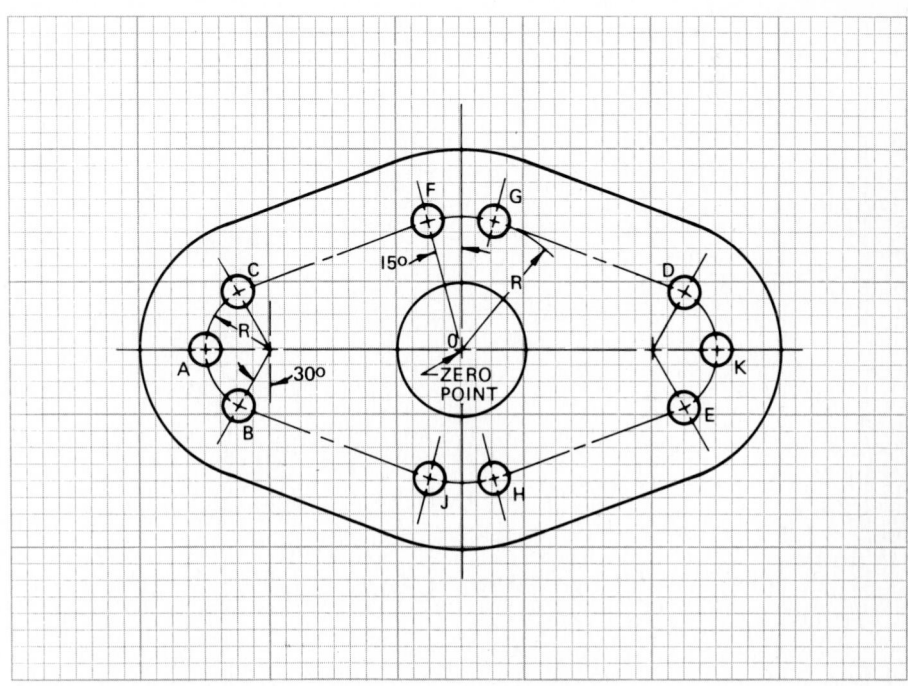

NOTE: GRID 8 X 8 TO THE INCH

Fig. 16-1-A   Cover plate.

2. On an A3- or B-size sheet, prepare a chart listing the X and Y (coordinate) locations and the quadrant for the points A to V shown on Fig. 16-1-B. The grid is 10 × 10 to the centimeter.

POINT	X AXIS	Y AXIS	QUADRANT
A			
B			
C			

Fig. 16-1-B Chart.

Fig. 16-1-C Cover plate.

NOTE: GRID 10 X 10 TO THE INCH

3. On a B- or A3-size sheet, prepare two drawings of the cover plate shown in Fig. 16-1-C. One drawing is to use point-to-point dimensioning for the holes; the other drawing is to use datum or coordinate dimensioning. Work with customary or metric dimensions as directed by your instructor. Only the dimensions locating the holes need be shown. Below each drawing prepare a chart listing each hole and their X and Y coordinates. The letters shown at the holes indicate the sequence in which they are to be drilled. Note the location of the zero point. Scale is full or 1:1.

Assignments for Unit 16-2, Three-Axis Control Systems
4. On a B- or A3-size sheet make a two-view drawing of the end plate shown in Fig. 16-2-A. Point-to-point dimensioning is to be used, and only the dimensions locating the holes need be shown. Below the drawing prepare a chart listing each hole and its X, Y, and Z coordinates. The letters on the holes show the sequence in which they are to be drilled. Calculating the Z coordinate is to be done in the same manner as for the part shown in Fig. 16-2-3. Scale is full or 1:1.

Note: Programming will be for the tap-drill holes and the six through holes shown. Zero point for the X and Y coordinates is the center of the end plate.

.250-20UNC-2B X .50 DEEP
4 HOLES EQ SPACED ON Ø1.500

Ø2.25

Ø.75

NOTE: HOLE D NOT SHOWN

Ø4.00

Z REFERENCE PLANE

.75

+Z

0 -Z

.75

1.12

ROUNDS AND FILLETS R.12

Ø.250
Ø.500 CBORE X .12 DEEP
6 HOLES EQ SPACED ON Ø3.000

Fig. 16-2-A End plate.

CHART I

LOCATE POINTS PI TO PI0 ON CHART

NOTE: ALL COORDINATES ARE +

POINT	X AXIS	Y AXIS	Z AXIS
PI	10	20	60
P2	20	70	70
P3	0	20	30
P4	100	60	75
P5	40	30	20
P6	40	60	10
P7	70	20	0
P8	50	50	50
P9	85	65	30
PI0	60	65	15

CHART 2

LOCATE COORDINATES AND RECORD IN TABLE

NOTE: ALL COORDINATES ARE +

POINT	X AXIS	Y AXIS	Z AXIS
PI	20		
P2			55
P3		60	
P4	30		
P5		30	
P6			0
P7			40
P8	65		
P9		30	
PI0			55

Fig. 16-2-B   Chart.

5. On a B- or A3-size sheet, locate points
   P1 to P10 on the coordinate chart 1 shown
   in Fig. 16-2-B. On chart 2 points P1 to P10
   are located, but only one of their coordinates is known. Accurately lay out the
   missing coordinates on the chart and
   record their values in the table. Scale—as
   shown.

6. On a B- or A3-size sheet make a one-
   view drawing of the terminal board
   shown in Fig. 16-2-C. Point-to-point programming is to be used to\locate each
   hole. Below the drawing prepare a chart
   listing each hole and its X and Y coordinates. The letters on the holes show the
   sequence in which they are to be drilled.
   Zero point for the X and Y coordinates is
   the center of the ϕ .438 hole.

Fig. 16-2-C   Terminal board

MATL—.12 THK FIBER

PART 4

Power Transmissions

CHAPTER 17

Belts, Chains, and Gears

UNIT 17-1
Belt Drives

FLAT BELTS

Flat-belt drives offer flexibility, shock absorption, efficient power transmission at high speeds, resistance to abrasive atmospheres, and comparatively low cost. The belts can operate on relatively small pulleys and can be spliced or connected for endless operation. However, because they require high tension, they also impose high bearing loads. They are sometimes noisier than other belt drives, will slip, and have comparatively low efficiency at moderate speeds. See Fig. 17-1-1.

Flat belts for power transmission can be divided into three classes:

1. *Conventional*: plain flat belt without teeth, grooves, or serrations.
2. *Grooved or serrated*: basic flat belt modified to provide the advantages of another type of transmission product, e.g., V-belts.
3. *Positive drive*: basic flat belt modified to eliminate the need for frictional force for power transmission.

Conventional belts are available in two types: *reinforced*, which utilize a tensile member to obtain strength, and *nonreinforced*, which depend upon the tensile strength of the basic material for its strength.

Longitudinally grooved or serrated belts use a flat belt as the tensile section and a series of adjacent V-shaped

Fig. 17-1-1　Flat-belt drives.

grooves for compression and tracking. These are generally known as *poly-V belts*. See Fig. 17-1-2.

Positive-drive belts use a flat belt as the tensile section and a series of evenly spaced teeth on the bottom surface. These teeth engage a similarly grooved pulley to achieve positive mesh. Positive-drive belts are also known as *timing belts*. See Fig. 17-1-3.

Fig. 17-1-2　Grooved or poly-V belt. (Raybestos-Manhattan [Canada])

Fig. 17-1-3   Timing belt. (Morse Chain Co.)

Fig. 17-1-4   Flat belt—rubberized-fabric type. (American Beltrite Rubber Co.)

Conventional Flat Belts

Conventional flat belts are available either as endless belts or as belting which can be spliced to make a needed length. An endless belt is best since it has no weak points caused by splices or connectors, and will usually operate much more smoothly. Either a vulcanized splice or mechanical fastener may be used.

Conventional belts are normally available in five basic materials.

1. Leather
2. Rubberized fabric or cord
3. Nonreinforced rubber or plastic
4. Reinforced leather
5. Fabric

Leather Most leather belts are made of plies of belting bonded together. They provide excellent coefficient of friction, flexibility, and long life and are easily repaired. On the other hand, their initial cost is high, they must be cleaned, and they require belt dressing. They also stretch and shrink, depending on atmospheric conditions.

This type of belt is used primarily for slow to moderate speeds, with a maximum of 6000 feet per minute (ft/min) [30 meters per second (m/s)] for medium to heavy loads, up to 500 horsepower (hp) [375 kilowatts (kW)], ratios of 16:1 are normally possible. It has good shock-absorbing qualities.

Rubberized Fabric or Cord Many types and grains of rubberized belting are presently available. Almost all are moisture-, acid- and alkali-resistant.

Rubberized Fabric This is the least expensive type of flat belting. See Fig. 17-1-4. It is made up of plies of cotton or synthetic duck, impregnated with rubber. Rubberized-fabric belts transmit less power per width for the same thickness and have a shorter life than leather belts.

Rubberized Cord These belts consist of a series of plies of rubber-impregnated cords. They offer high tensile strength for a modest size and mass.

Nonreinforced Rubber or Plastic For light-duty applications, flat belts are available in a number of unreinforced materials.
Rubber Basically a simple strip of rubber, these belts are available in various compounds. They are designed specifically for low-horsepower (kW), low-speed drives. They are especially useful for fixed-center drives since they can be simply stretched into place over their pulleys.
Plastic Unreinforced plastic belts transmit higher power loads than rubber belts. They are available in a number of plastic compounds.

Reinforced Leather These belts consist of a plastic tensile member, generally reorientated nylon, and leather top and bottom layers.

Fabric All-fabric belts may consist of a single piece of cotton or duck folded and sewn with rows of longitudinal stitches. Others are woven into endless forms.

Fabric belts are either made plain or treated with a chemical or rubber solution to improve their coefficient of friction.

The major advantage of all-fabric belts is their ability to track uniformly and to operate at high speeds. Capacity depends on the number of plies of fabric, size of thread, and belt width. They are used typically in check-sorting machines.

Grooved Belts

These are basically flat belts with a longitudinally ribbed underside. The *flat* belt section serves as the load-carrying component, and the ribs provide traction in the sheave grooves.

This type of belt, although it bears a resemblance to the conventional V-belt, operates on a different principle. Rather than depending on wedging action to transmit power, it depends solely on friction between sheave and belt. Power capacity depends on belt width; only a single belt, with a varying number of ribs, is used for each drive.

Positive-Drive Belts

Another variation of the flat belt is the positive-drive belt, commonly known as the *timing belt*. Basically a flat belt with a series of evenly spaced teeth on the inside circumference, it combines the advantages of the flat belt with the positive-grip features of chain and gears. The belt has high-strength steel or glass tensile members, with nylon-jacketed neoprene teeth. The belts are available in five stock pitches and in various widths. Special widths and pitches are available.

Positive-drive belts have many advantages. There is no slippage or speed variation, and a wide range of speed ratios is possible. Required belt tension is minimal, so that bearing loads are low.

These belts are not recommended where pulleys are misaligned. Also, high-speed operation may cause some noise, but this is not generally a problem at normal operating speeds.

Pulleys for Flat Belts

Different types of pulleys are used for flat, ribbed, and positive-drive belts.

Flat-Belt Pulleys These are generally made of cast iron. However, they are also available in steel and in various rim and hub combinations. They may have solid, spoked, or split hubs as well as other modifications of the basic pulley.

Crowning All power-transmission pulleys should be crowned or flanged. See Fig. 17-1-5.

Other Types Pulleys for ribbed and positive-drive belts are available in a variety of stock sizes and widths.

At least one pulley in a timing-belt drive must be flanged in order to keep the belt on the drive. For long-center drives, flanging both pulleys is recom-

Fig. 17-1-5 Crown on pulley.

mended but not required. Idler pulleys should not be crowned.

V-BELTS

V-belts are presently available in a wide variety of standardized sizes and types, for transmitting almost any amount of load power.

Normally, V-belt drives operate best at belt speeds between 1500 to 6000 ft/min (8 and 30 m/s). For standard belts, ideal (peak capacity) speed is approximately 4500 ft/min (23 m/s). Narrow V-belts, however, will operate up to 10 000 ft/min (50 m/s). A summary of belt characteristics is given in Fig. 17-1-6. For most drive applications, the maximum satisfactory speed ratio is approximately 7:1.

Advantages V-belt drives permit large speed ratios and provide long life (3 to 5 years). They are easily installed and removed, quiet, and low in maintenance; and they provide shock absorption between driver and driven shafts.

Limitations Because they are subject to a certain amount of creep and slip, V-belts should not be used where synchronous speeds are required.

Standard Dimensions

Cross Sections Industrial and agricultural V-belts are always made to standard cross sections. See Fig. 17-1-7.

Fig. 17-1-7 V-belt and pulley.

Industrial These are made in two types: heavy-duty (conventional, narrow) and light-duty. Conventional belts are available in A, B, C, D, and E sections. See Fig. 17-1-8. Narrow belts are made in 3V, 5V, and 8V sections. Light-duty belts come in 2L, 3L, 4L, and 5L sections.

Open-end belting is available in A, B, C, and D sections. Link-V belting, which is not covered by a standard, is made in A, B, C, D, and E sections, and in some sizes for low-horsepower (kilowatt) applications.

Wide-range V-belts, used for variable-speed drives, are available in Q, P, R, T, and W sections.

Agricultural These belts are made in the same sections as conventional belts. They are designated HA, HB,

Fig. 17-1-8 Industrial V-belts.

HC, HD, and HE; in double-V sections HAA, HBB, HCC, and HDD are available. Agricultural belts differ from industrial belts mainly in construction.

Automotive Belts for automotive applications are made in six SAE-designated cross sections identified by the nominal top widths .38, .50, .69, .75, .88, and 1.00 in. (10, 12, 17, 19, 22, and 25 mm).

Length Although endless V-belts can be manufactured in any length within a fairly wide range, manufacturers have standardized on certain lengths which are produced for stock.

Type of Belt	Maximum Power		Belt Speed for Max. Power		Maximum Speed		Max. Speed Ratio	Shock Absorption
	(hp)	(kW)	(ft/min)	(m/s)	(ft/min)	(m/s)		
Constant-Speed								
Light duty	7.5	5.6	3500	18	5000	25	8	Poor
Standard	350	260	4500	23	6000	30	7	Good
Super	500	375	5000	25	6000	30	7	Very good
Cogged	500	375	5000	25	6000	30	8	Very good
Steel cable	500	375	5000	25	8000	40	7	Poor
Narrow	270	200	7500	38	10000	50	7	Very good
Variable-Speed								
Conventional	300	225	—	—	6000	30	—	Good
Wide-range	75	55	—	—	6000	30	—	Good

*Stock items. Drives available to 1500 hp (1100 kW).

Fig. 17-1-6 V-belt characteristics.

Belt-Size Designation

For the different types of V-belts, the same basic method is used to designate belt size. Belt sizes are specified by a code designation consisting of symbols representing belt cross section followed by a designation of length. For conventional and light-duty belts, the length designation is in inches; for narrow belts the number represents tenths of an inch.

For example, a conventional V-belt designated B23 has a B cross section and a 23 in. standard length designation; a narrow belt designated 5V350 has a 5V cross section and a belt with a 35 in. effective outside length; and a light-duty V-belt designated 2L080 has a 2L cross section and an effective outside length of 80 in.

There are no standard methods for designating automotive belts. Variable-speed belts are designated by a code where the first two numbers denote the nominal belt width in sixteenths of an inch, the next two numbers denote the angle of the pulley groove, followed by the letter V, with numbers after that letter indicating length in tenths of an inch.

Basically, a V-belt consists of five component sections (Fig. 17-1-9)

1. Tensile members or load-carrying section
2. Low-durometer cushion section surrounding tensile members
3. Flexible top section
4. Bottom compression section
5. Cover or jacket

Fig. 17-1-9 Basic V-belt construction. (American Beltrite Rubber Co.)

Sheaves and Hubs

Most sheaves (the grooved wheels of pulleys) are made of cast iron, which is economical and stable and which provides long groove life. For light duty, sheaves may be of formed steel, cast

(A) SLIDING

(B) CRADLE

(C) SPRING TENSION

(D) PIVOTED

(E) APPLICATION OF A SLIDING MOTOR BASE

Fig. 17-1-10 Common types of motor bases. (T. B. Wood's Sons Co.)

(C) SINGLE DRIVE

Fig. 17-1-11 Single- and multiple-belt drives.

(A) SINGLE PULLEY

(B) DOUBLE PULLEY

(D) MULTIPLE DRIVE

iron, or plastic. Formed-steel sheaves are used primarily in automotive and agricultural applications. For special applications they may be made of steel or aluminum alloy. Typical V-belt applications are shown in Figs. 17-1-10 and 17-1-11.

Cast-iron sheaves are generally limited to 6500-ft/min (33-m/s) rim speeds. For speeds up to 10 000 ft/min (50 m/s), aluminum, steel, and ductile iron are used.

Sheaves are made with either regular or deep grooves. A deep-groove sheave is generally used when the V-belt enters the sheave at an angle, for example, in a quarter-turn drive, on vertical shaft drives, or wherever belt vibration may be a problem.

Formed-steel sheaves usually have an integral hub, but are sometimes available with removable bushings for various bore sizes.

The Use of Idler Pulleys

Idler pulleys are grooved sheaves or flat pulleys which do not serve to transmit power. Usually they are used as belt tighteners when it is not possible to move either shaft for belt installation and take-up, as between two line shafts, for example. It is better, and more economical in the long run, to provide movement for one of the shafts where it is at all possible to do so, rather than use idlers as belt tighteners. However, if idlers must be used, they are perfectly acceptable for multiple V-belt drives. Idler pulleys may also be used when the belt must be passed around some obstruction. Such a pulley may be run inside the drive or on the outside. An inside idler may be either a grooved sheave or a flat, uncrowned pulley, but an outside idler must be flat and without any crown.

An inside idler pulley invariably decreases the arc of contact of the belts on each loaded sheave of the drive. It should be at least as large as the small loaded sheave and located, preferably, on the slack side of the drive. See Fig. 17-1-12A.

(A) INSIDE IDLER PULLEY, AT LEAST AS LARGE AS THE SMALL SHEAVE, ON THE SLACK SIDE OF THE DRIVE

(B) OUTSIDE IDLER PULLEY, AT LEAST 1.3 LARGER THAN THE SMALL SHEAVE

(C) OUTSIDE IDLER PULLEY ON THE TIGHT SIDE OF THE DRIVE

(D) INSIDE IDLER PULLEY ON THE TIGHT SIDE OF THE DRIVE

Fig. 17-1-12 Location of idler pulleys.

An outside idler pulley invariably increases the arc of contact of the belts on each loaded sheave of the drive, but the amount of take-up that can be obtained is definitely limited by the belts as they run on the opposite side. If you are designing a V-belt drive and cannot provide movement for one of the shafts for belt installations and take-up, an outside idler pulley should be used. It should be at least one-third larger than the small loaded sheave and located, preferably, on the slack side of the drive. See Fig. 17-1-12B.

A flat idler pulley, whether used inside or outside the drive, should be located as close as possible to the place where the belts leave the sheave. On the slack side of the drive, which is the preferred location, this means as close as possible to the driver sheave. See Figs. 17-1-12A and B. On the tight side of the drive, this means as close as possible to the driven sheave. See Figs. 17-1-12C and D.

HOW TO SELECT A LIGHT-DUTY V-BELT DRIVE

The proper selection of V-belt drives for light machinery has been simplified and condensed into three steps. Complete selection involves the proper choice of

1. V-pulley size for driver shaft and belt cross section
2. V-pulley size for driven shaft
3. Belt length for required center distance

Proper duty classification helps to ensure maximum drive life. The following are typical duty classifications

- **Light Duty**
  Household washers
  Household ironers
  Dishwashers
  Fans and blowers
  Centrifugal pumps
- **Normal Duty**
  Oil burners
  Buffers
  Heating and ventilating fans
  Meat slicers
  Speed-up drives
  Drill presses
  Generators
  Power lawn mowers

- **Heavy Duty**
  Gasoline engine drives
  Metalworking machines
  Sanding machines
  Stokers
  Spray equipment
  Woodworking machines
  Lathes
  Industrial machines
  Refrigerators
  Compressors
  Piston or plunger pumps
  Grinders

The horsepower (kilowatt) ratings listed in Fig. 17-1-13 are suitable for normal-duty applications. For light duty, multiply the normal-duty rating by 1.2. For heavy duty, multiply the normal rating by 0.85.

Follow These Three Easy Steps

Step 1: Selecting Driver V-Pulley Diameter and Belt Cross Section First, classify the application and apply the proper service factor, as explained above. Refer to Fig. 17-1-13 for driver V-pulley diameter and belt cross section.

Step 2: Choosing Driven V-Pulley Diameter Refer to Fig. 17-1-14 for the speed of the motor. Locate desired driven speed in driver V-pulley column; read driven V-pulley diameter in the first column.

Step 3: Finding Belt Length and Center Distance Add the diameter of driver and driven V-pulley and refer to Fig. 17-1-15. Locate the sum of V-pulley diameters at the top of the chart, read down to the required centers, and read the belt length in the belt length column.

Although the amount of stretch in V-belts is relatively small, some adjustment between centers of pulleys is necessary to compensate for stretch and side wear on the belts and sheaves.

To design a belt drive, the following information should be known

1. The speed [revolutions per minute (r/min)] and horsepower (kilowatts) of the motor or driver unit
2. The speed (r/min) at which the driven shaft is to turn
3. The space available for the drive

RPM of Small Pulley	Outside Diameter of Small V-Pulley—Inches														
	1.50	1.75	2.00	2.25	2.50	2.75	3.00	3.25	3.50	3.75	4.00	4.25	4.50	4.75	5.00
200	...	...	...	...	...	...	...	...	...	...	0.18	0.22	0.24	0.28	0.29
400	...	...	0.06	0.08	0.12	0.15	0.18	0.22	0.25	0.31	0.35	0.42	0.46	0.52	0.56
600	0.04	0.07	0.08	0.12	0.18	0.22	0.27	0.32	0.36	0.44	0.51	0.58	0.66	0.73	0.81
800	0.05	0.08	0.11	0.15	0.22	0.28	0.34	0.41	0.45	0.55	0.64	0.74	0.81	0.93	1.00
1000	0.06	0.10	0.12	0.18	0.26	0.33	0.42	0.48	0.55	0.64	0.75	0.86	0.99	1.10	1.21
1160	0.07	0.11	0.15	0.21	0.29	0.38	0.46	0.54	0.62	0.69	0.84	0.98	1.07	1.23	1.35
1400	0.08	0.12	0.17	0.23	0.33	0.43	0.53	0.64	0.74	0.84	0.96	1.10	1.25	1.42	1.55
1600	0.08	0.14	0.19	0.25	0.36	0.48	0.58	0.69	0.80	0.90	1.02	1.20	1.36	1.53	1.68
1750	0.08	0.15	0.20	0.25	0.38	0.51	0.63	0.74	0.85	0.96	1.08	1.25	1.43	1.61	1.78
2000	0.09	0.16	0.22	0.28	0.41	0.55	0.68	0.81	0.92	1.05	1.17	1.35	1.54	1.73	1.90
2200	0.09	0.17	0.24	0.31	0.44	0.58	0.72	0.86	0.99	1.12	1.25	1.41	1.61	1.80	1.99
2400	0.10	0.18	0.25	0.32	0.45	0.61	0.76	0.91	1.05	1.19	1.32	1.45	1.65	1.86	2.02
2600	0.10	0.19	0.26	0.35	0.47	0.64	0.79	0.96	1.09	1.24	1.38	1.48	1.69	1.89	2.09
2800	0.11	0.19	0.28	0.36	0.48	0.66	0.83	0.99	1.14	1.28	1.42	1.48	1.71	1.91	2.11
3000	0.11	0.21	0.29	0.39	0.49	0.68	0.85	1.02	1.18	1.32	1.46	1.48	1.69	1.89	2.08
3200	0.11	0.21	0.30	0.39	0.51	0.70	0.88	1.05	1.20	1.36	1.50	1.50	1.67	1.86	2.03
3450	0.12	0.22	0.32	0.41	0.51	0.71	0.90	1.07	1.23	1.38	1.52	1.52	1.61	1.78	1.94
3600	0.12	0.22	0.33	0.42	0.52	0.72	0.91	1.09	1.25	1.40	1.54	1.54	1.54	1.71	1.85
3800	0.12	0.22	0.33	0.42	0.52	0.72	0.92	1.09	1.25	1.41	1.54	1.54	1.54	1.59	1.72
4000	0.12	0.22	0.34	0.44	0.53	0.72	0.92	1.10	1.26	1.40	1.52	1.52	1.52	1.52	1.55

FOR [ ] BACKGROUND USE A   FOR [ ] BACKGROUND USE A   FOR [ ] BACKGROUND USE A

10 / .38   6 / .22

12 / .50   A   8 / .31

16 / .66   B   10 / .41

MILLIMETERS
INCHES

NOTE: THIS TABLE INCORPORATES A SERVICE FACTOR OF 1.3.
FOR HEAVY DUTY, MULTIPLY NORMAL DUTY RATING BY .85.
FOR LIGHT DUTY, MULTIPLY NORMAL DUTY RATING BY 1.20.

NOTE: SIZES SHOWN ARE INCH SIZES SOFT CONVERTED TO MILLIMETERS.

KILOWATT RATINGS (METRIC)

R/min of Small Pulley	Outside Diameter of Small V-Pulley—Millimeters														
	38	44	51	57	64	70	76	83	89	95	102	108	114	121	127
200											0.13	0.16	0.18	0.21	0.22
400			0.04	0.06	0.09	0.11	0.13	0.16	0.19	0.23	0.26	0.31	0.34	0.39	0.42
600	0.03	0.05	0.06	0.09	0.13	0.16	0.20	0.24	0.27	0.33	0.38	0.43	0.49	0.54	0.60
800	0.04	0.06	0.08	0.11	0.16	0.21	0.25	0.31	0.34	0.41	0.48	0.55	0.60	0.69	0.75
1000	0.04	0.07	0.09	0.13	0.19	0.25	0.31	0.36	0.41	0.48	0.56	0.64	0.74	0.82	0.90
1160	0.05	0.08	0.11	0.16	0.22	0.28	0.34	0.40	0.46	0.51	0.63	0.73	0.80	0.92	1.01
1400	0.06	0.09	0.13	0.17	0.25	0.32	0.40	0.48	0.55	0.63	0.72	0.82	0.93	1.06	1.16
1600	0.06	0.10	0.14	0.19	0.27	0.36	0.43	0.51	0.60	0.67	0.76	0.90	1.01	1.14	1.25
1750	0.06	0.11	0.15	0.19	0.28	0.38	0.47	0.55	0.63	0.72	0.81	0.93	1.07	1.20	1.33
2000	0.07	0.12	0.16	0.21	0.31	0.41	0.51	0.60	0.69	0.78	0.87	1.01	1.15	1.29	1.42
2200	0.07	0.13	0.18	0.23	0.33	0.43	0.54	0.64	0.74	0.84	0.94	1.05	1.20	1.34	1.48
2400	0.07	0.13	0.19	0.24	0.34	0.46	0.57	0.68	0.78	0.89	0.98	1.08	1.23	1.39	1.51
2600	0.07	0.14	0.19	0.26	0.35	0.48	0.59	0.72	0.81	0.93	1.03	1.10	1.26	1.41	1.56
2800	0.08	0.14	0.21	0.27	0.36	0.49	0.62	0.74	0.85	0.95	1.06	1.10	1.28	1.42	1.57
3000	0.08	0.16	0.22	0.29	0.37	0.51	0.63	0.76	0.88	0.98	1.09	1.10	1.26	1.41	1.55
3200	0.08	0.16	0.22	0.29	0.38	0.52	0.66	0.78	0.90	1.01	1.12	1.12	1.25	1.39	1.51
3450	0.09	0.16	0.24	0.31	0.38	0.53	0.67	0.80	0.92	1.03	1.13	1.13	1.20	1.33	1.45
3600	0.09	0.16	0.25	0.31	0.39	0.54	0.68	0.81	0.93	1.04	1.15	1.15	1.15	1.28	1.38
3800	0.09	0.16	0.25	0.31	0.39	0.54	0.69	0.81	0.93	1.05	1.15	1.15	1.15	1.19	1.28
4000	0.09	0.16	0.25	0.33	0.40	0.54	0.69	0.82	0.94	1.05	1.15	1.15	1.13	1.13	1.16

Fig. 17-1-13 Calculating Pulley Diameter of Driven Shaft and Belt Cross Section. (T. B. Wood's Sons Co.)

EXAMPLE 1 A .5-hp (0.38-kW), 1750-r/min motor is to operate a drill press having a spindle speed of approximately 1200 r/min. The center distance between the motor shaft and spindle is approximately 19.5 in. (500 mm). The type of drive required is V-belt.

Solution Since drill press operations come under the classification of normal duty, no adjustment needs to be made to the horsepower (kilowatt) rating.

Step 1: Selecting Driver V-Pulley Diameter and Belt Cross Section (Fig. 17-1-13) Read down the extreme left column to the r/min figure nearest that of the speed of the motor, which is 1750 r/min. Read across this line to the figure closest to the design horsepower or kilowatts of the drive. The closest horsepower (kilowatt) rating is found to be .51 (0.38). Read up from the .51 hp (0.38-kW) figure. The figure at the top of the column is the outside diameter of the motor pulley in inches (millimeters). The .51 hp (0.38-kW) figure is in the white area. The reference at the bottom of the chart indicates the size of the belt required.

Pulley size for motor = 2.75 in. (70 mm)

Belt section = .50 in. (12 mm) wide × .31 in. (8 mm) thick

Step 2: Choosing Driven V-Pulley Diameter (Fig. 17-1-14) Refer to the table for driven speeds for 1750-r/min motors. Read across the top of the table to the figure nearest the small pulley size. Column 2.75 (70) corresponds exactly with the small pulley diameter. Read down this column to the figure nearest the desired speed (1200-r/min) of the driven shaft. The nearest figure is 1168. By reading to the left of this figure, the driven-pulley diameter is found to be 4.00 in. (102 mm).

Step 3: Finding Belt Length and Center Distance (Fig. 17-1-15) Add the diameter of the pulleys and select the number in the top row that is nearest to this sum.

Motor pulley diameter = 2.75 in. (70 mm)

Spindle pulley diameter = 4.00 in. (102 mm)

Sum of diameter = 6.75 in. (172 mm)

The exact sum of the diameters is not shown on the top row; use 7.00 in.

DriveN V-Pulley OD in./mm	DriveR V-Pulley OD—in./mm												
	1.50 / 38	1.75 / 44	2.00 / 51	2.25 / 57	2.50 / 64	2.75 / 70	3.00 / 76	3.25 / 83	3.50 / 89	3.75 / 95	4.00 / 102	4.25 / 108	4.50 / 114

DRIVEN SPEEDS FOR 1160 R/MIN MOTORS

DriveN V-Pulley OD in./mm	1.50 / 38	1.75 / 44	2.00 / 51	2.25 / 57	2.50 / 64	2.75 / 70	3.00 / 76	3.25 / 83	3.50 / 89	3.75 / 95	4.00 / 102	4.25 / 108	4.50 / 114
1.5/ 38	1160	1392	1625	1855	2085	2325	2550	2785	3015	3250	3480	3715	
2.0/ 51	829	995	1160	1325	1490	1658	1825	1988	2150	2315	2485	2650	
2.5/ 64	645	774	903	1031	1160	1290	1418	1546	1675	1805	1933	2032	2190
3.0/ 76	528	634	739	845	950	1057	1160	1266	1370	1475	1580	1685	1793
3.5/ 89	447	536	625	715	804	894	982	1071	1160	1248	1340	1428	1518
4.0/102	387	465	542	620	696	775	851	929	1008	1082	1160	1238	1315
4.5/114	341	409	477	545	614	682	750	819	886	955	1022	1091	1160
5.0/127	305	366	427	488	549	610	671	732	794	854	915	976	1039
5.5/140	277	332	381	442	497	553	608	663	718	774	829	884	939
6.0/152	253	302	353	404	454	505	555	605	655	706	756	806	857
7.0/178	215	258	301	344	388	430	474	516	560	602	648	688	732
8.0/203	187	224	262	297	337	374	411	449	486	524	561	599	636
10.0/254	149	179	208	238	268	298	328	357	387	417	446	477	506
12.0/305	123	148	173	197	222	247	272	296	321	346	370	395	420

DRIVEN SPEEDS FOR 1750 R/MIN MOTORS

DriveN V-Pulley OD in./mm	1.50 / 38	1.75 / 44	2.00 / 51	2.25 / 57	2.50 / 64	2.75 / 70	3.00 / 76	3.25 / 83	3.50 / 89	3.75 / 95	4.00 / 102	4.25 / 108	4.50 / 114
1.5/ 38	1750	2100	2450	2800	3150	3500	3850						
2.0/ 51	1250	1500	1750	2000	2250	2500	2750	3000	3250	3500	3750	4000	
2.5/ 64	974	1167	1360	1555	1750	1945	2140	2330	2530	2725	2915	3110	3305
3.0/ 76	797	955	1113	1272	1431	1590	1750	1910	2070	2225	2385	2545	2700
3.5/ 89	674	808	942	1077	1210	1346	1480	1615	1750	1885	2020	2155	2290
4.0/102	584	700	817	935	1050	1168	1283	1400	1518	1634	1750	1865	1985
4.5/114	516	618	720	824	926	1030	1131	1235	1339	1440	1543	1650	1750
5.0/127	462	554	646	737	830	922	1013	1105	1198	1290	1382	1473	1568
5.5/140	417	500	584	667	750	834	917	1000	1082	1167	1250	1333	1417
6.0/152	381	456	533	610	685	760	837	913	990	1065	1140	1217	1290
6.5/165	350	420	490	560	630	700	771	840	910	980	1050	1120	1190
7.0/178	324	389	454	518	584	648	713	778	843	907	973	1039	1102
8.0/203	282	339	394	451	507	564	620	676	734	789	845	902	959
9.0/229	250	300	350	400	450	500	550	600	650	700	750	800	850
10.0/254	224	270	315	360	405	450	495	540	585	630	675	720	765
11.0/279	203	244	285	326	366	407	448	488	530	570	610	652	692
12.0/305	186	224	261	298	336	373	410	446	485	522	560	596	634

DRIVEN SPEEDS FOR 3500 R/MIN MOTORS

DriveN V-Pulley OD in./mm	1.50 / 38	1.75 / 44	2.00 / 51	2.25 / 57	2.50 / 64	2.75 / 70	3.00 / 76	3.25 / 83	3.50 / 89	3.75 / 95	4.00 / 102	4.25 / 108	4.50 / 114
1.5/ 38	3500	4200	4900	5600	6300	7000	7700						
2.0/ 51	2500	3000	3500	4000	4500	5000	5500	6000	6500	7000	7500	8000	
2.5/ 64	1948	2324	2720	3110	3500	3890	4280	4660	5060	5450	5830	6220	6610
3.0/ 76	1594	1910	2236	2544	2862	3180	3500	3820	4140	4450	4770	5090	5400
3.5/ 89	1348	1616	1884	2154	2420	2692	2960	3230	3500	3770	4040	4310	4580
4.0/102	1168	1400	1634	1870	2030	2336	2566	2800	3036	3268	3500	3730	3970
4.5/114	1032	1236	1440	1648	1852	2060	2262	2470	2678	2880	3086	3300	3500
5.0/127	924	1108	1292	1474	1660	1844	2026	2210	2396	2580	2764	2946	3136
5.5/140	834	1000	1168	1334	1500	1668	1834	2000	2164	2334	2500	2666	2834
6.0/152	762	912	1066	1220	1370	1520	1774	1826	1980	2130	2280	2434	2580
6.5/165	700	840	980	1120	1260	1400	1542	1680	1820	1960	2100	2240	2380
7.0/178	648	778	908	1036	1168	1296	1426	1556	1686	1814	1946	2078	2204
8.0/203	564	678	788	902	1014	1128	1240	1352	1468	1578	1690	1804	1918
9.0/229	500	600	700	800	900	1000	1100	1200	1300	1400	1500	1600	1750
10.0/254	448	540	630	720	810	900	990	1080	1170	1260	1350	1440	1530
11.0/279	406	488	570	652	732	814	896	976	1060	1140	1220	1304	1384
12.0/305	372	448	522	596	672	746	820	892	970	1044	1120	1192	1268

Fig. 17-1-14 Calculating revolutions per minute and diameter of driven pulley. (T. B. Wood's Sons Co.)

(180 mm). Read down this column to the figure indicated below the shaded area. The 19.5 in. (500 mm) distance is the ideal center distance. Other figures in this column indicate alternative center distances. Use 19.4 in. (493 mm) since the approximate center distance required is 19.5 in. (500 mm). Follow along this line to the left to column Market Belt Length to obtain a belt length of 50 in. (1270 mm).

References and Source Material

1. *Machine Design*, Mechanical drives reference issue, 1979.
2. The Gates Rubber Company, Denver, Colorado.
3. T. B. Wood's Sons Company.

ASSIGNMENTS

See Assignments 1 and 2 for Unit 17-1 on page 361.

Review for Assignments

Unit 6-4   Bill of Material

SUM OF BOTH V-BELT PULLEY DIAMETERS—DIMENSIONS IN INCHES

Installation Allowance		Belt Length	4	4.5	5	5.5	6	6.5	7	7.5	8	8.5	9	9.5	10	10.5	11	11.5	12	12.5	13	13.5	14	14.5	15	15.5	16
.50	.50	16	4.9	4.5	4.1																						
.62	.50	18	5.9	5.5	5.1	4.6																					
.62	.50	20	6.9	6.5	6.1	5.6	5.2																				
.62	.50	22	7.9	7.5	7.1	6.6	6.2	5.8																			
.62	.50	24	8.9	8.5	8.1	7.6	7.2	6.8	6.3	5.8																	
.62	.50	26	9.9	9.5	9.1	8.6	8.2	7.8	7.3	6.9	6.5																
.62	.50	28	10.9	10.5	10.1	9.6	9.2	8.8	8.4	7.9	7.6	7.1	6.6														
.62	.50	30	11.9	11.5	11.1	10.6	10.2	9.8	9.4	8.9	8.6	8.1	7.7	7.3													
.62	.50	32	12.9	12.5	12.1	11.6	11.2	10.8	10.4	10.0	9.6	9.1	8.7	8.4	8.0												
.62	.50	34	13.9	13.5	13.1	12.7	12.2	11.8	11.4	11.0	10.6	10.2	9.7	9.4	9.0	8.6											
.62	.50	36	14.9	14.5	14.1	13.7	13.2	12.8	12.4	12.0	11.6	11.2	10.7	10.4	10.0	9.6	9.0										
.62	.50	38	15.9	15.5	15.1	14.7	14.2	13.8	13.4	13.0	12.6	12.2	11.8	11.4	11.0	10.6	10.0	9.7	9.1								
.75	.50	40	16.9	16.5	16.1	15.7	15.3	14.8	14.4	14.0	13.6	13.2	12.8	12.4	12.0	11.6	11.1	10.7	10.1	9.8							
.75	.50	42	17.9	17.5	17.1	16.7	16.3	15.8	15.4	15.0	14.6	14.2	13.8	13.4	13.1	12.6	12.1	11.7	11.2	10.8	10.2						
.75	.50	44	18.9	18.5	18.1	17.7	17.3	16.8	16.4	16.0	15.6	15.2	14.8	14.4	14.1	13.6	13.1	12.8	12.2	11.9	11.2	10.9					
.75	.50	46	19.9	19.5	19.1	18.7	18.3	17.9	17.4	17.0	16.6	16.2	15.8	15.4	15.1	14.6	14.1	13.8	13.2	12.9	12.3	12.0	10.9	10.5			
.75	.50	48	20.9	20.5	20.1	19.7	19.3	18.9	18.4	18.0	17.7	17.2	16.8	16.4	16.1	15.6	15.1	14.8	14.3	13.9	13.3	13.0	12.0	11.6	11.3		
.75	.50	50	21.9	21.5	21.1	20.7	20.3	19.9	19.4	19.0	18.7	18.2	17.8	17.4	17.1	16.7	16.2	15.8	15.3	14.9	14.4	14.0	13.1	12.7	12.4	12.1	11.7
.75	.50	52	22.9	22.5	22.1	21.7	21.3	20.9	20.4	20.0	19.7	19.2	18.8	18.4	18.1	17.7	17.2	16.8	16.3	15.9	15.4	15.0	14.1	13.8	13.5	13.1	12.8
.75	.50	54	23.9	23.5	23.1	22.7	22.3	21.9	21.4	21.0	20.7	20.2	19.8	19.4	19.1	18.7	18.2	17.8	17.3	17.0	16.4	16.1	15.2	14.8	14.5	14.2	13.8
.75	.50	56	24.9	24.5	24.1	23.7	23.3	22.9	22.4	22.0	21.7	21.2	20.8	20.4	20.1	19.7	19.2	18.8	18.3	18.0	17.4	17.1	16.2	15.9	15.6	15.2	14.9
.75	.50	58	25.9	25.5	25.1	24.7	24.3	23.9	23.4	23.0	22.7	22.2	21.8	21.4	21.1	20.7	20.2	19.8	19.3	19.0	18.5	18.1	17.3	16.9	16.6	16.3	15.9
.88	.75	60	26.9	26.5	26.1	25.7	25.3	24.9	24.5	24.0	23.7	23.2	22.8	22.4	22.1	21.7	21.2	20.8	20.4	20.0	19.5	19.1	18.3	18.0	17.6	17.3	17.0
.88	.75	62	27.9	27.5	27.1	26.7	26.3	25.9	25.5	25.0	24.7	24.3	23.8	23.4	23.1	22.7	22.2	21.8	21.4	21.0	20.5	20.1	19.4	19.0	18.7	18.3	18.0
.88	.75	64	28.9	28.5	28.1	27.7	27.3	26.9	26.5	26.0	25.7	25.3	24.8	24.4	24.1	23.7	23.2	22.9	22.4	22.0	21.5	21.1	20.4	20.0	19.7	19.4	19.0
.88	.75	66	29.9	29.5	29.1	28.7	28.3	27.9	27.5	27.0	26.7	26.3	25.9	25.4	25.1	24.7	24.2	23.9	23.4	23.0	22.5	22.2	21.4	21.1	20.7	20.4	20.0
.88	.75	68	30.9	30.5	30.1	29.7	29.3	28.9	28.5	28.1	27.7	27.3	26.9	26.4	26.1	25.7	25.2	24.9	24.4	24.0	23.5	23.2	22.4	22.1	21.7	21.4	21.0
.88	.75	70	31.9	31.5	31.1	30.7	30.3	29.9	29.5	29.1	28.7	28.3	27.9	27.4	27.1	26.7	26.2	25.9	25.4	25.0	24.5	24.2	23.5	23.1	22.8	22.4	22.1
.88	.75	72	32.9	32.5	32.1	31.7	31.3	30.9	30.5	30.1	29.7	29.3	28.9	28.4	28.1	27.7	27.2	26.9	26.4	26.0	25.5	25.2	24.5	24.1	23.8	23.4	23.1
.88	.75	74	33.9	33.5	33.1	32.7	32.3	31.9	31.5	31.1	30.7	30.3	29.9	29.4	29.1	28.7	28.2	27.9	27.4	27.0	26.5	26.2	25.5	25.1	24.8	24.4	24.1
.88	.75	76	34.9	34.5	34.1	33.7	33.3	32.9	32.5	32.1	31.7	31.3	30.9	30.4	30.1	29.7	29.2	28.9	28.4	28.0	27.6	27.2	26.5	26.2	25.8	25.5	25.1
.88	.75	78	35.9	35.5	35.1	34.7	34.2	33.9	33.5	33.1	32.7	32.3	31.9	31.4	31.1	30.7	30.2	29.9	29.4	29.0	28.6	28.2	27.5	27.2	26.8	26.5	26.1
.88	.75	80	36.9	36.5	36.1	35.7	35.3	34.9	34.5	34.1	33.7	33.3	32.9	32.4	32.1	31.7	31.3	30.9	30.4	30.0	29.6	29.2	28.6	28.2	27.9	27.5	27.1
.88	.75	82	37.6	37.1	36.7	36.3	35.9	35.5	35.1	34.7	34.3	33.9	33.5	33.0	32.7	32.3	31.9	31.5	31.0	30.7	30.2	29.8	29.2	28.8	28.5	28.1	27.8
.88	.75	84	38.9	38.5	38.1	37.7	37.3	36.9	36.5	36.1	35.7	35.3	34.9	34.4	34.1	33.7	33.3	32.9	32.4	32.1	31.6	31.2	30.6	30.2	29.9	29.5	29.2

SUM OF BOTH V-BELT PULLEY DIAMETERS—DIMENSIONS IN MILLIMETERS (METRIC)

Installation Allowance	Take-Up	Belt Length	125	140	150	165	180	190	200	215	230	240	255	265	280	290	305	320	330	345	355	370
12	10	410	105																			
15	10	460	130	117																		
15	10	510	155	142	132																	
15	10	560	180	168	157	147																
15	10	610	205	193	183	173	160	147														
15	10	660	231	218	208	198	185	175	165													
15	10	710	257	247	234	224	213	201	193	180	168											
15	10	760	282	272	259	249	239	226	218	205	196	185										
15	10	810	308	295	284	274	264	254	244	231	221	213	203									
15	10	860	333	323	310	300	290	280	269	259	246	239	228	218								
15	10	910	358	348	335	325	315	305	295	284	272	264	254	244	229							
15	10	960	384	373	361	351	340	330	320	310	300	290	279	269	254	246	231					
20	10	1010	409	399	389	376	366	356	345	335	325	315	305	295	282	272	257	249				
20	10	1070	434	424	414	401	391	381	371	361	351	340	333	320	307	297	284	274	259			
20	10	1120	460	450	439	427	417	406	396	386	376	366	358	345	333	325	310	302	284	277		
20	10	1170	485	475	465	455	442	432	422	411	401	391	384	371	358	351	335	328	312	304	277	266
20	10	1220	511	500	490	480	467	457	450	437	427	417	409	396	384	376	363	353	338	330	304	295
20	10	1270	536	526	516	505	493	483	475	462	452	442	434	424	411	401	389	378	366	358	333	323
20	10	1320	562	551	541	531	518	508	500	488	478	467	460	450	437	427	414	404	391	381	358	351
20	10	1370	587	577	566	556	544	534	526	513	503	493	485	475	462	452	439	431	417	410	386	376
20	10	1420	612	602	592	582	569	559	551	538	528	518	511	500	488	478	465	457	442	434	411	404
20	10	1470	638	627	617	607	594	584	577	564	554	544	536	526	513	503	490	482	470	460	439	429
22	15	1520	663	653	643	632	622	610	602	589	579	569	561	551	538	528	518	508	495	485	465	457
22	15	1570	688	678	668	658	648	635	627	617	605	594	587	577	564	554	544	533	520	511	493	483
22	15	1630	714	704	693	683	673	660	653	643	630	620	612	602	589	582	569	559	546	536	518	508
22	15	1680	740	729	719	709	699	686	678	668	658	645	638	627	615	607	594	584	571	564	544	536
22	15	1730	765	754	744	734	724	714	704	693	683	671	663	653	640	632	620	610	597	589	569	561
22	15	1780	790	780	770	759	749	739	729	719	709	696	688	678	665	653	645	635	622	615	597	587
22	15	1830	815	805	798	785	765	765	754	744	734	721	714	704	691	683	671	660	647	640	622	612
22	15	1880	841	831	820	810	800	790	780	770	759	747	739	729	716	709	696	685	673	665	648	638
22	15	1930	866	856	846	836	826	815	805	795	785	772	765	754	742	734	721	711	701	691	673	665
22	15	1980	892	881	871	861	851	841	831	820	810	798	790	780	767	759	747	736	726	716	698	691
22	15	2030	917	907	897	886	876	866	856	846	836	823	815	805	795	785	772	762	752	742	726	716
22	15	2080	932	922	912	902	892	881	871	861	851	838	831	820	810	800	787	780	767	757	742	732
22	15	2130	968	958	947	937	927	917	907	897	886	874	866	856	846	836	823	815	803	792	777	767

Fig. 17-1-15 Determining V-belt length from pulley diameters and center distances. (T. B. Wood's Sons Co.)

UNIT 17-2
Chain Drives

Nearly all types of power-transmission chains have two basic components: side bars or link plates, and pin and bushing joints. The chain articulates at each joint to operate around a toothed sprocket. The *pitch* of the chain is the distance between centers of the articulating joints.

Power-transmission chains have several advantages: relatively unrestricted shaft center distances, compactness, ease of assembly, elasticity in tension with no slip or creep, and ability to operate in relatively high-temperature atmospheres. A typical application can be seen in Fig. 17-2-1.

BASIC TYPES

There are six major types of power-transmission chains, with numerous modifications and special shapes for specific applications. A seventh type, the bead chain, is often used for light-duty applications. Figure 17-2-2 shows basic characteristics of four of the major types.

Detachable

The malleable detachable chain is made in a range of sizes from .902 to 4.063 in. (23 to 103 mm) pitch and ultimate strength from 700 to 17 000 lb/in.^{2} [5 to 110 megapascals (MPa)].

Of the same type is the steel detachable chain, covered by ANSI B29.6. This chain is made in sizes from .904 in. (23 mm) to just under 3.00 in. (76 mm) in pitch, with ultimate strength from 760 to 5000 lb/in.^{2} (35 to 240 MPa).

The ends of the detachable link are referred to as the *bar end* and the *hook end*.

Pintle

For slightly higher speeds [to about 450 ft/min (2.2 m/s)] and heavier loads, pintle chains are used. *Pintle chains* are made up of individual cast links having a full, round barrel end with offset sidebars. These links are intercoupled with steel pins. The ends of pintle chain links are referred to as the *barrel end* and the *open end*.

Many of these chains have been designed to operate over sprockets intended for detachable chain. Therefore, chains range from just over 1.00 in. (25 mm) up to 6.00 in. (150 mm) in pitch with ultimate strengths from 3600 to 30 000 lb/in.^{2} (25 to 200 MPa).

Offset-Sidebar

Steel offset-sidebar chains are used extensively as drive chains on construction machinery. They operate at speeds to 1000 ft/min (5 m/s) and transmit loads to approximately 250 hp (185 kW).

(A) PINTLE

(B) OFFSET

(C) ROLLER

(D) INVERTED TOOTH (SILENT)

(E) BEAD OR SLIDER

Fig. 17-2-2 Basic chain types. (A and D—Link Belt Ltd.; B and C—Dodge Mfg. Corp.; E—*Machine Design*)

Fig. 17-2-1 Chain drives.

Each link has two offset sidebars, one bushing, one roller, one pin, and, if the chain is detachable, a cotter pin. Some offset-sidebar chains are made without rollers.

Roller

Transmission roller chain (Fig. 17-2-3) is available in pitches from .25 to 3.00 in. (6 to 75 mm). In the single-width roller, the ultimate strength ranges from 925 to 130 000 lb/in.² (6 to 900 MPa). It is also available in multiple widths. Small-pitch sprockets can operate at speeds as high as 10 000 r/min, and 1000- to 1200-hp (750- to 900-kW) drives are not unusual.

These chains are assembled from roller links and pin links. If the chain is detachable, cotter pins are used in the chain pin holes.

ANSI B29.1 also covers a number of special types of roller chains. One is equipped with oil-impregnated, sintered, powdered metal bushings, for self-lubrication. This chain handles lighter loads at reduced speeds and is limited in application because it does not use rollers. Instead, it uses bushings of the same outside diameter as normal rollers. See Fig. 17-2-4.

(A) CHAIN TERMINOLOGY

SINGLE STEEL DOUBLE STEEL DOUBLE CAST IRON

(B) SPROCKETS

Fig. 17-2-3 Chain terminology and sprockets.

THIS TYPE USES OIL-IMPREGNATED SINTERED METAL BUSHING IN PLACE OF ROLLERS.

Fig. 17-2-4 Self-lubricating chain.

Another approach to self-lubrication has been the use of special roller chains with plastic sleeves between the chain rivets and bushings. The plastic reduces joint friction.

Double-Pitch

These are basically the same as roller chains, except that the pitch is twice as long. Roller and double-pitch chains have the same diameter pins and rollers, the same width rollers, and the same thickness of link plates.

Inverted-Tooth Silent

These are high-speed chains, used predominantly for prime-mover, power takeoff drives, such as on power cranes or shovels, machine tools, and pumps. Drives transmitting up to 1200 hp (900 kW) are in use.

These chains are made up of a series of tooth links, alternately assembled with either pins or a combination of joint components in such a way that the joint articulates between adjoining pitches. Center-guide chain has guide links which engage a groove or grooves in the sprocket, while the side-guide chain has guides which engage the sides of the sprocket.

Bead or Slider

Bead chains are used as manually controlled or slow-speed drives in numerous products such as television tuners, radio tuners, computing devices, time recorders, air conditioners, toys, display drives, ventilator controls, and venetian blinds.

SPROCKETS

Basic sprocket types used with precision steel roller chains conform to ANSI standards.

Used for mounting on flanges, hubs, or other devices, the plate sprocket is a flat, hubless sprocket.

Small- and medium-size hub sprockets are turned from bar stock or forgings or are made by welding a bar-stock hub to a hot-rolled plate. For small, low-load applications, only one hub extension may be needed. Large-diameter sprockets normally have two hub projections equidistant from the center plane of the sprocket.

Materials Although normally machined from gray-iron castings, sprockets are also available in cast steel or welded hub construction.

Sprockets made of sintered powdered metal, and from nylon and other plastics, have become economical in large quantities. These sprockets offer many advantages. For example, plastic sprockets require minimum lubrication and are widely used where cleanliness is essential.

DESIGN OF ROLLER CHAIN DRIVES

The design of a roller chain drive consists, primarily, of the selection of the chain and sprocket sizes. It also includes the determination of chain length, center distance, method of lubrication, and, in some cases, the arrangement of chain casings and idlers.

Unlike belt drives, which are based on lineal speeds in fpm or m/s, the limiting factor of chain drives is based on the rotative speed, or revolutions per minute, of the smaller sprocket, which in most installations is the driven member.

Design of chain drives is based not only on horsepower (kilowatts) and speeds but also on the following factors relative to broad service conditions

1. Average horsepower (kilowatts) to be transmitted. See Fig. 17-2-5.
2. The revolutions per minute of the driving and driven members.
3. Shaft diameter.
4. Permissible diameters of sprockets.
5. Load characteristics, whether smooth and steady, pulsating, heavy-starting, or subject to peaks.
6. Lubrication, whether periodic, occasional, or copious. Where chains are exposed to dust, dirt, or inju-

| Chain Speed | | Power | | Type of |
fpm	m/s	hp	kW	Chain
350	1.0	20	15	Detachable
450	2.2	40	30	Pintle
1000	5	250	190	Offset-sidebar
2500	12.5	1500	1100	Roller
4000	20	2500	1850	Silent

Fig. 17-2-5 Tentative selection factors for chain drives.

rious foreign matter, chain cases should be used.

7. Life expectancy: the amount of service required, or total life. It is much better to *overchain* than to skimp on the size of the chain used.

In designing chain drives, it is of the utmost importance to consider and study the pitch or size of the chain used. The number of revolutions per minute and the size of the smaller or faster-moving sprocket determine the pitch of chain that should be used.

Smaller-pitch chains in single or multiple width are adaptable for elevated-speed drives, and also for any speed drives where smoother and quieter performance is essential.

Large-pitch chains are adaptable for slow- and medium-speed drives.

Multiple-width roller chains are becoming increasingly popular. They not only solve the problem of transmitting greater power at higher speeds, but because of their smoother action they also substantially reduce the noise factor. See Fig. 17-2-6.

Size of Sprockets It is general practice to use a minimum size sprocket of 17

Fig. 17-2-6 Multiple-roller chain drive.

teeth in order to obtain smooth operation at high speeds. Nineteen- or 21-tooth sprockets should be considered from a standpoint of greater life expectancy, and smoother operation, because of the lessening of tooth impact. On slow-speed and special-purpose installations or where space limitations are involved, sprockets smaller than 17-tooth can be used. The normal maximum number of teeth is 120.

Ordinary practice indicates that the ratio of driver to driven sprockets should be no more than 6:1. The recommended chain wrap on driver is 120°.

Center Distances Center distances must be more than one-half the diameter of the smaller sprocket, plus one-half of the diameter of the larger sprocket; otherwise the sprocket teeth will touch. (When necessary, drives may be operated with a small amount of clearance between sprockets.) However, best results are obtained by using a center distance of 30 to 50 times the pitch of the chain used. Eighty times the pitch is considered maximum.

Chain Tension Chains should never run with both sides tight. Adjustable centers should be provided when possible to permit proper initial slack and to allow for periodic adjustment necessitated by natural chain wear. The chain sag should be equivalent to approximately 2 percent of the center distance. See Fig. 17-2-7.

Idler sprockets should be used as a means of taking up chain slack where it is not possible to provide adjustable centers.

Chain Length Chain length is a function of the number of teeth in both sprockets and of the center distance. In addition, the chain must consist of an integral number of pitches, with an even number preferable, in order to avoid the use of an offset link.

Chain Length Formula For simplicity it is customary to compute the chain length in terms of chain pitches and then to multiply the result by the chain pitch to obtain the length in inches (millimeters). The following formula is a quick and convenient method of finding the chain length in pitches. See Fig. 17-2-8.

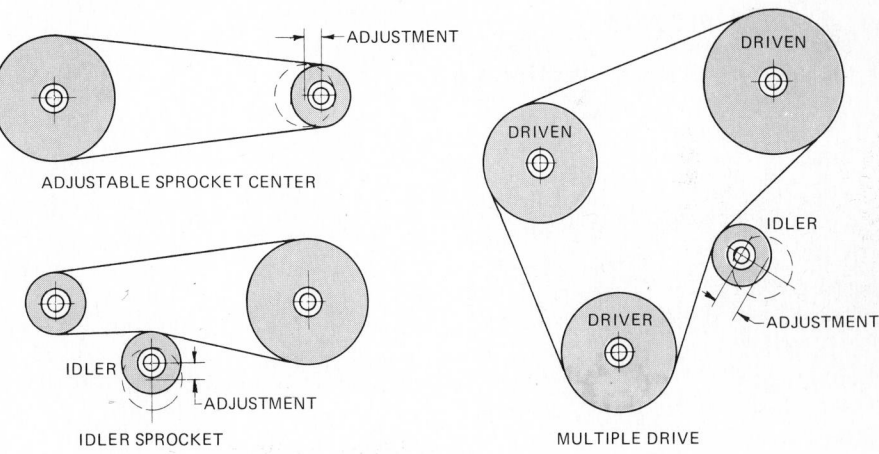

(A) METHOD OF CHAIN ADJUSTMENT

COUNTERSHAFT ADDED – ONE OR MORE DEPENDING ON DISTANCE

(B) CHAIN DRIVE WITH LONG CENTER DISTANCE

Fig. 17-2-7 Chain drives.

F	S	F	S	F	S	F	S	F	S	F	S
1	0.03	32	25.94	63	100.54	94	223.82	125	395.79	156	616.44
2	0.10	33	27.58	64	103.75	95	228.61	126	402.14	157	624.37
3	0.23	34	29.28	65	107.02	96	233.44	127	408.55	158	632.35
4	0.41	35	31.03	66	110.34	97	238.33	128	415.01	159	640.38
5	0.63	36	32.83	67	113.71	98	243.27	129	421.52	160	648.46
6	0.91	37	34.68	68	117.13	99	248.26	130	428.08	161	656.59
7	1.24	38	36.58	69	120.60	100	253.30	131	434.69	162	664.77
8	1.62	39	38.53	70	124.12	101	258.39	132	441.36	163	673.00
9	2.05	40	40.53	71	127.69	102	263.54	133	448.07	164	681.28
10	2.53	41	42.58	72	131.31	103	268.73	134	454.83	165	689.62
11	3.06	42	44.68	73	134.99	104	273.97	135	461.64	166	698.00
12	3.65	43	46.84	74	138.71	105	279.27	136	468.51	167	706.44
13	4.28	44	49.04	75	142.48	106	284.67	137	475.42	168	714.92
14	4.96	45	51.29	76	146.31	107	290.01	138	482.39	169	723.46
15	5.70	46	53.60	77	150.18	108	295.45	139	489.41	170	732.05
16	6.48	47	55.95	78	154.11	109	300.95	140	496.47	171	740.60
17	7.32	48	58.36	79	158.09	110	306.50	141	503.59	172	749.37
18	8.21	49	60.82	80	162.11	111	312.09	142	510.76	173	758.11
19	9.14	50	63.33	81	166.19	112	317.74	143	517.98	174	766.90
20	10.13	51	65.88	82	170.32	113	323.44	144	525.25	175	775.74
21	11.17	52	68.49	83	174.50	114	329.19	145	532.57	176	784.63
22	12.26	53	71.15	84	178.73	115	334.99	146	539.94	177	793.57
23	13.40	54	73.86	85	183.01	116	340.84	147	547.36	178	802.57
24	14.59	55	76.62	86	187.34	117	346.75	148	554.83	179	811.61
25	15.83	56	79.44	87	191.73	118	352.70	149	562.36	180	820.70
26	17.12	57	82.30	88	196.16	119	358.70	150	569.93	181	829.85
27	18.47	58	85.21	89	200.64	120	364.76	151	577.56	182	839.04
28	19.86	59	88.17	90	205.18	121	370.86	152	585.23	183	848.29
29	21.30	60	91.19	91	209.76	122	377.02	153	592.96	184	857.58
30	22.80	61	94.25	92	214.40	123	383.22	154	600.73	185	866.93
31	24.34	62	97.37	93	219.08	124	389.48	155	608.56	—	—

STEP 1	Divide center distance which is given in inches or millimeters by pitch of chain used obtaining C.
STEP 2	Add number of teeth in smaller sprocket to number of teeth in larger sprocket obtaining M.
STEP 3	Subtract number of teeth in smaller sprocket from number of teeth in larger sprocket which gives F in table above. Use Corresponding Constant S.
STEP 4	Chain length in pitches $= 2C + \dfrac{M}{2} + \dfrac{S}{C}$
STEP 5	Mulitply number of pitches by chain pitch used in order to get chain length in inches or millimeters.

Fig. 17-2-8 Determining chain length.

1. Divide center distance in inches (millimeters) by pitch of chain, obtaining C.
2. Add number of teeth in small sprocket to number of teeth in large sprocket, obtaining M.
3. Subtract number of teeth in small sprocket from number of teeth in large sprocket, obtaining value F to obtain the corresponding value of S.
4. Chain length in pitches equals

$$2C + \frac{M}{2} + \frac{S}{C}$$

A chain cannot contain the fractional part of a pitch. It is therefore necessary to increase the pitch to the next higher whole number, preferably an even number. The center distance must then be corrected.

5. Multiply number of pitches by chain pitch used in order to get chain length in inches (millimeters).

Drive Selection

The horsepower (kilowatt) ratings relate to the speed of the smaller sprocket, and drive selections are made on this basis, whether the drive is speed-reducing or speed-increasing. In making drive selections, consideration is given to the loads imposed on the chain by the type of input power and the type of equipment to be driven. Service factors are used to compensate for these loads, and the required horsepower (kilowatt) rating of the chain is determined by the following equation:

Required horsepower (kilowatt) rating (Fig. 17-2-9)

$$= \frac{\text{hp (kW) to be transmitted} \times \text{service factor}}{\text{multiple-strand factor}}$$

Figures 17-2-10 to 17-2-13 show the horsepower (kilowatt) ratings for just a few of the many roller chains available. For additional information refer to manufacturers' catalogs.

The horsepower (kilowatt) rating charts (Figs. 17-2-14 and 17-2-15) provide a quick means of determining the probable chain requirements.

Chain Drive Design

EXAMPLE 1 Select an electric motor-driven roller chain drive to transmit 5 hp (3.7 kW) from a countershaft to the main shaft of a wire drawing machine. The countershaft is 1.5 in. (38 mm) in diameter and operates at 1200 r/min. The main shaft is also 1.5 in. (38 mm) and must operate between 378 and 382 r/min. Shaft centers, once established,

SERVICE FACTOR FOR SINGLE CHAINS Horsepower or Kilowatt Ratings for Multiple-Strand Chains Equal Single-Strand Ratings Multiplied by Multiple-Strand Factor				MULTIPLE-STRAND FACTOR	
	Type of Input Power				
Type of Driven Load	Internal Combustion Engine with Hydraulic Drive	Electric Motor or Turbine	Internal Combustion Engine with Mechanical Drive	Number of Strands	Multiple-Strand Factor
Smooth	1.0	1.0	1.2	2	1.7
Moderate shock	1.2	1.3	1.4	3	2.5
Heavy shock	1.4	1.5	1.7	4	3.3

Fig. 17-2-9 Service- and multiple-strand factors for chain drives.

No. of Teeth Small Spkt.	.25 PITCH	NO. 25 ASA STANDARD ROLLER CHAIN Revolutions Per Minute—Small Sprocket															
	100	500	900	1200	1800	2500	3000	3500	4000	4500	5000	5500	6000	6500	7000	7500	8000
17	.086	.37	.62	.81	1.16	1.56	1.84	2.11	2.38	2.28	1.95	1.69	1.48	1.31	1.18	1.06	0.96
18	.092	.39	.66	.86	1.23	1.66	1.95	2.25	2.53	2.49	2.12	1.84	1.62	1.43	1.28	1.16	1.05
19	.097	.41	.70	.91	1.31	1.76	2.07	2.38	2.69	2.70	2.30	2.00	1.75	1.55	1.39	1.25	1.14
20	.103	.44	.74	.96	1.38	1.86	2.19	2.52	2.84	2.91	2.49	2.16	1.89	1.68	1.50	1.35	1.23
21	.108	.46	.78	1.01	1.46	1.96	2.31	2.65	2.99	3.13	2.68	2.32	2.04	1.80	1.61	1.46	1.32
22	.114	.48	.82	1.06	1.53	2.06	2.43	2.79	3.15	3.36	2.87	2.49	2.18	1.93	1.73	1.56	1.42
23	.119	.51	.86	1.12	1.61	2.16	2.55	2.93	3.30	3.59	3.07	2.66	2.33	2.07	1.85	1.67	1.51
24	.125	.53	.90	1.17	1.69	2.26	2.67	3.07	3.46	3.83	3.27	2.83	2.48	2.20	1.97	1.78	1.61
25	.131	.56	.94	1.22	1.76	2.37	2.79	3.20	3.61	4.02	3.48	3.01	2.64	2.34	2.10	1.89	1.72
28	.148	.63	1.07	1.38	1.99	2.67	3.15	3.62	4.08	4.54	4.12	3.57	3.13	2.78	2.49	2.24	2.04
30	.159	.68	1.15	1.49	2.14	2.88	3.39	3.90	4.40	4.89	4.57	3.96	3.47	3.08	2.76	2.49	2.26
32	.170	.73	1.23	1.60	2.30	3.09	3.64	4.18	4.71	5.24	5.03	4.36	3.83	3.39	3.04	2.74	2.49
35	.188	.80	1.36	1.76	2.53	3.40	4.01	4.61	5.19	5.78	5.76	4.99	4.38	3.88	3.48	3.13	2.85
40	.217	.92	1.57	2.03	2.93	3.93	4.63	5.32	6.00	6.67	7.04	6.10	5.35	4.75	4.25	3.83	3.48
45	.246	.05	1.78	2.31	3.32	4.46	5.26	6.04	6.81	7.58	8.33	7.28	6.39	5.66	5.07	4.57	4.15

No. of Teeth Small Spkt.	.38 PITCH	NO. 35 ASA STANDARD ROLLER CHAIN Revolutions Per Minute—Small Sprocket															
	100	500	900	1200	1800	2500	3000	3500	4000	4500	5000	5500	6000	6500	7000	7500	8000
17	.29	1.25	2.12	2.75	3.95	5.31	5.63	4.47	3.66	3.06	2.62	2.27	1.99	1.77	1.58	1.42	1.29
18	.31	1.33	2.25	2.92	4.20	5.65	6.13	4.87	3.98	3.34	2.85	2.47	2.17	1.92	1.72	1.55	1.41
19	.33	1.41	2.39	3.10	4.46	5.99	6.65	5.28	4.42	3.62	3.09	2.68	2.35	2.09	1.87	1.68	1.53
20	.35	1.49	2.53	3.27	4.71	6.33	7.18	5.70	4.67	3.91	3.34	2.90	2.54	2.25	2.02	1.82	1.65
21	.37	1.57	2.66	3.45	4.97	6.68	7.73	6.13	5.02	4.21	3.59	3.11	2.73	2.42	2.17	1.96	1.77
22	.39	1.65	2.80	3.63	5.22	7.02	8.27	6.58	5.38	4.51	3.85	3.34	2.93	2.60	2.33	2.10	1.90
23	.41	1.73	2.94	3.81	5.48	7.37	8.68	7.03	5.75	4.82	4.12	3.57	3.13	2.78	2.49	2.24	2.03
24	.43	1.81	3.08	3.98	5.74	7.71	9.09	7.49	6.13	5.14	4.39	3.80	3.34	2.96	2.65	2.39	2.17
25	.44	1.89	3.21	4.16	6.00	8.06	9.50	7.97	6.52	5.47	4.67	4.05	3.55	3.15	2.82	2.54	2.31
28	.50	2.14	3.63	4.71	6.78	9.11	10.7	9.44	7.73	6.48	5.53	4.80	4.21	3.73	3.34	3.01	2.73
30	.54	2.31	3.91	5.07	7.30	9.81	11.6	10.5	8.57	7.18	6.14	5.32	4.67	4.14	3.70	3.34	3.03
32	.58	2.47	4.20	5.44	7.83	10.5	12.4	11.5	9.44	7.91	6.76	5.86	5.14	4.56	4.08	3.68	3.34
35	.64	2.72	4.62	5.99	8.63	11.6	13.7	13.2	10.8	9.06	7.73	6.70	5.88	5.22	4.67	4.21	3.82
40	.74	3.15	5.34	6.92	9.96	13.4	15.8	16.1	13.2	11.1	9.45	8.19	7.19	6.37	5.70	5.14	4.67
45	.84	3.57	6.06	7.85	11.3	15.2	17.9	19.2	15.8	13.2	11.3	9.77	8.57	7.60	6.80	6.14	0

No. of Teeth Small Spkt.	.50 PITCH	NO. 40 ASA STANDARD ROLLER CHAIN Revolutions Per Minute—Small Sprocket															
	50	200	400	600	900	1200	1800	2400	3000	3500	4000	4500	5000	5500	6000	6500	7000
17	.37	1.29	2.40	3.45	4.98	6.45	8.96	5.82	4.17	3.31	2.71	2.27	1.94	1.68	1.47	1.31	1.17
18	.39	1.37	2.55	3.68	5.30	6.86	9.76	6.34	4.54	3.60	2.95	2.47	2.11	1.83	1.60	1.42	1.27
19	.42	1.45	2.71	3.90	5.62	7.27	10.5	6.88	4.92	3.91	3.20	2.68	2.29	1.98	1.74	1.54	1.38
20	.44	1.53	2.86	4.12	5.94	7.69	11.1	7.43	5.31	4.22	3.45	2.89	2.47	2.14	1.88	1.67	1.49
21	.46	1.62	3.02	4.34	6.26	8.11	11.7	7.99	5.72	4.54	3.71	3.11	2.66	2.30	2.02	1.79	1.60
22	.49	1.70	3.17	4.57	6.58	8.52	12.3	8.57	6.13	4.87	3.98	3.34	2.85	2.47	2.17	1.92	1.72
23	.51	1.78	3.33	4.79	6.90	8.94	12.9	9.16	6.55	5.20	4.26	3.57	3.05	2.64	2.32	2.06	1.84
24	.54	1.87	3.48	5.02	7.23	9.36	13.5	9.76	6.99	5.54	4.54	3.80	3.25	2.81	2.47	2.19	1.96
25	.56	1.95	3.64	5.24	7.55	9.78	14.1	10.4	7.43	5.89	4.82	4.04	3.45	2.99	2.63	2.33	0
28	.63	2.20	4.11	5.93	8.54	11.1	15.9	12.3	8.80	6.99	5.72	4.79	4.09	3.55	3.11	2.76	0
30	.68	2.38	4.43	6.38	9.20	11.9	17.2	13.6	9.76	7.75	6.34	5.31	4.54	3.93	3.45	0	
32	.73	2.55	4.75	6.85	9.86	12.8	18.4	15.0	10.8	8.54	6.99	5.86	5.00	4.33	3.80	0	
35	.81	2.80	5.24	7.54	10.9	14.1	20.3	17.2	12.3	9.76	7.99	6.70	5.72	4.96	0		
40	.93	3.24	6.05	8.71	12.5	16.3	23.4	21.0	15.0	11.9	9.76	8.18	6.99	0			
45	1.06	3.68	6.87	9.89	14.2	18.5	26.6	25.1	17.9	14.2	11.7	9.76	0				

Fig. 17-2-10   HP ratings for .25, .38, and .50 inch pitch single-strand roller chain. (American Sprocket Chain Manufacturers Association)

No. of Teeth Small Spkt.	.62 PITCH					NO. 50 ASA STANDARD ROLLER CHAIN Revolutions Per Minute—Small Sprocket										
	50	100	300	500	900	1200	1500	1800	2100	2400	2700	3000	3300	3500	4000	4500
17	.72	1.34	3.60	5.69	9.70	12.6	14.3	10.7	8.48	6.95	5.83	4.98	4.32	3.96	3.23	2.71
18	.77	1.43	3.83	6.05	10.3	13.4	15.6	11.7	9.24	7.58	6.35	5.42	4.70	4.31	3.52	2.95
19	.81	1.51	4.06	6.42	10.9	14.2	16.9	12.7	10.0	8.22	6.89	5.88	5.10	4.68	3.82	3.20
20	.86	1.60	4.30	6.78	11.6	15.0	18.2	13.7	10.8	8.87	7.44	6.35	5.51	5.05	4.12	3.45
21	.90	1.69	4.53	7.15	12.2	15.8	19.3	14.7	11.6	9.55	8.01	6.83	5.93	5.44	4.44	3.71
22	.95	1.77	4.76	7.52	12.8	16.6	20.3	15.8	12.5	10.2	8.59	7.33	6.36	5.83	4.76	3.98
23	1.00	1.86	5.00	7.89	13.4	17.4	21.3	16.9	13.3	10.9	9.18	7.83	6.79	6.23	5.08	4.26
24	1.04	1.95	5.23	8.26	14.1	18.3	22.3	18.0	14.2	11.7	9.78	8.34	7.24	6.64	5.42	4.54
25	1.09	2.04	5.47	8.63	14.7	19.1	23.3	19.1	15.1	12.4	10.4	8.88	7.70	7.06	5.76	4.83
28	1.20	2.30	6.18	9.76	16.6	21.6	26.3	22.7	17.9	14.7	12.3	10.5	9.13	8.37	6.83	0
30	1.33	2.42	6.66	10.5	17.9	23.2	28.4	25.1	19.9	16.3	13.7	11.7	10.1	9.28	7.57	0
32	1.42	2.66	7.14	11.3	19.2	24.9	30.4	27.7	21.9	18.0	15.0	12.9	11.1	10.2	8.34	0
35	1.57	2.93	7.86	12.4	21.2	27.4	33.5	31.7	25.1	20.5	17.2	14.7	12.8	11.7	9.55	0
40	1.81	3.38	9.08	14.3	24.4	31.1	38.7	38.7	30.6	25.1	21.0	18.0	15.6	14.3	0	
45	2.06	3.84	10.3	16.3	27.8	36.0	43.9	46.2	36.5	29.9	25.1	21.4	18.6	0		

No. of Teeth Small Spkt.	.75 PITCH					NO. 60 ASA STANDARD ROLLER CHAIN Revolutions Per Minute—Small Sprocket										
	50	100	200	500	700	900	1200	1400	1600	1800	2000	2200	2400	2600	2800	3000	3500
17	1.24	2.30	4.31	9.81	13.3	16.7	21.7	18.2	14.8	12.5	10.6	9.18	8.06	7.15	6.40	5.75	4.57
18	1.32	2.45	4.58	10.4	14.1	17.8	23.0	19.8	16.1	13.6	11.5	10.0	8.78	7.79	6.97	6.27	4.98
19	1.40	2.60	4.86	11.1	15.0	18.8	24.4	21.5	17.5	14.7	12.5	10.9	9.52	8.45	7.56	6.80	5.40
20	1.48	2.75	5.13	11.7	15.9	19.9	25.8	23.2	18.9	15.9	13.5	11.7	10.3	9.12	8.17	7.34	5.83
21	1.56	2.89	5.41	12.3	16.7	21.0	27.2	24.9	20.3	17.1	14.5	12.6	11.1	9.82	8.79	7.90	6.27
22	1.64	3.04	5.69	13.0	17.6	22.1	28.6	26.7	21.8	18.4	15.6	13.5	11.9	10.5	9.42	8.47	6.73
23	1.72	3.19	5.97	13.6	18.4	23.2	30.0	28.6	23.3	19.6	16.7	14.4	12.7	11.3	10.1	9.06	7.19
24	1.80	3.34	6.25	14.2	19.3	24.3	31.4	30.4	24.8	20.9	17.7	15.4	13.5	12.0	10.7	9.65	7.66
25	1.88	3.49	6.53	14.9	20.2	25.4	32.9	32.4	26.4	22.3	18.9	16.4	14.4	12.8	11.4	10.3	8.15
28	2.12	3.95	7.38	16.8	22.8	28.7	37.1	38.4	31.3	26.4	22.4	19.4	17.0	15.1	13.5	12.2	9.66
30	2.29	4.25	7.95	18.1	24.6	30.9	40.0	42.6	34.7	29.2	24.8	21.5	18.9	16.8	15.0	13.5	0
32	2.45	4.56	8.53	19.4	26.3	33.1	42.9	46.9	38.2	32.2	27.3	23.7	20.8	18.5	16.5	14.9	0

No. of Teeth Small Spkt.	1.00 PITCH					NO. 80 ASA STANDARD ROLLER CHAIN Revolutions Per Minute—Small Sprocket										
	25	50	100	200	300	400	500	700	900	1000	1200	1400	1600	1800	2000	2200	2400
17	1.55	2.88	5.38	10.0	14.5	18.7	22.9	31.0	38.9	37.6	28.6	22.7	18.6	15.6	13.3	11.5	10.1
18	1.64	3.07	5.72	10.7	15.4	19.9	24.4	33.0	41.3	41.0	31.2	24.8	20.3	17.0	14.5	12.6	11.0
19	1.74	3.25	6.07	11.3	16.3	21.1	25.8	35.0	43.8	44.5	33.9	26.9	22.0	18.4	15.7	13.6	12.0
20	1.84	3.44	6.42	12.0	17.2	22.3	27.3	37.0	46.4	48.1	36.6	29.0	23.8	19.9	17.0	14.7	12.9
21	1.94	3.62	6.76	12.6	18.2	23.6	28.8	39.0	48.9	51.7	39.4	31.2	25.6	21.4	18.3	15.9	13.9
22	2.04	3.81	7.11	13.3	19.1	24.8	30.3	41.0	51.4	55.5	42.2	33.5	27.4	23.0	19.6	17.0	14.9
23	2.14	4.00	7.46	13.9	20.0	26.0	31.7	43.0	53.9	59.2	45.1	35.8	29.3	24.6	21.0	18.2	15.9
24	2.24	4.19	7.81	14.6	21.0	27.2	33.3	45.0	56.4	62.0	48.1	38.1	31.2	26.2	22.3	19.4	17.0
25	2.34	4.38	8.17	15.2	21.9	28.4	34.8	47.0	59.0	64.9	51.1	40.6	33.2	27.8	23.8	20.6	8.34
28	2.65	4.94	9.23	17.2	24.8	32.1	39.3	53.2	66.6	73.3	60.6	48.1	39.4	33.0	28.2	24.4	0
30	2.85	5.33	9.94	18.5	26.7	34.6	42.3	57.3	71.8	78.9	67.2	53.3	43.6	36.6	31.2	24.5	0
32	3.06	5.71	10.7	19.9	28.6	37.1	45.3	61.4	77.0	84.7	74.0	58.7	48.1	40.3	34.4	0	

Fig. 17-2-11   HP ratings for .62, .75, and 1.00 inch pitch single-strand roller chain. (American Sprocket Chain Manufacturers Association)

No. of Teeth on Small Spkt.	6.0 PITCH						NO. 25 ASA STANDARD ROLLER CHAIN										
	Revolutions Per Minute of Small Sprocket																
	100	500	900	1200	1800	2500	3000	3500	4000	4500	5000	5500	6000	6500	7000	7500	8000
17	.064	0.3	0.5	0.6	0.9	1.2	1.4	1.6	1.8	1.7	1.5	1.3	1.1	1.0	0.9	0.8	0.7
18	.069	0.3	0.5	0.6	0.9	1.2	1.5	1.7	1.9	1.9	1.6	1.4	1.2	1.1	1.0	0.9	0.8
19	.072	0.3	0.5	0.7	1.0	1.3	1.5	1.8	2.0	2.0	1.7	1.5	1.3	1.2	1.0	0.9	0.9
20	.077	0.3	0.6	0.7	1.0	1.4	1.6	1.9	2.1	2.2	1.9	1.6	1.4	1.3	1.1	1.0	0.9
21	0.1	0.3	0.6	0.8	1.1	1.5	1.7	2.0	2.2	2.3	2.0	1.7	1.5	1.3	1.2	1.1	1.0
22	0.1	0.4	0.6	0.8	1.1	1.5	1.8	2.1	2.3	2.5	2.1	1.9	1.6	1.4	1.3	1.2	1.1
23	0.1	0.4	0.6	0.8	1.2	1.6	1.9	2.2	2.5	2.7	2.3	2.0	1.7	1.5	1.4	1.2	1.1
24	0.1	0.4	0.7	0.9	1.3	1.7	2.0	2.3	2.6	2.9	2.4	2.1	1.9	1.6	1.5	1.3	1.2
25	0.1	0.4	0.7	0.9	1.3	1.8	2.1	2.4	2.7	3.0	2.6	2.2	2.0	1.7	1.6	1.4	1.3
28	0.1	0.5	0.8	1.0	1.5	2.0	2.3	2.7	3.0	3.4	3.1	2.7	2.3	2.1	1.9	1.7	1.5
30	0.1	0.5	0.9	1.1	1.6	2.1	2.5	2.9	3.3	3.6	3.4	3.0	2.6	2.3	2.1	1.9	1.7
32	0.1	0.5	0.9	1.2	1.7	2.3	2.7	3.1	3.5	3.9	3.8	3.3	2.9	2.5	2.3	2.0	1.9
35	0.1	0.6	1.0	1.3	1.9	2.5	3.0	3.4	3.9	4.3	4.3	3.7	3.3	2.9	2.6	2.3	2.1
40	0.2	0.7	1.2	1.5	2.2	2.9	3.5	4.0	4.5	5.0	5.3	4.6	4.0	3.5	3.2	2.9	2.6
45	0.2	0.8	1.3	1.7	2.5	3.3	3.9	4.5	5.1	5.7	6.2	5.4	4.8	4.2	3.8	3.4	3.1

No. of Teeth on Small Spkt.	10 PITCH						NO. 35 ASA STANDARD ROLLER CHAIN										
	Revolutions Per Minute of Small Sprocket																
	100	500	900	1200	1800	2500	3000	3500	4000	4500	5000	5500	6000	6500	7000	7500	8000
17	0.2	0.9	1.6	2.1	2.9	4.0	4.2	3.3	2.7	2.3	2.0	1.7	1.5	1.3	1.2	1.1	1.0
18	0.2	1.0	1.7	2.2	3.1	4.2	4.6	3.6	3.0	2.5	2.1	1.8	1.6	1.4	1.3	1.2	1.1
19	0.2	1.1	1.8	2.3	3.3	4.5	5.0	3.9	3.1	2.7	2.3	2.0	1.8	1.6	1.4	1.3	1.1
20	0.3	1.1	1.9	2.4	3.5	4.7	5.4	4.3	3.5	2.9	2.5	2.2	2.0	1.7	1.5	1.4	1.2
21	0.3	1.2	2.0	2.6	3.7	5.0	5.8	4.6	3.7	3.1	2.7	2.3	2.0	1.8	1.6	1.5	1.3
22	0.3	1.2	2.1	2.7	3.9	5.2	6.2	4.9	4.0	3.4	2.9	2.5	2.2	1.9	1.7	1.6	1.4
23	0.3	1.3	2.2	2.8	4.1	5.5	6.5	5.2	4.3	3.6	3.1	2.7	2.3	2.1	1.9	1.7	1.5
24	0.3	1.4	2.3	3.0	4.3	5.8	6.8	5.6	4.6	3.8	3.3	2.8	2.5	2.2	2.0	1.8	1.6
25	0.3	1.4	2.4	3.1	4.5	6.0	7.1	5.9	4.9	4.1	3.5	3.0	2.6	2.3	2.1	1.9	1.7
28	0.4	1.6	2.7	3.5	5.1	6.8	8.0	7.0	5.8	4.8	4.1	3.6	3.1	2.8	2.5	2.2	2.0
30	0.4	1.7	2.9	3.8	5.4	7.3	8.7	7.8	6.4	5.4	4.6	4.0	3.5	3.1	2.8	2.5	2.3
32	0.4	1.8	3.1	4.1	5.8	7.8	9.3	8.6	7.0	5.9	5.0	4.4	3.8	3.4	3.0	2.7	2.5
35	0.5	2.0	3.4	4.5	6.4	8.7	10.2	9.8	8.1	6.8	5.8	5.0	4.4	4.0	3.5	3.1	2.8
40	0.6	2.3	4.0	5.2	7.4	10.0	11.8	12.0	9.8	8.3	7.0	6.1	5.4	4.8	4.3	3.8	3.5
45	0.7	2.7	4.5	5.9	8.4	11.3	13.4	14.3	11.8	9.8	8.4	7.3	6.4	5.7	5.1	4.6	

No. of Teeth on Small Spkt.	13 PITCH						NO. 40 ASA STANDARD ROLLER CHAIN										
	Revolutions Per Minute of Small Sprocket																
	50	200	400	600	900	1200	1800	2400	3000	3500	4000	4500	5000	5500	6000	6500	7000
17	0.3	0.9	1.8	2.6	3.7	4.8	6.7	4.3	3.1	2.5	2.0	1.7	1.4	1.3	1.1	1.0	0.9
18	0.3	1.0	1.9	2.7	4.0	5.1	7.3	4.7	3.4	2.7	2.2	1.8	1.6	1.4	1.2	1.1	0.9
19	0.3	1.1	2.0	2.9	4.1	5.4	7.8	5.1	3.8	2.9	2.4	2.0	1.7	1.5	1.3	1.1	1.0
20	0.3	1.1	2.1	3.0	4.4	5.7	8.3	5.5	4.0	3.1	2.6	2.1	1.8	1.6	1.4	1.2	1.1
21	0.3	1.2	2.3	3.2	4.7	6.1	8.7	6.0	4.3	3.4	2.8	2.3	2.0	1.7	1.5	1.3	1.2
22	0.4	1.3	2.4	3.4	4.9	6.4	9.2	6.4	4.6	3.6	3.0	2.5	2.1	1.8	1.6	1.4	1.3
23	0.4	1.3	2.5	3.6	5.1	6.7	9.6	6.8	4.9	3.9	3.2	2.7	2.3	2.0	1.7	1.5	1.4
24	0.4	1.4	2.6	3.7	5.4	7.0	10.1	7.3	5.2	4.1	3.4	2.8	2.4	2.1	1.8	1.6	1.5
25	0.4	1.5	2.7	3.9	5.6	7.3	10.5	7.8	5.5	4.4	3.6	3.0	2.6	2.2	2.0	1.7	
28	0.5	1.6	3.0	4.4	6.4	8.3	11.9	9.2	6.6	5.2	4.3	3.6	3.1	2.6	2.3	2.1	
30	0.5	1.8	3.3	4.8	6.9	8.9	12.8	10.1	7.3	5.8	4.8	3.1	3.4	2.9	2.6		
32	0.5	1.9	3.5	5.1	7.4	9.5	13.7	11.2	8.1	6.4	5.2	4.4	3.7	3.2	2.8		
35	0.6	2.1	3.9	5.6	8.1	10.5	15.1	12.8	9.2	7.3	6.0	5.0	4.3	3.7			
40	0.7	2.4	4.5	6.5	9.0	12.2	17.5	15.7	11.2	8.9	7.3	6.1	5.2				
45	0.7	2.7	5.1	7.4	10.6	13.8	19.8	18.7	13.4	10.6	8.8	7.3					

Fig. 17-2-12 Kilowatt ratings for 6-, 10-, and 13-pitch single-strand roller chain.

No. of Teeth on Small Spkt.	16 PITCH	NO. 50 ASA STANDARD ROLLER CHAIN														
	Revolutions Per Minute of Small Sprocket															
	50	100	300	500	900	1200	1500	1800	2100	2400	2700	3000	3300	3500	4000	4500
17	0.54	1.00	2.7	4.2	7.2	9.3	10.7	8.0	6.3	5.2	4.3	4.0	3.2	3.0	2.4	2.0
18	0.57	1.07	2.9	4.5	7.7	10.0	11.6	8.7	6.9	5.7	4.7	4.0	3.6	3.2	2.7	2.2
19	0.60	1.13	3.0	4.8	8.1	10.5	12.6	9.4	7.5	6.1	5.1	4.4	3.9	3.4	2.9	2.4
20	0.64	1.2	3.2	5.1	8.7	11.2	13.6	10.2	8.1	6.6	5.6	4.8	4.1	3.8	3.1	2.6
21	0.67	1.3	3.4	5.3	9.1	11.8	14.4	11.0	8.7	7.1	6.0	5.1	4.4	4.1	3.3	3.0
22	0.71	1.3	3.6	5.6	9.5	12.4	15.1	11.7	9.3	7.6	6.4	5.5	4.8	4.3	3.6	3.0
23	0.75	1.4	3.7	5.9	10.0	13.0	15.9	12.6	10.0	8.1	7.0	5.9	5.1	4.7	3.8	3.2
24	0.78	1.5	3.9	6.1	10.5	13.6	16.6	13.4	10.6	8.7	7.2	6.2	5.4	5.0	4.0	3.4
25	0.81	1.5	4.1	6.4	11.0	14.2	17.4	14.2	11.3	9.3	7.8	6.7	5.8	5.2	4.3	3.6
28	0.90	1.7	4.6	7.3	12.4	16.1	19.6	17.0	13.4	11.0	9.1	7.9	6.9	6.2	5.1	0
30	0.99	1.8	5.0	7.8	13.4	17.3	21.2	18.7	14.8	12.2	10.2	8.8	7.6	7.0	5.7	
32	1.06	2.0	5.3	8.4	14.3	18.6	22.7	20.7	16.3	13.4	12.0	10.0	8.2	8.0	6.2	
35	1.17	2.2	5.9	9.2	15.8	20.4	25.0	23.7	18.7	15.3	13.0	11.0	10.0	9.0	7.1	
40	1.35	2.5	6.8	10.6	18.2	23.2	28.9	28.9	22.8	18.7	15.7	13.4	11.7	11.0	0	
45	1.54	2.9	7.7	12.2	20.7	26.9	32.7	34.4	27.2	22.3	19.0	16.0	14.0	0		

No. of Teeth on Small Spkt.	20 PITCH	NO. 60 ASA STANDARD ROLLER CHAIN														
	Revolutions Per Minute of Small Sprocket															
	50	100	200	500	700	900	1200	1400	1600	1800	2000	2200	2400	2600	2800	3000
17	0.9	1.7	3.2	7.3	10.0	12.5	16.2	13.6	11.0	9.3	7.9	6.9	6.0	5.3	4.8	4.3
18	1.0	1.8	3.4	7.8	10.5	13.3	17.2	14.8	12.0	10.1	8.6	7.5	6.5	5.8	5.2	4.7
19	1.0	1.9	3.6	8.3	11.2	14.0	18.2	16.0	13.1	11.0	9.3	8.1	7.1	6.3	5.7	5.1
20	1.1	2.1	3.8	8.8	11.9	14.9	19.2	17.3	14.1	11.9	10.1	8.7	7.7	6.8	6.1	5.5
21	1.2	2.2	4.0	9.2	12.5	15.7	20.3	18.6	15.1	12.8	10.8	9.4	8.3	7.3	6.6	5.9
22	1.2	2.3	4.2	9.7	13.1	16.5	21.3	20.0	16.3	13.8	11.6	10.1	8.9	7.8	7.0	6.3
23	1.3	2.4	4.5	10.1	13.7	17.3	22.4	21.3	17.4	14.7	12.5	10.7	9.5	8.4	7.5	6.8
24	1.3	2.5	4.7	10.6	14.4	18.1	23.4	22.7	18.5	15.6	13.2	11.5	10.1	9.0	8.0	7.2
25	1.4	2.6	4.9	11.1	15.1	19.0	24.6	24.2	19.7	16.7	14.1	12.2	10.8	9.5	8.5	7.7
28	1.6	3.0	5.5	12.6	17.0	21.4	27.7	28.7	23.3	19.7	16.8	14.5	12.7	11.3	10.1	9.1
30	1.7	3.2	6.0	13.6	18.4	23.1	30.0	31.8	25.9	21.8	18.6	16.0	14.1	12.5	11.2	10.1
32	1.8	3.4	6.4	14.5	19.7	24.7	32.0	35.0	28.5	24.0	20.4	17.0	15.6	13.8	12.3	11.1

No. of Teeth on Small Spkt.	25 PITCH	NO. 80 ASA STANDARD ROLLER CHAIN														
	Revolutions Per Minute of Small Sprocket															
	25	50	100	200	300	400	500	700	900	1000	1200	1400	1600	1800	2000	2200
17	1.2	2.1	4.0	7.5	10.8	14.0	17.1	23.1	29.0	28.0	21.3	16.9	13.9	11.6	9.9	8.6
18	1.2	2.3	4.3	8.0	11.5	14.8	18.2	24.6	30.8	30.6	23.3	18.5	15.1	12.7	10.8	9.4
19	1.3	2.4	4.5	8.4	12.2	15.7	19.2	26.1	32.7	33.2	25.3	20.1	16.4	13.7	11.7	10.1
20	1.4	2.6	4.8	9.0	12.8	16.6	20.4	27.6	34.6	35.9	27.3	21.6	17.8	14.8	12.7	11.0
21	1.4	2.7	5.0	9.4	13.6	17.6	21.5	29.1	36.5	38.6	29.4	23.3	19.1	16.0	13.7	11.9
22	1.5	2.8	5.3	9.9	14.2	18.5	22.6	30.6	38.3	41.4	31.5	25.0	20.4	17.2	14.6	12.7
23	1.6	3.0	5.6	10.4	14.9	19.4	23.6	32.1	40.2	44.2	33.6	26.7	21.9	18.4	15.7	13.6
24	1.7	3.1	5.8	10.9	15.7	20.3	24.8	33.6	42.1	46.3	35.9	28.4	23.3	19.5	16.6	14.5
25	1.7	3.3	6.1	11.3	16.3	21.2	26.0	35.1	44.0	48.4	38.1	30.3	24.8	20.7	17.8	15.4
28	2.0	3.7	6.9	12.8	18.5	23.9	29.3	39.7	49.7	54.7	45.2	35.9	29.4	24.6	21.0	18.2
30	2.1	4.0	7.4	13.8	19.9	25.8	31.6	42.7	53.6	58.9	50.1	39.8	32.5	27.3	23.3	18.3
32	2.3	4.3	8.0	14.8	21.3	27.7	33.8	45.8	57.4	63.2	55.2	43.8	35.9	30.1	25.7	

Fig. 17-2-13 Kilowatt ratings for 16-, 20-, and 25-pitch single-strand roller chain.

are fixed and by initial calculations must be approximately 22.5 in. (570 mm). The load on the main shaft is uneven and presents peaks which place it in the heavy shock load category.

Solution

Step 1: Service Factor The corresponding service factor from Fig. 17-2-9 for heavy shock load and electric motor is 1.5.

Step 2: Design Horsepower (Kilowatts) Design horsepower is $5 \times 1.5 = 7.5$ hp. (Design kilowatts is $3.7 \times 1.5 = 5.6$ kW).

Step 3: Tentative Chain Selection On the horsepower rating chart (Fig. 17-2-14) the suggested selection using a design of 7.5 hp and a 1200-r/min sprocket is a no. 40 (.50 in. pitch) chain. On the kilowatt rating chart (Fig. 17-2-15) the suggested selection using a design of

5.6 kW and a 1200-r/min sprocket is a no. 40 (13 mm pitch) chain.

If a multiple-strand chain has been selected, determine the required horsepower (kilowatt) rating per strand from the following equation:

Required horsepower (kilowatt) rating =

$$\frac{\text{design hp (kW)}}{\text{multiple-strand factor}}$$

Refer to the RH columns shown in Fig. 17-2-9.

NOTE: The Maximum Horsepower Rating specified in each of the Strand Columns is not limiting for Chain Drives. Consult chain manufacturers on those applications which are above the horsepower range of the chart.

Fig. 17-2-14 HP rating chart.

Step 4: Final Selection of Chain and Small Sprocket On the horsepower rating table for a no. 40 chain, Fig. 17-2-10 at 1200 r/min, the computed design of 7.5 hp is realized with a 20-tooth sprocket. Use the same process on the kilowatt rating table (Fig. 17-2-11) for a no. 40 chain at 1200 r/min with a design of 5.6 kW to also find a 20-tooth sprocket. Follow down the column headed by the speed of the small sprocket (1200 r/min) and find the nearest value to the design horsepower. Follow this line horizontally to the left to find the number of teeth for the small sprocket.

For intermediate speeds or sprocket sizes not tabulated, interpolate between the appropriate columns or lines. Check the maximum bore for the selected sprocket. See Fig. 17-2-16. If the selected sprocket will not accommodate the shaft, use a larger sprocket or make a new sprocket and chain selection from the rating table for the next larger chain number. In this problem, the 20-tooth sprocket will accommodate the 1.5-in. (38-mm) shaft.

Step 5: Selection of the Large Sprocket Since the driver is to operate at 1200 r/min and the driven at a minimum of 378 r/min, the speed ratio = 1200/378 = 3.17:1 minimum. Therefore, the large sprocket should have 20 × 3.17 teeth = 63.4 teeth. Since the standard sprocket sizes near this number of teeth is either 60 or 70 teeth (Fig. 17-2-17), it may be more economical and time-saving to try to use a combination of standard sprockets. In rechecking the smaller sprocket, the 19-tooth sprocket would also be acceptable. This would require a large sprocket of 19 × 3.17 teeth = 60.2 teeth (use 60 teeth). Since the 19- and

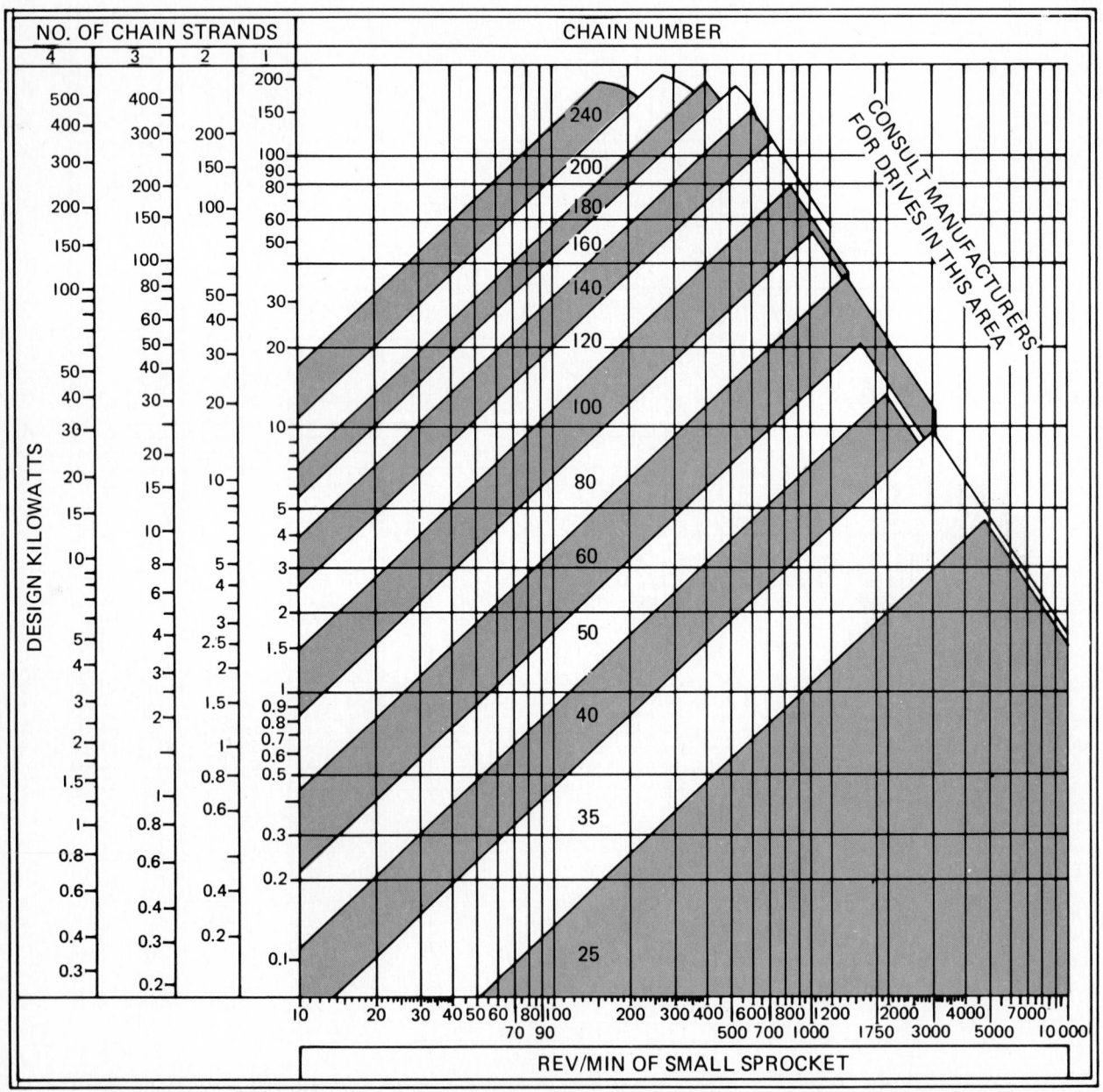

NOTE: THE MAXIMUM KILOWATT RATING SPECIFIED IN EACH OF THE STRAND COLUMNS IS NOT LIMITING FOR CHAIN DRIVES. CONSULT CHAIN MANUFACTURERS ON THOSE APPLICATIONS WHICH ARE ABOVE THE KILOWATT RANGE OF THE CHART.

Fig. 17-2-15 Kilowatt rating chart.

60-tooth sprockets are acceptable and standard, it would be more economical to use this combination.

Step 6: Chain Length in Pitches

Since 19- and 60-tooth sprockets are to be placed on 22.5 in. (570 mm) centers, calculations are as follows to determine chain length:

Chain length in pitches =

$$2C + \frac{M}{2} + \frac{S}{C}$$

where C = center distance ÷ pitch
= 22.5 ÷ .5 = 45
(= 570 ÷ 13 = 43.8 = 44)

M = total number of teeth on both sprockets
= 19 + 60 = 79

S = value obtained from table (Fig. 17-2-8)
F = 60 − 19 = 41
S = 42.58

Substituting values for C, M, and S, we get

Chain length in pitches =
$$2 \times 45 + \frac{79}{2} + \frac{42.58}{45} = 130.44$$

Since the chain is to couple to an even number of pitches, we will use 130 pitches since the leeway on the 22.5 in. (570 mm) centers is not critical.

Step 7: Chain Length in Inches (Millimeters)

Length of chain = number of pitches × pitch
= 130 × .5 = 65 in.
(= 128 × 13 = 1664 mm)

References and Source Material
1. *Machine Design*, Mechanical drives reference issue, 1979.
2. American Chain Association.

ASSIGNMENTS

See Assignments 3 to 8 for Unit 17-2 on page 362.

U.S. CUSTOMARY (INCH)										
No. of Teeth	.38 in. Pitch		.50 in. Pitch		.62 in. Pitch		.75 in. Pitch		1.00 in. Pitch	
	Maximum Bore	Maximum Hub Dia.	Maximum Bore	Maximum Hub Dia.	Maximum Bore	Maximum Hub Dia.	Maximum Bore	Maximum Hub Dia.	Maximum Bore	Maximum Hub Dia.
11	.59	.86	.78	1.17	.97	1.47	1.25	1.77	1.62	2.38
12	.62	.98	.88	1.33	1.16	1.67	1.28	2.02	1.78	2.70
13	.75	1.11	1.00	1.50	1.28	1.88	1.50	2.25	2.00	3.02
14	.84	1.23	1.16	1.66	1.31	2.08	1.75	2.50	2.28	3.34
15	.88	1.36	1.25	1.81	1.53	2.28	1.78	2.75	2.41	3.67
16	.97	1.47	1.28	1.98	1.69	2.48	1.97	2.98	2.72	3.98
17	1.09	1.59	1.38	2.14	1.78	2.69	2.22	3.22	2.81	4.31
18	1.22	1.72	1.53	2.30	1.88	2.89	2.28	3.47	3.12	4.64
19	1.25	1.84	1.69	2.45	2.06	3.08	2.44	3.70	3.31	4.95
20	1.28	1.95	1.78	2.62	2.25	3.28	2.69	3.95	3.50	5.28
21	1.31	2.08	1.78	2.78	2.28	3.48	2.81	4.19	3.75	5.59
22	1.44	2.20	1.94	2.94	2.44	3.69	2.94	4.44	3.88	5.92
23	1.56	2.31	2.09	3.09	2.62	3.89	3.12	4.67	4.19	6.23
24	1.69	2.44	2.25	3.27	2.81	4.08	3.25	4.91	4.56	6.56
25	1.75	2.56	2.28	3.42	2.84	4.28	3.38	5.16	4.69	6.88

METRIC (MILLIMETER)										
No. of Teeth	10 Pitch No. 25		13 Pitch No. 40		16 Pitch No. 50		20 Pitch No. 60		25 Pitch No. 80	
	Maximum Bore	Maximum Hub Dia.	Maximum Bore	Maximum Hub Dia.	Maximum Bore	Maximum Hub Dia.	Maximum Bore	Maximum Hub Dia.	Maximum Bore	Maximum Hub Dia.
11	15	22	20	30	25	37	32	45	41	50
12	16	25	22	34	29	42	33	51	45	69
13	20	28	25	38	33	47	38	57	51	76
14	22	32	30	42	33	53	45	64	58	85
15	22	35	32	46	39	58	45	70	61	93
16	25	38	32	50	43	63	50	76	69	101
17	28	40	35	54	45	68	56	82	71	109
18	31	44	39	58	48	73	58	88	79	118
19	32	47	43	62	52	78	62	94	84	126
20	33	50	45	67	57	83	68	100	89	134
21	34	53	45	71	58	88	71	106	95	142
22	36	56	49	75	62	94	75	111	98	150
23	40	59	53	79	67	99	79	119	106	158
24	43	62	57	83	71	104	83	124	116	167
25	45	65	58	87	72	109	86	131	119	175

Fig. 17-2-16 Maximum bore and hub diameters. (American Sprocket Chain Manufacturers Association)

NUMBER OF TEETH ON SPROCKET												
No. 25	No. 35		No. 40			No. 50			No. 60		No. 80	
9	9	48	8	34	59	9	34	59	9	34	9	34
10	10	54	9	35	60	10	35	60	10	35	10	35
11	11	60	10	36	70	11	36	70	11	36	11	36
12	12	70	11	37	72	12	37	72	12	37	12	37
13	13	72	12	38	80	13	38	80	13	38	13	38
14	14	80	13	39	84	14	39	84	14	39	14	39
15	15	84	14	40	96	15	40	96	15	40	15	40
16	16	96	15	41	112	16	41	112	16	41	16	41
17	17	112	16	42		17	42		17	42	17	42
18	18		17	43		18	43		18	43	18	43
19	19		18	44		19	44		19	44	19	44
20	20		19	45		20	45		20	45	20	45
21	21		20	46		21	46		21	46	21	46
22	22		21	47		22	47		22	47	22	47
24	23		22	48		23	48		23	48	23	48
25	24		23	49		24	49		24	49	24	54
26	25		24	50		25	50		25	50	25	60
28	26		25	51		26	51		26	51	26	
30	28		26	52		27	52		27	52	27	
32	30		27	53		28	53		28	53	28	
36	32		28	54		29	54		29	54	29	
40	35		30	55		30	55		30	60	30	
45	36		31	56		31	56		31	70	31	
48	40		32	57		32	57		32	72	32	
52	45		33	58		33	58		33	80	33	
60									84			

Fig. 17-2-17 Stock sprockets.

strength, wear, and material selection. Normally a drafter selects a gear from a catalog. Most gears are made of cast iron or steel, but brass, bronze, and plastic are used when factors such as wear or noise must be considered.

SPUR GEARS

Spur gear proportions and the shape of gear teeth are standardized, and the definitions, symbols, and formulas are given in Figs. 17-3-3 to 17-3-6.

Gears are used to transmit motion and power at constant angular velocity. The specific form of the gear which best produces this constant angular velocity is the *involute*.

Classically, the involute is described as the curve traced by a point on a taut string unwinding from a circle. This circle is called the *base circle*.

Every involute gear has only one base circle from which all the involute surfaces of the gear teeth are generated. This base circle is not a physical part of the gear and cannot be measured directly. The contact between mating involutes takes place along a line which is always tangent to and crosses between the two base circles. This is the *line of action*.

UNIT 17-3
Gear Drives

The function of a gear is to transmit motion, rotating or reciprocating, from one machine part to another and where necessary reduce or increase the revolutions of a shaft. *Gears* are rolling cylinders or cones having teeth on their contact surfaces to ensure positive motion. See Figs. 17-3-1 and 17-3-2.

There are many kinds of gears, and they may be grouped according to the position of the shafts that they connect. *Spur gears* connect parallel shafts, *bevel gears* connect shafts whose axes intersect, and *worm gears* connect shafts whose axes do not intersect. A spur gear with a rack converts rotary motion to reciprocating or linear motion. The smaller of two gears is known as the *pinion*.

Gear design is very complicated, dealing with such problems as

Fig. 17-3-1 Gears. (Boston Gear Works.)

MODULE FOR METRIC SIZE GEARS	DIAMETRAL PITCH FOR INCH SIZE GEARS	14.5°	20°
6.35	4		
5.08	5		
4.23	6		
3.18	8		
2.54	10		
2.17	12		
1.59	16		
1.27	20		
1.06	24		
0.79	32		

NOTE: MODULE SIZES SHOWN ARE CONVERTED INCH SIZES.

Fig. 17-3-2 Gear-teeth sizes.

Fig. 17-3-3 Gear-teeth terms.

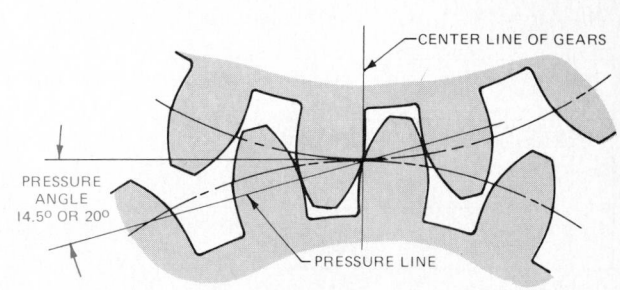

Fig. 17-3-4 Meshing of gear teeth.

The 14.5°-pressure angle has been used for many years and remains useful for duplicate or replacement gearing or in situations where the control of backlash is of primary importance.

The 20°-pressure angle has become the standard for new gearing because of the smoother and quieter running characteristics, greater load-carrying ability, and the fewer number of teeth affected by undercutting.

Standard spur gears having a 14.5°-pressure angle should have a minimum of 16 teeth with at least 40 teeth in a mating pair. Gears with 20°-pressure angle should have a minimum of 13 teeth with at least 26 teeth in a mating pair.

The formulas for the 14.5°- and the 20°-full-depth teeth are identical.

The 20°-stub tooth differs from the 20°-standard tooth depth. The stub tooth is shorter and stronger and therefore is preferred where maximum power transmission is required.

Drawing Gear Teeth

The teeth on a gear are not normally shown on the working drawings. Instead, they are represented by solid, broken, and hidden lines which will be discussed under working drawings. However, presentation or display drawings normally require the teeth to be shown.

Since the exact form of an involute tooth would require too much time to draw, approximate methods are used. The two most common methods are shown in Fig. 17-3-6. To draw the teeth using the approximate representation of involute spur-gear teeth, lay out the root, pitch, and outside circles. On the pitch circle mark off the circular thick-ness. Through the pitch point on the pitch circle draw the pressure line at an angle of 14.5° with the line tangent to the pitch circle for the 14.5°-involute tooth (use 15° for convenience) or 20° for the 20°-involute tooth. Draw the base circle tangent to this pressure line. With the compass set to a radius equal to one-eighth the pitch diameter

TERM AND SYMBOL	DEFINITION	FORMULA	
		METRIC GEARS	INCH GEARS
Pitch diameter—PD	The diameter of an imaginary circle on which the gear tooth is designed	$PD = MDL \times N$	$PD = N \div DP$
Number of teeth—N	The number of teeth on a gear	$N = PD \div MDL$	$N = PD \times DP$
Module—MDL	The length of pitch diameter per tooth	$MDL = PD \div N$	
Diametral pitch—DP	A ratio equal to the number of teeth on a gear for every inch of pitch diameter		$DP = N \div PD$
Addendum—ADD	The radial distance from the pitch circle to the top of the tooth	14.5° or 20° $ADD = MDL$ 20° stub $ADD = 0.8 \times MDL$	14.5° or 20° $ADD = 1 \div DP$ 20° stub $ADD = 0.8 \div DP$
Dedendum—DED	The radial distance from the pitch circle to the bottom of the tooth	14.5° or 20° $DED = 1.157 \times MDL$ 20° stub $DED = MDL$	14.5° or 20° $DED = 1.157 \div DP$ 20° stub $DED = 1 \div DP$
Whole depth—WD	The overall height of the tooth	14.5° or 20° $WD = 2.157 \times MDL$ 20° stub $WD = 1.8 \times MDL$	14.5° or 20° $WD = 2.157 \div P$ 20° stub $WD = 1.8 \div DP$
Clearance—CL	The radial distance between the bottom of one tooth and the top of the mating tooth	14.5° or 20° $CL = 0.157 \times MDL$ 20° stub $CL = 0.2 \times MDL$	14.5° or 20° $CL = 0.157 \div DP$ 20° stub $CL = 0.2 \div DP$
Outside diameter—OD	The overall diameter of the gear	14.5° or 20° $OD = PD + 2ADD$ $= PD + 2\,MDL$ 20° stub $OD = PD + 2ADD$ $= PD + 1.6\,MDL$	14.5° or 20° $OD = PD + 2ADD = (N + 2) \div PD$ 20° stub $OD = PD + 2ADD$ $= (N + 1.6) \div PD$
Root diameter—RD	The diameter at the bottom of the tooth	14.5° or 20° $RD = PD - 2DED$ $= PD + 2.314\,MDL$ 20° stub $RD = PD - 2DED$ $= PD + 2MDL$	14.5° or 20° $RD = PD - 2DED$ $= (N - 2.314) \div PD$ 20° stub $RD = PD - 2DED$ $= (N - 2) \div PD$
Base circle—BC	The circle from which the involute curve of the tooth is formed	$BC = PD\ Cos\ PA$	$BC = PD\ Cos\ PA$
Pressure angle—PA	The angle between the direction of pressure between contacting teeth and a line tangent to the pitch circle	14.5° or 20°	14.5° or 20°
Backlash	The clearance between the teeth of two meshing gears	——	——
Circular pitch—CP	The distance measured from the point of one tooth to the corresponding point on the adjacent tooth on the circumference of the pitch diameter	$CP = 3.1416\,PD \div N$ $= 3.1416\,MDL$	$CP = 3.1416\,PD \div N$ $= 3.1416 \div DP$
Circular thickness—T	The thickness of a tooth or space measured on the circumference of the pitch diameter	$T = 3.1416\,PD \div 2N$ $= 1.57\,PD \div N$ $= 1.57\,MDL$	$T = 3.1416\,PD \div 2N$ $= 1.57 \div DP$
Chordal thickness—Tc	The thickness of a tooth or space measured along a chord on the circumference of the pitch diameter	$Tc = PD\ sin\ (90° \div N)$	$Tc = PD\ sin\ (90° \div N)$
Chordal addendum—Ac	Chordal addendum, also known as Corrected addendum, is the perpendicular distance from chord to outside circumference of gear	$ADDc = ADD + T^2 \div 4PD$	$ADDc = ADD + T^2 \div 4PD$
Working depth—WKG DP	The depth of engagement of two gears. The sum of two addendums	$WKG\ DP = 2ADD$	$WKG\ DP = 2ADD$

Fig. 17-3-5 Spur gear definitions and formulas.

(A) EXACT FORM OF AN INVOLUTE SPUR GEAR TOOTH

(B) APPROXIMATE REPRESENTATION OF INVOLUTE SPUR GEAR TEETH

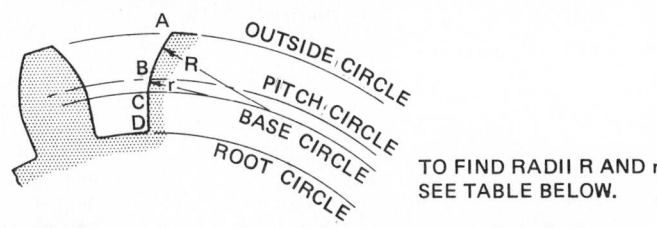

TO FIND RADII R AND r
SEE TABLE BELOW.

INCH GEARS Radii in Inches			METRIC GEARS Radii in Millimeters	
RADIUS R Divide No. by P	RADIUS r Divide No. by P	No. of Teeth	RADIUS R Multiply No. By MDL	RADIUS r Multiply No. by MDL
2.51	0.96	12	2.51	0.96
2.62	1.09	13	2.62	1.09
2.72	1.22	14	2.72	1.22
2.82	1.34	15	2.82	1.34
2.92	1.46	16	2.92	1.46
3.02	1.58	17	3.02	1.58
3.12	1.69	18	3.12	1.69
3.22	1.79	19	3.22	1.79
3.32	1.89	20	3.32	1.89
3.41	1.98	21	3.41	1.98
3.49	2.06	22	3.49	2.06
3.57	2.15	23	3.57	2.15
3.64	2.24	24	3.64	2.24
3.71	2.33	25	3.71	2.33
3.78	2.42	26	3.78	2.42
3.85	2.50	27	3.85	2.50
3.92	2.59	28	3.92	2.59
3.99	2.67	29	3.99	2.67
4.06	2.76	30	4.06	2.76
4.13	2.85	31	4.13	2.85
4.20	2.93	32	4.20	2.93
4.27	3.01	33	4.27	3.01
4.33	3.09	34	4.33	3.09
4.39	3.16	35	4.39	3.16
4.20	3.23	36	4.20	3.23
4.45	4.20	37-40	4.45	4.20
4.63	4.63	41-45	4.63	4.63
5.06	5.06	46-51	5.06	5.06
5.74	5.74	52-60	5.74	5.74
6.52	6.52	61-70	6.52	6.52
7.72	7.72	71-90	7.72	7.72
9.78	9.78	91-120	9.78	9.78
13.38	13.38	121-180	13.38	13.38
21.62	21.62	181-360	21.62	21.62

Fig. 17-3-6   Methods of drawing involute spur gear teeth.

and the compass point on the base circle, draw arcs passing through the circular thickness points established on the pitch diameter, starting at the base circle and ending at the top of the teeth. The part of the tooth profile below the base circle is drawn as a radial line ending in a small fillet at the root circle.

For a closer approximation of the involute tooth profile, the Grant's odontograph method is used. Lay out the outside, pitch, root, and base circles and circular thickness in the same manner as used in the approximation method. The top portion of the tooth profile from point A to point B is drawn with the radius R, and the portion of the tooth profile from point B to point C is drawn with the radius r.

The values of the radii R and r are found by dividing the numbers found in the table by the diametral pitch for inch-size gears or by multiplying the numbers in the table by the module for metric-size gears. The lower portion of the tooth from points C to D is drawn as a radial line ending in a small fillet at the root circle.

Working Drawings of Spur Gears

The working drawings of gears which are normally cut from blanks are not complicated. A sectional view is sufficient unless a front view is required to show web or arm details. Since the teeth are cut to shape by cutters, they need not be shown in the front view. See Figs. 17-3-7 and 17-3-8.

ANSI recommends the use of phantom lines for the outside and root circles and a center line for the pitch circle. In the section view, the root and outside circles are shown as solid lines.

The dimensioning for the gear is divided into two groups, because the finishing of the gear blank and the cutting of the teeth are separate operations in the shop. The gear-blank dimensions are shown on the drawing while the gear tooth information is given in a table.

The only differences in terminology between inch-size and metric-size gear drawings are the terms *diametral pitch* and *module*.

For inch-size gears, the term *diame-*

(A) PLAIN STYLE			

(A) PLAIN STYLE

(B) WEBBED STYLE

(C) WEBBED WITH CORED HOLES

(D) SPOKED STYLE

Fig. 17-3-7 Stock spur gear styles.

ROUNDS AND FILLETS R.10

Fig. 17-3-8 Working drawing of a spur gear.

CUTTING DATA	
NUMBER OF TEETH	30
PITCH DIAMETER	6.000
DIAMETRAL PITCH	5
PRESSURE ANGLE	14.5°
WHOLE DEPTH	.431
CHORDAL ADDENDUM	.204
CHORDAL THICKNESS	.300
CIRCULAR THICKNESS	.314
WORKING DEPTH	.400

tral pitch is used instead of the term module. The diametral pitch is a ratio of the number of teeth to a unit length of pitch diameter

Diametral pitch = DP = $\dfrac{N}{PD}$

Module is the term used on metric gears. It is the length of pitch diameter per tooth measure in millimeters

Module = MDL = $\dfrac{PD}{N}$

From these definitions it can be seen that the module is equal to the reciprocal of the diametral pitch and thus is not its metric dimensional equivalent. If the diametral pitch is known, the module can be obtained

Module = 25.4 ÷ diametral pitch

Gears presently being used in North America are designed in the inch sys-

tem and have a standard diametral pitch instead of a preferred standard module. Therefore it is recommended that the diametral pitch be referenced beneath the module when gears designed with standard inch pitches are used. For gears designed with standard modules, the diametral pitch need not be referenced on the gear drawing. The standard modules for metric gears are 0.8, 1, 1.25, 1.5, 2.25, 3, 4, 6, 7, 8, 9, 10, 12, and 16.

Spur-Gear Calculations

Center Distance The center distance between the two shaft centers is determined by adding the pitch diameter of the two gears and dividing the sum by 2.

EXAMPLE 1 A 12 DP, 36-tooth pinion mates with a 90-tooth gear. Find the center distance.

Pitch diameter
= number of teeth ÷ diametral pitch
= 36 ÷ 12 = 3.00 in. (pinion)
= 90 ÷ 12 = 7.50 in. (gear)

Sum of the two pitch diameters
= 3.00 in. + 7.50 in. = 10.50 in.

Center distance = ½ sum of the two pitch diameters
= $\dfrac{10.50}{2}$ = 5.25 in.

EXAMPLE 2 A 3.18-module, 24-tooth pinion mates with a 96-tooth gear. Find the center distance.

Pitch diameter (PD)
= number of teeth × module
= 24 × 3.18 = ⌀76.3 (pinion)
= 96 × 3.18 = ⌀305.2 (gear)

Sum of the two pitch diameters
= 76.3 + 305.2 = 381.5 mm
Center distance = ½ sum of the two pitch diameters =

$\dfrac{381.5}{2}$ = 190.75 mm

Ratio The ratio of gears is a relationship between any of the following:

1. Revolutions per minute of the gears
2. Number of teeth on the gears
3. Pitch diameter of the gears

The ratio is obtained by dividing the larger value of any of the three by the corresponding smaller value.

EXAMPLE 3 A gear rotates at 90 r/min and the pinion at 360 r/min.

Ratio $= \dfrac{360}{90} = 4$ or ratio $= 4:1$

EXAMPLE 4 A gear has 72 teeth, the pinion 18 teeth.

Ratio $= \dfrac{72}{18} = 4$ or ratio $= 4:1$

EXAMPLE 5 A gear with a pitch diameter of 8.500 in. meshes with a pinion having a pitch diameter of 2.125 in.

$$\text{Ratio} = \frac{\text{PD of gear}}{\text{PD of pinion}}$$

$$= \frac{8.500}{2.125} = 4$$

or   Ratio $= 4:1$

Determining the Pitch Diameter and Outside Diameter The pitch diameter of a gear can readily be found if the number of teeth and the diametral pitch or module are known. The OD is equal to the pitch diameter plus two addendums. The addendum for a 14.5°-spur-gear tooth is equal to $1 \div DP$ (U.S. Customary) or the module (metric).

EXAMPLE 6 A 14.5°-spur gear has a diametral pitch of 5 and 40 teeth.

$$\begin{aligned}
\text{Pitch diameter} &= N \div DP \\
&= 40 \div 5 \\
&= 8.00 \text{ in.}
\end{aligned}$$

$$\begin{aligned}
OD = PD + 2\,ADD &= 8.00 + 2\left(\frac{1}{DP}\right) \\
&= 8.00 + \frac{2}{5} \\
&= 8.40 \text{ in.}
\end{aligned}$$

EXAMPLE 7 A 14.5°-spur gear has a module of 6.35 and 34 teeth.

$$\begin{aligned}
\text{Pitch diameter} &= N \times MDL \\
&= 34 \times 6.35 \\
&= 216 \text{ mm}
\end{aligned}$$

$$\begin{aligned}
OD = PD + 2\,ADD &= 216 + 2(6.35) \\
&= 228.7 \text{ mm}
\end{aligned}$$

ASSIGNMENTS

See Assignments 9 through 11 for Unit 17-3 on page 362.

UNIT 17-4
Power-Transmitting Capacity of Spur Gears

Gear drives are required to operate under such a wide variety of conditions that it is very difficult and expensive to determine the best gear set for a particular application.

The most economical procedure is to select standard stock gears with an adequate load rating for the application.

Approximate horsepower (kilowatt) ratings for spur gears of various sizes (numbers of teeth), at several operating speeds (revolutions per minute), are given in catalogs with the spur-gear listings.

Ratings for gear sizes and/or speeds not listed may be estimated from the values shown in Fig. 17-4-2.

Pitch-line velocities exceeding 1000 ft/min (5 m/s) for 14.5° PA (pressure angle), or 1200 ft/min (6 m/s) for 20° PA, are not recommended for metallic spur gears. Ratings are listed for speeds below these limits.

The ratings given (or calculated) should be satisfactory for gears used under normal operating conditions, that is, when they are properly mounted and lubricated, carrying a smooth load (without shock) for 8 to 10 hours a day.

The charts shown in Fig. 17-4-2 indicate the approximate horsepower (kilowatt) ratings of 16- and 20-tooth steel spur gears of several tooth sizes operating at various speeds. They may be used to determine the approximate diametral pitch or module of a 16- or 20-tooth steel pinion that will carry the horsepower (kilowatts) required at the desired speed. The intersection of the lines representing values of revolutions per minute and horsepower (kilowatts) indicates the approximate gear diametral pitch (module) required.

The number of teeth normally should not be less than 16 to 20 in a 14.5° pinion, nor less than 13 in a 20° pinion.

Ratings shown for spur gears in catalogs normally are for class 1 service. For other classes of service, the service factors in Fig. 17-4-1 should be used.

Selecting the Spur-Gear Drive

1. Determine the class of service.
2. Multiply the horsepower (kilowatts) required for the application by the service factor.
3. Select spur-gear pinion with a catalog rating equal to or greater than the horsepower (kilowatts) determined in step 2.
4. Select driven spur gear with a catalog rating equal to or greater than the horsepower (kilowatts) determined in step 2.

EXAMPLE 1 Select a pair of 20° spur gears which will drive a machine at 150 r/min. Size of driving motor = 25 hp, 600 r/min. Service factor = 1.

Solution Since the service factor is 1, we do not need to increase or decrease the design horsepower. Refer to the charts shown in Fig. 17-4-2A, which show design data for 20°-spur gears having 16 and 20 teeth. Selecting a pinion having 16 teeth and reading vertically on the column showing 600 r/min to horsepower rating of 25, we find that the required DP is 4. (Select

SERVICE CLASS	OPERATING CONDITONS	SERVICE FACTOR
Class I	Continuous 8 to 10 hr per day duty, with smooth load (no shock)	1.0
Class II	Continuous 24-hr duty, with smooth load, or 8 to 10 hr per day, with moderate shock	1.2
Class III	Continuous 24-hr duty with moderate shock load	1.3
Class IV	Intermittent duty, not over 30 min per hr, with smooth load (no shock)	0.7
Class V	Hand operation, limited duty, with smooth load (no shock)	0.5

Heavy shock loads and/or severe conditions require the use of higher service factors. Such conditions may require Factors of 1.5 to 2.0 or greater than required for Class I service.

Fig. 17-4-1 Service class and factor for spur gears.

14.5°   16 AND 20 TOOTH STEEL   SPUR GEARS

20°   16 AND 20 TOOTH STEEL   SPUR GEARS

BASED ON
LEWIS FORMULA

REVOLUTIONS PER MINUTE
14.5° PRESSURE ANGLE

REVOLUTIONS PER MINUTE
20° PRESSURE ANGLE

(A) DIAMETRAL PITCH SELECTION CHARTS

BASED ON
LEWIS FORMULA

REVOLUTIONS PER MINUTE
14.5° PRESSURE ANGLE

REVOLUTIONS PER MINUTE
20° PRESSURE ANGLE

(B) MODULE SELECTION CHARTS

Fig. 17-4-2   Module and diametral pitch selection charts for 16 and 20 tooth spur gears. (Boston Gear Works.)

the DP equal to or greater than the horsepower required.)

- Pinion: N = 16, DP = 4, PD = 16 ÷ 4 = 4.000 in.
- Ratio 4:1
- Gear: N = 16 × 4 = 64, DP = 4, PD = 64 ÷ 4 = 16.000 in.

EXAMPLE 2 A 5 hp, 1200 r/min motor is used to drive a machine that runs 8 hours a day under moderate shock. If the machine is to run at 200 r/min and at the capacity of the motor, what spur gears would you select?

Solution The operating conditions of the machine are such that the machine fits into the service class 2 and requires a service factor of 1.2. Therefore

Horsepower required for design purposes = 5 hp × 1.2 = 6 hp

The pinion will be mounted on the motor and runs at 1200 r/min. Selecting a 20° pinion having 16 teeth, refer to Fig. 17-4-1B to find the required DP. Reading vertically on the 1200 r/min line and horizontally at 6 hp we find that the required DP is 8.

- Pinion: N = 16, DP = 8, PD = 16 ÷ 8 = 2.000 in.
- Gear: N = 16 × 6 = 96, DP = 8, PD = 96 ÷ 8 = 12.000 in.

EXAMPLE 3 A 7.5 kW, 900 r/min motor is attached by means of 14.5°-spur gears to a punch that operates 24 hours a day. The reduction in revolutions per minute is 4:1. Select a gear and pinion assuming that the punch is being operated at motor capacity.

Solution The operating conditions of the machine are such that the machine fits into the service class 3 and requires a service factor of 1.3. Therefore

Kilowatts required for design purposes = 7.5 × 1.3 = 9.75 kW

The pinion is mounted on the motor and runs at 900 r/min. Refer to Fig. 17-4-2B. Reading vertically on the 900 r/min line and horizontally at 9.75 kW, we may select either a pinion having a module of 5.08 and N of 20 or a module of 6.35 and N of 16. First, using a module of 5.08, we have

- Pinion: N = 20, MDL = 5.08, PD = 101.6 mm. Gear travels at 225 r/min, or one-fourth of pinion revolutions per minute.

- Gear: N = 4 × 20 = 80, MDL = 5.08, PD = 80 × 5.08 = 406.4 mm

Second, using a module of 6.35, we have

- Pinion: N = 16, MDL = 6.35, PD = 101.6 mm
- Gear: N = 64, MDL = 6.35, PD = 406.4 mm

Since both sets of gears are of the same diameter, the overall size is not a factor. Checking on the cost per set, we find that there would be considerable savings by selecting the gear and pinion having the module of 5.08. Since the extra strength of the gear set having a module of 6.35 is not required in this instance, we would recommend the gear and pinion having a module of 5.08.

Reference and Source Material
1. Boston Gear Works.

ASSIGNMENTS

See Assignments 12 and 13 for Unit 17-4 on page 363.

Rack and Pinion

A *rack* is a straight bar having teeth which engage the teeth on a spur gear. See Fig. 17-5-1. In theory, it is a spur gear having an infinite pitch diameter. Therefore, all circular dimensions become linear. The addendum, dedendum, and tooth thickness are the same as the mating spur gear. To draw the teeth on a rack, lay out the addendum and dedendum distances from the pitch line. Divide the pitch line into lineal pitch distances equal in size to the circular pitch on the gear. Divide each of these spaces in half to get the lineal thickness. Through these points draw the tooth faces at angles of 14.5 or 20° from the vertical lines. Darken in the top and bottom lines of the teeth and add the tooth fillets. The specifications for the teeth of the rack are given in the same manner as for spur gears. See Fig. 17-5-2.

ASSIGNMENT

See Assignment 14 for Unit 17-5 on page 364.

Review for Assignment
Unit 17-3 Gear Formulas

Fig. 17-5-1 Racks. (Gear Specialties, Inc.)

Fig. 17-5-2   Rack and pinion.

are transferred to the front view, and the profiles for the teeth are drawn. Radial lines from these points are taken, and the small end of the tooth is developed. The teeth on the side or section view are now drawn by projecting the teeth from the front view. Cast iron is normally used for large gears and small gears that are not subject to heavy duty. Often a gear and pinion are made of different materials for efficiency and durability. The pinion is made of a stronger material because the teeth on the pinion come into contact more times than the teeth on the gear. Common combinations are steel and cast iron, and steel and bronze.

For hp or kW ratings of bevel gears, refer to manufacturers' catalogs.

UNIT 17-6
Bevel Gears

Bevel gears are used to transmit power between two shafts whose axes intersect. The axes may intersect at any angle, but the most common is 90°. They are similar to rolling cones having the same apex. The teeth are the same shape as spur-gear teeth but taper toward the cone apex. Therefore, many spur-gear terms may apply to bevel gears. *Miter gears* are bevel gears having the same diametral pitch or module, pressure angle, and number of teeth.

Figures 17-6-1 and 17-6-2 show bevel gear definitions and formulas.

Working Drawings of Bevel Gears

The working drawings of bevel gears, like those of spur gears, give only the dimensions of the bevel gear blank. The cutting data for the teeth are given in a note or table. A single section view is normally used, unless a second view is required to show such details as spokes. Sometimes both the bevel gear and pinion are drawn together, showing their relationship. The dimensions and cutting data will depend on the method used in cutting the teeth, but the information shown in Fig. 17-6-3 is commonly used.

The actual gear teeth are often shown on assembly or display drawings. One of the most common conventions used for drawing the teeth is the Tredgold method which is shown in Fig. 17-6-4.

An arc whose radius is taken on the back cone is used as the pitch circle, and a tooth is developed using standard spur-gear formulas. Tooth sizes taken on the OD and pitch diameter

ASSIGNMENTS

See Assignments 15 and 16 for Unit 17-6 on page 364.

Review for Assignments

TERM	FORMULA
Addendum, dedendum, whole depth, pitch diameter, module, diametral pitch, number of teeth, circular pitch, chordal thickness, circular thickness	Same as for spur gears
Pitch cone radius	$\dfrac{PD}{2 \times \text{sin of pitch angle}}$
Pitch cone angle	Tan pitch angle $= \dfrac{PD \text{ of gear}}{PD \text{ of pinion}} = \dfrac{N \text{ of gear}}{N \text{ of pinion}}$
Addendum angle	Tan addendum angle $= \dfrac{\text{Addendum}}{\text{Pitch cone radius}}$
Dedendum angle	Tan dedendum angle $= \dfrac{\text{Dedendum}}{\text{Pitch cone radius}}$
Face angle	Pitch cone angle plus addendum angle
Cutting angle	Pitch cone angle minus dedendum angle
Back angle	Same as pitch cone angle
Angular addendum	Cos of pitch cone angle × addendum
Outside diameter	Pitch diameter plus two angular addendums
Crown height	Divide ½ the outside diameter by the tangent of the face angle
Face width	1½ to 2½ times the circular pitch
Chordal addendum	Addendum $+ \dfrac{\text{circular thickness}^2 \times \text{cos pitch cone angle}}{4PD}$

Fig. 17-6-1   Bevel-gear formulas.

Fig. 17-6-2 Bevel-gear nomenclature.

CUTTING DATA	
NO. OF TEETH	20
DIAMETRAL PITCH	5
TOOTH FORM	14.5° INV
CUTTING ANGLE	40°-25'
WHOLE DEPTH	.431
CHORDAL ADDENDUM	.204
CHORDAL THICKNESS	.314

Fig. 17-6-3 Working drawing of a bevel gear.

TOP VIEW

PROJECT TOOTH
LINES DOWN
FROM TOP VIEW

BACK CONE

Fig. 17-6-4 Bevel gear assembly or display drawing.

UNIT 17-7
Worm and Worm Gears

Worm gears are used to transmit power between two shafts that are at right angles to each other and are non-intersecting. The teeth on the worm are similar to the teeth on the rack, and the teeth on the worm gear are curved to conform with the teeth on the worm. Thread terms such as *pitch* and *lead* are used on the worm.

Since a single-thread worm gear in one revolution advances the worm gear only one tooth and space, a large reduction in velocity is obtained. Another feature of worm gearing is the high mechanical advantage acquired. The ratio of worm gear speed to the worm speed is the ratio between the number of teeth on the worm gear and the number of threads on the worm. A worm gear with 33 teeth and a worm with a multiple thread of three has a ratio of 11:1.

About 50:1 is the maximum ratio recommended. Since a single-thread worm has a low lead angle, it is inefficient and consequently not used to transmit power. The lead (or helix) angle should be between 25 and 45° for efficiency in transmitting power; as a result, multithread worms are used.

Fig. 17-7-1 How to identify a left-hand and right-hand worm gear.

Fig. 17-7-3 Worm gear and worm nomenclature.

Fig. 17-7-2 Location of bearings to absorb thrust load on worm and worm gear.

Fig. 17-7-4 Assembly drawing of a worm and worm gear assembly. (Dimensions not shown.)

The number of threads on a worm may vary from one to eight.

Figures 17-7-1 to 17-7-5 supply data on worm gear drawings and formulas.

Working Drawings of Worm and Worm Gears (Fig. 17-7-6)

These are similar to working drawings of other gears. A one-view section drawing is normally used for the worm gear. When a second view is required, the throat and root circles are shown as solid lines, and the outside circle is not shown on this view. As for the worm drawing, the root and the outside diameter are shown as solid lines, and a second view is not normally required.

When a worm and worm gear appear as an assembly drawing, both views are drawn and the conventional solid line for the OD of the worm and the throat diameter of the worm gear is shown as broken lines where the teeth mesh.

Term	Symbol	Formula	Definition
Pitch diameter of worm	PDw	$PDw = 2C - PDg$	
Pitch diameter of gear	PDg	$PDg = 2C - PDw$ or $\dfrac{NP}{\pi}$	
Pitch	P	$P = \dfrac{L}{T}$ $P = \dfrac{(2C - PDw) \times \pi}{N}$	The distance from one tooth to the corresponding point on the next tooth measured parallel to the worm axis. It is equal to the circular pitch on the worm gear
Lead	L	$L = \pi PDg \div R$ $L = P \times T$ $L = \text{Tan La} \times \pi PDw$	The distance the thread advances axially in one revolution of the worm
Threads	T	$T = \dfrac{L}{P}$	The number of threads or starts on worm; e.g., 2 for double thread, 3 for triple thread
Gear teeth	N	$N = \dfrac{\pi PDg}{P}$	Number of teeth on worm gear
Ratio	R	$R = \dfrac{N}{T}$	Divide number of gear teeth by number of worm threads
Center distance	C	$C = \dfrac{PDw + PDg}{2}$	
Addendum	ADD	$ADD = 0.318P$	Single and double threads
		$ADD = 0.286P$	Triple and quadruple threads
Whole depth	WD	$WD = 0.686P$	Single and double threads
		$WD = 0.623P$	Triple and quadruple threads
Outside diameter, worm	ODw	$ODw = PDw + 2ADD$	
Outside diameter, gear	ODg	$ODg = TD + 0.4775P$	Single and double threads
		$ODg = TD + 0.3183P$	Triple and quadruple threads
Throat diameter	TD	$TD = PDg + 2ADD$	
Face width, gear	F	$F = 2.38P + .25$ (inch) $F = 2.38P + 6$ (metric)	Single and double threads
		$F = 2.15P + .20$ (inch) $F = 2.15P + 5$ (metric)	Triple and quadruple threads
Face length, worm	FL	$FL = 6 \times P$	
Lead angle	La	$\text{Tan La} = \dfrac{L}{PDw \times 3.1416}$	Divide lead by circumference of pitch diameter of worm. Quotient is tangent of lead angle.
Throat radius	R_t	$R_t = \dfrac{PDw}{2} - ADD$	Subtract addendum from half of pitch diameter of worm
Rim radius	R_r	$R_r = \dfrac{PDw}{2} + P$	

Fig. 17-7-5 Worm and worm gear formulas.

CUTTING DATA	
NO. OF TEETH	36
ADDENDUM	.159
WHOLE DEPTH	.343
NO. OF THREADS	2
PITCH (AXIAL)	.500
PRESSURE ANGLE	14.5°
LEAD ANGLE	7°-53'
LEAD—RH	

CUTTING DATA	
NO. OF THREADS	2
PITCH	.500
PRESSURE ANGLE	14.5°
LEAD ANGLE	7°-53'
LEAD—RH	
WHOLE DEPTH	.343
ADDENDUM	.159

Fig. 17-7-6 Working drawing of a worm and worm gear.

ASSIGNMENTS

See Assignments 17 and 18 for Unit 17-7 on page 364.

Review for Assignments
Unit 19-1   Bearings
Unit 19-4   Oil Seals
Unit 7-5   Assemblies in Section
Unit 6-4   Bill of Material
Unit 9-2   Pin Fasteners

UNIT 17-8
Comparison of Chain, Gear, and Belt Drives

Chains, gears, and belts are used for power drives between rotating shafts that cannot be directly coupled. In this unit the characteristics of these media are compared, and the conditions favorable to the use of each type of drive are discussed.

CHAINS

A *chain drive* consists of an endless chain whose links mesh with toothed wheels, called *sprockets*, which are keyed to the shafts of the driving and driven mechanisms.

Roller Chains The unique feature of a roller chain is its freedom of joint action during its engagement with the sprocket. This is accomplished by articulation of the pins of the bushings, while the rollers turn on the outside of the bushings, thus eliminating rubbing action between the rollers and the sprocket teeth.

Silent Chains Comparable ease of joint action occurs in the engagement of the silent chain with the sprocket.

GEARS

A *simple gear drive* consists of a toothed driving wheel meshing with a similar driven wheel. Tooth forms are designed to ensure uniform angular rotation of the driven wheel during tooth engagement. Gears are available with precision-cut teeth or with unfinished teeth.

BELTS

A *belt drive* consists of an endless flexible belt that connects two wheels, or pulleys. Belt drives depend on friction between the belt and the pulley surfaces for the transmission of power.

In the case of V-belts, the friction for the transmission of the driving force is increased by wedging the belt into the grooves on the pulley.

V-belt drives are available in single or multiple strands for varying power-transmission requirements.

Another type of belt has shallow teeth molded on the inside of the driving face. The pulleys have teeth for engagement with the belt teeth.

CHAIN DRIVES COMPARED WITH GEAR DRIVES

Advantages of Chains
Shaft center distances for chain drives are relatively unrestricted, whereas with gears, the center distance must be such that the pitch surfaces of the gears are tangent. This advantage often will result in a simpler, less costly, and more practical design.

Chains are easily installed. While all drive media require proper installation, the assembly tolerances for chain drives are not as restricted as those for gears; and the resultant savings in the time of installation may be an important item in meeting the production schedule required of the driven machine.

The ease of chain installation is a definite advantage where later changes in design, such as speed ratio, capacity, and centers, are anticipated.

Advantages of Gears
When space limitations require the shortest possible distance between shaft centers, a gear drive is usually preferable to a chain drive.

The maximum speed ratio for satisfactory operation of a gear drive is usually greater than that for chain drive.

Gears can be operated at higher rotative speeds than chain drives.

CHAIN DRIVES COMPARED WITH BELT DRIVES

Advantages of Chains
Chain drives do not slip or creep as do belt drives. As a result, chains maintain a positive speed ratio between the driving and the driven shafts, and they are more efficient since no power is lost because of slippage.

Chain drives are more compact than belt drives. For a given capacity, a chain will be narrower than a belt, and sprockets will be smaller in diameter than pulleys; thus the chain drive will occupy less overall space.

Chains are easy to install. A chain can be installed by wrapping it around the sprockets and then slipping the pins of a connecting link into position.

The required minimum arc of contact is smaller for chains than for belts. This advantage becomes more pro-nounced as the speed ratio increases and thus permits chain drives to operate on much shorter shaft center distances.

Where several shafts are to be driven from a single shaft, positive speed synchronism between the driven shafts is usually imperative. For such applications, chains are more suitable.

Chains do not deteriorate with age; nor are they affected by sun, oil, and grease. Chains can operate at higher temperatures. Chain drives are more practical for low speeds.

Chain elongation resulting from normal wear is a slow process; the chain, therefore, requires infrequent adjustment. Belt stretch, however, necessitates frequent tightening by shaft adjustment, by idlers, or by shortening the belt.

Advantages of Belts
Since no metal-to-metal contact occurs between a belt and pulleys, belts require no lubrication, although leather belts need periodic applications of belt dressing to preserve their flexibility.

Generally speaking, a belt drive operates with less noise than a chain drive.

Flat belt drives can be used where extremely long center distances would make chain drives impractical.

In the extremely high-speed ranges, flat belts can be operated to better advantage than chains.

CONCLUSION

No one type of power drive is ideal for all types of service. This unit has discussed the relative merits of chain, gear, and belt drives, and should provide a guide to the selection of the best type for a given application.

Reference and Source Material
1. American Chain Association.

ASSIGNMENT

See Assignment 19 for Unit 17-8 on page 364.

Review for Assignment
Unit 17-1   Belt Drives
Unit 17-2   Chain Drives
Unit 17-3   Gear Drives

ASSIGNMENTS for Chapter 17

Assignments for Unit 17-1, Belt Drives

1. *V-Belt Drive Problems.* Do any three.
   (a) A 0.33 hp (0.25 kW), 1750 r/min motor is to operate a furnace blower having a shaft speed of approximately 765 r/min. The center distance between the motor and blower shafts is approximately 13.5 in. (340 mm). The type of drive required is V-belt.

   (b) A 0.50 hp (0.37 kW), 1160 r/min motor is used to operate a drill press. The spindle speed is 520 r/min, ±5 r/min. The center distance between the motor and blower shafts is approximately 22 in. (550 mm). Select a suitable V-belt.

   (c) A 1.5 hp (1.1 kW), 1750 r/min motor is to operate a band saw whose flywheel turns at approximately 800 r/min. A pulley attached to the flywheel shaft connects, by means of a V-belt, to the pulley on the motor shaft. Center-to-center distance of shafts is 13.5 in. (340 mm). Calculate the size of the V-belt required.

   (d) A 0.50 hp (0.37 kW), 1750 r/min motor drives a power hacksaw. The shaft on the hacksaw is to run at approximately 750 r/min, and the center-to-center distance of the shafts is 15.5 in. (400 mm). Calculate the size of V-belt required.

   (e) A 0.75 hp (0.6 kW), 1750 r/min motor is used to drive a punch machine whose flywheel turns at approximately 500 r/min. A pulley is attached to the flywheel shaft and connects to the motor pulley by means of a V-belt. Center-to-center distance is 17 in. (430 mm). Calculate the size of V-belt required.

2. *V-Belt Motor Drive.* On a B- or A3-size sheet, lay out a 0.25 horsepower (hp) or 0.2 kW motor to drive shaft A between 815 and 835 r/min by means of a V-belt drive. Refer to Fig. 17-1-A or 17-1-B. Design for normal duty. Other details can be seen in Fig. 17-1-10. Draw top and front views. From manufacturers' catalogs select the belt and pulleys and call for them in a bill of material. Scale is 1:4 inch or 1:5 metric.

Assignments for Unit 17-2, Chain Drives

3. A tumbler barrel is to be driven at approximately 40 r/min by a speed reducer powered by a 5 hp (3.7 kW) electric motor. The reducer output speed is 100 r/min, and the output shaft is 1.75 in. (44 mm) in diameter. The shaft diameter of the tumbling barrel is 2 in. (50 mm). The shaft center distance is approximately 36 in. (900 mm). Select a single chain (heavy shock).

4. This is the same as Assignment 3 except that a double chain is to be used.

5. The head shaft of an apron conveyor, which handles rough castings from a shakeout, operates at 66 r/min and is driven by a gear motor whose output is 7.5 hp (5.6 kW) at 100 r/min. The head shaft has a 2-in. (50-mm) diameter, and the gear motor shaft has a 1.75-in. (44-mm) diameter. Shaft center distance should not exceed 42 in. (1055 mm). Select a multiple chain (moderate shock).

6. A gear-type lubrication pump located in the base of a large hydraulic press is to be driven at 860 r/min from a 1.25-in. (32-mm) diameter shaft operating at 1000 r/min. The pump is rated at 3 hp (2.4 kW) and has a 1.375-in. (35-mm) diameter shaft. Shaft center distance must not be less than 10 in. (250 mm).

7. A 10-hp (7.5-kW), 480-r/min electric motor is to drive a line shaft, which is subject to light service, at 160 r/min. The motor shaft will be in approximately the

same horizontal plane as the line shaft. The diameters of the motor shaft and line shaft are 1.69 and 1.75 in. (42 and 44 mm), respectively. A shaft distance of 48 to 60 in. (1220 to 1520 mm) will be acceptable. Select a triple chain.

8. A centrifugal fan is to be driven at 2800 r/min by a 10-hp (7.5-kW) electric motor. The motor speed is 1800 r/min, and the shaft has a 1.375-in. (35-mm) diameter. The compressor shaft has a 1.25-in. (32-mm) diameter. The center distance is to be approximately 20 in. (500 mm). The overall drive must not exceed a 5-in. (125-mm) radius on the motor or a 3-in. (75-mm) radius at the fan.

Assignments for Unit 17-3, Gear Drives

9. On a B- or A3-size sheet, make working drawings for the two gears described in either Fig. 17-3-A or Fig. 17-3-B. Gear 1 will require one view only and be drawn to full scale or 1:1. Gear 2 will require

HP	C	L
.12	10.75	4.62
.16	10.75	4.62
.25	11.25	5.12
.33	11.75	5.62
.50	12.62	6.50
.75	13.50	7.38

115 VOLT   1750 RPM
MOTOR DIMENSIONS

Fig. 17-1-A   V-belt drive problems.

kW	C	L
0.1	274	118
0.12	274	118
0.20	286	130
0.25	286	144
0.37	320	166
0.56	344	188

115 VOLT   1750 REV/MIN
MOTOR DIMENSIONS

METRIC

Fig. 17-1-B   V-belt drive problems.

GEAR #1	GEAR #2
Tooth form—14.5°	Tooth form—20°
PD—6.00	N—50
DP—5	DP—5
Face width—1.00	Face width—1.75
Web—40	Shaft—∅1.75
Shaft—∅1.10	Hub—∅3.00 × 2.75
Hub—∅1.90 × 1.50 Lg	(total length)
Matl—MI	6 Spokes—60 Thk,
	1.50 wide, tapered
	to 1.10 wide
	Matl—MI

Fig. 17-3-A Single spur gears.

GEAR #1	GEAR #2
Tooth form—14.5°	Tooth form—20°
PD—127	N—44
Module—6.35	Module—6.35
Face width—26	Face width—46
Web—10	Shaft—∅45
Shaft—∅28	Hub—∅76 × 70.6
Hub—∅50 × 40 Lg	(total length)
Matl—MI	6 Spokes—16 Thk,
	40 wide, tapered
	to 30 wide
	Matl—MI

Fig. 17-3-B Single spur gears.

GEAR DATA
Tooth form—14.5°
Center-to-center distance—6.00
Gear—N—36
Face width—1.10
Shaft—∅1.25
Web—.40
Hub—∅2.10 × 1.50 Lg
Matl—MI
Pinion—N—24
Shaft—∅1.10
Matl—Steel

Fig. 17-3-C Meshing spur gears.

GEAR DATA
Tooth form—14.5°
Center-to-center distance—127
Gear—N—24
Face width—30
Shaft—∅32
Web—10
Hub—∅54 × 38 Lg
Matl—MI
Pinion—N—16
Shaft—∅30
Matl—Steel

Fig. 17-3-D Meshing spur gears.

GEAR	PD	N	DP	DIRECTION	R/MIN	CENTER DISTANCE
A	7.00	28	4	C'WISE	300	
B	3.00	12				
C	6.00					
D		12	3			

GEAR	PD	N	DP	DIRECTION	R/MIN	CENTER DISTANCE
A		54			160	
B	3.00		6			
C			5			
D	2.40			C'WISE	2400	

GEAR	PD	N	DP	DIRECTION	R/MIN	CENTER DISTANCE
A	7.50		4	A'WISE	240	
B		18				
C	10.00					
D	3.20		16			
E	8.00		6			
F		40				

Fig. 17-3-E Gear-train calculations.

GEAR	PD	N	MDL	DIRECTION	R/MIN	CENTER DISTANCE
A	228.0		6.35	C'WISE	300	
B		12				
C	76.32					
D		12	3.18			

GEAR	PD	N	MDL	DIRECTION	R/MIN	CENTER DISTANCE
A		54			160	
B			6			
C			5			
D				C'WISE	2400	

GEAR	PD	N	MDL	DIRECTION	R/MIN	CENTER DISTANCE
A	95.4			A'WISE	240	
B		18				
C	203					
D	67.68	18				
E	42.3		4.23			
F		40				

Fig. 17-3-F Gear-train calculations.

two views and will be drawn to half scale or 1:2. Select a proper key size and use your judgment for dimensions not given. Include with each gear drawing a cutting data block.

10. On a B- or A3-size sheet, prepare a working drawing of two gears in mesh from the information found in Fig. 17-3-C or 17-3-D. Show two views with three or four teeth shown in mesh. Add suitable keys and use your judgment for dimensions not given. Include cutting data for each gear. Scale is full or 1:1.

11. On a B- or A3-size sheet, complete the missing information on the gear-train problems shown in Fig. 17-3-E or 17-3-F.

Assignments for Unit 17-4, Power-Transmitting Capacity of Spur Gears

12. Show your calculations for the designs of suitable pairs of spur gears to operate the equipment described in a and b or c and d below.
    (a) A 1200-r/min motor drives, by means of a spur gear and pinion, a machine rated at 8 hp and operating under moderate shock 12 hours a day. The reduction in r/min is 4:1. Select a suitable pair of spur gears to transmit the power required.
    (b) A punch press rated at 22-hp, 900-r/min is to be driven by a 30-hp, 1200-r/min motor. The punch, which is subjected to moderate shock, will be in operation 16 hours a day. Select a suitable pair of spur gears to transmit the power required.
    (c) A 1200-r/min motor drives, by means of a spur gear and pinion, a machine rated at 7.5 kW and operating under moderate shock 8 hours a day. The reduction in r/min is 3:1. Select a suitable pair of spur gears to transmit the power required.
    (d) An 1800-r/min motor drives a machine which is rated at 2 kW and runs at 450 r/min under moderate shock 18 hours a day. Select a suitable pair of spur gears to transmit the power required.

13. Show your calculations for the design of suitable pairs of spur gears to operate the equipment described in a or b below.
    (a) A machine which works under smooth operating conditions is used twice a day for about 10 minutes. It is manually operated and is rated at 7 hp (6 kW) and runs at 800 r/min. Two motors are in stock at the plant; one rated at 7 hp (6 kW) and 1200 r/min, the other at 5 hp (4 kW) and 750 r/min. Spur gears are to be used.

Make a report to the plant engineer on the selection of motor and gears which you would recommend.

(b) A 900-r/min motor drives an air compressor which operates between 15 to 20 minutes every hour. The compressor which is rated at 5 kW (7.5 hp) runs at 600 r/min, under smooth operating conditions. Select a suitable pair of spur gears to transmit the power required.

Assignment for Unit 17-5, Rack and Pinion

14. On a B- or A3-size sheet, make a working assembly drawing of the gear and rack shown in Fig. 17-5-A or 17-5-B. Use your judgment for dimensions not given. Show four or five teeth in mesh. Scale is full or 1:1.

GEAR AND RACK DATA
Gear—N—36
DP—5
Tooth form—14.5°
Web—.50
Shaft—⌀1.25
Hub—⌀2.25 × 1.75 Lg
Face width—1.25
Matl—MI
Rack—Matl—Steel

Fig. 17-5-A   Gear and rack.

GEAR AND RACK DATA
Gear—N—30
MDL—5.08
Tooth form—14.5°
Web—11
Shaft—⌀35
Hub—⌀58 × 45 Lg
Face width—32
Matl—MI
Rack—Matl—Steel

Fig. 17-5-B   Gear and rack.

Assignments for Unit 17-6, Bevel Gears

15. On a B- or A3-size sheet, make a working drawing of a bevel gear from the data shown in Fig. 17-6-A or 17-6-B. Use your judgment for dimensions not given. Scale is full or 1:1.

16. On a B- or A3-size sheet, make an assembly working drawing of the gears described in Fig. 17-6-C or 17-6-D. Add to the drawing the cutting data for the gears. Use your judgment for dimensions not given. Scale is full or 1:1.

GEAR DATA
PD—4.50
Pitch cone angle—45°
DP—4
Tooth form—14.5°
Face width—1.25
Shaft—⌀1.00
Hub—⌀1.75 × 1.50 Lg
Web thickness—.62

Fig. 17-6-A   Single bevel gear.

GEAR DATA
PD—114
Pitch cone angle—45°
Module—6.35
Tooth form—14.5°
Face width—32
Shaft—⌀24
Hub—⌀44 × 32 Lg
Web thickness—16

Fig. 17-6-B   Single bevel gear.

GEAR DATA
DP—4
Face width—1.10
N—22
Shaft—⌀1.00
Hub—⌀1.90 × 1.50 Lg
Web—.75
Matl—MI
PINION DATA
N—14
Shaft—⌀.75
Hub—1.25 Lg
Matl—Steel

Fig. 17-6-C   Bevel gear assembly.

GEAR DATA
Module—6.35
Face width—30
N—22
Shaft—⌀25
Hub—⌀48 × 38 Lg
Web—20
Matl—MI
PINION DATA
N—14
Shaft—⌀20
Hub—32 Lg
Matl—Steel

Fig. 17-6-D   Bevel gear assembly.

Assignments for Unit 17-7, Worm and Worm Gears

17. On an A3- or B-size sheet, make a working drawing of a worm and the mating worm gear from the data given in a or b below. Use your judgment for dimensions not given. Include on the drawing the cutting data. Scale is full or 1:1.

(a) A worm and worm gear have a pitch of .5236 in. The gear, made of cast iron, has 30 teeth; shaft dia = .88, hub dia = 1.75, hub length = 1.90, face width = 1.00, web thickness = .40. The worm, made of hardened steel, is 3.50 long on a ⌀.88 shaft, pitch dia = 2.12, single thread, RH lead.

(b) A worm and worm gear have a pitch of 13.3. The gear, made of cast iron, has 30 teeth; shaft dia = 22, hub dia = 44, hub length = 48, face width = 25, web thickness = 10. The worm, made of hardened steel, is 88 long on a ⌀22 shaft, pitch dia = 54, single thread, RH lead.

18. On an A3 or B-size sheet, make a two-view detail assembly drawing of a worm and worm gear from the data given in (a) or (b) below. Use your judgment for dimensions not given. Include the cutting data on the drawing. Scale is half or 1:2.

(a) A worm and worm gear have a pitch of .75. The gear, made of phosphor bronze, has 24 teeth, shaft dia = 1.25, hub dia = 2.25, hub length = 2.50, web thickness = .50. The worm, made of steel, has a pitch dia = 2.50, left-hand double thread, shaft dia = 1.00.

(b) A worm and worm gear have a pitch of 19. The gear, made of phosphor bronze, has 24 teeth, shaft dia = 32, hub dia = 58, hub length = 64, web thickness = 13. The worm, made of steel, has a pitch dia = 64, left-hand double thread, shaft dia = 26.

Assignment for Unit 17-8, Comparison of Chain, Gear, and Belt Drives

19. Your supervisor has asked you to submit a report recommending the type of power transmission best suited for the machinery layouts shown in either Fig. 17-8-A or Fig. 17-8-B. Make a detailed report that specifies the power transmission parts required to properly operate the equipment.

Fig. 17-8-A Power-transmission drive.

METRIC

Fig. 17-8-B Power-transmission drive.

CHAPTER 18

Couplings, Clutches, Brakes, and Speed Reducers

UNIT 18-1
Couplings and Flexible Shafts

COUPLINGS

Couplings, as the name implies, are used to couple or join shafts. There are two types of couplings: permanent couplings and clutches.

Permanent couplings are not normally disconnected except for assembly or disassembly purposes, while clutches permit shafts to be connected or disconnected at will.

Permanent Couplings

Permanent couplings can be divided into three main categories: solid, flexible, and universal.

Solid Couplings Solid couplings should be used only when driving and driven shafts are mounted on a common rigid base, so that shafts can be perfectly aligned and will stay that way in service. If two shafts are not in exact alignment and are connected by a rigid coupling, excess bearing wear may occur on the bearing supporting the shaft. The steel sleeve coupling and the flanged coupling shown in Fig. 18-1-1 are solid couplings.

Flexible Couplings These are intended to compensate for unintentional misalignments or transient misalignments such as those caused by thermal expansion or vibration. They also pre-

(A) SLEEVE

(B) FLANGED

Fig. 18-1-1 Solid couplings.

vent shock from being transferred from one shaft to another and are recommended where several power machines are connected on one shaft. See Figs. 18-1-2 and 18-1-3.

Fig. 18-1-2 Flexible coupling. (Oldham Principle)

There are many types of flexible couplings, but all are similar in operation. There are two hubs, one on each shaft, connected by an intermediate part which may be flexible, floating, or both.

Flexible couplings may also be divided into three main categories: those that use mechanical movement, those that depend on the flexing of materials, and those that combine mechanical movement with flexing.

The table shown in Fig. 18-1-4 lists the most common types of couplings and their main qualities. It should be used only as a guide, since special materials can affect qualities considerably; and, of course, there are exceptions to every rule. For most jobs, any one of several couplings may be suitable; then cost determines the final selection.

To aid in selecting the correct-size coupling, most manufacturers rate power transmitted in horsepower per 100 revolutions per minute (kW per r/min) and give maximum permissible revolutions per minute. The rating can be determined by the simple formula

$$\text{hp per 100 rev/min} =$$

$$\frac{\text{driving hp} \times 100 \times \text{service factor}}{\text{coupling revolutions per minute}}$$

or

$$\text{Kilowatts per 100 rev/min} =$$

$$\frac{\text{driving kilowatts} \times \text{service factor}}{\text{coupling revolutions per minute}}$$

(A) ROLLER CHAIN

Silent

(B) SILENT CHAIN

Roller

(C) MORFLEX

(D) EXPLODED ASSEMBLY OF A RUBBER BALL COUPLING

(F) APPLICATION

(E) SURE-FLEX COUPLING

PARALLEL MISALIGNMENT

SHOCK AND VIBRATION

END FLOAT

ANGULAR MISALIGNMENT

(G) 4-WAY FLEXACTION

Fig. 18-1-3   Flexible couplings. (A, B, and C—Morse Chain Co.; D, F, and G—Commonwealth Mfg.; E—T. B. Wood's Sons Co.)

The service factor depends on the source of driving power and the type of duty. For smooth power sources such as an electric motor driving smooth loads like a centrifugal compressor, the factor is 1. It can be as high as 5 for reciprocating gasoline or diesel engines coupled to loads with cyclic torque variations such as a single-cylinder compressor without a flywheel.

yors yoke

Universal Couplings Commonly called universal joints, universal couplings are for applications where angular displacement of shafts is a design requirement. It is easier to select universal couplings than flexible couplings because there are fewer types of them. Most common is the Hooke's joint, which has a cross-type trunnion connected to driving and driven shafts by U-shaped endpieces. See Fig. 18-1-5. Its main disadvantage is that because the trunnion is always at right angles to the driven shaft, it gives a sinusoidal variation in angular velocity between shafts. Other disadvantages are that it cannot compensate for out-of-parallel alignments and it does not compensate for changing distances between driving and driven points when the angle between shafts changes. These disadvantages disappear when two universals are used, one with a sliding spline, as in automotive systems using the

(A) SINGLE JOINT

(B) DOUBLE JOINT

(C) CORRECT ARRANGEMENT ANGLES MUST BE EQUAL

Fig. 18-1-5   Universal joints—Hooke's type. (Boston Gear Works)

		Chain	Straight gear	Curved gear	Sliding insert	Metal ball	Metal disk	Plastic disk	Bellows	Helical spring	Rubber tire	Plastic insert	Rubber insert	Rubber ball	Radial spring	Grid spring
Angular misalignment	—high							*	*	*	*		*	*	*	
	—average		*	*	*	*						*			*	*
	—low	*	*													
Out-of-parallel misalignment	—high						*			*		*	*			
	—average				*	*			*			*			*	*
	—low	*	*	*		*				*						
End float	—high		*	*										*	*	*
	—average	*							*	*	*	*				
	—low									*						
r/min	—high	*	*	*			*			*						
	—average				*	*		*	*		*	*	*	*	*	*
	—low															
Torsional resilience	—high							*			*	*				
	—average						*		*		*			*	*	
	—low	*	*	*	*	*	*		*							
Lubrication required		*	*	*	*	*										

Fig. 18-1-4   Basic features of common flexible couplings.

Hotchkiss drive. Here, the transmission and differential pinion shafts are parallel, so rotational fluctuations are canceled out. When two joints are used in this manner, the U-shaped fittings on the drive-shaft ends must be parallel, or else the rotational fluctuations will be increased instead of canceling out.

If constant velocity is essential with only one universal, a special constant-velocity universal must be used. Most of these have some type of ball drive, where the driving points of contact bisect the driving angle. They are more complex than the Hooke's type and more expensive. The universal coupling shown in Fig. 18-1-6 is designed to transmit a constant velocity. The drive is through steel balls in races, designed so that the plane of contact between the balls and races always bisects the shaft angle. Flexible shafts also give constant velocity but are limited to transmitting relatively low power.

Fig. 18-1-6   Constant velocity universal coupling. (The Bendix Corp.)

FLEXIBLE SHAFTS

Flexible shafts are used to transmit power around corners and at various angles when driving and driven elements are not aligned. Speedometers, tachometers, and indicating and recording instruments are typical applications.

Flexible shafts are constructed of helically wound wire and designed for transmission of rotary power and motion between two points located so that their relative positions preclude the use of solid shafts.

Principal parts are shown in Fig. 18-1-7. They are

1. *Flexible shaft*, the bare working element without end fittings, sometimes called the *core* or *cable*.

(A) CABLE

(B) CASING

(I) LOOSE MALE NUT

(2) LOOSE FEMALE NUT

(C) CASING END FITTINGS

(D) END FITTING

Fig. 18-1-7   Principal parts of a flexible shaft.

2. *Shaft end fittings* fastened to the ends of a flexible shaft to permit connection to driving and driven elements.
3. *Flexible casing*, a flexible, tubelike covering which acts as a runway or guide for the flexible shaft, protects it from dirt and injury, and helps retain lubrication.
4. *Casing end fittings*, parts fastened to allow connection or coupling to the housing of the driving and driven elements.

References and Source Material

1. *Machine Design*, Mechanical drives reference issue, 1979.

ASSIGNMENTS

See Assignments 1 and 2 for Unit 18-1 on page 373.

Review for Assignments

Unit 6-4   Assembly Drawings
Unit 7-1   Full Sections
Unit 8-3   Drawing Nuts and Bolts
Unit 9-1   Keys

UNIT 18-2
Clutches and Brakes

CLUTCHES

The simplest use of a clutch is to start and stop a machine or rotating element without starting and stopping the prime mover. Clutching devices have been designed to do a variety of other jobs, such as to maintain constant speed, torque, and power or to limit torque. They are also used for automatic disconnection, quick starts and stops, gradual starts, and nonreversing and overrunning functions.

Three types of clutches currently in general use are mechanical, electric, and hydraulic. These categories break down into numerous subtypes providing special characteristics and functional capabilities.

Mechanical Clutches

Mechanical clutches are basically of two types: positive and friction. Positive clutches operate by meshing the metal teeth or jaws of the driving member with corresponding elements on the driven member. Friction clutches press one or more driving members against corresponding driven members, such as disks, bands, or shoes. Positive clutches will break but cannot slip; friction clutches slip but do not break.

Although not used as commonly as friction clutches, positive clutches have important applications. For example, they are used in many special synchronized drives. They are used extensively in machine tools, business machines, and household appliances. In integral sizes they are used in power presses, construction machinery, and automotive transmissions.

Positive clutches are available in three basic types, in addition to numerous special designs.

The *square-jaw clutch*, also called the *claw clutch* (Fig. 18-2-1A), originally used teeth in the as-cast condition.

Spiral-jaw clutches (Fig. 18-2-1B) have eliminated many of the undesirable features of the square-jaw type.

The spiral-jaw clutch will engage at higher speeds than the square-jaw, but

(A) SQUARE JAW

(B) SPIRAL JAW

(C) MULTIPLE TOOTH

Fig. 18-2-1   Positive clutches. (A and B—Link Belt Ltd.; C—The Bendix Corp.)

will shock-load the driver at engagement if load inertia is high. It drives in one direction only and has a tendency to freewheel.

Multiple-tooth clutches (Fig. 18-2-1C) have multiple teeth modified in various ways to provide strength or high-speed capability.

Friction clutches usually fall into two principal classifications: rim clutches, in which the contact pressure is applied, normally, to the shaft axis against a rim or drum; and axial clutches, in which the contact pressure is applied by shift movement along the shaft, as in cone or disk clutches.

One of the earliest of the axial-type clutches was the familiar cone clutch. See Fig. 18-2-2A.

In the disk type of clutch, one or more friction disks are clamped to metal plates. See Fig. 18-2-2B.

(A) CONE   (B) PLATE OR DISC

Fig. 18-2-2   Axial-type friction clutches.

Overrun clutches drive in one direction only and overrun or freewheel in the other direction. The operation of the overrun clutch is very simple. As the shaft and inner race rotate in the direction of the arrow shown in Fig. 18-2-3, the rollers roll up to the high spots of the inner race, thus jamming the inner and outer races as one, causing the outer race to rotate.

However, if the inner race should travel at a lower speed than the outer race, stop, or reverse its direction, then the outer race would not turn with the inner race because the rollers would not engage the outer race.

(A) ROLLERS JAMMING THE INNER AND OUTER RACES AS ONE

(B) ROLLERS NOT MAKING CONTACT WITH OUTER RACE

Fig. 18-2-3   Overrun clutch.

Electric Clutches

Electric clutches perform the same functions as mechanical clutches but are controlled electromagnetically rather than mechanically.

All electric clutches have one thing in common—a magnetic field that determines torque. Remote control is thus both feasible and economical.

An electric clutch acts as either an on-off or a continuous-slip device. Whenever a machine operation involves starting or stopping a motor more than approximately 12 times per minute (or 4 to 5 times per minute for an enclosed motor), an electric clutch should be used to clutch the load in and out, thus allowing the motor to run continuously.

The simplest and most commonly used electric clutch is the single-disk friction clutch. See Fig. 18-2-4. Its action is basically that of an electromagnet. When the electromagnet is energized, an armature is drawn into direct contact with the face of the electromagnet. In the single-disk clutch, the electromagnet is shaped like a doughnut, and the coil which creates the field is located inside the doughnut shell. Friction material is added to one side of the shell to form one of the clutch faces. The other face is that of the armature, a segmented iron disk.

Fig. 18-2-4   Electric clutch.

Hydraulic Clutches (Fluid Couplings)

Fluid couplings are similar to torque converters, except they do not have a stator between the impeller and the turbine. See Fig. 18-2-5. They simply transmit input torque, whereas the torque converter multiplies input torque. Some fluid couplings can be used as variable-speed drives.

Fig. 18-2-5   Fluid coupling. (Twin Disc Clutch Co.)

BRAKES

Basically, any brake is simply an extension of a clutch, in which one member is held stationary. Brakes may be used as on-off devices or as drags. Brakes may be classified into two major types: mechanical and electric.

Mechanical Brakes

A mechanical brake is a frictional device that converts kinetic energy to heat and dissipates it into the atmosphere. It can be actuated mechanically, pneumatically, hydraulically, or electrically.

Most commercial brakes are band, drum, or disk types.

Band Brakes These are among the oldest types of brakes. Basically, a band brake consists of a flexible steel band lined with friction material. See Fig. 18-2-6.

(A) AUXILIARY BRAKE FOR A TRUCK USUALLY MOUNTED ON THE TRANSMISSION

(B) COMBINATION BAND-DISC BRAKE USED MOSTLY IN FARM MACHINERY

Fig. 18-2-6   Band brakes. (The Bendix Corp.)

Drum Brakes Drum brakes are perhaps the best known type of mechanical brake, mainly because of their widespread use in automobiles. See Fig. 18-2-7. They are available in two basic configurations, external-contracting and internal-expanding.

Electric Brakes

Basically, an electric brake is similar to an electric clutch, except that one element is rigidly held. Like clutches, electric brakes are available in four

Fig. 18-2-7   Drum brakes. (The Bendix Corp.)

(A) SPRING-ACTUATED ELECTRICALLY RELEASED

(B) ELECTRICALLY ACTUATED

Fig. 18-2-8   Electromagnet friction-disk brakes. (A—Dings Brakes Co.; B—Warner Electric Brake & Clutch Co.)

basic types: magnetic-particle, eddy-current, hysteresis, and friction.

The most widely used type of electric brake is the electromagnetic friction-disk type, in which a friction unit is electrically actuated or released. See Fig. 18-2-8.

References and Source Material

1. *Machine Design*, Mechanical drives reference issue, 1979.

ASSIGNMENTS

See Assignments 3 and 4 for Unit 18-2 on page 373.

Review for Assignments

Unit 6-4   Assembly Drawings
Unit 7-1   Full Sections
Unit 7-2   Half Sections
Unit 8-3   Drawing Nuts and Bolts
Unit 9-1   Keys

UNIT 18-3
Adjustable-Speed Drives and Speed Reducers

PACKAGED ADJUSTABLE-SPEED DRIVES

Packaged mechanical adjustable-speed drives range in availability from small machine tools to trucks. They may provide only a few selected speeds or be able to vary speed infinitely over a wide range. Efficiency is generally high; for many units, efficiency will be over 90 percent depending on the type of drive. An example of a variable-speed drive is shown in Fig. 18-3-1.

Adjustable-speed drives can be classified in three major categories

1. Stepped
2. Stepless, limited range
3. Stepless, infinite range

GEAR DRIVES

Multispeed gear transmissions provide exact shaft speed at high efficiency. They are used in machine tools, mobile equipment, and other applications

Fig. 18-3-1 Variable-speed drive for 14 in. (35mm) lathe. (Rockwell Mfg. Co.)

(A) SLIDING GEAR

(B) CONSTANT MESH

(C) IDLER GEAR

(D) PLANETARY

Fig. 18-3-2 Gear-type adjustable-speed drives.

Fig. 18-3-3 Steel-belt variable drive. (Link Belt Ltd.)

where selective control of a number of fixed speeds is necessary.

Two broad categories of general transmissions are available. In one type, the speed is selected manually; in the other, speed changes occur automatically at predetermined points.

Basic Types

The most common selective-speed transmissions are parallel-axis arrangements. These can be broadly classified into four types.

Sliding Gears Speed adjustment is effected by sliding gears on one or more intermediate parallel shafts. See Fig. 18-3-2A. Shifting is generally accomplished by disengaging the input shaft.

Constant-Mesh Several gears of different sizes mounted rigidly to one shaft mesh with mating gears free to rotate on the other shaft. See Fig. 18-3-2B.

Idler One shaft carries several different-size gears that are rigidly mounted. See Fig. 18-3-2C. Speed adjustment is through an adjustable arm which carries an idler gear to connect with fixed or sliding gears on the other shaft. This arrangement is used to provide

stepped speeds in small increments and is frequently found in machine tools.

Planetary This type of gearing (Fig. 18-3-2D) is the most versatile and compact gear arrangement for a given ratio range and torque capacity. At the same time, it is also the most expensive, because of the clutching and braking elements necessary to control the operation of the unit. In addition, practical ratios available with planetary sets are limited.

BELT AND CHAIN DRIVES

Packaged belt and chain drives convert constant input speed into output that is steplessly variable within a certain range. They may contain integral motors and built-in gear reducers to obtain low output speeds. Electric motors are the most commonly used power-input devices.

All-Metal Belt In this unit, power is transmitted by means of a self-forming laminated steel belt or chain which engages radial teeth on conical pulley flanges. See Fig. 18-3-3.

V-Belt A highly compact belt drive for wide speed variation has been achieved through use of a modified form of the compound floating-sheave principle. Speed adjustment is accomplished by changing the position of the movable flange on the bottom input sheave.

FRICTION AND TRACTION DRIVES

Friction or *traction drives* transmit rotary motion by friction generated at the point or line of contact. Speeds are changed by moving the contact point or line relative to the centers of rotation of the driving and driven members. The amount of friction between the parts determines the transmittable power.

These drives operate best at constant load, with no sudden shocks to cause destructive slippage at the friction point.

Contact of metallic surfaces provides one practical method of obtaining stepless speed adjustment. See Fig. 18-3-4.

IMPULSE DRIVES

Adjustable impulse drives provide infinitely adjustable output speeds, usually in low-speed ranges. They provide

Fig. 18-3-4 Friction-type adjustable-speed drive. (Cone Drive Gears)

high-ratio speed reductions in a compact package. See Fig. 18-3-5.

The principle of operation is continuous indexing. The driving member engages the driven, moves it a predetermined distance, and then disengages.

Fig. 18-3-5 Impulse-type variable-speed drive. (Zero-Max Ind. Inc.)

(A) TYPICAL SINGLE STAGE ROTATING-HOUSING CONVERTER

(B) TYPICAL FLOW PATTERN IN CONVERTER

Fig. 18-3-6 Torque converter adjustable-speed drive.

TORQUE CONVERTERS

The hydraulic torque converter provides infinitely variable torque within its limits, solely in response to load variation. It makes use of the kinetic energy of fluid in motion. The torque converter is similar to a fluid coupling except for the addition of a stator.

Figure 18-3-6A shows the schematic blade arrangement of a typical single-stage, rotating-housing converter. Figure 18-3-6B shows the typical flow pattern among the impeller, turbine, and stator at stall, when maximum torque multiplication occurs, and at racing, when no torque is transmitted.

SPEED REDUCERS

Any device which reduces the speed of the driving unit is a speed reducer.

However, this section considers only those packaged units—primarily gear-type—whose prime function is the reduction of speed.

Base-Mounted Reducers

Base-mounted speed reducers are available in a number of gear types: helical, double-helical spur, spiral bevel, straight bevel, worm, double-enveloping worm, and herringbone.

In addition to the common geared reducers, many unconventional designs are available. Figure 18-3-7 shows a worm-type speed reducer.

Shaft-Mounted Reducers

A shaft-mounted speed reducer is an enclosed gear unit mounted on and supported by the input shaft of the driven machine. To prevent rotation of the housing, it is anchored by a torque-reaction member to a suitable rigid support.

These reducers are available as helical, herringbone, and spur-gear units, of single or multiple reduction stages (Fig. 18-3-8) with the output hub either concentric with or parallel to the input shaft of the speed reducer.

References and Source Materials

1. *Machine Design*, Mechanical drives reference issue, 1979.

ASSIGNMENTS

See Assignments 5 and 6 for Unit 18-3 on page 373.

Fig. 18-3-7 Base-mounted speed reducer. (Winsmith Div. of U.M.C.)

Fig. 18-3-8 Shaft-mounted reducer. (Dodge Mfg. Corp.)

ASSIGNMENTS for Chapter 18

Assignments for Unit 18-1, Couplings and Flexible Shafts

1. On a B- or A3-size sheet, lay out the motor-to-gear box drive unit shown in Fig. 18-1-A. A flexible coupling is required to connect the shafts, and the type of coupling specified is shown in the figure. Call for the correct-size coupling on the drawing. Scale is as specified.

2. On a B- or A3-size sheet, lay out the motor-to-pump drive assembly shown in Fig. 18-1-B. Flexible couplings are required to connect the shafts, and the type of coupling specified is shown in the figure. Call for the proper couplings on the drawing. Scale is as specified.

Assignments for Unit 18-2, Clutches and Brakes

3. On a B- or A3-size sheet, make a full-section assembly drawing of the friction clutch shown in Fig. 18-2-A. Scale is 1:1. Use your judgment for dimensions not shown.

4. On a B- or A3-size sheet, make a half-section assembly drawing of a gear mounted on the outer race hub of the overrunning clutch, model 12, shown in Fig. 18-2-B. Scale is full or 1:1. Use your judgment for dimensions not given.

Assignments for Unit 18-3, Adjustable-Speed Drives and Speed Reducers

5. On a B- or A3-size sheet, make a full-section assembly drawing of the sliding-gear speed reducer shown in Fig. 18-3-A. Support the gears on journal bearings. The gears are held to the shafts by set-screws and keys. Gears C and D are combined into one part and slide on the countershaft. See Fig. 18-3-2A. Complete the chart on the figure showing the two speeds available (when gears C and E mesh and when gears F and D mesh) for the different motor inputs. Use your judgment for dimensions not shown. Scale is 1:1.

6. On a B- or A3-size sheet, make a drawing of the speed reduction assembly shown in Fig. 18-3-B. The motor and worm-gear reducer are mounted on a table. The coupling FC15 joins the two. A steel sprocket, mounted directly on the reducer shaft, is to move a chain at an approximate rate of 40 ft per hour. Call out on the assembly drawing the catalog numbers for the coupling and sprocket selected. Scale is half size or 1:2. Show the coupling and sprocket in full section.

NO.	SHAFT Ø	Ø A	B	F	Ø G	L	Ø M
37 FN	.375	1.50	.70	VARIES WITH DIMENSION L	.88	TO SUIT YOUR REQUIREMENTS	.375
50 FN	.50	1.62	1.00		1.00		.50
62 FN	.62	1.75	1.10		1.38		.62
75 FN	.75	2.00	1.20		1.50		.75

FLEXIBLE COUPLING DATA

Fig. 18-1-A   Motor to gearbox drive.

METRIC

PART NO.	SHAFT SIZE	MAX R/MIN	A	B	C	D	H	F	G	N	L
163 A	Ø 40	5000	120	50	100	20	30	200	60	10	3
163 B	Ø 40	5000	120	50	115	20	30	215	60	10	3
163 C	Ø 40	5000	120	50	130	20	30	230	60	10	3
163 D	Ø 40	5000	120	50	145	20	30	245	60	10	3
163 E	Ø 40	5000	120	50	160	20	30	260	60	10	3

COUPLING DATA

Fig. 18-1-B   Motor to pump drive.

MODEL	A	B	C	D	E	F	G MAX	H MAX
F32N	90	85	43	21	60	25	32	25
F41N	116	100	50	25	66	32	40	25
F54N	140	125	55	35	90	32	45	50

METRIC

Fig. 18-2-A   Friction clutch.

SET SCREW
NO. 60
.75 PITCH ROLLER
CHAIN SPROCKET

CATALOG NO.	P.D.	TEETH	HUB DIA	L
KS8	1.96	8	1.50	1.38
KS9	2.19	9	1.65	1.38
KS10	2.43	10	1.94	1.38
KS11	2.66	11	2.10	1.25
KS12	2.90	12	2.10	1.25

SPEED REDUCTION ASSEMBLY

A FULL COMPLEMENT OF SPRAGS BETWEEN CONCENTRIC INNER AND OUTER RACES TRANSMITS POWER FROM ONE RACE TO THE OTHER BY WEDGING ACTION OF THE SPRAGS WHEN EITHER RACE IS ROTATED IN THE DRIVING DIRECTION. ROTATION IN THE OPPOSITE DIRECTION FREES THE SPRAGS AND CLUTCH IS DISENGAGED OR "OVER-RUNS."

F-S SERIES MODEL NUMBER	5	6	8	10	12	14
STANDARD BORE SIZE	.500 .625	.750	.875 1.000	1.125 1.250	1.375 1.500	1.625 1.750
KEYSEAT	.10x.05 .20x.10	20x.10	24x.12	30x.15	.30x.15 40x.20	44x.22
STANDARD HUB KEYWAY	.20x.10	20x.10	24x.12	30x.15	40x.20	44x.22
A	2.75	3.20	3.30	3.60	3.90	4.40
B	2.20	2.90	3.25	3.75	4.45	5.25
C	1.250 1.249	1.375 1.374	1.750 1.749	2.250 2.249	2.500 2.499	2.875 2.874
D	1.25	1.45	1.60	1.75	2.00	2.25
E	1.00	1.25	1.60	2.00	2.40	2.90
F	1.00	1.30	1.30	1.45	1.45	1.60
K	.048 .068	.048 .068	.056 .076	.056 .076	.056 .076	.056 .076
M	.900 .905	1.215 1.220	1.215 1.220	1.340 1.345	1.340 1.345	1.525 1.530
R	1.45	1.55	1.70	1.80	2.10	2.35
S	.55	1.00	.90	1.00	1.40	1.55
OIL HOLE	.25-28	.25-28	.25-28	.25-28	.25-28	.25-28

Fig. 18-2-B   Overrun clutch.

GEAR DATA: 20º SPUR GEAR, PITCH – 4, PITCH DIA – 6.00,
FACE WIDTH – 1.00, HUB   PROJECTION ONE SIDE ONLY – .25,
HUB DIA – 3.50, SHAFT DIA – 1.375.

COUPLING

CATALOG NO.	HOLE DIA	HOLE LENGTH	O.D.	HUB DIA	PROJ	LENGTH
FC 12	.50	.84	1.25	1.00	.60	2.30
FC 15	.50	1.00	1.50	1.25	.75	2.75
FC 20	.50	1.40	2.00	1.75	1.10	3.70

7.00   2.50

Ø .4995 .4990

4.25

MOTOR 1750 RPM

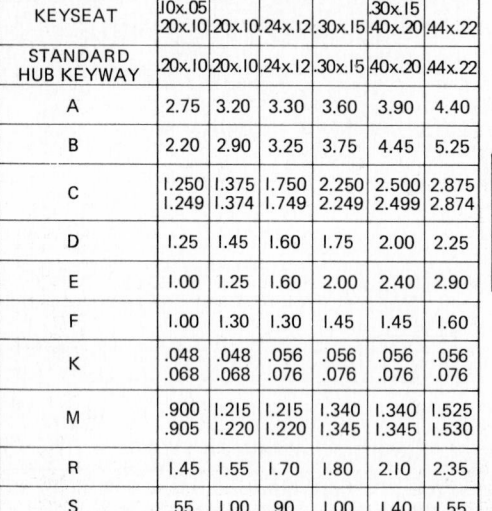

150

50   50

INPUT SHAFT   OUTPUT SHAFT

195   79.375

60

COUNTERSHAFT

GEAR	NUMBER OF TEETH
A	20
B	30
C	20
D	24
E	30
F	26

GEAR DATA:
SHAFT DIA – 20
FACE WIDTH – 15
MODULE – 3.175

METRIC

INPUT R/MIN	OUTPUT R/MIN
1150	
1750	

Fig. 18-3-A   Sliding-gear speed reducer.

9.90

KEY □ .18 X 1.40 LG

5.00   4.90

2.06   1.80

KEY □ .12 X 1.00 LG

OUTPUT   INPUT

4.38   4.25

Ø .8745 .8740   Ø .4995 .4990

.56   4.00   .50

RATIO 1740 : 1
WORM GEAR REDUCER

Fig. 18-3-B   Speed reduction assembly.

CHAPTER 19

Bearings, Lubricants, and Seals

UNIT 19-1
Bearings

Bearings permit smooth, low-friction movement between two surfaces. The movement can be either rotary (a shaft rotating within a mount) or linear (one surface moving along another).

Bearings can employ either a sliding or a rolling action. Bearings based on rolling action are called *rolling-element bearings*. Those based on sliding action are called *plain bearings*.

The basic principles of design and application of antifriction bearings were conceived many centuries ago. They originated for one purpose only —to lessen friction. Through the ages people wanted to move heavy objects across the earth's surface. As far back as 1100 B.C., we know that such friction was reduced by the insertion of rollers between the object and the surface over which it was being moved. The Assyrians and Babylonians used rollers to move enormous stones for their monuments and palaces. Down through history are recorded many similar examples of people's war on friction.

PLAIN BEARINGS

A plain bearing is any bearing that works by sliding action, with or without lubricant. This group encompasses essentially all types other than rolling-element bearings.

Plain bearings are often referred to as *sleeve bearings* or *thrust bearings*, terms that designate whether the bearing is loaded axially or radially.

Lubrication is critical to the operation of plain bearings, so their application and function are also often referred to according to the type of lubrication principle used. Thus, terms such as *hydrodynamic*, *fluid-film*, *hydrostatic*, *boundary-lubricated*, and *self-lubricated* are designations for particular types of plain bearings.

Although some materials have an inherent lubricity or can be lubricated by virtue of a film of slippery solid, most bearings operate with a fluid film —generally oil but sometimes a gas.

By far the largest number of bearings is oil-lubricated. The oil film can be maintained through pumping by a pressurization system, in which case the lubrication is termed *hydrostatic*. Or it can be maintained by a squeezing or wedging of lubricant produced by the rolling action of the bearing itself; this is termed *hydrodynamic* lubrication. The designs shown in Fig. 19-1-1 illustrate simple, effective arrangements for providing supplementary lubrication.

Bearing Types

Journal or Sleeve Bearings These are cylindrical or ring-shaped bearings designed to carry radial loads. See Fig. 19-1-2. The terms *sleeve* and *journal* are used more or less synonymously since *sleeve* refers to the general configuration while *journal* pertains to any

(A) OIL HOLE IN SHAFT

(B) OIL GROOVE IN BEARING

Fig. 19-1-1 Common methods of lubricating plain bearings.

Fig. 19-1-2 Journal or sleeve bearing.

portion of a shaft supported by a bearing. In another sense, however, the term *journal* may be reserved for two-piece bearings used to support the journals of an engine crankshaft.

The simplest and most widely used types of sleeve bearings are cast-bronze and porous-bronze (powdered-metal) cylindrical bearings. Cast-bronze bearings are oil- or grease-lubricated. Porous bearings are impregnated with oil and often have an oil reservoir in the housing.

Plastic bearings are being used increasingly in place of metal. Originally, plastic was used only in small, lightly loaded bearings where cost saving was the primary objective. More recently, plastics are being used because of functional advantages, including resistance to abrasion, and they are being made in large sizes.

Thrust Bearings This type of bearing differs from a sleeve bearing in that loads are supported axially rather than radially. See Fig. 19-1-3. Thin, disklike thrust bearings are called *thrust washers*.

Fig. 19-1-3 Thrust bearings.

Bearing Materials

Babbitts Tin and lead-base babbitts are among the most widely used bearing materials. They have an ability to embed dirt and have excellent compatibility properties under boundary-lubrication conditions.

In bushings for small motors and in automotive engine bearings, babbitt is generally used as a thin coating over a steel strip. For larger bearings in heavy-duty equipment, thick babbitt is cast on a rigid backing of steel or cast iron.

Bronzes and Copper Alloys Dozens of copper alloys are available as bearing materials. Most of these can be grouped into four classes: copper-lead, leaded-bronze, tin-bronze, and aluminum-bronze.

Aluminum Aluminum bearing alloys have high wear resistance, load-carrying capacity, fatigue strength, and thermal conductivity; excellent corrosion resistance; and low cost. They are used extensively in connecting-rods and main bearings in internal-combustion engines; in hydraulic gear pumps, in oil-well pumping equipment, in roll-neck bearings in steel mills; and in reciprocating compressors and aircraft equipment.

Porous Metals Sintered-metal self-lubricating bearings, often called *powdered-metal bearings*, are simple and low in cost. They are widely used in home appliances, small motors, machine tools, business machines, and farm and construction equipment.

Common methods used when supplementary lubrication for oil-impregnated bearings is needed are shown in Fig. 19-1-4.

References and Source Material
1. SKF Company Ltd.
2. *Machine Design*, Mechanical drives reference issue, 1979.

Fig. 19-1-4 Supplementary lubrication for oil-impregnated bearings.

ASSIGNMENTS

See Assignments 1 and 2 for Unit 19-1 on page 392.

Review for Assignments
Unit 7-5   Assembly Drawings in Section
Unit 17-3   Spur Gears
Appendix   Bolts, Setscrews

UNIT 19-2
Antifriction Bearings

Ball, roller, and needle bearings are classified as antifriction bearings since friction has been reduced to a minimum. They may be divided into two main groups: radial bearings and thrust bearings. Except for special designs, ball and roller bearings consist of two rings, a set of rolling elements, and a cage. The cage separates the rolling elements and spaces them evenly around the periphery (circumference of the circle). The nomenclature of an antifriction bearing is given in Fig. 19-2-1.

BEARING LOADS

Radial Load Loads acting perpendicular to the axis of the bearing are called *radial loads*. See Fig. 19-2-2. Although radial bearings are designed primarily for straight radial service, they will withstand considerable thrust loads when deep ball tracks in the raceway are used.

Thrust Load Loads applied parallel to the axis of the bearing are called *thrust loads*. Thrust bearings are not designed to carry radial loads.

Combination Radial and Thrust Loads When loads are exerted both parallel and perpendicular to the axis of the bearings, a combination radial and thrust bearing is used. The load ratings listed in the manufacturers' catalogs for this type of bearing are for either pure thrust loads or a combination of both radial and thrust loads.

BALL BEARINGS

Ball bearings fall roughly into three classes: *radial*, *thrust*, and *angular-*

Fig. 19-2-1　Antifriction bearing nomenclature. (SKF Company)

(A) RADIAL　　　(B) THRUST　　　(C) COMBINATION RADIAL AND THRUST

Fig. 19-2-2　Types of bearing loads.

DEEP GROOVE　SELF–ALIGNING　DOUBLE ROW　ANGULAR CONTACT　THRUST

Fig. 19-2-3　Ball bearings. (SKF Company)

contact. Angular-contact bearings are used for combined radial and thrust loads and where precise shaft location is needed. Uses of the other two types are described by their names: radial bearings for radial loads and thrust bearings for thrust loads. See Fig. 19-2-3.

Radial Bearings

Deep-groove bearings are the most widely used ball bearings. In addition to radial loads, they can carry substantial thrust loads at high speeds, in either direction. They require careful alignment between shaft and housing.

Self-aligning bearings come in two types: internal and external. In internal bearings, the outer-ring ball groove is ground as a spherical surface.

Externally self-aligning bearings have a spherical surface on the outside of the outer ring, which matches a concave spherical housing.

Double-row, deep-groove bearings embody the same principle of design as single-row bearings.

Double-row bearings can be used where high radial and thrust rigidity is needed and space is limited. They are about 60 to 80 percent wider than comparable single-row, deep-groove bearings, and they have about 50 percent more radial capacity.

Angular-contact thrust bearings can support a heavy thrust load in one direction, combined with a moderate radial load. High shoulders on the inner and outer rings provide steep contact angles for high thrust capacity and axial rigidity.

Thrust Bearings

In a sense, thrust bearings can be considered to be 90° angular-contact bearings. They support pure thrust loads at moderate speeds, but for practical purposes their radial load capacity is nil. Because they cannot support radial loads, ball thrust bearings must be used together with radial bearings.

Flat-race bearings consist of a pair of flat washers separated by the ball complement and a shaft-piloted retainer, so load capacity is limited. Contact stresses are high, and torque resistance is low.

One-directional, grooved-race bearings have grooved races very similar to those in radial bearings.

Two-directional, groove-race bearings consist of two stationary races, one rotating race, and two ball complements.

ROLLER BEARINGS

The principal types of roller bearings are *cylindrical, needle, tapered,* and *spherical*. In general, they have higher load capacities than ball bearings of the same size and are widely used in heavy-duty, moderate-speed applications. However, except for cylindrical bearings, they have lower speed capabilities than ball bearings. See Fig. 19-2-4.

Cylindrical Bearings

Cylindrical roller bearings have high radial capacity and provide accurate guidance to the rollers. Their low friction permits operation at high speed, and thrust loads of some magnitude can be carried through the flange-roller end contacts.

Unlike ball bearings, cylindrical roller bearings are generally lubricated with oil; most of the oil serves as a coolant.

Needle Bearings

Needle bearings are roller bearings with rollers that have high length-to-diameter ratios.

Compared with other roller bearings, needle bearings have much smaller rollers for a given bore size.

Loose-needle bearings are simply a full complement of needles in the annular space between two hardened machine components, which form the bearing raceways. They provide an effective and inexpensive bearing assembly with moderate speed capability, but they are sensitive to misalignment.

Caged assemblies are simply a roller complement with a retainer, placed between two hardened machine elements that act as raceways. Their speed capability is about 3 times higher than that of loose-needle bearings, but the smaller complement of needles reduces load capacity for the caged assemblies.

Thrust bearings are caged bearings with rollers assembled like the spokes of a wheel in a waferlike retainer.

Tapered Bearings

Tapered roller bearings are widely used in roll-neck applications in rolling mills, transmissions, gear reducers, geared shafting, steering mechanisms, and machine-tool spindles. Where speeds are low, grease lubrication suffices, but high speeds demand oil lubrication—and very high speeds demand special lubricating arrangements.

Spherical Bearings

Spherical roller bearings offer an unequaled combination of high load capacity, high tolerance to shock loads, and self-aligning ability, but they are speed-limited.

Single-row bearings are the most widely used tapered roller bearings. They have a high radial capacity and a thrust capacity about 60 percent of radial capacity.

Two-row bearings can replace two single-row bearings mounted back to back or face to face when the required capacity exceeds that of a single-row bearing.

BEARING SELECTION

Machine designers have a large variety of bearing types and sizes from which to choose. Each of these types has characteristics which make it best for a certain application. Although selection may sometimes present a complex problem requiring considerable experience, the following considerations are listed to serve as a general guide for conventional applications.

1. Generally, ball bearings are the less expensive choice in the smaller sizes and lighter loads, while roller bearings are less expensive for the larger sizes and heavier loads.
2. Roller bearings are more satisfactory under shock or impact loading than ball bearings.
3. If there is misalignment between housing and shaft, either a self-aligning ball or spherical roller bearing should be used.
4. Ball thrust bearings should be subjected to pure thrust loads only. At high speeds, a deep-groove or angular-contact ball bearing will usually be a better choice even for pure thrust loads.
5. Self-aligning ball bearings and cylindrical roller bearings have very low friction coefficients.
6. Deep-groove ball bearings are available with seals built into the bearings so that the bearing can be prelubricated and thus operate for long periods without attention.

BEARING CLASSIFICATIONS

Because of standardization of boundary dimensions, it is possible to

(A) CYLINDRICAL   (B) TAPERED   (C) SPHERICAL

LOOSE   CAGED

(D) NEEDLE

Fig. 19-2-4   Roller bearings. (The Torrington Co. & Orange Roller Bearing Co.)

replace a bearing by another bearing produced by a different manufacturer without any modification to the existing assembly.

Ball and roller bearings are classified into various series: rigid ball journals, self-aligning ball journals, rigid roller journals, etc. Each series is subdivided into types—extra light, light, medium, and heavy—to meet varying load requirements. Each type is manufactured to a range of standard sizes which are usually represented by the diameter of the bore. Therefore, when a bearing is ordered, the series, type, and size are specified.

Figure 19-2-5 shows a range of bearings to a common bore at A and to a common outside diameter at B. A selection can therefore be made for a given shaft size or for a given housing diameter, and the series selected will depend on the load which is applied to the bearing.

(A) COMMON BORE DIAMETER

(B) COMMON OUTSIDE DIAMETER

Fig. 19-2-5   Standard bearing sizes.

SHAFT AND HOUSING FITS

If a ball or roller bearing is to function satisfactorily, both the fit between the inner ring and the shaft and the fit between the outer ring and the housing must be suitable for the application. The desired fits can be obtained by selecting the proper tolerances for the shaft diameter and the housing bore.

Bearings may be mounted directly on the shaft or on tapered adapter sleeves. When the bearing is mounted directly on the shaft, the inner ring should be located against a shaft shoulder of proper height. This shoulder must be machined square with the bearing seat, and a shaft fillet should be used. The radius of the fillet must clear the corner radius of the inner ring. See Fig. 19-2-6. This also applies when the outer ring is mounted in the housing.

(A) STANDARD   (B) RELIEVED

Fig. 19-2-6   Correct shaft and housing fillet radii.

To hold the bearing inner ring axially on the shaft, a locknut and lockwasher are commonly used. See Fig. 19-2-7. Not only is this method effective and convenient, but nuts and washers specially made for the purpose are also readily obtainable. A tab in the bore of the lockwasher engages a slot in the shaft, and one of the many tabs on the periphery of the washer is bent over into one of the slots in the nut OD.

Instead of a nut, a retaining ring fitted into a groove in the shaft can be used for simple bearing arrangements. See Fig. 19-2-8.

If another machine component, such as a gear or pulley, is fitted alongside the bearing, the inner ring is often secured by means of a spacing sleeve. A sleeve is often used for spacing the inner rings when the bearings are located reasonably close together.

Some bearings are merely mounted against a shoulder without other means of securing the inner ring axially. This is particularly the case where there are no axial forces tending to displace the bearings on the shaft. The housings for the two bearings are rigidly connected, and when thrust occurs, the bearing taking the load is pressed against its shoulder.

On long standard shafting it is impractical to apply bearings, with an interference fit, directly on the shaft. Therefore, they are applied with

(A) LOCKWASHER LOCKNUT

(B) ADAPTOR SLEEVE

(C) WITHDRAWAL SLEEVE

Fig. 19-2-7   Locking devices.

tapered adapter sleeves. The outer surface of the sleeve is tapered to match the tapered bore of the bearing inner ring. This will provide the required tight fit between the inner ring and the shaft. The adapter sleeve is slotted to permit easy contraction and is threaded at the small end to fit a locknut. When the sleeve is drawn up tight between the bearing and the shaft, a press fit is provided at both the shaft and the inner ring.

(A) LOCKNUT

(B) FLOATING SNAP RING

(C) FIXED SPACING SLEEVE

(D) SPACING SLEEVE

(E) SHOULDER MOUNTING

(F) ADAPTOR SLEEVE

(G) WITHDRAWAL SLEEVE

Fig. 19-2-8   Axial mounting of inner rings.

(A) FIXED

(B) FLOATING

Fig. 19-2-9   Outer ring mountings.

If the operating conditions are such that the outer rings can be mounted with a push fit in the housing and *closed bearings* (bearings capable of carrying thrust load in either direction) are used, axial location may be controlled, as shown in Fig. 19-2-9A. The outer ring of the held bearing has a clearance of only .001 to .002 in. (0.05 to 0.1 mm) with the housing shoulders, while the floating bearings (Fig. 19-2-9B) have a free displacement axially in the housing.

One of the most critical factors affecting bearing operation is the mounting fit of the bearing on the shaft and in the housing. If there is any clearance or looseness between the shaft and the bore of the inner ring, the shaft, as it rotates, will roll along the bore of the inner ring. The rolling shaft in the bearing bore will cause the shaft to rapidly wear and become progressively looser. Soon it will become too sloppy for further operation. The best way to prevent this rolling action and wear is to press-fit the inner ring on the shaft.

Similar reasoning applies to a bearing subject to a load which rotates in space with the inner ring. Here, if the outer ring has a clearance in the housing bore, it will roll around the housing bore and wear loose. In this case it would be necessary to have the outer ring press-fit in the housing.

In all cases, it is necessary to press-fit the bearing ring which has relative rotation with respect to the direction of the radial load.

Seals for Grease Lubrication

In order that ball or roller bearings may operate properly, they must be

(A) SHIELD ONLY

(B) SHIELD AND WASHER

Fig. 19-2-10   Bearing seals for grease lubrication.

protected against loss of lubricant and entrance of dirt and dust on the bearing surfaces. In its simplest and least space-requiring form, this is accomplished in some types of bearings by the use of a thin steel shield on one or both sides of the bearing, fastened in a groove in the outer ring and reaching almost to the inner ring as illustrated in Fig. 19-2-10. All other types of bearings require a seal between the bearing housing and the shaft; the types and designs of which are shown in Fig. 19-2-11. Other types of seals are explained in Units 19-4 and 19-5.

Seals for Oil Lubrication

With oil lubrication, the sealing devices have the double function of protecting the bearing against contamination and retaining the lubricant in the housing. Protection is obtained by means of friction seals or flingers, as when grease lubrication is used. The

essential feature for retaining the oil is a groove in the rotating shaft, or a rotating ring or collar from whose edges the oil is thrown by centrifugal force. The oil-groove seal shown in Fig. 19-2-12A retains the oil effectively but should be used only in dry and dust-free places where there is little danger of contamination. Figure 19-2-12B shows examples of labyrinth seals which retain the oil and protect against contamination.

BEARING SYMBOLS

Simplified Representation The simplified representation (general symbol) of rolling bearings (see Fig. 19-2-13) should be used in all types of technical drawings, wherever it is not necessary to show the exact form or size of the rolling bearings or details of their inner design.

Where it is desirable to show the

(A) GENERAL SYMBOL   (B) APPLICATION

(C) WHEN IT IS DESIRABLE TO SHOW CONTOUR FORM

Fig. 19-2-13 Simplified representation of ball and roller bearings.

functional principle of the set of rolling elements, symbols for the appropriate type of rolling element and raceway surface are added. See Fig. 19-2-13C.

Pictorial Representation Pictorial representation of bearings, as shown in Fig. 19-2-14A, is used chiefly in catalogs and magazines. It is not recommended for production drawings because of the extra drafting time required.

Schematic Representation Designers and engineers frequently use schematic layouts in their initial design layout. The schematic diagrams of bearing types and their application are shown in Figs. 19-2-14C and 19-2-15.

(A) FELT RING   (B) GREASE GROOVES   (C) CUFF SEAL   (D) LABYRINTH SEALS

Fig. 19-2-11 Housing seals for grease lubrication.

(A) OIL GROOVES   (B) LABYRINTH SEALS

Fig. 19-2-12 Housing seals for oil lubrication.

References and Source Material
1. *Machine Design*, Mechanical drives reference issue, 1979.
2. SKF Company, LTD.

ASSIGNMENTS

See Assignments 3 through 5 for Unit 19-2 on page 393.

Review for Assignments
Unit 17-3   Spur Gears
Unit 9-3   Retaining Rings
Unit 9-1   Keys

BALL BEARINGS				ROLLER BEARINGS			THRUST BEARINGS		NEEDLE BEARINGS	
RADIAL DEEP-GROOVE	ANGULAR CONTACT	RADIAL DOUBLE ROW	SELF-ALIGNING DOUBLE ROW	CYLINDRICAL	SPHERICAL SELF ALIGNING	TAPERED	BALL	ROLLER	RADIAL	AXIAL

(A) PICTORIAL

(B) SIMPLIFIED

(C) SCHEMATIC

Fig. 19-2-14   Representation of bearings on drawings.

Fig. 19-2-15   Schematic representation of bearings.

UNIT 19-3
Premounted Bearings

Premounted bearing units consist of a bearing element and a housing, usually assembled to permit convenient adaptation to a machinery frame. All components are incorporated within a single unit to ensure proper protection, lubrication, and operation of the bearing. Both plain and rolling-element bearing units are available in a variety of housing designs and for a wide range of shaft sizes, as shown in Figs. 19-3-1 and 19-3-2.

Provision for lubrication is made within the units, and sealing elements retain the lubricant and exclude foreign materials. Some types are prelubricated and sealed at the factory.

Rigid and Self-Aligning Types Rigid premounted units require accurate alignment with the shaft.

Fig. 19-3-1   Adjustable shaft support with journal bearings.

(A) FLANGED HOUSING SELF-ALIGNING-SEALED

(B) PILLOW BLOCK SELF-ALIGNING-SEALED

Fig. 19-3-2   Premounted bearing units.

Self-aligning units compensate for minor misalignment in mounting structures, shaft deflection, and changes which may occur after installation. Self-alignment in sleeve and in some rolling types is accomplished by the use of separate inner housings into which the bearing element is assembled.

Expansion and Nonexpansion Types
Expansion bearings permit axial shaft movement. The principal application for expansion units is in equipment where shafts become heated and increase in length at a greater rate than the structure on which the bearings are mounted.
Nonexpansion bearings restrict shaft movement relative to the mounting structure and keep shaft and attached components accurately positioned. These bearings also serve as thrust bearings within their capacity. Nonexpansion sleeve bearings usually

require collars attached to the shaft at both ends of the housing.
Pillow blocks provide a convenient means of mounting shafts parallel to the surface of a supporting structure. Bolt holes are provided, usually elongated, to permit alignment; and dowel holes are sometimes predrilled for use in maintaining final position on the supporting member. Pillow blocks are available with rigid or self-aligning bearings of expansion or nonexpansion types and with either sleeve or rolling bearings. Housings are either split or solid.

Reference and Source Material
1. *Machine Design*, Mechanical drives reference issue, 1979.

ASSIGNMENTS

See Assignments 6 and 7 for Unit 19-3 on page 394.

Review for Assignments
Unit 2-6   Phantom Outlines
Unit 6-4   Assembly Drawings
Unit 7-1   Full Sections

UNIT 19-4
Lubricants and Radial Seals

LUBRICANTS
There are two main reasons why lubricants are used in any bearing: (1) to reduce friction between rubbing surfaces and (2) as coolants to carry off heat which may be generated in bearings. Either or both of these functions may be required of a lubricant on a particular bearing.

As friction reducers, lubricants can be considered from two aspects. When a hydrodynamic bearing is started, for instance, metal-to-metal contact occurs. Here, the actual oiliness of the lubricant lowers the coefficient of friction between the two sliding surfaces. In slider bearings operating on full fluid-film lubrication, the lubricant separates the two sliding surfaces completely, and shearing of the lubricant is substituted for sliding friction.

Any system of rolling elements, like a ball bearing, should theoretically reduce friction radically. If balls and rollers were perfectly smooth and inelastic, friction would be very low. But materials deform, and rolling elements slip under load. Also, uncaged balls or rollers tend to rub or slide against one another. When a separator or cage is present, the rolling elements slide against it, and the cage itself rubs against any guiding flange surfaces. Because of this sliding, lubrication is needed to minimize wear and friction.

All lubricants can be grouped roughly into three general types: oils, greases, and solid-film lubricants.

Oils and Greases
Whether to use oil or grease and what kind of oil or grease to use are questions that, for slider bearings, must usually be decided early in design phases, since bearing design depends on the lubricant and the type of lubrication selected.

Oils are slippery, hydrocarbon liquids. Grease is a semisolid, combining a fluid lubricant with a thickening agent, usually a soap. In the past, the soaps in greases were considered as storehouses for the oil. Pressure and temperature squeezed out the oil to lubricate bearing surfaces. This is probably only partly true. Soap molecules are attracted to metal surfaces. The long hydrocarbon chain molecule sticks separate the rubbing metal surfaces.

Both oil and grease are used to lubricate rolling- and sliding-contact bearings. In fact, either type of lubricant can be used in some applications, but each type has peculiar assets that equip it for certain types of applications.

Assets of Oil Some of the advantages of oils are:

1. Oil is easier to drain and refill. This is important if lubricating intervals are close together. It is also easier to control the fill volume of the oil in the housing or reservoir.
2. An oil lubricant for a bearing might also be usable at many other points in the machine, even eliminating the need for a second grease-type lubricant.
3. Oil is more effective than grease in carrying heat away from bearing and housing surfaces. In addition, oils are available for a greater range of operating speeds and temperatures than greases.
4. Oil readily feeds into all areas of contact and can carry away dirt, water, and the products of wear.

Assets of Grease Some of the advantages of greases are:

1. Grease does not flow as readily as oil, so it can be more easily retained in a housing. Since grease is easily contained, leakproof designs are unnecessary.
2. Less maintenance is required. There is no oil level to maintain; regreasing is infrequent.
3. Grease has better sealing abilities than oil. This asset may help to keep dirt and moisture out of the housing.

Solid-Film Lubricants
Even the smoothest machined surfaces have microscopic roller-coaster profiles. When one such surface slides on another, the surface irregularities complicate lubrication. Under hydrodynamic conditions, a lubricating liquid may be interposed between the surfaces to prevent them from scraping one peak against another.

When a pure lubricant or a mixture of lubricants is applied as dry powder, grease, or an oil suspension, it is called a *solid lubricant*. When the lubricant is applied in a uniformly thin layer, confining a high concentration of lubricant to a given area, it is called a *bonded dry film*. Inorganic or organic binders and solvents provide the vehicle for these films.

Lubricating Devices
Available lubricating devices range from simple fittings to completely automatic systems. Lubricating devices may be classified as internal or external. The bearings they serve can be lubricated individually or as a group.

Internal devices, or devices that use reservoirs contained within the housing of the bearing or group of bearings, are part of the original equipment design. External devices, or devices that rely on reservoirs in a separate housing, are often proprietary. Frequently, external devices are applied to bearings containing internal lubrication methods to provide additional lubricant.

Individual bearing devices include oil cups, hydraulic grease fittings, and drip oilers. Group methods generally supply lubricant under pressure through a distribution system to a number of bearings.

Hand Lubrication *Hand lubrication* refers to the manual use of any portable or semiportable lubrication equipment for bearing-by-bearing application.

For hand lubrication from portable devices, accessibly located fittings must be provided. For oil-lubricated bearings, the simplest provision is a drilled hole into which fluid lubricant is dripped. To avoid plugging or contamination, a tube or cup with a spring-loaded hinged lid is usually installed.

Fittings must be accessible not only to the coupling on a portable lubricating device, but also for possible field installation of other types of lubrication equipment. Pipe-thread connections are the most universal.

Individual Bearings When life expectancy is satisfactory, certain bearings require no lubrication maintenance—prelubricated, permanently sealed bearings (rolling-element or plain), solid or dry-film lubricated plain bearings, and porous bushings. But the capabilities of even these three types of bearings can be improved by adding internal reservoirs or external lubricating devices.

Internal Reservoirs The oldest method uses the direct contact of grease stored in the cavity of a rolling-element bearing or in the grooves of a plain bearing.

For oil lubrication, felt wicks or wool-waste packings are used to retain the oil and to transfer it to the moving surface by direct contact.

External Reservoirs These devices for individual bearings include drop-feed, constant-level, thermal-expansion, bottle, wick-feed oilers, and pressure grease cups. Only the drop-feed or gravity oiler (Fig. 19-4-1) is in fairly common use.

Fig. 19-4-1 External reservoir lubricating system.

Multiple Bearings A suitable housing or enclosure is required for all internal-reservoir methods for lubricating multiple bearings. This enclosure maintains proper lubricant level and prevents loss of lubricant from within the internal complex. The enclosure must also prevent the entrance of contaminants. Three common types of internal lubricating systems are shown in Fig. 19-4-2.

(A) BATH LUBRICATION (B) SPLASH LUBRICATION (C) SPLASH LUBRICATION WITH PRESSURE LUBRICATING SYSTEM

Fig. 19-4-2 Internal lubricating systems.

GREASE AND OIL SEALS

Factors in the Selection of Oil Seals

Efficient oil seals are available today for every application. But if modern sealing techniques and advancements are to be utilized, oil seals must be selected, inspected, and installed correctly. There are a number of factors to be considered in carrying out these operations.

The environment in which the seal will operate is the most important factor in the selection of materials to be used for the seal. In particular, the user must consider

1. Fluid to be sealed in
2. Fluid to be sealed out
3. Mean temperature of the environment
4. The shaft on which the sealing element runs
5. The sealing element designs for which tooling has been developed

The fluids to be sealed in are usually lubricants. The composition of fluid differs greatly, even within one classification.

The medium to be sealed out is usually air containing varying amounts of dust, gravel, water, slate, etc. Dirt can radically shorten seal life and often dictates the selection of the oil-seal compound. Dirt and dust get under the sealing lip and cause heavy wear and scoring of both the shaft and seal lip. Water and salt cause rusting of the shaft surface, with consequent pitting of the shaft and rapid wear of the seal lip.

The mean temperature of the environment has a radical effect on the seal life and strongly influences the choice of seal compound.

The shaft on which the seal must ride has some influence on the selection of the seal compound. If the shaft is hard and has a surface finish of better than 20 μin. (0.5 μm), any of the compounds can be used. On rougher shafts, the acrylics and silicones wear too rapidly to be used.

Sealing Materials The type of lubricant and the mean operating temperature usually govern the choice of the elastomer (any elastic substance resembling rubber) to be used for the seal compound. Since mean operating temperatures seldom exceed 220°F (105°C), nitrile rubber compounds are the most widely used sealing materials. They wear best, are easiest to mold, and are low in cost.

Silicone compounds are preferred for some applications. Not all silicone compounds are safe to use, however. Most will disintegrate rapidly in many automatic-transmission fluids and in some engine oils.

New fluorelastomer compounds, such as Viton, have a long life at very high temperatures in almost any lubricant. Their cost is high, however. They get quite stiff, but not brittle, at low temperatures.

RADIAL SEALS

Typical seal designs being used today feature both single- and dual-lip sealing elements bonded securely to metal cases that add strength and rigidity to the seals. The bonding of the sealing element to the case eliminates internal leakage resulting from clamping.

For several years, extensive tests and investigations have been conducted to evaluate the effect of various cross-sectional shapes for sealing elements. Several conditions must be satisfied in developing proper shapes, some of which conflict with others.

The sealing element must be flexible enough to follow shaft runout but stiff enough to prevent collapse under operating conditions.

The combination of angles between the trim surface and what is called the *approach angle* is critical. This is particularly true of the angle toward the oil. If an angle is too acute, an otherwise well-designed seal will perform poorly.

Installation The bore must be round and smooth. It must have a proper lead-in chamfer, with a minimum of tool leads and marks, and no tool-return grooves. A bottom should be designed into the bore, and the bore should be concentric with the bearing retention surface.

The shaft should have a chamfer, and, generally speaking, the surface should be approximately 20 μin. (0.5 μm) with above-C45 Rockwell hardness in abrasive applications and above-B80 Rockwell hardness when abrasive conditions are absent.

Felt Radial Seals

Felt is a built-up fabric made by interlocking fibers through a suitable combination of mechanical work, chemical action, moisture, and heat, without spinning, weaving, or knitting. It may consist of one or more classes of fibers—wool, reprocessed wool, or reused wool—which are used alone or combined with animal, vegetable, and synthetic fibers.

Felt has long been used as an important material for sealing purposes. The main reasons are oil wicking, oil absorption, filtration, resiliency, low friction, polishing action, and cost. See Fig. 19-4-3.

Radial Positive-Contact Seals

Radial positive-contact seals are dynamic rubbing seals. Operational effectiveness of a dynamic seal installation was once measured by an easy standard; if it did not leak too much too soon, it was a good seal. Today's operational concepts require sealing effectiveness with absolute minimal leakage over wide service parameters.

A radial positive-contact seal is a device which applies a sealing pressure to a mating cylindrical surface to retain fluids and, in some cases, to

(A) FELT RING IN RECESS. REMOVAL OF SHAFT FOR SEAL REPLACEMENT

(B) FELT RING HELD BY PLATE. EASILY REPLACED

(C) UNIT ASSEMBLY RECOMMENDED WHERE SPACE IS CRITICAL

(D) CUPPED FELT RING. GOOD AGAINST GRIT AND DUST

(E) MACHINED-CARRIER MOUNTING. WIDELY USED WITH BALL AND ROLLER BEARING

Fig. 19-4-3 Felt seal designs.

Fig. 19-4-4 Metal-cased radial seal nomenclature.

(A) LABYRINTH SEAL

(B) BUSHING AND RING SEAL

Fig. 19-4-5 Clearance seals.

exclude foreign matter. Although this definition fits almost all dynamic contact seals, including packings and felt rubbing seals, attention is given in this chapter to the types of seals more commonly known as *oil seals* or *shaft seals*. See Fig. 19-4-4.

The rotating-shaft application of the radial seal is most common. However, the radial seal is also used where shaft motion is oscillating or reciprocating.

Among the factors which recommend a shaft seal over other possible sealing media are ease of installation and small space allocation necessary in design of equipment, relative low cost for high effectiveness, and ability to handle simultaneously a wide range of variables while providing a positive sealing effect throughout.

Types Because of the variety of applications, the radial seal is manufactured in numerous types and sizes. They are generally categorized as

1. Cased seals, where the leather or synthetic sealing element is retained in a precision-manufactured metal case
2. Bonded seals, with the synthetic element permanently bonded to a

flat washer or to a formed-metal case

Seals of both categories can be provided with spring-tension elements, either garter-spring or finger type, for sealing low-viscosity fluids or where either shaft speed or eccentricity demands higher seal contact pressures.

Clearance Seals

Clearance seals limit leakage by closely controlling the angular clearance between the rotating or reciprocating shaft and the relatively stationary housing. There are two basic clearance-seal types: *labyrinth*, and *bushing* or *ring*. These seal types are employed when a small loss in efficiency because of leakage can be permitted. Clearance seals are used when pressure differentials are beyond the design limitations of contact seals (face and circumferential).

Labyrinths Some advantages of labyrinth seals are reliability, simplicity, and flexibility in material selection. They are used mainly in heavy industrial, power, and aircraft applications where relatively high leakage rates may be tolerated and where design simplicity is an absolute necessity.

A labyrinth seal consists of one or more thin strips or knives which are attached to either the stationary housing or the rotating shaft. A simple labyrinth is shown in Fig. 19-4-5A.

Bushing and Ring Seals The bushing-type seal is a close-fitting stationary sleeve within which the shaft rotates. Leakage from a high-pressure station at one end of the bushing to a region of low pressure at the other end is controlled by the restricted clearance between shaft and bushing. Ideally, the bushing and shaft are perfectly concentric, and no rubbing takes place. See Fig. 19-4-5B.

Split-Ring Seals

Split rings are used for a large number of seal applications. See Fig. 19-4-6. Expanding split rings (piston rings) are used in compressors, pumps, and internal-combustion engines. Applications for straight-cut and seal-joint rings are common in industrial and aerospace hydraulic and pneumatic cylinders (linear actuators), where the ruggedness of piston rings is advantageous and where various degrees of leakage can be tolerated.

Axial Mechanical Seals

By convention, the term *axial mechanical seal*, or *end-face seal*, designates a sealing device which forms a running seal between flat, precison-finished surfaces. Used for rotating shafts, the sealing surfaces usually are located in a plane at a right angle to the shaft. Forces which hold the rubbing faces in contact are parallel to the shaft.

(A) ACTION OF MEDIUM ON SPLIT RING SEAL

(B) STRAIGHT-CUT SEAL RING

(C) STEP SEAL RING

Fig. 19-4-6 Split-ring seals.

Axial mechanical seals replace conventional stuffing boxes where a fluid must be contained in spite of a substantial pressure head. These seals have many advantages, such as

1. Reduced friction and power losses
2. Elimination of wear on shaft or shaft sleeve
3. Zero or controlled leakage over a long service life
4. Relative insensitivity to shaft deflection or endplay
5. Freedom from periodic maintenance

Axial mechanical seals do have disadvantages. As precision components they demand careful handling and installation.

While differing in design detail, all mechanical seals make use of the following elements

1. Rotating seal rings
2. Stationary seal rings
3. Spring-loading devices
4. Static seals

The rotating seal ring and the stationary seal ring are spring-loaded together by the spring-loading apparatus, and sealing takes place on the surfaces of these two rings, which rub together. The static seal component of an axial mechanical seal stops leakage

of fluid past the juncture of the rotating seal ring and the shaft.

Since the rotating seal ring is stationary with respect to the turning shaft, sealing at their junction point is accomplished easily through the use of gaskets, O-rings, V-rings, cups, and so forth.

End-Face Seals

The main advantage of an end-face seal is its low leakage rate. For example, the ratio of leakage between mechanical packings and end-face seals averages about 100:1. In addition, the end-face seal causes little wear of the sleeve or shaft on which it seals. Dynamic sealing is created on the seal faces in a vertical plane to the shaft.

The basic development of an end-face seal is shown in Fig. 19-4-7. A shaft with a simple O-ring as its sealing member is provided with a housing that incorporates one of the sealing faces. The housing encloses the O-rings and effects a preload on the shaft, thereby ensuring its sealing. A spring assembly is added to energize the end-face member axially, providing spring pressure against the end-face member to keep the faces together during periods of shutdown or lack of hydraulic pressure in the unit. To complete the

basic seal, a stationary member is incorporated in the end cap of the unit.

A complete seal consists basically of two elements: the seal-head unit, which incorporates the housing, the end-face member, and the spring assembly; and the seal seat, which is the mating member that completes the precision-lapped face combination.

Shaft Sealing Shaft sealing elements include the O-ring, V-ring, U-cup, wedge, and bellows. See Fig. 19-4-8.

The first four of these elements constitute one category—the *pusher-type* seal. As the face wears, these sealing elements are pushed forward along the shaft to maintain the seal.

Pusher-Type Elements For the case of the O-ring, the hydraulic pressure and a mechanical preloading factor provide the sealing effect.

For the V-ring, U-cup, and wedge, the sealing function is created by mechanical and hydraulic means. Mechanical preloading to the shaft is provided by spring action incorporated in the seal design and by hydraulic pressure in the stuffing box.

The V-ring and U-cup designs seal at the shaft surface and at the mating surface of the housing. The sealing action is obtained from both spring force and hydraulic pressure acting against the spreader element, which reacts against

Fig. 19-4-7 Basic end-face seal design.

(A) O-RING (B) V-RING (C) U-CUP (D) WEDGE (E) ELASTOMER (F) CONVOLUTION

Fig. 19-4-8 Shaft seal configurations.

the wings of the seal, spreading it in both directions.

Bellows-Type Elements The bellows-shaped sealing member differs from the pusher type in that it forms a static seal between itself and the shaft. Hence, all axial movement is taken up by bellows flexure.

Molded Packings

Molded packings are often called *automatic*, *hydraulic*, or *mechanical* packings. As a general group, these packings usually do not require any gland adjustment after installation. The fluid being sealed supplies the pressure needed to produce the force for sealing the packings against the wearing surface.

This general classification of packings can be subdivided into two categories: lip and squeeze types.

Lip-Type Packings Lip-type packings of the flange, cup, U-cup, U-ring, and V-ring configurations are used almost exclusively for dynamic applications. Although rotary motions are encountered, the packings discussed here are used primarily for sealing during reciprocating motion. Hence, all the recommendations and designs mentioned apply to reciprocating service. See Fig. 19-4-9.

(A) FLANGED

(B) CUP

(C) U-CUP

(D) U-RING

(E) V-RING

Fig. 19-4-9 Lip-type packings.

Flanged Packings The flange, sometimes called the *hat*, is the least popular of all the lip-type packings.

Cup Packings Leather cup packings, one of the oldest types of lip or mechanical packings, are used in large volume for both hydraulic and pneumatic service at low and high pressures.

Squeeze-Type Packings Squeeze-type molded packings are made in a variety of sizes and shapes (see Fig. 19-4-10), but nearly all of them offer these advantages

D-RING

D-SHAPED RING MAKES GOOD ROD SEAL FOR RECIPROCATING MOTION. PERFORMS EQUALLY WELL IN HYDRAULIC OR PNEUMATIC APPLICATIONS.

DELTA-RING

TRIANGLE-SHAPED RING. SOLVES TWISTING PROBLEM OF O-RING, BUT SINCE FRICTION IS GREATER, EXPECTED LIFE IS RELATIVELY SHORT. HAS LIMITED APPLICATIONS.

O-RING

O-RING IS MOST COMMON FORM OF SQUEEZE PACKING. SEALS IN BOTH DIRECTIONS. HAS LOW INITIAL COST. USED FOR RECIPROCATING, OSCILLATING AND ROTATING MOTION.

T-RING

T-SHAPED RING IS NOT SUSCEPTIBLE TO SPIRAL FAILURES. USED AS ROD OR PISTON SEAL FOR RECIPROCATING MOTION, CAN BE USED FOR OSCILLATING MOTION UNDER LOW PRESSURES.

LOBED RING

SQUARE-SHAPED RING WITH FOUR ROUNDED LOBES. CAN BE USED IN CONVENTIONAL O-RING GROOVES FOR RECIPROCATING, ROTATING, AND OSCILLATING MOTION. SUPERIOR TO O-RING IN MOST ROTATING APPLICATIONS.

Fig. 19-4-10 Squeeze packings.

1. Low initial cost
2. Adaptability to limited space
3. Ease of installation
4. High efficiency
5. No need for adjustment
6. Tolerance to wide ranges of pressure, temperature, and fluids
7. Sealing in both directions
8. Relatively low friction

Squeeze-type packings are generally fitted to a rectangular groove, machined in a hydraulic or pneumatic mechanism. Nomenclature for the dimensions of a squeeze-packing seal groove is identified as that part which applies the *squeeze* to the cross section of the O-ring.

O-Rings O-ring seals work under the principle of controlled deformation. Some slight deformation is given the elastic O-ring in the form of diametral squeeze when it is installed. See Fig. 19-4-11. But it is the pressure from the confined fluid that produces the deformation which causes the elastic O-ring to seal.

There are three types of applications for dynamic O-rings:

1. Reciprocating, where the sealing action is that of a piston ring or a seal around a piston rod.
2. Oscillating, where the seal rotates back and forth through a limited number of degrees or several complete turns. This may be combined with very short reciprocating strokes. The main difference between oscillation and rotation is the amount of motion involved.
3. Rotating, where a shaft turns inside the ID of the O-ring.

Squeeze-type packings are economical and easy to install and can be used whenever conditions permit.

(A) DIAMETRAL SQUEEZE

(B) UNDER PRESSURE

O-RINGS ARE FITTED INTO RECTANGULAR GROOVES IN HYDRAULIC MECHANISMS, AND SEALED BY BEING FORCED, BY PRESSURE, INTO CREVICES.

Fig. 19-4-11 O-rings.

One of the ideal applications of an O-ring is as a piston seal in a hydraulic-actuating cylinder. Another common application uses the O-ring as a valve seat or as a valve stem packing.

(A) GENERAL SYMBOL   (B) APPLICATION

Fig. 19-4-12   Simplified representation of seals.

SEAL SYMBOLS

The simplified representation of seals as shown in Fig. 19-4-12 is recommended for use on drawings, wherever it is not necessary to show the exact form and size of seals.

Where it is desirable to show the functional principle of the seal, symbols for the appropriate type of seal are added. See Fig. 19-4-13.

References and Source Material
1. *Machine Design*, Mechanical drives reference issue, 1979.

ASSIGNMENTS

See Assignments 8 and 9 for Unit 19-4 on page 394.

Review for Assignments
Unit 19-2   Antifriction Bearings
Unit 9-3   Retaining Rings
Appendix   Cap Screws

UNIT 19-5
Static Seals and Sealants

O-RING SEALS

All static O-ring seals are classified as gasket-type seals. Static O-ring seals are generally easier to design into a unit than dynamic O-ring seals. Wider tolerances and rougher surface finishes are allowed on metal mating members. The amount of squeeze applied to the O-ring cross section can also be increased. This type of non-moving seal is used in flanges, flange fittings, flange unions and cylinder end caps, valve covers, plugs, etc.

Groove Design A rectangular groove is the most common for O-rings used as flange gaskets.

The rectangular groove can be machined half in the face plate and half in the flange, or the entire groove can be cut in one member. Various flange-type sealing designs and applications are shown in Fig. 19-5-1. In some flange-gasket designs, a triangular groove can be used to provide ease of machining and consequent reduced cost. Round-bottom grooves are also used.

FLAT NONMETALLIC GASKETS

A gasket creates and maintains a tight seal between separable members of a mechanical assembly. Although a seal may be obtained without a gasket, the gasket promotes an efficient initial seal and prolongs the useful life of an assembly.

Basic flange joints (Fig. 19-5-2) are suitable for all kinds of flat gaskets, plain or jacketed. For moderate pressures up to 200 lb/in.² (1400 kPa) the simple flange joint is applicable.

Metal-to-metal joints are particularly suitable for truly compressible materials, such as cork composition

SHAFT—SEALS		GROOVE—SEALS	
1	ONE TONGUE, LEFT SIDE, SEALING INSIDE.	1	SEALING OUTSIDE, LEFT.
2	ONE TONGUE, RIGHT SIDE, SEALING INSIDE.	2	SEALING OUTSIDE, RIGHT.
3	ONE TONGUE, LEFT SIDE, SEALING OUTSIDE.	3	SEALING OUTSIDE, LEFT, WITH BACKRING.
4	ONE TONGUE, RIGHT SIDE, SEALING OUTSIDE.	4	SEALING OUTSIDE, RIGHT, WITH BACKRING.
5	ONE TONGUE, LEFT SIDE, WITH DUST TONGUE, SEALING INSIDE.	5	SEALING INSIDE, LEFT.
6	ONE TONGUE, RIGHT SIDE, WITH DUST TONGUE, SEALING INSIDE.	6	SEALING INSIDE, RIGHT.
7	ONE TONGUE, RIGHT SIDE, WITH DUST TONGUE, SEALING OUTSIDE.	7	SEALING INSIDE, LEFT, WITH BACKRING.
8	ONE TONGUE, LEFT SIDE, WITH DUST TONGUE, SEALING OUTSIDE.	8	SEALING INSIDE, RIGHT, WITH BACKRING.
9	TONGUES, LEFT AND RIGHT, SEALING OUTSIDE.	9	SEALING INSIDE AND OUTSIDE.
10	TONGUES, LEFT AND RIGHT, SEALING INSIDE.	10	SEALING LEFT AND RIGHT.

Fig. 19-4-13   Functional representation of seals.

APPLICATIONS OF AN O-RING TO CYLINDER HEAD COVERS, BLIND FLANGES, ETC. THE HIGHER THE PRESSURE, THE TIGHTER THE SEAL. THIS DESIGN AUTOMATICALLY PRE-LOADS THE O-RING IN THE GROOVE. NUTS ARE TIGHTENED ONLY ENOUGH TO MAINTAIN METAL-TO-METAL SURFACE CONTACT. THIS TYPE OF INSTALLATION WILL SEAL HIGH PRESSURES WITHOUT THE EXCESSIVE BOLT STRESS NECESSARY WITH CONVENTIONAL GASKETED JOINTS.

TWO O-RINGS SIZES USED FOR SEALING A RECTANGULAR PRESSURE CHAMBER. THE OUTSIDE O-RING IS STRETCHED IN A GROOVE AND THE CAPSCREW HEADS ARE SEALED BY SMALL O-RINGS IN COUNTER-BORE. THIS DESIGN IS A SIMPLE AND EFFECTIVE METHOD OF SEALING X-RAY HEADS, GEAR PUMP END-PLATES AND OTHER APPLICATIONS WITHOUT REQUIR-ING LAPPED SURFACES.

O-RING GASKET AS USED ON A FLANGE UNION. THE O-RING MAKES A TIGHT SEAL WHEN THE UNION IS SCREWED DOWN FINGER-TIGHT. THE ROUND-BOTTOM GROOVE HAS THE SAME DIAMETER AS THE ACTUAL CROSS SECTION OF THE O-RING. THE O-RING PROTRUDES .02 TO .03 IN. (0.39/0.79 mm) ABOVE THE FACE OF THE FLANGE. THE VOLUME OF THE GROOVE IS CALCULATED TO BE THE SAME AS THE MINIMUM VOLUME OF THE O-RING.

A MODIFICATION OF THE FLANGE-TYPE GROOVE. THIS TYPE OF TRIANGULAR GROOVE IS USED WHERE EASE OF MACHIN-ING AND REDUCED COST ARE IMPORTANT. THIS TYPE OF DESIGN MAKES AN EFFECTIVE SEAL. HOWEVER, THE O-RING IS PERMANENTLY DEFORMED. PRESSURES ARE LIMITED ONLY BY THE CLEARANCE BETWEEN THE MATING METAL SURFACES AND THE STRENGTH OF THE METAL ITSELF.

Fig. 19-5-1   Flange-type static O-ring seal designs.

and cork-and-rubber. These joints bear a great similarity to joints designed for rubber O-rings. Gasket design considerations are shown in Fig. 19-5-3.

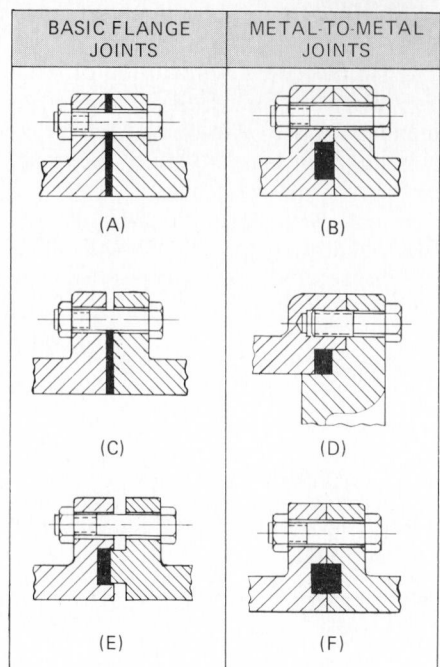

BASIC FLANGE JOINTS	METAL-TO-METAL JOINTS
(A)	(B)
(C)	(D)
(E)	(F)

Fig. 19-5-2   Flat gasket joints.

METALLIC GASKETS

Metallic gaskets are used for high pressures and for temperature extremes that cannot be handled by nonmetallic gaskets.

Solid metal gaskets generally require thick flanges. Thinner flanges can be used with metal O-rings. The rings are made of thin-walled metal tube, bent and welded to form a continuous circle.

SEALANTS

Sealants are used to exclude dust, dirt, moisture, and chemicals or to contain a liquid or gas. They can also protect against mechanical or chemical attack, exclude noise, improve appearance, and act as an adhesive. See Fig. 19-5-4.

Sealants are generally used for less severe conditions of temperature and pressure than gaskets. Sealants are categorized as hardening and nonhardening.

DETAIL	WHY FAULTY	SUGGESTED REMEDY
BOLT HOLES CLOSE TO EDGE	CAUSES BREAKAGE IN STRIPPING AND ASSEMBLY	PROJECTION OR "EAR" / NOTCH INSTEAD OF HOLE
VERY SMALL BOLT HOLES OR NONCIRCULAR OPENINGS	REQUIRE HAND PICKING...EASY TO MISS	AVOID HOLE SIZES UNDER 0.06 DIA. IF SMALL HOLE IS FOR LOCATING OR INDEXING, CHANGE TO NOTCH
TEAR-AWAY PARTS WITH OPEN SLOTS AT ATTACHED EDGES	SLOTS REQUIRE HAND PICKING, COSTLY DIES AND DIE MAINTENANCE	SIMPLE PERFORATION
THIN WALLS, DELICATE CROSS SECTION IN RELATION TO OVER-ALL SIZE	HIGH SCRAP LOSS; STRETCHING OR DISTORTION IN SHIPMENT OR USE. RESTRICTS CHOICE TO HIGH TENSILE STRENGTH MATERIALS	HAVE THE GASKET IN MIND DURING EARLY DESIGN STAGES
METALWORKING TOLER-ANCES APPLIED TO GASKET THICKNESS, DIAMETERS, LENGTH, WIDTH, ETC. ∅ 2.002 / 1.998	RESULTS IN PERFECTLY USABLE PARTS BEING REJECTED AT IN-COMING INSPECTION. REQUIRES TIME AND CORRESPONDENCE TO REACH AGREEMENT ON PRACTICAL LIMITS, INCREASES COST OF PARTS AND TOOLING. DELAYS DELIVERIES	MOST GASKET MATERIALS ARE COM-PRESSABLE. MANY ARE AFFECTED BY HUMIDITY CHANGES. TRY STANDARD OR COMMERCIAL TOL-ERANCES BEFORE CONCLUDING THAT SPECIAL ACCURACY IS REQUIRED
TRANSFERENCE OF FILLETS, RADII, ETC., FROM MATING METAL PARTS TO GASKET	UNLESS PART IS MOLDED, SUCH FEATURES MEAN EXTRA OPERATION AND HIGHER COST	MOST GASKET STOCKS WILL CON-FORM TO MATING PARTS WITHOUT PRESHAPING. BE SURE RADII, CHAMFERS, ETC., ARE FUNCTIONAL, NOT MERELY COPIED FROM METAL MEMBERS
LARGE GASKETS MADE IN SECTIONS WITH BEVELLED JOINTS	EXTRA OPERATION TO SKIVE. EXTRA OPERATION TO GLUE. DIFFICULT TO OBTAIN SMOOTH, EVEN JOINTS WITHOUT STEPS OR TRANSVERSE GROOVES	DIE-CUT DOVETAIL JOINT

Fig. 19-5-3   Common faults in gasket design and suggested remedies.

(A)　　　　　　　　　(B)　　　　　　　　　(C)

BUTT JOINT: USE SEALANT IF THICKNESS OF PLATE IS SUFFICIENT (A), OR BEAD SEALED (B), IF PLATES ARE THIN. TAPE CAN ALSO BE USED, (C). IF JOINT MOVES DUE TO DYNAMIC LOADS OR THERMAL EXPANSION AND CONTRACTION, A FLEXIBLE SEALANT WITH GOOD ADHESION MUST BE SELECTED. SELECT FLEXIBLE TAPE FOR BUTT JOINT IF MOVEMENT IS ANTICIPATED.

(A)　　　　　　　　　(B)　　　　　　　　　(C)

LAP JOINT: SANDWICH SEALANT BETWEEN MATING SURFACES, AND RIVET, BOLT, OR SPOT WELD SEAM TO SECURE JOINT (A). THICK PLATES CAN BE SEALED WITH A BEAD OF SEALANT (B), AND TAPE CAN ALSO BE USED (C), IF SUFFICIENT OVERLAP IS PROVIDED AS A SURFACE TO WHICH THE TAPE CAN ADHERE.

(A) t　　　(B)　　　　(C)　　　　(D)

POOR　　　GOOD　　　BETTER　　　BEST

ANGLE JOINT: SIMPLE BUTT JOINT CAN BE SEALED AS SHOWN IN (A), IF MATERIAL THICKNESS (t) IS SUFFICIENT. BUT BETTER CHOICE IS BEAD OF SEALANT SHOWN IN (B), WHICH IS INDEPENDENT OF MATERIAL THICKNESS. SUPPORTED ANGLE JOINTS WITH BEAD (C), OR SANDWICH SEAL (D), ARE BETTER CHOICES.

Fig. 19-5-4　Common methods for sealing joints.

(A) WIPER TYPE

(B) SCRAPER TYPE

(C) AXIAL TYPE

SEWN LEATHER　　　　SYNTHETIC RUBBER

(D) BOOT TYPE

Fig. 19-5-5　Exclusion seals.

Hardening sealants may be either rigid or flexible, depending on their composition. Nonhardening types are characterized by plasticizers that come to the surface continually, so that the sealant stays "wet" after application.

EXCLUSION SEALS

Exclusion seals are used to prevent the entry of foreign material into the moving parts of machinery. See Fig. 19-5-5. This protection is necessary because foreign material entry contaminates the lubricant and accelerates wear and corrosion. Static joints are easily sealed by tight fits and gaskets. Sealing between parts having relative motion, such as between a housing and a moving shaft, is more difficult.

Sometimes seals designed only for inclusion are used to perform the inclusion and exclusion functions simultaneously. This is inadvisable, except under very light service conditions. Inclusion seals usually do a poor exclusion job and are damaged by even small amounts of abrasive material.

Exclusion seals can be classified into four general groups: wiper, scraper, axial, and boot seals.

References and Source Material

1. *Machine Design*, Mechanical drives reference issue, 1979.

ASSIGNMENTS

See Assignments 10 and 11 for Unit 19-5 on page 397.

Review for Assignments
Unit 8-3　Spotfacing
Appendix　Cap Screws

ASSIGNMENTS for Chapter 19

Assignments for Unit 19-1, Bearings

1. On a B- or A3-size sheet, complete the assembly drawings shown in Fig. 19-1-A or 19-1-B. The shaft, shown in the right view, is supported by two plain bearings press-fitted into the shaft support and lubricated by means of an oil fitting. Select suitable bolts and fasten the shaft support to the mounting plate.

The shafts for the gear box are supported by two plain bearings press-fit-ted into the gear box. Setscrew collars, as shown in the Appendix, are mounted on the shafts to prevent lateral movement. Select suitable 16 DP (1.59 DPL), 20°-spur gears from manufacturers' catalogs which revolve the smaller shaft 4 times as fast as the larger shaft. Lock the gears to the shafts using setscrews and flats on the shaft. Scale is full or 1:1.

2. On a B- or A3-size sheet, complete the assembly drawings shown in Fig. 19-1-C or 19-1-D. For the assembly shown on the left, the largest portion of the shaft is positioned in the housing by a combination journal and thrust bearing. Complete the internal design of the housing to properly locate and support the bearing.

The shaft shown in the assembly on the right rests on a thrust bearing. Complete the housing detail, and show the bearing in position.

	A	B	C
19—1—A INCH	1.875	.375	.625
19—1—B METRIC	47.63	10	16

Fig. 19-1-A, B   Journal bearings.

	A	B	C	D
19—1—C INCH	2.00	.75	1.00	2.50
19—1—D METRIC	50	20	25	60

Fig. 19-1-C, D   Thrust and journal bearings.

Assignments for Unit 19-2, Antifriction Bearings

3. On a B- or A3-size sheet, complete the gearbox assembly drawing shown in Fig. 19-2-A or 19-2-B. Gears are mounted on shafts *A* and *B* and are positioned and held to the shafts by Woodruff keys and setscrews. The shafts are supported by radial ball bearings which are positioned on the shafts with retaining rings. The bearings are to be positioned and held to the housing by internal shoulders on the castings and by cover plates bolted to the housing. Each shaft will have one floating and one fixed outer ring mounting. The bearings will be purchased with seals on one side. From the information given, select suitable keys, bearings, retaining rings, and gears from the Appendix or manufacturers' catalogs. *Note*: Shaft *A* must be able to be removed from the housing with the gear in position. Scale is full or 1:1.

4. On a B- or A3-size sheet, complete the gearbox assembly drawing shown in Fig. 19-2-C or 19-2-D. Gear 1 and the shaft are cast as a single unit. Gears 2, 3, 6, and 7 are fastened to their respective shafts by keys and are held in location by retaining rings. Gears 4, 5, and 8 are formed as one part which slides along the lay-shaft meshing with gear 3, 6, or 7. Retaining rings located at each end of this sliding-gear assembly locate it in the three positions, and a key locks the assembly to the shaft. Radial ball bear-

Fig. 19-2-A, B Ball bearings.

ings are positioned at points *A* and *B* on each shaft. Each shaft will have one floating and one fixed outer ring mounting. The gear end of the primary shaft must be designed to house bearing *A* of the mainshaft. Refer to manufacturers' catalogs or the Appendix for standard parts. Scale is full or 1:1.

	⌀ A	⌀ B	C	GEAR DATA
19-2-A INCH	.785	1.00	3.000	DP 10 SHAFT A N=15 SHAFT B N=45
19-2-B METRIC	20	25	76.2	MODULE 2.54 SHAFT A N=15 SHAFT B N=15

ASSIGNMENT	SHAFT	A	B	GEAR DATA							
19-2-C INCH	PRIMARY	.394	.591	PITCH=10 20° SPUR GEAR FACE WIDTH=.60							
	MAINSHAFT	.394	.787	GEAR 1	2	3	4	5	6	7	8
	LAYSHAFT	.591	.591	N	20	40	25	35	30	30	35	25
19-2-D METRIC	PRIMARY	10	15	MODULE=2.54 20° SPUR GEAR FACE WIDTH=16							
	MAINSHAFT	10	20	GEAR 1	2	3	4	5	6	7	8
	LAYSHAFT	15	15	N	20	40	25	35	30	30	35	25

Fig. 19-2-C, D Gearbox.

Fig. 19-2-E   Lathe. (Timken Roller Bearing Co.)

Fig. 19-2-F   Honing gearbox. (Timken Roller Bearing Co.)

5. On a B- or A3-size sheet, make a schematic presentation of bearings, gears, and belts, similar to Fig. 19-2-15, of one of the assemblies shown in Fig. 19-2-E or 19-2-F. Scale is to suit.

Assignments for Unit 19-3, Premounted Bearings

6. On a B- or A3-size sheet, make a two-view (front- and side-view) assembly drawing of the adjustable shaft support shown in Fig. 19-3-A. Draw the front view in full section. Include on your drawing a bill of material. Scale is full or 1:1.

7. On a B- or A3-size sheet, make a one-view assembly drawing of the adjustable shaft support shown in Fig. 19-3-B. Show the bearing housing in its lowest position and a phantom outline of the bearing housing in its top position. Show only those dimensions that would be used for catalog purposes. Scale is full or 1:1.

Assignments for Unit 19-4, Lubricants and Radial Seals

8. On a B- or A3-size sheet, complete the two assemblies shown in Fig. 19-4-A or 19-4-B given the following information.

Radial Oil-Seal Assembly. The inner ring of the tapered roller bearing is held laterally on the shaft by the shaft shoulder. A cover plate which is bolted to the housing by four socket-head cap screws has a stepped shoulder the same diameter as that of the outside diameter of the bearing. This shoulder serves two purposes: it locks the outer ring of the bearing in position, and it locates the cover plate radially on the shaft. The cover plate has a recess to accommodate a metal-cased radial seal. The shaft diameter for the oil seal should be slightly smaller than the diameter of the shaft for the bearing. The outside face of the cover plate and the oil seal should be flush. Scale is full or 1:1.

Oil Ring Seal Assembly. A magnetized ring press-fitted into the housing firmly holds the mating ring on the shaft element by magnetic force. The carbon ring in the face of the mating ring, in balanced contact with the lapped surface of the magnet, forms a permanent, self-adjusting face seal. O-rings in both the mating ring and the magnetic shaft element (between the element and housing) prevent leakage of confined fluids. Scale is twice size or 2:1.

Fig. 19-3-A Adjustable shaft support.

PT 4 BEARING – X 4012
PT 5 GUIDE PIN MATL – .18 DRILL ROD X 1.00 LG, 2 REQD
PT 6 BASE PIN MATL – .18 DRILL ROD X 1.25 LG, 2 REQD
PT 7 OIL CAP – XGF–D5
PT 8 BOLT, HEX HD REGULAR – .25 – 20 NC X 1.62 LG, 2 REQD
PT 9 NUT, HEX HD REGULAR – .25 – 20 UNC, 2 REQD

BEARING AND HOUSING DETAIL

HOLE DETAIL

PT 2 BEARING HOUSING MATL–CI 1 REQD

.06 X 45 CHAMFER BOTH ENDS

PT 3 SHAFT MATL–CRS 2 REQD

PT 1 BASE MATL–CI 1 REQD

ROUNDS AND FILLETS R 3

PT 6 SET SCREW SLOTTED HEADLESS, CONE POINT M10 X 30 LG 3 REQD

PT 8 HEX HD JAM NUT M10 2 REQD

PT 2 VERTICAL SHAFT MATL–STEEL 1 REQD

PT 3 YOKE MATL–CI 1 REQD

φ 6 X 90° CSK 3 HOLES SPACED AT 90°

φ 20 SLIDE FIT FOR PT 2

ROUNDS AND FILLETS R 3

φ 8 SLOTS

PT 1 BASE MATL–CI 1 REQD

PT 5 BEARINGS MATL–BRONZE 2 REQD

φ 20 +0.02 0

φ 25 PRESS FIT IN PT 4

PT 4 BEARING HOUSING MATL–STEEL 1 REQD

PT 7 SET SCREW SLOTTED HEADLESS HEADER POINT M10 X 15 LG 2 REQD

NOTE – DIAMETERS SHOWN FOR SLIDE AND PRESS FITS ARE NOMINAL DIMENSIONS

METRIC

Fig. 19-3-B Adjustable shaft support.

	A	B	C	D
19–4–A INCH	.75	1.25	1.50	.75
19–4–B METRIC	20	30	38	19

INCH
0   .50   1.00

MILLIMETERS
0   10   20

HOUSING

ØA

ØB

INCH
0   .50   1.00   1.50

MILLIMETERS
0   10   20   30   40

RADIAL OIL SEAL ASSEMBLY

ØC

ØD

MATING RING

MAGNETIC SHAFT ELEMENT

HOUSING

OIL RING SEAL ASSEMBLY

Fig. 19-4-A, B Oil seals.

NOTE – ALL DIMENSIONS SHOWN ARE NOMINAL: CLEARANCES, TYPES OF FIT, O-RINGS, RETAINING RINGS, PACKING, SEALS, ETC. TO BE SELECTED BY STUDENT

(A) BRONZE BUSHING WITH FELT RING PACKING

(C) SQUEEZE PACKING (O-RING) HELD IN GROOVE IN PISTON

(B) THREE GROOVES IN PISTON SURFACE TO ACCOMODATE SPLIT-RING SEALS

(D) THREE INTERNAL RETAINING RINGS HELD IN ONE GROOVE IN HOUSING TO HOLD CYLINDER HEAD AGAINST SHOULDER

Fig. 19-4-C   Hydraulic cylinder.

	Ø A	Ø B	Ø C	Ø D	Ø E	F	G	Ø H	Ø J	K	L	M	N	Ø P	Ø R
19—5—A INCH	2.90	2.00	1.00	3.75	4.50	.25	1.80	4.75	3.50	.375	.60	.40	1.50	2.00	3.50
19—5—B METRIC	73	50	25	95	110	M6	45	120	90	M10	15	10	40	50	90

(K) HEX HD CAP SCREWS 4 EQUALLY SPACED

(F) HEX BOLTS & NUTS 4 EQUALLY SPACED

FLANGED PIPE COUPLING

CYLINDER-HEAD CAP

Fig. 19-5-A, B   Flanged pipe coupling and cylinder-head cap.

9. On a B- or A3-size sheet, complete the hydraulic cylinder assembly shown in Fig. 19-4-C. The sealing and fastener requirements are listed with the drawing. Add a bill of material calling out the standard sealing and fastener parts. Scale is full or 1:1.

Assignments for Unit 19-5, Static Seals and Sealants

10. On a B- or A3-size sheet, complete the two assemblies shown in Fig. 19-5-A or 19-5-B, given the following information.

Flanged Pipe Coupling. The flanges are fastened by hex bolts, nuts, and lockwashers. Alignment is accomplished by a tongue-and-groove joint, similar to that shown in Fig. 19-5-2E, and a gasket positioned in the groove provides the seal.

Cylinder Head Cap. The cylinder head cap is fastened to the cylinder head by four hex-head cap screws equally spaced. Locking is accomplished by lockwashers. Spotfacing on the cast head cap is required because of the rough finish of the casting. An O-ring provides the seal.

11. On a B- or A3-size sheet, make a detail drawing of the gasket used with the domed cover shown in Fig. 19-5-C. Material is neoprene. Scale is 1:2 or half size.

Fig. 19-5-C   Domed cover.

CHAPTER 20

Cams, Linkages, and Actuators

UNIT 20-1
Cams and Cam Motions

A *cam* is a machine element designed to generate a desired motion in a follower by means of direct contact. Cams are generally mounted on rotating shafts, although they can be used so that they remain stationary and the follower moves about them. Cams may also produce oscillating motion, or they may convert motions from one form to another.

The shape of a cam is always determined by the motion of the follower. The cam is actually the end product of a desired follower movement. From the standpoint of engineering alone, cams have many decided advantages over the fundamental kinematic four-bar linkages. See Fig. 20-1-1. Once they are understood, cams are easier to design and the action produced by them can be more accurately forecast. For example, to cause the follower system to remain stationary during a portion of a cycle is very difficult when linkages are used. With a cam this is accomplished by a contour surface which runs concentric with the center of rotation. To produce a given motion, velocity, or acceleration during a specific portion of a cycle is very difficult to do with linkages, whereas it

Fig. 20-1-1   Cam application. (Manifold Machinery Co.)

is comparatively easy with standard cam motions, especially when the design is achieved with the aid of a computer. See Fig. 20-1-2.

By far the most popular types of cams are the OD or plate cam and the drum or cylinder cam. In the case of the OD cam, the body of the cam is usually shaped like a disk with the cam contour developed along its circumference. With these cams the line of action of a follower is generally perpendicular to the cam axis. With the drum cam, the cam track is generally machined around the circumference of the drum. In this type of cam the line of

action is normally parallel to the cam axis. The level-winding mechanism on a fishing reel is an example of a drum cam. Other popular types of cams include the conjugate cam (multiple cams joined together); the face cam, in which the cam track is cut into the face of the disk; and the index cam, which is similar to a drum cam except that the motion of the follower passes in an arc over the cam itself.

As machine speeds increase, the need for properly designed quality cams becomes more evident. The essential specifications necessary to produce a cam of optimum quality are:

(A) OD OR PLATE CAM

(B) BARREL (DRUM OR CYLINDER) CAM

(C) CONJUGATE CAM

(D) FACE CAM

(E) COMBINATION DRUM AND PLATE CAM

(F) INDEX CAM

Fig. 20-1-2 Common cams.

1. Proper dynamic design which considers the velocity, acceleration, and jerk characteristics of the follower system. These include vibration and shaft torque analysis.
2. Proper material selection which takes into account cost, wear, and surface stresses produced by the system.

CAM NOMENCLATURE

(Refer to Figs. 20-1-3 and 20-1-4)

1. *Follower displacement* is generally defined as the position of the follower mechanism from a specific zero or rest position in relation to time or some fraction of the machine cycle (cam displacement) measured in degrees or inches (millimeters).

2. *Cam displacement*, measured in degrees or inches (millimeters), is the cam motion measured from a specific zero or rest position and relates to the follower mechanism as defined above.
3. *Cam profile* is the actual working surface contour of cam.
4. *Base circle* is the smallest circle drawn to the cam profile.
5. *Trace point* is the center line of the follower roller or its equivalent. When a flat follower is used, the cam profile is the envelope of successive positions of the flat follower.
6. *Pitch curve* is the locus of successive positions of the trace point as cam displacement takes place.
7. *Prime circle* is the smallest circle drawn in the pitch curve from the cam center. It is related to the base circle by the roller radius.
8. *Pressure angle* is the angle between the normal to the pitch curve and the instantaneous direction of motion of the follower.
9. *Pitch point* is the position on the pitch curve where the pressure angle is maximum.
10. *Pitch circle* is the circle which passes through the pitch point.
11. *Transition point* is the position of maximum velocity where acceleration changes from plus to minus (force on follower changes direction). In a closed cam this is sometimes referred to as the *crossover point*, where, because of the reversing acceleration, the follower roller leaves one cam profile and crosses over to the opposite (or conjugate) one.

Figure 20-1-5 illustrates typical cam and follower combinations used in machine design.

Fig. 20-1-3 Cam nomenclature.

Fig. 20-1-4 Cam displacement diagram.

FOLLOWER
MOTION

BASE
CIRCLE

FLAT FACE FOLLOWER

FOLLOWER MOTION

FOLLOWER

(A) RADIAL

FOLLOWER MOTION

OFFSET

FOLLOWER

(B) OFFSET RADIAL

SWINGING FOLLOWER

RADIAL FOLLOWERS

NO. I ROLLER AND CAM

NO. 2 ROLLER AND CAM

FOLLOWER
MOTION

CONJUGATE RADIAL
DUAL ROLLER FOLLOWERS

CLOSED CAM FOLLOWER

SPRING LOADED
CONJUGATE CAM ROLLERS

FOLLOWER

NO. I CAM

NO. 2 CAM

CONJUGATE SWING ARM
DUAL ROLLER FOLLOWERS

INDEX CAM FOLLOWER

Fig. 20-1-5 Typical cam and follower combinations.

CAM FOLLOWERS

The common types of cam followers are shown in Fig. 20-1-6. The roller follower is more suitable where high speeds, heat (friction), and wear are factors. See Fig. 20-1-7.

FOLLOWER MOTION

ROUND   FLAT   ROLLER   OFFSET ROLLER

Fig. 20-1-6 Types of cam followers.

CAM MOTIONS

In the early phases of the development of the cam mechanism, it is customary to work with only center lines to establish the desired motions. It is obvious that some data have been specified or determined from related parts of the design to establish the cam and linkage requirements and to provide base points from which to start the cam linkage design. These data will usually be the motion requirements and timing relationships of a particular part of the machine such as a feed slide, a folding mechanism, or a label applicator.

The choice of motion that the cam must produce will depend, first, on the cycle timing and, second, on the system or machine dynamics. For the purpose of showing cam layout techniques, cams producing the following motions will be discussed

1. Uniform motion
2. Parabolic motion
3. Harmonic motion
4. Cycloidal motion
5. Modified trapezoidal motion
6. Modified sine motion
7. Synthesized, modified sine-harmonic motion

Fig. 20-1-7 Cam-roller followers.

(A) APPLICATION

(B) YOKE MOUNTINGS FOR ROLLER BEARINGS

(C) ROLLER FOLLOWER

(D) ROLLER BEARINGS

time. The construction of the parabolic curve as shown in Fig. 20-1-8C is found in the same manner as detailed in Fig. 4-5-2.

By using the uniformly accelerated and retarded method of construction for this motion, the divisions will increase and decrease by a ratio of 1:3:5:5:3:1. For instance, a follower is to rise 2.25 in. in 180°. Plotting points every 30° and using six proportional divisions of 1:3:5:5:3:1, we find in the first 30° the follower rises one-eighteenth of the total rise of 2.25 in., or .125 in.; in the next 30° the follower rises three-eighteenths of the rise of 2.25 in., or .375 in., and in the third 30° the follower rises five-eighteenths of the rise of 2.25 in., or .625 in.; the fourth, fifth, and last rises being .625, .375, and .125 in., respectively. This motion would produce a jerk if used in connection with a cam having a dwell.

The first four are illustrated in Fig. 20-1-8.

Uniform Motion (Constant Velocity Motion)

Uniform motion is used when the follower is required to rise and drop at a uniform rate of speed. If a follower is to rise 1.50 in. in one-half of a revolution, or 180° of the cam, then for every 30° of cam rotation the follower would rise one-sixth of 1.50 in., or .25 in.

This curve is also referred to as a *straight-line motion*, and it is most commonly used in connection with screw machines to control the feed of a cutting tool. If it were used with a dwell area in a cam, there would be a jerk at the start and stop of the motion.

Since this kind of motion starts and ends abruptly, it is often modified slightly to reduce the shock on the follower. A radius is used at the beginning and end of the motion, and a line tangent to these arcs is drawn. The size of the radius varies between one-third and full-rise height depending on how sharp the rise is. This motion is known as *modified uniform motion*. Since this type of motion is not desirable for high speeds, motions that start and end slowly, reaching their maximum speed in the center, are used.

Parabolic Motion

Parabolic motion, commonly referred to as *uniformly accelerated and retarded motion*, or *constant acceleration*, is described by a curve found by taking the distance traveled and making it proportional to the square of the

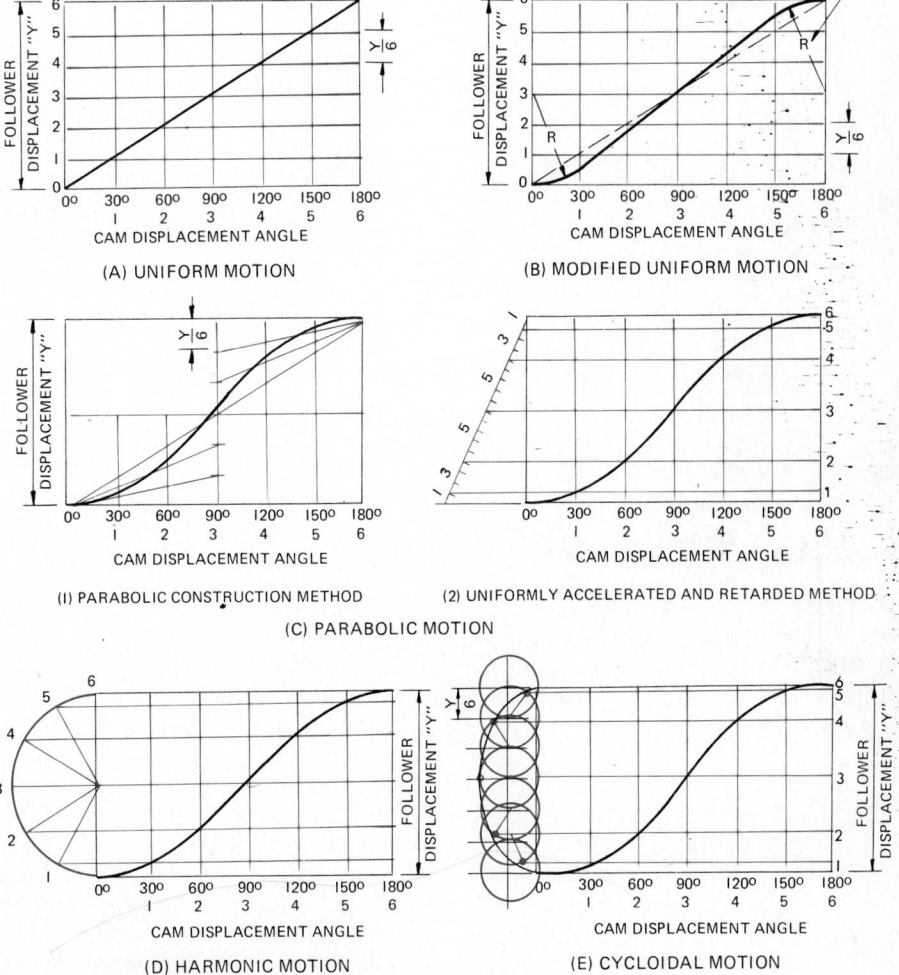

(A) UNIFORM MOTION

(B) MODIFIED UNIFORM MOTION

(I) PARABOLIC CONSTRUCTION METHOD

(2) UNIFORMLY ACCELERATED AND RETARDED METHOD

(C) PARABOLIC MOTION

(D) HARMONIC MOTION

(E) CYCLOIDAL MOTION

Fig. 20-1-8 Cam motions.

Harmonic Motion

This motion, often referred to as *crank motion*, is produced by a true eccentric operating against a flat follower whose surface is normal to the direction of linear displacement. Figure 20-1-9 illustrates this type of cam. However, it is more frequently necessary to produce a simple harmonic displacement with less than 360° of rotation of the cam, as illustrated in Fig. 20-1-10, and the ordinates for the cam pitch curve can then be determined as shown in Fig. 20-1-8D. It may be impossible to use a flat follower since the harmonic pitch curve usually has a reentrant or reversing curve and a flat follower would just *bridge* the hollow part. Since a roller follower is the most practical and reliable type, the development of the cam profile with this type of follower is shown. This motion would also produce a jerk if used in connection with a cam having a dwell.

To illustrate the effect of cam displacement for a given cam size and follower displacement on the pressure angle, the return, or fall, curve has been shown with a much larger angle. Note that the maximum pressure angle has been considerably reduced.

Cycloidal Motion

Figure 20-1-8E illustrates the graphic method of laying out a cycloidal profile using a rolling circle, as shown on the left end of the illustration. This curve, when generated accurately, produces a very smooth jerk-free motion when used in a cam having a dwell. This curve is best suited for light loading at high speeds.

Modified Trapezoidal Motion

The *modified trapezoid* is made by combining the cycloid and the constant-acceleration curve. Manufacturing accuracy requirements are less critical with the modified trapezoid than with the cycloidal curve. An advantage over the cycloid is lower acceleration, which means lower forces on output members (the follower system). High inertias can be handled more satisfactorily with the modified trapezoid than with the cycloid. This curve also is jerk-free when used in a cam having a dwell.

Modified Sine-Curve Motion

The *modified sine curve* is a combination of cycloidal and harmonic curves. This curve will absorb more errors than the modified trapezoid or the cycloidal curve. The torque change from positive to negative is 0.2 in the modified trapezoid and 0.4 in the modified sine curve. This means that the modified sine curve can stand a more flexible, or elastic, input drive than the modified trapezoid. This curve (modified sine) is ideal for high inertia, as well as for reasonably high speed.

Synthesized, Modified Sine-Harmonic Motion

Because of the complex makeup of the profiles of this curve, only the information shown in Fig. 20-1-11 is covered in this text.

SIMPLIFIED METHOD OF LAYING OUT A CAM MOTION

The method shown in Fig. 20-1-11B is a quick and accurate method for laying out a cam motion. The divisions shown on the lines in Fig. 20-1-11A are accurately divided into the proper divisions for the various cam motions. For example, it is required to construct a 2.00 in. parabolic rise in 120° of cam rotation.

Method

1. Draw two parallel horizontal lines 2.00 in. apart representing the rise.
2. Select a suitable distance for the cam displacement and divide the 120° into 10 equal parts (12°, 24°, 36°, etc.).
3. Using the edge of a sheet of paper, mark on it the divisions for the parabolic motion shown in Fig. 20-1-11A.

(A) FOLLOWER IN LOWEST POSITION (B) FOLLOWER IN HIGHEST POSITION (C) CAM ROTATED 30°

Fig. 20-1-9 Eccentric plate cam.

Fig. 20-1-10 Cam displacement diagram.

UNIFORM MOTION SCALE

HARMONIC MOTION SCALE

PARABOLIC MOTION SCALE

CYCLOIDAL MOTION SCALE

TRAPEZOID MOTION SCALE

MODIFIED SINE MOTION SCALE

SYNTHESIZED MODIFIED SINE-HARMONIC MOTION SCALE

(A) COMMON CAM MOTION SCALES

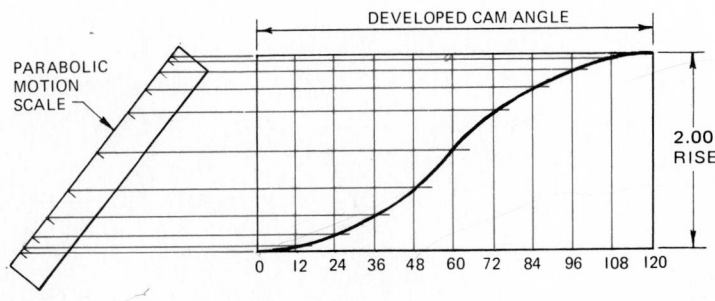

(B) SCALE APPLICATION

Fig. 20-1-11 Simplified method of laying out cam motion.

4. Using this marked paper as the scale, lay the scale between the base line and the top of the 2.00 in. rise, as shown in Fig. 20-1-11B, and transfer the points from the scale to the drawing.

5. Project these points horizontally to their respective cam divisions and draw the curve.

CAM DISPLACEMENT DIAGRAMS

In preparing cam drawings, a cam displacement diagram is drawn first to plot the motion of the follower. The curve on the drawing represents the path of the follower, not the face of the cam. The diagram can be any convenient length, but often it is drawn equal to the circumference of the base circle of the cam, and the height is drawn equal to the follower displacement. The lines drawn on the motion diagram are shown as radial lines on the cam drawing, and sizes are transferred from the motion diagram to the cam drawing. Figure 20-1-10 shows a cam displacement diagram having three different types of motion plus three dwell periods. Most cam displacement diagrams have cam displacement angles of 360°.

References and Source Material
1. Eonic Cams.
2. Commercial Cam and Machine Co.

ASSIGNMENT

See Assignment 1 for Unit 20-1 on page 415.

Review for Assignment
Unit 4-1   Dividing a Line into Equal Parts

UNIT 20-2
Plate Cams

In preparing cam drawings, the radial ordinates should be laid out in the opposite direction to that in which the cam rotates.

In drawing plate cams, the prime circle is constructed first. This circle represents the face of a flat follower or the

NOTE: DISTANCES A TO L (SHOWN IN COLOR) ON THE CAM DRAWING ARE TRANSFERRED TO DISPLACEMENT DIAGRAM.

CAM DRAWING

PRIME CIRCLE

PATH OF FOLLOWER (PITCH CIRCLE)

DISTANCE F = MAXIMUM FOLLOWER DISPLACEMENT = 2 X OFFSET

PATH OF FOLLOWER

FOLLOWER DISPLACEMENT

BASE LINE

ONE COMPLETE REVOLUTION OF CAM

DISPLACEMENT DIAGRAM

Fig. 20-2-1 Eccentric plate cam.

diagram is drawn first. The ordinate lines constructed on the motion or displacement diagram are drawn on the cam drawing as radial lines, and the corresponding distances from the base line to the motion curve are transposed to the cam drawing, locating the path of the follower. With cams using a roller follower, the roller diameter is then drawn in several positions along the path of the follower in order to construct the profile of the cam face. When the follower has a flat surface, the paths of the follower and the cam face are one.

Figure 20-2-2 shows a plate cam which produces a simple harmonic displacement with less than 360° of rotation of the cam. The ordinates for the cam pitch curve are constructed as shown in Fig. 20-1-8D. It may be impossible to use a flat follower since the harmonic pitch curve usually has a reentrant, or reversing, curve and a flat follower would just bridge the hollow part. Since a roller follower is the

center line of a roller follower, whichever is used, in it lowest position. It also represents the base line on the motion diagram.

One of the simplest cams to produce is the eccentric-plate cam, as illustrated in Fig. 20-2-1. The shape of the cam is a perfect circle, and the offset distance for the camshaft is equal to one-half the follower displacement. Radial lines are marked off on the cam drawing, with the center of the shaft as center. The distance between the prime circle and the center of the roller follower on the radial lines is transposed to the displacement diagram. The length of the displacement diagram can be any convenient size, but often the circumference of the prime circle is chosen in order to keep it in the same scale as the follower displacement height. Since this type of cam does not provide a dwell period, it has limited applications.

Since most cams combine motions and dwells in their design, the drawing sequence is different than that for eccentric cams. The cam displacement

PRESSURE ANGLE

PRIME CIRCLE

PITCH CURVE

PRESSURE ANGLE

CAM DRAWING

FOLLOWER DISPLACEMENT

RISE HARMONIC — DWELL — DROP HARMONIC — DWELL

DISPLACEMENT DIAGRAM

Fig. 20-2-2 Simple plate cam with harmonic motion.

most practical and reliable type, the development of the cam profile with this type of follower is shown.

CONJUGATE CAMS

Conjugate cams are used when a desired motion cannot be obtained with a single cam. See Fig. 20-2-3. Many indexing mechanisms use conjugate cams to obtain the necessary indexing. A displacement diagram is required for each cam.

Fig. 20-2-3   Conjugate cam.

TIMING DIAGRAMS

A convenient method of relating the movement of various machine members which are activated by cams is by the use of a timing diagram. Figure 20-2-4 shows the timing relationship for three cams. If displacements are plotted to scale, the diagram can be used for checking interferences. It can also be used for specifying the various types of transitions. If zero displacement is used to denote the prime circle radius, the timing diagram can be used by most manufacturers to produce finished cam data. The only additional data required would be a detailed drawing of the cam blank.

DIMENSIONING CAMS

The old method of developing cam contours on the drawing board by layout has been outdated. In the past, a detailed cam was developed from an enlarged layout, using swung arcs and straight lines. However, it is difficult to make a quality cam that has been developed by this method.

Fig. 20-2-4   Timing diagram.

To produce a master cam or a single cam, a table of cam radii with corresponding cam angles must be supplied. The cam is then cut on a milling machine, or some other suitable machine tool, by point settings. The result is a surface with a series of ridges which must be filed down to a smooth profile. The cam radius, cutting radius, and frequency of machine setting determine the extent of filing and the final accuracy of the profile. For accurate master cams, settings must often be in 0.5° increments, calculated to seconds. The preparation of this table may require the solution of six or eight equations for each of these machine settings.

If a cam has been developed by layouts and has a profile as shown in Fig. 20-2-5, it may appear that the easiest way to describe this contour is to scale the angle R_1 from 0° and scale the displacement D from the center of the cam to the profile surface. Admittedly, this method would define the surface or cam profile.

However, the only method of manufacturing that could be used is to broach the cam with a very small point cutter. If a cutter radius is added to the displacement value and a cut is made, the adjacent contour is undercut. To

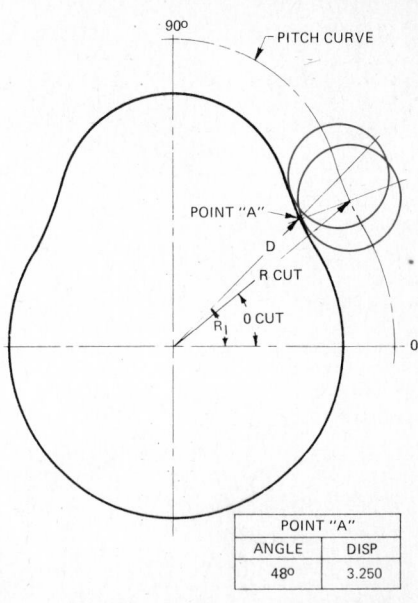

POINT "A"	
ANGLE	DISP
48°	3.250

Fig. 20-2-5   Dimensioning the cam profile.

properly cut the point A and maintain the contour, a new set of coordinates must be established. These are shown as R_{cut} and O_{cut}. To produce such data becomes a laborious and expensive task.

In describing a profile, *always* dimension to the pitch curve produced by the follower center. This holds true whether the cam is developed by layout or analytically. The actual follower location requires two physical dimensions: radial displacement and angular displacement. *Radial displacement* is expressed as a distance from the cam center or as a displacement from the prime circle. *Angular displacement* is measured in degrees from some zero reference, such as a keyseat, a dowel hole, or a timing hole, as shown in Fig. 20-2-6.

The easiest method of presenting these data is in tabular form, rather than dimensioning the detailed cam. Data should be given in at least 1° increments, although 0.5° increments are preferred. Increments of 0.5° allow the manufacturer to use discretion in selecting the intervals required to produce the finished cam.

Standard practice on tolerancing polar (angular) data is to hold the angles basic (zero tolerance) and to place all tolerance on the displacement value. The only angle that is toleranced is the one that relates the zero reference to some other point on the

FROM CENTER OR **FROM PRIME CIRCLE**

(A) DIMENSIONING RADIAL DISPLACEMENT

FROM KEYSEAT OR **FROM TIMING HOLE**

(B) DIMENSIONING ANGULAR DISPLACEMENT

Fig. 20-2-6 Dimensioning point "A" on the cam profile.

cam, such as a keyseat or dowel hole. Figure 20-2-7 shows this method, plus one of the ways in which to tolerance the pitch curve contour.

Figure 20-2-8 is an alternate method of establishing the pitch-curve tolerances. Both these examples contain three fundamental specifications

1. Tolerance on basic cam size
2. Tolerance on total transition
3. Tolerance on the pitch curve over some increment of cam angle

It is the third specification which ensures smooth continuity of the cam surface.

By far the simplest method of describing the contour is by denoting the type of transition. In this case, the type of dynamic curve chosen is called out on the detailed print, such as cycloidal, harmonic, modified-sine, etc. Most cam manufacturers are capable of producing their own incremental data. In most cases, the charge for this service is nominal. Figure 20-2-9 is an illustration of this type of cam detail drawing.

Variations from the smooth curve representing the tabulated values shall not exceed the limits as follows:

A — Tabulated values are to the center of a .2500 ±.0002 diameter cutter.

B — The base circle, as established by the initial cut, shall establish a reference. This initial cut shall not vary by more than .001 from the tabulated values.

C — The values of all errors, in the interval from 0° through 178°, shall lie within bands allowing either cyclic or random variations not exceeding ±.0002 from the (straight) center line of the band. The center line shall start at the initial cut point and shall end at 178° within .001 from the initial cut point.

D — In the interval from 178° through 360°, all errors shall lie within bands allowing either cyclic or random variations not exceeding .0005 from the (straight) center line of the band. The center line of this interval shall start at the actual end point of the first interval center line at 178° and end at the initial cut position at 360°.

Fig. 20-2-8 Alternate method of establishing pitch-curve tolerances.

Figure 20-2-10 illustrates a detailed cam drawing used by cam manufacturers. The radial and displacement dimensions for the motions are shown in tabular form.

CAM SIZE

Cam size depends primarily on three factors: the pressure angle, the curvature of profile, and the camshaft size. Secondary factors which affect size and design are cam-follower stresses, available cam material, and available space.

If a layout is made, such as shown in Fig. 20-2-11, it becomes obvious that the maximum pressure angle for a given cam and follower displacement becomes smaller as the cam-pitch circle becomes larger. It is advisable to limit this maximum pressure angle to 30 or 35°.

Figure 20-2-11 also shows how cam curvature is related to cam size. For a given displacement H, cam rotation B, and roller radius r, the larger cam with pitch-curve radius Rp_2 has a much easier curve to manufacture. Note how the smaller cam with radius Rp_1 has much smaller radii of curvature near the high end of the displacement.

TOLERANCES:
1–TABULATED VALUES ARE TO THE CENTER OF A Ø .500 FOLLOWER
2– ALL ANGLES ARE BASIC AND IN RELATION TO TIMING HOLE.
3–TOTAL TRANSITION DWELL TO DWELL MAY VARY ±.002 FROM TABULATED VALUES.
4–THE RELATIVE DIFFERENCE BETWEEN ANY TWO ADJACENT DEGREES MUST NOT EXCEED .0002 FROM THE DIFFERENCE ESTABLISHED BY THE TABULATED DATA.
5–DWELLS TO BE CONCENTRIC TO ID OF CAM WITHIN .0005 T.I.R. (TOTAL (INDICATOR) READING).

Fig. 20-2-7 Tolerancing polar data.

Fig. 20-2-9　Plate cam drawing.

ANGULAR DISPLACEMENT	RADIAL DISPLACEMENT FROM PRIME CIRCLE
0°	0
180°	1.250
210°	1.250
330°	0

TOLERANCE ON RADIAL DISPLACEMENT ± .001
TOLERANCE ON ANGULAR DISPLACEMENT ±0.5°

DISPLACEMENT DIAGRAM

NOTE: ANGULAR AND RADIAL DISPLACEMENT DIMENSIONS FOR MOTIONS SUPPLIED BY CAM MANUFACTURER.

ROTATION	DEG.	MIN.	SEC.	DISPLACEMENT	ROTATION	DEG.	MIN.	SEC.	DISPLACEMENT
120.000	119	48	49	8.650467	280.000	279	59	60	5.250001
121.000	120	48	49	8.650446	281.000	280	59	60	5.250014
122.000	121	48	49	8.650310	282.000	281	59	59	5.250105
123.000	122	48	48	8.649938	283.000	282	59	58	5.250350
......	....	...	...	....	......	....	...	...	....
182.000	181	6	2	6.856167	316.000	315	34	43	5.636968
183.000	182	6	3	6.799722	317.000	316	33	22	5.662956
......	....	...	...	....	......	....	...	...	....
208.000	207	32	54	5.671810	347.000	346	6	2	6.823881
209.000	208	34	30	5.641242	348.000	347	6	3	6.872336
......	....	...	...	....	......	....	...	...	....
239.000	238	59	60	5.250022	419.000	58	48	49	8.650454
240.000	239	59	60	5.250001	420.000	59	48	49	8.650467

Fig. 20-2-10　Cam manufacturing data.

Fig. 20-2-11　Increasing the cam size decreases the pressure angle.

Figure 20-2-11 also shows the effect of roller diameter on the shape and accuracy of the cam profile. Always use the smallest possible roller consistent with the load it has to carry.

Another factor affecting cam size is the cutting away of a previously generated cam profile by virtue of too large a cam roll. This is illustrated in Fig. 20-2-12. Basically, the cam-follower-roll radius must be less than the pitch-curve radii at any point along the pitch curve.

References and Source Material
1. Eonic Inc.

ASSIGNMENTS
See Assignments 2 through 4 for Unit 20-2 on page 416.

Review for Assignments
Unit 4-1　Dividing a Line into Equal Parts
Unit 9-1　Keys
Unit 20-1　Cam Motions

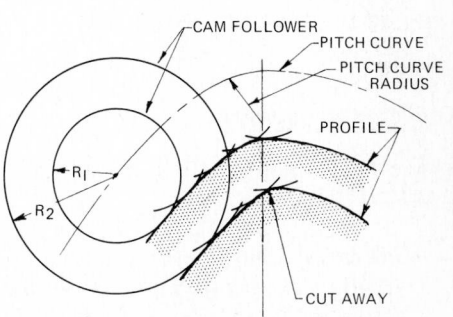

CAM FOLLOWER RADIUS MUST BE LESS THAN PITCH CURVE RADIUS AT ANY POINT TO AVOID CUT AWAY.

Fig. 20-2-12　Factors affecting cam size.

UNIT 20-3
Positive-Motion Cams

To ensure positive motion of the follower in both directions, positive-motion cams are employed. Two types are *face cams* and *cams with yoke-type followers*. Face cams are similar to plate cams, except that the follower engages a groove on the face of the cam rather than on the outside edge of the cam. One disadvantage to this type of cam is that the outer edge of the cam groove tends to rotate the roller in the direction opposite to that of the inner edge, resulting in wear in both the cam and the roller. However, this is not serious at slow speeds. Yoke-type followers are used for operating light mechanisms. The follower surface is flat or tangent to the curvature of the cam. With this type of cam, only one-half of the cam displacement diagram need be drawn since the other half of the cam is identical to the first half. See Figs. 20-3-1 and 20-3-2.

CAM — 360° CAM ROTATION

Fig. 20-3-1 Cam with yoke follower.

ASSIGNMENTS

See Assignments 5 through 8 for Unit 20-3 on page 417.

Review for Assignments

ANGULAR DISPLACEMENT FROM KEYSEAT	RADIAL DISPLACEMENT FROM PRIME CIRCLE
0°	0
30°	0
210°	.938
240°	.938

TOLERANCE ON RADIAL DISPLACEMENT ± .001
TOLERANCE ON ANGULAR DISPLACEMENT ± 0.5°
NOTE: ANGULAR AND RADIAL DISPLACEMENT DIMENSIONS FOR MOTIONS SUPPLIED BY CAM MANUFACTURER

Fig. 20-3-2 Face cam drawing.

UNIT 20-4
Drum Cams

The layout of a drum or cylinder cam starts, as with any cam, with the decision as to what profile and follower type will be used. Many cylinder cams are used with straight in-line followers so that the follower moves in a path parallel to the axis of the cam. The pitch surface is developed and shown as a rectangle (Fig. 20-4-1A), and the follower displacement is plotted with rectangular coordinates.

Theoretically, a tapered follower with its cone center on the cam axis should give the best results. See Fig. 20-4-1B. Actually, straight rollers give excellent results as long as the roller length and diameter are not too large in relation to the cam cylinder diameter. Swinging followers are used on indexing-type cylindrical cams, as shown in Fig. 20-1-5.

For drum or cylindrical groove cams, the displacement diagram is replaced by the developed surface of the cam, as shown in Fig. 20-4-2. The groove shown in the front view of the cam is found by projection.

(A) CAM DISPLACEMENT DIAGRAM FOR DRUM CAM

(B) TAPERED FOLLOWER

Fig. 20-4-1 Drum cam details.

Points from the developed surface of the cam and their corresponding points on the top view are projected to the front view, as shown by the letter *A* at position 210°.

ASSIGNMENTS

See Assignments 9 and 10 for Unit 20-4 on page 417.

Review for Assignments
Unit 4-1   Dividing a Line into Equal Parts
Unit 20-1   Cam Motions

UNIT 20-5
Indexing

Indexing is the conversion of a constant-speed rotary-input motion to an intermittent rotary-output motion. Press-feed tables, packaging machines, machine tools, switch gear, and feeding devices are but a few of the many machines found in industry that require indexing or intermittent motion.

Fig. 20-4-2 Drum cam drawing.

ANGULAR DISPLACEMENT FROM TIMING HOLE	DISPLACEMENT FROM BASE LINE
0°	0
210°	1.250
300°	0

TOLERANCE ON RADIAL DISPLACEMENT FROM BASE LINE   ± .001
TOLERANCE ON ANGULAR DISPLACEMENT FROM BASE LINE ±0.5°

NOTE: ANGULAR DISPLACEMENT AND DISPLACEMENT FROM BASE LINE SUPPLIED BY CAM MANUFACTURER.

DISPLACEMENT DIAGRAM

In recent years, advances in the design and manufacture of positive intermittent-motion devices have improved significantly the smoothness and speed of indexing motions possible with Geneva-type drives.

The manifold indexing mechanism (Fig. 20-5-1A) consists of two basic elements: a *cam* attached to the input shaft and a *turret* attached to the output shaft. The input and output shafts are at right angles but do not lie in the same plane. The cam is of concave globoidal form with a track that engages roller followers which project radially from the edge of the turret

disk. Part of the cam track is straight, so that when the roller followers engage it, no movement of the turret can take place. The angle of the cam occupied by this part is called the *dwell angle*.

The remaining part of the cam track progresses along the cam axis in helical fashion, thus rotating the turret. Before a roller leaves one end of the cam track, another roller enters the other end to maintain continuity of movement. The angle of the cam occupied by this part of the track is called the *cam index angle*. Thus, one revolution of the cam represents one

indexing cycle, during which the turret indexes from one station to the next and dwells for a specific period. The number of times that this takes place in one revolution of the turret is called the *number of stops*.

The tangent drive (Figs. 20-5-1C and 20-5-1E) consists of a constantly rotating driver and a driven wheel. The wheel may have four, five, six, or eight precision-machined radial slots. A matching cam follower, mounted on needle bearings on the driver, engages one of the slots on each revolution of the driver, thereby indexing the wheel. The concave section between the slots is precisely machined to mate with the locking hub of the driver to prevent movement of the wheel during dwell.

The tangent drive indexes over an angle equal to 360° divided by the number of slots or stations in its wheel. For example, each index of a four-station tangent drive is 90°; each index of a five-station drive is 72°. See Fig. 20-5-2.

The time ratio of a tangent drive is expressed by the arc (in degrees) of each revolution of the driver that the wheel is being indexed and the arc of each revolution that the wheel is at rest or dwell. The time ratio refers to each revolution of the driver and, therefore, remains constant, regardless of the driver speed. The actual speed of indexing is a function of the driver speed and is directly proportional to it.

The indexing application shown in Fig. 20-5-1F employs an overrun clutch and a rack and gear. The input or driver shaft is connected to a rack which converts rotary motion into reciprocating motion. The gear which is attached to an overrun clutch rotates in both directions. The overrun clutch drives the shaft in one direction but overruns or freewheels on the shaft in the other direction, producing an intermittent rotary motion.

(A) BARREL CAM
(Manifold Machinery Co. Ltd.)

(B) BARREL CAM
(Commercial Cam & Machine Co.)

(C) 6-STATION DRIVE
(Geneva Motions Corp.)

(D) CONJUGATE CAMS
(Commercial Cam & Machine Co.)

(E) 4 -STATION DRIVE
(Geneva Motions Corp.)
Fig. 20-5-1 Indexing mechanisms.

(F) OVERRUN CLUTCH, CRANK, AND GEAR AND RACK

References and Source Material

1. Manifold Machinery Company, Limited.
2. Geneva Motions Corporation.

ASSIGNMENT

See Assignment 11 for Unit 20-5 on page 417.

DIMENSIONS		MODEL	A	B	C	D	DRIVER					WHEEL		
							E	F	G	H	K	L	M	N
4 STATIONS STATIONS TIME RATIO: 90° INDEX - 270° DWELL	INCH	4S5	5.00	8.00	12.59	2.06	4.00	.50	1.50	.75	.88	.62	2.00	1.00
		4S5.5	5.50	8.62	13.75	2.06	4.40	.50	1.50	.75	.88	.62	2.00	1.00
		4S6	6.0	9.50	15.06	2.19	4.75	.50	1.50	.75	1.00	.75	2.00	1.00
	METRIC	4S75	75	124	194	48	62	12	40	16	16	12	45	20
		4S90	90	150	224	48	75	12	40	16	16	12	45	20
		4S100	100	164	256	50	82	12	40	20	20	16	48	22
		4S115	115	190	287	50	95	12	40	20	20	16	48	22
5 STATIONS STATIONS TIME RATIO: 108° INDEX - 252° DWELL	INCH	5S5	5.00	8.19	12.47	2.06	3.38	.50	1.50	.75	.88	.62	2.00	1.00
		5S5.5	5.50	9.00	13.69	2.06	3.69	.50	1.50	.75	.88	.62	2.00	1.00
		5S6	6.00	9.78	14.94	2.19	4.03	.50	1.50	.75	1.00	.75	2.00	1.00
	METRIC	5S75	75	124	190	48	53	12	40	16	16	12	45	20
		5S90	90	150	225	48	60	12	40	16	16	12	45	20
		5S100	100	164	252	50	70	12	40	20	20	16	48	22
		5S115	115	190	287	50	77	12	40	20	20	16	48	22
6 STATIONS STATIONS TIME RATIO: 120° INDEX - 240° DWELL	INCH	6S5	5.00	8.75	12.31	1.62	2.94	.50	1.50	.75	.88	.62	2.00	1.00
		6S5.5	5.50	9.62	13.50	1.62	3.14	.50	1.50	.75	.88	.62	2.00	1.00
		6S6	6.00	10.50	14.75	1.75	3.50	.50	1.50	.75	1.00	.75	2.00	1.00
	METRIC	6S75	75	134	188	35	46	12	40	16	16	12	45	20
		6S90	90	160	222	35	52	12	40	16	16	12	45	20
		6S100	100	180	250	38	60	12	40	20	20	16	48	22
		6S115	115	204	274	38	67	12	40	20	20	16	48	22
8 STATIONS STATIONS TIME RATIO: 135° INDEX - 225° DWELL	INCH	8S5	5.00	9.31	11.94	1.62	2.28	.50	1.50	.75	.75	.62	2.00	1.00
		8S5.5	5.50	10.22	13.12	1.62	2.50	.50	1.50	.75	.75	.62	2.00	1.00
		8S6	6.00	11.16	14.34	1.75	2.75	.50	1.50	.75	.75	.75	2.00	1.00
	METRIC	8S75	75	144	193	35	46	12	40	16	12	12	45	20
		8S90	90	170	215	35	40	12	40	16	12	12	45	20
		8S100	100	190	243	38	48	12	40	20	16	16	48	22
		8S115	115	214	274	38	52	12	40	20	16	16	48	22

Fig. 20-5-2 Indexing drives.

UNIT 20-6
Linkages

One of the ever-present problems in machine design is the mechanization of various interrelated motions. These motions are usually preassigned, often quite arbitarily, and specify the relationships of moving parts or simply the end motion of a single part. Illustrations are present in all types of machinery. Typical examples are seen in textile machinery, packaging machinery, printing presses, valve mechanisms in steam locomotives, machine tools, automotive equipment, household articles, instruments, computing devices, and many other common mechanisms. Upon closer observation, it will be noticed that all these devices are simply combinations and arrangements of basic mechanical elements such as gear trains, cam actuators, cranks and links, sliders, bolts and pulleys, and other rotating and sliding parts. Combinations of the crank, link, and sliding elements are commonly termed *bar linkages*.

Locus of a Point

The *locus* of a point in a linkage or mechanism is the path traced by that point as it moves according to certain controlled conditions. The study of loci is important in machine design to determine

1. The position of various links and joints in the cycle of the mechanism
2. The relative speeds of different parts
3. Forces exerted in the mechanism using applied machines in conjunction with diagrammatic layouts

The designer is often called upon to make these diagrammatic layouts to assist in the design of the container for linkages that will be of the most economical and space-saving nature, as well as to ensure that parts of adjacent linkages will not foul one another at any point of the movement of the machine.

Cams versus Linkages

The best-known solution for a function generator is the cam: *flat cams* for functions of single variables and *barrel cams* for functions of two variables.

As computing devices, linkage mechanisms enjoy a number of advantages over cams, with the one exception that the functions must be continuous. *Linkages* are essentially straight members joined together. Only a small number of dimensions need to be held closely. The joints make use of standard bearings, and the links in effect form a solid chain and are not subject to undue acceleration limitations.

The harmonic transformer and the four-bar linkage shown in Fig. 20-6-1 are the two bar linkages most commonly used as function generators. The harmonic transformer consists of crank, connecting link, and slider. It may be driven from the crank end when a rotary input and linear output

(A) PIVOT BEARING JOINT

(B) RADIAL–BEARING JOINT

(A) HARMONIC TRANSFORMER

(B) FOUR-BAR LINKAGE

Fig. 20-6-1 Linkages used as function generators.

(C) SELF–ALIGNING JOINT

Fig. 20-6-2 Linkage joints.

are desired or from the slide end when a linear input and rotary output are required. Two cranks and a connecting link form the four-bar linkage, whose input and output are both rotary. By assignment of correct values to the various parameters, these linkages will mechanize many single-variable functions. The selection of these values is termed a *linkage layout*. Typical linkage joints are shown in Fig. 20-6-2.

STRAIGHT-LINE MECHANISM

A *straight-line mechanism* is a linkage device used to guide a given point in an approximate straight line. Several such mechanisms use five or more links to produce exact straight-line motion of a given point. A four-link (or four-bar) mechanism, using finite links, can only approximate a straight line.

The four elements of the four-bar linkage (Fig. 20-6-3) are:

1. A link which will be caused to move so that one point on it, called the *indicator*, travels along the desired path. This link is composed of two rigidly connected parts: the *connecting link*, joining the drive point and the control point, and the *indicator link*, connecting the indicator to the drive point.
2. A *drive crank*, to which a turning torque is applied to move the mechanism and which is connected to the link at the drive point.
3. A *control crank*, which serves only to guide the link control point in the proper path.
4. The *base* of the machine, to which the two cranks are pivotally attached.

For identification purposes, *indicator path* is the term used to describe the approximate straight-line path through which the indicator travels, and *straight line* refers to the desired theoretical straight line. The indicator path and the straight line will coincide at three or four places.

SYSTEMS HAVING LINKAGES AND CAMS

A cam is of no value and can perform no useful function without a follower linkage. A *simple follower* is generally

Fig. 20-6-3 Terminology of a four-bar straight-line mechanism.

CONVERTING ROTARY MOTION INTO RECIPROCATING MOTION

CONVERTING ROTARY MOTION INTO OSCILLATING MOTION

TABLE MOVES FOUR TIMES CRANK OFFSET

CRANK WITH GEARS AND RACK

CONVERTING ROTARY MOTION INTO RECIPROCATING MOTION

Fig. 20-6-4 Converting rotary motion into oscillating or reciprocating motion.

not thought of as a linkage since it is usually a slide or plunger, such as an automotive valve assembly in a simple L-head engine. A *linkage* is generally considered to be a group of levers and links. See Fig. 20-6-4.

Figure 20-6-5 shows a typical cam linkage composed of a fairly heavy slide, a short link, and a bell crank. If the slide, which has the largest mass in the linkage, is to be moved with the most favorable accelerating forces, its displacement-to-time relationship must govern the shape of the cam profile. Since the bell crank and link swing about fixed and instantaneous centers during the stroke, the displacement increments of the slide and the cam profile do not bear the same relationship to the cam displacement. It is therefore necessary to plot the cam-

pitch curve from the slide displacement at each increment of cam displacement. Figure 20-6-6 shows how this is done.

For extra accuracy and for high-speed machinery, these displacements should be calculated.

References and Source Material
1. *Machine Design*.
2. Eonic, Inc.

ASSIGNMENTS

See Assignments 12 through 15 for Unit 20-6 on page 417.

Review for Assignments
Unit 9-4   Springs
Unit 20-1   Cam Motions

Fig. 20-6-6 Plotting the pitch curve from the slide displacement for the cam shown in Fig. 20-6-5.

Fig. 20-6-5 Cam linkage.

(A) EXTERNAL RATCHET

(B) U-SHAPED PAWL

(C) DOUBLE–ACTING ROTARY RATCHET

(D) INTERNAL RATCHET

(E) FRICTION RATCHET

(F) SHEET-METAL RATCHET AND PAWL

(G) JACK

(H) RATCHET WRENCH

Fig. 20-7-1 Ratchet and pawl application.

UNIT 20-7
Ratchet Wheels

Ratchet wheels are used to transform reciprocating or oscillatory motion into intermittent motion, to transmit motion in one direction only, or as an indexing device.

When a motion is to be transmitted at intervals rather than continuously and the loads are light, ratchets are ideal because of their low cost.

Common forms of rachets and pawls are shown in Fig. 20-7-1. The teeth in the ratchet engage the teeth in the pawl, permitting rotation in one direction only. The pawl, which fits into the ratchet teeth as shown in Fig. 20-7-2, is pivoted at one end. A spring or counterweight is normally used to maintain contact between the wheel and pawl. In Fig. 20-7-1(G) and (H), lever or pawl balls are used to shift the pawl to the alternative position so that the ratchet will work in reverse.

In friction ratchets [Fig. 20-7-1(E)], balls are used between the ratchet and the follower. As the ratchet rotates in the direction of the arrow shown, the balls roll up on the high spots on the teeth, wedging the ratchet and outer race together. If the direction of the

Fig. 20-7-2 Designing a ratchet wheel and pawl.

ratchet is reversed, the balls roll to the low points on the teeth and disengage the outer roller. This principle is used on overrunning clutches.

Ratchet wheels and pawls are also used widely to control drum rotation in hoisting equipment.

In designing a ratchet wheel and pawl, lay out points A, B, and C, as shown in Fig. 20-7-2, on the same circle to ensure the smallest forces are acting on the system.

Another method to increase the number of stops made by the ratchet wheel without increasing the number of teeth is the use of multiple pawls.

PAWLS LOCK RATCHET WHEEL EVERY 11.25°

Fig. 20-7-3 Ratchet with two pawls.

See Fig. 20-7-3. Adding another pawl of different length doubles the number of indexing positions.

ASSIGNMENTS

See Assignments 16 and 17 for Unit 20-7 on page 418.

Review for Assignments
Unit 9-4 Springs
Unit 9-1 Keys

ASSIGNMENTS for Chapter 20

Assignment for Unit 20-1, Cams and Cam Motions

1. On a B- or A3-size sheet, draw displacement diagrams for each of the two cams shown in Fig. 20-1-A or Fig. 20-1-B. Scale is full or 1:1.

CAM I	CAM 2
—Rise 2.00 in 150° with harmonic motion	—Rise 1.50 in 120° with uniform motion
—Dwell for 45°	—Dwell for 60°
—Drop 2.00 in 120° with uniformly accelerated and retarded motion	—Rise .50 in 60° with parobolic motion
—Dwell remainder	—Drop 2.00 with harmonic motion for remainder
—Displacement diagram 2.00 high	—Displacement diagram 2.00 high X 12.00 long
—X 12.00 long	

Fig. 20-1-A Cam displacement problem.

Assignments for Unit 20-2, Plate Cams

2. On a B- or A3-size sheet, design a plate cam that will activate a ∅.50-in. roller follower as follows:

> Rise 1.50 in. in 180° with harmonic motion
> Dwell 30°
> Drop 1.50 in. in 120° with modified uniform motion
> Dwell for remainder

Prime circle = 3.00 in., plate thickness = .50 in., shaft = ∅1.00 in., hub = ∅1.75 in. × 1.50 in. long, keyseat to suit. Add a chart to the drawing showing the angular and radial displacements every 15°, taking the radial measurements from the prime circle. Scale is full or 1:1.

3. On a B- or A3-size sheet, design a plate cam that will activate a ∅10-mm roller follower as follows:

> Rise 40 mm in 150° with uniformly accelerated and retarded motion
> Dwell for 45°
> Drop 40 mm in 120° with modified uniform motion
> Dwell for remainder

CAM I	CAM 2
—Rise 50 mm in 120° with cycloidal motion —Dwell for 60° —Drop 10 mm in 90° with uniform motion —Drop 40 mm in 90° with uniformly accelerated and retarded motion —Displacement diagram 50 mm high X 300 mm long	—Rise 30 mm in 90° with modified uniform motion —Dwell for 60° —Rise 20 mm in 45° with harmonic motion —Drop 50 mm in 120° with parobolic motion —Displacement diagram 50 mm high X 300 mm long

Fig. 20-1-B   Cam displacement problem.

Prime circle = 70 mm, plate thickness = 10 mm, shaft = ∅26 mm, hub = ∅44 × 32 mm long, keyseat to suit. Add a chart to the drawing showing the angular and radial displacements every 15°, taking the radial measurements from the prime circle. Scale is 1:1.

4. On a B- or A3-size sheet, lay out the parallel-drive indexing unit shown in Fig. 20-2-A or Fig. 20-2-B with the timing hole rotated to position B. Use your judgment for dimensions not shown. The angular and radial displacement values locate the center of the roller. Scale is half size or 1:2.

0°	3.74	25°	4.08	50°	5.51	75°	4.86	95°	3.47
5°	3.75	30°	4.32	55°	5.46	80°	4.59	100°	3.61
10°	3.76	35°	4.64	60°	5.37	85°	4.26	105°	3.68
15°	3.80	40°	4.95	65°	5.24	90°	3.78	110°	3.73
20°	3.90	45°	5.24	70°	5.06	93°	3.41	115° AND 120°	3.74

ANGULAR DISPLACEMENT AND RADIAL DISPLACEMENT FOR CAM "A". CAM "B" IS OPPOSITE HAND TO CAM "A". ANGULAR DIMENSIONS SHOWN ON DRAWING ARE FOR CAM "A".

Fig. 20-2-A   Parallel-drive indexing unit.

0°	94.9	25°	103.6	50°	139.9	75°	123.4	95°	88.1
5°	95.2	30°	109.7	55°	138.6	80°	116.5	100°	91.6
10°	95.5	35°	117.8	60°	136.3	85°	108.2	105°	93.4
15°	96.5	40°	125.7	65°	133.0	90°	96.0	110°	94.7
20°	99.0	45°	133.0	70°	128.5	93°	86.6	115° AND 120°	94.9

ANGULAR DISPLACEMENT AND RADIAL DISPLACEMENT FOR CAM "A". CAM "B" IS OPPOSITE HAND TO CAM "A". ANGULAR DIMENSIONS SHOWN ON DRAWING ARE FOR CAM "A".

Fig. 20-2-B   Parallel-drive indexing unit.

Assignments for Unit 20-3, Positive-Motion Cams

5. On a B- or A3-size sheet, make a two-view drawing of a face cam from the following information:

   Rise 1.20 in. in 150° with harmonic motion

   Dwell 30°

   Drop 1.20 in. in 120° with parabolic motion

   Dwell for remainder

   Roller = ∅.50 in., prime circle = 3.00 in., OD of face cam = 6.50 in., cam thickness = 1.00 in., groove depth = .38 in., shaft = ∅1.00 in., hub = ∅1.75 in. × 1.50 in. long. Add a suitable keyseat. Prepare a chart showing angular and radial displacement for every 15°, taking the radial measurements from the prime circle. Scale is full (1:1).

6. On a B- or A3-size sheet, make a two-view drawing of a face cam from the following information:

   Rise 24 mm in 120° with parabolic motion

   Dwell 45°

   Drop with cycloidal motion for the remainder

   Roller = ∅12 mm, prime circle = ∅80 mm, OD of face cam = 160 mm, cam thickness = 25 mm, groove depth = 12 mm, shaft = ∅24 mm, hub = ∅42 × 28 mm long. Add a suitable keyseat. Prepare a chart showing angular and radial displacement for every 15°, taking the radial measurements from the prime circle. Scale is 1:1.

7. Make a two-view drawing of a yoke cam that will raise the yoke 1.40 in. The cam is an eccentric cam having a dia. of 3.50 in. and a plate thickness of .75 in. Shaft = ∅1.06 in., hub = ∅1.75 in. × 1.10 in. long having the extension on one side only. The yoke is .40 in. thick and has a wall width of .75 in. A .25 in. × 1.25-in. steel guide bar is welded to the top and bottom of the yoke.

8. Make a two-view drawing of a yoke cam that will raise the yoke 35 mm. The cam is an eccentric cam having a dia. of 90 mm and a plate thickness of 20 mm. Shaft = ∅28 mm, hub = ∅44 mm × 30 mm long with the extension on one side only. The yoke is 10 mm thick and has a wall width of 20 mm. A 6 mm × 30 mm steel guide bar is welded to the top and bottom of the yoke.

Assignments for Unit 20-4, Drum Cams

9. On a B- or A3-size sheet, make a two-view drawing of a drum cam from the following information:

   Rise 1.50 in. in 120° with parabolic motion

   Dwell for 60°

   Drop 1.50 in. in 150° with modified sine motion

   Dwell for remainder

   Roller follower = ∅.50 in., cam = ∅3.00 in. × 3.50 in. long, follower groove = .40 in. deep. Use your judgment for dimensions not given. Show the full development of the cam, which will serve as a motion diagram. Prepare a chart showing the angular displacement from a timing hole located at 0°, and the displacement from the base line for every 15°. Scale is full (1:1).

10. On a B- or A3-size sheet, make a two-view drawing of a drum cam from the following information:

    Rise 32 mm in 150° with harmonic motion

    Dwell for 45°

    Drop 32 mm in 120° with trapezoid motion

    Dwell for remainder

    Roller follower = ∅14 mm, cam = ∅70 mm × 64 mm long, follower groove = 10 mm deep. Use your judgment for dimensions not given. Show the full development of the cam, which will serve as a motion diagram. Prepare a chart showing the angular displacement from a timing hole located at 0°, and the displacement from the base line for every 15°. Scale is 1:1.

Assignment for Unit 20-5, Indexing

11. On a B- or A3-size sheet, make a two-view drawing of the indexing drive 6S5 or 6S75 shown in Fig. 20-5-12. Use your judgment for dimensions not shown. Draw an angular displacement diagram, plotting points every 5° on the index cycle. Add suitable keyseats. Scale is full (1:1).

Assignments for Unit 20-6, Linkages

12. *Simple Crank Mechanism.* On a B- or A3-size sheet, lay out the two linkages shown in Fig. 20-6-A and plot the paths at 15° intervals of the points indicated. Scale is full (1:1).

13. On a B- or A3-size sheet lay out the two linkages shown in Fig. 20-6-B and plot the paths at 15° intervals of point *C* in (A) and *E* in (B). Points *C* and *E* are located at midpoint of links. Scale is full (1:1).

SIMPLE CRANK

CRANK WITH SLIDING ROD

Fig. 20-6-A   Simple crank mechanism.

(A) CROSS LINKED CRANK

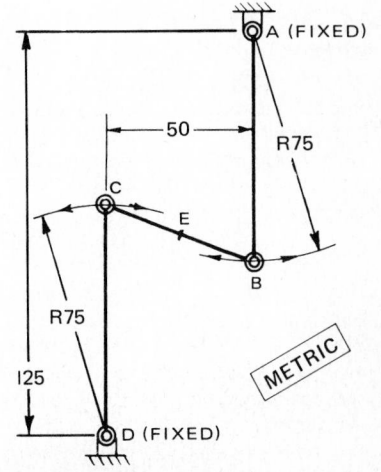

(B) WATT'S APPROXIMATE STRAIGHT LINE MECHANISM

Fig. 20-6-B   Linkage problems.

(A) SHAPER SHOWING QUICK RETURN MECHANISM

(B) MOTION DISPLACEMENT DIAGRAM

Fig. 20-6-C Shaper using Whitworth quick-return mechanism stroke.

14. On a B- or A3-size sheet, lay out the shaper shown in Fig. 20-6-C and complete the motion displacement diagram for two complete strokes. Take positions every 30° of trunnion rotation starting at position 240°. Motion displacement diagram size is 4.00 × 9.00 in. Scale is 1:4. Repeat the above by changing the 2.50 in. trunnion radius to 4.00 in.

15. On a B- or A3-size sheet lay out the two linkages shown in Fig. 20-6-D. For the Peaucellier's mechanism plot the path taken by point C. Plot points by moving point D every .25 in. For the toggle linkage, plot the distance X for every 15° of rotation of point A. AB = 2.50 in., BC = 1.75 in., BD = 2.25 in. Scale is full (1:1).

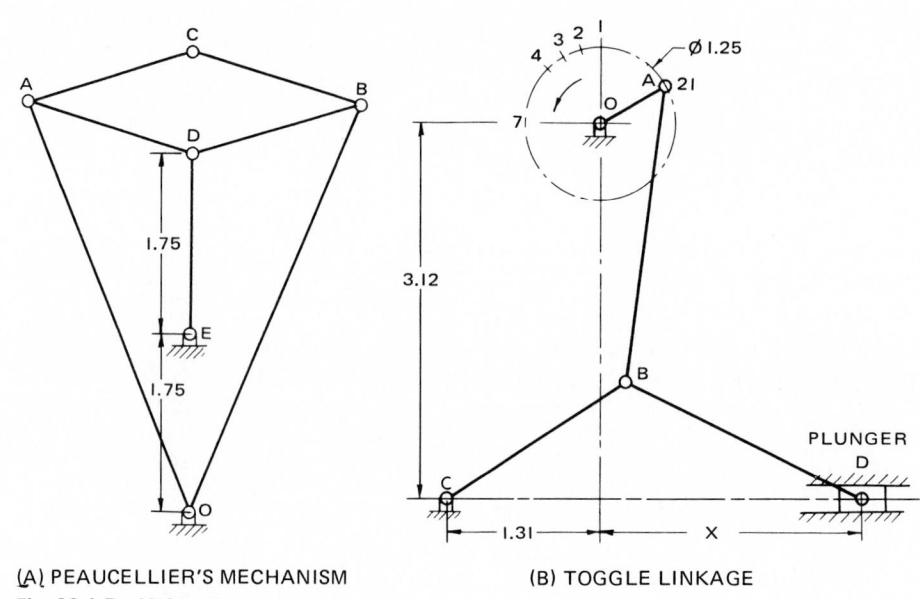

(A) PEAUCELLIER'S MECHANISM

(B) TOGGLE LINKAGE

Fig. 20-6-D Linkages.

Assignments for Unit 20-7, Ratchet Wheels

16. On a B- or A3-size sheet lay out a brake assembly using a U-shaped pawl with a ratchet wheel. The ratchet wheel is to have 24 teeth; OD of Ø146 mm; hub Ø48 mm; shaft Ø32.5 mm; keyseat to suit; width of teeth 12 mm; depth of teeth 10 mm; and the hub is to extend 16 mm on one side. Scale is 1:1. Show two views.

17. *Ratchet and Crank Mechanism.* On a B- or A3-size sheet, lay out a one-view drawing of the ratchet design shown in Fig. 20-7-A. Two pawls are used, a drive pawl as shown and a holding pawl held in position by a spring. Using crank rotation positions every 22.5°, plot the path of the end of the drive pawl. Use your judgment for dimensions not shown. Scale is 1:1.

Fig. 20-7-A Ratchet and crank mechanism.

CHAPTER 21

Fluid Power

UNIT 21-1
Hydraulics

The more complex industry becomes, the more vital becomes the role played by fluids in the industrial machine. One hundred years ago, water was the only important fluid which was conveyed from one point to another by pipe. Today, almost every conceivable fluid is handled in pipes during its production, processing, transportation, or utilization. The age of atomic energy and rocket power has added fluids such as liquid metals—for example, sodium, potassium, and bismuth—as well as liquid oxygen, nitrogen, etc., to the list of more common fluids such as oil, water, acids, and liquors that are being transported in pipes today. Nor is the transportation of fluids the only phase of hydraulics which warrants attention now. Hydraulic and pneumatic mechanisms are used extensively for the controls of modern aircraft, seagoing vessels, automotive equipment, machine tools, earthmoving and road-building machines, and even in scientific laboratory equipment where precise control of fluid flow is required.

So extensive are the applications of hydraulics and pneumatics that almost every designer has found it necessary to be familiar with at least the elementary laws of fluid flow.

BASIC PRINCIPLES

The science of hydraulics dates back several thousand years when water wheels, dams, and sluice gates were used to control the flow of water for irrigation and domestic use. Today, however, the term *hydraulics* commonly refers to *power hydraulics* in which fluid is used under controlled pressure to do work.

A fluid is infinitely flexible, yet as unyielding as steel. It can readily change its shape; it can be divided into parts to do work in different locations; it can move rapidly in one place and slowly in another; and it can transmit a force in any or all directions. No other medium except fluid combines the same degree of positiveness, accuracy, force, and pressure.

It is interesting and important to note that a hydraulic pump does not *pump* pressure. The pump merely produces flow. Pressure is generated only when a cylinder, motor, valve, or constriction tends to resist fluid flow. If the flow were to encounter only negligible resistance, the developed pressure would be slight.

Both force and pressure are primarily measures of effort. *Work*, however, is a measure of accomplishment. It describes the application of a force through a distance and flexibility of control, with the ability to transmit a maximum of power in a minimum of mass.

Although this unit does not aim at an in-depth treatment of hydraulic components and circuits, it will nevertheless be helpful to start by considering some of the fundamental principles involved.

Force, Pressure, Work, and Power

Force is defined as any cause which tends to produce or modify motion. To move an object, such as a machine tool head, force must be applied to it. The amount of force required depends on the object's inertia. In the foot-pound or U.S. Customary system, force can be expressed in any of the units of weight, but it is commonly expressed in pounds. In the metric system force is measured in newtons (N) for light forces, kilonewtons (kN) for intermediate forces, and meganewtons (MN) for strong forces. (A newton is approximately one-quarter of a pound-force.)

In the U.S. Customary system, pressure is usually expressed in terms of pounds per square inch (psi). In the metric system, pressure is measured in pascals. A pascal (Pa) is the pressure produced when a force of one newton is applied to an area of one square meter (Pa = N/m^2).

The pascal is a very small unit of measure, equivalent to approximately 0.001 pounds per square inch (lb/in.^{2}).

It is used for very low-pressure applications. In most instances the kilopascal (kPa) and megapascal (MPa) are used. In fluid power applications the kilopascal is the recommended unit of pressure. A 1-kPa pressure is equal to approximately 20.9 lb/ft², or 6.895 lb/in.²

The Earth's atmosphere provides an example of the relationship between force and pressure. The blanket of air enveloping the Earth's surface is of such volume that its total mass could be measured in U.S. Customary tons (2000 lb), or metric tons (megagrams). However, the force exerted by the mass of a column of air 1 in.² in cross-sectional area is only 14.7 lb/in.² (1 m² in cross-sectional area is only 101.325 kPa) at sea level. Thus atmospheric pressure at sea level is 14.7 lb/in.² (101.325 kPa), but for low-accuracy work 15 lb/in.² (100 kPa) is used. The relationship between force, pressure, and area is expressed mathematically as

Force = pressure × area
(U.S. Customary)
Force in lb = lb/in.² × in.²
(Metric)   Newtons = pascals × meters²
N = Pa × m²

As mentioned previously, work is a measure of accomplishment, describing the application of a force moving through a distance. In the U.S. Customary system, work is expressed in terms of in.-lb or ft-lb. Thus, if a force of 1200 lb moves a ram 4 in., the work accomplished equals 4800 in.-lb, or 400 ft-lb. In the metric system, work is expressed in joules. A joule (J) is the energy required or expended to move a force of one newton a distance of one meter. The joule is the special name given the derived unit newtonmeter (N • m). The kilojoule (kJ) is the preferred unit. The joule (J) is used to express small quantities of energy, megajoules (MJ) are used for large quantities. Thus, if a force of 500 N moves a ram 3 m, the work accomplished equals 1500 J, or 1.5 kJ. The formula is

Work = force × distance
ft-lb = lb × ft
J = N × m

The concept of work makes no allowance for the time factor. *Power* is work per unit of time. In the U.S. Customary system, *horsepower* (hp) is the standard unit of power measurement. One horsepower is that amount of power necessary to raise 33 000 lb one ft in one minute or 550 lb one ft in one second. In the metric system the watt (W) is the standard unit for all forms of power: heat, electric, mechanical, etc. A *watt* is energy per unit of time measured in joules per second (J/s). The preferred units are watt (W) for small values, *kilowatt* (kW) for intermediate values, and *megawatt* (MW) for large values. One watt is that amount of power necessary to move one newton one meter in one second.

Behaviors and Mechanics of Fluids

In the 17th century, Pascal formulated the fundamental law which forms the basis of modern hydraulics. Pascal's law states: *Pressure at any one point in a static liquid is the same in every direction and exerts equal force on equal areas.* Figure 21-1-1 illustrates this principle.

Fig. 21-1-1 Pascal's law.

Since fluids are practically incompressible, mechanical forces may be transmitted, multiplied, or controlled by means of hydraulic fluids under pressure. This is demonstrated in Fig. 21-1-2. Assume that the masses of pistons *P* and *W* are equal. Assume also that the face of piston *P* has an area of 1 in.², that the face of piston *W* has an area of 50 in.², and that the two communicating containers are filled with liquid.

If we were to exert a downward force of 2 lb on piston *P*, this force would be transmitted undiminished in every direction, acting with equal force on equal areas. Since the area of

Fig. 21-1-2 Transmission of mechanical forces.

piston *P* is 1 in.², the pressure at every point in the system is 2 psi. When a pressure of 2 psi is applied to the lower surface of piston *W*, the resultant upward force is 2 × 50, or 100 lb, because the area of piston *W* is 50 in.² and force equals pressure times area.

In either case, we can multiply force only by sacrificing distance proportionally. If we move piston *P* downward a distance of 50 in., we have forced 50 in.³ of liquid to the underside of piston *W*. Since piston *W* has an area of 50 in.², 50 in.³ of liquid can raise piston *W* only 1 in.

Characteristics of Flow Pascal's law neglects the factor of friction because it deals with static fluids. When a fluid flows in a hydraulic circuit, friction results and heat is produced. Thus some of the energy being transferred is lost in the form of heat energy.

Although friction can never be eliminated entirely, it can be controlled to some extent. The four main causes of excess friction in hydraulic lines are excessive length of lines, excessive number of bends and fittings or improper (too sharp) bends, excessive fluid velocity caused by undersized lines, and excessive viscosity of fluid.

Figure 21-1-3 illustrates the effect of friction upon pressure. Since pressure is a result of resistance to flow, the pressure at point *B* is zero. Assume that the mass of the fluid and the diameter of tube *A* are such that a static pressure of 10 psi is created at point *C*. Then the flow from point *C* to point *B*

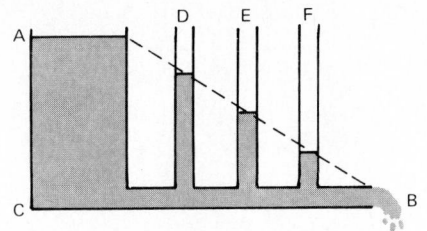

Fig. 21-1-3 Illustrating effect of friction upon pressure.

has resulted in a pressure drop of 10 psi.

The potential energy at point *C* is completely dissipated in moving the fluid to point *B*. This potential energy has been converted into the heat energy produced by the friction of the fluid moving through the tubes.

The height of the fluid in tubes *D*, *E*, and *F* illustrates the action of friction in producing a pressure drop. In a moving fluid, pressure drop tends to increase and pressure tends to decrease as the distance from the source of pressure increases.

HELPFUL INFORMATION

Some of the points already mentioned are repeated below, along with several additional facts that may help in understanding machinery hydraulics.

1. Oil is the most commonly used hydraulic fluid because it serves as a lubricant for hydraulic components and is practically incompressible.
2. The mass of oil varies considerably with change in viscosity. However, 55 to 58 lb/ft³ (2.6 to 2.8 kPa) covers the viscosity range of common hydraulic fluids.
3. Pressure at the bottom of a 1-ft column of oil will be approximately 0.4 psi. To find the approximate pressure at the bottom of any oil column, multiply the height in feet by 0.4.
4. There must be a pressure drop (pressure difference) across an orifice or constriction to cause flow through it. Conversely, if there is no flow, there will be no pressure drop.
5. A fluid is pushed into a pump. Atmospheric pressure supplies this push (in an unsupercharged pump) at 14.7 lb/in.² (101.3 kPa) at sea level.

6. The force exerted by a cylinder is dependent on the pressure applied and the piston area.
7. The speed of a cylinder is dependent on its piston area and the rate of fluid flow into it.
8. Te flow velocity through a pipe varies inversely with the square of the inside diameter (ID). Doubling the ID increases the area 4 times.

HYDRAULIC CIRCUITS

All *hydraulic circuits* are essentially the same regardless of the application (machine tools, airplanes, farm equipment, boats, etc.). There are four basic components required: a tank (reservoir) to hold the fluid, a pump to force the fluid through the system (the pump is driven by an electric motor or other power source), valves to control fluid pressure and flow, and an actuator (a cylinder for linear or a motor for rotary motion) to convert the energy of fluid movement into mechanical force to do the work.

The complexity of hydraulic systems will vary, of course, depending on the application.

HYDRAULIC EQUIPMENT

Reservoirs

Although it primarily serves as a supply source for hydraulic system fluid, a

well-designed reservoir serves other useful purposes. Its construction assists in the separation of air and contaminants from the fluid and helps to dissipate heat generated within the system.

A reservoir which conforms to Joint Industry Conference (JIC) specifications is illustrated in Fig. 21-1-4. The tank is constructed of hot-rolled steel plates with welded joints. Extensions of the tank ends support the tank and may be bolted to the floor. The tank bottom is concave and has a drain plug in the middle.

Strainers and Filters

To ensure long life and trouble-free performance of hydraulic components, it is important to keep the hydraulic fluid clean.

Filters, strainers, and magnetic plugs can be used to remove foreign particles from the hydraulic fluid and are most effective as safeguards against contamination. Typical filters and strainers are shown in Fig. 21-1-5.

Hydraulic Fluids

Hydraulic fluid characteristics have an important effect on equipment performance and maintenance.

In addition to serving as a power transmitting medium, hydraulic fluid must keep wear to a minimum by providing good lubrication.

Fig. 21-1-4 Reservoir.

(A) GEAR

(B) VANE

(C) TYPICAL OIL INTAKE STRAINER INSTALLATIONS

Fig. 21-1-5 Filters and strainers.

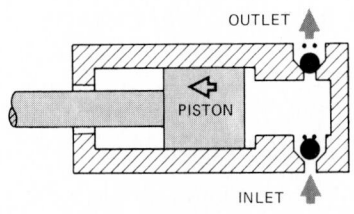

(C) RECIPROCATING

Fig. 21-1-6 Positive-displacement pumps.

Hydraulic Pumps

Hydraulic pumps are devices for converting mechanical energy into hydraulic energy. When a hydraulic pump is operated, it performs two functions. First, its mechanical action creates a partial vacuum at the pump inlet which enables atmospheric pressure in the reservoir to force liquid through the inlet line into the pump. Second, its mechanical action delivers this liquid to the pump outlet and forces it into the hydraulic system.

Pumps are broadly classified as either non-positive-displacement or positive-displacement units.

A *non-positive-displacement pump* produces a continuous flow. However, because of its design, there is no positive internal seal against leakage, and its output varies considerably as pressure varies.

Practically all pumps used in power hydraulic systems—whether on industrial machinery, mobile vehicles, or aircraft—are of the positive-displacement type.

A *positive-displacement pump* produces a pulsating flow, but since it provides a positive internal seal against leakage, its output is relatively unaffected by variations in system pressure. See Fig. 21-1-6.

Actuators

Hydraulic actuators and hydraulic pumps have opposite functions; the actuators convert hydraulic energy back to mechanical energy to perform useful work. In a typical circuit, the actuator is mechanically linked to the workload, and is actuated by fluid from the pump so that thrust or torque is transferred to the work.

Actuators can be classified broadly as either linear or rotary. A *linear actuator* such as a ram or cylinder is used for such operations as clamping, pressing, or traverse and feed motions. *Rotary actuator* applications include chucking, indexing, and turning.

Linear Actuators The simplest linear actuator is the single-acting cylinder or ram (Fig. 21-1-7A), which applies force in only one direction. Fluid directed into the housing displaces the rod hydraulically; the retracting force can be gravity or some mechanical means, such as a spring.

A double-acting cylinder (Fig. 21-1-7B) permits application of hydraulic pressure on either side of the piston to control linear motion in either of two opposite directions. The type shown is called a *differential cylinder*

(A) SINGLE ACTING

(B) DOUBLE ACTING DIFFERENTIAL TYPE

(C) DOUBLE ACTING NON-DIFFERENTIAL TYPE

Fig. 21-1-7 Cylinders.

because the piston area at the left is larger, providing a slower, more powerful work stroke when fluid pressure is applied to the LH side. The return stroke will be faster due to the smaller piston area. Reciprocating motions of this type are required on machine tools such as shapers. If equal forces in both directions are required, the piston rod is designed to extend through both ends of the cylinder, as in Fig. 21-1-7C. This is a *non-differential-type cylinder*.

Rotary Actuators Rotary actuators or motors, like rotary pumps, are of either the gear, vane, or piston design. The piston design is further divided into radial or axial types. Actually, many hydraulic pumps can be used as motors with little or no modification.

Valves

Valves are used in hydraulic circuits to control pressure, as well as the direction and rate of fluid flow. They can be classified as pressure controls, directional controls, and flow controls.

Valves, like pipes, are generally rated according to size and pressure. The valve name may be based on its usual function (relief valve) or a feature of its construction (gate valve).

Pressure Control Valves The most common type of pressure control valve is the *relief valve*. A relief valve may be used to provide overload protection for circuit components or to limit the force or torque exerted by a linear actuator or rotary motor. With a relief valve the flow is diverted back to the reservoir.

A relief valve of the simple type is shown in Fig. 21-1-8A. The valve is installed so that one port is connected to the pressure line, the other to the reservoir. The ball is held on its seat by thrust of the spring. Spring thrust can be changed by turning the adjusting screw.

When pressure at the valve inlet is insufficient to overcome spring force, the ball remains on its seat and the valve is closed as shown. The position of the ball prevents flow through the valve.

When pressure at the valve inlet exceeds the adjusted spring force, the ball is forced off its seat and the valve is opened. Liquid flows from the pressure line through the valve to the reservoir. This diversion of flow prevents further pressure increase in the pressure line. When pressure decreases below the valve setting, the spring reseats the ball and the valve is again closed.

Sequence valves, as shown in Fig. 21-1-8B, are used on machines requiring operations which must occur in a proper sequence. Sequence valves will not operate until the pressure of one unit has reached a certain level. Sequence valves differ from relief valves in that when a sequence valve is used the flow is diverted to another portion of the system to perform work.

Pressure-reducing valves are used to block or modulate flow at a preset pressure. Unlike the two valves discussed so far, reducing valves are normally open, the most common being a spool valve, which is shown in Fig. 21-1-8C.

Directional Control Valves Although all *directional control valves* have a common function of controlling direction of fluid flow, they vary considerably in construction and operation.

Directional controls can be classified on the basis of principal characteristics: the type of internal valving element, the method of actuating the valving element, and the number of positions of the valving element as well as the flow paths created in the various positions.

Check valves are the simplest of all directional control valves. They permit fluid to flow in one direction only. See Fig. 21-1-9.

Fig. 21-1-9 Check valves.

Multiple-way valves are known as two-way, three-way, or four-way, depending on the number of primary parts the valve contains. See Figs. 21-1-10 through 21-1-12.

Many variations are possible when classifying valves according to the number of positions or flow paths. They may be of the simple on-off variety or have a wide selection of flow paths through them.

(A) PRESSURE RELIEF

(B) SEQUENCE VALVE

(C) PRESSURE REDUCING

Fig. 21-1-8 Pressure-control valves.

Fig. 21-1-10 Two-way, two-position valves.

The operation of a two-way sliding-spool valve is shown in Fig. 21-1-10.

Usually containing a sliding spool, two-way valves provide a choice of two flow paths exclusive of the valve's center position. In either shifted position, the pressure port is open to one cylinder port, but the opposite cylinder port is not open to the tank. (In a four-way valve, the opposite cylinder port would be open to the tank, making two flow paths in either shifted position.)

Flow Control Valves The widespread use of hydraulic circuitry in machine tool applications stems in part from the ease with which traverse and feed rates can be controlled through the use of various types of flow controls.

The speed of an actuator, whether it be a hydraulic cylinder or a motor, is determined by its displacement and the amount of fluid available to it.

NORMAL POSITION

ACTUATED LEFT

TO TANK

Fig. 21-1-11 Four-way, two-position valve.

Fig. 21-1-12 Four-way, three-position valve.

Changing the displacement would change the operating speed and would have an inverse effect on the force or torque output. For this and other reasons, such as the impracticability of changing the bore of a cylinder, speed control usually is accomplished by controlling the flow rate.

With full pump delivery directed into a cylinder, the actuator travels at its maximum speed. A larger pump would move it faster; a smaller one would move it more slowly. Logically, then, the use of a variable delivery pump would provide a relatively simple means of controlling actuator speed. Since operating pressure would vary with the workload, the volumetric efficiency of the pump would determine feed-rate accuracy.

A restrictor-type flow control valve (Fig. 21-1-13) has a compensator spool, which is held in position by a light spring, and pressure sensor beyond the throttle. In this type, the compensator spool is normally held in the open position, and tends to shut off, not permitting fluid in excess of the throttle setting to enter the valve. The balance of the pump delivery is then available for other purposes.

Fig. 21-1-13 Flow-control valve.

(A) PICTORIAL DIAGRAM

(B) CUTAWAY DIAGRAM

Fig. 21-1-14   Comparison between pictorial and cutaway diagrams.

FLUID CIRCUIT DIAGRAMS

An accurate diagram of the fluid circuit is one of the most important pieces of literature accompanying a machine. The information shown in the circuit diagram is essential for an understanding of the operation of the machine and for installation or troubleshooting.

Not all circuit diagrams are complete. For instance, a diagram which is to be used only for piping would not need to show the sequence of operations or sizes and ratings of the components.

Fluid circuit diagrams are of four types, namely, pictorial, cutaway, graphical, and combination. The type

Fig. 21-1-15   Typical pictorial diagrams.

used depends on the amount and kind of information required. Pictorial and cutaway diagrams are illustrated in Fig. 21-1-14.

Pictorial Diagrams

A pictorial diagram is primarily used to show the piping arrangement of a circuit. The symbols are outline drawings showing the actual external shape of the components, with the piping shown to the various parts of the units. Because they show nothing of the internal construction or function of the components, pictorial diagrams have little value for instruction or troubleshooting. Pictorial symbols are difficult to standardize on a functional basis. Typical pictorial symbols are shown in Figs. 21-1-14A and 21-1-15.

Cutaway Diagrams

Cutaway diagrams contain much information about the operation of a circuit and the construction and operation of its components. These diagrams are ideally suited for instruction and are widely used for that purpose. Because of the time and cost involved, they are seldom made for other purposes. Cutaway symbols are difficult to draw, and the part functions are not readily apparent. Figures 21-1-5 and 21-1-14B illustrate typical cutaway symbols for common hydraulic parts.

Arrangement of Symbols

Symbols are arranged in the diagram to facilitate the use of straight interconnecting lines. Where components have definite mechanical, functional, or otherwise important relationships to one another, their symbols should be so placed in the diagram. The conductors (interconnecting lines) are drawn straight for clarity. They are not intended to indicate actual installation of pipelines and are drawn as single or double lines depending on the type of diagram. Single lines are used on pictorial diagrams, double lines on cutaway diagrams.

References and Source Materials
1. Crane Canada, Limited.
2. Vickers, Incorporated.

ASSIGNMENTS

See Assignments 1 and 2 for Unit 21-1 on page 433.

UNIT 21-2
Graphical Diagrams

Graphical diagrams[1] are preferred by most application and service engineers for designing and troubleshooting hydraulic circuits. The graphical symbols are combinations of simple geometric figures which are easy to work with and which clearly define the types and functions of the components. No attempt is made to show the shape or internal construction of a component. Note the differences in the diagrams shown in Fig. 21-2-1.

Some of the standard graphical symbols specified by the Joint Industry Conference (JIC) are shown in Fig. 21-2-2. The symbols can be combined in any form that is necessary to correctly depict a composite unit. Unless multiple diagrams are furnished showing the various phases of operation, the symbols will be shown in a diagram in their normal or neutral position. The reservoir symbol, like that of the electrical ground symbol, may be repeated several times on a graphical diagram.

(A) GRAPHICAL DIAGRAM

(B) CUTAWAY DIAGRAM

(C) PICTORIAL DIAGRAM

(D) COMBINATION DIAGRAM

LIST OF COMPONENTS
A—RESERVOIR
B—ELECTRIC MOTOR
C—PUMP
D—MAXIMUM PRESSURE (RELIEF) VALVE
E—DIRECTIONAL VALVE
F—FLOW CONTROL VALVE
G—RIGHT ANGLE CHECK VALVE
H—CYLINDER

Fig. 21-2-1 Fluid circuit diagrams.

LINE, WORKING (MAIN)

LINE, PILOT (FOR CONTROL)

LINE, LIQUID DRAIN

(A) CONDUCTORS

ONE DIRECTION   BOTH DIRECTIONS

(B) FLOW

FILTERING ELEMENT

(C) FILTER STRAINER

(D) MOTOR

FIXED DISPLACEMENT   VARIABLE DISPLACEMENT

(E) PUMPS

SINGLE ACTING

SINGLE END ROD   DOUBLE END ROD
DOUBLE ACTING

(F) CYLINDERS

VENTED   PRESSURIZED

ABOVE FLUID LEVEL

*SHOW LINE ENTERING OR EXITING BELOW RESERVOIR ONLY WHEN BOTTOM CONNECTION IS ESSENTIAL TO CIRCUIT FUNCTION.

BELOW FLUID LEVEL

BELOW   ABOVE
SIMPLIFIED

(G) RESERVOIR WITH CONNECTING LINES

Fig. 21-2-2 Graphical symbols.

(A) MISCELLANEOUS VALVES

CHECK

2-WAY MANUAL

(B) FLOW CONTROL VALVES

FIXED

VARIABLE

ADJUSTABLE WITH BYPASS

(C) PRESSURE CONTROL VLAVES

RELIEF

RELIEF WITH VENT

SEQUENCE

REDUCING

XX DENOTES PRESSURE SETTING

SINGLE

MULTIPLE

ENVELOPES

PORTS

PORTS CLOSED OR BLOCKED

PORTS OPEN

ENVELOPES AND PORTS

SYMBOL APPLICATION

SPRING

MANUAL – GENERAL SYMBOL WITHOUT INDICATING TYPE

PUSH BUTTON

LEVER

PEDAL OR TREADLE

SOLENOID

MECHANICAL

ACTUATORS AND CONTROLS

NORMALLY CLOSED

NORMALLY OPEN

INFINITE POSITION

TWO-WAY VALVES

NORMALLY CLOSED

TO CYLINDER

INLET (PRESSURE) TO TANK OR EXHAUST

UNACTUATED

TO CYLINDER

PUSH

INLET (PRESSURE) TO TANK OR EXHAUST

ACTUATED

NORMALLY OPEN

TO CYLINDER

TO TANK OR EXHAUST INLET (PRESSURE)

UNACTUATED

TO CYLINDER

PUSH

TO TANK OR EXHAUST INLET (PRESSURE)

ACTUATED

THREE-WAY TWO-POSITION VALVE

NORMAL

ACTUATED

T TO TANK

T TO TANK

TWO POSITION

NORMAL

ACTUATED RIGHT

ACTUATED LEFT

T TO TANK

T TO TANK

THREE POSITION

FOUR-WAY VALVES

(D) DIRECTIONAL CONTROL VALVES

Fig. 21-2-3 Graphical valve symbols.

Directional valves, because of various types of controls and flow paths, require a number of different symbols. A basic directional valve symbol consists of an envelope (square) for each position of the valve. The envelopes are drawn side by side in their correct order with each showing the flow path for its position. The method of actuating a valve is also shown symbolically. See Fig. 21-2-3.

SYMBOL RULES

1. Symbols show connections, flow paths, and function of the component represented. They do not indicate conditions occurring during transition from one flow-path arrangement to another. Further, they do not indicate construction or values such as pressure, flow-rate, and other component settings.
2. Symbols do not indicate the location of ports, direction of shifting of spools, or position of control elements on actual components.
3. Symbols may be rotated or reversed without altering their meaning.
4. Line width does not alter meaning of symbols.
5. Symbols may be drawn any suitable size. Size may be varied for emphasis or clarity.
6. Letter combinations used as parts of graphical symbols are not necessarily abbreviations.
7. Where flow lines cross, a loop is used, except within a symbol envelope. A loop may be used in this case if clarity is improved by its use.
8. In multiple envelope symbols, the flow condition shown nearest a control symbol takes place when that control is caused or permitted to actuate.
9. Each symbol is drawn to show the normal or neutral condition of the component unless several circuit diagrams are furnished showing various phases of circuit operation.
10. Arrows should be used within symbol envelopes to show the direction of flow path in the component as used in the application represented. A double-ended arrow is used to indicate reversing flow.
11. External ports are where flow lines connect to the basic symbol, except where component enclosure is used.
12. External ports are at the intersections of flow lines and the component enclosure symbol when enclosure is used.

HYDRAULIC CIRCUITS

EXAMPLE 1 Figure 21-2-4 is a typical example of a simple hydraulic circuit. When the directional control valve (two-position, three-way, spring-operated, push-button control) is in the position shown in Fig. 21-2-4A, the cylinder piston extends. When the piston reaches the end of its stroke, the pump flow is then diverted through the pressure control (relief) valve.

When the push button is operated on the directional control valve, the port to the pump is blocked, as shown in Fig. 21-2-4B. The weight (mass) of the piston forces the hydraulic fluid from the cylinder through the directional control valve to the reservoir, and the piston retracts. Since pump flow through the directional control valve is blocked, the fluid is diverted through the pressure control (relief) valve back to the reservoir.

EXAMPLE 2 Figure 21-2-5 shows a simple hydraulic circuit using a three-position, four-way valve controlling the movement of a double-acting cylinder. Figure 21-2-5A shows the control valve in its neutral position. Since pump flow through the directional control valve is blocked, the fluid is diverted through the pressure control (relief) valve back to the reservoir.

Fig. 21-2-4 Simple hydraulic circuit.

Fig. 21-2-5 Graphic diagram showing a three-position, four-way valve controlling a cylinder.

When the operator shifts the control valve to the right, fluid is directed to the head of the cylinder (Fig. 21-2-5B). The fluid at the other end of the cylinder is directed back to the reservoir. The pump continues to pump oil after the cylinder rod has completed its stroke. This excess oil is returned via the relief valve to the reservoir.

At the end of the forward stroke, the operator shifts the control valve to the left which causes the cylinder rod to retract (Fig. 21-2-5C). Fluid is directed to the rod end of the cylinder and the fluid from the head end returns to the reservoir. When the operator releases the control valve handle, the valve returns to its neutral position.

Reference and Source Material
1. Vickers, Incorporated.

ASSIGNMENTS

See Assignments 3 and 4 for Unit 21-2 on page 434.

UNIT 21-3
Pneumatics

Compressed air has been used as a means of transmitting power for a long time, but in recent years pneumatics has taken on a new sophistication. With the availability of miniaturized valves and fluidic devices, air can now supply the nerves and brains, as well as the muscles, of complex automated equipment.

What is air? *Air* may be defined as a colorless, odorless, tasteless gas, principally composed of nitrogen, oxygen, carbon dioxide, and water vapor. Its molecules are widely separated and in constant motion, traveling in straight lines. At sea level, air exists at a pressure of 14.7 lb/in.^{2} (101.3 kPa). Air mixes readily with other fluids.

BASIC LAWS

There are several simple basic laws governing the behavior of air that are important to pneumatic system design.

Boyle's Law This law defines the relationship between volume and pressure and states that if the temperature is constant, the volume of a given mass of gas varies inversely as its absolute pressure. It can be written

$$P_1V_1 = P_2V_2$$

where P_1 = the initial pressure (absolute)
  V_1 = the initial volume
  P_2 = the final pressure (absolute)
  V_2 = the final volume

Charles' Law This law refers to the behavior of a gas when changes in temperature take place and states that when there is no change in volume, the pressure of a gas varies directly with its absolute temperature.

Pascal's Law This law states that the pressure of a gas in a container is transmitted undiminished in all directions and acts at right angles to the surfaces of the container. See Fig. 21-1-1.

Fig. 21-3-1 Receiver critical back pressure is 53 percent of supply pressure.

Airflow
Air will only flow when a difference of pressure exists and flows *toward* the lower pressure. The rate of flow depends on the initial pressure, the difference in pressure, and the size, shape, and smoothness of the orifice. The shape and smoothness of the orifice are important. A straight or smoothly curved pipe assists airflow, and smooth inner walls reduce friction. Rough surfaces slow adjacent layers of air and reduce the effective area of the pipe.

Critical Back-Pressure Ratio
In an enclosed air system, one other factor must be considered. Figure 21-3-1 shows a supply tank at a constant pressure of 100 psi connected through a stop valve to a receiver tank at atmospheric pressure (0 psi). If the stop valve is opened, air will flow from A to B at a rate dependent on the orifice and pipe characteristics. Continuing flow will reduce the differential pressure between the supply and receiver tanks, but it is an observed phenomenon in gas behavior that the rate of flow does not change until the receiver tank pressure has reached 53 percent of the supply pressure, in this case, 53 psi. Conversely, if the pressure in the receiver falls below 53 psi, any further decrease in pressure will not increase airflow. This is called the *critical back-pressure ratio*, and can be important when selecting the size of a valve.

PNEUMATIC EQUIPMENT

Air Supply Installation
Before any pneumatic equipment can operate, it must have a supply of air at the correct pressure, in sufficient volume, and properly conditioned. See Fig. 21-3-2. The supply starts at the compressor which has a rated output pressure in psi or pascals and a rated volume in standard cubic feet per minute (SCFM) or cubic meters per second. This is often known as *free air delivered* (FAD). A standard cubic foot (ft^3) of air is the weight of air contained in 1 ft^3 of space at 70°F and 14.7 psi absolute. A standard cubic meter of air is the mass of air contained in 1 m^3 of space at 21°C and 101.3 kPa.

The compressor obeys Boyle's law by pushing air at atmospheric pressure into a smaller volume to increase the

Fig. 21-3-2 Basic pneumatic components.

pressure. Unfortunately, Charles' law dictates that as pressure increases, so does temperature; and warm air contains more water vapor than cool air.

To decrease water content, a filter should be fitted immediately downstream of the compressor where moisture can be precipitated and drained off.

Although there is a definite trend toward oil-free compressors, industrial compressors are still predominantly of the oil-flooded type. Most have internal oil separators that are of quite high efficiency, but a certain amount of oil vapor does get through. This oil can contain wear particles from compressor parts; it also tends to be burned due to the heat of compression. To prevent contamination of the air system, it is recommended that an efficient oil separator be fitted between the compressor discharge point and the air receiver.

Air Distribution System
The air distribution system will be considered as beginning at the storage tank where a supply of air is available with most of the oil vapor and solids already removed. Between the tank outlet and the feeds to the individual systems, the biggest problem is how to get rid of water. A certain amount of precipitation will take place in the tank, so a drain is normally provided there. Methods of removing water in other parts of a pneumatic circuit are shown in Fig. 21-3-3.

Conditioning the Air
So far, the compression of the air, the removal of the moisture from the air, and the piping of the air to the supply points have been discussed. The next concern is to condition the air before it enters the tool, control, or power elements to do some useful work.

Filters The average filter available (Fig. 21-3-4) is designed to do two jobs: remove moisture and remove dirt. When air enters the filter, its path is changed abruptly to a rotary direction, so centrifugal force hurls any particles of water outward. Here they collect on the sides of the filter bowl and gravitate to the bottom where a baffle prevents air turbulence. This area is called the *quiet zone* and is drained by an automatic or manual valve.

Dirt exists to a greater or lesser extent in various forms in any plant system and is intercepted in most filters by a cartridge element. These ele-

Fig. 21-3-4 Typical air filter.

ments are rated by the size, in microns, of particles they will intercept. A micron is .000 254 in. or 0.001 mm.

For work or power air, a filter of 50- to 60-micron rating is normally sufficient, since this air comes in contact with contamination from cylinder or shaft packings. Control air is not normally subject to this, so should be passed through a secondary filter of about 5-micron rating. This finer filtration will add appreciably to the efficiency and life of the control section of the circuit.

Pressure Regulators (Fig. 21-3-5)
The three main reasons for regulating air pressure are

1. To keep wasted air to a minimum
2. To achieve maximum consistency in circuit performance
3. To maintain an optimum balance between work output of components and their wear rate

A regulator must be chosen to give the correct working range, usually 0 to 125 lb/in.^{2} (0 to 1000 kPa).

In locating the regulator, always try to get it upstream of a lubricator, since air can be contaminated by the interaction of some oils on the regulator diaphragm.

Fig. 21-3-5 Simple spring-type regulator.

Air Lubricators
The standard-type lubricator (Fig. 21-3-6) operates by creating a pressure differential between the lubricant container and a metering chamber. This causes the oil to enter the metering

Fig. 21-3-3 Air distribution diagram.

Fig. 21-3-6 *Air-lubricator oil-fog type.*

chamber and disperse into the pipeline as fog. Lubricators are designed to operate with a certain size pipe and airflow.

Common types of lubricators in use are the oil-fog type and the micro-fog type. The *oil-fog type* disperses relatively large drops of oil which have a tendency to fall out early, the normal carrying distance before fallout being about 15 ft (5 m) in a straight length of pipe.

The *micro-fog type* disperses much smaller oil particles which remain in suspension more easily. The suggested carrying distance before fallout for this type is 25 ft (8 m) in a straight pipe.

In choosing the lubricant, it is preferable to use the grade recommended by the component manufacturers.

AIR CIRCUIT COMPONENTS

Air circuit elements can be considered under the three separate functions they perform: power, control, and signal.

Power Elements

Cylinders The most common power elements are cylinders, and many factors influence their choice.

If work is to be performed in one direction only, then a *single-acting cylinder* (Fig. 21-1-7A) may be used. This type is retracted by an internal spring, an external load or gravity, or bucking air. If work is to be done in both directions, i.e., if the return load exceeds cylinder friction, a *double-acting cylinder* (Fig. 21-1-7B) is needed.

The size of the cylinder will depend on the magnitude of load and the distance it must be moved. The bore can be found from the formula $A = F/P$,

where F is the load and P is the air pressure available. The stroke should equal the distance to be moved.

After it is decided which cylinder will be used, its air consumption must be calculated. This is the volume found by calculating the volume of the free end of the cylinder, adding the volume of the piston rod end if the piston is double-acting, and multiplying by the frequency in strokes per minute.

This calculation is important, but it is often neglected. It must be done not only to decide the size of lubricator required, but also to check if the plant system has the capacity to operate the circuit.

Air Motors *Air motors* are used to convert the energy of compressed air into continuous torque. They are not the most efficient means of producing torque, since the average motor needs about 5 hp (4 kW) at the compressor to produce 1.25 hp (1 kW) at the motor, but they still have a lot in their favor.

With the growing popularity of pneumatic controls, air motors can make some machines or systems independent of electricity, particularly if the compressor is driven by an internal combustion engine. They cannot be harmed by stalling; they do not overheat, since the air expanding through the motor has a cooling effect; they have a very high power-to-mass ratio; and they are very reliable (seldom break down, only wear out slowly with lots of warning).

There are three main types. See Fig. 21-3-7. The two *piston types* are the workhorses, giving high power at lower speeds. *Vane types* are the racehorses, suitable for ligher loads at higher speeds, and they are more compact. The majority of low-kilowatt (horsepower) motors in use are of this type.

In selecting an air motor, it must be remembered that power is propor-

tional to revolutions per minute and pressure. This is the pressure at the motor inlet, so air lines must be big enough to pass the required volume of air. If several motors are operating intermittently from the same air system, surge tanks or air reservoirs should be provided. The motor chosen should give the required power and revolutions per minute at about one-half the maximum pressure, so that there will be no power loss under adverse conditions.

Control Elements

Power Valves *Power valves* direct airflow to and from the work elements.

There are various designs of valve action, and it is helpful if the action requires low thrust to actuate the valve. More important, the required thrust should be constant and unaffected by variations in pressure or airflow through the valve or by changes in friction within the valve.

To eliminate unnecessary air wastage as the valve shifts, the action should block the air supply from the connecting flow paths in the valve as it moves through midposition.

In most applications, the valve action should be detented to keep it in its selected position in the event of failure in the controlling medium (electricity or air).

A good valve action provides a variety of flow paths by accepting and valving air at any port. This allows the use of the same valve for different circuit functions, such as normally open, normally closed, two-way, three-way, dual-pressure, and others.

There are three methods commonly used to actuate a power valve: mechanical, electrical, and air. In a two-position valve, any combination of these may be used.

The selection of power valves is governed by the airflow required, the

(A) AXIAL PISTON

(B) RADIAL PISTON

(C) VANE TYPE

Fig. 21-3-7 **Air motors.**

flow paths needed, and the method of actuation.

Valves are termed by the number of flow paths they provide, either two- or three-position.

Two-position means that two flow conditions exist that are relative to the position of the valve.

Three-position is similar but has a third flow condition when the valve mechanism is centered.

One other factor must be known when describing a power valve. That is the number of *ways* it may be used and refers basically to the number of ports or connections in the valve body.

Figure 21-3-8 shows two types of two-way, two-position valves. Air connected to one port on the right valve will only flow out of the other port when the valve is actuated. This is known as a normally closed, two-way, two-position valve. On the left is a normally open, two-way, two-position valve, since air flows until the valve is actuated.

PNEUMATIC CIRCUIT DIAGRAMS

Pictorial and Cutaway Diagrams The same types of symbols as those shown in the hydraulic pictorial and cutaway diagrams also apply to pneumatic circuits. See Unit 21-1.

Graphic Symbols With the exception of the symbols shown in Fig. 21-3-9, pneumatic graphic symbols are identical to those used in hydraulic circuits. See Figs. 21-2-2 and 21-2-3.

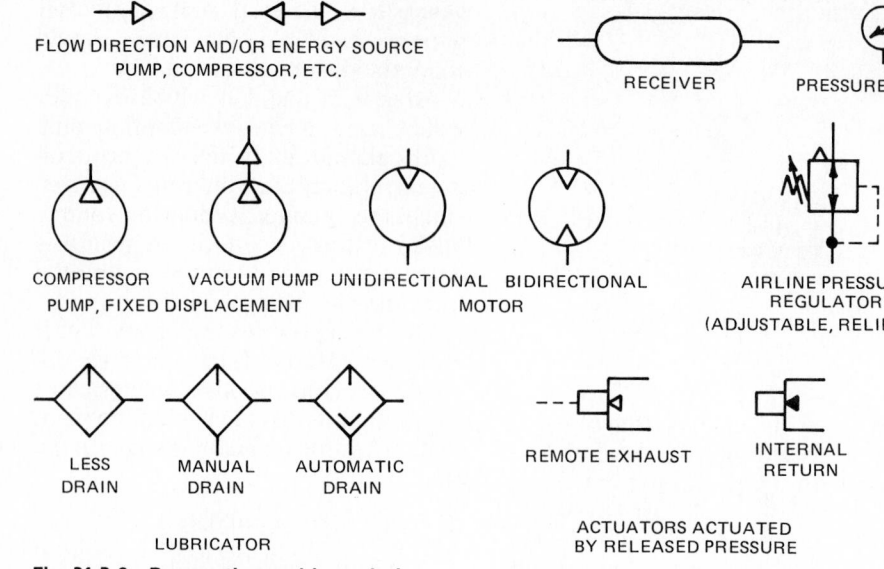

Fig. 21-3-9 Pneumatic graphic symbols.

Pneumatic Circuits

EXAMPLE 1 See Fig. 21-3-10. When the operator shifts the control valve to the right, air is directed to the head end of the cylinder. The return air is directed back through the control valve to the atmosphere. At the end of the forward stroke, the operator shifts the control valve to the left. Air is directed to the rod end of the cylinder, and the air from the head end is directed to the atmosphere by way of the control valve. The pump continues to pump air after the cylinder rod has completed its stroke. This excess air is disposed of by the pressure regulator valve to the atmosphere. When the operator releases the control-valve handle, the spring-centered valve returns to neutral. The pressure on the cylinder is relieved while the pressure inlet remains closed, preventing the valve from draining the compressor. With the pressure removed from both ends of the cylinder, the cylinder rod can be moved readily. This action is called *floating* and is used in both air and hydraulic circuitry.

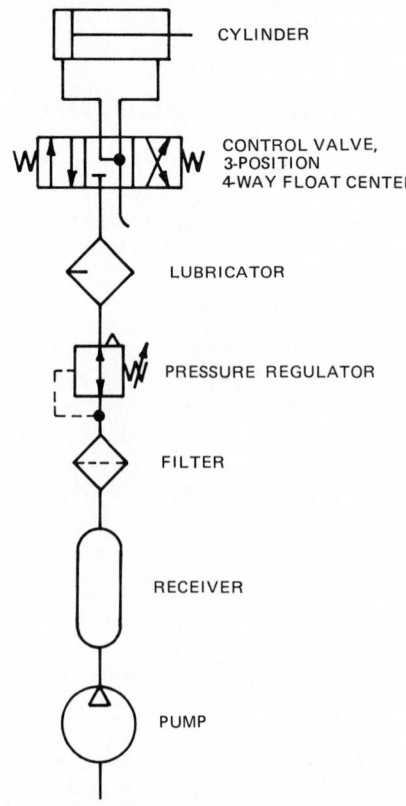

Fig. 21-3-10 Simple pneumatic circuit.

Fig. 21-3-8 Two-way, two-position valves.

Fig. 21-3-11 Sequence circuit.

WORK CYLINDER

WORK

SEQUENCE VALVE 1

SEQUENCE VALVE 2

CLAMP CYLINDER

CONTROL VALVE 2-POSITION, 4-WAY

LUBRICATION

PRESSURE REGULATOR

FILTER

RECEIVER

PUMP

EXAMPLE 2 A typical sequence circuit is shown in Fig. 21-3-11. The sequence of operation is (1) extend clamp cylinder, (2) extend work cylinder, (3) retract work cylinder, and (4) retract clamp cylinder. When the control valve is shifted, air is directed into the head of the clamp cylinder extending the piston and clamping the part. Air also flows to sequence valve 1, but no flow occurs through the valve because of the spring-loaded ball. After the clamp cylinder has completed its stroke, pressure builds up in the line and overcomes the spring tension in the valve, permitting air to pass through the valve to the head of the work cylinder. The piston of the work cylinder extends to perform the work. The pressure in the line is controlled by the pressure regulator valve.

When the control valve is released, air pressure flows into the rod end of the work cylinder, retracting the piston. The air also flows to sequence valve 2 but is blocked by the spring-loaded valve. The air from the work cylinder head is forced through the sequence and control valves to the atmosphere.

After the work cylinder is fully retracted, the pressure again builds up and overcomes the spring tension in sequence valve 2, forcing air into the rod end of the clamp cylinder and thus retracting the piston and releasing the work. The air in the head of the work cylinder is forced out of the cylinder through the sequence and control valves to the atmosphere.

References and Source Material
1. J. Mooney, "Course on Basic Pneumatics," *Design Engineering*.
2. Holman Bros., Limited, Maxam-Nopak Division.

ASSIGNMENTS

See Assignments 5 through 7 for Unit 21-3 on page 435.

ASSIGNMENTS for Chapter 21

Assignments for Unit 21-1, Hydraulics

1. On a B- or A3-size sheet, make a pictorial diagram of the automobile brake hydraulic system shown in Fig. 21-1-A. Label the parts. Scale is to suit.

2. On a B- or A3-size sheet, make a cutaway diagram of the hydraulic circuit shown in Fig. 21-1-B. Label the parts. Scale is to suit.

Fig. 21-1-A Brake hydraulic system.

MASTER CYLINDER

BRAKE PEDAL

FRONT DISK BRAKE

ALARM INDICATOR SWITCH

REAR WHEEL CYLINDER

BRAKE LINES

REAR WHEEL CYLINDER

FRONT DISK BRAKE

Fig. 21-1-B Hydraulic circuit with one double-acting differential-type cylinder.

CHECK VALVE

DIRECTIONAL VALVE

FLOW CONTROL VALVE

PRESSURE RELIEF VALVE

CYLINDER

PUMP

MOTOR

RESERVOIR

Assignments for Unit 21-2, Graphical Diagrams

3. On a B- or A3-size sheet, make a graphical diagram of the hydraulic circuit shown in Fig. 21-2-A.

4. On a B- or A3-size sheet, make a graphical diagram of the hydraulic circuit shown in Fig. 21-2-B.

(1) MID STROKE EXTENDING
Solenoid B of valve F is held energized during the extending stroke. Vent line from valve D is blocked at valve G. Delivery of pump B is directed through F into head end of cylinder H. Discharge from rod end of H flows to tank through valves F and C.

LEGEND
A – RESERVOIR
B – STRAINER
C – ELECTRIC MOTOR
D – FLEXIBLE COUPLING
E – PUMP
F – MAXIMUM PRESSURE
    (RELIEF) VALVE WITH VENT
G – DIRECTIONAL VALVE MANUAL
    –3 POSITION, 4 CONNECTION
H – CHECK VALVE
J – VALVE (SEQUENCE)
    SINGLE FLOW PATH
K – CHECK VALVE
L – CYLINDER NO. I
M – CYLINDER NO. 2

POSITION OF DIRECTIONAL VALVE
FOR EXTENDING PISTON

POSITION OF DIRECTIONAL
VALVE IN NEUTRAL POSITION

POSITION OF DIRECTIONAL
VALVE FOR RETRACTING PISTON

LEGEND
A – RESERVOIR
B – PUMP
C – CHECK VALVE
D – RELIEF VALVE WITH VENT
E – CHECK VALVE
F – 4-WAY 3-POSITION DIRECTIONAL
    CONTROL VALVE
G – 3-WAY 2-POSITION DIRECTIONAL
    CONTROL VALVE
H – CYLINDER
J – STRAINER

(2) MID STROKE RETRACTING
At end of extension stroke, cam on cylinder H contacts limit switch LS. This causes solenoid B of valve F to be de-energized. F shifts to the spring offset position and directs delivery of pump B into rod end of H. Discharge from head end of H flows to tank through valves F and C.

(3) AUTOMATIC STOP
At end of retraction stroke, cam on cylinder H depresses valve G. Valve D is now vented through valves E, G, F, and C. Delivery of pump B returns to tank over valve D at low pressure. Pressure drop through C assures pilot pressure for operation of F.

TO SEQUENCE THE ADVANCE OF TWO CYLINDERS, A SINGLE FLOW SEQUENCE VALVE IN THE SUPPLY LINE OF NO.2 CYLINDER (M) IS SET TO OPEN AT A PRESSURE IN EXCESS OF THAT REQUIRED TO ADVANCE NO.1 CYLINDER (L). AS THE DIRECTIONAL VALVE IS SHIFTED IN POSITION FOR EXTENDING PISTONS, FLOW IS DIRECTED FROM PORT (a). WHEN CYLINDER (L) COMPLETES ITS STROKE, SYSTEM PRESSURE BUILDS UP OPENING THE SEQUENCE VALVE (J) AND ALLOWING FLOW TO ADVANCE NO.2 CYLINDER (M). BY SHIFTING THE DIRECTIONAL CONTROL VALVE TO NEUTRAL, THE CYLINDERS CAN BE STOPPED AND HELD IN ANY POSITION WHILE PERMITTING THE PUMP OUTPUT TO RESERVOIR THROUGH PORT (d). WHEN THE DIRECTIONAL CONTROL VALVE IS SHIFTED TO THE RETRACTING POSITION FLOW IS DIRECTED FROM PORT (p) TO PORT (b). THE BY-PASS CHECK VALVE (K) ALLOWS NO.2 CYLINDER (M) TO RETRACT FREELY.

SEQUENCING THE ADVANCED STROKE OF TWO HYDRAULIC CYLINDERS

Fig. 21-2-A   Sequencing the advanced stroke of two hydraulic cylinders.

(4) PUSHBUTTON START
Depressing a push button causes solenoid B of valve F to be held energized. F shifts to connect head end of cylinder H to pump B, and rod end of H to tank. Pilot flow from vent of D stops and check valve F closes. Pressures equalize through balance hole in hydrostat of D causing it to start to close. Acceleration of H takes place during the closing of the hydrostat of D.

AUTOMATIC VENTING OF RELIEF VALVE AT END OR EACH CYCLE BY CAM-OPERATED PILOT VALVE

Fig. 21-2-B   Automatic venting of a relief valve at end of each cycle by a cam-operated pilot valve.

Assignments for Unit 21-3, Pneumatics

5. On a B- or A3-size sheet, make a graphical diagram of the pneumatic circuit shown in Fig. 21-3-A. The circuit operations are as follows.

All air entering the circuit passes through the lubrication-control unit, then it is delivered at a constant controllable pressure to the inlet port of valves 1, 3, 5, 6, 7, 8, 10, 12, 13, 14, and 15.

The operation of this automatic circuit depends entirely on the operation of valve 2, which is depressed only when a bar of sufficient length is properly positioned in the machine.

With the bar in place, the operator depresses valve 1. Valve 1 passes a feed through valve 2 to the bottom end of valve 3, which passes air unrestricted to the bottom end of cylinder *A*, thus extending the piston rod to lift the saw.

Toward the end of the upward stroke, valve 5 is momentarily tripped, passing air to the bottom end of valve 6, which operates, passing air to the top end of cylinder *B*, thus retracting the piston rod to open the vise. When nearing its retracted position, the piston rod momentarily trips valve 7 to pass air to the bottom end of valve 8, which operates, passing air to the top end of cylinder *C*, thus retracting the piston rod to open the clamp. When fully retracted, a pressure buildup operates sequence valve 9, which allows air to pass into the RH end of the cylinder *D*, thus retracting the piston rod. At an appropriate point valve 10 is momentarily tripped to pass air to the top end of valve 8, which operates, passing air to the bottom end of cylinder *C*, thus extending the piston rod to clamp the bar. When fully extended, a pressure buildup operates sequence valve 11, which allows air to pass to the LH end of cylinder *D*, thus extending the piston rod to feed the required length of bar into the vise.

At a predetermined point, the piston rod of cylinder *D* momentarily trips valve 12 to pass air to the top end of valve 6, which operates, passing air to the bottom end of cylinder *B*, thus extending the piston rod to close the vise. As the vise closes, valve 13 is tripped to pass air to the top end of valve 3, which operates, exhausting the line to the bottom of cylinder *A*. The saw, therefore, descends under its own mass at a speed controlled by adjustment of restrictor valve 4. When the bar has been cut through, the saw trips valve 14 to pass air to the RH end of valve 15, which operates, passing air via valve 2 to the bottom end of valve 3, which operates to recommence the cycle.

6. On a B- or A3-size sheet, make a graphical diagram of the pneumatic circuit shown in Fig. 21-3-B.

Fig. 21-3-A   Control circuit for an automatic sawing machine.

Fig. 21-3-B   Operating a series of cylinders in sequence.

PART 5

Special Fields of Drafting

CHAPTER 22

Development and Intersections

UNIT 22-1
Surface Developments and Intersections

SURFACE DEVELOPMENTS

Many objects, such as cardboard and metal boxes, tin cans, funnels, cake pans, furnace pipes, elbows, ducts, and roof gutters, are made from flat sheet material that, when folded, formed, or rolled, will take the shape of an object. Since a definite shape and size are desired, a regular orthographic drawing of the object, such as shown in Fig. 22-1-1, is made first; then a development drawing is made to show the complete surface or surfaces laid out in a flat plane.

Sheet-Metal Development

Surface development drawing is sometimes referred to as *pattern drawing*, because the layout, when made on heavy cardboard, thin metal, or wood, is used as a pattern for tracing out the developed shape on flat material. Such patterns are used extensively in sheet-metal shops.

When making a development drawing of an object which will be constructed of thin metal, such as a tin can or a dust pan, the drafter must be concerned not only with the developed

Fig. 22-1-1 Sheet-metal application. (Johnson & Johnson)

surfaces but also with the joining of the edges of these surfaces and with exposed edges. An allowance must be made for the additional material necessary for such seams and edges. The drafter must also indicate where the material is bent. Several commonly used methods of marking bend lines are shown in Fig. 22-1-2. If the finished part is not shown with the development drawing, instructions such as bend up 90°, bend down 180°, bend up 45°, are shown beside each bend line. Figure 22-1-3 shows a number of common methods for seaming and edging. Seams are used to join edges. The

Fig. 22-1-2 Common methods used to identify fold or bend lines on development drawings.

EDGES

ALLOWANCE = 2.5 X WIRE Ø
WIRED

SINGLE
ALLOWANCE = A

DOUBLE
ALLOWANCE = 2A

HEMMED

ALLOWANCE
SIDE = A
BOTTOM = 2A
CUP JOINT

ALLOWANCE
SIDE = 2A + B
BOTTOM = A
PITTSBURGH CORNER LOCK

ALLOWANCE
BOTTOM = 3A

(1) BEADED DOVETAIL

RIVET
ALLOWANCE
BOTTOM = A
CONNECTOR = A + B

(2) FLANGED DOVETAIL

ALLOWANCE
BOTTOM = A

(3) PLAIN DOVETAIL

SIDE OUTLET JOINTS

ALLOWANCE = A
SINGLE-LAP SOLDERED

ALLOWANCE = A
SINGLE-LAP RIVETED

ALLOWANCE = 3A
FLAT LOCK

ALLOWANCE
SIDES = A
CONNECTOR = 2A
CAP-STRIP CONNECTOR

ALLOWANCE
SIDES = A
S-HOOK = 3A
S-HOOK SLIP JOINT

SEAMS

Fig. 22-1-3 Joints, seams, and edges.

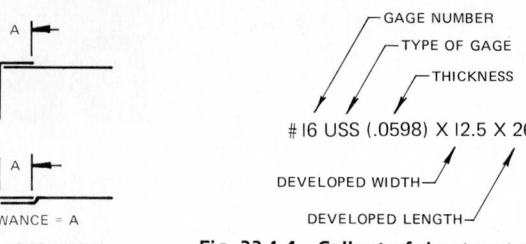

GAGE NUMBER
TYPE OF GAGE
THICKNESS

16 USS (.0598) X 12.5 X 26

DEVELOPED WIDTH
DEVELOPED LENGTH

Fig. 22-1-4 Callout of sheet-metal sizes.

segments are then stretched and sewed together to give the desired shape.

Sheet-Metal Sizes

Metal thicknesses up to .25 in. (6 mm) are usually designated by a series of gage numbers, the more common gages being shown in Table 153 of the Appendix. Metal .25 in. and over is given in inch or millimeter sizes. In calling for the material size of sheet-metal developments, customary practice is to give the gage number, type of gage, and its inch or millimeter equivalent in brackets followed by the developed width and length. See Fig. 22-1-4.

STRAIGHT-LINE DEVELOPMENT

This is the term given to the development of an object that has surfaces on a flat plane of projection. The true size

seams may be fastened together by lock seams, solder, rivets, adhesive, or welds. Exposed edges are folded or wired to give the edge added strength and to eliminate the sharp edge.

A surface is said to be *developable* if a thin sheet of flexible material, such as paper, can be wrapped smoothly about its surface. Objects that have plane, or flat, surfaces, or single-curved surfaces are developable; but if a surface is double-curved or warped, approximate methods must be used to develop the surface. The development of a spherical shape would thus be approximate, and the material would be stretched to compensate for small inaccuracies. For example, the coverings for a football or a basketball are made in segments, each segment cut to an approximate developed shape; the

SAFE EDGE
(HEMMED EDGE)

LAP SEAM

SIDE

END   BOTTOM   END

SIDE

5.20

7.70

DEVELOPMENT

Fig. 22-1-5 Development of a rectangular box.

of each side of the object is known, and these sides can be laid out in successive order. Figure 22-1-5 shows the development of a simple rectangular box having a bottom and four sides. Note that in the development of the box, an allowance is made for lap seams at the corners and for a folded edge. The fold lines are shown as thin, unbroken lines. Note also that all lines for each surface are straight.

Figure 22-1-6 shows a development drawing with a complete set of folding instructions. Figure 22-1-7 shows a letter box development drawing where the back is higher than the front surface.

ASSIGNMENT

See Assignment 1 for Unit 22-1 on page 453.

UNIT 22-2
The Packaging Industry

Packaging, which involves the principles of surface development, is one of the largest and most diversified industries in the world. Most products are packaged in metal, plastic, or cardboard containers. Many products, from candy-coated gum to large television sets, are packaged in cardboard containers. See Figs. 22-2-1 and 22-2-2. Such containers, often referred to as *cartons*, in many instances must be attractive as well as functional. They may be designed for sales appeal as well as for protection against contamination, shipping, and handling. They are also designed for temporary or permanent use.

Fig. 22-2-1 Typical commercial containers. (Photo: Fred Newman)

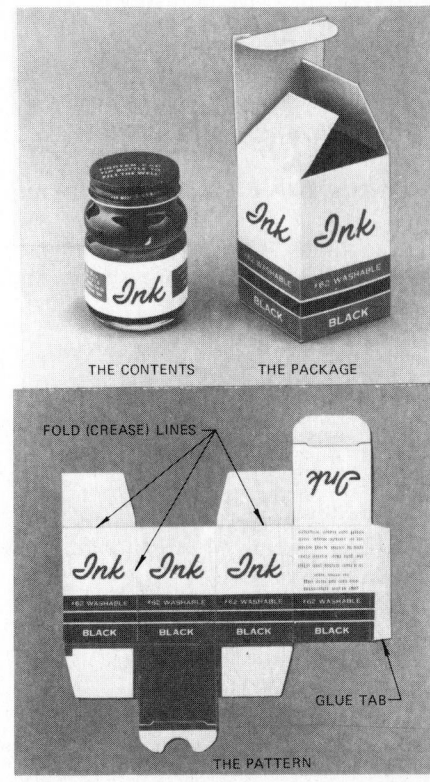

Fig. 22-2-2 A familiar container made by cutting and folding a flat sheet.

Fig. 22-1-6 Development drawing with a complete set of folding instructions.

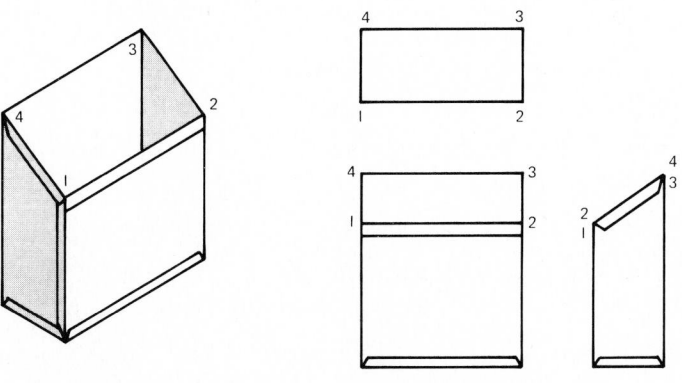

Fig. 22-1-7 Development drawing of a letter box.

DEVELOPMENT (OUTSIDE SURFACE SHOWN)

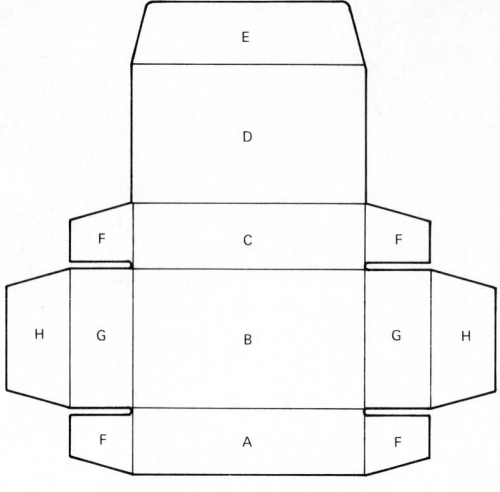

Fig. 22-2-3 Development of a one-piece carton with fold-down corners.

ASSIGNMENTS

See Assignments 2 through 4 for Unit 22-2 on page 453.

Review for Assignments
Unit 22-1   Surface Developments
Unit 4-3   Constructing a Polygon

Many cartons are printed, cut, creased, and sent to the customer in a flat position. See Fig. 22-2-3. They take less space to store and ship and are readily assembled. Locking devices such as tabs hold each box together. This type of container is used extensively by food chain operators such as Dunkin' Donuts, McDonald's, and Taco Bell. Other shapes, such as hexagons and octagons, as shown in Figs. 22-2-4 and 22-2-5, are becoming popular because of their novel form.

UNIT 22-3
Radial Line Development of Flat Surfaces

Development of a Right Pyramid with True Length of Edge Lines Shown See Fig. 22-3-1. A right pyramid is a pyramid having all the lateral edges (from vertex to the base) of equal length. Since the true length of the lateral edges is shown in the front view (line 0-1 or 0-3) and the top view shows the true lengths of the edges of the base (lines 1-2, 2-3, etc.), the development may be constructed as follows: with 0 as center (corresponding to the apex) and with a radius equal to the true length of the lateral edges (line 0-1 in the front view), draw an arc as shown. Drop a perpendicular from 0 to intersect the arc at point 3. With a radius equal to the length of the edges of the base (line 1-2 on the top view), start at point 3 and step off the distances 3-2, 2-1, 3-4, and 4-1 on the large arc. Join these points with straight lines. These points are then connected to point 0 by straight lines to complete the development. Lines 0-2, 0-3, and 0-4 are the lines on which the development is folded to shape the pyramid. The base and seam allowances have been omitted for clarity.

Fig. 22-2-4 Development of a truncated hexagon.

NOTE—ALLOWANCES FOR SEAMS AND JOINTS NOT SHOWN

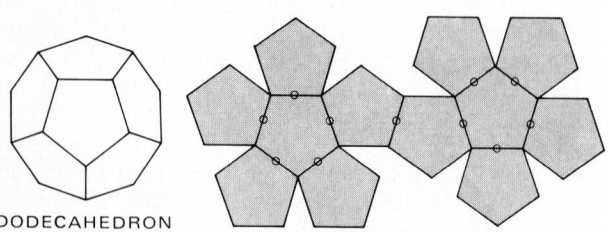

DODECAHEDRON

Fig. 22-2-5 Twelve- and twenty-sided shapes.

ICOSAHEDRON

Fig. 22-3-1 Development of a right pyramid with true length of edge lines shown.

Fig. 22-3-2 Development of a right pyramid with true length edge lines not shown.

In developing a truncated pyramid of this type, the procedure is the same as described above, except that only a portion of lines 0-1, 0-2, 0-3, and 0-4 is required. The positions of points B and D in the top view are found by projecting lines horizontally from points B and D in the front view to intersect the true-length line 0-3 at B_1. Project a vertical line from point B_1 to intersect point B_2 in the top view. Rotate B_2 90° from point 0 to intersect line 0-2 at B and 0-4 at D. It will be noted that only lines A-1 and C-3 appear as their true length in the front view. The true length of lines B-2 and D-4 may be found by projecting a horizontal line from points B and D to point E on the true-length line 0-1.

To complete the development, step off distances 1-A on line 1-0, 2-B on line 2-0, 3-C on line 3-0, and 4-D on line 4-0. Join points A, B, C, D, A with straight lines. The top surface of the truncated pyramid may be added to the development as follows: With A as center and

with a radius equal to distance AC in the front view, swing an arc. With B as center and with a radius equal to line BC on the development, swing an arc intersecting the first arc at C_1. Join point B to point C_1 with a straight line. With A as center and with a radius equal to line AB in the development, swing an arc. With B as center and with a radius equal to distance BD in the top view, swing an arc intersecting the first arc at D_1. Join AD_1 and D_1C_1 with straight lines.

Development of a Right Pyramid with True Length of Edge Lines not Shown See Fig. 22-3-2. In order to construct the development, the true length of the edge lines 0-1, 0-2, etc., must first be found. The true length of the edge lines would be equal to the hypotenuse of a right triangle having one leg equal in length to the projected edge line in the top view and the other leg equal to the height of the projected edge line in the

front view. Since only one true-length line is required, it may be developed directly on the front view rather than by making a separate true-length diagram. With 0 in the top view as center and radius equal to distance 0-1 in the top view, swing an arc from point 1 until it intersects the center line at point 1_1. Project a vertical line down to the front view, intersecting the base line at point 1_1. Line 0-1_1 is the true length of the edge lines. The development may now be constructed in a similar manner to that outlined in the previous development.

In developing a truncated pyramid of this type, the procedure is the same except only the truncated edge lines are required. The true length of the truncated edge lines is required and may be found by projecting lines horizontally from points A, B, C, and D in the front view to intersect the true-length line 0-1_1 at points F and E, respectively. Line F-1_1 is the true length of the truncated edge lines $A1$

and $B1$, and line $E1_1$ is the true length of the remaining truncated edge lines $C3$ and $D4$. The sides of the truncated pyramid may now be constructed in the development view. The top surface of the truncated pyramid may be added to the development as follows: With points A and B on the development as centers and with a radius equal in length to line BC on the development, swing light arcs. With a radius equal in length to the true distance between points A and C or B and D (this is found on the true-length diagram constructed to the left of the front view) and with center B, swing an arc intersecting the first arc at C. Repeat, using point A as center and intersecting the other arc at point C. Join points B, C, and A with straight lines to complete the top surface. The base and seam lines have been omitted for clarity.

Development of an Oblique Pyramid See Fig. 22-3-3, an oblique pyramid having all its lateral edges of unequal length. The true length of each of these edges must first be found as shown in the true-length diagram. The development may now be constructed as follows: Lay out base line 1-2 in the development view equal in length to the base line 1-2 found in the top view. With point 1 as center and radius equal in length to line 0-1 in the true-length diagram, swing an arc. With point 2 as center and radius equal in length to line 0-2 in the true-length diagram, swing

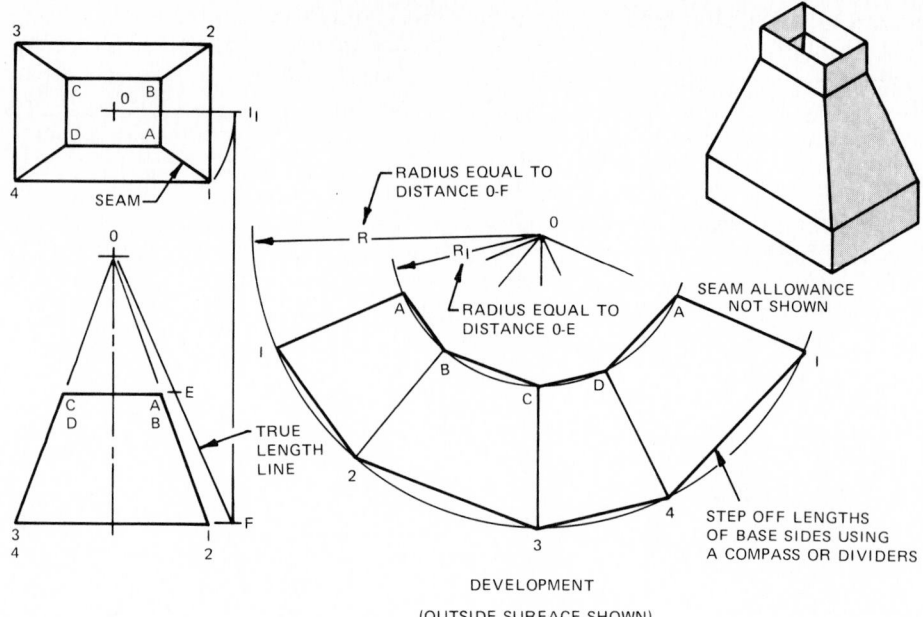

Fig. 22-3-4 Development of a transition piece.

an arc intersecting the first arc at 0. With point 0 as center and radius equal in length to line 0-3 in the true-length diagram, swing an arc. With point 2 as center and radius equal in length to base line 2-3 found in the top view, swing an arc intersecting the first arc at point 3. Locate point 4 and point 1 in a similar manner, and join these points, as shown, with straight lines. The base and seam lines have been omitted on the development drawing.

Development of a Transition Piece See

Fig. 22-3-4. The development of the transition piece is created in a similar manner to that of the development of the right pyramid (Fig. 22-3-2).

ASSIGNMENTS

See Assignments 5 and 6 for Unit 22-3 on page 454.

Review for Assignments
Unit 22-1   Surface Developments

UNIT 22-4
Parallel Line Development

The lateral, or curved, surface of a cylindrically shaped object, such as a tin can, is developable since it has a single-curved surface of one constant radius. The development technique used for such objects is called *parallel line development*. Figure 22-4-1A shows the development of the lateral surface of a simple hollow cylinder. The width of the development is equal to the height of the cylinder, and the length of the development is equal to the circumference of the cylinder plus the seam allowance. Figure 22-4-1B

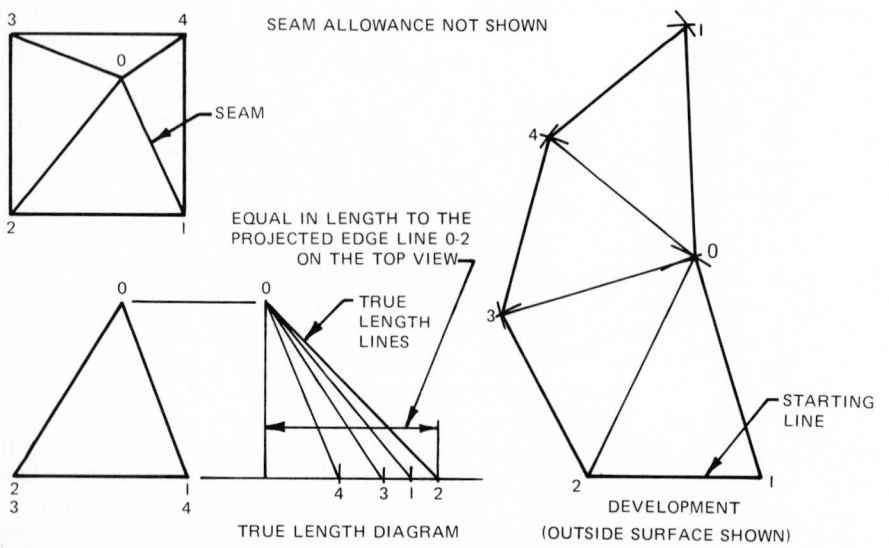

Fig. 22-3-3   Development of an oblique pyramid by triangulation.

shows the development of a cylinder with the top truncated at a 45° angle (one-half of a two-piece 90° elbow). Points of intersection are established to give the curved shape on the development. These points are derived from the intersection of a length location, representing a certain distance around the circumference from a starting point, and the height location at that same point on the circumference. The closer the points of intersection are to one another, the greater the accuracy of the development. An irregular curve is used to connect the points of intersection.

Figure 22-4-1C shows the development of the surface of a cylinder with both the top and the bottom truncated at an angle of 22.5° (the center part of a three-piece elbow). It is normal practice in sheet-metal work to place the seam on the shortest side. In the development of elbows, however, this prac-tice would result in considerable waste of material, as illustrated by Fig. 22-4-2A. To avoid this waste and to simplify cutting the pieces, the seams are alternately placed 180° apart, as illustrated by Fig. 22-4-2B for a two-piece elbow and by Fig. 22-4-2C for a three-piece elbow. Refer to Figs. 22-4-3 and 22-4-4 for complete devel-opments of two- and four-piece elbows.

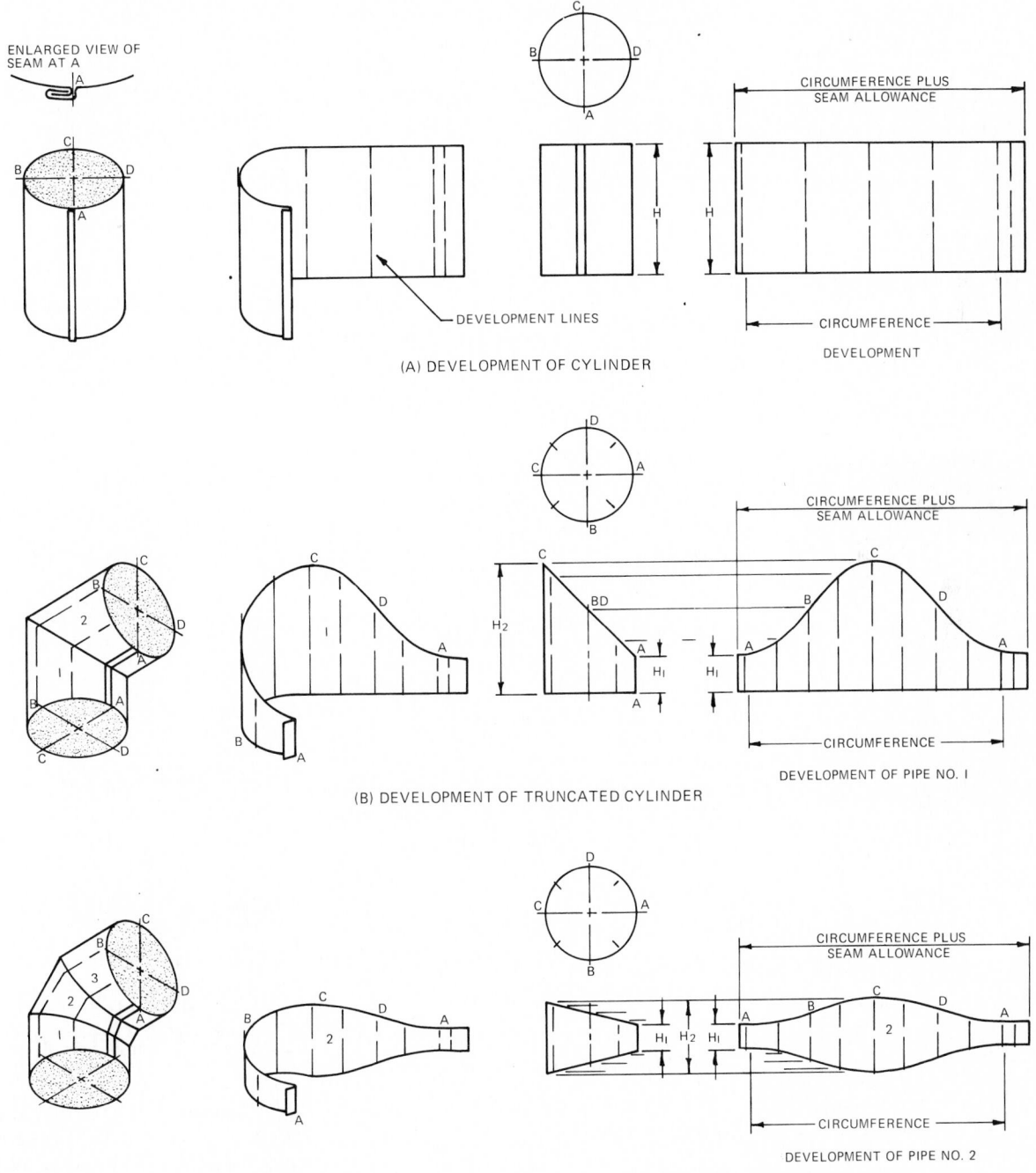

(A) DEVELOPMENT OF CYLINDER

(B) DEVELOPMENT OF TRUNCATED CYLINDER

(C) DEVELOPMENT OF A CYLINDER WITH THE TOP AND BOTTOM TRUNCATED

Fig. 22-4-1 Development of cylinders.

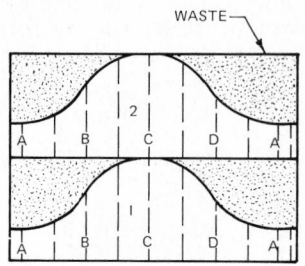

(A) DEVELOPMENT OF A 2-PIECE ELBOW WITH BOTH SEAMS ON LINE A

(B) DEVELOPMENT OF A 2-PIECE ELBOW WITH SEAMS ON LINES A AND C

(C) DEVELOPMENT OF A 3-PIECE ELBOW WITH SEAMS ALTERNATED ON LINES A AND C

Fig. 22-4-2   Location of seams on elbows.

ALLOWANCES FOR SEAMS AND JOINTS NOT SHOWN

DEVELOPMENT OF UPPER PART

CIRCUMFERENCE = DIA X 3.1416

DEVELOPMENT OF LOWER PART

Fig. 22-4-3   Development of a two-piece elbow.

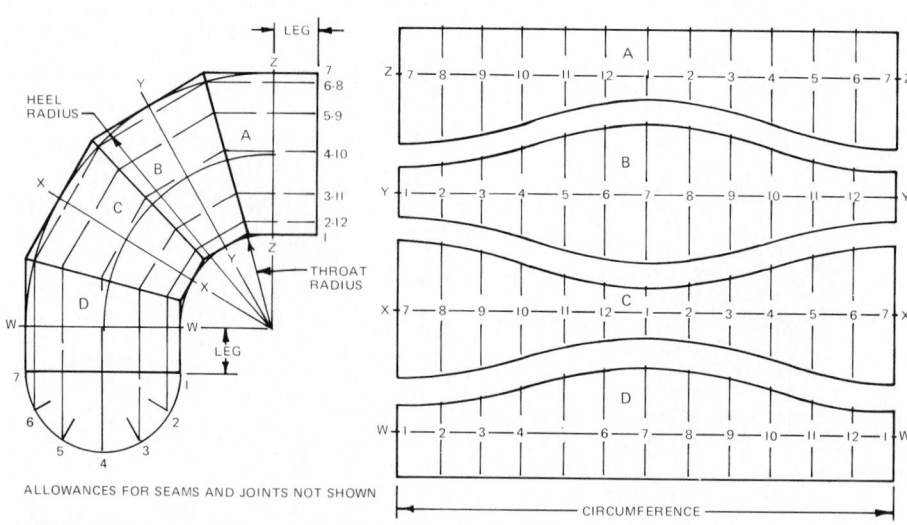

ALLOWANCES FOR SEAMS AND JOINTS NOT SHOWN

CIRCUMFERENCE

Fig. 22-4-4   Development of a four-piece elbow.

ASSIGNMENTS

See Assignments 7 and 8 for Unit 22-4 on page 455.

Review for Assignments

Unit 22-1   Surface Developments

UNIT 22-5
Radial Line Development of Conical Surfaces

Development of a Cone   The surface of a cone is developable, because a thin sheet of flexible material can be wrapped smoothly about it. The two dimensions necessary to make the

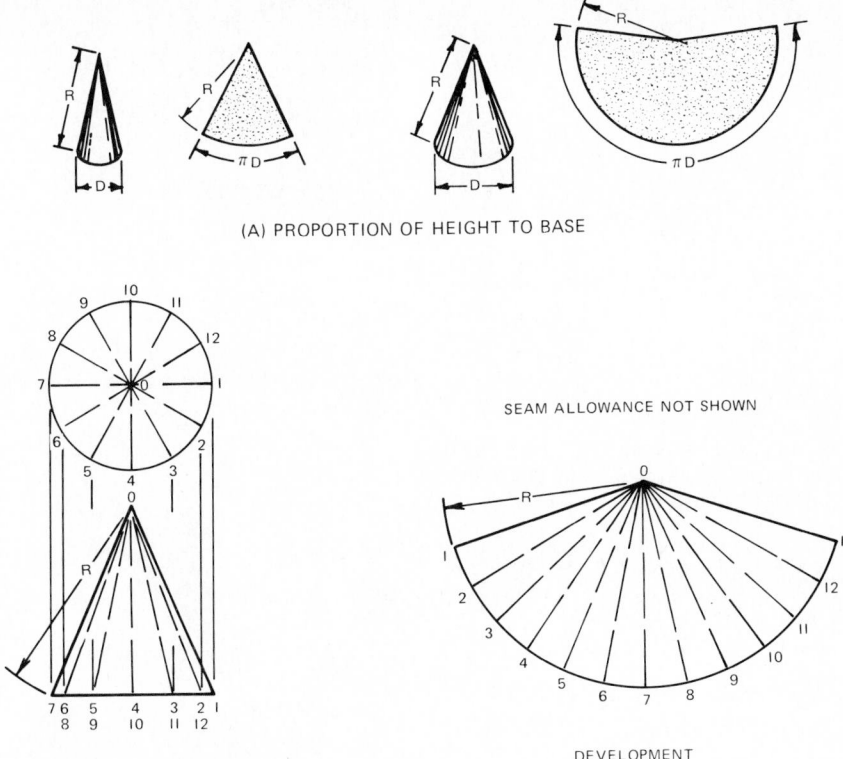

(A) PROPORTION OF HEIGHT TO BASE

SEAM ALLOWANCE NOT SHOWN

DEVELOPMENT

(B) DEVELOPMENT PROCEDURE

Fig. 22-5-1 Development of a cone.

APEX

FRUSTUM

DEVELOPMENT

DEVELOPMENT

(A) PROPORTION OF HEIGHT TO BASE

SEAM ALLOWANCE NOT SHOWN

DEVELOPMENT

(B) DEVELOPMENT PROCEDURE

Fig. 22-5-2 Development of a truncated cone.

development of the surface are the slant height of the cone and the circumference of its base. For a right circular cone (symmetrical about the vertical axis), the developed shape is a sector of a circle. The radius for this sector is the slant height of the cone, and the length around the perimeter of the sector is equal to the circumference of the base. The proportion of the height to the base diameter determines the size of the sector, as illustrated by Fig. 22-5-1A.

Figure 22-5-1B shows the steps in the development of a cone. The top view is divided into a convenient number of equal divisions, in this instance 12. The chordal distance between these points is used to step off the length of arc on the development. The radius R for the development is seen as the slant height in the front view. If a cone is truncated at an angle to the base, the inside shape on the development no longer has a constant radius; that is, it is an ellipse, which must be plotted by establishing points of intersection. The divisions made on the top view are projected down to the base of the cone in the front view. Element lines are drawn from these points to the apex of the cone. These element lines are seen in their true length only when the viewer is looking at right angles to them. Thus the points at which they cross the truncation line must be carried across, parallel to the base, to the outside element line, which is seen in its true length. The development is first made to represent the complete surface of the cone. Element lines are drawn from the step-off points about the circumference to the center point. True-length settings for each element line are taken from the front view and marked off on the corresponding element lines in the development. An irregular curve is used to connect these points of intersection, giving the proper inside shape.

Development of a Truncated Cone The development of a frustum of a cone is the development of a full cone less the development of the part removed, as shown in Fig. 22-5-2. Note that, at all times, the radius setting, either R_1 or R_2, is a slant height, a distance taken on the surface of the cone.

When the top of a cone is truncated at an angle to the base, the top surface

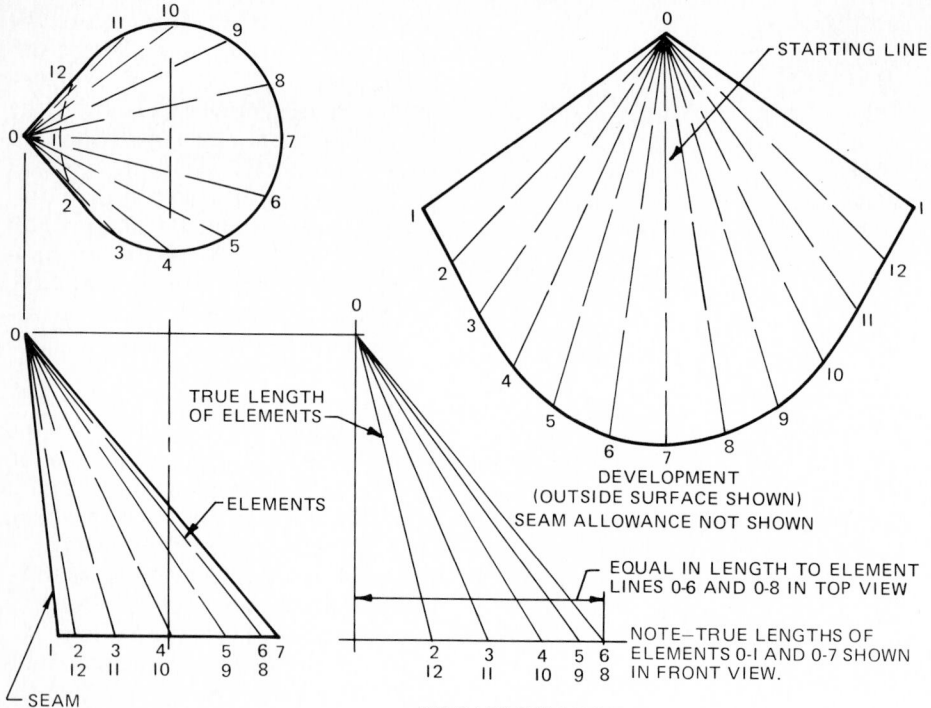

Fig. 22-5-3 Development of an oblique cone.

(labels in figure:)
STARTING LINE

TRUE LENGTH OF ELEMENTS

ELEMENTS

DEVELOPMENT
(OUTSIDE SURFACE SHOWN)
SEAM ALLOWANCE NOT SHOWN

EQUAL IN LENGTH TO ELEMENT LINES 0-6 AND 0-8 IN TOP VIEW

NOTE—TRUE LENGTHS OF ELEMENTS 0-1 AND 0-7 SHOWN IN FRONT VIEW.

SEAM

TRUE LENGTH DIAGRAM

UNIT 22-6
Development of Transition Pieces

TRIANGULATION

Nondevelopable surfaces can be developed approximately by assuming the surface to be made from a series of triangular surfaces laid side by side to form the development. This form of development is known as triangulation. Refer to Figs. 22-6-1 and 22-6-2.

Fig. 22-6-1 Forming a square-to-round transition piece.

will not be seen as a true circle. This shape must also be plotted by establishing points of intersection. True radius settings for each element line are taken from the front view and marked off on the corresponding element line in the top view. These points are connected with an irregular curve to give the correct oval shape for the top surface. If the development of the sloping top surface is required, an auxiliary view of this surface shows its true shape.

Development of an Oblique Cone See Fig. 22-5-3. The development of an oblique cone is generally accomplished by the triangulation method. The base of the cone is divided into a convenient number of equal parts and elements; 0-1, 0-2, etc., are drawn in the top view and projected down and drawn in the front view. The true lengths of the elements are not shown in either the top or front view but would be equal in length to the hypotenuse of a right-angle triangle having one leg equal in length to the projected element in the top view and the other leg equal to the height of the projected element in the front view.

When it is necessary to find the true length of a number of edges, or elements, then a true-length diagram is drawn adjacent to the front view. This prevents the front view from being cluttered with lines.

Since the development of the oblique cone will be symmetrical, the starting line will be element 0-7. The development is constructed as follows: With 0 as center and radius equal to the true length of element 0-6, draw an arc. With 7 as center and radius equal to distance 6-7 in the top view, draw a second arc intersecting the first at point 6. Draw element 0-6 on the development. With 0 as center and the radius equal to the true length of element 0-5, draw an arc. With 6 as center and the radius equal to distance 5-6 in the top view, draw a second arc intersecting the first at point 5. Draw element 0-5 on the development. This is repeated until all the element lines are located on the development view. No seam allowance is shown on the development.

ASSIGNMENT

See Assignment 9 for Unit 22-5 on page 456.

Review for Assignment
Unit 22-3   Radial Line Development of Flat Surfaces

Fig. 22-6-2 Transition pieces.

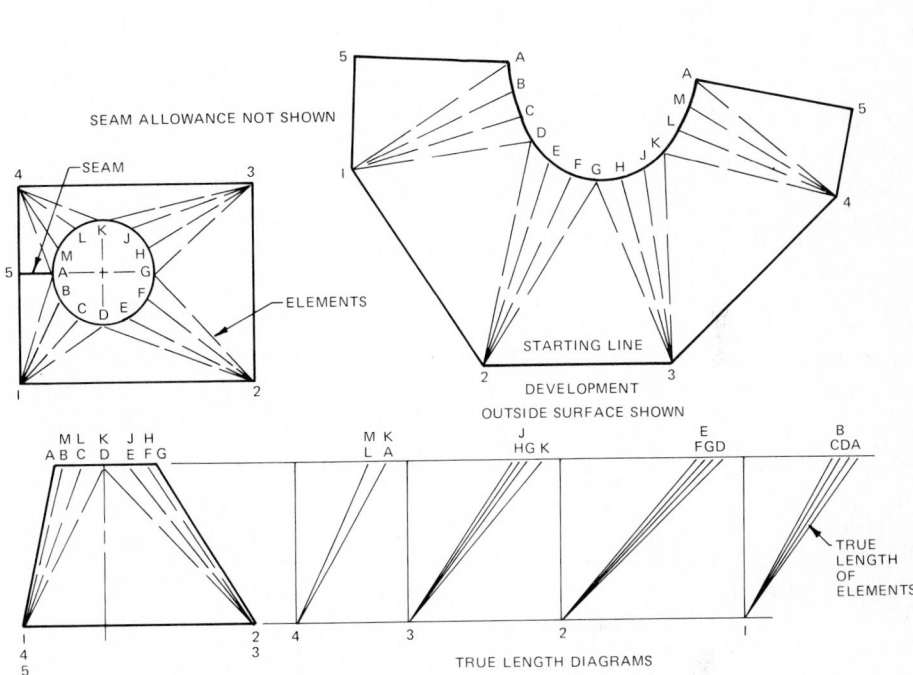

Fig. 22-6-3 Development of a transition piece—square to round.

Fig. 22-6-4 Development of an offset transition piece—rectangular to round.

Development of a Transition Piece—Square to Round

See Fig. 22-6-3. The transition piece shown is used to connect round and square pipes. It can be seen from both the development and the pictorial drawings that the transition piece is made of four isosceles triangles whose bases connect with the square duct and four parts of an oblique cone having the circle as the base and the corners of the square pipe as the vertices. To make the development, a true-length diagram is drawn first. When the true length of line 1A is known, the four equal isosceles triangles can be developed. After the triangle G-2-3 has been developed, the partial developments of the oblique cone are added until points D and K have been located. Next the isosceles triangles D-1-2 and K-3-4 are added, then the partial cones, and, last, half of the isosceles triangle placed at each side of the development.

Development of an Offset Transition Piece—Rectangular to Round

See Fig. 22-6-4. The development of the transition piece shown is constructed in the same manner as the one previously developed, except that all the elements are of different lengths. To avoid confusion, four true-length diagrams are drawn and the true-length lines are clearly labeled.

Transition Piece Connecting Two Circular Pipes—Parallel Joints

See Fig. 22-6-5. The development of a transition piece connecting two circular pipes is similar

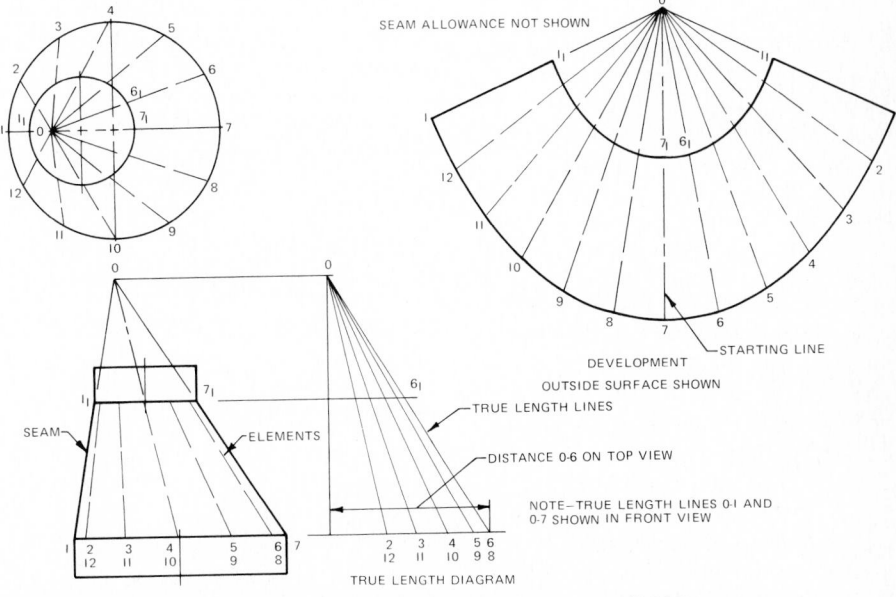

Fig. 22-6-5 Transition piece connecting two circular pipes—parallel joints.

to the development of an oblique cone (Fig. 22-5-3), except that the cone is truncated. The apex of the cone, 0, is located by drawing the two given pipe diameters in their proper positions and extending the radial lines 1-1$_1$ and 7-7$_1$ to intersect at point 0. First the development is made to represent the complete development of the cone, and then the top portion is removed. Radius settings for distances 0-2$_1$ and 0-3$_1$ on the development are taken from the true-length diagram.

Transition Piece Connecting Two Circular Pipes—Oblique Joints See Fig. 22-6-6. When the joints between the pipe and transition piece are not perpendicular to the pipe axis, then the transition piece may be developed as shown. Since the top and bottom of the transition piece will be elliptical in shape, a partial auxiliary view is required to find the true length of the chords between the end points of the elements. The development is then constructed in a manner similar to that outlined for Fig. 22-6-5.

ASSIGNMENT

See Assignment 10 for Unit 22-6 on page 456.

Review for Assignment
Unit 22-5   Development of a Cone

Fig. 22-7-1   Development of a sphere—gore method.

PARTIAL DEVELOPMENT

Fig. 22-7-2   Development of a sphere—zone method.

Fig. 22-6-6   Transition piece connecting two circular pipes—oblique joints.

UNIT 22-7
Development of a Sphere

Since the surface of a sphere is double-curved, it is not developable. However, the surface may be approximately developed by either the gore or the zone method. In the gore method (Fig. 22-7-1) the surface is divided into a number of equal sections, each section being considered as a section of a cylinder. Only one section need be developed, for it will serve as a pattern for the others. In the zone method (Fig. 22-7-2), the sphere is divided into horizontal zones and each zone is developed as a frustum of a cone.

ASSIGNMENTS

See Assignments 11 and 12 for Unit 22-7 on page 457.

UNIT 22-8
Intersection of Flat Surfaces—Lines Perpendicular

Whenever two surfaces meet, there is a line common to both called the *line of intersection*. In making the orthographic drawing of objects that comprise two or more intersecting parts, the lines of intersection of these parts must be plotted on the orthographic views. Figures 22-8-1 and 22-8-2 illustrate this plotting technique for the intersection of flat-sided prisms. Figure 22-8-1 shows the development of the parts. A numbering technique is very valuable in plotting lines of intersection. In the illustrations shown, the lines of intersection appear in the front view. The end points for these lines are established by projecting the height position from the right side view to intersect the corresponding length

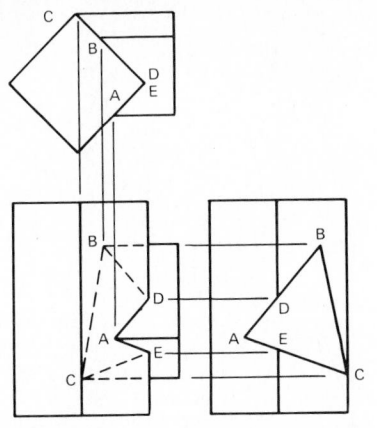

Fig. 22-8-2 Intersecting prisms at right angles.

position projected from the top view. When the prisms are flat-sided, the lines of intersection are straight, and the lines in the development are straight.

Intersecting Prisms—Triangle and Pyramid See Fig. 22-8-3. In drawing the intersection of these two prisms, the points of intersection in the top view are found by projecting the points of intersection from the front view. If a development of the pyramid is required, true-length lines, which do not appear in either the top or the front view, must be found. Since only a few true-length lines are unknown, line 0-D

Fig. 22-8-3 Intersecting prisms—triangle and pyramid.

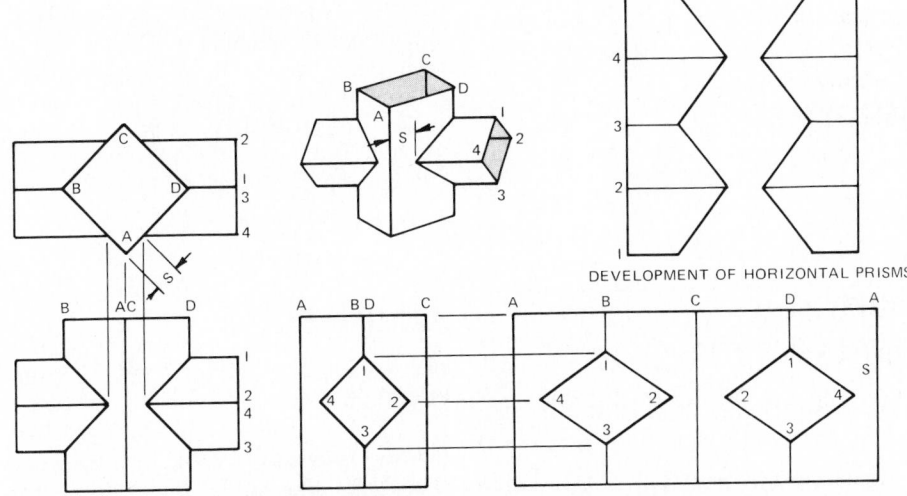

Fig. 22-8-1 Plotting lines of intersection and making development drawings of intersecting prisms.

on the front view serves as a true-length diagram. Lines $0\text{-}2_1$, $0\text{-}1_1$, and $0\text{-}5_1$ on line 0-D are the true lengths of lines 0-2, 0-1, and 0-5, respectively. Point 1 on surface 0-E-F and 0-B-D is located on the development as follows: Draw a straight line through points 0 and 1 in the front view to intersect the base at point 7. Transfer distance C7 in the front view to the development view. Join points 0 and 7 with a straight line. With center 0 and radius equal to distance $0\text{-}1_1$ shown on the front view, swing an arc intersecting line 0-7 at point 1.

Points on the development of the triangular prism are found by projecting lines from the top view to the development view and transferring distances between points (numbers) and lines (letters) from the front view to corresponding points on the development.

Intersecting Prisms Not at Right Angles— Hexagon and Rectangle See Fig. 22-8-4. Often a partial auxiliary view is drawn to locate points of intersection such as points *A* and *B* on line *L*.

Fig. 22-8-4   Intersecting prisms not at right angles—hexagon and rectangle.

Intersecting Prisms Not at Right Angles— Hexagon and Triangle See Fig. 22-8-5. An auxiliary view is required to locate points of intersection, such as points *A*, *B*, and *C*. To complete the side view, ends of lines *D*, *E*, and *F* and points of intersection *A*, *B*, and *C* are projected from the top view. Distances between the lines of the hexagon and triangle are transferred from either the top view or the auxiliary view.

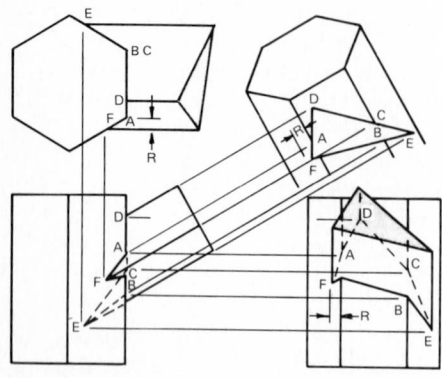

Fig. 22-8-5   Intersecting prisms not at right angles—hexagon and triangle.

Fig. 22-8-6   Intersecting prisms—triangle and pyramid.

Intersecting Prisms—Triangle and Pyramid See Fig. 22-8-6. Another method commonly used to locate points of intersection of lines and surfaces is the use of vertical cutting planes located on the edges piercing the surface. Thus section *R-R* locates point *C*, section *S-S* locates point *A*, and section *T-T* locates point *B*. The sectional views shown are for illustrative purposes only and need not be drawn. To establish point *C*, extend line *LC* in the top view to intersect line 0-3 at C_2 and line 1-3 at C_1. Project vertical lines from C_1 and C_2 down the front view, locating points C_1 on base line 1-3 and C_2 on line 0-3. Join points C_1 and C_2 at point *C*. Extend a vertical line up from point *C* to the top view, intersecting line C_2L at point *C*. Repeat for points *A* and *B*.

ASSIGNMENT

See Assignment 13 for Unit 22-8 on page 457.

UNIT 22-9
Intersection of Cylindrical Surfaces

90° Reducing Tee See Fig. 22-9-1. This figure illustrates the plotting technique for the intersection of cylinders.

Because there are no edges on the cylinders, element lines of reference are established about the cylinders in their orthographic views. In the top view, the element lines for the small cylinder are drawn to touch the surface of the large cylinder; for example, line 2 touches at *T*. This point location is then projected down to the front view to intersect the corresponding element line, establishing the height at that point. The points of intersection thus established are connected by an irregular curve to produce the line of intersection. The same points of reference used to establish the line of intersection are used to draw the development.

45° Reducing Tee See Fig. 22-9-2. This figure illustrates the intersection of a small pipe at an angle of 45° to a large pipe. The same techniques of plotting reference points are used as were previously described for a 90° reducing tee.

ASSIGNMENT

See Assignment 14 for Unit 22-9 on page 459.

Review for Assignment
Unit 22-1   Parallel Line Development
Unit 22-8   Intersection of Flat
             Surfaces

Fig. 22-9-1 Plotting lines of intersection and making development drawings for a 90° reducing tee.

Fig. 22-9-2 Plotting lines of intersection making development drawings for a 45° reducing tee.

Fig. 22-10-1 Intersecting prisms—hexagon and cone.

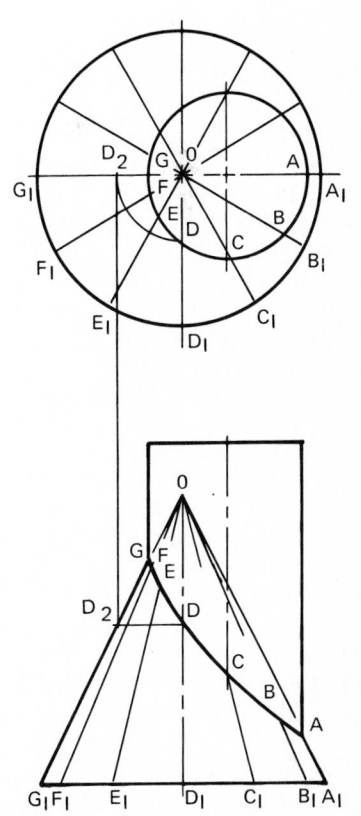

Fig. 22-10-2 Intersecting prisms—cone and cylinder.

UNIT 22-10
Intersecting Prisms

Intersecting Prisms—Hexagon and Cone
See Fig. 22-10-1. The lines of intersection between the hexagon and cone are developed as follows. Divide each side of the hexagon into four parts (lines every 15°). Through point A in the top view, swing an arc intersecting horizontal center line 0-4 at A_1. From point A_1 drop a vertical line down to the front view, intersecting line 0-D at point A_1. Draw a light horizontal line through A_1 in the front view. Drop ver-

tical lines down from point A in the top view, intersecting this line at point A. Repeat, locating point B. Point C may be located in the front view by extending a light horizontal line from point C on line 0-D. Join points A, B, and C with a line, which forms a hyperbolic curve.

Intersecting Prisms—Cone and Cylinder
See Fig. 22-10-2. The intersections of the cone and cylinder elements in the top view are first found and are then projected down to the corresponding elements in the front view. A smooth curve is drawn through these points to produce the line of intersection.

Fig. 22-10-3 Intersecting prisms—cone and cylinder.

SECTION TAKEN ON PLANE 3

Fig. 22-10-4 Intersecting prisms—pyramid and cylinder.

Intersecting Prisms—Cone and Cylinder See Fig. 22-10-3. The line of intersection between the cone and the cylinder is found by assuming the front view to have a series of horizontal cutting planes passing through points 2, 3, 4, 5, and 6. The cutting-plane line passes through the intersection of the cone and cylinder. Each point on the line of intersection is developed in a manner similar to that for Fig. 22-10-2.

Intersecting Prisms—Pyramid and Cylinder See Fig. 22-10-4. Cutting-plane lines taken horizontally through points 2, 3, 4, 5, and 6 in the front view are used to locate the lines of intersection. Point 5 on the line of intersection is located as follows. Draw a horizontal line through point 5 on the half circle in the front view to intersect line 0-*A* at point 5_1. Extend a vertical line from point 5_1 to the top view to intersect line 0-*A* at point 5_1. Extend a line from 5_1 in the top view at an angle of 45° to intersect line 0-*D*. From this intersection, extend a line at an angle of 45° in the direction of line 0-*C* to intersect a horizontal line passing through point 5 on the half circle in the top view. The intersection of these lines is point 5. To locate point 5 in the front view, drop a vertical line from point 5 in the top view to intersect the horizontal line passing through point 5 of the half circle. Locate the other points in a similar manner, and connect them with a smooth, curved line.

Intersecting Prisms—Cone and Oblique Cylinder See Fig. 22-10-5. The cylinder shown in the auxiliary view is divided into 12 equal parts. Element lines are drawn from the apex of the cone, point 0, through points 2 to 6 to the base of the cone, establishing points 2_1 to 6_1 inclusive. These points are projected to corresponding points in the front view and the element lines are drawn.

The element lines in the top view are located by projecting vertical lines up from points 2_1 to 6_1 in the front view to intersect the base circle in the top view. The lines of intersection are then found by projecting lines from the circle to meet their corresponding element lines.

ASSIGNMENT

See Assignment 15 for Unit 22-10 on page 459.

Review for Assignment
Unit 22-3 Radial Line Development of Flat Surfaces
Unit 22-4 Parallel Line Development
Unit 22-5 Radial Line Development of Conical Surfaces

AUXILIARY VIEW

Fig. 22-10-5 Intersecting prisms—cone and oblique cylinder.

ASSIGNMENTS for Chapter 22

Assignment for Unit 22-1, Surface Developments and Intersections

1. On a B- or A3-size sheet, make a development drawing complete with bending instructions of one of the parts shown in Figs. 22-1-A to 22-1-D. Dimension the development drawing, showing the distance between bend lines, and show the overall sizes. Scale is full (1:1).

Fig. 22-1-A Nail box.

Fig. 22-1-B Wall tray.

Fig. 22-1-C Cake tray.

Assignments for Unit 22-2, the Packaging Industry

2. On a B- or A3-size sheet, make a development drawing of one of the boxes shown in Figs. 22-2-A or 22-2-B. Scale is full or 1:1. After the development drawing has been checked by your instructor, add suitable seams and joint allowances. Then cut out the development, score on the bend lines, and form and glue the box together.

U.S. CUSTOMARY W = 1.75 H = 2.88 All seams .25 in. wide, placed on the inside and glued. Material is .02-in. cardboard.

METRIC W = 44 H = 75 All seams 6 mm wide, placed on the inside and glued. Material is 0.5-mm cardboard.

Fig. 22-1-D Memo pad holder.

Fig. 22-2-A Hexagon box.   Fig. 22-2-B Octagon box.

3. On a B- or A3-size sheet, make a development drawing of either the pencil or swim goggle boxes shown in Figs. 22-2-C and 22-2-D. On the exterior surface of the box, lay out a design which has eye appeal (color can be used) and which contains in the design the name of the item being sold, a company name, a slogan, and any other feature you believe would improve the salability of the article. Cut out the development, score on the bend lines, and glue together. *Note:* With reference to the swim goggle box, the box is completely sealed and must be broken to remove the goggles. Scale is full (1:1).

4. Many containers are designed for a dual purpose. The main purpose is to accommodate the article being sold. The secondary purpose is to use the container as a novelty item after the article is removed. This next product is such a container. A company which produces cat and dog food wishes to have a dodecahedron- (12-sided) or icosahedron- (20-sided) shaped container having illustrations of different animals on the sides. The container can be used to hold small articles or as an art object (mobile) which can be hung from the ceiling.

On a B- or A3-size sheet, lay out one of the containers shown in Figs. 22-2-E and 22-2-F. One of the sides forms a lid. On the exterior surface of the container add a suitable design. Cut out the development, score on the bend lines, and glue together. Scale is full (1:1).

U.S. CUSTOMARY W = 1.75 in. All seams .20 in. wide, placed on the inside and glued. Material is .02-in. cardboard.

METRIC W = 45 All seams 6 mm wide, placed on the inside and glued. Material is 0.5-mm cardboard.

Assignments for Unit 22-3, Radial Line Development of Flat Surfaces

5. On a B- or A3-size sheet, make a development drawing of one of the concentric pyramids shown in Figs. 22-3-A to 22-3-D. Add suitable seams. Scale is full (1:1).

6. On a B- or A3-size sheet, make a development drawing of the eccentric pyramid shown in Fig. 22-3-E. Add suitable seams. Scale is full (1:1).

MATL - 0.5 CARDBOARD

Fig. 22-2-C   Pencil box.

NOTES:
BOX TO OPEN FROM FRONT ONLY.

MATL - .02 CARDBOARD.

Fig. 22-2-D   Swim goggle box.

ICOSAHEDRON

Fig. 22-2-E   Food container box.

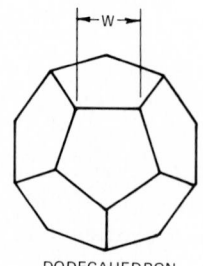

DODECAHEDRON

Fig. 22-2-F   Food container box.

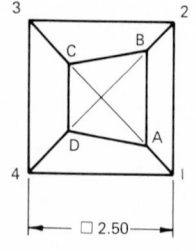

NOTES:
SEAM IS AT A–I.
TOP IS HINGED AT A–B.
BOTTOM IS HINGED AT I–2.

MATL – .01 CARDBOARD.

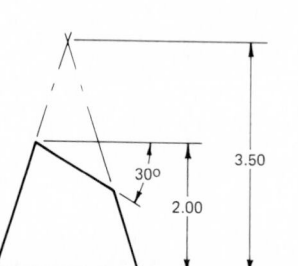

Fig. 22-3-A   Truncated concentric pyramid.

NOTES:
SEAM IS AT A-I.
TOP IS HINGED AT A-B.
BOTTOM IS HINGED AT I-2.

MATL - 0.3 CARDBOARD

Fig. 22-3-B   Truncated concentric pyramid.

NOTES:
SEAM IS AT A–I.
TOP IS HINGED AT A–B.
BOTTOM IS HINGED AT I–2.
MATL – .02 CARDBOARD.

Fig. 22-3-C Truncated concentric pyramid.

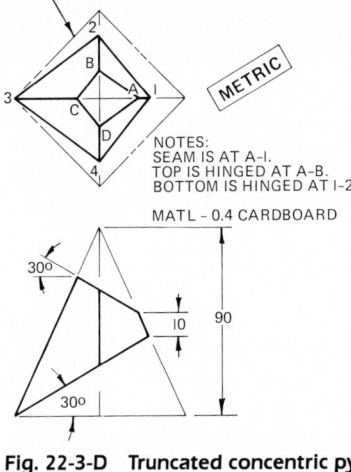

METRIC

NOTES:
SEAM IS AT A–I.
TOP IS HINGED AT A–B.
BOTTOM IS HINGED AT I–2.

MATL – 0.4 CARDBOARD

Fig. 22-3-D Truncated concentric pyramid.

NOTES:
SEAM IS AT A–I.
TOP IS HINGED AT A–D.
BOTTOM IS HINGED AT I–4.
MATL – .02 CARDBOARD.

Fig. 22-3-E Truncated eccentric pyramid.

Assignments for Unit 22-4, Parallel Line Development

7. On a B- or A3-size sheet, make a two-view and a development drawing of one of the elbows shown in Figs. 22-4-A to 22-4-C. Add suitable seams. Scale is full (1:1).

8. On a B- or A3-size sheet, make a two-view and a development drawing of one of the parts shown in Figs. 22-4-D or 22-4-E. Add suitable seams. Scale is full (1:1).

NOTES:
ALL SEAMS .25 WIDE.
MATL – 18 USS

Fig. 22-4-A Three-piece elbow.

NOTES:
ALL SEAMS 6 WIDE.
MATL – 18 USS

Fig. 22-4-B Two-piece elbow.

NOTES:
ALL SEAMS .25 WIDE
MATL – 18 USS

Fig. 22-4-C Four-piece elbow.

Fig. 22-4-D Sugar scoop.

Fig. 22-4-E Planter.

Assignment for Unit 22-5, Radial Line Development of Conical Surfaces

9. On a B- or A3-size sheet, make development drawings of one of the assembled parts shown in Figs. 22-5-A to 22-5-E. Use your judgment for the other views required and add suitable seams. Scale is full (1:1) for Figs. 22-5-A to 22-5-D, and half size or 1:2 for Fig. 22-5-E.

Assignment for Unit 22-6, Development of Transition Pieces

10. On a B- or A3-size sheet, make a two-view drawing plus a development drawing of the transition piece of one of the parts shown in Figs. 22-6-A to 22-6-D. Use your judgment for size and location of seams. Scale is full (1:1).

MATL – 24 USS
Fig. 22-5-A Funnel.

METRIC
MATL – 24 USS
Fig. 22-5-B Offset funnel.

MATL – 22USS
Fig. 22-5-C Truncated cone.

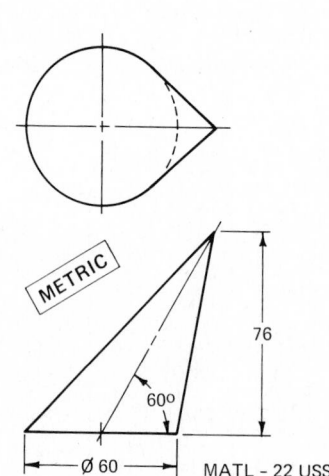
METRIC
MATL – 22 USS
Fig. 22-5-D Oblique cone.

MATL – 22 USS
Fig. 22-5-E Measuring can.

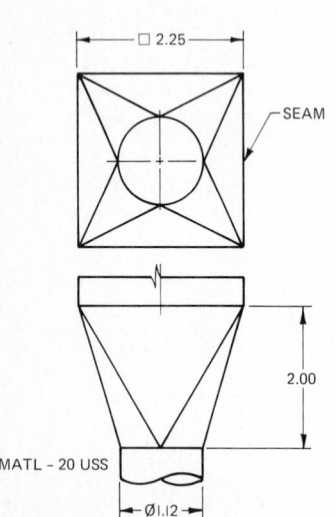
MATL – 20 USS
Fig. 22-6-A Concentric transition piece—square to round.

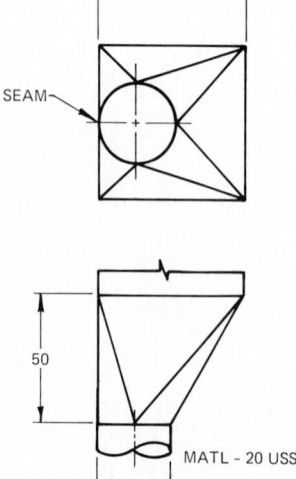
MATL – 20 USS
Fig. 22-6-B Offset transition piece—square to round.

MATL – 20 USS
Fig. 22-6-C Transition piece—square to round.

MATL – 20 USS
METRIC
Fig. 22-6-D Transition piece—round to round.

Assignments for Unit 22-7, Development of a Sphere

11. On a B- or A3-size sheet, make a development drawing of the ball shown in Fig. 22-7-A. Use your judgment for the views required. Either the gore or zone method may be used. Scale is full (1:1). Ball diameter is 3.00 in. or 80 mm.

12. On a B- or A3-size sheet, make development drawings for the three parts of the thermos bottle liner shown in Fig. 22-7-B. Seam allowances need not be shown. Scale is half size (1:2).

Assignment for Unit 22-8, Intersection of Flat Surfaces— Lines Perpendicular

13. On a B- or A3-size sheet, select one of the assembled parts shown in Figs. 22-8-A to 22-8-D, complete the views and make development drawings of the vertical and horizontal parts.

A

B

Fig. 22-8-A Intersecting prisms.

Fig. 22-7-A Ball.

A

B

C

D

Fig. 22-8-B Intersecting prisms.

Fig. 22-7-B Thermos bottle liner.

Fig. 22-8-C Intersecting prisms.

Fig. 22-8-D Intersecting prisms.

METRIC

METRIC

Fig. 22-9-A Intersecting prisms.

A

B

Assignment for Unit 22-9, Intersection of Cylindrical Surfaces

14. On a B- or A3-size sheet, select one of the intersections shown in Figs. 22-9-A to 22-9-D. Draw all assemblies and complete the lines of intersection on all views. Scale is full (1:1).

(A) 90º REDUCING TEE

(B) 45ºREDUCING TEE

Fig. 22-9-C   Tees and laterals.

Assignment for Unit 22-10, Intersecting Prisms

15. On a B- or A3-size sheet, select one of the intersections shown in Figs. 22-10-A to 22-10-C. Complete the lines of intersection on the partially completed views, and make a development drawing of the vertical part(s). Scale is full (1:1).

(A) 90º OFFSET REDUCING TEE

(B) 45ºOFFSET REDUCING TEE

Fig. 22-9-D   Tees and laterals.

A

Fig. 22-10-C
Intersecting prisms.

C

Fig. 22-9-B   Intersecting prisms.

Fig. 22-10-A
Intersecting prisms.

Fig. 22-10-B
Intersecting prisms.

CHAPTER 23

Pipe Drawings

UNIT 23-1
Pipes

One hundred years ago, water was the only important fluid which was conveyed from one point to another in pipes. Today almost every conceivable fluid is handled in pipes, during its production, processing, transportation, or utilization. The age of atomic energy and rocket power has added fluids such as liquid metals, oxygen, and nitrogen to the list of more common fluids such as oil, water, gases, and acids that are being transported in pipes today. Nor is the transportation of fluids the only phase of hydraulics which warrants attention now. Hydraulic and pneumatic mechanisms are used extensively for the control of machinery and numerous other types of equipment. Piping is also used as a structural element in columns and handrails. It is for these reasons that drafters and engineers should become familiar with pipe drawings.

KINDS OF PIPE

Steel and Wrought-Iron Pipe
This type of pipe carries water, steam, oil, and gas and is commonly used where high temperatures and pressures are encountered. Standard steel and cast-iron pipe is specified by the nominal diameter, which is always less than the actual inner diameter (ID) of

the pipe. This pipe was available up to recent times in only three wall thicknesses—standard, extra-strong, and double extra-strong. See Fig. 23-1-1. In order to use common fittings with these different wall thicknesses of pipe, the outer diameter (OD) of each remained the same, and the extra metal was added to the ID to increase the wall thickness of the extra-strong and double extra-strong pipe.

(A) STANDARD SCHEDULE 40   (B) EXTRA STRONG SCHEDULE 80   (C) DOUBLE EXTRA STRONG

Fig. 23-1-1   A comparison between steel pipe.

The nominal size of pipe is given in inch sizes, but the inside and outside diameters and wall thickness are given in millimeter sizes in the metric system.

Because of the demand for a greater variety of pipe for increased pressure and temperature uses, ANSI has now made available 10 different wall thicknesses of pipe, each designated by a schedule number. Standard pipe is now referred to as *schedule 40 pipe* and extra-strong pipe as *schedule 80*. Pipe over 12 in. is referred to as *OD pipe*, and the nominal size is the OD of the pipe.

Cast-Iron Pipe
Cast-iron pipe is often installed underground to carry water, gas, and sewage. It is also used for low-pressure steam connections. Cast-iron pipe joints are normally of the flanged type or the bell-and-spigot type.

Seamless Brass and Copper Pipe
These types of pipe are used extensively in plumbing because of their ability to withstand corrosion. They have the same nominal diameter as steel or iron pipe, but they have thinner wall sections.

Copper Tubing
This pipe is used in plumbing and heating and where vibration and misalignment are factors, such as in automotive, hydraulic, and pneumatic design.

Plastic Pipe
This pipe or tubing, because of its corrosion and chemical resistance, is used extensively in the chemical industry. It is flexible and readily installed but is not recommended where heat or pressure is a factor.

PIPE JOINTS AND FITTINGS

Parts that are used to join pipe are called *fittings*. They may be used to change size or direction and to join or provide branch connections. They fall into three general classes; screwed, welded, and flanged. See Fig. 23-1-2.

(A) FLANGED

(B) BELL AND SPIGOT

(C) SCREWED

(D) SOLDERED

(E) WELDED

Fig. 23-1-2　Common types of pipe joints.

A = PITCH DIAMETER AT END OF PIPE D−(0.05D + 1.1)P
B = PITCH DIAMETER AT GAGING NOTCH (A + 0.0625F)
E = EFFECTIVE THREAD (0.8D + 6.8)P
F = NORMAL ENGAGEMENT BY HAND
P = PITCH
DEPTH OF THREAD =0.8P

Fig. 23-1-5　American standard pipe thread.

Other methods are used for cast-iron pipe, copper, and plastic tubing.

Pipe fittings are specified by the nominal pipe size, the name of the fitting, and the material. Some fittings, such as tees, crosses, and elbows, are used to connect different sizes of pipe. These are called *reducing fittings*, and their nominal pipe sizes must be specified. The largest opening of the through run is given first, followed by the opposite end and the outlet. Figure 23-1-3 illustrates the method of designating sizes of reducing fittings.

Screwed Fittings

Screwed fittings, as shown in Fig. 23-1-4, are generally used on small pipe design of 2.50 in. or less. Common practice is to use a pipe compound (a mixture of lead and oil) on the

TEE　　TEE　　CROSS　　CROSS

CROSS　　LATERAL　　LATERAL

Fig. 23-1-3　Order of specifying the openings of reducing fittings.

90° ELBOW　　45° ELBOW　　45° STREET ELBOW

90° STREET ELBOW　　TEE　　SERVICE TEE

45° Y-BEND　　REDUCER　　COUPLING

RETURN BEND　　CAP　　CROSS

Fig. 23-1-4　Screwed fittings.

threaded connection to provide a lubricant and to seal any irregularities.

The American standard pipe thread is of two types—tapered or straight. The tapered thread, which is the more common, employs a 1:16 taper on the diameter of both the external and internal threads. See Fig. 23-1-5. This fixes the distance to which the pipe enters the fitting and ensures a tight joint.

Straight threads are used for special applications which are listed in the ANSI handbook.

Both the taper and straight pipe threads have the same number of threads per inch of nominal pipe size, and a pipe with a tapered thread may thread into a fitting having a straight thread, resulting in a tight seal.

Tapered threads are designated on drawings as NPT (American Standard

EXTERNAL THREAD

END VIEW　　SECTION VIEW

(A) SCHEMATIC REPRESENTATION

TAPER SHOWN　　TAPER NOT SHOWN
EXTERNAL THREAD

END VIEW　　SECTION VEIW

(B) SIMPLIFIED REPRESENTATION

Fig. 23-1-6　Pipe thread conventions.

Pipe Taper Thread) and may be drawn with or without the taper, as shown in Fig. 23-1-6. When drawn in tapered form, the taper is exaggerated. Straight pipe threads are designated on drawings as NPS (American Standard Pipe Straight Thread), and standard thread symbols are used. All pipe threads are assumed to be tapered unless otherwise specified.

Welded Fittings

Welded fittings are used where connections are to be permanent, and on high-pressure and high-temperature

lines. Other advantages over flanged or screwed fittings are that welded pipes are easier to insulate, they may be placed closer together, and they are lighter in weight (mass). The ends of the pipe and pipe fittings are normally beveled, as shown in Fig. 23-1-7, to accommodate the weld. Joint rings may be used when welded pipe must be disassembled periodically.

Flanges

Flanges provide a quick means of disassembling pipe. Flanges are attached to the pipe ends by welding, screwing,

SCREWED   WELDED

Fig. 23-1-9   Flanges.

or lapping. The flange faces are then drawn together by bolts, the size and spacing being determined by the size and working pressure of the joint. See Figs. 23-1-8 and 23-1-9.

VALVES

Valves are used in piping systems to stop or to regulate the flow of fluids and gases. A few of the more common types are described here.

Gate Valves

Gate valves are used to control the flow of liquids. The wedge, or gate, lifts to allow full, unobstructed flow and lowers to stop it completely. See Fig. 23-1-10A. These valves are normally used where operation is infrequent and are not intended for throttling or close control.

Globe Valves

These valves are used to control the flow of liquids or gases. The design of the globe valve requires two changes in the direction of flow, which slightly reduces the pressure in the system. The globe valve shown in Fig. 23-1-10B is installed so that the pressure is on the disk which assists the spring in the cap to make a tight closure. This type of valve is recommended for the control of air, steam, gas, or other compressibles where instantaneous on-and-off operation is essential. Figure 23-1-10C is recommended for the control of liquids such as hot or cold water, gasoline, oil, or solvents, where the sudden closure of a valve might cause objectionable and destructive water hammer. The cap is fitted with a spring-loaded piston dashpot arrange-

90° SHORT RADIUS ELBOW	90° LONG RADIUS ELBOW	REDUCING ELBOW	45° ELBOW	STRAIGHT CROSS
180° SHORT RADIUS RETURN	180° LONG RADIUS RUN		STRAIGHT TEE	REDUCING TEE
STRAIGHT LATERAL	CONCENTRIC REDUCER	ECCENTRIC REDUCER	FLANGE	CAP

Fig. 23-1-7   Welded fittings. (Tube Turns)

| 90° ELBOW | 90° REDUCING ELBOW | 45° STRAIGHT ELBOW | TEE STRAIGHT | TEE REDUCING |
| 45° LATERAL STRAIGHT | TAPER REDUCER | ECCENTRIC REDUCER | CROSS STRAIGHT | SIDE OUTLET ELBOW STRAIGHT |

Fig. 23-1-8   Flanged fittings.

(A) GATE VALVE   (B) GLOBE VALVE   (C) GLOBE VALVE   (D) CHECK VALVE

Fig. 23-1-10   Common valves. (Jenkins Bros. Ltd.)

(A) DOUBLE-LINE DRAWING

(B) SINGLE-LINE DRAWING

(C) FORMER SINGLE-LINE DRAWING SYMBOLS

Fig. 23-1-11   Piping drawing symbols.

ment which retards closure times and helps eliminate shock.

Check Valves

As the name implies, check valves permit flow in one direction, but check all reverse flow. They are operated by the pressure and velocity of line flow alone, and they have no external means of control or operation. See Fig. 23-1-10D.

PIPING DRAWINGS

The purpose of piping drawings is to show the size and location of pipes, fittings, and valves. Since these items may be purchased, a set of symbols has been developed to portray these features on a drawing.

There are two types of piping drawings in use—single-line and double-line drawings. See Fig. 23-1-11. Double-line drawings take more time to draw and therefore are not recommended for production drawings. They are, however, suitable for catalogs and other applications where the visual appearance is more important than the extra drafting time taken to make the drawing.

Single-Line Drawings

Beyond dispute, single-line drawings, also known as *simplified representations*, of pipelines are able to provide substantial savings without loss of clarity or reduction of comprehensiveness of information. As such, the simplified method is used whenever possible.

Single-line piping drawings, as the name implies, use a single line to show the arrangement of the pipe and fittings. The center line of the pipe, regardless of pipe size, is drawn as a thick line to which the valve symbols are added. The size of the symbol is left to the discretion of the drafter. When the pipelines carry different liquids, such as cold or hot water, a coded line symbol is often used.

Drawing Projection Two methods of projection are used—orthographic and pictorial. Orthographic projection, as shown in Fig. 23-1-12, is recommended for the representation of single pipes either straight or bent in one plane only. However, this method is also used for more complicated piping.

Fig. 23-1-12   Single-line orthographic piping drawing.

Fig. 23-1-15   Pipe connection.

Pictorial projection, as shown in Fig. 23-1-13, is recommended for all pipes bent in more than one plane and for assembly and layout work, because the finished drawing is easier to understand.

Crossings Crossing of pipes without connections are normally to be depicted without interrupting the line representing the hidden line (Fig. 23-1-14); but when it is desirable to indicate that one pipe must pass behind the other, the line representing the pipe farthest from the viewer will be shown with a break, or interruption, where the other pipe passes in front of it. For microform purposes, the break

(A) CROSSING OF PIPE SHOWN WITHOUT INTERRUPTING THE PIPE PASSING BEHIND THE NEAREST PIPE.

(B) USING AN INTERRUPTED LINE TO INDICATE PIPE FURTHEST AWAY.

Fig. 23-1-14   Crossing of pipes.

should not be less than 10 times the line thickness.

Connections Permanent connections or junctions, whether made by welding or other processes such as gluing and soldering, are to be indicated on the drawing by a heavy dot (Fig. 23-1-15). A general note or specification may describe the process used.

Fig. 23-1-13   Single-line pictorial piping drawing.

Detachable connections or junctions may be shown by a single thick line instead of a heavy dot, as shown in Fig. 23-1-16. The specifications, a general note, or the bill of materials will list the types of fittings such as flanges, union, or coupling and whether the fittings are flanged or threaded.

Fig. 23-1-16   Adjoining apparatus.

Fittings If no specific symbols are standardized, fittings like tees, elbows, crosses, etc., are not specially drawn, but are represented, like pipe, by a continuous line. The circular symbol for a tee or elbow may be used when it is necessary to indicate whether the piping is coming toward or going away from the viewer, as shown in Fig. 23-1-17. Elbows on isometric drawings may be shown without the radius. However, the change of direction that the piping takes should be quite clear if this method is used.

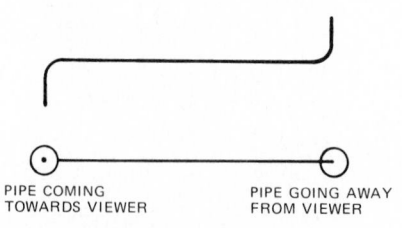

(A) PIPE LINE WITHOUT FLANGES CONNECTED TO ENDS OF PIPE

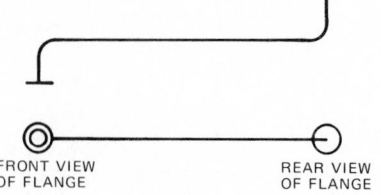

(B) FLANGES CONNECTED TO ENDS OF PIPE LINE

Fig. 23-1-17   Indicating ends of pipe lines.

Adjoining Apparatus If needed, adjoining apparatus, such as tanks, machinery, etc., not belonging to the piping itself may be shown by an outline drawn with a thin phantom line. See Fig. 23-1-16.

Dimensioning

- Dimensions for pipe and pipe fittings are always given from center to center of pipe and to the outer face of the pipe end or flange (Fig. 23-1-18).
- Pipe lengths are not normally shown on the drawings, but left to the pipe fitter.
- Pipe and fitting sizes and general notes are placed on the drawing beside the part concerned or, where space is restricted, with a leader.
- A bill of materials is usually provided with the drawing.
- Pipes with bends are dimensioned from vertex to vertex.
- Radii and angles of bends may be dimensioned as shown in Fig. 23-1-18B. Whenever possible, the smaller of the supplementary angles is to be specified.
- The outer diameter and wall thickness of the pipe may be specified on the line representing the pipe or elsewhere (parts list, general note, specification, etc.).

SHOWING THE RADIUS OF ELBOW OPTIONAL

(A) LINEAR DIMENSIONS

(B) RADII AND ANGLES OF BENDS

(C) PIPE SIZE INDICATED ON DRAWINGS

Fig. 23-1-18 Dimensioning piping drawings.

Orthographic Piping Symbols

Pipe Symbols If flanges are not attached to the ends of the pipelines when orthographic projections are drawn in, pipeline symbols indicating the direction of the pipe are required. If the pipeline is coming toward the front (or viewer), it will be shown by two concentric circles, the smaller one being solid. See Fig. 23-1-17A. If the pipeline is going toward the back (or away from the viewer), it will be shown by one circle. No extra lines are required on the other views.

Flange Symbols See Fig. 23-1-17B. Flanges are to be represented, irrespective of their type and sizes, by two concentric circles in the front view, by one circle in the rear view, and by a short stroke in the side view, while lines of equal thickness, as chosen for the representation of pipes, are used.

Valve Symbols Symbols representing valves are drawn with continuous thin lines (as opposed to thick lines for piping and flanges). The valve spindles should be shown only if it is necessary to define their positions. It will be assumed that unless otherwise specified, the valve spindle is in the position shown in Fig. 23-1-19.

Source Material
- Crane Limited
- Jenkins Bros. Limited
- ANSI handbook

ASSIGNMENTS

See Assignments 1 and 2 for Unit 23-1 on page 468.

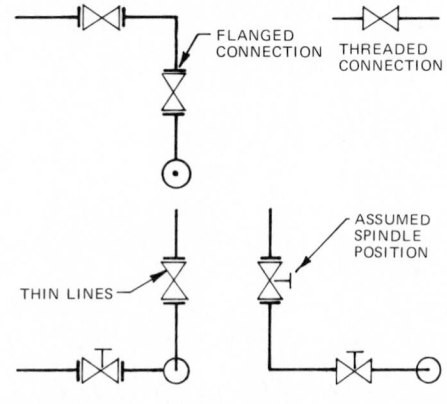

NOTE: WHEN VALVE SPINDLES NOT SHOWN IT WILL BE ASSUMED THAT THEY WILL BE IN THE POSITIONS INDICATED ABOVE.

Fig. 23-1-19 Valve symbols.

UNIT 23-2
Isometric Projection of Piping Drawings

The scale of the drawing applies to the dimension taken along the coordinate axes (isometric axes). In drawing to the isometric projection method, the following rules should be followed.

- Parts of pipe that run parallel to the coordinate axes are drawn without any special indication of being parallel to the isometric axes. See Figs. 23-2-1 and 23-2-2.
- With reference to calculations or programming for computer drafting, it will probably be necessary to indicate the x, y, and z axes (coordinates) on the drawing.

ISOMETRIC COORDINATE AXES

POSITIONING PIPING ON COORDINATE AXES

Fig. 23-2-1 Coordinate axes for piping drawings.

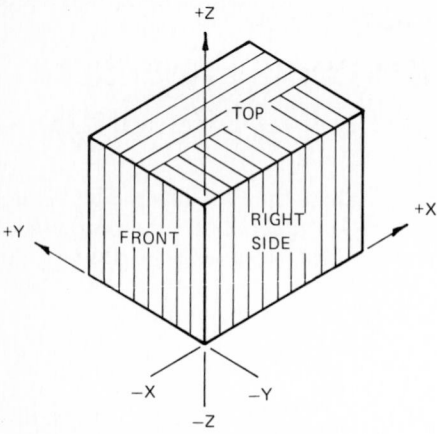

Fig. 23-2-2 Coordinates for piping drawings.

Flanges

Flanges are to be represented, irrespective of their type and sizes, by short strokes of equal-thickness lines as chosen for the representation of the pipes. See Fig. 23-2-3.

Flanges at the ends of vertical pipe parts should preferably be drawn to an angle of 30° to horizontal and flanges at the ends of horizontal pipe parts in a vertical direction.

(A) FLANGES FOR VERTICAL PIPE

(B) FLANGES FOR HORIZONTAL PIPE

Fig. 23-2-3 Flange positioning for isometric drawings.

Valves

For isometric drawings it will be For isometric drawings it will be assumed that unless otherwise specified, the valve spindle is in the position shown in Fig. 23-2-4. Valve spindles should be drawn only if it is necessary to define their positions. Deviations from these positions can be described by specifying the angle to which the valve is rotated in a clockwise, or right-hand, direction when looking in the direction of the positive x, y, or z axes (Fig. 23-2-5).

(A) VALVES WITH THREADED CONNECTIONS

(B) VALVES WITH FLANGE CONNECTIONS

Fig. 23-2-4 Valve symbols.

Dimensioning

The preferred method of dimensioning isometric pipe drawings is the unidirectional system because of the ease in execution and reading. See Fig. 23-2-6.

Source Material
• Crane Limited
• Jenkins Bros. Limited

ASSIGNMENTS

See Assignments 3 and 4 for Unit 23-2 on page 470.

Review for Assignments
Unit 14-1   Pictorial Drawings

| DRAWING CALLOUT | MEANS THIS |

Fig. 23-2-5 Deviations from normal position of valve spindle.

Fig. 23-2-6 Unidirectional dimensioning.

UNIT 23-3
Supplementary Piping Information

Direction of Flow The direction of flow may be indicated by an arrowhead on the line representing the piping, as shown in Fig. 23-3-1.

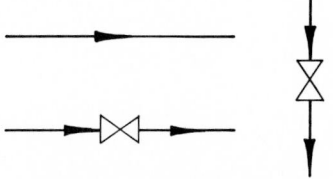

Fig. 23-3-1   Indicating direction of flow.

Level Indicators Level indications in lieu of linear measurements may be used to show the height of pipelines and fittings. The preferred method of indicating levels is shown in Fig. 23-3-2.

Fig. 23-3-2   Level indicators.

Specifying Slope on Pipes The direction of slope is indicated by an arrow located above the pipe pointing from the higher to the lower level. The amount of slope may be specified by either a general note on the drawing or one of the methods shown in Fig. 23-3-3. However, with long piping runs, it may be useful to specify the slope by reference to a datum and level indication as shown in Fig. 23-3-3C.

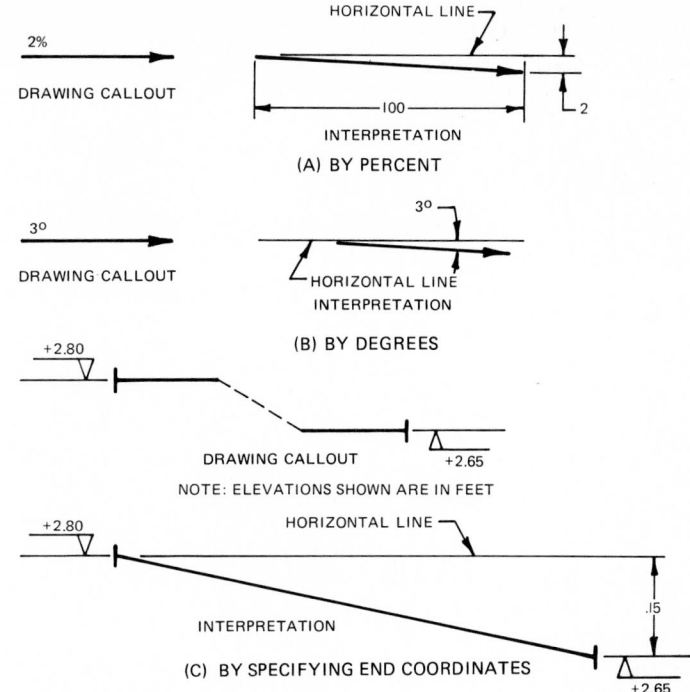

Fig. 23-3-3   Specifying slope of pipes.

Support and Hangers Support and hangers are to be represented by their appropriate symbols, as shown in Fig. 23-3-4A. The representation of repetitive accessories may be simplified, as shown in Fig. 23-3-4B.

Fig. 23-3-4   Supports and hangers.

Transition Pieces Transition pieces for changing the cross section are indicated by the symbols shown in Fig. 23-3-5. The relevant nominal sizes are indicated above the symbols.

Fig. 23-3-5   Transition pieces.

Pipe Runs not Parallel with Coordinate Axes

Deviations from the directions of the coordinate axes are to be indicated by means of hatched planes, as follows.

1. For a part of a pipe situated in a plane parallel to one of the vertical projection planes, vertical hatching lines are drawn to indicate the vertical projection plane (Fig. 23-3-6A).

2. For a part of a pipe situated in a plane parallel to the horizontal coordinate plane, horizontal hatching lines are drawn to indicate the horizontal projection plane (Fig. 23-3-6B).

3. For a part of a pipe not running parallel to any of the coordinate planes, both vertical and horizontal hatching lines are drawn to indicate the vertical and horizontal projection planes (Fig. 23-3-6C).

   If desired, in additon to the coordinate planes, the prism of which the pipe part forms the diagonal may be shown in thin lines (Fig. 23-3-6D).
   If such hatching is not convenient, for instance when using automated drafting equipment, it may be omitted but should be replaced with the thin-line rectangle or parallelopiped whose

(A) (B)

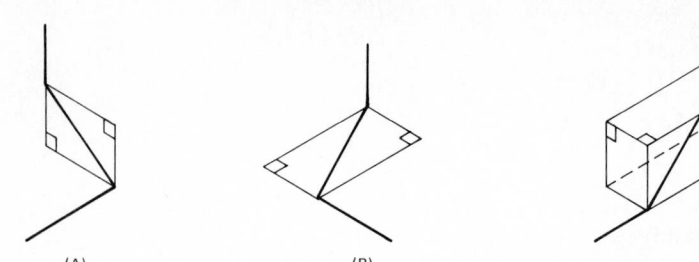

(A) (B) (C)

Fig. 23-3-7 Alternate method of indicating pipe runs not in the direction of the coordinate axes.

(C) (D)

Fig. 23-3-6 Indication of pipe runs not in the direction of the coordinate axes.

diagonal coincides with the pipe. See Fig. 23-3-7. Application of projection planes is shown in Fig. 23-3-8.

Source Material
• Jenkins Bros. Limited

ASSIGNMENTS

See Assignments 5 and 6 for Unit 23-3 on page 471.

FRONT

(A) ISOMETRIC

Fig. 23-3-8 Application of projection planes.

NOTE—RADIUS TO INDICATE DIRECTION OF RUN

(B) ORTHOGRAPHIC

ASSIGNMENTS for Chapter 23

Assignments for Unit 23-1, Pipes

1. On a B- or A3-size sheet, make a three-view drawing of the fuel-oil supply system shown in Fig. 23-1-A. Include with the drawing a bill of materials calling for all the pipe fittings and valves. The following valves are used: (1) relief valves, (2) globe valves, and (3) check valves. The numbers shown on the assignment correspond with the numbers listed above. Unions are used above the fuel oil pumps as detachable connections for ease of assembling and disassembling. The gages and temperature-sensing element have

1.00-in. pipe connections and, as such, necessitate the use of reducing tees. Scale is ½ in. = 1 ft (U.S. customary) or 1:20 (metric).

2. Heated tanks must be provided for the storage of industrial heating oil in most plants using this fuel for boiler furnaces generating heat or power or for processing furnaces.

To ensure uninterrupted service when cleaning, or in the event of a breakdown of one of the systems, a duplicate installation of tanks is shown in this layout. Since circulation must be provided to keep the oil fluid, a return line as well as a suction

line from the tanks is shown. A valve is provided directly to the suction line. Connections are provided for both high and low suction. High suction guards against difficulties from sediment, while the low suction is necessary when the fuel oil supply is extremely low.

The free blow shown on the steam-line connections to the heating coils in the tanks is important for testing for the presence of oil in the steam return line, since oil would indicate a leak.

Extra-heavy globe valves of the re-grind-renew type are recommended on the oil lines to ensure maximum safety in

ALL PIPE AND FITTINGS 2.00 INCH
PIPE SIZE EXCEPT WHERE NOTED
IN ASSIGNMENT.

Fig. 23-1-A Fuel-oil-supply system.

the transmission of hazardous fluid and to meet code requirements. Outside screw and yoke gate valves are suggested because they show at a glance if the valve is opened or closed.

On a B- or A3-size sheet, make a three-view drawing of the fuel-oil storage connections with heating coil as shown in Fig. 23-1-B. Include with the drawing a bill of materials calling for all the pipe fittings and valves. The oil lines are 1.50-in. pipe, and the steam line is 2.00-in. pipe. For the scale, see the drawing.

CODE	VALVES	SERVICE
A	GLOBE-REGRIND-RENEW	OIL LINES
B	GLOBE-BEVEL SEAT	STEAM LINES
C	GATE	STEAM LINES

Fig. 23-1-B Fuel-oil-storage connections with heating coils. (Jenkins Bros. Ltd.)

Assignments for Unit 23-2, Isometric Projection of Piping Drawings

3. For starting diesel engines, the most dependable and widely used method is an air system of the type illustrated in Fig. 23-2-A. With this starting system hooked up to diesel installations generating power and heat for such buildings as factories, hotels, large apartment houses, and stores, interruptions such as might occur through failure of electric supply or storage cells are avoided.

Safety valves are provided for the compressor and the air storage tanks. Check valves are installed on the air storage tank feed lines and the compressor discharge lines to prevent accidental discharge of the tanks.

Piping is arranged so that the compressor will either fill the storage tanks and/or pump directly to the engines. Any of the three storage tanks may be used for starting, and pressure gages indicate their readiness. The engines are fitted with quick-opening valves to admit air quickly at full pressure and shut it off the instant rotation is obtained. A bronze globe valve is installed to permit complete shutdown of the engine for repairs and for regulation of the air flow. Drains are provided at low points to remove condensate from the air storage tanks, lines, and engine feeds.

Globe valves are recommended throughout this hook-up except on the main shutoff lines where gate valves are used because of infrequent operation.

All valves connected to horizontal pipelines 6 ft (1800 mm) or higher above the floor will have their spindles located on the underside for ease of operation. Other horizontally positioned valves will have their spindles in the upright position.

All valves connected to vertical pipes will have their valve spindles oriented to the front of the drawing.

Flanges are to be located on the top pipeline near the three starting air tanks and near the air compressor for assembly and disassembly purposes. Flanges are located on the starting diesel engines.

On a B- or A3-size sheet, make an isometric piping drawing for the diesel engine starting system shown in Fig. 23-2-A. Scale is ¼ in. = 1 ft (U.S. customary) or 1:50 (metric). Include on your drawing a bill of materials calling for the pipe fittings and valves. All the fittings are threaded, and 1.50-in. pipe is used throughout.

4. In the piping layout of a boiler room shown in Fig. 23-2-B, boilers 1, 3, and 5 are connected to supply the hot water to rooms located on the first floor. Check valves are placed on the cold-water lines

Fig. 23-2-A Diesel-engine starting air system.

CODE	VALVE	SERVICE
A	BRONZE GLOBE	AIR STORAGE TANK FEED LINES
B	BRONZE GLOBE	AIR STORAGE TANK DISCHARGE LINES
C	BRONZE GLOBE	DIESEL ENGINE SHUTOFF CONTROL
D	BRONZE GLOBE	AIR COMPRESSOR DISCHARGE
E	BRONZE GLOBE	PRESSURE GAGE SHUTOFF
F	BRONZE GLOBE	DRAIN VALVES
G	SPINDLE GATES	MAIN LINE SHUTOFF
H	BRONZE CHECK	AIR COMPRESSOR CHECK
J	BRONZE CHECK	AIR STORAGE TANK FEED LINES

Fig. 23-2-B Piping layout of boiler room. (Jenkins Bros. Ltd.)

adjacent to each boiler to prevent the hot water from backing into the cold-water lines. Gate valves are used to shut off the main hot- and cold-water supply lines. Globe valves are placed near the boilers on the hot-water lines. Flanged connections are used at the boilers for ease of assembly and disassembly. The size of pipe is indicated on the plane view, and the scale is shown on the drawing.

On a B- or A3-size sheet, make an isometric drawing of the piping layout of a boiler room. Include on the drawing a bill of materials calling for the pipe fittings and valves. All fittings are of flanged type.

Assignments for Unit 23-3, Supplementary Piping Information

5. The one-story, taxpayer-type building (Fig. 23-3-A) has been developed and steadily improved as a result of the movement of shopping centers to suburban areas. This type of building, which is multiplying rapidly, is constructed either with or without basement. It houses retail stores, service establishments, amusement centers, restaurants, and offices.

Heat and plumbing services in such buildings are usually provided by the owner or operator; and for this reason, he or she might give careful consideration to low-cost and trouble-free installations. An oil- or gas-fired steam boiler with automatic control, and a separate gas-fired heater for hot-water supply, will generally meet these requirements.

The two-pipe heating system, located in the basement in the installation illustrated, utilizes unit heaters with individual thermostatic controls. Valuable extra floor space is made available for tenants' use because the heaters are hung from the ceiling. Since the heaters in each store or each section of a store are automatically controlled, fuel savings are effected and even heating is ensured.

On a B- or A3-size sheet, make an isometric drawing of the piping layout shown. Include with the drawing a bill of materials, calling for all the pipe fittings and valves. Pipe hangers are required for every 8 ft (2400 mm) of piping. Direction of flow, level indicators for horizontal piping using the basement floor as zero, indication of pipe runs not in the direction of the coordinate axes (see Fig. 23-3-6), and a drainage slope of 1:20 are to be shown on the drawing. Scale is 1/8 in. = 1 ft (U.S. customary) and 1:100 (metric). Use 1.50-in. pipe.

CODE	VALVE	SERVICE
A	BRONZE GATE	WATER SERVICE SHUTOFF
B	BRONZE GATE	DISTRIBUTION SHUTOFF
C	BRONZE GLOBE	WATER SUPPLY TO BOILER
D	BRONZE SWING CHECK	PREVENT BOILER BACKFLOW
E	BRONZE GLOBE	EMERGENCY BOILER FILL
F	BRONZE GLOBE	DRAINS SHUTOFF
G	BRONZE GLOBE	STEAM SUPPLY TO HEATERS
H	BRONZE GATE	CONDENSATE DRAIN SHUTOFFS
J	BRONZE GLOBE	STEAM MAIN TRAP CONNECTION
K	BRONZE GLOBE	TRAP BYPASS
L	BRONZE GATE	HOT WATER HEATER SHUTOFF
M	BRONZE SWING CHECK	PREVENT WATER HEATER BACKFLOW
N	BRONZE GATE	WATER SUPPLY SHUTOFFS
P	I.B.B.M. GATE	STEAM MAIN SHUTOFF
R	BRONZE GATE	RETURN SHUTOFF

1. FLOAT AND THERMO TRAP
2. WATER METER
3. HOT WATER TO STORE
4. COLD WATER TO STORE
5. DRAIN

Fig. 23-3-A Piping connections for plumbing and heating in a small building. (Jenkins Bros. Ltd.)

6. Light and medium fuel oils, numbers 1, 2, 3, and 5 (cold), which do not require preheating can be handled in a layout such as shown in Fig. 23-3-B. Since the expense of a preheater installation can be eliminated, this relatively simple system is economical and easy to operate. Similar systems are often installed in hotels, apartment houses, office buildings, large residences, and small industrial plants.

Fuel oil, stored in an unheated tank, flows through a large-mesh, twin-type strainer to a motor-driven pump which provides the necessary oil pressure for satisfactory operation. The oil then passes through a fine-mesh strainer which removes any small particles that might clog the burner. Oil flow to the burner is controlled by a burner control valve which opens or shuts according to the boiler pressure.

Although one fuel oil pump can adequately handle the maximum boiler demands, two are recommended to provide a second pump for standby service in case of breakdown. Each pump is provided with a pressure relief valve as a protection against excessive oil pressure, which might become high enough to cause leaks in the oil piping. Check valves in the relief lines prevent relieved oil from entering the idle standby pump.

Bronze valves are recommended throughout and must be of the appropriate pressure rating. The plug-type globe valve, recommended for the important individual burner shutoff, ensures positive tightness when closed and extremely close regulation of oil flow, both of which are essential to good oil burner operation.

The swing check valve indicated in this layout is exceptionally serviceable for the nonreturn control of steam, oil, water, and gas. It is generally used in connection with a gate valve, offering comparable full, free flow.

On a B- or A3-size sheet, make a single-line isometric drawing of the piping drawing shown. Design your own symbols for the indicators (gages, strainers, etc.) for which there are no standard symbols. The coding of these items should be shown clearly off the main drawing. Show the direction of flow and indicate on the drawing that all horizontal pipes require a slope of 1:20 for drainage purposes. Using the floor as zero elevation, scale the drawing and show by means of level indicator symbols the height of all horizontal pipelines. Include on your drawing a bill of materials, listing all the valves and fittings. Scale for Fig. 23-3-B is ¼ in. = 1 ft (U.S. customary) and 1:50 (metric).

CODE	VALVES	SERVICE
A	FIG. 470 GATE	SUCTION STRAINER SHUTOFF
B	FIG. 2058 GLOBE	SUCTION STRAINER BY PASS
C	FIG. 470 GATE	PUMP SUCTION SHUT OFF
D	FIG. 962 SWING CHECK	PUMP DISCHARGE PREVENT BACK FLOW
E	FIG. 530-A GLOBE	PUMP DISCHARGE SHUTOFF
F	FIG. 962 SWING CHECK	PREVENT BACKFLOW IN PRESSURE RELIEF LINES
G	FIG. 280 GATE	DISCHARGE STRAINER SHUT OFF
H	FIG. 530-A GLOBE	DISCHARGE STRAINER SHUT OFF
J	FIG. 530-A GLOBE	OIL BURNER SHUT OFF
K	FIG. 592 GLOBE	OIL BURNER CONTROL
L	FIG. 703 NEEDLE	PRESSURE GAGE CONTROL

SCALE

10 8 6 4 2 0 1 2
FEET

SCALE

3 2 1 0
METERS

Fig. 23-3-B Oil-burner piping for light oils. (Jenkins Bros. Ltd.)

CHAPTER 24

Structural Drafting

UNIT 24-1
Structural Drafting

The training of the structural steel drafter is of vital importance to the engineering profession, the construction industry, and every structural steel fabricator.

THE BUILDING PROCESS

The steps through which a building proceeds from conceptual planning to finished products are, generally speaking, as follows.

1. An owner with the appropriate financing establishes the requirements for a building to fulfill some particular function.

2. A design team (usually an architect and an engineer) studies the owner's needs in reference to set standards and conventions. The following factors may influence the preliminary design: available materials, construction costs, building codes, zoning, health requirements, local bylaws, land condition, fire protection, finance, and setbacks.

3. With these parameters and the owner's requirements, the consultant or design team prepares sketches of the finished building, floor plans, and cost estimates, which are submitted to the owner for approval.

4. The design team then takes over the job. The structural design group designs the building frame, taking into consideration factors

Fig. 24-1-1   Erecting fabricated steel supports for a building.

which influence the type and location of structural members.

5. When the structural arrangements have been finalized, layout drawings are made. These give distances from center line to center line, size and location of structural components, and other specifics of the design. When the layout drawings have been completed, checked, and approved, they are sent to steel fabricators for tendering. *Tendering* involves quoting a price (usually a price for detailing, supply, fabrication, and erection of the steel members).

6. When the contract has been received by the steel fabricator, he or she makes a list of material required so that the basic shapes can be ordered from the steel producer. The fabricator also begins to detail (draw the individual building members). These are referred to as *shop drawings*.

7. As the shop drawings are completed, they are sent to the shop in order that parts may be fabricated. It is usually during this period that the fabricator will make the erection drawings in conjunction with the structural design group.

8. As the steel is fabricated, it is either stored in the yard or sent to the construction site if it is required immediately.

9. At the site, the steel is erected using the erection drawings. See Fig. 24-1-1.

SYMBOL	WWF	W	M	S	C	MC	WWT	WT OR MT	L
SHAPE									
NAME	WELDED WIDE–FLANGE SHAPES	WIDE–FLANGE SHAPES	MISCELLANEOUS SHAPES	STANDARD BEAMS	STANDARD CHANNELS	MISCELLANEOUS CHANNELS	STRUCTURAL TEES		EQUAL LEG / UNEQUAL LEG / ANGLES

Fig. 24-1-2 Common structural steel shapes.

STRUCTURAL STEEL—PLAIN MATERIAL

It is important to remember that the steel produced at the rolling mills and shipped to the fabricating shop comes in a wide variety of shapes (approximately 600) and forms. At this stage it is called *plain material*.

Many of these materials are shown in Fig. 24-1-2. They can be classified and designated as follows:

1. S shapes (formerly called standard beams or I beams) are rolled in many sizes 3 to 20 in. (75 to 500 mm).
2. C shapes (formerly called standard channels) are available in sizes ranging from 3 to 18 in. (80 to 450 mm).
3. W shapes (formerly called wide-flange shapes) and welded wide-flange (WWF) beams and columns. W shapes are available in sizes ranging from 6 to 36 in. (150 to 900 mm) WWF shapes, sometimes referred to as H shapes, range in size from 14 to 48 in. (350 to 1200 mm).

4. M shapes (formerly called joists and light beams) are similar in contour to the W shapes. They are available in sizes ranging from 6 to 16 in. (150 to 400 mm).
5. Structural tees are produced by splitting S or W shapes, usually through the center of their webs, thus forming two T-shaped pieces from each beam.
6. L shapes, or angles, consisting of two legs set at right angles, are available in sizes ranging from 3 to 8 in. (75 to 200 mm).
7. Hollow structural sections (HSS) consist of round, square, and rectangular sections.
8. Plates, and round and rectangular bars.

Fig. 24-1-3 Structural steel callouts.

Shape	U.S. Customary Examples See Note 1		Metric Size Examples See Note 2
	New Designation	Old Designation	
Welded Wide Flange Shapes (WWF Shapes)			
—Beam	WWF48 × 320	48WWF320	WWF1000 × 244
—Columns			WWF350 × 315
Wide Flange Shapes (W Shapes)	W24 × 76	24WF76	W600 × 114
	W14 × 26	14B26	W160 × 18
Miscellaneous Shapes (M Shapes)	M8 × 18.5	8M18.5	M200 × 56
	M10 × 9	10JR9.0	M160 × 30
Standard Beams (S Shapes)	S24 × 100	24I100	S380 × 64
Standard Channels (C Shapes)	C12 × 20.7	12C20.7	C250 × 23
Structural Tees			
—cut from WWF Shapes	WWT24 × 160	ST24WWF160	WWT280 × 210
—cut from W Shapes	WT12 × 38	ST12WF38	WT130 × 16
—cut from M Shapes	MT4 × 9.25	ST4M9.25	MT100 × 14
Bearing Piles (HP Shapes)	HP14 × 73	14BP73	HP350 × 109
Angles (L Shapes) (leg dimensions × thickness)	L6 × 6 × .75	L6 × 6 × ¾	L75 × 75 × 6
	L6 × 4 × .62	16 × 4 × ⅝	L150 × 100 × 13
Plates (width × thickness)	20 × .50	20 × ½	500 × 12
Square Bar (side)	⌀ 1.00	Bar 1 ⌀	⌀ 25
Round Bar (diameter)	Ø1.25	Bar 1¼ Ø	Ø30
Flat Bar (width × thickness)	250 × .25	Bar 2½ × ¼	60 × 6
Round Pipe (type of pipe × OD × wall thickness)	12.75 OD × .375	12¾ × ⅜	XS 102 OD × 8
Square and Rectangular Hollow Structural Sections (outside dimensions × wall thickness)	HSS4 × 4 × .375	4 × 4RT × ⅜	HSS102 × 102 × 8
	HSS8 × 4 × .375	8 × 4RT × ⅜	
Steel Pipe Piles (OD × wall thickness)			320 OD × 6

Note 1—Values shown are nominal depth (inches) × weight per foot length (pounds).
Note 2—Values shown are nominal depth (millimeters) × mass per meter length (kilograms).
Note 3—Metric size examples shown are not necessarily the equivalents of the inch size examples shown.

Fig. 24-1-4 Abbreviations for shapes, plates, bars, and tubes.

(A) BEAM DESIGNATION PRIOR TO 1972

(B) BEAM DESIGNATION FROM 1972 ON

(C) METRIC BEAM DESIGNATION AND DIMENSIONING

Fig. 24-1-5 Building floor framing plan. (partial view)

Fig. 24-1-7 Slopes and dimensions of flanges.

When steel shapes are designated on drawings, a standard method of abbreviating should be followed that will identify the size and shape of the steel part. See Figs. 24-1-3 and 24-1-4. However, the method for calling for these standard shapes has changed over the last few years. When called upon to revise or modify existing drawings, the drafter must do so in the same convention used previously on that drawing. Therefore, it is important that the drafter not only have the most up-to-date knowledge, but also be familiar with previous standards still in use on old drawings. See Fig. 24-1-5.

Besides having to know the type of shapes available and their drawing designation, one must also be familiar with framing construction terms and where these shapes are used. See Fig. 24-1-6.

The abbreviations shown are intended only for use on design drawings. When lists of materials are being prepared for ordering from the mills, the requirements of the respective mills from which the material is to be ordered should be observed.

All S, C, and MC shapes have a 16.67 percent slope on the inside faces of the flanges. This is equivalent to 9°28' or a bevel of 1:6. W-shaped beams and columns are rolled with parallel face flanges or with a 5 percent slope (2°51') on the inside of the flange. See Fig. 24-1-7.

In structural steel shape tables, dimensions such as K and mean thickness of sloping flanges are given. Since the mean thickness of the sloping flange is given, these dimensions may also be used for all flange shapes.

If it is necessary to have the exact dimensions of a particular shape, they must be obtained from the individual mill's structural shape catalog. These catalogs also give the pertinent radial dimensions in question. See Fig. 24-1-8.

It is customary, on details made to a scale of 1:8, 1:10 or smaller, for the curve indicating the toes of angles and of flanges, the interior fillets between legs of angles, and the interior fillets between web, or stem, and flanges to be omitted in the drawing. It is usual to exaggerate on detail drawings the thickness of the leg, stem, web, or flange.

1. Anchors or hangers for open-web steel joists
2. Anchors for structural steel
3. Bases of steel and iron for steel or iron columns
4. Beams, purlins, girts
5. Bearing plates for structural steel
6. Bracing for steel members or frames
7. Brackets attached to the steel frame
8. Columns, concrete-filled pipe, and struts
9. Conveyor structural steel frame work
10. Steel joists, open-web steel joists, bracing, and accessories supplied with joists
11. Separators, angles, tees, clips, and other detail fittings
12. Floor and roof plates (raised pattern or plain (connected to steel frame)
13. Girders
14. Rivets and bolts
15. Headers or trimmers for support of open-web steel joists where such headers or trimmers frame into structural steel members
16. Light-gage cold-formed steel used to support floor and roofs
17. Lintels shown on the framing plans or otherwise scheduled
18. Shelf angles

Fig. 24-1-6 Structural steel terms.

W SHAPES   S SHAPES   C SHAPES

Steel Grades

There are hundreds of grades of steel produced in mills today. However, only a few of those are suitable for structural applications. The most common structural grade used in the United States is ASTM A36. All structural members discussed in this chapter will be assumed to be fabricated from ASTM A36, while the bolts are made from A307 or A325 depending on the strength required.

Mill Tolerances

There are certain permissible deviations brought about in the manufacture of structural steel that the drafter should understand. These permissible deviations from the published dimensions and contours (as listed in the AISC manual, in mill catalogs, and from the lengths specified by the purchaser) are referred to as *mill tolerances*. See Figs. 24-1-9 to 24-1-11.

The factors which contribute to the necessity for a mill tolerance are as follows:

1. The high speed of the rolling operations required to prevent the metal from cooling before the process has been completed.
2. The varying skill of operators in squeezing together the rolls for successive passes of the metal, particularly the final pass. See Fig. 24-1-12.
3. The springing and wearing of the rolls, and other mechanical factors.
4. The warping of the steel in the process of cooling.
5. The subsequent shrinkage in the length of a shape which was cut while the metal was still hot.

Under rolling tolerances (Fig. 24-1-9) note that the maximum overall depth (C) can be ¼ in. over the nominal depth. For example, a W24 × 94 beam (Fig. 24-1-8) is shown as having a 24¼-in. depth. However, its finished actual depth at C after rolling could be 24½ in. The depth at center line A could be either 24⅜ in. (24¼ + ⅛) or 24⅛ in. (24¼ − ⅛). The width of flange B could be 9¼ in. (9 + ¼) or 8¹³⁄₁₆ in. (9 − ³⁄₁₆) in place of the 9 in. width.

Suppose the W24 × 94 is ordered cut to length from the mill as a 55-ft long piece. It might be received by the fabricator either as 55′-¾ (55 ft + ¾ in.) or 54′-11⅝ (55 ft − ⅜ in.) long

Inch Designation	Depth d	Flange Width b	Flange Thickness t	Web Thickness w	k	Metric Designation	Depth d	Width b
W24 × 94	24¼	9	⅞	½	1⅝	W610 × 140	617	230
W24 × 76	23⅞	9	¹¹⁄₁₆	⁷⁄₁₆	1⁷⁄₁₆	W610 × 113	608	228
W18 × 114	18½	11⅞	1	⅝	1¹¹⁄₁₆	W460 × 177	482	286
W18 × 105	18⅜	11¾	¹⁵⁄₁₆	⁹⁄₁₆	1⅝	W460 × 158	476	284
W18 × 96	18⅛	11¾	¹³⁄₁₆	½	1½	W460 × 144	472	283
W18 × 60	18¼	7½	¹¹⁄₁₆	⁷⁄₁₆	1³⁄₁₆	W460 × 89	463	192
W16 × 78	16⅜	8⅝	⅞	½	1⅝	W410 × 114	420	261
W16 × 40	16	7	½	⁵⁄₁₆	1⅛	W410 × 60	407	178
W14 × 74	14¼	10⅛	¹³⁄₁₆	⁷⁄₁₆	1½	W360 × 110	357	255
W14 × 48	13¾	8	⁹⁄₁₆	⁵⁄₁₆	1¼	W360 × 72	350	204
W12 × 58	12¼	10	⅝	⅜	1⅜	W310 × 86	310	254
W12 × 50	12¼	8⅛	⅝	⅜	1⅜	W310 × 74	310	205
W12 × 36	12¼	6⅝	⁹⁄₁₆	⁵⁄₁₆	1¹⁄₁₆	W310 × 52	317	167
W12 × 27	12	6½	⅜	¼	¹⁵⁄₁₆	W310 × 39	310	165
W10 × 49	10	10	⁹⁄₁₆	⁵⁄₁₆	1⅛	W250 × 73	253	254
W10 × 33	9¾	8	⁷⁄₁₆	⁵⁄₁₆	1	W250 × 49	247	202
W10 × 25	10⅛	5¾	⁷⁄₁₆	¼	1	W250 × 39	262	147
W10 × 15	10	4	¼	¼	¹³⁄₁₆	W250 × 22	254	102
W8 × 35	8⅛	8	½	⁵⁄₁₆	1	W200 × 52	206	204
W8 × 28	8	6½	⁷⁄₁₆	⁵⁄₁₆	¹⁵⁄₁₆	W200 × 46	203	203
W8 × 20	8⅛	5¼	⅜	¼	⅞	W200 × 42	205	166
W8 × 15	8⅛	4	⁵⁄₁₆	¼	¹³⁄₁₆	W200 × 36	201	165
S24 × 100	24	7¼	⅞	¾	1¾	S610 × 149	610	184
S24 × 90	24	7⅛	⅞	⅝	1¾	S610 × 134	610	181
S20 × 95	20	7¼	¹⁵⁄₁₆	¹³⁄₁₆	1⅞	S510 × 141	508	183
S20 × 75	20	6⅜	¹³⁄₁₆	⅝	1⅝	S510 × 112	508	162
S18 × 70	18	6¼	¹¹⁄₁₆	¹¹⁄₁₆	1½	S460 × 104	457	159
S15 × 50	15	5⅝	⅝	⁹⁄₁₆	1⅜	S380 × 74	381	143
S12 × 50	12	5½	¹¹⁄₁₆	¹¹⁄₁₆	1⁷⁄₁₆	S310 × 74	305	139
S12 × 35	12	5⅛	⁹⁄₁₆	⁷⁄₁₆	1³⁄₁₆	S310 × 52	305	129
S10 × 35	10	5	½	⅝	1⅛	S250 × 52	254	126
S8 × 23	8	4⅛	⁷⁄₁₆	⁷⁄₁₆	1	S200 × 34	203	106
C15 × 50	15	3¾	⅝	¹¹⁄₁₆	1⁷⁄₁₆	C380 × 74	381	94
C15 × 40	15	3½	⅝	½	1⁷⁄₁₆	C380 × 60	381	89
C12 × 30	12	3⅛	½	½	1⅛	C310 × 45	305	80
C12 × 20.7	12	3	½	⁵⁄₁₆	1⅛	C310 × 31	305	74
C10 × 30	10	3	⁷⁄₁₆	¹¹⁄₁₆	1	C250 × 45	254	76
C10 × 20	10	2¾	⁷⁄₁₆	⅜	1	C250 × 30	254	69

Fig. 24-1-8   Properties of common structural steel shapes.

ROLLING TOLERANCES (Inches)

Nominal Size	Depth A		Width of Flange B		Out of Square T or T₁	Web Off Center E	Maximum Depth of Any Cross Section C
	Over	Under	Over	Under	Max.	Max.	Over Nominal Size
12 in. and under	1/8	1/8	1/4	3/16	3/16	3/16	1/4
Over 12 in.	1/8	1/8	1/4	3/16	1/4	3/16	1/4

CUTTING TOLERANCES (Inches)

W Shapes Nominal Depth	Variation from Specified Length for Lengths Given									
	To 30 Incl		Over 30 To 40 Incl		Over 40 To 50 Incl		Over 50 To 65 Incl		Over 65	
	Over	Under	Over	Under	Over	Under	Over	Under	Over	Under
Beams 24 in. and under	3/8	3/8	1/2	3/8	5/8	3/8	3/4	3/8	1	3/8
Beams over 24 in. All Columns	1/2	1/2	5/8	1/2	3/4	1/2	1	1/2	1 1/8	1/2

Fig. 24-1-9 Rolling and cutting tolerances for W shapes.

ROLLING TOLERANCES (Inches)

Shape	Nominal Specified Size	Depth A		Flange Width B		Out of Square T or T₁	Out of Square C – D
		Over	Under	Over	Under	B B	B
S	3 to 7 incl.	1/16	1/16	1/8	1/16	0.03	0.03
	Over 7 to 14 incl.	1/8	1/16	1/8	1/8	0.03	0.03
	Over 14 to 24 incl.	3/16	1/8	3/16	3/16	0.03	0.03
C	3 to 7 incl.	1/16	1/16	1/16	1/16	0.03	0.03
	Over 7 to 14 incl.	1/8	1/16	1/8	1/8	0.03	0.03
	Over 14	3/16	1/8	1/8	3/16	0.03	0.03

CUTTING TOLERANCES (inches)

Shape	Variation from Specified Lengths									
	To 30 incl.		Over 30 to 40 Incl.		Over 40 to 50 Incl.		Over 50 to 65 Incl.		Over 65	
	Over	Under	Over	Under	Over	Under	Over	Under	Over	Under
S and C Shapes	1/2	1/4	3/4	1/4	1	1/4	1 1/8	1/4	1 1/4	1/4

Fig. 24-1-10 Rolling and cutting tolerances for S and C shapes.

S SHAPES

C SHAPES

Fig. 24-1-11 Error between mating shapes.

CHANNEL

SMALL STRUCTURAL SHAPES MAY BE
FORMED BY A WIDE VARIETY OF
PASSING PROCEDURES.

Fig. 24-1-12 The making of a C shape.

piece. The fabricators have standards for ordering the plain material that take into consideration these cutting tolerances.

Although this variation of length would not be tolerated in the shop, it is essential that the detailer be aware of its possible occurrence, so that when he or she specifies the required stock, the material obtained will fulfill the purpose for which it was ordered. Another example of mill tolerances is shown in Fig. 24-1-13.

Detailers can usually disregard mill tolerances when detailing light- and medium-mass trusses, standard beams, standard channels, struts, and most plate girders. However, consideration must be given to the tolerances for all wide-flange beams and other heavy parts.

MINIMUM LENGTH OF JOINED BEAM=

$(42'\text{-}0 - \frac{3}{8}) + (35'\text{-}0 - \frac{3}{8}) = 76'\text{-}11\frac{1}{4}$

MAXIMUM LENGTH OF JOINED BEAM =

$(42'\text{-}0 + \frac{3}{8}) + (35'\text{-}0 + \frac{3}{8}) = 77'\text{-}1\frac{1}{8}$

REFER TO FIGURE 24-1-9

Fig. 24-1-13 Calculating minimum and maximum length of joined beam.

The handbooks prepared by the American Institute of Steel Construction (AISC) lists all the available structural shapes, their properties, and dimensions. However, for the convenience of the student, all the dimensions which are required for examples and problems are reproduced in this text.

STRUCTURAL DRAWING PRACTICES

Figure 24-1-14 is a table indicating the scale and the type of structural drafting in which the scale is most frequently used.

Dimensioning

The general practice, in dimensioning structural drawings, is to use the aligned method for dimensions and to place the dimensions above the dimension lines. Otherwise, the same general guidelines as used in mechanical drafting will apply. All dimensions shown in this chapter will be in feet and inches or inches. Metric (millimeters) measurements could also have been used.

Dimensions should be arranged in a manner most convenient to all who must use the drawing. They should not crowd the sketch and should cross the fewest possible number of other lines. The longest and overall dimensions should be farthest away from the views to which they apply. Dimensioning and descriptions of components (billing), in general, should be placed outside the picture. Dimensions should be given to the center lines of beams, to the backs of angles, and as explained later, to the backs of channels. They should be given to the top or bottom of beams and channels (whichever level is to be held), but never to both top and bottom, because of a possible overrun or underrun, in the depth, resulting from rolling.

As shown in Fig. 24-1-15, when four or more equal spaces between bolts are required, it is recommended that the information be given as 4 @ 2 = 8 instead of repeating 2 four times. This reduces the possibility of error, both in

Fig. 24-1-15 Dimensioning structural drawings.

Scales		Principal Drawings
Inch and Foot	Millimeter and Meter	Used
Full	1:1	Layout
3 = 1'-0	1:5	Layout
1½ = 1'-0	1:10	Layout or detail
¾ = 1'-0	1:20	Detail
⅜ = 1'-0		Erection or design
3/16 = 1'-0		Erection or design
3/32 = 1'-0		Erection or design
1 = 1'-0		Detail
½ = 1'-0		Detail or erection
¼ = 1'-0	1:50	Erection or design
⅛ = 1'-0	1:100	Erection or design

Fig. 24-1-14 Scales for structural drawings.

reading the drawing and in layout of the work in the shop. Do not include in such an equation the distance locating the group itself from some reference point, even though the distance may happen to be the same as the increment of spacing.

Elevation detail dimensions, known as *levels*, are generally furnished by a note on the drawing. When it is desirable to show the level or vertical distance above some established reference point (usually ground level), the value is given in inches (or millimeters) and placed above the level symbol, as shown in Fig. 24-1-15. A plus or minus precedes the value, indicating that the level specified is higher or lower than the reference point.

Another dimensioning practice is to enclose bolt and hole sizes in diamond-shaped frames, as shown in Fig. 24-1-15. This helps to differentiate the circular sizes from the linear dimensions. Bolt symbols are shown in Fig. 24-1-16.

SHOP   FIELD

Fig. 24-1-16   Bolt symbols.

The neatness, and hence the legibility, of shop drawings is enhanced by lining up notes and dimensions which have the same common purpose. Thus if the 5-in. cut instructions (Fig. 24-1-15) were required at both ends of the beam, they would be shown at the same elevation, even though the dimension lines were not drawn from end to end of the sketch. Attention paid to these features, resulting in an orderly and systematic presentation of the necessary information, does much to enhance the finished appearance of a shop drawing.

Source Material
• American Institute of Steel Construction

ASSIGNMENT

See Assignment 1 for Unit 24-1 on page 493.

UNIT 24-2
Beams

As a rule, each beam in a system of floor or roof framing makes a convenient erection unit. Hence, the required shop fabrication for each beam is shown on a shop drawing, which provides complete information for that beam. Such a drawing seldom pictures any part of the adjacent members to which this beam will later be joined in the field. However, in the preparation of the beam detail drawing, all the features that have a bearing on the later installation of the beam into its proper location in the frame, as indicated on the design drawing, must be investigated.

The location of the open holes to be provided in the beam for its field connection must match the location of similar holes in the supporting members. Proper clearances must be provided so that the beam can be swung into position after its supporting members have been erected. Any possible interference must be eliminated by cutting away the excess material.

The various fabricating shop drafting rooms do not always agree among themselves on a standard way of making shop drawings. In this text, details will be presented in a manner that all shops could use.

The two principal kinds of beam connections generally used are the framed and the seated types. In the framed type, the beam is connected by means of fittings (generally a pair of

SEATED
CONNECTION

FRAMED
CONNECTION

Fig. 24-2-1   Beam to column connections.

short angles) attached to its web. With seated connections, the end of the beam rests on a ledge, or seat, which receives the load from the beam just as if the end of the beam rested upon a wall. See Fig. 24-2-1.

It should be noted that the depth of the beam, dimensions in relation to the depth of the beam, end connections, cuts, and spacing of holes are drawn to scale. Copes, blocks, and cuts are shown in Fig. 24-2-2. It is the practice of the structural detailer to draw the depth dimensions to scale so that the relation of detail is correct and so that the fabricator can interpret the relation of holes to bolts or holes more readily.

C

NOTE I   "CUT NOT CHIP"

NOTE 2   "CUT AND CHIP"

NOTE I : PREFERRED
NOTE 2: USE IF SURFACE C
MUST BE FLUSH WITH WEB

COPE   BLOCK   CUT

Fig. 24-2-2   Copes, blocks, and cuts.

The length of beam and dimensions in relation to the length can be drawn to scale but are usually foreshortened. The reason that the length is usually foreshortened is that the scale length would, in most cases, take more space on a drawing than is economical and would have no practical value to the fabricator. However, foreshortening the length so much that the holes in the web or some of the detail will appear crowded or ambiguous should be avoided.

The structural detailer does not always draw to the exact scale, but exaggerates the drawings to clarify details. An example is the two lines that would show the thickness of the top or bottom flange of a beam.

ASSEMBLY CLEARANCES

In order for members to assemble readily, clearances are required between beams and columns or beams and beams. It may also be necessary to

NOTE — PRACTICE IS TO MAKE Q_1 AND Q_2 DIMENSIONS MULTIPLES OF $\frac{1}{4}$ IN.

Fig. 24-2-3   Assembly clearances.

cut or shape the ends of beams for mating parts to fit properly. The recommended clearances are shown in Fig. 24-2-3.

SIMPLE SQUARE-FRAMED BEAMS

The information—such as member length, size, and type; number of bolts; or type of fastener—required by the structural detailer is obtained from the design drawing. These drawings usually describe the type of construction, end loads or loads at support if not normal type and size of bolts, member shape and size, and any other data that would be required by the detailer.

Figure 24-2-4 represents part of a design drawing for a steel-framed floor system as viewed from above. With its notes, it contains, with the exception of connection-angle detail, all the necessary information required by the shop detailer to detail the W18 × 60 beam. Unless otherwise shown by dimensions or notes, members shown on the design drawing are presumed to be parallel or at right angles to one another, with their webs in a vertical plane, and to be in a level position from end to end. Elevation detail dimensions of beams are generally furnished by a note on the drawing. In Fig. 24-2-4A the vertical distance, or elevation, is placed above the level symbol and is shown as +98'-6, for the W21 × 73 beam and +98'-9 for the W24 × 76 beam. It might have been given by a note reading ALL STEEL FLUSH,

ELEVATION TOP OF STEEL SHOWN THUS: (+98'-6)

METHOD A

ELEVATION: ALL STEEL FLUSH, TOP AT ELEV. +98'-6

– CONNECTIONS: TWO ANGLES 4 X 3 X $\frac{1}{4}$ X 9 AT EACH END OF BEAM

METHOD B

Fig. 24-2-4   Partial design drawing.

TOP AT ELEV. +98'-6, as shown by Fig. 24-2-4B. Note that the top elevation of W24 × 76 is designated by (+3), meaning that the top of the beam is 3 in. above the reference elevation of +98'-6, or (+98'-9) as presented in Fig. 24-2-4A.

Before starting the drawing, the detailer should first establish what the beam detail is going to look like. This is

achieved by making sketches of the connections at both ends. The detailer first investigates the connection at one end, for this example the north end of the W18 × 60 of Fig. 24-2-4. A sketch, shown in Fig. 24-2-5A, is then made of the W18 framing into the W24. This section represents what would be seen if a viewer looked at the connection from the west side of the W18. A sketch is then made of the south end connection of W18. From the sketches the necessary detail requirements can be obtained. Of importance are the number of bolts or size of fillet welds required, and the size of the connecting angles. In this unit, the connecting-angle sizes are given. In Unit 24-3 the calculations for angle size and connections are covered in detail.

Note that the south end flange of the W18 × 60 is flush with that of the supporting W21 × 73 flange. Figure 24-2-5B is produced similarly to Fig. 24-2-5A for the north end, except we find that the flange of the W18 will interfere with that of the W21. Thus it becomes necessary to notch out or cope the W18.

Some shops would not dimension such a cut, but would give it a standard mark. Others would simply note on the drawing COPE TO W21 × 73 and let the shop work out its proper shape and size.

Note that the intersection of the horizontal and vertical is not a sharp corner (reentrant-cut), but is cut to a small radius to provide a fillet at this point. However, since the shop has been trained to provide these fillets, they are not generally shown on the detail drawing.

(A) NORTH–END BEAM CONNECTION

(B) SOUTH–END BEAM CONNECTION

Fig. 24-2-5   Detail of W18 x 60 beam.

Since the two beams are flush on top, the minimum depth of the cut Q is made equal to or greater than the K distance for the W21 × 73 beam.

K Distance for the W21 Beam = 1⅜
Following the guidelines shown in Fig. 24-2-3, the Q_1 dimension for this beam is 1½ in. The length of the cut, Q_2, as measured from the backs of the connecting angles, should allow for a minimum ½ in. clearance between the toe of the supporting beam and the flange of the supported beam. To determine the length of dimension Q_2, add ½ in. to half the flange width of the W21 beam. From this value subtract half the web thickness of the W21 beam and ¹⁄₁₆ in. The ¹⁄₁₆-in. dimension is the clearance allowed between the web face and the outer face of the connecting angles.

Therefore Q_2, = ½ + 4⅛ − ⁷⁄₃₂ − ¹⁄₁₆ = 4¹¹⁄₃₂. As previously used for Q_1, the dimension Q_2 should be raised to the nearest length evenly divisible by ¼ in. Thus the 4¹¹⁄₃₂ in. dimension (Q_2) is raised to 4½ in.

With the exception of the connecting-angle data, Fig. 24-2-6 is the completed shop drawing of the W18 × 60 beam. Note the following points.

1. The minus dimensions (− ⁵⁄₁₆), shown outside and opposite the dimension line for the back-to-back distance of the end connection angles (25′-5⅜), are the distances from the center lines of the supporting beams to the back of each connecting angle. For a beam framing to other shapes, the minus dimension (setback distance) is equal to half the web thickness of the supporting member plus ¹⁄₁₆ in., rounded off to the nearest ¹⁄₁₆ in.

2. The center-to center distance of 25′-6 between the two supporting beams is shown for reference purposes.

3. The actual or ordered length of the W18 × 60 should be such that its ends are about ½ in. short of the backs of the connecting angles. This is to allow for inaccurate cutting to specified length at the mill or in the shop and therefore eliminate any extra expense caused by recutting or trimming during fabrication.

4. No top or bottom views are necessary because no holes are required in either flange. In general, the shop should not be required to look at views which convey no necessary instructions.

5. The end connection angles are shown, but not detailed. Information on detailing these angles is given in Unit 24-3. Note the end view of the angles is shown but the W18 beam is not drawn.

6. The complete beam is given a shipping or erection mark, B15, to identify it in the office, shop, and field. There are many systems currently in use for establishing the shipping mark. One of the most common methods is to use a capital letter followed by a sheet number. Each separate shipping piece, detailed on one sheet, has the same number preceded by a different letter. In this example the detail drawing of the W18 × 60 is the second sketch on sheet 15; the third would be C15, the fourth D15; and so on.

7. The connection angles are given assembly or template marks, usually lowercase letters. This is done for two reasons:
   (a) It saves the detailing of these angles again, when they are used on the same piece (as on the south end of the beam in this example) or on other beams on the same sheet.
   (b) The angles will be punched on a different machine from the one used for the beam. The assembly mark is a guarantee that the correct angle will be assembled on the correct beam. On the given detail, the material required to fabricate only one complete shipping piece is listed, or billed. When duplication of the shipping pieces is

required, the shop multiplies the billing for one complete piece by the total number of assemblies required.

Source Material
- American Institute of Steel Construction

ASSIGNMENTS
See Assignments 2 and 3 for Unit 24-2 on page 494.

Review for Assignment
Unit 7-14   Conventional Breaks
Unit 24-1   Structural Steel—Plain Material

UNIT 24-3
Standard Connections

Standard framed-beam connections are used for framing structural steel. Since riveting is almost nonexistent in most fabricating plants today, rivets will not be considered in this context. Standard connection angles are shop-welded or bolted to the beam web and field-bolted to their supporting member. Only high-strength bolted connections of the friction type will be considered in this chapter.

When detailing individual members, the shop detailer should bear in mind that each individual member must be joined to other members. The place or location of one member's attachment shape or plate along with the means of fastening is called the *connection plate*, or the *connection* for short.

BOLTED CONNECTIONS

Bolts are placed on standard lines or gages. The distance between bolt holes is referred to as the *bolt pitch*, or *pitch*. The gage and the pitch for a multiple bolt connection detail must be sufficiently large to allow for the wrench clearance when a bolt adjacent to a previously installed bolt or adjacent to another part of the shape being joined is tightened. Figure 24-3-1 shows the recommended gages to be used for structural shapes.

NOTE: DETAIL FOR THE CONNECTING ANGLES NOT SHOWN ON THIS DRAWING

Fig. 24-2-6   Detail drawing of W18 x 60 beam.

LEG SIZE (INCHES)

GAGE	8	6	5	4	3½	3	2½	2	1¾	1⅜	1
G_1	4½	3½	3⅜	2½	2⅜	1¾	1⅜	1 3/16	1	⅞	⅝
G_2	3⅛	2⅜	1¾								
G_3	3⅛	2½	2								

(A) INCH SIZES

S SHAPES

FLANGE WIDTH	G
2⅜ TO 2¾	1½
3 TO 3⅜	1¾
3½ TO 4	2
4 TO 4¾	2⅛
5 TO 5¾	3
6 TO 7¼	3½
7½ TO 8	5

C SHAPES

FLANGE WIDTH	G
1⅜ TO 1½	¾
1¾	1
1⅜ TO 2	1 3/16
2⅛	1 3/16
2⅛ TO 2⅜	1⅜
2½ TO 2⅝	1⅜
2⅞ TO 3⅛	1¾
3⅜ TO 3⅝	2⅛
3¾ TO 4⅜	2½

W AND M SHAPES

FLANGE WIDTH	G
4	2⅛
5 TO 5¾	3
6 TO 7⅛	3½
7½ AND UP	5⅛

EXTRA GAGES FOR W COLUMNS: 2 / 2¾ / 3½ and 2⅛ / 3 / 3½

DEPTH OF W AND M COLUMN: G, 2⅛, 3¾, 5 — for 6, 8, 10 TO 14

LEG SIZE (MILLIMETERS)

GAGE	200	150	125	100	90	75	65	50	45	35	25
G_1	115	90	80	65	60	45	35	30	25	22	16
G_2	80	60	45								
G_3	80	65	50								

(B) METRIC SIZES

S SHAPES

FLANGE WIDTH	G
60 TO 70	40
75 TO 85	45
90 TO 100	50
100 TO 120	55
125 TO 145	75
150 TO 185	95
190 TO 200	125

C SHAPES

FLANGE WIDTH	G
35 TO 40	20
45	25
45 TO 50	30
55	30
55 TO 60	35
65 TO 70	40
75 TO 80	45
85 TO 90	55
95 TO 110	65

W AND M SHAPES

FLANGE WIDTH	G
100	55
125 TO 145	75
150 TO 180	95
190 AND UP	130

EXTRA GAGES FOR W COLUMNS: 50 / 70 / 85 and 55 / 75 / 90

DEPTH OF W AND M COLUMN: G, 55, 95, 125 — for 150, 200, 250 TO 350

Fig. 24-3-1 Recommended gages.

Of prime importance is consistency of detail; for example, gages on an individual member should not vary throughout the length of the member. If a connection plate, made from a sheared piece of steel, as shown in Fig. 24-3-2, is to have three holes across its width of 10 in. for ⅝-in. bolts, then the gage could be 3¾ in. with edge distances of 1¼ in. If the fasteners to be used are ¾-in. diameter bolts, then the edge distance of 1¼ in. would have to be adjusted. The minimum distance from the center of a bolt hole to any edge should not be less than that given by Fig. 24-3-3. For the ¾ in. bolt, the

(A) ⅝ – BOLTED CONNECTION

MIN EDGE DISTANCE = 1⅛ (SEE FIG. 24·3·3), Ø⅝ BOLTS, GAGE — 1¼, 3¾, 3¾, 1¼ (total 10)

(B) ⅝ – BOLTED CONNECTION

MIN EDGE DISTANCE = 1⅜ (SEE FIG. 24·3·3), Ø¾ BOLTS, GAGE — 1⅜, 3⅝, 3⅝, 1⅜ (total 10)

Fig. 24-3-2 Establishing gage sizes from edge distance.

	BOLT DIAMETER	AT SHEARED EDGE	AT ROLLED OR GAS CUT EDGE
U.S. CUSTOMARY (INCHES)	½	1	¾
	⅝	1⅛	⅞
	¾	1⅜	1
	⅞	1½	1⅛
	1	1⅝	1 3/16
	1⅛	1¾	1⅜
	1¼	2	1½
	1⅜	2½	1 13/16
METRIC (MILLIMETERS)	14	26	20
	16	28	22
	20	34	26
	22	38	28
	24	42	30
	27	46	34
	30	52	38
	36	64	46

Fig. 24-3-3 Minimum edge distance for bolt holes.

		SOCKETS		MIN. CLEAR.	
	BOLT SIZE	A	B	E	F
U. S. CUST. (IN.)	5/8	2 3/8	1 3/4	5/8	1 1/4
	3/4	3	2 1/4	1 1/8	1 1/4
	1	3 1/2	2 5/8	1 1/8	1 7/8
	1 1/4	4	3 1/8	1 3/8	1 13/16
	1 3/8	4 1/4	3 1/4		
METRIC (mm)	16	70	55	28	32
	20	80	58	29	34
	24	90	65	33	36
	30	110	75	38	42
	36	130	85	43	48

		BOLT SIZE	C	D
U. S. CUST. (IN.)	LIGHT WRENCHES	5/8 TO 1	13 1/4 TO 14	2
	HEAVY WRENCHES	1 TO 1 3/8	15 1/4 TO 17 1/4	2
METRIC (mm)	LIGHT WRENCHES	16 TO 24	337 TO 356	54
	HEAVY WRENCHES	24 TO 36	375 TO 438	64

Fig. 24-3-4 Minimum erection clearances.

minimum edge distance to the sheared edge is 1⅜ in. Since the plate is 10 in. wide and a minimum edge distance of 1⅜ in. is required, the gage required would be (10 − 2 × 1⅜) ÷ 2, or 3⅝ in. If the connection plate had been an angle, the same reasoning as used above would still pertain to the connection detail. Another consideration is wrench clearance, which is illustrated in Fig. 24-3-4.

Occasionally, a gage is too small for both holes to be placed adjacent to one another at right angles. When this happens, staggered centers are used as illustrated in Fig. 24-3-5.

EXAMPLE 1 Given a flat bar 4½ in. in width, which is to have a double line of Ø⅝ in. bolts and a gage of 2 in., calculate the pitch of the bolts. See Fig. 24-3-6.

Solution From Fig. 24-3-4 the clearance E required for a ⅝ in. bolt is 1³⁄₃₂. The minimum recommended distance between holes is 2E, or 2³⁄₁₆ in., which is greater than the gage of 2 in. Therefore, staggered holes will be required.

To determine the minimum pitch, refer to Fig. 24-3-5. Read down the 2 in. gage column to find where the dimension 2³⁄₁₆ (2.18) will fall. If the dimension is not exact, use the next larger number (2.25). To the extreme left of the 2.25 value is the pitch required. For this problem, the pitch is 1.

EXAMPLE 2 Figure 24-3-7 shows a partial design drawing similar to Fig. 24-2-4 except that it includes information concerning the connection. It represents part of a design drawing for a

Pitch (in.)	Gage (inches)													
	1	1¼	1½	1¾	2	2¼	2½	2¾	3	3¼	3½	3¾	4	4¼
¼	1.00	1.25	1.50	1.75	2.00	2.25	2.50	2.75	3.00	3.25	3.50	3.75	4.00	4.25
½	1.10	1.35	1.60	1.80	2.05	2.30	2.55	2.80	3.05	3.30	3.55	3.80	4.05	4.30
¾	1.25	1.45	1.70	1.90	2.15	2.35	2.60	2.85	3.10	3.35	3.60	3.85	4.10	4.35
1	1.40	1.60	1.80	2.00	2.25	2.45	2.70	2.95	3.15	3.40	3.65	3.90	4.15	4.40
1¼	1.60	1.75	1.95	2.15	2.35	2.55	2.80	3.00	3.25	3.50	3.70	3.95	4.20	4.45
1½	1.80	1.95	2.10	2.30	2.50	2.70	2.90	3.15	3.35	3.60	3.80	4.05	4.30	4.50
1¾	2.05	2.15	2.30	2.50	2.65	2.85	3.05	3.25	3.50	3.70	3.90	4.15	4.40	4.60
2	2.25	2.35	2.50	2.65	2.85	3.00	3.20	3.40	3.60	3.80	4.05	4.25	4.50	4.70
2¼	2.45	2.55	2.70	2.85	3.00	3.20	3.35	3.55	3.75	3.95	4.15	4.40	4.60	4.85
2½	2.70	2.80	2.90	3.05	3.20	3.35	3.55	3.70	3.90	4.10	4.30	4.50	4.75	4.95
2¾	2.95	3.00	3.15	3.25	3.40	3.55	3.70	3.90	4.10	4.25	4.45	4.65	4.85	5.10
3	3.15	3.25	3.35	3.50	3.60	3.75	3.90	4.10	4.25	4.40	4.60	4.80	5.00	5.20
3¼	3.40	3.50	3.60	3.70	3.80	3.95	4.10	4.25	4.40	4.60	4.80	5.00	5.20	5.35
3½	3.65	3.70	3.80	3.90	4.05	4.15	4.30	4.45	4.60	4.80	4.95	5.15	5.35	5.50
3¾	3.90	3.95	4.05	4.15	4.25	4.40	4.50	4.65	4.80	5.00	5.15	5.30	5.50	5.70

(U.S. CUSTOMARY (in.))

Pitch (mm)	Gage (millimeters)																	
	25	30	35	40	45	50	55	60	65	70	75	80	85	90	95	100	105	110
5	25	30	35	40	45	50	55	60	65	70	75	80	85	90	95	100	105	110
10	27	32	36	41	46	51	56	61	66	71	76	81	86	91	96	100	105	110
15	29	34	38	43	47	52	57	62	67	72	76	81	86	91	96	101	106	111
20	32	36	40	45	49	54	59	63	68	73	78	82	87	92	97	102	107	112
25	35	39	43	47	51	56	60	65	70	74	79	84	89	93	98	103	108	113
30	39	42	46	50	54	58	63	67	72	76	81	85	90	95	100	104	109	114
35	43	46	49	53	57	61	65	69	74	78	83	87	92	97	101	106	111	115
40	47	50	53	57	60	64	68	72	76	81	85	89	94	98	103	108	112	117
45	51	54	57	60	64	67	71	75	79	83	87	92	96	101	105	110	114	119
50	56	58	61	64	67	71	74	78	82	86	90	94	99	103	107	112	116	121
55	60	63	65	68	71	74	78	81	85	89	93	97	101	105	110	114	119	123
60	65	67	69	72	75	78	81	85	88	92	96	100	104	108	112	117	121	125
65	70	72	74	76	79	82	85	88	92	96	99	103	107	111	115	119	123	128
70	74	76	78	81	83	86	89	92	96	99	103	106	110	114	118	122	126	130
75	79	81	83	85	87	90	93	96	99	103	106	110	113	117	121	125	129	133
80	84	85	87	89	92	94	97	100	103	106	110	113	117	120	124	128	132	136
85	89	90	92	94	96	99	101	104	107	110	113	117	120	124	127	131	135	139
90	93	95	97	98	101	103	105	108	111	114	117	120	124	127	131	135	138	142

(METRIC (mm))

Fig. 24-3-5 Staggered fasteners.

Fig. 24-3-6 Gage and pitch layout.

Fig. 24-3-8 Indicating beam reactions on drawings.

steel-framed floor system as viewed from above. With its notes, it contains all the necessary information required by the shop detailer to detail the W18 × 60 beam. With the exception of the connection-angle detail, all information

ELEVATION TOP OF STEEL SHOWN THUS (+98'-6)

NOTES:
ALL HOLES Ø $\frac{13}{16}$
ALL CONNECTIONS TO DEVELOP FULL LENGTH
    UNLESS OTHERWISE SPECIFIED
BOLTS: $\frac{3}{4}$ A325
CONNECTING ANGLES WELDED TO BEAM,
    BOLTED TO SUPPORT

Fig. 24-3-7 Partial design drawing.

tion pertaining to this beam was covered in Unit 24-2. Therefore, this unit will deal only with the connection-angle detail.

Solution The problem is to select a connection for the W18 beam which will be able to carry the reaction values. Often the reaction values are calculated and given on the design drawing, as shown in Fig. 24-3-8. If the reaction values are not given, the connection is designed to support half the total uniform load capacity. In this case, since the reaction values are not given and there is no indication on the design drawing to indicate other than a uniformly distributed load when the building is complete, we consult Fig. 24-3-9, which states the maximum allowable load for a W18 × 60 beam having a span of 25'-6 lies between the 126 and 118 kip values (approximately 120 kips). One kip equals 1000 lb. Connection design load is one-half of this value or 60 kips.

Referring to Fig. 24-3-10 under the column Bolt Capacity in Kips Friction Connections, we find that four bolts

are required to support this load. The length of the connecting angles shown for four bolts is 11½ in. for which the minimum and maximum depth recommendations for beams are 15 and 24 in. Since these limits bracket the actual depth of the W18 beam, it is acceptable.

Referring to the column Web Framing Leg with Welds, we find the maximum weld capacity of a four-bolt per vertical line angle with a ³⁄₁₆ in. weld is 140 kips for a 3 in. angle width and 134 kips for a 2½ in. angle width. Both are greater than the 60 kips allowable load and thus are acceptable. Therefore, the smaller angle width of 2½ in. is selected for the welded framing leg. Practice dictates that the angle thickness should be ¹⁄₁₆ in. greater than the weld size; hence the minimum required angle thickness is ³⁄₁₆ + ¹⁄₁₆ = ¼ in.

Another check for the minimum thickness of the angle and the minimum permissible web thickness for the beam must be made. In order to determine the minimum angle thickness, refer to the section Minimum Required Web Thickness and Angles Where Bolted of Fig. 24-3-10. Since connection angles are assumed to be material which has a yield strength of 44 000 psi (see note 1 located below the table), the minimum web and angle thickness specified is .34 in. Note 2 states that if the connection is used for outstanding angle legs, then the value as specified can be halved. Therefore, the minimum thickness of angle that can be used is .17 in. This is less than the ¼ in. required for welding; and as such, the ¼ in. angle thickness is selected. The web thickness of the W18 × 60 beam is ⁷⁄₁₆ in. which is acceptable.

BEAM LOAD TABLE IN KIPS (1000 lb)								BEAM LOAD TABLE IN KILONEWTONS (kN)								
	W18 Beam				W12 Beam				W460 Beam				W310 Beam			
Span	Pounds per Foot				Pounds per Foot				Span	Mass per Meter				Mass per Meter			
Feet	60	55	50	45	40	36	31	27	mm	89	82	74	67	60	52	45	39
16	177	160	145	130	81	74	63	55	5000	864	782	708	631	395	361	308	266
18	161	146	132	118	74	67	57	50	5500	785	711	644	573	359	328	280	242
20	148	134	121	108	67	62	53	46	6000	720	652	590	526	329	301	257	222
22	136	123	112	100	62	57	49	42	6500	665	601	545	485	304	278	237	205
24	126	115	104	92	58	53	45	39	7000	617	559	506	451	282	258	220	190
26	118	106	97	86	54	50	42	36	7500	576	521	472	420	263	241	205	177
28	111	100	91	81	51	46	40	34	8000	540	489	443	394	247	226	193	166
30	104	94	85	76	48	44	37	32	8500	508	460	417	371	232	212	181	157

Fig. 24-3-9 Beam load tables.

WEB FRAMING LEGS **OUTSTANDING LEGS**

EITHER LEG WITH 7/8-A325 BOLTS				WEB FRAMING LEG WITH E70XX WELDS				
Angle Width and Gage	Bolts Per Vertical Line	Bolt Capacity in Kips Friction Connections	Bearing Connections	Weld Capacity in Kips — Fillet Size, D (in.)				Angle Width (in.)
				3/16	1/4	5/16		
W = 3½ G = 5½ G₁ = 2¼	2	38	47	81	108	135		
	3	54	70	114	152	190		
	4	72	94	140	186	232		
	5	90	118	165	220	275		W = 3
	6	108	142	190	252	315		
	7	126	165	214	286	358		
	8	144	188	240	320	400		
W = 3 G = 4 G₁ = 1¾	2	38	47	81	108	135		
	3	54	70	108	144	180		
	4	72	94	134	178	222		
	5	90	118	158	210	262		W = 2½
	6	108	142	182	243	304		
	7	126	165	207	276	345		
	8	144	188	232	310	388		
Minimum Required Web Thickness and Angles Where Bolted (inches)								
			.34	.27	.40	.54	.68	

EITHER LEG WITH M20-A325 BOLTS				WEB FRAMING LEG WITH E-480XX WELDS				
Angle Width and Gage	Bolts Per Vertical Line	Bolt Capacity in Kilonewtons Friction Connections	Bearing Connections	Weld Capacity in Kilonewtons — Fillet Size, D (mm)				Angle Width mm
				4	6	8	10	
W = 90 G = 130 G₁ = 60	2	169	210	324	484	646	808	
	3	240	314	452	678	904	1130	
	4	320	419	551	826	1100	1380	
	5	400	524	649	974	1300	1620	W = 75
	6	480	629	748	1120	1500	1870	
	7	560	734	846	1270	1690	2120	
	8	640	839	945	1420	1890	2360	
W = 75 G = 100 G₁ = 45	2	169	210	330	484	646	808	
	3	240	314	428	641	855	1070	
	4	320	419	526	789	1050	1320	
	5	400	524	625	937	1250	1560	W = 65
	6	480	629	723	1080	1450	1810	
	7	560	734	822	1230	1640	2050	
	8	640	839	920	1380	1840	2300	
Minimum Required Web Thickness and Angles Where Bolted								
FY = 300			8.6	6.9	10.3	13.8	17.2	

OUTSTANDING LEG WITH E70XX WELDS				LENGTH OF CONNECTING ANGLES			
Angle Width (in.)	Weld Capacity in Kips — Fillet Size D (in.)			Bolts Per Vertical Line	Suggested Beam Depth Limits (in.) Min. Max.		Angle Length L (in.)
	3/16	1/4	5/16				
W = 3½	24	32	40	2	8 – 12		5½
	56	74	92	3	12 – 18		8½
	94	126	158	4	15 – 24		11½
	125	167	208	5	18 – 30		14½
	150	200	250	6	22 – 36		17½
	176	234	292	7	24 – 42		20½
	200	268	335	8	28 – 48		23½
W = 3	30	40	50	2	8 – 12		5½
	68	91	114	3	12 – 18		8½
	98	130	162	4	15 – 24		11½
	125	167	208	5	18 – 32		14½
	150	200	250	6	22 – 36		17½
	176	234	292	7	24 – 44		20½
	200	268	335	8	28 – 48		23½

OUTSTANDING LEG WITH E480XX WELDS					LENGTH OF CONNECTING ANGLES			
Angle Width mm	Weld Capacity in Kilonewtons — Fillet Size D (mm)				Bolts Per Vertical Line	Suggested Beam Depth Limits (mm) Min. Max.		Angle Lenth L (mm)
	4	6	8	10				
W = 90	98	141	186	232	2	200 – 300		150
	228	329	438	546	3	300 – 450		230
	384	563	755	952	4	380 – 600		310
	490	743	1000	1263	5	450 – 800		390
	589	891	1200	1510	6	550 – 900		470
	687	1040	1400	1760	7	600 – 1100		550
	786	1190	1590	2000	8	700 – 1200		630
W = 75	125	180	238	297	2	200 – 300		150
	276	405	544	681	3	300 – 450		230
	386	581	789	1004	4	380 – 600		310
	490	743	1000	1263	5	450 – 800		390
	589	891	1200	1510	6	550 – 900		470
	687	1040	1400	1760	7	600 – 1100		550
	786	1190	1590	2000	8	700 – 1200		630
Minimum Required Web or Flange Thickness*								
FY = 300	3.4	5.2	6.9	8.6				

To be used for educational purposes only.

*Thickness listed is for supporting material with beams attached to one side only. If beams are attached to both sides of the supporting material, use double the minimum thickness listed.

Note 1. Connection angles are assumed to be material with minimum yield strength of 44 000 psi (300 Mpa).
Note 2. For connections with outstanding legs bolted, the minimum required thickness of the supporting material is one-half the thickness listed above, if beams are attached to one side of the supporting material.

Fig. 24-3-10 Double angle beam connections for .75 in. (M20)-A325 bolts and E 480xx fillet welds.

The next step is to select the gage (distance between bolt centers on the two outstanding legs). The recommended gage and angle widths shown are 5½ and 3½, or 4 and 3, respectively. Both the 3½- and 3-in. wide angles are acceptable as far as load capacities are concerned. The clearances shown in Fig. 24-3-11 would be a factor in deciding to use either the 3- or 3½-in. wide outstanding length. In this problem, the 3-in. leg was chosen because the use of universal joints is now a widely accepted practice.

The G_1 distance of 1¾ in. shown in Fig. 24-3-1 should be used instead of the actual 1²⁵/₃₂-in. dimension shown in Fig. 24-3-11, since a hole clearance of 1/16 in. (13/16 in. diameter holes) is used for the Ø¾ bolts.

EXAMPLE 3 With reference to Fig. 24-3-12 select a connection for the W12 beam which will be able to carry the reaction value shown.
Solution The reaction value for the connection is 50 kips. Referring to Fig. 24-3-10 under the column Bolt Capac-

ALL HOLES Ø 13/16 IN.
BOLTS: Ø 3/4 IN. —ASTM A325
CONNECTING ANGLES WELDED TO BEAM, BOLTED TO SUPPORT.

Fig. 24-3-12   Partial design drawing.

ity in Kips Friction Connections, we find that three bolts are required to support this load. The length of the connecting angles shown for three bolts is 8½ in., for which the minimum and maximum depth recommendations for beams are 1'-0 and 1'-6. Since

these limits bracket the actual depth of the W12 beam, it is acceptable.

Referring to the column Web Framing Leg with Welds in Fig. 24-3-10, we find the maximum weld capacity of a three bolt per vertical line with a 3/16 in. weld is 114 kips for a 3-in. angle width. Practice dictates that the angle thickness should be 1/16 in. greater than the weld size; hence the minimum required angle thickness is 3/16 + 1/16 = ¼ in.

Another check for the minimum thickness of the angle and the minimum permissible web thickness for the beam must be made. In order to determine the minimum angle thickness, refer to the section Minimum Required Web Thickness and Angles Where Bolted. Since connection angles are assumed to be material which has a yield strength of 44 000 lb/in.[2] (see note 1 located below the table), the minimum web and angle thickness specified is .34 in. Note 2 states that if the connection is used for outstanding angle legs, then the value specified can be halved. Therefore the minimum angle that can be used is .34 ÷ 2 = .17 in. This is less than the ¼ in. required for welding, and as such, the ¼ in. angle thickness is selected. The web thickness of the W12 × 36 beam is .31 in., which is acceptable.

The next step is to select the gage (distance between bolt centers on the two outstanding legs). The recommended gage and angle widths shown are 5½ and 3½ in., or 4 and 3 in., respectively. Both the 3½- and the 3-in. wide angle legs are acceptable as far as load capacities are concerned. Therefore the 3-in. wide leg is chosen.

The G_1 distance of 1¾ in. shown in Fig. 24-3-1 and the gage distance of 4 in. are selected since a hole clearance of 1/16 in. (13/16-in. diameter holes) is used for the ¾-in. diameter bolts.

Some fabricators on occasion choose to use high-strength bolts in the shop to fasten the connection angles to the web of the beam. The combination of the shop and field high-strength bolts in one connection presents other problems, such as additional clearance requirements for entering and tightening the high-strength bolts. These requirements mean larger connection angles, larger gages, and larger spread. An alternate solution to the increased sizes would be to stagger the bolt cen-

TWO - L3 X 2½ X ¼

GAGE 4

WEB OF W18 X 60

7/16

1 25/32

3

1 7/32

Ø 2 5/8

Ø 2 3/8

2 3/16

ERECTION CLEARANCE USING UNIVERSAL JOINT

TWO - L3 X 2½ X ¼

FOR WRENCH SIZES AND CLEARANCES SEE FIG. 28-3-4

GAGE 5½

WEB OF W18 X 60

7/16

2 17/32

15/32

3

2 3/16

NOTE: INSUFFICIENT EDGE DISTANCE. SEE FIG. 24-3-3

ERECTION CLEARANCE WITHOUT USING UNIVERSAL JOINT

Fig. 24-3-11   How erection clearances control gage and connecting angle sizes.

Fig. 24-3-13   Detail of north end of W18 x 60 beam shown in Fig. 24-2-7.

ters of the outstanding leg. In order not to overcomplicate the subject at this point, all connections discussed in this unit will be angles shop-welded to the web and field-bolted to the supporting members.

The location of the connection angles on the horizontal beam must now be set. Practice is to set the distance from the uppermost bolt hole to the top of the beam equal to the pitch distance. From the detailer's sketch, shown in Fig. 24-3-13, the complete detail drawing of the beam (Fig. 24-3-14) is made. Note the following points.

1. The end connection angles are completely detailed. Following their own individual shop standards, many shops could do this with less information.
2. The web leg is 2½ in.
3. The outstanding leg of each connection angle is 3 in. and the recom-

mended gage for this leg size is 1¾ in.

4. The open holes, for connection to the web of the supporting beam, are spaced 4 in. apart, center to center. This distance is called the *spread*. It is $2 \times G_1 +$ web thickness of the W18 × 60 beam = $2 \times 1\frac{3}{4} + \frac{7}{16} = 3\frac{15}{16}$. Use 4 in.
5. The pitch, or distance, bolt to bolt, along any gage line is 3 in.
6. The end distance is equal to half of the remainder left after subtracting the sum of all bolt spaces from the length of the angles. In the case of the four-row connection, it equals $(11\frac{1}{2} - 3 - 3 - 3) \div 2 = 1\frac{3}{4}$ in.
7. Instead of noting them on each individual detail, as in Fig. 24-3-14, it is usual practice to call for the sizes of bolts and holes once on each sheet in a general note. Such a note covers all the bolts and holes on the sheet, with exceptions noted on the individual details where they occur.

Source Material

• American Institute of Steel Construction

ASSIGNMENTS

See Assignments 4 and 5 for Unit 24-3 on page 497.

Review for Assignment

Unit 24-1   Structural Drawing Practices
Unit 24-2   Assembly Clearances

UNIT 24-4
Sectioning

In many instances the cross-hatching of sections may be omitted on shop drawings because its use is not needed to make the drawing any clearer. In structural detailing, the practice is to omit all lines on a drawing that serve no significant purpose.

BOTTOM VIEWS

In shop detail drawings, the bottom flange or face of a shape is never viewed from below; that is, the drafter does not stand below the object and look squarely up at it. Rather, she or he cuts a section such as A-A in Fig. 24-4-1 and views the bottom flange by looking squarely down on the top side of it.

SECTION A-A — USED AS BOTTOM VIEW IN STRUCTURAL DRAFTING. CUTTING PLANE LINE AND CROSSHATCHING NOT DRAWN

NORMAL BOTTOM VIEW – NOT USED IN STRUCTURAL DRAFTING

Fig. 24-4-1   Bottom view in structural drafting.

The reason for substituting a bottom section for a bottom view is to obtain a better correlation between it and the top view. For example, it will be more apparent whether a connection on one side of the top flange and a connection on the bottom flange are on the same side or opposite sides of the member. Note that the cross-hatching is omitted, as previously mentioned.

ELIMINATION OF TOP AND BOTTOM VIEWS

In the preceding examples, we found that a top and bottom view were not necessary because no holes were

TWO BEAMS – B15

Fig. 24-3-14   Complete beam detail of W18 x 60 from partial design drawing, Fig. 24-3-7.

Fig. 24-4-2 Detail of beam.

Fig. 24-4-3 Elimination of top and bottom views.

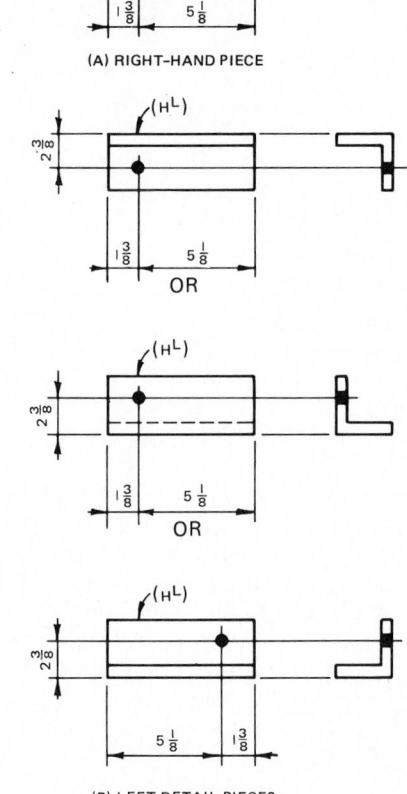

Fig. 24-4-4 Right and left detail pieces.

required in either flange. Let us now look at an example where there are holes in the bottom and top flanges and see if top and bottom views are required.

In Fig. 24-4-2, the detailing shows a top view and a picture of the bottom flange, taken as a section looking down. Note that the dash line in the top view and the solid lines in the bottom view that depict the web are not drawn continuously across the length of the member. Neither is the cut section of the web blackened or cross-hatched. Yet, the drawings is complete, readable, and understandable to the fabricator. Remember, use as few lines as possible to describe the object and the shop fabrication.

Detailing in Fig. 24-4-3 eliminates these views. In order to eliminate views of the top and bottom flanges, instructions (including necessary dimensions) for cutting these flanges at the right-hand end have been covered in the note on the web view concerning cutting. The transverse distance between gage lines on the flanges is covered by the note GA = 5. In both cases, symmetry about the center line of the beam web is understood. These notes must be explicit in showing what fabrication, if any, is required on each flange. Both methods of presentation are common practice.

RIGHT- AND LEFT-HAND DETAILS

Very frequently detail material, such as connection angles and other fittings, is used under conditions where one piece must be the exact opposite of another. In such cases, both the RH (right-hand) and the LH (left-hand) pieces are fabricated from the same sketch; that is, from the detail of the RH piece.

The piece which is made like the drawing is identified by use of the letter R, added to its assembly mark, thus: H^R for right. The one which is made opposite-hand has the letter L added to its assembly mark, thus: H^L for left. No assembly mark should be marked R unless there is an exact opposite, or LH detail piece, needed on the sheet, because all detail pieces are assumed to be RH as shown on the drawing unless otherwise noted. Likewise, no assembly mark should be marked L unless there is also a corresponding right.

If a drawing is placed in front of a mirror, the required RH detail would appear as represented by the drawing, and the required LH detail would appear as reflected in the mirror.

An understanding of rights and lefts, if not innate, may be gotten from Fig. 24-4-4. Note that the two views of H^L to the right of the drawing, even in their rotated position, still picture a fitting which is opposite hand to H^R.

Pieces which in their assembled position may appear to be rights and lefts often really are alike. Thus the fittings, shown on the left side of Fig.

24-3-14; can be turned upside down and used on the right side of the beam web.

When rights and lefts of whole shipping pieces are encountered, it is the practice in some fabricating shops to note the RH piece AS SHOWN and the LH piece OPPOSITE HAND in the required list. If two shipping pieces are involved, one the exact opposite of the other, the required listing under the single sketch might read (Fig. 24-4-5):

ONE BEAM A150^R—AS SHOWN
ONE BEAM A150^L—OPPOSITE HAND

If two shipping pieces are involved, one practically but not exactly the opposite of the other, and they are detailed on the same sketch, the required listing under the single sketch would read:

ONE BEAM A150—AS SHOWN AND NOTED
ONE BEAM B150—OPPOSITE HAND AND NOTED

ONE-CHANNEL-A 150ᴿ — AS SHOWN
ONE· *do* ·A 150ᴸ — OPP HAND

ONE-CHANNEL-A 150 — AS SHOWN AND NOTED
ONE· *do* ·B 150 — OPP HAND AND NOTED

(A) IF THESE ARE THE DRAWINGS THE SHOP GETS TO WORK FROM

A 150ᴿ AS FABRICATED

A 150 AS FABRICATED

A 150ᴸ AS FABRICATED

B 150 AS FABRICATED

(B) THIS IS WHAT THE SHOP FURNISHES (MAKES)

Fig. 24-4-5 Right and left shipping pieces.

In the case of exact RH and LH shipping pieces, the shipping mark may be the same except for the R and L notation; in the case of combined but different RH and LH shipping pieces, the shipping marks are always different. In both cases only the RH or AS SHOWN shipping piece is detailed. It is the shop which does the reversing, according to the notation OPPOSITE HAND, in the required list.

Before an attempt is made to detail pieces involving combinations of rights and lefts, Fig. 24-4-5 should be studied. If differences between pieces are minor, it is common practice to combine the details of two or more different pieces in a single sketch, by noting the differences, for example, in the case of the two web holes required in B150 but not required in A150.

Source Material

• American Institute of Steel Construction

ASSIGNMENTS

See Assignments 6 and 7 for Unit 24-4 on page 497.

Review for Assignments

Unit 24-1 Structural Steel—Plain Material
Unit 24-2 Beams and Assembly Clearances
Unit 24-3 Standard Connections

UNIT 24-5
Seated Beam Connections

Seated beam connections are used to connect beams to column webs or flanges. There are two types: unstiffened seat connections and stiffened seat connections. Only the unstiffened (angles) will be covered in this unit.

The following procedure is suggested for choosing a seated beam connection. Assume that a W12 × 27 beam is to be placed between two columns, as shown in Fig. 24-5-1. As in

Fig. 24-5-1 Partial design drawing.

the example for the framed beam. Unit 24-3, the beam reactions must first be established. If they are not shown on the drawing, they must be calculated. To do this, the length of the beam must be known. This is found by subtracting half of the nominal depth of each of the two supporting columns from the center-to-center distance. The nominal depth of the columns can be found from Fig. 24-1-8. In this example, the length of the span is $16'-0 - 10 = 15'-2$ in. From the beam load tables in Fig. 24-3-9 for a W12 × 27 beam having a span of 16 ft, the total allowable uniform load is 55 kips. The reaction at the end of the beam at each support is half the total load, or 27.5 kips. The length of the seated beam is now determined. From Fig. 24-1-8, the flange width b for the W12 × 27 beam is 6½ in. Note that Fig. 24-5-2 gives tables for a supporting angle length L of 6 and 8 in. Since the flange width of 6½ in. is greater than the L (length) of 6 in., the angle length of $L = 8$ in. for the seat angle will have to be used. The seat angle thickness must now be calculated. The web thickness of the W12 × 27 beam is ¼ in. Refer to Fig. 24-5-2, under the heading Outstanding Leg Capacity—kips, $L = 8$ in., and Beam Web Thickness = ¼ in. Read across until a leg capacity of 27½ kips or greater is found. This occurs at the kip value of 34 where the angle thickness is ⅝ in.

It is preferable for most fabricators to shop-weld the seat angle to the column, since the seat will provide support for the beam during erection. Under the heading Vertical Leg Weld Capacity, the angle thickness as previously determined was found to be ⅝ in.; therefore the maximum permissible weld size would be ⅝ − 1/16, or 9/16 in. (use ½ in.). From the table a ½ in. fillet weld will resist a force of 31 kips when a 4 × 4 angle size is used. To the right of this column, the angle thickness range of ⅜ to ⅝ in. is specified. The required angle thickness range as determined previously was ⅝ in.; therefore the dimensions for the seat angle are 4 × 4 × ⅝ × 8. The next step in the process is to make a sketch of the detail, as shown in Fig. 24-5-3.

The beam will be fastened to the seat angles using ∅⅝—A325 bolts. In order to determine their gage, reference should be made to Fig. 24-3-1. The

SEATED BEAM CONNECTIONS*
METRIC BOLTS E480XX FILLET WELDS**
U.S. Customary (in.)

Outstanding Leg Capacity—Kips
(Based on 3½ or 4 in. Outstanding Leg)

Angle Length		L = 6 in.						L = 8 in.					
Angle Thickness		3/8	1/2	5/8	3/4	7/8	1	3/8	1/2	5/8	3/4	7/8	1
Beam Web Thickness	3/16	15	20	25	31	35	39	17	22	28	35	39	44
	1/4	18	24	30	38	43	48	20	27	34	42	47	53
	5/16	22	29	36	46	52	57	24	32	40	50	57	64
	3/8	26	33	43	54	61	68	29	38	47	59	66	74
	7/16	—	37	46	58	66	73	—	41	51	64	71	80
	1/2	—	42	52	66	74	82	—	46	57	72	81	90
	9/16	—	48	59	74	83	93	—	52	65	81	91	101

*To be used for educational purposes only.
**Welding resistances have been soft converted.

U.S. Customary (in.)							
Vertical Leg Weld Capacity (Kips)							
Fillet Weld Size D-E70XX Electrodes						†Angle Sizes	Angle Thickness
1/4	5/16	3/8	7/16	1/2	5/8		
22	26	28	29	31	32	4 x 4	3/8-5/8
34	40	43	45	48	50	5 x 3½	3/8-3/4
43	51	54	56	59	64	6 x 4	3/8-7/8
72	88	95	99	103	113	8 x 4	1/2-1

†Long vertical leg.

SEATED BEAM CONNECTIONS*
METRIC BOLTS E480XX FILLET WELDS**
Metric (mm)

Outstanding Leg Capacity (kN)
(Based on 90 or 100 mm Outstanding Leg)

Angle Length		L = 150 mm						L = 200 mm					
Angle Thickness		10	13	16	20	#25	##30	10	13	16	20	#25	##30
Beam Web Thickness	4	60.2	80.4	101	128	161	195	68.3	90.8	113	143	181	219
	5	70.4	93.7	117	148	187	226	79.4	105	131	166	209	252
	6	81.0	107	133	168	212	256	90.7	120	149	188	236	284
	7	89.1	118	147	186	234	282	99.5	132	164	206	260	313
	8	96.9	129	160	203	255	308	108	143	178	224	283	341
	9	106	141	175	221	278	335	118	156	194	244	307	370
	10	116	153	190	239	301	362	128	169	209	263	331	398
	11		166	205	258	323	389		182	226	283	355	427
	12		179	221	277	346	416		196	244	303	379	455

*To be used for educational purposes only.   #100 mm Outstanding Leg Only.
**Welding resistances have been soft converted.   ##125 mm Outstanding Leg Only.

Metric (mm)							
Vertical Leg Weld Capacity (kN)							
Fillet Weld Size D—E480XX Electrodes						*Angle Sizes	Angle Thickness
6	8	10	12	14	16		
97	115	123	130	133	142	100 x 100	6:16
151	178	189	199	208	221	125 x 90	8:20
192	228	241	255	266	282	150 x 100	8:25
323	393	423	450	470	503	200 x 125	10:30

*Long vertical leg.

Fig. 24-5-2   Seated beam connections.

recommended gage for a W12 × 27 which has a flange width of 6½ in. is 3¾ in. Also from Fig. 24-3-3, the minimum distance for an ∅5/8 in. bolt to a rolled edge is 7/8 in. Therefore, the end distance (center of bolt to end of beam) is 4 − 7/8 − ½ = 2 5/8. Use 2½ in.

A top, or cap, angle is used to provide lateral support at the top of the beam. Since it is not required to resist any calculated moment at the end of the beam, this angle can be relatively small. For the top angle 4 × 4 × ¼ × 4 is recommended. In this example, no limitations have been specified by the design drawing as to the top angles, and therefore the angle can be placed as shown. However, if the top clearance had been critical, the angle could have been placed in the optional position on either side of the web, whichever provided the most convenient position for field erection purposes.

The top angle is welded to the column and bolted to the beam with two ∅5/8 bolts having a gage of 2 1/8 in. as recommended in Fig. 24-3-1. The length of the beam required is equal to the center-to-center distance of the columns minus half of each of the column depths (or the column depth if both columns are the same) minus the ½-in. nominal clearance at each end. For this example, the length of the beam is 16′-0 − 10 − 2(½) or 15′-1 in. The detail drawing of the W12 beam is shown in Fig. 24-5-4.

Source Material
1. American Institute of Steel Construction

ASSIGNMENTS

See Assignments 8 and 9 for Unit 24-5 on page 497.

Review for Assignment
Unit 24-1   Structural Steel—Plain Material
Unit 24-2   Beams and Assembly Clearances
Unit 24-4   Elimination of Top and Bottom Views

Fig. 24-5-3 Sketch of seated beam connection for partial design drawing—Fig. 24-5-1.

person laying out the beam details who first marks the locations of the center line for a group of holes on the beam and then centers a template at this point, by which he or she can center-punch the location of all the holes required in the group.

In the detail drawing (Fig. 24-6-1), the fabricator preferred to dimension to the center line of the channel webs rather than the backs of the channels. The direction in which the flanges of these channels are to be placed has been indicated on the drawing by the channel symbol, which has been drawn with the web parallel to the line showing the members. When they are installed, the flanges of these channels must point in the same direction as the flanges of the symbol.

Fig. 24-5-4 Detail drawing of W12 beam shown in Fig. 24-5-1.

Fig. 24-6-1 Dimensioning to center line of channel webs.

UNIT 24-6
Dimensioning

Dimensioning techniques were discussed in Units 24-1 and 24-3. The following are additional items for consideration.

Note that longitudinal dimensions along the beams shown in Fig. 24-6-1 are given to the center line of the groups of open holes required for field connections. This practice can serve two useful purposes. First, it simplifies the dimensioning work for the drafter and later for the checker, since the distances to the center lines of the beams are the dimensions given on the design drawing and the erection plan. Second, it is a convenience to the

Fig. 24-6-2 Dimensioning from the left end of beam.

ONE - BEAM — A3
ONE - do   — B3

Fig. 24-6-3   Dimensioning to the backs of channels.

BEAM W 42

PARTS NOT SHOWN
SAME AS FOR W 42

BEAM X 42

PARTS NOT SHOWN
SAME AS FOR W 42

BEAM Y 42

Fig. 24-6-4   Partial view detailing.

In some shops, however, in addition to locating groups of holes, as noted above, a method of using extension, stub, or running dimensions is employed. This consists of specifying the overall dimension from the left end of the beam to the center line of each group of holes, as shown in Fig. 24-6-2. Note that this practice was also followed in Fig. 24-6-1, to the first line of holes, but here it was done for reference only. In either case, it lessens the shop layout person's work by eliminating the need for calculations and therefore reduces the possibility of an error being made, especially in a shop that uses automated punching equipment.

If the fabricator had preferred to dimension to the backs of the chan-

nels, the dimensions would be as shown in Fig. 24-6-3. Dimensions and notes not shown are the same as in Fig. 24-6-1, except that no note is required to identify the dimension reference lines locating the groups of open holes in the beam web. The dimensions $2\frac{5}{8}$ in. and $2\frac{7}{8}$ in. to each side of the reference lines provide the clue that these open holes are to receive the connection for a channel. It would be understood that the back of the channel will be located toward the smaller of these two dimensions.

Another time-saving device commonly used in the drafting room when pieces are alike, except for some end or intermediate detail, is partial view detailing.

Instead of completely redetailing each piece or trying to combine too many dissimilar details on one sketch by the use of notes indicating to which piece each detail applies, completely detail one piece first. Then, for the second piece, only the difference between it and the first piece needs to be detailed. This partial view is supplemented by a note stating that the parts not shown are THE SAME or possibly OPPOSITE HAND TO the first piece detailed. Figure 24-6-4 represents this practice.

Bills of Materials

From the bills of materials, or material bills, the workers in the yard, where the structural shapes are stocked, cut the

DATE TO P.A. 26											CONT. NO.					
FG'D BY	DATE										BILL NO. M 3a					
CHK'D BY	DATE															
CUSTOMER				INSP. AT MILLS			LUMP SUM POUND PRICE COST PLUS				MADE BY	DATE				
DESCRIPTION				SPECN'S			KIND OF MAT'L				CHK'D BY	DATE				

				BILL OF STRUCTURAL MATERIAL							SHIPPING LIST					
ITEM NO.	NO. PIECES	ASSEMBLY MARK	DESCR.	SIZE	LENGTH	P BILL REFERENCE	EST. WEIGHT			P. A.	NO. PCS.	DESCRIPTION	MARK		ACTUAL WEIGHT	SHIPPING RECORD
1	2		W	W24 X 36	20'-10		1	5 0 0			1	BEAM	A3			783
2	8	a	L	3 X $2\frac{1}{2}$ X $\frac{5}{16}$	1'-$5\frac{1}{2}$			6 6			1	BEAM	B3			783
3							1	5 6 6								
4																
5																

Fig. 24-6-5   Sample bill of material.

material to length, cut the number of pieces shown on the bills, and send the material into the fabricating shop. From the bills of materials the shipping department tallies the number of pieces to be shipped. Therefore, it is extremely important that the drafter include in the bill of materials all the material that is shown on the drawing.

The sample shop bill of materials (Fig. 24-6-5) shows a typical form that is used in billing the material on shop drawings. The first items are the billing for beam A3 and beam B3 that are shown detailed in Fig. 24-6-1. Note that beam A3 is different from B3 only in the longitudinal spacing of holes in the web, and that the material for the two beams is identical. When the material is the same, the billing of members on a shop drawing is grouped to avoid repetition.

Calculation of Weights (Masses)

After the billing operation, it may be necessary to figure the weight or mass of the material on the bill of materials. The weight (mass) of the materials is very important in that the basis of payment for the fabricated steel may be a price per pound (kilogram). For that reason, the weight (mass) must be accurate to the nearest pound (kilogram). Also, the shop uses the calculated weight (mass) of a member to avoid overloading the cranes or other transporting machinery. The shipping department uses the calculated weights (masses) for making up loads and as a basis of payment for shipping. The erection department is interested in weights (masses) of members to plan erection procedure and equipment. The weights (masses) of structural

steel bolts and common structural steel shapes are found in the Appendix.

Source Material
• American Institute of Steel Construction

ASSIGNMENTS

See Assignments 10 and 11 for Unit 24-6 on page 498.

Review for Assignment
Unit 24-1   Structural Steel—Plain Material
Unit 24-2   Beams and Assembly Clearances
Unit 24-3   Standard Connections
Unit 24-4   Elimination of Top and Bottom Views
Unit 24-5   Seated Beam Connections

ASSIGNMENTS for Chapter 24

Assignment for Unit 24-1, Structural Drafting

1. Calculate the limits, tolerances, and sizes of the beams shown in either Fig. 24-1-A, or Fig. 24-1-B. Refer to Fig. 24-1-8 and structural steel handbooks for sizes of structural steel shapes.

DIM	NOMINAL	TOLERANCE		MINIMUM	MAXIMUM
A		+	−		
B		+	−		
C		+	−		

DIM	NOMINAL	TOLERANCE		MINIMUM	MAXIMUM
A		+	−		
B		+	−		
C		+	−		

DIM	NOMINAL	TOLERANCE		MINIMUM	MAXIMUM
A		+	−		
B		+	−		
C		+	−		

Fig. 24-1-A   Beam sizes.

DIM	TOLERANCE		MINIMUM	MAXIMUM
A	+	−		
B	+	−		
C				

Fig. 24-1-B Beam sizes.

Assignments for Unit 24-2, Beams

2. On a B- or A3-size sheet, make detail drawings of the two connections shown on either Fig. 24-2-A or Fig. 24-2-B. Refer to Fig. 24-1-8 and structural steel manuals. The connection angles are welded to the beam web, and the out-standing angles are bolted to the con-necting beam. The bolts and holes need not be shown on these drawings. Scale is 1:8 (U.S. Customary) or 1:10 (metric).

3. On a B- or A3-size sheet, make sketches of the connections of both ends of the center beam shown on either Fig. 24-2-C or Fig. 24-2-D. After the beam connec-tion sketches have been completed and approved by your instructor, prepare a working drawing of the beam from the sketches and information shown on the drawing. The bolt holes on the out-standing legs of the connection angles need not be shown. Use a conventional break to shorten the length of the beam.

Figure	Connection Sketches	Detail Drawing of Beam
24-2-C	1:8	1.4
24-2-D	1:10	1:5

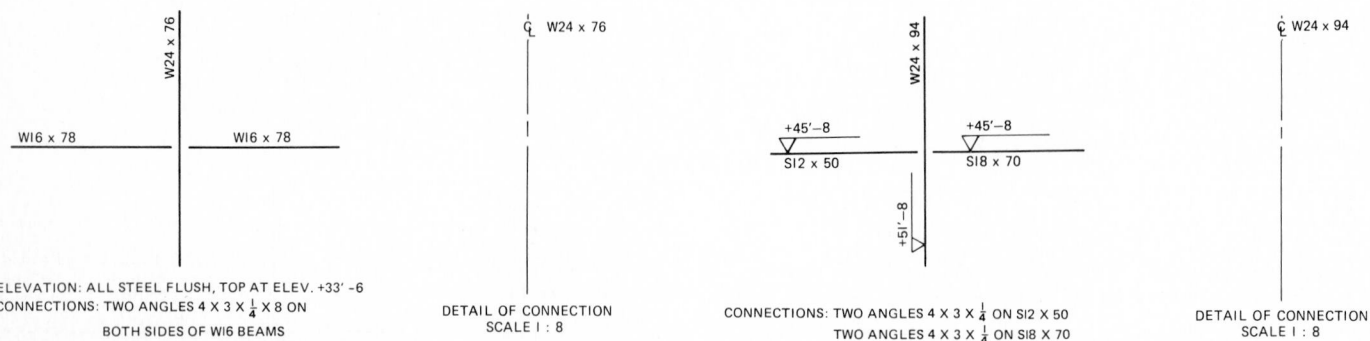

Fig. 24-2-A Beam and connection details.

ELEVATION: ALL STEEL FLUSH, TOP AT ELEV. +10 200
CONNECTIONS: TWO ANGLES 100 x 75 x 6 x 200 ON BOTH
SIDES OF W410 X 114 BEAM

DETAIL OF CONNECTION
SCALE 1 : 10

CONNECTIONS: TWO ANGLES 100 x 75 x 6 x 140 ON S310 X 74 BEAM
TWO ANGLES 100 x 75 x 6 x 250 ON S460 X 104 BEAM

DETAIL OF CONNECTION
SCALE 1 : 10

Fig. 24-2-B   Beam and connection details.

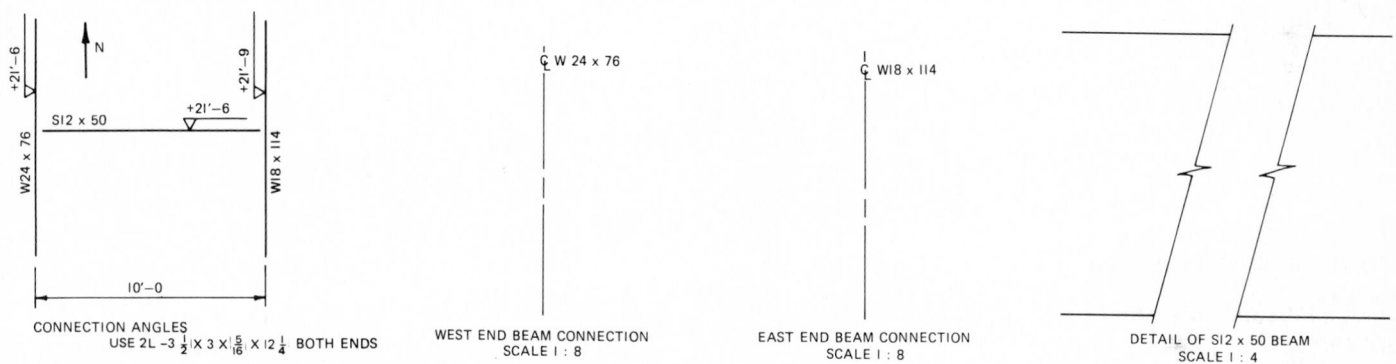

CONNECTION ANGLES
USE 2L –3 ½ X 3 X $\frac{5}{16}$ X 12 ¼ BOTH ENDS

WEST END BEAM CONNECTION
SCALE 1 : 8

EAST END BEAM CONNECTION
SCALE 1 : 8

DETAIL OF S12 x 50 BEAM
SCALE 1 : 4

Fig. 24-2-C   Beam and connection details.

CONNECTION ANGLES
USE 2L–90 x 75 x 8 x 310–BOTH ENDS

WEST END BEAM CONNECTION
SCALE 1 : 10

EAST END BEAM CONNECTION
SCALE 1 : 10

DETAIL OF S380 X 74 BEAM
SCALE 1 : 5

Fig. 24-2-D   Beam and connection details.

ELEVATION: TOP OF STEEL TO BE (+50'-3") UNLESS OTHERWISE SHOWN
NOTES: ALL HOLES Ø .81   BOLTS Ø .75

Fig. 24-3-A   Beam connections.

NORTH END CONNECTION
OF BEAM E3
SCALE 1 : 8

SOUTH END CONNECTION
OF BEAM K3
SCALE 1 : 8

WEST END CONNECTION
OF BEAM N3
SCALE 1 : 8

ELEVATIONS: TOP OF STEEL TO BE (+15 240) UNLESS OTHERWISE SHOWN
NOTES: ALL HOLES Ø 22   BOLTS M20 — A325

Fig. 24-3-B   Beam connections.

METRIC

NORTH END CONNECTION
OF BEAM E3
SCALE 1 : 10

SOUTH END CONNECTION
OF BEAM K3
SCALE 1 : 10

WEST END CONNECTION
OF BEAM N3
SCALE 1 : 10

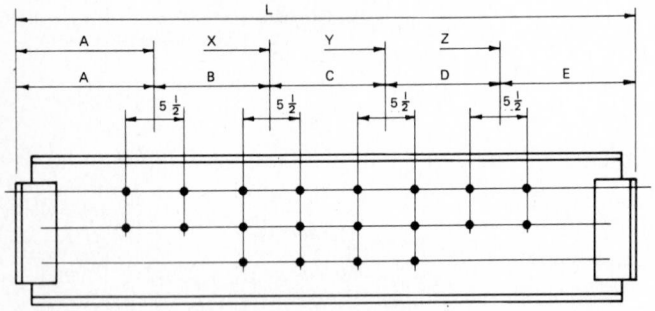

USING THE BEAM SHOWN ABOVE AND DIMENSIONS A–E CALCULATE DIMENSIONS L TO Z
AND COMPLETE THE CHART SHOWN BELOW.

USING THE BEAM SHOWN ABOVE AND DIMENSIONS A–E CALCULATE DIMENSIONS L TO Z
AND COMPLETE THE CHART SHOWN BELOW.

	PROBLEM					
DIM	1	2	3	4	5	6
A	5'-6$\frac{3}{4}$	3'-6	4'-1$\frac{1}{2}$	2'-10$\frac{3}{4}$	8'-0	4'-10
B	3'-6$\frac{3}{8}$	3'-5$\frac{7}{8}$	4'-7$\frac{1}{4}$	3'-10$\frac{1}{2}$	8'-9$\frac{1}{2}$	3'-11$\frac{1}{2}$
C	2'-7$\frac{7}{8}$	3'-6$\frac{1}{2}$	4'-1	4'-10	7'-9$\frac{3}{4}$	2'-11$\frac{1}{2}$
D	6'-10$\frac{1}{2}$	3'-6$\frac{1}{4}$	3'-9$\frac{1}{4}$	5'-9$\frac{3}{4}$	8'-7$\frac{1}{2}$	3'-11
E	3'-10$\frac{1}{2}$	3'-6$\frac{1}{4}$	4'-11$\frac{3}{4}$	6'-9$\frac{1}{2}$	7'-8	4'-11
L						
X						
Y						
Z						

Fig. 24-3-C   Calculation of dimensions.

	PROBLEM		
DIM	7	8	9
A	3'-6$\frac{1}{2}$	4'-6	7'-7
B	6'-1$\frac{1}{4}$	9'-0$\frac{1}{2}$	13'-7$\frac{1}{2}$
C	8'-4$\frac{3}{4}$	14'-2$\frac{1}{2}$	19'-6$\frac{3}{4}$
D	11'-4	18'-3	27'-6$\frac{1}{2}$
E	13'-11	22'-6$\frac{3}{4}$	131'-9
L			
V			
W			
X			
Y			
Z			

Assignments for Unit 24-3, Standard Connections

4. On a B- or A3-size sheet, sketch the following beam connections from the design sketch shown in Fig. 24-3-A or Fig. 24-3-B.
(A) North end connection of beam E3.
(B) South end connection of beam K3.
(C) West end connection of beam N3.
Scale is 1:8 or 1:10.
5. On a B- or A3-size sheet, calculate the missing dimensions from the charts and drawings shown in Fig. 24-3-C.

Assignments for Unit 24-4, Sectioning

6. On a B- or A3-size sheet, prepare one drawing for the two beams shown in Fig. 24-4-A (W16 × 40) or 24-4-B (W400 × 60). Eliminate the top and bottom views. Scale is 1:8 or 1:10.
7. On a B- or A3-size sheet, make a complete working drawing of beams B8^R, B8^L, and C6^R, shown on Fig. 24-4-A or 24-4-B. Scale is 1:8 or 1:10.

Assignments for Unit 24-5, Seated Beam Connections

8. On a B- or A3-size sheet, make detail sketches of the seated beam connections at both ends of beam A shown in Fig. 24-5-A or 24-5-B. Scale is 1:8 or 1:10.
9. On a B- or A3-size sheet, make a one-view detail drawing of beam A in assignment 8. Scale is to suit.

NOTES:
- TOP OF ALL MEMBERS AT ELEVATION +70'−3 EXCEPT WHERE NOTED
- SHOP CONNECTIONS WELDED
- FIELD CONNECTIONS .75−A325 FRICTION TYPE BOLTS
- USE DOUBLE−ANGLE BEAM CONNECTIONS

Fig. 24-4-A   One-view beam drawing.

NOTES:
- TOP OF ALL MEMBERS AT ELEVATION +21 400 EXCEPT WHERE NOTED
- SHOP CONNECTIONS WELDED
- FIELD CONNECTIONS M20−A325 FRICTION TYPE BOLTS
- USE DOUBLE−ANGLE BEAM CONNECTIONS

Fig. 24-4-B   One-view beam drawing.

DETAIL OF WEST END OF W18 x 60 BEAM CONNECTION
SCALE 1 : 8

DETAIL OF EAST END OF W18 x 60 BEAM CONNECTION
SCALE 1 : 8

Fig. 24-5-A   Sketches of seated connection.

DETAIL OF WEST END OF W460 X 89 BEAM CONNECTION
SCALE 1 : 10

PLAN
TOP OF BEAMS AT ELEV. +28 000
Ø 20 A325 FRICTION BOLTS

METRIC

DETAIL OF EAST END OF W460 X 89 BEAM CONNECTION
SCALE 1 : 10

Fig. 24-5-B   Sketches of seated connections.

Assignments for Unit 24-6, Dimensioning

10. On a B- or A3-size sheet, prepare a complete detail drawing of beams D3, E3, G3, K3, M3, N3, C3, and F3 shown on Fig. 24-6-A or 24-6-B. Dimension to the center line of channel webs. The connection angles are welded to the beams, and the outstanding angles are bolted to the connecting beams. Make sketches of the beam connections. Scale is 1:8 or 1:10.

11. On a B- or A3-size sheet, prepare a bill of materials and a shipping list for the beams in assignment 10.

ELEVATION: TOP OF STEEL TO BE (+50'-3) UNLESS OTHERWISE SHOWN
NOTES: ALL HOLES Ø .81   BOLTS Ø .75–A325
MAKE A FULLY DIMENSIONED DETAIL DRAWING OF
BEAM C3 SHOWN ON THE FLOOR PLAN ABOVE.

Fig. 24-6-A   Detail drawings.

ELEVATIONS: TOP OF STEEL TO BE (+15 240) UNLESS OTHERWISE SHOWN
NOTES: ALL HOLES Ø 22   BOLTS M20 – A325
MAKE A FULLY DIMENSIONED DETAIL DRAWING OF
BEAM C3 SHOWN ON THE FLOOR PLAN ABOVE.

METRIC

Fig. 24-6-B   Detail drawings.

CHAPTER 25

Electrical and Electronics Drawings

UNIT 25-1
Pictorial Drawings

Mechanical drafters and technicians can no longer be isolated from electrical and electronics drawings. With the steady increase in automation and electronics equipment, they are now required to either produce or understand electrical and electronics drawings.

In addition to the standard detail and assembly drawings used to manufacture and assemble electrical components, electrical diagrams are also used to show how to connect the wires and to explain how the circuits operate.

Although there are many types of electrical and electronics drawings and diagrams, only the most widely used will be covered in this chapter. These are

1. Pictorial drawings
2. Connection diagrams
3. Elementary (schematic) diagrams
4. Printed circuit (PC) drawings
5. Block and logic diagrams

Although electrical drawings for residential and commercial buildings are also widely used, the authors feel that this type of drafting should be dealt with in architectural texts.

PICTORIAL DRAWINGS

Pictorial drawings, as the name implies, are drawings of pictures of parts or components, showing how they are electrically and mechanically connected. See Fig. 25-1-1. They are prepared mainly for the use of persons who are not trained in reading technical drawings, and as such, the parts drawn should be as realistic as possible. Photographs as well as line drawings (plan, isometric, oblique, and perspective) may be used, the method which best illustrates the subject being recommended. Additional information such as notes, color coding, and arrows are added whenever the clarity of the drawing can be improved. The addition of shading and screens usually makes the drawing more realistic and easier to read.

Examples of pictorial drawings used by assembly workers of electrical

Fig. 25-1-1   Pictorial drawing of a charging system. (Ford Motor Company)

(A) PLAN VIEW—VHF-TO-UHF BLOCK CONVERTER

(B) PERSPECTIVE

(C) ISOMETRIC

(D) USING SCREENS, ARROWS, AND NOTES

Fig. 25-1-2 Pictorial drawings. (Heath Company)

equipment and by companies supplying assembly instructions for do-it-yourself kits are shown in Fig. 25-1-2.

ASSIGNMENTS

See Assignments 1 and 2 for Unit 25-1 on page 514.

UNIT 25-2
Connection Diagrams

In this era of mass production of electronics equipment by nontechnical personnel and with the publication of an increased number of repair manuals and building kits for the do-it-yourself enthusiast, a *connection diagram* is required to show the proper electrical connections. This connection diagram, formerly known as a wiring diagram, simply shows the external connections of the various components in the electrical system. The internal connections of the components are usually omitted to avoid confusion.

As electrical and electronic symbols would be meaningless to the persons connecting the components, the components are represented pictorially and the connection points are shown clearly. However, symbols may be used in diagrams specifically intended for professional assemblers or repairers. Connection diagrams may or may not be drawn to scale, but the individual parts are placed in their relative positions and are drawn with either solid or broken lines.

The connecting wires are drawn as straight horizontal or vertical solid lines, except when several lines are joined at one connection or the line will interfere with a component. In these instances the lines may be angled to improve the clarity of the connection.

In order to reduce assembly and repair time and lessen the chance of error when there are many connecting wires, color coding is often used. Each wire is covered with a different color of insulation, and the color name is placed beside or on the wire on the drawing. Alternately a color code listing the color of insulation and its designation may be placed on the drawing; the color designation appears at each terminal, both on the drawing and the actual part.

When the position of the wire is not a factor, the lines representing the wires are spaced to provide a clear and well-balanced layout. The ends of the wires terminate at components. This type of drawing is known as a point-to-

Fig. 25-2-1   Point-to-point connection diagram of a boat's electrical system.

point connection diagram. See Fig. 25-2-1.

Often when there are several wires close together, as in a conduit, or held together by a harness, one thick line, called a *highway*, is used instead of the several separate lines. When clarity is required to show the direction a wire takes when it enters the highway, an arc or a 45° line is used to indicate the direction of travel. Several highways may be desirable on a drawing because of electrical, physical, or other factors. This type of drawing is called a high-

way-type connection diagram. See Fig. 25-2-2.

To minimize the cost of an electrical assembly, many of the wires are joined together prior to the final assembly. Each wire is cut to the required length, positioned on a board or jig and taped together to form a wiring harness. See Fig. 25-2-3. The length of each wire extending beyond the harness, its breakout point, and its color must be known. A pictorial drawing showing a wiring harness installation is shown in Fig. 25-2-4.

Fig. 25-2-2   Highway-type connection diagram of a clothes dryer. (Frigidaire)

COLOR	SYMBOL
BLACK	B, B-I
WHITE	W, W-I, W-2
RED	R, R-I
GREEN	G (GROUND)
YELLOW	Y
BLUE	BU
BROWN	BR
PURPLE	P
ORANGE	OR
WHITE-BLACK	W-B
BLACK-WHITE	B-W
RED-WHITE	R-W

△ TERMINAL
▪ SPLICE CONNECTOR BLACK
▫ SPLICE CONNECTOR WHITE

WIRING AS VIEWED FROM REAR SERVICE POSITION

WIRE		
COLOR	SYMBOL	LENGTH
YELLOW	Y	91
PURPLE	P	80
BROWN	BR	65
BLACK	B	91
ORANGE	O	54
WHITE	W	59

STRIP ENDS OF WIRE FOR .50

NOTE: DIMENSIONS SHOWN ARE IN INCHES

Fig. 25-2-3   Harness drawing.

Fig. 25-2-4   Dash panel-to-engine wiring harness. (General Motors Corp.)

ASSIGNMENTS

See Assignments 3 to 6 for Unit 25-2 on page 515.

UNIT 25-3
Elementary (Schematic) Diagrams

Elementary diagrams, formerly called schematic diagrams, show the connection and function of a circuit in its simplest form, using graphical symbols. They do not show the physical relationship of the components, nor mechanical connections as shown in Fig. 25-2-2. The components may be repositioned or inverted to simplify and improve the clarity of the drawing. The elementary diagram for the clothes dryer (Fig. 25-2-2) is shown in Fig. 25-3-1. The elementary diagram is the drawing most frequently used by engineers, designers, and electronics personnel as they are interested mostly in the design and function of the equipment. See Fig. 25-3-2 for an example.

NOTE— JUNCTION TERMINAL "W" AND TIMER MOTER NOT SHOWN WITH TIMER. WIRES "B" AND "B-I" FROM TERMINAL CONNECTOR (NOT SHOWN) ARE SHOWN AS ONE WIRE ON ELEMENTARY DIAGRAM AS THEY ARE CONNECTED IN PARALLEL.

Fig. 25-3-1   Elementary diagram for the clothes dryer shown in Fig. 25-2-2. (Frigidaire)

Fig. 25-3-2   Partial elementary diagram of a receiver.

The connecting wires are drawn as straight horizontal or vertical lines. Unlike connection diagrams, wire connections may be made at any convenient location on the diagram and the connection normally is shown as a small solid circle. This symbol must be clearly visible, otherwise a connection could be mistaken for a crossover. Ground symbols are used frequently on schematic diagrams instead of wire connections. The ground symbol is usually drawn so that the lines are horizontal and the symbol tapers towards the bottom of the drawing.

The following points should be used as a guide when laying out an elementary diagram:

1. Keep lines to a minimum.
2. Avoid crossovers where possible.
3. Components may be rotated.
4. Maintain a uniform symbol size.
5. Parts may be repositioned for clarity.
6. Allow space for component identification.
7. The diagram does not need to be to scale.
8. Alignment of similar components where feasible makes a more pleasing and professional looking drawing.
9. The use of a grid background on the drawing eliminates the scaling of many components and their spacing, thus saving many hours of drafting time.

(A) COMMON BASE   (B) COMMON EMITTER   (C) COMMON COLLECTOR

NOTE: IF PNP TRANSISTORS ARE USED, THE BATTERY POLARITIES SHOULD BE REVERSED

Fig. 25-3-3   Three methods of connecting bipolar NPN transistors as amplifiers.

(A) RECTANGLE

(B) TRIANGLE

Fig. 25-3-4 Symbols for integrated circuits (composite assemblies).

SYMBOLS FOR ELEMENTARY DIAGRAMS

Standard symbols, as shown in the Appendix, have been prepared for use on electrical and electronics diagrams. Although symbols may be drawn to any convenient size, electrical and electronic symbol templates are available for use by drafters and engineers. In conjunction with the graphical symbol all replaceable parts should be referenced with a number and letter to indicate the value and type of the component represented by the symbol. The reference may be placed above, below, or on either side of the part. Numbering of the components (R1, R2, R3, etc.) is normally done from left to right.

Transistor and Other Semiconductors

The symbols for transistors and other more commonly used semiconductors are shown in the Appendix. The line with an arrow pointing in the direction of the conventional current flow is the emitter, and the line without the arrow is the collector. The emitter and collector lines touch the base line.

Three common methods of connecting bipolar NPN transistors as amplifiers are shown in Fig. 25-3-3.

Integrated Circuit Symbols

One of the most recent far-reaching developments in the field of electronics is that of microelectronics, which started with the transistor and has evolved at a very rapid pace.

The symbols already discussed for elementary diagrams have been for individual parts. Integrated circuits (IC), which may contain hundreds of components, may be used as a part of a larger circuit.

Two symbols are presently used to depict an integrated circuit, the rectangle and the equilateral triangle, as shown in Fig. 25-3-4. The triangle symbol is used for ICs that are amplifiers. The symbol normally lists the manufacturer's number and function, and the pin numbers. Simplification of the overall diagram normally dictates the orientation of the rectangle.

Although there are no particular sizes or rules governing the drawing of these symbols, generally they should be of one size, with a maximum of two sizes for one drawing.

A typical elementary diagram incorporating ICs is shown in Fig. 25-3-5.

ASSIGNMENTS

See Assignments 7 through 11 for Unit 25-3 on page 516.

Fig. 25-3-5 Elementary diagram for a melody organ.

UNIT 25-4
Printed Circuits

Practically all mass-produced electronics equipment (radios, televisions, etc.) are now using printed circuits (PC) instead of wire-lead connections. The advantages are many: uniformity in production, elimination of practically all assembly wiring errors, reduced cost, and miniaturization are a few. Printed circuits or etched circuits are thin ribbons of copper or other metals placed on a plastic board and shaped to replace the wire leads formerly used to connect the components. See Figs. 25-4-1 and 25-4-2. There are three types of PC boards: single-sided, with conductors on one side; double-sided, with conductors on both sides; and multilayer, with conductors at different levels. Only single-sided PC boards will be discussed in this unit.

For many applications, manual PC drafting is still used but the trend is to computer graphics and photoplotting. Both techniques have proved efficient and cost-effective.

A printed circuit is made from an accurately drawn layout. The layout is normally drawn on a grid to an enlarged scale. A polyester film is laid over the layout and traced using pressure sensitive symbols and tapes. The symbols center the pads and holes in the layout, and are connected with tapes to form circuit paths. The drawing is then photographically reduced to the desired size onto a copper foil

Fig. 25-4-1   Printed circuit used on the amplifier shown in Fig. 25-4-2. (General Electric)

(A) ELECTRONIC COMPONENTS WITH PRINTED CIRCUIT.

Fig. 25-4-2   Electrical drawings for a simple amplifier. (General Electric)

(B) ELEMENTARY DIAGRAM.

UNLESS OTHERWISE NOTED:
All voltages shown are typical voltages with no signal applied to circuit.
Measurements shown may deviate ±10%.
A - denotes ampere connectors.
Resistors shown are 1/2 W   K = 1000   MΩ = 1 000 000
Capacitor values less than 1 in microfarad; capacitor values
   more than 1 in mmf.
Arrows on controls indicate clockwise rotation.
DC voltages measured from B-(gnd) with 20 000 Ω/V meter.
Line voltage maintained at 120 VAC, 60 cycles.
◄█ -Bowties indicate cut points on conductor pattern
   for circuit testing with ohmmeter.

TERMINAL VIEW OF   TERMINAL VIEW OF
TRI, TR2, TR5, & TR6   TR3, TR4, TR7, & TR8

* - REFER TO SERVICE MANUAL COVERING SPECIFIC MODEL.
** - REFER TO RECORD CHANGER SERVICE MANUAL.

Fig. 25-4-2 (Cont'd)   Electrical drawings for a simple amplifier. (General Electric)

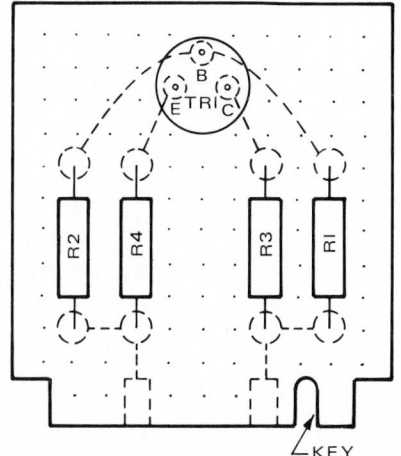

(A) ELEMENTARY (SCHEMATIC) DIAGRAM

(B) TOP VIEW OF PRINTED CIRCUIT BOARD
(COMPONENT SIDE)

Fig. 25-4-3   Layout of a simple printed circuit board.

which is bonded to a plastic board. See
Fig. 25-4-3. Holes are drilled at various
locations in the board and leads from
the components are inserted into circular conductors, called *lands*, located
on the circuit side of the board. When
all of the components to be connected
are in position, the circuit side of the
board is then dipped into molten solder, making all of the soldered connections at one time.

(C) BOTTOM VIEW OF PRINTED CIRCUIT BOARD
(CIRCUIT SIDE)

BATTERY

(A) RUNNING CONDUCTOR AROUND COMPONENT

(B) REPOSITIONING COMPONENT

(C) USING AN INSULATED JUMPER (CROSSOVER)

THE PROBLEM:

TO JOIN TOP OF CI
TO BASE (TERMINAL B)
OF TRANSISTOR (TRI)

Fig. 25-4-4 Crossovers.

Crossovers should be avoided whenever possible. Running the conductor around the land of the interfering component, or repositioning the component, as shown in Figs. 25-4-4A and B, should be given priority over crossovers. If however, a crossover cannot be avoided when laying out a printed circuit, then a jumper wire (crossover) may be used, as shown in Fig. 25-4-4C. The jumper wire is treated in the same manner as other components and placed on the component side of the circuit board.

Printed Circuit Drawings The following points are a guide for the preparation of printed circuit drawings:

1. Establish the physical size of each component and where the leads are located. Physical requirements may determine the position of the components on the board. Transistor leads are shown in reverse when viewed from the component side of the circuit board. In most cases the emitter (E) is close to the tab on the transistor, Fig. 25-4-5. Check the orientation of components such as diodes and capacitors, Fig. 25-4-6. Some capacitors are reversible, while others are not. The polarity is shown when the capacitor is not reversible.

(A) METAL – CASE TYPE

(B) METAL – CASE TYPE

(C) PLASTIC – CASE TYPE

Fig. 25-4-5 Location of leads on typical transistors.

2. Study the elementary drawing. What components have common connections?
3. Prepare a rough trial wiring diagram, with the components drawn to scale, prior to starting the final diagram. Remember, in most cases, an enlarged scale is desirable.
4. Locate the parts on the wiring layout avoiding crossovers where possible.
5. All holes should be located at an intersection of the grid lines. Standard grid spacing should be used.
6. Conductors should be kept to a minimum length (See Fig. 25-4-7), be at least .10 in. (2.5 mm) from the edge of the circuit board, have a minimum width of .04 in. (1 mm), and be

spaced a minimum of .04 in. (1 mm) apart.
7. The minimum center-to-center spacing for component holes should be component length plus .12 in. (3 mm).
8. Long leads, such as grounds, are best located near or around the edge of the circuit board.
9. When using integrated circuits on PC boards, such as shown in Fig. 25-4-8, a socket base to hold the IC is connected to the circuit board. Typical base sizes are shown in Fig. 25-4-9.

ASSIGNMENTS

See Assignments 12 through 15 for Unit 25-4 on page 518.

Fig. 25-4-6   Orientation of capacitors and diodes.

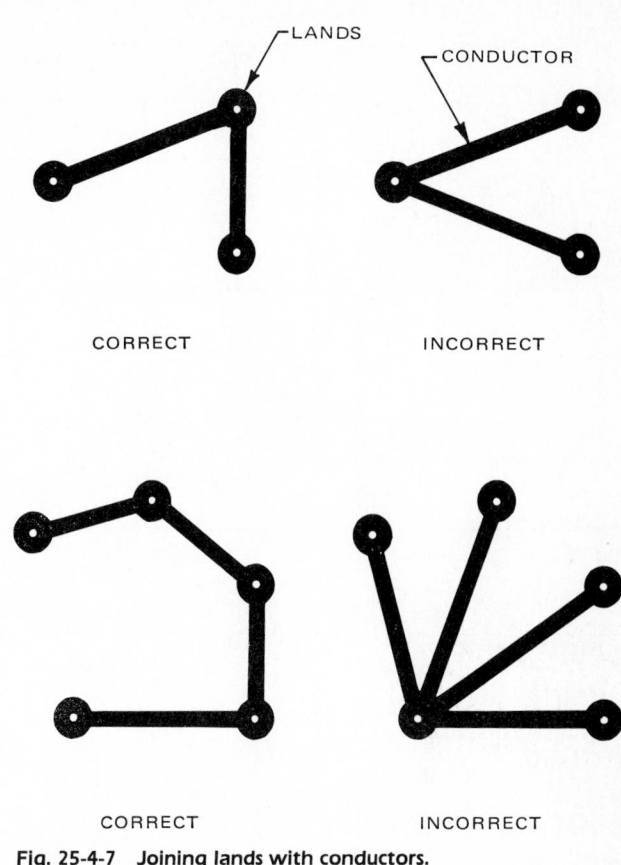

Fig. 25-4-7   Joining lands with conductors.

Fig. 25-4-8   A PC board with IC's. (Quadram Corp.)

Fig. 25-4-9   IC socket bases.

UNIT 25-5
Block and Logic Diagrams

BLOCK DIAGRAMS

Block diagrams in electrical and electronics drafting are used to simplify the understanding of circuits. Their utter simplicity enables you to tell at a glance the relative position and function of each part of the circuit. Block diagrams are used by designers in the early stages of planning a project. You should note that a block diagram shows only the relationship between the components, and does not show the electrical connections.

A *block diagram*, as its name suggests, consists of a series of blocks (or boxes), connected by straight lines.

The identities of the respective units are placed inside or adjacent to the blocks, in abbreviated form where necessary.

Each block in the diagram represents a stage or subcircuit within a circuit. These blocks are usually drawn as squares, rectangles, or triangles and are uniform in size, shape, and spacing, regardless of the physical size they represent. Certain components such as antennas, speakers, and microphones are shown by means of a symbol rather than a block. See Fig. 25-5-1. Block diagrams for alternative or future components are indicated by a broken line, the broken line being the same weight as the lines showing the solid blocks.

Fig. 25-5-2   Logic flow diagram.

Fig. 25-5-1   Block diagram of a radio-phonograph combination.

The blocks are joined by a single line, which indicates the signal path from block to block. The signal path is normally shown from left to right and the connecting line may be lighter or darker than the blocks, depending on which is being emphasized. When arrows are used on the connecting line to show the signal path, it is referred to as a flow diagram.

LOGIC DIAGRAMS

A logic diagram is a diagram representing the logic elements and their implementations without necessarily expressing construction or engineering details. A logic symbol is the graphic representation of a logic function.

Computer design has been largely responsible for the steady growth of logic functions which can be performed by basic circuits. See Fig. 25-5-2.

Graphic Symbols

The symbols representing logic functions shown in Fig. 25-5-2 are the symbols approved by ANSI/Y32.14 (IEEE 91)—Graphic Symbols for Logic Diagrams. The distinctive symbols shown in Fig. 25-5-3, being easier to understand, are the symbols more commonly used.

The following guidelines should be observed when using graphical symbols on logic diagrams.

1. In most cases, the meaning of a symbol is defined by its form. The size and line thickness does not affect the meaning of a symbol. In some cases, it may be desirable to use different sizes of symbols: (a) to emphasize certain aspects, or (b) to

Fig. 25-5-4 Recommended symbol proportions for distinctive-shape logic symbols.

facilitate the inclusion of additional information.

2. Logic symbol size should be governed by the space necessary for internal annotations and the length of the side needed to accommodate input and output lines and pin numbers at an acceptable spacing.

3. Graphic symbols may be drawn to any proportional size that suits a particular diagram, provided that the selection of size takes into account the anticipated reduction or enlargement. Standard proportions for these symbols are shown in Fig. 25-5-4.

Fig. 25-5-3 Partial layout of shift right, shift left logic diagram. (Heath Co.)

(A) ADDING INPUTS AND OUTPUTS

INPUT LINES

X X X

OUTPUT LINE

(B) SYMBOL EXTENSION TO ACCOMODATE INPUTS

EXTENSION

A1-3
A2-3
A3-3

3

D1-3

MEANS

A1
A2
A3

D1

A1
A2
A3

D2

A1
A2
A3

D3

(C) ONE SYMBOL MAY BE USED TO REPRESENT SEVERAL IDENTICAL LOGIC SYMBOLS

NEGATION INDICATOR SYMBOL

LOGIC SYMBOL

INPUT SIDE

OUTPUT SIDE

(D) NEGATION INDICATOR SYMBOL

POLARITY INDICATOR SYMBOL

LOGIC SYMBOL

INPUT SIDE

OUTPUT SIDE

(E) POLARITY INDICATOR SYMBOL

DYNAMIC INDICATOR SYMBOL

LOGIC SYMBOL

INPUT SIDE

(F) ADDING, DYNAMIC INDICATOR SYMBOL, TO INPUT LINE

NONLOGIC INDICATOR SYMBOL

LOGIC SYMBOL

INPUT SIDE

OUTPUT SIDE

(G) ADDING NONLOGIC INDICATOR SYMBOL

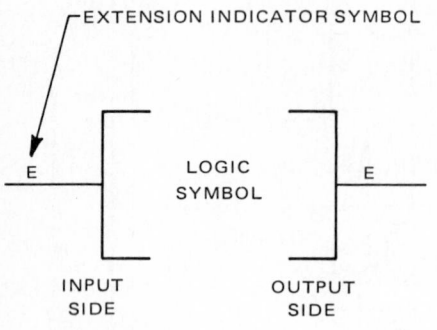

EXTENSION INDICATOR SYMBOL

E

LOGIC SYMBOL

E

INPUT SIDE

OUTPUT SIDE

(H) EXTENDER – CONNECTION INDICATOR SYMBOL

INHIBITING – INPUT INDICATOR SYMBOL

LOGIC SYMBOL

INPUT SIDE

(J) INHIBITING – INPUT INDICATOR SYMBOL

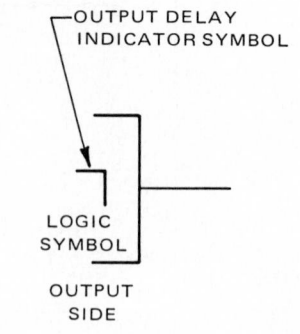

OUTPUT DELAY INDICATOR SYMBOL

LOGIC SYMBOL

OUTPUT SIDE

(K) OUTPUT – DELAY INDICATOR SYMBOL

Fig. 25-5-5 Adding additional logic symbols to basic symbol.

4. It is preferred that all text be readable from the bottom, that is, right side up.

5. Input and output lines are preferably placed on opposite sides of the symbol and should join the outline of the symbol at right angles. Input lines are normally shown on the left side, output lines on the right side. See Fig. 25-5-5.

6. Pin numbers, when used, are shown outside the graphic symbol. See Fig. 25-5-6.

7. Recommended arrangements of internal information are shown in Fig. 25-5-7.

Diagram Preparation The following guidelines should be observed when preparing a logic diagram:

1. The parts should be spaced to provide an even balance between blank spaces and lines.

2. The logic diagram should use a layout which follows the circuit, signal, or transmission path either from input to output, source to load, or in order of functional sequence.

3. The layout should be such that the principal flow of information is from left to right of the diagram, or from top to bottom.

4. Functionally related symbols should be grouped and placed close to one another.

5. Lines should be drawn horizontally or vertically except in those cases where oblique lines aid in the clarity of the digram.

6. Highways can be used to simplify logic diagrams where groups of similar signals are encountered.

ASSIGNMENTS

See Assignments 16 to 18 for Unit 25-5 on page 520.

EXAMPLE I

(A) DISTINCTIVE SHAPED SYMBOLS

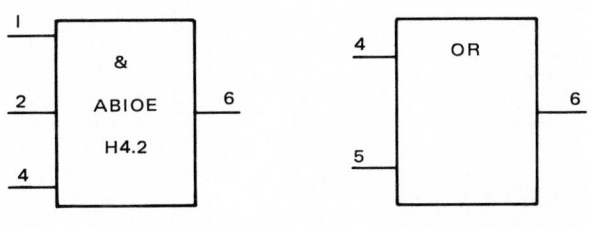

(B) RECTANGULAR SHAPED SYMBOLS

Fig. 25-5-6 Adding pin numbers to basic symbols.

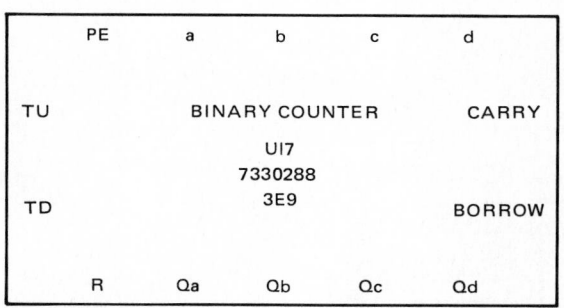

EXAMPLE 2

Fig. 25-5-7 Recommended arrangement of internal information.

ELECTRICAL AND ELECTRONICS DRAWINGS **513**

ASSIGNMENTS for Chapter 25

Assignments for Unit 25-1, Pictorial Drawings

1. Prepare a pictorial (oblique) drawing of the transformer shown in Fig. 25-1-A showing the following set of instructions with reference to the transformer leads:

   Black lead to extend 6.5 in. beyond side A of transformer.
   Red lead to extend 5.25 in. beyond side A of transformer.
   Black-red lead to extend 3.5 in. beyond side A of transformer.
   Black-green lead to extend 5.5 in. beyond side A of transformer.
   .25 in. of insulation to be removed from the end of all leads.

2. Figure 25-1-B shows a schematic diagram and the electrical components for part of a power supply. On a B- or A3-size sheet prepare a pictorial drawing showing the layout and connection of the parts. The leads from the components pass through the holes on the circuit board and are connected by wires on the back side. The holes, which can be enlarged to suit, serve as mounting holes for the bolt-down components.

CIRCUIT BOARD (LENGTH AND WIDTH TO SUIT)

Ø.06 HOLES SPACED .25 IN. CENTER TO CENTER

.25   .25

TRANSFORMER

0.38A FUSE
Ø.10
.25

DO-41
4 REQUIRED
Ø.10
.25

CAPACITOR 2000 μF
Ø.90
1.90

CAPACITOR 10 μ F
Ø.25
.75

(A) COMPONENTS
NOTE: LEADS ON COMPONENTS MAY BE BENT TO SUIT

IC11 INTEGRATED CIRCUIT

TO GROUND
IN   OUT
.38
1.00
.75
1.25

6-32UNC
HEX NUTS
.38
CIRCUIT BOARD
TERMINAL CONNECTOR

Fig. 25-1-A   Transformer.

BLACK
BLACK-RED
WINDINGS
2.50 1.10
.70
BLACK-GREEN
RED

Fig. 25-1-B   Part of a power supply.

TRANSFORMER
117 V AC
DO-41
0.38A FUSE
CAPACITOR 2000 μF, 50 V
I C11
TERMINAL
+15
CAPACITOR 10 μF, 50 V
(B) SCHEMATIC DIAGRAM

Assignments for Unit 25-2, Connection Diagrams

3. On a B- or A3-size sheet make a highway-type connection diagram of the outboard electrical system shown in Fig. 25-2-1.

4. On a B- or A3-size sheet make a point-to-point connection diagram of the clothes dryer shown in Fig. 25-2-2.

5. On a B- or A3-size sheet make a point-to-point connection diagram of one of the appliances shown in Figs. 25-2-A or 25-2-B.

6. On a B- or A3-size sheet prepare harness drawings for one of the appliances shown in Figs. 25-2-2, 25-2-A or 25-2-B. Use the scale of 1:4 (U.S. Customary) or 1:5 (metric) to measure the cable lengths and positions, rounding off the measurements to the nearest .50 in. or 10 mm. Strip the ends of wires for .50 in. (U.S. Customary) or 15 mm (metric). Use rectangular coordinate dimensioning. Scale is half size (1:2).

Fig. 25-2-A Connection diagram for dishwasher. (Frigidaire)

Fig. 25-2-B Connection diagram for a wall oven. (Frigidaire)

Assignments for Unit 25-3, Elementary (Schematic) Diagrams

7. On a B- or A3-size sheet make an elementary diagram from one of the connection diagrams shown in Fig. 25-2-A or 25-2-B. For electrical symbols not shown in the Appendix, use the symbols shown on the assignment drawings.

8. On a B- or A3-size sheet make an elementary diagram of one of the circuits shown in Figs. 25-3-A to 25-3-C. Refer to the Appendix for symbols.

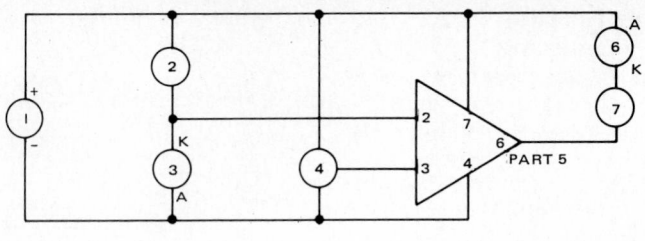

PART		PART	
1	12-V battery	5	741 operational amplifier (IC)
2	27 000-Ω resistor	6	Light-emitting diode
3	6.2-V zener diode	7	1000-Ω resistor
4	100 000-Ω potentiometer (adjustable contact resistor)		

Fig. 25-3-A Low battery indicator.

PART		PART	
1	Phototransistor (NPN-type)	6	100 000-Ω resistor
2	100 000-Ω resistor	7	100 000-Ω resistor
3	50 000-Ω variable resistor (rheostat)	8	555 timer IC
4	741 operational amplifier IC	9	0.1-μF capacitor
5	0.1-μF capacitor	10	0.01-μF capacitor
		G	Ground

Fig. 25-3-B Fiber optic receiver.

PART		PART	
1	1-MΩ potentiometer	12	0.1-μF Mylar capacitor
2	Amplifier 4069 IC	13	100 000-Ω potentiometer
3	Amplifier 4069 IC4	14	amplifier LM386 IC4
4	1-μF capacitor	15	0.1-μF Mylar capacitor
5	0.47-μF capacitor	16	0.047-μF Mylar capacitor
6	0.1-μF capacitor	17	10-Ω resistor
7	0.05-μF capacitor	18	1000-Ω resistor
8	500-μF capacitor	19	Light-emitting diode
9	100-μF capacitor	20	250-μF capacitor
10	TL507C IC (integrated circuit)	21	8-Ω speaker
11	100 000-Ω potentiometer	G	Ground

Fig. 25-3-C Sound-effects generator.

9. On a C- or A2-size sheet make an elementary diagram of the audio amplifier shown in Fig. 25-3-D. Refer to the Appendix for symbols.

10. In a B- or A3-size sheet make an elementary diagram of the mailbox sentry shown in Fig. 25-3-E. Add the values to the symbols. *Note:* Parts may be repositioned and similar numbers are electrically joined together.

PART		PART		PART		PARTS	
R1	10 000 Ω	R9	Adjustable Contact	C1	8 μF	T1	1K/1K C.T. transformer with magnetic core
R2	150 000 Ω		10 000 Ω	C2	.50 μF	T2	100 C.T./V.C. transformer with magnetic core
R3	6800 Ω		Taper Audio V.C.	C3	.02 μF		
R4	Adjustable Contact	R10	220 000 Ω	C4	.20 μF	TR1, TR2, TR3 GE2N323	
	50 000 Ω	R11	2200 Ω	C5	.005 μF		
	Bass Linear	R12	4700 Ω	C6	.10 μF	TR4, TR5 GE2N321 (with clip-on heatsink)	
R5	1000 Ω	R13	33 000 Ω	C7	10 μF		
R6	10 000 Ω	R14	47 000 Ω	C8	10 μF	S1 speaker	
R7	100 000 Ω	R15	1500 Ω	C9	50 μF		
R8	Adjustable Contact	R16	330 Ω	C10	50 μF	SW switch	
	50 000 Ω	R17	220 Ω	C11	50 μF		
	Treble Linear	R18	1200 Ω			B1 battery, 6V	
		R19	33 Ω				
						Ⓖ ground	

Fig. 25-3-D 5-transistor audio amplifier.

PARTS LIST
B1 9-V battery
C1 10-μF, 25-V electrolytic
C2, C3 1-10-μF, 25-V electrolytic
C4 0.001 μF, 25-V ceramic disc capacitor
C5 .01-μF, 25-V ceramic disc capacitor
C6, C7, C8 0.01-μF, 25-V ceramic disc capacitor
IC 4001 quad NOR gate
IC 555 timer

PARTS LIST
LED1—Red light-emitting diode
Q1 2N2222 NPN silicon transistor
The following are ¼-W, 10% resistors:
R1, R2 22 kΩ
R3 1 kΩ
R4 4.7 kΩ
R5 10 kΩ
R6 47 kΩ
R7 100 kΩ
R8 2.2 MΩ

PARTS LIST
S1 Normally open microswitch, magnetic reed switch
S2 through S4 Normally open pushbutton switch, panel mount
SPKR1 8-Ω, 2-in. speaker

Fig. 25-3-E Mailbox sentry.

Fig. 25-3-F Keyboard/display circuit using IC's.

11. On a C- or A2-size sheet make an elementary diagram of the keyboard display circuit shown in Fig. 25-3-F. Show the values with the symbols. *Note:* Parts may be repositioned, similar numbers are electrically joined together, number 2 is ground, and identify each component, i.e., R1 − 1000 Ω, R2 − 1000 Ω,....

Assignments for Unit 25-4, Printed Circuits

12. On a B- or A3-size sheet draw the top and bottom views of the circuit board shown in either Fig. 25-4-A or 25-4-B. Complete the conductor connections in the top view from the information shown in the elementary diagram. The grid lines shown on the printed circuit board are .125 in. or 4 mm on centers. Conductor size and minimum clearance to be .06 in. or 1.5 mm. Land diameter is .12 in. or 3 mm. Scale is twice size (2:1).

13. On a B- or A3-size sheet make an elementary diagram from the printed circuit and the component location diagram for one of the circuits shown in either Fig. 25-4-C or 25-4-D. Include on the drawing a bill of material calling for the capacitors, resistors, and transistors.

(A) ELEMENTARY DIAGRAM

(B) PARTIALLY COMPLETED TOP VIEW (COMPONENT SIDE)

Fig. 25-4-A Printed circuit board 1.

(A) ELEMENTARY DIAGRAM

(B) PARTIALLY COMPLETED TOP VIEW (COMPONENT SIDE)

Fig. 25-4-B   Printed circuit board 2.

CRYSTAL MONAURAL CARTRIDGE

CHANGER MOTOR

SPEAKER

4-WAY RECTIFIER

(B) ELEMENTARY DIAGRAMS FOR CONNECTING COMPONENTS

Fig. 25-4-D   Phonograph amplifier.

RIGHT CHANNEL INPUT TO AMPLIFIER

PREAMPLIFIER DC VOLTAGE −20 V DC

LEFT CHANNEL INPUT TO AMPLIFIER

CARTRIDGE GROUND

LEFT CARTRIDGE INPUT

RIGHT CARTRIDGE INPUT

(A) COMPONENT LOCATION DIAGRAM

TRANSISTOR   STEREO CARTRIDGE

(B) ELEMENTARY DIAGRAMS

CAPACITORS	RESISTORS
C1 0.05 μF	R1 10m Ω
C2 0.01 μF	R2 10m Ω
	R3 22k Ω
TRANSISTORS	R4 22k Ω
TR1 NPN	R5 8.2k Ω
TR2 NPN	R6 8.2k Ω
	R7 1m Ω

(C) COMPONENTS FOR AMPLIFIER

Fig. 25-4-C   Preamplifier. (General Electric)

VOLUME   TONE

BLACK   GREEN   RED   BLACK   BROWN

TO CARTRIDGE GROUND

TO SPEAKER

TO CARTRIDGE

TO SPEAKER

TO CHANGER MOTOR

(A) COMPONENT LOCATION DIAGRAM

CAPACITORS	TRANSISTORS	RESISTORS	
C1 400 μF, 25 V	TR1 2N2716	R2 680 Ω	R10 56 Ω
C2 150 μF, 15 V	TR2 2N527	R3 1200 Ω	R14 100 kΩ
C5 10 μF, 20 V	TR3 2N2713	R5 68 kΩ	R15 56 Ω
C7 .005 μF	TR4 2N525	R6 56 kΩ	R16 M VAR.
C8 330 μF		R8 100 kΩ	R17 10 M VAR.
	FUSES	R9 180 Ω	R18 4.7 Ω
	F1 1¼ A		R30 10Ω

(C) COMPONENTS FOR PHONOGRAPH AMPLIFIER

14. The three amplifiers in the elementary diagram shown in Fig. 25-4-E are to be replaced by the LM348 IC. Amplifier 4 (leads 12, 13, and 14) is not used. On a B- or A3-size sheet redraw the elementary diagram.

15. On a B- or A3-size sheet design a printed circuit board for the redesigned circuit in problem 14. Draw the top and bottom views. Use Ø.12 in. or Ø3 mm lands and .06 in. or 1:5 mm conductor widths. Scale is twice size (2:1).

REPLACEMENT PART
FOR THE THREE AMPLIFIERS

Fig. 25-4-E Partial layout for an AC/DC converter.

PARTS LIST
R1 10 kΩ
R2 20 kΩ
R3, R4, R5 20 kΩ, 1%
R6 10 kΩ, 1%
R7 30 kΩ

PARTS LIST
R8 50 kΩ
R9 10-kΩ trimming pot
R10 100 kΩ
C1 0.68-μF, 15-V electrolytic
C2, C3 47-μF, 15-V electrolytic
C4 10-μF, 15-V electrolytic
D1, D2 IN914

Assignments for Unit 25-5, Block and Logic Diagrams

16. Prepare a block diagram of a home intercom system from Fig. 25-5-A and the following information:

1. Four intercoms (located at the front door, the back door, the work shop, and the recreation room) connect to a multiplexer which in turn connects to two units, a speech synthesizer, and speech recognition hardware.
2. The last two units in (1) connect (two-way flow) to the CPU module.
3. Two sensors, one indoor and one outdoor, feed into the CPU module.
4. The power to operate the system feeds into a line voltage monitor and battery backup system, which in turn, is connected to the CPU module.
5. The line voltage monitor and backup system also connects to an RCT unit, which in turn, is connected (two-way flow) to the CPU module.
6. A 16-channel ac remote control transmitter links the power control receivers located throughout the house to the CPU module.

17. On a B- or A3-size sheet prepare a block diagram with arrows showing direction flow of a hybrid TV set from the following information:

1. Antenna connects to a TV tuner/mixer

2. TV tuner/mixer connects to a broadband 1F amplifier
3. Broadband IF amplifier connects to two units: video bandpass filter and audio IF filter.
4. Video bandpass filter connects to detector
5. Detector connects to video amplifier
6. Video amplifier connects to CRT

7. Audio IF filter connects to mixer/detector
8. Mixer/detector connects the two amplifiers 5.5 MHz.
9. Each amplifier connects to an FM detector
10. One FM detector connects to channel 1
11. Other FM detector connects to channel 2

Fig. 25-5-A Home intercom system.

18. On a B- or A3-size sheet prepare a logic diagram from the information shown in Fig. 25-5-B. Convert to distinctive-shape logic symbols and add the missing connections listed on the drawing.

RI TO R6
2.2K Ω

I/O PORT SELECT

I/O PORT SELECTOR

IC 14 PIN 16

IC 15 PIN I

IC 14 PINS 4, 16

IC 14 PIN 23

LOGIC SYMBOL	ICs
AMPLIFIER	IC2, IC7, IC9, IC10
OR	IC3, IC4, IC5, IC6 IC11, IC13
AND	IC8, IC12
NEGATION INDICATOR	IC2 PORT 8 IC10 PORT 10 IC7 PORT 2 IC 12 PORT 8 IC8 PORT 6

CONNECTIONS TO BE MADE:
IC3 TERM. 11 TO IC4 TERM. 4 AND IC11 TERM. 12
IC4 TERM. 5 TO IC8 TERM. 6
IC5 TERM. 10 TO IC7 TERM. 1, IC10 TERM. 10, AND IC12 TERM. 9
IC6 TERM. 1 TO IC7 TERM. 2
IC9 TERM. 12 TO IC10 TERM. 11
IC11 TERM. 11 TO IC13 TERM. 10
IC12 TERM. 8 TO IC13 TERM. 9
GROUND TERM. 7 ON IC1, IC3, IC9, AND IC13
GROUND TERM. 8 ON IC1

Fig. 25-5-B   Partial logic diagram for a remote control house wiring.

CHAPTER 26

Jigs and Fixtures

UNIT 26-1
Jig and Fixture Design

With mass production and interchangeable assembly being used extensively in industry, it is imperative that components be machined and sized to identical standards. To do this, devices called *jigs* and *fixtures* are employed to hold and locate the work or to guide the tools while machining operations are being performed. Jigs and fixtures also cut down machining time, thus lowering production costs.

A jig is a device which holds the work and locates the path of the tool. See Fig. 26-1-1. Generally, a jig may readily be moved about or repositioned. An example of this would be a drill jig which may reposition the work several times when many holes are required in the workpiece, the drill being located each time by a drill bushing located on the jig. Jigs are used extensively for drilling, reaming, tapping, and counterboring operations. A fixture, as the name implies, is fixed to the worktable of the machine and locates the work in an exact position relative to the cutting tool. The fixture does not guide the tool. Fixtures are often employed when milling, grinding, welding, and honing operations are required.

Fig. 26-1-1   Drill jig. (Ex-Cell-O Corp. Ltd.)

JIGS

There are two main types of jigs: those used for machining purposes and those used for assembly purposes.

When a jig is used in conjunction with a machine tool, its function is to locate the component, hold it firmly, and guide the cutting tool during its operation. The jig need not be secured to the machine. The term thus used usually refers to drilling, reaming, tap-

ping, and boring operations. See Fig. 26-1-2. The size of the jig in this case is limited by the proportions of the machine and the handling characteristic required of the jig. This type of jig is normally moved about frequently and is stored when not in use.

When a jig is used for assembly purposes, its function is to locate separate component parts and hold them rigidly in their correct relative positions to each other while they are being connected. These parts usually form large structural frameworks from which accurate locators are taken.

The Design of Jigs
The design of a jig is governed by five major factors

1. The machining operation or operations involved
2. The number of parts to be produced
3. The degree of accuracy of the component
4. The state of the component
5. Any other relevant factors such as portability requirements and external locations

Machining Operation(s) Involved As stated earlier, the term *jig* usually refers to a drilling, reaming, tapping, or boring device. More often than not, a jig may perform a combination of these functions—such as drill and

(A) LOADING THE WORKPIECE

(B) TAPPING THE WORKPIECE

**Fig. 26-1-2 Drilling and tapping jig.
(Northwestern Tools)**

ream, drill and tap, drill, ream and counterbore, etc. Drilling, reaming, and tapping jigs and their combinations are usually similar in construction since all these operations are performed on the one machine, the drill press.

Number of Parts to Be Produced The number of parts to be produced has an important effect on design. For example, in very large quantity production, the cost of an expensive clamping device may be recovered many times over, as a result of time saved through its use. Of course, in small quantity production, the cost of the device might not be recovered; thus a cheaper device should be used.

Degree of Accuracy of the Component It is logical, of course, that if a component is required to be very accurate, the tool producing it will have to be even more accurate.

Stage of the Component The designer must know the stage of manufacturing of the component so that he or she can use any available machined faces for location purposes.

Any Other Relevant Factors Sometimes it is necessary to bolt a jig to the table of the machine. For example, when a large hole is being drilled or reamed, the designer must know what facilities for clamping may be available on the machine.

Before designing a jig, the designer must have or be able to find all information such as that given above. The information is given to the designer in the form of a working drawing of the component, a process sheet showing the sequence of operations on the component, and general information usually available in the department.

Having progressed this far, there are several principles that must also be considered before the designer can finally decide on the design of the jig. The designer must consider

1. The machine on which the operation is to be performed
2. Loading and unloading of the component: (*a*) clearances necessary for locating the part; (*b*) methods of guarding against improper loading
3. Rapid methods of clamping the work
4. Chip clearance and chip removal
5. Allowance for observation of operation where possible
6. Safety in operation

There are many other considerations, of course, but these are the major ones.

If the following questions are asked and answered before the jig design is started, much time and money (time *is* money) will be saved and trouble will be avoided.

Jigs in General

1. Can a component be inserted and withdrawn without difficulty?
2. Should the component be located to ensure symmetry or balance, i.e., optical balance or material balance?
3. Have the best points of location been chosen with regard to the accuracy of location and the function of the component?
4. Are hardened location points provided where necessary?
5. Can the locating points be adjusted where required to make allowance for wear of forging dies or patterns?
6. Are locations clear of flash or burrs?
7. Will the locating devices permit a commercial variation in the machining of the component without affecting the accuracy of location or causing it to bind in the jig?
8. Can the jig be easily cleared of metal shavings and grit, particularly on the locating faces?
9. Are all the clamps strong enough?
10. Will any clamp-operating lever or nut be in a dangerous position, i.e., near the cutter?
11. Are all clamps and clamping screws in the most accessible and natural positions?
12. Can wrenches be eliminated by the use of ball or eccentric levers?
13. Will one wrench fit all clamp-operating bolts and nuts?
14. Is the component well supported against the actions of pressure of the cutter?
15. Is the jig foolproof? That is, can the component, tools, or bushings, etc., be wrongly inserted or used?
16. Has the operator an unobstructed view of the component, particularly at the points of location, clamping, and cut?
17. Is the jig as light as possible, consistent with the desired strength?
18. Can coolant, if used, reach the point of cut?
19. Have loose parts been eliminated wherever possible?
20. Have standard parts been used where circumstances allow?
21. Can the jig be designed to hold RH and LH or other similar or complementary components which may be required?
22. Where will burrs be formed, and is clearance for them arranged?
23. Are all corners and sharp edges that are likely to cut the operator shown well radiused on the drawing?
24. Are locating and other working faces and holes protected as far as possible from dirt and cuttings?
25. Will the jig as designed produce components within the required degree of accuracy?

Drilling and Boring Jigs

26. Do drills, tools, etc., enter the component at the face which

directly adjoins the face of the component to which it fits?

27. Have all slip bushings necessary for reaming, spotfacing, tapping, counterboring, seating, etc., been arranged?

Milling Jigs

28. Have clamps, etc., been designed to permit the use of the smallest possible diameter of cutter(s)?
29. Will the cutter mandrel clear all parts of the jig when it passes over?
30. Have means for setting the cutter(s) in the correct position been provided?

Drill Jigs

Drill jigs are of two general types: open jigs and closed or box-type jigs. Open jigs are often referred to as *plate* or *template jigs*. Closed jigs will be discussed in Unit 26-5.

Open Jigs The simplest tool used to locate holes for drilling is the plate jig or drill template. It consists of a plate with holes to guide the drills, and it has locating pins that locate the workpiece on the jig; or, the workpiece may be nested on the jig and then both are turned over for the drilling operation. Jigs of this type are generally without clamping devices. They are used where the cost of more elaborate tools would not be justified.

A separate base is often used with the template or top plate, thus forming the sandwich type of jig. The base may have holes or grooves to provide clearance for the end of the drill as it breaks through the work.

In the drill jig shown in Fig. 26-1-3, the component is not clamped into or onto the jig. The jig rests upon the component. Since the center-to-center distance between the holes is probably more critical and held to closer tolerances than the distance between the holes and the edge of the part, a locking pin is used to ensure the center-to-center hole accuracy. After the first hole is drilled, the locking pin is inserted into the drill jig and workpiece.

Drill Bushings

These are precision tools that guide cutting tools such as drills and reamers into precise locations in a workpiece.

(A) THE WORKPIECE

(B) PLACING JIG OVER WORKPIECE AND DRILLING FIRST HOLE

(C) DRILLING SECOND HOLE AFTER PLACING LOCKING PIN IN HOLE

Fig. 26-1-3 A simple plate jig.

Assembled in a jig or fixture, drill bushings are capable of producing duplicate parts to extremely close tolerances in regard to location and hole size.

A variety of bushings have been developed for a wide range of portable or machine drilling, reaming, and tapping operations. They include headless and head press-fit bushings, slip and fixed renewable bushings, headless and head liners, thin-wall bushings, and a number of embedment bushings for plastic or castable tooling, soft materials, and special applications.

Each type of bushing has its preferred use. Only proper selection can give the service that the manufacturer has built into the product.

To select the proper bushing, it is necessary to consider not only the function of the jig but also the quantity of production. Life of the average bushing is no more than 5000 to 10 000 pieces.

Press-Fit Bushings Press-fit bushings are available in two basic styles: headless (type P) and headed (type H). See Fig. 26-1-4. These bushings are permanently pressed into the jig plate or fixture. Press-fit bushings are recommended for use in limited production runs where replacement due to wear is not anticipated during the life of the tooling, and where a single operation, such as drilling only or reaming only, is performed. Headless press-fit bushings offer two advantages: they can be installed flush with the jig plate without counterboring the mounting hole, and they can be mounted closer together than headed bushings. However, where space permits, the use of headed press-fit bushings is preferable in any application where heavy axial loads may eventually force a headless bushing out of the jig plate. Typical

(A) HEADLESS   (B) HEAD TYPE

Fig. 26-1-4 Press-fit drill bushings. (American Drill Bushing Co.)

Fig. 26-1-5 Chip and burr clearance.

(A) RECOMMENDED CLEARANCE BETWEEN WORKPIECE AND BUSHING

CUTTING TOOL WITH NORMAL BACK TAPER
GUIDING EFFECT OF BUSHING REDUCED
JIG
ID
WORKPIECE
NO CLEARANCE (MAXIMUM PRECISION DRILLING ONLY)
EQUAL TO ONE HALF OF DRILL DIA (SMALL CHIPS)
ONE TO ONE AND ONE HALF OF DRILL DIA (LONG STRINGY CHIPS)
NORMAL CHIP CLEARANCE
EXCESSIVE CHIP CLEARANCE
EXCESSIVE CHIP CLEARANCE
BURR CLEARANCE
SECONDARY BURR
(B) BURR CLEARANCE

(A) BUTTON CYLINDRICAL   (B) PRESS-FIT CYLINDRICAL   (C) SCREW-TYPE HEXAGONAL

(D) CAST   (E) WELDED   (F) SCREW-TYPE CYLINDRICAL

Fig. 26-2-2 Typical jig feet.

sizes of standard press-fit drill jig bushings are shown in the appendix.

Renewable bushings are covered in Unit 26-3.

Installation

Chip Control See Fig. 26-1-5. Sufficient clearance should be provided between the bushing and the workpiece to permit the removal of chips. The exception to this rule occurs in drilling operations requiring maximum precision where the bushings should be in direct contact with the workpiece. However, suitable chip clearance should be provided in most applications because the abrasive action of metal particles will accelerate bushing wear.

Burr Clearance See Fig. 26-1-5B. Burr clearance should be provided between the bushing and the workpiece when wiry metals such as copper are drilled. Metals of this type tend to produce secondary burrs around the top of the drilled holes; the burrs act to lift the jig from the workpiece and to cause difficulty in the removal of workpieces from side-loaded jigs. The recommended burr clearance is one-half the bushing ID.

Reference
1. American Drill Bushing Co.

ASSIGNMENT

See Assignment 1 for Unit 26-1 on page 536.

Review for Assignment
Unit 7-5   Assemblies in Sections

UNIT 26-2
Drill Jig Components

Jig Body
The frame which holds the various parts of a jig assembly is called the *jig body*. It may be in one piece or bolted or welded together. Rigid construction is necessary because of the accuracy required, yet the jig should be light enough to provide ease in handling. Sharp edges or burrs which may harm the operator should be removed. Supporting legs—a minimum of four being recommended—should be provided on the opposite side of each drilling surface. Standard shapes have been designed for jig bodies and are generally more economical than fabricating units in the shop. See Figs. 26-2-1 and 26-2-2.

Cap Screws and Dowel Pins
The purpose of cap screws in jig design is to hold together fabricated parts.

Fig. 26-2-1 Machined cast iron and aluminum sections for construction of jigs and fixtures. (Standard Parts Co.)

Dowel pins provide the necessary alignment between the parts, a minimum of two being recommended. See Fig. 26-2-3. Wherever possible, cap screws should be recessed and have socket fillister heads. This type of cap screw can be tightened, with a greater amount of pressure providing better holding power. When thin stock is to be fastened together and counterboring is not possible, a hexagon-head cap screw is used.

Dowel pins may be tapered or straight, the latter being used more frequently. A press fit into the two parts ensures the proper alignment required in jig design.

DOWEL PINS—HARDENED AND GROUND. CLOSE SLIP FIT IN "A", FORCE FIT IN "B"

DOWEL PINS—USED TO ALIGN PARTS. 2 MINIMUM SOCKET SCREWS USED TO HOLD PARTS TOGETHER

Fig. 26-2-3 Dowel pins and cap screws.

Locating Devices
The shape of the object determines the type of location best suited for the part. Pins, pads, and recesses are the more common methods used to locate the workpiece on the jig.

Internal Locating Devices A machined recess in the jig plate (Fig. 26-2-4A), and a nesting ring (Fig. 26-2-4B), attached to the plate are two methods used to locate a part having a circular projection. The latter method is preferred because the part can be machined more readily and can be replaced when worn. Dowel pins—normally two—position the ring while fastening screws (the number being determined by the size of the ring)

Fig. 26-2-4 Common locating devices.

(A) MACHINED RECESS IN JIG BODY
(B) NESTING RING
(C) HEADLESS BUSHING

INTERNAL LOCATING DEVICES

(D) STRAIGHT STUD—PRESS FIT SHANK
(E) STRAIGHT STUD—THREADED SHANK
(F) DISK LOCATOR

EXTERNAL LOCATING DEVICES

secure it to the plate. For small cylindrical extensions, a headless bushing mounted flush with the locating surface may be used, provided that the shoulder of the workpiece rests on the locating surface, as shown in Fig. 26-2-4C. An example of a drilling jig that has an internal locating device is shown in Fig. 26-2-5.

External Locating Devices Locating studs (Fig. 26-2-4D) provides an excellent means of locating workpieces with circular holes. When it is desirable to

clamp the workpiece to the stud, the stud should be lengthened and fastened in place by a nut and washer (Fig. 26-2-4E). This secures the stud to the jig body and also provides for the interchanging of studs when necessary. Disk-type locators (Fig. 26-2-4F) are used when the locating diameter is over 2 in. (50 mm). Dowels and fastening screws, the number determined by the size of the disk, locate and secure the disk to the plate.

Stops When the workpiece cannot be located by recesses or projections as outlined above, locating stops are used. They are classified as either fixed or adjustable. See Fig. 26-2-6.

The commonest types of fixed stops are the stop pin, flattened shoulder plug, crowned shoulder plug, and stop pads. While stop pins (dowels) are the most economical, their main disadvantages are rapid wear and marring of the finished surface of the workpiece. Shoulder plugs, with one side of the head flattened, provide a greater bearing surface and will not wear as readily.

Centralizers Circular workpieces, or flat workpieces with rounded or angled ends, may be located or centered by centralizers, as shown in Figs. 26-2-7 and 26-2-8.

Workpiece Supports The workpiece must be supported so as to avoid distortion caused by either clamping or machining. See Fig. 26-2-9. The surfaces supporting the workpiece are called *workpiece supports* and are classified as either fixed or adjustable. They should be located, as nearly as possible, directly opposite the clamping force. It is recommended that four

Fig. 26-2-5 Plate drill jig for drilling holes in flange.

NOTE: A LOCKING PIN TO BE INSERTED IN FIRST DRILLED HOLE TO INSURE PROPER ALIGNMENT OF HOLES

FEET
LOCKING NUT
C WASHER
Ø "A" SLIDE FIT BETWEEN STUD AND WORKPIECE
PRESS FIT BETWEEN STUD AND PLATE

BURR RELIEF
WORKPIECE
STOP PIN

WORKPIECE
FLATTENED SHOULDER PLUG

WORKPIECE
CROWNED SHOULDER PLUG

WORKPIECE
CAST PADS

(A) FIXED STOPS

WORKPIECE
SIDE LOCKING WITH JAM NUT

WORKPIECE
TOP LOCKING WITH SETSCREW

(B) ADJUSTABLE STOPS

Fig. 26-2-6 Fixed and adjustable stops.

WORKPIECE
45°
(A) INTERNAL V—TYPE

HARDENED DOWEL PINS
WORKPIECE
45°
(B) DOWEL PINS

45°
WORKPIECE
(C) EXTERNAL V—TYPE

Fig. 26-2-7 Centralizers.

Fig. 26-2-8 V-bushing drill jig. (Acme Industrial Co.)

small work support areas be used in lieu of one large area, because the latter may produce a rocking condition.

The jig body with metal cut away and steel blocks, called *rest buttons*, are the more common types of fixed supports used.

(1) JIG BODY   (2) REST BUTTON

(A) FIXED WORKPIECE SUPPORTS

(1) SIDE LOCK   (2) TOP LOCK

(B) ADJUSTABLE WORKPIECE SUPPORTS (JACK SCREWS)

(C) ADJUSTABLE WORKPIECE SUPPORT (JACK PIN)

Fig. 26-2-9 Workpiece supports.

(A) LONG-TRAVEL CAM-LOCK CLAMP

(B) TOGGLE SCREW CLAMP

(C) SPRING-LOADED HOOK CLAMP

(D) CONICAL-POINT SETSCREW CLAMP

(E) TWO-DIRECTION CLAMP

(F) HINGED CAM-ASSEMBLY CLAMP

Fig. 26-2-10 Common clamping devices.

Clamping Devices

The clamping components must be designed to securely hold the workpiece but not distort it, to be quickly and easily locked and unlocked, and to swing out of the way during loading and unloading. Some of the more common types of clamps are shown in Fig. 26-2-10.

Screw clamps are commonly used because they do not tend to loosen under vibration and they provide adequate clamping force. One of the simplest types of screw clamps is the cone-point setscrew. The incline on the screw tends to push the workpiece against the locating pads as well as against the stops. It is best suited for clamping unfinished surfaces such as castings because the point of the setscrew will mar the workpiece surface.

The toggle-head type clamp provides a larger contact surface with the workpiece, thereby reducing the possibility of marring. It is also ideally suited for clamping workpieces having side drafts. Where only moderate clamping pressures are required, a knurled knob, lever nuts, or a thumbscrew may be used.

The two-direction clamp provides both side and top clamping. As pressure is exerted on the end of the screw thread, the clamp is pivoted about the pivot pin, producing a downward pressure at the top of the workpiece.

Since screw-thread-type clamps are relatively slow, they are often used in combination with other devices to speed up the clamping and unclamping operations. The travel cam lock assembly and the hinged cam assembly clamps are two such devices.

KNURLED-HEAD SCREW

ADJUSTABLE FIXED STOP

C WASHER

T-SLOT NUTS

FLANGED NUT

QUARTER-TURN SCREW

LOCATING PIN

SWING C WASHER

SWING BOLT

HAND KNOB SCREW

SHOULDER SCREW

SPHERICAL WASHER

FIXTURE KEY

SURE-LOCK FIXTURE KEY

HAND KNOB SCREW
WITH SWIVEL TOGGLE

LEVER NUT

TORQUE SCREW

BALL-HANDLE KNOB

SWING CLAMP

ADJUSTABLE GOOSE-NECK CLAMP

Fig. 26-2-11   Standard jig and fixture parts. (Standard Parts Co.)

Locking Pins

A locking pin is used in jig design to lock or hold the workpiece securely to the jig plate while the second or subsequent holes are being drilled. After the first hole is drilled, the locking pin is inserted through the drill bushing into the drilled hole in the workpiece, locking the drill jig and workpiece together. When more than two holes are drilled, a second locking pin is used to maintain proper alignment. The use of a locking pin is illustrated in Fig. 26-1-3C.

Miscellaneous Standard Parts

Figure 26-2-11 shows some of the more common standard jig components. The designer should, wherever possible, use standard parts in the design in order to simplify the work and reduce the manufacturing cost.

Design Examples

EXAMPLE 1 An alternative drill jig for the workpiece shown in Fig. 26-1-3 is

shown in Fig. 26-2-12. This jig employs a lever arm and a knurled head screw which applies pressure on two sides of the workpiece, forcing it against the

DRILL

JIG FEET (MINIMUM 4). THE TOP OF THE SHOULDER SCREW
IS GROUND FLUSH WITH THE OTHER THREE FEET TO ACT
AS THE FOURTH FOOT

TORQUE
SCREW

SHOULDER
SCREW
(PIVOT PIN)

LEVER ARM

Fig. 26-2-12   Alternative plate jig for workpiece shown in Figure 26-1-3.

locating pins. This jig not only locates but also holds the workpiece in position.

EXAMPLE 2 For drilling a series of bolt holes in a flange, a drill jig, as shown in Fig. 26-2-5, may be used. The base surface of the flange and the diameter A of the workpiece, which were previously machined, are used as locating surfaces. The workpiece slips over the stud and rests on the jig plate. The C-washer is then inserted over the workpiece, and the locking nut is screwed down to securely clamp the parts together. The size of the locknut is selected to clear diameter A. The body of the jig is designed to protect the threads on the stud from being damaged. The position shown is the loading and unloading position. For drilling, the jig must be inverted and the side walls which act as feet must be machined to level and true-up the jig. Notice that part of the sides has been machined away, leaving only the four small surfaces to act as jig feet.

Fig. 26-2-13 Screw latch clamp jig.

EXAMPLE 3 The screw latch clamp jig, similar to the one shown in Fig. 26-2-13, is frequently used because of its simple design and fast clamping action. All the parts shown, with the exception of the clamp plate, are standard items which can be purchased.

ASSIGNMENTS

See Assignments 2 and 3 for Unit 26-2 on page 536.

Review for Assignment
Unit 7-5   Assemblies in Section
Unit 26-1   The Design of Jigs

UNIT 26-3
Renewable Bushings

Renewable bushings are designed to be easily replaced and are available in two styles—fixed renewable (type F) and slip renewable (type S). See Fig. 26-3-1. Both types are installed as slip fits in liners, are designed for long production runs, and are intended to remain fixed in the jig or fixture until worn out. In most applications, the liners (head or headless types) are mounted flush with the jig plate; however, projected mounting is sometimes used for head liners when the jig plate is too thin to accept a suitable counterbore.

Lock screws are suitable only for use with flush-mounted liners, usually

in light-duty applications. For heavier-duty applications, clamps provide a better means of locking the bushing against the effects of vibration and torque produced by drill rotation.

Replacement of a fixed renewable bushing can be accomplished simply by removing its lock screw or clamp, without removing the jig from the production line. Slip renewable bushings are recommended for production runs of any length where more than one operation is performed in a hole, such as drilling and then reaming or counterboring. This type is especially designed for fast change by merely turning and lifting the bushing. Slip renewable bushings may be interchanged in the same liner without affecting the centering accuracy. For

example, suppose you wish to produce a hole in a part between .5000 and .5003 in. in diameter. Since a drill alone cannot come that close, the hole must be reamed. First, drill through a .4844-in. ID slip renewable bushing. After drilling the hole, remove the bushing and replace it with a .5000-in. slip renewable bushing for the reaming operation. Both bushings fit perfectly in the same .75-in. ID jig hole liner.

Typical installations of fixed and slip renewable bushings are shown in Fig. 26-3-2.

LINERS

Liners are permanently pressed into the jig plate or fixture both to provide precision mounting holes or correct slip fit for renewable bushings and to prevent wear of soft jig plates caused by frequent replacement of renewable bushings. Liners are available in two basic styles: headless (type L) and headed (type HL). Headless and headed liners are similar to headless and headed bushings in their advantages and limitations. See the appendix.

Installation

Multiple Operations See Fig. 26-3-3A. In performing multiple operations such as drilling and reaming, slip renewable bushings of different lengths may be used to obtain the combined advantages of adequate chip

UN-A-LOK LINERS PROVIDE HEAD LINER AND SHOULDER LOCK IN ONE. THEY ARE USED WITH SLIP-RENEWABLE BUSHINGS.

(A) LINER

SLIP-RENEWABLE BUSHINGS ADD FLEXIBILITY, PERMIT MULTIPLE OPERATIONS AND CAN BE CHANGED QUICKLY.

SLIP RENEWABLE

FIXED-RENEWABLE BUSHINGS ARE FOR LONG PRODUCTION RUNS. THEY ARE HELD IN PLACE BY LOCK SCREW.

FIXED RENEWABLE

(B) BUSHINGS

LINER
SLIP RENEWABLE

LOCK SCREW
FIXED RENEWABLE

SLIP TYPE

HEADLESS LINER
FIXED TYPE

(C) INSTALLATION

Fig. 26-3-1 Renewable bushings. (American Drill Bushing Co.)

FIXED RENEWABLE BUSHINGS		SLIP RENEWABLE BUSHINGS
FLUSH MOUNTED (HEAD OR HEADLESS LINERS)		
LOCK SCREW • RECOMMENDED FOR LIGHT DRILLING APPLICATIONS • SMALL HEAD DIAMETER PERMITS CLOSE BUSHING PLACEMENT	**ROUND CLAMP** • PROVIDES BETTER SECURITY THAN LOCK SCREW IN HEAVIER DUTY APPLICATIONS • DIAMETER SAME AS LOCK-SCREW HEAD FOR CLOSE BUSHING PLACEMENT	**LOCK SCREW** • RECOMMENDED FOR LIGHT DRILLING OPERATIONS • SMALL HEAD DIAMETER PERMITS CLOSE BUSHING PLACEMENT
ROUND END CLAMP • FOR LOCKING STANDARD FIXED-RENEWABLE BUSHINGS • PROVIDES LARGE BEARING SURFACE AGAINST JIG PLATE	**FLAT CLAMP** • FOR TYPE FX BUSHINGS ONLY • PROVIDES MAXIMUM SECURITY AGAINST VIBRATION AND TORQUE	**ROUND END CLAMP** • FOR HEAVY DUTY APPLICATIONS • PROVIDES LARGE BEARING SURFACE AGAINST JIG PLATE
PROJECTED MOUNTED (HEAD LINERS ONLY)		
ROUND END CLAMP • FOR LOCKING STANDARD FIXED-RENEWABLE BUSHINGS IN PROJECTED MOUNTED LINERS	**FLAT CLAMP** • FOR TYPE FX BUSHINGS ONLY • PROVIDES MAXIMUM SECURITY AGAINST VIBRATION AND TORQUE	**ROUND END CLAMP** • FOR ANY APPLICATION USING SLIP-RENEWABLE BUSHINGS INSTALLED IN PROJECTED MOUNTED HEAD LINERS

Fig. 26-3-2 Typical installations of renewal bushings. (American Drill Bushing Co.)

removal and precise accuracy. The slip renewable bushing should be short enough to provide proper chip clearance during the drilling operation, while the reamer bushing may be long enough to contact or closely approach the workpiece, thus providing maximum guiding effect during the reaming operation.

Close-Hole Patterns See Fig. 26-3-3B. For many applications requiring close center-to-center placement of bushings, thin-wall and miniature head series will prove helpful; however, for especially difficult close-hole patterns, it may be necessary to grind flats on the bushing ODs and/or heads to achieve minimum spacing.

Irregular Work Surfaces See Fig. 26-3-3C. When bushings are adapted to suit applications involving irregular work surfaces, the ends of the bushings should be formed to the contour of the workpiece. In many applications of this nature, the drill point does not enter perpendicular to the work surface and has a tendency to skid or wander. For this reason, the distance between the bushing and the workpiece must be held to a minimum so that the full guiding effect of the bushing can be obtained.

Design Example

The jig shown in Fig. 26-3-4 is similar in design to the one shown in Fig. 26-2-5 except for the feet and bushings. Holes in the jig plate are provided for the feet, which are purchased items. The bushings are the fixed renewable type, designed for long production runs.

Reference

1. American Drill Bushing Co.

ASSIGNMENT

See Assignment 4 for Unit 26-3 on page 536.

Fig. 26-3-3 Installation considerations.

NOTE: A LOCKING PIN TO BE INSERTED IN FIRST DRILLED HOLE TO INSURE PROPER ALIGNMENT OF HOLES

Fig. 26-3-4 Open jig. (Standard Parts Co.)

UNIT 26-4
Dimensioning Jig Drawings

The finished detail drawing for the simple plate jig (Fig. 26-1-3) is shown in Fig. 26-4-1. A brief explanation of why the various dimensions were chosen is as follows.

1. *Distance between Holes.* The dimension between the ∅.328-in. holes on the workpiece is 1.498–1.502 in. Therefore, the tolerance allowed on this dimension is .004 in. The distance between the drill bushings on the jig plate must be kept to a closer tolerance because of bushing wear. A .002-in. tolerance was chosen for the center distance between the bushings, and the limits were placed midway between the workpiece limits. Thus the center-to-center distance was established at 1.499–1.501 in.

2. *Size of Bushing Holes.* In tool design it is general practice to show on the drawing only a note listing the nominal diameter of the hole and the part number of the mating part. It is the job of the machinist to select the proper diameters to ensure a press fit.

3. *Size of Dowel Pinholes.* Dowels are commercially available, at low cost, in a wide range of standard sizes. Standard commercial dowels are finished to .0002 in. (0.006 mm) larger than the nominal diameter with a tolerance of ± .0001 in. (0.003 mm). The size of the dowel pins used is ∅.25 (∅.2501–.2503) × .75 in. long. One end of the dowel pin has a chamfer to facilitate pressing it into the jig plate and to permit easy loading onto the workpiece.

   As mentioned above, a note calling for the nominal diameter and the part number of the mating part covers all the information required.

4. *Center Distance between Dowels.* Since the width and length of the workpiece are shown in nominal dimensions, the tolerance permitted on these dimensions is ± .005 in. Thus the size of the largest workpiece which would be permissible is

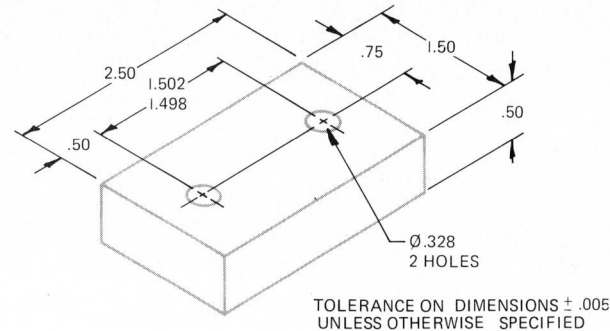

TOLERANCE ON DIMENSIONS ± .005
UNLESS OTHERWISE SPECIFIED

(A) CALCULATING DISTANCES BETWEEN DOWEL PINS

(B) DIMENSIONING JIG PLATE SHOWN IN FIGURE 26-1-3

Fig. 26-4-1 Dimensioning jig drawings.

1.505 × 2.505 in. These are the workpiece sizes which are used in calculating the center-to-center distance between dowel pins. A clearance of .001 to .003 in. was decided on between the maximum workpiece size and the largest dowel pin. Thus the center-to-center distances between dowels were calculated to be 1.7563–1.7583 and 2.7563–2.7583 in.

5. *Center Distance. Bushing and Dowel.* The maximum limits were calculated by taking half the difference between the maximum limits of 2.758 and 1.501 in. for length and half the limit 1.758 in. for width. An allowance of −.001 in. was given to these dimensions.

ASSIGNMENT

See Assignment 5 for Unit 26-4 on page 536.

Review for Assignment
Unit 26-1   The Design of Jigs
Unit 26-2   Drill Jig Components
Unit 26-3   Renewable Bushings

UNIT 26-5
Closed Jigs

The closed or box type of drill jig, within which the work is clamped, is usually used when holes have to be drilled in several directions. To firmly support the jig, sets of supporting legs or feet must be provided on the side of the box opposite each of the drilling faces. The jig is normally opened by swinging back a leaf or cover. The part to be drilled is placed within the box and accurately located and clamped with devices which are, as a rule, permanently attached to the jig body. The jig body frame, shown in Figs. 26-5-1 and 26-5-2, is typical of the type of jig body readily purchased.

Design Example
The box-type jig shown in Fig. 26-5-2 is designed to permit drilling from two

CASTING

AN INTEGRAL CASTING OF ALUMINUM THAT FLOWS THROUGH THE CAST IRON LEGS. ALL WALLS OF THE CASTING, INCLUDING THE LEAF, ARE ENGINEERED TO TAKE STANDARD DRILL JIG BUSHINGS. ALL SIX SIDES CAN BE USED FOR DRILLING. THE UNIQUE CONSTRUCTION OF THIS JIG ALLOWS MACHINING OUT FOR CHIP CLEARANCE WITHOUT DANGER OF DISTORTION.

INTERLOCKING LEG JOINT

"POP-UP LEAF"

ONE-HAND OPERATION. THUMB SCREW RELEASES LEAF INSTANTLY. LEAF IS MADE OF STEEL. THE SOLID HINGE PIN GOES ALL THE WAY THROUGH. PRECISION FITTED TO CAST-IRON LEGS. NO FERROUS METAL MOVING PARTS RUB AGAINST ALUMINUM—THEREFORE NO GALLING.

ONE-HAND OPERATION

SPRING PLUNGER ACTION CAUSES LEAF TO "POP-UP" INSTANTLY. ALL OF WALL OF JIG MAY BE USED FOR BUSHINGS EXCEPT .50 in DEPTH ILLUSTRATED.

LEGS

LEGS ARE CAST IRON. ENTIRE UNIT IS SQUARED TO TOLERANCE ±.00I. WHERE BOTTOM OF JIG IS NOT USED FOR BUSHINGS, LEGS MAY BE CUT OFF AT THE TOP FOR LESS OPERATING INTERFERENCE.

Fig. 26-5-1   Tumble box-type jig body. (Standard Parts Co.)

HEAD-TYPE PRESS FIT BUSHING

CLAMP SCREW

THUMB SCREW HOLDS DOWN POP-UP LEAF

FOOLPROOF PIN

TUMBLE BOX JIG

POP-UP LEAF

REST BUTTONS

ADJUSTABLE FIXED STOP

Fig. 26-5-2   Closed or box-type drill jig.

sides. The complete design is made up of standard parts, and the job of the toolroom is assembling rather than manufacturing and assembling. The use of adjustable stops reduces machining time and permits adjustment and reuse at a later date.

ASSIGNMENT

See Assignment 6 for Unit 26-5 on page 537.

Review for Assignment
Unit 26-1   The Design of Jigs
Unit 26-2   Drill Jig Components
Unit 26-3   Renewable Bushings

UNIT 26-6
Fixtures

A *fixture* is a device which supports, locates, and holds a workpiece securely in position while machining operations are being performed. It should be noted that the accuracy of the machining depends on the quality of the machine and tools used.

Milling Fixtures

The most common type of fixture used is the milling fixture. See Figs. 26-6-1 and 26-6-2. It may be clamped to the milling machine table or held in the milling machine vise. Before a milling fixture is designed, information such as the size and spacing of the T slots, crossfeed, and horizontal traverse of

the table must be known. Most drafting offices have this information tabulated in the form of a chart, and the designer may select the most suitable milling machine.

In laying out a fixture, the drawing should be checked to see that no part of the fixture will interfere with the milling arbor or arbor supports. The standard practice of many designers is to show the cutter and arbor on the assembly drawing.

FIXTURE COMPONENTS

Fixture Base
Fixture components and the workpiece are usually located on a base, which is securely fastened to the milling machine table with clamping lugs or slots. See Fig. 26-6-3. The size of the lug opening corresponds to the T slot width on the milling machine table. In addition, the base is usually provided with keys or tongues which sit on the table T slots, aligning the fixture so that the workpiece is perpendicular to the cutter arbor axis and parallel to the sides of the cutter. Standard fixture bases in a wide variety of sizes are at the disposal of the designer and are shown in the appendix.

Clamps
In milling fixture design, forces resulting from the feed of the table and the rotation of the cutter are encountered. These forces are normally counteracted by the clamp forces. For this reason, fixture clamps must be of heavier design than jig clamps and must be properly located. See Figs. 26-6-4 and 26-6-5.

Fig. 26-6-3   Milling fixture base. (Standard Parts Co.)

Fig. 26-6-4   Toggle clamps.

Fig. 26-6-1   Vertical milling fixture for slotting a hole clamped in a milling vise. (Ex-Cell-O Corp. Ltd.)

Fig. 26-6-2   Typical milling fixture. (Standard Parts Co.)

| Clamp Assembly | Thread | | | | | Travel | | | Capacity |
	A	B	C	D	E	F	G	H	J
20005M	10	16	50	M6	24	7	12	25	22
20010M	12	25	82	M8	32	9	46	38	28
20020M	16	32	100	M10	48	11	25	58	42
20030M	16	32	125	M12	48	13	38	58	42
20040M	22	38	125	M16	50	17	38	64	45
20050M	25	45	165	M20	66	21	45	82	50

Fig. 26-6-5 Strap clamp assemblies. (American Drill Bushing Co.)

Set Blocks

Cutter set blocks are mounted on the fixture to properly position the milling cutter in relation to the workpiece. See Figs. 26-6-6 and 26-6-7. The locating surfaces of the set blocks are offset from the finished surfaces on the workpiece that are to be machined. Feeler gages the same thickness as the offset are placed on the located surfaces of the set block, and the position of the milling fixture is adjusted until the cutter touches the feeler gage. The space between the cutter and set block ensures clearance between the cutter and set block during the machining operation. Set blocks are normally fastened to the fixture body with cap screws and dowel pins.

USE SINGLY TO SUPPORT STEP CLAMP STRAP.

USE IN PAIRS TO SUPPORT PLAIN CLAMP STRAP.

MATCHED SERRATIONS ON ALL STEP BLOCKS PERMIT USE OF MIXED SIZES.

Fig. 26-6-6 Application of holding components. (American Drill Bushing Co.)

STEP BLOCK

U.S. Customary (in.)

Width	Height B	Capacity A
1.00	1.18	.75-1.50
1.38	1.75	1.25-2.50
1.75	3.50	2.50-6.00

PLAIN

STEP

CLAMP STRAPS

Metric (mm)

Width	Height B	Capacity A
25	30	20-40
35	45	30-60
45	90	60-150

U.S. Customary (in.)

Part	C	D	E	F
A-930	.38 to .50	1.00	.50	2.50
A-940	.50	1.19	.75	6.00
A-950	.62	1.19	1.00	6.00
A-960	.75	1.19	1.19	8.00
A-970	1.00	2.00	1.38	10.00

Metric (mm)

Part	C	D	E	F
A-930M	10 to 12	25	12	60
A-940M	12	30	20	150
A-950M	18	30	25	150
A-960M	20	30	30	200
A-970M	25	50	35	250

Fig. 26-6-7 Holding components. (American Drill Bushing Co.)

Fixture Design Considerations

1. Is the fixture foolproof? Does the design permit only one way of loading?
2. Does the fixture permit rapid loading and unloading?
3. Is ample chip clearance provided?
4. Is the fixture kept as low as possible to avoid chatter and springing of the work?
5. Are the cutting forces taken on the base rather than on the clamp?
6. Does any part of the fixture interfere with the milling arbor or supports during the machining operation?
7. Are the clamps located in front of the workpiece?

Sequence in Laying Out a Fixture

See Fig. 26-6-8. The following sequence is recommended in laying out a fixture.

1. Draw the necessary views of the workpiece. Leave sufficient room for drawing in the fixture details.
2. Draw the locating devices.
3. Draw the cutter and arbor.
4. Draw the clamping arrangement.
5. Draw the set blocks, if required.
6. Draw the fixture base and keying arrangement.

ASSIGNMENT

See Assignment 7 for Unit 26-6 on page 537.

Review for Assignment
Unit 7-5   Assemblies in Section

(A) DRAW 3 VIEWS OF THE WORKPIECE AND ADD SUITABLE LOCATING DEVICES

NOTE POSITION OF DIAMOND LOCATING PIN

LOCATING PIN

SLOT TO BE MILLED

(B) ADD CLAMPS AND SHOW LOCATION OF CUTTER AND ARBOR

ADJUSTING HEEL-CLAMP ASSEMBLY

CUTTER

ARBOR

(C) DRAW IN SET BLOCK AND BASE DETAIL

CLAMPING LUGS PROVIDED TO CLAMP FIXTURE TO MILLING MACHINE TABLE

STANDARD FIXTURE BASE

DIRECTION OF TABLE FEED OPTIONAL

FEELER GAGE THICKNESS

FEELER GAGE THICKNESS

SET BLOCK

STANDARD FIXTURE KEYS SET IN TABLE T-SLOTS

Fig. 26-6-8   Sequence in drawing a simple milling fixture.

ASSIGNMENTS for Chapter 26

Assignments for Unit 26-1, Jig and Fixture Design

1. On a B- or A3-size sheet, design a simple plate jig for drilling the holes in one of the parts shown in Figs. 26-1-A to 26-1-D. Scale is full (1:1). State the sequence of operations and the time at which locking pins are employed.

Assignments for Unit 26-2, Drill Jig Components

2. On a B- or A3-size sheet, design a jig for drilling the small holes in the part in either Fig. 26-2-A or Fig. 26-2-B. The large center hole and finished base should be the features used for locating the part in the jig. A locking pin is recommended for alignment after the first hole is drilled. Standard components should be used wherever possible.

3. On a B- or A3-size sheet, design a drill jig for the smaller of the two holes shown in Fig. 26-2-C or 26-2-D. The hole in the hub of the part and the finished surfaces should be the features used for locating the part in the jig. A locking pin for the part shown in Fig. 26-2-D is optional, depending on the design. Standard components should be used wherever possible. Scale is full (1:1).

Assignments for Unit 26-3, Renewable Bushings

4. On a B- or A3-size sheet, design a jig using slip renewable bushings for one of the parts shown in Fig. 26-3-A or 26-3-B. For the jig for Fig. 26-3-B, only two drilling operations are required—a tap drill hole and a clearance hole for the shank of the cap screw. The clearance hole for the head of the cap screw and the threading of the holes will each be done as a separate operation. Show the drill bushing for the largest hole in the assembly, and show a detail drawing of the slip bushings used for the smallest or smaller holes. Only nominal sizes need to be shown. Scale is full (1:1).

Assignments for Unit 26-4, Dimensioning Jig Drawings

5. On a B- or A3-size sheet, design a simple plate jig for drilling the holes in one of the parts shown in Fig. 26-4-A or 26-4-B. The size of the dowel pins used in the design is .2502 ± .0001 in. or 6.006 ± 0.003 mm. After your overall design has been approved by your instructor, dimension the jig plate as per procedures outlined in this unit. Scale is full (1:1).

Fig. 26-1-A   Connector.

Fig. 26-1-B   Spacer.

Fig. 26-1-C   Washer.

Fig. 26-1-D   Cover plate.

Fig. 26-2-A   Plate.

Fig. 26-2-B   Flanged bracket.

Fig. 26-2-C   Link.

Fig. 26-2-D   Connector.

Ø.34
CBORE Ø.50 X .31 DEEP
4 HOLES EVENLY SPACED ON
Ø2.38 BOLT CIRCLE

.750– 12 UNC

.88

1.62

.62

Ø3.50

MATL – SAE 1110

Fig. 26-3-A   End plate.

Ø.38
2 HOLES

45°

1.50

3.00

1.50

1.50

.75

45°

.75

1.50

3.00

MATL – BRASS .38 THICK
TOLERANCE ON DIMENSIONS ±.005
UNLESS OTHERWISE SPECIFIED

Fig. 26-4-A   Locking plate.

1.00

1.50

.50

.50

R .90

Ø1.00
3 HOLES

1.50

2.75

R 1.00

Ø2.00

R .88

1.75

Ø.75
2 HOLES

.50

MATL – MALLEABLE IRON

Fig. 26-5-B   Swivel hanger.

Ø.18
CBORE Ø.28
2 HOLES

Ø.501

Ø1.12

.40

2.00

.44

1.12

.18

.06

8–32 UNC–2B
2 HOLES

.94

MATL – SAE 1110

Fig. 26-3-B   Coupling.

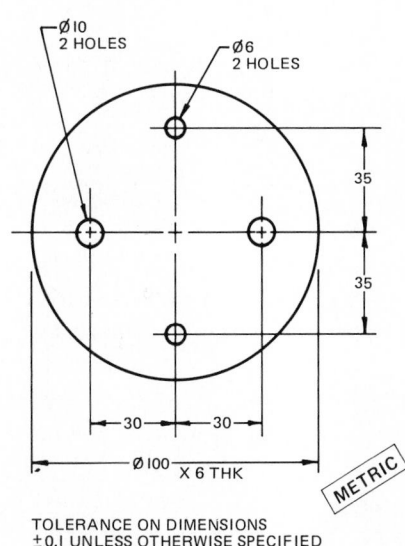

Ø10
2 HOLES

Ø6
2 HOLES

35

35

30

30

Ø100
X 6 THK

METRIC

TOLERANCE ON DIMENSIONS
±0.1 UNLESS OTHERWISE SPECIFIED

Fig. 26-4-B   Spacer.

Ø.750

1.06

Ø2.12

.44

1.25

.44

Ø1.38

MATL – SAE 1020

Fig. 26-6-A   Drive link.

Assignments for Unit 26-5, Closed Jigs

6. On a B- or A3-size sheet, design a box-type jig for drilling all the holes in one of the parts shown in Fig. 26-5-A or 26-5-B. Select a suitable tumble-box jig from the appendix. Use standard components wherever possible. Scale is full (1:1).

Assignments for Unit 26-6, Fixtures

7. On a B- or A3-size sheet, design a simple milling fixture to mill out the two outside portions on the top of the part in Fig. 26-6-A or the slot in Fig. 26-6-B. Use standard components and refer to manufacturers' catalogs wherever possible. Draw the workpiece in red. Scale is half size (1:2).

Ø20.5

Ø14.5

66

Ø35

Ø25

Ø5.1

6

5

Ø3.1

25

25

10

MATL – MALLEABLE IRON

METRIC

Fig. 26-5-A   Link.

16

Ø 20.6
20.5

16

45°

Ø 48

2

18

MATL – SAE 1030

METRIC

Fig. 26-6-B   Sleeve.

CHAPTER 27

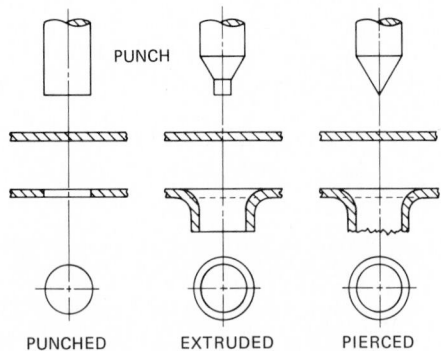

STOCK STRIP

PIECE PART (BLANK)

PUNCH OPENING

PIECE PART (BLANK)

SLUG

Die Design

UNIT 27-1
Die Design

Because of the wide use of sheet metal in the construction of many products, die design and its associated tooling have become indispensable to the engineering industry.

A casual examination of the goods on sale in any hardware store will make it clear that presswork or stamping is the foundation of the mass-production industry. Also, designers will often seek to replace expensive castings and forgings with parts of equal strength constructed from sheet metal for one or more of the following reasons:

1. Faster production, because of the speed at which the presses can operate.
2. Cheaper costs, since many press operations can be carried out by unskilled labor.
3. Lighter construction by skillful arrangement of the sheet-metal components.

STAMPING

Stamping is the art of pressworking sheet metal to change its shape by the use of punches and dies. It may involve punching out a hole or the product itself from a sheet of metal. It may also involve bending or forming, drawing, and coining. See Figs. 27-1-1 and 27-1-2.

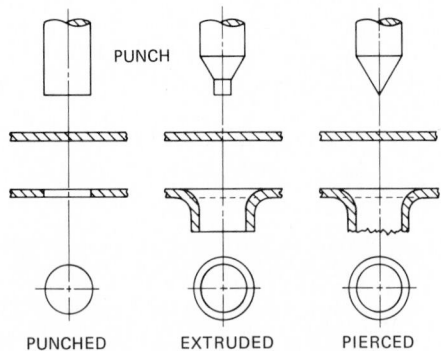

PUNCH

PUNCHED   EXTRUDED   PIERCED

Fig. 27-1-1   Methods of producing holes.

Stamping may be divided into two general classifications: forming and shearing.

Forming *Forming* includes stampings made by forming sheet metal to the shape desired without cutting or shearing the metal.

Shearing *Shearing* includes stampings made by shearing the sheet metal either to change the outline or to cut holes in the interior of the part.

Blanking is the process of shearing or cutting out the size and shape of a flat piece necessary to produce the desired finish part.

Punching forms a hole or opening in the part.

PRINCIPLES OF BLANKING DIES

Before beginning the study of die design, you must have a clear understanding of the following terms.

PUNCH
DIE SHOE
PRESS BED

Fig. 27-1-2   Punch press used to punch and form metal. (Whitney Metal Tool Co.)

Piece Part A *piece part* is the product of a die.

Stock Material This is the general term for any of the various materials from which the piece part is made.

Die The word *die* has several definitions. This book utilizes two: (1) a complete production tool, the purpose of which is to produce piece parts consistently to required specifications, and (2) the female part of a complete die.

Punch A *punch* is the male member of a complete die which mates, or acts in conjunction with, the female die to produce a desired effect upon the material being worked.

Parts Produced by Blanking

The piece part in Fig. 27-1-3 is produced by blanking. It is shown in relation to the stock material from which it was blanked. The terms used to describe the various aspects of the work are indicated. The *stock strip* is stock material of the required thickness (*T*) which has been cut to strips of suitable width for the particular job. The stock strip is fed through the die from right to left, as indicated by the feed direction arrow. The *advance* is the distance the stock must be fed (advanced) to allow a clean blanking operation at each press stroke. The *scrap bridge* is the scrap remaining between the openings in the stock strip after blanking. The advance distance is equal to the width of the piece part plus the scrap bridge. The back scrap is so called because it is located toward the back of the press (away from the operator). The front scrap is on the side toward the front of the press (toward the operator). The *lead* end of the stock strip is the end that is fed into the die first. The opposite end of the strip is called the *tail*.

The piece part shown in Fig. 27-1-4 is called a *blank* because further work is required to complete it. The slug is the material that is cut out of the blank or piece part by a punching operation.

Whether a die is a blanking die or a punching die depends solely upon its intended use. It is called a *blanking die* if it is meant to produce blanks (B), as shown in Fig. 27-1-3, or a desired contour and size by cutting them out of the required type of material, called the *stock strip*. The blanks are the desired product (piece parts) made by the die.

Fig. 27-1-4 Relationship of stock material, piece part, and slug.

Description of Blanking Die

Figures 27-1-3 and 27-1-5 show the basic elements of a standard parts blanking die. These elements are the die block, in which the proper female die opening has been made, an adaptor (may or may not be required depending on the size of hole being made), the punch, and the stripper. In addition, guide and stopping devices are required to control the stock. The die block and stripper are secured to the die shoe, which in turn is fastened to the press bed of the punch press. The punch is locked into the punch holder, which in turn is secured to the press ram which moves up and down. Standard punch and die parts are shown in Fig. 27-1-6.

Press Cycle during Blanking See Fig. 27-1-5A. The at-rest position of the press ram is at the top of the stroke. The ram is then at its greatest distance from the bed of the press. When the press is tripped, the clutch allows engagement of the crankshaft with the press flywheel. The crankshaft then rotates through 360°, or one full turn. During the first half of the cycle, the ram is driven toward the press bed. During the last half of the cycle, the ram moves away from the press bed. The distance traveled by the ram during the half cycle (top of stroke to bottom of stroke) is called the *press stroke*.

(A) RELATIONSHIP OF PIECE PART AND STOCK STRIP

(B) PUNCH AND DIE ASSEMBLY

Fig. 27-1-3 Simple blanking die.

Fig. 27-1-5   Simple blanking operation using standard components shown in Fig. 27-1-6.

(A) PUNCH IN "AT REST" POSITION

(B) PUNCH SHEARS THROUGH STOCK AND BLANKS OUT PIECE PART

(C) STRIPPER PREVENTS STOCK FROM RISING WITH PUNCH ON RETURN STROKE

STRIPPER

PT 1—DIE SHOE WITH T SLOTS
PT 2—DIE SHOE WITHOUT T SLOTS

SLOT DETAIL

Ø.25 SET SCREW

Ø.25 RADIAL LOCKING PIN

ADJUSTABLE GUIDE BLOCK

PUNCHES

PUNCH SIZE

DIE ADAPTOR

PUNCH HOLDER

DIE BLOCK

HOLE SIZE AND SHAPE (SEE PUNCHES)

HOLE SIZE + .03 in. CLEARANCE PER SIDE

PUNCH SHAPE	AVAILABLE SIZES
CIRCLES	.12 TO 1.00 in. IN $\frac{1}{32}$ in. INCREMENTS
SQUARES	.12 TO .75 in. IN $\frac{1}{32}$ in. INCREMENTS
RECTANGLES	.12 X .75 .16 X .75 .19 X .75 .25 X .75 .31 X .75 .38 X .75 .50 X .75
OVALS	.12 X 1.00 .16 X 1.00 .19 X 1.00 .25 X 1.00 .31 X 1.00 .38 X 1.00 .50 X 1.00

Fig. 27-1-6   Standard punch and die parts. (Whitney Metal Tool Co.)

Action of Blanking or Punching a Die
Refer to Fig. 27-1-5B. The stock material (A) is fed or loaded in the proper position on the top surface of the die block. When the press is tripped, the ram drives the punch through the stock strip (A) into the die opening, thereby producing an opening in the stock material by cutting out the blank or slug (B).

Stripping After the blanking has taken place, the punch is returned to the open or at-rest position by the press ram as it completes its cycle. The stock material clings to the punch and will remain on the punch unless something is done to prevent it. This is the function of the stripper: it keeps the stock material from traveling with the punch on the return stroke (Fig. 27-1-5C).

Reaction of the Stock Material—Shearing Action

The result of the forces imposed on the stock material by the working of the blanking die is a shearing action. This shearing action may be considered in three stages, which are important to the die maker because of their direct relationship to the dimensional qualities and appearance of piece parts. They are also related to the effective working and life of the die. These stages showing the reaction of the stock material are illustrated progressively in Fig. 27-1-7.

Critical Stages of Shearing Action of Metal *First Stage: Plastic Deforma-*

(A) FIRST STAGE—PLASTIC DEFORMATION

(B) SECOND STAGE—PENETRATION (SHEAR)

(C) THIRD STAGE—FRACTURE

Fig. 27-1-7   Critical stages of shearing action on metal.

tion. The stock material has been placed on the die, the press has been tripped, and the punch is being driven toward the die. The punch contacts the stock material and exerts pressure upon it. When the elastic limit of the stock material is exceeded, plastic deformation takes place.

Second Stage: Penetration. As the driving force of the ram continues, the punch is forced to penetrate the stock material, and the blank or slug is displaced into the die, which opens a corresponding amount. This is the true shearing portion of the cutting cycle, from which the term *shearing action* is derived.

Third Stage: Fracture. Further continuation of the punching pressure then causes fractures to start at the cutting edges of the punch and the die.

These three stages of shear action are responsible for the characteristic appearance of piece parts produced by blanking.

Cutting Clearance

Proper cutting clearance is necessary to the life of the die and the quality of the piece part. Excessive cutting clearance results in objectionable piece-part characteristics; insufficient cutting clearance causes undue stress and wear on the cutting members of the tool because of the greater punching effort required.

Determining Cutting Clearance The physical properties and the thickness of the stock material are the factors that determine the amount of cutting clearance. The thickness is easily measured, but the physical properties in relation to cutting clearance are not. See Fig. 27-1-8.

Cutting clearance should be expressed in terms of percentage of stock material thickness per side.

The percentage varies with the properties of the material. A suggested list

Fig. 27-1-8   Optimum cutting clearance.

of percentages for various materials is given in Fig. 27-1-9.

There is another result of the necessity for cutting clearance that must be studied and thoroughly understood by the tool and die maker. This is the effect of cutting clearance on the actual dimensions of the piece part, as shown in Fig. 27-1-10.

CUTTING CLEARANCE PER SIDE (Percentage of Stock Thickness)

Material	Irregular Contours	Round
Aluminum		
Soft, less than .06 in. (1.5 mm) thick	3%	2%
Soft, more than .06 in. (1.5 mm) thick	5%	3%
Hard	5–8%	4–6%
Brass & Copper		
Soft	3%	2%
Half Hard	4%	3%
Hard	5–6%	4%
Steel		
Low Carbon Soft	3%	2%
Half Hard	4%	2%
Hard	5%	3%
Silicon Steel	4–5%	3%
Stainless Steel	5–8%	4–6%

Fig. 27-1-9   Cutting clearances.

Fig. 27-1-10   Piece part sizes in relation to punch and die size.

RELATIONSHIP OF PIECE-PART SIZES TO PUNCH AND DIE SIZES

When blanked piece parts are measured, the measurement is made at the cut band. The blank or slug measures larger than the opening in the stock material from which it was punched. The reasons for this are obvious upon examination of Fig. 27-1-10. The actual cutting of the blank or slug is done by

CLEARANCE HOLE IN STRIPPER — PUNCH — STRIPPER PLATE — DIE BLOCK

(A) PLAIN PIN

(B) SHOULDER PIN   (C) HEADED PIN

Fig. 27-1-11   Solid pin stops.

the cutting edge of the die opening. Therefore, the die opening determines the size of a blank or slug. The actual cutting of the opening in the stock material is done by the punch. Therefore, the size of a punched opening is determined by the punch.

A careful dimensional check of a blank or slug will often reveal that its overall dimensions are slightly larger than the die opening that produced it. This is because the cutting action caused it to be compressed in the die opening. After it passed through the die, the pressure was released; therefore, it expanded a slight amount. Conversely, a punched opening will often be slightly smaller than the punch which produced it. When the punch was withdrawn from the stock material during the stripping stage, the opening closed slightly. Most materials react this way. Satisfactory compensation for this condition is usually achieved by making the punch .0005 to .001 in. (0.01 to 0.02 mm) larger overall than the desired opening to be punched. If the blank is the desired product, the die opening is then made .0005 to .001 in. (0.01 to 0.02 mm) smaller overall than the desired blank.

DIE STOPS

In the majority of applications, stops are installed on dies for the purpose of arresting the feeding movement of the stock strip.

Figure 27-1-11 illustrates three types of solid stop pins. A mounting hole is provided at the desired location in the required die component. The stop pin shown in view A is mounted in a die block. The pin is a light drive fit in the mounting hole. The mounting hole is generally made to suit the stop pin so that a standard pin size can be used. A clearance hole for the pin should be provided in the die shoe.

The stock strip shown in Fig. 27-1-12 is located by a simple stop pin pressed into the die shown. In operation, the edge of the hole in the stock strip butts against the stop pin, holding the stock strip at this position until the piece part is punched out. The stock strip is then

4 GUIDE LOCK ASSEMBLIES

STOP PIN   STOCK STRIP

DIE SHOE WITH T SLOTS

(A) USING STOP PIN

4 GUIDE PINS (RADIAL LOCKING PINS—SOLID OR HOLLOW SPRING)
I RADIAL LOCKING PIN USED AS A STOP PIN
STOCK STRIP

DIE SHOWN WITHOUT T SLOTS

(B) USING RADIAL PIN

Fig. 27-1-12   Stopping the stock strip.

4 GUIDE PINS   DIE ADAPTOR   SETSCREW   GUIDE BLOCK
GUIDE PIN PRESS FIT IN GUIDE BLOCK
GUIDE BLOCK RUNNING FIT IN T—SLOT LOCKED IN POSITION BY SETSCREW
DIE SHOE   STOCK STRIP

(A) 4 GUIDE PINS LOCATE STOCK STRIP

ROUND-HEAD SCREW WITH SQUARE NUT
GUIDE PINS
ADJUSTABLE GUIDE STRIP (BACK GAGE)
DIE ADAPTOR
DIE SHOE
STOCK STRIP
RESTING BAR ATTACHED TO GUIDE STRIP

(B) GUIDE STRIP AND PINS

Fig. 27-1-13   Guiding the stock strip.

raised and moved along so that the next hole drops over the stop pin. Two simple ways of stopping the stock strip are shown. The operator should be able to see the stop action. As such it may be necessary to cut away part of the stripper plate so that the stop pin may be seen. See Fig. 27-1-13.

Guide Strip and Pins

The stock strip must be guided into the die. This is accomplished by guide pins and guide strips. The two shown in Fig. 27-1-13 are standard parts, adjustable, and, as such, can be used on more than one stock strip size. The same screws which hold the stripper plate can be used for the guide strip.

If four guide pins are used, such as shown in Fig. 27-1-13A, then a spacer or shim will be required under the stripper plate in order that the stock strip may be raised sufficiently to clear the stop or radial locking pin. Other forms of stops, such as automatic or spring stops, are also used but will not be covered at this time. See Fig. 27-1-14.

Fig. 27-1-14 Cutting notch in stripper plate so operator may see stop action.

NEST GAGES

Nest gages are commonly used on dies which perform secondary operations, such as piercing, forming, etc. A typical fixed-pin nest gage is shown in Fig. 27-1-15.

The purpose of a nest gage is to locate and position the workpiece properly in the die.

There are four important conditions which must be satisfied in order to pro-

Fig. 27-1-15 Typical fixed-pin nest gage.

vide suitable nest gaging: accuracy, loading ease, unloading ease, and foolproofing.

Reference

1. From *Basic Diemaking* by the National Tool, Die, and Precision Machining Association. Copyright, 1963. McGraw-Hill Book Company. Used by permission.

ASSIGNMENTS

See Assignments 1 and 2 for Unit 27-1 on page 552.

Review for Assignment
Unit 6-4   Assembly Drawings
Appendix   Bolt and Nut Sizes

UNIT 27-2
Nesting of Blanks

In die design, one of the first considerations is the layout of the piece parts on the stock strip. Since over 50 percent of the cost of a stamping is for material, the designer must take care to utilize the raw stock to full advantage. The blanks must be positioned so as to utilize the maximum area of the stock strip. See Fig. 27-2-1.

To accomplish this, a layout of the stock strip is made showing the exact location of each blank. Only a portion of the strip need be drawn if the remaining part of the strip is repetitive.

Fig. 27-2-1 Nesting of blanks.

To reduce the amount of scrap, it may be desirable to alternate or reverse the position of every other blank. However, with this layout method the strip must be fed through the die twice, which involves additional handling costs to produce the same number of parts. Figure 27-2-2 shows how, by modifying the original design slightly, a stock saving could be achieved.

Fig. 27-2-2 Stock saving.

DESIGN CONSIDERATIONS

In designing parts which are to be stamped, certain design considerations should be adhered to. Some of the more basic design rules follow.

Specifying Cutoffs There are several correct cutoffs that can be used for economical stampings once the material has been sheared to the correct width. If sharp corners are permissible, the square corner is most economical. Corners along the edge of the strip should be sharp, and those not adjacent to the edge should be round. See Figs. 27-2-3 and 27-2-4.

Fig. 27-2-3   Specifying cutoffs.

Fig. 27-2-4   Corner design.

Minimum Sections Tabs or slots should never be less than 1.5 to 2 times material thickness in width and never less than .04 in. (1.0 mm). Its length should not be greater than 5 times its width. See Fig. 27-2-5.

Grain Direction If the metal grain runs in a direction contrary to the strength requirement of the part, the strength of a stamping may be reduced considerably. Therefore, in designing highly

W MIN = 1.5 THICKNESS
L MIN = 5W

Fig. 27-2-5   Minimum practical sections.

Fig. 27-2-6   Specifying grain direction.

stressed parts, the direction of the grain of the metal should be considered and, when necessary, specified on the drawing. See Fig. 27-2-6.

Hole Sizes and Tolerances For general economy, a hole diameter should not be less than the thickness of the material. If the hole is less than the material thickness or less than .04 in. (1 mm), the hole should be drilled and the burr removed at added cost.

Unless specified, tolerances shown on hole diameters are considered to apply to the punch side only. See Fig. 27-2-7.

Fig. 27-2-7   Punched holes.

Holes near Blank Edge A hole can be punched without causing a bulge if the web is a minimum of the stock thickness. See Fig. 27-2-8. A bulge will result whenever the web is less than the stock thickness.

Distance between Holes The distance between holes or between a hole and the edge of a part should be large enough to prevent tearing of metal and excessive die wear. Recommended minimum distances are shown in Fig. 27-2-9.

Fig. 27-2-8   Specifying hole opening near blank edge.

METAL THICKNESS		"A" MINIMUM
INCHES	MILLIMETERS	
UP TO .06 INC.	UP TO 1.6 INC.	.12 [3]
OVER .06	OVER 1.6	TWO TIMES METAL THICKNESS

Fig. 27-2-9   Distance between holes.

Notches with Vertices Notches in highly stressed parts should be specified with a radius at the vertex because a sharp vee might provide the starting point of a tear. The radius should be a minimum of twice the metal thickness. A sharp vertex is allowed on lower stress parts when it will aid in lowering design costs. See Fig. 27-2-10.

Reference
General Motors Drafting Standards

ASSIGNENT

See Assignment 3 for Unit 27-2 on page 552.

MINIMUM RADIUS = 2 X METAL THICKNESS. PREFERRED FOR HIGHLY STRESSED PARTS.

GOOD

SHARP VERTEX PERMITTED FOR LOW-STRESSED PARTS.

POOR

Fig. 27-2-10   Notches with vertices.

UNIT 27-3
Die Sets and Components

The punches and dies discussed in Unit 27-1 are relatively simple and inexpensive and have limited use. Many parts produced by punches and dies are more complicated than those already discussed and, as such, require custom-designed die parts.

DIE SETS

A *die set* (Fig. 27-3-1) can be defined as a subpress unit consisting of a lower shoe and an upper shoe, together with the guideposts and bushings by which the shoes are aligned.

The following terms are either directly pertinent or closely related to die sets.

Die Shoe The die-set base is called the *die shoe* (or *die holder*).

Punch Holder The die-set top member is called the *punch holder* (or *punch shoe*).

Shank Most punch holders in the smaller sizes are made with a shank which fits the clamping hole in the lower end of the punch press ram. The shank is used to center the die set in the press and to secure the punch holder to the ram.

Guideposts *Guideposts* (also called *leader pins* or *guide pins*) are cylindrical pins which provide a means of alignment for the die set.

Guidepost Bushings These are installed in the opposing shoe and engage the guideposts with a close sliding fit. The guideposts and bushings, acting together, align the die set.

Flange This is a ledge which is flush with the bottom surface of a die shoe or the top surface of a punch holder.

Fig. 27-3-2   Headless punch.

The ledge extends beyond the die area to provide a means for clamping the shoe member to the bolster plate or press ram, whichever is appropriate.

Die Area This is the area available on the top surface of the die shoe and the lower surface of the punch holder for the mounting of punch and die components.

Shut Height of Die This is the distance from the bottom of the die shoe to the top of the punch holder when the die is in its closed working position.

PUNCHES

As previously defined, a punch is a male member of a complete die. It supplements or complements the female die in order to produce a desired effect upon the material being worked.

Punches Mounted in Punch Plates

A punch plate is so called because its function is to retain and/or position a punch or punches. Punches assembled into punch plates are, for many applications, the most practical method of punch mounting.

One method of assembling a punch and punch plate combination is shown in Fig. 27-3-2.

Perforator-Type Punches

A perforator-type punch can be described as a cutting punch 1.00 in. (25 mm) or less in diameter, if it is round. If it is not round, then its contour may be circumscribed by a circle whose diameter is 1.00 in. (25 mm) or less.

For the sake of convenience, these punches are commonly called *perforators* whether or not their function

Fig. 27-3-1   Components of typical die set.

PUNCH HOLDER
SHANK
GUIDE POST BUSHING
GUIDE POST
DIE SHOE
FLANGE
BOLT SLOT

CLEARANCE — .001—.002 in. (0.025—0.050mm)
(A) OPEN SLOT KEYSEAT

CLEARANCE — .001—.002 in. (0.025—0.050mm)
(B) POCKET-TYPE KEYSEAT

.001—.002 in. (0.025—0.050mm) — CLEARANCE
(C) SEPARATE KEYSEAT

Fig. 27-3-3   Typical key arrangement for punches.

is strictly one of perforating. As a rule, perforators are mounted in punch plates. Common types of key arrangements are shown in Fig. 27-3-3.

Step-Head Perforators For general application, the step-head perforator shown in Fig. 27-3-4 is probably the type most widely used.

Fig. 27-3-4　Step-head perforator.

MOUNTING SCREWS AND DOWELS

Any construction method that securely holds a die component in assembly and simultaneously ensures accurate positioning of the component throughout the life of the die may be said to be satisfactory.

It is common procedure to use a combination of screws and dowels to provide for proper mounting (assembling) of the various components that make up a die. The function of the dowels is to provide and maintain accurate positioning of the component. The function of the screws is to retain the component securely in the doweled position.

DIE BLOCKS

The term *die block* refers to the complete unit component which mates with the punch or punches in the manner required to produce the desired effect upon the stock material.

Among the simplest of die block constructions is the one-piece ring shown in Fig. 27-3-5. Dowel pins are used to locate and position the die ring. The screw and dowel locations shown are conventionally typical for smaller rings of this kind. Bolt circle diameter C should be such that the screw and dowel positions will not be too close to the wall of the die-shoe clearance

Fig. 27-3-5　One-piece die ring—dowel mounted.

opening. For rings with minimal cross section W, this usually requires that dimension F be somewhat larger than E. For rings with wider cross sections W, the bolt circle diameter may be made to divide W equally, that is, E = F.

Note that one dowel pin B is offset. This is a foolproofing procedure; if the dowel were not offset, it would be possible to mount this ring upside down.

Another common construction for round work is shown in Fig. 27-3-6. Here, the die ring is set in the die shoe.

STRIPPERS

Stripping is the act of removing the work from the punch or punches. A *stripper* is a device for stripping.

There are two basic stripping actions, depending upon whether the punch is moving or stationary. With a

Fig. 27-3-6　One-piece die ring—set in.

moving punch, the stripper acts to arrest the stock material, permitting the punch to withdraw from it. This action is depicted in Figs. 27-3-7 and 27-3-8. Figure 27-3-7 illustrates the arresting action of a fixed stripper.

View A The die is in a closed position, with the punch entered, at the bottom of the press stroke.

Fig. 27-3-7　Stripping action—fixed stripper, punch moving with ram.

VIEW A

VIEW B

STRIPPER SCREW
PUNCH HOLDER
SPRING
SPRING PILOT
STRIPPER
PUNCH
WORK
(STOCK MATERIAL)
DIE BLOCK
DIE SHOE

VIEW C

Fig. 27-3-8 *Stripping action—pressure-pad stripper, punch moving with ram.*

View B The punch is ascending on the return stroke. The stock material (work) clings to the punch, moving upward with it until it contacts the intervening stripper.

View C The punch, as it continues upward, is withdrawn from the stock material, permitting the stock material to drop toward the die surface.

View D The working cycle is completed. The punch has returned to its up position. The workpiece has dropped back to the die face and is ready to be removed.

Figure 27-3-8 depicts the arresting action of a pressure-pad–type stripper. The stripper shown is spring-actuated. Gages, spring pins, etc., have been omitted for purposes of clarification.

View A The die is in a closed position, punch entered, at the bottom of the press stroke. The potential stripper travel for the stripper action is indicated at *M*.

View B The punch assembly is ascending on the return stroke. The stripper is

still in the position shown in view A. The springs have maintained this stripper position, forcing the ascending punch to withdraw from the stock material, as shown. Stripper travel *M* relative to the punch is noted.

View C The working cycle is completed at the top of the press stroke. The punch holder has carried the punch and stripper assembly with it to the up position. The workpiece has been left behind, lying on the die surface to be retrieved or ejected as required.

ANGULAR CLEARANCE

Angular clearance is a draft or taper applied to the sidewalls of a die opening (Fig. 27-3-9) in order to relieve the internal pressure of the blank or slug as it passes through the opening.

Specifying Angular Clearance Angular clearance should be expressed in terms of the amount of clearance per side, not as an overall or included-angle figure.

Maximum and Minimum Angular Clearance With few exceptions, an angle of 2° per side can be considered a maximum desirable clearance angle. A practical minimum clearance angle is 0.13° per side (1:250 taper).

Methods of Providing Angular Clearance Figure 27-3-9 illustrates the application of angular clearance to die openings. In view A, the opening is

PUNCH
CUTTING FACE OF DIE BLOCK
CUTTING LAND
ANGULAR CLEARANCE
DIE BLOCK

(A) WITH CUTTING LAND

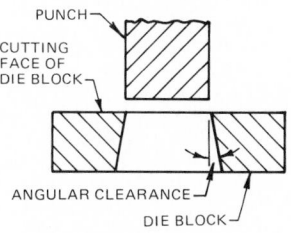

PUNCH
CUTTING FACE OF DIE BLOCK
ANGULAR CLEARANCE
DIE BLOCK

(B) WITHOUT CUTTING LAND

Fig. 27-3-9 Angular clearance.

made with a cutting land contiguous to the cutting edge and extending into the die opening. The height of the cutting land should be equal to the thickness of the stock material, but should not be less than .08 in. (2 mm). A height of .16 in. (4 mm) is considered good average practice for cutting material less than .16 in. thick.

View B shows the angular clearance beginning at the cutting edge. This method is used when it is necessary or desirable to relieve the compression stored within the blanked part as soon as possible.

BLANK AND SLUG OPENINGS IN DIE SHOES

After a blank or slug has passed through its die opening, it falls through the clearance opening in the die shoe. The following requirements must be met: (1) the blanks or slugs must fall freely, without interference; (2) the contour of the clearance opening in the die shoe should be made as simple as possible; (3) the opening should not weaken the die shoe any more than necessary; and (4) the contour of the die shoe opening must be such that it provides adequate support for the die block.

Various typical die-shoe clearance openings are shown in Fig. 27-3-10.

Straight and Tapered Die-Shoe Openings At times it is difficult to decide whether die-shoe clearance openings should be made straight or tapered (Fig. 27-3-10C and D).

The tapered opening is unquestionably the safest. The straight opening, however, is often easier to make, especially when the die shoe is large and/or thick.

Amount of Taper The amount of taper on the sidewalls of the drop-through opening in a die shoe is not critical. An angle of 0.5 to 2° will satisfy most cases.

Amount of Offset The top edge of the die-shoe opening is offset from the bottom edge of the die opening. The amount of offset is not critical.

ARRANGEMENT OF VIEWS

In laying out the design of a punch and die, an approved arrangement of views

(A) ROUND OPENINGS FOR BLANK OR SLUG PASSAGE THROUGH DIE SHOE

(B) DIE-SHOE CLEARANCE OPENING CONTOURED TO PROVIDE SUPPORT FOR DIE-OPENING CONTOUR

(C) STRAIGHT WALL BLANK OR SLUG CLEARANCE OPENING IN DIE SHOE

(D) TAPERED BLANK OR SLUG CLEARANCE OPENING IN DIE SHOE

Fig. 27-3-10   Die-shoe openings.

is recommended over the theoretical arrangement of views, as illustrated in Fig. 27-3-11. The stock strip is normally drawn in red or shown by phantom or broken lines on the assembly drawing. Standard die sets are shown in Table 86 of the Appendix.

BLANKING AND PUNCHING DIES

Blanking Dies

Figure 27-3-12 is an assembly drawing of the die which produces the blank for the piece part illustrated. This die does the blanking operation only. The holes in the piece part are punched in a subsequent punching die.

Each component of the die is identified by detail number and name. The number required of each component or detail is also shown. Socket-head cap screws are indicated by the abbreviation SHCS.

Operation of Die The stock material is furnished in strip form 2.69 in. wide. It is fed across the die face from right to left until the lead end contacts the stop, at detail 6. The press is then tripped, causing the first piece to be blanked out. After the blanking, the scrap bridge is lifted over the stop. As soon as the bridge passes over the stop, the strip is dropped back to the die face. The head of the stop is now within the previously blanked-out opening. The stock strip is then advanced until the edge of the blanked-out opening contacts the stop. This

process is repeated at every press stroke until each strip is used up.

Fixed Stop If a plain pin (no head) type of fixed stop were used, it would have to be installed at a safe distance from the cut edge in order to avoid weakening the die block. As a result, the scrap bridge would be wider than necessary, which would mean, of course, that stock material was being wasted. In order to minimize the amount of scrap while still maintaining adequate die-block strength, the die shown has a headed fixed-pin stop. The diameter of the head is made rather large in proportion to the shank diameter, which enables the mounting hole to be located a safe distance from the edge of the die opening. The thickness or height of the stop head should be held to a practical minimum in order to make the feeding of the stock strip as easy as possible. The stop is shown in detail in Fig. 27-3-13.

Stripping A gap-type fixed stripper is used. The gap is made relatively high for ease of feeding. The height of the gap and the open front provide visibility and accessibility for the press operator.

Blanking Die Details

The die details are shown in Fig. 27-3-13. For the sake of clarity, only the overall dimensions and the dimensions pertinent to producing the piece part are shown. Abbreviations used are OHTS for oil-hardening tool steel, CRS for cold-rolled steel, and RC60 to

(A) THEORETICAL ARRANGEMENT OF VIEWS

(B) APPROVED ARRANGEMENT OF VIEWS

Fig. 27-3-11   Arrangement of views for punch and die drawings.

Fig. 27-3-12 Blank-through die assembly.

THE PIECE PART
18 (.048) U.S. GAGE

DET	REQ	DESCRIPTION
13	2	DOWEL
12	2	S.H.C.S.
11	1	STOCK REST
10	3	S.H.C.S.
9	1	DOWEL
8	2	S.H.C.S.
7	2	DOWEL
6	1	STOP
5	1	PUNCH
4	1	STRIPPER
3	1	BACK RAIL
2	1	DIE BLOCK
1	1	DIE SET

SCRAP BRIDGE

DIRECTION OF FEED

STOCK WIDTH

④ STRIPPER
1 REQD MATL—CRS

③ BACK GAGE
1 REQD MATL—CRS

⑪ STOCK REST
1 REQD MATL—CRS

② DIE BLOCK 1 REQD
MATL—OHTS RC60—62

⑥ STOP 1 REQD
MATL—CRS
CYANIDE HDN

⑤ PUNCH 1 REQD
MATL—OHTS RC60—62

HOLE DATA "A" HOLES—REAM FOR DOWELS (3)
"B" HOLES—CLEARANCE FOR CAP SCREWS (2)
"C" HOLES—TAPPED FOR MOUNTING SCREWS (3)

Fig. 27-3-13 Details of blank-through die.

NOTE: ONLY DIMENSIONS PERTINENT TO DIE
DESIGN SHOWN ON DETAIL DRAWINGS

62 to indicate the hardness of the hardened two-steel components as measured on the C scale of a Rockwell tester.

Die Block The die block, part 2, is for a blanking die. Therefore, the piece parts produced will be sized by the die opening. The piece part specifies a width of 1.500 + .000, − .005 in. The tolerance is negative (minus). This gives the width dimension a range of 1.495 to 1.500 in. The mean dimension is 1.4975 in. With this in mind, a figure of 1.4975 in. was chosen for the die opening width.

The length of the piece part is 2.500 + .005 − .000 in. This tolerance is positive (plus). The range is 2.500 to 2.505 in., and the mean dimension is, obviously, 2.5025. The die opening figure selected for this dimension is 2.502 in.

The figures chosen give a die opening size of 1.497 × 2.502 in. These figures are the optimum die opening sizes for the given conditions. A die opening made to these figures will yield a blank (or piece part) that will measure very nearly the exact mean of the dimensions specified on the piece-part drawing.

Tolerances are specified on the piece-part drawing. The purpose of the tolerances is to specify a range of acceptable sizes: all piece parts produced within the range are acceptable. This does not mean that the die opening can be made to the same tolerance range as the piece part.

A good choice for a minimum acceptable die opening size is 1.4955 × 2.5005 in. The maximum acceptable die opening should be 1.498 × 2.503 in.

Punch The punch sizes shown on part 5 are 1.493 × 2.498 in., with a corner radius of .123 in. These cutting figures are the result of deducting the cutting clearance from the optimum die opening dimensions. It is standard die design procedure to show punch and die dimensions in this manner.

Back Gage The back gage, part 3, is 1.09 in. wide. This dimension is determined as follows. The stock material width is specified as 2.69 in. One-half of this stock width is 1.345 in. The distance from the center line of the die opening to the back of the die block is 2.44 in.

Subtracting 1.345 from 2.44 leaves a difference of 1.095 in. from the edge of the stock strip to the back of the die block. Using 1.09-in. as the dimension for the width of the back gage aids the mounting of the gage of the die block.

Stripper The stripper opening, part 4, measures 1.56 × 2.56 in. These dimensions give a clearance space of approximately .03 in. on each side between the punch and the stripper.

Punch Die

Figure 27-3-14 is an assembly drawing of a die to punch two holes in the piece part illustrated. This die performs the punching of the holes only; the blank was produced by the blank-through die shown in Fig. 27-3-12. For convenience, the piece part has been shown on each die assembly drawing.

The punch die is set up in the same manner as the blank-through die assembly (Fig. 27-3-12).

Spring Stripper This die has a moving stripper actuated by springs. The designer elected to use this type of stripper for reasons of accessibility for the press operator. Using a spring stripper for this die also ensures a flatter piece part after piercing, because the stripper pressure holds the part against the face of the die block while the cutting action is taking place.

Instruction Plate Note that the drawing calls for an instruction plate. The plate, which reads LOAD PART BURR SIDE DOWN, must be clearly lettered and located at the front of the die where it can be easily seen by the operator. The reason for the plate is that the piece-part drawing specifies a burr side, and because the blank is symmetrical, there is no way to foolproof the loading of the blank in the die. As a result, the press operator must be instructed to feed the part into the die in such a manner that the burr

side of the blank and the burr side of the pierced holes will agree, as specified.

Locating and Feeding Proper location of the pierced holes in the piece part is achieved by a nest area created by the nest pins, part 9. When the blank is placed in the nest area, it is confined between the nest pins, which are correctly located with respect to the die openings and punches.

Cutting Clearance When the cutting clearance for the blank-through die was determined, the clearance was found to be .002 in. per side. However, for the punching die, owing to the fact that the punchings are round, a cutting clearance of .0015 in. per side was chosen.

Punch Die Details

Figure 27-3-15 is the detail sheet for the punch die. Major overall sizes and dimensions associated directly with the piece-part requirements are shown.

Punches Refer to Fig. 27-3-15, parts 7 and 8. One of the openings in the piece part is dimensioned .502 + .003, − .000 in. diameter. For this opening, a punch diameter of .504 in. was decided upon. A punch which has a point or working diameter of .504 in. will produce an opening in the piece part of .5035 to .504 in., which is optimum for this particular opening.

The other opening is specified as .688 ± .005 in. diameter. Since the mean dimension for this figure is .688 in., a punch diameter of .688 in. is used.

It is an accepted practice to make the shank diameters of round punches in .062-in. increments up to .50-in. diameter and in .125 in. increments for diameters larger than .50 in. Using diameters of these size increments means that the punch-plate openings can be sized by reamers that are common toolroom equipment. Following this practice led to a choice of .625- and .750-in. diameters for the respective punch shanks.

The punch shanks are fitted to the punch plate. The proper fit is a light tap fit, not a press fit.

Die Block Since this die block, part 2, is for a punching die, the size of the die openings is derived from the punches.

DET	REQD	DESCRIPTION
13	1	SHEDDER PIN
12	2	S.H.C.S.
11	4	DOWEL
10	3	S.H.C.S
9	5	NEST PIN
8	1	PUNCH
7	1	PUNCH
6	1	PUNCH PLATE
5	4	SPRING
4	4	S.H.S.S.
3	1	STRIPPER
2	1	DIE BLOCK
1	1	DIE SET

THE PIECE PART

MATL— 18 (.048) HARD BRASS

BURR SIDE

LOAD PART BURR SIDE DOWN

Fig. 27-3-14 Punch die assembly.

Fig. 27-3-15 Punch die details.

HOLE DATA:

A—REAM FOR NEST PINS (5)
B—REAM FOR DOWELS (2)
C—TAP FOR MOUNTING SCREWS (3)
D—CLEARANCE HOLES FOR NEST PINS (5)
E—CLEARANCE HOLE FOR TOP OF SHREDDER PIN (1)

The punch diameters are .688 and .504 in. A cutting clearance of .0015 in. per side is used. The die opening diameters are, therefore, .691 and .507 in.

The center-to-center distance between the die openings is taken directly from the piece-part specifications, which give 1.000 + .005, − 000 in. Thus the dimensional range is 1.000 to 1.005 in. For this range the exact mean dimension is 1.0025 in.

After the optimum die opening size and location are decided, the next consideration is to ensure proper location of the pierced openings in relation to the contour of the piece part. This is the function of the nest pins, part 9. In order to serve their purpose, the nest pins must be accurately located. The nest must fit the blank closely enough to ensure the required accuracy of location. Some clearance, however, should exist between the nest pins and the blank to permit ease of loading and unloading. The amount of clearance to allow, of course, depends upon the accuracy requirements of the piece part. In this particular case, a clearance of .0005 to .001 in. between the edges of the blank and the ∅.250 in. nest pins is satisfactory. Thus the center-to-center distances between the nest pins are 1.751 (1.500 + .250

+ .001) and 2.756 (2.505 + .250 + .001) in.

Punch Plate Refer to Fig. 27-3-15·part 6. The .750- and .625 in. diameters are chosen because a light tap fit is used.

The center-to-center distance is picked up or transferred from the hardened die block. This procedure is followed to eliminate the possibility of minor alignment discrepancies which might result from slight dimensional changes caused by the die-block heat treatment.

The counterbores which receive the heads of the punches are made larger than the punch heads. Clearing for the heads in this way eliminates the possibility of misalignment resulting from contact between the head of the punch and the sidewall of the counterbore.

Foolproofing is accomplished by using dowels of differing diameters.

A shedder or spring pin, part 13, is provided in the stripper. It is a standard purchased item, the length of which cannot be altered.

Stripper The diameters of the stripper openings (Fig. 27-3-15, part 3) are .562 and .750 in., respectively. These figures give a clearance of approximately .03 in. per side between the punches and the stripper. So much clearance is

permissible because of the stock material thickness and is desirable from the die maker's viewpoint because it eliminates the need for precise location and sizing of the stripper openings.

Spring strippers such as this one must have shedder pins to prevent the possibility of a piece part adhering to the stripper. In this stripper, the shedder pin, part 13, is a standard purchased item. It contains a pin which is spring-loaded. A suitable tapped hole is provided in the stripper to receive this unit, which is shown in assembly with the stripper.

Reference

1. From *Basic Diemaking* by the National Tool, Die, and Precision Machining Association. Copyright, 1963. McGraw-Hill Book Company. Used by permission.

ASSIGNMENTS

See Assignments 4 through 6 for Unit 27-3 on page 552.

Review for Assignments

Unit 8-3   Cap Screws
Unit 9-2   Pin Fasteners
Unit 9-4   Springs

ASSIGNMENTS for Chapter 27

Assignments for Unit 27-1, Die Design

1. On a B- or A3-size sheet, lay out a standard die assembly for the part shown in Fig. 27-1-A or 27-1-B. Wherever it is practical, use standard parts as shown in Fig. 27-1-6. Select suitable stops and guides. The punch need not be shown. Show the stock strip slug and piece part in a second color. Scale is half size (1:2).
2. On a B- or A3-size sheet, lay out a standard punch and die assembly for the piece part shown in Fig. 27-1-C. The piece part has already been punched out of the stock strip, and the part must be positioned for the hole to be punched. A nest gage similar to the type shown in Fig. 27-1-15 is recommended. Show the piece part in a second color. Scale is full (1:1).

Assignment for Unit 27-2, Nesting of Blanks

3. On a B- or A3-size sheet, design stock strips for the parts shown on Fig. 27-2-A or 27-2-B. Dimension the width of the stock strip. Scale is full (1:1).

Assignments for Unit 27-3, Die Sets and Components

4. On a B- or A3-size sheet, design suitable punches for the blanking and the punching for the two parts shown in Fig. 27-3-A or 27-3-B. The die set to be used is B1-361 or B1-361M, as shown in Table 86 in the Appendix. Draw only the front and bottom views of the punch. Scale is full (1:1).
5. On a B- or A3-size sheet, design a suitable die and stripper plate for one of the two parts shown in Fig. 27-3-C or 27-3-D. The die set to be used is B1-361 or B1-361M, as shown in Table 86 in the Appendix. Draw the front, top, and side views of the die and stripper plate. Scale is full (1:1).
6. On a B- or A3-size sheet, draw the details for the blank-through die in assignment 5. Show all the details, but only those dimensions pertinent to the blanking design need be shown on the detail drawings. Scale is to suit.

Fig. 27-1-A Stop.

Fig. 27-1-C Nest gage.

Fig. 27-1-B Guide.

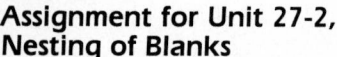

Fig. 27-2-A Nesting of blanks.

.15 THICK   .12 THICK   .10 THICK

3.04 THICK

1.9 THICK

2.66 THICK

Fig. 27-2-B Nesting of blanks.

Fig. 27-3-A   Punch and blank designs.

Fig. 27-3-C   Die and stripper plate design.

METRIC

Fig. 27-3-B   Punch and blank designs.

METRIC

Fig. 27-3-D   Die and stripper plate design.

PART 6

Advanced Drafting Design

CHAPTER 28

Applied Mechanics

UNIT 28-1
Forces

Weight (mass) and density have many applications in industry and construction. A few examples are

- Defining quantities of materials, such as bags of mortar and tons of steel.
- Defining physical characteristics of materials like steel beams (W24 × 94), 210-lb asphalt shingles, and 3.5-lb density board.
- Defining load capacity for supports, bridges, cranes, and elevators.

A *force* is that which changes, or tends to change, the state of rest or uniform motion of a body.

U.S. Customary Units
In the U.S. Customary System, the force values are given in pounds, kips, and tons. The methods for solving problems in either metric or U.S. Customary units are identical.

Metric Units
The term mass (not weight) is used to refer to the quantity of matter in an object (rather than to the force of gravity acting on it). Mass is always measured in terms of the kilogram, gram, or some related unit; that is, a multiple or submultiple of the gram. Whenever a quantity is specified in such a unit, mass, not force, is the quantity under consideration.

The force of gravity, pulling downwards, acts on each mass on the Earth. This "mass" and the "force of gravity," while having entirely different characteristics, are linked. The word "weight" in the metric system is avoided, since it is ambiguous; it sometimes means "mass" and sometimes "force of gravity," depending on the context.

Force is measured in newtons. The preferred units are newtons (N) for small forces, kilonewtons (kN) for intermediate forces and meganewtons (MN) for large forces. One newton is approximately one-quarter of a one-pound force.

Using a spring scale, it would take about 10 N to lift a 1-kg stone, perhaps 4 N to pull it along the floor, and maybe 6.5 N to pull it up a ramp.

In this chapter, the force of gravity acting on the part, rather than the mass of the part is given in the examples and problems.

GRAPHIC REPRESENTATION OF FORCES

The graphical method of solving mechanical problems involving forces is often used because it is quick and accurate. The force is shown graphically. To describe completely the force, the following particulars must be given

1. Its magnitude
2. Its point of application
3. Its direction
4. Its sense, i.e., whether it is pushing or pulling

A line is drawn to a given length to represent the magnitude of the force. The direction of this line is parallel to the direction of the force. The sense of the force is indicated by an arrow on the line indicating whether it is acting toward or away from the point of application. The graphical representation of the force is called a *vector*.

Thus a pull of 6 tons (T) acting at a point *A* at 45° to the horizontal would be represented by the vector *AB*, as shown in Fig. 28-1-1. Using the scale .25 in. = 1 T, the length of the vector would be 1.50 in.

Fig. 28-1-1   A vector.

A body is said to be in equilibrium if the forces acting at a point balance one another. If two equal and opposite forces act at a point in a straight line, the body is in equilibrium. Examples are tie bars, which are bars under pull or tension, and struts or columns, which are bars under push or compression (Fig. 28-1-2).

TWO FORCES ACTING AT A POINT

Two or more forces acting at a point may be replaced by one force that will

Fig. 28-1-2 Types of forces acting on supports.

FORCE   FORCE   FORCE

REACTION OF
SUPPORTS ARE
EQUAL BUT OPPOSITE
IN DIRECTION TO FORCE

FORCE   FORCE   FORCE

TIE BAR   STRUT   COLUMN
(TENSION)   (COMPRESSION) (COMPRESSION)

arrows representing the direction of the forces are pointing the same way around the triangle.

MORE THAN TWO FORCES ACTING AT A POINT

Resultants or equilibrants may be found for any number of forces acting at a point and in one plane. Let A, B, C, and D represent forces acting at a point O, shown in Fig. 28-1-6A.

Using the parallelogram of forces method shown in Fig. 28-1-6B, we find the resultant R_1 for forces A and B and resultant R_2 for forces C and D. Using resultants R_1 and R_2 in Fig. 28-1-6C instead of the forces A, B, C, and D, we find the resultant R of the four forces. The equilibrant or force required to keep the forces A, B, C, and D in equilibrium would be equal to R but would act in the opposite direction.

POLYGON OF FORCES

Using the polygon of forces method shown in Fig. 28-1-7, which is the

produce the same effect. This force is called the *resultant* of the forces. If two opposite forces of 8 and 5 T act at a point O in a straight line, as in Fig. 28-1-3, a resultant force of 3 T acting in the same direction as the 8 T force could replace the two original forces.

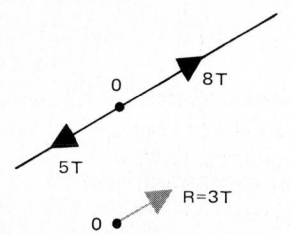

Fig. 28-1-3 Resultant when two forces act in a straight line.

If two opposite forces F_1 and F_2 act at point O at angles of 120° to each other, as in Fig. 28-1-4, the resultant force R may be found by drawing the two forces to scale and completing the parallelogram. The diagonal OC would be the resultant, and the magnitude of the force could be measured. This method is called the *parallelogram of forces*. Accuracy of direction and distance is important in laying out forces.

Another way of finding the resultant is the *triangle of force* method. The known force vectors are laid end to end with the forces traveling in the same direction. The resultant R is found by joining the beginning of the first vector to the end of the last vector, as shown in Fig. 28-1-5A, and the

direction of the resultant force is in the combined direction of the other two forces.

If a force equal to the resultant of forces F_1 and F_2, but acting in the opposite direction, was to act at O, as shown in Fig. 28-1-5B, the object would be in equilibrium, since the forces acting at point O tend to balance one another. This force balancing the other forces is known as the *equilibrant*.

The equilibrant is found in a similar manner to the resultant, by using the triangle of force method. Note that the

Fig. 28-1-4 Parallelogram of forces.

(A) TRIANGLE OF FORCE METHOD FOR FINDING RESULTANT

(B) FORCES IN EQUILIBRIUM

(C) TRIANGLE OF FORCE METHOD FOR FINDING EQUILIBRANT

Fig. 28-1-5 Resultant and equilibrant of two forces acting at a point using the triangle of force method.

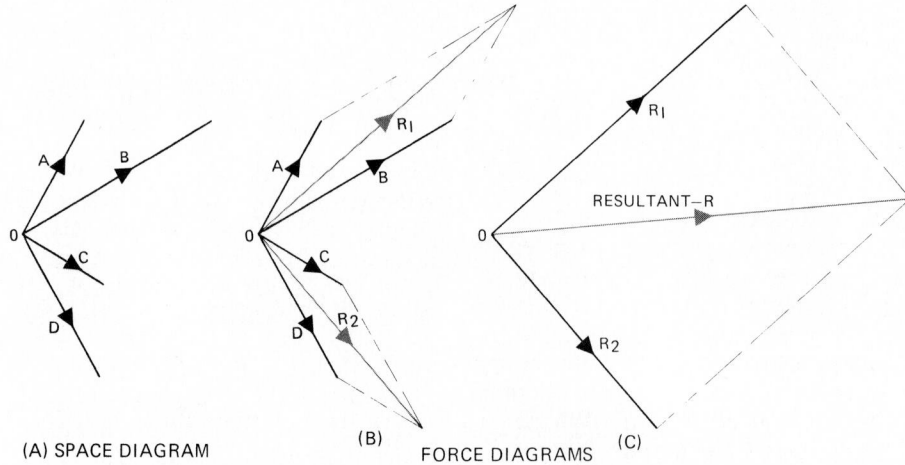

(A) SPACE DIAGRAM   (B) FORCE DIAGRAMS   (C)

Fig. 28-1-6 Parallelogram of forces method of finding resultant of R for more than two forces acting at a point.

(A) SPACE DIAGRAM   (B) FORCE DIAGRAM

Fig. 28-1-7   Polygon of forces method for finding the equilibrant for more than two forces acting at a point.

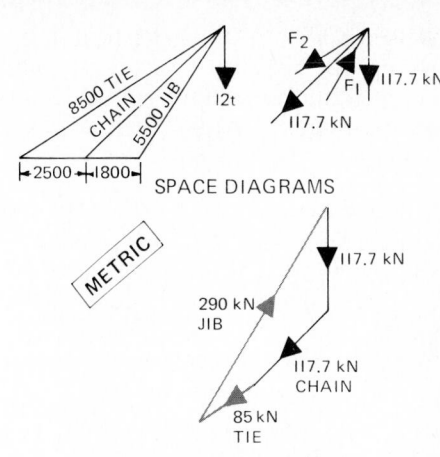

SPACE DIAGRAMS

METRIC

FORCE DIAGRAM

Fig. 28-1-10   Solution to Example 3 by the force diagram method.

extension of the triangle method, join the forces A, B, C, and D end to end to form a polygon. Be careful to keep the arrows pointing the same way around. The line joining the beginning of the first force and the end of the last force is the equilibrant.

The following examples illustrate how vector diagrams are applied to practical problems.

EXAMPLE 1 A crane lifts a steel boiler by means of a chain sling. The sling makes angles of 30 and 45° with the boiler. Find the forces acting on the sling if the weight of the boiler is 8 tons.

Solution The force on the crane's chain supporting the sling is equal to the force created by the weight of the boiler. The angles on the sling indicate the direction of the sling forces F_1 and F_2, as shown in the space diagram in Fig. 28-1-8. The force diagram is then drawn to a suitable scale and the lengths of lines F_1 and F_2 are measured to find the magnitude of the forces acting on the sling.

EXAMPLE 2 In Fig. 28-1-9, a simple wall crane has a 400 lb load applied at the end of the jib. Neglecting the weight of the crane parts, calculate the forces acting on the tie and jib.

Solution A space diagram is drawn first to find the direction of the force acting on the tie. With the direction of the three forces and the magnitude of one force W known, a force diagram is then drawn to a suitable scale. The length of lines F_1 and F_2 can now be measured to find the magnitude of the forces acting on the jib and tie.

EXAMPLE 3 In Fig. 28-1-10, a machine having a mass of 12 tonnes (t) is lifted by a jib crane. The length of the jib is 5500 mm and an 8500-mm tie is fastened to a point 4300 mm behind the base of the jib. The chain that lifts the machine passes over a pulley at the top of the jib and connects to a winch located 1800 mm behind the base of the jib. Find the forces acting on the jib and tie.

Solution A 1-t mass creates a gravitational force of 9806.65 N. The 12-t

mass exerts a force of 12×9806.65 N $= 117\ 700$ N or 117.7 kN on the chain. A space diagram is drawn first to find the direction of the forces acting on the tie, jib, and chain connected to the winch. With the direction of the four forces and the magnitude of two forces known, a force diagram is then drawn to a suitable scale. The length of lines F_1 and F_2 can now be measured to find the magnitude of the forces acting on the jib and tie.

EXAMPLE 4 In Fig. 28-1-11, a simple roof truss has a force W applied at the top. Find the forces acting on the rafters, tie bar, and walls.

Solution A space diagram is drawn first to find the direction of the force acting on the rafters. A force diagram

SPACE DIAGRAMS

FORCE DIAGRAM

Fig. 28-1-8   Soluiton to Example 1 by the force diagram method.

SPACE DIAGRAMS

FORCE DIAGRAM

Fig. 28-1-9   Solution to Example 2 by the force diagram method.

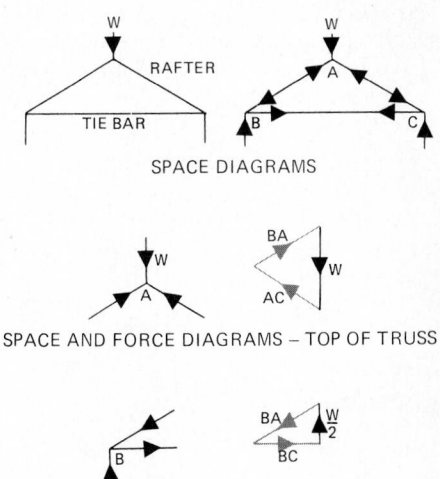

SPACE DIAGRAMS

SPACE AND FORCE DIAGRAMS – TOP OF TRUSS

SPACE AND FORCE DIAGRAMS AT TOP OF WALL

Fig. 28-1-11   Solution to Example 4 by the force diagram method.

taken at point *A* is drawn as the rafters *AB* and *AC* support the load *W*. The forces acting on rafters *AB* and *AC* may be found by making a force diagram taken at point *A*, drawing *W* to scale, and measuring the values of *AB* and *AC*.

Since the force acting on the wall is equal to *W*/2 (the roof design and load being symmetrical) and the forces acting at points *B* and *C* being equal, only one force diagram need be drawn to find the force acting on the tie.

ASSIGNMENTS

See Assignments 1 through 3 for Unit 28-1 on page 566.

UNIT 28-2
Beams

BOW'S NOTATION

In the previous illustrations, the forces have been identified as F_1, F_2, *R*, etc. Another system of identifying forces, called *Bow's notation*, is helpful in solving force problems. In the space diagram (Fig. 28-2-1), a boldface capital letter, **A**, **B**, **C**, etc., is placed in the space between two forces and the force is referred to by the two boldface capital letters in the adjoining spaces. The force **AB** in the space diagram is represented by the vector *ab* in the force diagram, the letters *a* and *b* being placed at the beginning and end, respectively, of the vector. The letters in the space diagram are usually given in alphabetical order and in a clockwise direction.

EQUILIBRIUM POLYGON

Equilibrium or *funicular polygons* are used in the graphical solutions for find-

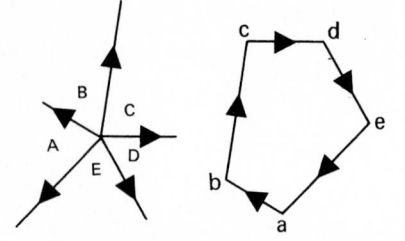

(A) SPACE DIAGRAM (B) FORCE DIAGRAM
Fig. 28-2-1 Bow's notation.

(A) SPACE DIAGRAM

NOTE: THE FORCES SHOWN MAY BE U. S. CUSTOMARY OR METRIC UNITS SUCH AS POUNDS, TONS, NEWTONS, KILONEWTONS, ETC.

RESULTANT PASSES THROUGH INTERSECTION OF LINES *oa* AND *of*

EQUILIBRIUM POLYGON

FINDING POSITION OF RESULTANT
(C) SPACE DIAGRAM
REPEATED FOR CLARITY

(I) FINDING RESULTANT

FORCE DIAGRAM BECOMES A STRAIGHT LINE WHEN FORCES ARE PARALLEL

FINDING MAGNITUDE AND DIRECTION OF RESULTANT
(B) FORCE DIAGRAM

SCALE LINES *ag* AND *gf* TO FIND MAGNITUDE OF REACTIONS

EQUILIBRIUM POLYGON

FINDING DIRECTION OF LINE *og*
(D) SPACE DIAGRAM

DRAW LINE *og* AFTER FINDING ITS DIRECTION ON THE EQUILIBRIUM POLYGON

(E) FORCE DIAGRAM

(II) FINDING REACTIONS

Fig. 28-2-2 Graphic method of finding resultant and reactions of vertical forces acting on a beam.

ing the magnitude, direction, and point of application of resultants, equilibrants, and reactions. They are also used to check whether or not a number of forces are in equilibrium.

Graphic Method of Finding Resultant and Reactions of Vertical Forces Acting on a Beam

EXAMPLE 1 A number of parallel forces are acting on a beam as shown in Fig. 28-2-2A. It is required to find, graphically, the magnitude and point

of application of the resultant and the magnitude of the reactions.

Solution First draw a space and force diagram using Bow's notation. Note that the force diagram is a straight line and not a polygon when all the forces are parallel. The magnitude of the resultant is found by measuring the distance from *a* to *f* on the force diagram. The direction of the resultant is also established, but its position with respect to the six forces is still unknown. To find its position, locate a point *o* anywhere on the force diagram and join *o* to each letter with a line.

NOTE: THE FORCES SHOWN MAY BE U. S. CUSTOMARY OR METRIC UNITS SUCH AS POUNDS, TONS, NEWTONS, KILONEWTONS, ETC.

(A) SPACE DIAGRAM

RESULTANT PASSES THROUGH INTERSECTION OF LINES oa AND oe

EQUILIBRIUM POLYGON

R

FINDING POSITION OF RESULTANT

(C) SPACE DIAGRAM
(REPEATED FOR CLARITY)

R=228

FINDING MAGNITUDE AND DIRECTION OF RESULTANT

(B) FORCE DIAGRAM

Fig. 28-2-3   Graphic method of finding resultant of nonparallel forces acting on a beam.

Draw a line anywhere in space **B** of the space diagram (Fig. 28-2-2C) parallel to line *ob* in the force diagram, until it intersects forces **AB** and **BC**. Next draw a line parallel to *oc* in space **C** but starting where line *ob* intersects force **BC**. This is continued until the equilibrium polygon is completed. The resultant *R* passes through the intersection of lines *oa* and *of*, thus determining its position. The polygon constructed in the space diagram is called an equilibrium or funicular polygon.

The magnitude of the reaction forces isd found by making an equilibrium polygon (see Fig. 28-2-2D), which includes reaction forces **AG** and **FG**. These forces were not needed to find the resultant. To construct the equilibrium polygon, draw a line *oa* anywhere in space **A** and parallel to line *oa* in the force diagram, as shown in Fig. 28-2-2E. Repeat for lines *ob*, *oc*, *od*, and *oe*, having each of these lines touch each other as shown.

Line *of* will not have any length on the equilibrium polygon since forces **EF** and **FG** act in the same line. Draw line *og* in the space diagram by joining the start of line *oa* to *of*. Now draw line *og* on the force diagram parallel to line *og* in the space diagram. The magnitude of the reactions (two vertical upward forces supporting the beam) **GA** of 122 units and **FG** of 150 units may be found by measuring lines *ag* and *gf* on the force diagram. Note, the units for the forces may be U.S. Customary or metric such as pounds, tons, newtons, kilonewtons, etc.

Graphic Method of Finding Resultant of Nonparallel Forces Acting on a Beam

EXAMPLE 2 A number of nonparallel forces are acting on a beam as shown in Fig. 28-2-3A. It is required to find, graphically, the magnitude and the point of application of the resultant.
Solution To find the resultant, first draw a space and force diagram and label the forces, using Bow's notation. The magnitude and the direction of the resultant are shown by line *ae* on the force diagram (Fig. 28-2-3B). The location of the resultant with respect to the four forces must now be found. Locate

a point *o* anywhere on the force diagram and join *o* to each letter with a line. Draw a line *ob* anywhere in space **B** of the space diagram (Fig. 28-2-3C), parallel to line *ob* in the force diagram. Next draw a line parallel to *oc* in space **C** but starting where line *ob* intersects force **BC**. This is continued until the equilibrium polygon is completed. The resultant *R* is then drawn through the intersection of lines *oa* and *oe*, thus determining its position.

ASSIGNMENTS

See Assignments 4 through 7 for Unit 28-2 on page 566.

UNIT 28-3
Truss Reactions When Loads Are Parallel

BRIDGE AND ROOF TRUSSES

The graphical solution offers a quick and convenient method of checking or determining truss calculations. Some of the more common types of roof and bridge trusses are shown in Fig. 28-3-1. The loads that the truss and abutments support are combinations of the mass of the truss and the mass of the materials placed on the roof or truss, for example, snow, wind, and live loads

(I) HOWE TRUSS

(2) WARREN TRUSS

(3) PRATT TRUSS

(4) HOWE TRUSS

(5) FINK TRUSS

(6) CRESCENT TRUSS

(7) FAN TRUSS

(8) CAMBERED FINK TRUSS

(9) BOW-STRING HOWE TRUSS

Fig. 28-3-1   Common types of trusses.

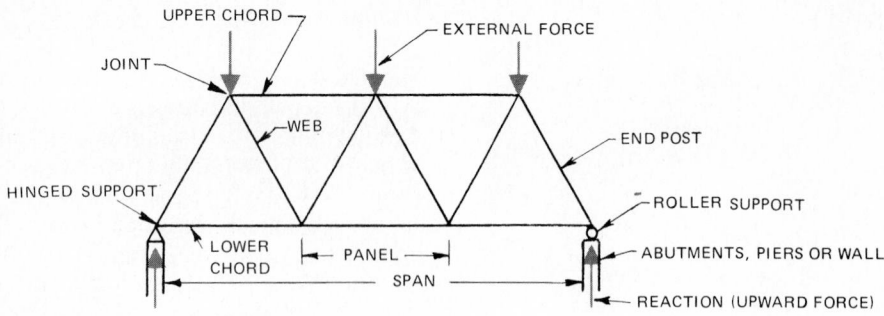

Fig. 28-3-2   Truss terminology.

such as cars and trains. The upward forces of the abutments or walls are called the *reactions*. Rollers are often used at one support of the truss to allow for expansion and contraction. Truss terminology is shown in Fig. 28-3-2.

GRAPHIC METHOD OF FINDING TRUSS REACTIONS WHEN LOADS ARE VERTICAL

Consider the truss in Fig. 28-3-3A. The forces acting on the truss are vertical downward forces, while the reaction forces are vertical upward forces, since they must be equal and opposite to the truss forces. The lines of action and directions of all forces are known, and only the magnitude of the reactions must be computed. Draw a space diagram and label the forces using Bow's notation. Next draw the force diagram. Note that the force diagram becomes a straight line when all the forces are parallel. The magnitudes of the reactions **FG** and **GA** are not yet known, but their combined magnitude is equal to 1100 units, the length of line *af* on the force diagram. Locate a point *o* anywhere on the force diagram and join *o* to points *a* and *f* with a broken line.

Draw a line *ob* parallel to line *ob* in the force diagram, anywhere in space **B** of the space diagram (Fig. 28-3-3D) until it intersects forces **AB** and **BC**. Draw a line parallel to *oc* in space **C**, but start where the line *ob* intersects force **BC**. Repeat for lines *od* and *oe*. Line *oa* will not have any length on the equilibrium polygen since forces **AB** and **GA** act in the same line. The same is true for line *of* since forces **EF** and **FG** act in the same line. Close the polygon by joining *oa* to *of* with line *og*. Draw line *og* on the force diagram par-

allel to the line *og* on the space diagram. The values of the reactions **GA** of 625 units and **FG** of 475 units may be found by measuring lines *ga* and *fg* on the force diagram.

GRAPHIC METHOD OF FINDING TRUSS REACTIONS WHEN LOADS ARE NOT VERTICAL

When the direction of the resultant of the roof and wind load is not vertical, the reactions are not normally parallel.

The solution for finding the magnitude and direction of the reactions when the truss has fixed supports is based on the assumption that the hori-

zontal components of the reactions are equal, since the truss is assumed to be rigid and rigidly held to the supports.

Consider the truss shown in Fig. 28-3-4A. The forces acting on the truss are inclined downward forces, while the reaction forces are inclined upward forces. The lines of action and direction of all the downward forces are known and only the magnitude and direction of the reactions need to be computed. Draw a space diagram and label the forces using Bow's notation. Next draw the force diagram as shown in Fig. 28-3-4C. The resultant of the downward forces, line *ae* on the force diagram, is equal but opposite to the resultant of the reaction forces.

The magnitude and direction of the reactions **EF** and **FA** are not yet known, but the magnitude of their resultant is equal to 950 units. Locate a point *o* anywhere on the force diagram and join *o* to points *a* and *e* with a line. Draw a line *ob* parallel to line *ob* on the force diagram anywhere in space **B** of the space diagram (Fig. 28-3-4D) until it intersects forces **AB** and **BC**.

Draw a line parallel to *oc* in space **C** starting where line *ob* intersects force **BC**. Repeat for lines *od* and *oe*. Line *oa* will not have any length on the equi-

NOTE: THE FORCES SHOWN MAY BE U.S. CUSTOMARY OR METRIC UNITS SUCH AS POUNDS, TONS, NEWTONS, KILONEWTONS, ETC.

(A) TO FIND MAGNITUDE AND DIRECTION OF TRUSS REACTION

(B) SPACE DIAGRAM SHOWING ALL THE EXTERNAL FORCES

(D) SPACE DIAGRAM REPEATED FOR CLARITY

FINDING DIRECTION OF LINE *og*

FORCE DIAGRAM BECOMES A STRAIGHT LINE WHEN FORCES ARE PARALLEL

DRAW LINE *og* AFTER FINDING ITS DIRECTION ON THE EQUILIBRIUM POLYGON

FINDING MAGNITUDE OF REACTIONS

(C) FORCE DIAGRAM

Fig. 28-3-3   Graphic method of finding truss reactions when loads are vertical.

WIND LOADS

(A) TO FIND MAGNITUDE AND DIRECTION OF TRUSS REACTIONS

FORCE DIAGRAM BECOMES A STRAIGHT LINE WHEN FORCES ARE PARALLEL

DRAW LINE of AFTER FINDING ITS DIRECTION ON THE EQUILIBRIUM POLYGON

FINDING RESULTANT OF REACTIONS

(C) FORCE DIAGRAM

NOTE: THE FORCES SHOWN MAY BE U. S. CUSTOMARY OR METRIC UNITS SUCH AS POUNDS, TONS, NEWTONS, KILONEWTONS, ETC.

EQUILIBRIUM POLYGON

FINDING DIRECTION OF LINE of

(D) SPACE DIAGRAM (REPEATED FOR CLARITY)

REACTION FORCE UNKNOWN

REACTION FORCE UNKNOWN

MAGNITUDE UNKNOWN, BUT ASSUME DIRECTION OPPOSITE TO WIND LOADS

(B) SPACE DIAGRAM SHOWING ALL THE EXTERNAL FORCES

Fig. 28-3-4   Graphic method of finding truss reactions when loads are not vertical.

FINDING MAGNITUDE AND DIRECTION OF REACTIONS ASSUMING HORIZONTAL COMPONENTS OF REACTIONS TO BE EQUAL

(E) FORCE DIAGRAM (REPEATED FOR CLARITY)

librium polygon since forces **AB** and **FA** act in the same line. Close the equilibrium polygon by joining *oa* and *oe* with line *of*. Draw line *of* on the force diagram parallel to line *of* on the space diagram. The values of 625 units and 325 units may be found by measuring lines *fa* and *ef* on the force diagram. These values are the individual resultants of reactions **EF** and **FA**. Since the horizontal components of the reactions are assumed to be equal, the magnitude and direction of reactions **FA** and **EF** can be found by completing the force diagram as shown in Fig. 28-3-4E.

Distance *X* represents the horizontal component of the combined reaction forces. Since the truss is rigidly held to the supports, we may assume that each support will take half the horizontal force. Therefore *X*/2 represents the horizontal component of each reaction. From point *f* on line *ae*, extend a horizontal line until it intersects the vertical line bisecting distance *X* at point f_1. The values of the reactions **FA** of 590 units and **EF** of 375 units may be found by measuring lines f_1a and ef_1 on the force diagram. The directions of the reaction forces will be parallel to lines f_1a and ef_1.

REACTION OF HINGED-PIN AND ROLLER SUPPORTS

Normally in bridge and roof truss design, one end of the truss is supported by a hinged pin and the other

end rests on a roller. This provides for changes in the length of the truss because of temperature changes. The reaction of the roller support is taken through the roller and is usually perpendicular to the path of the roller. The reaction of the hinged-pin support is equal to the direction and size of force required to keep the structure in equilibrium. If the forces acting on the bridge or roof are vertical, then we can assume that the reactions of the hinge and rollers are also vertical. If the resultant of the forces acting on the bridge or roof is inclined due to wind loads and the reaction at the roller support is vertical, then the reaction at the hinged-pin support must be inclined.

GRAPHIC METHOD OF FINDING TRUSS REACTIONS WHEN LOADS ARE PARALLEL FOR HINGED-PIN AND ROLLER SUPPORT

Consider the truss shown in Fig. 28-3-5A. The forces acting on the truss are inclined downward forces caused by wind loads, a vertical upward force through the center of the roller, and an inclined upward force at the hinged-pin support. It is required to find the magnitude of the reactions and the direction of the reaction at the hinged-pin support.

Draw a space diagram (Fig. 28-3-5B), and label all the forces using

Bow's notation. Draw wind forces **AB**, **BC**, **CD**, and **DE** on the force diagram. Since the direction of reaction force **EF** is known, its direction can be drawn on the force diagram. Its length or magnitude is not known. Reaction force **FA** cannot be drawn since both its magnitude and direction are not known, but its line of action passes through the point of support. Thus the equilibrium polygon is started at the pin support. Locate a point *o* anywhere on the force diagram and join *o* to points *a* and *e*. Draw a line *ob* parallel to line *ob* on the force diagram in space **B** of the space diagram (Fig. 28-3-5D) starting at the pin support until it intersects force **BC**. Draw a line parallel to *oc* in space **C** but starting where line *ob* intersects force **BC**. Repeat for lines *od* and *oe*. Line *oa* will not have any length on the equilibrium polygon since forces **AB** and **FA** act at the same point. Close the equilibrium polygon by joining *oa* to *oe* with line *of*. Draw line *of* on the force diagram parallel to line *of* on the space diagram until it intersects reaction force *ef* at point *f*. Close the force poly-

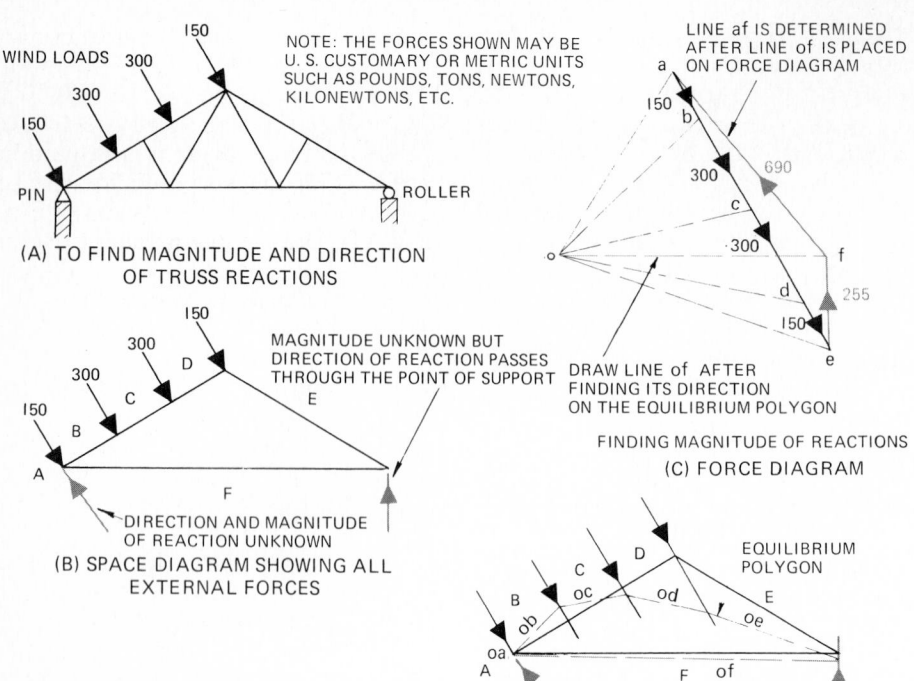

NOTE: THE FORCES SHOWN MAY BE U. S. CUSTOMARY OR METRIC UNITS SUCH AS POUNDS, TONS, NEWTONS, KILONEWTONS, ETC.

WIND LOADS

PIN   ROLLER

(A) TO FIND MAGNITUDE AND DIRECTION OF TRUSS REACTIONS

MAGNITUDE UNKNOWN BUT DIRECTION OF REACTION PASSES THROUGH THE POINT OF SUPPORT

DIRECTION AND MAGNITUDE OF REACTION UNKNOWN

(B) SPACE DIAGRAM SHOWING ALL EXTERNAL FORCES

LINE af IS DETERMINED AFTER LINE of IS PLACED ON FORCE DIAGRAM

DRAW LINE of AFTER FINDING ITS DIRECTION ON THE EQUILIBRIUM POLYGON

FINDING MAGNITUDE OF REACTIONS

(C) FORCE DIAGRAM

EQUILIBRIUM POLYGON

FINDING DIRECTION OF LINE of

(D) SPACE DIAGRAM (REPEATED FOR CLARITY)

Fig. 28-3-5 Graphic method of finding truss reaction when loads are parallel for hinged-pin and roller support.

the direction of the reaction at the hinged support can be found. The lines of action of the resultant wind loads and the vertical reaction force are extended until they intersect at point *O* (Fig. 28-3-6B). Since the lines of action of all three forces must pass through point *O*, a line joining point *O* to the point of intersection at the hinged support determines the direction of the left reaction.

Knowing the direction of the three forces and the magnitude of the resultant, the magnitudes of the truss reactions can readily be found by making a force diagram and measuring lines *ca* and *bc*.

ASSIGNMENTS

See Assignments 8 and 9 for Unit 28-3 on page 567.

gon with line *fa*. The direction of the hinged reaction force **FA** is parallel to line *fa*. The values of the reactions **EF** of 255 units and **FA** of 690 units may be found by measuring lines *ef* and *fa* on the force diagram.

ALTERNATIVE GRAPHIC METHOD OF FINDING TRUSS REACTIONS WHEN LOADS ARE PARALLEL FOR HINGED-PIN AND ROLLER SUPPORT

An alternative method can be employed for finding the truss reactions when the reaction forces at the pin and roller and the load force are not parallel. Consider the forces acting in Fig. 28-3-6A. The four wind forces of 150, 300, 300, and 150 units can be replaced by a resultant force of 900 units acting midway on the truss since the loads are symmetrical. The direction of the reaction at the hinge support is not known, but the direction of the reaction at the roller support will be vertical. Since the lines of action of any three nonparallel forces in equilibrium intersect at a common point,

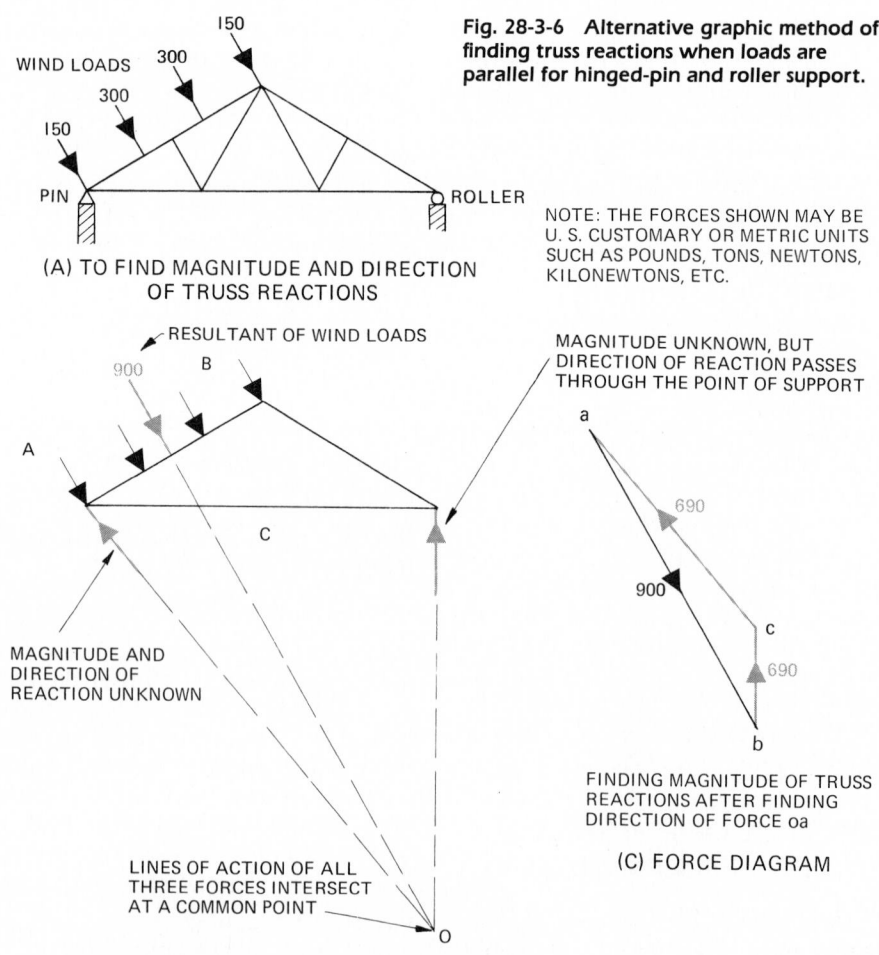

Fig. 28-3-6 Alternative graphic method of finding truss reactions when loads are parallel for hinged-pin and roller support.

WIND LOADS

PIN   ROLLER

(A) TO FIND MAGNITUDE AND DIRECTION OF TRUSS REACTIONS

NOTE: THE FORCES SHOWN MAY BE U. S. CUSTOMARY OR METRIC UNITS SUCH AS POUNDS, TONS, NEWTONS, KILONEWTONS, ETC.

RESULTANT OF WIND LOADS

MAGNITUDE AND DIRECTION OF REACTION UNKNOWN

LINES OF ACTION OF ALL THREE FORCES INTERSECT AT A COMMON POINT

(B) FINDING DIRECTION OF TRUSS REACTIONS

MAGNITUDE UNKNOWN, BUT DIRECTION OF REACTION PASSES THROUGH THE POINT OF SUPPORT

FINDING MAGNITUDE OF TRUSS REACTIONS AFTER FINDING DIRECTION OF FORCE oa

(C) FORCE DIAGRAM

UNIT 28-4
Truss Reactions When Loads Are Not Parallel

GRAPHIC METHOD OF FINDING TRUSS REACTIONS WHEN WIND AND TRUSS LOADS ARE NOT PARALLEL

Consider the truss shown in Fig. 28-4-1A. The forces acting on the truss are a vertical downward force, an inclined downward force, and the reaction forces, which are inclined upward forces and normally do not act in the same direction. Since the truss is rigid and rigidly held to the supports, we may assume that the horizontal components of the reactions are equal.

The lines of action and direction of the downward forces are known, and only the magnitude and direction of the reactions need be computed. Draw a space diagram and label the forces using Bow's notation. Next draw a force diagram (Fig. 28-4-1B). The combined resultant of the two reaction forces is equal to *ac*. The individual magnitudes and directions of the reaction forces **CD** and **DA** are not yet known, but the magnitude of their combined resultant is equal to 970 units. Locate a point *o* anywhere on the force diagram and join *o* to points *a*, *b*, and *c* with a line, as shown in Fig. 28-4-1D. Draw a line *oa* parallel to line *oa* on the force diagram anywhere in space **A** of the space diagram (Fig. 28-4-1C) until it intersects forces **DA** and **AB**. Draw a line parallel to *ob* in space **B** but start where line *oa* intersects force **AB**. Repeat for line *oc*. Close the equilibrium polygon with line *od*.

Draw line *od* on the force diagram (Fig. 28-4-1D) parallel to line *od* on the equilibrium polygon. The values of 560 units and 410 units may be found by measuring lines *ad* and *dc* on the force diagram.

These values are the individual resultants of reactions **CD** and **DA**. Since the horizontal components of the reactions are assumed to be equal, the magnitude and direction of reactions **CD** and **DA** can be found by drawing the force diagram as shown in Fig. 28-4-1F. Distance *X* represents the horizontal component of the combined reaction forces, therefore *X*/2 represents the horizontal component of each reaction. From point *d* on line *ac*, extend a horizontal line until it intersects the vertical line bisecting distance *X* at point d_1. The values of the reactions **CD** of 425 units and **DA** of 550 units may be found by measuring lines cd_1 and d_1a on the force diagram. The directions of the reaction forces will be parallel to lines cd_1 and d_1a.

GRAPHIC METHOD OF FINDING TRUSS REACTIONS, ROLLER AT ONE END, WHEN WIND AND TRUSS LOADS ARE NOT PARALLEL

Consider the truss shown in Fig. 28-4-2A. The forces acting on the truss are a vertical downward force, an inclined downward force, a vertical upward force through the center of the roller, and an inclined upward force at the hinged-pin support. Draw a space diagram and label all the forces using Bow's notation. Next, partially draw the force diagram showing the known forces **AB** and **BC**. Force **CD** is a vertical upward force, but its magnitude is not known. Locate a point *o* anywhere on the force diagram and join *o* to points *a*, *b*, and *c*. Draw a line *oa* paral-

RESULTANT OF WIND LOADS
RESULTANT OF TRUSS LOADS
540
460

NOTE: THE FORCES SHOWN MAY BE U. S. CUSTOMARY OR METRIC UNITS SUCH AS POUNDS, TONS, NEWTONS, KILONEWTONS, ETC.

TO FIND MAGNITUDE AND DIRECTION OF TRUSS REACTIONS

(A) SPACE DIAGRAM

a
540
970
b
460
c

FINDING COMBINED RESULTANT OF REACTION FORCES

(B) FORCE DIAGRAM

460
540
B
C
A
D

(C) SPACE DIAGRAM SHOWING DIRECTION OF ALL EXTERNAL FORCES. MAGNITUDE AND DIRECTION OF REACTIONS UNKNOWN

DRAW LINE od AFTER FINDING ITS DIRECTION ON THE EQUILIBRIUM POLYGON

a
560
540
o
d
b
410
460
c

FINDING INDIVIDUAL RESULTANTS OF REACTION FORCES

(D) FORCE DIAGRAM (REPEATED FOR CLARITY)

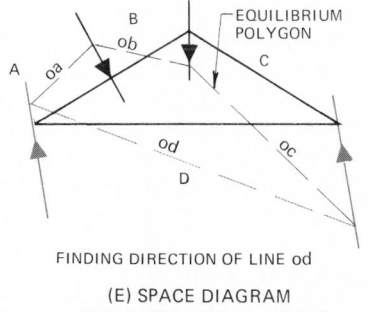

EQUILIBRIUM POLYGON
B
ob
C
A
oa
od
oc
D

FINDING DIRECTION OF LINE od

(E) SPACE DIAGRAM (REPEATED FOR CLARITY)

Fig. 28-4-1 Graphic method of finding truss reactions when wind and mass loads are not parallel.

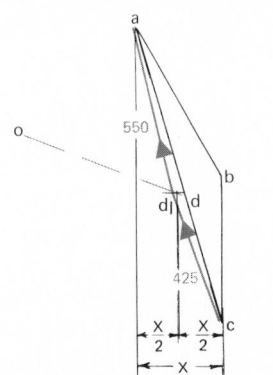

a
o
550
b
d_1
d
425
c
$\frac{X}{2}$ $\frac{X}{2}$
X

FINDING MAGNITUDE AND DIRECTION OF REACTIONS

(F) FORCE DIAGRAM (REPEATED FOR CLARITY)

RESULTANT OF WIND LOADS

RESULTANT OF TRUSS LOADS

560

320

NOTE: THE FORCES SHOWN MAY BE U. S. CUSTOMARY OR METRIC UNITS SUCH AS POUNDS, TONS, NEWTONS, KILONEWTONS, ETC.

PIN   ROLLER

TO FIND MAGNITUDE AND DIRECTION OF TRUSS REACTIONS

(A) SPACE DIAGRAM

a

320

510

b

560

o

d

360

c

DRAW LINE od AFTER FINDING ITS DIRECTION ON THE EQUILIBRIUM POLYGON

FINDING MAGNITUDES OF REACTIONS

(C) FORCE DIAGRAM

560

320

B

C

A

MAGNITUDE UNKNOWN, BUT DIRECTION OF REACTION PASSES THROUGH THE POINT OF SUPPORT

D

DIRECTION AND MAGNITUDE OF REACTION UNKNOWN

(B) SPACE DIAGRAM SHOWING ALL EXTERNAL FORCES

EQUILIBRIUM POLYGON

B

ob

A

oa

C

oc

od

D

FINDING DIRECTION OF LINE od

(D) SPACE DIAGRAM (REPEATED FOR CLARITY)

Fig. 28-4-2 Graphic method of finding truss reactions, roller at one end, when wind and mass loads are not parallel.

lel to line *oa* on the force diagram anywhere in space **A** of the space diagram (Fig. 28-4-2D), until it intersects forces **DA** and **AB**. Draw a line parallel to *ob* in space **B** but start where line *oa* intersects force **AB**. Repeat for line *oc*. Close the equilibrium polygon by joining *oa* and *oc* with line *od*. Draw line *od* on the force diagram parallel to line *od* on the space diagram until it intersects line *bc* at point *d*. Join *a* to *d* with a line which represents the direction of force **DA**. The values of the reactions **CD** of 360 units and **DA** of 510 units may be found by measuring lines *cd* and *da* on the force diagram.

ASSIGNMENTS

See Assignments 10 and 11 for Unit 28-4 on page 567.

UNIT 28-5
Internal Forces or Stresses in a Truss

GRAPHIC METHOD OF FINDING INTERNAL FORCES IN A ROOF TRUSS

The previous examples relating to roof trusses have been confined to finding the external forces acting on the roof and walls. Once the external forces have been calculated, the forces acting in the truss members can be determined graphically by two methods. Consider the truss shown in Fig. 28-5-1A. Since the forces are symmetrical, the reaction forces will be equal; namely 4000 units each. Using Bow's notation and stating all the forces, draw a space diagram to scale, as in Fig. 28-5-1B. Consider joint *ABHG*, the left support. Two of the four forces are known (Fig. 28-5-1C). Draw the force diagram (Fig. 28-5-1G) to a convenient scale, starting with vectors *ga* and *ab*. The next force in order is **BH**. Draw a line parallel to **BH** through point *b*. Point *h*, which is one

end of vector *bh*, has not yet been established. Draw a line parallel to the last force **HG** through point *g*. Since point *h* in on this line as well as on line *bh*, *h* must be their point of intersection. The arrows must travel in the same direction around the polygon, thus indicating the direction of the forces acting at the joint. We find that truss member *BH* is under compression and truss member *HG* is under tension. Next, consider joint *BCJH* in Fig. 28-5-1D. There are two known forces, **BC** of 2000 units and **HB** which was found to be 6000 units and under compression. The directions of forces **CJ** and **JH** are known, but their magnitude is not. The member *CJ* is under compression, but we do not know whether member *JH* is a tie or a strut. Draw the force diagram (Fig. 28-5-1H), to a convenient scale by first drawing the vectors *hb* and *bc*. Draw a line parallel to *CJ* through point *c*. Point *J*, which is one end of the vector, has not yet been established. Draw a line parallel to the last force **JH** through point *h*. Since point *j* is on this line as well as on line *cj*, *j* must be their point of intersection. The arrows must travel in the same direction around the polygon, thus indicating the direction of the force acting at the joint. We find that both truss members *CJ* and *JH* are under compression. Because of this, they are struts. Next consider joint *CDLKJ* (Fig. 28-5-1E). The magnitude of three of the five forces is known and the magnitude of the two unknown forces is equal since the loads are symmetrical about the center of the truss. Draw the force diagram (Fig. 28-5-1J) to a convenient scale by first drawing the vectors *jc*, *cd* and *dl*. The next force in order is **LK**. Draw a line parallel to **LK** through point *L*. Point *k*, which is one end of the vector, has not yet been established. Draw a line parallel to the last force **KL** through point *j*. Since point *k* is also on this line as well as on line *jk*, it must be at their point of intersection. The arrows must travel around the polygon in the same direction, thus indicating the direction of the forces acting at the joint. We find that truss members *LK* and *KJ* are under tension and are tie bars. Only one member remains to be calculated. It is not known whether the truss member *GK* is under compression or tension, nor is its magnitude known.

Fig. 28-5-1 Graphic method of finding internal forces in a roof truss.

From the space diagram shown in Fig. 28-5-1E, draw the force diagram in Fig. 28-5-1K. **GK** is found to have a magnitude of 3500 units and is under tension.

The second method for finding the forces in the truss members is much faster. A single diagram, called a *loading diagram*, that combines all the separate force polygons is used. The first step in constructing a loading diagram is to draw the force diagram of the external forces, as in Fig. 28-5-1L.

Now consider a joint where only one or two forces are unknown, such as joint *ABHG*. The force diagram *ab*, *bh*, *hg*, *ga* can be drawn on the stress diagram in a similar manner to that previously explained. Next, a force diagram similar to Fig. 28-5-1H for joint *BCJH* is constructed on the loading diagram, adding vectors *jh* and *cj*. The same procedure is used for joints *CDLKJ* and *GHJK* until the loading diagram is complete. Only the direc-

tions of the external forces are shown on the loading diagram, since the directions of the other forces would alternate at the different joints.

ASSIGNMENTS

See Assignments 12 through 14 for Unit 28-5 on page 568.

ASSIGNMENTS for Chapter 28

Fig. 28-1-A   Finding resultants.

Fig. 28-1-B   Finding equilibrants.

Fig. 28-1-C   Finding forces acting on the tie and jib.

Assignments for Unit 28-1, Forces

1. On a B- or A3-size sheet, lay out the six problems shown in Fig. 28-1-A, and find their resultant. Scale is to suit. The values shown may be either pounds or newtons.

2. On a B- or A3-size sheet, lay out the six problems shown in Fig. 28-1-B, and find their equilibrants. Scale is to suit. The values shown may be either pounds or newtons.

3. On a B- or A3-size sheet, lay out the four problems shown in Fig. 28-1-C, and find the forces acting on each member. Scale is to suit.

Assignments for Unit 28-2, Beams

4. On a B- or A3-size sheet, lay out the two beams shown in Fig. 28-2-A, and using the graphic method, find the resultant and reactions of the vertical forces. Scale is to suit.

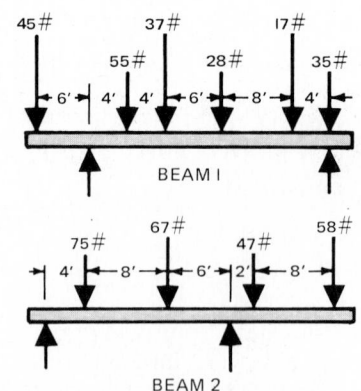

Fig. 28-2-A   Vertical forces acting on a beam.

5. On a B- or A3-size sheet, lay out the two beams shown in Fig. 28-2-B, and using the graphic method, find the resultant and reactions of the vertical forces. Scale is to suit.

6. On a B- or A3-size sheet, lay out the two beams shown in Fig. 28-2-C, and using

Fig. 28-2-B   Vertical forces acting on a beam.

the graphic method, find the resultant of the nonparallel forces. Scale is to suit.

7. On a B- or A3-size sheet, lay out the two beams shown in Fig. 28-2-D, and using the graphic method, find the resultant of the nonparallel forces. Scale is to suit.

BEAM 1

BEAM 2

Fig. 28-2-C   Angular forces acting on a beam.

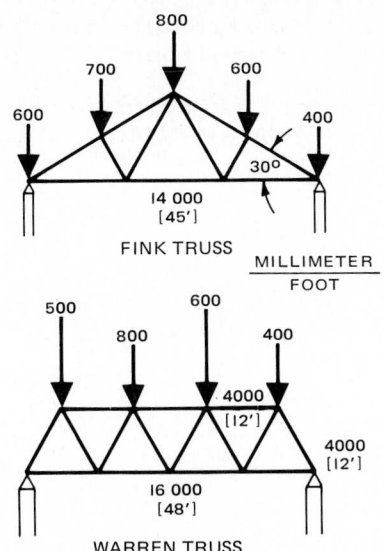

FINK TRUSS

MILLIMETER
FOOT

WARREN TRUSS

NOTE:
—FORCE VALUES ARE lb (U.S.
CUSTOMARY) OR kN (METRIC).
—LINEAR VALUES ARE ft (U.S.
CUSTOMARY) OR mm (METRIC).

Fig. 28-3-A   Vertical loads on truss.

FINK TRUSS

MILLIMETER
FOOT

FAN TRUSS

NOTE:
—FORCE VALUES ARE lb (U.S.
CUSTOMARY) OR kN (METRIC).
—LINEAR VALUES ARE ft (U.S.
CUSTOMARY) OR mm (METRIC).

Fig. 28-4-A   Nonparallel loads acting on truss.

BEAM 1

BEAM 2

Fig. 28-2-D   Angular forces acting on a beam.

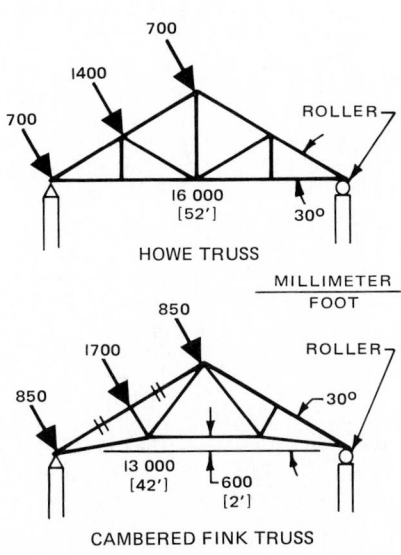

HOWE TRUSS

MILLIMETER
FOOT

CAMBERED FINK TRUSS

NOTE:
—FORCE VALUES ARE lb (U.S.
CUSTOMARY) OR kN (METRIC).
—LINEAR VALUES ARE ft (U.S.
CUSTOMARY) OR mm (METRIC).

Fig. 28-3-B   Parallel but not vertical loads on truss-hinged and roller supports.

CAMBERED FINK TRUSS

MILLIMETER
FOOT

HOWE TRUSS

NOTE:
—FORCE VALUES ARE lb (U.S.
CUSTOMARY) OR kN (METRIC).
—LINEAR VAUES ARE ft (U.S.
CUSTOMARY) OR mm (METRIC).

Fig. 28-4-B   Nonparallel loads acting on truss—roller at one end.

Assignments for Unit 28-3, Truss Reactions when Loads are Parallel

8. On a B- or A3-size sheet, lay out the two trusses shown in Fig. 28-3-A, and by the graphic method find the magnitude of the reactions. Scale is to suit.

9. On a B- or A3-size sheet, lay out the two trusses shown in Fig. 28-3-B, and by the graphic method find the magnitude and direction of the reactions. Scale is to suit.

Assignments for Unit 28-4, Truss Reactions when Loads are not Parallel

10. On a B- or A3-size sheet, lay out the two trusses shown in Fig. 28-4-A, and by the graphic method find the magnitude and direction of the reactions. Scale is to suit.

11. On a B- or A3-size sheet, lay out the two trusses shown in Fig. 28-4-B, and by the graphic method find the magnitude and direction of the reactions. Scale is to suit.

Assignments for Unit 28-5, Internal Forces or Stresses In a Truss

12. On a B- or A3-size sheet, lay out the fink truss shown in Fig. 28-5-A, and by the graphic method find the magnitude of the reactions and the internal forces acting on each member. Scale is to suit.

13. On a B- or A3-size sheet, lay out the fan truss shown in Fig. 28-5-B, and by the graphic method find the magnitude of the reactions and the internal forces acting on each member. Scale is to suit.

14. On a B- or A3-size sheet, lay out the crescent truss shown in Fig. 28-5-C, and by the graphic method find the magnitude and direction of the reactions and the internal forces acting on each member. Scale is to suit.

MILLIMETER / FOOT FINK TRUSS

NOTE:
-FORCE VALUES ARE tons (U.S. CUSTOMARY) OR kN (METRIC).
-LINEAR DIMENSIONS ARE ft (U.S. CUSTOMARY) OR mm (METRIC).

Fig. 28-5-A   Internal roof forces—fink truss.

MILLIMETER / FOOT FAN TRUSS

NOTE:
-FORCE VALUES ARE tons (U.S. CUSTOMARY) OR kN (METRIC).
-LINEAR DIMENSIONS ARE ft (U.S. CUSTOMARY) OR mm (METRIC).

Fig. 28-5-B   Internal roof forces—fan truss.

MILLIMETER / FOOT CRESCENT TRUSS

NOTE:
-FORCE VALUES ARE kN (METRIC) OR tons (U.S. CUSTOMARY).
-LINEAR DIMENSIONS ARE mm (METRIC) OR ft (U.S. CUSTOMARY).

Fig. 28-5-C   Internal roof forces—crescent truss.

CHAPTER 29

Strength of Materials

UNIT 29-1
Stresses and Strain

Relationship Between Mass and Force

Mass is the quantity of matter in a body. The mass of an object remains constant, regardless of its location on earth. Mass is measured in ounces, pounds, and tons (U.S. Customary), or grams, kilograms, and metric tons (metric).

Some examples of the use of these units of measurement are in

- Defining quantities of material, packaged in bulk, such as bags of mortar and tons of sand
- Defining physical characteristics of material such as 210-lb asphalt shingles and 18-oz, 24-oz, or 32-oz glass
- Defining load capacities for building elements, elevator cranes, hoists, bridges, roads, supports, and bearing surfaces
- Specifying application of materials such as 20-lb roofing asphalt per mopping per 100 ft² (U.S. Customary) or 10-kg roofing asphalt per mopping per 100 m² (metric)
- Establishing costs for materials, unit prices, and rates on an ounce, pound, or ton (U.S. Customary) or gram, kilogram, or metric ton basis (metric)

Force is the external agent which changes or tends to change the condition of rest of a body. Force is mea-

sured in ounce-force, pound-force, ton-force and kip-force (U.S. Customary) or in newtons (N) for light forces, kilonewtons (kN) for intermediate forces, and meganewtons (MN) for heavy forces (metric).

Forces related to the design and construction processes are numerous: bearing capacity, applied weight (mass under the influence of gravity) of live, dead, and mobile loads, connection load, etc. Force may be concentrated on a tiny spot or applied over an immense area.

To convert kilograms to a force value, multiply the mass value (in kilograms) by 9.806 65 to obtain the force in newtons.

STRESSES

When a force acts on a piece of material, internal resistance or forces are set up in the material to resist the external force. The resisting forces are called *stresses* and are measured in pounds per square inch or square foot (U.S. Customary) or pascals (metric). A *pascal* (Pa) is a pressure or stress produced when a force of one newton (N) is applied to an area of one square meter (m²).

$$Pa = \frac{N}{m^2}$$

The pascal is a very small unit of measure. It is used for very low-stress applications. In most instances the kilopascal (kPa) and megapascal (MPa) are used.

In solving stress problems the following formulas can be used

$$Stress = \frac{force}{area} \quad or \quad area = \frac{force}{stress}$$

$$Force = stress \times area$$

There are three kinds of stresses: *tension*, *compression*, and *shear*. Tension, or tensile stress, is caused by an external force that tends to pull apart or stretch the material. Tie bars supporting heating units or fans from ceiling members are examples of parts subject to tensile stress. See Fig. 29-1-1.

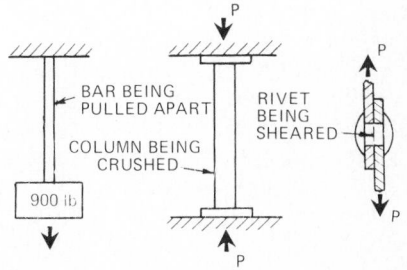

(A) TENSILE   (B) COMPRESSIVE   (C) SHEAR

(D) COMBINATION STRESSES FOUND IN A WOODEN BEAM

Fig. 29-1-1   Stresses.

Compression, or compressive stress, is caused by external forces that tend to crush or push the material together. Basement posts and walls are parts subject to compressive stress.

Shear stress is caused by external forces that tend to cause the particles within the material to slide past one another. Rivets holding metal plates together are subject to shear stress.

These stresses often appear in combination. In a simple beam supporting a load, all three stresses occur. There is a tensile stress along the bottom of the beam, a compressive stress along the top of the beam, and a shear stress at each side at the abutments.

The *ultimate strength* of a material is the highest unit of stress that the material can withstand without breaking.

LOADS

The external forces acting on a body, called *loads* and measured in pounds, tons, and kips (U.S. Customary) or newtons, kilonewtons, and meganewtons (metric), are classified according to the manner in which they are applied.

A *static* load is one that is applied gradually to a part and that remains practically constant once the maximum load is reached. The weight (mass) of a building acting on its foundation is an example of a static load. Static loads are also referred to as *dead* loads.

An *impact* or *shock load* is one that is applied suddenly on an object for a short time. When a nail is struck by a hammer or a train passes over a portion of track, the loads resulting from these actions are known as impact loads.

Repeated loads are loads that are alternately applied and removed many times. An example of a part that is subjected to this type of load is a connecting rod in an automobile engine.

Only static and impact loads will be dealt with in this text.

TYPES OF STRESS

Since there are three types of stresses, a material will have three different ultimate strengths. When machine parts are designed, it is not feasible to work precisely to the ultimate strength of the material since the addition of any shock or unforeseen load would cause breaking.

Therefore, when parts are designed, another stress known as the *safe working stress* (allowable unit stress) is used. This stress is obtained by dividing the ultimate strength of the material by a number called the *factor of safety*. Hence

Term	Symbol	Formula
Force or Load	F	$F = A \times S$
Area	A	$A = \dfrac{F}{S}$
Stress	S	$S = \dfrac{F}{A}$
Ultimate Strength	Su	$Su = S \times FS$
Factor of Safety	FS	$FS = \dfrac{Su}{S}$
Deformation (Unit Strain)	Du	$Du = \dfrac{Dt}{L} = \dfrac{S}{E}$
Deformation (Total Strain)	Dt	$Dt = Du \times L$
Coefficients of Linear Expansion	N	
Modulus of Elasticity	E	$E = \dfrac{S}{Du}$
Length of Part being Deformed	L	$L = \dfrac{Dt}{Du}$

Fig. 29-1-2   Common terms, symbols, and formulas.

Safe working stress (S) =

$$\dfrac{\text{ultimate strength (Su)}}{\text{factor of safety (FS)}}$$

Other common terms and formulas are shown in Fig. 29-1-2. The number used for the factor of safety varies according to the material, the proposed location of the part, and the type of force that it must withstand. For example, a wooden part that is subjected to a shock force would have a greater factor of safety than a steel part that is subjected to a dead load. Factors of safety are not used as frequently as they were in the past since many of today's codes for structures and machines list the allowable unit or working stresses to be used. However, in certain applications, such as aircraft design, the ultimate strength and factors of safety are often used. Allowable unit stresses for steel will be covered in greater detail later in the chapter.

Figure 29-1-3 shows the average values of the ultimate strengths of various materials.

EXAMPLE 1 A 2-ton weight is suspended from a 1.25 × 2.00 in. steel bar. What is the unit stress in pounds per square inch (psi)?

Solution The unit stress will be the force divided by the area

Material		Ultimate Strength			Allowable Unit Stress			Modulus of Elasticity Tension
		Tension	Com-pression	Shear	Tension	Com-pression	Shear	
Aluminum	10^3 lb/in.^{2}	**15**	**12**	**12**				**11 000**
	MPa	103	83	83				78 500
Brass	10^3 lb/in.^{2}	**21**	**30**	**36**				**9 000**
	MPa	145	207	248				62 000
Copper	10^3 lb/in.^{2}	**34**	**32**	**36**				**15 000**
	MPa	235	220	248				103 500
Cast Iron Gray	10^3 lb/in.^{2}	**21**	**90**	**24**		**15**	**4**	**14 000**
	MPa	145	620	165		103	28	96 500
Cast Iron Malleable	10^3 lb/in.^{2}	**31**	**46**	**40**	**5.2**	**7.5**	**6.6**	**25 000**
	MPa	214	317	275	35	52	45	172 000
Cast Iron Wrought	10^3 lb/in.^{2}	**48**	**48**	**40**	**12**	**12**	**10**	**28 000**
	MPa	330	330	275	83	83	69	193 000
Steel A572-50	10^3 lb/in.^{2}	**65**	**65**	**50**	**30**	**30**	**20**	**29 000**
A572M-350 (See Fig. 29-1-8)	MPa	450	450	345	210	210	140	200 000

Bold figures denote U.S. Customary system values.

Fig. 29-1-3   Physical properties of common materials.

Stress = $\dfrac{\text{force}}{\text{area}}$

$$S = \dfrac{F}{A}$$

$$= \dfrac{2 \times 2000}{1.25 \times 2.00}$$

$$= 1600 \text{ psi}$$

EXAMPLE 2 What tensile force would cause a $\phi 2.00$ A572-50 steel rod to fail?
Solution The cross-sectional area of the rod is equal to $\pi R^2 = 3.1416 \times 1.00 \times 1.00 = 3.1416 \text{ in.}^2$. From the table shown in Fig. 29-1-3, A572-50 steel has an ultimate strength of 65 000 psi. Therefore, the tensile force that would cause the rod to fail is

Area $\times$ ultimate strength =
$\qquad$ 3.1416 $\times$ 65 000 = 204 204 lb.

EXAMPLE 3 What is the maximum permissible tensile load an A572M-350 structural steel column having a cross-sectional area of 300 mm², can carry if the factor of safety is given as 5?
Solution From the table shown in Fig. 29-1-3, A572M-350 structural steel has an ultimate tensile strength of 450 MPa.

Working stress = $\dfrac{\text{ultimate strength}}{\text{factor of safety}}$

$$= \dfrac{450}{5} = 90 \text{ MPa}$$

Therefore, maximum allowable load = $S \times A$ = working stress $\times$ area = (90×10^6) Pa $\times$ (300×10^{-6}) = 27 kN.

EXAMPLE 4 A $10'' \times 10'' \times 8'\text{-}0$ Western hemlock construction-grade post supports a weight of 75 000 lb. Does this meet the minimum requirements as recommended by the Institute of Timber Construction (ITC)?
Solution From the table shown in Fig. 29-1-4 under the headings Carrying Load Independently, Compression, and Parallel to Grain, we find the allowable unit stress for Western hemlock, construction grade, is 1100 psi. Therefore

Maximum allowable load =
$\qquad A \times S = 10 \times 10 \times 1100$
$\qquad\qquad\qquad = 110\ 000 \text{ lb}$

The load is acceptable.

Next check for buckling, using the formula $l/d \leq 10$, where L = length in inches and d = least dimension of compression member in inches. $l/d = (8 \times 12) \div 10 = 9.6$. Therefore, the post carrying this load would meet ITC requirements.

EXAMPLE 5 What force is required to punch a 1.50 in. diameter hole in a no. 12 USS gage sheet?
Solution The sheared area will be equal to the circumference of the circle multiplied by the thickness of the sheet. Circumference of a 1.50 in. diameter hole = 4.71 in. Thickness of a no. 12 USS plate (see Appendix) = .109 in. Sheared area = 4.71 $\times$ .109 = .513 in.². Ultimate shear strength of steel (see Fig. 29-1-3) is 50 000 psi. Therefore the force required to punch the hole = $A \times S$ = .513 $\times$ 50 000 = 25 650 lb.

DEFORMATION

When an object is subjected to a load or force, the shape of the material is changed slightly. This change in length is called *strain*, or *deformation*. See Fig. 29-1-5. The length of an object is shortened by a compressive force or lengthened by a tensile force. The change in size is called *total elongation* and is normally measured in inches (U.S. Customary) or millimeters (metric) while the change in length per inch or millimeter is called *unit elongation* and is normally measured in inches per

Group	Grade of Lumber		Carrying Load Independently Working Stress				
			Bending		Compression		
			Stress at Extreme Fiber	Longi-tudinal Shear	Parallel to Grain $1/d \leq 10$	Perpen-dicular to Grain	Tension Parallel to Grain
Douglas Fir	Construction	lb/in.²	**1500**	**120**	**1200**	**415**	**1500**
		MPa	10	0.8	8	2.9	10
	Standard	lb/in.²	**1200**	**95**	**1000**	**390**	**1200**
		MPa	8	0.7	7	2.7	8
Western Hemlock	Construction	lb/in.²	**1500**	**100**	**1100**	**365**	**1500**
		MPa	10	0.7	8	2.5	10
	Standard	lb/in.²	**1200**	**80**	**1000**	**365**	**1200**
		MPa	8	0.6	7	2.5	8
Spruce (All)	Structural	lb/in.²	**1050**	**90**	**750**	**300**	**1050**
		MPa	7	0.6	6	2.1	9
	Construction	lb/in.²	**840**	**70**	**600**	**300**	**840**
		MPa	6	0.5	4	2.1	6
Red Cedar & Pine	Structural	lb/in.²	**900**	**80**	**750**	**260**	**900**
		MPa	6	0.6	5	1.8	8
	Construction	lb/in.²	**720**	**65**	**600**	**260**	**720**
		MPa	5	0.5	4	1.8	5

Note: Values shown are for teaching purposes only. Consult your local building codes for exact values.
$\qquad$ Bold figures denote U.S. Customary system values.

Fig. 29-1-4   Allowable unit stresses for sawn timber members.

(A) TENSILE LOAD $\qquad$ (B) COMPRESSIVE LOAD

(C) LOAD ON BEAM

Fig. 29-1-5   Deformation due to loads.

Fig. 29-1-6   Stress-strain diagram for A572-42 carbon steel.

inch (U.S. Customary), or millimeters per millimeter (metric). Normally the deformation is so small that it cannot be detected by the naked eye. The deformation or sag that occurs on a beam when a load is applied is called *deflection*.

Stress-Strain Diagram

The relationship between stress and strain for any material is best shown by a diagram; see Fig. 29-1-6. A piece of carbon steel 1.00 in. × 1.00 in. having an area of 1.00 in.² was subjected to a tensile load which was increased each time by 5000 lb, and the results were recorded. Up to point *A* on the graph, the elongation of the bar was proportional to the stress. Point *A*, which was recorded at 29 000 lb, was the elastic limit for that steel. After point *A*, the elongation increased at a faster rate. At a stress slightly higher than the elastic limit, deformation occurred without an increase in stress. This is known as the *yield point* of the material. As the tension increased, the bar elongated until point *B* was reached. This was the largest load applied, which was recorded at 65 000 lb. Beyond this point the bar continued to stretch or elongate with less tension. Point *B* was the *ultimate strength* of the material. The breaking point of the bar was point *C*, which was recorded at 48 000 lb.

The strength of any material may be plotted and calculated in a similar manner, although not all materials act in

the same way. An example of this would be a cast-iron part. Since the ultimate strength and the breaking point would be the same, the part would break at the maximum load.

If a force acting on an object is not great, the material will return to its original shape when the force is removed. This tendency to return to the original shape after being deformed is called *elasticity* and varies greatly in different materials. For example, lead is said to have little or no elasticity, while spring steel has a great amount. If, however, the material does not return to its original shape after it has been subjected to a force, it is said to be stressed beyond its elastic limit. Up to this elastic limit the deformation is proportional to the load; that is, the unit stress is proportional to the unit strain at any point in a material up to its elastic limit. This is known as *Hooke's law*. Beyond the elastic limit, the deformation ceases to be proportional to the load. The elastic limit of a material is difficult to determine accurately.

The *modulus of elasticity* of a material is defined as the ratio of unit stress to unit deformation (the stress in 1 in. divided by the deformation in 1 in.) and is denoted by the letter *E*. It may be used for finding the elongation per inch or millimeter caused by any given load.

In the metric system, the modulus of elasticity is the stress in pascals divided by the deformation in one millimeter.

EXAMPLE 6   A steel bar 10 ft long elongates .075 in. under a tensile force. Calculate the unit deformation.
Solution

Unit elongation (*Du*)

$$= \frac{\text{total strain or deformation } (Dt)}{\text{length of part } (L)}$$

$$= \frac{.075}{10 \times 12} = .000\ 63 \text{ in. per in.}$$

EXAMPLE 7   Find the unit deformation on a piece of steel produced by a stress of 45 000 psi.
Solution   From Fig. 29-1-3 we find the modulus of elasticity for steel is 29 000. Therefore,

$$Du = \frac{\mathbf{S}}{\mathbf{E}} = \frac{45\ 000}{29\ 000} = .001\ 552 \text{ in. per in.}$$

EXAMPLE 8   A .25 × 1.00 in. steel bar 15 ft in length supports a tensile load of 5000 lb. Find the total deformation.
Solution

$$\mathbf{S} = \frac{\mathbf{P}}{A} = \frac{5000}{.25 \times 1.00} = 20\ 000 \text{ psi}$$

The modulus of elasticity for steel (see Fig. 29-1-3) = 29 000. *Du* = **S**/**E** = 20 000 ÷ 29 000 000 = .000 69 in. per in. Therefore,

$$Dt = Du \times L = .000\ 69 \times 15 \times 12 = .1242 \text{ in.}$$

Temperature Stresses   When the temperature of a piece of metal is changed, the length of the metal will be either decreased or increased, depending on whether the temperature of the metal is lowered or raised. If, however, the part is rigidly held and is restrained from changing its length, stresses known as *temperature stresses* will result. The main factors concerning temperature stress are (1) amount of heat involved, (2) material undergoing temperature change (aluminum, iron, etc.), and (3) length of part. In order to avoid these stresses, trusses or girders of long spans frequently have one end placed on a roller or a sliding plate.

The linear change per inch or millimeter of length of a part for a degree of change in temperature is called the *coefficient of linear expansion or contraction*. The coefficients of common materials are shown in Fig. 29-1-7. Thus, the total deformation resulting from temperature change can be found as follows. Let total strain or deformation be *Dt*, the coefficient of linear expansion *Ce*, temperature change

Material	Ce Coefficient of Linear Expansion Inches per inch per °F [Millimeters per millimeter per °C]
Aluminum	.000 012 8 [0.000 023 0]
Brass	.000 010 4 [0.000 018 7]
Bronze	.000 010 1 [0.000 018 2]
Copper	.000 009 3 [0.000 016 7]
Iron—Cast	.000 006 2 [0.000 011 2]
Iron—Wrought	.000 006 8 [0.000 012 2]
Steel—Hard	.000 007 4 [0.000 013 3]
Steel—Medium	.000 006 7 [0.000 012 1]
Steel—Soft	.000 006 1 [0.000 011 0]

Bracketed figures denote metric values.

Fig. 29-1-7 Coefficients of expansion.

Description	Steel Standard	U.S. CUSTOMARY		Allowable Unit Stress—kips/in.²				Steel Standard	METRIC		Allowable Unit Stress—MPa			
		Ultimate Tensile Strength—kips/in.²	Yield Point—kips/in.²	Tension	Compression	Bending	Shear		Ultimate Tensile Strength—MPa	Yield Point—MPa	Tension	Compression	Bending	Shear
General Construction	A36	58	36	22	22	24	14.5	A36	400	250	150	150	165	100
	A572 -60	75	60	36	36	40	24	A572M -410	520	410	245	245	270	165
	-55	70	55	33	33	36	22	-380	480	380	230	230	250	150
	-50	65	50	30	30	33	20	-350	450	350	210	210	230	140
	-45	60	45	27	27	30	18	-310	410	310	185	185	205	125
	-42	60	42	25	25	28	16.8	-290	410	290	175	175	190	115

Notes 1. Metric designations and values were not available at time of printing. They are soft converted.
2. Values shown are for steel having a maximum thickness of 50 mm (2.00 in.).

Fig. 29-1-8 Allowable working stress for steel.

(°F) T, and length of part (in.) L. Then $Dt = Ce \times T \times L$. In the metric system, degrees Celsius (°C) is used.

EXAMPLE 9 A medium steel bar 100 in. long is raised from 70 to 170°F. How much does it expand?

Solution The coefficient of linear expansion for medium steel (see Fig. 29-1-7) is .000 006 7. Therefore

Total deformation $(Dt) = Ce \times T \times L$
= .000 006 7 × 100 × 100 = .067 in.

EXAMPLE 10 If the steel bar in Example 9 is restrained and is 2 in. square, what compressive stress is placed on the bar and what load is placed on the restraining members?

Solution

Total strain = .067 in.

Unit strain = $\dfrac{.067}{100}$
= .000 67 in. per in.

Du (unit strain) = $\dfrac{S}{E}$

where S = stress
E = modulus of elasticity. See Fig. 29-1-3.
Therefore
$S = Du \times E$
= .000 67 × 29 000 000
= 19 430 psi

Therefore

Load = stress × area
= 19 430 × 2.00 × 2.00
= 77 720 lb

UNIT STRESSES FOR STEEL

The increasing use of high-strength steels no longer permits the continuation of a standard design specification based on the exclusive use of one grade of steel. These high-strength steels afford as much as a 50 percent increase in strength as compared to common structural carbon steel.

To simplify matters, permissible unit stresses for the various grades of steel are given in terms of a percentage of a specified minimum yield point. These unit stresses are not to exceed 61 percent of the yield point. For steel having a yield point of 36 kips/in.², the permissible unit stress would be 22 kips/in.², which provides for a factor of safety of 1.64. Figure 29-1-8 lists the various grades of steels and their allowable unit stresses. In keeping with the inclusion of steels of several strength grades, a number of corresponding specifications for cast-steel forgings and other materials such as rivets, welding electrodes, and high-strength bolts have been introduced.

ASSIGNMENTS

See Assignments 1 through 32 for Unit 29-1 on page 596.

UNIT 29-2
Bolted and Riveted Joints

It is assumed that the reader has a full understanding of the many advantages of bolted and riveted construction and possesses a knowledge of this type of working drawing and terminology.

The factors of safety for fasteners used in tension are preferably based upon ultimate strength rather than yield point since ultimate strength is of much greater significance for fasteners. The permissible working stresses as shown in Fig. 29-2-1 represent working loads which are approximately one-third to one-half of the value of the ultimate loads observed in tests.

For greater convenience in the proportioning of the bolted connections, permissible stresses for bolts are now given in terms applicable to their normal body area, i.e., the area of the unthreaded shank.

The tensile stress permitted for A307 bolts and threaded parts of A36 steel is equivalent to 22 000 psi (pounds per square inch) or 150 MPa (metric) applied at the root area of the threads. See Fig. 29-1-8.

Permissible stresses for rivets are given in terms applicable to the nominal cross-sectional area of the rivet before driving. See Fig. 29-2-1.

TENSION										
	20 kips/in.² Rivet Size in Inches					140 MPa Rivet Size in Millimeters				
Rivet Dia. in. (mm)	.50	.625	.75	.875	1.00	12	16	20	22	25
Area in.² (mm²)	.196	.307	.442	.601	.785	113	201	314	380	491
Load kips (kN)	3.93	6.14	8.84	12.03	15.71	15.82	28.14	43.96	53.2	68.74

SHEAR
Check Below to Ensure that the Allowable Load is Not Governed by Bearing

	15 kips/in.² Rivet Size in Inches					100 MPa Rivet Size in Millimeters				
Rivet Dia. in. (mm)	.50	.625	.75	.875	1.00	12	16	20	22	25
Single Shear kips (kN)	2.94	4.60	6.63	9.02	11.78	11.3	20.1	31.4	38	49.1
Double Shear kips (kN)	5.89	9.20	13.25	18.04	23.56	22.6	40.2	62.8	76	98.2

BEARING
Single and Multiple Shear
Check to Ensure that the Allowable Load is Not Governed by Shear

Thickness of Material		45 kips/in.² Rivet Size in Inches					310 MPa Rivet Size in Millimeters				
in.	mm	.50	.625	.75	.875	1.00	12	16	20	22	25
.188	5	4.22	5.27	6.33	7.38		18.6	24.8	31	34.1	38.8
.250	6	5.62	7.03	8.44	9.84	11.25	22.3	29.8	37.2	40.9	46.5
.312	8	7.03	8.79	10.55	12.31	14.06	29.8	39.7	49.6	54.6	62
.375	10		10.55	12.66	14.77	16.88		49.6	62	68.2	77.5
.500	12				19.69	22.50				81.4	93
1.000	25	22.50	28.13	33.75	39.38	45.00	3.72	4.96	6.2	6.82	7.75

For material thickness other than those shown, the bearing value is the value for 1.00 in. (U.S. Customary) or 1 mm (metric) multiplied by the actual thickness. Use values shown in bottom line.

Fig. 29-2-1 Allowable load in kips per square inch (U.S. Customary) and kilonewtons (metric) for structural steel.

The most common methods of bolting or riveting plates together are by lapping or butting the plates, as shown in Fig. 29-2-2. There are many areas where a failure may occur in this type of connection (Fig. 29-2-3). In the lap joint the rivet may shear between the two plates. Since the rivet would shear in only one plate, it is said to be in *single shear*. The area that would shear would be the cross-sectional area of the rivet.

There is a possibility that the plate may fail by tearing away at its weakest point, the section through the rivet or bolt hole. This would be a tension failure. The area that would fail would be the area of the plate at the center of the hole less the area of the hole.

A third type of failure would be for the rivet or bolt to rip through or crush the plate directly beneath it. This is called a *bearing failure*, and the area that would fail in the plate would be equal to the diameter of the fastener times the plate thickness. If more than one bolt or rivet were used, the load would be divided equally on the fasteners. Since only two pieces of metal are joined, they are said to be in *single bearing*.

In the butt joint shown in Fig. 29-2-3, the rivet would have to be sliced into two sections if the joint were to fail by shear. The rivet is said to be in *double shear*, and twice the area of the rivet is used in the shear calculations.

Fig. 29-2-2 Plate connections.

Fig. 29-2-3 Stress areas in lap and butt joints.

If the joint fails by tension, that is, pulls or tears away, it will do so at its weakest point—the section through the hole. Since the two outside plates are pulling in one direction and the center plate in the other, the smaller of the two areas must be used in calculating the tensile strength of the joint.

In calculating for bearing failure, a greater allowable working stress is permissible for rivets and high-strength bolts over ordinary bolts.

Rivet Holes

In calculating the stresses in riveted and bolted joints, a distinction must be made between structural joints and joints in boilers, pipes, and tanks. In structural work, the steel members are generally punched and drilled .06 in. (1.5 mm) larger than the rivet in the shop and then taken to the site for assembly. In calculating the tensile stress in the joint, the size of the hole is taken as .12 in. (3 mm) greater than the nominal diameter of the rivet. This is to allow for any unseen damage that may occur around the hole when it is punched and assembled. Areas for shear and bearing are based on the nominal rivet diameter.

In the construction of boilers, etc., where leakage may be a problem, it is essential that the rivet holes line up. These holes are often reamed at assembly. As the finished rivet fills the hole (which is larger than the rivet) completely, the diameter of the hole is used for computing all the stresses.

Spacing of Rivets or Bolts

The minimum distance between the centers of fastener holes is 3 times the diameter of the fastener, but when possible, the distance shall be not less than shown in Fig. 29-2-4.

The maximum pitch of rivets or bolts in line with the stress of compression members composed of plates and shapes does not exceed 16 times the thickness of the thinnest outside plate or shape or 20 times the thickness of the thinnest enclosed plate or shape, with a maximum of 12 in. (300 mm). When two or more gage center lines are used with rivets and bolts staggered, the maximum pitch of rivets or bolts in the line of stress in each gage line shall not exceed 24 times the thickness of the thinnest plate or shape, with a maximum of 18 in. (450 mm).

The distance between lines of rivets or bolts measured at right angles to the line of stress shall not exceed 32 times the thickness of the thinnest plate or shape. The minimum distance from the center of any punched hole to any edge shall be that given in Fig. 29-2-4.

EXAMPLE 1 Lap joint. Two steel bars, .50 × 2.00 in. are lapped and joined by a .75-in. rivet. What is the allowable tensile load that could be applied to the joint? The holes for rivets are to be punched. Plate material is A36 steel.

Solution There are three areas that must be checked

1. The bars failing under a tensile load at the holes
2. The rivet shearing
3. The bearing on the bars directly below the rivet

As the holes are punched, the diameter of the hole will be taken as .12 in. larger than the rivet diameter for calculation purposes.

Area 1 *Bars failing under a tensile load.* Area of plate at center line of hole = .50 × (2.00 − .87) = .565 in.^{2}. Allowable unit stress = 22 kips/in.^{2}. See Fig. 29-1-8. Therefore

Allowable load = $S \times A$
= 22 000 × .565 = 12 430 lb

Area 2 *Rivet shearing.* See Fig. 29-2-1.

Single shear for Ø.75 rivet = 6.63 kips or 6 630 lb

Area 3 *Bearing on plate below rivet.* A Ø.75 rivet is bearing on .50-in. thick steel. Allowable load = (33.75 × .50) = 16 875 lb (see note at bottom of bearing table, Fig. 29-2-1).

The weakest area would be the shear on the rivet. Therefore, allowable tensile load that joint could support = 6 630 lb.

EXAMPLE 2 Single-riveted butt joint (Fig. 29-2-5). A boiler has a single-riveted butt joint. The boiler plate is .44 in., and the two cover plates are .31 in. thick. The rivets are Ø.75 in. and are spaced 3.00 in. apart. Calculate the main stresses that could safely be applied to this joint. Plate material is A36 steel.

Solution Since the pitch of the rivets is 3.00 in., it is assumed that the width of the section taken for calculation purposes is 3.00 in. As in Example 1, there are three areas to be checked

Fig. 29-2-5   Single-riveted butt joint on a boiler.

Fastener Diameter		In Sheared Edge		In Rolled Edge of Plates		In Rolled Edge of Structural Shapes		Minimum Spacing of Rivets or Bolts	
Inches	mm	Inches	mm	Inches	mm	Inches	mm	Inches	mm
.50	12	1.00	25	.90	23	.75	20	2.00	50
.625	16	1.10	28	1.00	25	.90	23	2.25	58
.75	20	1.25	32	1.10	28	1.00	25	2.50	65
.875	22	1.50	38	1.25	32	1.10	28	3.00	75
1.00	25	1.75	45	1.50	38	1.25	32	3.50	90
1.125	30	2.00	50	1.75	45	1.50	38	4.00	100
1.25	32	2.25	60	2.00	50	1.75	45	4.50	115

Fig. 29-2-4   Minimum edge distances and spacings for rivets and bolts.

1. The section failing under a tensile load
2. The rivet shearing
3. The bearing on the steel plate below the rivet.

As previously mentioned, for boiler-plate construction, the finished rivet is assumed to be the same size as the drilled hole, namely, .81 in.

Area 1 *Plate failing under a tensile load.* Since the area of the two outside plates is greater than the area of the middle plate, the middle plate will fail first. Diameter of rivet hole = .81 in. Area of middle plate = $(3.00 - .81) \times .44 = .959$ in.^{2}. Allowable unit stress = 22 kips. See Fig. 29-1-8. Therefore

Allowable load = $\mathbf{S} \times \mathbf{A}$
$= 22\,000 \times .959 = 21\,100$ lb

Area 2 *Rivet shear.* Since the rivet would have to shear in two places, it is considered to be in double shear.

Shear area =
$$\frac{\pi \times .81^2}{4} \times 2 = 1.03 \text{ in.}^2$$

Allowable shear stress = 14.5 kips. Therefore

Allowable load = $\mathbf{S} \times \mathbf{A}$
$= 14\,500 \times 1.03 = 14\,935$ lb

Area 3 *Bearing on plates.* Middle plate (double shear) area = $.44 \times .31 = .356$ in.^{2}. Outside plates (single shear) have an area equal to

Area = $2 \times .31 \times .31 = .502$ in.^{2}

The weaker area would be the middle plate failing under bearing. Allowable unit stress = 45 kips per in.^{2}. Allowable load = $\mathbf{S} \times \mathbf{A} = 45\,000 \times .356 = 16\,020$ lb.

Therefore, the weakest area of the three areas checked would be the rivet shearing. Allowable load on joint = 14 935 lb.

EXAMPLE 3 Roof truss (Fig. 29-2-6). A roof truss has loads of 75 and 64 kips acting on the upper and lower chord members. Calculate the number of ∅.75-in. rivets required to safely carry these loads.

Solution Since the .38 in. gusset is enclosed by two .31-in.-thick angles, the rivets are in double shear. In calculating the bearing stress, it will be noted that the two outer angles having a combined thickness of .62 in. (two .31-in.-thick angles) are stronger than

Fig. 29-2-6 Roof truss.

the .38-in.-thick gusset. Since the problem is one of determining the number of rivets required to carry the load, it can be assumed that the size of the steel is satisfactory for the applied loads. Refer to Fig. 29-2-1.

1. Number of ∅.75-in. rivets in double shear required for

Upper chord = $75 \div 13.25 = 6$ rivets
Lower chord = $64 \div 13.25 = 5$ rivets

2. Number of ∅.75-in. rivets bearing on .38-in. plate required for

Upper chord = $75 \div 12.66 = 6$ rivets
Lower chord = $64 \div 12.66 = 5$ rivets

Therefore, the minimum allowable rivets required for the upper and lower chords are 6 and 5 rivets, respectively.

EXAMPLE 4 Double-riveted butt joint (Fig. 29-2-7). A boiler has a double-riveted butt joint. The boiler plate is .38 in. thick, and the two cover plates are .25 in. The rivets are ∅.75 in. A section of the riveted joint is shown. Calculate the main stresses in the joint when the boiler plate is subject to a tensile strength of 6000 psi.

SECTION C–C
(A)

Fig. 29-2-7 Double-riveted butt joint on a boiler.

Solution The length of the repeated section is 6.00 in. Since both sides of the joint are the same, only one side of the joint (shown in Fig. 29-2-7B) is used in computing the stresses. There are two rivets in double shear and one rivet in single shear. The rivets are .75 in. in diameter and are placed in .81-in. drilled holes. As mentioned earlier in boiler work, finished rivets are assumed to be the same size as the drilled holes; thus for calculation purposes the rivets will be .81 in. in diameter. The total force exerted on the repeated section is

$\mathbf{F} = \mathbf{S} \times \mathbf{A}$
$= 6000 \times 6.00 \times .38 = 13\,680$ lb

There are two rivets in double shear and one rivet in single shear, comprising shear areas. It will be assumed that each shear area will carry one-fifth of the load. Shear force on each rivet = $13\,680 \div 5 = 2736$ lb. Therefore the unit shear stress on rivets is

$$S = \frac{F}{A} = \frac{2736}{[(\pi \times 81^2) \div 4]} = \frac{2736}{.515}$$
$$= 5313 \text{ psi}$$

SECTION D–D
(B)

The upper cover plate transmits two-fifths of the load. Therefore $\mathbf{F}_1 = .4 \times$ 13 680 = 5472 lb.

The lower cover plate transmits three-fifths of the load. Therefore $\mathbf{F}_2$ = .6 × 13 680 = 8208 lb. Stress on boiler plate taken at section A

$$S = \frac{\mathbf{F}}{A} = \frac{13\ 680}{(6.00 - .81) \times .38}$$

$$= \frac{13\ 680}{1.97} = 6944 \text{ psi}$$

Since one-fifth of the total load has been transmitted to the lower cover plate at section A, the load on the boiler plate at section B is .8 × 13 680 = 10 944 lb. Stress on .38-in. boiler plate taken at section B is

$$S = \frac{\mathbf{F}}{A} = \frac{10\ 944}{(6.00 - 1.62) \times .38} = \frac{10\ 944}{1.664}$$

$$= 6577 \text{ psi}$$

Since the lower cover plate transmits three-fifths of the total load, the largest stress on the two cover plates will occur on the lower cover plate at section B. Stress on the bottom cover plate at section B is

$$S = \frac{\mathbf{F}_2}{A} = \frac{8208}{(6.00 - 1.62) \times .25}$$

$$= \frac{8208}{1.095} = 7496 \text{ psi}$$

In calculating the bearing stresses, the bearing area for the rivet in single shear is the rivet diameter times the thickness of the thinner plate connected (the cover plate). The bearing area for the rivet in double shear is the rivet diameter times the thickness of the boiler plate. The rivets in double shear are subjected to twice the bearing load of those in single shear. Bearing stress at a rivet in single shear (section A) is

$$S = \frac{\mathbf{F}}{A} = \frac{2736}{.81 \times .25} = \frac{2736}{.20}$$

$$= 13\ 680 \text{ psi}$$

Bearing stress at a rivet in double shear (section B) is

$$S = \frac{\mathbf{F}_1}{A} = \frac{5472}{.81 \times .38} = \frac{5472}{.308}$$

$$= 17\ 766 \text{ psi}$$

Stresses in Thin-Wall Cylinders An important application of riveted and

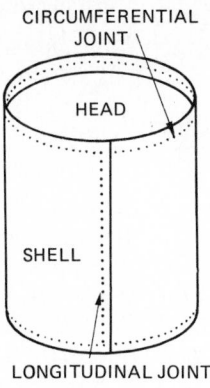

(A) TERMINOLOGY

Fig. 29-2-8 Thin-wall cylinder.

(B) HALF SECTION

welded joints is in the construction of boilers and tanks. The pressure of gases or liquids upon the walls of a tank acts outwardly in all directions and uniformly. Therefore, the cylinder shell on a thin-wall vessel is designed with the assumption that the stress is uniform throughout the wall thickness.

The tensile stress in the ends of the cylinder, caused by the pressure inside, is called *longitudinal stress*, or *tension*. The tensile stress acting in the circumferential direction is called *hoop stress*, or *tension*.

EXAMPLE 5 A tank of 48.00-in. diameter is made of .25-in. steel plate. The internal pressure is 150 psi. Calculate the size of rivets required if the pitch on the longitudinal and circumferential joints is 3.00 in.

Solution

1. Calculate rivet size for longitudinal seam. Figure 29-2-8 shows a half-section of the tank. The internal pressure of 150 psi acts on the shell surface at every point. The total force acting on the half of the tank shown would be equal to the area of the tank taken at its center times the pressure, or (48.00 × 72.00)in.² × 150 psi = 518 400 lb. The combined equal pressures of $\mathbf{F}_1$ and $\mathbf{F}_1$ acting on the tank wall are equal in magnitude to $\mathbf{P}$ but act in opposite directions.

Only the pitch distance of 3.00 in. the repeated section, need be used in calculating the size of the rivet along the joint. Therefore

$\mathbf{F}_1$ for repeated section =
$$\frac{48.00 \times 3.00 \times 150}{2}$$

$$= 10\ 800 \text{ lb}$$

or load acting on each rivet. As previously mentioned, in boiler construction the diameter of the rivet hole, which is .06 in. larger than the diameter of the rivet, is used in computing all the stresses. Refer to Fig. 29-2-1. The allowable stress in single shear is 15 kips per in.² The chart shows values of 9.02 and 11.78 kips for rivet sizes of ⌀.875 and ⌀1.00 in., respectively. Since the finished size of the .875-in. diameter rivet will be .938 in. in diameter, the allowable load will be computed on the final size. Therefore the allowable load for a .938-in. diameter rivet will be

Area × stress =
($\pi \times .938^2 \div 4$) in.² × 15 kips =
.69 in.² × 15 000 = 10 365 lb

Since this is less than the load acting on the rivet, the next size larger rivet must be used, and therefore the ⌀1.00-in. rivet is required. The allowable load in bearing for a ⌀1.00-in. rivet on .25-in. steel plate is 11 250 lb. Therefore the size of rivet required along the longitudinal seam is ⌀1.00 in.

2. Calculate rivet size for circumferential joint. Number of pitches or repeated sections on circumference equals

$$\frac{\pi \times \text{diameter}}{\text{pitch}} = \frac{\pi \times 48.00}{3.00} = 50.3$$

Use 51 rivets.

Pressure exerted on head of tank = pressure × area = 150 psi × ($\pi \times 24.00^2$) in.² = 271 434 lb. Therefore

$$\text{Load per rivet} = \frac{271\ 434}{51} = 5322 \text{ lb}$$

Note: When the pitches on the longitudinal and the circumferential joints are equal, then the load per pitch on the circumferential joint is one-half of the load per pitch on the longitudinal joint.

Refer to Fig. 29-2-1. The allowable load in single shear for a $\emptyset$.62-in. rivet (use .69 in. for calculations) is 5609 lb and the allowable load in bearing for a $\emptyset$.62-in. rivet (use .69 in. for calculations) on .25-in. steel is 7762 lb. Therefore the size of rivet required along circumferential joint is $\emptyset$.62 in.

BOLTS, SCREWS, AND STUDS

As mentioned at the beginning of this unit, the tensile stress permitted for A307 bolts and threaded parts of A36 steel is equivalent to 22 000 psi applied at the root area of the threads.

EXAMPLE 6 What force is required to strip the threads on a 1.000-8 UNC regular hex nut and bolt?
Solution The sheared area will be equal to the circumference of the root circle multiplied by the height of the nut. Root diameter of 1.000-8 UNC thread = .847 in., circumference = 2.66 in. Height of 1.000 in. regular hex nut (see Appendix) = .875 in. Shear area = 2.66 × .875 = 2.33 in.^{2}. Ultimate tensile strength of A36 steel (see Fig. 29-1-8) = 58 kips.
Therefore

Force required to strip threads =
S × A = 58 000 × 2.33 = 135 140 lb

EXAMPLE 7 A platform is supported by four A36 steel rods which are suspended from the ceiling. The end of the rods are threaded, and plate washers and nuts are attached. The platform is to support a load of 16 000 lb and any two of the four rods must be capable of supporting the entire load.

Solution The design load for each rod is 16 000 ÷ 2 = 8000 lb. Allowable unit stress (see Fig. 29-1-8) = 22 000 lb per in.^{2}.

$$A = \frac{P}{S} = \frac{8000}{22\ 000} = .364 \text{ in.}^2$$

Note that in example 6 the shear area for a 1.000 in. threaded nut is 2.33 in.^{2}. Therefore, a smaller threaded nut is required. Try a .375 UNC style 1 hex nut (see Appendix and use tap drill size for root diameter).

Area = circumference of root diameter × height of nut = (π × .312) × .328 = .321 in.^{2}. A greater root area is required. A style 2 hex nut which is thicker or a .438 in. nut will be needed. Area of a .375 UNC style 2 hex nut = (π × .312) × .406 = .398 in.^{2}. Therefore, $\emptyset$.375 rods with UNC threads and style 2 hex nuts meet the design requirements.

As explained in Chap. 8, metric threaded fasteners are defined by property class numbers which designate their strength. The first number of a two-digit symbol or the first two numbers of a three-digit symbol approximates 1 percent of the minimum tensile stress in megapascals.

The last numeral approximates one-tenth of the ratio expressed as a percentage between minimum yield stress and minimum tensile stress.

EXAMPLE 8 What mass can be supported by an M24 × 3 stud, property class 8.8, if a factor of safety of 4 is added to the requirements?
Solution Refer to Fig. 29-2-9. Under the 8.8 column, an M24 × 3 thread has a tensile strength (the maximum force permitted) of 293 kN. Adding a factor of safety to this value we find the permissible force is 293 ÷ 4 = 73.25 kN. 1N = 0.102 kg

Therefore, mass that can be supported = 73 250 × 0.102 = 7472 kg

References and Source Material
1. American Institute of Steel Construction

ASSIGNMENTS

See Assignments 33 and 34 for Unit 29-2 on page 596.

Review for Assignments
Unit 29-1 Stresses

UNIT 29-3
Welded Joints

In addition to riveting, welding is also employed in the joining of structural steel. Fabricated steel construction has also replaced many parts formerly made by casting because of the lower cost and the greater strength at a considerable reduction in size and weight, or mass.

The two types of welds most frequently used are fillet and butt welds. Thus only these types will be covered in this unit.

FILLET WELDS

The fillet weld is used to join two parts that either overlap or join at an angle, normally perpendicular, to each other. In the calculations of strength of fillet welds, the *effective area* is considered as the *effective length* of weld times the *effective throat thickness*, as illustrated in Fig. 29-3-1.

Thread	Tensile Stress Area mm^2	Thread Shear Area mm^2	Tensile Strength (kN) for Property Class						
			4.6	4.8	5.8	8.8	9.8	10.9	12.9
M10 × 1.5	58	15.6	23.2	24.4	30.2		52.2	60.3	70.8
M12 × 1.75	84	19	33.7	35.4	43.8		75.9	87.7	103
M14 × 2	115	22.4	46	48.3	59.8		104	120	140
M16 × 2	157	26.1	62.8	65.9	81.6	130	141	163	192
M20 × 2.5	245	33.3	98		127	203		255	299
M24 × 3	353	40.5	141		184	293		367	431
M30 × 3.5	561	51.6	224			466		583	684
M36 × 4	817	63.1	327			678		850	997

Fig. 29-2-9 Load capacities of threaded fasteners.

Fig. 29-3-1 Fillet weld nomenclature.

Electrodes			
U.S. Customary		Metric	
E60XX	E70XX	E410XX	E480XX
13 500 lb/in.²	15 700 lb/in.²	185 MPa	216 MPa

Fig. 29-3-2 Shear stress for electrodes.

For example, a .38 in. fillet weld 6.00 in. long (effective length) has an effective area of .38 × .7 × 6.00, or 1.596 in.².

For purposes of calculating the strength of welds in this unit, the shear stresses shown in Fig. 29-3-2 will be used. The strength of the previous weld would be stress times effective area.

With E60xx electrodes the weld strength is

13 500 × 1.596 = 21 546 lb

Using E70xx elecrodes, the weld strength is

15 700 × 1.596 = 25 057 lb

Another method of calculating the strength of welds is to multiply the leg size of the weld by effective length. The shear resistance factors (SRF) shown in Fig. 29-3-3 are based on the shear stresses shown in Fig. 29-3-2.

The strength of the previous weld using E70xx electrodes would be 4.2 × 6 = 25.2 kips or 25 200 lb.

The strength value for specified weld sizes is the more convenient method to use.

The following recommendations should be adhered to when welded joints are designed.

- Even-number-mm-size welds, such as shown in Fig. 29-3-3, should be used whenever possible.
- For metric length of welds use lengths evenly divisible by 5, such as 40, 50, 60, etc.
- Fillet welds should be at least .06 in. (2 mm) less than the thickness of the part being welded.
- Welds should be located on both sides of T joints evenly spaced around the line of action of the applied load.

EXAMPLE 1 Two .38 × 6.00 in. steel bars are welded with F70 electrodes as shown in Fig. 29-3-4. What size weld is required if a tensile load of 30 kips is applied?

| | Allowable Load per Inch Length in Kips | |
	Base Metal	Metal Electrodes
Fillet Weld Size Inches	A36[3] E60XX Electrode	A572 E70XX[4] Electrode
.18	1.8	2.1
.25	2.4	2.8
.31	3.0	3.5
.38	3.6	4.2
.44	4.2	4.9
.50	4.8	5.6
.62	6.0	7.0
.75	7.2	8.4

1. Based on 600 lb per .062 in. of weld thickness (13 500 lb/in.² shear stress).
2. Based on 700 lb per .062 in. of weld thickness (15 800 lb/in.² shear stress).

| | Allowable Load per mm Length in kN | |
	Base Metal	Electrodes
Fillet Weld Size mm	A36[1] E410XX Electrode	A572M[2] E480XX Electrode
4	0.5	0.6
6	0.8	0.9
8	1	1.2
10	1.3	1.5
12	1.6	1.8
16	2.1	2.4
20	2.6	3

3. Based on shear resistance factor of 0.131 kN per mm of weld for 1 mm of weld length (185 MPa shear stress).
4. Based on shear resistance factor of 0.153 kN per mm of weld for 1 mm of weld length (216 MPa shear stress).

Fig. 29-3-3 Strength of fillet welds.

Solution

$$\text{Weld size} = \frac{\text{load}}{\text{length} \times \text{SRF}}$$

$$\text{SRF} = \frac{30}{2 \times 6.00} = 2.5$$

Refer to Fig. 29-3-3.
A .25 in. fillet weld is required.

EXAMPLE 2 A .75 × 4.00 in. A572-50 bar is welded to a column. What are the size and length of the fillet welds required if a tensile load of 40 kips is applied?
Solution Since two fillet welds will be used, one on each side of the bar, each fillet weld will be designed to resist a force of 40 kips ÷ 2 = 20 kips.

Fig. 29-3-4 Lap joint.

With a weld running the entire length of 4.00 in., a weld having the strength of 20 ÷ 4 or 5 kips per inch of length is required.

Refer to Fig. 29-3-3: a .50 in. fillet weld is selected. With a .62 in. weld, a weld length of 20 ÷ 7.0 = 2.85 in. (use 3.00 in.) would be required.

EXAMPLE 3 A 250 × 10 mm A572M-380 steel plate is connected by a pair of fillet welds to the bottom flange of a beam, as shown in Fig. 29-3-5. The plate is subjected to a tensile load of 450 kN. What are the minimum size and length of weld recommended for this connection, if the maximum stress on the plate is 220 MPa?

Fig. 29-3-5 Bar fillet welded both sides.

Solution Before the weld size is chosen, the minimum plate area at the weld should be established. The minimum plate area, for calculating purposes, at the welded area (see Fig. 29-3-6) is width times plate thickness.

Load = stress × area.

Therefore

Minimum width of plate
$$= \frac{\text{load}}{\text{stress} \times \text{plate thickness in meters}}$$
$$= \frac{450\ 000\ \text{N}}{(220 \times 10^6)\ \text{Pa} \times 0.01\ \text{m}}$$
$$= 0.2045\ \text{m or } 204.5\ \text{mm}$$

Thus, the plate width of 250 mm is acceptable. If the minimum weld length is 210 mm per side (min. plate width), the load per mm of weld is 450 ÷ (2 × 210) = 1.07 kN.

Refer to Fig. 29-3-3. An 8-mm fillet weld is required. If the fillet weld were to run the entire length of the plate,

Fig. 29-3-6 Plate area at weld.

then the load per mm of weld = 450 ÷ (2 × 250) = 0.9 kN. This would permit a 6-mm weld to be used, which is more economical.

Intermittent Fillet Welds

Intermittent fillet welds may be used to transfer calculated stress across a joint when the strength required is less than that developed by a continuous fillet weld of the smallest permitted size, and to join components of built-up members. The effective length of any segment of intermittent fillet welding should be not less than 4 times the weld size, with a minimum of 1.50 in. (40 mm).

EXAMPLE 4 Calculate the size of the intermittent weld shown in Fig. 29-3-7 to safely carry a tensile load of 180 kips. Use F70xx electrodes.
Solution

Total length of welds
$$= 11 \times 2.00 \times 2 = 44.00 \text{ in.}$$

Load per inch length of weld
$$= \frac{180}{44} = 4.09 \text{ kips.}$$

Refer to Fig. 29-3-3. Size of fillet weld required is .38 in.

Fillet Welds for Angle Iron

When tension or compression members are connected by two side fillet welds as shown in the previous examples, the weld should be placed in the same line of action as the force being transmitted by the weld. For members having symmetrical cross sections, the length of weld on each side of the member should be equal. For members having unsymmetrical cross sections, as shown in Fig. 29-3-8 where an angle iron is welded to a steel plate, the lengths of welds are so proportioned that the line of action of the force transmitted by the weld will be along the axes of the two members. This is accomplished by assuming that the line of action on the angle member is on the *centroidal axis* (center of gravity) and by making the lengths of welds such that $L_1 \times A = L_2 \times B$.

EXAMPLE 5 A $5.00 \times 3.00 \times .38$ in. A572-45 angle welded to a steel plate transmits a load of 54 kips. Calculate the length of welds on each side of the angle so that the load acts along the centroidal axis of the angle. See Fig. 29-3-8.
Solution The maximum fillet weld for .38-in.-thick material is .31 in. The allowable load per inch length of a .31-in. weld is 3.5 kips. See Fig. 29-3-3. Therefore

Minimum permissible length of weld
$$= \frac{54}{3.5} = 15.43 \text{ in. (use 15.50 in.)}$$

Referring to the Appendix, we find the back of the short leg of the angle is 1.70 in. from the centroidal axis.

Consider the 54 kip load being transferred to the plate by loads P_1 and P_2 where $P_1 = 3.5 \times L_1$, $P_2 = 3.5 \times L_2$, and P_1 and P_2 both equal 54 kips. Taking moments about a point on length L_2 (refer to Fig. 29-4-3 for calculation of moments), we have

$$P_1 \times 5.00 = 54 \times 1.70$$
$$P_1 = 54 \times 1.70 \div 5.00$$
$$= 18.3 = 3.5 \times L_1$$

Therefore

$$L_1 = 5.23 \text{ in. (use 5.25 in.)}$$
$$L_1 + L_2 = 15.50 \text{ in.}$$
$$L_2 = 15.50 - 5.25 = 10.25 \text{ in.}$$

Therefore weld lengths of 5.25 (L_1) and 10.25 in. (L_2) are selected.

EXAMPLE 6 Use the same members and load as in the previous example, except that the fillet weld is welded on three sides, as shown in Fig. 29-3-9.
Solution The design calls for 15.50 in. of weld to be used: 5.00 in. of the weld lies along the back of the angle so that the remaining 10.25 in. of weld is equal to the combined length of welds L_1 and L_2.

The 15.50 in. of weld should be located so that the line of action of the force transmitted by the weld is along the centroidal axis of the angle.

Fig. 29-3-7 Intermittent weld.

Fig. 29-3-8 Fillet weld on both sides of angle iron.

Fig. 29-3-9 Fillet weld on sides and end of angle iron.

For calculation purposes, assume there are three welds, P_1, P_2, and P_3, whose combined loads equal 54 kips, and the allowable load per inch length of a .06 in. weld is 3.5 kips. P_1 lies along distance L_1 and is equal to $3.5L_1$ kips. P_2 lies along distance L_2 and is equal to $3.5L_2$ kips. P_3 lies midway along the 5.00-in. width, or 2.50 in. from P_1, and is equal to $5.00 \times 3.5 = 17.5$ kips. At a point on line L_2

Clockwise moments
$$= (P_1 \times 5.00) + (P_3 \times 2.50)$$
$$= (L_1 \times 3.5 \times 5.00) + (17.5 \times 2.50)$$
$$= 17.5L_1 + 43.75 \text{ in. kips}$$

Counterclockwise moments
$$= 1.70 \times 54 = 91.8 \text{ in. kips}$$

Clockwise moments = counterclockwise moments
$$17.5L_1 + 43.75 \text{ in. kips} = 91.8 \text{ in. kips}$$
$$17.5L_1 = 48.05 \text{ in. kips}$$
$$L_1 = 2.75 \text{ in.}$$
$$L_1 + L_2 = 10.50 \text{ in.}$$
$$L_2 = 10.50 - 2.75$$
$$= 7.75 \text{ in.}$$

Therefore weld lengths of 2.75 in. (L_1) and 7.75 in. (L_2) are used on the sides of the angle.

TYPE OF BUTT WELD		PARTIAL PENETRATION WELDED ONE SIDE	COMPLETE PENETRATION WELDED BOTH SIDES
SQUARE	FLUSH	T (MAX .12 in.) (MAX 3 mm) — EFFECTIVE THROAT THICKNESS = 0.5 T	T (MAX .31 in.) (MAX 8 mm)
	OPEN	T (MAX .25 in.) (MAX 6 mm) — EFFECTIVE THROAT THICKNESS = 0.75 T	EFFECTIVE THROAT THICKNESS = T
BEVEL V AND U		T (UNLIMITED) SINGLE BEVEL — T (UNLIMITED) DOUBLE BEVEL EFFECTIVE THROAT THICKNESS = T − .25 in.(6 mm)	T (UNLIMITED) — T (UNLIMITED) EFFECTIVE THROAT THICKNESS = T

Fig. 29-3-10 Strength of butt welds.

BUTT WELDS

The butt weld is used to join two pieces of metal that lie on the same plane. In the calculations of strength of butt joints, the effective area of butt welds shall be considered as the effective length of weld times the effective throat thickness. The effective throat thickness depends on the metal thickness, the gap between the adjoining parts, the type of butt weld, and whether the weld is on one or both sides. See Fig. 29-3-10.

EXAMPLE 7 An open-square butt weld, welded one side, is used to join two A36 steel plates .25 × 10.00 in. Compute the safe tensile load that can be applied to the joint.
Solution Refer to Fig. 29-3-10. The effective throat thickness for an open-square butt weld = $0.75T = 0.75 \times .25 = .188$ in. Next (refer to Fig. 29-1-8), the allowable unit tensile stress for A36 steel is 22 kips. Therefore

Safe load = area × unit stress
$$= (.188 \times 10.00) \times 22\,000$$
$$= 141.36 \text{ kips}$$

EXAMPLE 8 A ⌀36 in. boiler, made of A36 steel, has to withstand a steam pressure of 350 psi. A single-V butt joint, welded one side, is to be used. What is the thickness of boiler plate required?
Solution Figure 29-3-11 shows a half-section of the boiler. The total force P acting on the cylinder is the resultant pressure of the internal pressure of 350

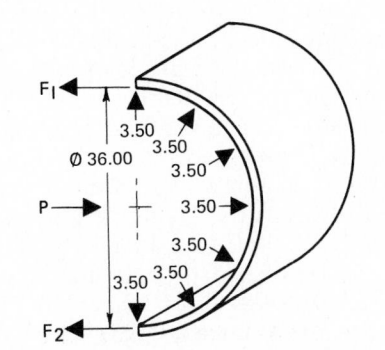

Fig. 29-3-11 Welded boiler section.

psi acting in all directions on the cylinder wall. It will be assumed that for thin-walled cylinders the resultant force P will equal the diameter of the cylinder in inches times the length of the cylinder in inches times the pressure acting within the cylinder.

The total force P is resisted by two equal forces F_1 and F_2. Taking a section of the tank 1 in. in length and calculating the forces P, F_1, and F_2, we have

$$P = S \times A$$
$$= 350 \times 1.00 \times 36.00$$
$$= 12\,600 \text{ lb}$$
$$F_1 = 12\,600 \div 2 = 6300 \text{ lb}$$

The single-V butt weld will have to withstand a force of 6300 lb for every inch of weld. Allowable unit stress for A36 steel is 22 000 psi. Weld stress equals plate stress. Therefore

Effective throat thickness of weld
$$= F_1 \div (S \times \text{length of section})$$
$$= 6300 \div (22\,000 \times 1.00)$$
$$= .28 \text{ in.}$$

Refer to Fig. 29-3-10; note that the

effective throat thickness of a single-V butt weld, welded one side, under tension is equal to $T - .25$ in., where T is equal to the thickness of plate. Therefore, minimum plate thickness must be $.25 + .28 = .53$ in. to safely carry the load. With a welded-both-sides joint, the plate thickness could be reduced to .28 in., which would be a considerable saving.

Reference and Source Material

1. American Institute of Steel Construction.

ASSIGNMENTS

See Assignments 35 through 58 for Unit 29-3 on page 597.

UNIT 29-4
Beams

A *beam* is a structural member or machine part which supports transverse (i.e., perpendicular) loads and reactions. Most beams are placed in a horizontal position with vertical forces acting on them. Examples are floor and ceiling joists, lintels, and floor beams. This unit covers the design of simple beams only where buckling and twisting are not factors and where the beams are of uniform size and shape for the entire length. Forces acting on the beams are assumed to be in the same plane.

Types of Beams Beams are classified according to the manner in which they are supported. Some of the more common types of beams are shown in Fig. 29-4-1. They are

1. Cantilever beam: a beam that has one fixed end
2. Simple beam: a beam that is supported at each end
3. Overhanging beam: a beam that has one or both ends projecting beyond its supports
4. Beams with both ends fixed
5. Beams fixed at one end and supported at the other end
6. Continuous beam: a beam supported at more than two points

(A) CANTILEVER BEAM

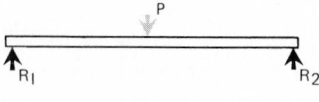

(B) SIMPLE BEAM—
SUPPORTED AT BOTH ENDS

(C) OVERHANGING BEAM

(D) BEAM WITH FIXED ENDS

(E) BEAM FIXED AT ONE END,
SUPPORTED AT OTHER END

(F) CONTINUOUS BEAM

Fig. 29-4-1   Common types of beams.

KINDS OF LOADS

Two types of loads commonly occur on beams: *concentrated* and *uniformly distributed* loads. A concentrated load extends over a short length of the beam and for calculation purposes is considered as acting at one point. It is usually represented by a line with an arrow indicating its direction of force and the letter P as shown in Fig. 29-4-2A. Concentrated loads are generally expressed in pounds or kips (U.S. Customary), or kilonewtons or meganewtons (metric). One kip equals 1000 pounds. A uniformly distributed

(A) THREE METHODS OF INDICATING
CONCENTRATED LOADS

(B) UNIFORMLY DISTRIBUTED LOADS

(C) COMBINATION OF CONCENTRATED
AND UNIFORMLY DISTRIBUTED LOADS

Fig. 29-4-2   Representation of loads on beam drawings.

load is one in which the load is distributed uniformly over a given length or over the entire length of the beam. The weight or mass of the beam is an example of a uniformly distributed load. This type of load is generally expressed in pounds or kips per foot (U.S. Customary), or newtons per meter or kilonewtons per meter (metric). The load is represented in the figure by a rectangular block resting on the beam, as shown in Fig. 29-4-2B.

The upward forces, or supports, that hold the beam in a state of equilibrium are called the *reactions* and are designated by the letters R_1 (left side) and R_2 (right side). The sum of the reactions $R_1 + R_2$, known as the *forces acting upward*, are equal and opposite to the downward forces or loads.

MOMENTS

When a force acts upon an object at a distance from the object, as through a beam, the force is called a *moment*. See Fig. 29-4-3. A moment is the tendency of a force to cause rotation about a given point or axis. The magnitude of a moment is equal to the magnitude of the force times the perpendicular distance to the point. Since the force is measured in pounds and the distance in feet or inches, moments are measured in foot-pounds (ft·lb) or

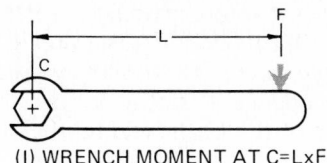

(I) WRENCH MOMENT AT C=L×F

(2) LEVER MOMENT AT
F=W$_1$ × L$_1$=W$_2$ × L$_2$

(3) CRANK MOMENT AT C=W×L

(4) CANTILEVER BEAM
MOMENT AT WALL=W×L

(5) SIMPLE BEAM MOMENT
AT W=R$_1$× L$_1$=R$_2$×L$_2$

Fig. 29-4-3 Application of moments.

inch-pounds (in.·lb). In the metric system the moments are measured in newton-meters (N·m).

If a number of forces acting on a point are in equilibrium, the sum of the moments of all the forces about that point is zero. Therefore, the sum of the moments of all forces that tend to produce clockwise moments about a given point is equal and opposite to the sum of all the forces that tend to produce counterclockwise moments at that given point. This *law of equilibrium* is very helpful in solving beam reaction.

EXAMPLE 1 A force of 20 lb is applied at the end of a wrench 12 in. from the center of the bolt which is being held by the wrench. Calculate the moment.
Solution The moment may be found by

multiplying the force times the distance: 20 × 12 = 240 in.·lb.

EXAMPLE 2 A cantilever beam supports a concentrated load of 500 lb located 12 ft from the support. Neglecting the mass of the beam, calculate the moment at the wall.
Solution The moment taken at the wall or support may be found by multiplying the force times the distance: 500 × 12 = 600 ft·lb.

EXAMPLE 3 A beam 15 ft long has a concentrated load of 1200 lb acting 5 ft away from the left reaction. Neglecting the mass of the beam, calculate the reaction forces. See Fig. 29-4-4.

MOMENT DIAGRAM ABOUT R$_1$

Fig. 29-4-4 Simple beam with concentrated load.

Solution Taking moments about reactions R$_1$ and R$_2$ we have

Clockwise moments using pounds and feet = 5 × 1200 = 6000 ft·lb.

Counterclockwise moments = 15 × R$_2$

$$\begin{matrix} \text{Clockwise} \\ \text{moments} \end{matrix} = \begin{matrix} \text{Counterclockwise} \\ \text{moments} \end{matrix}$$

6000 ft·lb = 15 × R$_2$

Therefore

$$R_2 = \frac{6000}{15} = 400 \text{ lb}$$

$$R_1 + R_2 = 1200 \text{ lb}$$

Thus

$$R_1 = 1200 - 400 = 800 \text{ lb}$$

EXAMPLE 4 Use the same data given in Example 3, but include the force of gravity acting on the beam, which is 40 lb/ft. See Fig. 29-4-5.
Solution Since the force of gravity acting on the beam is uniformly dis-

CONCENTRATED LOAD
REPLACING UNIFORMLY
DISTRIBUTED LOAD
OF 40 LB/FT

(C) MOMENT DIAGRAM ABOUT R$_1$

Fig. 29-4-5 Simple beam with uniformly distributed and concentrated loads.

tributed, a concentrated load of 600 lb located midway on the beam, as shown in Fig. 29-4-5B, would have the same effect on the reactions R$_1$ and R$_2$.

Therefore by substituting the 600-lb concentrated load 7′-6 from the reactions for the uniformly distributed load, the reaction forces can now be found.

Taking moments about reaction R$_1$ (Fig. 29-4-5C), we have

Clockwise moments using pounds and feet

$$\begin{aligned} &= 1200 \times 5 + 600 \times 7.5 \\ &= 6000 + 4500 \\ &= 10\,500 \text{ ft·lb} \end{aligned}$$

$$\begin{matrix} \text{Clockwise} \\ \text{moments} \end{matrix} = \begin{matrix} \text{Counterclockwise} \\ \text{moments} \end{matrix}$$

10 500 ft·lb = 15 × R$_2$

Therefore

$$R_2 = 10\,500 \div 15 = 700 \text{ lb}$$

$$R_1 + R_2 = 1800 \text{ lb}$$

Thus

$$R_1 = 1800 - 700 = 1100 \text{ lb}$$

ASSIGNMENTS

See Assignments 59 and 60 for Unit 29-4 on page 598.

UNIT 29-5
Shear Diagrams

When a beam supports a load, there is a tendency for the beam to fail by shear. In design work, it is essential to know what shear force a beam must resist at any section. The *vertical shear force* at a section of a beam is the algebraic sum of all the external forces acting on either side of the section. This can be further simplified by stating that the vertical shear at any section is equal to the product of the reaction minus the loads. The *section* is the name given to the cross section of the beam where the calculations are made. For simplification, only the forces acting to the left of the section will be calculated in this text.

Shear is designated as either *positive shear* or *negative shear*. When the sum of the vertical forces to the left of the section is upward, the shear is positive. When the sum of the vertical forces to the left of the section is downward, the shear is negative. See Fig. 29-5-1.

This information is represented in a shear force diagram which is normally drawn below the loading diagram of the beam. A horizontal zero base line is drawn to the same horizontal scale as the loading diagram, and the positive shear is shown above this line while the negative shear is drawn below it. The magnitude of the shear at each section is shown by vertical lines drawn to a convenient scale.

In order to identify the section at which the shear is taken, a symbol (the letter *V* followed by a number) is used. The letter *V* refers to the magnitude of the vertical shear, and the number refers to the horizontal distance from the left end of the beam. Thus V_4 refers to the shear force taken at a section 4 ft away from the left reaction (R_1) of a simple beam, or 4 ft away from the free end of a cantilever beam. The mass of the beam will not be considered unless specified in the examples or problems.

CANTILEVER BEAMS

Cantilever beams should be drawn with the support shown at the RH side.

EXAMPLE 1 Figure 29-5-2A represents a cantilever beam with a concentrated

(A) POSITIVE SHEAR

(B) NEGATIVE SHEAR

Fig. 29-5-1 Designation of positive and negative shear.

(A) WITH CONCENTRATED LOAD

(1) LOADING DIAGRAM
(2) SECTION TAKEN 3' FROM LOAD
(3) SECTION TAKEN 6' FROM LOAD
(4) SECTION TAKEN 9' FROM LOAD
(5) SECTION TAKEN AT WALL

(B) WITH UNIFORMLY DISTRIBUTED LOAD

(1) LOADING DIAGRAM
(2) SECTION TAKEN 9' FROM WALL
(3) SECTION TAKEN 6' FROM WALL
(4) SECTION TAKEN 3' FROM WALL
(5) SECTION TAKEN AT WALL

Fig. 29-5-2 Construction of shear diagram for cantilever beams.

load at the free end. Construct the shear diagram.

Solution Taking sections at various points along the beam and calculating V, the vertical shear to the left of the section, we have

$$V_0 = 0 - 300 = -300 \text{ lb}$$
$$V_3 = 0 - 300 = -300 \text{ lb}$$
$$V_6 = 0 - 300 = -300 \text{ lb}$$
$$V_9 = 0 - 300 = -300 \text{ lb}$$
$$V_{12} = 0 - 300 = -300 \text{ lb}$$

Since there is no reaction to the left of the section, the shear values are all negative and are drawn below the base line.

2.5 m
800 N/m
4 m

0   0
−2000 N

(A) UNIFORMLY DISTRIBUTED LOAD AT END OF BEAM

500 LB
9'
60 LB FT
16'

0   0
−500 LB
−1040 LB

(B) COMBINATION CONCENTRATED AND A PARTIAL UNIFORMLY DISTRIBUTED LOAD

650 LB
8'   10'
80 LB FT
18'

NEGATIVE SHEAR AS FORCE TO THE LEFT OF SECTION ACTING DOWNWARD

0   0
−640 LB
−1290 LB
−2090 LB

(C) COMBINATION CONCENTRATED AND UNIFORMLY DISTRIBUTED LOAD

Fig. 29-5-3 Shear diagrams for cantilever beams.

EXAMPLE 2 Figure 29-5-2B illustrates a cantilever beam with a uniformly distributed load. Construct the shear diagram.

Solution Taking sections at various points along the beam, starting at the free end, we have

$$V_0 = 0 - 0 = 0 \text{ lb}$$
$$V_3 = 0 - (800 \times 3) = -2400 \text{ lb}$$
$$V_6 = 0 - (800 \times 6) = -4800 \text{ lb}$$
$$V_9 = 0 - (800 \times 9) = -7200 \text{ lb}$$
$$V_{12} = 0 - (800 \times 12) = -9600 \text{ lb}$$

EXAMPLE 3 Figure 29-5-3A illustrates a cantilever beam with a uniformly distributed load at the end of the beam. Construct the shear diagram.

Solution Taking sections at various points along the beam, starting at the free end, we have

$$V_0 = 0 - 0 = 0 \text{ N}$$
$$V_1 = 0 - 800 = -800 \text{ N}$$
$$V_2 = 0 - 800 \times 2 = -1600 \text{ N}$$
$$V_{2.5} = 0 - 800 \times 2.5 = -2000 \text{ N}$$
$$V_3 = 0 - 800 \times 2.5 = -2000 \text{ N}$$
$$V_4 = 0 - 800 \times 2.5 = -2000 \text{ N}$$

EXAMPLE 4 Figure 29-5-3B shows a cantilever beam with a concentrated load at the free end of the beam and a uniformly distributed load at the fixed end. Construct the shear diagram.

Solution Taking sections at various points along the beam, starting at the free end, we have

$$V_0 = 0 - 500 = -500 \text{ lb}$$
$$V_4 = 0 - 500 = -500 \text{ lb}$$
$$V_8 = 0 - 500 - (1 \times 60) = -560 \text{ lb}$$
$$V_{12} = 0 - 500 - (5 \times 60) = -800 \text{ lb}$$
$$V_{16} = 0 - 500 - (9 \times 60) = -1040 \text{ lb}$$

EXAMPLE 5 Figure 29-5-3C shows a cantilever beam with a uniformly distributed load and a concentrated load acting in the middle section of the beam. Construct the shear diagram.

Solution Taking sections at various points along the beam, starting at the free end, we have

$$V_0 = 0 - 0 = 0 \text{ lb}$$
$$V_4 = 0 - (4 \times 80) = -320 \text{ lb}$$

$$V_8 = 0 - (8 \times 80) = -1290 \text{ lb}$$
$$V_{12} = 0 - (12 \times 80) = -1610 \text{ lb}$$
$$V_{18} = 0 - (18 \times 80) = -2090 \text{ lb}$$

SIMPLE BEAMS

In constructing the shear diagram for a simple beam, the magnitude of the reactions must be calculated first. The shear diagram is constructed in the same manner as for a cantilever beam. For calculation purposes V_0 will be considered as the section where the beam leaves the reaction R_1, and the shear section taken at R_2 will be considered as the section where the beam leaves the reaction R_2.

EXAMPLE 6 Figure 29-5-4A illustrates a simple beam with a concentrated load of 800 lb at the center of the span. Construct the shear diagram.

Solution From the figure it is apparent that the reactions are 400 lb each, because of the symmetrical loading. As previously mentioned, only the forces acting to the left of the section will be used for calculating the shear diagram. At the left end of the beam, the only force acting on the beam is the upward force of the reaction R_1. Therefore the shear force at $V_0 = 1400 - 0 = +400$ lb. Taking sections along the beam, we find that

$$V_4 = 400 - 0 = +400 \text{ lb}$$
$$V_7 = 400 - 0 = +400 \text{ lb}$$

Up to and including the section just to the left of the center of the beam, no new forces are encountered; therefore from V_0 to $V_{7.99}$ the shear force is $+400$ lb. At the center of the beam the downward force of 800 lb occurs; thus

$$V_8 = 400 - 800 = -400 \text{ lb}$$

Since no new loads are encountered for the remainder of the beam unitl R_2 is reached,

$$V_{12} = 400 - 800 = -400 \text{ lb}$$
$$V_{14} = 400 - 800 = -400 \text{ lb}$$

Just before R_2 is reached, the shear force from V_8 to V_{16} is -400 lb.

Note that the shear diagram passes through zero, from $+400$ lb to -400 lb at the 800 lb load.

Fig. 29-5-4 (A) WITH CONCENTRATED LOAD

800 LB
8'   8'
16'
R_1 (I) LOADING DIAGRAM R_2

800 LB
4'
400 LB   400 LB
+400 LB   ←SHEAR FORCE AT SECTION
0   0
(2) SECTION TAKEN 4' FROM R_1

800 LB
8'
400 LB   400 LB
+400 LB   ←SHEAR FORCE JUST BEFORE SECTION
0   0
←SHEAR FORCE AT SECTION
−400 LB
(3) SECTION TAKEN 8' FROM R_1

800 LB
12'
400 LB   400 LB
+400 LB
0   0
SHEAR FORCE AT SECTION
−400 LB
(4) SECTION TAKEN 12' FROM R_1

800 LB
16'
400 LB   400 LB
+400 LB
0   0
SHEAR FORCE AT SECTION   −400 LB   COMPLETED SHEAR DIAGRAM
(5) SECTION TAKEN AT R_2

(A) WITH CONCENTRATED LOAD

(B) WITH UNIFORMLY DISTRIBUTED LOAD

400 N/m
6 m
R_1 (I) LOADING DIAGRAM R_2

400 N/m
1.5 m
1200 N   1200 N
+1200 N   ←SHEAR FORCE AT SECTION
+600 N
0
SHEAR FORCE AT R_1
(2) SECTION TAKEN 1.5 m FROM R_1

400 N/m
3 m
1200 N   1200 N
+1200 N   ←SHEAR FORCE AT SECTION
0   0
(3) SECTION TAKEN 3 m FROM R_1

400 N/m
4.5 m
1200 N   1200 N
+1200 N
0   SHEAR FORCE AT SECTION   0
−600 N
(4) SECTION TAKEN 4.5 m FROM R_1

400 N/m
6 m
1200 N   1200 N
+1200 N
0   0
SHEAR FORCE AT R_2 −1200 N
(5) SECTION TAKEN AT R_2

(B) WITH UNIFORMLY DISTRIBUTED LOAD

Fig. 29-5-4   Construction of shear diagram for simple beams.

EXAMPLE 7 Figure 29-5-4B illustrates a simple beam with a uniformly distributed load. Construct the shear diagram.
Solution Taking sections at intervals along the beam, we have

$$V_0 = 1200 - 0 = +1200 \text{ N}$$

$$V_{1.5} = 1200 - 1.5 \times 400 = +600 \text{ N}$$

$$V_3 = 1200 - 3 \times 400 = +0 \text{ N}$$

$$V_{4.5} = 1200 - 4.5 \times 400 = -600 \text{ N}$$

$$V_6 = 1200 - 6 \times 400 = -1200 \text{ N}$$

EXAMPLE 8 Figure 29-5-5 illustrates a simple beam with a partial, uniformly distributed load starting at reaction R_1. Construct the shear diagram.

Fig. 29-5-5   Simple beam with partial, uniformly distributed load.

Solution The values of reactions R_1 and R_2 must first be found. For calculation purposes, a concentrated load of 250×8, or 2000 lb, acting at the center of the uniformly distributed load will be used in place of the uniformly distributed load. Taking moments about R_1, we have

Clockwise moments
$$= 4 \times (8 \times 250) = 8000 \text{ ft·lb}$$

Counterclockwise moments $= 20 \times R_2$

$$R_2 = 8000 \div 20 = 400 \text{ lb}$$

$$R_1 = 2000 - 400 = 1600 \text{ lb}$$

Taking sections at intervals along the beam starting at reaction R_1, we have

$$V_0 = 1600 - 0 = +1600 \text{ lb}$$

$$V_4 = 1600 - (4 \times 250) = +600 \text{ lb}$$

$$V_8 = 1600 - (8 \times 250) = -400 \text{ lb}$$

$$V_{12} = 1600 - (8 \times 250) = -400 \text{ lb}$$

$$V_{20} = 1600 - (8 \times 250) = -400 \text{ lb}$$

From the shear diagram it can be seen that the shear passes from positive shear to negative shear between V_4 and V_8. How to locate the position of zero shear will be discussed in Unit 29-6.

EXAMPLE 9 Figure 29-5-6 illustrates a simple beam with two concentrated loads. Construct the shear diagram.
Solution The reactions must first be calculated. Taking moments about reaction R_1, we have

Fig. 29-5-6 Simple beam with two concentrated loads.

Fig. 29-5-7 Simple beam with uniformly distributed load and concentrated load.

Fig. 29-5-8 Overhanging beam.

Clockwise moments
$$= (4 \times 750) + (10 \times 600)$$
$$= 3000 + 6000 = 9000 \text{ ft·lb}$$

Counterclockwise moments $= 18 \times R_2$

$$\begin{array}{cc} \text{Clockwise} & \text{Counterclockwise} \\ \text{moments} & \text{moments} \end{array}$$

$18 \times R_2 = 9000 \text{ ft·lb}$

$$R_2 = \frac{9000}{18} = 500 \text{ lb}$$

$R_1 + R_2 = 750 + 600 = 1350 \text{ lb}$

Therefore

$$R_1 = 1350 - 500 = 850 \text{ lb}$$

Taking sections at intervals along the beam, we have

$V_0 = 850 - 0 = +850 \text{ lb}$

$V_4 = 850 - 750 = +100 \text{ lb}$

$V_{10} = 850 - 750 - 600 = -500 \text{ lb}$

$V_{18} = 850 - 750 - 600 = -500 \text{ lb}$

EXAMPLE 10 Figure 29-5-7 illustrates a simple beam with a uniformly distributed load and a concentrated load acting on it. Construct the shear diagram.
Solution The reactions must be calculated first. For calculation purposes, a concentrated load of 12 × 150, or 1800 lb, acting at the center of the beam, will be used in place of the uniformly distributed load. Taking moments about R_1, we have

Clockwise moments
$$= (4 \times 1200) + (6 \times 1800)$$
$$= 4800 + 10\ 800 = 15\ 600 \text{ ft·lb}$$

$$\begin{array}{cc} \text{Clockwise} & \text{Counterclockwise} \\ \text{moments} & \text{moments} \end{array}$$

$12 \times R_2 = 15\ 600 \text{ ft·lb}$

$R_2 = 15\ 600 \div 12 = 1300 \text{ lb}$

$R_1 + R_2 = 1200 + 1800 = 3000 \text{ lb}$

Therefore

$R_1 = 3000 - R_2 = 3000 - 1300 = 1700 \text{ lb}.$

Taking sections at intervals along the beam gives us

$V_0 = 1700 - 0 = +1700 \text{ lb}$

$V_2 = 1700 - (2 \times 150) = +1400 \text{ lb}$

$V_4 = 1700 - (4 \times 150) - 1200 = -100 \text{ lb}$

$V_8 = 1700 - (8 \times 150) - 1200 = -700 \text{ lb}$

$V_{10} = 1700 - (10 \times 150) - 1200 = -1000 \text{ lb}$

$V_{12} = 1700 - (12 \times 150) - 1200 = -1300 \text{ lb}$

EXAMPLE 11 Figure 29-5-8 illustrates an overhanging beam. Construct the shear diagram.
Solution First the reactions must be calculated. Taking moments about R_1, we have

Clockwise moments
$$= 1.4 \times 24 + 3.6 \times 35 + 7 \times 30$$
$$= 33.6 + 126 + 210$$
$$= 369.6 \text{ kN·m}$$

Counterclockwise moments $= 5 \times R_2$

$$\begin{array}{cc} \text{Clockwise} & \text{Counterclockwise} \\ \text{moments} & \text{moments} \end{array}$$

$5 \times R_2 = 369.6 \text{ kN·m}$

$R_2 = 369.6 \div 5 = 73.92 \text{ kN}$

$R_1 + R_2 = 24 + 35 + 30 = 89 \text{ kN}$

Therefore

$$R_1 = 89 - 73.92 = 15.08 \text{ kN}$$

Taking sections at intervals along the beam gives us

$V_0 = 15.08 - 0 = +15.08 \text{ kN}$

$V_{1.4} = 15.08 - 24 = -8.92 \text{ kN}$

$V_3 = 15.08 - 24 = -8.92 \text{ kN}$

$V_{3.6} = 15.08 - 24 - 35 = -43.92 \text{ kN}$

$V_5 = 15.08 - 24 - 35 + 73.92 = +30 \text{ kN}$

$V_7 = 15.08 - 24 - 35 + 73.92 = +30 \text{ kN}$

Conclusion

From the examples given, the following conclusions can be drawn for shear diagrams.

- Where there are concentrated loads, the shear lines are straight horizontal lines changing in value at the loads.
- Where there are uniformly distributed loads, the shear lines are straight inclined lines, the slope of the line being proportional to the load.
- At each concentrated load, including reactions, the shear line rises or drops vertically by an amount equal to the load at that section.

ASSIGNMENTS

See Assignments 61 and 62 for Unit 29-5 on page 598.

UNIT 29-6
Bending Moment Diagrams

As previously mentioned, when a load acts on a beam, the force tends to shear the beam. In addition to producing this shearing action, the load tends to *deflect* or *bend* the beam. To determine this deflection, which varies along the beam, the bending stresses must be calculated. Just as the shear diagram shows the shear at any section along the beam, a bending moment diagram is similarly constructed to show the bending moment at any point along the beam and also to indicate where the maximum bending occurs.

The *bending moment* at any section along the beam is equal to the sum of all the moments of the forces acting to the right or left of the beam. In drawing bending moment diagrams, the following points should be noted

1. Forces are taken to the left of the section.
2. Upward moments are considered positive and are shown above the base line on the bending moment diagram.
3. Downward moments are considered negative and are shown below the base line.
4. The bending moment diagram is drawn directly below the shear diagram and to the same scale.
5. Shear is equal to reaction minus loads.

6. Bending moments are equal to reaction moments minus load moments.
7. In calculating the bending moments at any given section along the beam, the capital letter M is used to designate the bending moments. It is followed by a subscript which indicates the distance from the LH end of the beam. Thus, M_4 indicates the bending moments 4 ft from the LH end of the beam.

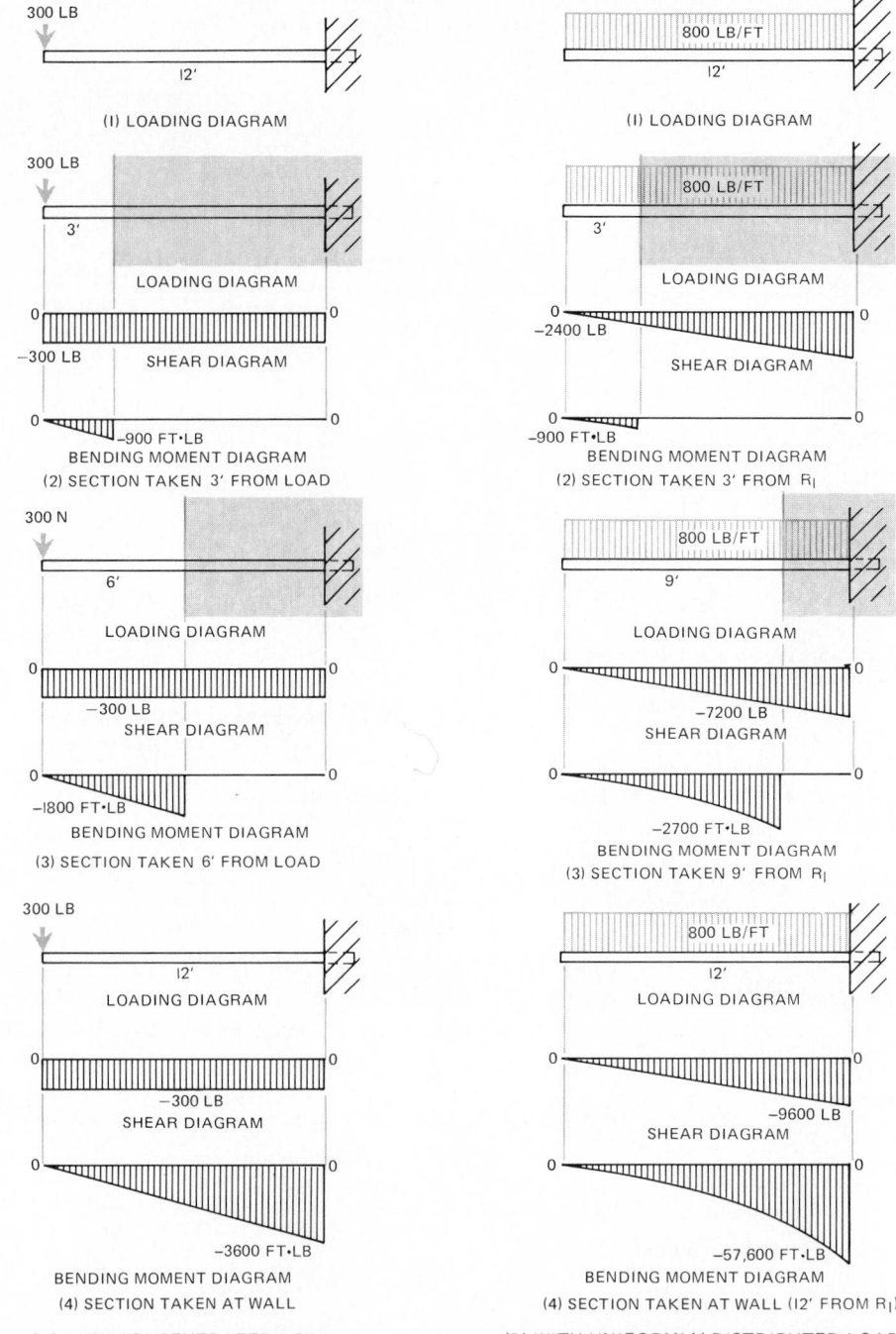

Fig. 29-6-1 **Construction of bending moment diagram for cantilever beams.**

EXAMPLE 1 Figure 29-6-1A shows a cantilever beam with a concentrated load applied to the free end. Figure 29-5-2A shows the shear diagram development for this beam. Construct the bending moment diagram.

Solution Taking moments at intervals along the beam starting at the LH end, we find the moments will be negative as the force is acting downward. Thus we have

$$M_0 = -300 \times 0 = 0$$
$$M_3 = -300 \times 3 = -900 \text{ ft·lb}$$
$$M_6 = -300 \times 6 = -1800 \text{ ft·lb}$$
$$M_9 = -300 \times 9 = -2700 \text{ ft·lb}$$
$$M_{12} = -300 \times 12 = -3600 \text{ ft·lb}$$

EXAMPLE 2 Figure 29-6-1B shows a cantilever beam with a uniformly distributed load. Figure 29-5-2B shows the shear diagram development for this beam. Construct the bending moment diagram.

Solution The bending moment at M_0 is zero. The 3 ft section to the right of R_1 weighs 3×800 or 2400 lb. The force of any uniform load can be considered as acting at its center of gravity. Thus the 2400 lb load can be considered as acting 1.5 ft away from R_1. Therefore, we have

$$M_3 = -(800 \times 3) \times 1.5 = -3600 \text{ ft·lb}$$
$$M_6 = -(800 \times 6) \times 3 = -14\,400 \text{ ft·lb}$$
$$M_9 = -(800 \times 9) \times 4.5 = -32\,400 \text{ ft·lb}$$
$$M_{12} = -(800 \times 12) \times 6 = -57\,600 \text{ ft·lb}$$

EXAMPLE 3 Figure 29-6-2 shows a cantilever beam with a uniformly distributed load at the end of the beam. The shear diagram development was explained in Unit 29-5, Example 3, Fig. 29-5-3A. Construct the bending moment diagram.

Solution Taking moments at intervals along the beam starting from the LH end, we have

$$M_0 = 0$$
$$M_1 = -800 \times 1 \times 0.5 = -400 \text{ N·m}$$
$$M_2 = -800 \times 2 \times 1 = -1600 \text{ N·m}$$
$$M_{2.5} = -800 \times 2.5 \times 1.25 = -2500 \text{ N·m}$$
$$M_3 = -800 \times 2.5 \times (3 - 1.25) = -3500 \text{ N·m}$$

Note that at M_3 the distance from the section to the center of gravity of the load is $3 - 1.25 = 1.75$ m, since the center of gravity is 1.25 m from the LH end.

$$M_4 = -800 \times 2.5 \times (4 - 1.25) = -5500 \text{ N·m}$$

Note that from $M_{2.5}$ to M_4 the line on the bending moment diagram is straight.

EXAMPLE 4 Figure 29-6-3 shows a cantilever beam with a combination concentrated and a partial, uniformly distributed load. The shear diagram development was explained in Unit 29-5, Example 4, Fig. 29-5-3B. Construct the bending moment diagram.

Solution The bending moments from M_0 to M_{16} are calculated in the same manner as in Example 1.

$$M_0 = -500 \times 0 = 0$$
$$M_1 = -500 \times 1 = -500 \text{ ft·lb}$$
$$M_4 = -500 \times 4 = -2000 \text{ ft·lb}$$
$$M_7 = -500 \times 7 = -3500 \text{ ft·lb}$$

$$M_9 = -(500 \times 9) - (60 \times 2 \times 1) = -4620 \text{ ft·lb}$$
$$M_{12} = -(500 \times 12) - (60 \times 5 \times 2.5) = -6750 \text{ ft·lb}$$
$$M_{16} = -(500 \times 16) - (60 \times 9 \times 4.5) = -10\,430 \text{ ft·lb}$$

EXAMPLE 5 Figure 29-6-4 shows a cantilever beam with a concentrated and uniformly distributed load. The shear diagram development was explained in Unit 29-5, Example 5, Fig. 29-5-3C. Construct the bending moment diagram.

Solution The bending moments from M_0 to M_8 are calculated in the same manner as in Example 2.

$$M_0 = 0$$
$$M_2 = -80 \times 2 \times 1 = -160 \text{ ft·lb}$$
$$M_4 = -80 \times 4 \times 2 = -640 \text{ ft·lb}$$
$$M_8 = -(80 \times 8 \times 4) - (650 \times 0) = -2560 \text{ ft·lb}$$
$$M_{10} = -(80 \times 10 \times 5) - (650 \times 2) = -5300 \text{ ft·lb}$$
$$M_{12} = -(80 \times 12 \times 6) - (650 \times 4) = -8360 \text{ ft·lb}$$
$$M_{14} = -(80 \times 14 \times 7) - (650 \times 6) = -11\,740 \text{ ft·lb}$$
$$M_{18} = -(80 \times 16 \times 8) - (650 \times 6) = -14\,140 \text{ ft·lb}$$

Fig. 29-6-2   Cantilever beam with uniformly distributed load at end of beam.

Fig. 29-6-3   Cantilever beam with concentrated load and partial, uniformly distributed loads.

Fig. 29-6-4   Cantilever beam with concentrated and uniformly distributed loads.

EXAMPLE 6 Figure 29-6-5 shows a simple beam with a partial, uniformly distributed load. The development of the shear diagram was explained in Unit 29-5, Example 8, Fig. 29-5-5. Construct the bending moment diagram.

Solution Taking moments at intervals along the beam, starting at reaction R_1, gives us

$M_0 = +1600 \times 0 = 0$

$M_4 = +1600 \times 4 - 250 \times 4 \times 2$
$\quad = +3400$ ft·lb

$M_6 = +1600 \times 6 - 250 \times 6 \times 3$
$\quad = +5100$ ft·lb

$M_8 = +1600 \times 8 - 250 \times 8 \times 4$
$\quad = +4800$ ft·lb

From the bending moment calculations it can be seen that the maximum bending moment occurs somewhere between M_6 and M_8 where the zero shear takes place.

The distance between R_1 and zero shear can be found as follows. Let the distance from R_1 to the point at zero shear be X. Thus, we have

$V_X = +1600 - 250 \times X = 0$

$X = 1600 \div 250 = 6.4$ ft

Maximum bending moment occurs at $M_{6.4}$

$M_{6.4} = +1600 \times 6.4 - 250 \times 6.4 \times 3.2$
$\quad = +5120$ ft·lb

Fig. 29-6-5 Simple beam with partial, uniformly distributed load.

For calculating the moments for the remaining sections, the uniformly distributed load will be considered as a 2000 lb concentrated load acting 4 ft from R_1.

$M_{12} = +1600 \times 12 - 2000 \times 8$
$\quad = +3200$ ft·lb

$M_{16} = +1600 \times 16 - 2000 \times 12$
$\quad = +1600$ ft·lb

$M_{20} = +1600 \times 20 - 2000 \times 16 = 0$

EXAMPLE 7 Figure 29-6-6A shows a simple beam with a concentrated load acting in the center of the beam. The development of the shear diagram was explained in Unit 29-5, Example 6, Fig. 29-5-4A. Construct the bending moment diagram.

Solution Taking moments at intervals along the beam, starting at reaction R_1 which is acting upward, we have

(A) WITH CONCENTRATED LOAD

(B) WITH UNIFORMLY DISTRIBUTED LOAD

Fig. 29-6-6 Construction of bending moment diagram for simple beam.

$M_0 = +400 \times 0 - 0$

$M_2 = +400 \times 2 = +800 \text{ ft·lb}$

$M_4 = +400 \times 4 = +1600 \text{ ft·lb}$

$M_8 = +400 \times 8 - 800 \times 0$
$= +3200 \text{ ft·lb}$

$M_{12} = +400 \times 12 - 800 \times 4$
$= +1600 \text{ ft·lb}$

$M_{14} = +400 \times 14 - 800 \times 6$
$= +800 \text{ ft·lb}$

$M_{16} = +400 \times 16 - 800 \times 8 = 0$

Note: The maximum bending moment occurs at the point where shear passes through zero.

EXAMPLE 8 Fig. 29-6-6B shows a simple beam with a uniformly distributed load. The development of the shear diagram was explained in Unit 29-5, Example 7, Fig. 29-5-4B. Construct the bending moment diagram.

Solution The bending moment at M_0 is zero. The 1-m section to the right of R_1 creates a force of 400 N. The force of any uniform load can be considered as acting at its center of gravity. Thus, the 400-N load can be considered as acting 0.5 m away from R_1. Therefore, we have

$M_1 = +1200 \times 1 - 400 \times 1 \times 0.5$
$= +1000 \text{ N·m}$

$M_2 = +1200 \times 2 - 400 \times 2 \times 1$
$= +1600 \text{ N·m}$

$M_3 = +1200 \times 3 - 400 \times 3 \times 1.5$
$= +1800 \text{ N·m}$

$M_4 = +1200 \times 4 - 400 \times 4 \times 2$
$= +1600 \text{ N·m}$

$M_5 = +1200 \times 5 - 400 \times 5 \times 2.5$
$= +1000 \text{ N·m}$

$M_6 = +1200 \times 6 - 400 \times 6 \times 3 = 0$

Note that the maximum bending moment of +1800 N·m occurs at zero shear.

EXAMPLE 9 Figure 29-6-7 shows a simple beam with two concentrated loads. The development of the shear diagram was explained in Unit 29-5, Example 9, Fig. 29-5-6. Construct the bending moment diagram.

Solution Taking moments at intervals along the beam, starting at reaction R_1, gives us

$M_0 = +850 \times 0 = 0$

$M_2 = +850 \times 2 = +1700 \text{ ft·lb}$

$M_4 = +850 \times 4 - 750 \times 0$
$= +3400 \text{ ft·lb}$

Fig. 29-6-7   Simple beam with two concentrated loads.

$M_5 = +850 \times 5 - 750 \times 1$
$= +3500 \text{ ft·lb}$

$M_8 = +850 \times 8 - 750 \times 4$
$= +3800 \text{ ft·lb}$

$M_{10} = +850 \times 10 - 750 \times 6$
$- 600 \times 0 = +4000 \text{ ft·lb}$

$M_{12} = +850 \times 12 - 750 \times 8$
$- 600 \times 2 = +3000 \text{ ft·lb}$

$M_{14} = +850 \times 14 - 750 \times 10$
$- 600 \times 4 = +2000 \text{ ft·lb}$

$M_{18} = +850 \times 18 - 750 \times 14$
$- 600 \times 8 = 0$

EXAMPLE 10 Figure 29-6-8 shows a simple beam with a uniformly distributed load and a concentrated load. The development of the shear diagram was explained in Unit 29-5, Example 10, Fig. 29-5-7. Construct the bending moment diagram.

Solution Taking moments at intervals along the beam, starting at reaction R_1, we have

$M_0 = +1700 \times 0 = 0$

$M_2 = +1700 \times 2 - 150 \times 2 \times 1$
$= +3100 \text{ ft·lb}$

$M_4 = +1700 \times 4 - 150 \times 4 \times 2$
$- 1200 \times 0 = +5600 \text{ ft·lb}$

$M_6 = +1700 \times 6 - 150 \times 6 \times 3$
$- 1200 \times 2 = +5100 \text{ ft·lb}$

$M_8 = +1700 \times 8 - 150 \times 8 \times 4$
$- 1200 \times 4 = +4000 \text{ ft·lb}$

$M_{10} = +1700 \times 10 - 150 \times 10 \times 5$
$- 1200 \times 6 = +2300 \text{ ft·lb}$

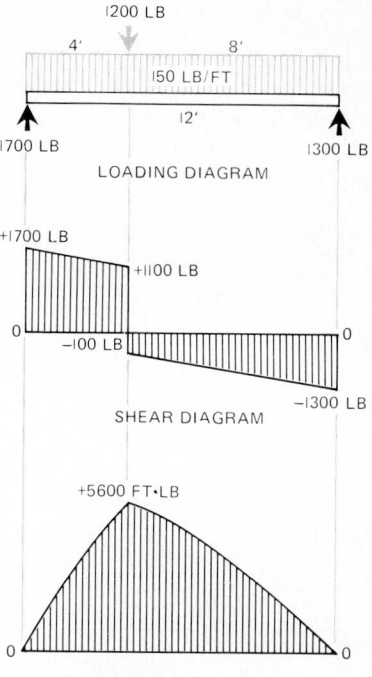

Fig. 29-6-8   Simple beam with uniformly distributed and concentrated loads.

$M_{12} = +1700 \times 12 - 150 \times 12 \times 6$
$- 1200 \times 8 = 0$

EXAMPLE 11 Figure 29-6-9 shows an overhanging beam with three concentrated loads. The development of the shear diagram was explained in Unit 29-5, Example 11, Fig. 29-5-8. Construct the bending moment diagram.

Solution Taking moments at intervals along the beam, starting at reaction R_1, we have

$M_0 = +15.08 \times 0 = 0$

$M_1 = +15.08 \times 1 = +15.08 \text{ kN·m}$

$M_{1.4} = +15.08 \times 1.4 - 24 \times 0$
$= +21.1 \text{ kN·m}$

$M_2 = +15.08 \times 2 - 24 \times 0.6$
$= +16.12 \text{ kN·m}$

$M_3 = +15.08 \times 3 - 24 \times 1.6$
$= +6.84 \text{ kN·m}$

$M_{3.6} = +15.08 \times 3.6 - 24 \times 2.2$
$- 35 \times 0 = +1.39 \text{ kN·m}$

$M_4 = +15.08 \times 4 - 24 \times 2.6 - 35 \times$
$0.4 = -16.08 \text{ kN·m}$

$M_5 = +15.08 \times 5 - 24 \times 3.6 - 35 \times$
$1.4 + 73.92 \times 0 = -60 \text{ kN·m}$

$M_6 = +15.08 \times 6 - 24 \times 4.6 - 35 \times$
$2.4 + 73.92 \times 1 = -30 \text{ kN·m}$

$M_7 = +15.08 \times 7 - 24 \times 5.6 - 35 \times$
$3.4 + 73.92 \times 2 = 0$

LOADING DIAGRAM

24 kN   35 kN   30 kN
1.4 m ↓ 2.2 m ↓   ↓
        5 m   2 m
15.08 kN   73.92 kN

+30 kN

+15.08 kN

−8.92 kN

−43.92 kN

SHEAR DIAGRAM

+21.1 kN·m

+1.39 kN

3.63

−60 kN·m

BENDING MOMENT DIAGRAM

Fig. 29-6-9 Overhanging beam.

From the bending moment diagram it can be seen that zero bending moment occurs to the right of the 35-kN load. Its exact location can be found as follows.

Let the distance between R_1 and the point where zero takes place be X. Then

$$M_X = 0$$
$$M_X = +15.08 \times X - 24(X - 1.4)$$
$$\quad - 35(X - 3.6) = 0$$
$$\quad = +15.08X - 24X + 33.6$$
$$\quad - 35X + 126$$
$$43.92X = 159.6$$
$$X = 3.63 \text{mm}$$

EXAMPLE 12 Figure 29-6-10 shows a simple beam with a partial, uniformly distributed load and a concentrated load. Find the position and magnitude of the maximum bending moment.

Solution Reactions R_1 and R_2 must be calculated first. Taking moments about R_1, we have

Clockwise moments
$$= 1.25 \times 2 \times 2.5 + 3.2 \times 3$$
$$= 15.85 \text{ kN·m}$$

LOADING DIAGRAM

2.5 m   3 kN
2 kN/m   ↓   1.8 m
        5 m
R_1 R_2

CALCULATING REACTIONS

2.5   3 kN
2 kN/m   ↓
        5 m
4.83 kN   3.17 kN

+4.83 kN

0   0
            −0.17 kN   0
X
                    −3.17 kN

SHEAR DIAGRAM

+5.83 kN·m

0   0

BENDING MOMENT DIAGRAM

Fig. 29-6-10 Simple beam with a partial, uniformly distributed load and a concentrated load.

Counterclockwise moments $= 5 \times R_2$
$$R_2 = 15.85 \div 5 = 3.17 \text{ kN}$$
$$R_1 = 2.5 \times 2 + 3 - 3.17 = 4.83 \text{ kN}$$

Next construct the shear diagram, taking sections at intervals along the beam, starting at reaction R_1

$$V_0 = 4.83 - 0 = +4.83 \text{ kN}$$
$$V_1 = 4.83 - 1 \times 2 = +2.83 \text{ kN}$$
$$V_2 = 4.83 - 2 \times 2 = +0.83 \text{ kN}$$
$$V_{2.5} = 4.83 - 2.5 \times 2 = -0.17 \text{ kN}$$
$$V_{3.2} = 4.83 - 2.5 \times 2 - 3 = -3.17 \text{ kN}$$
$$V_5 = 4.83 - 2.5 \times 2 - 3 = -3.17 \text{ kN}$$

From the shear diagram, it is noted that the section having zero shear lies somewhere between R_1 and the end of the 2.5-m uniformly distributed load. Its exact position, X distance from R_1, may be found by taking the shear at V_X, which is zero. Thus, $V_X = 4.83 - X \times 2 = 0$. Therefore, $X = 4.83 \div 2 = 2.415$m.

The maximum bending moment will occur where the shear passes through

zero or 2.415 m from R_1. Thus the maximum bending moment is

$$M_{2.415} = 4.83 \times 2.415 - 2 \times 2.415$$
$$\times 1.2075 = 5.83 \text{ kN·m}$$

EXAMPLE 13 A simple beam 8 m long carries a 3600-N concentrated load 2 m from the left abutment. Calculate the maximum bending moment and shear.
Solution The maximum bending moment for a simple beam with a concentrated load at any point (see Fig. 29-6-11) is

$$\frac{FAB}{L} = \frac{3600 \times 2 \times 6}{8} = 5400 \text{ N·m}$$

Maximum shear $=$

$$\frac{FB}{L} = \frac{3600 \times 6}{8} = 2700 \text{ N}$$

Conclusion

From the examples given, the following conclusions can be drawn from bending moment diagrams.

1. Where there are no loads on a part of a beam, the bending moment line is a straight, sloping line.
2. Where there is a uniformly distributed load, the bending moment line is a curve.
3. The maximum bending moment occurs at a section on the beam at which the shear passes through zero.

Standard beam formulas are shown in Fig. 29-6-11.

ASSIGNMENTS

See Assignments 63 and 64 for Unit 29-6 on page 598.

UNIT 29-7
Beam Design

It has been found from experience that beams normally fail at the section where the bending moment is maximum, rather than by shearing at the supports. Therefore, in beam design, it is customary first to select a suitable beam size to withstand the bending forces and then to check it for shear and deflection. The ability of a beam to resist bending depends on such factors as the material used, the shape of its

BEAM AND LOADING L IN METERS	MAXIMUM BENDING MOMENTS	MAXIMUM SHEAR	MAXIMUM DEFLECTION
(cantilever, point load F)	FL	F	$\dfrac{FL^3}{3EI}$
(cantilever, distributed load N/m OR kN/m)	$\dfrac{NL^2}{2}$	NL	$\dfrac{NL^4}{8EI}$
(simple beam, central point load F, L/2 + L/2)	$\dfrac{FL}{4}$	$\dfrac{F}{2}$	$\dfrac{FL^3}{48EI}$
(simple beam, off-center load F; A, B)	$\dfrac{FAB}{L}$	WHEN B IS GREATER THAN A $\dfrac{FB}{L}$	$\dfrac{FA^2B^2}{3EIL}$
(simple beam, distributed load N/m OR kN/m)	$\dfrac{NL^2}{8}$	$\dfrac{NL}{2}$	$\dfrac{5NL^3}{384EI}$

E = MODULUS OF ELASTICITY   I = MOMENT OF INERTIA

Fig. 29-6-11 Maximum bending moments, shear, and deflection for commonly occurring loads on beams.

pascals (metric), respectively. The stress in pascals is divided by 10^6 to obtain the stress per square millimeter. The section modulus for certain regular sections can be found from the formulas given in Fig. 29-7-1. The values of Z for structural-steel shapes and many common circular and rectangular sizes are tabulated in most engineers' handbooks.

The letter S is frequently used in textbooks to designate section modulus. However, to avoid confusion with the letter S for stress, the letter Z will be used to designate section modulus throughout this chapter.

Structural shapes may be placed in two general positions, as shown in Fig. 29-7-2. Since the resistance to bending will depend on the position of the beam with regard to its neutral axis, two section modulus values are generally shown in engineering tables. One value is used when the beam is in the upright position, as shown in Fig. 29-7-2C(1) where the X-X axis is the neutral axis; the other is used when the beam is in the flat position, as shown in Fig. 29-7-2C(2), where the Y-Y axis is the neutral axis. The *neutral axis* is defined as the axis which passes

cross section, and the way the cross section is turned with respect to the load. To illustrate this last point, one may bend a flat steel rule across its thin axis; but if the steel rule is set on its edge, then it is virtually impossible to bend the rule in the direction of its width. This resistance to bending can be measured in terms of a quantity called the *section modulus* of the section concerned. The theory and the mathematics behind the development of the section modulus of beams and shapes will not be covered in this text.

Thus the ability of any beam to resist bending is directly related to its section modulus, which is expressed in cubic inches (U.S. Customary) or cubic millimeters (metric) and is denoted Z in calculations. The bending stress S, the bending moment M_0, and the section modulus are related by the formula $M_0 = Z \times S \div 10^6$ in which the quantities are inch-pounds, cubic inches, and pounds (U.S. Customary) and newton-millimeters, cubic millimeters, and

$$Z = \frac{BD^2}{6}$$

$$Z = \frac{BD^3 - bd^3}{6D}$$

$$Z = \frac{\pi\,D^3}{32}$$

$$Z = \frac{\pi\,(D^4 - d^4)}{32D}$$

Fig. 29-7-1 Formulas for section moduli for common shapes.

(A) LOCATION OF NEUTRAL AXIS

(B) NEUTRAL AXES OF RECTANGLES

(C) NEUTRAL AXES OF BEAMS

Fig. 29-7-2 Neutral axis.

through the centroid of the cross-sectional area.

The majority of engineering handbooks show only one illustration of the structural shape with both the X-X and Y-Y axes shown as illustrated in Fig. 29-7-2C(3).

SHEARING STRESSES IN BEAMS

In designing beams for vertical shear, it is customary to consider only the full height of the webs of S, C, WT, and WWT beams to carry the full load; the flanges are not considered.

EXAMPLE 1 A cantilever beam 10 ft long supports a 5000 lb load at the end of the beam. What size A36 beam is required?

Solution First select a beam to withstand the bending forces. Refer to Fig. 29-6-11; maximum bending moments $= FL = 5000 \times 10 \times 12 = 600\,000$ in.·lb. The allowable bend stress for A36 steel (see Fig. 29-1-8) is 24 kips/in.^{2}.

Section modulus required $= Z$

$= M \div S = 600\,000 \div 24\,000$
$= 25$ in.^{3}

Refer to Fig. 29-7-3. A W8 × 31 has a section modulus of 27.5 in.^{3}, which is acceptable. If the depth of the beam is not an important design factor, then the W10 × 29 beam, which is lighter and has a section modulus of 30.9 in.^{3}, would be the most economical.

Next the beam must be checked for vertical shear. Maximum shear force $= 5000$ lb. Web area of a W8 × 31 beam (refer to the structural-steel handbook) $= 8 \times 31 = 2.48$ in.^{2}. Vertical shear stress $= 5000 \div 2.48 = 2016$ lb or 2 kips/in.^{2}. Permissible shear stress for steel (see Fig. 29-1-8) is 14.5 kips/in.^{2}. Therefore the W8 × 31 beam is acceptable.

EXAMPLE 2 A simple beam 6 m long supports a uniformly distributed load of 6 kN·m. Neglecting the mass of the beam, select the lightest A572M-310 beam to safely carry this load.

Solution First select a beam to withstand the bending forces. Refer to Fig. 29-6-11. Maximum bending moments $= NL^2 \div 8 = 6000 \times 6^2 \div 8 = 27\,000$ N·m, or 27×10^6 N·mm. The allowable

U.S. CUSTOMARY			METRIC*		
Section Modulus In.^{3}	Shape	Moment of Inertia = I In.^{4**}	Section Modulus 10^3 mm^3	Shape	Moment of Inertia = I 10^6 mm^{4***}
157	W18 × 85	1440	3050	W460 × 128	637
151	W16 × 88	1220	2850	W410 × 132	538
151	W21 × 73	1600	2830	W530 × 109	667
142	W18 × 77	1290	2670	W460 × 113	556
140	W21 × 68	1480	2620	W530 × 101	617
131	W14 × 84	928	2560	W360 × 134	415
125	W12 × 92	789	2420	W310 × 143	348
121	W14 × 78	851	2270	W360 × 122	365
118	W18 × 64	1050	2180	W460 × 97	445
116	W16 × 71	941	2160	W310 × 129	308
116	W12 × 85	723	2130	W410 × 100	398
112	W14 × 74	797	2060	W360 × 110	331
107	W12 × 79	663	1950	W310 × 118	275
105	W16 × 64	836	1830	W460 × 82	370
97.5	W12 × 72	597	1770	W310 × 107	248
94.3	W16 × 58	748	1730	W410 × 85	315
92.2	W14 × 61	641	1680	W360 × 91	267
88	W12 × 65	533	1590	W310 × 97	222
80.9	W16 × 50	657	1510	W410 × 74	275
73.6	W10 × 66	382	1400	W250 × 101	164
70.6	W12 × 53	426	1280	W310 × 79	177
64.8	W12 × 50	395	1190	W310 × 74	165
64.8	S15 × 50	486	1140	S380 × 64	187
62.7	W14 × 43	429	1140	W360 × 64	178
60.5	W10 × 54	306	1140	S380 × 64	187
59.6	S15 × 42.9	447	1090	W250 × 80	126
54.7	W14 × 38	386	1010	W360 × 57	161
51.9	W12 × 40	310	1010	S310 × 74	128
50.8	S12 × 50	305	941	W310 × 60	129
49.2	W10 × 45	249	901	W250 × 67	104
43.3	W8 × 48	184	803	W200 × 71	76.6
41.8	W14 × 30	290	779	W360 × 45	122
36.3	S12 × 31.8	218	690	S310 × 47	91.1
35.1	W10 × 33	171	633	W250 × 49	70.6
30.9	W10 × 29	158	602	W250 × 45	71.1
27.5	W8 × 31	110	496	W200 × 46	45.5
24.8	S10 × 25.4	124	465	S250 × 38	51.4
24.3	W8 × 28	97.8	446	W200 × 42	40.9
21.6	W10 × 21	107	424	W250 × 33	48.9
20.8	W8 × 24	82.5	380	W200 × 36	34.4
16.2	S8 × 23	64.9	316	S200 × 34	27.0
14.2	W8 × 17	56.6	279	W200 × 27	25.8
13.4	W6 × 20	41.5	244	W150 × 30	17.2
12	M12 × 11.8	71.9	232	M310 × 17.6	29.7
10.1	W6 × 16	31.7	192	W150 × 24	13.4
7.23	W6 × 12	21.7	136	W150 × 18	9.16
5.43	W4 × 13	11.3	103	W100 × 19	4.76
5.08	W6 × 85	14.8	103	W150 × 14	6.87

*Soft Converted.   **Taken at X-X axis.

Fig. 29-7-3 Section modulus and moment of inertia for shapes used as beams.

bending stress for A572M-310 steel (see Fig. 29-1-8) is 205 MPa. Section modulus required $= Z = M \div (S \div 10^6) = 27 \times 10^6 \times 10^6 \div (205 \times 10^6) = 131\ 700\ mm^3$.

Referring to Fig. 29-7-3, we find that a W150 × 18 beam has a section modulus of 136 000 mm³, which is acceptable.

Next the beam must be checked for vertical shear.

Maximum shear force (Fig. 29-6-11)
$$= \frac{NL}{2} = \frac{6\ kN \times 6}{2} = 18\ kN$$

Web area of a W150 × 18 beam (refer to the structural-steel handbook) $= 153 \times 6 = 918\ mm^2$. Average vertical shear stress $= 18\ 000 \div 0.000\ 918 = 19.6$ MPa. The permissible vertical shear stress for A572M-310 steel (see Fig. 29-1-8) is 125 MPa. Therefore the W150 × 18 beam is acceptable.

EXAMPLE 3 A simple beam 16 ft long supports a 30 000 lb concentrated load 4 ft from the left abutment. What size A36 beam is required?

Solution First select a beam to withstand the bending forces. Refer to Fig. 29-6-11. Maximum bending moments $= FAB \div L = 30\ 000 \times 4 \times 12 \div 16 = 90\ 000$ ft·lb, or 1 080 000 in.·lb. Allowable bending stress for A36 steel (see Fig. 29-1-8) is 24 kips/in.².

Section modulus required $= Z$
$= M \div S$
$= 1\ 080\ 000 \div 24\ 000$
$= 45\ in.^3$

Referring to Fig. 29-7-3, we find that a W10 × 45 beam has a section modulus of 49.2.

Next the beam must be checked for vertical shear.

Maximum shear force
$$= \frac{30\ 000 \times 12}{16} = 22\ 500\ lb$$

Web area of a W10 × 45 beam (refer to the structural-steel handbook) = 10.12 × .38 = 3.85 in.².

Average vertical shear stress =
$$\frac{22\ 500}{3.85} = 5844\ lb$$

The permissible vertical shear stress for A36 steel (see Fig. 29-1-8) is 14.5 kips/in.². Therefore the W10 × 45 beam is acceptable.

EXAMPLE 4 A floor 5 m wide has a uniformly distributed load of 3000 N·m. The floor joists are 38 mm wide and are spaced 400 mm center-to-center. If the allowable bending stress is not to exceed 9600 kPa, what depth of floor joists must be used?

Solution On each floor joist, the uniformly distributed load is 3000 × (400 ÷ 1000) = 1200 N·m. For a simple beam with a uniformly distributed load, the maximum bending moment $= FL^2 \div 8 = (1200 \times 5 \times 5) \div 8 = 3750$ N·m, or 375×10^4 N·mm. Allowable bending stress = 10 MPa. Therefore

Section modulus required $= Z$
$$= M \div \frac{S}{10^6}$$
$$= 375 \times 10^4 \times \frac{10^6}{10 \times 10^6}$$
$$= 375\ 000\ mm^3$$

The section modulus for the joist $= bd^2 \div 6$ where b = width = 38 mm and d = depth, which is unknown. Therefore

$$d = \sqrt{\frac{375\ 000 \times 6}{38}} = 243\ mm$$

Since standard joist sizes are 38 × 184, 38 × 235, and 38 × 286, the joist size of 38 × 286 is selected.

DEFLECTION OF BEAMS

The vertical distance a horizontally placed beam moves when it bends under an applied load is called *deflection*. Since deflection may cause cracking in plastered ceilings or buckling of floors, the limitations placed on the deflection of the beam may be the governing factor in its selection. In building construction the maximum deflection of beams is limited to 1/360 of the span, or, for cantilever beams, 1/180 of the span. After the beam is selected to withstand the bending and shearing stresses, it must then be checked for deflection. The theory and the mathematics behind the development of beam defection will not be covered in this text. The formulas using the *double-integration method* for finding the deflection of simple beams are shown in Fig. 29-6-11.

EXAMPLE 5 A W8 × 28 cantilever beam 8 ft long has a concentrated load of 5000 lb at its free end. Is the deflection excessive?

Solution Refer to Fig. 29-6-11. The maximum deflection for a cantilever beam having a concentrated load at its free end is $FL^3 \div 3EI$, where $F = 5000$ lb, $L = 96$ in., $E = 29 \times 10^6$ lb/in.² (see Fig. 29-1-3), and $I = 97.8$ in.⁴ (Fig. 29-7-3). Therefore

Maximum deflection
$$= \frac{5000 \times 96^3}{3 \times 97.8 \times 29 \times 10^6} = .52\ in.$$

Allowable deflection = 1/180 of the span (for cantilever beams) = 96 ÷ 180 = .53 in. Therefore, since the maximum deflection is less than the allowable, the W8 × 28 beam is acceptable.

EXAMPLE 6 A W310 × 60 simple beam has a concentrated load of 27 kN acting at the center of the beam. The beam span is 6 m. Check for deflection.

Solution Refer to Fig. 29-6-11. The maximum deflection for a simple beam with a concentrated load is $FL^3 \div 48EI$ where $F = 27$ kN, $L = 6000$ mm, $E = 200\ 000$ MPa (Fig. 29-1-3), and $I = 129 \times 10^6$ mm⁴ (Fig. 29-7-3). Therefore

Maximum deflection
$$= \frac{27\ 000 \times 6000^3}{48 \times 200\ 000 \times 129 \times 10^6} = 4.7\ mm$$

Allowable deflection = span ÷ 360 = 6000 ÷ 360 = 16.7 mm. Since the maximum deflection is less than the allowable, the W310 × 60 beam is acceptable.

ASSIGNMENTS

See Assignments 65 through 88 for Unit 29-7 on page 599.

ASSIGNMENTS for Chapter 29

Assignments for Unit 29-1, Stresses and Strains

U.S. CUSTOMARY ASSIGNMENTS

1. A load of 40 000 lb is suspended from a steel rod. What is the area of the bar if the allowable unit stress is 12 000 pounds per square inch (lb/in.²)? What is the minimum stock size that can be used (a) of a round rod, (b) of a square bar?
2. What may safely be placed on an 8 × 8 in. wooden post if the allowable unit stress is 880 lb/in.²?
3. What is the unit stress on a 4 × 4 × .50 in. angle iron if the angle is subjected to a 60 000-lb load? Rounds and fillets need not be considered for calculation purposes.
4. What must be the diameter of an A36 steel rod that safely lifts a load of 15 000 lb? The factor of safety is 5.
5. What is the maximum tensile load that can be carried by an A36 steel strut having a cross-sectional area of 1.75 in.²? The factor of safety is 3. Refer to Fig. 29-1-8.
6. A load of 40 000 lb is hung from an A36 steel bar 4 × .375 in. What is the factor of safety?
7. A machine of 12 t rests on four structural-steel legs. If the factor of safety is 5, what is the cross-sectional area of each leg if the load is equally proportioned on each leg? Use A572-50 steel.
8. What is the maximum load a 6-in. schedule 40 pipe can support using the working stress shown in Fig. 29-1-3?
9. A steel bar 16 ft long elongates .124 in. under a tensile force. Calculate the unit deformation.
10. What deformation will occur in a 1.00-in.-diameter steel tie rod, 8 ft long, when a tensile load of 8 tons is applied?
11. What load will elongate a .375 × 1 in. × 10 ft long steel tie rod .125 in.?
12. What unit deformation is produced by a compressive stress of 34 000 lb in a 1.80²-in. steel bar?
13. A 1 × 1 × .125 in. angle iron 12 ft long supports a tensile load of 4000 lb. Find the total deformation. Ignore the rounds and fillets for calculation purposes.
14. Calculate the allowable load that may be placed on a short 6 × 6 in. post. The material is Eastern spruce, building grade.
15. A flat mild steel bar of rectangular section 3 × .75 in. carries an axial pull of 12 tons. Find the tensile stress and the elongation in inches if the unloaded length is 20 ft.
16. Calculate the load required to punch a 2-in.-diameter hole in a piece of mild steel plate .25 in. thick, assuming that the ultimate shear strength of the steel is 60 000 lb/in.².

METRIC ASSIGNMENTS

17. A load of 350 kN is suspended from a steel rod. What is the area of the bar if the allowable unit stress is 90 MPa? What is the minimum stock size that can be used (a) of a round rod, (b) of a square bar?
18. What may safely be placed on a 150 × 150 wooden post if the allowable unit stress is 8 MPa?
19. What is the unit stress on a 75 × 75 × 10 mm angle iron if the angle is subjected to a 250-kN load? Rounds and fillets need not be considered for calculation purposes.
20. What must be the diameter of an A36 steel rod that safely lifts a load of 90 kN? The factor of safety is 4.
21. What is the maximum tensile load that can be carried by an A36 steel strut having a cross-sectional area of 900 mm²? The factor of safety is 4. Refer to Fig. 29-1-8.
22. A load of 200 kN is hung from an A36 steel bar 75 × 12. What is the factor of safety?
23. A machine of 8 t rests on four structural steel legs. If the factor of safety is 4, what is the cross-sectional area of each leg if the load is equally proportioned on each leg? Use A572M-350 steel.
24. What is the maximum load a 4-in. schedule 40 pipe can support using the working stress shown in Fig. 29-1-3?
25. A steel bar 5000 mm long elongates 3.15 mm under a tensile force. Calculate the unit deformation.
26. What deformation will occur in a 25 mm diameter steel tie rod, 2400 mm long, when a tensile load of 70 kN is applied?
27. What load will elongate a 10 × 25 × 3000 mm long steel tie rod 3 mm?
28. What unit deformation is produced by a tensile load of 150 kN in a 25² mm steel bar?
29. A 25 × 25 × 3 mm angle iron 3600 mm long supports a tensile load of 18 kN. Find the total deformation. Ignore the rounds and fillets for calculation purposes.
30. Calculate the allowable load that may be placed on a short 150 × 150 mm post. The material is Eastern spruce, construction grade. Refer to Fig. 29-1-4.
31. A flat, mild steel bar of rectangular section 75 × 20 mm supports a mass of 12 t. Find the tensile stress and the elongation if the bar length is 6000 mm.
32. Calculate the load required to punch a 50-mm-diameter hole in a piece of mild steel plate 6 mm thick, assuming that the ultimate shear strength of the steel is 415 MPa.

Assignments for Unit 29-2, Bolted and Riveted Joints

33. On a B- or A3-size sheet, show the calculations for the single-riveted butt joint and the riveted roof truss shown in Fig. 29-2-A or 29-2-B. For the single-riveted butt joint, calculate the main stresses that could safely be applied to the joint. Plate material is A36 steel. For the riveted roof truss, the roof truss has loads as shown acting on the upper and lower chord members. Calculate the number of rivets required to carry these loads.
34. On a B- or A3-size sheet, show the calculations for the double-riveted butt joint and the riveted lap joint shown in Fig. 29-2-C or 29-2-D. For the double-riveted butt joint, calculate the main stresses in the boiler joint when the boiler plate is subjected to the tension stress shown. In the case of the riveted

SINGLE RIVETED BUTT JOINT

RIVETED ROOF TRUSS

Fig. 29-2-A   Riveted joints.

SINGLE RIVETED BUTT JOINT

RIVETED ROOF TRUSS

Fig. 29-2-B Riveted Joints.

COVER PLATES 5 THICK

STRESS = 70 MPa

DOUBLE RIVETED BUTT JOINT

COVER PLATES .25 THICK

STRESS 8000 PSI
DOUBLE RIVETED BUTT JOINT

TENSILE LOAD = 18 000 LB
RIVETED LAP JOINT

Fig. 29-2-C Riveted Joints.

TENSILE LOAD = 100 kN
RIVETED LAP JOINT

Fig. 29-2-D Riveted Joints.

lap joint, a structural joint is fastened by four rivets. Calculate the main stresses when the tensile load shown is applied.

Assignments for Unit 29-3, Welded Joints

U.S. CUSTOMARY ASSIGNMENTS

35. Calculate the safe load in tension on an open-square butt weld, welded one side, joining two .25 × 6 in. A36 steel plates.

36. Do the same as in Problem 35 except use an open square, welded both sides.

37. A .50 × 10 in. A36 steel plate, welded both sides, is connected to a steel plate. Calculate the length of the two .375-in. side fillet welds required if the plate is subject to a tensile load of 15 000 lb.

38. Two A572-55 structural-steel plates 1.50 in. thick are connected by a double-U butt weld. Calculate the allowable load per inch the weld will support.

39. A Ø2 in. A572-42 steel shaft is welded to a steel plate by a 25-in. fillet weld. Calculate the safe tensile load that could be supported by the weld.

40. A 4 × 4 × .625 in. A36 steel angle supporting a load of 90 kips is welded to a steel plate. Calculate the length of side

welds required if .44-in. fillet welds are used and the line of action on the angle is taken through the centroidal axis.

41. Do the same as in Problem 40 except the weld is on the sides and end of the angle.

42. A boiler 42 in. in diameter, made of A36 steel, .50 in. thick is welded by a single-V butt weld. What is the allowable pressure in pounds per square inch of the boiler?

43. Do the same as in Problem 42 except use a single-V butt, welded both sides.

44. A boiler 42 in. in diameter made of A36 steel, has to withstand a steam pressure of 325 lb/in.². A single-bevel butt joint, welded one side, is used. What is the required thickness of boiler plate?

45. An intermittent fillet weld is used on both sides of a welded T A572-55 steel section 16 ft long. The weld length is 3 in. on a pitch of 12 in. If a tensile load of 600 kips is uniformly distributed along the welded section, what size fillet weld is required?

46. What is the maximum tensile load that can be applied to an A36 steel welded section 8 ft long? A .38-in. intermittent fillet weld 3 in. long on a 12-in. pitch is used on each side of the plate.

METRIC ASSIGNMENTS

47. Calculate the safe load in tension on an open-square butt weld, welded one side, joining two 6 × 150 mm A36 steel plates.

48. Do the same as in Problem 47 except use an open square, welded both sides.

49. A 12 × 250 mm A36 steel bar, welded both sides, is connected to a steel plate. Calculate the length of the two 10-mm side fillet welds required if the plate is subjected to a tensile load of 130 kN.

50. Two A572M-380 structural-steel plates 40 mm thick are connected by a double-U butt weld. Calculate the allowable load the weld will support in a 100-mm wide section.

51. A Ø50 A572M-290 steel shaft is welded to a steel plate by a 6-mm fillet weld. Calculate the safe tensile load that could be supported by the weld.

52. A 100 × 100 × 13 mm A36 steel angle supporting a load of 400 kN is welded to a steel plate. Calculate the length of side welds required using the largest permissible fillet welds. The line of action on the angle is taken through the centroidal axis.

53. Do the same as in Problem 52 except that the weld is on the sides and end of the angle.

54. A boiler 1000 mm in diameter, made of A36 steel 12 mm thick, is welded by a single-V butt weld. What is the maximum allowable pressure in the boiler?

55. Do the same as in Problem 54 except use a single-V butt, welded both sides.

56. A boiler 1000 mm in diameter is made of A36 steel and has to withstand a steam pressure of 2.2 MPa. A single-bevel butt joint, welded one side, is used. What is the required thickness of boiler plate?

57. An intermittent fillet weld is used on both sides of a welded T A572M-380 steel section 5000 mm long. The weld length is 50 mm on a 200-mm pitch. If a tensile load of 1.3 MN is uniformly distributed along the welded section, what size fillet weld is required?

58. What is the maximum tensile load that can be applied to an A36 steel welded section 2400 mm long? A 12-mm intermittent fillet weld 75 mm long on a 300-mm pitch is used on each side of the plate.

Assignments for Unit 29-4, Beams

59. On a B- or A3-size sheet, calculate the moments and loads for the problems shown in Fig. 29-4-A or 29-4-B.

60. On a B- or A3-size sheet, calculate the moments and reactions for the problems shown in Fig. 29-4-C or 29-4-D.

Assignments for Unit 29-5, Shear Diagrams

61. On a B- or A3-size sheet, draw the four beams shown in Fig. 29-5-A or 29-5-B, showing the loading diagram, the shear diagram, and the calculations. Scale is to suit.

62. On a B- or A3-size sheet, draw the four beams shown in Fig. 29-5-C or 29-5-D, showing the loading diagram, calculated reaction diagram, and the shear diagram plus the calculations. Scale is to suit.

Assignments for Unit 29-6, Bending Moment Diagrams

63. On a B- or A3-size sheet, draw the bending moment diagrams for the four cantilevered beams shown in Fig. 29-5-

Fig. 29-4-A Calculating moments and loads.

Fig. 29-4-B Calculating moments and loads.

Fig. 29-4-C Calculating moments and reactions.

Fig. 29-4-D Calculating moments and reactions.

598 ADVANCED DRAFTING DESIGN

Fig. 29-5-A Shear problems. **Fig. 29-5-B** Shear problems. **Fig. 29-5-C** Shear problems. **Fig. 29-5-D** Shear problems.

A or 29-5-B. Below each diagram show the calculations. Scale is to suit.

64. On a B- or A3-size sheet, draw the bending moment diagrams for the four concentrated loaded beams shown in Fig. 29-5-C or 29-5-D. Below each diagram show the bending moment calculations.

Assignments for Unit 29-7, Beam Design

U.S. CUSTOMARY ASSIGNMENTS

65. Select the lightest beam from Fig. 30-7-3 that can be used to safely carry a 4000-lb load suspended from the end of a 9-ft cantilever beam. Use A572-45 steel.

66. A simple beam 30 ft long supports a uniformly distributed load of 300 lb/ft. Neglecting the weight of the beam, select the lightest A36 steel beam that will safely carry this load.

67. A simple beam 24 ft long supports a 44 000-lb concentrated load 10 ft from the left abutment. Neglecting the mass of the beam, select the lightest A36 steel beam that will safely carry this load.

68. A cantilever beam 16 ft long carries a uniformly distributed load of 500 lb/ft. Neglecting the weight of the beam, select the lightest A572-50 steel beam that will safely carry this load.

69. What is the maximum uniformly distributed load that can be safely placed on a simple W12 × 40 beam 20 ft long made of A572-45 steel?

70. A simple beam 20 ft long has a uniformly distributed load of 300 lb/ft and a 2400-lb concentrated load 4 ft from the left abutment. Select the lightest A36 steel beam that will safely carry this load.

71. How many 2 × 12 in. (1⅝ × 11⅝ in. actual size) wood beams must be joined to form a simple beam that will carry safely a concentrated load of 6000 lb

acting at the center of the beam? The beam is to be 12 ft long, and the allowable bending stress must not exceed 1200 lb/in.².

72. A simple W10 × 66 beam, 20 ft long, is made of A36 steel. Calculate the largest concentrated load the beam can support at the middle of the span and the largest uniformly distributed load the beam can support?

73. A W10 × 29 beam has a 20-ft span. What is the largest concentrated load, located at the center, that the beam can carry if the maximum deflection is the span ÷ 360?

74. Do the same as in Problem 73 except substitute a uniformly distributed load for a concentrated load.

75. An S12 × 31.8 cantilever beam 12 ft long has a concentrated load of 8000 lb at its free end. Is the deflection excessive?

76. A 10-in.-wide × 16-in.-deep simple wooden beam has a span of 16 ft and a uniformly distributed load of 1000 lb/ft. Check the beam for deflection, using values of 1 320 000 for E and 2948 for I.

METRIC ASSIGNMENTS

77. Select the lighest beam from Fig. 29-7-3 that can be used to safely carry an 18-kN load suspended from the end of a 3-m cantilever beam. Use A572M-310 steel.

78. A simple beam 10 m long supports a uniformly distributed load of 5 kN·m. Neglecting the mass of the beam, select the lighest A36 steel beam that will safely carry this load.

79. A simple beam 8 m long supports a 200-kN concentrated load 3 m from the left abutment. Neglecting the mass of the beam, select the lighest A36 beam that will safely carry this load.

80. A cantilever beam 5 m long carries a uniformly distributed load of 8 kN·m. Neglecting the mass of the beam, select the lightest A572M-350 beam that will safely carry this load.

81. What is the maximum uniformly distributed load that can be safely placed on a simple W300 × 60 beam 6 m long made of A572M-310 steel?

82. A simple beam 6 m long has a uniformly distributed load of 5 kN·m and a 10-kN concentrated load 1.2 m from the left abutment. Select the lightest A36 steel beam from Fig. 29-1-8 that will safely carry this load.

83. How many 38 × 286-mm wood beams must be joined to form a simple beam that will carry safely a concentrated load of 25 kN acting at the center of the beam? The beam is to be 3.5 m long, and the allowable bending stress must not exceed 8000 kPa.

84. A simple W250 × 98 beam 6 m long is made of A36 steel. Calculate the largest concentrated load the beam can support at the middle of the span and the largest uniformly distributed load the beam can support?

85. A W250 × 43 beam has a 6-m span. What is the largest concentrated load, located at the center, that the beam can carry if the maximum deflection is the span ÷ 360?

86. Do the same as in Problem 85 except substitute a uniformly distributed load for the concentrated load.

87. An S300 × 47 cantilever beam 4 m long has a concentrated load of 35 kN at its free end. Is the deflection excessive?

88. A 250-mm-wide × 400-mm-deep simple wooden beam has a span of 5 m and a uniformly distributed load of 14 kN·m. Check the beam for deflection using values of 9400 for E and 1226 for I.

CHAPTER 30

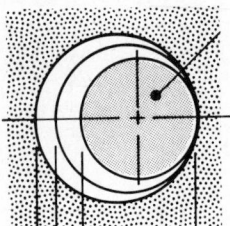

Engineering Tolerancing

UNIT 30-1
Modern Engineering Tolerancing

An engineering drawing of a manufactured part is intended to convey information from the designer to the manufacturer and inspector. It must contain all information necessary for the part to be correctly manufactured. It must also enable an inspector to make a precise determination of whether the parts are acceptable.

Therefore each drawing must convey three essential types of information

1. The material to be used
2. The size or dimensions of the part
3. The shape or geometric characteristics

The drawing must also specify permissible variations for each of these aspects, in the form of tolerance or limits.

Materials are usually covered by separate specifications or supplementary documents, and the drawings need only make reference to these.

Size is specified by linear and angular dimensions. Tolerances may be applied directly to these dimensions or may be specified by means of a general tolerance note.

Shape and geometric characteristics, such as orientation and position, are described by views on the drawing, supplemented to some extent by dimensions.

In the past tolerances were often shown for which no precise interpretation existed, such as on dimensions which originated at nonexistent center lines. Specification of datum features was often omitted, resulting in measurements being made from actual surfaces when datums were intended. There was confusion concerning the precise effect of various methods of expressing tolerances and of the number of decimal places used. While tolerancing of geometric characteristics was sometimes specified in the form of simple notes, no precise methods or interpretations were established. Straight or circular lines were drawn, without specifying how straight or round they should be. Square corners were drawn without specifying how much the 90° angle could vary.

Modern systems of tolerancing, which include geometric and positional tolerancing, use of datum and datum targets, and more precise interpretations of linear and angular tolerances, provide designers and drafters with a means of expressing permissible variations in a very precise manner. Furthermore, the methods and sym-bols are international in scope and therefore help break down language barriers.

It is not necessary to use geometric tolerances for every feature on a part drawing. In most cases it is to be expected that if each feature meets all dimensional tolerances, form variations will be adequately controlled by the accuracy of the manufacturing process and equipment used.

This chapter covers the application of modern tolerancing methods on drawings.

National and International Standards References are made to technical drawing standards published by United States and ISO standardizing bodies. These bodies are generally referred to by their acronyms, as shown in Fig. 30-1-1.

Most of the symbols in all these standards are identical, but there are some variations. These are chiefly in the methods of indicating datum features and of applying the symbols to drawings. In view of the exchange of drawings among the United States and other countries, it would be advan-

ACRONYM	STANDARDIZING BODY	STANDARD FOR DIMENSIONING AND TOLERANCING
ANSI	AMERICAN NATIONAL STANDARDS INSTITUTE	ANSI Y14.5
ISO	INTERNATIONAL ORGANIZATION FOR STANDARDIZATION	ISO R1101

Fig. 30-1-1 **Standardizing bodies.**

tageous for drafters and designers to become acquainted with these different symbols.

For this reason whenever differences between United States and ISO standards occur, two methods are shown in some of the illustrations, and each is labeled with the acronym of the appropriate standardizing body, ANSI or ISO (ISO, CSA, and BSI [British Standards Institute] all use identical symbols). However, differences in symbols or methods of application do not in any way affect the principles or interpretation of tolerances, unless specifically noted.

Illustrations

Most of the drawings in this chapter are not complete working drawings. They are intended only to illustrate a principle. Therefore, to avoid distraction from the information being presented, most of the details that are not essential to explain the principle have been omitted.

DEFINITIONS OF BASIC TERMS

Definitions of some of the basic terms used in dimensioning and tolerancing of drawings follow. While these terms are not new, their exact meanings warrant special attention in order that there be no ambiguity in the precise interpretation of tolerancing methods described in this chapter.

Dimension

A *dimension* is a geometric characteristic, of which the size is specified, such as diameter, length, angle, location, or center distance. The term is also used for convenience to indicate the size or value of a dimension, as specified on a drawing. See Fig. 30-1-2.

Tolerance

The *tolerance* on a dimension is the total permissible variation in its size,

Fig. 30-1-2   Dimensions of a part.

(A) TOLERANCE SIZE

(B) TYPE OF TOLERANCE

Fig. 30-1-3   Tolerances.

which is equal to the difference between the limits of size. The plural term *tolerances* is sometimes used to denote the permissible variations from the specified size when the tolerance is expressed bilaterally.

For example, in Fig. 30-1-3A the tolerance on the center distance dimension $1.50 \pm .04$ is .08 in., but in common practice the values $+.04$ and $-.04$ are often referred to as the tolerances.

A *bilateral tolerance* is a tolerance which is expressed as plus and minus values, where neither is zero, to denote permissible variations in both directions from the specified size.

A *unilateral tolerance* is one which applies in only one direction from the specified size, so that the permissible variation in the other direction is zero.

Size of Dimensions

In theory, it is impossible to produce a part to an exact size, because every part, if measured with sufficient accuracy, would be found to be a slightly different size. However, for purposes of discussion and interpretation, a number of distinct sizes for each dimension have to be recognized.

Actual Size In ordinary practice, *actual size* simply means the measured size of an individual part.

Nominal Size The *nominal size* is the designation of size used for purposes of general identification.

The nominal size is used in referring to a part in an assembly drawing stocklist, in a specification, or in other such documents. It is very often identical to the basic size but may differ widely; for example, the diameter of a .50 in. steel pipe is 0.84 in. (21.34 mm). The nominal size is .50 in.

Specified Size This is the size specified on the drawing when the size is associated with a tolerance. The specified size is usually identical to the design size or, if no allowance is involved, to the basic size.

Basic Size The *basic size* of a dimension is the theoretical size from which the limits for that dimension are derived by the application of the allowance and the tolerance.

Figure 30-1-4 shows two mating features with the tolerance and allowance zones exaggerated, to illustrate the sizes, tolerances, and allowances. This figure also illustrates the origin of tolerance block diagrams, as shown in Fig. 30-1-5, which are commonly used to show the relationships among part limits, gage or inspection limits, and gage tolerances.

Design Size The *design size* of a dimension is the size in relation to which the

Fig. 30-1-4   Sizes of mating parts.

Fig. 30-1-5 Tolerance block diagram.

tolerance for that dimension is assigned.

Theoretically, it is the size on which the design of the individual feature is based, and therefore it is the size which should be specified on the drawing. For dimensions of mating features it is derived from the basic size by the application of the allowance, but when there is no allowance, it is identical to the basic size.

Deviations

The differences between the basic, or zero, line and the maximum and minimum sizes are called the *upper* and *lower deviations*, respectively.

Thus in Fig. 30-1-6 the upper deviation of the external part is − .001, and the lower deviation is − .003. For the hole diameter, the upper deviation is + .002, and the lower deviation is + .001, whereas for the length of the pin the upper and lower deviations are + .02 and − .02, respectively.

Fig. 30-1-6 Deviations.

Exact Dimensions

True-Position Dimension *True-position* is the theoretical exact location established by basic dimensions.

Datum Dimension A *datum dimension* is a dimension which establishes the true position of a datum or datum target.

Basic Dimension A *basic dimension* represents the theoretical exact size,

profile, orientation, or location of a feature or datum target.

These terms are called *exact dimensions*. See Fig. 30-1-7. They are shown without direct tolerances, and each is enclosed in a rectangular frame to indicate that the tolerances in the general tolerance note do not apply.

Fig. 30-1-7 Exact dimensions.

Feature

A *feature* is a specific, characteristic portion of a part, such as a surface, hole, slot, screw thread, or profile.

While a feature may include one or more surfaces, the term is generally used in geometric tolerancing in a more restricted sense, to indicate a specific point, line, or surface. Some examples are the axis of a hole, the edge of a part, or a single flat or curved surface, to which reference is being made or which forms the basis for a datum.

Axis

An *axis* is a theoretical straight line about which a part or circular feature revolves or could be considered to revolve. See Fig. 30-1-8.

Fig. 30-1-8 Divergence of axis and center line when part is deformed.

INTERPRETATION OF DRAWINGS AND DIMENSIONS

It should not be necessary to specify the geometric shape of a feature, unless some particular precision is required. Lines which appear to be straight imply straightness; those that appear to be round imply circularity; those that appear to be parallel imply parallelism; those that appear to be square imply perpendicularity; center lines imply symmetry; and features that appear to be concentric about a common center line imply concentricity.

Therefore it is not necessary to add angular dimensions of 90° to corners of rectangular parts nor to specify that opposite sides are parallel.

However, if a particular departure from the illustrated form is permissible, or if a certain degree of precision of form is required, it must be specified. If a slight departure from the true geometric form or position is required, it should be exaggerated pictorially in order to show clearly where the dimensions apply. Figure 30-1-9 shows some examples. Dimensions which are not to scale should be underlined freehand.

Fig. 30-1-9 Exaggeration of small dimensions.

Point-to-Point Dimensions

When datums are not specified, linear dimensions are intended to apply on a point-to-point basis, either between opposing points on the indicated surfaces or directly between the points marked on the drawing.

The examples shown in Fig. 30-1-10 should help to clarify this principle of point-to-point dimensions.

Location Dimensions with Datums

When location dimensions originate

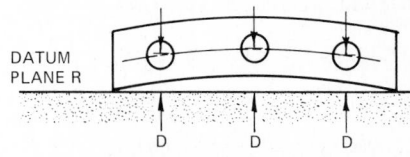

Fig. 30-1-11 Dimension referred to a datum.

Fig. 30-1-10 Point-to-point dimensions when datums not used.

from a feature or surface specified as a datum, measurement is made from the theoretical datum, not from the actual feature or surface of the part.

There will be many cases where a curved center line, as shown in Fig. 30-1-10F, would not meet functional requirements or where the position of the hole in Fig. 30-1-10H would be required to be measured parallel to the base. This can easily be specified by referring the dimension to a datum feature, as shown in Fig. 30-1-11. This will be more fully explained in Unit 30-7, where the interpretation of coordinate tolerances is compared with geometric and positional tolerances.

Assumed Datums

There are often cases where the basic rules for measurements on a point-to-point basis cannot be applied, because the originating points, lines, or surfaces are offset in relation to the features located by the dimensions. It is then necessary to assume a suitable datum, which is usually the theoretical extension of one of the lines or surfaces involved.

The following general rules cover three types of dimensioning procedures commonly encountered.

1. If a dimension refers to two parallel edges or planes, the larger edge or

surface is assumed to be the datum feature. For example, if the surfaces of the part shown in Fig. 30-1-12A were not quite parallel, as shown in the lower view, dimension D would be acceptable if the top surface were within limits when measured at a and b, but need not be within limits if measured at c.

2. If only one of the extension lines refers to a straight edge or surface, the extension of that edge or surface is assumed to be the datum. Thus in Fig. 30-1-12B measurement of dimension A is made to a datum surface as shown at a in the bottom view.

3. If both extension lines refer to offset points rather than to edges or surfaces, generally it should be assumed that the datum is a line running through one of these points and parallel to the line or surface to which it is dimensionally related. Thus in Fig. 30-1-12C dimension A is measured from the center of hole D to a line through the center of hole C which is parallel to the base line, as at a.

Permissible Form Variations

The actual size of a feature must be within the limits of size, as specified on the drawing, at all points of measurement. This means that each measurement, made at any cross section of the feature, must be not greater than the maximum limit of size nor smaller than the minimum limit of size. See Fig. 30-1-13.

Fig. 30-1-12 Assumed datums.

Fig. 30-1-13 Deviations permitted by toleranced dimensions.

By themselves, toleranced linear dimensions, or limits of size, do not give specific control over many other variations of form, orientation, and, to some extent, position, such as errors of squareness of related features or deviations caused by bending of parts, lobing, eccentricity, and the like. Therefore features may actually cross the boundaries of perfect form at the maximum material size and at the minimum material size.

In order to meet functional requirements, it is often necessary to control such deviations. This is done to ensure that parts are not only within their limits of size but also within specified limits of geometric form, orientation, and position. In the case of mating parts, such as holes and shafts, it is usually necessary to ensure that they do not cross the boundary of perfect form at the maximum material size, by reason of being bent or otherwise deformed. This condition is shown in Fig. 30-1-14, where features do not cross the maximum material boundary but are permitted to cross the boundary of perfect form at the minimum material condition.

If only size tolerances or limits of size are specified for an individual fea-

ture and no geometric tolerance is given, no element of the feature would extend beyond the maximum material boundary of perfect form. Examples are shown in Fig. 30-1-15. According to ANSI rules, all parts are expected to have perfect form of individual features at the maximum material condition.

Reference and Source Material

1. ANSI Y 14.5M Dimensioning and Tolerancing

ASSIGNMENTS

See Assignments 1 and 2 for Unit 30-1 on page 645.

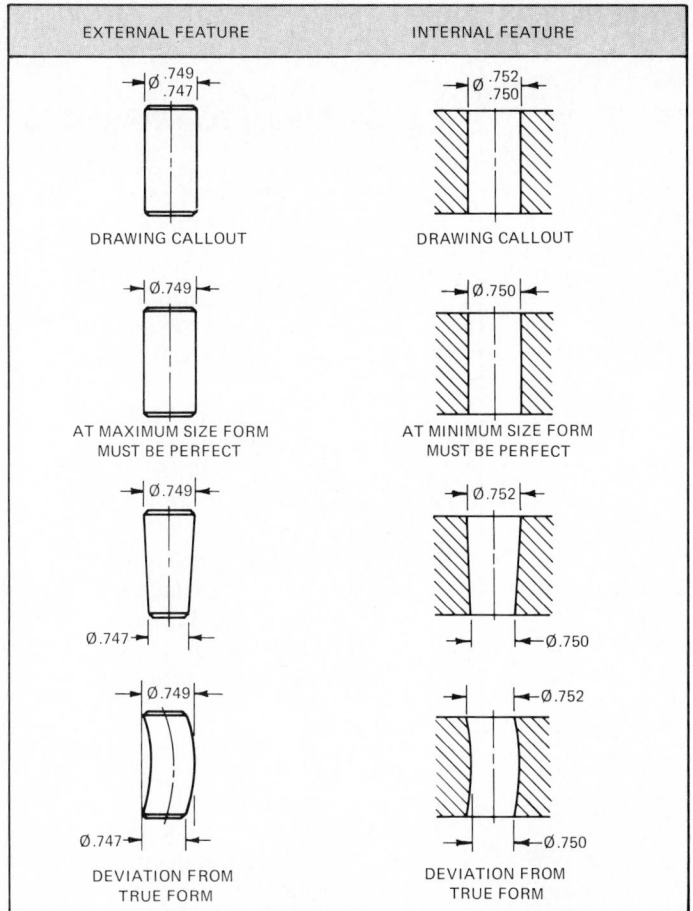

EXTERNAL FEATURE	INTERNAL FEATURE

DRAWING CALLOUT · Ø .749/.747 | DRAWING CALLOUT · Ø .752/.750

AT MAXIMUM SIZE FORM MUST BE PERFECT · Ø.749 | AT MINIMUM SIZE FORM MUST BE PERFECT · Ø.750

DEVIATION FROM TRUE FORM · Ø.749 · Ø.747 | DEVIATION FROM TRUE FORM · Ø.752 · Ø.750

Fig. 30-1-14 Examples of deviation of form when perfect form at the maximum material condition is required.

DRAWING CALLOUT	POSSIBLE DEVIATION ACCEPTED BY LIMIT GAGES

(A) EXTERNAL FEATURE (SHAFT) · LENGTH · DIA

(B) INTERNAL FEATURE (HOLE) · LENGTH · DIA

RING GAGE · PART · MIN · MAX

PART · MAX · MIN · PLUG GAGE

Fig. 30-1-15 Form variations accepted by limit gages.

UNIT 30-2
Geometric Tolerancing

A geometric tolerance is the maximum permissible variation of form, orientation, or location of a feature from that indicated or specified on the drawing. The tolerance value represents the width or diameter of the tolerance zone, within which the point, line, or surface of the feature shall lie.

From this definition it follows that a feature would be permitted to have any variation of form, or take up any position, within the specified geometric tolerance zone.

For example, a line controlled by a straightness tolerance of .006 in. must be contained within a tolerance zone .006 in. wide. See Fig. 30-2-1.

POINTS, LINES, AND SURFACES

The production and measurement of engineering parts deals, in most cases, with surfaces of objects. These surfaces may be flat, cylindrical, conical, or spherical or have some more or less irregular shape or contour.

Fig. 30-2-1 Tolerance zone for straightness of a line.

Measurement, however, usually has to take place at specific points. A line or surface is evaluated dimensionally by making a series of measurements at various points along its length.

Therefore, geometric tolerances are chiefly concerned with points and lines, while surfaces are considered to be composed of a series of line elements running in two or more directions.

Points have position but no size, and therefore position is the only characteristic that requires control. Lines and surfaces have to be controlled for form, orientation, and location. Therefore geometric tolerances provide for control of the characteristics shown in Fig. 30-2-2.

FEATURE	TYPE OF TOLERANCE	CHARACTERISTIC	SYMBOL	SEE UNIT
INDIVIDUAL FEATURES	FORM	STRAIGHTNESS	—	30-2, 30-4
		FLATNESS	▱	30 – 4
		CIRCULARITY (ROUNDNESS)	○	30 – 9
		CYLINDRICITY	⌭	
INDIVIDUAL OR RELATED FEATURES	PROFILE	PROFILE OF A LINE	⌒	30 – 10
		PROFILE OF A SURFACE	⌓	
RELATED FEATURES	ORIENTATION	ANGULARITY	∠	30 – 6
		PERPENDICULARITY	⊥	
		PARALLELISM	//	
	LOCATION	POSITION	⊕	30 – 7
		CONCENTRICITY	◎	
	RUNOUT	CIRCULAR RUNOUT	↗	30 – 11
		TOTAL RUNOUT	↗↗	

Fig. 30-2-2   Geometric characteristic symbol.

FEATURE CONTROL FRAME

Some geometric tolerances have been used for many years in the form of notes such as "PARALLEL WITH SURFACE A WITHIN .001" and "STRAIGHT WITHIN .12." While such notes are now obsolete, the reader should be prepared to recognize them on older drawings.

The modern method is to specify geometric tolerances by means of the *feature control frame*. A feature control frame for an individual feature is divided into compartments containing the geometric characteristic symbol followed by the tolerance. See Fig. 30-2-3. Where applicable, the tolerance is preceded by the diameter symbol (see Unit 30-4) and is followed by a material condition symbol (see Unit 30-3).

When necessary, other compartments are added to contain datum references, as explained in Unit 30-5.

Geometric characteristic symbols used for form of a line are shown in Fig. 30-2-2. Other symbols will be

introduced as required, but all are shown for reference purposes.

APPLICATION TO DRAWINGS

The feature control frame is related to the feature by one of the following methods (shown in Fig. 30-2-4):

1. Locating the frame below the leader-directed callout or dimension pertaining to the feature.
2. Running a leader from the frame to the feature.
3. Attaching a side or end of the frame to an extension line extending from a plane surface feature.
4. Attaching a side or end of the frame to an extension of the dimension line pertaining to a feature of size. ISO practice is to attach the dimension line to the feature control frame and place the feature of size above or below the frame.

APPLICATION TO SURFACES

The arrowhead of the leader from the feature control frame should touch the surface of the feature or the extension line of the surface.

The leader from the feature control frame should be directed at the feature in its characteristic profile. Thus, in Fig. 30-2-5 the straightness tolerance is

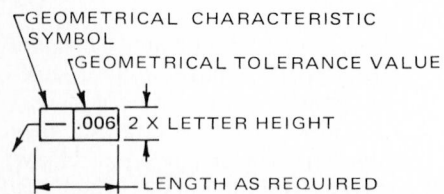

GEOMETRICAL CHARACTERISTIC SYMBOL
GEOMETRICAL TOLERANCE VALUE

— .006   2 × LETTER HEIGHT
LENGTH AS REQUIRED

Fig. 30-2-3   Feature control frame for an individual feature.

Fig. 30-2-4   Placement of feature control frame.

STRAIGHTNESS TOLERANCE   CIRCULARITY TOLERANCE

Fig. 30-2-5 Preferred location of feature control symbol.

directed to the side view, and the circularity tolerance to the end view. This may not always be possible, and a tolerance connected to an alternative view, such as a circularity tolerance connected to a side view, is acceptable.

When two or more feature control frames apply to the same feature, they are drawn together with a single leader and arrowhead, as shown in Fig. 30-2-6.

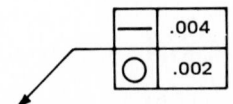

Fig. 30-2-6 Two controls of one surface.

CIRCULAR TOLERANCE ZONES

When the resulting tolerance zone is circular or cylindrical, such as when straightness of the center line of a cylindrical feature is specified, a diameter symbol precedes the tolerance value in the feature control frame. See Fig. 30-2-7.

STRAIGHTNESS

Lines Straightness is fundamentally a characteristic of a line, such as the edge of a part or a line scribed on a surface. A straightness tolerance is specified on a drawing by means of a feature control frame, which is directed by a leader to the line requiring control, as shown in Fig. 30-2-8. It states in symbolic form that the line shall be straight within .006 in. This means that the line shall be contained within a tolerance zone consisting of the area between two parallel straight

lines in the same plane, separated by the specified tolerance.

Theoretically, straightness could be measured by bringing a straightedge into contact with the line and determining that any space between the straightedge and the line does not exceed the specified tolerance.

(A) DRAWING CALLOUT

(B) INTERPRETATION

Fig. 30-2-7 Circular tolerance zone.

(A) DRAWING CALLOUT

(B) STRAIGHTNESS TOLERANCE ZONE

(C) CHECKING WITH A STRAIGHTEDGE

Fig. 30-2-8 Straightness symbol and application.

Cylindrical Surfaces For cylindrical parts, or curved surfaces which are straight in one direction, the feature control frame should be directed to the side view, where line elements appear as a straight line, as shown in Figs. 30-2-9 and 30-2-10.

(A) DRAWING CALLOUT

REFERS TO LINE ELEMENTS ON SURFACE

BENDING ERROR

CONCAVE ERROR

CONVEX ERROR

(B) INTERPRETATION

NOTE – NO PART OF THE CYLINDRICAL SURFACE MAY LIE OUTSIDE THE LIMITS OF SIZE

Fig. 30-2-9 Straightness errors in surface elements of a cylindrical surface.

(A) DRAWING CALLOUT

REFERS TO LINE ELEMENTS
ON SURFACE

.004 TOLERANCE ZONE

(B) INTERPRETATION

Fig. 30-2-10   Straightness of surface line elements.

(A) DRAWING CALLOUT

DRAWING CALLOUT
REFERS TO EACH
LINE ELEMENT

.004 TOLERANCE ZONE

(B) INTERPRETATION

Fig. 30-2-11   Straightness of a conical surface.

DRAWING CALLOUT
MEANS STRAIGHT WITHIN .003
MEASURED IN DIRECTION
OF ARROWS

INTERPRETATION

(A) STRAIGHTNESS IN ONE DIRECTION

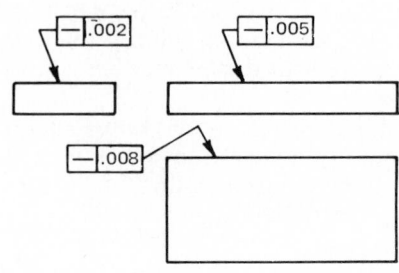

DRAWING CALLOUT

STRAIGHT WITHIN .002 MEASURED
IN DIRECTION OF ARROWS

STRAIGHT WITHIN .005
MEASURED IN DIRECTION
OF ARROWS

STRAIGHT WITHIN
.008 MEASURED IN
DIRECTION OF
ARROWS

INTERPRETION

(B) STRAIGHTNESS IN SEVERAL
DIRECTIONS

(C) THREE STRAIGHTNESS TOLERANCES
ON ONE VIEW

Fig. 30-2-12   Three straightness tolerances on one view.

A straightness tolerance thus applied to the surface controls surface elements only. Therefore it would control bending or a wavy condition of the surface or a barrel-shaped part, but it would not necessarily control the straightness of the center line or the conicity of the cylinder.

Straightness of a cylindrical surface is interpreted to mean that each line element of the surface shall be contained within a tolerance zone consisting of the space between two parallel planes, separated by the width of the specified tolerance, when the part is rolled along one of the planes. All circular elements of the surface must be within the specified size tolerance. No error in straightness would be permitted if the diameter were at its maximum material size. The straightness tolerance must be less than the size tolerance.

Conical Surfaces A straightness tolerance can be applied to a conical surface in the same manner as for a cylindrical surface, as shown in Fig. 30-2-11, and will ensure that the rate of taper is uniform. The actual rate of taper, or the taper angle, must be separately toleranced.

Flat Surfaces A straightness tolerance applied to a flat surface indicates straightness control in one direction only and must be directed to the line on the drawing representing the surface to be controlled and the direction in which control is required, as shown in Fig. 30-2-12A. It is then interpreted to mean that each line element on the surface in the indicated direction shall lie within a tolerance zone.

Different straightness tolerances may be specified in two or more directions when required, as shown in Fig. 30-2-12B. However, if the same straightness tolerance is required in two coordinate directions on the same surface, a flatness tolerance rather than a straightness tolerance is used.

If it is not otherwise necessary to draw all three views, the straightness tolerances may all be shown on a single view by indicating the direction with short lines terminated by arrowheads, as shown in Fig. 30-2-12C.

Straightness in a Specified Length It is often desirable on long parts to specify a straightness tolerance over a specific length, either with or without a maximum overall tolerance. For example,

a certain amount of bow may be quite acceptable if spread over the entire length, but quite undesirable if concentrated at some point along the length.

This requirement is specified on the drawing by including the specified

Fig. 30-2-13 Tolerance in a specified length.

length with the tolerance in the feature control frame and separating them with a diagonal line, as shown in Fig. 30-2-13. The expression .01/2.50 means .01 in. in any 2.50 in. long portion of the part.

It can be readily shown that for a part which is uniformly bowed in a circular arc, a bow of .01 in. in 10 in. would be .04 in. in 20 in., or .16 in. in 40 in. etc., as illustrated in Fig. 30-2-14. When a maximum overall tolerance is

Fig. 30-2-14 Resultant in specifying straightness per unit length with no specified total.

to be combined with a tolerance in a specified length, the tolerances are shown in a double feature control symbol, as in Fig. 30-2-15. An example is: STRAIGHT WITHIN .01 IN. FOR THE FULL LENGTH, BUT NOT TO EXCEED .003 IN. IN ANY 1.00 IN. LENGTH.

Fig. 30-2-15 Overall tolerance combined with a tolerance in a specified length.

Reference and Source Material
1. ANSI Y 14.5M Dimensioning and Tolerancing

ASSIGNMENT
See Assignment 3 for Unit 30-2 on page 647.

UNIT 30-3
Relationship to Feature of Size

DEFINITIONS

Maximum Material Condition (MMC) This term refers to that limit of size of a feature which results in the part containing the maximum amount of material. Thus it is the maximum limit of size for an external feature, such as a shaft, or the minimum limit of size for an internal feature, such as a hole. See Fig. 30-3-1.

Virtual Condition (Size) *Virtual condition* refers to the overall envelope of perfect form within which the feature would just fit. For an external feature such as a shaft, it is the maximum measured size plus the effect of actual form variations, such as straightness, flatness, or roundness. For an internal feature such as a hole, it is the minimum measured size minus the effect of such form variations

Least Material Condition (LMC) This term refers to that size of a feature which results in the part containing the minimum amount of material. Thus it is the minimum limit of size for an

external feature, for example, a shaft, and the maximum limit of size for an internal feature, such as a hole.

Regardless of Feature Size (RFS) This term indicates that a form or positional tolerance applies to any size of a feature which lies within its size tolerance.

FEATURES OF SIZE

Geometric tolerances so far considered concern only lines, line elements, and single surfaces. These are features having no diameter or thickness, and tolerances applied to them cannot be affected by feature size.

Features of size are features which do have diameter or thickness. These may be cylinders, such as shafts and holes. They may also be slots, tabs, or rectangular or flat parts, where two parallel, flat surfaces are considered to form a single feature.

If freedom of assembly of mating parts is the chief criterion for establishing a geometric tolerance for a feature of size, the least favorable assembly condition exists when the parts are made to the maximum material condition. Further geometric variations can then be permitted, without jeopardizing assembly, as the features approach their least material condition.

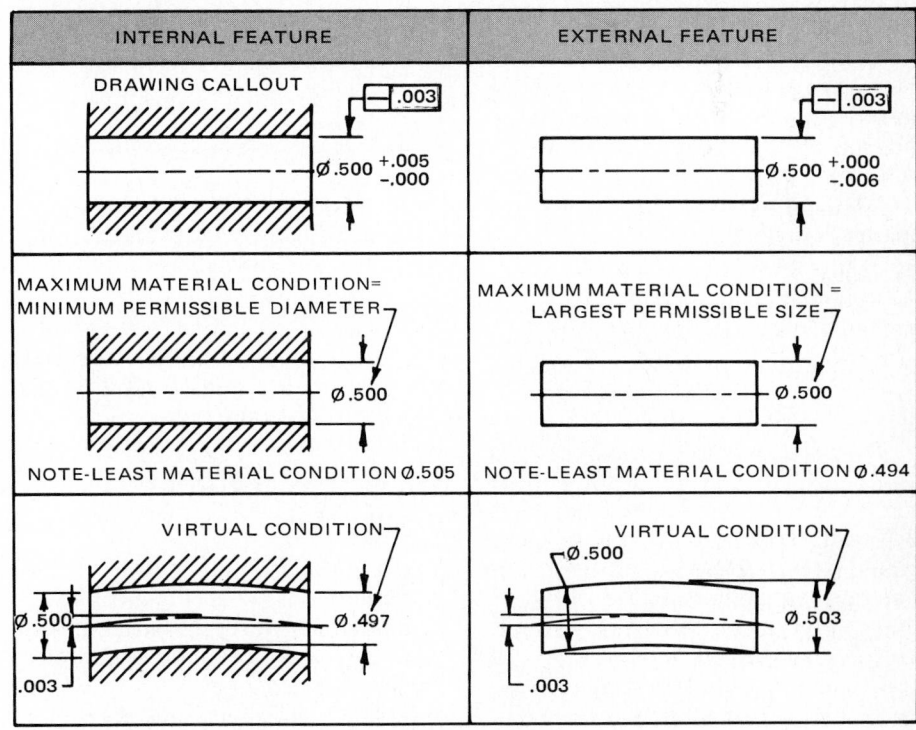

Fig. 30-3-1 Maximum material and virtual condition.

Fig. 30-3-2   Effect of form variations.

EXAMPLE 1 The effect of a form tolerance is shown in Fig. 30-3-2, where a cylindrical pin of ∅.307–.312 in. is intended to assemble into a round hole of ∅.312–.316 in. If both parts are at their maximum material condition of ∅.312 in., it is evident that both would have to be perfectly round and straight in order to assemble. However, if the pin was at its least material condition of ∅.307 in., it could be bent up to .005 in. and still assemble in the smallest permissible hole.

EXAMPLE 2 Another example, based on the location of features, is shown in Fig. 30-3-3. This shows a part with two projecting pins required to assemble into a mating part having two holes at same center distance.

The worst assembly condition exists when the pins and holes are at their maximum material condition, which is ∅.250 in. Theoretically, these parts would just assemble if their form, orientation (squareness to the surface), and center distances were perfect. However, if the pins and holes were at their least material condition of ∅.247 and ∅.253 in., respectively, it is evident that one center distance could be

Fig. 30-3-3   Effect on location.

increased and the other decreased by .003 in. without jeopardizing the assembly condition.

MAXIMUM MATERIAL CONDITION

The symbol for maximum material condition is shown in Fig. 30-3-4. The symbol dimensions are based on percentages of the recommended letter height of dimensions.

If a geometric tolerance is required to be modified on an MMC basis, it is

H= LETTER HEIGHT OF DIMENSIONS

Fig. 30-3-4   MMC symbol.

Fig. 30-3-5   Application of MMC symbol.

specified on the drawing by including the symbol ⓜ immediately after the tolerance value in the feature control frame as shown in Fig. 30-3-5.

A form tolerance modified in this way can be applied only to a feature of size; it cannot be applied to a single surface. It controls the boundary of the feature, such as a complete cylindrical surface, or two parallel surfaces of a flat feature. This permits the feature surface or surfaces to cross the maximum material boundary by the amount of the form tolerance. If it is required that the virtual condition be kept within the maximum material boundary, the form tolerance must be specified as zero at MMC, as shown in Fig. 30-3-6.

U. S. CUSTOMARY   METRIC

Fig. 30-3-6   MMC symbol with zero tolerance.

Application with Maximum Value It is sometimes necessary to ensure that the geometric tolerance does not vary over the full range permitted by the size variations. For such applications a maximum limit may be applied to the geometric tolerance, in addition to the tolerance permitted at the maximum material condition, as shown in Fig. 30-3-7.

REGARDLESS OF FEATURE SIZE (RFS)

When MMC is not specified with a geometric tolerance for a feature of

EXAMPLE I

.002 (M) .005 MAX

EXAMPLE 2

Fig. 30-3-7   Tolerance with a maximum value.

size, no relationship is intended to exist between the feature size and the geometric tolerance. In other words, the tolerance applies regardless of feature size.

In this case, the geometric tolerance controls the form, orientation, or location of the center line, axis, or median plane of the feature.

The regardless of feature size symbol shown in Fig. 30-3-8 is shown only with a tolerance of position. See Unit 30-7.

Fig. 30-3-8   Application of RFS symbol.

Reference and Source Material
1. ANSI Y 14.5M Dimensioning and Tolerancing

ASSIGNMENTS

See Assignments 4 through 7 for Unit 30-3 on page 647.

UNIT 30-4
Straightness of Features and Flatness

In applying a geometric tolerance to a feature of size, it is often desirable to control the feature as a whole, rather than merely its surface elements. Such control may be intended to control the center line or median plane, in which

case it applies regardless of feature size (RFS). Alternatively it may be intended to control the bounding surfaces of the feature, in which case the tolerance is modified on an MMC basis. Straightness tolerances using both of these methods are described in this unit.

The same straightness symbol is used in the feature control symbol as for straightness of surface elements in Unit 30-3. However, when not modified by MMC, the feature control symbol may be directed to extension lines from the diameter or thickness, as shown in Fig. 30-4-1.

Figures 30-4-1 and 30-4-2 show examples of cylindrical features where all circular elements of the surface are to be within the specified size tolerance; however, the boundary of perfect form at MMC may be violated. This violation is permissible when the feature control frame is associated

with the size dimension or attached to an extension of the dimension line. In this instance, a diameter symbol precedes the tolerance value as the tolerance zone is circular at any given point, and the tolerance is applied on either an RFS or MMC basis. The straightness tolerance may be greater than the size tolerance where necessary. The collective effect of size and form variation can produce a virtual condition equal to the MMC size plus the straightness tolerance. When applied on an RFS basis, as in Fig. 30-4-1, the maximum straightness tolerance is the specified tolerance. When applied on an MMC basis, as in Fig. 30-4-2, the maximum straightness tolerance is the specified tolerance plus the amount the feature departs from its MMC size. The axis or center line of the actual feature must lie within the specified cylindrical tolerance zone.

(A) DRAWING CALLOUT

FEATURE SIZE	DIAMETER TOLERANCE ZONE ALLOWED
.615	.015
.614	.015
.613	.015
↓	↓
.606	.015
.605	.015

(B) INTERPRETATION

Fig. 30-4-1   Specifying straightness—RFS.

(A) DRAWING CALLOUT

FEATURE SIZE	DIAMETER TOLERANCE ZONE ALLOWED
.615	.015
.614	.016
.613	.017
↓	↓
.606	.024
.605	.025

(B) INTERPRETATION

Fig. 30-4-2   Specifying straightness—MMC.

A straightness tolerance, not modified by MMC, may be applied to parts or features of any size or shape, provided they have a center line or median plane which is intended to be straight in the direction indicated. Examples are parts having a cross section which is hexagonal, square, or rectangular. See Fig. 30-4-3.

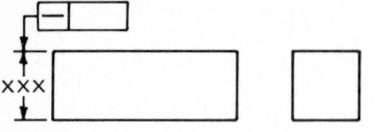

(A) SQUARE AND RECTANGULAR PARTS

(B) REGULAR POLYGONS

Fig. 30-4-3 Straightness of the center line—RFS.

Tolerances directed in this manner apply to straightness of the center line or center plane between all opposing line elements of the surfaces in the longitudinal direction to which the control is directed. The width of the tolerance zone is in the direction of the arrowhead. If the cross section forms a regular polygon, such as a hexagon or square, the tolerance applies to the center lines between each pair of sides, without its being necessary to so state on the drawing. If the cross section is circular, the tolerance zone becomes circular and a diameter symbol then precedes the tolerance, as shown in Fig. 30-4-4.

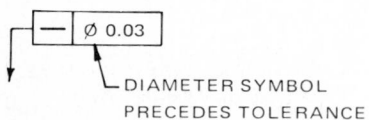

Fig. 30-4-4 Diameter symbol added when tolerance zone is circular or cylindrical.

CONTROL IN SPECIFIC DIRECTIONS

As already stated, straightness of a center line or median plane applies

only to center lines that run in the direction of the line or line elements to which the straightness tolerance is directed. If there could be some ambiguity, a note should be added, such as THIS DIRECTION ONLY, as shown in Fig. 30-4-5A. If the part is circular and it is intended that the tolerance apply in all directions, a diameter symbol should precede the tolerance value, as shown in Fig. 30-4-5B.

Fig. 30-4-5 Direction of application of straightness.

If a tolerance is shown in two directions, it is measured in these two directions, and the tolerance zone is then a parallelepiped, as shown in Fig. 30-4-6.

Fig. 30-4-7 Straightness—0 MMC.

FEATURE SIZE	PERMISSIBLE STRAIGHTNESS ERROR
.624	.000
.623	.001
.622	.002
.621	.003
.620	.004
.619	.005
.618	.006

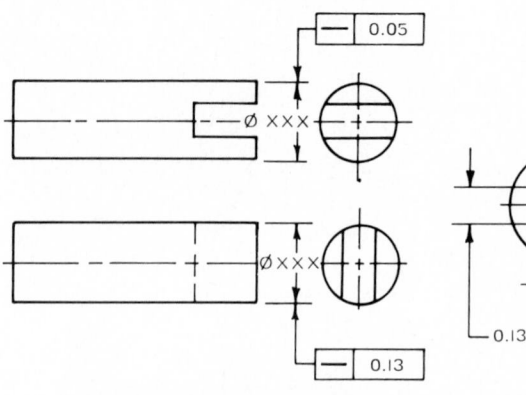

(A) DRAWING CALLOUT

Fig. 30-4-6 Straightness in two directions.

(B) TOLERANCE ZONE

STRAIGHTNESS—ZERO MMC

It is quite permissible to specify a geometric tolerance of zero MMC, which means that the virtual condition coincides with the maximum material size. See Fig. 30-4-7. Therefore, if a feature is at its maximum material limit everywhere, no errors of straightness are permitted.

Straightness on an MMC basis can be applied to any part or feature having straight-line elements in a plane which includes the diameter or thickness. This includes practically all the parts already shown on an RFS basis. However, it should not be used for features which do not have a uniform cross section.

STRAIGHTNESS WITH A MAXIMUM VALUE

If it is desired to ensure that the straightness error does not become too great when the part approaches the least material condition, a maximum value may be added, as shown in Fig. 30-4-8.

FLATNESS

The symbol for flatness is a parallelogram, with angles of 60° as shown in Fig. 30-4-9. The length and height are based on a percentage of the height of the lettering used on the drawing.

Flatness of a Surface

Flatness of a surface is a condition in which all surface elements are in one

H = RECOMMENDED LETTER HEIGHT

Fig. 30-4-9 Flatness symbol.

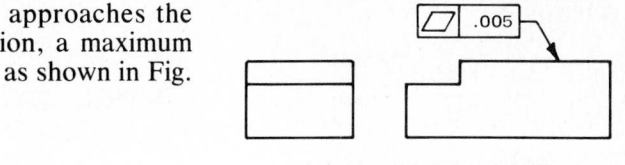

(A) DRAWING CALLOUT

TOLERANCE ZONE IS SPACE BETWEEN PARALLEL PLANES .005 APART

(B) INTERPRETATION

Fig. 30-4-10 Flatness of a surface.

(A)   (B)

Fig. 30-4-11 Controlling flatness on two or more surfaces.

plane. On such a surface all line elements in two or more directions are straight.

A flatness tolerance is applied to a line representing the surface of a part by means of a feature control frame, as shown in Fig. 30-4-10.

A flatness tolerance means that all points on the surface shall be contained within a tolerance zone consisting of the space between two parallel planes which are separated by the specified tolerance. These planes may be oriented in any manner to contain the surface; that is, they are not necessarily parallel to the base.

If the same control is desired on two or more surfaces, a suitable note may be added instead of repeating the symbol, as shown in Fig. 30-4-11.

Reference and Source Material
1. ANSI Y 14.5M Dimensioning and Tolerancing

ASSIGNMENTS
See Assignments 8 through 12 for Unit 30-4 on page 647.

Fig. 30-4-8 Straightness of a hole and shaft with a maximum value.

SHAFT	HOLE

Ⓜ Ø.000 Ⓜ .002 MAX

Ø .998 / .994

DRAWING CALLOUT

Ⓜ Ø.000 Ⓜ .001 MAX

Ø 1.003 / 1.000

DRAWING CALLOUT

Ø.998 EXTREME VIRTUAL CONDITION

Ø1.000 EXTREME VIRTUAL CONDITION

FEATURE SIZE	DIAMETER TOLERANCE ZONE ALLOWED
.998	.000
.997	.001
.996	.002
.995	.002
.994	.002

PERMISSIBLE VARIATIONS

FEATURE SIZE	DIAMETER TOLERANCE ZONE ALLOWED
1.000	.000
1.001	.001
1.002	.001
1.003	.001

PERMISSIBLE VARIATIONS

UNIT 30-5
Datums and the Three-Plane Concept

DATUMS

Datum A *datum* is a point, line, plane, or other geometric surface from which dimensions are measured when so specified or to which geometric tolerances are referenced. A datum has an exact form and represents an exact or fixed location, for purposes of manufacture or measurement.

Datum Feature A *datum feature* is a feature of a part, such as an edge, a surface, or a hole, which forms the basis for a datum or is used to establish its location.

DATUMS FOR GEOMETRIC TOLERANCING

As defined, datums are exact geometric points, lines, or surfaces, each based on one or more datum features of the part. Surfaces are usually either flat or cylindrical, but other shapes are used when necessary. The datum features, being physical surfaces of the part, are subject to manufacturing errors and variations. For example, a flat surface of a part, if greatly magnified, will show some irregularity. If brought into contact with a perfect plane, it will touch only at the highest points, as shown in Fig. 30-5-1. The true datums are theoretical but are considered to exist in the form of locating surfaces of machines, fixtures, and gaging equipment on which the part rests or with which it makes contact during manufacture and measurement.

THREE-PLANE SYSTEM

Geometric tolerances, such as straightness and flatness, refer to unrelated lines and surfaces and do not require the use of datums.

Orientation and locational tolerances refer to related features; that is, they control the relationship of features to one another or to a datum or datum system. Such datum features must be properly identified on the drawing.

Fig. 30-5-1 Magnified section of a flat surface.

Usually only one datum is required for orientation purposes, but positional relationships may require a datum system consisting of two or three datums. These datums are designated as *primary*, *secondary*, and *tertiary*. When these datums are plane surfaces that are mutually perpendicular, they are commonly referred to as a *three-plane datum system*, or a *datum reference frame*.

Primary Datum If the primary datum feature is a flat surface, it could lie on a suitable plane surface, such as the surface of a gage, which would then become a primary datum, as shown in Fig. 30-5-2. Theoretically, there will be a minimum of three high spots on the flat surface which will come in contact with the surface of the gage.

Secondary Datum If the part, while lying on this primary plane, is brought into contact with a secondary plane, it will theoretically touch at a minimum of two points.

Tertiary Datum The part can now be slid along, while maintaining contact with the primary and secondary planes, until it contacts a third plane. This plane then becomes the tertiary datum, and the part will theoretically touch it at only one point.

These three planes constitute a datum system from which measurements can be taken. They will appear on the drawing, as shown in Fig. 30-5-3, except that the datum features will be identified in their correct sequence by the methods described later in this unit.

It must be remembered that the majority of parts are not of the simple rectangular shape, and considerably more ingenuity may be required to establish suitable datums for more complex shapes.

Identification of Datums
Datum symbols are required to serve two purposes:

(C) TERTIARY DATUM

Fig. 30-5-2 The datum planes.

1. To locate the datum surface or feature on the drawing
2. To identify the datum feature so that it can easily be referred to in other requirements.

There are two methods of datum symbolization in general use for such purposes: one is shown and used in ANSI standards; the other, the ISO method, is used in most other countries of the world.

Fig. 30-5-3   Three-plane datum system.

ANSI Datum Symbolization

In the ANSI system, every datum feature is identified by a capital letter, enclosed in a rectangular box.

A dash is placed before and after the letter, to identify it as applying to a datum feature, as shown in Fig. 30-5-4.

Fig. 30-5-4   Datum feature identification— ANSI.

This identifying symbol may be directed to the datum feature in any one of the following ways.

1. By attaching a side, end, or corner of the symbol frame to an extension line from the feature
2. By running a leader with arrowhead from the symbol frame to the feature

3. By adding the symbol to a note, a dimension, or a feature control frame pertaining to the feature.

These methods are illustrated in Fig. 30-5-6.

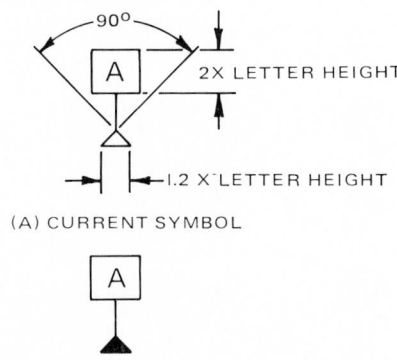

Fig. 30-5-5   ISO datum feature indicator symbol.

ISO Datum Symbolization

The ISO method is used in Canadian, British, and most other national standards. The ISO datum feature symbol is a right-angled triangle, with a leader projecting from the 90° apex, as shown in Fig. 30-5-5. The base of the triangle should be slightly greater than the height of the lettering used on the drawing. The triangle was formerly filled in, but because of microforming and cost reduction it was changed to a hollow triangle.

The datum is identified by a capital letter placed in a square frame and connected to the leader.

Application of Datum Identification to Drawings

The datum feature symbol is place on or directed to the datum feature in one of the following ways:

1. The datum feature symbol is placed on or directed to the outline of the feature when the datum feature is the surface itself or line elements of the surface. This is the preferred method.
2. It is on an extension line from the datum feature surface, but not in line with the dimension. This method is useful when the datum feature surface is very small or if, for other reasons, it is difficult to place the symbol on the line representing the surface.

3. It is on an extension line, in line with the feature dimension, when the datum is the datum axis of the cylindrical surface represented by the extension line, especially when applied on an MMC basis.

4. It is on two extension lines, using two datum symbols on the same leader, when two surfaces form a single datum, such as two sides of a slot or of a rectangular part.

5. It is on a leader line to a small feature, such as a hole, when the surface is too small to place the symbol on the surface and extension lines are not used.

All these methods are illustrated in Fig. 30-5-7. The symbol should not be placed on a hidden-feature line, but on the extension line to it or on the surface line in a sectional view.

Association with Geometric Tolerances

The datum letter is placed in the feature control symbol frame by adding an extra compartment for the datum reference, as shown in Fig. 30-5-8.

If two or more datum references are involved, then additional frames are added and the datum references are placed in these frames in the correct order, that is, primary, secondary, and tertiary datums, as shown in Fig. 30-5-9.

Multiple Datum Features

If a single datum is established by two datum features, such as two ends of a shaft, the features are each identified by separate letters. Both letters are then placed in the same compartment of the feature control symbol, with a dash between them, as shown in Fig. 30-5-10B. The datum, in this case, is the common line between the two datum features.

Datums Based on Features of Size

When a feature of size is specified as a datum feature, the datum has to be established from the full surface of a cylindrical feature or from two opposing surfaces of other features of size. However, the true datum is a datum axis, center line, or median plane of the feature.

There are two methods of establishing such datums: on an MMC basis or without MMC.

Fig. 30-5-6   Placement of datum symbol for single features.

ANSI CALLOUT	ISO CALLOUT

(A) SYMBOL PLACED ON OR DIRECTED TO THE EXTENSION
LINE FROM ONE DATUM SURFACE IN LINE WITH THE DIMENSION.
DATUM IS TWO PARALLEL PLANES SEPARATED BY THE MMC.

(B) DATUM SYMBOL IS PLACED IN LINE WITH THE DIMENSION.

(C) DATUM IS A CYLINDER WITH A DIAMETER EQUAL TO MMC

(D) VARIATIONS HAVING DATUM SYMBOL IN LINE WITH DIMENSION

Fig. 30-5-7 Placement of datum symbol for feature of size.

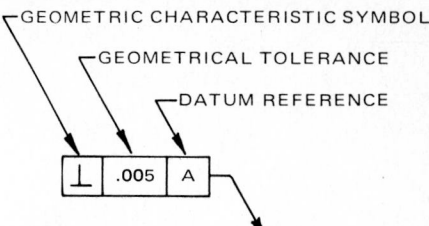

Fig. 30-5-8 Feature control symbol referenced to a datum.

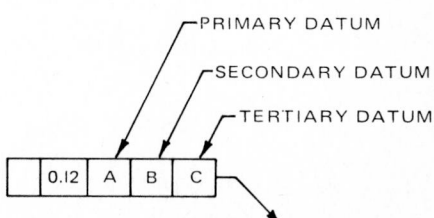

Fig. 30-5-9 Multiple datum references.

DRAWING CALLOUT

INTERPRETATION
(A) COPLANAR DATUM FEATURES

DRAWING CALLOUT

INTERPRETATION
(B) COAXIAL DATUM FEATURES

Fig. 30-5-10 Two datum features for one datum.

DATUM FEATURES—
WITHOUT MMC

When a feature of size is specified as a datum feature and is not modified on an MMC basis, the datum axis or median plane is established from the virtual size of each individual part.

When there are other features on a common center line, the symbol should be directed to extension lines and be associated with or in line with the dimension, as shown in Fig. 30-5-11.

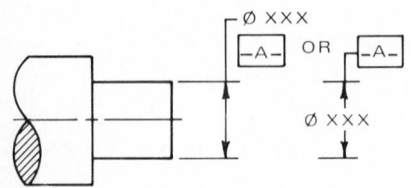

Fig. 30-5-11 Symbols on extension lines when center line is common to other features.

External Cylindrical Datum Features If the axis of an external cylindrical feature is designated as a primary datum feature, the datum is the axis of an imaginary perfect cylinder. This is the smallest such cylinder which can be circumscribed around the feature so that it just contacts the highest points on the surface. Geometric tolerances which are referenced to such a datum refer theoretically to the axis of this datum cylinder, as shown in Fig. 30-5-12.

Cylindrical Features as Secondary Datums It is often desirable to specify an end face or flange of a cylindrical feature as the primary datum, using the cylindrical feature as the secondary datum. In such cases the secondary datum symbol may be directed to the cylindrical surface rather than to the center line.

The primary datum is then a perfect plane on which the part would normally rest. The secondary datum is still the axis of an imaginary perfect cylinder, but also one that is perpendicular to the primary datum. This cylinder would theoretically touch the feature at only two points. This is illustrated in Fig. 30-5-13, where the part has been purposely drawn out of square to show the effect of such deviations.

(A) DRAWING CALLOUT

DATUM CYLINDER · DATUM AXIS · PART

EXAMPLE I

DATUM AXIS · DATUM CYLINDER

EXAMPLE 2

(B) INTERPRETATION

Fig. 30-5-12 Datum passed on center line of part.

(A) DRAWING CALLOUT

DATUM AXIS · SECONDARY DATUM CYLINDER B · PRIMARY DATUM PLANE A

(B) INTERPRETATION

Fig. 30-5-13 Cylindrical feature as secondary datum.

Internal Cylindrical Datum Features If an internal cylindrical feature, such as a hole, is specified as a primary datum feature, the datum is again the axis of an imaginary perfect cylinder. In this

(A) DRAWING CALLOUT

DATUM AXIS · DATUM CYLINDER · DATUM AXIS · DATUM CYLINDER

(B) INTERPRETATION

Fig. 30-5-14 Internal cylindrical feature as datum.

case it is the largest such cylinder which can be inscribed within the feature. It will therefore contact the inner surface of the feature at the lowest points on the surface, as shown in Fig. 30-5-14.

DATUM FEATURES—MMC

The measurement of geometric tolerances, which refer to datums that are features of size, is greatly simplified if the datums are specified on an MMC basis. This permits the use of fixed "go" gages.

Datum features on an MMC basis always apply at their virtual condition. *Virtual condition* refers to the potential boundary of a feature, as specified on a drawing, derived from the collective effect of the maximum material limit of size and the specified form or orientation tolerance. These are added for external features, such as shafts, and subtracted for internal features, such as holes and slots.

If no form or orientation tolerance is specified, it is assumed, for datum reference purposes, that the form tolerance is zero MMC.

The fact that a datum applies on an MMC basis is indicated in the feature control frame by the addition of the MMC symbol Ⓜ immediately following the datum reference, as shown in Fig. 30-5-15. When there is more than one datum reference, the MMC sym-

GEOMETRICAL CHARACTERISTIC
SYMBOL
GEOMETRICAL TOLERANCE
DATUMS

.004 Ⓜ A Ⓜ B

THIS INDICATES THAT DATUM A
APPLIES ON MMC BASIS WHILE
DATUM B DOES NOT

Fig. 30-5-15   Reference to datums—MMC.

bol must be added for each datum where this modification is required.

The datum identification symbol could be shown in any of the ways used for data that are not on an MMC basis. However, it is convenient to direct the symbol in line or in association with the feature size dimension. It may also be coupled to the form tolerance if one is used, as shown in Fig. 30-5-16. This is permissible because an MMC datum is based on the maximum material size of the feature and the applicable form tolerance.

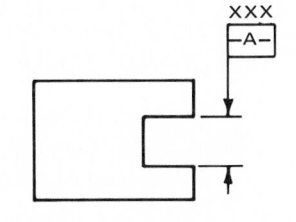

(A) APPLIED ON MMC BASIS

(B) COUPLED WITH FORM TOLERANCES

Fig. 30-5-16   Specifying datums.

External Features with Regular Cross Sections—MMC For external features having a cross section which is circular or which comprises a regular polygon, the datum will be of the same shape as the datum feature. The width or diameter of the datum will be equal to the

maximum material limit of size, plus the specified form tolerance.

In Fig. 30-5-17, because no form tolerance is specified, the cylindrical gaging elements are made to the maximum material size of .565 in. This provides an exact location of the part when it is made to the maximum material condition. However, it allows a deviation of .003 in. in any direction from true position if the part is everywhere at the minimum material size of .559 in. and there are no form errors.

Fig. 30-5-17   Gage element for circular datum.

Rectangular Features—MMC If the specified datum feature consists of two flat surfaces and the cross section is not a regular polygon, then the datum consists of two parallel planes. These are separated by a distance equal to the maximum material condition, plus the specified form tolerance.

Figure 30-5-18 shows a flatness tolerance which applies separately to

(A) DRAWING CALLOUT

(B) LOCATING ELEMENT OF GAGE

Fig. 30-5-18   Gage element for rectangular datum feature.

both datum feature surfaces. It should be noted that in such cases the form tolerance is not doubled in calculating the virtual condition. This is because if the part is everywhere at its maximum material condition, as shown in the gaging position, then a convex point on one side can only be offset by a concave point on the other side. If this were not true, the size dimension would be exceeded.

It should be noted that Fig. 30-5-18B shows the element of a gage which locates on the datum on an MMC basis. It does not check the flatness requirement.

Internal Features—MMC The same rules apply to internal features, except that the maximum material condition is the minimum limit of size, and the specified form tolerance is subtracted from this limit to obtain the virtual condition. The size of the locating element of the gage is identical to the virtual condition, as shown in Fig. 30-5-19.

(A) ANSI DRAWING CALLOUT

(B) LOCATING ELEMENT OF GAGE

Fig. 30-5-19   Gage element for internal datum feature.

Reference and Source Material

1. ANSI Y 14.5M Dimensioning and Tolerancing

ASSIGNMENTS
See Assignments 13 through 16 for Unit 30-5 on page 648.

UNIT 30-6
Orientation

ANGULAR RELATIONSHIPS OF FLAT SURFACES

Orientation refers to the angular relationship which exists between two or more lines, surfaces, or other features.

The general geometric characteristic for orientation is termed *angularity*. This term may be used to describe angular relationships, of any angle, between straight lines or surfaces with straight-line elements, such as flat or cylindrical surfaces. For two particular types of angularity special terms are used. These are *perpendicularity*, or squareness, for features related by a 90° angle, and *parallelism* for features related by an angle of 0°.

An orientation tolerance indicates a relationship between two or more features. Whenever possible, the feature to which the controlled feature is related should be designated as a datum.

SYMBOLS FOR GEOMETRIC CHARACTERISTICS

There are three geometric symbols for these characteristics, as shown in Fig. 30-6-1. The proportions are based on the height of the lettering used on the drawing.

For geometric tolerancing of angularity, the angle between the datum and the controlled feature should be stated as a basic angle. Therefore it should be enclosed in a rectangular frame, as shown in Fig. 30-6-2, to indicate that the general tolerance note does not apply. However, the angle need not be stated for either perpendicularity (90°) or parallelism (0°).

Fig. 30-6-2 Orientation tolerances with identified datum.

The datum is identified with a capital letter, and the same letter is then used in the feature control symbol, as shown in Fig. 30-6-2.

For the tolerancing of angularity the same characteristic symbols are used for both lines and surfaces.

ORIENTATION TOLERANCING OF FLAT SURFACES

Figure 30-6-3 shows three simple parts in which one flat surface is designated as a datum feature and another flat surface is related to it by one of the orientation tolerances.

Each of these tolerances is interpreted to mean that the designated surface shall be contained within a tolerance zone consisting of the space between two parallel planes, separated by the specified tolerance (.002 in.) and related to the datum by the basic angle specified (30, 90, or 0°).

Control in Two Directions

The measuring principles for angularity indicate the method of aligning the part prior to making angularity measurements. Proper alignment ensures that line elements of the surface perpendicular to the angular line elements are parallel to the datum.

For example, the part in Fig. 30-6-4 will be aligned so that line elements running horizontally in the left-hand view will be parallel to the datum. However, these line elements will bear a proper relationship with the sides, ends, and top faces only if these surfaces are true and square with the primary datum.

It may be functionally more important to measure the angle in a direction parallel to a side or perpendicular to the front or back face. In this case, one side or face must be chosen as a secondary datum, as shown in Fig. 30-6-4.

Under these circumstances, the part is aligned on the angle plate so that the secondary datum *B* is exactly parallel to the side of the angle plate.

LINES RELATED TO SURFACES

Internal Cylindrical Features

Figure 30-6-5 shows some simple parts in which the axis or center line of a hole is related by an orientation tolerance to a flat surface. The flat surface is designated as the datum feature.

The center line of the hole must be contained within a tolerance zone consisting of the space between two parallel planes. These planes are separated by a specified tolerance of .006 in. and are related to the datum by one of the basic angles 45, 90, or 0°. Figure 30-6-6 clearly illustrates the tolerance zone for angularity.

When the tolerance is one of perpendicularity, the tolerance-zone planes can be revolved around the feature axis without the angle being affected. The tolerance zone therefore becomes a cylinder. This cylindrical zone is perpendicular to the datum and has a diameter equal to the specified tolerance, as shown in Fig. 30-6-7.

Control of Direction of Angularity The angularity tolerance in Fig. 30-6-5 controls angularity of the hole in the direction in which it has its greatest slope, that is, the minimum angle. In some cases, it may be functionally more important to control the angle in a specific direction, such as parallel to a side. In this case, a secondary datum should be added, as shown in Fig. 30-6-8.

ANGULARITY
H = LETTER HEIGHT

PERPENDICULARITY
(SQUARENESS)

PARALLELISM

Fig. 30-6-1 Orientation symbols.

ANGULARITY   PERPENDICULARITY   PARALLELISM

(A) ANSI CALLOUT

OR

(B) INTERNATIONAL CALLOUT

TOLERANCE ZONE .002 WIDE

30°

DATUM PLANE

TOLERANCE ZONE .002 WIDE

90°

TOLERANCE ZONE .002 WIDE

(C) INTERPRETATION

Fig. 30-6-3   Orientation tolerancing.

Fig. 30-6-4   Angularity referred to a datum system.

(A) ANGULARITY

Fig. 30-6-5   Orientation of lines and surfaces.

(B) PERPENDICULARITY

(C) PARALLELISM

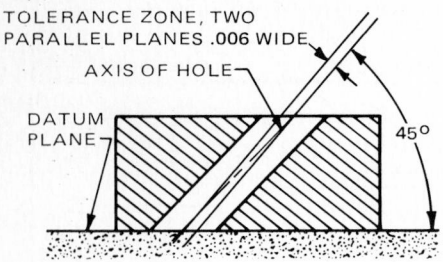

TOLERANCE ZONE, TWO
PARALLEL PLANES .006 WIDE

AXIS OF HOLE

DATUM
PLANE

45°

Fig. 30-6-6   Tolerance zone for angularity—Fig. 30-6-5.

PARALLEL PLANES CAN BE REVOLVED,
THUS TOLERANCE ZONE BECOMES
A CYLINDER

TOLERANCE ZONES
.006

DATUM
PLANE A

90°

Fig. 30-6-7   Tolerance zone for perpendicularity—Fig. 30-6-5.

| // | 0.12 | A | B |

45°

SECONDARY DATUM

Fig. 30-6-8   Angularity referred to two datums.

Control of Parallelism in Two Directions
The feature control symbol shown in Fig. 30-6-5C controls parallelism with a base only. If control with a side is also required, the side should be designated as a secondary datum, as shown in Fig. 30-6-9. The tolerance zone will be a parallelepiped, and two separate measurements will have to be made.

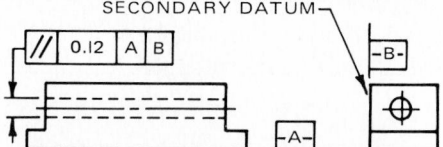

SECONDARY DATUM

| // | 0.12 | A | B |

-B-

-A-

Fig. 30-6-9   Parallelism controlled in two directions.

Controls on MMC Basis   Since a hole is a feature of size, any of the tolerances shown in Fig. 30-6-5 can be modified on an MMC basis. This is specified by adding the symbol Ⓜ after the tolerance. Figure 30-6-10 shows an example.

Ø .250 ±.002

| ⊥ | Ø .006 Ⓜ | A |

-A-

Fig. 30-6-10   Feature controlled on MMC basis.

Control of Center Lines   Tolerances intended to control orientation of the center line of a feature are applied to drawings as shown in Fig. 30-6-11.

The center line of the cylindrical feature must be contained within a tolerance zone consisting of the space between two parallel planes separated by the specified tolerance. The parallel planes are related to the datum by the basic angle of 75, 90, or 0°.

Since the tolerance planes for perpendicularity can be revolved around the feature axis, the tolerance zone effectively becomes a cylinder. The diameter of this cylinder is equal to the specified tolerance.

Tolerances Applied to Line Elements of the Surface

In some cases it may be more important to control line elements of the cylindrical surface, instead of its axis or center line. For this purpose the tolerance is directed to the surface with a single arrowhead, as shown in Fig. 30-6-12.

Except for perpendicularity, such controls apply to only two line elements of the surface. These are the lines on opposite sides of the feature,

Ø XXX

| ∠ | 0.1 | A |

75°

-A-

(A) ANGULARITY

Ø XXX

| ⊥ | Ø 0.1 | A |

-A-

(B) PERPENDICULARITY

| // | 0.1 | A |

-A-

(C) PARALLELISM

Fig. 30-6-11   Orientation of external cylindrical features.

which lie in a plane perpendicular to the datum and which include the axis of the cylindrical feature. There are two tolerance zones, each of which may be considered to be the area between two parallel lines, which lie in this measurement plane. These lines are separated by the specified tolerance and are related to the datum by the basic angle. Figure 30-6-13 shows these zones for a parallelism tolerance. Figure 30-6-14 illustrates perpendicularity and parallelism on an MMC basis.

For perpendicularity the measurement plane can be revolved around the axis without altering its relationship to the datum. Therefore, unless other-

(A) ANGULARITY

(B) PERPENDICULARITY

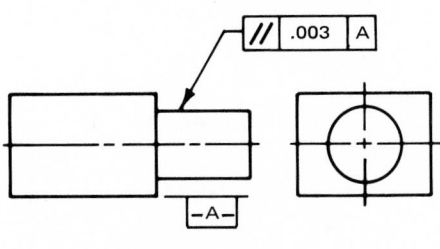

(C) PARALLELISM

Fig. 30-6-12 Orientation of surface elements.

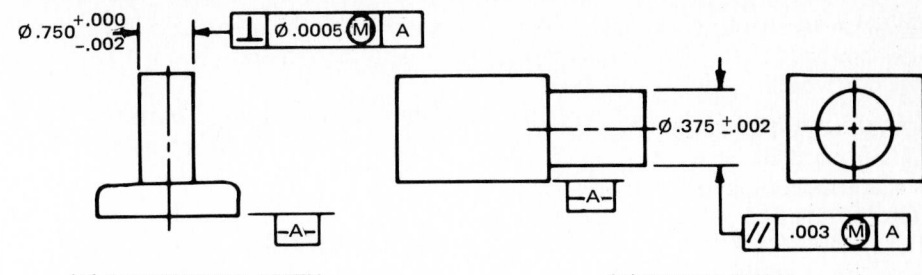

(A) PERPENDICULARITY

(B) PARALLELISM

Fig. 30-6-14 Perpendicularity and parallelism—MMC.

(A) DRAWING CALLOUT

(B) TOLERANCE ZONE

Fig. 30-6-15 Angularity of flat surface with an external cylinder.

(A) DRAWING CALLOUT

(B) TOLERANCE ZONE

Fig. 30-6-16 Angularity of flat surface with an internal cylinder.

wise specified, perpendicularity applies to line elements in all positions of the measuring plane.

Orientation Using Lines as Datums

The following examples of orientation tolerancing involve internal or external cylindrical datum features. These features establish datums which are either line elements of the datum fea-

ture surface or the axes of datum cylinders. Many of the methods are also applicable to the center lines of noncylindrical features, such as those having square or hexagonal cross sections.

When MMC Is Not Specified The tolerance zone is always the space between two parallel planes separated by the specified tolerance and related to the datum by the basic angle. The controlled feature, whether it be a flat surface, the center line of the feature, or a line element of the surface, must be contained within this tolerance zone.

EXAMPLE 1 Figure 30-6-15 shows a flat surface related to an external cylindrical datum feature. Since the MMC is not specified, the real datum is the axis of a datum cylinder, which is the smallest perfect cylinder that can be circumscribed around the feature. The tolerance zone is the space between two parallel planes .004 in. apart, which are related to the datum axis by the basic angle.

EXAMPLE 2 Figure 30-6-16 is similar to Fig. 30-6-15 except that the flat surface is related to an internal cylindrical feature. The datum is therefore the axis of the largest perfect cylinder which will fit within the center hole of the part. The tolerance zone is the space

Fig. 30-6-13 Tolerance zones for parallelism in Fig. 30-6-12.

between two parallel planes, .004 in. apart, related to this datum axis by the specified basic angle.

EXAMPLE 3 Figure 30-6-17 shows a simple parallelism requirement of a flat surface in relation to a cylindrical hole. The tolerance zone is the space between two parallel planes, 0.1 mm apart and parallel to the datum axis.

(A) DRAWING CALLOUT

(B) TOLERANCE ZONE

Fig. 30-6-17 Parallelism of a flat surface with a cylindrical hole.

EXAMPLE 4 Figure 30-6-18 shows a requirement for perpendicularity of the axis of a hole with line elements of a cylindrical surface. There are actually two line elements of the cylindrical surface perpendicular to the hole, rep-

(A) DRAWING CALLOUT

(B) TOLERANCE ZONE

Fig. 30-6-18 Perpendicularity of hole with line elements of surface.

resented by the upper and lower solid lines in the illustration.

Reference and Source Material
1. ANSI Y 14.5M Dimensioning and Tolerancing

ASSIGNMENTS

See Assignments 17 through 21 for Unit 30-6 on page 649.

UNIT 30-7
Tolerancing for Location of Features

The location of features is one of the most frequently used applications of dimensions on technical drawings. Tolerancing may be accomplished either by *coordinate tolerances* applied to the dimensions or by *geometric (positional) tolerancing*.

Positional tolerancing is especially useful when applied on an MMC basis to groups or patterns of holes or other small features in the mass production of parts. This method meets functional requirements in most cases and permits assessment with simple gaging procedures.

Most examples in this unit are devoted to the principles involved in the location of small, round holes, because they represent the most commonly used applications. The same principles apply, however, to the location of other features, such as slots, tabs, bosses, and noncircular holes.

TOLERANCING METHODS

A single hole is usually located by means of rectangular coordinate dimensions, extending from suitable edges or other features of the part to the axis of the hole. Other dimensioning methods, such as polar coordinates, may be used when circumstances warrant.

There are two standard methods of tolerancing the location of holes, as illustrated in Fig. 30-7-1:

1. *Coordinate tolerancing*, which refers to tolerances applied directly to the coordinate dimensions or to

(A) COORDINATE TOLERANCING

(B) POSITIONAL TOLERANCING – RFS

(C) POSITIONAL TOLERANCING – MMC

Fig. 30-7-1 Comparison of tolerancing methods.

applicable tolerances specified in a general tolerance note.

2. (*a*) *Positional tolerancing*, RFS (regardless of feature size).
(*b*) *Positional tolerancing*, MMC basis. These positional tolerancing methods are part of the system of geometric tolerancing.

Any of these tolerancing methods can be substituted one for the other, although with differing results. It is

(A) EQUAL TOLERANCES

(B) UNEQUAL TOLERANCES

(C) POLAR TOLERANCES

Fig. 30-7-2 Tolerance zones for coordinate tolerances.

Fig. 30-7-4 Square tolerance zone.

For the examples shown in Fig. 30-7-2, the tolerance zones are shown in Fig. 30-7-5, and the maximum tolerance values are as shown in the following examples.

necessary, however, to first analyze the widely used method of coordinate tolerancing in order to explain and understand the advantages and disadvantages of the positional tolerancing methods.

COORDINATE TOLERANCING

Coordinate dimensions and tolerances may be applied to the location of a single hole, as shown in Fig. 30-7-2. They locate the hole axis and result in a rectangular or wedge-shaped tolernace zone within which the axis of the hole must lie.

If the two coordinate tolerances are equal, the tolerance zone formed will be a square. Unequal tolerances result in a rectangular tolerance zone. Where one of the locating dimensions is a radius, polar dimensioning gives a circular ring section tolerance zone. For simplicity, square tolerance zones are used in the analyses of most of the examples in this section.

It should be noted that the tolerance zone extends for the full depth of the hole, that is, the whole length of the axis. This is illustrated in Fig. 30-7-3 and explained in more detail in a later unit. In most of the illustrations, tolerances will be analyzed as they apply at the surface of the part, where the axis is represented by a point.

Fig. 30-7-3 Tolerance zone extending through part.

Maximum Permissible Error

The actual position of the feature axis may be anywhere within the rectangular tolerance zone. For square tolerance zones, the maximum allowable variation from the desired position occurs in a direction of 45° from the direction of the coordinate dimensions. See Fig. 30-7-4.

For rectangular tolerance zones this maximum tolerance is the square root of the sum of the squares of the individual tolerances, or expressed mathematically

$$\sqrt{X^2 + Y^2}$$

EXAMPLE 1 $\sqrt{.010^2 + .010^2} = .014$

EXAMPLE 2 $\sqrt{.010^2 + .010^2} = .0224$

EXAMPLE 3

Fig. 30-7-5 Tolerance zones for parts shown in Fig. 30-7-2.

EXAMPLE 1

$$\sqrt{.010^2 + .010^2} = .014$$

EXAMPLE 2

$$\sqrt{.010^2 + .020^2} = .0224$$

EXAMPLE 3 For polar coordinates the extreme variation is

$$\sqrt{A^2 + T^2}$$

where $A = R \tan a$
$T = $ tolerance on radius
$R = $ mean radius
$a = $ angular tolerance

Thus, the extreme variation in the third example is

$$\sqrt{(1.25 \times .017\,45)^2 + .020^2} = .03$$

Note: Mathematically, A in the above formula should be $2R \tan a/2$, instead of $R \tan a$, and T should be $T \cos A/2$; but the difference in results is quite insignificant for the tolerances normally used.

Some values of tan A for commonly used angular tolerances are as follows:

A	tan a	A	tan a	A	tan a
0° 5′	0.00145	0°25′	0.00727	0°45′	0.01309
0°10′	0.00291	0°30′	0.00873	0°50′	0.01454
0°15′	0.00436	0°35′	0.01018	0°55′	0.01600
0°20′	0.00582	0°40′	0.01164	1° 0′	0.01745

Use of Chart

A quick and easy method of finding the maximum positional error permitted with coordinate tolerancing, without having to calculate squares and square roots, is by use of a chart like that shown in Fig. 30-7-6.

In the first example shown in Fig. 30-7-2, the tolerance in both directions is .010 in. The extensions of the horizontal and vertical lines of .010 in the chart intersect at point A, which lies between the radii of .013 and .014 in. When interpolated and rounded to three decimal places, the maximum permissible variation of position is .014 in.

In the second example shown in Fig. 30-7-2, the tolerances are .010 in. in one direction and .020 in. in the other. The extensions of the vertical and horizontal lines at .010 and .020 in., respectively, in the chart intersect at point B, which lies between the radii of .022 and .023 in. When interpolated and rounded to three decimal places, the maximum variation of position is .022 in. Figure 30-7-6 also shows a chart for use with tolerances in millimeters.

Advantages of Coordinate Tolerancing

The advantages claimed for direct coordinate tolerancing are as follows:

1. It is simple and easily understood, and, therefore, it is commonly used.
2. It permits direct measurements to be made with standard instruments and does not require the use of special-purpose functional gages or other calculations.

Disadvantages of Coordinate Tolerancing

There are a number of disadvantages to the direct tolerancing method:

1. It results in a square or rectangular tolerance zone within which the axis must lie. For a square zone this permits a variation in a 45° direction of approximately 1.4 times the specified tolerance. This amount of variation may necessitate the specification of tolerances which are only 70 percent of those that are functionally acceptable.
2. It may result in an undesirable accumulation of tolerances when several features are involved, especially when chain dimensioning is used.
3. It is more difficult to assess clearances between mating features and components than when positional tolerancing is used, especially when a group or a pattern of features is involved.

Fig. 30-7-6 Chart for calculating maximum tolerance using coordinate tolerancing.

4. It does not correspond to the control exercised by fixed functional "go" gages, often desirable in mass production of parts. This becomes particularly important in dealing with a group of holes. With direct coordinate tolerancing, the location of each hole has to be measured separately in two directions, whereas with positional tolerancing on an MMC basis one functional gage checks all holes in one operation.

POSITIONAL TOLERANCING

Positional tolerancing is part of the system of geometric tolerancing. In this system the location of features is shown either by coordinate dimensions or by polar and angular dimensions, except that the dimensions are shown without direct tolerances. These dimensions represent the basic sizes and are commonly known as *true-position* or *basic* dimensions. Each such dimension is enclosed in a rectangular frame to indicate that it represents an exact value, to which tolerances shown in the general tolerance note do not apply. See Fig. 30-7-7. The frame size need not be any larger than that necessary to enclose the dimension. Permissible deviations from the basic dimension are then given by a positional tolerance as described in this unit.

Fig. 30-7-7 Identifying true-position dimensions.

Fig. 30-7-8 Position symbol.

SYMBOL FOR POSITION

The geometric characteristic symbol for position is a circle with two solid center lines, as shown in Fig. 30-7-8. This symbol is used in the feature control frame in the same manner as for other geometric tolerances.

Positional tolerances may be specified on an MMC, RFS, or LMC (least material condition) basis.

When specified on the MMC basis, the specified tolerance is independent of the size of the feature. Where the actual size of the feature has departed from MMC, an increase in the tolerance is allowed equal to the amount of such departure.

When specified on an RFS basis, the specified tolerance is independent of the size of the feature, and the tolerance is limited to the specified value regardless of the actual size of the feature.

As positional tolerance controls the position of the axis of the hole, the feature control frame is normally attached to the size of the feature, as shown in Fig. 30-7-9.

The positional tolerance represents the diameter of a cylindrical tolerance zone, located at true position as determined by the true-position dimensions on the drawing, within which the axis or center line of the hole must lie.

Except for the fact that the tolerance zone is circular instead of square, a positional tolerance on this basis has exactly the same meaning as direct coordinate tolerancing with equal tolerances in both directions.

It has already been shown that with rectangular coordinate tolerancing the maximum permissible error in location is not the value indicated by the horizontal and vertical tolerances, but rather is equivalent to the length of the

(A) DRAWING CALLOUT

Fig. 30-7-9 Positional tolerancing—regardless of feature size.

diagonal between the two tolerances. For square tolerance zones this amount is 1.4 times the specified tolerance values. The specified tolerance can therefore be increased to an amount equal to the diagonal of the coordinate tolerance zone without affecting the clearance between the hole and its mating part.

It is quite practical, however, to replace coordinate tolerances with a positional tolerance having a value equal to the diagonal of the coordinate tolerance zone. This does not affect the clearance between the hole and its mating part, yet it offers 57 percent more tolerance area, as shown in Fig. 30-7-10. Such a change would most likely result in a reduction in the number of parts rejected for positional errors.

A simpler method is to make coordinate measurements and evaluate them on a chart, as shown in Fig. 30-7-11. For example, if measurements of four parts are as shown in the table below, only two are acceptable. These positions are shown on the chart.

Part	Measurements		Acceptability
A	.565	.752	Rejected
B	.562	.754	Accepted
C	.557	.753	Accepted
D	.556	.754	Rejected

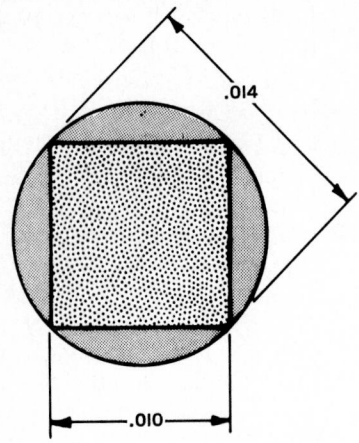

AREA OF CIRCUMSCRIBED
CIRCULAR ZONE = 157 % OF
SQUARE TOLERANCE ZONE

Fig. 30-7-10   Relationship of tolerance zones.

POSITIONAL TOLERANCING— MAXIMUM MATERIAL CONDITION

The problems of tolerancing for the position of holes are simplified when positional tolerancing is applied on an MMC basis. This overcomes the disadvantages listed for the coordinate method. Positional tolerancing simplifies measuring procedures because it permits the use of functional "go" gages. It also permits an increase in positional variations as the size departs from the maximum material size without jeopardizing free assembly of mating features.

A positional tolerance on an MMC basis is specified on a drawing, on either the front or the side view, as shown in Fig. 30-7-12. The MMC sym-

Fig. 30-7-12   Positional tolerancing—MMC.

bol Ⓜ is added in the feature control frame immediately after the tolerance.

A positional tolerance applied to a hole on an MMC basis means that the boundary of the hole must fall outside a perfect cylinder having a diameter equal to the maximum material condition of the feature minus the positional tolerance. This cylinder is located with its axis at true position. The hole must, of course, meet its diameter limits.

The effect is illustrated in Fig. 30-7-13, where the gage cylinder is shown at true position and the minimum and maximum diameter holes are drawn to show the extreme permissible variations in position in one direction.

Therefore, if a hole is at its maximum material condition (minimum diameter), the position of its axis must lie within a circular tolerance zone having a diameter equal to the specified tolerance. If the hole is at its maximum diameter (least material condition), the diameter of the tolerance zone is increased by the amount of the feature tolerance. The greatest deviation in one direction from true position is therefore

$$\frac{H + P}{2}$$

where H = hole diameter tolerance
P = positional tolerance

An example is illustrated in Fig. 30-7-14. The hole at its least material condition has a .543-in. diameter.

$H = .004$
$P = .006$

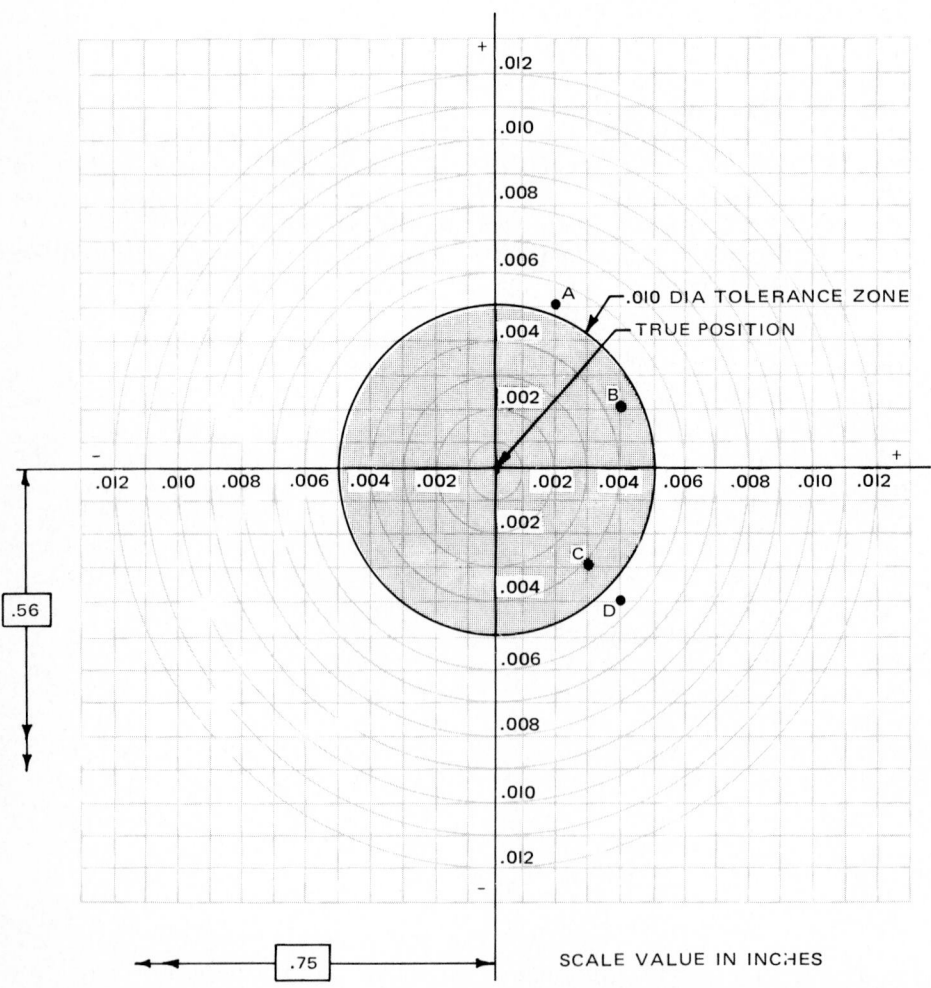

NOTE – DARK CIRCULAR AREA REPRESENTS ⌀ .010 TOLERANCE ZONE

Fig. 30-7-11   Chart for evaluating positional tolerancing—RFS.

Fig. 30-7-13 Positional variations for tolerancing—MMC.

Therefore, the greatest deviation in any one direction is

$$\frac{H + P}{2} = \frac{.004 + .006}{2} = .005$$

It must be emphasized that positional tolerancing, even on an MMC basis, is not a cure-all for positional tolerancing problems; each method of tolerancing has its own area of usefulness. In each application a method must be selected which best suits the particular case. The preferred methods for various types of applications are as follows.

Positional Tolerancing on an MMC Basis This method is preferred when production quantities warrant the provision of functional "go" gages, because gaging is then limited to one simple operation, even when a group of holes is involved. This method also facilitates manufacture by permitting larger variations in position when the diameter departs from the maximum material

condition. It cannot be used when it is essential that variations in location of the axis be observed regardless of feature size.

Coordinate Tolerancing Method This method is preferred in many applications where it is not economical to provide functional gages. Coordinate tolerancing may require a reduction in the specified tolerance from the estimated permissible variation in order to compensate for a square tolerance zone. Also, it does not permit an increase in the permissible variation as the size approaches the least material condition.

Positional Tolerancing, RFS This method is very similar to the coordinate tolerancing method except that it results in a round, instead of a square, tolerance zone. The specified tolerance value can therefore be increased by 40 percent without affecting free assembly of mating parts. However, it is probably the least used method since measurement entails the use of mathematics or a chart.

Projected Tolerance Zone
The application of this concept is recommended where the variation in perpendicularity of threaded or press-fit holes could cause fasteners, such as

(A) DRAWING CALLOUT

(B) INTERPRETATION

Fig. 30-7-14 Hole with an MMC positional tolerance.

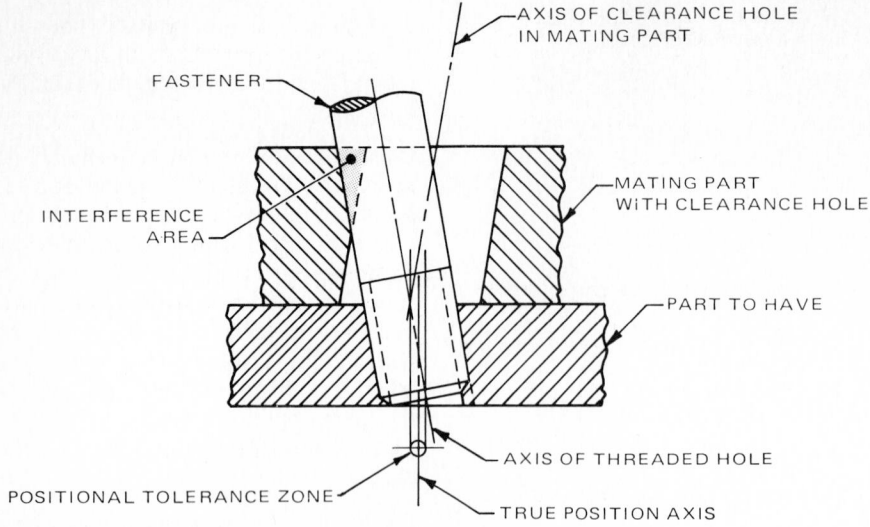

Fig. 30-7-15   Illustrating how fastener can interfere with mating part.

(A) DRAWING CALLOUT   (B) INTERPRETATION

Fig. 30-7-16   Specifying a projected tolerance zone.

In the examples given so far, the position of the axis of the hole was established by using dimensions from the actual surfaces of the part. These surfaces were not designated as datums. The true or mean position of the axis was therefore a line parallel to a surface line element on each of the surfaces from which the dimensions were drawn, as illustrated in Fig. 30-8-1.

(A) DRAWING CALLOUT

(B) INTERPRETATION OF TRUE POSITION

Fig. 30-8-1   Lines from which measurements are made.

screws, studs, or pins, to interfere with mating parts, as shown in Fig. 30-7-15. Interference can occur where a positional tolerance is applied to the depth of a hole and the hole axis is inclined within allowable limits. Figure 30-7-16 illustrates the application of a positional tolerance using a projected tolerance zone. The specified value for the projected tolerance zone is the minimum value and represents the

maximum permissible mating part thickness.

Reference and Source Material
1. ANSI Y 14.5M Dimensioning and Tolerancing

ASSIGNMENTS

See Assignments 22 through 27 for Unit 30-7 on page 650.

If these sides are off-square with one another or with other surfaces of the part, the true position of the axis would be similarly off-square, as shown in a somewhat exaggerated format in Fig. 30-8-2.

In some applications this may be the desired requirement, but in most cases it is preferable to have the hole either related to other surfaces or features or

DIMENSIONS APPLY TO THE SHORTEST DISTANCE FROM THE AXIS OF THE HOLE TO THE SIDE

Fig. 30-8-2 Results when sides are off-square.

related to a full side rather than a line on the surface. It is then necessary to specify the desired datum feature or features in the required order of priority.

The first consideration in such applications is to decide on the primary datum feature. The usual course of action is to specify as the primary datum the surface into which the hole is produced. This will ensure that the true position of the axis is perpendicular to this surface or at the basic angle, if it is other than 90°. Secondary and tertiary datum features are then selected and identified, if required.

Figure 30-8-3 shows a part similar to that shown in Fig. 30-8-1 but with the addition of a primary datum feature and the MMC modifier. Figure 30-8-4 shows the same part with three datum features specified.

(A) DRAWING CALLOUT (B) INTERPRETATION OF TRUE POSITION

Fig. 30-8-4 Part with three datum features specified.

Fig. 30-8-5 Datum system for a long hole.

Long Holes It is not always essential to have the true position of a hole perpendicular to the face into which the hole is produced. It may be functionally more important, especially with long holes, to have it parallel to one of the sides. Figure 30-8-5 is a case in point. In this example the sides are designated as primary and secondary datums. A tertiary datum is not required.

Circular Datums Circular features, such as holes or external cylindrical features, can be used as datums just as readily as flat surfaces. In the simple

part shown in Fig. 30-8-6, it is quite evident that the true position of the small hole is established from the axis of the large hole. In cases like these, it may not be necessary to specify one of the holes as a datum, although it would facilitate gaging if the datum were specified on an MMC basis.

In other cases, such as that shown in Fig. 30-8-7, it is essential to specify the datum; otherwise, the origin of the true-position dimension would be left in doubt. It could be either the axis of the hole or the axis of the outside cylindrical surface.

Fig. 30-8-3 Part with one datum feature specified.

Fig. 30-8-6 Part where specification of the datum feature may not be required.

Fig. 30-8-7 Position referenced to a circular datum.

Multiple-Hole Datums The axis of holes is sometimes specified as a datum feature with MMC being specified for the datum reference. On an MMC basis, any number of holes or similar features which form a group or pattern may be specified as a single datum. All features forming such a datum must be related with a positional tolerance on an MMC basis. See Fig. 30-8-8.

DATUM TARGETS

The full feature surface was used to establish a datum for the features so far designated as datum features. This may not always be practical for the following reasons:

1. The surface of a feature may be so large that a gage designed to make contact with the full surface may be too expensive or too cumbersome to use.

2. Functional requirements of the part may necessitate the use of only a portion of a surface as a datum feature, for example, the portion which contacts a mating part in assembly.

3. A surface selected as a datum feature may not be sufficiently true, and a flat datum feature may rock when placed on a datum plane, so that accurate and repeatable measurements from the surface would not be possible. This is particularly so for surfaces of castings, forgings, weldments, and some sheet-metal and formed parts.

A useful technique to overcome such problems is the datum target method. In this method certain points, lines, or small areas on the surfaces are selected as the bases for establishment of datums. For flat surfaces, this usually requires three target points or

areas for a primary datum, two for a secondary datum, and one for a tertiary datum.

It is not necessary to use targets for all datums. It is quite logical, for example, to use targets for the primary datum and other surfaces or features for secondary and tertiary datums if required; or to use a flat surface of a part as the primary datum and to locate fixed points or lines on the edges as secondary and tertiary datums.

Datum targets should be spaced as far apart from each other as possible to provide maximum rigidity for making measurements.

Identification of Targets

Each datum target is shown on a view of the part in its desired location by means of a datum target symbol. These symbols are shown in Fig. 30-8-9.

Each datum target is then identified by means of a datum target identification symbol. This symbol consists of a circle with a diameter approximately 3.5 times the height of the lettering used on the drawing. This circle is divided by a horizontal line, as shown in Fig. 30-8-10. The upper half contains the target area size, where applicable, and the lower half contains a number and letter which identifies that particular target in the datum system. Targets should be numbered consecutively; for example, in a three-plane, six-point datum system, if the datums are A, B, and C, the datum targets would be A_1, A_2, A_3, B_4, B_5, and C_6.

Each datum target symbol is connected to its datum target by a leader. The use of a solid leader line indicates

Fig. 30-8-8 Group of holes forming a single datum.

DATUM POINT	A CROSS ON THE SURFACE	✕
DATUM LINE	OR A PHANTOM LINE ON THE SURFACE	
	OR A CROSS ON THE PROFILE, (WHERE THE LINE APPEARS AS A POINT ON THE SURFACE)	
DATUM AREA	A SECTION-LINED AREA ON THE SURFACE ENCLOSED BY PHANTOM LINES	

Fig. 30-8-9   Symbols for datum targets.

(A)

(B)

Fig. 30-8-10   Datum target identification symbol.

Fig. 30-8-11   Symbol for a datum target point.

(C)

Fig. 30-8-13   Location of part on datum target points.

that the datum target is on the near (visible) surface. The use of a dashed leader line indicates that the datum target is on the far (hidden) surface. Arrowheads or dots are not used on these leaders.

The datum surface may also be identified in the normal manner.

Dimensioning for Target Location

The location of datum targets is shown by means of datum dimensions. Each dimension is shown, without tolerances, enclosed in a rectangular frame, indicating that the general tolerance does not apply. Dimensions locating a set of datum targets should be dimensionally related or have a common origin.

Target Points

Each target point is shown on the surface, in its desired location, by means of a cross, drawn at approximately 45° to the coordinate dimensions. The cross is twice the height of the lettering used, as shown in Fig. 30-8-11. Figure

30-8-12 illustrates two such points as they would appear on a drawing.

Target points may be represented on tools, fixtures, and gages by spherically ended pins, as shown in Fig. 30-8-13.

Fig. 30-8-12   Datum target points on a surface.

Targets not in the Same Plane

In most applications datum target points which form a single datum are all located on the same surface, as shown in Fig. 30-8-12. However, this is not essential. They may be located on different surfaces, to meet functional requirements, as shown, for example, in Fig. 30-8-14. In some cases the

Fig. 30-8-14   Datum target points on different planes.

datum plane may be located in space, that is, not actually touching the part, as shown in Fig. 30-8-15. In such applications the controlled features must be dimensioned from the specified datum, and the position of the datum from the datum targets must be shown by means of exact datum dimensions. For example, in Fig. 30-8-15 datum *B* is positioned by means of datum dimensions .38, .50, and 1.06. The top surface is controlled from this datum by means of a toleranced dimension, and the hole is positioned by means of a true-position dimension 1.00 and a positional tolerance.

Fig. 30-8-15 Datum outside of part profile.

Target Lines

A datum target line is identified by the symbol "*X*" on the edge view of the surface, a phantom line on the surface view, or both (Fig. 30-8-16).

Fig. 30-8-16 Datum target line

Datum Target Areas

Datum target areas are indicated by drawing the datum target boundary on the plan view in phantom lines and section-lining the area, as shown in Fig. 30-8-17. The datum target symbol is directed to the area by an arrowless leader. The diameter of the circular target area is shown in the upper half of the datum target symbol.

Fig. 30-8-17 Datum target areas.

Datum target areas may have any desired shape, a few of which are shown in Fig. 30-8-18. Target areas should be kept as small as possible, consistent with functional requirements, to avoid having large, section-lined areas on the drawing.

Fig. 30-8-18 Typical target areas.

Reference and Source Material

1. ANSI Y 14.5M Dimensioning and Tolerancing

ASSIGNMENTS

See Assignments 28 and 29 for Unit 30-8 on page 650.

UNIT 30-9
Circularity (Roundness) and Cylindricity

CIRCULARITY

Circularity refers to a condition of a circular line or the surface of a circular feature where all points on the line, or on the periphery of a plane cross section of the feature, are equidistant from a common center point.

Examples of circular features would include disks, spheres, cylinders, and cones. The measurement plane for a sphere is any plane which passes through a section of maximum diameter. For a cylinder, cone, or other nonspherical feature, the measurement plane is any plane perpendicular to the axis or center line.

Errors of Circularity

Errors of circularity (out-of-roundness) of a circular line or the periphery of a cross section of a circular feature may occur as *ovality*, where differences appear between the major and minor axes; as *lobing*, where the diametral values may be constant or nearly so; or as *random irregularities* from a true circle. All these errors are illustrated in Fig. 30-9-1. The geometric characteristic symbol for circularity is simply a circle, having a diameter equal to 1.5 times the height of letters on the drawing, as shown in Fig. 30-9-2.

(A) OVALITY

(B) LOBING

(C) IRREGULAR

Fig. 30-9-1 Common types of circularity errors.

Fig. 30-9-2 Circularity symbol.

Circularity Tolerance—RFS

A circularity tolerance may be specified by using this symbol in the feature control frame. It is expressed on an RFS basis.

A circularity tolerance specifies the width of an annular tolerance zone, bounded by two concentric circles in the same plane, within which the circular line or the periphery of the feature in that plane shall lie, as shown in Fig. 30-9-3. A circularity tolerance cannot be modified on an MMC basis since it controls surface elements only. The circularity tolerance must be less than the size tolerance.

(A) DRAWING CALLOUT

(B) TOLERANCE ZONE

Fig. 30-9-3 Circularity tolerance.

Circularity of Noncylindrical Parts

Noncylindrical parts refer to conical parts and other features which are circular in cross section but which have variable diameters, such as those shown in Fig. 30-9-4. Since many sizes of circles may be involved, it is usually best to direct the circularity tolerance to the longitudinal surfaces as shown.

PART I

PART 2

Fig. 30-9-4 Circularity tolerance on noncylindrical parts.

CYLINDRICITY

Cylindricity refers to a condition of a surface which forms a cylinder where the surface elements in cross sections parallel to the axis are straight and parallel and in cross sections perpendicular to the axis are round. Cylindricity thus combines in one term geometric form tolerances for circularity, straightness, and parallelism of the surface elements.

Cylindricity tolerances can be applied only to cylindrical surfaces, such as round holes and shafts. No specific geometric tolerances have been devised for other circular forms, which require the use of several geometric tolerances. A conical surface, for example, must be controlled by a combination of tolerances for circularity, straightness, and angularity.

Errors of cylindricity may be caused by out-of-roundness, like ovality or lobing, by errors of straightness caused by bending or by diametral variation, by errors of parallelism like conicity or taper, and by random irregularities from a true cylindrical form.

The geometric characteristic symbol for cylindricity consists of a circle with two tangent lines at 60°, as shown in Fig. 30-9-5.

Cylindricity Tolerance—RFS

The cylindricity tolerance symbol is used in a feature control frame in the usual manner, and it is directed to the cylindrical surface, in either the side or end view, as shown in Fig. 30-9-6.

Since cylindricity is a form tolerance controlling surface elements only, it cannot be modified on an MMC basis. The cylindricity tolerance must be less than the size tolerance.

Fig. 30-9-5 Cylindricity symbol.

(A) DRAWING CALLOUT

(B) PERMISSIBLE FORM ERRORS

Fig. 30-9-6 Cylindricity tolerance directed to either view.

Reference and Source Material

1. ANSI Y 14.5M Dimensioning and Tolerancing

ASSIGNMENTS

See Assignments 30 through 33 for Unit 30-9 on page 651.

UNIT 30-10
Profile Tolerancing

PROFILES

A *profile* is the outline form or shape of a line or surface. A *line profile* may be the outline of a part or feature as depicted in a view on a drawing. It may represent the edge of a part, or it may refer to line elements of a surface in a single direction, such as the outline of cross sections through the part. In contrast, a *surface profile* outlines the form or shape of a complete surface.

The elements of a line profile may be straight lines, arcs, or other curved lines. The elements of a surface profile may be flat surfaces, spherical surfaces, cylindrical surfaces, or surfaces composed of various line profiles in two or more directions.

A profile tolerance may be applied to a single, independent line or surface, or any part of a complex line or surface. It may be applied to a feature of size, in which case MMC is specified. It may be related to a datum, in which case it also controls orientation and, in some cases, position of the line or surface.

PROFILE SYMBOLS

There are two geometric characteristic symbols for profiles, one for lines and one for surfaces. Separate symbols are required, because it is often necessary to distinguish between line elements of a surface and the complete surface itself. The symbol for profile of a line consists of a semicircle with a diameter equal to twice the lettering size used on the drawing. The symbol for profile of a surface is identical except that the semicircle is closed by a straight line at the bottom, as shown in Fig. 30-10-1.

Fig. 30-10-1 Profile symbols.

All other geometric tolerances of form and orientation are merely special cases of profile tolerancing.

Profile tolerances are used to control the position of lines and surfaces which are neither flat nor cylindrical.

PROFILE OF A LINE

A profile-of-a-line tolerance may be directed to a line of any length or shape. If a datum is not referenced and the line is intended to be straight, the profile tolerance is identical to a straightness tolerance in meaning and interpretation. If the line is intended to be circular, the profile tolerance is identical to a circularity tolerance. The form of lines of any other shape or combinations of straight and curved lines must be controlled by a profile tolerance.

Similarly, a profile-of-a-surface tolerance may be directed to a surface of any size or shape. If a datum is not referenced and the surface is intended to be flat, the profile tolerance is identical to a flatness tolerance. If the surface is intended to be cylindrical, the profile tolerance is identical to a cylindricity tolerance. To control the form of all other surfaces, or combinations of straight, flat, and curved surfaces, a profile tolerance must be used.

A profile-of-a-line tolerance is specified in the usual manner, by including the symbol and tolerance in a feature control frame directed to the line to be controlled, as shown in Fig. 30-10-2.

If the line on the drawing to which the tolerance is directed represents a surface, the tolerance applies to all line elements of the surface parallel to the plane of the view on the drawing, unless otherwise specified.

The tolerance indicates a tolerance zone consisting of the area between two parallel lines, separated by the specified tolerance, which are themselves parallel to the basic form of the line in a plane parallel to the view on the drawing.

Bilateral and Unilateral Tolerances

The profile tolerance zone is normally equally disposed about the basic profile in a form known as a *bilateral tolerance zone*. The width of this zone is always measured normal to the profile

(A) DRAWING CALLOUT

(B) BILATERAL TOLERANCE ZONE

Fig. 30-10-2 Simple profile with profile tolerance that is bilateral.

surface. The tolerance zone may be considered to be bounded by two lines enveloping a series of circles, each having a diameter equal to the specified profile tolerance, with their centers on the theoretical, basic profile, as shown in Fig. 30-10-2.

Occasionally it is desirable to have the tolerance zone wholly on one side of the basic profile instead of equally divided on both sides. Such zones are called *unilateral tolerance zones*. They are specified by showing a thick chain or zone line close to the profile surface. The tolerance is directed to this line, as shown in Fig. 30-10-3.

Method of Dimensioning

The true or mean profile is established by means of basic dimensions, each of which is enclosed in a rectangular

(A) TOLERANCE ZONE ON OUTSIDE OF TRUE PROFILE

(B) TOLERANCE ZONE ON INSIDE OF TRUE PROFILE

Fig. 30-10-3 Unilateral tolerance zones.

Fig. 30-10-4 Position and form as separate requirements.

frame to indicate that the tolerance in the general tolerance note does not apply.

When the profile tolerance is not intended to control the position of the profile, there must be a clear distinction between dimensions which control the position of the profile and those which control the form or shape of the profile.

Any convenient method of dimensioning may be used to establish the basic profile. Examples are chain or common-point dimensions, dimensioning to points on a surface or to the intersection of lines, dimensioning located on tangent radii, and angles.

To illustrate, the simple part in Fig. 30-10-4 shows a dimension of .90 ± .01 controlling the height of the profile. This dimension must be separately measured. The radius of 1.500 in. is a basic dimension, and it becomes part of the profile. Therefore the profile tolerance zone has radii of 1.497 and 1.503, but is free to flat in any direction

Fig. 30-10-5 Position and radius separate from form.

within the limits of the positional tolerance zone in order to enclose the curved profile.

If the radius were shown as a toleranced dimension, without the rectangular frame, as in Fig. 30-10-5, it would become a separate measurement.

Figure 30-10-6 shows a more complex profile, where the profile is located by a single toleranced dimension. There are, however, five basic dimensions defining the true profile.

In this case, the tolerance on the height indicates a tolerance zone .06 in. wide extending the full length of the profile. This is because the profile is established by basic dimensions. No other dimension exists to effect the orientation or height. The profile tolerance specifies a .008-in.-wide tolerance zone, which may lie anywhere within the .06 in. tolerance zone.

Extent of Controlled Profile

The profile is generally intended to extend to the first abrupt change or sharp corner. For example, in Fig. 30-10-6 it extends from the upper left-to the upper right-hand corners, unless otherwise specified. If the extent of the

(A) DRAWING CALLOUT

TOLERANCE ZONE FOR FORM OF PROFILE .008 WIDE SHOWN IN AN EXTREME POSITION

TOLERANCE ZONE FOR POSITION OF PROFILE

(B) PROFILE TOLERANCE ZONE

Fig. 30-10-6 Profile defined by basic dimensions.

(A) DRAWING CALLOUT

(B) TOLERANCE ZONE

Fig. 30-10-7 Specifying extent of profile.

(A) DRAWING CALLOUT

TOLERANCE ZONES

BASIC PROFILE

(B) TOLERANCE ZONE

Fig. 30-10-9 Dual tolerance zones.

Fig. 30-10-11 Profile tolerance required for all around the surface.

profile is not clearly identified by sharp corners or by basic profile dimensions, it must be indicated by a note under the feature control symbol, such as FROM A TO B as shown in Fig. 30-10-7.

If the controlled profile includes a sharp corner, the corner represents a discontinuity of the tolerance boundary, and the boundary is considered to extend to the intersection of the boundary lines, as shown in Fig. 30-10-8.

If different profile tolerances are required on different segments of a surface, the extent of each profile tolerance is indicated by the use of reference letters to identify the extremities. See Fig. 30-10-9.

All-Around Profile Tolerances
Where a profile tolerance applies all around the profile of a part, the symbol used to designate "all around" is placed on the leader from the feature control frame. See Figs. 30-10-10 and 30-10-11.

Profile Tolerance on an MMC Basis
A profile tolerance may be specified on an MMC basis for any profile that includes opposing surfaces. This method is especially suitable when the tolerance applies all around the part.

Specification on an MMC basis is particularly useful for parts of simple or regular cross sections, such as square, rectangular, hexagonal, or round bars, as shown in Fig. 30-10-12.

(A)

(B) (C)

Fig. 30-10-12 Profile tolerance—MMC.

PROFILE OF A SURFACE

As already stated, a profile-of-a-line tolerance, when directed to a line on a drawing which represents a surface, applies to the profile of all cross sections parallel to the view on the drawing, unless otherwise specified. This is illustrated by the cross sections *AA* and *BB* in Fig. 30-10-13.

If the profile in a plane normal to the plane of the drawing requires tolerancing, a separate tolerance may be added

BASIC PROFILE

WIDTH OF TOLERANCE ZONE

TOLERANCE ZONE EXTENDS TO THIS POINT

Fig. 30-10-8 Tolerance zone at a sharp corner.

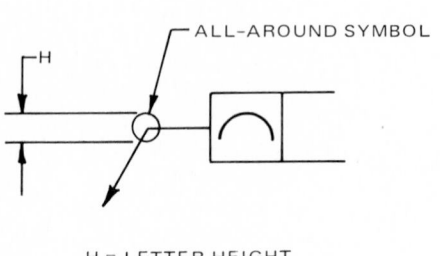

H = LETTER HEIGHT

Fig. 30-10-10 All-around symbol.

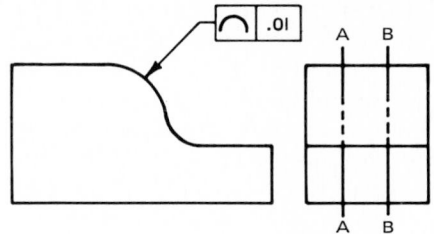

Fig. 30-10-13   Tolerance applies to profile of cross section.

(A) DRAWING CALLOUT

to the side view. Frequently this profile will be straight, in which case a straightness tolerance may be substituted, as shown in Fig. 30-10-14.

If the same tolerance is intended to apply over the whole surface, instead of to lines or line elements in specific directions, the profile of a surface symbol is used, as shown in Fig. 30-10-15. While the profile tolerance may be directed to the surface in either view, it is usually directed to the view showing the more characteristic profile.

The profile-of-a-surface tolerance indicates a tolerance zone having the same form as the basic surface, with a uniform width equal to the specified tolerance within which the entire surface must lie.

Fig. 30-10-14   Profile and straightness tolerance to same surface.

(A) DRAWING CALLOUT

(B) TOLERANCE ZONE

Fig. 30-10-15   Profile of a surface.

(B) TOLERANCE ZONE

Fig. 30-10-16   Profile referenced to a datum.

Orientation

In the examples so far, the profile tolerances have been shown without reference to a datum, and it has been permissible to move evaluating charts in any direction in an effort to encompass the line or surface profile. More often than not, requirements call for the profile to be oriented to some other surface or feature of the part. This is specified simply by indicating suitable datums. Figure 30-10-16 shows a simple part where the base is designated as a datum.

Position of Lines and Surfaces

Position of features, such as holes and bosses, is normally controlled by a positional tolerance. Such a tolerance automatically controls the orientation and form of the feature, since all points on the feature surface must lie within the positional tolerance zone.

Positional tolerances may also be applied to surfaces, such as the position of the flat surface in Fig. 30-10-17. The surface must be dimensioned from the datum feature or features by means of basic dimensions, as shown. The tolerance zone is then established at an exact position from the datum, and all points on the surface must lie within this tolerance zone without further adjustment.

Positional tolerancing may also be applied to lines or surfaces of other

(A) DRAWING CALLOUT

(B) TOLERANCE ZONE

Fig. 30-10-17   Position of a surface.

shapes. It is usually confined, however, to straight lines, circular lines, flat or cylindrical surfaces, or sometimes regular shapes such as hexagons. For example, in Fig. 30-10-18, it is immaterial whether the geometric symbol is position or profile—the interpretation is identical.

The criterion which distinguishes a profile tolerance as applying to position or to orientation is whether the profile is related to the datum by a basic dimension or by a toleranced dimension. This is illustrated by Fig. 30-10-19.

(A) DRAWING CALLOUT

(B) TOLERANCE ZONE

Fig. 30-10-18   Profile tolerance for position.

(A) PROFILE TOLERANCE CONTROLS
FORM OF PROFILE ONLY

(B) PROFILE TOLERANCE CONTROLS
FORM AND ORIENTAION OF PROFILE

(C) PROFILE TOLERANCE CONTROLS
FORM, ORIENTATION, AND POSITION
OF PROFILE

Fig. 30-10-19   Comparison of profile tolerances.

BASIC VALUES	
DEGREES	RADIUS
0	1.000
30	1.000
60	.936
90	.896
120	.858
150	.819
180	.780
210	.741
240	.702
270	.663
290	.546
340	.546

Fig. 30-10-20   Profile tolerances control form and size of cam profile.

Cylindrical Data

Profile tolerances controlling position are very useful for parts which can be revolved around a cylindrical datum feature, such as screw-machine cams. The position or size as well as the form can easily be assessed by revolving the part on the datum axis in conjunction with a dividing head, with an indicator gage to make direct measurements on the periphery. Figure 30-10-20 shows an example.

Reference and Source Material
1. ANSI Y 14.5M Dimensioning and Tolerancing.

ASSIGNMENTS

See Assignments 34 through 37 for Unit 30-10 on page 652.

UNIT 30-11
Correlative Tolerances

Correlative geometric tolerancing refers to tolerancing for the control of two or more features intended to be correlated in position or attitude. Examples of such correlated tolerancing include coplanarity, for control of two or more flat surfaces; positional tolerance at MMC for symmetrical relationships, such as control of features equally disposed about a center line; concentricity and coaxiality, for control of features having common axes or center lines; and runout, for control of surfaces related to an axis. These are all tolerances of location, for which the positional symbol and positional tolerances could be used. Special symbols, however, have been

(A) DRAWING CALLOUT

(B) TOLERANCE ZONE

Fig. 30-11-3   Surface referenced to a datum system.

(A) DRAWING CALLOUT

.003 WIDE TOLERANCE ZONE
(B) TOLERANCE ZONE

Fig. 30-11-1   Specify profile or a surface for coplanar surfaces.

provided for some of them to clarify and simplify drawing callout requirements.

When position is to be separately controlled, other form or orientation tolerances may be applied to control the correlation of features.

COPLANARITY

Coplanarity refers to the relative position of two or more flat surfaces, which are intended to lie in the same geometric plane. A profile of a surface tolerance may be used where it is desirable to treat two or more surfaces as a single interrupted or noncontinuous surface. See Fig. 30-11-1. In this case, a control similar to a flatness tolerance is provided. Both surfaces must lie within the specified limits of size.

Figure 30-11-2 shows a case where coplanar surfaces are also required to be accurately located and parallel to another surface of a part, which is then designated as a datum feature. In this case, the datum and the controlled surfaces must be associated by a basic dimension.

It is sometimes necessary to reference surfaces to a datum system instead of to a single datum surface. This occurs when it is necessary to

control coplanar surfaces perpendicular to a datum instead of parallel to it. Figure 30-11-3 shows a case where the coplanar surfaces are required to be perpendicular to the axis of a hole.

TOLERANCING SYMMETRICALLY LOCATED FEATURES

Where it is required that a feature be located symmetrically with respect to the center plane of a datum feature, positional tolerancing is used. A symmetrical relationship may be controlled by specifying a positional tolerance at MMC as illustrated in Figs. 30-11-4 and 30-11-5.

CONCENTRICITY

Concentricity is a condition in which two or more features, such as circles, spheres, cylinders, cones, or hexagons, share a common center or axis. An example would be a round hole through the center of a cylindrical part.

A concentricity tolerance is a particular case of a positional tolerance. It controls the permissible variation in position, or eccentricity, of the center line of the controlled feature in relation to the axis of the datum feature when

(A) DRAWING CALLOUT

Fig. 30-11-2   Coplanar surfaces parallel to a datum.

(B) TOLERANCE ZONE

7.8–8.2

(A) DRAWING CALLOUT

16.8
16.6

─B─

─A─

THESE DIMENSIONS ARE EQUAL
REGARDLESS OF THEIR SIZE

MEDIAN PLANE OF DATUM A

0.12 WIDE TOLERANCE ZONE
WHEN SLOT IS AT MMC (7.8)

(B) TOLERANCE ZONE

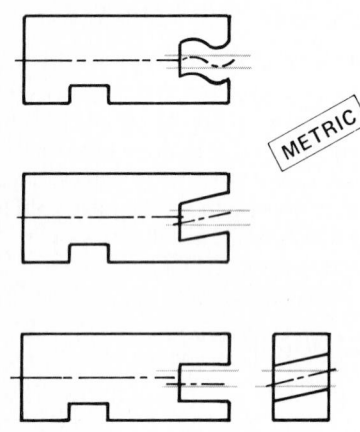

METRIC

(C) PERMISSIBLE VARIATIONS

Fig. 30-11-4   Positional tolerancing at MMC for symmetry.

DRAWING CALLOUT   GAGING PRINCIPLE

EXAMPLE I

DRAWING CALLOUT   GAGING PRINCIPLE

EXAMPLE 2

Fig. 30-11-5   Symmetry tolerance—MMC.

1.5 H   H

H =LETTER HEIGHT

Fig. 30-11-6   Concentricity symbol.

Concentricity—RFS

Concentricity tolerance, because of its unique characteristic, is always used on an RFS basis.

Figure 30-11-7 shows a common type of part where the outer diameter is required to be concentric with the center hole, which is designated as a datum feature.

the feature and datum are intended to be concentric or coaxial.

The tolerance zone for circles is a circle; for other circular features it is a cylinder, concentric with the datum axis, having a diameter equal to the specified tolerance. The center of all cross sections normal to the axis of the controlled feature must lie within this tolerance zone.

The geometric characteristic symbol used for concentricity consists of two concentric circles, having diameters equal to the actual height (1:1) and 1.5 times the height of lettering used on the drawing. See Fig. 30-11-6.

Figure 30-11-8 shows an example of a part where two cylindrical portions are intended to be coaxial. This figure also illustrates the extreme errors of eccentricity and parallelism that the concentricity tolerance would permit.

A concentricity tolerance may be referenced to a datum system, instead of to a single datum, to meet certain functional requirements. Figure 30-11-9 gives an example in which the tolerance zone is perpendicular to datum A and also concentric with the axis of datum B in the plane of datum A.

Where it is necessary to control coaxiality of related features within their limits of size, a zero positional tolerance at MMC is specified. The datum feature is normally specified on an MMC basis (Fig. 30-11-10).

Fig. 30-11-7   Cylindrical part with cylindricity tolerance.

(A) DRAWING CALLOUT

Ø .004 TOLERANCE ZONE

.002

(B) EXTREME ECCENTRICITY

AXIS OF DATUM A
Ø .004 TOLERANCE ZONE

AXIS OF SMALL DIAMETER

(C) EXTREME ANGULAR VARIATION

Fig. 30-11-8   Concentricity of cylindrical features.

Runout

Runout is a composite tolerance used to control the functional relationship of one or more features of a part to a datum axis. The types of features controlled by runout tolerances include those surfaces constructed around a datum axis and those constructed at right angles to a datum axis. See Fig. 30-11-11.

Each feature must be within its runout tolerance when rotated about

(A) DRAWING CALLOUT

AXIS OF DATUM FEATURE "A" AT MMC

FEATURE DIAMETER SIZE	DIAMETER TOLERANCE ZONE ALLOWED
1.000	.000
.999	.001
.998	.002
.997	.003
.996	.004
.995	.005

(B) TOLERANCE ZONES

Fig. 30-11-10   Zero positional tolerancing at MMC for coaxiality.

the datum axis. The tolerance specified for a controlled surface is the total tolerance or full indicator movement (FIM) in inspection and international terminology.

There are two types of runout control, circular runout and total runout. The type used is dependent upon design requirements and manufacturing considerations. The geometric characteristic symbols for runout are shown in Fig. 30-11-12.

H = LETTER HEIGHT

Fig. 30-11-12   Runout symbols.

Circular Runout

Circular runout provides control of circular elements of a surface. The tolerance is applied independently at any circular measuring position as the part is rotated 360°. See Fig. 30-11-13. Where applied to surfaces constructed around a datum axis, circular runout controls variations such as circularity and coaxiality. Where applied to surfaces constructed at right angles to the datum axis, circular runout controls wobble.

Where a runout tolerance applies to a specific portion of a surface, a chain line is drawn adjacent to the surface profile to show the desired length. See Fig. 30-11-13.

Fig. 30-11-9   Concentricity referenced to datum system.

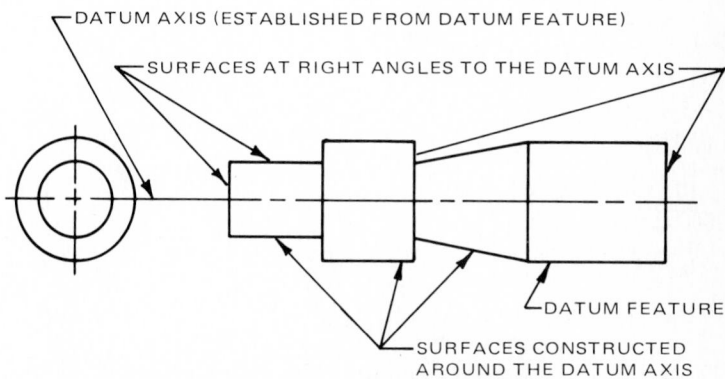

Fig. 30-11-11   Features applicable to runout tolerancing.

(A) DRAWING CALLOUT

(B) METHOD OF MEASURING

Fig. 30-11-13   Specifying circular runout relative to a datum diameter.

Total Runout

Total runout concerns the runout of a complete surface, not merely the runout of each circular element. For measurement purposes the checking indicator must traverse the full length or extent of the surface while the part is revolved about its datum axis. Measurements are made over the whole surface without resetting the indicator. Total runout is the difference between the lowest indicator reading in any position and the highest reading in that or in any other position on the same surface. Thus in Fig. 30-11-14 the tolerance zone is the space between two concentric cylinders separated by the specified tolerance and coaxial with the datum axis.

Figure 30-11-15 illustrates the application of runout tolerances where two datum diameters act as a single datum axis to which the features are related.

Establishing Datums

In many examples the datum axis has been established from centers drilled in the two ends of the part, in which case the part is mounted between centers for measurement purposes. This is an ideal method of mounting and revolving the part when such centers have been provided for manufacturing purposes. When centers are not provided, any cylindrical or conical surface may be used to establish the datum axis if it is chosen on the basis of the functional requirements of the part.

Figure 30-11-16 shows a simple, external cylindrical feature specified as the datum feature.

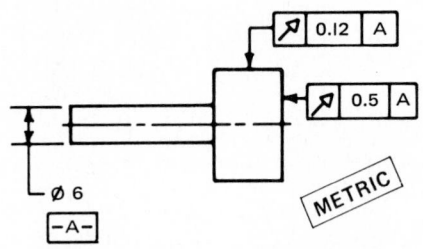

Fig. 30-11-16   Cylindrical datum feature.

Reference and Source Material

1. ANSI Y 14.5M Dimensioning and Tolerancing

ASSIGNMENTS

See Assignments 38 through 42 for Unit 30-11 on page 653.

(A) DRAWING CALLOUT

(B) TOLERANCE ZONE

Fig. 30-11-14   Tolerance zone for total runout.

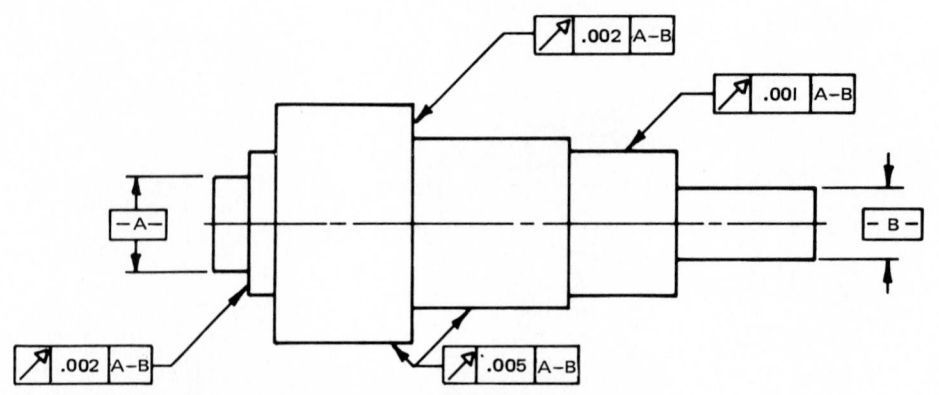

Fig. 30-11-15   Specifying runout relative to two datum diameters.

ASSIGNMENTS for Chapter 30

Assignments for Unit 30-1, Modern Engineering Tolerancing

1. Parts may deviate from true form and still be acceptable provided the measurements lie within the limits of size. Show by means of a sketch with dimensions two acceptable form variations for each part shown in Fig. 30-1-A.

2. On a B- or A3-size sheet, prepare sketches from the drawings shown in Fig. 30-1-B or 30-1-C and the following information.

    (a) Using illustration (A) make a tolerance block diagram similar to Fig. 30-1-5. Show the deviations and limits of size.

    (b) Draw illustration (B) and shade in and dimension the tolerance zone.

    (c) The exaggeration of sizes is used when it improves the clarity of the drawing. Draw illustration (C) and exaggerate the sizes which would improve the readability of the drawing. Dimension the exaggerated features.

    (d) With reference to illustration (D), is the part acceptable? State your reason.

    (e) With reference to the drawing callout shown in illustration (E), what parts would pass inspection?

    (f) In the drawing callout in illustration (F), what parts in illustration (E) would pass inspection?

(A)

(B)

Ø .188 HOLE AT THIS POINT TO BE 1° RIGHT OF CENTER MEASURED AT HOLE A

.02 STEP AT THIS POINT

A

Ø .016 HOLE AT THIS POINT

(C)

PART 1

.24

.02

PART 2

.26

.02

PART 3

.25

.01

.02

PART 4

.24

.02

SIZE OF PARTS

(E)

.60 ± .02

2.00 ± .04

1.20 ± .01

(A)

2.00 ± .02

Ø .784 / .780

(B)

.70 ± .02   .625 ± .008

(C)

1.00 ± .02

2.00 ± .02

.600 ± .010

Fig. 30-1-A   Assignments.

1.00 ± .01

3.00 ± .02

DRAWING CALLOUT

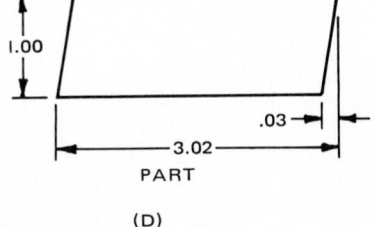

1.00

3.02

.03

PART

(D)

Fig. 30-1-B   Assignments.

.25 ± .01

DATUM FEATURE

(F)

(A)

Ø 35.013 / 34.088

Ø 35.041 / 35.000

35 ± 1

DRAWING CALLOUT

(B)

R 25 / 24

60° / 59°

12.5 / 12.0

(C)

0.5 SAWCUTS
10 mm WIDE
CENTER SECTION

Ø 1
4 HOLES

90°

189°

HORIZONTAL

SLOPED

STARTING OF ANGLE

(D)

20 ± 0.5

50 ± 0.5

DRAWING CALLOUT

0.5

0.5

20.5

19.5

50

PART

METRIC

Fig. 30-1-C Assignments.

(E)

CURVED

PART 1

34

2

FLAT

PART 2

34

2

FLAT

PART 3

36

34

2

CURVED

PART 4

36

2

SIZE OF PARTS

(F)

35 ± 1

PART 5

DATUM FEATURE

Assignment of Unit 30-2, Geometric Tolerancing

3. With reference to Fig. 30-2-A and the information given below, add the feature control frames to the following parts:

Part 1. Surface A to have a straightness tolerance of .004 in.

Part 2. Surface M to have a straightness tolerance of .006 in.
Surface N to have a straightness tolerance of .008 in.

Part 3. Surface R to be straight within .006 in. for the full length, but not to exceed .002 in. for any 2.00 in. length.

Part 4. With straightness specified as shown, what is the maximum permissible deviation from straightness if the radius is (a) .496 in., (b) .501 in., (c) .504 in.?

Part 5. Eliminate the bottom view and place the feature control frames on the front and side views.

PART 1

A

Ø .755 / .745   Ø .505 / .495   M   N   Ø 1.005 / .995

PART 2

R

PART 3

- .002

R .500 ± .004

PART 4

- .002   - .005
- .004
- .001

PART 5

Fig. 30-2-A Assignments.

Assignments for Unit 30-3, Relationship to Feature of Size

4. With reference to Fig. 30-3-A, what is the maximum deviation permitted from straightness for the surface of the diameter if the shaft was (*a*) at MMC, (*b*) at LMC, (*c*) Ø.621?

5. With reference to Fig. 30-3-B, calculate the MMC, LMC, and extreme virtual condition of the hole and shaft.

6. With reference to Fig. 30-3-B, if the hole was straight and at its MMC, how much could the shaft be bent and still assemble if the shaft diameter was (*a*) at MMC, (*b*) at LMC, (*c*) at Ø17.94?

7. With reference to Fig. 30-3-C, calculate the limits for the distances between the holes when the pins and holes are at (*a*) MMC, (*b*) LMC, and (*c*) the distance between the pins is exactly 3.500 in.

Fig. 30-3-A Assignment.

Fig. 30-3-B Assignment.

Fig. 30-3-C Assignment.

Assignments for Unit 30-4, Straightness of Features and Flatness

8. With reference to Fig. 30-4-A are parts A to E acceptable?

9. If Ⓜ was added to the straightness tolerance shown in Fig. 30-4-A, what parts would be acceptable?

10. Dimension the ring and snap gage shown in Fig. 30-4-B to check the pins shown. The ring gage should be of such a size as to check the entire length of pin. The two open ends of the snap gage should measure the minimum and maximum acceptable pin diameters.

PART	FEATURE SIZE	STRAIGHTNESS DEVIATION	ACCEPTABLE
A	.747	.001	
B	.741	.004	
C	.742	.005	
D	.740	.006	
E	.740	.003	

Fig. 30-4-A Assignment.

RING GAGE   SNAP GAGE

Fig. 30-4-B Assignment.

11. In Fig. 30-4-C, part 1 is required to fit into part 2 so that there will not be any interference and the maximum clearance will never exceed .005 in. Show suitable flatness tolerances for both parts and a size with the largest tolerance for part 2.

12. Show the tolerance zones and widths, for the three parts shown in Fig. 30-4-D. Determine the extreme virtual conditions for parts 2 and 3.

PART 1

PART 2

Fig. 30-4-C Assignment.

PART 1

PART 2

PART 3

Fig. 30-4-D Assignment.

Assignments for Unit 30-5, Datums and the Three-Plane Concept

13. Draw the top and front views of Fig. 30-5-A and show the following information:

- Surface A is datum A.
- Surface B is datum B.
- Surfaces C and D are datum features which form a single datum.
- Add the geometric tolerances from the information shown on the drawing.

14. With reference to Fig. 30-5-B:

- Pins 1, 2, and 3 are used to establish the secondary and tertiary datums for the part shown.
- Make a two-view drawing of the part shown and identify the primary, secondary, and tertiary datum planes as A, B, and C, respectively.
- Place the flatness tolerance specified for the back of the slot on the drawing.
- With reference to the slot tolerance and assuming that the top surface is smooth and flat, are the three parts shown acceptable?

15. With reference to Fig. 30-5-C: diameter M is to be used as datum A, the end face of diameter N is to be used as datum B, and the two sides of the slot in diameter N are to be used as datum C. Prepare two drawings, one with ANSI drawing standards the other with ISO drawing standards, which will identify these datums.

16. Prepare a three-view drawing showing the datums and feature control frames for Fig. 30-5-D.

-DATUM A IS TO BE STRAIGHT WITHIN .008 FOR THE 4.00 LENGTH BUT THE STRAIGHTNESS ERROR NOT TO EXCEED .002 FOR ANY 1.00 LENGTH
-DATUM B IS TO BE FLAT WITHIN .004
-SURFACE B IS TO BE PARALLEL TO DATUM C-D WITHIN .006

Fig. 30-5-A   Assignment.

Ø 10 PINS
PIN 2
PIN 1
PIN 3
BASE PLATE

25 ± 0.25
PART
50.1
20
METRIC

THE BACK OF THE SLOT IS TO HAVE A FLATNESS TOLERANCE OF 0.2

25
PART 1
25.03
0.02

24.75
PART 2
25.4
0.3

25
PART 3
25.3
0.1

Fig. 30-5-B   Assignment.

Ø N
Ø M
Ø L
END FACE

Fig. 30-5-C   Assignment.

DATUM B Ø 1.856 ± .004
STRAIGHTNESS TOLERANCE OF .002 AT MMC
DATUM A
1.62 ± .02
2.25 ± .02
3.50 +.02 -.00
.870 .854
DATUM D
Ø 2.500 +.000 -.006
STRAIGHTNESS TOLERANCE OF .003 AT MMC
BOTH SIDES FLAT WITHIN .015 MMC
BASE FLAT WITHIN .01

Fig. 30-5-D   Assignment.

Assignments for Unit 30-6, Orientation

17. Today's drafters must be capable of interpreting and preparing drawings for use in other countries as well as in their own locality. From the information shown in Fig. 30-6-A prepare a three-view sketch showing the datums and geometric tolerancing for use in the United States and with ISO standards.

18. The surfaces shown in Fig. 30-6-B are required to be controlled in the following manner. Surfaces A, B, C, and D are datums A, B, C, D, respectively. Prepare a three-view drawing showing the datums and feature control symbols from the information supplied.

19. In Fig. 30-6-C it is required that the hole be parallel with datum A within ± 0.5°. Given that the tangent of 0.5° is 0.0087, show the drawing callout and describe the tolerance zone.

20. In Fig. 30-6-D it is functionally necessary that the shaft portion of the part not depart from perpendicularity with the holes by more than the tolerance specified. Show the drawing callout for this, and indicate the shape and size of the tolerance zone.

21. Sketch the basic elements of a gage to check the perpendicularity requirement in Fig. 30-6-E. If the large hole were round and straight and at the size measured, what would be the greatest angularity error of the hole axis?

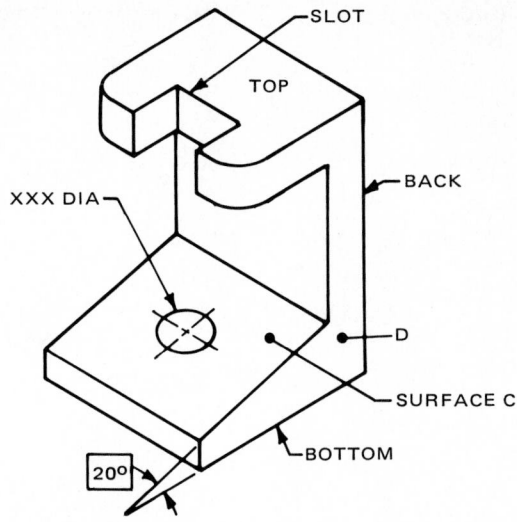

SLOT

TOP

XXX DIA

BACK

D

SURFACE C

BOTTOM

20°

BRACKET
- BOTTOM TO BE DATUM A
- BACK TO BE DATUM B
- HOLE TO BE PERPENDICULAR TO BOTTOM WITHIN .003
- BACK TO BE PERPENDICULAR TO BOTTOM WITHIN .004
- TOP TO BE PARALLEL WITH BOTTOM WITHIN .005
- SURFACE C TO HAVE AN ANGULARITY TOLERANCE OF .006
  WITH THE BOTTOM. SURFACE D TO BE THE SECONDARY
  DATUM FOR THIS FEATURE.
- THE SIDES OF THE SLOT TO BE PARALLEL WITH EACH
  OTHER WITHIN .002

Fig. 30-6-A Bracket.

DATUM SURFACE A

Ø .250 $^{+.002}_{-.000}$

Ø .898 $\pm$ .001

Fig. 30-6-C Assignment.

Ø 1.004 $^{+.000}_{-.002}$

1.25

Ø 502 $^{+.001}_{-.000}$

MAXIMUM PERPENDICULARITY TOLERANCE BETWEEN
HOLES AND SHAFT 0.05 IN 1.00 IN.

Fig. 30-6-D Assignment.

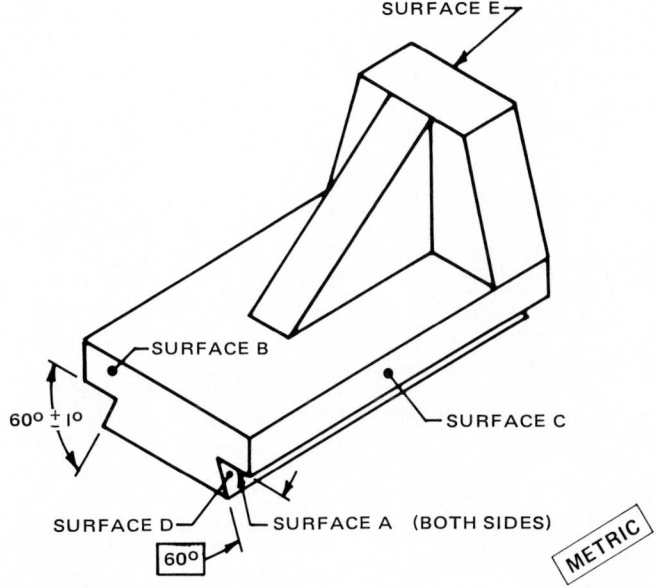

SURFACE E

SURFACE B

SURFACE C

60° $\pm$ 1°

SURFACE D

SURFACE A   (BOTH SIDES)

60°

METRIC

DOVETAIL SLIDE
- SURFACE D OF THE DOVETAIL MUST HAVE AN ANGULARITY
  TOLERANCE OF 0.05mm WITH DATUM A.
- SURFACE C SHOULD BE PERPENDICULAR TO DATUM A WITHIN
  0.03mm.
- SURFACE E MUST BE PERPENDICULAR TO DATUMS A AND D
  WITHIN 0.02mm.

Fig. 30-6-B Dovetail slide.

−A−

36 $^{0}_{-0.1}$

⊥ Ø 0.05 Ⓜ A

16.5 $^{+0.12}_{0}$

METRIC

MEASURED SIZE OF LARGE HOLE Ø 16.62

Fig. 30-6-E Assignment.

Assignments for Unit 30-7, Tolerancing for Location of Features

22. If coordinate tolerances as shown in Fig. 30-7-A are given, what are the shapes of the tolerance zones and the distance between extreme permissible positions of the holes?

23. In Fig. 30-7-B add the largest equal tolerances so that if two such parts are assembled with the edges aligned, the distance between their hole centers could never be more than that shown.

24. If a tolerance shown in Fig. 30-7-C is specified for the vertical dimension, what tolerance should be added to the horizontal dimension to meet the same requirement as that required in Assignment 23?

25. In order to assemble correctly, the hole in the part in Fig. 30-7-D must not vary from its true position by more than that shown on the drawing when the hole is at its smallest size.

    (a) Show suitable tolerancing to achieve this:

- By means of coordinate tolerancing
- By positional tolerancing without MMC
- By positional tolerancing on an MMC basis

    (b) What would be the maximum permissible departure from true position if the hole were at its maximum diameter, using positional tolerancing RFS?

    (c) What would be the maximum permissible departure from true position with a maximum-diameter hole and a positional tolerance on an MMC basis?

26. The part shown in Fig. 30-7-E is set on a revolving table, adjusted so that the part revolves about the true-position center of the large hole.

    (a) If both indicators give identical readings and the results shown are obtained, which parts are acceptable?

    (b) What is the positional error for each part?

27. With reference to Fig. 30-7-D a projected tolerance zone of .60 in. is required for the ∅.502 in. hole. Show how this would be added to the feature control frame.

Assignments for Unit 30-8, Datums for Positional Tolerancing

28. On a B- or A3-size sheet, make a three-view drawing of the bearing housing shown in Fig. 30-8-A showing the datum features. Only the dimensions related to the datums need be shown. Scale is half size or 1:2.

29. On a B- or A3-size sheet, make a two-view drawing of the part shown in Fig. 30-8-B showing the datum features.

Only the dimensions related to the datums need be shown. Scale is 1:1. Datum information is as follows:

- *Primary datum* (three points). Points A_1 and A_2 are located on center of surface M, one-fifth the height distance from the top and bottom, respectively. Point A_3 is located on the center of surface N midway between the center of the hole and the right end.
- *Secondary datum* is a datum line located on the center of surface D.
- *Tertiary datum* is a datum point located on the center of surface E.

Fig. 30-7-C Assignment.

(A)

(B)

Fig. 30-7-A Assignment.

MAXIMUM VARIATION OF HOLE
IN ANY DIRECTION = .0014

Fig. 30-7-D Assignment.

CALCULATE TOLERANCES
TO NEAREST 0.02
MAXIMUM DISTANCE BETWEEN
HOLE CENTERS = 0.5

Fig. 30-7-B Assignment.

PART NO.	SIZE OF MANDREL	HIGHEST READING	LOWEST READING
1	8.00	1.54	1.32
2	8.06	0.18	-0.07
3	7.96	1.87	1.59
4	7.94	1.72	1.48
5	8.00	1.95	1.85
6	8.05	1.24	1.02

Fig. 30-7-E Assignment.

ROUNDS & FILLETS R.20

DATUM AND LOCATION				
DATUM DESCRIPTION		LOCATION FROM		
		PRIMARY DATUM PLANE	SECONDARY DATUM PLANE	TERTIARY DATUM PLANE
DATUM A TARGET AREAS Ø .50	A1		.60	1.40
	A2		.60	5.00
	A3		3.60	3.20
DATUM B TARGET LINES	B4			.80
	B5			5.60
DATUM C TARGET POINT	C6	.60	1.80	

Fig. 30-8-A   Assignment.

Fig. 30-8-B   Assignment.

Fig. 30-9-A   Assignment.

Fig. 30-9-B   Assignment.

Fig. 30-9-C   Assignment.

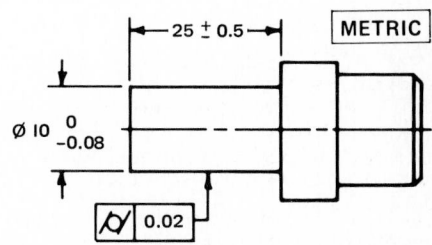

Fig. 30-9-D   Assignment.

Assignments for Unit 30-9, Circularity (Roundness) and Cylindricity

30. Sketch the tolerance zone for the circularity tolerance in Fig. 30-9-A. If measurements made at cross sections AA, BB, and CC, indicate that all points on the periphery fall within the annular rings shown, would you conclude that the part met the specified circularity tolerance? If not, which cross section is not acceptable?

31. Add circularity tolerances to the diameters shown in Fig. 30-9-B. The circularity tolerances are to be one-fifth of the size tolerances for each diameter.

32. Show on each part in Fig. 30-9-C a cylindrical tolerance. The size of the cylindrical tolerance is to equal one-quarter the size tolerance for each diameter.

33. Sketch the tolerance zone for the cylindrical tolerance in Fig. 30-9-D indicating its size and shape for a part shown.

Assignments for Unit 30-10, Profile Tolerancing

34. In Fig. 30-10-A it is required to have the form of the indented portion controlled by the line profile tolerance of .006 in. Show the tolerance and sketch the resulting tolerance zone on the drawing.

35. It is required to control the profile in Fig. 30-10-B with the tolerance described on the drawing. Add the line profile tolerance to the drawing and sketch the resulting tolerance zone.

36. Draw the tolerance zone, showing its relationship to the datums, for the profile tolerance shown in Fig. 30-10-C.

37. A cam is dimensioned as shown in Fig. 30-10-D. If parts are measured with an indicator which was set to zero and the following readings were obtained, which parts shown in the chart would be acceptable? Of the nonacceptable parts, which could be made acceptable by regrinding?

Fig. 30-10-C   Assignment.

Fig. 30-10-A   Assignment.

CONTROL THE PROFILE A TO B WITH A LINE PROFILE TOLERANCE OF .003 EXCEPT THAT THE .546 STRAIGHT PORTION CAN BE PERMITTED TO VARY VERTICALLY BY ± .01.

Fig. 30-10-B   Assignment.

ANGLE	C RAD
0	24.5
30	27.5
60	27.5
90	25
120	22
150	20.5
180 TO 270	20
300	20.5
330	22

ANGLE	DISTANCE FROM CENTER			
	PART 1	PART 2	PART 3	PART 4
0	24.6	24.7	24.5	24.4
30	27.8	27.7	27.4	27.4
60	27.6	27.6	27.25	27.5
90	25.1	25	24.6	25.1
120	22.4	21.9	21.7	22.1
150	20.6	20.4	20.2	20.5
180 TO 270	20.2 TO 20.3	19.8 TO 19.75	19.6 TO 20.3	20.1 TO 20
300	20.4	20.3	20.2	20.4
330	21.9	22	21.7	21.9

Fig. 30-10-D   Assignment.

Assignments for Unit 30-11, Correlative Tolerances

38. In Fig. 30-11-A show tolerances which will ensure that features are symmetrical at their maximum material conditions as shown on drawings.

39. It is required to have the three flat surfaces in Fig. 30-11-B coplanar with one another and perpendicular with the axis of the center hole within the tolerances specified. Add suitable geometric tolerances to the drawing. Show the tolerance zones for the part.

40. It is required to have the top diameter in Fig. 30-11-C concentric with the bottom diameter, when resting on surface A. Show how this should be specified on the drawing on an RFS basis.

41. Show the size and form of the tolerance zone for the concentricity tolerance shown in Fig. 30-11-D.

42. The part shown in Fig. 30-11-E is intended to function by rotating with the two end diameters supported in bearings. The two larger diameters are required to run true within the tolerance specified. Show how this would be toleranced.

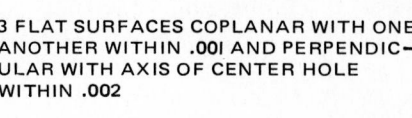

3 FLAT SURFACES COPLANAR WITH ONE ANOTHER WITHIN .001 AND PERPENDICULAR WITH AXIS OF CENTER HOLE WITHIN .002

Ø .546 $^{+.008}_{-.000}$

.60 ± .02

Fig. 30-11-B   Assignment.

Ø 35 ± 0.05

Ø 15 ± 0.05

Ø 15 CONCENTRIC WITH Ø 35 WITHIN 0.03 WHEN RESTING ON SURFACE A

METRIC

SURFACE A

Fig. 30-11-C   Assignment.

PART 1

Ø .156 $^{+.008}_{-.000}$ 2 HOLES

.240 ± .006

RECTANGULAR PROJECTION TO BE SYMMETRICAL WITH HOLES WITHIN .002

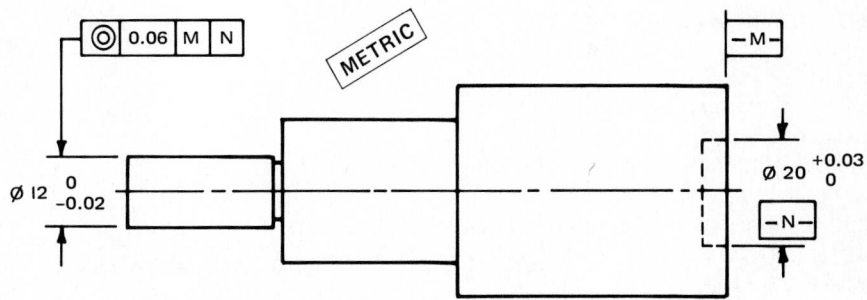

⌖ | 0.06 | M | N

METRIC

Ø 12 $^{0}_{-0.02}$

Ø 20 $^{+0.03}_{0}$

—M—

—N—

Fig. 30-11-D   Assignment.

PART 2

Ø .156 $^{+.008}_{-.000}$

Ø .546 $^{+.000}_{-.004}$

Ø.156 HOLE TO BE SYMMETRICAL WITH Ø.546 WITHIN .001

PART 3

.312 ± .004

1.000 ± .004

THE 2 SLOTS TO BE SIMULTANEOUSLY SYMMETRICAL WITH THE 1.000 WIDTH WITHIN ZERO TOLERANCE

Fig. 30-11-A   Assignment.

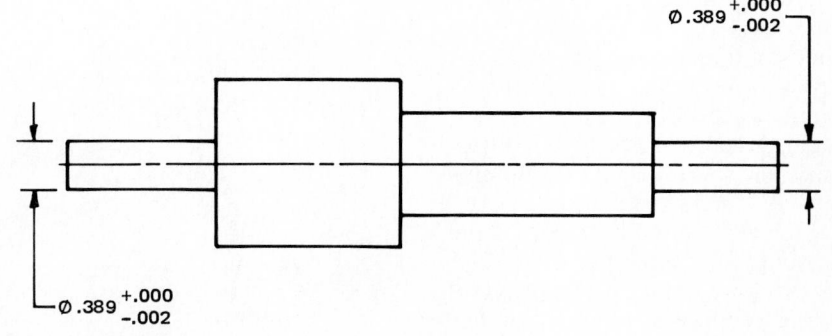

Ø .389 $^{+.000}_{-.002}$

Ø .389 $^{+.000}_{-.002}$

Fig. 30-11-E   Assignment.

CHAPTER 31

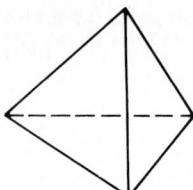

Descriptive Geometry

UNIT 31-1
Graphic Solutions

A major problem in technical drawing and design is the creation of projections for finding the true views of lines and planes. The following is a brief review of the principles of descriptive geometry involved in the solution of such problems. The designer working along with an engineering team can solve problems graphically with geometric elements. Structures that occupy space have three-dimensional forms made up of a combination of geometric elements (Fig. 31-1-1). The graphic solutions of three-dimensional forms require an understanding of the space relations that points, lines, and planes share in forming any given shape. Problems which many times require mathematical solutions can often be solved graphically with an accuracy that will allow manufacturing and construction. *Basic descriptive geometry* is one of the designer's methods of thinking through and solving problems. In the 18th century a French mathematician, Gaspard Monge, developed the principles of graphically solving spatial problems related to military structures. Descriptive geometry was introduced to the U.S. Military Academy at West Point by Claude Crozet in 1816. The Mongean method of presentation has changed, but the basic principles are

Fig. 31-1-1   Geometric space-frame structure, Franklin Park Mall, Toledo, Ohio. (Unistrut Corp.)

still taught in engineering schools throughout the world. The visual studies required in descriptive geometry assist in the development of the reasoning powers used in graphically solving problems.

In this chapter a graphic technique is used to analyze all geometric elements. The visual examination of geometric elements will assist in the description of structures of every possible shape. The basic shape of most structures designed by human beings is rectangular. This is a result of the relatively simple planning needed to form this type of structure. Figure 31-1-2 shows the basic geometric elements and some of the common and unusual geometric features of engineering designs.

REFERENCE PLANES

Unfolding of the reference planes forms a two-dimensional surface which a drafter uses to construct and solve problems (Fig. 31-1-3). The planes are labeled so that T represents the top or horizontal reference plane, F represents the front or vertical reference plane, and S represents the side (end) or profile reference plane. Thus a point 1 on the part, line, or plane would be identified as 1_F on the front reference plane, 1_T on the top reference plane, and 1_S on the side reference plane.

The folding lines shown on the box are referred to as *reference lines* on the drawing. Other reference planes and reference lines are drawn and labeled as required.

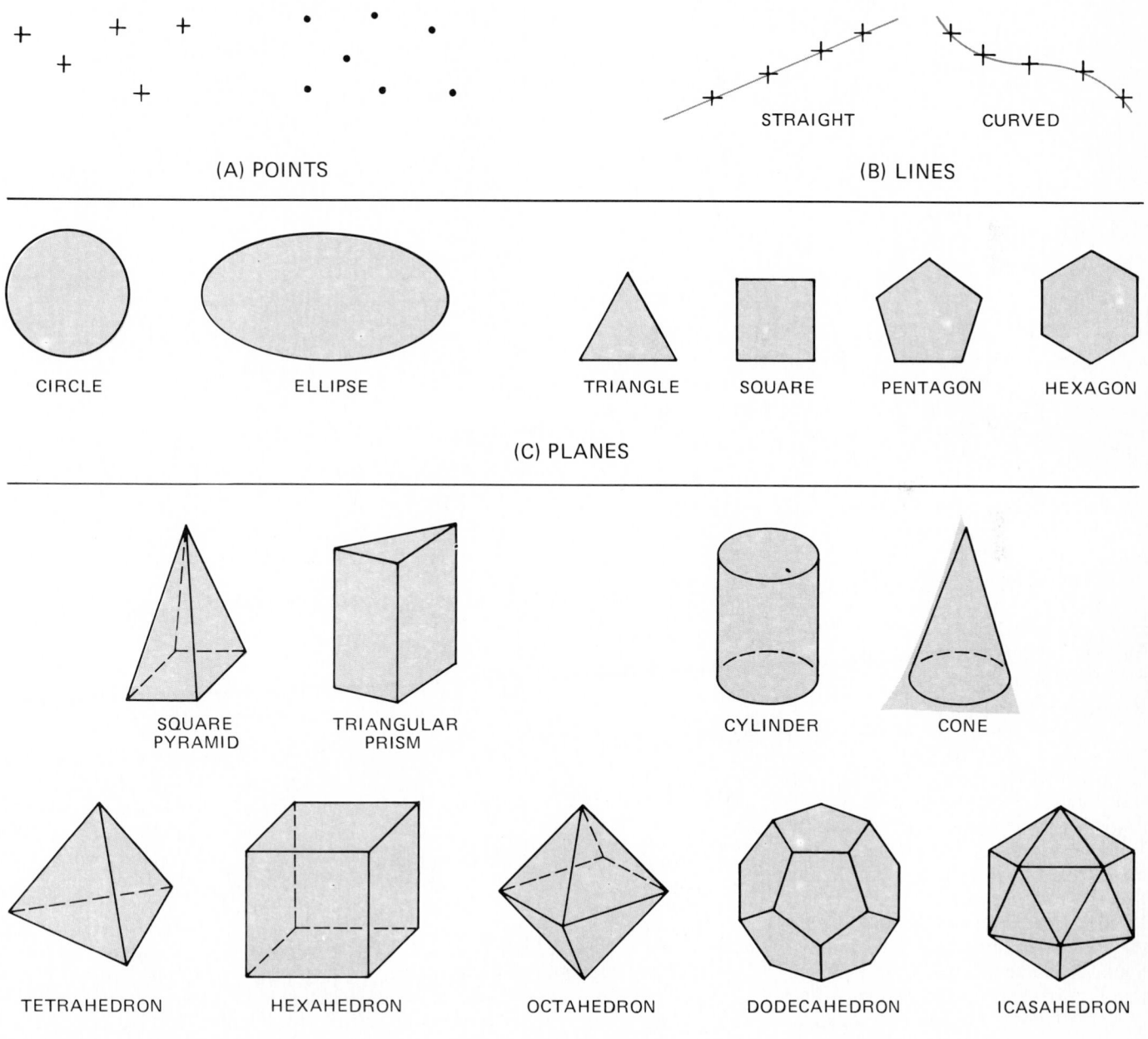

(A) POINTS (B) LINES

CIRCLE   ELLIPSE   TRIANGLE   SQUARE   PENTAGON   HEXAGON

(C) PLANES

SQUARE PYRAMID   TRIANGULAR PRISM   CYLINDER   CONE

TETRAHEDRON   HEXAHEDRON   OCTAHEDRON   DODECAHEDRON   ICASAHEDRON

(D) SOLIDS

Fig. 31-1-2 Basic geometric elements and shapes.

(A) PICTORIAL VIEW OF REFERENCE PLANES

(A) POINTS A AND B IDENTIFIED ON
UNFOLDED REFERENCE PLANE

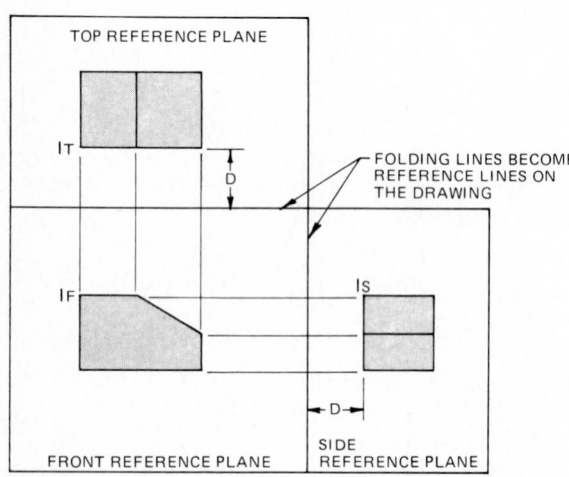

(B) UNFOLDING OF THE THREE REFERENCE PLANES

Fig. 31-1-3 Reference planes.

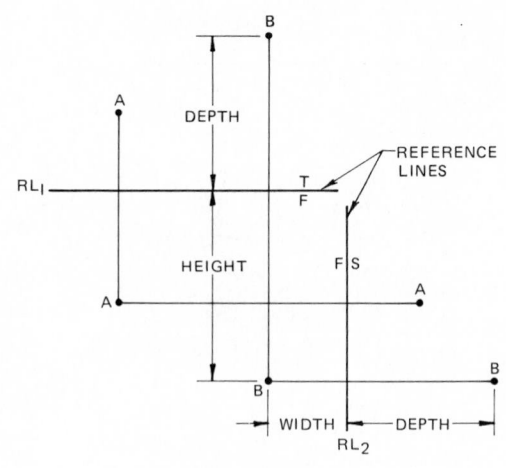

(B) POINTS A AND B IDENTIFIED TO
REFERENCE LINES

Fig. 31-1-4 Points in space.

POINTS

A point can be considered physically real and can be located by a small dot or a small cross. It is normally identified by two or more projections. In Fig. 31-1-4A points A and B are located on all three reference planes. Notice that the unfolding of the three planes forms a two-dimensional surface with the fold lines remaining. The folding lines are labeled as shown to indicate that F represents the front view, T represents the top view, and S represents the profile or right-side view. In Fig. 31-1-4B the planes are replaced with reference lines RL_1 and RL_2, which

are placed in the same position as the fold lines in Fig. 31-1-4A.

LINES IN SPACE

Lines in descriptive geometry are grouped into three classes depending on how they are positioned in relation to the reference lines.

Normal Lines A line which is perpendicular to the reference plane will project as a point on that plane. In Fig. 31-1-5A, line AB is perpendicular to the front reference plane. As such, it is shown as a point ($A_F B_F$) in the front view and as a true-length line in the top

and side views (lines $A_T B_T$ and $A_S B_S$, respectively).

Inclined Lines A line that appears inclined in one plane, as shown in Fig. 31-1-5B, and is parallel to one on the other two principal planes will appear foreshortened in the other two views. The inclined line shown in the front view will be the true length of line AB.

Oblique Lines A line that appears inclined in all three views is an oblique line. It is neither parallel nor perpendicular to any of the three planes. The true length of the line is not shown in any of these views. See Fig. 31-1-5C.

(A) NORMAL LINE

(B) INCLINED LINE

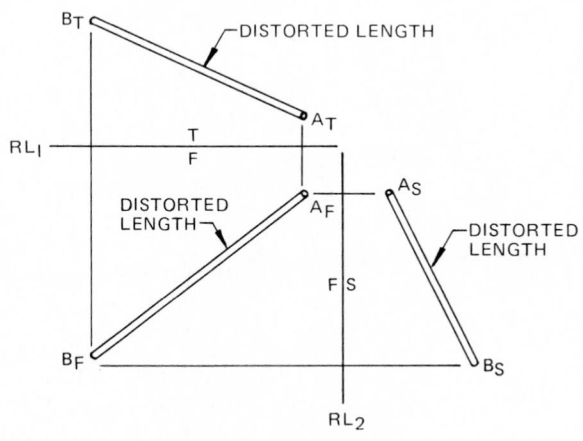

(C) OBLIQUE LINES

Fig. 31-1-5 Lines in space.

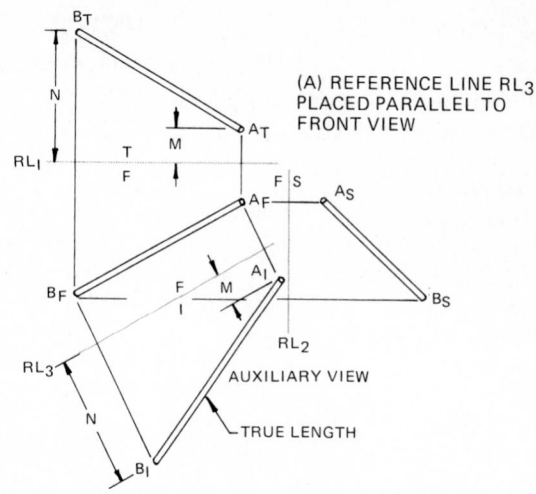

(A) REFERENCE LINE RL₃ PLACED PARALLEL TO FRONT VIEW

(B) REFERENCE LINE RL₃ PLACED PARALLEL TO SIDE VIEW

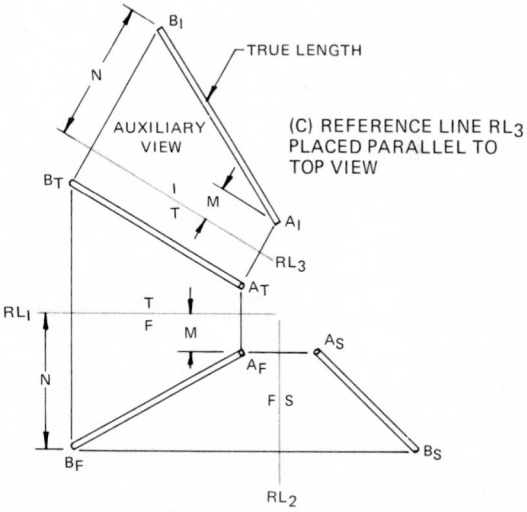

(C) REFERENCE LINE RL₃ PLACED PARALLEL TO TOP VIEW

Fig. 31-1-6 True length of an oblique line by auxiliary projection.

TRUE LENGTH OF AN OBLIQUE LINE BY AUXILIARY PROJECTION

The true length of a normal line and the inclined line have line projections which are parallel to a principal plane of projection. Thus, a line parallel to a plane of projection shows true length in that projection. Since an oblique line is not parallel to any of the three principal reference planes, an auxiliary reference line RL_3 can be placed parallel to any one of the oblique lines, as shown in Fig. 31-1-6. Transfer distances designated as M and N shown in the regular views to the auxiliary view locating points A_1 and B_1, respectively. Join points A_1 and B_1 with a line, obtaining the true length of line AB.

POINT ON A LINE

The line $A_F B_F$ in the front view of Fig. 31-1-7A contains a point C. To place point C on the line in the other two views, it is necessary to project construction lines perpendicular to the reference lines RL_1 and RL_2, as shown in Fig. 31-1-7. The construction lines are projected to line $A_T B_T$ in the top view and line $A_S B_S$ in the side view, locating point C on the line in these views.

If point C was to be located on the true length of line AB, then another reference line, such as RL_1, is required, and the distances L and M in the front view are then used to locate the true length of line $A_1 B_1$ in the auxiliary view. Position C is then projected perpendicular to line $A_S B_S$ in the side view to locate C on the true-length line.

POINT-ON-POINT VIEW OF A LINE

When the front view $A_F B_F$ and top view $A_T B_T$ are given, as in Fig. 31-1-8, and the point-on-point view of line AB is required, the procedure is as follows:

- Draw a true-length view of AB by the method described in True Length of an Oblique Line by Auxiliary Projection (Fig. 31-1-6).
- Draw a reference line perpendicular to the true length of line $A_1 B_1$, and label it RL_3.
- The next adjacent, second auxiliary view $A_2 B_2$ will be a point-on-point view of line AB.

ASSIGNMENTS

See Assignments 1 and 2 for Unit 31-1 on page 666.

UNIT 31-2
Planes in Space

Planes for practical studies are considered to be without thickness and can be extended without limit. A plane may be represented or determined by intersecting lines, two parallel lines, a line and a point, three points, or a triangle.

The three basic planes, referred to as the *normal plane*, *inclined plane*, and *oblique plane*, are identified by their relationship to the three principal reference planes. Figure 31-2-1 illustrates the three basic planes, each plane being triangular in shape.

Normal Plane This is a plane whose surface, in this case a triangular surface, appears in its true shape in the front view and as a line in the other two views.

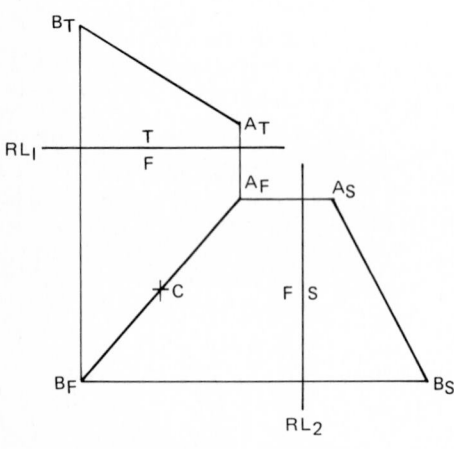

(A) TO LOCATE POINT C ON LINE A–B IN OTHER VIEWS

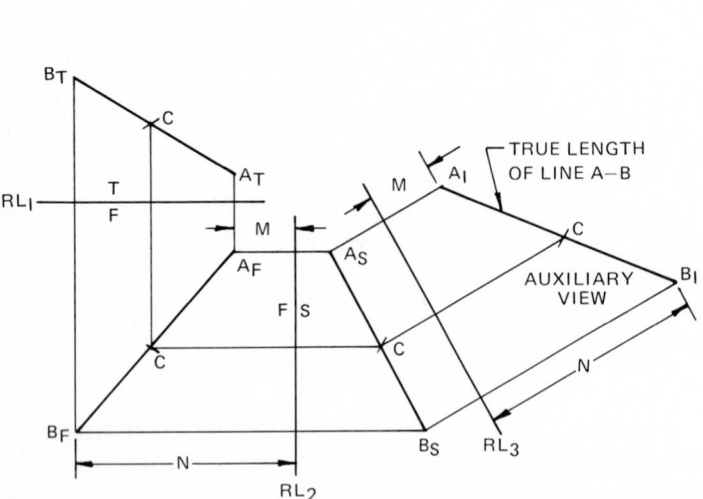

Fig. 31-1-7   Point on a line.

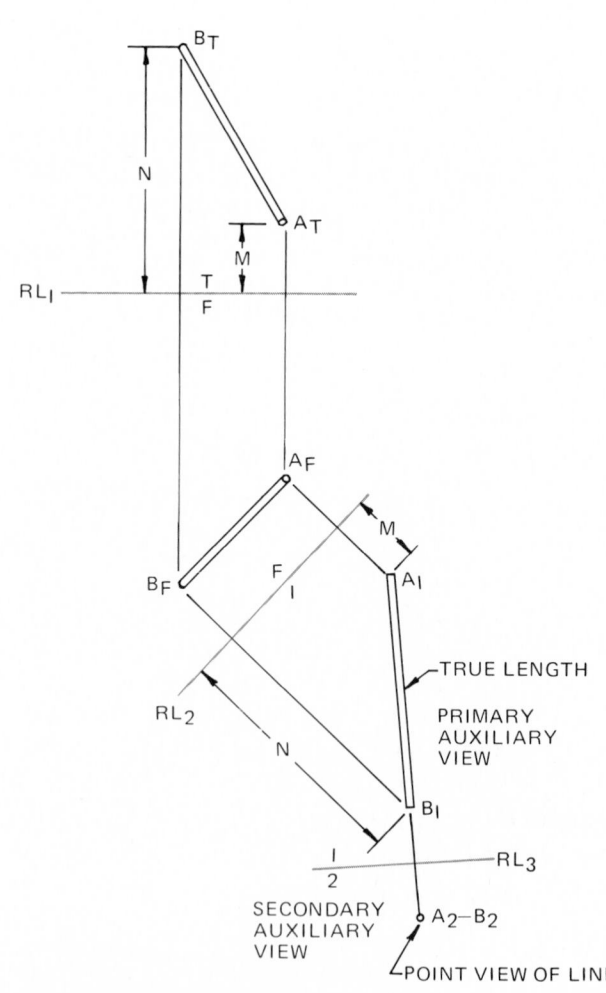

Fig. 31-1-8   Point-on-point view of a line.

(A) NORMAL PLANE

(B) INCLINED PLANE

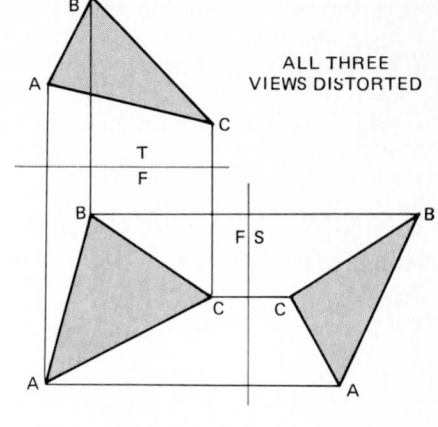

(C) OBLIQUE PLANE

Fig. 31-2-1 Planes in space.

Inclined Plane This results when the shape of the triangular plane appears distorted in two views and as a line in the other view.

Oblique Plane This is a plane whose shape appears distorted in all three views.

LOCATING A LINE IN A PLANE

The top and front views shown in Fig. 31-2-2A show a triangular plane *ABC* and lines *RS* and *MN*, each located in one of the views. To find their location in the other views, refer to Fig. 31-2-2B and the following procedures.

To locate line RS in the front view:

- Line $R_T S_T$ crosses over lines $A_T B_T$ and $A_T C_T$ at points D_T and E_T, respectively.

- Project points D_T and E_T to front view, locating points D_F and E_F.
- Extend a line through points D_F and E_F.
- The length of the line can be found by projecting points R_T and S_T to the front view, locating the end points R_F and S_F.

To locate line MN in the top view:

- Extend line $M_F N_F$ in the front view, locating points H_F and G_F on lines $A_F B_F$ and $A_F C_F$, respectively.
- Project points H_F and G_F to the top view, locating points H_T and G_T.
- Draw a line through points H_T and G_T.
- Project points M_F and N_F to top view, locating points on line $M_T N_T$.

LOCATING A POINT ON A PLANE

The top and front views shown in Fig. 31-2-3A show a triangular plane *ABC* and points *R* and *S* each located in one of the views. To find their location in the other views, refer to Fig. 31-2-3B and the following procedures.

To locate point R in the front view:

- Draw a line from A_T passing through point R_T to a point M_T on line $B_T C_T$.
- Project point M_T to front view, locating point M_F.
- Join points A_F and M_F with a line.
- Project point R_T to front view, locating point R_F.

To locate point S in the top view:

- Draw a line between points B_F and S_F, locating point N_F on line $A_F C_F$.

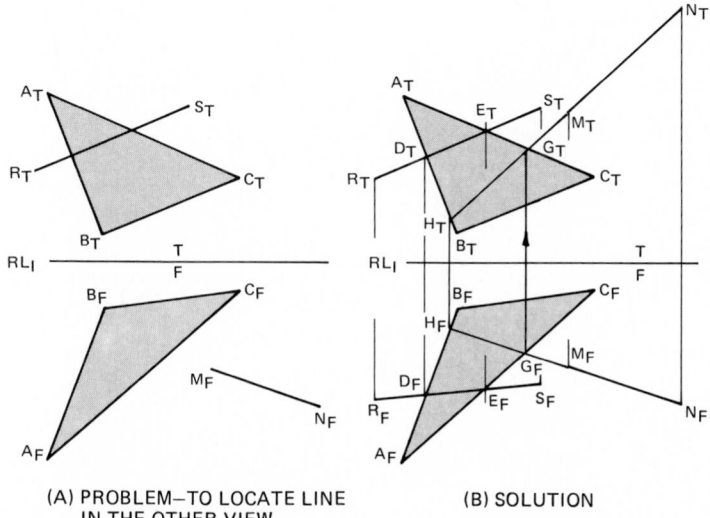

(A) PROBLEM—TO LOCATE LINE IN THE OTHER VIEW

(B) SOLUTION

Fig. 31-2-2 Locating a line in a plane.

(A) PROBLEM—TO LOCATE POINT IN OTHER VIEW

(B) SOLUTION

Fig. 31-2-3 Locating a point in a plane.

- Project point N_F to top view, locating point N_T.
- Draw a line through points B_T and N_T.
- Project point S_F to top view, locating point S_T.

LOCATING THE PIERCING POINT OF A LINE AND A PLANE— CUTTING-PLANE METHOD

The top and front views shown in Fig. 31-2-4 show a line UV passing somewhere through plane ABC. The piercing point of the line through the plane is found as follows:

- Locate points D_T and E_T in the top view.
- Project points D_T and E_T to the front view, locating points D_F and E_F.
- The intersection of lines $D_F E_F$ and $U_F V_F$ is the piercing point, labeled O_F.
- Project point O_F to top view, locating point O_T.

LOCATING THE PIERCING POINT OF A LINE AND A PLANE— AUXILIARY-VIEW METHOD

The top and front views in Fig. 31-2-5 show a line UV passing somewhere through plane ABC. The piercing point of the line through the plane is found as follows:

- Draw line $A_T D_T$ in the top view parallel to reference line RL_1.
- Project point D_T to front view, locating point D_F.
- Draw reference line RL_2 perpendicular to a line intersecting points A_F and D_F in the front view.
- Draw an auxiliary view which shows the edge view of the plane. The intersection of the plane and line locates the piercing point O_1.
- Project the piercing point to front and top views.

ASSIGNMENTS

See Assignments 3 and 4 for Unit 31-2 on page 666.

UNIT 31-3
Establishing Visibility of Lines in Space

VISIBILITY OF SKEW LINES BY TESTING

In the example of the two nonintersecting pipes shown in Fig. 31-3-1A, it is not apparent which pipe is nearest the viewer at the crossing points in the two views. To establish which of the pipes lies in front of the other, the following procedure is used.

To establish the visible pipe at the crossing shown in the top view (Fig. 31-3-1B):

- Label the crossing of lines $A_T B_T$ and $C_T D_T$ as ①, ②.
- Project the crossing point to the front view, establishing points ① and ②.
- Point ① is closer to reference line RL_1, which means that line $A_T B_T$ is nearer when the top view is being observed and thus is visible.

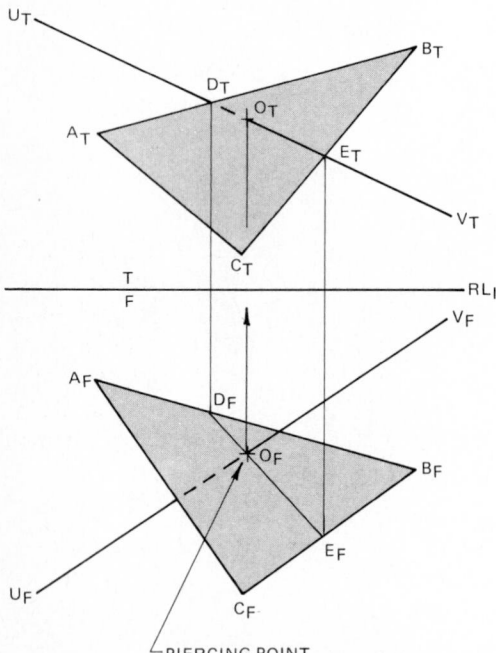

Fig. 31-2-4 Locating piercing point of a line and a plane, cutting plane method.

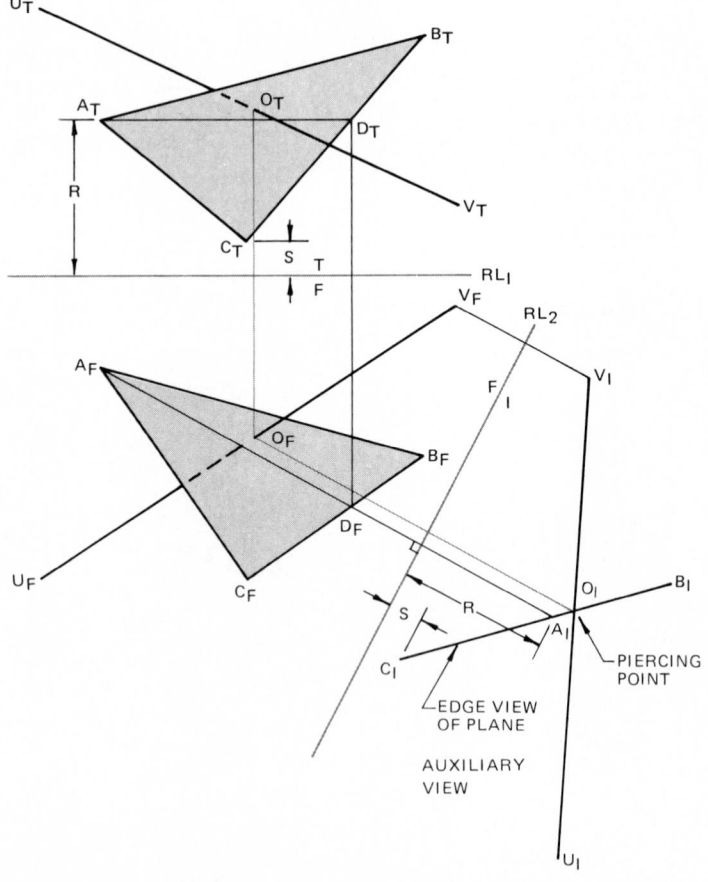

Fig. 31-2-5 Piercing point of a line and a plane, auxiliary view method.

(A) PROBLEM

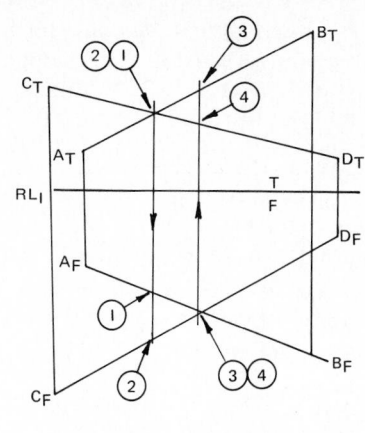

(B) ESTABLISHING LINES WHICH ARE
CLOSER TO OBSERVER

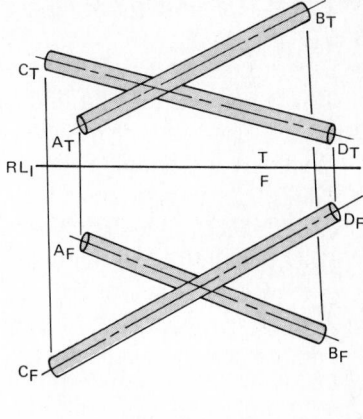

(C) SOLUTION

Fig. 31-3-1   Visibility of skew lines by testing.

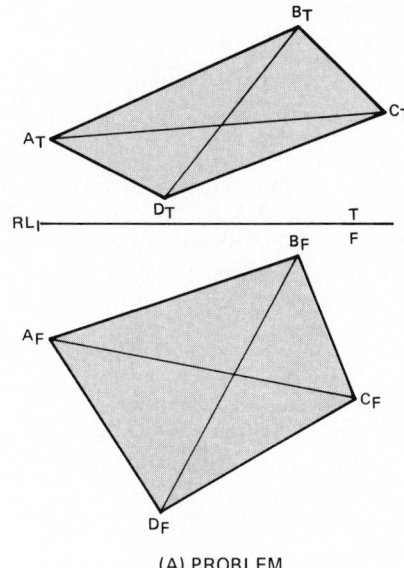

(A) PROBLEM

(B) ESTABLISHING LINES WHICH ARE
CLOSER TO OBSERVER

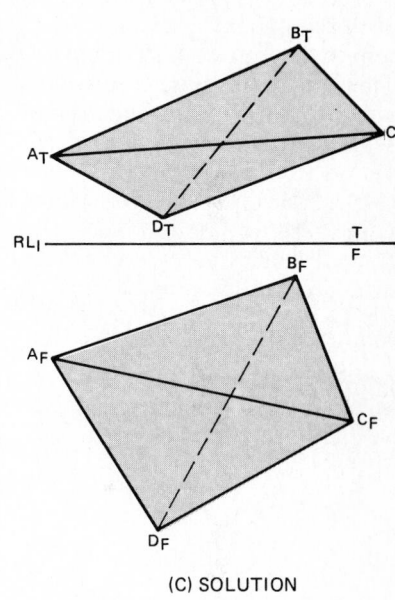

(C) SOLUTION

Fig. 31-3-2   Establishing visibility of lines and surfaces by testing.

To establish the visible pipe at the crossing shown in the front view:

- Label the crossing of lines A_FB_F and C_FD_F as ③, ④.
- Project the crossing point to top view, establishing points ③ and ④.
- Point ④ is closer to reference line RL_1, which means that line C_FD_F is nearer when the front view is being observed and thus is visible. Figure 31-3-1C shows the correct crossings of the pipes.

VISIBILITY OF LINES AND SURFACES BY TESTING

In cases where points or lines are approximately the same distance away from the viewer, it may be necessary to graphically check the visibility of lines and points, as in Fig. 31-3-2.

To check visibility of lines A_TC_T and B_TD_T in the top view:

- Label intersection of lines A_TC_T and B_TD_T as ①, ②.
- Project the point of intersection to front view, establishing point ① on line A_FC_F and point ② on line B_FD_F.

- Point ① is closer to reference line RL_1, which means that line A_TC_T is nearer when the top view is being observed and thus is visible.
- Point ② is farther away from reference line RL_1, which means that line B_TD_T would not be seen when one is viewing from the top.

To check visibility of lines A_FC_F and B_FD_F in the front view:

- Label the intersection of lines A_FC_F and B_FD_F as ③, ④.
- Project the point of intersection to top view, establishing point ③ on

line $A_T C_T$ and point ④ on line $B_T D_T$.

- Point ③ is closer to reference line RL_1, which means that line $A_F C_F$ is nearer when the front view is being observed and thus is visible.
- Point ④ is farther away from reference line RL_1, which means that line $B_F D_F$ would not be seen when one is viewing from the front. Figure 31-3-2C shows the completed top and front views of the part.

VISIBILITY OF LINES AND SURFACES BY OBSERVATION

In order to fully understand the shape of an object, it is necessary to know which lines and surfaces are visible in each of the views. Determining their visibility can, in most cases, be done by inspection. With reference to Fig. 31-3-3A, the outline of the part is obviously visible. However, the visibility of lines and surfaces within the outline must be determined. This is accomplished by determining the position of O_F in the front view. Since position O is the closest point to the reference line RL_1 in the front view, it must be the point which is closest to the observer when viewing the top

view. Thus it can be seen, and the lines converging to point O_T are visible.

With reference to determining the visibility of the lines in the front view, see the top view. Plane $O_T C_T D_T$ is closest to reference line RL_1. Therefore it must be the closest surface when the observer is looking at the front view, and it must be visible. Since point B_T in the top view is farthest away from reference line RL_1, it is the point which is farthest away when the observer is looking at the front view. Since it lies behind surface $O_F C_F D_F$, it cannot be seen.

From this example it may be stated that lines or points closest to the observer will be visible, and lines and points farthest away from the viewer but lying within the outline of the view will be hidden.

ASSIGNMENTS

See Assignments 5 and 6 for Unit 31-3 on page 669.

Distances Between Lines and Points

DISTANCE FROM A POINT TO A LINE

When the front and side views are given, as in Fig. 31-4-1, and the shortest distance between line AB and point P is required, the procedure is as follows:

- Draw reference line RL_2 parallel to line $A_S B_S$ in the side view.
- Transfer distances designated as R, S, and U in the front view to the primary auxiliary view. The resulting line $A_1 B_1$ in the auxiliary view is the true length of line AB.
- Next draw reference line RL_3 perpendicular to line $A_1 B_1$.
- Transfer distances designated as V and W in the side view to the secondary auxiliary view, establishing points P_2 and $A_2 B_2$, the latter being the point view of line AB.
- The shortest distance between point P and line AB is shown in the secondary auxiliary view.

Figure 31-4-2 illustrates the application of the point-on-point view of a line to determine the clearance between a hydraulic cylinder and a clip on the wheel housing.

SHORTEST DISTANCE BETWEEN TWO SKEW LINES

When the front and top views are given, as in Fig. 31-4-3, and the shortest distance between the two lines AB and CD is required, the procedure is as follows:

- Draw reference line RL_2 parallel to line $A_F B_F$ in the front view.
- Transfer the distances designated as R, S, U, and V in the top view to the primary auxiliary view. The resulting line $A_1 B_1$ in the auxiliary view is the true length of line AB.
- Next draw reference line RL_3 perpendicular to line $A_1 B_1$.
- Transfer the distances designated as L, M, and N in the front view to the secondary auxiliary view. The resulting line $C_2 D_2$ is the true length

(A) PROBLEM—TO DETERMINE VISIBILITY OF LINES

(B) SOLUTION

Fig. 31-3-3 Visibility of lines and surfaces by observation.

Fig. 31-4-1 Distance from a point to a line.

Fig. 31-4-2 Design application of distance from a point to a line.

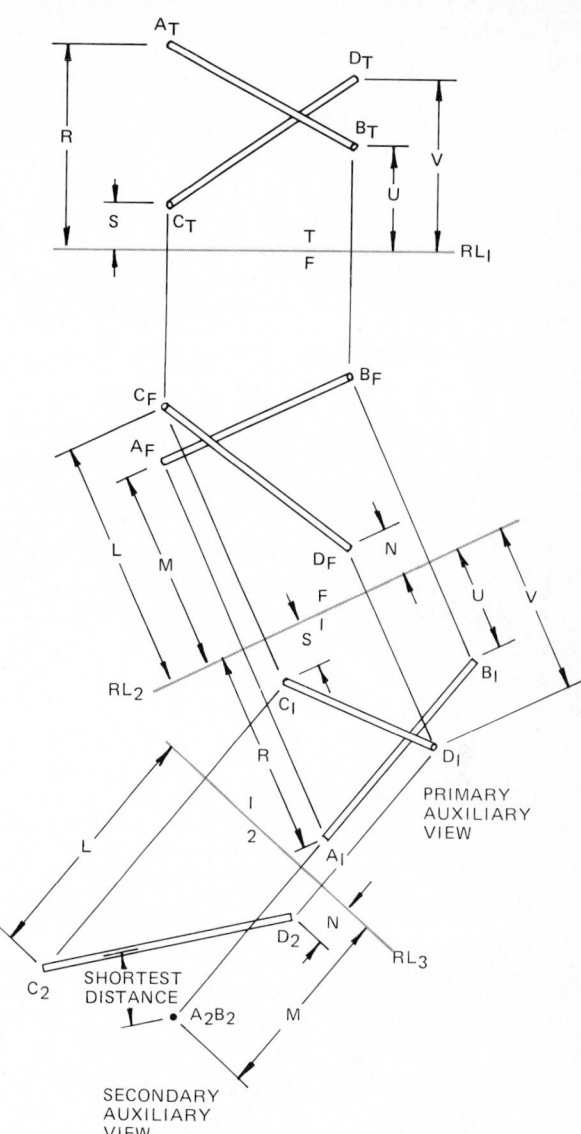

Fig. 31-4-3 Shortest distance between skew lines.

of line *CD*. Point A_2B_2 is a point view of line *AB*.

- The shortest distance between the two lines *AB* and *CD* is shown in the secondary auxiliary view.

ASSIGNMENTS

See Assignments 7 and 8 for Unit 31-4 on page 669.

UNIT 31-5
Edge and True View of Planes

The primary planes of projection are *horizontal*, *vertical* (or *frontal*), and *profile*. A plane that is not parallel to a primary plane is not visible in its true dimensions. To show a plane in true view, it must be revolved until it is parallel to a projection plane. Figure 31-5-1 shows an oblique plane *ABC* in the top and front views. The object is to find the true view of this plane. When the top and front views are examined carefully, no line is parallel to the reference line in either view. However, a line C_TD_T can be drawn on the plane parallel to the reference line RL_1 and projected to the front view for its true length D_FC_F. Next:

- Draw reference line RL_2 perpendicular to line D_FC_F in the front view.

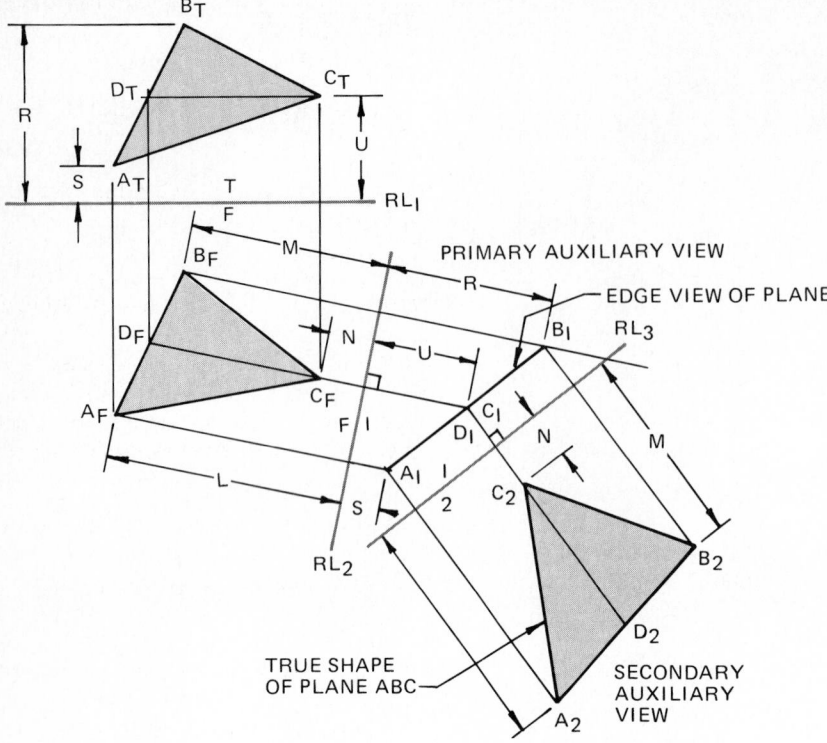

Fig. 31-5-1   True view of a plane.

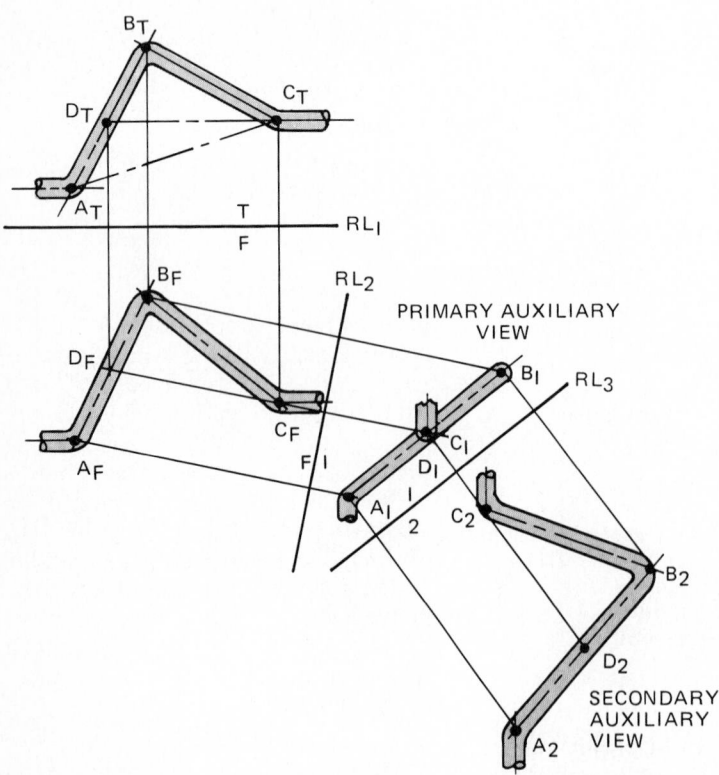

Fig. 31-5-2   Design application of true view of a plane (Fig. 31-5-1).

- Transfer the distances designated R, S, and U in the top view to the primary auxiliary view. The resulting line A_1B_1 is the edge view of the plane.
- Draw reference line RL_3 parallel to line A_1B_1.
- Transfer the distances L, M, and N in the front view to construct the true shape of plane ABC in the secondary auxiliary view.

Design Application   Figure 31-5-2 shows the application of the procedure followed for Fig. 31-5-1. Points A, B, C, and D correspond in both drawings, but line AC is omitted in Fig. 31-5-2 since it serves no practical purpose in the design.

PLANES IN COMBINATION

Figure 31-5-3 demonstrates a solution where a combination of planes is involved. Note that $A_TB_TC_T$ and $A_FB_FC_F$ form one plane while $B_TC_TD_T$ and $B_FC_FD_F$ form another. Also line BC is common to both planes. The objective in the problem is to find the true bends at the angles ABC and BCD. The procedure is as follows:

- Construct the primary auxiliary view which shows the true length of line BC.
- Construct a secondary auxiliary view which shows BC as a point-on-point view. The result is the edge view of both planes ABC and BCD in the secondary auxiliary view.
- Since any view adjacent to a point-on-point view of a line must show the line in its true length, BC will be in true length in secondary auxiliary views 2 and 3. Therefore, projecting perpendicularly from the edge views in the secondary auxiliary view 1 to the secondary auxiliary views 2 and 3 gives not only BC in true length in auxiliary views 2 and 3, but also the true angles ABC and BCD.

ASSIGNMENTS

See Assignments 9 and 10 for Unit 31-5 on page 670.

Fig. 31-5-3 Use of planes in combination.

UNIT 31-6
Angles Between Lines and Planes

THE ANGLE A LINE MAKES WITH A PLANE

The top and front views in Fig. 31-6-1 show a line UV passing somewhere through plane ABC. The true angle between the line and the plane will be shown in the view which shows the edge view of the plane and the true length of the line. This view is found as follows:

- Draw line $A_T D_T$ in the top view parallel to reference line RL_1.
- Project point D_T to front view, locating point D_F.
- Draw reference line RL_2 perpendicular to a line intersecting points A_F and D_F in the front view.
- Draw the primary auxiliary view which shows the edge view of the plane but distorted length of line UV. The point of intersection between the line and edge view of the plane is established.
- Draw reference line RL_3 parallel to edge view of the plane shown in the auxiliary view.
- Draw the secondary auxiliary view 1 which shows the true view of the plane and location of the piercing point.

Fig. 31-6-1 The angle a line makes with a plane.

- Draw reference line RL_4 parallel to line V_2U_2.
- Draw the secondary auxiliary view 2 which shows the true length of line UV and the true angle between line and edge view of plane.

EDGE LINES OF TWO PLANES

Figure 31-6-2 shows a line of intersection AB made by two planes, triangles ABC and ABD. When the top and front views are given, the point-on-point view of line AB and the true angle between the planes are found as follows:

- Draw reference line RL_2 parallel to line A_FB_F in the front view.
- Transfer the distances designated R, S, and U in the top view to the primary auxiliary view. The resulting line A_1B_1 in the auxiliary view is the true length of line AB.
- Next draw reference line RL_3 perpendicular to line A_1B_1.
- Transfer the distances designated as L, M, and N in the front view to the secondary auxiliary view.
- Point A_2B_2 is a point-on-point view of line AB. The true angle between the two planes is seen in the secondary auxiliary view.

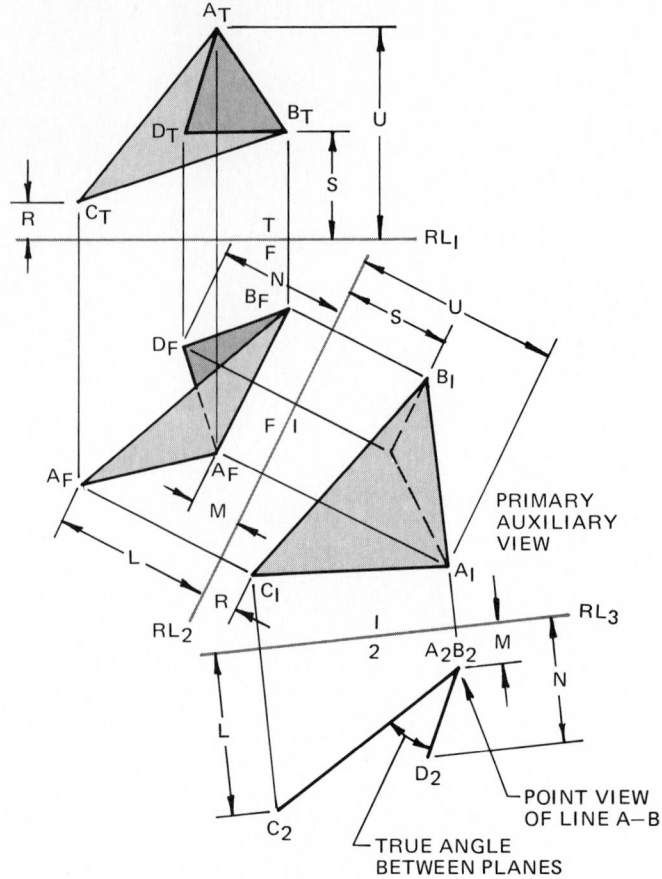

Fig. 31-6-2   Edge lines of two planes.

ASSIGNMENTS

See Assignments 11 through 13 for Unit 31-6 on page 671.

ASSIGNMENTS for Chapter 31

Assignments for Unit 31-1, Graphic Solutions
1. Divide a B- or A3-size sheet into four parts. With the use of reference lines, locate the points in the third view for the drawings shown in Fig. 31-1-A. Scale to suit.
2. Divide a B- or A3-size sheet into four parts. With the use of reference lines, locate the lines in the other views for the drawings shown in Fig. 31-1-B. Scale to suit.

Assignments for Unit 31-2, Planes In Space
3. *Locating a Plane or a Line in Space*. On a B- or A3-size sheet, complete the three drawings shown in Fig. 31-2-A. Scale to suit.
4. *Locating a Point in Space and the Piercing Point of a Line and a Plane*. On a B- or A3-size sheet, complete the drawings shown in Fig. 31-2-B. Scale to suit.

(1) LOCATE POINTS IN THE TOP VIEW

(2) LOCATE POINTS IN THE SIDE VIEW

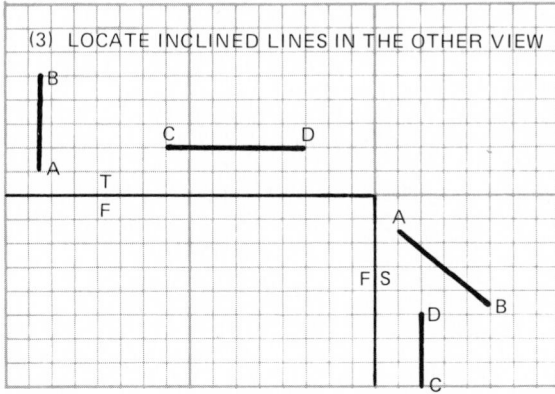

(3) LOCATE INCLINED LINES IN THE OTHER VIEW

(4) LOCATE NORMAL AND INCLINED LINES IN THE OTHER VIEW

Fig. 31-1-A Assignments.

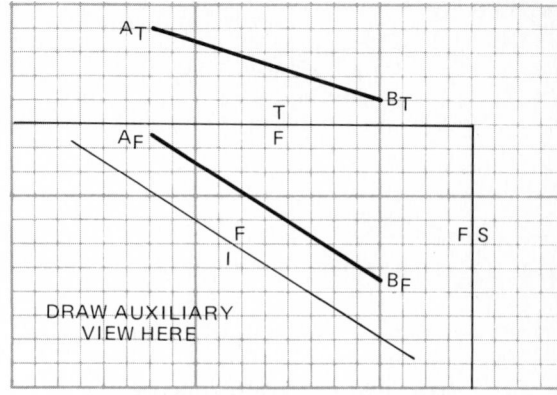

DRAW AUXILIARY VIEW HERE

(1) DRAW SIDE AND AUXILIARY VIEWS
(INDICATE TRUE LENGTH OF LINE A—B)

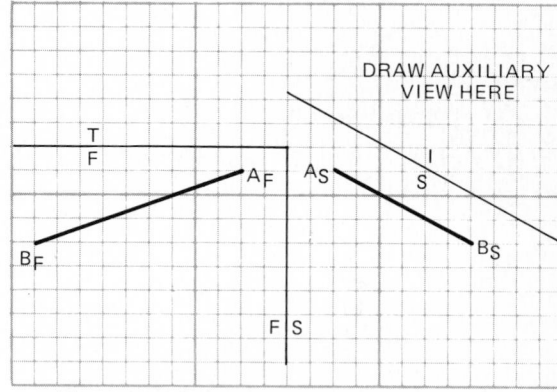

DRAW AUXILIARY VIEW HERE

(2) DRAW TOP AND AUXILIARY VIEWS
(INDICATE TRUE LENGTH OF LINE A—B)

PROJECT AUXILIARY VIEW FROM SIDE VIEW SHOWING TRUE LENGTH OF LINE A—B. LOCATE POINT C ON LINE A—B IN ALL VIEWS. PROJECT FROM THE PRIMARY AUXILIARY VIEW THE SECONDARY AUXILIARY VIEW SHOWING THE POINT VIEW OF LINE A—B

(3) DRAW SIDE VIEW

(4) DRAW TOP VIEW. PROJECT AUXILIARY VIEW FROM FRONT VIEW SHOWING TRUE LENGTH OF LINE A—B. LOCATE POINT C ON LINE A—B IN ALL VIEWS. PROJECT FROM THE PRIMARY AUXILIARY VIEW THE SECONDARY AUXILIARY VIEW SHOWING THE POINT VIEW OF LINE A—B.

Fig. 31-1-B Assignments.

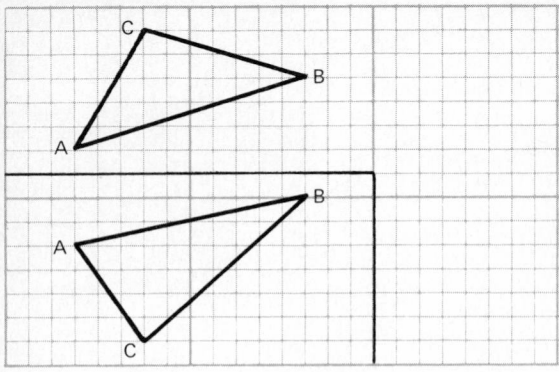

(1) DRAW THE SIDE VIEW OF TRIANGLE ABC

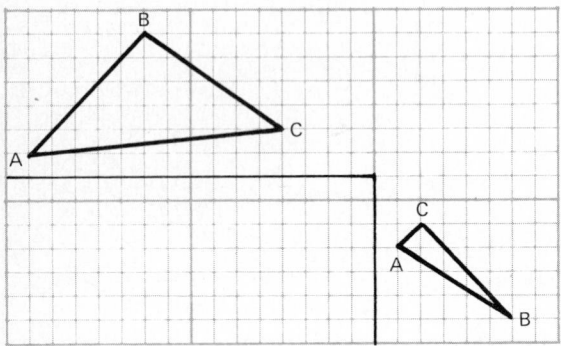

(2) DRAW THE FRONT VIEW OF TRIANGLE ABC

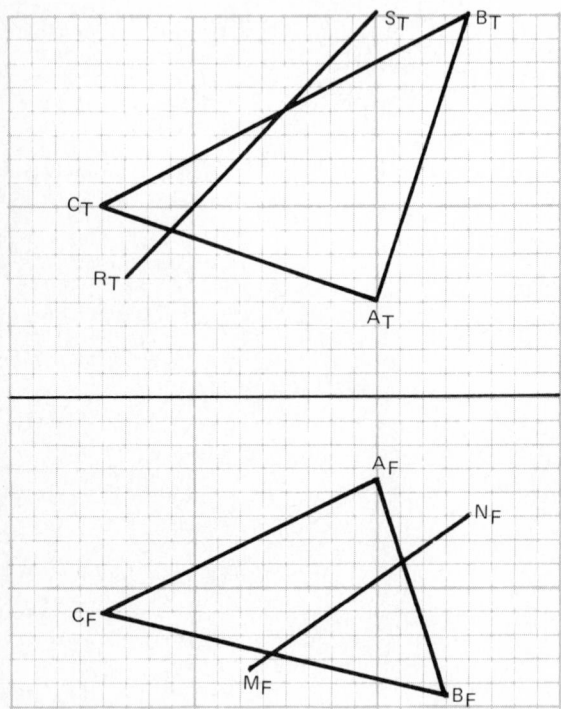

(3) LOCATING A LINE IN THE OTHER VIEW

Fig. 31-2-A Assignments.

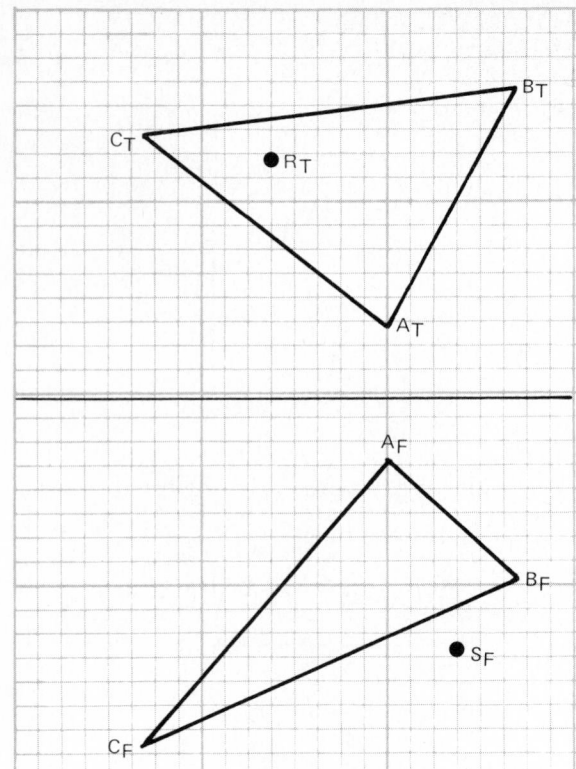

(1) LOCATING A POINT IN SPACE

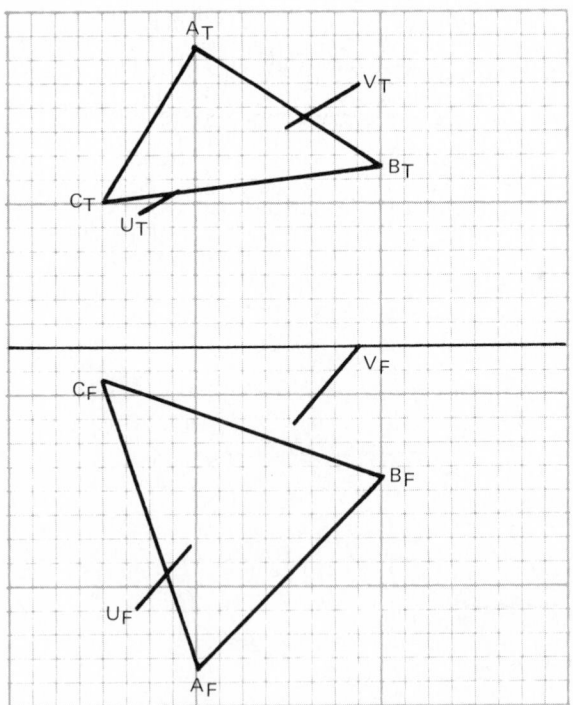

(2) LOCATING PIERCING POINT OF A LINE AND A PLANE

Fig. 31-2-B Assignments.

Assignments for Unit 31-3, Establishing Visibility of Lines In Space

5. *Visibility of Lines and Surfaces by Observation and Testing.* On a B- or A3-size sheet, lay out the drawings shown in Fig. 31-3-A. By observation, sketch the circular pipes (drawings 1 and 2) in a manner similar to that shown in Fig. 31-3-1, showing the direction in which the pipes are sloping and which pipe is closer to the observer in the two views. By testing in a manner similar to that shown in Fig. 31-3-2, establish the visibility of lines and surfaces of drawings 3 and 4. Scale to suit.

6. *Establishing Visibility of Lines and Surfaces by Testing.* Divide a B- or A3-size sheet into four parts, as shown in Fig. 31-3-B. By testing in the manner shown in Fig. 31-3-2, complete the two-view drawings, showing the visible and hidden lines. Scale to suit.

Assignments for Unit 31-4, Distances Between Lines and Points

7. *Finding Distance from a Point to a Line.* Divide a B- or A3-size sheet into two parts. With the use of reference lines and auxiliary views, find the distance from a point to a line in the two problems shown in Fig. 31-4-A. Scale to suit.

8. *Finding the Shortest Distance between Skew Lines.* Divide a B- or A3-size sheet into two parts. With the use of reference lines and auxiliary views, find the shortest distance between the skew lines for the two problems shown in Fig. 31-4-B. Scale to suit.

Fig. 31-3-A   Assignment.

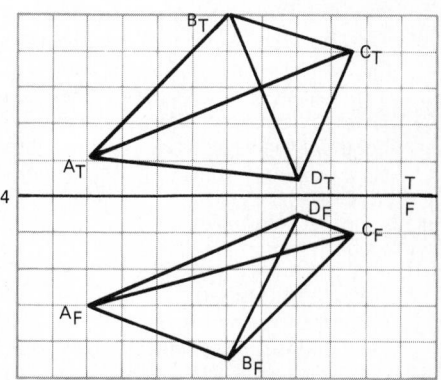

Fig. 31-3-B   Assignment.

Fig. 31-4-A Assignment.

1

2

Fig. 31-4-B Assignment.

1

2

Assignments for Unit 31-5, Edge and True View of Planes

9. *Finding True Angles of Intersecting Planes.* Divide a B- or A3-size sheet into two parts. With the use of reference lines and auxiliary views, find the true angles of the intersecting planes for the two problems shown in Fig. 31-5-A. Scale to suit.

10. *True View of a Plane.* Divide a B- or A3-size sheet into two parts. With the use of reference lines and auxiliary views, find the true view of the plane for the two problems shown in Fig. 31-5-B. Scale to suit.

Fig. 31-5-A Assignment.

1

2

Assignments for Unit 31-6, Angles Between Lines and Planes

11. *The Angle a Line Makes with a Plane.* On a B- or A3-size sheet, find the angle the line makes with a plane for the layout shown in Fig. 31-6-A. Scale to suit.

12. *The Angle a Line Makes with a Plane.* On a B- or A3-size sheet, find the angle the line makes with a plane for the layout shown in Fig. 31-6-B. Scale to suit.

13. *Edge Line of Two Planes.* Divide a B- or A3-size sheet into two parts. With the use of reference lines and auxiliary views, find the edge line of the two planes shown in the two problems in Fig. 31-6-C. Scale to suit.

Fig. 31-6-A   Assignment.

Fig. 31-6-B   Assignment.

Fig. 31-5-B   Assignment.

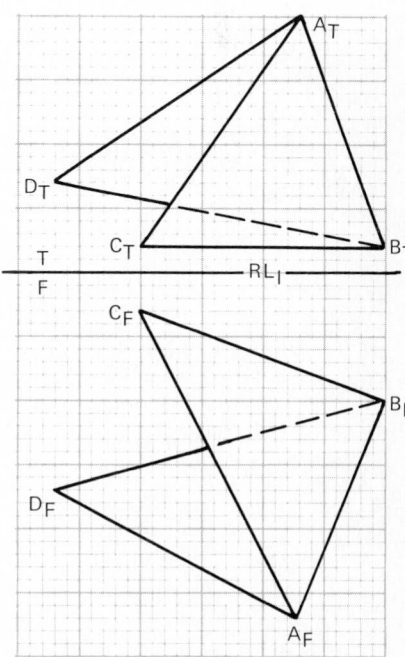

Fig. 31-6-C   Assignment.

CHAPTER 32

Computer-Aided Design and Drafting

UNIT 32-1
Introduction

This chapter describes the concepts and system equipment of computer-aided design and drafting (CAD).

Computer-aided manufacturing (CAM) is covered in Chap. 16, "Drawings for Numerical Control."

COMPUTER-AIDED DRAFTING

The use of the computer in design and drafting is the most significant devel-

opment to occur recently in these fields. The way in which drawings are prepared has been revolutionized. Producing engineering drawings on a computer is a process known as *computer-aided drafting*, commonly referred to as CAD. See Fig. 32-1-1. If information is to be sent directly to the

Fig. 32-1-1   CAD drafting station. (Cascade Graphics Development)

which are displayed on the screen in a line-by-line format. Each function or instruction is displayed. Operation and output is determined by the information originally built into the computer. For illustration purposes only, a simple partial program is shown in Fig. 32-1-3A. This program is used to con-

```
100   HOME
110   PAGE
120   INPUT   A, B
130   INPUT   C, D
140   INPUT   E, F
150   MOVE    A, B
160   DRAW    C, D
170   DRAW    E, F
180   DRAW    A, B
190   END
```

(A) LINE PROGRAM INSTRUCTIONS

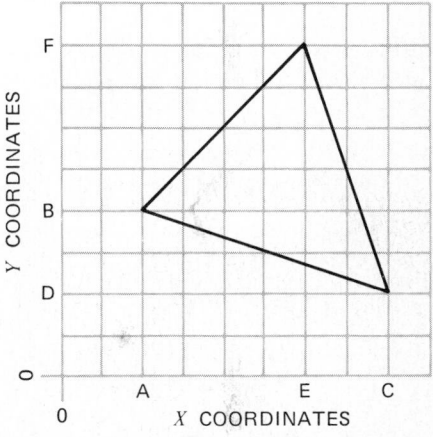

(B) GRAPH DISPLAY FROM INSTRUCTIONS

Fig. 32-1-3   Programming the computer to draw a triangle.

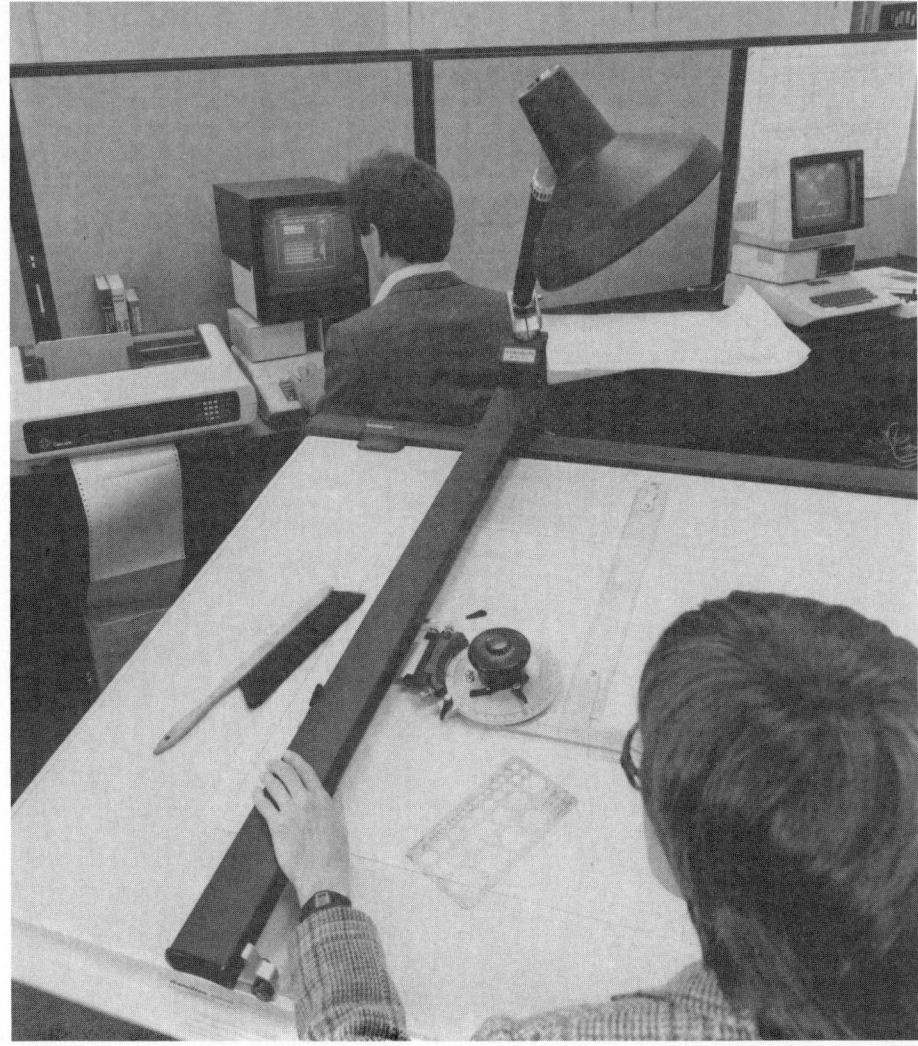

Fig. 32-1-2   The present and future state of the art in drafting. (Cascade Graphics Development)

fabricating machinery, the process is referred to as *computer-aided drafting–computer-aided manufacturing* (CAD-CAM). With the advent of CAD, the traditional drafting table has been, or is being, replaced with a station similar to that shown in the background of Fig. 32-1-2.

Commercial computer-aided drafting was introduced in 1964. Only recently, however, has the dramatic impact of this new technical tool been felt.

Whether preparing manual (traditional) or automated (computer-aided) drawings, the drafter must be familiar with the principles of drafting. Only the skill requirements and drafting equipment are different.

With manual drafting, skills in drawing lines and lettering are mandatory, and equipment such as drafting tables, drafting machines, parallel edges, scales, triangles, templates, and pencils are required.

With automated drafting, skills in drawing lines and lettering are not required. A cathode-ray terminal, a processing unit, a digitizer, and a plotter replace the manual drafting equipment and make the above-mentioned skills redundant.

The computer, at first, appears to be a mysterious machine. It is actually an electronic device with no brain. Its capability is limited to basic logical functions. These functions must be determined and fed into the computer by human beings.

A computer's memory enables the drafter to program a drawing. The *program* is a set of detailed instructions

struct a triangle which is displayed graphically in Fig. 32-1-3B. Instructions may be presented in one of several recognized standard languages. The language known as BASIC is used in Fig. 32-1-3.

The advancement in commercially produced microcomputing equipment has led the way in widespread CAD implementation. The term *microminiaturization* refers to the technology of the integrated-circuit (IC) chip. Discrete-component printed-circuit (PC) boards in computers have been replaced by microprocessors. Another innovation contributing to the widespread use of CAD was the development of the graphics display station using a cathode-ray tube (CRT). The CRT allows an image (drawing) to be projected on a television-type screen. The computer user interacts with the system through the monitor as shown in Fig. 32-1-4. Changes or additions to the drawing can be made. This interface between the computer and the computer user is known as *interactive computer graphics*.

Computer-aided drafting systems can relieve the drafter and designer from many tedious chores such as redrawing. It cannot, however, replace the skilled drafter. A CAD system cannot think. It should be thought of as an additional tool at the drafter's disposal. It is like a template which helps the drafter draw more accurately and quickly. Thus, it is an eco-nomically sound investment for aid in preparing and revising drawings. Often, a large initial financial outlay is required for computer equipment. To be cost-effective, the system demands greater efficiency, and CAD offers the possibility of increased productivity.

Reducing drafting time is of prime importance. The drafting part of a project is often considered a bottleneck. Traditional drafters spend approximately two-thirds of their time "laying lead." Only one-third is spent for all other jobs combined, including design. The use of CAD systems can change this. Drawings and design changes can be accomplished much more rapidly, resulting in a quicker turnaround time. Consequently, projects flow more quickly through production. The traditional drafting bottleneck is eliminated.

Besides increasing the speed with which a job is done, a CAD system can perform many of the tedious and repetitive tasks ordinarily required of a drafter. This includes such tasks as lettering and differentiation of line weights. CAD thus frees the drafter to be more creative while it quickly performs the mundane tasks of drafting. It has proved to be, conservatively speaking, at least a 30 percent improvement in production in terms of time spent on drawing.

A CAD system by itself cannot create. A drafter must create the drawing, and thus a strong design and drafting background remains essential. A parallel example is the use of hand-held calculators in solving math problems. Knowledge of math is still required to solve the problem. What has been eliminated is the tedious task of manually working out the calculations.

It may not be practical to handle all the workload in a design or drafting office on a CAD system. Although most design and drafting work most certainly can benefit from it, some functions will continue to be performed by traditional means. For example, certain drawings in the construction and electronics fields can be designed more quickly on a drafting board. Other factors may depend on product line, company standards, and refinement of the state of the art. Thus some companies use CAD for only a portion of the workload. Still others use CAD almost exclusively. Whatever the percentage of CAD use, one fact is certain. It has had, and will continue to have, a dramatic effect on design and drafting careers.

Once a CAD system has been installed, the required personnel must be hired or trained. Trained personnel generally originate from one of three popular sources: educational institutions, CAD equipment manufacturer training courses, and individual company programs. Regardless of how an individual is instructed in the use of a CAD system, such training is best accomplished by an experienced designer. To be competent requires three to six months of using the equipment. This time period will vary depending on the individual, the type of CAD equipment, and the level of sophistication desired.

Fig. 32-1-4 Interactive Graphics Computer. (Cascade Graphics Development)

UNIT 32-2
CAD Systems

A CAD system—whether small, medium, or large—may consist of various combinations of equipment. The specific combination largely depends on the needs of the user. Various types of drawings, referred to as *hard copy* may be preferred by certain companies. Other companies may not require any drawings. Thus one company will choose a piece of equipment that prepares a drawing one way, and another company will select equipment that uses another method. Still a third company may not utilize any copy equipment, and will instead send instructions by computer directly to the shop.

There are four basic operations involved in computer-aided drafting: input, storage, manipulation, and output.

INPUT

Information is required to create a drawing. It may incorporate old drawings, new design sketches, and verbal and/or written instructions. Commonly used input devices are an alphanumeric keyboard (*alpha* meaning "letter" and *numeric* meaning "number") and a tablet using an electronic pen or puck.

STORAGE

Information in the form of drawing symbols and drawings is stored on disks or in the computer and can be called up at any time to revise or create a new drawing. All the symbols used in mechanical, electrical, or architectural drawing are available.

MANIPULATION

Stored data may be called out to produce a drawing. For schematic diagrams symbols are sized, rotated, enlarged, or reduced, and they may be positioned in such a manner as to produce an image on the display screen. Commands are given through the keyboard to activate a drawing, turn on or off a selected grid (dot pattern), or select the type of line required.

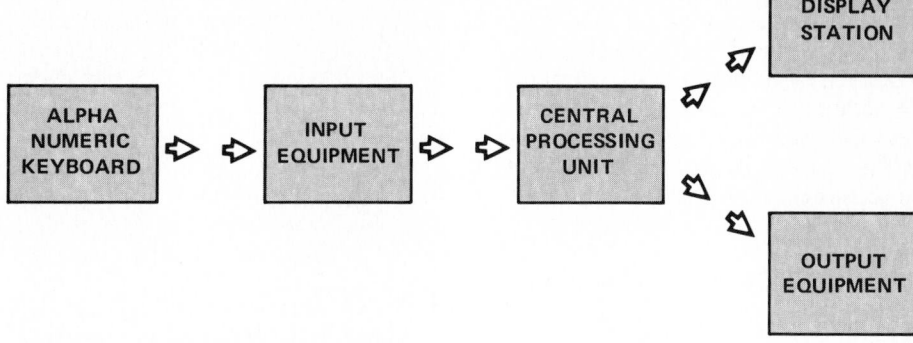

Fig. 32-2-1   A typical system arrangement.

OUTPUT

Once the desired drawing is completed on the screen, a print can be produced from a plotter, copier, or printer. Prints of what is shown on the screen may also be made at any time during the design and construction stages of the drawing.

Figure 32-2-1 shows a diagram of a typical complete system arrangement. A graphic display station and the central processing unit are considered part of the processing equipment. An alphanumeric keyboard, which is used to input data manually, is normally attached to the graphic display station to constitute one unit. This combination is commonly referred to as a *ter-minal*. A central processing unit may be reserved for a single purpose and may be attached to the terminal. This latter combination is commonly referred to as a *computer*.

The typical system arrangement is *interactive*. This means human interaction occurs between the central processing unit and graphic display station. An alphanumeric keyboard or other input equipment may aid this process. After the design and/or drawing on a graphic display station has been completed, the information may be transferred to various output devices.

Figure 32-2-2 shows a typical system. The design console is positioned at the left. It includes a graphic display

Fig. 32-2-2   System equipment setup. (Cascade Graphics Development)

station, processor, and keyboard. Also being operated by the drafter is an input device called a *graphics tablet* or *digitizer*. The unit located to the left of the keyboard is the central processing unit; it includes the programs and the means to process them. The piece of equipment shown at the right is an output device called a *plotter*.

GRAPHIC DISPLAY STATION

The purpose of the graphic display station is to project an image onto a screen. The image will display data either in alphanumeric (written) or graphical (pictorial) form. The primary use of the graphic display station in CAD application is for graphical feedback—the user can view a picture of the design as it is being entered into the system.

Cathode-Ray Tube

The most popular graphic display station is the *cathode-ray tube*, abbreviated and referred to as a CRT. Different types of CRTs are available.

The *vector-writing* CRT is drawn on by means of an *XY* coordinate system. It first locates points, then connects them with lines.

The *raster* CRT differs from the vector CRT in how displayed data are represented. It uses a grid network to display the image in a manner similar to the television screen display. Each grid is either a dark or light image that falls within a square area that appears on the screen as a dot. Each dot is known as a *pixel*. The resolution of a raster-developed image will depend on the closeness of lines forming the grid pattern. The smaller the pixels, the more resolution (or clearness) the image has. There is almost an unlimited number of bits of picture information available with raster scan. This can eliminate uneven jagged lines (known as "jaggies" or stair-stepping). An example of high and low resolution is shown in Fig. 32-2-3. Note the difference in resolution. The intensity of light in each pixel of a raster system can be controlled. By varying the intensity, shading can be produced for a more realistic image.

Raster display is quickly accomplished by the *refresh method*. The picture is refreshed, or scanned, from

(A) HIGH RESOLUTION

(B) LOW RESOLUTION

Fig. 32-2-3 CRT displays. (Lexidata Corp.)

left to right and from top to bottom very rapidly. The complete image is redrawn at a rate of speed beyond human comprehension. When compared with the vector method, "raster is faster."

A *storage tube* allows a line to be plotted without using the refresh method. A phosphor surface on the tube is bombarded by electrons to produce the image.

The above types of CRTs will produce an image in one color, similar to the conventional black and white television screen image. Any raster refresh system can be color-enhanced quite readily by using three electron guns rather than one. This is similar to

the common color television screen. Each gun will emit one of the primary colors: red, green, or blue. From this, any color pattern may be achieved.

KEYBOARD

The keyboard allows communication with the microprocessor. It is used to input data manually, primarily for nongraphical work. The keyboard resembles that of a standard typewriter, as shown in the lower portion of Fig. 32-2-4. *Alpha* refers to the keys that input letters of the alphabet. *Numeric* refers to the other keys, each of which inputs a number. The user may type in an alphanumeric instruction.

The alphanumeric keyboard is normally combined with a CRT. This terminal enables the operator to view immediately on the screen each manual instruction.

CENTRAL PROCESSING UNIT

The central processing unit (CPU) is the computing portion of the system. A multitude of integrated-circuit (IC) chips are combined to allow the performance of fundamental computations. The number of computations, or capacity of the unit, is designated by the number of bytes. *Byte* is the base term used to describe a character of memory containing 8 bits. A *bit* is a *bi*nary digi*t* based upon the two-digit binary code of 0 and 1. All messages are sent to a microprocessor by means of the binary code. The digit 1 indicates the presence of a signal (on). The digit 0 indicates the lack of a signal (off). This two-digit signal is transmitted by the operator when a key is pressed on the alphanumeric keyboard. A typical computer comprised of a CPU, CRT,

and keyboard is shown in Fig. 32-2-4.

The memory capacity of the CPU will determine the size of the CAD system. A *micro* unit is a small system such as a home computer. A *mini* unit is medium-sized but has large capability, and a *mainframe* unit is a large system.

UNIT 32-3
CAD Software

PROGRAM LANGUAGE

Software includes written sets of instructions referred to as programs, which are used to input information into the system. Programs are written in a variety of languages, the most common being FORTRAN. Another common language is BASIC, which is popular because it uses English and math phrases that are easy to under-

stand. The notations are comprised of statements rather than sentences. Two other program languages are PASCAL and C. No matter which language is used, drafters and designers need not develop a knowledge of them. The drafter and designer will normally serve as the users of programs rather than the program developers.

STORAGE

Small and medium-sized CAD systems will store programs on a disk or cassette. The primary storage medium is the disk, on which programs are stored magnetically on a plastic-like (Mylar) surface. A common disk known as a *floppy* disk is shown in Fig. 32-3-1. A cassette, while much less popular, may also be used. The larger CAD units consisting of a mainframe utilize a different format for software. Data may be keyed directly to the computer for storage. This is referred to as *on-line data entry*. As with all computer instructions the language of electrical impulses (the binary system) is used.

Fig. 32-2-4   Desktop computer. (Cascade Graphics Development)

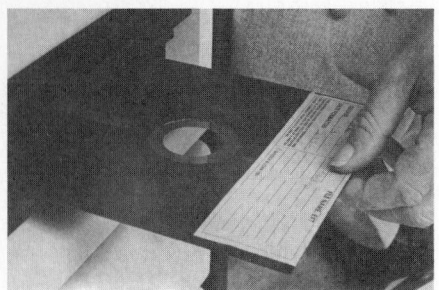

Fig. 32-3-1 Floppy disk
with protective cover.
(Tektronix, Inc.)

PROCESSING MEMORY

Additional devices are required to allow input to or output from the processing equipment. Among those required are the drives and memories. The systems which use disk or tape cassettes require disk and tape drive equipment to run the software. Typical floppy and hard disk drives are shown in the background (right) in Fig. 32-2-4.

Computers must have memory systems to store programs and data. The memory may be either permanent or temporary. Permanent memory is referred to as *read-only memory* (ROM). Temporary memory is referred to as *random-access memory* (RAM). It provides temporary storage locations for entries made by an input device, thus allowing a program to be developed. Data flow to and from RAM is controlled by the CPU. The complete set of statements (or program), up to system capacity, are stored in RAM. They will be executed sequentially whenever proper direction is given. The result will be displayed on the CRT. The contents of a program on a disk can also be loaded in RAM.

UNIT 32-4
Input and Output Peripherals

CAD system capability is enhanced by several peripheral input and output devices. A *peripheral* is an additional

Fig. 32-4-1 Graphics input tablet with stylus. (Cascade Graphics Development)

piece of equipment that may be used in conjunction with, but is not part of, the computer.

GRAPHICS TABLET

A graphics tablet, shown in Fig. 32-4-1, is an input device. In terms of graphics, it is far more important than the keyboard. Also, it has many purposes. One use is for quick and accurate graphic conversion. A rough sketch, for example, can be converted to a finished drawing. It is done by transferring point and line locations to the CRT screen. This method is commonly known as *digitizing*.

Several pieces of equipment are used in conjunction with the graphics tablet. These may include a stylus or pen, a push-button cursor or puck, a power module or console, and menus. The tablet itself is a flat surface and is available in a range of sizes. Many horizontal and vertical wires beneath the tablet surface comprise an electronic grid network which is used to sense the instruction given by the user.

Puck and Stylus

The position of each desired point of a drawing or sketch on the tablet is sensed by a puck or stylus. Several styles of puck and stylus are available (see Fig. 32-4-2A). Pucks have fine black crosshairs that are used for positioning. Pressing the appropriate button on the puck or pressing down on the stylus will digitize the point and send horizontal (X) and vertical (Y) data to the computer. The result will be displayed on the CRT screen as a bright mark (cursor). All the data may be digitized in this manner. It may be repeated as often as necessary to complete a drawing.

Tablet Menu

Most graphic tablets are provided with a *menu*. The menu contains a selection of commands that may be executed.

LIGHT PEN

A light pen is used as an input pointing device. Attached to one end of the pen is a cable through which an electric signal is transmitted. The other end of the pen may be positioned by hand to a desired screen location. After positioning, the tip of the pen is touched to the screen. Depressing it causes the pen to activate. Light spots are sensed. A signal is sent to the system indicating the pen position. The pen may be moved about the display screen. To indicate the current position under consideration, the position is illuminated. Ways to show this include a blinking character (rectangle, arrow, dot, or crosshair) or an extra bright spot. This mark on the screen is referred to as a *cursor*.

JOYSTICK

Direction Control

A *joystick* is another type of device used to control the cursor. It can be added to many systems, further enhancing CAD capability. The joystick shown in Fig. 32-4-4 is simply an extended version of the type commonly seen with video games. An electrical connection to the computer is made with a cable. The joystick steers a lighted cursor on the console screen. This is done by tilting the joystick lever in the desired direction. The cursor is correspondingly repositioned in the same direction.

Speed Control

The rate of speed the cursor is moved on the screen is variable. It is proportional to the distance that the stick is moved, or tilted, from the vertical position.

Use

A joystick is a handy device for certain functions, for example, creation of general symbol shapes which may be too cumbersome to perform with other processing controls. With a joystick they may be rapidly created. It is not useful, however, for highly-accurate applications. If a symbol drawn accurately to size must be generated, the joystick should not be used. In such a case, each point coordinate should be keyed in with the alphanumeric keyboard.

Fig. 32-4-2 (A) Graphics pucks and styli. (Summagraphics). (B) Application. (Cascade Graphics Development)

The commands may appear on the screen or, more likely, be placed directly on the tablet. The main purpose of all menus is for rapid data conversion. A symbol, for example, may be generated from a menu by the use of the stylus or puck. This is accomplished by pointing to the menu item as shown on Fig. 32-4-3 and touching down (digitizing). Next, signal the desired position on the drawing area of the graphics tablet as shown in Fig. 32-4-2B. The result is the item or symbol displayed on the CRT screen. The process is repeated until the drawing, as displayed on the CRT, is finished.

Fig. 32-4-3   Making a selection from the command menu. (Cascade Graphics Development)

media. Because of the digital movement, a plotter is considered a vector device. The main use of the plotter is to produce an image that already appears on the CRT graphic display. This can be any combination of lines and alphanumerics. If a CAD system is an automated drafting machine, the plotter is the part replacing laying lead.

Ink pens normally are used to produce a permanent copy of a drawing. Various types of pens can be used, such as wet ink, felt tip, or liquid ball. They may be a single color or multicolored. The pens will draw on various types of media such as vellum and Mylar. The drawings produced are of high quality, uniform, precise, and expensive. Although there are faster output devices, it is the most dominant means of producing a final CAD drawing.

PRINTER

A printer is an output device that duplicates the CRT display quickly and conveniently. The primary advantage of the technique is that of speed. It produces output much more quickly

PEN PLOTTER

A line-type digital plotter is an electromechanical graphics output device. These units, shown in Fig. 32-4-5 and 32-4-6, are capable of two-dimensional movement between a pen and paper

Fig. 32-4-4   Joystick. (Cascade)

Fig. 32-4-5   Flatbed pen plotter. (Bausch & Lomb, Inc.)

Fig. 32-4-6 Microgrip pen plotter. (Auto-trol Technology Corp.)

the information is transmitted directly from one data base to another data base. The NC tape is actually the drawing on a different medium.

CAM accomplishes process planning and processing among other tasks. Process planning is the sequence of production steps. Mechanical processing has to do with tool path creation. Tool path movement is controlled by a numerical control tape. Output of the CRT is punched onto a tape which is prepared by selectively placing holes in paper like that shown in Fig. 32-4-7. The punched tapes convert the image on the screen to recorded binary coded information. This describes each desired movement of the machine. Thus, numerically controlled machining is computer-controlled machining. A machine operator is not required. The punched tape is an output of the CRT. The finished part is an output of the tape. Note that NC eliminates the need for a hard-copy engineering drawing. The drawing is stored in the data base known as a *geometrical data base.*

Modern NC equipment allows three-axis (X, Y, Z) generation or movement. This includes forward-backward, left-right, and up-down. The machines which are contolled are the common types found in industry such as lathes, drilling machines, and milling machines.

than pen plotting. Complex graphic screen displays may be copied by the touch of a button. This includes any combination of graphic and non-graphic (text and characters) display. The copy, however, does not approach the level of quality produced by the pen plotter. Thus, it is used primarily for preliminary check prints rather than final copy. It is, for example, very useful for a quick preview at various intermediate steps of a design project.

COMPUTER-AIDED MANUFACTURING

Numerical Control

Computer-aided manufacturing (CAM) uses the result of a computer-aided design. Combining CAD and CAM has had the effect of radically increasing productivity and accuracy. This is accomplished by the preparation of the design of a product using CAD. The instructions for the manufacture or preparation of that design are sent directly to the factory. One method of transmitting the information is known as *numerical control* (NC).

Numerical-control tapes and equipment can store the designs. These may be used with a variety of production-related processes. This may be accomplished without the need for an actual engineering drawing. In other words,

Fig. 32-4-7 Numerical control tape.

Robotics

The newest feature of computer-aided manufacturing is known as *robotics*. Robot machinery appears similar to the CRT display shown in Fig. 32-4-8. It differs from numerical-control machinery in that movement is now the prime duty. Automatic manipulators are used to perform a variety of materials-handling functions. The robot manipulators are arms and hands. They will grasp, operate, assemble, and handle. This is especially useful in environments intolerable to human beings. They may be delicate enough to pick up an egg, yet strong enough to exert a great force on large steel products.

As stated earlier, NC equipment has three-axis movement. Robots have the ability to move about in the X, Y, or Z direction. This includes six degrees of freedom. A degree of freedom is a constrained or unconstrained movement, for example, the ability to turn or rotate. On a robot these are known as waist, shoulder, elbow, wrist rotation, wrist bending, and flange. These modes allow essentially any positioning to a target position. A simplistic version of this control is similar to the two-axis (XY) pen-plotter output. Robots also may be controlled by digitizing or a light pen. An instruction will be sent to the arm.

The ultimate aim of many in industry is to organize every step of manufacturing around computer automation. When this occurs, the factory will be controlled by personnel in the engineering and design office. They will have the ability both to design and produce products. This creates the potential for the totally automated industrial facility.

Fig. 32-4-8 Robot display on a CRT. (Evans & Southerland Computer Corp.)

UNIT 32-5
Developing a Drawing by Use of a Grid System

The remaining units in Chap. 32 will cover the procedures required to produce a basic engineering drawing using CAD. Each procedure will be referred to as a command or function. Many are actually not commands but are objects of commands. The term *command*, however, is often used to describe any action. The result of a function or command is universal. The procedure, however, will vary somewhat from system to system. Thus, to be consistent throughout the remaining units in this chapter, only one CAD system, the Cascade II, will be used to explain the procedures.

MENUS

The master menu for a CAD system is similar in nature to the menus found in restaurants. Where the restaurant menus list the foods under different categories—seafood, meat, wines, desserts, etc.—the CAD master menu lists the major drafting symbols and/or tasks (see Figs. 32-5-1 and 32-5-2). Each master menu item will have an auxiliary menu. This allows the selection of different types of graphics. Where required the auxiliary menus are further subdivided into secondary auxiliary menus (see Fig. 32-5-3).

CGD ©1982 MASTER MENU

SELECT FROM MASTER MENU

50 USER MENU	67 FASTENERS	84 DIMENSION
51 TYPEWRTR	68 ALIGNMENT	85 LINE WIDN
52 TEXT	69 WELD SYMBOLS	86 DRILL HOLES
53 LINES	70 CREA GRP	87 SPRINGS
54 ARCS	71 FILE GRP	88 ZOOM
55 FILLETS	72 GET GROUP	89 PAN
56 FLOW LINE	73 GRP LIBED	90 FULLWINDW
57 POLYARC	74 GRP EDIT	
58 CONST LINES	75 PLACE GRP	
59 CURVES	76 DIG/MENST	93 PLOT
60 MOVCPYDEL	77 CROSHATCH	94 DEACTIVAT
61 ARC EDIT	78 AREA/PERM	95 FAST PLOT
62 ANSI SYM	79 DRAW DEF	96 CNVT DRAW
63 ROT SCALE	80 TEXT EDIT	
64 COPY REPT	81 EASY EDIT	98 DIAGNSTIC
65 LINE EDIT	82 PROPERTY	99 REDRAW
66 BALLOONS	83 MECH DIM	

Fig. 32-5-2 CAD master menu. (Cascade Graphics Development)

To display the master menu on the monitor simply type a two-key code on the keyboard of the computer console.

To activate or display an auxiliary menu, simply enter the two-digit number preceding the desired task shown on the master menu.

For example, suppose the CAD operator wishes to add center lines to a drawing being worked on. To do this the following commands are given.

● Type 53

(See master menu.) This activates and displays the auxiliary menu LINES. Center lines do not appear on this menu but are covered in a secondary auxiliary menu 12 LINE-STYLE (Fig. 32-5-3).

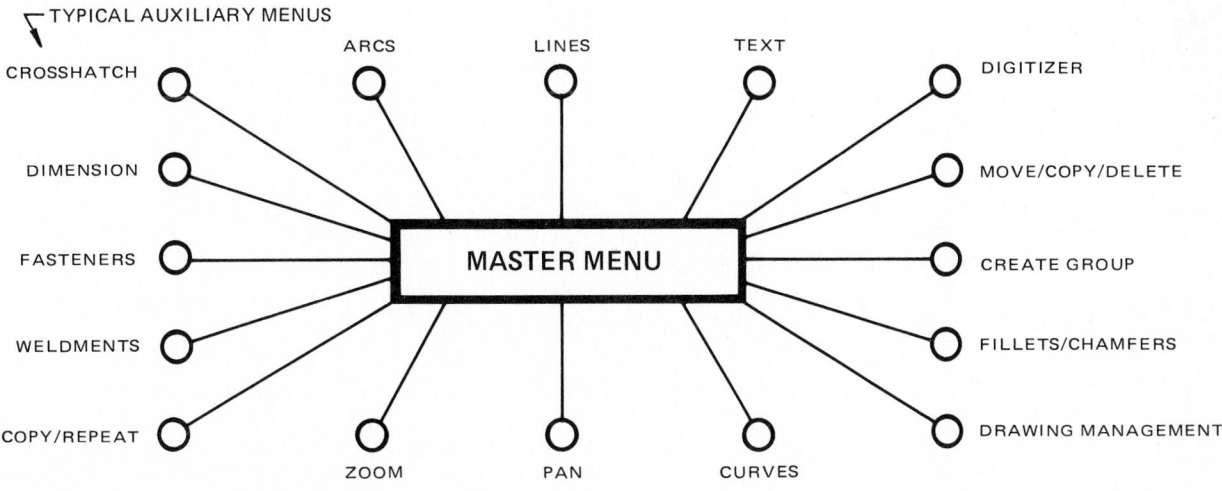

Fig. 32-5-1 Available graphics from master menu.

TYPICAL AUXILIARY MENUS

CROSSHATCH · ARCS · LINES · TEXT · DIGITIZER

DIMENSION · MOVE/COPY/DELETE

FASTENERS · MASTER MENU · CREATE GROUP

WELDMENTS · FILLETS/CHAMFERS

COPY/REPEAT · DRAWING MANAGEMENT

ZOOM · PAN · CURVES

CGD ©1982 GRID NON-AXIS L 1 P 1 I 3

PICK START OF LINE

* * LINESTYLE * *
11 SOLID LINE
12 LINESTYLE

15 NEW START

38 MULTI-SEGMENT

24 DRAW TO POINT   41 RECTANGLE
25 MOVE TO POINT   42 PARALLELOGRAM

CGD ©1982 GRID NON-AXIS L 1 P 1 I 3

ENTER LINESTYLE

12 DASH
13 CENTER
14 PHANTOM
15 SOLIDRAW

18 REFERENCE

20 EXIT

AUXILIARY MENU
53 LINES

SECONDARY AUXILIARY MENU
12 LINESTYLE (FROM AUXILIARY MENU 53 LINES)

Fig. 32-5-3   Auxiliary and secondary auxiliary menus. (Cascade Graphics Development)

- Type 12   The menu LINE-STYLE now appears on the monitor. The number preceding the line CENTER is 13.

- Type 13   Before any center lines can be drawn, the system must be returned to the auxiliary menu LINES. Select item 20 EXIT to return to the auxiliary menu.

- Type 20   This returns us to the auxiliary menu 53. The system is now programmed to draw center lines. The instruction on top of the screen reads PICK START OF LINE. In order to do this the drawing must now be displayed on the screen.

- Type ESC/G This displays the drawing. Select the starting point by moving the stylus across the graphics input tablet. The end point of the stylus appears as a cursor (a mobile +) on the screen. When the cursor is in position

where the center line should start, press down on the stylus. This locks in the starting position. Now move the stylus to locate the position where the center line should stop. When the cursor is in position, press down on the stylus. This locks in the end of the center line. The center line will then appear on the drawing. To draw the next center line simply press down on the stylus where the start and end of the next center line should appear. Continue this sequence until all the required center lines are drawn.

SELECTING GRID SIZE

Auxiliary menu 79 DRAWING DEFINITIONS allows the operator to define the grid size and select other drawing parameters such as drawing size, scale, and hit window size.

When designing a drawing using the grid system, the designer must first decide on whether the grid should be

in metric or in U.S. Customary units (inches) and the size of grid to be selected.

Once the grid size is determined, the grid may be displayed or removed from the CRT by typing / (slash) and G on the keyboard or by pressing the stylus on the grid sensing element on the command menu of the graphics input tablet.

STYLUS INPUT

To locate a desired point for stylus input, move the stylus across the tablet without pressing the tip and observe the movement of the cursor on the graphics screen. When the cursor is at the desired location, simply press the stylus to input the point coordinates.

Four stylus input modes are available. The user has a choice of either grid pick or free pick in conjunction with either axis lock or non-axis lock. For example, a line is to be drawn between the two cursor marks shown in Fig. 32-5-4A. In grid pick–non-axis lock, stylus inputs are "snapped" to the closest grid point (Fig. 32-5-4B); whereas in free pick–non-axis lock, stylus inputs remain exactly where they are placed (Fig. 32-5-4C). In the grid pick–axis lock mode, stylus inputs are snapped to the closest X or Y axis (Fig. 32-5-4D); whereas in free pick–axis lock mode, this does not happen (Fig. 32-5-4E). In both cases, however, the axis lock mode will produce

(A) STYLUS INPUT FOR LINE TO BE DRAWN

RESULT
(1) Snap to grid point.
(2) Horizontal, vertical, and inclined lines.

(B) GRID PICK—NON-AXIS LOCK

RESULT
(1) Points chosen anyplace on screen.
(2) Horizontal, vertical, and inclined lines.

(C) FREE PICK—NON-AXIS LOCK

RESULT
(1) Snap to grid point.
(2) Horizontal or vertical lines only.

(D) GRID PICK—AXIS LOCK

RESULT
(1) Points chosen anyplace on screen.
(2) Horizontal or vertical lines only.

(E) FREE PICK—AXIS LOCK

Fig. 32-5-4   Stylus input modes.

(A) DRAWING TO BE MADE

(B) DRAWING FIRST LINE

(C) DRAWING SECOND LINE

(D) DRAWING THIRD LINE

Fig. 32-5-5   Developing the drawing by use of a grid system.

an exact horizontal or vertical line (whichever is closest).

The two-digit stylus input mode numbers, shown below, can be entered in response to typing the two-digit number shown below. Other systems may vary. Note that these modes work only in conjunction with the stylus; they do not adjust absolute, relative, or polar coordinate values entered via the keyboard.

The modes currently in effect appear on the top line of the auxiliary menu screen.

Stylus Input Modes	Number
Grid pick	00
Free pick	01
Non-axis lock	03
Axis lock	04

An example of drawing a part on a CAD system is shown in Fig. 32-5-5. The two-view sketch shown in Fig. 32-5-5A shows that the drawing is in inches and that a grid size of 1 in. would place all the lines on the grids. The stylus input modes selected would be grid pick–axis lock.

Before any graphical data can be displayed on the CRT screen, the system must be turned on using the manufacturer's start-up procedure, a new or existing drawing stored in the drawing file must be logged into the computer, and the graphics tablet must be digitized.

By selecting auxiliary menu 53 LINES, 11 SOLID LINE is automatically selected (see Fig. 32-5-3) and the placing of lines on the CRT can now begin. Move the stylus across the drawing area surface of the graphics input tablet until the cursor is at or near point 1, starting point A. Press down on the stylus to lock in the start of the line. Next locate point 2 (5 grid spaces of 1 in.) to the right of point 1 and press down on the stylus. A solid line will show on the screen joining points 1 to 2. Next locate point 3, which is 2 spaces (2 in.) above point 2, and press down on the stylus. Line segment 2-3 will appear. Continue until the front view is complete. The lines joining points 1 to 6 to 1 is defined as one line bent in different directions. Note that in the grid pick–axis lock mode the cursor need only be near the grid dot and not exactly on it, which enables the operator to draw the line segments very quickly.

To start a new line, the computer must be given the command 15 NEW START. This is accomplished by pressing the stylus down on square 15 of the command menu on the graphics input tablet, or by typing 15 on the keyboard. The start of the second line is located 1 space (1 in.) above point 6 of line 1. Locate line segments 7-8, 8-9, 9-10, and 10-7 in the same manner as described for line 1-6. Before starting the third line press the stylus down on square 15. The start of the third line is point 11 located on line segment 7-8. Locate point 12 and complete the line.

If desired, the grid can now be removed from the CRT by typing /G.

ASSIGNMENT

See Assignment 1 for Unit 32-5 on page 691.

UNIT 32-6
Developing a Drawing by Coordinate Input

Point coordinates that are entered via the keyboard can be specified using relative, absolute, or polar coordinates. Relative (point-to-point) coordinates are relative to the last coordinates that were picked or typed. Absolute coordinates are always relative to the drawing origin. Polar coordinates are referenced to the last input coordinates, where the distance is a given length and the angle is expressed in degrees and measured counterclockwise relative to the X axis.

RELATIVE COORDINATE INPUT

1. Pick reference point (drawing origin $X = 0$, $Y = 0$) with stylus.
2. Type the first set of coordinates, which must be enclosed in parentheses: (X, Y).
3. As necessary, continue to enter coordinates.

(A) STARTING FIRST LINE

(B) STARTING SECOND AND THIRD LINES

RELATIVE COORDINATE INPUT
1. Pick point A with stylus (coordinates 0,0)
2. Type (5,0)
3. Type (0,2)
4. Type (-2,0)
5. Type (0,1)
6. Type (-3,0)
7. Type (0,-3)
8. Type or digitize 15 (NEW START)
9. Type (0,1) point B
10. Type (5,0)
11. Type (0,2)
12. Type (-5,0)
13. Type (0,-2)
14. Type or digitize 15 (NEW START)
15. Type (3,0) point C
16. Type (0,2)
17. Type or digitize 15 (FINISH)

ABSOLUTE COORDINATE INPUT
1. Pick point A with stylus (coordinates 0,0)
2. Type [5,0]
3. Type [5,2]
4. Type [3,2]
5. Type [3,3]
6. Type [0,3]
7. Type [0,0]
8. Type or digitize 15 (NEW START)
9. Type [0,4] point B
10. Type [5,4]
11. Type [5,6]
12. Type [0,6]
13. Type [0,4]
14. Type or digitize 15 (NEW START)
15. Type [3,4] point C
16. Type [3,6]
17. Type or digitize 15 (FINISH)

Fig. 32-6-1   Developing a drawing (Fig. 32-5-5) by coordinate input.

ABSOLUTE COORDINATE INPUT

1. Type the absolute coordinates enclosed in brackets: [X, Y].
2. Type the second set of coordinates enclosed in brackets: [1.5, 2.5].
3. As necessary, continue to enter coordinates.

POLAR COORDINATE INPUT

1. Pick a reference point with the stylus.
2. Type the distance and the angle surrounded by a less-than symbol and a greater-than symbol: <3.875, 45D>. The D after the angle indicates degrees. If it is omitted, radians are assumed. To indicate 45° 30′ 10″ type 45D30M10S.

The same drawing shown in Fig. 32-5-5 can be developed by relative or absolute coordinates as shown in Fig. 32-6-1.

ASSIGNMENTS

See Assignments 2 to 7 for Unit 32-6 on page 691.

UNIT 32-7
Circles and Arcs

Menu 54 ARCS provides two methods for constructing arcs and circles; see Fig. 32-7-1. Arcs are always drawn clockwise. For many drawings it may be desirable to place the center lines on the CRT first.

CIRCLES

Method 1: Using Center and Radius (Fig. 32-7-2A)

When menu 54 is selected, line style 11 SOLID, 18 CIRCLE, and 24 CENTER/RADIUS parameters are automatically selected as these are most commonly used. Line styles may be changed by keying in the number preceding the desired line style.

CGD ©1982 GRID NON-AXIS L 1 P 1 I 3

PICK CENTER OF CIRCLE OR REJECT

* * LINESTYLE * *
11 SOLID
12 DASH
13 CENTER
14 PHANTOM

17 ARC
18 CIRCLE
19 CON CIRC

24 CENTER/RADIUS
25 DIAMETER ENDPOINTS

(A) AUXILIARY MENU: CIRCLES

CGD ©1982 GRID NON-AXIS L 1 P 1 I 3

PICK ORIGIN OF ARC OR REJECT

* * LINESTYLE * *
11 SOLID
12 DASH
13 CENTER
14 PHANTOM

16 DIRECTION
17 ARC
18 CIRCLE
19 CON CIRC

21 NEW CENTER

24 CENTER/RADIUS

(B) SECONDARY AUXILIARY MENU: CONCENTRIC CIRCLES

CGD ©1982 GRID NON-AXIS L 1 P 1 I 3

PICK ORIGIN OF ARC OR REJECT

* * LINESTYLE * *
11 SOLID
12 DASH
13 CENTER
14 PHANTOM

16 DIRECTION
17 ARC
18 CIRCLE
19 CON CIRC

24 3-POINT ARC
25 START/END/RADIUS

(C) SECONDARY AUXILIARY MENU: ARCS

Fig. 32-7-1 Menu 54. Circles and arcs.

(A) CENTER AND RADIUS

(B) DIAMETER ENDPOINTS

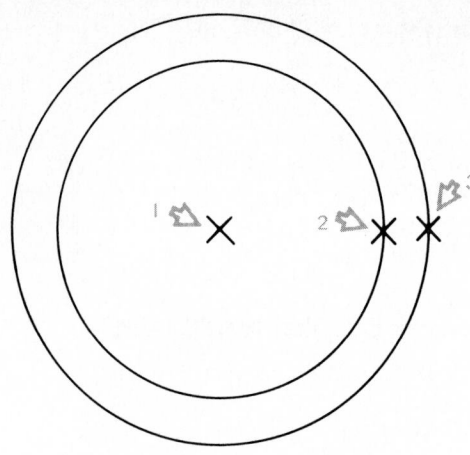

(C) CONCENTRIC CIRCLES

Fig. 32-7-2 Drawing circles.

1. Pick center of circle. Move the stylus across the drawing area surface of the graphics input tablet until the cursor is positioned where the center of the circle is to be located, and press down on the stylus.
2. Pick point on circle. Move the stylus until the cursor rests on any point on this new circle radius and press down on the stylus. The circle will appear on the CRT.

Method 2: Using Diameter Endpoints (Fig. 32-7-2B)

Type 25 DIAMETER ENDPOINTS (Fig. 32-7-1A).

1. Pick point 1, one end of the diameter.
2. Pick point 2, the opposite end of the diameter. The circle will appear on the CRT.

Method 3: Drawing Concentric Circles (Fig. 32-7-2C)

Type 19 CON CRC (Fig. 32-7-1A). A secondary auxiliary menu appears on the CRT.
Type 24 CENTER/RADIUS (Fig. 32-7-1B).

1. Pick center point.
2. Pick point (2) on diameter of first circle. The circle will appear on the CRT.
3. Pick point (3) on diameter of second circle. The second circle will appear on the CRT.

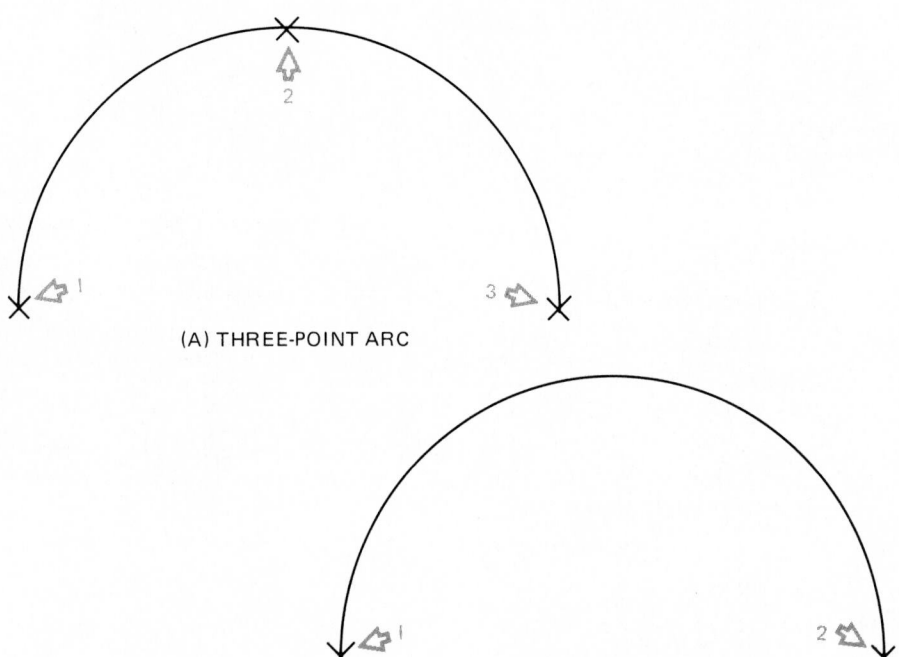

(A) THREE-POINT ARC

(B) START/END/RADIUS

Fig. 32-7-3 Drawing arcs.

ARCS

Method 1: Three-Point Arc (Fig. 32-7-3A)

Type 17 ARC (Fig. 32-7-1A). A secondary auxiliary menu appears on the CRT.
Type 24 3-POINT ARC (Fig. 32-7-1C).

1. Pick an endpoint.
2. Pick a second point.
3. Pick the other endpoint. The arc will appear on the CRT.

Method 2: Start/End/Radius (Fig. 32-7-3B)

Type 17 ARC (Fig. 32-7-1A). A secondary auxiliary menu appears.
Type 25 START/END/RADIUS (Fig. 32-7-1C).

1. Pick an endpoint.
2. Pick the other endpoint.
3. Type RADIUS. The arc will appear on the CRT.

ASSIGNMENT

See Assignment 8 for Unit 32-7 on page 693.

UNIT 32-8
Fillets and Chamfers

Menu 55 FILLETS/CHAMFERS allows the user to add fillets and chamfers to a drawing. See Fig. 32-8-1. Since the computer is programmed to place an arc between two intersecting lines, the term *fillet* in CAD refers to both fillets (inside arcs) and rounds (outside arcs) in manual drafting.

FILLETS

Select menu item 10 FILLETS to draw an arc that is tangent to two picked lines. To specify any radius for the arc, select menu item 24 from the auxiliary menu shown in Fig. 32-8-1A. Type in the desired radius and press RETURN. Select menu item 29 to erase the lines between the intersection point of the picked lines and the endpoints of the arc. Select menu item 30 to erase the lines between the endpoints of the arc and the nonintersecting endpoints of the picked lines.

EXAMPLE 1. (Fig. 32-8-2A)

Type 24   RADIUS.

1. Type radius value (press RETURN).

Type 29   TRIM A.
Type 10   FILLETS.

2. Pick first line.
3. Pick second line.

(Repeat steps 2 and 3 for five additional sets of lines.)

Type 42   CONTINUE.

4. Result of task (Fig. 32-8-2B).

EXAMPLE 2 (Fig. 32-8-3A)

Type 24   RADIUS.

1. Type radius value (press RETURN).

Type 30   TRIM B.
Type 10   FILLETS.

2. Pick first line.
3. Pick second line.

Type 42   CONTINUE.

4. Result of task (Fig. 32-8-3B).

CGD ©1982 GRID NON-AXIS L 1 P 1 I 3

PICK FIRST LINE

10 FILLETS   ** TRIM **
11 CHAMFERS   28 NO TRIM
                              29 TRIM A
                              30 TRIM B

24 RADIUS=0.1250   42 CONTINUE

(A) AUXILIARY MENU: FILLETS

CGD ©1982 GRID NON-AXIS L 1 P 1 I 3

PICK FIRST LINE

10 FILLETS   * DISTANCE ALONG *
11 CHAMFERS   28 1ST LINE=0.0000
                              29 2ND LINE=0.0000

24 ANGLE=45   42 CONTINUE

(B) SECONDARY AUXILIARY MENU: CHAMFERS

Fig. 32-8-1   Menu 55: Fillets and chamfers.

(A) EXISTING DRAWING

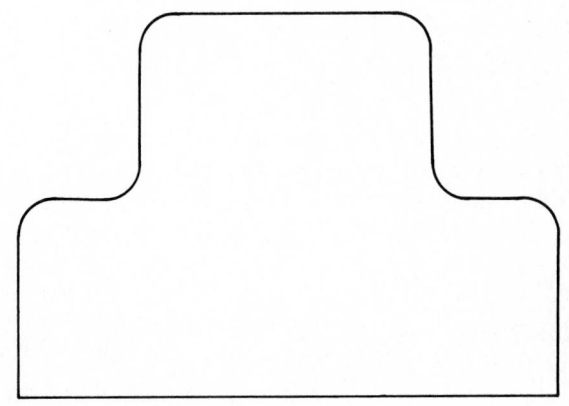

(B) RESULT OF TASK

Fig. 32-8-2   Adding fillets and rounds to a drawing.

(A) EXISTING DRAWING

Fig. 32-8-3 Adding a fillet to a drawing.

(B) RESULT OF TASK

Type 11 CHAMFERS.

3. Pick first line.
4. Pick second line.

Type 42 CONTINUE.

5. Pick first line.
6. Pick second line.

Note: Order of picks is important.

Type 42 CONTINUE.

7. Result of task.

CHAMFERS

Select menu item 11 CHAMFERS to construct a chamfer on a right angle. The secondary auxiliary menu shown in Fig. 32-8-1B appears on the CRT. The selected lines cannot be members of a group and must form a right angle (90°). Select menu items 28 and 29 to enter the distance on the picked lines. Select menu item 24 to change the angle of the chamfer relative to the first line picked. This angle will be effective only if the distance along one of the picked lines is set at zero. If both angles are nonzero, this angle will be determined by the given distances.

EXAMPLE 1 (Fig. 32-8-4A)

Type 11 CHAMFERS (Fig. 32-8-1).
  A secondary auxiliary menu appears.
Type 24 ANGLE.

1. Type angle value 30.

Type 28 1ST LINE.

2. Type line length value of 10 mm.

Note: The second line length value should be set to zero.

Type 11 CHAMFERS.

3. Pick first line.
4. Pick second line.

Type 42 CONTINUE.

5. Pick first line.
6. Pick second line.

Note: Order of picks is important.

Type 42 CONTINUE.

7. Result of task.

EXAMPLE 2 (Fig. 32-8-4B)

Type 11 CHAMFERS (Fig. 32-8-1).
  A secondary auxiliary menu appears.

Type 28 1ST LINE.

1. Type line length value of 10 mm.

Type 29 2ND LINE.

2. Type line length value of 5 mm.

EXISTING DRAWING

RESULT OF TASK

(A) GIVEN ANGLE PLUS LENGTH OF ONE SIDE

EXISTING DRAWING

RESULT OF TASK

(B) GIVEN LENGTHS OF BOTH SIDES

Fig. 32-8-4 Adding chamfers to a drawing.

ASSIGNMENT

See Assignment 9 or Unit 32-8 on page 693.

ASSIGNMENTS for Chapter 32

Assignment for Unit 32-5
Developing a Drawing by Use of a Grid System

1. Using the commands given in Unit 32-5 lay out the sequence of commands or functions required to produce a three-view drawing of an appropriate grid for any three parts shown in Fig. 32-5-A. Use graph paper and a different-colored pencil to show each of the plotted lines. Indicate the starting point and direction of each line. Allow one grid space between views.

Assignments for Unit 32-6
Developing a Drawing by Coordinate Input

2. Select one of the parts shown in Fig. 32-6-A and plot the relative coordinates of each of the line intersections. The bottom left-hand corner is the drawing origin or starting point. Move in a clockwise direction. Scale: One grid square equals 1 in. or 25 mm.
3. Same as Assignment 2 except use absolute coordinates.
4. Select any two problems shown in Fig. 32-5-A and lay out the sequence of commands required to produce a three-view CAD drawing. Scale: One grid square equals 1 in. or 25 mm. Allow one grid space between views and use relative coordinate dimensioning. The bottom left-hand corner of the front view is the drawing origin or starting point.
5. Same as Assignment 4 except use absolute coordinates.
6. Using absolute coordinates plot Figs. 32-6-B and 32-6-C on a B- or A3-size sheet. The bottom left-hand corner of the drawing is the starting point.
7. Using relative coordinates plot Figs. 32-6-D and 32-6-E on a B- or A3-size sheet. The bottom left-hand corner of the drawing is the starting point.

FRONT

FRONT

FRONT

FRONT

FRONT

FRONT

Fig. 32-5-A   Drawing assignment.

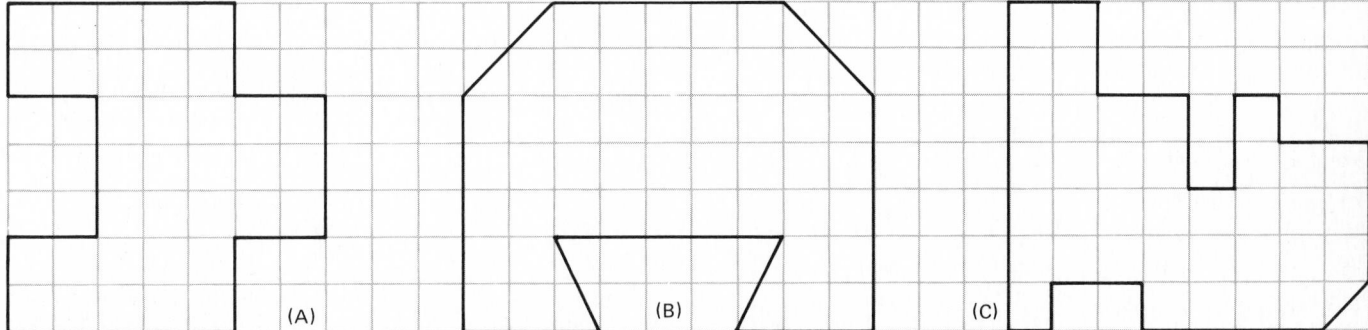

(A)   (B)   (C)

Fig. 32-6-A   Drawing assignment.

ABSOLUTE COORDINATES (INCHES)

Point	X Axis	Y Axis
1	0	0
2	7.00	0
3	8.50	1.00
4	8.50	2.25
5	7.00	3.00
6	8.50	3.75
7	8.50	5.25
8	7.00	6.25
9	5.50	6.25
10	5.50	4.50
11	4.75	4.50
12	4.75	6.25
13	3.50	6.25
14	3.50	5.50
15	1.50	5.50
16	1.50	6.25
17	1.50	6.25
18	.75	6.25
19	0	5.5
20	0	0
New Start		
21	.75	1.50
22	2.25	3.50
23	.75	3.50
24	.75	1.50
New Start		
25	3.00	2.25
26	5.75	2.25
27	5.75	3.75
28	3.00	3.75
29	3.00	2.25
New Start		
30	2.75	.75
31	6.00	.75
32	5.25	1.50
33	3.50	1.50
34	2.75	.75

Fig. 32-6-B   Absolute coordinates, inches.

ABSOLUTE COORDINATES (METRIC)

Point	X Axis	Y Axis
1	0	0
2	50	0
3	50	20
4	120	20
5	120	0
6	150	0
7	180	30
8	220	30
9	220	100
10	160	100
11	160	130
12	140	130
13	140	160
14	110	160
15	120	140
16	90	140
17	70	100
18	40	120
19	60	160
20	40	160
21	0	140
22	0	80
23	20	40
24	0	40
25	0	0
New Start		
26	40	50
27	160	50
28	120	90
29	40	70
30	40	50

Fig. 32-6-C   Absolute coordinates, metric.

RELATIVE COORDINATES (INCHES)

Point	X Axis	Y Axis
1	0	0
2	4.50	0
3	0	.75
4	−.75	0
5	0	.75
6	−.75	0
7	0	.75
8	−3.00	0
9	0	−2.25
New Start—Solid		
10	0	.75
11	3.75	0
New Start		
12	−.75	.75
13	−3.00	0
New Start—Solid		
14	0	1.50
15	4.50	0
16	0	2.25
17	−4.50	0
18	0	−2.25
New Start—Solid		
19	0	.75
20	3.75	0
21	0	1.50
New Start—Solid		
22	−.75	0
23	0	−.75
24	−3.00	0
New Start—Solid		
25	5.25	−4.50
26	2.50	0
27	0	2.25
28	−1.00	0
29	0	−.75
30	−.75	0
31	0	−.75
32	−.75	0
33	0	−.75
New Start—Solid		
34	.75	.75
35	1.75	0
New Start—Solid		
36	0	.75
37	−1.00	0

Fig. 32-6-D   Relative coordinates, inches.

RELATIVE COORDINATES (METRIC)

Point	X Axis	Y Axis
1	0	0
2	40	0
3	0	20
4	40	0
5	0	− 20
6	40	0
7	0	60
8	− 20	0
9	0	− 20
10	− 80	0
11	0	20
12	− 20	0
13	0	− 60
	New Start—Solid	
14	140	0
15	60	0
16	0	60
17	− 60	0
18	0	− 60
	New Start—Solid	
19	− 140	80
20	120	0
21	0	60
22	− 120	0
23	0	− 60
	New Start—Solid	
24	20	0
25	0	60
	New Start—Solid	
26	80	C
27	0	− 60
	New Start—Hidden	
28	− 10	0
29	0	60
	New Start—Hidden	
30	− 40	0
31	0	− 60
	New Start—Hidden	
32	100	− 40
33	60	0
	New Start	
34	0	− 20
35	− 60	0

Fig. 32-6-E Relative coordinates, metric.

Assignment for Unit 32-7
Circles and Arcs

8. Using the commands given in Units 32-5 to 32-7 lay out the sequence of commands or functions required to produce a three-view drawing on an appropriate grid for one of the parts shown in Fig. 32-7-A or 32-7-B. Use graph paper and different-colored pencils to show each of the plotted lines. Indicate the starting point and direction of each line. Allow one grid space between views.

Assignment for Unit 32-8
Fillets and Chamfers

9. Using the commands given in Units 32-5 to 32-8 lay out the sequence of commands or functions required to produce a two-view drawing on an appropriate grid for one of the parts shown in Fig. 32-8-A or 32-8-B. Use graph paper and different-colored pencils to show each of the plotted lines. Indicate the starting point and direction of each line. Allow two grid spaces between views.

Fig. 32-7-A Link.

Fig. 32-7-B Cradle bracket.

Fig. 32-8-A Shaft support.

Fig. 32-8-B Bracket.

CHAPTER 33

Design Concepts

UNIT 33-1
The Design Process

The purpose of any design department is to create a product which not only will function efficiently but also will be a financial success. Although most designs are more complicated than the examples in this chapter, the main steps in designing a product follow a similar pattern.

In the design process, considerations should be given to each of the steps shown in Fig. 33-1-1. The design can be treated as a process in which the input is the problem and the output is the solution.

A detailer may work from a complete set of instructions such as a complete assembly drawing or may have a free hand in the design of the part. If a detailer extracts the information from a complete assembly drawing, then many of the decisions have already been made by the designer or engineer. If the designer or engineer has made the decisions, there are several factors the detailer must consider before starting the detail drawing. Normally, the final design is a compromise of many factors.

THE ENGINEERING APPROACH TO SUCCESSFUL DESIGN

Each field of engineering has its techniques and rules—and its standards for the use of the construction materials peculiar to that field.

The steps from idea to production are based on logical and well-known design principles. These principles apply equally to the manufacture of gears, optical systems, industrial components, consumer articles, or rockets—and regardless of the material of construction. These steps are not necessarily in order, but all are essential for a successful application.

Defining the End-Use Requirements

As an initial step, the product designer must anticipate the conditions of use and the performance requirements of the product. Consideration must be given to such things as environment, load, speed of production, life expectancy, optimum size, maintenance, shape, color, strength, and stiffness.

These end-use requirements can be ascertained through market analyses, surveys, examinations of similar products, testing, general experience, and, frequently, material suppliers.

A clear definition of product requirements leads directly to choice of the construction material.

Selecting the Material

Material is a very important factor to consider in designing a part. Perhaps plastics are the better choice of material over wood or metal. Would one choice of metal be better than another? What elements will come into contact with water? If the part is to be immersed in an oil solution, will the choice of plastics be minimized or ruled out? Is strength a factor? If so, what materials will meet the stresses required? What material is in stock or easily obtainable? Is the material *the* correct choice if a plating or coating is required?

There are thousands of engineering materials available, yet no single one will exhibit all desired properties in their proper relationships. Therefore, a compromise among properties, cost, and manufacturing process determines the construction material. Even within one series, materials differ because of varying formulations. Just as steel compositions vary—tool steel and

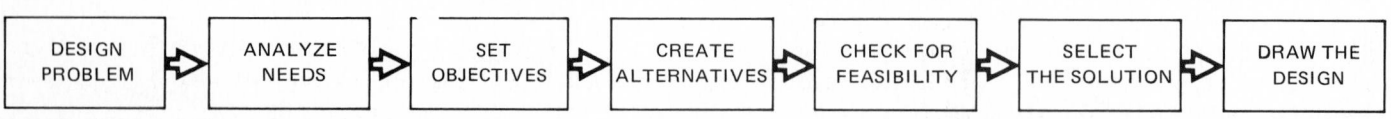

Fig. 33-1-1 Steps in the design process.

stainless steel, for example—so do the plastics.

The designer needs a firm set of properties and engineering data upon which to base the first design. The data can come from handbooks or, more likely, from the published literature provided by the manufacturers of materials.

Drafting the Preliminary Design
The designer blends the end-use requirements and the properties of the selected material into a preliminary design.

Engineering techniques and formulas are used to achieve the three requirements of design success:

- Economic feasibility
- Functional feasibility
- Attractive appearance

The production method to be used will often set limitations on design. The designer should be aware of the strengths and weaknesses of the method selected. The material supplier and the processor, with their experience in hundreds of applications, can assist here.

Prototyping the Design
The prototype is the opportunity for the designer to see the product as a three-dimensional object. This, too, is the first opportunity for checking the engineering design. The quality of the prototype is quite important. The method used in producing the prototype may not be the same as that planned for the final production line, but the design must be identical to that expected on the production line—otherwise tests may be misleading and analysis false.

If the search for the right material has been narrowed to only two or three, prototyping will help spotlight one.

Testing the Design
Every design should be given an actual or simulated service test while in the prototype stage to ensure that the obvious is not overlooked and that the not-so-obvious is taken into account. The end-use requirements dictate the design testing program. An engine part might be given temperature, vibration, and hydrocarbon-resistance tests; a

luggage fixture might be subjected to abrasion tests; and a toaster knob might be checked for electrical and heat insulation. Other tests, such as field testing or consumer reactions, are part of the necessary procedure for completely evaluating any design.

Taking a Second Look
The second look at the design provides an answer to the basic question, "Is the product doing the right job at the right price?"

At this point, most products can be improved by redesigning for better production economies or for important functional or aesthetic changes. Weak sections can be strengthened, colors changed, and new features added. Substantial changes in design will require retesting. Now is the time to set up production. The first step is to write the specification.

Writing Meaningful Specifications
The purpose of the specification is to eliminate any variations in the product that will not satisfy the functional, aesthetic, or economic requirements. The specifications are a complete set of written requirements which the part must meet. The specifications for the part should include such things as the material of construction by brand and generic name, method of fabrication, dimensions, color, surface finish, packaging, printing, and every other detail of production to which there could be more than one answer.

Setting Up Production
How many parts are required? When a large quantity is required, the number of methods of producing the parts increases. Perhaps a casting, a forging, or a stamping may be the most sound choice. If only a few parts are required, then prefabricating or machining may be the better choice.

Production Should the part be manufactured in the plant, or sent out to be produced? In many cases company policy may be to produce the part within the plant. If this is the case, then the production choices are limited to the methods available within the plant.

After the specification is written but before the production line can start, tooling must be designed, built, and

integrated with the processing equipment. (In some cases, dies and molds can be started while testing is in progress.)

Production efficiency and economy can be realized through proper design of tools. The processor is an important source of aid in this area.

Time Factor
In some instances, such as when a breakdown of a machine is holding up production within the plant, the best method of producing a part may take second place if it involves too much time.

Workforce
This factor ties in with time. A machine breaks down at 2:00 P.M. It is essential that the machine be back in operation by 8:00 A.M. the following day. What personnel are available to produce the part, with overtime, or is there a night shift?

Controlling Quality
Good inspection practice requires a checklist to maintain a consistently good product. The inspection checklist, for the most part, will conform to the end-use requirements set forth in the specifications.

Here, too, it is beneficial to consult with the supplier or the molder, who knows the processing and finishing characteristics of the material chosen.

PART SPECIFICATIONS
All material applications start out as ideas in someone's mind. From this point the idea must be developed into a production item. The transition is accomplished in a series of logical steps. Ensuring the quality of the final production item starts with the writing of a set of specifications.

The specifications are a complete set of written requirements; the purpose is to ensure that the finished part will perform as intended. The scope of the specification depends on the performance required of the part. In general, as specifications become more complex, the cost of the part increases.

Let's take a look at a typical, although hypothetical, part (shown in Fig. 33-1-2). Assume that it has been developed through the steps of the

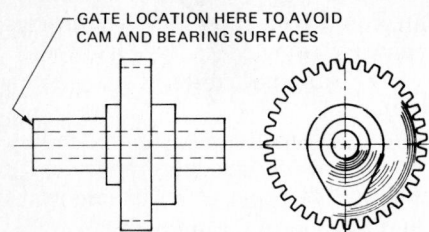

GATE LOCATION HERE TO AVOID
CAM AND BEARING SURFACES

Fig. 33-1-2 Combined gear and cam.

engineering approach and is ready for a meaningful specification.

Any good specifications should contain three basic portions: (1) the raw material, (2) the design of the part, and (3) the performance of the part in use.

Raw Material

The raw material for any part is selected for its physical properties, with due regard for economic and engineering requirements. The section of specifications dealing with raw material should be divided into two major parts: identification and quality.

Design of the Part

The second major portion of the specifications involves the design of the part. The design specification includes tolerances and surface finish. If the part is going to be cast, then parting lines, flash, gate location, and warpage must be considered.

Dimensional Tolerances The dimensional tolerances should be as close as required for functioning. When tolerances become tighter than necessary, the cost of both tooling and fabrication rises very rapidly (Fig. 33-1-3). Such tight dimensional tolerances require very close control of the processing variables and necessitate extra inspection, which contributes to a high unit cost.

Fig. 33-1-3 Cost of part increases rapidly as tolerances become smaller.

Other important considerations are:

1. Critical dimensions should be identified with specific tolerances; let overall tolerances control less important dimensions.
2. If parts are to be machined, allow generous tolerances in these areas.

Surface Finish The type and degree of surface finish required should be clearly indicated. If a highly polished finish is necessary, it should be specified. On other surfaces, finish need not be covered except in general terms. Surfaces which must be clear of imperfections such as tool marks, sinks, blisters, and flow lines should be clearly indicated.

Parting Line The location of a parting line of a mold may be influenced by flash, part appearance, and structural design. In such cases, parting-line location should be specified on the drawings. Experienced processors can assist in altering part design to simplify tooling or molding.

The parting line in the gear-cam part could be located in several places. Placing the parting line on one end of the gear simplifies the tooling requirement.

Flash Where flash is undesirable, the drawing should so indicate. The length of allowable flash may be given in a measurable dimension.

Gating The gate should be located where it will cause the least difficulty. The specification on gating should define areas to avoid, such as cam or bearing surfaces. To permit maximum concentricity in the hypothetical part, the gate should be located as shown in Fig. 33-1-2.

Warpage The allowable warpage of parts after molding should be specified, even though all dimensions may be met. When minimum warpage is desired, the cooling of parts in a jig or fixture may be necessary. It should be remembered that post-molding operations (annealing, moisture conditioning, etc.) may cause warpage.

Minimum warpage depends on a proper balance of several factors. These include uniform part thickness, location of knockout pins, and optimum molding conditions.

PERFORMANCE SPECIFICATIONS

The third major portion of the specifications concerns the quality of the part. Earlier work on design, material selection, and end-use testing has given assurance of part performance. Performance specifications are concerned with two factors: the first is visual inspection, and the second is simulated service tests.

Visual Tests This kind of test is concerned with color, weld lines, etc. The test is established by considering the effect of each of these on the final performance of the part. Not all the factors may be necessary for every part.

Simulated Service Test The second test for a part should simulate the ultimate use for the part. Care must be taken to use a meaningful test. Excessive speeds, loads, or impacts well beyond ultimate requirement can frequently cause rejection of good parts.

A simulated service test on a part like the cam-gear assembly of Fig. 33-1-2 might consist of impact and/or torque loading.

DOs AND DON'Ts FOR DESIGNERS

Designers want reliably functioning parts that are dependably procurable at the lowest installed cost. They can best meet their needs by consulting with vendors and understanding custom-metal-part manufacturing and pricing. Here is a checklist to use in the design of a part or assembly.

Don'ts

1. Don't specify tolerances tighter than essential for mechanism functioning.
2. Don't specify every dimension as mandatory; mark noncritical ones as reference only.
3. Don't specify material that is too good (too expensive) for the service.
4. Don't specify material that is available only on special purchase unless there is no alternative. If in doubt, ask your vendor.

Do's

1. Do leave adequate space for assembly, i.e., bolt clearance, finger grip, etc.
2. Do consider manufacturing economics.
3. Do consider utilizing stock items when you need only a small quantity of parts. Your savings in design time, procurement costs, and delivery time may be appreciable.
4. Do realize that for small quantities or one part, the cost of raw material is not important; material availability and minimum-quantity purchase restrictions are important.
5. Do realize that for large-quantity purchases, precise specification of raw material can be extremely important.
6. Do realize that the total cost of a custom part is not the purchase cost but the installed cost.
7. Do consider, in your product reliability, the relation between part cost, part reliability, and the cost of replacement of a broken part, including lost production time.

References and Source Material

1. E. I. duPont de Nemours & Co. (Inc.)
2. The Wallace Barnes Company Ltd.

ASSIGNMENTS

See Assignments 1 through 6 for Unit 33-1 on page 703.

UNIT 33-2
Assembly Considerations

An *assembly* is a combination of two or more parts which are joined by any of a number of different methods. A *subassembly* is made to facilitate the production of a larger assembly.

This unit describes various methods of attaching component parts to produce an assembly and some of the problems concerning assembly cost, tools, and practicability.

Although various examples of assembly methods and attachments are presented, they are not to be considered the only methods nor are they to have any preference over other means. Product design, volume, cost, and facilities are the determining factors influencing the need for an assembly.

The quality of the finished product as an assembly depends on effective attachment or fastening methods, regardless of the quality of the individual parts.

All assemblies, regardless of size, shape, or design, should be given the following considerations, since, in many cases, an analysis will dictate changes in design which will effect a cost savings.

COST OF ASSEMBLY

The following points should be carefully noted in determining the least expensive method of assembly.

Product Volume

Careful design will reduce costs of assembly at any volume level. However, many times the greatest savings are realized when the volume is high enough to justify the capital expenditure necessary for time-saving equipment that could not be justified at lower volumes. In many cases, it can be readily demonstrated that the procurement of highly specialized machinery may be justified by the increased efficiency made possible by such equipment.

Product Design

The cost of any part or assembly is the responsibility of the design engineer and engineering management. To effectively control costs, it is essential that the design, fabrication, and assembly costs be continuously foremost in the minds of all responsible personnel. It is generally true that the simpler the design, the lower the cost of producing the finished product. In considering new designs or methods, all factors involved must be taken into account in calculating the savings or increased costs, including equipment obsolescence.

There is no general solution for any assembly problem. For example, an assembly may be made in one plant following a set sequence, while in another plant it may be made quite differently—the difference being due to established plant practices, equipment and facilities, tooling fixture design, volume, and labor costs.

Methods of assembly also have an important bearing on cost. Figure 33-2-1 shows a typical cost analysis

Operation or Material	Cost Factor	Qty & Pos	COST COMPARISONS						
			Spot Welds	Proj Welds	Rivets	Arc Welds	Bolts & Nuts	Bolts & Tapping Plate	Blind Rivets
Method			1	2	3	4	5	6	7
Spotwelding	100	3B	300					200	
Projection Welding	100	3B		300					
Forming Weld Proj	89	3B		267					
Punching Hole	89	4A			356		356	356	356
Rivet	70	2A			140				
Driving Rivets	96	2A			192				192
Arc Welding (1 in.)	250	3C				750			
Bolt	115	2A					230	230	
Nut	106	2A					212		
Lockwasher	18	2A					36	36	
Assembling Bolts	136	2A					272	272	
Tapping Plate (Matl)	321	1A						321	
Drilling Hole	89	2A						178	
Tapping Hole	89	2A						178	
Blind Rivet	742	2A							1484
TOTAL COST			300	567	688	750	1106	1771	2032

The above table is for illustrative purposes only and its application should be adjusted to costs prevailing at the time of its use. Cost comparisons are based on spotwelding as Unit 100. The table is not intended to indicate that the least costly method is the best; function and strength of assembly must also be considered.

Fig. 33-2-1 Assembly methods cost analysis chart with reference to Figure 33-2-2.

Fig. 33-2-2   Bracket assembly.

covering seven possible methods for assembling a simple bracket to its carrying member, as illustrated in Fig. 33-2-2.

Sometimes methods of attachment are designed into the product without careful consideration of the economics involved. Obviously there is considerable difference in the effectiveness of the various attaching means. However, with cost comparisons at hand similar to those shown in Fig. 33-2-1, it becomes only a question of simple economics to choose the least expensive method that will do the job satisfactorily.

Ease of Assembly

The cost of assembly labor and equipment, as well as space requirements for assembly operations and equipment, is directly dependent on the ease and speed with which the assembly can be made.

Intricate assemblies require careful, slow hand fitting or expensive jigs and fixtures, or both. If the assembly is made from simple components which can be rapidly assembled, the cost will be lower.

Hole tolerances should be as liberal as is commensurate with the functional requirements of the assembly in order to facilitate the assembly operations. In many instances, a redesign may be justified to eliminate tight fits and unnecessarily close clearances which slow up the operation and make the assembly difficult. Not only should each subassembly be reviewed from this standpoint, but accessibility of all parts used to attach the subassembly to the main assembly should also be investigated to determine whether the tools normally used by the production department can reach the points of attachment.

Quality

Thought must be given to the finished appearance, the functional limitations, and the sales appeal of the completed product. Lack of attention to refinements in the assembly will otherwise completely offset the closest attention to the details.

Service

One of the most frequently heard criticisms of assembled products is the difficulty and cost of removing and replacing some minor part of the unit. Often the labor cost of replacing a bearing, gasket, or minor assembly exceeds the cost of the replaced parts. The cost of replacing parts can be minimized if consideration is given during the design period to providing for rapid and easy disassembly of functioning parts.

It must also be remembered that automobiles, trucks, industrial engines, airplanes, locomotives, etc., have to function in all types of weather, and parts may be subject to moisture and rust. Likewise, products made to handle corrosive vapors and liquids must be given special consideration, and the designer should use fastenings which will be least affected by such exposure.

ATTACHMENTS

Attaching methods used in assemblies are broadly divided into three categories: permanent, semipermanent, and quickly detachable or connectable. Each has an important function in the assembly of component parts.

Permanent Attachments

Permanent attachments include welding, brazing, soldering, riveting, peening, staking, crimping, spinning, stapling, stitching, pressing, and shrinking. Welding is the most popular because of its satisfactory attachment and because it can be accomplished by many different processes.

Welding *Welding* is the process of joining metallic parts by fusing them at their junction with heat and with or without pressure. For a more complete discussion, see Chap. 11 on welding. In considering resistance welded attachments, select electrode shapes from a clearance standpoint (Fig. 33-2-3).

Fig. 33-2-3   Resistance welding.

Brazing *Brazing* is the process of joining metallic parts by heating them at the junction points to a suitable temperature and using a nonferrous filler metal that has a melting point below that of the base metals.

Soft Soldering *Soft soldering* is the process of joining metal parts by melting into their heated joints an alloy of nonferrous metal. The silver brazing alloys, which are often called *hard solders*, have a much higher melting point and fall within the field of brazing.

Riveting *Rivets* are a permanent type of fastener for attaching parts of an assembly. The most common types are solid, blind, split, and tubular.

Solid rivets are used in assemblies that are not intended to be taken apart.

Blind rivets are designed for use where it is impossible to have access to both ends of the rivet, such as riveting a bracket to a box section. In other applications, they may also be used in place of solid rivets; however, they are more costly than solid rivets. The cost of installation time plus the unit price of both methods should be considered before a process is chosen.

Tubular rivets do not make as strong joints as solid rivets, but they can be easily installed by either a spinning or a squeezing process. Although spinning is considered more desirable from a strength standpoint, the squeezing operation is used more generally because of the simplicity of equipment.

Split rivets are somewhat limited in their applications. They are usually installed with the same type of squeezer equipment used for tubular rivets. An application of tubular and split rivets is shown in Fig. 33-2-4.

Fig. 33-2-4 Tubular and split rivet applications.

Impact riveting, known also as *peening*, is used to secure a shoulder pin or rivet in an assembly of two or more parts. Impact riveting can be used to advantage where stock thickness or hardness of parts varies; the operator can control the force and number of blows required to produce a secure assembly. In impact riveting, a round shaft is often swaged into contact with the sides of a hexagonal hole to solidly lock the pin and eliminate any possibility of rotation of the pin in the part (Fig. 33-2-5).

Fig. 33-2-5 Impact riveting.

Spin riveting results in a better head-bearing surface than impact or squeeze riveting, and has less tendency to cause shaft distortion than squeeze riveting. However, it is usually slower in operation, and the tool cost is normally much higher. Spin riveting can be used to advantage where one of the assembled parts must be free to move.

Squeeze riveting can be used to advantage in fastening two or more parts where the holes for the rivet may be slightly mismatched, as well as in true matched holes. Squeeze riveting

of pins into parts tends to distort the pin below the riveting point and, by so doing, fill the hole in the plates even though they may be slightly mismatched. This method of riveting can be done with either hot or cold rivets (Fig. 33-2-6).

Fig. 33-2-6 Squeeze riveting.

Crimping This method is used to secure two or more pieces of metal in assembly by folding over the metal of one part to squeeze or clinch the other part or parts.

In crimping, the part must be designed to allow enough extruded metal on the crimped part to fold over in complete contact with its assembly mates but without excess metal which may be forced out from under the crimping punch. Successful crimping requires a die or tool designed for the specific crimping operation, as illustrated in Fig. 33-2-7. Crimping is less expensive than riveting or welding and can be used when the metal of one part is ductile enough to allow folding over without cracking.

Stitching This method is used to secure metal to metal, fabric to metal, rubber to metal, etc., as shown in Fig. 33-2-8.

Fig. 33-2-7 Crimping.

Fig. 33-2-8 Stitching.

Press Fit The term *press fit* applies to the assembling of a part, such as a shaft, into a hole which is slightly smaller in diameter than the shaft (Fig. 33-2-9). The degree of interference depends on the size of the hole, the mass of material around the hole, and the kind and quality of material.

Fig. 33-2-9 Press fit.

Shrink Fit This method is a modification of a press fit, adapted particularly to large diameters. Diameters that would provide sufficient interference to hold the two parts together permanently could not be pressed together cold. In these cases, the ring is preheated, then slipped over the shaft or wheel and allowed to cool in place. As the ring cools, it shrinks to its normal diameter, thus producing a pressure on the shaft sufficient to hold it.

Cementing Cementing or bonding with a suitable adhesive agent is another method used in production to make permanent or semipermanent assemblies.

Semipermanent Attachments
Semipermanent attachments include threaded fastenings, such as bolts, screws, studs, and nuts as well as washers, nails, and pins. Many factors must be taken into consideration when a fastener selection is made, such as strength, appearance, permanence, corrosion resistance, materials to be

joined, cost, assembling, and disassembling.

Bolts The proper diameter for a bolt is usually determined by design requirement and controlled by the engineer or designer. The factors which govern this decision are the strength requirement of the assembled unit and the material and heat treatment of the bolt. The type of head is also determined by design requirements such as unit pressure exerted by the bolt head, space limitations, and driving torque.

Hexagonal bolts are in most general use. They have a washer face, or the underside of the head is chamfered. They may be used in a threaded hole or with a nut. A typical application of a hexagonal bolt is shown in Fig. 33-2-10.

Fig. 33-2-10 Hexagon bolt application.

Flanged hex-head bolts are usually specified where a bolt is to be used against a material that has a relatively low compressive strength, such as aluminum. The flanged head is also advantageous where an oversized hole or slotted hole is necessary.

Round-head bolts are made with variously shaped necks under the head for such specific purposes as embedding in wood or metal to prevent rotation or as a means of retention in thin metal, as shown in Fig. 33-2-11.

Square-head bolts are better adapted to heavy machinery, conveyors, and fixtures.

In selecting thread pitches, the bolt material strength and internal thread material strength must be considered, since a coarse pitch produces a stronger internal thread and a fine pitch produces a stronger external thread.

In general, coarse threads should be used in materials which have relatively low shear strength, such as castings

Fig. 33-2-11 Round-head square-neck bolt assembly.

and soft metals, and for applications requiring rapid assembly or disassembly. Fine threads should be used where fine adjustment is necessary and where thin walls may be encountered.

Studs These are sometimes called *stud bolts*. Studs have threads on both ends, to be screwed permanently into a fixed part at one end and receive a nut on the exposed end. They are made of different materials depending on their use. Normally, they are made with coarse threads on the stud end and fine threads on the nut end, as shown in Fig. 33-2-12.

Fig. 33-2-12 Stud application.

Machine Screws Machine screws generally differ from bolts in range of diameters, head shapes, and driver provisions. Their use is restricted to light assemblies such as instrument-panel mountings, moldings, and wire and pipe clips.

The *flat head* is used where a flush surface is required. The *oval head* is generally used for reasons of appearance. Other head types are used for functional reasons; for example, *pan* and *truss heads* are used to cover large clearance holes and elongated holes.

The *hexagonal heads* are preferable from a driving standpoint; however, they are not suitable for appearance in many locations. For appearance, the *cross-recess head* is popular.

Setscrews Setscrews are used extensively in tools, jigs and fixtures, control knobs, hand wheels, cam levers, and collars. In order to avoid accidents to operators, setscrews with the end protruding above the hole should never be used on power-rotated or oscillating parts. A typical setscrew installation is shown in Fig. 33-2-13.

Fig. 33-2-13 Setscrew application.

Screw-and-Washer Assemblies A preassembled screw and washer comprise a unit assembly, as shown in Fig. 33-2-14. The washer is free to rotate relative to the screw and is held in place under the head of the screw by the threads, which are rolled after the washer is assembled. Screw-and-

(A) SCREW AND WASHER ASSEMBLIES

(B) APPLICATION

Fig. 33-2-14 Screw-and-washer assemblies.

washer assemblies result in a labor savings, since only one part need be handled. In addition, they ensure that a washer will be included in the assembly. Procurement and stock control are also simplified. These are factors which bear consideration in specifying screws and washers for attachments and should be weighed against the added unit cost for screw-and-washer assemblies.

Drive Screws for Metal Hardened metallic drive screws provide a permanent fastening for heavy sheet metal, castings, plastics, etc., and may be used in place of tapping screws or machine screws. Drive screws are hammered or otherwise forced into holes of suitable size. The unthreaded pilot guides the drive screw in straight, and the hardened spiral thread, which extends to the head, forms the required mating thread in the hole.

The thickness of metal into which the screw is to be driven must be at least approximately the same as the outside diameter of the drive screw to ensure adequate thread engagement. An advantage in using these screws in place of machine screws is the elimination of tapped holes; however, a pilot hole is necessary (Fig. 33-2-15).

Fig. 33-2-15 Metallic drive screw.

Tapping Screws These screws were developed primarily to eliminate tapping operations or nuts in certain assemblies of sheet-metal parts, plastics, and soft castings.

Nuts Many types of nuts are available for specific requirements. It is desirable to minimize the use of special designs in favor of the more commonly used nuts.

Slotted nuts with cotter pins or wire can be used to help retain the nut on the bolt.

Jam nuts are used where height is restricted or as a means of locking the working nut, if assembled as shown in Fig. 33-2-16.

Fig. 33-2-16 Jam nut application.

A *locknut* is a nut having a special means for gripping an externally threaded member so that relative back-off rotation between the nut and the companion member is impeded.

Prevailing-torque-type locknuts employ a self-contained locking feature such as deformed or undersize threads, variable lead angle, plastic or fiber washers, or plug inserts. This type of nut resists screwing on, as well as unscrewing, and does not depend on bolt load for locking (Fig. 33-2-17).

Fig. 33-2-17 Prevailing torque locknut.

Free-running locknuts develop their locking action after the nut has been seated by reactive spring force against the threads or by friction against the bearing surface.

Spring nuts are made of thin spring metal and have arched prongs or formed embossments to fit a single lead of a mating screw thread. Spring nuts are used extensively for sheet-metal construction where relatively high torques and strength are not required.

Another type of spring nut is available which can be pushed on over rivets, tubing, nails, or other unthreaded parts and provides a positive bite that grips securely even on very smooth surfaces. Figure 33-2-18 shows a typical application.

Stamped nuts are usually fabricated from thin spring steel and have arched prongs formed to fit a single lead of a mating screw thread. They have the

Fig. 33-2-18 Push-on spring nut.

same functional usage as a spring nut, with the additional advantage of provisions for turning the nut (Fig. 33-2-19).

Crown nuts are generally used where it is desirable to cover the end of the externally threaded part for purposes of appearance or protection from sharp edges.

Fig. 33-2-19 Stamped nuts.

Wing nuts, as the name implies, are provided with two wings to facilitate hand tightening and loosening. They are used where high torque is not required and where the nuts are to be disassembled and reassembled frequently.

Barrel and *sleeve nuts* are usually made to resemble a screw head at the outer or exposed end. They are used in assemblies where any other type of nut would present a less favorable appearance.

Clinch nuts were developed for sheet-metal assemblies where the nut is inaccessible for wrenching. They are provided on one side with a shoulder and smaller pilot, which is inserted into a preformed hole in the sheet metal, and are permanently attached by spinning or staking the portion of pilot extending through the hole (Fig. 33-2-20).

Fig. 33-2-20   Assembled clinch nuts.

Weld nuts are similar to the clinch nuts in function. However, they are supplied with weld projections and are either spot- or projection-welded to the sheet metal instead of being peened.

Washers   The four basic types of washers are flat (plain) washers, conical-spring washers, helical-spring lockwashers, and tooth lockwashers.

Flat washers are used under the head of a screw or bolt, or under a nut, for four principal purposes:

1. To spread the load over a greater area
2. To reduce frictional variations during assembly
3. To provide bearing surface over large clearance holes or slots
4. To prevent marring of parts during assembly

Conical-spring washers are made of steel that has been hardened and tempered. The relatively high supporting load and spring return make this washer effective where bolt tension

may be lost because of such factors as thermal expansion or compression set of gaskets.

Helical-spring lockwashers are usually used as a hardened thrust washer or as a spacer.

Pins   Cotter pins, spring pins, groove pins, taper pins, and clevis pins are used to retain parts of an assembly in relative position.

Cotter pins are used for retaining slotted nuts, movable links or rods, etc., as shown in Fig. 33-2-21.

Fig. 33-2-21   Clevis-pin application.

Groove pins are straight pins having longitudinal grooves rolled or pressed into the body which provide a reactive expansion effect when the pin is driven into a drilled hole.

Spring pins provide a spring effect which serves to retain the pin when it is driven into a drilled hole of diameter slightly smaller than that of the pin. These types of pins eliminate reaming or peening and can be disassembled a number of times without serious loss of holding power.

Taper pins serve the same functional purpose as groove pins. However, they require taper-reamed holes at assembly and are retained only by taper lock, which can totally disengage when minor displacement occurs. Assemblies using taper pins are more costly than grooved pins.

The primary purpose of *clevis pins* is to attach clevises to rod ends and levers and to serve as bearings. They are held in place by cotter pins, as shown in Fig. 33-2-21.

Quickly Detachable Attachments   Quickly detachable attachments include clips or certain kinds of snap fasteners. These attaching means are convenient and time-saving by providing for quick assembly and disassembly.

DESIGN CHECKLIST

The following design checklist will serve as a helpful guide in reviewing a design.

1. Keep the number of separate pieces in a subassembly as low as practicable by combining single components into one assembly if functional requirements permit.
2. Check particularly left- and right-hand parts to determine whether they can be made identical and so avoid stocking an extra part.
3. Check advisability of using lockwashers assembled to bolts as purchased assemblies.
4. Check to determine if similar parts between models can be standardized.
5. If subassemblies require alignment, design the parts to secure such alignment without the use of special jigs by providing tabs, shoulders, notches, or contour locators.
6. Provide adequate clearances between parts for assembly tools.
7. Provide clearances between parts to allow for tolerance stackup variations.
8. Specify the simplest, most effective, and cheapest type of attachment practicable and commensurate with the functional requirements.
9. Avoid blind-riveting operations wherever possible.
10. Where rivets are used, provide sufficient clearances between parts and from flanges to permit use of a standard riveting gun.
11. Avoid use of slotted nuts and cotter pins wherever possible.
12. Standardize as far as practicable bolt and thread sizes. Hold the number of bolt lengths to a minimum and recheck frequently to reduce number of lengths in use.
13. Try to avoid riveting or welding operations in shop areas where this type of equipment is not normally used.

Reference and Source Materials
1. General Motors

ASSIGNMENTS

See Assignments 7 through 12 for Unit 33-2 on page 703.

ASSIGNMENTS for Chapter 33

Assignments for Unit 33-1, The Design Process

1. Your drafting supervisor has assigned to you the responsibility of designing an attractive single toggle switch plate for use in kitchens and bathrooms. The toggle plate and clearance hole requirements are shown in Fig. 33-1-A. The production run will exceed 25 000 and four different color plates are required. On a B- or A3-size sheet, lay out the design of the plate and include on the drawing the production and specification data which you would submit with your design.

2. Design a suitable handle for a metal storage cabinet from the following data and the information shown in Fig. 33-1-B. The latch is to be released with a quarter turn (anticlockwise direction) of the square latch shaft. The shaft is .38 in. square and protrudes beyond the cover by 1.00 in. The lock and shaft are securely fastened to the inside face of the cover and as such do not require any support on the handle side of the cover plate. A .62-in. diameter hole is punched in the .062-in.-thick cover plate. Quantity 5000. On a B- or A3-size sheet, lay out the design of the handle and include on the drawing the production and specification data which you would submit with the design.

3. Fig. 33-1-C. Your drafting supervisor has asked you to design a table-mounted holder for a cassette microphone. The material is to be of such quality as to have an attractive finish as well as support the mass of the microphone. A $\varnothing$20-mm tapered hole (large end) with a 3-mm wide slot on top is required for attaching the microphone to the holder. On a B- or A3-size sheet, lay out the holder and show the microphone in phantom lines.

4. Design a personalized key holder for a recreation vehicle that is used occasionally. When not in use it must hang on a wall or key rack. The material is to be lightweight and colorful.

5. Due to increased drafting room costs, your supervisor has asked you to design a holder to use up the short ends of wooden pencils that have been too small to hold in the hand. The material you choose should have good characteristics for grip and color. A quantity of 1000 is required.

6. Design a small tic-tac-toe board that could be given away as a souvenir or award. Your school crest or name must appear on the board or be an intrinsic part of the design. The material selected should be one that could be produced in one of your shops on campus.

Assignments for Unit 33-2, Assembly Considerations

7. Fig. 33-2-A. Design a combined pencil holder and masking tape dispenser which can be mounted on the edge of a drafting table. It should hold four pencils and one roll of masking tape. The tape has a 3-in.-diameter center hole and is 1-in. wide. Prepare an assembly drawing and include a bill of material. The material selected should be able to be used in mass production methods.

8. A 20-mm-diameter steel rotary shaft must be supported at 2-meter intervals along a ceiling. Split journal bearings, 40 mm long, are recommended by the engineering department. A maximum of eight brackets are required. On a B- or A3-size sheet, design a suitable bracket complete with dimensions.

9. Same as Assignment 8 except the quantity requied is 2000.

10. Two vertical .312-in.-diameter electrical conductors are to be supported on the inside wall of an oil-filled metal tank. The conductors, made of copper, have no insulation on them. The voltage they carry is such that there must be a minimum distance of 2.00 in. between conductors or any other metal when supported by nonconductive material, such as plastic, wood, etc. Although there are no external forces acting on these conductors, they should be supported every 24 in. On a B- or A3-size sheet, prepare an assembly drawing showing the conductors, support, and tank wall. Include a bill of material. On a second sheet prepare the details of the parts required. Scale to suit.

11. Design a container to store 12 tape cassettes. This container should have a lid or cover and look attractive when installed on the bottom edge of an automobile instrument panel.

12. Design a napkin holder to sit on the table. The napkins can be stored either flat, 6.00 in. square, or folded in half 3.00 × 6.00 in. Their position for dispensing is your choice as long as one napkin can be taken at a time. The holder must hold a minimum of 12 napkins and come in a variety of colors or a selection of wood grains to complement the modern kitchen.

PLATE MOUNTING AND TOGGLE SWITCH CLEARANCE

Fig. 33-1-A   Assignment.

Fig. 33-1-B   Assignment.

Fig. 33-1-C   Assignment.

Fig. 33-2-A   Assignment.

APPENDIX

ANSI	Y14.1	Drawing Sheet Size and Format
ANSI	Y14.2M	Line Conventions and Lettering
ANSI	Y14.3	Projections
ANSI	Y14.4	Pictorial Drawing
ANSI	Y14.5M	Dimensioning and Tolerancing for Engineering Drawings
ANSI	Y14.6	Screw Thread Representation
ANSI	Y14.7	Gears, Splines, and Serrations
ANSI	Y14.7.	Gear Drawing Standards—Part 1, for Spur, Helical, Double Helical and Rack
ANSI	Y14.9	Forgings
ANSI	Y14.10	Metal Stampings
ANSI	Y14.11	Plastics
ANSI	Y14.14	Mechanical Assemblies
ANSI	Y14.15	Electrical and Electronics Diagrams
ANSI	Y14.15A	Interconnection Diagrams
ANSI	Y14.17	Fluid Power Diagrams
ANSI	Y14.36	Surface Texture Symbols
ANSI	Y32.2	Graphic Symbols for Electrical and Electronics Diagrams
ANSI	Y32.9	Graphic Electrical Wiring Symbols for Architectural and Electrical Layout Drawings
ANSI	B1.1	Unified Screw Threads
ANSI	B4.2	Preferred Metric Limits and Fits
ANSI	B18.2.1	Square and Hex Bolts and Screws
ANSI	B18.2.2	Square and Hex Nuts
ANSI	B18.3	Socket Cap, Shoulder, and Setscrews
ANSI	B18.6.2	Slotted-Head Cap Screws, Square-Head Setscrews, Slotted-Headless Setscrews
ANSI	B18.6.3	Machine Screws and Machine Screw Nuts
ANSI	B17.2	Woodruff Key and Keyslot Dimensions
ANSI	B17.1	Keys and Keyseats
ANSI	B18.21.1	Lock Washers
ANSI	B27.2	Plain Washers
ANSI	B46.1	Surface Texture
ANSI	B94.6	Knurling
ANSI	B94.11M	Twist Drills
ANSI	Z210.1	Metric Practice

Table 1 ANSI publications.

Quantity	Metric Unit	Symbol	Metric to Inch-Pound Unit	Inch-Pound to Metric Unit
Length	millimeter	mm	1 mm = 0.0394 in.	1 in. = 25.4 mm
	centimeter	cm	1 cm = 0.394 in.	1 ft. = 30.5 cm
	meter	m	1 m = 39.37 in. = 3.28 ft	1 yd. = 0.914 m = 914 mm
	kilometer	km	1 km = 0.62 mile	1 mile = 1.61 km
Area	square millimeter	mm²	1 mm² = 0.001 55 sq. in.	1 sq. in. = 6 452 mm²
	square centimeter	cm²	1 cm² = 0.155 sq. in.	1 sq. ft. = 0.093 m²
	square meter	m²	1 m² = 10.8 sq. ft. = 1.2 sq. yd.	1 sq. yd. = 0.836 m²
Mass	milligram	mg	1 g = 0.035 oz.	1 oz. = 28.3 g
	gram	g	1 kg = 2.205 lb.	1 lb. = 0.454 kg
	kilogram	kg	1 tonne = 1.102 tons	1 ton = 907.2 kg
	tonne	t		= 0.907 tonnes
Volume	cubic centimeter	cm³	1 mm³ = 0.000 061 cu. in.	1 fl. oz. = 28.4 cm³
	cubic meter	m³	1 cm³ = 0.061 cu. in.	1 cu. in. = 16.387 cm³
	milliliter	m	1 m³ = 35.3 cu ft. = 1.308 cu. yd.	1 cu. ft. = 0.028 m³
			1 mℓ = 0.035 fl. oz.	1 cu. yd. = 0.756 m³
Capacity	liter	L	U.S. Measure 1 pt. = 0.473 L 1 qt. = 0.946 L 1 gal = 3.785 L Imperial Measure 1 pt. = 0.568 L 1 qt. = 1.137 L 1 gal = 4.546 L	U.S. Measure 1 L = 2.113 pt. = 1.057 qt. = 0.264 gal. Imperial Measure 1 L = 1.76 pt. = 0.88 qt. = 0.22 gal.
Temperature	Celsius degree	°C	$°C = \frac{5}{9}(°F-32)$	$°F = \frac{9}{5} \times °C + 32$
Force	newton	N	1 N = 0.225 lb (f)	1 lb (f) = 4.45N
	kilonewton	kN	1 kN = 0.225 kip (f) = 0.112 ton (f)	= 0.004 448 kN
Energy/Work	joule	J	1 J = 0.737 ft · lb	1 ft · lb = 1.355 J
	kilojoule	kJ	1 J = 0.948 Btu	1 Btu = 1.055 J
	megajoule	MJ	1 MJ = 0.278 kWh	1 kWh = 3.6 MJ
Power	kilowatt	kW	1 kW = 1.34 hp	1 hp (550 ft · lb/s) = 0.746 kW
			1 W = 0.0226 ft · lb/min.	1 ft · lb/min = 44.2537 W
Pressure	kilopascal	kPa	1 kPa = 0.145 psi = 20.885 psf = 0.01 ton-force per sq. ft.	1 psi = 6.895 kPa 1 lb-force/sq. ft. = 47.88 Pa 1 ton-force/sq. ft. = 95.76 kPa
	*kilogram per square centimeter	kg/cm²	1 kg/cm² = 13.780 psi	
Torque	newton meter	N · m	1 N · m = 0.74 lb · ft	1 lb · ft = 1.36 N · m
	*kilogram meter	kg/m	1 kg/m = 7.24 lb · ft	1 lb · ft = 0.14 kg/m
	*kilogram per centimeter	kg/cm	1 kg/cm = 0.86 lb · in	1 lb · in = 1.2 kg/cm
Speed/Velocity	meters per second	m/s	1 m/s = 3.28 ft/s	1 ft/s = 0.305 m/s
	kilometers per hour	km/h	1 km/h = 0.62 mph	1 mph = 1.61 km/h

*Not SI units, but included here because they are employed on some of the gages and indicators currently in use in industry.

Table 2   Metric conversion tables.

ONE HUNDREDTH OF AN INCH INCREMENTS TO ONE INCH

Inch	.00	.01	.02	.03	.04	.05	.06	.07	.08	.09
.00	0.00	0.25	0.51	0.76	1.02	1.27	1.52	1.78	2.03	2.29
.10	2.54	2.79	3.05	3.30	3.56	3.81	4.06	4.32	4.57	4.83
.20	5.08	5.33	5.59	5.84	6.10	6.35	6.60	6.86	7.11	7.37
.30	7.62	7.87	8.13	8.38	8.64	8.89	9.14	9.40	9.65	9.91
.40	10.16	10.41	10.67	10.92	11.18	11.43	11.68	11.94	12.19	12.45
.50	12.70	12.95	13.21	13.46	13.72	13.97	14.22	14.48	14.73	14.99
.60	15.24	15.49	15.75	16.00	16.26	16.51	16.76	17.02	17.27	17.53
.70	17.78	18.03	18.29	18.54	18.80	19.05	19.30	19.56	19.81	20.07
.80	20.32	20.57	20.83	21.08	21.34	21.59	21.84	22.10	22.35	22.61
.90	22.86	23.11	23.37	23.62	23.88	24.13	24.38	24.64	24.89	25.15

ONE TENTH OF AN INCH INCREMENTS TO TWENTY INCHES

Inches	0	.10	.20	.30	.40	.50	.60	.70	.80	.90
0	0.0	2.5	5.1	7.6	10.2	12.7	15.2	17.8	20.3	22.9
1	25.4	27.9	30.5	33.0	35.6	38.1	40.6	43.2	45.7	48.3
2	50.8	53.3	55.9	58.4	61.0	63.5	66.0	68.6	71.1	73.7
3	76.2	78.7	81.3	83.8	86.4	88.9	91.4	94.0	96.5	99.1
4	101.6	104.1	106.7	109.2	111.8	114.3	116.8	119.4	121.9	124.5
5	127.0	129.5	132.1	134.6	137.2	139.7	142.2	144.8	147.3	149.9
6	152.4	154.9	157.5	160.0	162.6	165.1	167.6	170.2	172.7	175.3
7	177.8	180.3	182.9	185.4	188.0	190.5	193.0	195.6	198.1	200.7
8	203.2	205.7	208.3	210.8	213.4	215.9	218.4	221.0	223.5	226.1
9	228.6	231.1	233.7	236.2	238.8	241.3	243.8	246.4	248.9	251.5
10	254.0	256.5	259.1	261.6	264.2	266.7	269.2	271.8	274.3	276.9
11	279.4	281.9	284.5	287.0	289.6	292.1	294.6	297.2	299.7	302.3
12	304.8	307.3	309.9	312.4	315.0	317.5	320.0	322.6	325.1	327.7
13	330.2	332.7	335.3	337.8	340.4	342.9	345.4	348.0	350.5	353.1
14	355.6	358.1	360.7	363.2	365.8	368.3	370.8	373.4	375.9	378.5
15	381.0	383.5	386.1	388.6	391.2	393.7	396.2	398.8	401.3	403.9
16	406.4	408.9	411.5	414.0	416.6	419.1	421.6	424.2	426.7	429.3
17	431.8	434.3	436.9	439.4	442.0	444.5	447.0	449.6	452.1	454.7
18	457.2	459.7	462.3	464.8	467.4	469.9	472.4	475.0	477.5	480.1
19	482.6	485.1	487.7	490.2	492.8	495.3	497.8	500.4	502.9	505.5
20	508.0	510.5	513.1	515.6	518.2	520.7	523.2	525.8	528.3	530.9

Table 3   Conversion of decimals of an inch to millimeters.

IN.	0	1/16	1/8	3/16	1/4	5/16	3/8	7/16	1/2	9/16	5/8	11/16	3/4	13/16	7/8	15/16
0	.0	1.6	3.2	4.8	6.4	7.9	9.5	11.1	12.7	14.3	15.9	17.5	19.1	20.6	22.2	23.8
1	25.4	27.0	28.6	30.2	31.8	33.3	34.9	36.5	38.1	39.7	41.3	42.9	44.5	46.0	47.6	49.2
2	50.8	52.4	54.0	55.6	57.2	58.7	60.3	61.9	63.5	65.1	66.7	68.3	69.9	71.4	73.0	74.6
3	76.2	77.8	79.4	81.0	82.6	84.1	85.7	87.3	88.9	90.5	92.1	93.7	95.3	96.8	98.4	100.0
4	101.6	103.2	104.8	106.4	108.0	109.5	111.1	112.7	114.3	115.9	117.5	119.1	120.7	122.2	123.8	125.4
5	127.0	128.6	130.2	131.8	133.4	134.9	136.5	138.1	139.7	141.3	142.9	144.5	146.1	147.6	149.2	150.8
6	152.4	154.0	155.6	157.2	158.8	160.3	161.9	163.5	165.1	166.7	168.3	169.9	171.5	173.0	174.6	176.2
7	177.8	179.4	181.0	182.6	184.2	185.7	187.3	188.9	190.5	192.1	193.7	195.3	196.9	198.4	200.0	201.6
8	203.2	204.8	206.4	208.0	209.6	211.1	212.7	214.3	215.9	217.5	219.1	220.7	222.3	223.8	225.4	227.0
9	228.6	230.2	231.8	233.4	235.0	236.5	238.1	239.7	241.3	242.9	244.5	246.1	247.7	249.2	250.8	252.4
10	254.0	255.6	257.2	258.8	260.4	261.9	263.5	265.1	266.7	268.3	269.9	271.5	273.1	274.6	276.2	277.8
11	279.4	281.0	282.6	284.2	285.8	287.3	288.9	290.5	292.1	293.7	295.3	296.9	298.5	300.0	301.6	303.2
12	304.8	306.4	308.0	309.6	311.2	312.7	314.3	315.9	317.5	319.1	320.7	322.3	323.9	325.4	327.0	328.6
13	330.2	331.8	333.4	335.0	336.6	338.1	339.7	341.3	342.9	344.5	346.1	347.7	349.3	350.8	352.4	354.0
14	355.6	357.2	358.8	360.4	362.0	363.5	365.1	366.7	368.3	369.9	371.5	373.1	374.7	376.2	377.8	379.4

Table 4   Conversion of fractions of an inch to millimeters.

Fraction	Decimal	Fraction	Decimal
$\frac{1}{64}$	0.015625	$\frac{33}{64}$	0.515625
$\frac{1}{32}$	0.03125	$\frac{17}{32}$	0.53125
$\frac{3}{64}$	0.046875	$\frac{35}{64}$	0.546875
$\frac{1}{16}$	0.0625	$\frac{9}{16}$	0.5625
$\frac{5}{64}$	0.078125	$\frac{37}{64}$	0.578125
$\frac{3}{32}$	0.09375	$\frac{19}{32}$	0.59375
$\frac{7}{64}$	0.109375	$\frac{39}{64}$	0.609375
$\frac{1}{8}$	0.1250	$\frac{5}{8}$	0.6250
$\frac{9}{64}$	0.140625	$\frac{41}{64}$	0.640625
$\frac{5}{32}$	0.15625	$\frac{21}{32}$	0.65625
$\frac{11}{64}$	0.171875	$\frac{43}{64}$	0.671875
$\frac{3}{16}$	0.1875	$\frac{11}{16}$	0.6875
$\frac{13}{64}$	0.203125	$\frac{45}{64}$	0.703125
$\frac{7}{32}$	0.21875	$\frac{23}{32}$	0.71875
$\frac{15}{64}$	0.234375	$\frac{47}{64}$	0.734375
$\frac{1}{4}$	0.2500	$\frac{3}{4}$	0.7500
$\frac{17}{64}$	0.265625	$\frac{49}{64}$	0.765625
$\frac{9}{32}$	0.28125	$\frac{25}{32}$	0.78125
$\frac{19}{64}$	0.296875	$\frac{51}{64}$	0.796875
$\frac{5}{16}$	0.3125	$\frac{13}{16}$	0.8125
$\frac{21}{64}$	0.328125	$\frac{53}{64}$	0.828125
$\frac{11}{32}$	0.34375	$\frac{27}{32}$	0.84375
$\frac{23}{64}$	0.359375	$\frac{55}{64}$	0.859375
$\frac{3}{8}$	0.3750	$\frac{7}{8}$	0.8750
$\frac{25}{64}$	0.390625	$\frac{57}{64}$	0.890625
$\frac{13}{32}$	0.40625	$\frac{29}{32}$	0.90625
$\frac{27}{64}$	0.421875	$\frac{59}{64}$	0.921875
$\frac{7}{16}$	0.4375	$\frac{15}{16}$	0.9375
$\frac{29}{64}$	0.453125	$\frac{61}{64}$	0.953125
$\frac{15}{32}$	0.46875	$\frac{31}{32}$	0.96875
$\frac{31}{64}$	0.484375	$\frac{63}{64}$	0.984375
$\frac{1}{2}$	0.5000	1	1.0000

Table 5 Decimal equivalents of common inch fractions.

ANGLE	SINE	COSINE	TAN	COTAN	ANGLE
0°	.0000	1.0000	.0000	θ	90°
1°	0.0175	0.9998	0.0175	57.290	89°
2°	0.0349	0.9994	0.0349	28.636	88°
3°	0.0523	0.9986	0.0524	19.081	87°
4°	0.0698	0.9976	0.0699	14.301	86°
5°	0.0872	0.9962	0.0875	11.430	85°
6°	0.1045	0.9945	0.1051	9.5144	84°
7°	0.1219	0.9925	0.1228	8.1443	83°
8°	0.1392	0.9903	0.1405	7.1154	82°
9°	0.1564	0.9877	0.1584	6.3138	81°
10°	0.1736	0.9848	0.1763	5.6713	80°
11°	0.1908	0.9816	0.1944	5.1446	79°
12°	0.2079	0.9781	0.2126	4.7046	78°
13°	0.2250	0.9744	0.2309	4.3315	77°
14°	0.2419	0.9703	0.2493	4.0108	76°
15°	0.2588	0.9659	0.2679	3.7321	75°
16°	0.2756	0.9613	0.2867	3.4874	74°
17°	0.2924	0.9563	0.3057	3.2709	73°
18°	0.3090	0.9511	0.3249	3.0777	72°
19°	0.3256	0.9455	0.3443	2.9042	71°
20°	0.3420	0.9397	0.3640	2.7475	70°
21°	0.3584	0.9336	0.3839	2.6051	69°
22°	0.3746	0.9272	0.4040	2.4751	68°
23°	0.3907	0.9205	0.4245	2.3559	67°
24°	0.4067	0.9135	0.4452	2.2460	66°
25°	0.4226	0.9063	0.4663	2.1445	65°
26°	0.4384	0.8988	0.4877	2.0503	64°
27°	0.4540	0.8910	0.5095	1.9626	63°
28°	0.4695	0.8829	0.5317	1.8807	62°
29°	0.4848	0.8746	0.5543	1.8040	61°
30°	0.5000	0.8660	0.5774	1.7321	60°
31°	0.5150	0.8572	0.6009	1.6643	59°
32°	0.5299	0.8480	0.6249	1.6003	58°
33°	0.5446	0.8387	0.6494	1.5399	57°
34°	0.5592	0.8290	0.6745	1.4826	56°
35°	0.5736	0.8192	0.7002	1.4281	55°
36°	0.5878	0.8090	0.7265	1.3764	54°
37°	0.6018	0.7986	0.7536	1.3270	53°
38°	0.6157	0.7880	0.7813	1.2799	52°
39°	0.6293	0.7771	0.8098	1.2349	51°
40°	0.6428	0.7660	0.8391	1.1918	50°
41°	0.6561	0.7547	0.8693	1.1504	49°
42°	0.6691	0.7431	0.9004	1.1106	48°
43°	0.6820	0.7314	0.9325	1.0724	47°
44°	0.6947	0.7193	0.9657	1.0355	46°
45°	0.7071	0.7071	0.0000	1.0000	45°
ANGLE	COSINE	SINE	COTAN	TAN	ANGLE

Table 6 Trigonometric functions.

Table 7 Function of numbers.

NUMBER	SQUARE	SQUARE ROOT	CIRCUMFERENCE OF CIRCLE	AREA OF CIRCLE
1	1	1	3.14	0.78
2	4	1.41	6.28	3.14
3	9	1.73	9.43	7.07
4	16	2.00	12.57	12.57
5	25	2.34	15.71	19.64
6	36	2.4495	18.85	28.27
7	49	2.6458	21.99	38.48
8	64	2.8284	25.13	50.27
9	81	3.0000	28.27	63.62
10	100	3.1623	31.46	78.54
11	121	3.3166	34.56	95.03
12	144	3.4641	37.70	113.09
13	169	3.6056	40.84	132.73
14	196	3.7417	43.98	153.94
15	225	3.8730	47.12	176.72
16	256	4.0000	50.27	201.06
17	289	4.1231	53.41	226.98
18	324	4.2426	56.55	254.47
19	361	4.3589	59.69	283.53
20	400	4.4721	62.83	314.16
21	441	4.5826	65.97	346.36
22	484	4.6904	69.12	380.13
23	529	4.7958	72.26	415.48
24	576	4.8990	75.39	452.39
25	625	5.0000	78.54	490.87
26	676	5.0990	81.68	530.93
27	729	5.1962	84.82	572.56
28	784	5.2915	87.97	615.75
29	841	5.3852	91.11	660.52
30	900	5.4772	94.25	706.86
31	961	5.5678	97.39	754.77
32	1024	5.6569	100.53	804.25
33	1089	5.7446	103.67	855.30
34	1156	5.8310	106.81	907.92
35	1225	5.9161	109.96	962.113
36	1296	6.0000	113.10	1017.88
37	1369	6.0828	116.24	1075.21
38	1444	6.1644	119.38	1134.11
39	1521	6.2450	122.52	1194.59
40	1600	6.3246	125.66	1256.64
41	1681	6.4031	128.81	1320.25
42	1764	6.4807	131.95	1385.44
43	1849	6.5574	135.09	1452.20
44	1936	6.6332	138.23	1520.53
45	2025	6.7082	141.37	1590.43
46	2116	6.7823	144.51	1661.90
47	2209	6.8557	147.65	1734.94
48	2304	6.9282	150.80	1809.56
49	2401	7.0000	153.94	1885.74
50	2500	7.0711	157.08	1963.50
51	2601	7.1414	160.22	2042.82
52	2704	7.2111	163.36	2123.72
53	2809	7.2801	166.50	2206.18
54	2916	7.3485	169.65	2290.22
55	3025	7.4162	172.79	2375.83
56	3136	7.4833	175.93	2463.01
57	3249	7.5498	179.07	2551.76
58	3364	7.6158	182.21	2642.08
59	3481	7.6811	185.35	2733.97
60	3600	7.7460	188.50	2827.43
61	3721	7.8102	191.64	3922.47
62	3844	7.8740	194.78	3019.07
63	3969	7.9373	197.92	3117.25
64	4096	8.0000	201.06	3216.99
65	4225	8.0623	204.20	3318.31
66	4356	8.1240	207.35	3421.19
67	4489	8.1854	210.49	3525.65
68	4624	8.2462	213.63	3631.68
69	4761	8.3066	216.77	3739.28
70	4900	8.3666	219.91	3848.50
71	5041	8.4261	223.05	3959.19
72	5184	8.4853	226.19	4071.50
73	5329	8.5440	229.34	4185.39
74	5476	8.6023	232.48	4300.84
75	5625	8.6603	235.62	4417.88
76	5776	8.7178	238.76	4536.47
77	5929	8.7750	241.90	4656.64
78	6084	8.8318	245.04	4778.37
79	6241	8.8882	248.19	4901.68
80	6400	8.9443	251.33	5026.56
81	6561	9.0000	254.47	5183.01
82	6724	9.0554	257.61	5281.03
83	6889	9.1104	260.75	5410.62
84	7056	9.1652	263.89	5541.78
85	7225	9.2200	267.04	5674.52
86	7396	9.2736	270.18	5808.82
87	7569	9.3274	273.32	5944.69
88	7744	9.3808	276.46	6082.14
89	7921	9.4340	279.60	6221.15
90	8100	9.4868	282.74	6361.74
91	8281	9.5393	285.89	6503.90
92	8464	9.5917	289.03	6647.63
93	8649	9.6437	292.17	6792.92
94	8836	9.6954	295.31	6939.79
95	9025	9.7468	298.45	7088.24
96	9216	9.7979	301.59	7238.25
97	9409	9.8489	304.74	7389.83
98	9604	9.8995	307.88	7542.98
99	9801	9.9509	311.02	7697.71
100	10 000	10.000	314.16	7854.00
101	10 201	10.0499	317.30	8011.87
102	10 404	10.0995	320.44	8171.30
103	10 609	10.1489	323.58	8332.31
104	10 816	10.1980	326.73	8494.89
105	11 025	10.2470	329.87	8659.04

And	&	Machined	√ or √
Across Flats	A/F	Machine Steel	MS or MACH ST
American National Standards Institute	ANSI	Material	MATL
Angular	ANG	Maximum	MAX
Approximate	APPROX	Maximum Material Condition	Ⓜ or MMC
Assembly	ASSY	Meter	m
Basic	BSC	Metric Thread	M
Bill of Material	B/M	Micrometer	μm
Bolt Circle	BC	Millimeter	mm
Brass	BR	Minimum	MIN
Brown and Sharpe Gage	B & S GA	Minute (Angle)	MIN
Bushing	BUSH	Newton	N
Canada Standards Institute	CSI	Nominal	NOM
Casting	CSTG	Not to Scale	—
Cast Iron	CI	Number	NO
Centimeter	cm	On Center	OC
Center Line	₵	Outside Diameter	OD
Center to Center	C to C	Parallel	PAR
Chamfered	CHAM	Pascal	Pa
Circularity	CIR	Perpendicular	PERP
Cold-Rolled Steel	CRS	Pitch	P
Concentric	CONC	Pitch Circle Diameter	PCD
Counterbore	⌴ or CBORE	Pitch Diameter	PD
Countersink	∨ or CSK	Plate	PL
Cubic Centimeter	cm³	Radian	rad
Cubic Meter	m³	Radius	R
Datum	DATUM	Reference or Reference Dimension	( ) or REF
Deep	↓	Regardless of Feature Size	Ⓡ
Degree (Angle)	° or DEG	Revolutions per Minute	rev/min
Diameter	⌀ or DIA	Right Hand	RH
Diametral Pitch	DP	Second (Arc)	(″)
Dimension	DIM	Second (Time)	SEC
Drawing	DWG	Section	SECT
Eccentric	ECC	Slotted	SLOT
Figure	FIG	Socket	SOCK
Finish All Over	FAO	Spherical	SPHER
Gage	GA	Spotface	⌴ or SFACE
Heat Treat	HT TR	Square	□ or SQ
Head	HD	Square Centimeter	cm²
Heavy	HVY	Square Meter	m²
Hexagon	HEX	Steel	STL
Hydraulic	HYD	Straight	STR
Inside Diameter	ID	Symmetrical	‡ or SYM
International Organization for Standardization	ISO	Thread	THD
Iron Pipe Size	IPS	Through	THRU
Kilogram	kg	Tolerance	TOL
Kilometer	km	True Profile	TP
Large End	LE	Undercut	UCUT
Left Hand	LH	U.S. Sheet-Metal Gage	USS GA
Liter	L	Watt	W
		Wrought iron	WI

ISO ANSI
Ⓐ -A-

Table 8 Abbreviations and symbols used on technical drawings.

NUMBER OR LETTER SIZE DRILL	SIZE		NUMBER OR LETTER SIZE DRILL	SIZE		NUMBER OR LETTER SIZE DRILL	SIZE		NUMBER OR LETTER SIZE DRILL	SIZE	
	mm	INCHES		mm	INCHES		mm	INCHES		mm	INCHES
80	0.343	.014	50	1.778	.070	20	4.089	.161	K	7.137	.281
79	0.368	.015	49	1.854	.073	19	4.216	.166	L	7.366	.290
78	0.406	.016	48	1.930	.076	18	4.305	.170	M	7.493	.295
77	0.457	.018	47	1.994	.079	17	4.394	.173	N	7.671	.302
76	0.508	.020	46	2.057	.081	16	4.496	.177	O	8.026	.316
75	0.533	.021	45	2.083	.082	15	4.572	.180	P	8.204	.323
74	0.572	.023	44	2.184	.086	14	4.623	.182	Q	8.433	.332
73	0.610	.024	43	2.261	.089	13	4.700	.185	R	8.611	.339
72	0.635	.025	42	2.375	.094	12	4.800	.189	S	8.839	.348
71	0.660	.026	41	2.438	.096	11	4.851	.191	T	9.093	.358
70	0.711	.028	40	2.489	.098	10	4.915	.194	U	9.347	.368
69	0.742	.029	39	2.527	.100	9	4.978	.196	V	9.576	.377
68	0.787	.031	38	2.578	.102	8	5.080	.199	W	9.804	.386
67	0.813	.032	37	2.642	.104	7	5.105	.201	X	10.084	.397
66	0.838	.033	36	2.705	.107	6	5.182	.204	Y	10.262	.404
65	0.889	.035	35	2.794	.110	5	5.220	.206	Z	10.490	.413
64	0.914	.036	34	2.819	.111	4	5.309	.209			
63	0.940	.037	33	2.870	.113	3	5.410	.213			
62	0.965	.038	32	2.946	.116	2	5.613	.221			
61	0.991	.039	31	3.048	.120	1	5.791	.228			
60	1.016	.040	30	3.264	.129	A	5.944	.234			
59	1.041	.041	29	3.354	.136	B	6.045	.238			
58	1.069	.042	28	3.569	.141	C	6.147	.242			
57	1.092	.043	27	3.658	.144	D	6.248	.246			
56	1.181	.047	26	3.734	.147	E	6.350	.250			
55	1.321	.052	25	3.797	.150	F	6.528	.257			
54	1.397	.055	24	3.861	.152	G	6.629	.261			
53	1.511	.060	23	3.912	.154	H	6.756	.266			
52	1.613	.064	22	3.988	.157	I	6.909	.272			
51	1.702	.067	21	4.039	.159	J	7.036	.277			

Table 9   Number and letter-size drills.

METRIC DRILL SIZES		Reference Decimal Equivalent (Inches)	METRIC DRILL SIZES		Reference Decimal Equivalent (Inches)	METRIC DRILL SIZES		Reference Decimal Equivalent (Inches)
Preferred	Available		Preferred	Available		Preferred	Available	
—	0.40	.0157	—	2.7	.1063	14	—	.5512
—	0.42	.0165	2.8	—	.1102	—	14.5	.5709
—	0.45	.0177	—	2.9	.1142	15	—	.5906
—	0.48	.0189	3.0	—	.1181	—	15.5	.6102
0.50	—	.0197	—	3.1	.1220	16	—	.6299
—	0.52	.0205	3.2	—	.1260	—	16.5	.6496
0.55	—	.0217	—	3.3	.1299	17	—	.6693
—	0.58	.0228	3.4	—	.1339	—	17.5	.6890
0.60	—	.0236	—	3.5	.1378	18	—	.7087
—	0.62	.0244	3.6	—	.1417	—	18.5	.7283
0.65	—	.0256	—	3.7	.1457	19	—	.7480
—	0.68	.0268	3.8	—	.1496	—	19.5	.7677
0.70	—	.0276	—	3.9	.1535	20	—	.7874
—	0.72	.0283	4.0	—	.1575	—	20.5	.8071
0.75	—	.0295	—	4.1	.1614	21	—	.8268
—	0.78	.0307	4.2	—	.1654	—	21.5	.8465
0.80	—	.0315	—	4.4	.1732	22	—	.8661
—	0.82	.0323	4.5	—	.1772	—	23	.9055
0.85	—	.0335	—	4.6	.1811	24	—	.9449
—	0.88	.0346	4.8	—	.1890	25	—	.9843
0.90	—	.0354	5.0	—	.1969	26	—	1.0236
—	0.92	.0362	—	5.2	.2047	—	27	1.0630
0.95	—	.0374	5.3	—	.2087	28	—	1.1024
—	0.98	.0386	—	5.4	.2126	—	29	1.1417
1.00	—	.0394	5.6	—	.2205	30	—	1.1811
—	1.03	.0406	—	5.8	.2283	—	31	1.2205
1.05	—	.0413	6.0	—	.2362	32	—	1.2598
—	1.08	.0425	—	6.2	.2441	—	33	1.2992
1.10	—	.0433	6.3	—	.2480	34	—	1.3386
—	1.15	.0453	—	6.5	.2559	—	35	1.3780
1.20	—	.0472	6.7	—	.2638	36	—	1.4173
1.25	—	.0492	—	6.8	.2677	—	37	1.4567
1.3	—	.0512	—	6.9	.2717	38	—	1.4361
—	1.35	.0531	7.1	—	.2795	—	39	1.5354
1.4	—	.0551	—	7.3	.2874	40	—	1.5748
—	1.45	.0571	7.5	—	.2953	—	41	1.6142
1.5	—	.0591	—	7.8	.3071	42	—	1.6535
—	1.55	.0610	8.0	—	.3150	—	43.5	1.7126
1.6	—	.0630	—	8.2	.3228	45	—	1.7717
—	1.65	.0650	8.5	—	.3346	—	46.5	1.8307
1.7	—	.0669	—	8.8	.3465	48	—	1.8898
—	1.75	.0689	9.0	—	.3543	50	—	1.9685
1.8	—	.0709	—	9.2	.3622	—	51.5	2.0276
—	1.85	.0728	9.5	—	.3740	53	—	2.0866
1.9	—	.0748	—	9.8	.3858	—	54	2.1260
—	1.95	.0768	10	—	.3937	56	—	2.2047
2.0	—	.0787	—	10.3	.4055	—	58	2.2835
—	2.05	.0807	10.5	—	.4134	60	—	2.3622
2.1	—	.0827	—	10.8	.4252			
—	2.15	.0846	11	—	.4331			
2.2	—	.0866	—	11.5	.4528			
—	2.3	.0906	12	—	.4724			
2.4	—	.0945	12.5	—	.4921			
2.5	—	.0984	13	—	.5118			
2.6	—	.1024	—	13.5	.5315			

Table 10   Metric twist drill sizes.

SIZE INCHES		Graded Pitch Series						Constant Pitch Series					
		Coarse UNC		Fine UNF		Extra Fine UNEF		8 UN		12 UN		16 UN	
Number or Fraction	Deci-mal	Threads per Inch	Tap Drill Dia.	Threads per Inch	Tap Drill Dia.	Threads per Inch	Tap Drill Dia.	Threads per Inch	Tap Drill Dia.	Threads per Inch	Tap Drill Dia.	Threads per Inch	Tap Drill Dia.
0	.060	—	—	80	$3/64$	—	—	—	—	—	—	—	—
2	.086	56	No. 50	64	No. 49	—	—	—	—	—	—	—	—
4	.112	40	No. 43	48	No. 42	—	—	—	—	—	—	—	—
5	.125	40	No. 38	44	No. 37	—	—	—	—	—	—	—	—
6	.138	32	No. 36	40	No. 33	—	—	—	—	—	—	—	—
8	.164	32	No. 29	36	No. 29	—	—	—	—	—	—	—	—
10	.190	24	No. 25	32	No. 21	—	—	—	—	—	—	—	—
$1/4$	.250	20	7	28	3	32	.219	—	—	—	—	—	—
$5/16$	.312	18	F	24	1	32	.281	—	—	—	—	—	—
$3/8$	.375	16	.312	24	Q	32	.344	—	—	—	—	UNC	—
$7/16$	.438	14	U	20	.391	28	Y	—	—	—	—	16	V
$1/2$	.500	13	.422	20	.453	28	.469	—	—	—	—	16	.438
$9/16$	.562	12	.484	18	.516	24	.516	—	—	UNC	—	16	.500
$5/8$	.625	11	.531	18	.578	24	.578	—	—	12	.547	16	.562
$3/4$	.750	10	.656	16	.688	20	.703	—	—	12	.672	UNF	—
$7/8$	.875	9	.766	14	.812	20	.828	—	—	12	.797	16	.812
1	1.000	8	.875	12	.922	20	.953	UNC	—	UNF	—	16	.938
$1 1/8$	1.125	7	.984	12	1.047	18	1.078	8	1.000	UNF	—	16	1.062
$1 1/4$	1.250	7	1.109	12	1.172	18	1.188	8	1.125	UNF	—	16	1.188
$1 3/8$	1.375	6	1.219	12	1.297	18	1.312	8	1.250	UNF	—	16	1.312
$1 1/2$	1.500	6	1.344	12	1.422	18	1.438	8	1.375	UNF	—	16	1.438
$1 5/8$	1.625	—	—	—	—	18	—	8	1.500	12	1.547	16	1.562
$1 3/4$	1.750	5	1.562	—	—	—	—	8	1.625	12	1.672	16	1.688
$1 7/8$	1.875	—	—	—	—	—	—	8	1.750	12	1.797	16	1.812
2	2.000	4.5	1.781	—	—	—	—	8	1.875	12	1.922	16	1.938
$2 1/4$	2.250	4.5	2.031	—	—	—	—	8	2.125	12	2.172	16	2.188
$2 1/2$	2.500	4	2.250	—	—	—	—	8	2.375	12	2.422	16	2.438
$2 3/4$	2.750	4	2.500	—	—	—	—	8	2.625	12	2.672	16	2.688
3	3.000	4	2.750	—	—	—	—	8	2.875	12	2.922	16	2.938
$3 1/4$	3.250	4	3.000	—	—	—	—	8	3.125	12	3.172	16	3.188
$3 1/2$	3.500	4	3.250	—	—	—	—	8	3.375	12	3.422	16	3.438
$3 3/4$	3.750	4	3.500	—	—	—	—	8	3.625	12	3.668	16	3.688
4	4.000	4	3.750	—	—	—	—	8	3.875	12	3.922	16	3.938

Note: The tap diameter sizes shown are nominal. The class and length of thread will govern the limits on the tapped hole size.

Table 11   Inch screw threads.

Table 12 ISO metric screw threads.

NOMINAL SIZE DIA (mm) Preferred	SERIES WITH GRADED PITCHES				SERIES WITH CONSTANT PITCHES																	
	COARSE		FINE		4		3		2		1.5		1.25		1		0.75		0.5		0.35	
	Thread Pitch	Tap Drill Size	Thread Pitch	Tap Drill Size	Thread Pitch	Tap Drill Size	Thread Pitch	Tap Drill Size	Thread Pitch	Tap Drill Size	Thread Pitch	Tap Drill Size	Thread Pitch	Tap Drill Size	Thread Pitch	Tap Drill Size	Thread Pitch	Tap Drill Size	Thread Pitch	Tap Drill Size	Thread Pitch	Tap Drill Size
1.6	0.35	1.25																				
1.8	0.35	1.45																				
2	0.4	1.6																				
2.2	0.45	1.75																				
2.5	0.45	2.05																			0.35	2.15
3	0.5	2.5																			0.35	2.65
3.5	0.6	2.9																			0.35	3.15
4	0.7	3.3																	0.5	3.5		
4.5	0.75	3.7																	0.5	4.0		
5	0.8	4.2																	0.5	4.5		
6	1	5.0															0.75	5.2				
8	1.25	6.7	1	7.0											1	7.0	0.75	7.2				
10	1.5	8.5	1.25	8.7									1.25	8.7	1	9.0	0.75	9.2				
12	1.75	10.2	1.25	10.8							1.5	10.5	1.25	10.7	1	11						
14	2	12	1.5	12.5							1.5	12.5	1.25	12.7	1	13						
16	2	14	1.5	14.5							1.5	14.5			1	15						
18	2.5	15.5	1.5	16.5					2	16	1.5	16.5			1	17						
20	2.5	17.5	1.5	18.5					2	18	1.5	18.5			1	19						
22	2.5	19.5	1.5	20.5					2	20	1.5	20.5			1	21						
24	3	21	2	22					2	22	1.5	22.5			1	23						
27	3	24	2	25					2	25	1.5	25.5			1	26						
30	3.5	26.5	2	28					2	28	1.5	28.5			1	29						
33	3.5	29.5	2	31					2	31	1.5	31.5										
36	4	32	3	33					2	34	1.5	34.5										
39	4	35	3	36					2	37	1.5	37.5										
42	4.5	37.5	3	39	4	38	3	39	2	40	1.5	40.5										
45	4.5	39	3	42	4	41	3	42	2	43	1.5	43.5										
48	5	43	3	45	4	44	3	45	2	46	1.5	46.5										

	U.S. CUSTOMARY (INCHES)										METRIC (MILLIMETERS)										
Nominal Size	Hexagon Head		Socket Head		Flat Head		Fillister Head		Round or Oval Head		Nominal Size	Hexagon Head		Socket Head		Flat Head		Fillister Head		Pan Head	
	A	H	A	H	A	H	A	H	A	H		A	H	A	H	A	H	A	H	A	H
.250	.44	.17	.38	.25	.50	.14	.38	.17	.44	.19	M3	5.5	2	5.5	3	5.6	1.6	6	2.4	5.6	1.9
.312	.50	.22	.47	.31	.62	.18	.44	.20	.56	.25	4	7	2.8	7	4	7.5	2.2	8	3.1	7.5	2.5
.375	.56	.25	.56	.38	.75	.21	.56	.25	.62	.27	5	8.5	3.5	9	5	9.2	2.5	10	3.8	9.2	3.1
.438	.62	.30	.66	.44	.81	.21	.62	.30	.75	.33	6	10	4	10	6	11	3	12	4.6	11	3.8
.500	.75	.34	.75	.50	.88	.21	.75	.33	.81	.35	8	13	5.5	13	8	14.5	4	16	6	14.5	5
.625	.94	.42	.94	.62	1.12	.28	.88	.42	1.00	.44	10	17	7	16	10	18	.5	20	7.5	18	6.2
.750	1.12	.50	1.12	.75	1.38	.35	1.00	.50	1.25	.55	12	19	8	18	12						
											14	22	9	22	14						
											16	24	10	24	16						

Table 13 Common cap screws.

U.S. CUSTOMARY (INCHES)			METRIC (MILLIMETERS)		
Nominal Bolt Size	Width Across Flats F	Thickness T	Nominal Bolt Size and Thread Pitch	Width Across Flats F	Thickness T
.250	.438	.172	M5 x 0.8	8	3.9
.312	.500	.219	M6 x 1	10	4.7
.375	.562	.250	M8 x 1.25	13	5.7
.438	.625	.297			
.500	.750	.344	M10 x 1.5	15	6.8
.625	.938	.422	M12 x 1.75	18	8
.750	1.125	.500	M14 x 2	21	9.3
.875	1.312	.578	M16 x 2	24	10.5
1.000	1.500	.672	M20 x 2.5	30	13.1
1.125	1.688	.750	M24 x 3	36	15.6
1.250	1.875	.844	M30 x 3.5	46	19.5
1.375	2.062	.906	M36 x 4	55	23.4
1.500	2.250	1.000			

Table 14 Hexagon-head bolts and cap screws.

| NOMINAL BOLT SIZE | HEAD SIZES | | |
AND THREAD PITCH	F	A	H
M5 x 0.8	9.4	5.9	5
M6 x 1	11.8	7.4	6.3
M8 x 1.25	15	9.4	8
M10 x 1.5	18.6	11.7	10
M12 x 1.75	22.8	14	12
M14 x 2	26.4	16.3	14
M16 x 2	30.3	18.7	16
M20 x 2.5	37.4	23.4	20

Table 15   Twelve-spline flange screws.

SETSCREW HEADS

SETSCREW POINTS

U.S. CUSTOMARY (INCHES)		METRIC (MILLIMETERS)	
Nominal Size	Key Size	Nominal Size	Key Size
.125	.06	M1.4	0.7
.138	.06	2	0.9
.164	.08	3	1.5
.190	.09	4	2
.250	.12	5	2
.312	.16	6	3
.375	.19	8	4
.500	.25	10	5
.625	.31	12	6
.750	.38	16	8

Table 16   Setscrews.

WASHER FACE
REGULAR
STYLE 1

THICK
STYLE 2

U.S. CUSTOMARY (INCHES)			
	Distance Across Flats F	Thickness Max.	
Nominal Nut Size		Style 1 H	Style 2 H₁
.250	.438	.218	.281
.312	.500	.266	.328
.375	.562	.328	.406
.438	.625	.375	.453
.500	.750	.438	.562
.562	.875	.484	.609
.625	.938	.547	.719
.750	1.125	.641	.812
.875	1.312	.750	.906
1.000	1.500	.859	1.000
1.125	1.688	.969	1.156
1.250	1.875	1.062	1.250
1.375	2.062	1.172	1.375
1.500	2.250	1.281	1.500

METRIC (MILLIMETERS)			
	Distance Across Flats F	Thickness Max.	
Nominal Nut Size and Thread Pitch		Style 1 H	Style 2 H₁
M4 x 0.7	7	—	3.2
M5 x 0.8	8	4.5	5.3
M6 x 1	10	5.6	6.5
M8 x 1.25	13	6.6	7.8
M10 x 1.5	15	9	10.7
M12 x 1.75	18	10.7	12.8
M14 x 2	21	12.5	14.9
M16 x 2	24	14.5	17.4
M20 x 2.5	30	18.4	21.2
M24 x 3	36	22	25.4
M30 x 3.5	46	26.7	31
M36 x 4	55	32	37.6

Table 17   Hexagon-head nuts.

STYLE 1

STYLE 2

METRIC (MILLIMETERS)							
Nominal Nut Size and Thread Pitch	Width Across Flats F	Style 1				Style 2	
		H	J	K	M	H	J
M6 x 1	10	5.8	3	1	14.2	6.7	3.7
M8 x 1.25	13	6.8	3.7	1.3	17.6	8	4.5
M10 x 1.5	15	9.6	5.5	1.5	21.5	11.2	6.7
M12 x 1.75	18	11.6	6.7	2	25.6	13.5	8.2
M14 x 2	21	13.4	7.8	2.3	29.6	15.7	9.6
M16 x 2	24	15.9	9.5	2.5	34.2	18.4	11.7
M20 x 2.5	30	19.2	11.1	2.8	42.3	22	12.6

Table 18   Hex flange nuts.

HEX NUTS

SIZE, SHAPE, AND LOCATION OF THE PREVAILING-TORQUE ELEMENT OPTIONAL

HEX FLANGE NUTS

SIZE, SHAPE, AND LOCATION OF THE PREVAILING-TORQUE ELEMENT OPTIONAL

NOMINAL NUT SIZE AND THREAD PITCH	WIDTH ACROSS FLATS F	HEX NUTS				HEX FLANGE NUTS					
		Style 1		Style 2		Style 1				Style 2	
		H max.	J max.	H max.	J max.	H	J	K	M	H	J
M5 × 0.8	8.0	6.1	2.3	7.6	2.9						
M6 × 1	10	7.6	3	8.8	3.7	7.6	3	1	14.2	8.8	3.7
M8 × 1.25	13	9.1	3.7	10.3	4.5	9.1	3.7	1.3	17.6	10.3	4.5
M10 × 1.5	15	12	5.5	14	6.7	12	5.5	1.5	21.5	14	6.7
M12 × 1.75	18	14.2	6.7	16.8	8.2	14.4	6.7	2	25.6	16.8	8.2
M14 × 2	21	16.5	7.8	18.9	9.6	16.6	7.8	2.3	29.6	18.9	9.6
M16 × 2	24	18.5	9.5	21.4	11.7	18.9	9.5	2.5	34.2	21.4	11.7
M20 × 2.5	30	23.4	11.1	26.5	12.6	23.4	11.1	2.8	42.3	26.5	12.6
M24 × 3	36	28	13.3	31.4	15.1						
M30 × 3.5	46	33.7	16.4	38	18.5						
M36 × 4	55	40	20.1	45.6	22.8						

Table 19   Prevailing-torque insert-type nuts.

SLOTTED FLAT COUNTERSUNK HEAD

SLOTTED OVAL COUNTERSUNK HEAD

PAN HEAD

HEX HEAD

HEX WASHER HEAD

U.S. CUSTOMARY (INCHES)

NOMINAL SIZE		SLOTTED FLAT COUNTERSUNK HEAD		SLOTTED OVAL COUNTERSUNK HEAD		PAN HEAD		HEX HEAD		HEX WASHER HEAD		
No.	DIA.	A	H	A	H	A	H	A	H	A	B	H
2	.086	.17	.05	.17	.05	.17	.05	.12	.05	.12	.17	.05
4	.112	.23	.07	.23	.07	.22	.07	.19	.06	.19	.24	.06
6	.138	.28	.08	.28	.08	.27	.08	.25	.09	.25	.33	.09
8	.164	.33	.10	.33	.10	.32	.10	.25	.11	.25	.35	.11
10	.190	.39	.17	.39	.17	.37	.11	.31	.12	.31	.41	.12

METRIC (MILLIMETERS)

NOMINAL SIZE	SLOTTED FLAT COUNTERSUNK HEAD		SLOTTED OVAL COUNTERSUNK HEAD		PAN HEAD			HEX HEAD		HEX WASHER HEAD		
							Slot Recess					
	A	H	A	H	A	H	H	A	H	A	B	H
2	3.6	1.2	3.6	1.2	3.9	1.4	1.6	3.2	1.3	3.2	4.2	1.3
2.5	4.6	1.5	4.6	1.5	4.9	1.7	2	4	1.4	4	5.3	1.4
3	5.5	1.8	5.5	1.8	5.8	1.9	1.3	5	1.5	5	6.2	1.5
3.5	6.5	2.1	6.5	2.1	6.8	2.3	2.5	5.5	2.4	5.5	7.5	2.4
4	7.5	2.3	7.5	2.3	7.8	2.6	2.8	7	2.8	7	9.2	2.8
5	9.5	2.9	9.5	2.9	9.8	3.1	3.5	8	3	8	10.5	3
6	11.9	3.6	11.9	3.6	12	3.9	4.3	10	4.8	10	13.2	4.8
8	15.2	4.4	15.2	4.4	15.6	5	5.6	13	5.8	13	17.2	5.8
10	1.9	5.4	19	5.4	19.5	6.2	7	15	7.5	15	19.8	7.5
12	22.9	6.4	22.9	6.4	23.4	7.5	8.3	18	9.5	18	23.8	9.5

Table 20   Tapping screws.

Table 21

KIND OF MATERIAL	THREAD-FORMING								THREAD-CUTTING			SELF DRILLING	
	Type A	Type B	Type AB	HEX HEAD B	SWAGE FORM*	SWAGE FORM* B	Type U	Type 21	Type F*	Type L	Type B-F*	DRIL-KWICK	TAPITS*
SHEET METAL .0156 to .0469in. thick (0.4 to 1.2mm) (Steel, Brass, Aluminum, Monel, etc.)	●	●	●	●	●	●		●				●	●
SHEET STAINLESS STEEL .0156 to .0469in thick (0.4 to 1.2mm)	●	●		●	●	●		●	●			●	●
SHEET METAL .20 to .50in. thick (1.2 to 5mm) (Steel, Brass, Aluminum, etc.)		●	●	●	●	●	●	●	●			●	
STRUCTURAL STEEL .20 to .50in. thick (1.2 to .5mm)				●	●		●		●				
CASTINGS (Aluminum, Magnesium, Zinc, Brass, Bronze, etc.)		●	●	●	●	●	●		●				
CASTINGS (Grey Iron, Malleable Iron, Steel, etc.)					●		●		●				
FORGINGS (Steel, Brass, Bronze, etc.)					●		●		●				
PLYWOOD, Resin Impregnated: Compreg, Pregwood, etc. **NATURAL WOODS**	●	●	●	●					●		●	●	●
ASBESTOS and other compositions: Ebony, Asbestos, Transite, Fiberglas, Insurok, etc.	●	●	●	●		●						●	●
PHENOL FORMALDEHYDE: **Molded:** Bakelite, Durez, etc. **Cast:** Catalin, Marblette, etc. **Laminated:** Formica, Textolite, etc.		●	●	●		●	●		●		●		
UREA FORMALDEHYDE: **Molded:** Plaskon, Beetle, etc. **MELAMINE FORMALDEHYDE:** Melantite, Melamac						●	●				●		
CELLULOSE ACETATES and NITRATES: Tenite, Lumarith, Plastacele Pyralin, Celanese, etc. **ACRYLATE & STYRENE RESINS:** Lucite, Plexiglas, Styron, etc.		●	●	●	●	●	●				●		
NYLON PLASTICS: Nylon, Zytel					●	●	●			●			

Table 21 Selector guide to thread cutting screws.

Table 22 Common washer sizes.

U.S. CUSTOMARY (INCHES)

Bolt Size	Flat Washers Type A–N Id	Od	Thick	Lockwashers Regular Id	Od	Thick
#6	.156	.375	.049	.141	.250	.031
#8	.188	.438	.049	.168	.293	.040
#10	.219	.500	.049	.194	.334	.047
#12	.250	.562	.065	.221	.377	.056
.250	.281	.625	.065	.255	.489	.062
.312	.344	.688	.065	.318	.586	.078
.375	.406	.812	.065	.382	.683	.094
.438	.469	.922	.065	.446	.779	.109
.500	.531	1.062	.095	.509	.873	.125
.562	.594	1.156	.095	.572	.971	.141
.625	.656	1.312	.095	.636	1.079	.156
.750	.812	1.469	.134	.766	1.271	.188
.875	.938	1.750	.134	.890	1.464	.219
1.000	1.062	2.000	.134	1.017	1.661	.250
1.125	1.250	2.250	.134	1.144	1.853	.281
1.250	1.375	2.500	.165	1.271	2.045	.312
1.375	1.500	2.750	.165	1.398	2.239	.344
1.500	1.625	3.000	.165	1.525	2.430	.375

METRIC (MILLIMETERS)

Bolt Size	Flat Washers Id	Od	Thick	Lockwashers Id	Od	Thick	Spring Lockwashers Id	Od	Thick
2	2.2	5.5	0.5	2.1	3.3	0.5			
3	3.2	7	0.5	3.1	5.7	0.8			
4	4.3	9	0.8	4.1	7.1	0.9	4.2	8	0.3 / 0.4
5	5.3	11	1	5.1	8.7	1.2	5.2	10	0.4 / 0.5
6	6.4	12	1.5	6.1	11.1	1.6	6.2	12.5	0.5 / 0.7
7	7.4	14	1.5	7.1	12.1	1.6	7.2	14	0.5 / 0.8
8	8.4	17	2	8.2	14.2	2	8.2	16	0.6 / 0.9
10	10.5	21	2.5	10.2	17.2	2.2	10.2	20	0.8 / 1.1
12	13	24	2.5	12.3	20.2	2.5	12.2	25	0.9 / 1.5
14	15	28	2.5	14.2	23.2	3	14.2	28	1.0 / 1.5
16	17	30	3	16.2	26.2	3.5	16.3	31.5	1.2 / 1.7
18	19	34	3	18.2	28.2	3.5	18.3	35.5	1.2 / 2.0
20	21	36	3	20.2	32.2	4	20.4	40	1.5 / 2.25
22	23	39	4	22.5	34.5	4	22.4	45	1.75 / 2.5
24	25	44	4	24.5	38.5	5			
27	28	50	4	27.5	41.5	5			
30	31	56	4	30.5	46.5	6			

Table 23 Belleville washers.
(Wallace Barnes Co. Ltd.)

U.S. CUSTOMARY (INCHES)

Outside Diameter Max.	Inside Diameter Min.	Stock Thickness T	H Approx.
.187	.093	.010	.015
.250	.125	.009 / .013	.017 / .020
.281	.138	.010 / .015	.020 / .023
.312	.156	.011 / .017	.022 / .025
.343	.164	.013 / .019	.024 / .028
.375	.190	.015 / .020	.027 / .030
.500	.255	.018 / .025	.034 / .038
.625	.317	.022 / .032	.042 / .048
.750	.380	.028 / .040	.051 / .059
.875	.442	.031 / .045	.059 / .067
1.000	.505	.035 / .050	.067 / .075
1.125	.567	.038 / .056	.073 / .084
1.250	.630	.040 / .062	.082 / .092
1.375	.692	.044 / .067	.088 / .101

METRIC (MILLIMETERS)

Outside Diameter Max.	Inside Diameter Min.	Stock Thickness T	H Approx.
4.8	2.4	0.16 / 0.25	0.33 / 0.38
6.4	3.2	0.22 / 0.34	0.44 / 0.51
7.9	4	0.27 / 0.42	0.55 / 0.64
9.5	4.8	0.38 / 0.51	0.69 / 0.76
12.7	6.5	0.46 / 0.64 / 0.97	0.86 / 0.97 / 1.20
15.9	8.1	0.56 / 0.81	1.07 / 1.22
19.1	9.7	0.71 / 1.02 / 1.42	1.3 / 1.5 / 1.8
22.2	11.2	0.79 / 1.14	1.5 / 1.7
25.4	12.8	0.89 / 1.27 / 1.85	1.7 / 1.9 / 2.3
28.6	14.4	0.97 / 1.42	1.9 / 2.1
31.8	16	1.02 / 1.58	2.1 / 2.3
34.9	17.6	1.12 / 1.70	2.2 / 2.6
38.1	19.2	1.14 / 1.83	2.4 / 2.7
44.5	22.4	1.45 / 2.16	2.9 / 3.3
50.8	25.4	1.65 / 2.46	3.3 / 3.7
63.5	31.8	2.03 / 3.05	4.1 / 4.6

U.S. CUSTOMARY (INCHES)						METRIC (MILLIMETERS)					
Diameter of Shaft		Square Key Nominal Size		Flat Key Nominal Size		Diameter of Shaft		Square Key Nominal Size		Flat Key Nominal Size	
From	To	W	H	W	H	Over	Up To	W	H	W	H
.500	.562	.125	.125	.125	.094	6	8	2	2		
.625	.875	.188	.188	.188	.125	8	10	3	3		
.938	1.250	.250	.250	.250	.188	10	12	4	4		
						12	17	5	5		
1.312	1.375	.312	.312	.312	.250	17	22	6	6		
1.438	1.750	.375	.375	.375	.250	22	30	7	7	8	7
1.812	2.250	.500	.500	.500	.375	30	38	8	8	10	8
						38	44	9	9	12	8
						44	50	10	10	14	9
						50	58	12	12	16	10

Table 24 Square and flat stock keys.

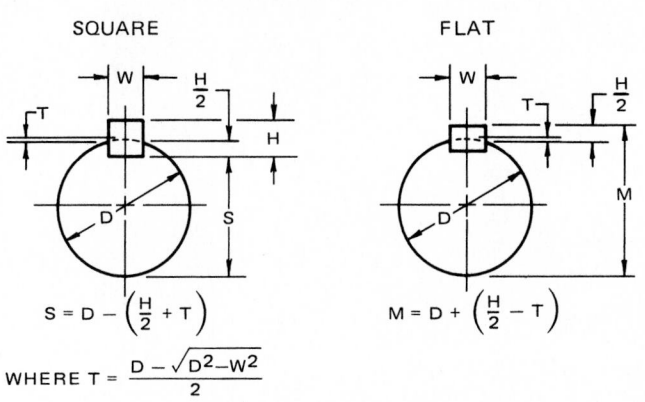

SQUARE

FLAT

$$S = D - \left(\frac{H}{2} + T\right)$$

$$M = D + \left(\frac{H}{2} - T\right)$$

$$\text{WHERE } T = \frac{D - \sqrt{D^2 - W^2}}{2}$$

WOODRUFF KEYS

U.S. CUSTOMARY (INCHES)						METRIC (MILLIMETERS)					
Nominal Size	Key			Keyseat	Key No.	Nominal Size	Key			Keyseat
A × B	E	C	D	H		A × B	E	C	D	H
.062 × .500	.047	.203	.194	.172	204	1.6 × 12.7	1.5	5.1	4.8	4.2
.094 × .500	.047	.203	.194	.156	304	2.4 × 12.7	1.3	5.1	4.8	3.8
.094 × .625	.062	.250	.240	.203	305	2.4 × 15.9	1.5	6.4	6.1	5.1
.125 × .500	.049	.203	.194	.141	404	3.2 × 12.7	1.3	5.1	4.8	3.6
.125 × .625	.062	.250	.240	.188	405	3.2 × 15.9	1.5	6.4	6.1	4.6
.125 × .750	.062	.313	.303	.251	406	3.2 × 19.1	1.5	7.9	7.6	6.4
.156 × .625	.062	.250	.240	.172	505	4.0 × 15.9	1.5	6.4	6.1	4.3
.156 × .750	.062	.313	.303	.235	506	4.0 × 19.1	1.5	7.9	7.6	5.8
.156 × .875	.062	.375	.365	.297	507	4.0 × 22.2	1.5	9.7	9.1	7.4
.188 × .750	.062	.313	.303	.219	606	4.8 × 19.1	1.5	7.9	7.6	5.3
.188 × .875	.062	.375	.365	.281	607	4.8 × 22.2	1.5	9.7	9.1	7.1
.188 × 1.000	.062	.438	.428	.344	608	4.8 × 25.4	1.5	11.2	10.9	8.6
.188 × 1.125	.078	.484	.475	.390	609	4.8 × 28.6	2.0	12.2	11.9	9.9
.250 × .875	.062	.375	.365	.250	807	6.4 × 22.2	1.5	9.7	9.1	6.4
.250 × 1.000	.062	.438	.428	.313	808	6.4 × 25.4	1.5	11.2	10.9	7.9

NOTE: METRIC KEY SIZES WERE NOT AVAILABLE AT THE TIME OF PUBLICATION. SIZES SHOWN ARE INCH-DESIGNED KEY-SIZES SOFT CONVERTED TO MILLIMETERS. CONVERSION WAS NECESSARY TO ALLOW THE STUDENT TO COMPARE KEYS WITH SLOT SIZES GIVEN IN MILLIMETERS.

Table 25 Woodruff keys.

L (MIN) = 4W
L (MAX) = 16W

L (MIN) = 2W

U.S. CUSTOMARY (INCHES)

Shaft Diameter	Square Type					Flat Type				
	W	H	C	D	E	W	H	C	D	E
.500–.562	.125	.125	.250	.219	.156	.125	.094	.188	.125	.125
.625–.875	.188	.188	.312	.281	.219	.188	.125	.250	.188	.156
.938–1.250	.250	.250	.438	.344	.344	.250	.188	.312	.250	.188
1.312–1.375	.312	.312	.562	.406	.406	.312	.250	.375	.312	.250
1.438–1.750	.375	.375	.688	.469	.469	.375	.250	.438	.375	.312
1.812–2.250	.500	.500	.875	.594	.625	.500	.375	.625	.500	.438
2.312–2.750	.625	.625	1.062	.719	.750	.625	.438	.750	.625	.500
2.875–3.250	.750	.750	1.250	.875	.875	.750	.500	.875	.750	.625

METRIC (MILLIMETERS)

Shaft Diameter	Square Type					Flat Type				
	W	H	C	D	E	W	H	C	D	E
12–14	3.2	3.2	6.4	5.4	4	3.2	2.4	5	3.2	3.2
16–22	4.8	4.8	10	7	5.4	4.8	3.2	6.4	5	4
24–32	6.4	6.4	11	8.6	8.6	6.4	5	8	6.4	5
34–35	8	8	14	10	10	8	6.4	10	8	6.4
36–44	10	10	18	12	12	10	6.4	11	10	8
46–58	13	13	22	15	16	13	10	16	13	11
60–70	16	16	27	19	20	16	11	20	16	13
72–82	20	20	32	22	22	20	13	22	20	16

Note: Metric standards governing key sizes were not available at the time of publication. The sizes given in the above chart are "soft conversion" from current standards and are not representative of the precise metric key sizes which may be available in the future. Metric sizes are given only to allow the student to complete the drawing assignment.

Table 26 Square and flat gib-head keys.

U.S. CUSTOMARY (INCHES)

Key No.	L	W	H	D
2	.500	.094	.141	.094
4	.625	.094	.141	.094
6	.625	.156	.234	.156
8	.750	.156	.234	.156
10	.875	.156	.234	.156
12	.875	.234	.328	.219
14	1.00	.234	.328	.234
16	1.125	.188	.281	.188
18	1.125	.250	.375	.250
20	1.250	.219	.328	.219
22	1.375	.250	.375	.250
24	1.50	.250	.375	.250
26	2.00	.188	.281	.188
28	2.00	.312	.469	.312
30	3.00	.375	.562	.375
32	3.00	.500	.750	.500
34	3.00	.625	.938	.625

METRIC (MILLIMETERS)

Key No.	L	W	H	D
2	12	2.4	3.6	2.4
4	16	2.4	3.6	2.4
6	16	4	6	4
8	20	4	6	4
10	22	4	6	4
12	22	6	8.4	7
14	25	6	8.4	6
16	28	5	7	5
18	28	6.4	10	6.4
20	32	7	8	5
22	35	6.4	10	6.4
24	38	6.4	10	6.4
26	50	5	7	5
28	50	8	12	8
30	75	10	14	10
32	75	12	20	12
34	75	16	24	16

Table 27 Pratt and Whitney keys.

STANDARD

MITRE END

PRONG SQUARE CUT

HAMMER LOCK

EXTENDED MITRE END

BEVEL POINT

CHISEL POINT

U.S. CUSTOMARY (INCHES)				METRIC (MILLIMETERS)			
Nominal Bolt or Thread Size Range	Nominal Cotter-Pin Size (A)	Cotter-Pin Hole	Min. End Clearance	Nominal Bolt or Thread-Size Range	Nominal Cotter-Pin Size (A)	Cotter-Pin Hole	Min. End Clearance*
.125	.031	.047	.06	–2.5	0.6	0.8	1.5
.188	.047	.062	.08	2.5–3.5	0.8	1.0	2.0
.250	.062	.078	.11	3.5–4.5	1.0	1.2	2.0
.312	.078	.094	.11	4.5–5.5	1.2	1.4	2.5
.375	.094	.109	.14	5.5–7.0	1.6	1.8	2.5
.438	.109	.125	.14	7.0–9.0	2.0	2.2	3.0
.500	.125	.141	.18	9.0–11	2.5	2.8	3.5
.562	.141	.156	.25	11–14	3.2	3.6	5
.625	.156	.172	.40	14–20	4	4.5	6
1.000–1.125	.188	.203	.40	20–27	5	5.6	7
1.250–1.375	.219	.234	.46	27–39	6.3	6.7	10
1.500–1.625	.250	.266	.46	39–56	8.0	8.5	15
				56–80	10	10.5	20

*End of bolt to center of hole

Table 28   Cotter pins.

COTTER PIN HOLE F

U.S. CUSTOMARY (INCHES)						METRIC (MILLIMETERS)					
Pin Dia					Drill Size	Pin Dia.					Drill Size
A	B	C	D	E	Min. F	A	B	C	D	E	Min. F
.188	.31	.06	.59	.11	.078	4	6	1	16	2.2	1
.250	.38	.09	.80	.12	.078	6	10	2	20	3.2	1.6
.312	.44	.09	.97	.16	.109	8	14	3	24	3.5	2
.375	.50	.12	1.09	.16	.109	10	18	4	28	4.5	3.2
.500	.62	.16	1.42	.22	.141	12	20	4	36	5.5	3.2
.625	.81	.20	1.72	.25	.141	16	25	4.5	44	6	4
.750	.94	.25	2.05	.30	.172	20	30	5	52	8	5
1.000	1.19	.34	2.62	.36	.172	24	36	6	66	9	6.3

Table 29   Clevis pins.

U.S. CUSTOMARY (INCHES)

NUMBER	7/0	6/0	5/0	4/0	3/0	2/0	0	1	2	3	4	5	6	7	8	9
SIZE (LARGE END)	.062	.078	.094	.109	.125	.141	.152	.172	.193	.219	.250	.289	.314	.409	.492	.591
LENGTH .375	x	x														
.500	x	x	x	x	x	x	x									
.625	x	x	x	x	x	x	x									
.750		x	x	x	x	x	x	x	x	x						
.875						x	x	x	x	x						
1.000			x	x	x	x	x	x	x	x	x	x				
1.250						x	x	x	x	x	x	x	x			
1.500							x	x	x	x	x	x	x			
1.750								x	x	x	x	x	x			
2.000								x	x	x	x	x	x	x	x	
2.250									x	x	x	x	x	x	x	
2.500									x	x	x	x	x	x	x	
2.750										x	x	x	x	x	x	x

METRIC (MILLIMETERS)

SIZE (LARGE END)	1.6	2	2.4	2.8	3.2	3.6	4	4.4	4.9	5.6	6.4	7.4	8	10.4	12.5	15
LENGTH 10	x	x														
12	x	x	x	x	x	x	x									
16	x	x	x	x	x	x	x									
20		x	x	x	x	x	x	x	x	x						
22						x	x	x	x	x						
25			x	x	x	x	x	x	x	x	x	x				
30						x	x	x	x	x	x	x	x			
40							x	x	x	x	x	x	x			
45								x	x	x	x	x	x			
50								x	x	x	x	x	x	x	x	
55									x	x	x	x	x	x	x	
65									x	x	x	x	x	x	x	
70										x	x	x	x	x	x	x

Table 30　Taper pins.

Table 31　Spring pins.

PIN DIAMETER (INCHES)								PIN DIAMETER (MILLIMETERS)									
Length	.062	.094	.125	.156	.188	.250	.312	Length	1.5	2	2.5	3	4	5	6	8	10	12
.250	x	x						5	x	x								
.375	x	x	x					10	x	x	x	x						
.500	x	x	x	x	x			15	x	x	x	x	x	x				
.625	x	x	x	x	x	x		20	x	x	x	x	x	x	x			
.750	x	x	x	x	x	x	x	25	x	x	x	x	x	x	x	x		
.875	x	x	x	x	x	x		30		x	x	x	x	x	x	x	x	x
1.00	x	x	x	x	x	x	x	35		x	x	x	x	x	x	x	x	x
1.250		x	x	x	x	x	x	40		x	x	x	x	x	x	x	x	x
1.500		x	x	x	x	x	x	45			x	x	x	x	x	x	x	x
1.750			x	x	x	x	x	50			x	x	x	x	x	x	x	x
2.000			x	x	x	x	x	55				x	x	x	x	x	x	x
2.225				x	x	x	x	60				x	x	x	x	x	x	x
2.500			x		x	x	x	70							x	x	x	x
3.000						x	x	75							x	x	x	x
3.500						x	x	80							x	x	x	x

| | U.S. CUSTOMARY (INCHES) | | | | | | | | METRIC (MILLIMETERS) | | | | | | | |
| | PIN DIAMETER | | | | | | | | PIN DIAMETER | | | | | | | |
Length	.09	.125	.188	.250	.312	.375	.500	Length	2	3	4	5	6	8	10	12
.250	x	x						5	x	x	x					
.375	x	x	x					10	x	x	x	x	x			
.500	x	x	x	x				15	x	x	x	x	x	x		
.625	x	x	x	x	x			20	x	x	x	x	x	x	x	
.750	x	x	x	x	x	x		25	x	x	x	x	x	x	x	x
.875	x	x	x	x	x	x		30	x	x	x	x	x	x	x	x
1.000	x	x	x	x	x	x	x	35		x	x	x	x	x	x	x
1.250	x	x	x	x	x	x	x	40			x	x	x	x	x	x
1.500		x	x	x	x	x	x	45			x	x	x	x	x	x
1.750			x	x	x	x	x	50			x	x	x	x	x	x
2.000		x		x	x	x	x	55					x	x	x	x
2.250		x		x	x	x	x	60					x	x	x	x
2.275				x	x	x	x	65					x	x	x	x
3.000				x	x	x	x	70							x	x
								75							x	x

Note: Metric size pins were not available at the time of publication. Sizes were soft converted to allow students to complete drawing assignments.

Table 32 Groove pins.

WIDELY USED FOR FASTENING BRACKETS

ATTACHING NAMEPLATES, INSTRUCTION PANELS

| | U.S. CUSTOMARY (INCHES) | | | | | | | | | | METRIC (MILLIMETERS) | | | | | | | |
| STUD NUMBER | SHANK DIA. | DRILL SIZE | HEAD DIA. | STANDARD LENGTHS | | | | | | STUD NUMBER | SHANK DIA. | DRILL SIZE | HEAD DIA. | STANDARD LENGTHS | | | | | |
				.125	.188	.250	.312	.375	.500					4	6	8	10	12	14
0	.067	51	.130	•	•	•				0	1.7	1.7	3.3	•	•	•			
2	.086	44	.162	•	•	•				2	2.2	2.2	4.1	•	•	•			
4	.104	37	.211		•	•	•			4	2.6	2.6	5.4		•	•	•		
6	.120	31	.260			•	•	•		6	3.0	3.0	6.6			•	•	•	
7	.136	29	.309				•	•	•	7	3.4	3.4	7.8				•	•	•
8	.144	27	.309					•	•	8	3.8	3.8	7.8					•	•
10	.161	20	.359					•	•	10	4.1	4.1	9.1					•	•
12	.196	9	.408						•	12	5.0	5.0	10.4						•
14	.221	2	.457						•	14	5.6	5.6	11.6						•
16	.250	¼	.472						•	16	6.3	6.3	12						•

Note: Metric size studs were not available at the time of publication. Sizes were soft converted to allow students to complete drawing assignments.

Table 33 Grooved studs. (Drive-Lok)

Southco Lion Fasteners

Use these columns first to locate your correct GRIP LENGTH

Grip = Total thickness of all sheets fastened together

Minimum Grip	Nominal Grip	Maximum Grip	Length Under Head L
1	2	3	5
2	3	4	6
4	5	6	8
5	6	7	10
7	8	9	12
9	10	11	14
11	12	13	16
13	14	15	18
15	16	17	20
19	20	21	24
23	24	25	28
27	28	29	32

.125in. (3mm) Dia — Part Numbers

Length Under Head L	Universal Head	100° Csk Head	Full Brazier Head
5	✓	✓	✓
6	✓	✓	✓
8	✓	✓	✓
10	✓	✓	✓
12	✓	✓	✓
14	✓	✓	✓
16	✓		✓
18			
20			
24			
28			
32			

.156in. (4mm) Dia — Part Numbers

Length Under Head L	Universal Head	100° Csk Head
5	✓	
6	✓	✓
8	✓	✓
10	✓	✓
12	✓	✓
14	✓	✓
16	✓	✓
18	✓	✓
20		
24		
28		
32		

.188in. (5mm) Dia — Part Numbers

Length Under Head L	Universal Head	100° Csk Head	Full Brazier Head	All Purpose Liner Head
5	✓			✓
6	✓			✓
8	✓	✓		✓
10	✓	✓		✓
12	✓	✓	✓	✓
14	✓	✓	✓	✓
16	✓	✓	✓	✓
18	✓	✓	✓	
20	✓		✓	✓
24	✓			
28	✓		✓	
32	✓			✓

.250in. (6mm) Dia — Part Numbers

Length Under Head L	Universal Head	100° Csk Head	Full Brazier Head
5	✓	✓	✓
6	✓	✓	✓
8	✓	✓	✓
10	✓	✓	✓
12	✓	✓	✓
14	✓	✓	✓
16	✓	✓	✓
18	✓	✓	✓
20	✓	✓	✓
24	✓	✓	✓
28	✓	✓	✓
32	✓	✓	✓

Note: Metric drive rivets were not available at the time of publication. Sizes were soft converted to allow students to complete drawing assignments.

HIT THE PIN — Drive pin flush with rivet head

Expanding prongs clinch sheets tightly, eliminating gaps.

Metal and wood pulled tightly together. Nothing protrudes through wood.

IN METAL — GRIP — Grip Length = Total thickness of sheets to be fastened.

IN WOOD — Use "L" Dimension (length under head) instead of grip length. L = M (thickness of metal) + D (hole depth in wood).

Table 34 Aluminum drive rivets.

DRILL HOLE SLIGHTLY UNDERSIZE

REAM FULL SIZE

DRIVE OR PRESS LOK DOWELS INTO PLACE

LOK DOWELS LOCK SECURELY AND PARTS SEPARATE EASILY

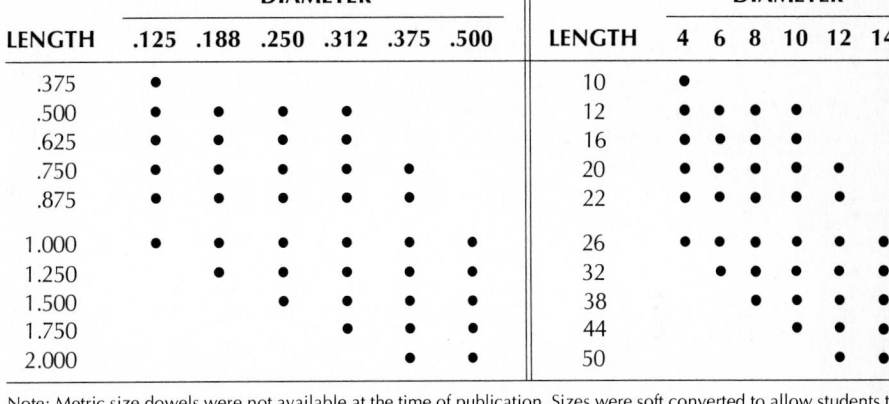

U.S. CUSTOMARY (INCHES)							METRIC (MILLIMETERS)						
	DIAMETER							DIAMETER					
LENGTH	.125	.188	.250	.312	.375	.500	LENGTH	4	6	8	10	12	14
.375	•						10	•					
.500	•	•	•	•			12	•	•	•	•		
.625	•	•	•	•			16	•	•	•	•		
.750	•	•	•	•	•		20	•	•	•	•	•	
.875	•	•	•	•	•		22	•	•	•	•	•	
1.000	•	•	•	•	•	•	26	•	•	•	•	•	•
1.250		•	•	•	•	•	32		•	•	•	•	•
1.500			•	•	•	•	38			•	•	•	•
1.750				•	•	•	44				•	•	•
2.000					•	•	50					•	•

Note: Metric size dowels were not available at the time of publication. Sizes were soft converted to allow students to complete drawing assignments.

Table 35 Lok dowels. (Drive-Lok)

(A) SEMITUBULAR

(B) SPLIT

	D	A	B	C	E	MIN. LENGTH
U.S. CUSTOMARY (INCHES)	.062	.125	.031	.062	—	.062
	.094	.156	.031	.062	.031	.078
	.109	.188	.031	.078	—	.078
	.125	.218	.047	.109	.049	.109
	.141	.250	.047	.125	.049	.125
	.188	.312	.062	.141	.062	.156
	.219	.438	.062	.188	.062	.188
	.250	.500	.078	.219	.094	.219
	.312	.562	.109	.250		.250
METRIC (MILLIMETERS)	1.5	2.8	0.4	1.2		1.6
	2.2	3.7	0.6	1.6	0.8	2.0
	2.5	4.7	0.7	2.0		2.0
	3.1	5.5	0.9	2.4	1.0	2.4
	3.6	5.9	1.0	3.2	1.2	3.2
	4.7	7.9	1.5	3.9	1.6	4.0
	5.4	11.1	1.7	4.8	1.8	4.8
	6.3	12.7	2.0	5.6	2.2	5.6
	7.7	14.3	2.4	6.2		6.4

Table 36 Semi-tubular and split rivets.

U.S. CUSTOMARY (INCHES)				METRIC (MILLIMETERS)			
HOLE DIA.	PANEL RANGE	HEAD Dia.	HEAD Height	HOLE DIA.	PANEL RANGE	HEAD Dia.	HEAD Height
.125	.031–.140	.188	.047	3.18	0.8– 3.6	4.8	1.2
	.031–.125	.218	.062		0.8– 3.2	5.5	1.5
.156	.250–.375	.218	.047	4.01	5.9– 9.4	5.5	1.3
.188	.062–.156	.375	.125	4.75	1.6– 4.0	9.5	3.2
	.156–.281	.438	.094		4.0– 7.1	11.1	1.9
.219	.062–.125	.375	.094	5.54	1.6– 3.2	9.5	2.4
	.094–.312		.078		2.4– 8.0		2.0
.250	.094–.219	.625	.125	6.35	2.3– 5.6	16	3.2
	.125–.375	.750	.062		3.2– 9.5	19	1.3
.297	.140–.328	.500	.078	7.14	3.4– 8.1	12.3	1.9
.375	.250–.500	.438	.109	9.53	6.4–12.7	11.1	2.6
.500	.312–.375	.750	.109	12.7	8.1– 9.4	19	2.5

Table 37 Plastic rivets.

ENLARGED DETAIL
OF GROOVE PROFILE
AND EDGE MARGIN (Z)

SHAFT DIA.	EXTERNAL SERIES	RETAINING RING DIMENSIONS		GROOVE DIMENSIONS				MAX. CORNER RADII AND CHAMFER OF RETAINED PARTS		EDGE MARGIN	NOMINAL GROOVE DEPTH (REF)
				DIAMETER		WIDTH					
S	Size—No.	D	t	G	Tol.	W	Tol.	R Max.	Ch. Max.	z	d
U.S. CUSTOMARY (INCHES)											
.188	5100-18	.168	.015	.175	±.0015	.018	+.002	.014	.008	.018	.006
.250	5100-25	.225	.025	.230	±.0015	.029	+.003	.018	.011	.030	.010
.312	5100-31	.281	.025	.290	±.002	.029	+.003	.020	.012	.033	.011
.375	5100-37	.338	.025	.352	±.002	.029	+.003	.026	.015	.036	.012
.500	5100-50	.461	.035	.468	±.002	.039	+.003	.034	.020	.048	.016
.625	5100-62	.579	.035	.588	±.003	.039	+.003	.041	.025	.055	.018
.750	5100-75	.693	.042	.704	±.003	.046	+.003	.046	.027	.069	.023
.875	5100-87	.810	.042	.821	±.003	.046	+.003	.051	.031	.081	.027
1.000	5100-100	.925	.042	.940	±.003	.046	+.003	.057	.034	.090	.030
1.125	5100-112	1.041	.050	1.059	±.004	.056	+.004	.063	.038	.099	.033
1.250	5100-125	1.156	.050	1.176	±.004	.056	+.004	.068	.041	.111	.037
1.375	5100-137	1.272	.050	1.291	±.004	.056	+.004	.072	.043	.126	.042
1.500	5100-150	1.387	.050	1.406	±.004	.056	+.004	.079	.047	.141	.047
METRIC (MILLIMETERS)											
4	M5100-4	3.6	0.25	3.80	−0.08	0.32	+0.05	0.35	0.25	0.3	0.10
6	M5100-6	5.5	0.4	5.70	−0.08	0.5	+0.1	0.35	0.25	0.5	0.15
8	M5100-8	7.2	0.6	7.50	−0.1	0.7	+0.15	0.5	0.35	0.8	0.25
10	M5100-10	9.0	0.6	9.40	−0.1	0.7	+0.15	0.7	0.4	0.9	0.30
12	M5100-12	10.9	0.6	11.35	−0.12	0.7	+0.15	0.8	0.45	1.0	0.33
14	M5100-14	12.9	0.9	13.25	−0.12	1.0	+0.15	0.9	0.5	1.2	0.38
16	M5100-16	14.7	0.9	15.10	−0.15	1.0	+0.15	1.1	0.6	1.4	0.45
18	M5100-18	16.7	1.1	17.00	−0.15	1.2	+0.15	1.2	0.7	1.5	0.50
20	M5100-20	18.4	1.1	18.85	−0.15	1.2	+0.15	1.2	0.7	1.7	0.58
22	M5100-22	20.3	1.1	20.70	−0.15	1.2	+0.15	1.3	0.8	1.9	0.65
24	M5100-24	22.2	1.1	22.60	−0.15	1.2	+0.15	1.4	0.8	2.1	0.70
25	M5100-25	23.1	1.1	23.50	−0.15	1.2	+0.15	1.4	0.8	2.3	0.75
30	M5100-30	27.9	1.3	28.35	−0.2	1.4	+0.15	1.6	1.0	2.5	0.83
35	M5100-35	32.3	1.3	32.9	−0.2	1.4	+0.15	1.8	1.1	3.1	1.05
40	M5100-40	36.8	1.6	37.7	−0.3	1.75	+0.2	2.1	1.2	3.4	1.15
45	M5100-45	41.6	1.6	42.4	−0.3	1.75	+0.2	2.3	1.4	3.9	1.3
50	M5100-50	46.2	1.6	47.2	−0.3	1.75	+0.2	2.4	1.4	4.2	1.4

Table 38 Retaining rings—external. (© 1965, 1958 Waldes Koh-i-noor, Inc. Reprinted with permission.)

ENLARGED DETAIL
OF GROOVE PROFILE
AND EDGE MARGIN (Z)

| | HOUSING DIA. | INTERNAL SERIES | RETAINING RING DIMENSIONS | | GROOVE DIMENSIONS | | | | MAX. CORNER RADII AND CHAMFER OF RETAINED PARTS | | EDGE MARGIN | NOMINAL GROOVE DEPTH |
| | | | | | DIAMETER | | WIDTH | | | | | |
	S	Size—No.	D	t	G	Tol.	W	Tol.	R Max.	Ch. Max.	z	d
U.S. CUSTOMARY (INCHES)	.250	N5000-25	.280	.015	.268	±.001	.018	+.002	.011	.008	.027	.009
	.312	N5000-31	.346	.015	.330	±.001	.018	+.002	.016	.013	.027	.009
	.375	N5000-37	.415	.025	.397	±.002	.029	+.003	.023	.018	.033	.011
	.500	N5000-50	.548	.035	.530	±.002	.039	+.003	.027	.021	.045	.015
	.625	N5000-62	.694	.035	.665	±.002	.039	+.003	.027	.021	.060	.020
	.750	N5000-75	.831	.035	.796	±.002	.039	+.003	.032	.025	.069	.023
	.875	N5000-87	.971	.042	.931	±.003	.046	+.003	.035	.028	.084	.028
	1.000	N5000-100	1.111	.042	1.066	±.003	.046	+.003	.042	.034	.099	.033
	1.125	N5000-112	1.249	.050	1.197	±.004	.056	+.004	.047	.036	.108	.036
	1.250	N5000-125	1.388	.050	1.330	±.004	.056	+.004	.048	.038	.120	.040
	1.375	N5000-137	1.526	.050	1.461	±.004	.056	+.004	.048	.038	.129	.043
	1.500	N5000-150	1.660	.050	1.594	±.004	.056	+.004	.048	.038	.141	.047
METRIC (MILLIMETERS)	8	MN5000-8	8.80	0.4	8.40	+0.6	0.5	+0.1	0.4	0.3	0.6	0.2
	10	MN5000-10	11.10	0.6	10.50	+0.1	0.7	+0.15	0.5	0.35	0.8	0.25
	12	MN5000-12	13.30	0.6	12.65	+0.1	0.7	+0.15	0.6	0.4	1.0	0.33
	14	MN5000-14	15.45	0.9	14.80	+0.1	1.0	+0.15	0.7	0.5	1.2	0.40
	16	MN5000-16	17.70	0.9	16.90	+0.1	1.0	+0.15	0.7	0.5	1.4	0.45
	18	MN5000-18	20.05	0.9	19.05	+0.1	1.0	+0.15	0.75	0.6	1.6	0.53
	20	MN5000-20	22.25	0.9	21.15	+0.15	1.0	+0.15	0.9	0.7	1.7	0.57
	22	MN5000-22	24.40	1.1	23.30	+0.15	1.2	+0.15	0.9	0.7	1.9	0.65
	24	MN5000-24	26.55	1.1	25.4	+0.15	1.2	+0.15	1.0	0.8	2.1	0.70
	25	MN5000-25	27.75	1.1	26.6	+0.15	1.2	+0.15	1.0	0.8	2.4	0.80
	30	MN5000-30	33.40	1.3	31.9	+0.2	1.4	+0.15	1.2	1.0	2.9	0.95
	35	MN5000-35	38.75	1.3	37.2	+0.2	1.4	+0.15	1.2	1.0	3.3	1.10
	40	MN5000-40	44.25	1.6	42.4	+0.2	1.75	+0.2	1.7	1.3	3.6	1.20
	45	MN5000-45	49.95	1.6	47.6	+0.2	1.75	+0.2	1.7	1.3	3.9	1.30
	50	MN5000-50	55.35	1.6	53.1	+0.2	1.75	+0.2	1.7	1.3	4.6	1.55

Table 39   Retaining rings—internal. (© 1965, 1958 Waldes Koh-i-noor, Inc. Reprinted with permission.)

| | SHAFT DIA. | EXTERNAL SERIES 11-410 | RETAINING RING DIMENSIONS | | GROOVE DIMENSIONS | | | | MAXIMUM ALLOWABLE CORNER RADII AND CHAMFER OF RETAINED PARTS | | EDGE MARGIN | NOMINAL GROOVE DEPTH (REF) |
| | | | | | Diameter | | Width | | | | | |
	S	Size—No.	Y	t	G	Tol.	W	Tol.	R Max.	Ch. Max.	z	d
U.S. CUSTOMARY (INCHES)	.250	11-410-25	.311	.025	.222	−.004	.029	+.003	.023	.018	.030	.015
	.312	11-410-31	.376	.025	.278	−.004	.029	+.003	.024	.018	.036	.018
	.375	11-410-37	.448	.025	.337	−.004	.029	+.003	.026	.020	.040	.020
	.500	11-410-50	.581	.025	.453	−.006	.039	+.003	.030	.023	.050	.025
	.625	11-410-62	.715	.035	.566	−.006	.039	+.003	.033	.025	.062	.031
	.750	11-410-75	.845	.042	.679	−.006	.046	+.003	.036	.027	.074	.037
	.875	11-410-87	.987	.042	.792	−.006	.046	+.003	.040	.031	.086	.043
	1.000	11-410-100	1.127	.042	.903	−.006	.046	+.003	.046	.035	.100	.050
	1.125	11-410-112	1.267	.050	1.017	−.008	.056	+.004	.052	.040	.112	.056
	1.250	11-410-125	1.410	.050	1.130	−.008	.056	+.004	.057	.044	.124	.062
	1.375	11-410-137	1.550	.050	1.241	−.008	.056	+.004	.062	.048	.138	.069
	1.500	11-410-150	1.691	.050	1.354	−.008	.056	+.004	.069	.053	.150	.075
	1.750	11-410-175	1.975	.062	1.581	−.010	.068	+.004	.081	.062	.174	.087
	2.000	11-410-200	2.257	.062	1.805	−.010	.068	+.004	.091	.070	.200	.100
METRIC (MILLIMETERS)	8	11-410-080	10	0.6	7	−0.1	0.7	+0.15	0.6	0.45	1.5	0.5
	10	11-410-100	12.2	0.6	9	−0.1	0.7	+0.15	0.6	0.45	1.5	0.5
	12	11-410-120	14.4	0.6	10.9	−0.1	0.7	+0.15	0.6	0.45	1.7	0.5
	14	11-410-140	16.3	1	12.7	−0.1	1.1	+0.15	1	0.8	2	0.65
	16	11-410-160	18.5	1	14.5	−0.1	1.1	+0.15	1	0.8	2.3	0.75
	18	11-410-180	20.4	1.2	16.3	−0.1	1.3	+0.15	1.2	0.9	2.6	0.85
	20	11-410-200	22.6	1.2	18.1	−0.2	1.3	+0.15	1.2	0.9	2.9	0.95
	22	11-410-220	25	1.2	19.9	−0.2	1.3	+0.15	1.2	0.9	3.2	1.05
	24	11-410-240	27.1	1.2	21.7	−0.2	1.3	+0.15	1.2	0.9	3.5	1.15
	25	11-410-250	28.3	1.2	22.6	−0.2	1.3	+0.15	1.2	0.9	3.6	1.2
	30	11-410-300	33.7	1.5	27	−0.2	1.3	+0.2	1.5	1.15	4.5	1.5
	35	11-410-350	39.4	1.5	31.5	−0.25	1.6	+0.2	1.5	1.15	5.3	1.75
	40	11-410-400	45	1.5	36	−0.25	1.6	+0.2	1.5	1.15	6	2
	45	11-410-450	50.6	1.5	40.5	−0.25	1.6	+0.2	1.5	1.15	6.8	2.25
	50	11-410-500	56.4	2	45	−0.25	2.2	+0.2	2	1.5	7.5	2.5

Table 40 Retaining rings—radial assembly. (© 1965, 1958 Waldes Koh-i-noor, Inc. Reprinted with permission.)

MORSE TAPERS

No. of Taper	TAPER	
	inches per Foot	mm per 100 mm
0	.625	5.21
1	.599	4.99
2	.599	4.99
3	.602	5.02
4	.623	5.19
5	.631	5.26
6	.626	5.22
7	.624	5.20

BROWN AND SHARPE TAPERS

No. of Taper	TAPER	
	inches per Foot	mm per 100 mm
1	.502	4.18
2	.502	4.18
3	.502	4.18
4	.502	4.18
5	.502	4.18
6	.503	4.19
7	.502	4.18
8	.501	4.18
9	.501	4.18
10	.516	4.3
11	.501	4.18
12	.500	4.17
13	.500	4.17
14	.500	4.17
15	.500	4.17
16	.500	4.17

Table 41   Machine tapers.

PRODUCT	OUTSTANDING FEATURES	APPLICATION METHOD	COLOR
1357	A high performance adhesive with long bonding range, excellent initial strength. Meets specification requirements of MMM-A-121 (supersedes MIL-A-1154 C), MIL-A-5092 B, Type II, and MIL-A-21366. Bonds rubber, cloth, wood, foamed glass, paper honeycomb, decorative plastic laminates. Also used with metal-to-metal for bonds of moderate strength.	Spray or Brush	Gray/ Green or Olive
2210	Fast drying, exhibits aggressive tack that allows coated surfaces to knit easily under hand roller pressure. Excellent water and oil resistance. Meets specification requirements of MMM-A-121 (supersedes MIL-A-1154 C), MIL-A-21366, and MMM-A-00130a. Bonds a wide range of materials including rubber, leather, cloth, aluminum, wood, hardboard. Used extensively for bonding decorative plastic laminates.	Brush, Roller, or Trowel	Yellow
2215	Fast drying and has a rapid rate of strength build-up. Its aggressive tack permits adhesive coated surfaces to bond easily with moderate pressure. Bonds decorative plastic laminates to metal, wood, and particle board. Also used for general bonding of rubber, leather, cloth, aluminum, wood, hardboard, etc.	Spray	Light Yellow
2218	Offers rapid strength build-up, high-ultimate strength. Has a high softening point and excellent resistance to plastic flow. Adhesion to steel is especially good. Meets specification requirements of MMM-A-121 (supersedes MIL-A-1154 C). Bonds high density decorative plastic laminates to metal or wood. Widely used to fabricate honeycomb and sandwich-type building panels with various face sheets, including porcelain enamel steel.	Spray	Green
2226	Water dispersed, has high immediate bond strength, long bonding range. Changes color from blue to green while drying. Used to bond foamed plastics, plastic laminate, wood, rubber, plywood, wallboard, wood veneer, plaster, and canvas to themselves and to each other.	Spray or Brush	Wet: Lt. Blue Dry: Green
4420	High performance, fast-drying adhesive designed for application by pressure curtain coating and mechanical roll coating. Bonds decorative plastic laminates to plywood or particle board and is suitable for conveyor line production of laminated panels of various types such as aluminum to wood or hardboard.	Roll Coating	Yellow
4488	Lower viscosity version of Cement 4420 for use specifically with flow-over or Weir-type curtain coaters. Bonds decorative plastic laminates to plywood or particle board and is suitable for conveyor line production of laminated panels of various types, such as aluminum to wood or hardboard.	Curtain and Roll Coating	Yellow
4518	Designed for spray application with automatic or production line equipment. Dries very fast, requires pressure from a niproll (rotary press) or platen press to assure proper bonding. Bonds decorative plastic laminates to plywood or particle board on both flat work and postforming. Also used for conveyor line production of laminated panels of various types such as aluminum to wood or hardboard. Meets requirements of MIL-A-5092 B, Type II.	Spray	Green
4729	Similar to Cement 2218 except that it is formulated with a nonflammable solvent. Requires force drying to prevent blushing.	Spray	Red
5034	Water dispersed, has high immediate bond strength and long bonding range. Bonds foamed plastics, plastic laminate, wood, rubber, plywood, wallboard, wood veneer, plaster, and canvas to themselves and to each other.	Spray or Brush	Neutral

Table 42   Physical properties and application data of adhesives. (3M Company.)

EXAMPLE: RC2 SLIDING FIT FOR A
Ø 1.50 NOMINAL HOLE DIAMETER

Nominal Size Range Inches		Class RC1 Precision Sliding			Class RC2 Sliding Fit			Class RC3 Precision Running			Class RC4 Close Running		
		Hole Tol. GR5	Minimum Clearance	Shaft Tol. GR4	Hole Tol. GR6	Minimum Clearance	Shaft Tol. GR5	Hole Tol. GR7	Minimum Clearance	Shaft Tol. GR6	Hole Tol. GR8	Minimum Clearance	Shaft Tol. GR7
Over	To	−0		+0	−0		+0	−0		+0	−0		+0
0	.12	+0.15	0.1	−0.12	+0.25	0.1	−0.15	+0.4	0.3	−0.25	+0.6	0.3	−0.4
.12	.24	+0.2	0.15	−0.15	+0.3	0.15	−0.2	+0.5	0.4	−0.3	+0.7	0.4	−0.5
.24	.40	+0.25	0.2	−0.15	+0.4	0.2	−0.25	+0.6	0.5	−0.4	+0.9	0.5	−0.6
.40	.71	+0.3	0.25	−0.2	+0.4	0.25	−0.3	+0.7	0.6	−0.4	+1.0	0.6	−0.7
.71	1.19	+0.4	0.3	−0.25	+0.5	0.3	−0.4	+0.8	0.8	−0.5	+1.2	0.8	−0.8
1.19	1.97	+0.4	0.4	−0.3	+0.6	0.4	−0.4	+1.0	1.0	−0.6	+1.6	1.0	−1.0
1.97	3.15	+0.5	0.4	−0.3	+0.7	0.4	−0.5	+1.2	1.2	−0.7	+1.8	1.2	−1.2
3.15	4.73	+0.6	0.5	−0.4	+0.9	0.5	−0.6	+1.4	1.4	−0.9	+2.2	1.4	−1.4
4.73	7.09	+0.7	0.6	−0.5	+1.0	0.6	−0.7	+1.6	1.6	−1.0	+2.5	1.6	−1.6
7.09	9.85	+0.8	0.6	−0.6	+1.2	0.6	−0.8	+1.8	2.0	−1.2	+2.8	2.0	−1.8
9.85	12.41	+0.9	0.8	−0.6	+1.2	0.8	−0.9	+2.0	2.5	−1.2	+3.0	2.5	−2.0
12.41	15.75	+1.0	1.0	−0.7	+1.4	1.0	−1.0	+2.2	3.0	−1.4	+3.5	3.0	−2.2

Table 43   Running and sliding fits. (Values in thousandths of an inch.)

RUNNING AND SLIDING FITS

Class RC5 Medium Running			Class RC6 Medium Running			Class RC7 Free Running			Class RC8 Loose Running			Class RC9 Loose Running		
Hole Tol. GR8	Minimum Clearance	Shaft Tol. GR7	Hole Tol. GR9	Minimum Clearance	Shaft Tol. GR8	Hole Tol. GR9	Minimum Clearance	Shaft Tol. GR8	Hole Tol. GR10	Minimum Clearance	Shaft Tol. GR9	Hole Tol. GR11	Minimum Clearance	Shaft Tol. GR10
−0		+0	−0		+0	−0		+0	−0		+0	−0		+0
+0.6	0.6	−0.4	+1.0	0.6	−0.6	+1.0	1.0	−0.6	+1.6	2.5	−1.0	+2.5	4.0	−1.6
+0.7	0.8	−0.5	+1.2	0.8	−0.7	+1.2	1.2	−0.7	+1.8	2.8	−1.2	+3.0	4.5	−1.8
+0.9	1.0	−0.6	+1.4	1.0	−0.9	+1.4	1.6	−0.9	+2.2	3.0	−1.4	+3.5	5.0	−2.2
+1.0	1.2	−0.7	+1.6	1.2	−1.0	+1.6	2.0	−1.0	+2.8	3.5	−1.6	+4.0	6.0	−2.8
+1.2	1.6	−0.8	+2.0	1.6	−1.2	+2.0	2.5	−1.2	+3.5	4.5	−2.0	+5.0	7.0	−3.5
+1.6	2.0	−1.0	+2.5	2.0	−1.6	+2.5	3.0	−1.6	+4.0	5.0	−2.5	+6.0	8.0	−4.0
+1.8	2.5	−1.2	+3.0	2.5	−1.8	+3.0	4.0	−1.8	+4.5	6.0	−3.0	+7.0	9.0	−4.5
+2.2	3.0	−1.4	+3.5	3.0	−2.2	+3.5	5.0	−2.2	+5.0	7.0	−3.5	+9.0	10.0	−5.0
+2.5	3.5	−1.6	+4.0	3.5	−2.5	+4.0	6.0	−2.5	+6.0	8.0	−4.0	+10.0	12.0	−6.0
+2.8	4.5	−1.8	+4.5	4.0	−2.8	+4.5	7.0	−2.8	+7.0	10.0	−4.5	+12.0	15.0	−7.0
+3.0	5.0	−2.0	+5.0	5.0	−3.0	+5.0	8.0	−3.0	+8.0	12.0	−5.0	+12.0	18.0	−8.0
+3.5	6.0	−2.2	+6.0	6.0	−3.5	+6.0	10.0	−3.5	+9.0	14.0	−6.0	+14.0	22.0	−9.0

Table 43 (cont'd.) Running and sliding fits. (Values in thousandths of an inch.)

EXAMPLE: LC2 LOCATIONAL FIT
FOR A Ø1.50 NOMINAL HOLE DIAMETER

LOCATIONAL CLEARANCE FITS

Nominal Size Range Inches		Class LC1			Class LC2			Class LC3			Class LC4			Class LC5		
		Hole Tol. GR6	Minimum Clearance	Shaft Tol. GR5	Hole Tol. GR7	Minimum Clearance	Shaft Tol. GR6	Hole Tol. GR8	Minimum Clearance	Shaft Tol. GR7	Hole Tol. GR10	Minimum Clearance	Shaft Tol. GR9	Hole Tol. GR7	Minimum Clearance	Shaft Tol. GR6
Over	To	−0		+0	−0		+0	−0		+0	−0		+0	−0		+0
0	.12	+0.25	0	−0.15	+0.4	0	−0.25	+0.6	0	−0.4	+1.6	0	−1.0	+0.4	0.1	−0.25
.12	.24	+0.3	0	−0.2	+0.5	0	−0.3	+0.7	0	−0.5	+1.8	0	−1.2	+0.5	0.15	−0.3
.24	.40	+0.4	0	−0.25	+0.6	0	−0.4	+0.9	0	−0.6	+2.2	0	−1.4	+0.6	0.2	−0.4
.40	.71	+0.4	0	−0.3	+0.7	0	−0.4	+1.0	0	−0.7	+2.8	0	−1.6	+0.7	0.25	−0.4
.71	1.19	+0.5	0	−0.4	+0.8	0	−0.5	+1.2	0	−0.8	+3.5	0	−2.0	+0.8	0.3	−0.5
1.19	1.97	+0.6	0	−0.4	+1.0	0	−0.6	+1.6	0	−1.0	+4.0	0	−2.5	+1.0	0.4	−0.6
1.97	3.15	+0.7	0	−0.5	+1.2	0	−0.7	+1.8	0	−1.2	+4.5	0	−3.0	+1.2	0.4	−0.7
3.15	4.73	+0.9	0	−0.6	+1.4	0	−0.9	+2.2	0	−1.4	+5.0	0	−3.5	+1.4	0.5	−0.9
4.73	7.09	+1.0	0	−0.7	+1.6	0	−1.0	+2.5	0	−1.6	+6.0	0	−4.0	+1.6	0.6	−1.0
7.09	9.85	+1.2	0	−0.8	+1.8	0	−1.2	+2.8	0	−1.8	+7.0	0	−4.5	+1.8	0.6	−1.2
9.85	12.41	+1.2	0	−0.9	+2.0	0	−1.2	+3.0	0	−2.0	+8.0	0	−5.0	+2.0	0.7	−1.2
12.41	15.75	+1.4	0	−1.0	+2.2	0	−1.4	+3.5	0	−2.2	+9.0	0	−6.0	+2.2	0.7	−1.4

Table 44 **Locational clearance fits.** (Values in thousandths of an inch.)

Nominal Size Range Inches		Class LT1			Class LT2		
		Hole Tol. GR7	Maximum Interference	Shaft Tol. GR6	Hole Tol. GR8	Maximum Interference	Shaft Tol. GR7
Over	To	−0		+0	−0		+0
0	.12	+0.4	0.1	−0.25	+0.6	0.2	−0.4
.12	.24	+0.5	0.15	−0.3	+0.7	0.25	−0.5
.24	.40	+0.6	0.2	−0.4	+0.9	0.3	−0.6
.40	.71	+0.7	0.2	−0.4	+1.0	0.3	−0.7
.71	1.19	+0.8	0.25	−0.5	+1.2	0.4	−0.8
1.19	1.97	+1.0	0.3	−0.6	+1.6	0.5	−1.0
1.97	3.15	+1.2	0.3	−0.7	+1.8	0.6	−1.2
3.15	4.73	+1.4	0.4	−0.9	+2.2	0.7	−1.4
4.73	7.09	+1.6	0.5	−1.0	+2.5	0.8	−1.6
7.09	9.85	+1.8	0.6	−1.2	+2.8	0.9	−1.8
9.85	12.41	+2.0	0.6	−1.2	+3.0	1.0	−2.0
12.41	15.75	+2.2	0.7	−1.4	+3.5	1.0	−2.2

Table 45 **Transition fits.** (Values in thousandths of an inch.)

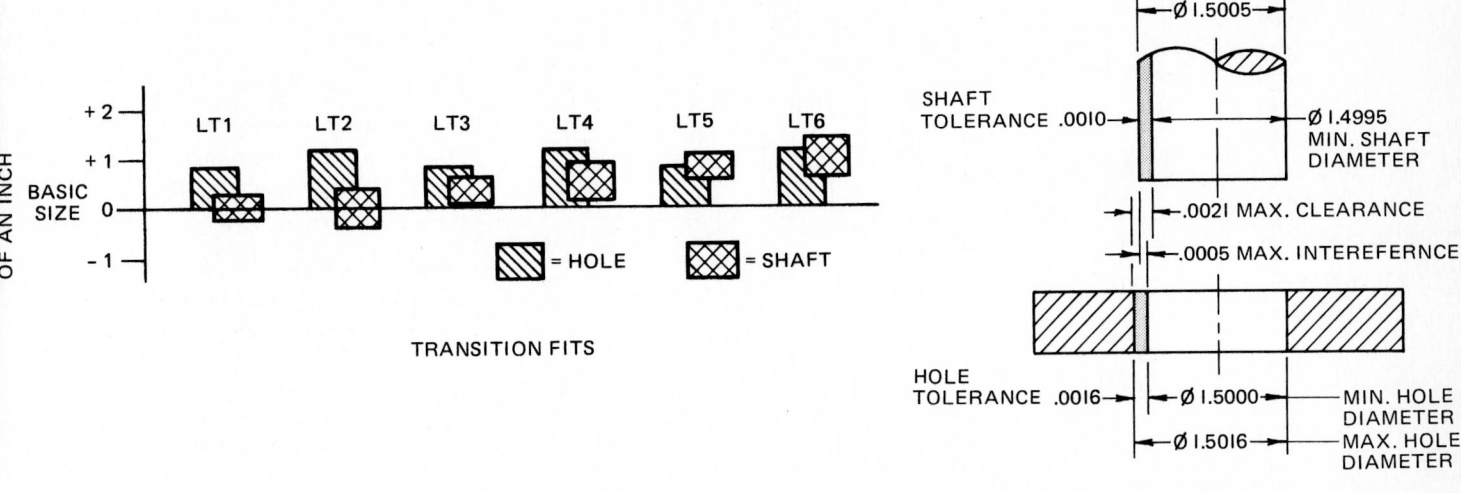

EXAMPLE: LT2 TRANSITION FIT
FOR A Ø1.50 NOMINAL HOLE DIAMETER

TRANSITION FITS

	Class LC6			Class LC7			Class LC8			Class LC9			Class LC10			Class LC11		
	Hole Tol. GR9	Minimum Clearance	Shaft Tol. GR8	Hole Tol. GR10	Minimum Clearance	Shaft Tol. GR9	Hole Tol. GR10	Minimum Clearance	Shaft Tol. GR9	Hole Tol. GR11	Minimum Clearance	Shaft Tol. GR10	Hole Tol. GR12	Minimum Clearance	Shaft Tol. GR11	Hole Tol. GR13	Minimum Clearance	Shaft Tol. GR12
	−0		+0	−0		+0	−0		+0	−0		+0	−0		+0	−0		+0
	+1.0	0.3	−0.6	+1.6	0.6	−1.0	+1.6	1.0	−1.0	+2.5	2.5	−1.6	+4.0	4.0	−2.5	+6.0	5.0	−4.0
	+1.2	0.4	−0.7	+1.8	0.8	−1.2	+1.8	1.2	−1.2	+3.0	2.8	−1.8	+5.0	4.5	−3.0	+7.0	6.0	−5.0
	+1.4	0.5	−0.9	+2.2	1.0	−1.4	+2.2	1.6	−1.4	+3.5	3.0	−2.2	+6.0	5.0	−3.5	+9.0	7.0	−6.0
	+1.6	0.6	−1.0	+2.8	1.2	−1.6	+2.8	2.0	−1.6	+4.0	3.5	−2.8	+7.0	6.0	−4.0	+10.0	8.0	−7.0
	+2.0	0.8	−1.2	+3.5	1.6	−2.0	+3.5	2.5	−2.0	+5.0	4.5	−3.5	+8.0	7.0	−5.0	+12.0	10.0	−8.0
	+2.5	1.0	−1.6	+4.0	2.0	−2.5	+4.0	3.6	−2.5	+6.0	5.0	−4.0	+10.0	8.0	−6.0	+16.0	12.0	−10.0
	+3.0	1.2	−1.8	+4.5	2.5	−3.0	+4.5	4.0	−3.0	+7.0	6.0	−4.5	+12.0	10.0	−7.0	+18.0	14.0	−12.0
	+3.5	1.4	−2.2	+5.0	3.0	−3.5	+5.0	5.0	−3.5	+9.0	7.0	−5.0	+14.0	11.0	−9.0	+22.0	16.0	−14.0
	+4.0	1.6	−2.5	+6.0	3.5	−4.0	+6.0	6.0	−4.0	+10.0	8.0	−6.0	+16.0	12.0	−10.0	+25.0	18.0	−16.0
	+4.5	2.0	−2.8	+7.0	4.0	−4.5	+7.0	7.0	−4.5	+12.0	10.0	−7.0	+18.0	16.0	−12.0	+28.0	22.0	−18.0
	+5.0	2.2	−3.0	+8.0	4.5	−5.0	+8.0	7.0	−5.0	+12.0	12.0	−8.0	+20.0	20.0	−12.0	+30.0	28.0	−20.0
	+6.0	2.5	−3.5	+9.0	5.0	−6.0	+9.0	8.0	−6.0	+14.0	14.0	−9.0	+22.0	22.0	−14.0	+35.0	30.0	−22.0

Table 44 (cont'd). Locational clearance fits. (Values in thousandths of an inch.)

	Class LT3			Class LT4			Class LT5			Class LT6		
	Hole Tol. GR7	Maximum Interference	Shaft Tol. GR6	Hole Tol. GR8	Maximum Interference	Shaft Tol. GR7	Hole Tol. GR7	Maximum Interference	Shaft Tol. GR6	Hole Tol. GR8	Maximum Interference	Shaft Tol. GR7
	−0		+0	−0		+0	−0		+0	−0		+0
	+0.4	0.25	−0.25	+0.6	0.4	−0.4	+0.4	0.5	−0.25	+0.6	0.65	−0.4
	+0.5	0.4	−0.3	+0.7	0.6	−0.5	+0.5	0.6	−0.3	+0.7	0.8	−0.5
	+0.6	0.5	−0.4	+0.9	0.7	−0.6	+0.6	0.8	−0.4	+0.9	1.0	−0.6
	+0.7	0.5	−0.4	+1.0	0.8	−0.7	+0.7	0.9	−0.4	+1.0	1.2	−0.7
	+0.8	0.6	−0.5	+1.2	0.9	−0.8	+0.8	1.1	−0.5	+1.2	1.4	−0.8
	+1.0	0.7	−0.6	+1.6	1.1	−1.0	+1.0	1.3	−0.6	+1.6	1.7	−1.0
	+1.2	0.8	−0.7	+1.8	1.3	−1.2	+1.2	1.5	−0.7	+1.8	2.0	−1.2
	+1.4	1.0	−0.9	+2.2	1.5	−1.4	+1.4	1.9	−0.9	+2.2	2.4	−1.4
	+1.6	1.1	−1.0	+2.5	1.7	−1.6	+1.6	2.2	−1.0	+2.5	2.8	−1.6
	+1.8	1.4	−1.2	+2.8	2.0	−1.8	+1.8	2.6	−1.2	+2.8	3.2	−1.8
	+2.0	1.4	−1.2	+3.0	2.2	−2.0	+2.0	2.6	−1.2	+3.0	3.4	−2.0
	+2.2	1.6	−1.4	+3.5	2.4	−2.2	+2.2	3.0	−1.4	+3.5	3.8	−2.2

Table 45 (cont'd). Transition fits. (Values in thousandths of an inch.)

EXAMPLE: LN2 LOCATIONAL INTERFERENCE FIT FOR A Ø1.50 NOMINAL HOLE DIAMETER

Nominal Size Range Inches		Class LN1 Light Press Fit			Class LN2 Medium Press Fit		
		Hole Tol. GR6	Maximum Interference	Shaft Tol. GR5	Hole Tol. GR7	Maximum Interference	Shaft Tol. GR6
Over	To	−0		+0	−0		+0
0	.12	+0.25	0.4	−0.15	+0.4	0.65	−0.25
.12	.24	+0.3	0.5	−0.2	+0.5	0.8	−0.3
.24	.40	+0.4	0.65	−0.25	+0.6	1.0	−0.4
.40	.71	+0.4	0.7	−0.3	+0.7	1.1	−0.4
.71	1.19	+0.5	0.9	−0.4	+0.8	1.3	−0.5
1.19	1.97	+0.6	1.0	−0.4	+1.0	1.6	−0.6
1.97	3.15	+0.7	1.3	−0.5	+1.2	2.1	−0.7
3.15	4.73	+0.9	1.6	−0.6	+1.4	2.5	−0.9
4.73	7.09	+1.0	1.9	−0.7	+1.6	2.8	−1.0
7.09	9.85	+1.2	2.2	−0.8	+1.8	3.2	−1.2
9.85	12.41	+1.2	2.3	−0.9	+2.0	3.4	−1.2
12.41	15.75	+1.4	2.6	−1.0	+2.2	3.9	−1.4

Table 46   Locational interference fits. (Values in thousandths of an inch.)

Nominal Size Range Inches		Class FN1 Light Drive Fit			Class FN2 Medium Drive Fit		
		Hole Tol. GR6	Maximum Interference	Shaft Tol. GR5	Hole Tol. GR7	Maximum Interference	Shaft Tol. GR6
Over	To	−0		+0	−0		+0
0	.12	+0.25	0.5	−0.15	+0.4	0.85	−0.25
.12	.24	+0.3	0.6	−0.2	+0.5	1.0	−0.3
.24	.40	+0.4	0.75	−0.25	+0.6	1.4	−0.4
.40	.56	+0.4	0.8	−0.3	+0.7	1.6	−0.4
.56	.71	+0.4	0.9	−0.3	+0.7	1.6	−0.4
.71	.95	+0.5	1.1	−0.4	+0.8	1.9	−0.5
.95	1.19	+0.5	1.2	−0.4	+0.8	1.9	−0.5
1.19	1.58	+0.6	1.3	−0.4	+1.0	2.4	−0.6
1.58	1.97	+0.6	1.4	−0.4	+1.0	2.4	−0.6
1.97	2.56	+0.7	1.8	−0.5	+1.2	2.7	−0.7
2.56	3.15	+0.7	1.9	−0.5	+1.2	2.9	−0.7
3.15	3.94	+0.9	2.4	−0.6	+1.4	3.7	−0.9

Table 47   Force and shrink fits. (Values in thousandths of an inch.)

EXAMPLE: FN2 MEDIUM DRIVE
FIT FOR A Ø 1.50 NOMINAL HOLE DIAMETER

	Class LN3 Heavy Press Fit			Class LN4			Class LN5			Class LN6		
Hole Tol. GR7	Maximum Interference	Shaft Tol. GR6	Hole Tol. GR8	Maximum Interference	Shaft Tol. GR7	Hole Tol. GR9	Maximum Interference	Shaft Tol. GR8	Hole Tol. GR10	Maximum Interference	Shaft Tol. GR9
−0		+0	−0		+0	−0		+0	−0		+0
+0.4	0.75	−0.25	+0.6	1.2	−0.4	+1.0	1.8	−0.6	+1.6	3.0	−1.0
+0.5	0.9	−0.3	+0.7	1.5	−0.5	+1.2	2.3	−0.7	+1.8	3.6	−1.2
+0.6	1.2	−0.4	+0.9	1.8	−0.6	+1.4	2.8	−0.9	+2.2	4.4	−1.4
+0.7	1.4	−0.4	+1.0	2.2	−0.7	+1.6	3.4	−1.0	+2.8	5.6	−1.6
+0.8	1.7	−0.5	+1.2	2.6	−0.8	+2.0	4.2	−1.2	+3.5	7.0	−2.0
+1.0	2.0	−0.6	+1.6	3.4	−1.0	+2.5	5.3	−1.6	+4.0	8.5	−2.5
+1.2	2.3	−0.7	+1.8	4.0	−1.2	+3.0	6.3	−1.8	+4.5	10.0	−3.0
+1.4	2.9	−0.9	+2.2	4.8	−1.4	+4.0	7.7	−2.2	+5.0	11.5	−3.5
+1.6	3.5	−1.0	+2.5	5.6	−1.6	+4.5	8.7	−2.5	+6.0	13.5	−4.0
+1.8	4.2	−1.2	+2.8	6.6	−1.8	+5.0	10.3	−2.8	+7.0	16.5	−4.5
+2.0	4.7	−1.2	+3.0	7.5	−2.0	+6.0	12.0	−3.0	+8.0	19	−5.0
+2.2	5.9	−1.4	+3.5	8.7	−2.2	+6.0	14.5	−3.5	+9.0	23	−6.0

Table 46 (cont'd.) Locational interference fits. (Values in thousandths of an inch.)

	Class FN3 Heavy Drive Fit			Class FN4 Shrink Fit			FN5 Heavy Shrink Fit		
Hole Tol. GR7	Maximum Interference	Shaft Tol. GR6	Hole Tol. GR7	Maximum Interference	Shaft Tol. GR6	Hole Tol. GR8	Maximum Interference	Shaft Tol. GR7
−0		+0	−0		+0	−0		+0
			+0.4	0.95	−0.25	+0.6	1.3	−0.4
			+0.5	1.2	−0.3	+0.7	1.7	−0.5
			+0.6	1.6	−0.4	+0.9	2.0	−0.6
			+0.7	1.8	−0.4	+1.0	2.3	−0.7
			+0.7	1.8	−0.4	+1.0	2.5	−0.7
			+0.8	2.1	−0.5	+1.2	3.0	−0.8
+0.8	2.1	−0.5	+0.8	2.3	−0.5	+1.2	3.3	−0.8
+1.0	2.6	−0.6	+1.0	3.1	−0.6	+1.6	4.0	−1.0
+1.0	2.8	−0.6	+1.0	3.4	−0.6	+1.6	5.0	−1.0
+1.2	3.2	−0.7	+1.2	4.2	−0.7	+1.8	6.2	−1.2
+1.2	3.7	−0.7	+1.2	4.7	−0.7	+1.8	7.2	−1.2
+1.4	4.4	−0.9	+1.4	5.9	−0.9	+2.2	8.4	−1.4

Table 47 (cont'd.) Force and shrink fits. (Values in thousandths of an inch.)

INTERNATIONAL TOLERANCE GRADES

Basic sizes		IT01	IT0	IT1	IT2	IT3	IT4	IT5	IT6	IT7	IT8	IT9	IT10	IT11	IT12	IT13	IT14	IT15	IT16
Over	Up to and including																		
0	3	0.0003	0.0005	0.0008	0.0012	0.002	0.003	0.004	0.006	0.010	0.014	0.025	0.040	0.060	0.100	0.140	0.250	0.400	0.600
3	6	0.0004	0.0006	0.001	0.0015	0.0025	0.004	0.005	0.008	0.012	0.018	0.030	0.048	0.075	0.120	0.180	0.300	0.480	0.750
6	10	0.0004	0.0006	0.001	0.0015	0.0025	0.004	0.006	0.009	0.015	0.022	0.036	0.058	0.090	0.150	0.220	0.360	0.580	0.900
10	18	0.0005	0.0008	0.0012	0.002	0.003	0.005	0.008	0.011	0.018	0.027	0.043	0.070	0.110	0.180	0.270	0.430	0.700	1.100
18	30	0.0006	0.001	0.0015	0.0025	0.004	0.006	0.009	0.013	0.021	0.033	0.052	0.084	0.130	0.210	0.330	0.520	0.840	1.300
30	50	0.0006	0.001	0.0015	0.0025	0.004	0.007	0.011	0.016	0.025	0.039	0.062	0.100	0.160	0.250	0.390	0.620	1.000	1.600
50	80	0.0008	0.0012	0.002	0.003	0.005	0.008	0.013	0.019	0.030	0.046	0.074	0.120	0.190	0.300	0.460	0.740	1.200	1.900
80	120	0.001	0.0015	0.0025	0.004	0.006	0.010	0.015	0.022	0.035	0.054	0.087	0.140	0.220	0.350	0.540	0.870	1.400	2.200
120	180	0.0012	0.002	0.0035	0.005	0.008	0.012	0.018	0.025	0.040	0.063	0.100	0.160	0.250	0.400	0.630	1.000	1.600	2.500
180	250	0.002	0.003	0.0045	0.007	0.010	0.014	0.020	0.029	0.046	0.072	0.115	0.185	0.290	0.460	0.720	1.150	1.850	2.900
250	315	0.0025	0.004	0.006	0.008	0.012	0.016	0.023	0.032	0.052	0.081	0.130	0.210	0.320	0.520	0.810	1.300	2.100	3.200
315	400	0.003	0.005	0.007	0.009	0.013	0.018	0.025	0.036	0.057	0.089	0.140	0.230	0.360	0.570	0.890	1.400	2.300	3.600
400	500	0.004	0.006	0.008	0.010	0.015	0.020	0.027	0.040	0.063	0.097	0.155	0.250	0.400	0.630	0.970	1.550	2.500	4.000
500	630	0.0045	0.006	0.009	0.011	0.016	0.022	0.030	0.044	0.070	0.110	0.175	0.280	0.440	0.700	1.100	1.750	2.800	4.400
630	800	0.005	0.007	0.010	0.013	0.018	0.025	0.035	0.050	0.080	0.125	0.200	0.320	0.500	0.800	1.250	2.000	3.200	5.000
800	1000	0.0055	0.008	0.011	0.015	0.021	0.029	0.040	0.056	0.090	0.140	0.230	0.360	0.560	0.900	1.400	2.300	3.600	5.600
1000	1250	0.0065	0.009	0.013	0.018	0.024	0.034	0.046	0.066	0.105	0.165	0.260	0.420	0.660	1.050	1.650	2.600	4.200	6.600
1250	1600	0.008	0.011	0.015	0.021	0.029	0.040	0.054	0.078	0.125	0.195	0.310	0.500	0.780	1.250	1.950	3.100	5.000	7.800
1600	2000	0.009	0.013	0.018	0.025	0.035	0.048	0.065	0.092	0.150	0.230	0.370	0.600	0.920	1.500	2.300	3.700	6.000	9.200
2000	2500	0.011	0.015	0.022	0.030	0.041	0.057	0.077	0.110	0.175	0.280	0.440	0.700	1.100	1.750	2.800	4.400	7.000	11.000
2500	3150	0.013	0.018	0.026	0.036	0.050	0.069	0.093	0.135	0.210	0.330	0.540	0.860	1.350	2.100	3.300	5.400	8.600	13.500

Table 48 **International Tolerance Grades.** (Values in millimeters.)

Legend: First choice tolerance zones encircled (ANSI B4.2 preferred)
Second choice tolerance zones framed (ISO 1829 selected)
Third choice tolerance zones open

TOLERANCE ZONES FOR INTERNAL DIMENSIONS (HOLES)

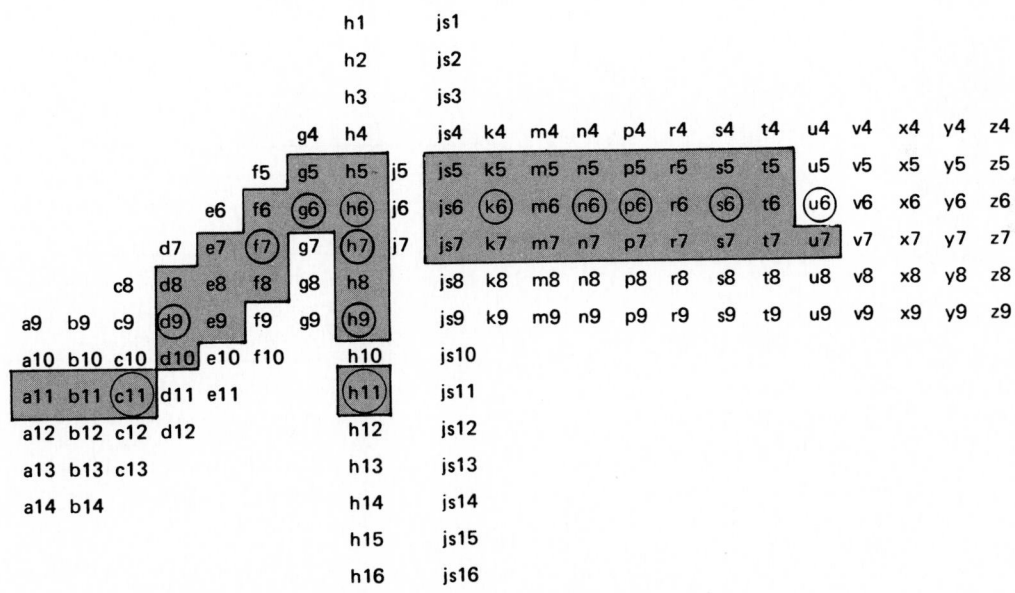

Legend: First choice tolerance zones encircled (ANSI B4.2 preferred)
Second choice tolerance zones framed (ISO 1829 selected)
Third choice tolerance zones open

TOLERANCE ZONES FOR EXTERNAL DIMENSIONS (SHAFTS)

Table 48 (cont'd.)   International Tolerance Grades

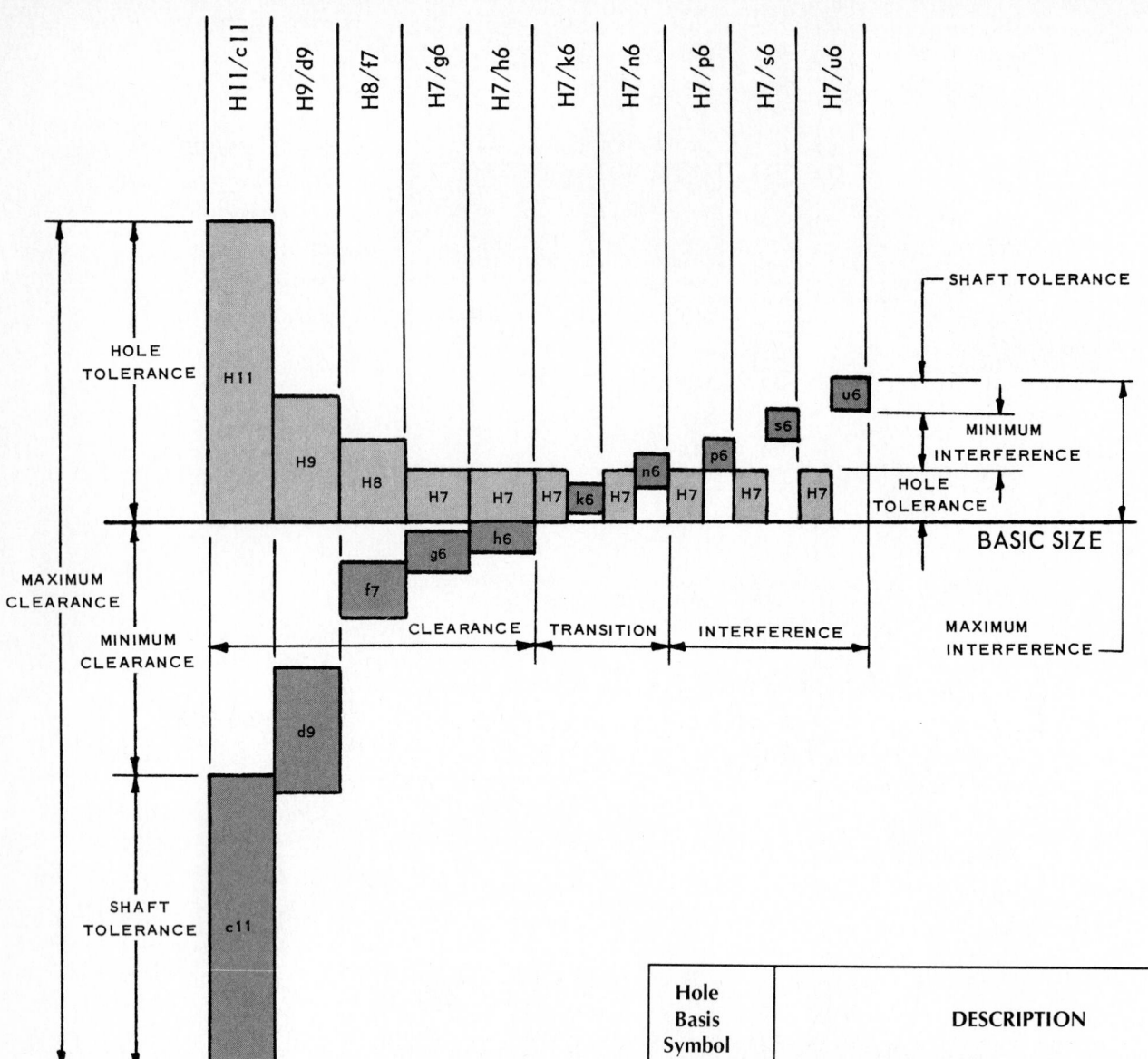

Hole Basis Symbol	DESCRIPTION
H11/c11	*Loose running* fit for wide commercial tolerances or allowances on external members.
H9/d9	*Free running* fit not for use where accuracy is essential, but good for large temperature variations, high running speeds, or heavy journal pressures.
H8/f7	*Close running* fit for running on accurate machines and for accurate location at moderate speeds and journal pressures.
H7/g6	*Sliding* fit not intended to run freely, but to move and turn freely and locate accurately.

Hole Basis Symbol	DESCRIPTION
H7/h6	*Locational clearance* fit provides snug fit for locating stationary parts; but can be freely assembled and disassembled.
H7/k6	*Locational transition* fit for accurate location, a compromise between clearance and interference.
H7/n6	*Locational transition* fit for more accurate location where greater interference is permissible.
H7/p6	*Locational interference* fit for parts requiring rigidity and alignment with prime accuracy of location but without special bore pressure requirements.
H7/s6	*Medium drive* fit for ordinary steel parts or shrink fits on light sections, the tightest fit usable with cast iron.
H7/u6	*Force* fit suitable for parts which can be highly stressed or for shrink fits where the heavy pressing forces required are impractical.

Table 49 Preferred hole basis fits description.

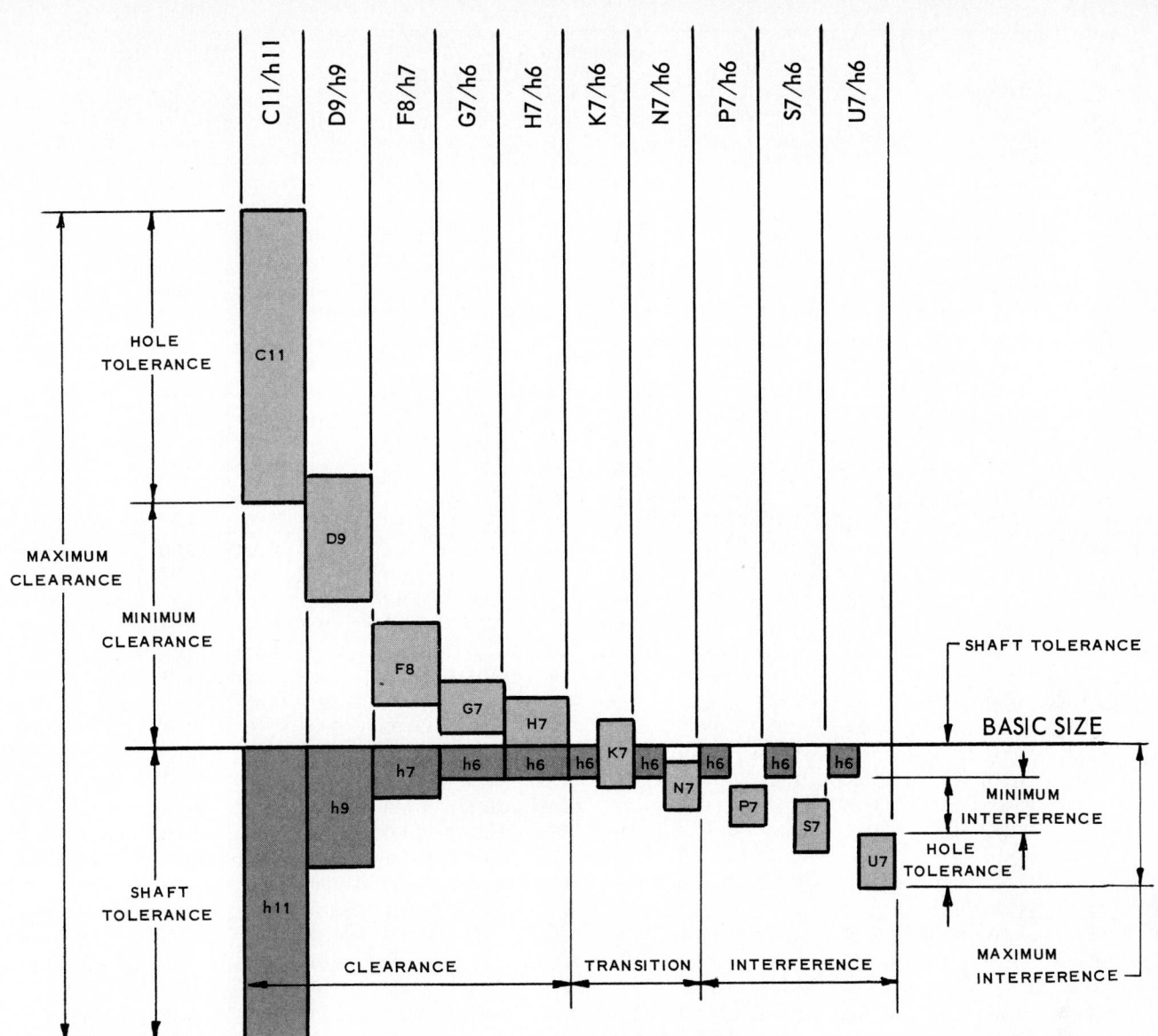

Shaft Basis Symbol	DESCRIPTION
C11/h11	*Loose running* fit for wide commercial tolerances or allowances on external members.
D9/h9	*Free running* fit not for use where accuracy is essential, but good for large temperature variations, high running speeds, or heavy journal pressures.
F8/h7	*Close running* fit for running on accurate machines and for accurate location at moderate speeds and journal pressures.
G7/h6	*Sliding* fit not intended to run freely, but to move and turn freely and locate accurately.
H7/h6	*Locational clearance* fit provides snug fit for locating stationary parts; but can be freely assembled and disassembled.

Shaft Basis Symbol	DESCRIPTION
K7/h6	*Locational transition* fit for accurate location, a compromise between clearance and interference.
N7/h6	*Locational transition* fit for more accurate location where greater interference is permissible.
P7/h6	*Locational interference* fit for parts requiring rigidity and alignment with prime accuracy of location but without special bore pressure requirements.
S7/h6	*Medium drive* fit for ordinary steel parts or shrink fits on light sections, the tightest fit usable with cast iron.
U7/h6	*Force* fit suitable for parts which can be highly stressed or for shrink fits where the heavy pressing forces required are impractical.

Table 50 Preferred shaft basis fits description.

PREFERRED HOLE BASIS CLEARANCE FITS

BASIC SIZE		LOOSE RUNNING Hole H11	Shaft c11	Fit	FREE RUNNING Hole H9	Shaft d9	Fit	CLOSE RUNNING Hole H8	Shaft f7	Fit	SLIDING Hole H7	Shaft g6	Fit	LOCATIONAL CLEARANCE Hole H7	Shaft h6	Fit
1	MAX	1.060	0.940	0.180	1.025	0.980	0.070	1.014	0.994	0.030	1.010	0.998	0.018	1.010	1.000	0.016
	MIN	1.000	0.880	0.060	1.000	0.955	0.020	1.000	0.984	0.006	1.000	0.992	0.002	1.000	0.994	0.000
1.2	MAX	1.260	1.140	0.180	1.225	1.180	0.070	1.214	1.194	0.030	1.210	1.198	0.018	1.210	1.200	0.016
	MIN	1.200	1.080	0.060	1.200	1.155	0.020	1.200	1.184	0.006	1.200	1.192	0.002	1.200	1.194	0.000
1.6	MAX	1.660	1.540	0.180	1.625	1.580	0.070	1.614	1.594	0.030	1.610	1.598	0.018	1.610	1.600	0.016
	MIN	1.600	1.480	0.060	1.600	1.555	0.020	1.600	1.584	0.006	1.600	1.592	0.002	1.600	1.594	0.000
2	MAX	2.060	1.940	0.180	2.025	1.980	0.070	2.014	1.994	0.030	2.010	1.998	0.018	2.010	2.000	0.016
	MIN	2.000	1.880	0.060	2.000	1.955	0.020	2.000	1.984	0.006	2.000	1.992	0.002	2.000	1.994	0.000
2.5	MAX	2.560	2.440	0.180	2.525	2.480	0.070	2.514	2.494	0.030	2.510	2.498	0.018	2.510	2.500	0.016
	MIN	2.500	2.380	0.060	2.500	2.455	0.020	2.500	2.484	0.006	2.500	2.492	0.002	2.500	2.494	0.000
3	MAX	3.060	2.940	0.180	3.025	2.980	0.070	3.014	2.994	0.030	3.010	2.998	0.018	3.010	3.000	0.016
	MIN	3.000	2.880	0.060	3.000	2.955	0.020	3.000	2.984	0.006	3.000	2.992	0.002	3.000	2.994	0.000
4	MAX	4.075	3.930	0.220	4.030	3.970	0.090	4.018	3.990	0.040	4.012	3.996	0.024	4.012	4.000	0.020
	MIN	4.000	3.855	0.070	4.000	3.940	0.030	4.000	3.978	0.010	4.000	3.988	0.004	4.000	3.992	0.000
5	MAX	5.075	4.930	0.220	5.030	4.970	0.090	5.018	4.990	0.040	5.012	4.996	0.024	5.012	5.000	0.020
	MIN	5.000	4.855	0.070	5.000	4.940	0.030	5.000	4.978	0.010	5.000	4.988	0.004	5.000	4.992	0.000
6	MAX	6.075	5.930	0.220	6.030	5.970	0.090	6.018	5.990	0.040	6.012	5.996	0.024	6.012	6.000	0.020
	MIN	6.000	5.855	0.070	6.000	5.940	0.030	6.000	5.978	0.010	6.000	5.988	0.004	6.000	5.992	0.000
8	MAX	8.090	7.920	0.260	8.036	7.960	0.112	8.022	7.987	0.050	8.015	7.995	0.029	8.015	8.000	0.024
	MIN	8.000	7.830	0.080	8.000	7.924	0.040	8.000	7.972	0.013	8.000	7.986	0.006	8.000	7.991	0.000
10	MAX	10.090	9.920	0.260	10.036	9.960	0.112	10.022	9.987	0.050	10.015	9.995	0.029	10.015	10.000	0.024
	MIN	10.000	9.830	0.080	10.000	9.924	0.040	10.000	9.972	0.013	10.000	9.986	0.005	10.000	9.991	0.000
12	MAX	12.110	11.905	0.315	12.043	11.950	0.136	12.027	11.984	0.061	12.018	11.994	0.035	12.018	12.000	0.029
	MIN	12.000	11.795	0.095	12.000	11.907	0.050	12.000	11.966	0.016	12.000	11.983	0.006	12.000	11.989	0.000
16	MAX	16.110	15.905	0.315	16.043	15.950	0.136	16.027	15.984	0.061	16.018	15.994	0.035	16.018	16.000	0.029
	MIN	16.000	15.795	0.095	16.000	15.907	0.050	16.000	15.966	0.016	16.000	15.983	0.006	16.000	15.989	0.000
20	MAX	20.130	19.890	0.370	20.052	19.935	0.169	20.033	19.980	0.074	20.021	19.993	0.041	20.021	20.000	0.034
	MIN	20.000	19.760	0.110	20.000	19.883	0.065	20.000	19.959	0.020	20.000	19.980	0.007	20.000	19.987	0.000
25	MAX	25.130	24.890	0.370	25.052	24.935	0.169	25.033	24.980	0.074	25.021	24.993	0.042	25.021	25.000	0.034
	MIN	25.000	24.760	0.110	25.000	24.883	0.065	25.000	24.959	0.020	25.000	24.980	0.007	25.000	24.987	0.000
30	MAX	30.130	29.890	0.370	30.052	29.935	0.169	30.033	29.980	0.074	30.021	29.993	0.041	30.021	30.000	0.034
	MIN	30.000	29.760	0.110	30.000	29.883	0.065	30.000	29.959	0.020	30.000	29.980	0.007	30.000	29.987	0.000
40	MAX	40.160	39.880	0.440	40.062	39.920	0.204	40.039	39.975	0.089	40.025	39.991	0.050	40.025	40.000	0.041
	MIN	40.000	39.720	0.120	40.000	39.858	0.080	40.000	39.950	0.025	40.000	39.975	0.009	40.000	39.984	0.000
50	MAX	50.160	49.870	0.450	50.062	49.920	0.204	50.039	49.975	0.089	50.025	49.991	0.050	50.025	50.000	0.041
	MIN	50.000	49.710	0.130	50.000	49.858	0.080	50.000	49.950	0.025	50.000	49.975	0.009	50.000	49.984	0.000
60	MAX	60.190	59.860	0.520	60.074	59.900	0.248	60.046	59.970	0.106	60.030	59.990	0.059	60.030	60.000	0.049
	MIN	60.000	59.670	0.140	60.000	59.826	0.100	60.000	59.940	0.030	60.000	59.971	0.010	60.000	59.981	0.000
80	MAX	80.190	79.850	0.530	80.074	79.900	0.248	80.046	79.970	0.106	80.030	79.990	0.059	80.030	80.000	0.049
	MIN	80.000	79.660	0.150	80.000	79.826	0.100	80.000	79.940	0.030	80.000	79.971	0.010	80.000	79.981	0.000
100	MAX	100.220	99.830	0.610	100.087	99.880	0.294	100.054	99.964	0.125	100.035	99.988	0.069	100.035	100.000	0.057
	MIN	100.000	99.610	0.170	100.000	99.793	0.120	100.000	99.929	0.036	100.000	99.966	0.012	100.000	99.978	0.000
120	MAX	120.220	119.820	0.620	120.087	119.880	0.294	120.054	119.964	0.125	120.035	119.988	0.069	120.035	120.000	0.057
	MIN	120.000	119.600	0.180	120.000	119.793	0.120	120.000	119.929	0.036	120.000	119.966	0.012	120.000	119.978	0.000
160	MAX	160.250	159.790	0.710	160.100	159.855	0.345	160.063	159.957	0.146	160.040	159.986	0.079	160.040	160.000	0.065
	MIN	160.000	159.540	0.210	160.000	159.755	0.145	160.000	159.917	0.043	160.000	159.961	0.014	160.000	159.975	0.000

Table 51 Preferred hole basis fits. (Dimensions in millimeters.)

		LOCATIONAL TRANSN.			LOCATIONAL TRANSN.			LOCATIONAL INTERF.			MEDIUM DRIVE			FORCE		
BASIC SIZE		Hole H7	Shaft k6	Fit	Hole H7	Shaft n6	Fit	Hole H7	Shaft p6	Fit	Hole H7	Shaft s6	Fit	Hole H7	Shaft u6	Fit
1	MAX	1.010	1.006	0.010	1.010	1.010	0.006	1.010	1.012	0.004	1.010	1.020	−0.004	1.010	1.024	−0.008
	MIN	1.000	1.000	−0.006	1.000	1.004	−0.010	1.000	1.006	−0.012	1.000	1.014	−0.020	1.000	1.018	−0.024
1.2	MAX	1.210	1.206	0.010	1.210	1.210	0.006	1.210	1.212	0.004	1.210	1.220	−0.004	1.210	1.224	−0.008
	MIN	1.200	1.200	−0.006	1.200	1.204	−0.010	1.200	1.206	−0.012	1.200	1.214	−0.020	1.200	1.218	−0.024
1.6	MAX	1.610	1.606	0.010	1.610	1.610	0.006	1.610	1.612	0.004	1.610	1.620	−0.004	1.610	1.624	−0.008
	MIN	1.600	1.600	−0.006	1.600	1.604	−0.010	1.600	1.606	−0.012	1.600	1.614	−0.020	1.600	1.618	−0.024
2	MAX	2.010	2.006	0.010	2.010	2.010	0.006	2.010	2.012	0.004	2.010	2.020	−0.004	2.010	2.024	−0.008
	MIN	2.000	2.000	−0.006	2.000	2.004	−0.010	2.000	2.006	−0.012	2.000	2.014	−0.020	2.000	2.018	−0.024
2.5	MAX	2.510	2.506	0.010	2.510	2.510	0.006	2.510	2.512	0.004	2.510	2.520	−0.004	2.510	2.524	−0.008
	MIN	2.500	2.500	−0.006	2.500	2.504	−0.010	2.500	2.506	−0.012	2.500	2.514	−0.020	2.500	2.518	−0.024
3	MAX	3.010	3.006	0.010	3.010	3.010	0.006	3.010	3.012	0.004	3.010	3.020	−0.004	3.010	3.024	−0.008
	MIN	3.000	3.000	−0.006	3.000	3.004	−0.010	3.000	3.006	−0.012	3.000	3.014	−0.020	3.000	3.018	−0.024
4	MAX	4.012	4.009	0.011	4.012	4.016	0.004	4.012	4.020	0.000	4.012	4.027	−0.007	4.012	4.031	−0.011
	MIN	4.000	4.001	−0.009	4.000	4.008	−0.016	4.000	4.012	−0.020	4.000	4.019	−0.027	4.000	4.023	−0.031
5	MAX	5.012	5.009	0.011	5.012	5.016	0.004	5.012	5.020	0.000	5.012	5.027	−0.007	5.012	5.031	−0.011
	MIN	5.000	5.001	−0.009	5.000	5.008	−0.016	5.000	5.012	−0.020	5.000	5.019	−0.027	5.000	5.023	−0.031
6	MAX	6.012	6.009	0.011	6.012	6.016	0.004	6.012	6.020	0.000	6.012	6.027	−0.007	6.012	6.031	−0.011
	MIN	6.000	6.001	−0.009	6.000	6.008	−0.016	6.000	6.012	−0.020	6.000	6.019	−0.027	6.000	6.023	−0.031
8	MAX	8.015	8.010	0.014	8.015	8.019	0.005	8.015	8.024	0.000	8.015	8.032	−0.008	8.015	8.037	−0.013
	MIN	8.000	8.001	−0.010	8.000	8.010	−0.019	8.000	8.015	−0.024	8.000	8.023	−0.032	8.000	8.028	−0.037
10	MAX	10.015	10.010	0.014	10.015	10.019	0.005	10.015	10.024	0.000	10.015	10.032	−0.008	10.015	10.037	−0.013
	MIN	10.000	10.001	−0.010	10.000	10.010	−0.019	10.000	10.015	−0.024	10.000	10.023	−0.032	10.000	10.028	−0.037
12	MAX	12.018	12.012	0.017	12.018	12.023	0.006	12.018	12.029	0.000	12.018	12.039	−0.010	12.018	12.044	−0.015
	MIN	12.000	12.001	−0.012	12.000	12.012	−0.023	12.000	12.018	−0.029	12.000	12.028	−0.039	12.000	12.033	−0.044
16	MAX	16.018	16.012	0.017	16.018	16.023	0.006	16.018	16.029	0.000	16.018	16.039	−0.010	16.018	16.044	−0.015
	MIN	16.000	16.001	−0.012	16.000	16.012	−0.023	16.000	16.018	−0.029	16.000	16.028	−0.039	16.000	16.033	−0.044
20	MAX	20.021	20.015	0.019	20.021	20.028	0.006	20.021	20.035	−0.001	20.021	20.048	−0.014	20.021	20.054	−0.020
	MIN	20.000	20.002	−0.015	20.000	20.015	−0.028	20.000	20.022	−0.035	20.000	20.035	−0.048	20.000	20.041	−0.054
25	MAX	25.021	25.015	0.019	25.021	25.028	0.006	25.021	25.035	−0.001	25.021	25.048	−0.014	25.021	25.061	−0.027
	MIN	25.000	25.002	−0.015	25.000	25.015	−0.028	25.000	25.022	−0.035	25.000	25.035	−0.048	25.000	25.048	−0.061
30	MAX	30.021	30.015	0.019	30.021	30.028	0.006	30.021	30.035	−0.001	30.021	30.048	−0.014	30.021	30.061	−0.027
	MIN	30.000	30.002	−0.015	30.000	30.015	−0.028	30.000	30.022	−0.035	30.000	30.035	−0.048	30.000	30.048	−0.061
40	MAX	40.025	40.018	0.023	40.025	40.033	0.008	40.025	40.042	−0.001	40.025	40.059	−0.018	40.025	40.076	−0.035
	MIN	40.000	40.002	−0.018	40.000	40.017	−0.033	40.000	40.026	−0.042	40.000	40.043	−0.059	40.000	40.060	−0.076
50	MAX	50.025	50.018	0.023	50.025	50.033	0.008	50.025	50.042	−0.001	50.025	50.059	−0.018	50.025	50.086	−0.045
	MIN	50.002	50.000	−0.018	50.000	50.017	−0.033	50.000	50.026	−0.042	50.000	50.043	−0.059	50.000	50.070	−0.086
60	MAX	60.030	60.021	0.028	60.030	60.039	0.010	60.030	60.051	−0.002	60.030	60.072	−0.023	60.030	60.106	−0.057
	MIN	60.000	60.002	−0.021	60.000	60.020	−0.039	60.000	60.032	−0.051	60.000	60.053	−0.072	60.000	60.087	−0.106
80	MAX	80.030	80.021	0.028	80.030	80.039	0.010	80.030	80.051	−0.002	80.030	80.078	−0.029	80.030	80.121	−0.072
	MIN	80.000	80.002	−0.021	80.000	80.020	−0.039	80.000	80.032	−0.051	80.000	80.059	−0.078	80.000	80.102	−0.121
100	MAX	100.035	100.025	0.032	100.035	100.045	0.012	100.035	100.059	−0.002	100.035	100.093	−0.036	100.035	100.146	−0.089
	MIN	100.000	100.003	−0.025	100.000	100.023	−0.045	100.000	100.037	−0.059	100.000	100.071	−0.093	100.000	100.124	−0.146
120	MAX	120.035	120.025	0.032	120.035	120.045	0.012	120.035	120.059	−0.002	120.035	120.101	−0.044	120.035	120.166	−0.109
	MIN	120.000	120.003	−0.025	120.000	120.023	−0.045	120.000	120.037	−0.059	120.000	120.079	−0.101	120.000	120.144	−0.166
120	MAX	160.040	160.028	0.037	160.045	160.052	0.013	160.040	160.068	−0.003	160.040	160.125	−0.060	160.040	160.215	−0.150
	MIN	160.000	160.003	−0.028	160.000	160.027	−0.052	160.000	160.043	−0.068	160.000	160.000	−0.125	160.000	160.190	−0.215

PREFERRED HOLE BASIS TRANSITION AND INTERFERENCE FITS

Table 51 (cont'd.) Preferred hole basis fits.

		LOOSE RUNNING			FREE RUNNING			CLOSE RUNNING			SLIDING			LOCATIONAL CLEARANCE		
BASIC SIZE		Hole C11	Shaft h11	Fit	Hole D9	Shaft h9	Fit	Hole F8	Shaft h7	Fit	Hole G7	Shaft h6	Fit	Hole H7	Shaft h6	Fit
1	MAX	1.120	1.000	0.180	1.045	1.000	0.070	1.020	1.000	0.030	1.012	1.000	0.018	1.010	1.000	0.016
	MIN	1.060	0.940	0.060	1.020	0.975	0.020	1.006	0.990	0.006	1.002	0.994	0.002	1.000	0.994	0.000
1.2	MAX	1.320	1.200	0.180	1.245	1.200	0.070	1.220	1.200	0.030	1.212	1.200	0.018	1.210	1.200	0.016
	MIN	1.260	1.140	0.060	1.220	1.175	0.020	1.206	1.190	0.006	1.202	1.194	0.002	1.200	1.194	0.000
1.6	MAX	1.720	1.600	0.180	1.645	1.600	0.070	1.620	1.600	0.030	1.612	1.600	0.018	1.610	1.600	0.016
	MIN	1.660	1.540	0.060	1.620	1.575	0.020	1.606	1.590	0.006	1.602	1.594	0.002	1.600	1.594	0.000
2	MAX	2.120	2.000	0.180	2.045	2.000	0.070	2.020	2.000	0.030	2.012	2.000	0.018	2.010	2.000	0.016
	MIN	2.060	1.940	0.060	2.020	1.975	0.020	2.006	1.990	0.006	2.002	1.994	0.002	2.000	1.994	0.000
2.5	MAX	2.620	2.500	0.180	2.545	2.500	0.070	2.520	2.500	0.030	2.512	2.500	0.018	2.510	2.500	0.016
	MIN	2.560	2.440	0.060	2.520	2.475	0.020	2.506	2.490	0.006	2.502	2.494	0.002	2.500	2.494	0.000
3	MAX	3.120	3.000	0.180	3.045	3.000	0.070	3.020	3.000	0.030	3.012	3.000	0.018	3.010	3.000	0.016
	MIN	3.060	2.940	0.060	3.020	2.975	0.020	3.006	2.990	0.006	3.002	2.994	0.002	3.000	2.994	0.000
4	MAX	4.145	4.000	.0220	4.060	4.000	0.090	4.028	4.000	0.040	4.016	4.000	0.024	4.012	4.000	0.020
	MIN	4.070	3.925	0.070	4.030	3.970	0.030	4.010	3.988	0.010	4.004	3.992	0.004	4.000	3.992	0.000
5	MAX	5.145	5.000	0.220	5.060	5.000	0.090	5.028	5.000	0.040	5.016	5.000	0.024	5.012	5.000	0.020
	MIN	5.070	4.925	0.070	5.030	4.970	0.030	5.010	4.988	0.010	5.004	4.992	0.004	5.000	4.992	0.000
6	MAX	6.145	6.000	0.220	6.060	6.000	0.090	6.028	6.000	0.040	6.016	6.000	0.024	6.012	6.000	0.020
	MIN	6.070	5.925	0.070	6.030	5.970	0.030	6.010	5.988	0.010	6.004	5.992	0.004	6.000	5.992	0.000
8	MAX	8.170	8.000	0.260	8.076	8.000	0.112	8.035	8.000	0.050	8.020	8.000	0.029	8.015	8.000	0.024
	MIN	8.080	7.910	0.080	8.040	7.964	0.040	8.013	7.985	0.013	8.005	7.991	0.005	8.000	7.991	0.000
10	MAX	10.170	10.000	0.260	10.076	10.000	0.112	10.035	10.000	0.050	10.020	10.000	0.029	10.015	10.000	0.024
	MIN	10.080	9.910	0.080	10.040	9.964	0.040	10.013	9.985	0.013	10.005	9.991	0.005	10.000	9.991	0.000
12	MAX	12.205	12.000	0.315	12.093	12.000	0.136	12.043	12.000	0.061	12.024	12.000	0.035	12.018	12.000	0.029
	MIN	12.095	11.890	0.095	12.050	11.957	0.050	12.016	11.982	0.016	12.006	11.989	0.006	12.000	11.989	0.000
16	MAX	16.205	16.000	0.315	16.093	16.000	0.136	16.043	16.000	0.061	16.024	16.000	0.035	16.018	16.000	0.029
	MIN	16.095	15.890	0.095	16.050	15.957	0.050	16.016	15.982	0.016	16.006	15.989	0.006	16.000	15.989	0.000
20	MAX	20.240	20.000	0.370	20.117	20.000	0.169	20.053	20.000	0.074	20.028	20.000	0.041	20.021	20.000	0.034
	MIN	20.110	19.870	0.110	20.065	19.948	0.065	20.020	19.979	0.020	20.007	19.987	0.007	20.000	19.987	0.000
25	MAX	25.240	25.000	0.370	25.117	25.000	0.169	25.053	25.000	0.074	25.028	25.000	0.041	25.021	25.000	0.034
	MIN	25.110	24.870	0.110	25.065	24.948	0.065	25.020	24.979	0.020	25.007	24.987	0.007	25.000	24.987	0.000
30	MAX	30.240	30.000	0.370	30.117	30.000	0.169	30.053	30.000	0.074	30.028	30.000	0.041	30.021	30.000	0.034
	MIN	30.110	29.870	0.110	30.065	29.948	0.065	30.020	29.979	0.020	30.007	29.987	0.007	30.000	29.987	0.000
40	MAX	40.280	40.000	0.440	40.142	40.000	0.204	40.064	40.000	0.089	40.034	40.000	0.050	40.025	40.000	0.041
	MIN	40.120	39.840	0.120	40.080	39.938	0.080	40.025	39.975	0.025	40.009	39.984	0.009	40.000	39.984	0.000
50	MAX	50.290	50.000	0.450	50.142	50.000	0.204	50.064	50.000	0.089	50.034	50.000	0.050	50.025	50.000	0.041
	MIN	50.130	49.840	0.130	50.080	49.938	0.080	50.025	49.975	0.025	50.009	49.984	0.009	50.000	49.984	0.000
60	MAX	60.330	60.000	0.520	60.174	60.000	0.248	60.076	60.000	0.106	60.040	60.000	0.059	60.030	60.000	0.049
	MIN	60.140	59.810	0.140	60.100	59.926	0.100	60.030	59.970	0.030	60.010	59.981	0.010	60.000	59.981	0.000
80	MAX	80.340	80.000	0.530	80.174	80.000	0.248	80.076	80.000	0.106	80.040	80.000	0.059	80.030	80.000	0.049
	MIN	80.150	79.810	0.150	80.100	79.926	0.100	80.030	79.970	0.030	80.010	79.981	0.010	80.000	79.981	0.000
100	MAX	100.390	100.000	0.610	100.207	100.000	0.294	100.090	100.000	0.125	100.047	100.000	0.069	100.035	100.000	0.057
	MIN	100.170	99.780	0.170	100.120	99.913	0.120	100.036	99.965	0.036	100.012	99.978	0.012	100.000	99.978	0.000
120	MAX	120.400	120.000	0.620	120.207	120.000	0.294	120.090	120.000	0.125	120.047	120.000	0.069	120.035	120.000	0.057
	MIN	120.180	119.780	0.180	120.120	119.913	0.120	120.036	119.965	0.036	120.012	119.978	0.012	120.000	119.978	0.000
160	MAX	160.460	160.000	0.710	160.245	160.000	0.345	160.106	160.000	0.146	160.054	160.000	0.079	160.040	160.000	0.065
	MIN	160.210	159.750	0.210	160.145	159.900	0.145	160.043	159.960	0.043	160.014	159.975	0.014	160.000	159.975	0.000

PREFERRED SHAFT BASIS CLEARANCE FITS

Table 52 Preferred shaft basis fits. (Dimensions in millimeters.)

		PREFERRED SHAFT BASIS TRANSITION AND INTERFERENCE FITS														
		LOCATIONAL TRANSN.			LOCATIONAL TRANSN.			LOCATIONAL INTERF.			MEDIUM DRIVE			FORCE		
BASIC SIZE		Hole K7	Shaft h6	Fit	Hole N7	Shaft h6	Fit	Hole P7	Shaft h6	Fit	Hole S7	Shaft h6	Fit	Hole U7	Shaft h6	Fit
1	MAX	1.000	1.000	0.006	0.996	1.000	0.002	0.994	1.000	0.000	0.986	1.000	−0.008	0.982	1.000	−0.012
	MIN	0.990	0.994	−0.010	0.986	0.994	−0.014	0.984	0.994	−0.016	0.976	0.994	−0.024	0.972	0.994	−0.028
1.2	MAX	1.200	1.200	0.006	1.196	1.200	0.002	1.194	1.200	0.000	1.186	1.200	−0.008	1.182	1.200	−0.012
	MIN	1.190	1.194	−0.010	1.186	1.194	−0.014	1.184	1.194	−0.016	1.176	1.194	−0.024	1.172	1.194	−0.028
1.6	MAX	1.600	1.600	0.006	1.596	1.600	0.002	1.594	1.600	0.000	1.586	1.600	−0.008	1.582	1.600	−0.012
	MIN	1.590	1.594	−0.010	1.586	1.594	−0.014	1.584	1.594	−0.016	1.576	1.594	−0.024	1.572	1.594	−0.028
2	MAX	2.000	2.000	0.006	1.996	2.000	0.002	1.994	2.000	0.000	1.986	2.000	−0.008	1.982	2.000	−0.012
	MIN	1.990	1.994	−0.010	1.986	1.994	−0.014	1.984	1.994	−0.016	1.976	1.994	−0.024	1.972	1.994	−0.028
2.5	MAX	2.500	2.500	0.006	2.496	2.500	0.002	2.494	2.500	0.000	2.486	2.500	−0.008	2.482	2.500	−0.012
	MIN	2.490	2.494	−0.010	2.486	2.494	−0.014	2.484	2.494	−0.016	2.476	2.494	−0.024	2.472	2.494	−0.028
3	MAX	3.000	3.000	0.006	2.996	3.000	0.002	2.994	3.000	0.000	2.986	3.000	−0.008	2.982	3.000	−0.012
	MIN	2.990	2.994	−0.010	2.986	2.994	−0.014	2.984	2.994	−0.016	2.976	2.994	−0.024	2.972	2.994	−0.028
4	MAX	4.003	4.000	0.011	3.996	4.000	0.004	3.992	4.000	0.000	3.985	4.000	−0.007	3.981	4.000	−0.011
	MIN	3.991	3.992	−0.009	3.984	3.992	−0.016	3.980	3.992	−0.020	3.973	3.992	−0.027	3.969	3.992	−0.031
5	MAX	5.003	5.000	0.011	4.996	5.000	0.004	4.992	5.000	0.000	4.985	5.000	−0.007	4.981	5.000	−0.011
	MIN	4.991	4.992	−0.009	4.984	4.992	−0.016	4.980	4.992	−0.020	4.973	4.992	−0.027	4.969	4.992	−0.031
6	MAX	6.003	6.000	0.011	5.996	6.000	0.004	5.992	6.000	0.000	5.985	6.000	−0.007	5.981	6.000	−0.011
	MIN	5.991	5.992	−0.009	5.984	5.992	−0.016	5.980	5.992	−0.020	5.973	5.992	−0.027	5.969	5.992	−0.031
8	MAX	8.005	8.000	0.014	7.996	8.000	0.005	7.991	8.000	0.000	7.983	8.000	−0.008	7.978	8.000	−0.013
	MIN	7.990	7.991	−0.010	7.981	7.991	−0.019	7.976	7.991	−0.024	7.968	7.991	−0.032	7.963	7.991	−0.037
10	MAX	10.005	10.000	0.014	9.996	10.000	0.005	9.991	10.000	0.000	9.983	10.000	−0.008	9.978	10.000	−0.013
	MIN	9.990	9.991	−0.010	9.981	9.991	−0.019	9.976	9.991	−0.024	9.968	9.991	−0.032	9.963	9.991	−0.037
12	MAX	12.006	12.000	0.017	11.995	12.000	0.006	11.989	12.000	0.000	11.979	12.000	−0.010	11.974	12.000	−0.015
	MIN	11.988	11.989	−0.012	11.977	11.989	−0.023	11.971	11.989	−0.029	11.961	11.989	−0.039	11.956	11.989	−0.044
16	MAX	16.006	16.000	0.017	15.955	16.000	0.006	15.989	16.000	0.000	15.979	16.000	−0.010	15.974	16.000	−0.015
	MIN	15.988	15.989	−0.012	15.977	15.989	−0.023	15.971	15.989	−0.029	15.961	15.989	−0.039	15.956	15.989	−0.044
20	MAX	20.006	20.000	0.019	19.993	20.000	0.006	19.986	20.000	−0.001	19.973	20.000	−0.014	19.967	20.000	−0.020
	MIN	19.985	19.987	−0.015	19.972	19.987	−0.028	19.965	19.987	−0.035	19.952	19.987	−0.048	19.946	19.987	−0.054
25	MAX	25.006	25.000	0.019	24.993	25.000	0.006	24.986	25.000	−0.001	24.973	25.000	−0.014	24.960	25.000	−0.027
	MIN	24.985	24.987	−0.015	24.972	24.987	−0.028	24.965	24.987	−0.035	24.952	24.987	−0.048	24.939	24.987	−0.061
30	MAX	30.006	30.000	0.019	29.993	30.000	0.006	29.986	30.000	−0.001	29.973	30.000	−0.014	29.960	30.000	−0.027
	MIN	29.985	29.987	−0.015	29.972	29.987	−0.028	29.965	29.987	−0.035	29.952	29.987	−0.048	29.939	29.987	−0.061
40	MAX	40.007	40.000	0.023	39.992	40.000	0.008	39.983	40.000	−0.001	39.966	40.000	−0.018	39.949	40.000	−0.035
	MIN	39.982	39.984	−0.018	39.967	39.984	−0.033	39.958	39.984	−0.042	39.941	39.984	−0.059	39.924	39.984	−0.076
50	MAX	50.007	50.000	0.023	49.992	50.000	0.008	49.983	50.000	−0.001	49.966	50.000	−0.018	49.939	50.000	−0.045
	MIN	49.982	49.984	−0.018	49.967	49.984	−0.033	49.958	49.984	−0.042	49.941	49.984	−0.059	49.914	49.984	−0.086
60	MAX	60.009	60.000	0.028	59.991	60.000	0.010	59.979	60.000	−0.002	59.958	60.000	−0.023	59.924	60.000	−0.057
	MIN	59.979	59.981	−0.021	59.961	59.981	−0.039	59.949	59.981	−0.051	59.928	59.981	−0.072	59.894	59.981	−0.106
80	MAX	80.009	80.000	0.028	79.991	80.000	0.010	79.979	80.000	−0.002	79.952	80.000	−0.029	79.909	80.000	−0.072
	MIN	79.979	79.981	−0.021	79.961	79.981	−0.039	79.949	79.981	−0.051	79.922	79.981	−0.078	79.879	79.981	−0.121
100	MAX	100.010	100.000	0.032	99.990	100.000	0.012	99.976	100.000	−0.002	99.942	100.000	−0.036	99.889	100.000	−0.089
	MIN	99.975	99.978	−0.025	99.955	99.978	−0.045	99.941	99.978	−0.059	99.907	99.978	−0.093	99.854	99.978	−0.146
120	MAX	120.010	120.000	0.032	119.990	120.000	0.012	119.976	120.000	−0.002	119.934	120.000	−0.044	119.869	120.000	−0.109
	MIN	119.975	119.978	−0.025	119.955	119.978	−0.045	119.941	119.978	−0.059	119.899	119.978	−0.101	119.834	119.978	−0.166
160	MAX	160.012	160.000	0.037	159.988	160.000	0.013	159.972	160.000	−0.003	159.915	160.000	−0.060	159.825	160.000	−0.150
	MIN	159.972	159.975	−0.028	159.948	159.975	−0.052	159.932	159.975	−0.068	159.875	159.975	−0.125	159.785	159.975	−0.215

Table 52 (cont'd.)   Preferred shaft basis fits.

NORTH AMERICAN GAGES / EUROPEAN GAGES

U.S. Standard (USS)			U.S. Standard (Revised) Formerly Manufactures Standard			American Standard or Brown and Sharpe (B & S)			United States Steel Wire Gage			Birmingham (BWG)			New Birmingham (BG)			Imperial Wire Gage Imperial Standard (SWG)		
Gage	in.	mm	Gage	in.	mm	Gage	in.	mm	Gage	in.	mm	Gage	in.	mm	Gage	in.	mm	Gage	in.	mm
			3	.240	6.01	3	.229	5.83												
4	.234	5.95	4	.224	5.70	4	.204	5.19	4	.225	5.72	4	.238	6.05	4	.250	6.35	4	.232	5.89
5	.219	5.56	5	.209	5.31	5	.182	4.62	5	.207	5.26	5	.220	5.59	5	.223	5.65	5	.212	5.39
6	.203	5.16	6	.194	4.94	6	.162	4.12	6	.192	4.88	6	.203	5.16	6	.198	5.03	6	.192	4.88
7	.188	4.76	7	.179	4.55	7	.144	3.67	7	.177	4.50	7	.180	4.57	7	.176	4.48	7	.176	4.47
8	.172	4.37	8	.164	4.18	8	.129	3.26	8	.162	4.11	8	.165	4.19	8	.157	3.99	8	.160	4.06
9	.156	3.97	9	.149	3.80	9	.114	2.91	9	.148	3.77	9	.148	3.76	9	.140	3.55	9	.144	3.66
10	.141	3.57	10	.135	3.42	10	.102	2.59	10	.135	3.43	10	.134	3.40	10	.125	3.18	10	.128	3.25
11	.125	3.18	11	.120	3.04	11	.091	2.30	11	.121	3.06	11	.120	3.05	11	.111	2.83	11	.116	2.95
12	.109	2.78	12	.105	2.66	12	.081	2.05	12	.106	2.68	12	.109	2.77	12	.099	2.52	12	.104	2.64
13	.094	2.38	13	.090	2.78	13	.072	1.83	13	.092	2.32	13	.095	2.41	13	.088	2.24	13	.092	2.34
14	.078	1.98	14	.075	1.90	14	.064	1.63	14	.080	2.03	14	.083	2.11	14	.079	1.99	14	.080	2.03
15	.070	1.79	15	.067	1.71	15	.057	1.45	15	.072	1.83	15	.072	1.83	15	.070	1.78	15	.072	1.83
16	.063	1.59	16	.060	1.52	16	.051	1.29	16	.063	1.63	16	.065	1.65	16	.063	1.59	16	.064	1.63
17	.056	1.43	17	.054	1.37	17	.045	1.15	17	.054	1.37	17	.058	1.47	17	.056	1.41	17	.056	1.42
18	.050	1.27	18	.048	1.21	18	.040	1.02	18	.048	1.21	18	.049	1.25	18	.050	2.58	18	.048	1.22
19	.044	1.11	19	.042	1.06	19	.036	0.91	19	.041	1.04	19	.042	1.07	19	.044	1.19	19	.040	1.02
20	.038	0.95	20	.036	0.91	20	.032	0.81	20	.035	0.88	20	.035	0.89	20	.039	1.00	20	.036	0.91
21	.034	0.87	21	.033	0.84	21	.029	0.72	21	.032	0.81	21	.032	0.81	21	.035	0.89	21	.032	0.81
22	.031	0.79	22	.030	0.76	22	.025	0.65	22	.029	0.73	22	.028	0.71	22	.031	0.79	22	.028	0.71
23	.028	0.71	23	.027	0.68	23	.023	0.57	23	.026	0.66	23	.025	0.64	23	.028	0.71	23	.024	0.61
24	.025	0.64	24	.024	0.61	24	.020	0.51	24	.023	0.58	24	.022	0.56	24	.025	0.63	24	.022	0.56
25	.022	0.56	25	.021	0.53	25	.018	0.46	25	.020	0.52	25	.020	0.51	25	.022	0.56	25	.020	0.51
26	.019	0.48	26	.018	0.46	26	.016	0.40	26	.018	0.46	26	.018	0.46	26	.020	0.50	26	.018	0.46
27	.017	0.44	27	.016	0.42	27	.014	0.36	27	.017	0.44	27	.016	0.41	27	.017	0.44	27	.016	0.42
28	.016	0.40	28	.015	0.38	28	.013	0.32	28	.016	0.41	28	.014	0.36	28	.016	0.40	28	.015	0.38
29	.014	0.36	29	.014	0.34	29	.011	0.29	29	.015	0.38	29	.013	0.33	29	.014	0.35	29	.014	0.35
30	.013	0.32	30	.012	0.31	30	.010	0.25	30	.014	0.36	30	.012	0.31	30	.012	0.31	30	.012	0.32
31	.011	0.28	31	.011	0.27	31	.009	0.23	31	.013	0.34	31	.010	0.25	31	.011	0.28			
32	.010	0.26	32	.010	0.25	32	.008	0.20	32	.013	0.33	32	.009	0.23				32	.011	0.27
33	.009	0.24	33	.009	0.23	33	.007	0.18	33	.012	0.30	33	.008	0.20	33	.009	0.22	33	.010	0.25
34	.009	0.22	34	.008	0.21	34	.006	0.16	34	.010	0.26	34	.007	0.18	34	.008	0.20	34	.009	0.23
									35	.010	0.24	35	.005	0.13	35	.007	0.18	35	.008	0.21
36	.007	0.18	36	.007	0.17	36	.005	0.13	36	.009	0.23	36	.004	0.10	36	.006	0.16			
									37	.008	0.22							37	.007	0.17
38	.006	0.16	38	.006	0.15	38	.004	0.10	38	.008	0.20				38	.005	0.12	38	.006	0.15
									39	.008	0.19									
									40	.007	0.18				40	.004	0.10	40	.005	0.12
									41	.007	0.17							42	.004	0.10

NORTH AMERICAN GAGES headings: Ferrous metals, such as galvanized steel, tin plate (U.S. Standard USS); U.S. Standard (Revised) Formerly Manufactures Standard; Nonferrous metals, such as copper, brass, aluminum (American Standard or Brown and Sharpe B & S); Steel and iron wire and bare copper piano wire (United States Steel Wire Gage).

EUROPEAN GAGES headings: Galvanized steel, tin plate, copper, strip steel and steel, copper and aluminum tubes (Birmingham BWG; New Birmingham BG); Nonferrous (Imperial Wire Gage Imperial Standard SWG).

Note: Metric standards governing gage sizes were not available at the time of publication. The sizes given in the above chart are "soft conversion" from current inch standards and are not meant to be representative of the precise metric gage sizes which may be available in the future. Conversions are given only to allow the student to compare gage sizes readily with the metric drill sizes.

Table 53   Wire and sheet-metal gages and thicknesses.

FLAT SHEET

U.S. CUSTOMARY (INCHES)

Manufacturers Standard Gauge	Decimal Inch Equivalent	Pounds Per Square Foot
8	.164	6.88
9	.149	6.25
10	.134	5.63
11	.120	5.00
12	.105	4.38
13	.090	3.75
14	.075	3.13
15	.067	2.81
16	.060	2.50
17	.054	2.25
18	.048	2.00
19	.042	1.75
20	.036	1.50
21	.033	1.38
22	.030	1.25
23	.027	1.13
24	.024	1.00
25	.021	.88
26	.018	.75
27	.016	.69
28	.015	.63
29	.014	.56
30	.012	.50
31	.011	.44
32	.010	.41

METRIC (MILLIMETERS)

Millimeters Thickness	Cold Rolled Sheet	Hot Rolled Sheet	Galvanized Steel Sheet	Aluminum	Coated Sheet	Mass in kg/m²
12						94.2
11						86.3
10		●				78.5
9		●				70.6
8		●				62.8
7.5		●				58.9
7		●				54.9
6.5		●				51.0
6		●				47.1
5.5		●				43.2
5		●				39.2
4.8		●				37.7
4.5		●				35.3
4.2		●				33.0
4		●	●			31.4
3.8		●	●			29.8
3.6		●	●			28.6
3.5		●	●			27.5
3.4		●	●			26.7
3.2	●	●	●			25.1
3	●	●	●			23.5
2.8	●	●	●			22.0
2.6	●	●	●			20.4
2.5	●	●	●			19.6
2.4	●	●	●			18.8
2.2	●	●	●			17.3
2.1	●	●	●			16.5
2	●	●	●			15.7
1.9	●	●	●			14.9
1.8	●	●	●			14.1
1.7	●	●	●			13.3
1.6	●	●	●			12.6
1.5	●		●		●	11.8
1.4	●		●		●	11.0
1.3	●		●		●	10.2
1.2	●		●		●	9.4
1.1	●		●		●	8.6
1.05	●		●		●	8.2
1.0	●		●		●	7.8
0.95	●		●		●	7.5
0.9	●		●		●	7.1
0.85	●		●		●	6.7
0.8	●		●		●	6.3
0.75	●		●		●	5.9
0.7	●		●		●	5.5
0.65	●		●		●	5.1
0.6	●		●		●	4.7
0.55	●		●		●	4.3
0.5	●		●		●	3.9
0.45	●		●			3.5
0.4	●		●			3.1
0.35						2.7

Size In.	Area Sq. In. Round	Square	Weight Lb. per Ft. Round	Square
0				
1/16	.0031	.0039	.010	.013
1/8	.0123	.0156	.042	.053
3/16	.0276	.0352	.094	.120
1/4	.0491	.0625	.167	.213
5/16	.0767	.0977	.261	.332
3/8	.1105	.1406	.376	.478
7/16	.1503	.1914	.511	.651
1/2	.1963	.2500	.668	.850
9/16	.2485	.3164	.845	1.076
5/8	.3068	.3906	1.043	1.328
11/16	.3712	.4727	1.262	1.607
3/4	.4418	.5625	1.502	1.913
13/16	.5185	.6602	1.763	2.245
7/8	.6013	.7656	2.044	2.603
15/16	.6903	.8789	2.347	2.988
1	.7854	1.0000	2.670	3.400
1 1/8	.9940	1.2656	3.380	4.303
1 1/4	1.2272	1.5625	4.172	5.313
1 3/8	1.4849	1.8906	5.049	6.428
1 1/2	1.7671	2.2500	6.008	7.650

Size mm	Area mm² Round	Square	Mass kg/m Round	Square
1.6	2.01	2.56	0.016	0.020
1.8	2.54	3.24	0.020	0.025
2	3.14	4	0.025	0.031
2.5	4.91	6.25	0.039	0.049
3	7.07	9	0.055	0.071
4	12.57	16	0.099	0.125
5	19.64	25	0.154	0.196
6	28.27	36	0.222	0.282
7	38.48	49	0.302	0.384
8	50.27	64	0.394	0.502
9	63.62	81	0.499	0.635
10	78.54	100	0.616	0.784
12	113.1	144	0.887	1.129
13	132.73	169	1.041	1.325
14	153.94	196	1.207	1.537
15	176.72	225	1.386	1.764
16	201.06	256	1.577	2.007
18	254.47	324	1.996	2.541
20	314.16	400	2.464	3.137
22	380.13	484	2.981	3.795
24	452.39	576	3.548	4.517
25	490.88	625	3.850	4.901

Table 54 Weight (mass) and areas of flat, square, and round steel.

| | U.S. CUSTOMARY (INCHES) | | | | METRIC (MILLIMETERS) | | | | |
| EQUAL ANGLES | | UNEQUAL ANGLES | | | EQUAL ANGLES | | UNEQUAL ANGLES | | |
Section Size	Centroidal Axis X	Section Size	Centroidal Axis X	Y	Section Size	Centroidal Axis X	Section Size	Centroidal Axis X	Y
8.00 × 8.00 × .75	2.28	8.00 × 4.00 × .75	.95	2.95	200 × 200 × 20	57.4	200 × 125 × 20	31.5	69
.62	2.23	.62	.91	2.91	16	55.9	16	30	67.5
.50	2.19	.50	.86	2.86	13	54.8	13	28.9	66.4
6.00 × 6.00 × .75	1.78	6.00 × 4.00 × .75	1.08	2.08	150 × 150 × 20	44.8	150 × 100 × 20	27.4	52.4
.62	1.73	.62	1.03	2.03	16	43.4	16	25.9	50.9
.50	1.68	.50	.99	1.99	13	42.3	13	24.9	49.9
.38	1.64	.38	.94	1.94	10	41.2	10	23.8	48.8
5.00 × 5.00 × .62	1.48	5.00 × 3.50 × .62	.95	1.70	125 × 125 × 16	37.1	125 × 90 × 16	24.7	42.2
.50	1.43	.50	.91	1.66	13	36	13	23.7	41.2
.38	1.39	.38	.86	1.61	10	34.9	10	22.6	40.1
.31	1.37	.31	.84	1.59	8	34.2	8	21.8	39.3
4.00 × 4.00 × .50	1.18	4.00 × 3.00 × .50	.83	1.33	100 × 100 × 13	29.8	100 × 75 × 13	20.9	33.4
.38	1.14	.38	.78	1.28	10	28.7	10	19.8	32.3
.31	1.12	.31	.76	1.26	8	28	8	19	31.5
.25	1.09	.25	.74	1.24	6	27.2	6	18.3	30.8
3.50 × 3.50 × .50	1.06	3.50 × 3.00 × .50	.88	1.13	90 × 90 × 13	27.2	90 × 75 × 16	22.8	30.3
.38	1.01	.44	.85	1.00	10	26.2	13	21.8	29.3
.31	.99	.38	.83	1.08	8	25.5	10	20.7	28.2
.25	.97	.31	.81	1.06	6	24.7	8	20	27.5
3.00 × 3.00 × .50	.93	.25	.79	1.04	75 × 75 × 13	23.5	6	19.3	26.8
.38	.89	3.50 × 2.50 × .50	.75	1.00	10	22.4	90 × 65 × 16	19.5	32
.31	.87	.44	.73	.98	8	21.7	13	18.4	30.9
.25	.84	.38	.71	.96	6	21	10	17.3	29.8
2.50 × 2.50 × .38	.76	.31	.68	.93	65 × 65 × 10	19.9	8	16.6	29.1
.31	.74	.25	.66	.91	8	19.2	6	15.9	28.4
.25	.72	3.00 × 2.00 × .50	.58	1.08	6	18.5	75 × 50 × 13	14.8	27.3
2.00 × 2.00 × .38	.64	.38	.54	1.04	50 × 50 × 10	16.1	10	13.7	26.2
.31	.61	.31	.52	1.02	8	15.4	8	13	25.5
.25	.59	.25	.49	.99	6	14.7	6	12.2	24.7
1.75 × 1.75 × .25	.53	2.50 × 2.00 × .38	.58	.83	4	14	65 × 45 × 10	12.9	22.9
.19	.51	.31	.56	.81	45 × 45 × 8	14.2	8	12.2	22.2
.12	.48	.25	.54	.79	6	13.4	6	11.4	21.4
1.50 × 1.50 × .25	.47	.19	.51	.76	4	12.7	4	10.7	20.7
.19	.44	2.00 × 1.50 × .25	.41	.66	35 × 35 × 8	11.6	50 × 35 × 8	10.1	17.6
.12	.42	.19	.39	.64	6	10.9	6	9.42	16.9
1.00 × 1.00 × .25	.34	.12	.37	.62	4	10.2	4	8.7	16.2
.19	.32	1.75 × 1.25 × .25	.35	.60	25 × 25 × 6	8.4	45 × 25 × 6	6.71	16.7
		.19	.33	.58	4	7.7	4	5.98	16

Table 55   Angle iron sizes

	Nominal Pipe Size Inches	Outside Diameter	Wall Thickness			Approx. Distance Pipe Enters Fitting L	Weight (lbs/ft)		
			Sched. 40 (Standard)	Sched. 80 (Extra Strong)	Sched. 160		Sched. 40 (Standard)	Sched. 80 (Extra Strong)	Sched. 160
U.S. CUSTOMARY (INCHES)	⅛ (.125)	.405	.068	.095	—	.188	.24	.31	—
	¼ (.250)	.540	.088	.119	—	.281	.42	.54	—
	⅜ (.375)	.675	.091	.126	—	.297	.57	.74	—
	½ (.500)	.840	.109	.147	.188	.375	.85	1.09	1.31
	¾ (.750)	1.050	.113	.154	.219	.406	1.13	1.47	1.94
	1.00	1.315	.133	.179	.250	.500	1.68	2.17	2.84
	1.25	1.660	.140	.191	.250	.549	2.27	3.00	3.76
	1.50	1.900	.145	.200	.281	.562	2.72	3.63	4.86
	2	2.375	.154	.218	.344	.578	3.65	5.02	7.46
	2.5	2.875	.203	.276	.375	.875	5.79	7.66	10.01
	3	3.500	.216	.300	.438	.938	7.58	10.25	14.31
	3.5	4.000	.226	.318	—	1.000	9.11	12.51	—
	4	4.500	.237	.337	.531	1.062	10.79	14.98	22.52
	5	5.563	.258	.375	.625	1.156	14.62	20.78	32.96
	6	6.625	.280	.432	.719	1.250	18.97	28.57	45.34
	8	8.625	.322	.500	.906	1.469	28.55	43.39	74.71

	Nominal Pipe Size Inches	Outside Diameter mm	Wall Thickness			Approx. Distance Pipe Enters Fitting L	Mass (kg/m)		
			Sched. 40 (Standard)	Sched. 80 (Extra Strong)	Sched. 160		Sched. 40 (Standard)	Sched. 80 (Extra Strong)	Sched. 160
METRIC (MILLIMETERS)	⅛ (.125)	10.3	1.7	2.4	—	5	0.36	0.46	—
	¼ (.250)	13.7	2.2	3.0	—	7	0.63	0.80	—
	⅜ (.375)	17.1	2.3	3.2	—	8	0.85	1.10	—
	½ (.500)	21.3	2.8	3.7	4.8	10	1.26	1.62	1.95
	¾ (.750)	26.7	2.9	3.9	5.6	11	1.68	2.19	2.89
	1.00	33.4	3.4	4.6	6.4	13	2.50	3.23	4.23
	1.25	42.1	3.6	4.9	6.4	14	3.38	4.46	5.60
	1.50	48.3	3.7	5.1	7.1	14	4.05	5.40	7.23
	2.00	60.3	3.9	5.5	8.7	15	5.43	7.47	11.10
	2.50	73	5.2	7.0	9.5	22	8.62	11.40	14.90
	3.00	88.9	5.5	7.6	11.1	24	11.28	15.25	21.30
	3.50	101.6	5.7	8.1	—	25	13.56	18.62	—
	4.00	114.3	6.0	8.6	13.5	27	16.06	22.30	33.51
	5.00	141.3	6.6	9.5	15.9	29	21.76	30.92	49.05
	6.00	168.3	7.1	11.0	18.3	32	28.23	42.52	67.47
	8.00	219	8.2	12.7	23.0	38	42.49	64.57	111.18

Table 56   American standard wrought steel pipe.

90° ELBOW   TEE   CROSS   45° ELBOW   LATERAL

Nominal Pipe Size in Inches	U.S. CUSTOMARY (INCHES)						METRIC (MILLIMETERS)					
	A	Min B	Min C	D	E	F	A	Min B	Min C	D	E	F
.25	.81	.38	.93	—	—	.73	21	10	24	—	—	19
.375	.95	.44	1.12	—	—	.80	24	11	28	—	—	20
.50	1.12	.50	1.34	2.50	1.87	.88	28	13	34	64	47	22
.75	1.31	.56	1.63	3.00	2.25	.98	33	14	41	76	57	25
1.00	1.50	.62	1.95	3.50	2.75	1.12	38	16	50	89	70	28
1.25	1.75	.69	2.39	4.25	3.25	1.29	44	18	61	108	83	33
1.50	1.94	.75	2.68	4.87	3.81	1.43	49	19	68	124	97	36
2.00	2.25	.84	3.28	5.75	4.25	1.68	57	21	83	146	108	43
2.50	2.70	.94	3.86	6.75	5.18	1.95	69	24	98	171	132	50
3.00	3.08	1.00	4.62	7.87	6.12	2.17	78	25	117	200	155	55
3.50	3.42	1.06	5.20	8.87	6.87	2.39	87	27	132	225	174	61
4.00	3.79	1.12	5.79	9.75	7.62	2.61	96	28	147	248	194	66
5.00	4.50	1.18	7.05	11.62	9.25	3.05	114	30	179	295	235	77
6.00	5.13	1.28	8.28	13.43	10.75	3.46	130	33	210	341	273	88
8.00	6.56	1.47	10.63	16.94	13.63	4.28	167	37	270	430	346	109
10.00	8.08	1.68	13.12	20.69	16.75	5.16	205	43	333	613	425	131

Table 57   American standard (125 lb) cast-iron screwed-pipe fittings.

Nominal Pipe Size in Inches	U.S. CUSTOMARY (INCHES)							METRIC (MILLIMETERS)						
	A	B	C	D	E	F	G	A	B	C	D	E	F	G
.125	.69	.20	.69	—	—	—	.96	18	5.0	18	—	—	—	24
.250	.81	.22	.84	—	—	.73	1.06	21	5	21	—	—	19	27
.375	.95	.23	1.02	1.93	1.43	.80	1.16	24	6	26	49	36	20	29
.500	1.12	.25	1.20	2.32	1.71	.88	1.34	28	6	30	59	43	22	34
.750	1.31	.27	1.46	2.77	2.05	.98	1.52	33	7	37	70	52	25	39
1.00	1.50	.30	1.77	3.28	2.43	1.12	1.67	38	8	45	83	62	28	42
1.25	1.75	.34	2.15	3.94	2.92	1.29	1.93	44	9	55	100	74	33	49
1.50	1.94	.37	2.43	4.38	3.28	1.43	2.15	49	9	62	111	83	36	55
2.00	2.25	.42	2.96	5.17	3.93	1.68	2.53	57	11	75	131	100	43	64
2.50	2.70	.48	3.59	6.25	4.73	1.95	2.88	69	12	91	159	120	50	73
3.00	3.08	.55	4.29	7.26	5.55	2.17	3.18	78	14	109	184	141	55	81
3.50	3.42	.60	4.84	—	—	2.39	3.43	87	15	123	—	—	61	87
4.00	3.79	.66	5.40	8.98		2.61	3.69	96	17	137	228	177	66	94
5.00	4.50	.78	6.58		6.97	3.05	—	114	20	167	—	—	77	—
6.00	5.13	.90	7.77			3.46	—	130	23	197	—	—	88	—

Table 58   American standard (150 lb) malleable-iron screwed-pipe fittings.

90° ELBOW   45° ELBOW   TEE   CROSS   LATERAL   COUPLING

90° ELBOW 90° LONG RADIUS ELBOW 45° ELBOW

REDUCING ELBOW SIDE OUTLET ELBOW TEE

SIDE OUTLET TEE CROSS

LATERAL REDUCER ECCENTRIC REDUCER

NOMINAL PIPE SIZE IN INCHES		A	B	C	D	E	F	G	H
U.S. CUSTOMARY (INCHES)	1.50	4.00	6.00	5.00	9.00	7.00	2.25	—	.56
	2.00	4.50	6.50	6.00	10.50	8.00	2.50	5.00	.62
	2.50	5.00	7.00	7.00	12.00	9.50	3.00	5.50	.69
	3.00	5.50	7.75	7.50	13.00	10.00	3.00	6.00	.75
	3.50	6.00	8.50	8.50	14.50	11.50	3.50	6.50	.81
	4.00	6.50	9.00	9.00	15.00	12.00	4.00	7.00	.94
	5.00	7.50	10.25	10.00	17.00	13.50	4.50	8.00	.94
	6.00	8.00	11.50	11.00	18.00	14.50	5.00	9.00	1.00
	8.00	9.00	14.00	13.50	22.00	17.50	5.50	11.00	1.12
	10.00	11.00	16.50	16.00	25.50	20.50	6.50	12.00	1.19
METRIC (MILLIMETERS)	1.50	102	152	127	229	178	57	—	14
	2.00	114	165	152	267	203	64	127	16
	2.50	127	178	178	305	241	76	140	18
	3.00	140	197	190	330	254	76	152	19
	3.50	153	216	216	368	292	89	165	21
	4.00	165	229	229	381	305	102	178	24
	5.00	190	260	254	432	343	114	203	24
	6.00	203	292	280	457	368	127	229	25
	8.00	229	356	343	559	445	140	279	28
	10.00	279	419	406	648	521	165	305	30

Table 59 American standard flanged fittings.

ELBOW SHORT RADIUS ELBOW LONG RADIUS 45° ELBOW LONG RADIUS

TEE CROSS

REDUCER ECCENTRIC REDUCER

NOMINAL PIPE SIZE IN INCHES		A	B	C	D	E	F
U.S. CUSTOMARY (INCHES)	1.50	1.50	2.25	1.12	2.25	2.25	2.50
	2.00	2.00	3.00	1.38	2.50	2.50	3.00
	2.50	2.50	3.75	1.75	3.00	3.00	3.50
	3.00	3.00	4.50	2.00	3.38	3.38	3.50
	3.50	3.50	5.25	2.25	3.75	3.75	4.00
	4.00	4.00	6.00	2.50	4.12	4.12	4.00
	5.00	5.00	7.50	3.12	4.89	4.89	5.00
	6.00	6.00	9.00	3.75	5.62	5.62	5.50
	8.00	8.00	12.00	5.00	7.00	7.00	6.00
	10.00	10.00	15.00	6.25	8.50	8.50	7.00
METRIC (MILLIMETERS)	1.50	38	57	28	57	57	64
	2.00	51	76	35	64	64	76
	2.50	64	95	44	76	76	89
	3.00	76	114	51	86	86	89
	3.50	89	133	57	95	95	102
	4.00	102	152	64	105	105	102
	5.00	127	190	79	124	124	127
	6.00	152	229	95	143	143	140
	8.00	203	305	127	178	178	152
	10.00	254	381	159	216	216	178

Table 60 American standard steel butt-welding fittings.

CHECK VALVES

LIFT   SWING

Nominal Pipe Size in Inches	Lift Check A Screwed	Lift Check A Flanged	B	Swing Check C Screwed	Swing Check C Flanged	D
U.S. CUSTOMARY (INCHES)						
2.00	6.50	—	3.50	6.50	8.00	4.25
2.50	7.00	—	4.25	7.00	8.50	4.81
3.00	8.00	9.50	5.00	8.00	9.50	5.06
3.50	—	—	—	9.00	10.50	5.81
4.00	10.00	11.50	6.25	10.00	11.50	6.19
5.00	—	13.00	7.00	11.25	13.00	7.19
6.00	—	14.00	8.25	12.50	14.00	7.50
8.00	—	—	—	—	19.50	10.19
10.00	—	—	—	—	24.50	12.12
METRIC (MILLIMETERS)						
2.00	165	—	89	165	203	108
2.50	178	—	108	178	216	122
3.00	203	241	127	203	241	129
3.50	—	—	—	229	267	148
4.00	254	292	159	254	292	157
5.00	—	330	178	286	330	183
6.00	—	356	210	318	356	190
8.00	—	—	—	—	495	259
10.00	—	—	—	—	622	308

GATE VALVES

Nominal Pipe Size in Inches	Nonrising Spindle E Screwed	Nonrising Spindle E Flanged	F	Rising Spindle G Screwed	Rising Spindle G Flanged	H
U.S. CUSTOMARY (INCHES)						
2.00	4.75	7.00	10.50	4.75	7.00	13.12
2.50	5.50	7.50	11.19	5.50	7.50	14.50
3.00	6.00	8.00	12.62	6.00	8.00	16.62
3.50	6.62	8.50	13.31	6.62	8.50	18.44
4.00	7.12	9.00	15.25	7.12	9.00	21.06
5.00	8.12	10.00	17.88	8.12	10.00	25
6.00	9.00	10.50	20.19	9.00	10.50	29.25
8.00	10.00	11.50	24.00	10.00	11.50	37.25
10.00	—	13.00	28.19	—	13.00	44.12
METRIC (MILLIMETERS)						
2.00	121	178	267	121	178	333
2.50	140	190	284	140	190	368
3.00	152	203	321	152	203	422
3.50	168	216	338	168	216	468
4.00	181	229	387	181	229	535
5.00	206	254	454	206	254	635
6.00	229	267	513	229	267	743
8.00	254	292	610	254	292	946
10.00	—	330	716	—	330	1121

GLOBE AND ANGLE VALVES

GLOBE   ANGLE

Nominal Pipe Size in Inches	Globe J Screwed	Globe J Flanged	K	Angle L Screwed	Angle L Flanged	M
U.S. CUSTOMARY (INCHES)						
2.00	4.75	7.00	9.44	3.50	3.88	10.38
2.50	5.50	7.50	11.06	3.88	4.50	12.06
3.00	6.00	8.00	12.38	4.69	4.62	12.69
3.50	6.62	8.50	13.19	5.00	5.38	13.62
4.00	7.12	9.00	15.25	6.00	5.88	14.88
5.00	8.12	10.00	17.25	6.31	6.50	17.69
6.00	9.00	10.50	18.81	8.00	8.00	19.06
8.00	10.00	11.50	22.12	—	9.25	22.75
10.00	—	13.00	24.75	—	10.62	24.94
METRIC (MILLIMETERS)						
2.00	121	178	240	89	98	264
2.50	140	190	281	98	114	281
3.00	152	203	314	119	118	346
3.50	168	216	335	127	136	348
4.00	181	229	387	152	149	378
5.00	206	254	438	160	165	449
6.00	229	267	478	203	203	484
8.00	254	292	562	—	235	578
10.00	—	330	629	—	270	633

Dimensions taken from manufacturer's catalogs for drawing purposes.

Table 61   Common valves.

FITTING		FLANGED	SCREWED	WELDED	BELL AND SPIGOT	SOLDERED
BUSHING						
CAP						
CROSS	STRAIGHT SIZE					
ELBOW	45 DEGREE					
	90 DEGREE					
	TURNED DOWN					
	TURNED UP					
	LONG RADIUS					
	STREET					
JOINT (COUPLING)	CONNECTING PIPE					
LATERAL						
PLUG	PIPE PLUG					
REDUCER	CONCENTRIC					
	ECCENTRIC					
TEE	STRAIGHT					
	OUTLET UP					
	OUTLET DOWN					
UNION						
VALVES	CHECK STRAIGHTWAY					
	GATE					
	GLOBE					
	ANGLE GLOBE ELEVATION					
	GLOBE PLAN					

Table 62   Graphic symbols formerly used for pipe fittings.

U.S. CUSTOMARY (INCHES)								METRIC (MILLIMETERS)					
	W = .071	W = .087	W = .125	W = .174	W = .209	W = .256		W = 1.5	W = 2	W = 3	W = 4	W = 5	W = 6
ID							ID	OD	OD	OD	OD	OD	OD
.124	x						3	6					
.158	x						4	7					
.132	x						5	8					
.248	x						6	9					
.280	x						7	10					
.295	x	x	x				8	11	12	14			
.354	x	x	x				9	12	13	15			
.394	x	x	x	x	x	x	10	13	14	16	18	20	
.465	x	x	x	x	x	x	12	15	16	18	20	22	24
.551	x	x	x	x	x	x	14	17	18	20	22	24	26
.591	x	x	x	x	x	x	15	18	19	21	23	25	27
.622	x	x	x	x	x	x	16	19	20	22	24	26	28
.630	x	x	x	x	x	x	18	21	22	24	26	28	30
.662	x	x	x	x	x	x	20	23	24	26	28	30	32
.669	x	x	x	x	x	x	22	25	26	28	30	32	34
.787	x	x	x	x	x	x	24	27	28	30	32	34	36
.850	x	x	x	x	x	x	25	28	29	31	33	35	37
1.024	x	x	x	x	x	x	26	29	30	32	34	36	38
1.102	x	x	x	x	x	x	28	31	32	34	36	38	40
1.181	x	x	x	x	x	x	30	33	34	36	38	40	42
1.220	x	x	x	x	x	x	32	35	36	38	40	42	44
1.299	x	x	x	x	x	x	34	37	38	40	42	44	46
1.319	x	x	x	x	x	x	35	38	39	41	43	45	47
1.496	x	x	x	x	x	x	36	39	40	42	44	46	48
1.524	x	x	x	x	x	x	38	41	42	44	46	48	50
1.575	x	x	x	x	x	x	40	43	44	46	48	50	52
1.654	x	x	x	x	x	x	42	45	46	48	50	52	54
1.732	x	x	x	x	x	x	44	47	48	50	52	54	56
1.772	x	x	x	x	x	x	45	48	49	51	53	55	57
1.811	x	x	x	x	x	x	46	49	50	52	54	56	58
1.890	x	x	x	x	x	x	48	51	52	54	56	58	60
2.008	x	x	x	x	x	x	50	53	54	56	58	60	62
2.087	x	x	x	x	x	x	52	55	56	58	60	62	64
2.165	x	x	x	.174	.209	.256	54	57	58	60	62	64	66
2.205	x	x	x	x	x	x	55	58	59	61	63	65	67
2.244	x	x	x	x	x	x	56	59	60	62	64	66	68
2.323	x	x	x	x	x	x	58	61	62	64	66	68	70
2.402	x	x	x	x	x	x	60	63	64	66	68	70	72
2.639	x	x	x	x	x	x	65	68	69	71	73	75	77
2.835	x	x	x	x	x	x	70	73	74	76	78	80	82
3.031	x	x	x	x	x	x	75	78	79	81	83	85	87
3.228	x	x	x	x	x	x	80	83	84	86	88	90	92

Table 63   O rings.

Inside Dia		Outside Dia		Width	
in.	mm	in.	mm	in.	mm
.375	10	.753	19	.25	6
.375	10	.840	21	.31	8
.438	11	1.003	26	.31	8
.438	11	1.128	28	.31	8
.500	12	1.003	26	.31	8
.500	12	1.128	28	.31	8
.500	12	1.254	32	.38	10
.562	14	1.003	26	.31	8
.562	14	1.128	28	.31	8
.625	16	1.250	32	.38	10
.625	16	1.128	28	.31	8
.625	16	1.250	32	.38	10
.688	18	1.379	35	.38	10
.688	18	1.128	28	.31	8
.688	18	1.254	32	.38	10
.750	20	1.379	35	.38	10
.750	20	1.254	32	.38	
.750	20	1.379	35		10
.750	20	1.503	38	.38	10
.8125	21	1.756	44	.44	12
.8125	21	1.254	32	.38	10
.8125	21	1.379	35	.38	10
.8125	21	1.503	38	.38	10
.875	22	1.756	44	.44	12
.875	22	1.379	35	.38	10
.875	22	1.503	38	.38	10
.875	22	1.628	42	.44	12
.938	24	1.756	44	.44	12
.938	24	1.503	38	.38	10
.938	24	1.628	42	.38	10
1.000	25	1.756	44	.44	12
1.000	25	1.503	38	.38	10
1.000	25	1.756	44	.44	12
1.000	25	1.878	48	.44	12
1.000	25	2.004	50	.44	12
1.062	26	1.503	38	.38	10
	26	1.628	42	.44	12
	26	1.756	44	.44	12
1.125	28	1.628	42	.44	12
	28	1.756	44	.44	12
	28	1.987	50	.50	12
1.188	30	1.832	46	.44	12
	30	1.987	50	.50	12
	30	2.254	58	.50	12
1.25	32	1.756	44	.44	12
	32	1.878	48	.44	12
	32	2.066	52	.50	12
1.312	34	2.060	52	.44	12
	34	2.254	58	.50	12
	34	2.378	60	.50	12
1.375	35	2.066	52	.44	12
	35	2.254	58	.50	12
	35	2.441	62	.50	12
1.438	36	2.254	58	.50	12
	36	2.506	64	.50	12
	36	2.627	66	.50	12
1.500	38	2.254	58	.38	10
	38	2.410	62	.50	12
	38	2.720	70	.50	12
1.562	40	2.441	62	.50	12
	40	2.690	68	.50	12
	40	2.879	74	.50	12
1.625	42	2.441	62	.38	10
	42	2.879	74	.38	10
	42	2.627	66	.50	12
	42	2.879	74	.50	12
	42	3.066	78	.50	12
1.75	44	2.254	58	.50	12
	44	2.441	62	.50	12
	44	2.506	64	.50	12

Table 64 Oil seals.

Shaft Dia		Outside Dia		Width		Size of Set Screw	
in.	mm	in.	mm	in.	mm	in.	mm
.375	10	.75	20	.40	10	.250	M6
.500	12	1.00	25	.44	10	.250	M6
.625	16	1.10	28	.50	12	.3125	M8
.750	20	1.20	30	.56	14	.3125	M8
.875	22	1.50	40	.56	14	.3125	M8
1.000	24	1.60	40	.60	16	.3125	M8

Table 65 Setscrew collars.

Figures show springs wound left-hand

Fig. 1
90° Deflection

Fig. 2
180° Deflection

Fig. 3
270° Deflection

Fig. 4
360 Deflection

\multicolumn{8}{c}{U.S. CUSTOMARY (INCHES)}								\multicolumn{8}{c}{METRIC (MILLIMETERS)}							
Wire Dia	Out-side Dia	Pos. of Ends Fig.	Def, Deg.	Torque in · lb	Radius R	E	Min. Axial Space L	Wire Dia	Out-side Dia	Pos. of Ends Fig.	Def, Deg.	Torque N · m	Radius R	E	Min. Axial Space L
	.124	1	90		.250	.50	.067		3.2	1	90		6.4	13	1.7
	.133	2	180		.250	.50	.105		3.4	2	180		6.4	13	2.7
.014	.124	3	270	.070	.250	.50	.158	0.35	3.2	3	270	0.008	6.4	13	4.0
	.194	2	180		.375	.75	.077		4.9	2	180		9.5	19	2.0
	.201	3	270		.375	.75	.102		5.1	3	270		9.5	19	2.6
	.204	4	360		.375	.75	.126		5.2	4	360		9.5	19	3.2
	.160	1	90		.250	.50	.081		4.1	1	90		6.4	13	2.1
	.172	2	180		.250	.50	.127		4.4	2	180		6.4	13	3.2
.017	.160	3	270	.117	.250	.50	.192	0.43	4.1	3	270	0.013	6.4	13	4.9
	.249	2	180		.375	.75	.094		6.3	2	180		9.5	19	2.4
	.259	3	270		.375	.75	.123		6.6	3	270		9.5	19	3.1
	.235	4	360		.375	.75	.170		6.0	4	360		9.5	19	4.3
	.191	1	90		.375	.75	.095		4.9	1	90		9.5	19	2.4
	.179	2	180		.375	.75	.170		4.6	2	180		9.5	19	4.3
.020	.175	3	270	.187	.375	.75	.245	0.51	4.5	3	270	0.021	9.5	19	6.2
	.242	2	180		.500	1.00	.130		6.2	2	180		12.7	25	3.3
	.268	3	270		.500	1.00	.165		6.8	3	270		12.7	25	4.2
	.254	4	360		.500	1.00	.250		6.5	4	360		12.7	25	6.4
	.204	1	90		.375	.75	.109		5.2	1	90		9.5	19	2.8
	.191	2	180		.375	.75	.196		4.9	2	180		9.5	19	5.0
.023	.187	3	270	.308	.375	.75	.282	0.59	4.8	3	270	0.035	9.5	19	7.2
	.259	2	180		.500	1.00	.150		6.6	2	180		12.7	25	3.8
	.251	3	270		.500	1.00	.213		6.4	3	270		12.7	25	5.4
	.271	4	360		.500	1.00	.253		6.9	4	360		12.7	25	6.5
	.267	1	90		.500	1.00	.133		6.8	1	90		12.7	25	3.4
	.249	2	180		.500	1.00	.238		6.3	2	180		12.7	25	6.0
.028	.245	3	270	.515	.500	1.00	.344	0.71	6.2	3	270	0.058	12.7	25	8.8
	.340	2	180		.500	1.00	.182		8.7	2	180		12.7	25	4.6
	.329	3	270		.500	1.00	.259		8.4	3	270		12.7	25	6.6
	.355	4	360		.500	1.00	.308		9.0	4	360		12.7	25	7.9

Table 66 Torsion springs. (Wallace Barnes Co. Ltd.)

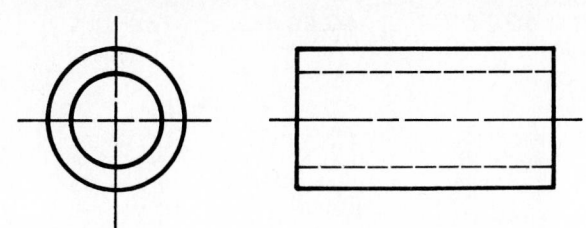

Inside Dia in.	Inside Dia mm	Outside Dia in.	Outside Dia mm	15 / .50	20 / .75	25 / 1.00	30 / 1.25	35 / 1.50	40 / 1.75	50 / 2.00	60 / 2.50	75 / 3.00
.375	10	.625	16	•	•	•	•					
		.750	20	•	•	•	•					
.500	12	.625	16	•	•	•	•	•				
		.750	20	•	•	•	•		•	•		
.625	16	.875	22	•	•	•	•	•	•	•		
		1.000	25	•	•	•	•	•	•	•		
.750	20	1.000	25	•	•	•	•	•	•	•		
		1.125	28	•	•	•	•	•	•	•		
.875	22	1.125	28		•	•	•	•	•	•	•	
		1.250	32		•	•	•	•	•	•	•	
1.000	25	1.250	32			•	•	•	•	•	•	•
		1.375	35			•	•	•	•	•	•	•

Table 67   Standard plain (journal) bearings.

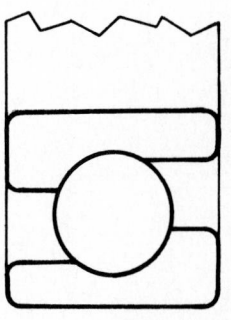

Series	Bearing Number	Bore d in.	Bore d mm	Outside Diameter D in.	Outside Diameter D mm	Width W in.	Width W mm	Basic Load Rating Kips	Basic Load Rating kN
Light	7205B	0.984	25	2.047	52	0.591	15	2.5	11.4
	7206B	1.181	30	2.441	62	0.630	16	3.4	15.6
	7207B	1.378	35	2.835	72	0.670	17	4.6	20.8
	7208B	1.575	40	3.150	80	0.709	18	5.5	24.5
	7209B	1.772	45	3.347	85	0.748	19	6.2	27.5
	7210B	1.969	50	3.543	90	0.787	20	6.4	28.5
	7211B	2.165	55	3.937	100	0.827	21	8.2	36
	7212B	2.362	60	4.331	110	0.866	22	9.7	43
Medium	7304B	0.787	20	2.047	52	0.591	15	3.1	13.4
	7305B	0.984	25	2.441	62	0.670	17	4.3	19
	7306B	1.181	30	2.835	72	0.748	19	5.4	24
	7307B	1.378	35	3.150	80	0.827	21	6.3	28
	7308B	1.575	40	3.543	90	0.906	23	7.8	34.5
	7309B	1.772	45	3.937	100	0.984	25	10	45
	7310B	1.969	50	4.331	110	1.063	27	11.6	52
	7311B	2.165	55	4.724	120	1.142	29	13.4	61
	7312B	2.362	60	5.118	130	1.221	31	15.6	69.5
Heavy	7405B	0.984	25	3.150	80	0.827	21	6.8	30.5
	7406B	1.181	30	3.543	90	0.906	23	8.3	36.5
	7407B	1.378	35	3.937	100	0.984	25	10.4	46.5
	7408B	1.575	40	4.331	110	1.063	27	12	54
	7409B	1.772	45	4.724	120	1.142	29	14.6	65.5
	7410B	1.969	50	5.118	130	1.221	31	16.6	73.5
	7411B	2.165	55	5.512	140	1.299	33	19	85
	7412B	2.362	60	5.906	150	1.378	35	20.4	91.5

Table 68   Angular contact ball bearings. (SKF Co. Ltd.)

| | | | | | | | | ONE SHIELD | TWO SHIELDS | ONE RED SEAL | TWO RED SEALS | SNAP RING | SNAP RING ONE SHIELD |

60   **60 Z**   **60 2Z**   **60 RS**   **60 2RS**   **60 NR**   **60 ZNR**

Bearing Number							Nominal Bearing Dimensions								
							d		D		B		F	M	S	H	T
							in.	mm	in.	mm	in.	mm	Millimeters				
6000	6000 Z	6000 2Z	6000 RS	6000 2RS			.3937	10	1.0236	26	.3150	8					
6001	6001 Z	6001 2Z	6001 RS	6001 2RS			.4724	12	1.1024	28	.3150	8					
6002	6002 Z	6002 2Z	6002 RS	6002 2RS	6002 NR	6002 ZNR	.5906	15	1.2598	32	.3543	9	30	36.5	3	2	1
6003	6003 Z	6003 2Z	6003 RS	6003 2RS			.6693	17	1.3780	35	.3937	10					
6004	6004 Z	6004 2Z	6004 RS	6004 2RS			.7874	20	1.6535	42	.4724	12					
6005	6005 Z	6005 2Z	6005 RS	6005 2RS			.9843	25	1.8504	47	.4724	12					
6006	6006 Z	6006 2Z	6006 RS	6006 2RS			1.1811	30	2.1654	55	.5118	13					
6007	6007 Z	6007 2Z	6007 RS	6007 2RS			1.3780	35	2.4409	62	.5512	14					
6008	6008 Z	6008 2Z	6008 RS	6008 2RS			1.5748	40	2.6772	68	.5906	15					
6009							1.7717	45	2.9528	75	.6299	16					
6010	6010 Z	6010 2Z	6010 RS	6010 2RS	6010 NR	6010 ZNR	1.9685	50	3.1496	80	.6299	16	76.8	86	4	2.5	1.6
6011	6011 Z	6011 2Z	6011 RS	6011 2RS	6011 NR	6011 ZNR	2.1654	55	3.5433	90	.7087	18	86.8	96.5	5	3	2.4
6012							2.3622	60	3.7402	95	.7087	18					
6013	6013 Z	6013 2Z			6013 NR	6013 ZNR	2.5591	65	3.9370	100	.7087	18	96.8	106	5	3	2.4
6014							2.7559	70	4.3307	110	.7874	20					
6015	6015 Z	6015 2Z			6015 NR	6015 ZNR	2.9528	75	4.5276	115	.7874	20	112	121	5	3	2.4

Note: When shield is required on same side as snap ring, order as ZNBR.

Table 69   Radial ball bearings. (SKF Co. Ltd.)

Bearing Number	Nominal External Sizes						U.S. CUSTOMARY (in.)						METRIC (mm)					
	Bore		Outside Dia		Width													
	in.	mm	in.	mm	in.	mm	F	C	S	P	L	O	F	C	S	P	L	O
4059	.590	15	1.375	35	.430	11	.089	.344	.781	.719	1.094	1.250	2	9	20	18	28	32
03062	.625	16	1.625	41	.562	14	.125	.438	.844	.781	1.312	1.469	3	11	21	20	33	37
5062	.625	16	1.850	46	.566	14	.129	.438	.938	.844	1.562	1.688	3	11	24	21	40	43
1749	.688	17	1.570	40	.545	14	.125	.420	.906	.781	1.312	1.438	3	11	23	20	37	36
9195	.695	17.6	1.938	49	.906	20	.219	.562	1.062	.969	1.625	1.750	6	14	27	24	41	44
9070	.695	17.6	1.938	49	.906	23	.219	.688	1.062	.969	1.562	1.750	6	17	27	24	39	44
1774	.748	19	2.240	57	.753	20	.138	.625	1.062	.969	1.875	2.031	4	16	27	24	48	51
1949	.750	20	1.781	45	.610	16	.135	.475	1.000	.922	1.500	1.625	4	12	25	59	38	41
9067	.750	20	1.938	50	.835	21	.148	.688	1.000	.922	1.531	1.750	4	17	25	23	39	44
9078	.750	20	1.938	50	.906	23	.219	.562	1.000	.922	1.625	1.750	6	17	25	23	39	44
7075	.750	20	2.000	51	.591	15	.091	.375	1.000	.938	1.719	1.844	3	12	25	24	44	47
7087	.857	22	1.969	50	.531	14	.156	.375	1.125	1.062	1.719	1.844	4	10	28	27	44	47
1380	.875	22	2.125	54	.762	20	.200	.562	1.125	1.094	1.750	1.906	5	15	28	28	44	48
1280	.875	22	2.250	54	.875	22	.188	.688	1.125	1.094	1.875	2.062	5	17	28	28	48	52
4643	1.000	25	1.980	48	.560	14	.140	.420	1.219	1.156	1.719	1.844	4	10	33	29	44	47
1780	1.000	25	2.240	57	.762	20	.781	.625	1.188	1.172	1.844	2.031	4	16	30	30	48	52

Table 70 Tapered roller bearings. (SKF Co. Ltd.)

Inside Dia		Thickness		Outside Dia					
in.	mm	in.	mm	in.	mm	in.	mm	in.	mm
.500	12	.12 .18	3 5	.80	20	1.10	25	1.25	30
.625	16	.12 .18	3 5	1.25	30	1.50	40	1.75	46
.750	20	.12 .18	3 5	1.30	34	1.60	40	2.00	50
.875	22	.12 .18	3 5	1.30	34	1.60	40	2.00	50
1.000	25	.12 .18 .25	3 5 6	1.60	40	2.00	50	2.25	60

Table 71 Thrust plain bearings.

BEARING NUMBER	Nominal Bearing Dimensions								Max. Fillet Radius* in.	Shoulder Diameter in.	
	d		d_2		D		H			Shaft Min.	Housing Max.
	in.	mm	in.	mm	in.	mm	in.	mm			
51100	.3937	10	.402	10.2	.9449	24	.354	9	.012	.750	.563
51101	.4724	12	.480	12.2	1.0236	26	.354	9	.012	.844	.656
51102	.5906	15	.598	15.2	1.1024	28	.354	9	.012	.938	.750
51103	.6693	17	.677	17.2	1.1811	30	.354	9	.012	1.031	.844
51104	.7874	20	.795	20.2	1.3780	35	.394	10	.012	1.188	.969
51105	.9843	25	.992	25.2	1.6535	42	.433	11	.024	1.438	1.219
51106	1.1811	30	1.189	30.2	1.8504	47	.433	11	.024	1.563	1.469
51107X	1.3780	35	1.386	35.2	2.0472	52	.472	12	.024	1.813	1.656
51108	1.5748	40	1.583	40.2	2.3622	60	.512	13	.024	2.063	1.875
51109	1.7717	45	1.780	45.2	2.5591	65	.551	14	.024	2.281	2.063
51110	1.9685	50	1.976	50.2	2.7559	70	.551	14	.024	2.500	2.250
51111	2.1654	55	2.173	55.2	3.0709	78	.630	16	.024	2.750	2.500
51112	2.3622	60	2.370	60.2	3.3465	85	.669	17	.039	3.000	2.719
51113	2.5591	65	2.567	65.2	3.5433	90	.709	18	.039	3.188	2.875
51114	2.7559	70	2.764	70.2	3.7402	95	.709	18	.039	3.375	3.063
51115	2.9528	75	2.961	75.2	3.9370	100	.748	19	.039	3.625	3.250
51116	3.1496	80	3.157	80.2	4.1339	105	.748	19	.039	3.813	3.438
51117	3.3465	85	3.354	85.2	4.3307	110	.748	19	.039	4.000	3.625

*The maximum fillet on the shaft or in the housing, which will be cleared by the bearing corner.

Table 72 Thrust roller bearings. (SKF Co. Ltd.)

STEEL AND IRON SPUR GEARS
14.5° PRESSURE ANGLE
(Will not operate with 20° Spurs)
Actual Tooth Size

SPUR GEARS STEEL AND IRON
20° PRESSURE ANGLE
(Will not operate with 14.5° Spurs)
Actual Tooth Size

Style	No. of Teeth	Inch Sizes 8 Pitch 1.25 Face				Metric Sizes (mm) 3.18 Module 30 Face			
		Pitch Dia	Hole	Hub Dia	Hub Proj.	Pitch Dia	Hole	Hub Dia	Hub Proj.
Steel Plain	12	1.500	.750	1.12	.75	38.2	20	28	20
	14	1.750	.750	1.38	.75	44.5	20	35	20
	15	1.875	.875	1.50	.75	47.7	22	40	20
	16	2.000	.875	1.62	.75	50.9	22	40	20
	18	2.250	.875	1.88	.75	57.2	22	48	20
	20	2.500	.875	2.12	.75	63.6	22	54	20
	22	2.750	.875	2.38	.75	70.0	22	60	20
Cast Iron Web	24	3.000	.875	2.12	1.00	76.3	22	54	25
	28	3.500	.875	2.25	1.00	89.0	22	56	25
	30	3.750	.875	2.25	1.00	95.4	22	56	25
	32	4.000	1.000	2.25	1.00	101.8	25	56	25
	36	4.500	1.000	2.50	1.00	114.5	25	64	25
	40	5.000	1.000	2.50	1.00	127.2	25	64	25
	42	5.250	1.000	2.50	1.00	133.6	25	64	25
	44	5.500	1.000	2.50	1.00	139.9	25	64	25
	48	6.000	1.000	2.50	1.00	152.6	25	64	25
	54	6.750	1.000	2.50	1.00	171.7	25	64	25
Cast Iron Spoke	56	7.000	1.000	2.50	1.00	178.1	25	64	25
	60	7.500	1.000	2.50	1.00	190.8	25	64	25
	64	8.000	1.000	2.50	1.00	203.5	25	64	25
	72	9.000	1.000	2.50	1.00	229.0	25	64	25
	80	10.000	1.125	3.00	1.12	254.4	28	76	28
	84	10.500	1.125	3.00	1.12	267.1	28	76	28
	88	11.000	1.125	3.00	1.12	279.8	28	76	28
	96	12.000	1.125	3.00	1.12	305.3	28	76	28

Note: Metric size gears were not available at time of publication. Values were soft converted for problem solving only.

Table 73 Eight-pitch (3.18 module) spur-gear data. (Boston Gear Works)

3.18 Module (8 Pitch)

Style	No. of Teeth	Inch Sizes 6 Pitch 1.50 Face				Metric Sizes (mm) 4.23 Module 40 Face			
		Pitch Dia	Hole	Hub Dia	Hub Proj.	Pitch Dia	Hole	Hub Dia	Hub Proj.
Steel Plain	12	2.000	1.00	1.50	.88	50.8	25	38	22
	14	2.333	1.00	1.81	.88	59.2	25	46	22
	15	2.500	1.00	2.00	.88	63.5	25	50	22
	16	2.667	1.00	2.16	.88	67.7	25	55	22
	18	3.000	1.00	2.50	.88	76.1	25	64	22
	20	3.333	1.00	2.84	.88	84.6	25	72	22
	21	3.500	1.00	3.00	.88	88.8	25	76	22
Cast Iron Web	24	4.000	1.12	2.50	1.00	101.5	28	64	25
	27	4.500	1.12	2.50	1.00	114.2	28	64	25
	30	5.000	1.12	2.50	1.00	126.9	28	64	25
	32	5.333	1.12	2.50	1.00	135.4	28	64	25
	33	5.500	1.12	2.50	1.00	139.6	28	64	25
	36	6.000	1.12	2.50	1.00	152.3	28	64	25
	40	6.667	1.12	2.50	1.00	169.2	28	64	25
	42	7.000	1.12	2.50	1.00	177.7	28	64	25
Cast Iron Spoke	48	8.000	1.12	2.50	1.00	203.0	28	64	25
	54	9.000	1.12	2.50	1.00	228.4	28	64	25
	60	10.000	1.25	3.00	1.25	253.8	30	76	30
	64	10.667	1.25	3.00	1.25	270.7	30	76	30
	66	11.000	1.25	3.00	1.25	279.2	30	76	30
	72	12.000	1.25	3.00	1.25	304.6	30	76	30
	84	14.000	1.25	3.25	1.25	355.3	30	82	30

Note: Metric size gears were not available at time of publication. Values were soft converted for problem solving only.

Table 74 Six-pitch (4.23 module) spur-gear data. (Boston Gear Works)

SPUR GEARS STEEL AND IRON
14.5° PRESSURE ANGLE
(Will not operate with 20° Spurs)
Actual Tooth Size

SPUR GEARS STEEL AND IRON
20° PRESSURE ANGLE
(Will not operate with 14.5° Spurs)
Actual Tooth Size

SPUR GEARS STEEL AND IRON
14.5° PRESSURE ANGLE
(Will not operate with 20° Spurs)

Actual Tooth Size

SPUR GEARS STEEL AND IRON
20° PRESSURE ANGLE
(Will not operate with 14.5° Spurs)

Actual Tooth Size

Style		No. of Teeth	Inch Sizes 5 Pitch 2.00 Face				Metric Sizes (mm) 5.08 Module 50 Face			
			Pitch Dia	Hole	Hub Dia	Hub Proj.	Pitch Dia	Hole	Hub Dia	Hub Proj.
Steel Plain		12	2.400	1.06	1.78	.88	61.0	26	45	22
		14	2.800	1.06	2.18	.88	71.1	26	55	22
		15	3.000	1.06	2.38	.88	76.2	26	60	22
		16	3.200	1.06	2.59	.88	81.3	26	65	22
		18	3.600	1.06	3.00	.88	91.4	26	75	22
		20	4.000	1.06	3.38	.88	101.6	26	85	22
Cast Iron	Web	24	4.800	1.06	3.00	1.25	121.9	26	75	30
		25	5.000	1.06	3.00	1.25	127.0	26	75	30
		30	6.000	1.06	3.00	1.25	152.4	26	75	30
		35	7.000	1.18	3.00	1.25	177.8	30	75	30
		40	8.000	1.18	3.00	1.25	203.2	30	75	30
		45	9.000	1.18	3.00	1.25	228.6	30	75	30
	Spoke	50	10.000	1.18	3.50	1.25	254.0	30	90	30
		55	11.000	1.18	3.50	1.25	279.4	30	90	30
		60	12.000	1.18	3.50	1.25	304.8	30	90	30
		70	14.000	1.18	3.50	1.25	355.6	30	90	30
		80	16.000	1.18	3.50	1.25	406.4	30	90	30
		90	18.000	1.18	3.50	1.25	457.2	30	90	30
		100	20.000	1.31	3.75	1.50	508.0	32	95	38
		110	22.000	1.31	3.75	1.50	558.8	32	95	38
		120	24.000	1.31	4.00	1.50	609.6	32	100	38

Note: Metric size gears were not available at the time of publication. Values were soft converted for problem solving only.

Table 75 Five-pitch (5.08 module) spur-gear data. (Boston Gear Works)

OVERALL LENGTH

FACE

HUB PROJ

PITCH DIA

HOLE

HUB DIA

5.08 Module (5 Pitch)

Style		No. of Teeth	Inch Sizes 4 Pitch 2.75 Face				Metric Sizes (mm) 6.35 Module 70 Face			
			Pitch Dia	Hole	Hub Dia	Hub Proj.	Pitch Dia	Hole	Hub Dia	Hub Proj.
Steel Plain		12	3.000	1.12	2.25	.88	76.2	28	58	22
		14	3.500	1.12	2.75	.88	88.9	28	70	22
		15	3.750	1.12	3.00	.88	95.3	28	76	22
		16	4.000	1.12	3.25	.88	101.6	28	82	22
		18	4.500	1.12	3.75	.88	114.3	28	96	22
		20	5.000	1.12	4.25	.88	127.0	28	108	22
		22	5.500	1.12	4.75	.88	139.7	28	120	22
Cast Iron	Web	24	6.000	1.12	3.50	1.50	152.4	28	90	38
		28	7.000	1.25	3.50	1.50	177.8	30	90	38
		30	7.500	1.25	3.50	1.50	190.5	30	90	38
		32	8.000	1.25	3.50	1.50	203.2	30	90	38
		36	9.000	1.25	3.50	1.50	228.6	30	90	38
		40	10.000	1.25	4.00	1.50	254.0	30	100	38
		42	10.500	1.25	4.00	1.50	266.7	30	100	38
		44	11.000	1.25	4.00	1.50	279.4	30	100	38
	Spoke	48	12.000	1.25	4.00	1.50	304.8	30	100	38
		54	13.500	1.25	4.00	1.50	342.9	30	100	38
		56	14.000	1.25	4.00	1.50	355.6	30	100	38
		60	15.000	1.25	4.00	1.50	381.0	30	100	38
		64	16.000	1.25	4.00	1.50	406.4	30	100	38
		72	18.000	1.25	4.00	1.50	457.2	30	100	38
		80	20.000	1.38	4.50	1.50	508.0	35	115	38

Note: Metric size gears were not available at the time of publication. Values were soft converted for problem solving only.

Table 76 Four-pitch (6.35 module) spur-gear data. (Boston Gear Works)

SPUR GEARS STEEL AND IRON
14.5° PRESSURE ANGLE
(Will not operate with 20° Spurs)

Actual Tooth Size

SPUR GEARS STEEL AND IRON
20° PRESSURE ANGLE
(Will not operate with 14.5° Spurs)

Actual Tooth Size

Module (Pitch)	Teeth	U.S. CUSTOMARY (in.)									METRIC (mm)								
		Pitch Dia	Face	OD Approx.	D	MD	Hub Dia	Hub Prov. Approx.	Hole	Keyseat	Pitch Dia	Face	OD Approx.	D	MD	Hub Dia	Hub Prov. Approx.	Hole	Keyseat
2.54 (10)	20	2.000	.44	2.15	1.36	2.00	1.62	.81	.500	.12 × .06	50	11.2	54	34	50	42	20	12	3 × 1.5
									.625	.18 × .09								16	5 × 2.5
									.750	.18 × .09								20	5 × 2.5
2.54 (10)	25	2.500	.55	2.65	1.62	2.44	2.00	.94	.750	.18 × .09	62	14	68	41	62	50	24	20	5 × 2.5
									.875	.18 × .09								22	5 × 2.5
									1.000	.25 × .12								25	6 × 3
3.18 (8)	24	3.000	.64	3.18	1.58	2.56	1.75	.81	.750	.18 × .09	76	16	80	40	65	45	20	20	5 × 2.5
					1.76	2.75	2.25	1.06	1.000	.25 × .12				45	70	58	27	25	6 × 3
					1.76	2.75	2.50	1.12	1.250	.25 × .12				45	70	64	28	30	6 × 3
3.18 (8)	28	3.500	.75	3.68	2.09	3.25	2.50	1.25	1.000		88	16	93	53	82	64	32	25	
									1.188	.25 × .12								30	6 × 3
									1.250									32	
4.23 (6)	24	4.000	.86	4.24	2.31	3.62	3.00	1.31	1.25	.25 × .12	100	22	108	58	92	76	33	32	6 × 3
	27	4.500	.96	4.74	2.62	4.12	3.25	1.50	1.50	.38 × .18	114	24	120	66	104	82	38	38	10 × 5
									1.50	.38 × .18								38	10 × 5
5.08 (5)	25	5.000	1.10	5.29	3.00	4.62	3.50	1.75	1.38	.31 × .16	128	28	134	76	117	88	44	35	8 × 4
									1.50	.38 × .18								38	10 × 5
									1.75	.38 × .18								44	10 × 5

NOTE: Metric size gears were not available at time of publication. Values were soft converted for problem solving only.

Table 77 Miter gears. (Boston Gear Works)

20° PRESSURE ANGLE

Module (Pitch)	Teeth	U.S. CUSTOMARY (in.)								METRIC (mm)							
		Pitch Dia	Face	Hole	D	MD	Hub Dia	Hub Proj. Approx.	OD Approx.	Pitch Dia	Face	Hole	D	MD	Hub Dia	Hub Proj. Approx.	OD Approx.
2.54 (10)	50	5.000	.70	.750	1.30	2.62	2.00	1.00	5.06	127	18	20	33.0	66.6	50	25	128.5
	25	2.500			1.55	3.38	2.00	.75	2.75	63.5			39.4	85.8	50	20	69.8
	60	6.000	.78	1.00	1.86	2.75	3.00	1.38	6.04	152.4	20	25	47.2	69.8	76	35	153.4
	20	2.000		.875	2.16	4.38	1.75	1.30	2.27	50.8		22	54.9	111.2	45	33	57.6
	60	6.000	.72	.875	1.62	2.25	2.50	1.12	6.03	152.4	18.3	22	41.1	57.2	64	28	153.2
	15	1.500		.625	1.60	3.88	1.44	.84	1.78	38.1		16	40.6	98.6	36	22	45.2
3.18 (8)	40	5.000	.82	1.000	1.84	2.88	3.00	1.25	5.08	127.2	20.8	25	46.7	67.6	76	32	129
	20	2.500			2.28	4.00	2.12	1.40	2.81	63.6			57.9	101.6	54	35	71.4
	48	6.000	.84	.875	1.62	2.38	2.75	1.00	6.05	152.6	21.4	22	41.2	60.4	70	25	153.7
	16	2.000		.750	2.08	4.25	1.75	1.88	2.35	50.9		20	52.8	108	45	48	59.7
	64	8.000	.84	1.000	1.88	2.75	2.75	1.25	8.04	203.5	21.4	25	47.7	69.8	70	32	204.2
	16	2.000		.875	2.09	5.25	1.88	1.22	2.36	50.9		22	53.1	133.4	48	30	60
4.23 (6)	40	5.000	.82	1.000	1.84	2.88	3.00	1.25	5.08	169.2	20.8	25	46.7	67.6	76	32	129
	20	2.500			2.28	4.00	2.12	1.40	2.81	84.6			57.9	101.6	54	35	71.4
	48	6.000	.84	.875	1.62	2.38	2.75	1.00	6.05	203	21.4	22	41.1	60.4	70	25	153.7
	16	2.000		.750	2.08	4.25	1.75	1.18	2.35	67.7		20	52.8	108	45	30	59.7
	64	8.000	.84	1.000	1.88	2.75	2.75	1.25	8.04	270.2	21.4	25	47.7	69.8	70	32	204.2
	16	2.000		.875	2.09	5.25	1.88	1.22	2.36	67.7		22	53.1	133.4	48	30	60
	36	6.000	1.06	1.125	2.25	3.50	3.25	1.50	6.10	152.3	27	28	57.2	88.9	82	38	155
	18	3.000			2.76	4.75	2.50	1.60	3.41	76.1			70.1	120.6	64	40	86.6
	45	7.500	1.06	1.125	2.12	3.00	3.25	1.25	7.57	190.4	27	28	53.8	76.2	82	32	192.3
	15	2.500		.875	2.56	5.25	2.12	1.44	2.95	63.4		22	65.0	133.4	54	36	75

NOTE: Metric size gears were not available at time of publication. Values were soft converted for problem solving only.

Table 78   Bevel gears. (Boston Gear Works)

TYPE P
HEADLESS PRESS FIT BUSHINGS

TYPE H
HEAD PRESS FIT BUSHINGS

	Inside Dia (A)		Outside Dia (B)		Bushing Length C											Technical Data	
	From	To	Max.	Min.	.25	.31	.38	.50	.75	1.00	1.38	1.75	2.12	2.50	3.00	F	G
U.S. CUSTOMARY (in.)	.125	.194	.3141	.3138	●	●	●	●	●	●	●	●				.42	.12
	.188	.257	.4078	.4075	●	●	●	●	●	●	●	●	●			.50	.16
	.188	.316	.5017	.5014	●	●	●	●	●	●	●	●	●			.60	.22
	.312	.438	.6267	.6264	●	●	●	●	●	●	●	●	●	●		.80	.22
	.312	.531	.7518	.7515	●	●	●	●	●	●	●	●	●	●	●	.92	.22
	.500	.656	.8768	.8765		●	●	●	●	●	●	●	●	●	●	1.10	.25
	.500	.766	1.0018	1.0015				●	●	●	●	●	●	●	●	1.24	.32
	.625	1.031	1.3772	1.3768				●	●	●	●	●	●	●	●	1.60	.38
	1.000	1.390	1.7523	1.7519					●	●	●	●	●	●	●	1.98	.38
					6	8	10	12	16	20	25	30	35	40	45	F	G
METRIC (mm)	3	5	7.978	7.971	●	●	●	●	●	●	●	●	●	●		10	3
	5	6.5	10.358	10.351	●	●	●	●	●	●	●	●	●	●		12	4
	5	8	12.972	12.736	●	●	●	●	●	●	●	●	●	●		15	5
	8	11	15.918	15.911	●	●	●	●	●	●	●	●	●	●		20	6
	8	13.5	19.096	19.088	●	●	●	●	●	●	●	●	●	●	●	24	6
	12	16	22.271	22.263		●	●	●	●	●	●	●	●	●	●	28	6
	12	20	25.446	25.438			●	●	●	●	●	●	●	●	●	32	8
	16	26	34.981	34.971			●	●	●	●	●	●	●	●	●	40	10
	25	35	44.508	44.498			●	●	●	●	●	●	●	●	●	50	10

Note: Metric size bushings were not available at time of publication. Values shown were soft converted for problem solving only.

Table 79   Press-fit drill jig bushings. (American Drill Bushing Co.).

TYPE S
SLIP RENEWABLE BUSHINGS

TYPE F
FIXED RENEWABLE BUSHINGS

NOTE: Type "FX" bushings (at the manufacturer's option) may be supplied with lock screw slot opposite the flat clamp milling.

	Inside Dia (A)		Outside Dia (B)		Bushing Length C										Technical Data								
	From	To	From	To	.38	.50	.62	.75	1.00	1.38	1.75	2.12	2.50	F	G	H	J	K	L	M	N	R
U.S. CUSTOMARY (in.)	.064	.089	.3125	.3123	●	●	●	●	●	●				.55	.38	.12	.17	.17	65°	.62	.68	.50
	.094	.194	.3125	.3123	●	●	●	●	●	●				.55	.38	.12	.17	.17	65°	.62	.68	.50
	.141	.188	.5000	.4998	●	●	●	●	●	●	●	●		.80	.44	.12	.30	.26	65°	.75	.78	.62
	.189	.344	.5000	.4998	●	●	●	●	●	●	●	●		.80	.44	.12	.30	.26	65°	.75	.78	.62
	.281	.562	.7500	.7498		●	●	●	●	●	●	●	●	1.05	.44	.12	.42	.39	50°	.88	.80	.75
	.469	.781	1.0000	.9998		●	●	●	●	●	●	●	●	1.42	.44	.18	.60	.50	35°	1.11	1.14	.92
	.719	1.062	1.3750	1.3747			●	●	●	●	●	●	●	1.80	.44	.18	.78	.69	30°	1.30	1.33	1.10
	.969	1.406	1.7500	1.7497			●	●	●	●	●	●	●	2.30	.62	.18	1.00	.88	30°	1.64	1.52	1.39
					10	12	16	20	25	30	40	50	60	F	G	H	J	K	L	M	N	R
METRIC (mm)	1.6	2.3	7.938	7.932	●	●	●	●	●	●				14	10	3	4.3	4.3	65°	16	17	13
	2.4	4.9	7.938	7.932	●	●	●	●	●	●				14	10	3	4.3	4.3	65°	16	17	13
	3.6	4.8	12.700	12.695	●	●	●	●	●	●	●	●		20	11	3	7.6	7.6	65°	20	20	16
	4.8	8.7	12.700	12.695	●	●	●	●	●	●	●	●		20	11	3	7.6	7.6	65°	20	20	16
	7.1	14.2	19.050	19.045		●	●	●	●	●	●	●	●	26	11	3	10.7	10.7	50°	22	20	20
	11.9	19.8	25.400	25.350		●	●	●	●	●	●	●	●	36	11	4.5	15.2	15.2	35°	28	29	24
	18.2	26.9	34.925	34.917			●	●	●	●	●	●	●	46	11	4.5	19.8	17.5	30°	33	34	28
	24.6	35.7	44.450	44.442			●	●	●	●	●	●	●	58	16	4.5	25.4	22.4	30°	42	38	35

Note: Metric size bushings were not available at time of publication. Values shown were soft converted for problem solving only.

Table 80   Renewable drill jig bushings. (American Drill Bushing Co.)

HEADLESS LINER   HEAD LINER

	Nominal Inside Dia	Outside Dia (B)		Liner Length C										Technical Data	
	(A)	Max.	Min.	.31	.38	.50	.62	.75	1.00	1.38	1.75	2.12	2.50	F	G
U.S. CUSTOMARY (in.)	.188	.3141	.3138	●	●	●	●	●						—	—
	.312	.5017	.5014	●	●	●	●	●	●	●				.62	.09
	.500	.7518	.7515	●	●	●	●	●	●	●	●	●		.88	.09
	.750	1.0018	1.0015			●	●	●	●	●	●	●	●	1.12	.12
	1.000	1.3772	1.3768			●	●	●	●	●	●	●	●	1.50	.12
	1.375	1.7523	1.7519					●	●	●	●	●	●	1.88	.18
				8	10	12	16	20	25	35	45	55	65	F	G
METRIC (mm)	5	7.978	7.971	●	●	●	●	●						—	—
	8	12.743	12.736	●	●	●	●	●	●	●				16	2.3
	12	19.096	19.088	●	●	●	●	●	●	●	●	●		22	2.3
	20	25.446	25.438			●	●	●	●	●	●	●	●	28	3.0
	25	34.981	34.971			●	●	●	●	●	●	●	●	38	3.0
	35	44.508	44.498				●	●	●	●	●	●	●	48	4.5

LINER FOR SLIP RENEWABLE BUSHINGS

	Nominal Inside Dia	Outside Dia (B)		Liner Length C						Technical Data						
	(A)	Max.	Min.	.31	.38	.50	.75	1.00	1.38	D	E	F	G	H	J	R
U.S. CUSTOMARY (in.)	.188	.3141	.3138	●	●	●	●			.23	.11	.25	.36	.06	.10	.16
	.312	.5017	.5014		●	●	●			.31	.18	.38	.56	.09	.14	.31
	.500	.7518	.7515			●	●	●		.31	.31	.50	.81	.09	.14	.31
	.750	1.0018	1.0015			●	●	●		.38	.44	.66	1.06	.12	.14	.31
	1.000	1.3772	1.3768			●	●	●	●	.38	.61	.88	1.44	.12	.20	.31
				8	10	12	20	25	35	D	E	F	G	H	J	R
METRIC (mm)	5	7.978	7.971	●	●	●	●			5.5	3	6	9	1.5	2.5	4
	8	12.743	12.736		●	●	●			8.1	1.5	10	14	2.3	3.5	8
	12	19.096	19.088			●	●	●		8.1	8	12	20	2.3	3.5	8
	20	25.446	25.438			●	●	●		9.5	11	16	27	3	3.5	8
	25	34.981	34.971			●	●	●	●	11	15	22	36	3	5	8

Note: Metric size drill bushing liners were not available at time of publication. Values shown were soft converted for problem solving only.

Table 81   Drill bushing liners. (American Drill Bushing Co.)

	Part No.	A	B	C	E	F	Thread
U.S. Customary (in.)	LS-1	.625	.375	.625	.250	.138	.312-18
	LS-2	.875	.375	.625	.375	.200	.312-18
	LS-3	1.000	.434	.750	.375	.200	.375-16
	LS-4	1.062	.434	.750	.434	.231	.375/16
Metric (mm)	LS-1M	16	10	16	6	2.5	M8
	LS-2M	22	10	16	10	3.5	M8
	LS-3M	25	11	20	10	5.0	M10
	LS-4M	27	11	20	11	5.8	M10

ROUND CLAMPS

	Part No.	A	B	D	E	F	G	H	Replaces Lock Screw
U.S. Customary (in.)	CL-1	.625	.312	.156	.531	.109	.125	.312UNC	LS-1
	CL-2	.625	.438	.219	.906	.125	.188	.312UNC	LS-2
	CL-3	.750	.500	.281	1.406	.156	.188	.375UNC	LS-3
	CL-4	.750	.531	.250	2.500	.156	.219	.375UNC	LS-4
Metric (mm)	CL-1M	16	8	4	14	2.7	3	M8	LS-1M
	CL-2M	16	12	5.5	23	3.2	4.5	M8	LS-2M
	CL-3M	20	12	7	35	4.0	4.5	M10	LS-3M
	CL-4M	20	13	6	64	10.5	4.5	M10	LS-4M

Table 82 Locking devices for renewable bushings. (American Drill Bushing Co.)

ROUND END CLAMPS

	Part No.	FOR FLUSH MOUNTING OF LINERS						
		A	B	D	E	F	G	H
U.S. Customary (in.)	RE-1	.438	.750	.625	.312	.188	.125	.250UNC
	RE-2	.500	.750	.625	.375	.250	.188	.250UNC
	RE-3	.625	1.000	.750	.375	.312	.188	.312UNC
Metric (mm)	RE-1M	11	20	16	8	5	3	M6
	RE-2M	12	20	16	10	6	4.5	M6
	RE-3M	16	25	20	10	7	4.5	M8

FLAT CLAMPS

	Part No.	FOR FLUSH MOUNTING OF LINERS						
		A	B	D	E	F	G	H
U.S. Customary (in.)	FC-1	.500	.875	.625	.312	.188	.125	.250UNC
	FC-2	.625	1.000	1.00	.375	.250	.188	.312UNC
	FC-3	.625	1.125	1.00	.375	.312	.188	.312UNC
Metric (mm)	FC-1M	12	22	16	8	5	3	M6
	FC-2M	16	25	25	10	6	4.5	M8
	FC-3M	16	28	25	10	7	4.5	M8

	Part No.	FOR PROJECTED MOUNTING OF SHOULDER LINERS						
		A	B	D	E	F	G	H
U.S. Customary (in.)	FC-10	.50	.875	.625	.375	.188	.219	.250UNC
	FC-10	.50	.875	.625	.375	.188	.250	.250UNC
Metric (mm)	FC-10M	12	22	16	10	5	5	M6
	FC-11M	12	22	16	10	5	6	M6

Note: Metric sizes were not available at time of publication. Values shown were soft converted for problem solving only.

Table 82 Continued.

	A	B	C	D	E	Cat. No.	A-B	C	Cat. No.	A-B	C	D
U.S. Customary (in.)	6.00	9.00	1.25	1.88	1.35	ET-300	3.00	.62	UT-300	3.00	.62	1.75
	8.00	12.00	1.38	2.00	1.62	ET-400	4.00	.75	UT-400	4.00	.75	2.62
	10.00	15.00	1.50	2.00	1.75	ET-500	5.00	.75	UT-500	5.00	.75	3.25
	12.00	18.00	1.75	2.25	2.00	ET-600	6.00	1.00	UT-600	6.00	1.00	3.50
						ET-800	8.00	1.25	UT-800	8.00	1.25	4.88
Metric (mm)	150	225	30	45	35	ET-75M	75	15	UT-75M	75	15	45
	200	300	35	50	40	ET-100M	100	20	UT-100M	100	20	65
	250	375	40	50	45	ET-125M	125	20	UT-125M	125	20	85
	300	450	45	55	50	ET-150M	150	25	UT-150M	150	25	90
						ET-200M	200	30	UT-200M	200	30	120

	A	B	C	D	E	Cat. No.	A-B	C	Cat. No.	A-B	C
U.S. Customary (in.)	6.00	9.00	1.25	1.25	1.25	U-300	3.00	.62	L-300	3.00	.62
	8.00	12.00	1.38	1.25	1.25	U-400	4.00	.75	L-400	4.00	.75
	10.00	15.00	1.50	1.38	1.38	U-500	5.00	.75	L-500	5.00	.75
						U-600	6.00	1.00	L-600	6.00	1.00
						U-800	8.00	1.25	L-800	8.00	1.25
Metric (mm)	150	225	30	30	30	U-75M	75	15	L-75M	75	15
	200	300	35	30	30	U-100M	100	20	L-100M	100	20
	250	375	40	35	35	U-125M	125	20	L-125M	125	20
						U-150M	150	25	L-150M	150	25
						U-200M	200	30	L-200M	200	30

Table 83   Fixture bases and microsections.

WITH OR WITHOUT POP-UP LEAF

U.S. CUSTOMARY (INCHES)

Nominal Capacity	Inside Dimensions			Outside Dimensions					
	W	L	D	A	B	C	T	E	F
2.00 × 2.00 × 1.50/2.00/2.50	2.62	2.50	1.50/2.00/2.50	3.88	3.75	3.06/3.56/4.06	.31	2.48	.50
2.00 × 3.00 × 1.50/2.00/2.50	2.62	3.50	1.50/2.00/2.50	3.88	4.75	3.06/3.56/4.06	.31	2.48	.50
2.00 × 4.00 × 1.50/2.00/2.50	2.62	4.50		3.88	5.75	3.06/3.56/4.06	.31	2.48	.50
3.00 × 3.00 × 1.50/2.00/3.00	3.62	3.50	1.50/2.00/3.00	5.00	4.88	3.25/3.75/4.25	.38	3.48	.56
3.00 × 4.00 × 2.00/2.50/3.00	3.62	4.50	2.00/2.50/3.00	5.00	5.88	3.75/4.25/4.75	.38	3.48	.56
3.00 × 5.00 × 2.00/2.50/3.00	3.62	5.50	2.00/2.50/3.00	5.00	6.88	3.75/4.25/4.75	.38	3.48	.56
3.00 × 6.00 × 2.00/2.50/3.00	3.62	6.50	2.00/2.50/3.00	5.00	7.88	3.75/4.25/4.75	.38	3.48	.56
4.00 × 4.00 × 2.00/3.00/4.00	4.62	4.50	2.00/3.00/4.00	6.25	6.12	4.00/5.00/6.00	.50	4.48	.56
4.00 × 5.00 × 2.00/3.00/4.00	4.62	5.50	2.00/3.00/4.00	6.25	7.12	4.00/5.00/6.00	.50	4.48	.56
4.00 × 6.00 × 2.00/3.00/4.00	4.62	6.50	2.00/3.00/4.00	6.25	8.12	4.00/5.00/6.00	.50	4.48	.56
4.00 × 8.00 × 2.00/3.00/4.00	4.62	8.50	2.00/3.00/4.00	6.25	10.12	4.00/5.00/6.00	.50	4.48	.56

METRIC (MILLIMETERS)

Nominal Capacity	Inside Dimensions			Outside Dimensions					
	W	L	D	A	B	C	T	E	F
50 × 50 × 40/50/60	65	60	40/50/60	100	95	80/90/100	8	60	15
50 × 75 × 40/50/60	65	85	40/50/60	100	120	80/90/100	8	60	15
50 × 100 × 40/50/60	65	110	40/50/60	100	145	80/90/100	8	60	15
75 × 75 × 40/50/75	90	85	40/50/75	130	125	85/95/120	10	85	15
75 × 100 × 50/60/75	90	110	50/60/75	130	150	95/105/120	10	85	15
75 × 125 × 50/60/75	90	135	50/60/75	130	175	95/105/120	10	85	15
75 × 150 × 50/60/75	90	160	50/60/75	130	200	95/105/120	10	85	15
100 × 100 × 50/75/100	115	110	50/75/100	160	155	100/125/150	12	110	15
100 × 125 × 50/75/100	115	135	50/75/100	160	180	100/125/150	12	110	15
100 × 150 × 50/75/100	115	160	50/75/100	160	205	100/125/150	12	110	15
100 × 200 × 50/75/100	115	210	50/75/100	160	255	100/125/150	12	110	15

Note: Metric sizes were not available at time of publication. Values were soft converted for problem solving only.

Table 84 Tumble box jig. (Standard Parts Co.)

Table 85   Graphic symbols for electrical and electronics diagrams.

PATH, TRANSMISSION (CONT'D)

GROUPING OF LEADS

OR

OR

POLARITY

POSITIVE +

NEGATIVE –

RECTIFIER OR DIODE

DIODE, GENERAL A—▷|—R

RECTIFIER, BRIDGE TYPE

RESISTOR

GENERAL

TAPPED

ADJUSTABLE CONTACT (OFTEN REFERRED TO AS POTENTIOMETER OR VOLUME CONTROL)

ADJUSTABLE OR CONTINUOUSLY ADJUSTABLE

SEMICONDUCTOR DEVICES
(TRANSISTORS, DIODES)

NAME OF TERMINAL	LETTER (NOT PART OF SYMBOL)
ANODE	A
BASE	B
COLLECTOR	C
DRAIN	D
EMITTER	E
GATE	G
CATHODE	K
SOURCE	S
MAIN TERMINAL	T

PHOTO DIODE (PHOTOEMISSIVE TYPE)

UNIDIRECTIONAL DIODE, VOLTAGE REGULATOR

SEMICONDUCTOR DIODE RECTIFIER DIODE — ANODE / CATHODE

PNP TYPE TRANSISTOR — C, B, E

NPN TYPE TRANSISTOR — C, B, E

UNIJUNCTION TRANSISTOR N-TYPE BASE — E, B2, BI

N-CHANNEL JUNCTION GATE — G, D, S

N-CHANNEL INSULATED GATE — G, D, U, S

P-CHANNEL JUNCTION GATE — G, D, S

P-CHANNEL INSULATED GATE — G, D, S

THYRISTOR; TRIAC — G, T, T

THYRISTOR DIODE BI-SWITCH — G, T, T

THYRISTOR, SEMICONDUCTOR CONTROLLED RECTIFIER — G, A, K

PHOTOTRANSISTOR, 3 TERMINAL PNP TYPE — E, C, P

PHOTOTRANSISTOR, 2 TERMINAL NPN TYPE — E, C

LIGHT SENSORS

SYMMETRICAL PHOTOCONDUCTIVE TRANSDUCER (RESISTIVE)

SWITCH

SINGLE THROW SINGLE POLE

DOUBLE THROW SINGLE POLE

PUSH BUTTON, CIRCUIT CLOSING (MAKE)

PUSH BUTTON, CIRCUIT OPENING (BREAK)

MULTIPOSITION (ANY NUMBER OF POSITIONS) OR

TRANSFORMERS, INDUCTORS, WINDINGS

TRANSFORMERS

GENERAL OR

MAGNETIC CORE OR

INDUCTORS AND WINDINGS

GENERAL OR

MAGNETIC CORE

ADJUSTABLE OR

Table 85   (Cont'd.)

	Width W	Length L	Thickness														
			Die T	Punch S	A	B	C	D	E	F	G	H	J	K	M	N
U.S. CUSTOMARY (in.)	3.00	4.00	1.25	1.00	3.00	5.00	1.12	2.00	1.88	6.50	7.50	1.50	.75	1.06	.62	.50
	3.00	6.00	1.50	1.25	5.00	5.25	1.25	2.00	2.00	8.50	10.00	1.68	.88	1.06	.75	.62
	3.00	8.00	1.50	1.25	7.00	5.25	1.25	2.00	2.00	10.50	12.00	1.68	.88	1.06	.75	.62
	3.00	10.00	1.50	1.25	9.00	5.44	1.38	2.00	2.06	12.50	14.00	1.81	1.00	1.06	.75	.62
	4.00	4.00	1.50	1.25	3.00	6.25	1.25	2.50	2.50	6.50	7.50	1.68	.88	1.06	.75	.62
	4.00	5.00	1.50	1.25	4.00	6.25	1.25	2.50	2.50	7.50	8.50	1.68	.88	1.06	.75	.62
	4.00	6.00	1.50	1.25	5.00	6.25	1.25	2.50	2.50	8.50	10.00	1.68	.88	1.06	.75	.62
	4.00	8.00	1.50	1.25	7.00	6.25	1.25	2.50	2.50	10.50	12.00	1.68	.88	1.06	.75	.62
METRIC (mm)	75	100	30	25	75	125	25	50	50	165	190	40	20	28	16	12
	75	150	40	30	125	135	35	50	50	215	250	45	22	28	20	16
	75	200	40	30	175	135	35	50	50	265	300	45	22	28	20	16
	75	250	40	30	225	140	40	50	50	320	350	45	25	28	20	16
	100	100	40	30	75	160	30	65	65	165	190	45	22	28	20	16
	100	125	40	30	100	160	30	65	65	190	215	45	22	28	20	16
	100	150	40	30	125	160	30	65	65	215	250	45	22	28	20	16
	100	200	40	30	175	160	30	65	65	265	300	45	22	28	20	16

Note: Metric size die sets were not available at time of publication. Values shown were soft converted for problem solving only.

Table 86   Die sets. (E.A. Baumbach Mfg. Co.)

Table 87   Form and proportion of geometric tolerancing symbols.

Table 88   Form and proportion of dimensioning symbols.

SYMBOL FOR:	ANSI Y14.5	ISO
STRAIGHTNESS	—	—
FLATNESS	▱	▱
CIRCULARITY	○	○
CYLINDRICITY	⌭	⌭
PROFILE OF A LINE	⌒	⌒
PROFILE OF A SURFACE	⌓	⌓
ALL AROUND—PROFILE	↙⟲	NONE
ANGULARITY	∠	∠
PERPENDICULARITY	⊥	⊥
PARALLELISM	∥	∥
POSITION	⊕	⊕
CONCENTRICITY/COAXIALITY	◎	◎
SYMMETRY	NONE	≡
CIRCULAR RUNOUT	*↗	↗
TOTAL RUNOUT	*↗↗	↗↗
AT MAXIMUM MATERIAL CONDITION	Ⓜ	Ⓜ
AT LEAST MATERIAL CONDITION	Ⓛ	NONE
REGARDLESS OF FEATURE SIZE	Ⓢ	NONE
PROJECTED TOLERANCE ZONE	Ⓟ	Ⓟ
DIAMETER	⌀	⌀
BASIC DIMENSION	50	50
REFERENCE DIMENSION	(50)	(50)
DATUM FEATURE	-A-	⟂ OR ⟂—Ⓐ
DATUM TARGET	⌀6/A1	⌀6/A1
TARGET POINT	✕	✕
DIMENSION ORIGIN	⊕→	NONE
FEATURE CONTROL FRAME	⊕ ⌀0.5Ⓜ A B C	⊕ ⌀0.5Ⓜ A B C
CONICAL TAPER	▷	▷
SLOPE	◁	◁
COUNTERBORE/SPOTFACE	⌴	NONE
COUNTERSINK	⌵	NONE
DEPTH/DEEP	↧	NONE
SQUARE (SHAPE)	□	□
DIMENSION NOT TO SCALE	<u>15</u>	<u>15</u>
NUMBER OF TIMES/PLACES	8X	8X
ARC LENGTH	⏜105	NONE
RADIUS	R	R
SPHERICAL RADIUS	SR	NONE
SPHERICAL DIAMETER	S⌀	NONE

*MAY BE FILLED IN

Table 89   Comparison of ANSI and ISO symbols.

INDEX